# Carbon Management Technology Conference (CMTC 2017)



## Global CCUS Innovation Nexus

Houston, Texas, USA
17-20 July 2017

Volume 1 of 2

ISBN: 978-1-5108-4815-3

**Printed from e-media with permission by:**

Curran Associates, Inc.
57 Morehouse Lane
Red Hook, NY 12571



Printed by Curran Associates, Inc. (2017)



AIChE
120 Wall Street, FL 23
New York, NY 10005-4020

Phone: (800) 242-4363
Fax: (203) 775-5177

www.aiche.org

**Additional copies of this publication are available from:**

Curran Associates, Inc.
57 Morehouse Lane
Red Hook, NY 12571 USA
Phone: 845-758-0400
Fax: 845-758-2633
Email: curran@proceedings.com
Web: www.proceedings.com

# TABLE OF CONTENTS

CARBON MANAGEMENT TECHNOLOGY CONFERENCE

# CMTC 2017: Global CCUS Innovation Nexus

July 17-20, 2017
Houston, Texas, USA

## CMTC 2017 Technical Program

### Monday, July 17, 2017

| | | |
|---|---|---|
| 9:00 AM | 5:00 PM | Registration |
| 9:30 AM | 12:00 PM | Petra Nova Tour, NRG & JX Nippon |
| 1:00 PM | 5:00 PM | Oyster Bayou Field Tour, Denbury Resources |
| 1:00 PM | 6:00 PM | Port Arthur Refinery Tour, Air Products & Valero |
| 7:00 PM | 8:00 PM | Welcoming Reception |

### Tuesday, July 18, 2017

| | | |
|---|---|---|
| 7:00 AM | 12:00 PM | Registration |
| *7:30 AM* | *8:00 AM* | *Coffee (sponsored by ExxonMobil)* |
| **8:00 AM** | **11:50 AM** | **Tuesday Plenary** |
| 8:00 AM | 8:35 AM | Keynote 1 - Greetings from Conference Chairs |
| 8:35 AM | 9:10 AM | Keynote 2 - Charles McConnell<br>Executive Director, Rice University Energy and Environmental Initiative<br>Pathways not Targets. Let's Get Serious. |
| 9:10 AM | 9:45 AM | Keynote 3 - Jeff Erikson<br>General Manager for the Americas, Global CCS Institute<br>Carbon Capture at a Crossroads |
| *9:45 AM* | *10:05 AM* | *Break (sponsored by ExxonMobil)* |
| 10:05 AM | 10:40 AM | Keynote 4- Janet Gellici<br>Chief Executive Officer, National Coal Council<br>Charting a Course for Coal's Future |
| 10:40 AM | 11:15 AM | Keynote 5 - Frederick Eames<br>Partner, Hunton & Williams<br>U.S. Direction on CCUS |
| 11:15 AM | 11:50 AM | Keynote 6 - Robert Williams<br>Senior Research Scientist, Andlinger Center for Energy and Environment, Princeton University<br>Negative Emissions Technologies and Strategies Based on Geological Storage of Photosynthetic CO2 |
| 11:50 AM | 1:00 PM | Lunch *(sponsored by Mitsubishi Heavy Industries) (Room 335)* |
| 1:00 PM | 1:35 PM | Keynote 7 - Jennifer Holmgren<br>Chief Executive Officer, LanzaTech<br>Carbon: The Final Frontier |
| **1:35 PM** | **2:20 PM** | **Keynote 8 - Panel Discussion: Opportunities that Exist for CCU Technologies**<br>Chair - Jennifer Holmgren, Chief Executive Officer, Lanzatech<br>Panelist - Matt Carr, Executive Director, Algae Biomass Organization (ABO)<br>Panelist - Robert Sturtz, Vice President of Business Development, World Fuel Services<br>Panelist - Greer Tidwell, Director Environmental Management, Bridgestone |
| *2:20 PM* | *2:40 PM* | *Break* |

| | | BOA - C | Room 330 | Room 329 |
|---|---|---|---|---|
| **2:40 PM** | **4:20 PM** | **Session 1 - Solvent-Based CO2 Capture Systems *(sponsored by Mitsubishi Heavy Industries)* Chair: Ron Munson, Global CCS Institute** | **Session 2 - EOR 1 Chair: Mehdi Zeidouni, Louisiana State University** | **Session 3 - Business Frameworks Chair: Dale Keairns, Deloitte Consulting** |
| 2:40 PM | 3:05 PM | Kevin O'Brien, Illinois Sustainable Technology Center, UIUC | Robert Balch, New Mexico Institute of Mining and Technology | Ramón Gil-Egui, Bureau of Economic Geology |
| | | CCUS As a Regional Economic Development Tool: Planning and Design Considerations | The Intersection of Environment and EOR: How Carbon Capture is Changing Tertiary Recovery | Matching Environmental and Economic Performance of CCUS Systems: An Approach to a Decision-Making Methodology for Sustainable Development |
| 3:05 PM | 3:30 PM | Jonathan V. Pelgen, University of Kentucky | Benjamin Cook, University of Wyoming | Sa'd Shannak, KAPSARC |
| | | Reducing Regeneration Energy and Capital Costs in an Advanced Pccc System | Electricity in the Oil Patch: The Power Costs of Production, CO2-EOR, Electric Lines and Onsite Generators | Could China Meet Its CO2 Reduction Goal By CO2-EOR? |

| | | | | |
|---|---|---|---|---|
| 3:30 PM | 3:55 PM | Avi Patkar, CCS USA | David Merchant, Merchant Consulting | Louis Fradette, CO2 Solutions Inc. |
| | | Carbon Clean Solutions Limited Cdrmax® CO2 Capture Technology | Enhanced Oil Recovery – the History of CO2 Conventional Wag Injection Techniques Developed from Lab in the 1950's to 2017 | Valorisation Carbone Quebec – a Unique CO2 Capture and Reuse Project |
| 3:55 PM | 4:20 PM | Andrew Sexton, Trimeric Corporation | Edward Ennin | Anna Sberegaeva, The National Academies of Sciences, Engineering, and Medicine |
| | | Pilot Plant NO2 Removal with Aqueous Sulfite | Optimizing Reservoir Oil Displacement and CO2 storage in the Texas Farnsworth Unit | Developing a Research Agenda for Carbon Removal, Sequestration, and Utilization – Two Studies at the National Academies of Sciences, Engineering, and Medicine |

| | | |
|---|---|---|
| *4:20 PM* | *4:40 PM* | *Break* |
| 4:40 PM | 5:40 PM | Rapid-Fire Poster Session |
| 5:40 PM | 6:40 PM | Poster Session + Light Reception |

## Wednesday, July 19, 2017

| | | |
|---|---|---|
| 7:00 AM | 12:00 PM | Registration |
| *7:30 AM* | *8:00 AM* | *Coffee* |
| **7:50 AM** | **8:00 AM** | **Remarks from the Chair** |
| **8:00 AM** | **11:50 AM** | **Wednesday Plenary** |
| 8:00 AM | 8:35 AM | Keynote 9 - Haroon Kheshgi<br>ExxonMobil Research and Engineering Company<br>The Potential Role of CCS in Electricity and Liquid Fuel Production Sectors |
| 8:35 AM | 9:10 AM | Keynote 10 - Mike Monea<br>President and CEO, The International CCS Knowledge Centre<br>Introduction to The International CCS Knowledge Centre |
| 9:10 AM | 9:45 AM | Keynote 11 - Bill Brown<br>Chief Executive Officer, NET Power, LLC |
| *9:45 AM* | *10:05 AM* | *Break* |
| 10:05 AM | 10:40 AM | Keynote 12 - David Greeson<br>Vice President, Development, NRG Energy, Inc.<br>Clean Coal is a Reality |
| 10:40 AM | 11:15 AM | Keynote 13 - TBD |
| 11:15 AM | 11:50 AM | Keynote 14 - Juan Anguiano<br>Oil and Gas Climate Initiative<br>The Oil and Gas Climate Initiative Works Scope in CCUS |
| 11:50 AM | 1:00 PM | Lunch (Room 335) |

| | | BOA - C | Room 330 | Room 329 |
|---|---|---|---|---|
| **1:00 PM** | **2:15 PM** | **Session 4 - Membrane-Based Capture Systems<br>Chair: Samuel Tam, US DOE Office of Fossil Energy** | **Session 5 - EOR 2<br>Chairs: Vida Gholami, Advanced Resources International, Inc. & Greeshma Gadikota, University of Wisconsin** | **Session 6 - Next Generation Capture Technologies 1<br>Chair: Richard Adamson, Adamson StrAdvisory** |
| 1:00 PM | 1:25 PM | Jay Kniep, Membrane Technology and Research, Inc. | William Ampomah, New Mexico Institute of Mining and Technology | JinWon Lee, Postech |
| | | Pilot Test Results from a PolarisTM Membrane 1 MWe CO2 Capture System | Study of CO2-EOR Performance in the Morrow Formation, Ochiltree County, Texas | Spray Towers Can Perform Better Than Packed Towers in Capturing CO2 |
| 1:25 PM | 1:50 PM | Shiguang Li, Gas Technology Institute | Hamed A. Hamed | Aaron Sayre, Sustainable Energy Solutions |
| | | Pilot-Scale Development of a PEEK Hollow Fiber Membrane Contactor Process for Post-Combustion CO2 Capture | Study of Applying Solar Energy in Enhanced Oil Recovery. | Field Testing of Cryogenic Carbon Capture |
| 1:50 PM | 2:15 PM | Shilu Fu, American Air Liquide | Junjie Yang, Baker Hughes | Soheil Khiavi, Inventys |
| | | Bench Scale Testing of Next Generation Hollow Fiber Membrane Modules | Miscible CO2 Flooding Using Horizontal Multi-Fractured Wells in the San Andres Formation, TX – a Feasibility Study | Can Engineered Structured Adsorbent Alter the Capital Cost Paradigm of Carbon Capture? |

| Start | End | | | |
|---|---|---|---|---|
| *2:15 PM* | *2:35 PM* | *Break* | | |
| **2:35 PM** | **4:15 PM** | **Session 7 - International Test Center Network Panel**<br>**Chair: Frank Morton, National Carbon Capture Center** | **Session 9 - EOR 3**<br>**Chairs: Jacqi Roueche, Lynxnet LLC, USGS Contractor & Greeshma Gadikota, University of Wisconsin** | **Session 10 - CO2 Storage Case Studies**<br>**Chair: David Riestenberg, Advanced Resources International, Inc.** |
| 2:35 PM | 3:00 PM | | Christian A Dalton, Louisiana State University<br>Bayesian Proxy-Based Optimization of Immiscible-CO2 Huff-n-Puff Flooding to Enhance Oil Recovery in Unconventional Shale Oil Reservoirs | Kyle Worth, PTRC<br>Aquistore: the World's Most Comprehensive Industrial-Scale Field Laboratory for the Investigation of CO2 Geological Storage |
| 3:00 PM | 3:25 PM | Panelist: Jesse Thompson, University of Kentucky<br>Panelist: Jon Gibbins, UK CCS Research Centre<br>Panelist: Abdul Qader, CO2CRC | Japan Trivedi, University of Alberta<br>Evaluating the Potential of CO2 Foam and Polymer Enhanced Foam for Heavy Oil Recovery and CO2 Sequestration | Vello A. Kuuskraa, Advanced Resources International<br>The Project ECO2S Phase 2 Carbonsafe Field Project |
| 3:25 PM | 3:50 PM | **Session 8 - CCUS Transition Issues: The Legal and Policy Landscape**<br>**Chair: Mary Ellen Ternes, Earth & Water Law LLC** | Wes Peck, University of North Dakota<br>Demonstration of Secure CO2 Geologic Storage Associated with Enhanced Oil Recovery in the PCOR Partnership Region. | Joseph Swisher, Electric Power Research Institute<br>Techno-economic analysis of extraction, treatment, and distribution of brines as a part of pressure management operations for CO2 storage in saline aquifers |
| 3:50 PM | 4:15 PM | Panelist: Frederick Eames, Hunton & William<br>Panelist: Charles McConnell, Rice University<br>Panelist: E. Dennis Griffith, Granherne<br>Panelist: Victor B. Flatt, University of North Carolina | Nick Jones, University of Wyoming<br>Wyoming Reservoir Information Tool - a Map and Query Based Web Application for Oil and Gas Reservoirs | Abdul Qader, CO2CRC<br>Carbon Capture and Storage Demonstration By CO2CRC |
| *4:15 PM* | *4:35 PM* | *Break* | | |
| **4:35 PM** | **5:50 PM** | **Session 11 - Offshore Storage**<br>**Chair: Alan Weakly, Society of Mining, Metallurgy and Exploration (SME)** | **Session 12 - Storage Integrity and Monitoring**<br>**Chair: Greeshma Gadikota, University of Wisconsin** | **Session 13 - CO2 Utilization**<br>**Chair: Kevin O'Brien, Illinois Sustainability Technology Center** |
| 4:35 PM | 5:00 PM | Vello A. Kuuskraa, Advanced Resources International<br>Establishing CO2 Utilization, Storage, and Pipeline Systems for Oil Fields in Shallow and Deep Waters of the Gulf of Mexico | Ahmed Abdulhamid Mahmoud, King Fahd University of Petroleum & Minerals<br>Investigation of Nano Clay-Based Cement Alteration and Associated Conductivity Change in CO2-Rich Environments Relevant to Geological Carbon Sequestration | Ferdinando Tessarolo<br>Carbon Farming associated with Solar Energy Dissipation |
| 5:00 PM | 5:25 PM | Jack C. Pashin, Oklahoma State University<br>Offshore CO2 Storage Potential of the Mafla Continental Shelf, Eastern Gulf of Mexico | Yilin Mao, Louisiana State University<br>Analytical Temperature Modeling for Early Detection and Rate Estimation of CO2 Wellbore Leakage | Andres Clarens, University of Virginia<br>Life Cycle Implications of Using CO2-Based Fracturing Fluids As a Substitute for Slickwater |
| 5:25 PM | 5:50 PM | Marina Rossi, Virginia Tech<br>Carbon Dioxide Storage Resource Assessment of the Mid-Atlantic Ocean Seaboard: Analysis of Geological and Geophysical Vintage Data | Muhammad Zulqarnain, Louisiana State University<br>Risk Based Approach to Identify the Leakage Potential of Wells in Depleted Oil and Gas Fields for CO2 Geological Sequestration | Brian Tang, Shockwave CO2, LLC<br>Shockwave CO2 |

5:50 PM 6:50 PM Poster Session

## Thursday, July 20, 2017

| | | |
|---|---|---|
| 7:00 AM | 9:00 AM | Registration |
| *7:30 AM* | *8:00 AM* | *Coffee* |
| **7:50 AM** | **8:00 AM** | **Remarks from the Chair** |
| **8:00 AM** | **10:20 AM** | **Thursday Plenary** |
| 8:00 AM | 8:35 AM | Keynote 15 - Frank Morton<br>National Carbon Capture Center<br>International Collaboration on CCUS R&D |
| 8:35 AM | 9:10 AM | Keynote 16 - Steve Carpenter<br>Director, Enhanced Oil Recovery Institute, School of Energy Resources, University of Wyoming<br>CCUS, Wyoming, & Non-OECD Economies: ISO Standards as the Common Denominator |
| 9:10 AM | 9:45 AM | Keynote 17 - Anthony Ku<br>Director of Advanced Technologies, National Institute of Clean and low-carbon Energy (NICE), Shenhua Group<br>Carbon Management at Shenhua: NICE R&D initiatives and CCUS Demonstrations |
| 9:45 AM | 10:20 AM | Keynote 18 - Jon Gibbins<br>Director, UK CCS Research Centre |
| *10:20 AM* | *10:40 AM* | *Break* |

| | | BOA - C | Room 330 | Room 329 |
|---|---|---|---|---|
| **10:40 AM** | **12:20 PM** | **Session 14 - Analysis and Modeling for CCUS Technologies**<br>**Chairs: Anne Oudinot, Advanced Resources International & Jacqi Roueche, Lynxnet LLC, USGS Contractor** | **Session 15 - Storage Characterization**<br>**Chair: Steven Carpenter, University of Wyoming** | **Session 16 - Next Generation Capture Technologies 2**<br>**Chair: Tim Merkel, Membrane Technology & Research** |
| 10:40 AM | 11:05 AM | Dale Keairns, Deloitte Consulting LLP<br>Utilizing Systems Analysis to Inform Carbon Capture Technology Development | Ashwin Pasumarti, Battelle Memorial Institute<br>Assessing the Effective CO2 Storage Capacity of a Reservoir Using a Geomechanical Framework: A Case Study of a Site in the Arches Province of the Midwest U.S. | WS Winston Ho, The Ohio State University<br>New Amine-Containing Membranes for CO2 Capture |
| 11:05 AM | 11:30 AM | Christina Reynolds, University of Michigan<br>Mobile Carbon Dioxide Removal: A Baseline Cost Estimate and Testing Protocol | Isis Fukai, Battelle Memorial Institute<br>Characterization of CO2 Storage Resource and Reservoir Feasibility of Deep Saline Cambrian-Ordovician Formations in Eastern Ohio | Shiguang Li, Gas Technology Institute<br>Graphene Oxide-PEEK Hybrid Membrane Process for CO2 Capture |
| 11:30 AM | 11:55 AM | Seyed Mostafa Safdarnejad, Intech Process Automation<br>Performance Comparison of a Cryogenic and Chemical Absorption Carbon Capture Systems in Response to a Dynamic Electricity Load Profile | David Riestenberg, Advanced Resources International, Inc.<br>Project ECO2s: Geologic Characterization and Preliminary Storage Capacity Estimates | Mustapha Soukri, RTI International<br>Hybrid MOF Solid Sorbent-Based CO2 Capture Process for Power and Industrial Applications |
| 11:55 AM | 12:20 PM | Michael Buckner, Colorado State University<br>Automotive CO2 Mitigation Via an Onboard Bosch Reactor System | Autumn Haagsma, Battelle Memorial Institute<br>Impact of Geologic Diversity on Static Earth Models in a CO2-EOR Reef Complex | Amanda Warriner, University of Kentucky<br>Evaluating the Effect of Aqueous Calcium Chloride Desiccant on Cooling Tower Performance in a Heat-Integrated Post-Combustion CO2 Capture System |

12:20 PM 12:30 PM Closing Remarks

**CMTC 2017**

# Spray Towers Can Perform Better than Packed Towers in Capturing CO2

2017. 7. 19

Minki Cho and Jin W. Lee (jwlee@postech.ac.kr)
Dep't of Mech. Eng., POSTECH
Pohang, KOREA

EnvTF Lab.

POSTECH

## Capture Performance versus Cost

1. chemical absorption using liquid absorbent
   - $1 - \eta \sim \exp(-\alpha\, h_m A_L/Q_g) \sim \exp[-\alpha\, (h_m A'')(Q_L/Q_g)(L/V_f)]$
   - factors affecting $\eta$ ; $h_m$ and $A_L[= A'L \sim A''Q_L(L/V_f)]$

2. common practice
   - increase $Q_L L/V_f$ → easy and direct → cost more money
   - $Q_L$ (solvent flow) → energy for flow and regeneration
   - L (tower height) → construction for structure and packing
   - $V_f$ (falling speed) → construction + maintenance for internal packing

3. $\eta$ enhancement via $h_m A''$ ?
   - A" (specific surface area per liquid volume); film → drop
   - efforts to enhance $h_m$ have not been very successful

EnvTF Lab.

POSTECH

# Possibility of Enhancing $h_m$

1. $h_m$ ; gas-interface-liquid
   - $h_m$ in the gas phase ; fast enough → no/little additional benefit
   - overall $h_m$ dominated by the internal dispersion within the liquid phase

2. $h_m$ for a single liquid drop or film
   - saturation($\Delta\Phi$) or effectiveness → $\tau_f/\tau_d \sim d^{-(2\sim2.5)}$
     $\tau_d$, time for internal diffusion; $\tau_f$, contact/falling time
   - smaller drops or thinner films always advantageous ; $\Phi \sim d^{-(2\text{-}2.5)}$
   - liquid size difficult to change ; $l \sim V_g^{1/2}$, $V_t \sim d_p^{1-1.5}$

3. mass-averaged $\underline{h}_m$
   - every non-uniformity always deteriorates performance
   - highly non-uniform ; drop-size, spatial distribution
   - * reduced non-uniformities → enhanced $\underline{h}_m$
     → even without any change on the single-drop level

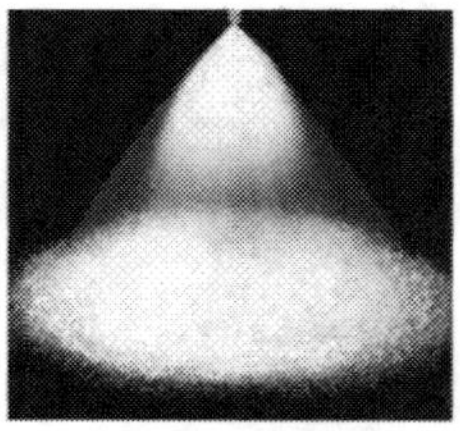

EnvTF Lab. POSTECH

# Effect of Drop-Size Non-uniformity

- example ; saturation($\Phi$) for $\sigma_g = 2.0$
  - drop size ; log-normal distribution
  - $\Phi(\underline{d}) \sim 70\%$
  - $\underline{\Phi} \sim 11\% \sim \Phi(2\underline{d})$
  - * $\underline{\Phi}(\sigma_g = 1.0) = \Phi(\underline{d}) \sim 70\%$

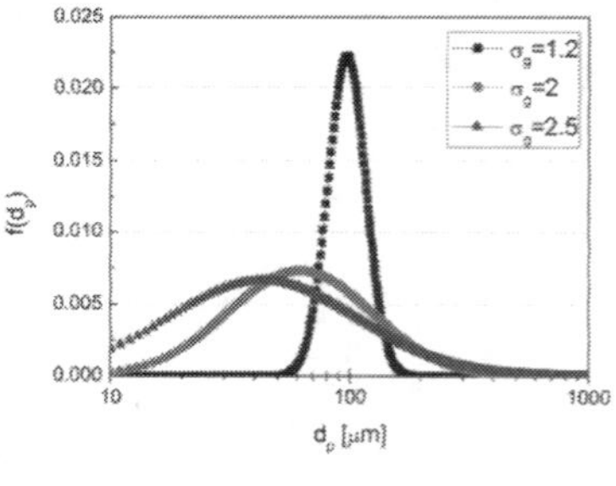


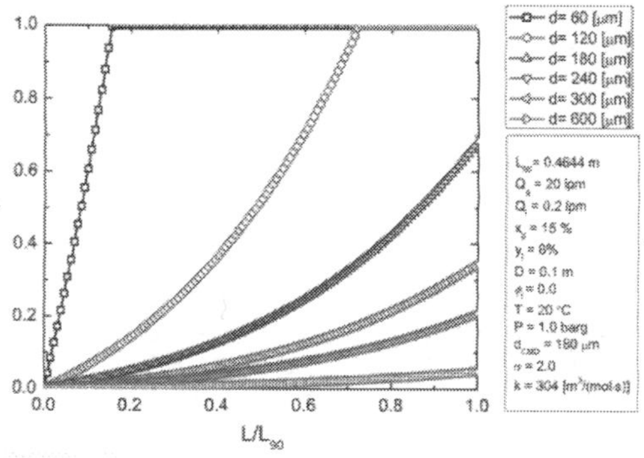


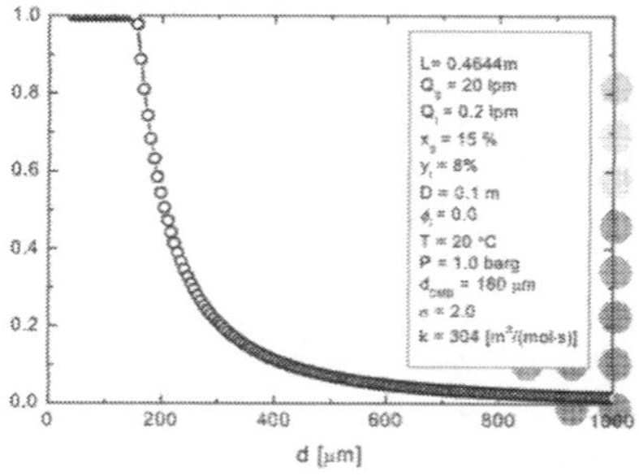


EnvTF Lab. POSTECH

## Spray Tower against Packed Tower

1. strong points
   - surface area (A") ; large and easy to control
   - diffusion in 3-D → shorter diffusion time
   - internal circulation → enhanced mixing
   - * theoretical performance → much better
   - simple structure and low cost ; no packing

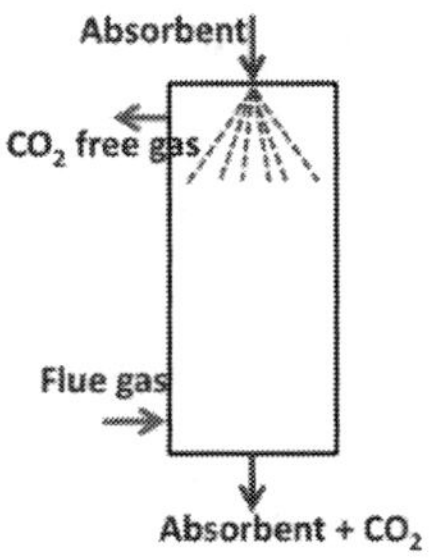


2. weak points
   - high falling speed → short capture time
   - spatial distribution ; highly non-uniform
   - drop-size non-uniform ; $\sigma_g > 2.0$
     - → residence/reaction time
     - → big drops lost to the wall
     - → small droplets lost by fly-back
     - * the bigger drops are the more ineffective

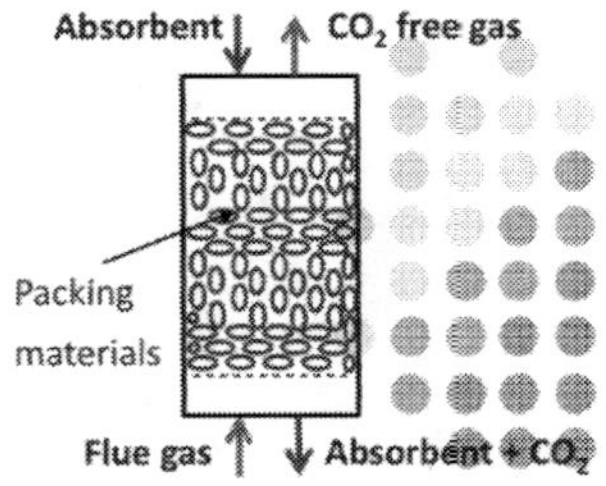


EnvTF Lab. POSTECH

## Room for Performance Improvement

1. packed tower
   - better capture ; mainly due to the slow fall
   - little room for further improvement ; film thickness ~ $f(Q_L, V_g)$
   - packing materials, heavy structure

2. spray tower
   - larger surface area + 3-directional diffusion + internal circulation
   - poorer capture ; wall loss + fly-back + various high non-uniformities
     - → spatial distribution, drop-size, gas velocity (higher porosity)
   - large room for performance improvement

3. chance for performance enhancement
   - via reducing the drop-size non-uniformity
   - negligible additional cost

EnvTF Lab. POSTECH

## Performance Simulation of Spray Tower

1. Non-uniform gas velocity, drop size and droplet spatial distribution

2. Poly-disperse drop size distribution ; log-normal

$$f(i) = \frac{1}{\sqrt{2\pi}\cdot d(i)\cdot \ln\sigma}\cdot \exp\left(-\frac{\left[\ln\left(d(i)/d_{CMD}\right)\right]^2}{2[\ln\sigma]^2}\right)$$

$$N(i) = N_{tot}\cdot f(i)\cdot \Delta d(i) = \frac{Q_l\cdot[f(i)\cdot\Delta d(i)]}{\sum_1^{i_{max}}[q(i)\cdot f(i)\cdot \Delta d(i)]}$$

$$n'''(i) = \frac{N(i)}{A_C\cdot V_f(i)} \qquad N_{tot} = Q_l/\overline{q}$$

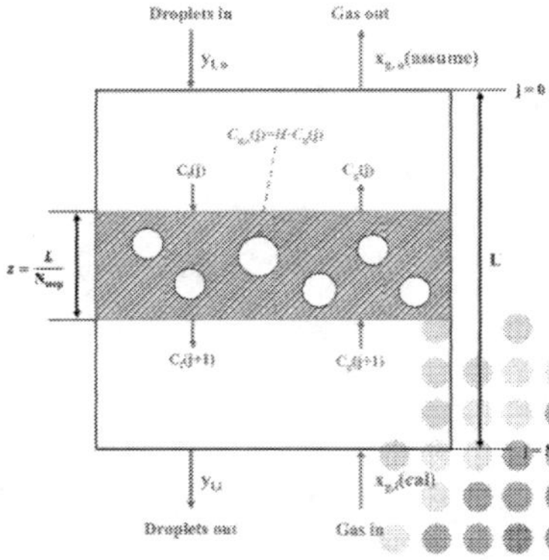


3. Mass balance equations

$$dM(i,j) = m''_g(i,j)\cdot A(i)\cdot[n'''(i)\cdot z\cdot A_C]$$

$$dM(j) = \sum_1^i dM(i,j) = \sum_1^i m''_g(i,j)\cdot\{\pi\cdot[d(i)]^2\}\cdot[n'''(i)\cdot z\cdot A_C]$$

EnvTF Lab.

POSTECH

## Effect of Drop-size Uniformity

- reduced size variation ($\sigma_g$) → $(Q_L/Q_g)L$ reduced (same η condition)
  - → 1/2 with $\sigma_g$ = 1.6 and 1/3 with $\sigma_g$ = 1.2 (relative to $\sigma_g$ = 2.0)
  - → system size (L) reduced ; same ratio (same $Q_L/Q_g$ condition)
- enhanced capture for any type of solvent
- enhancement factor ; independent of the system size L and $V_g$ (~ 1:10)

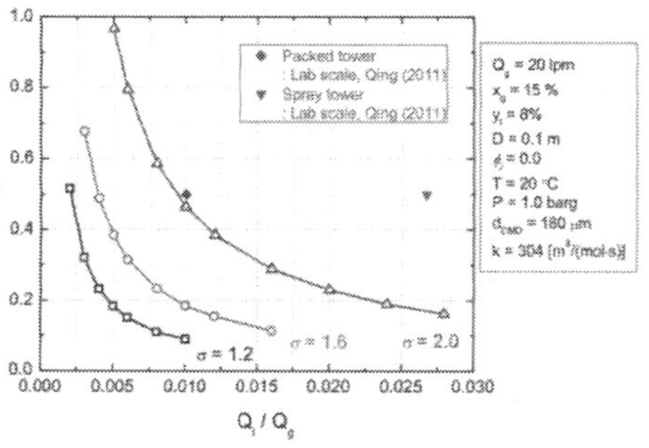


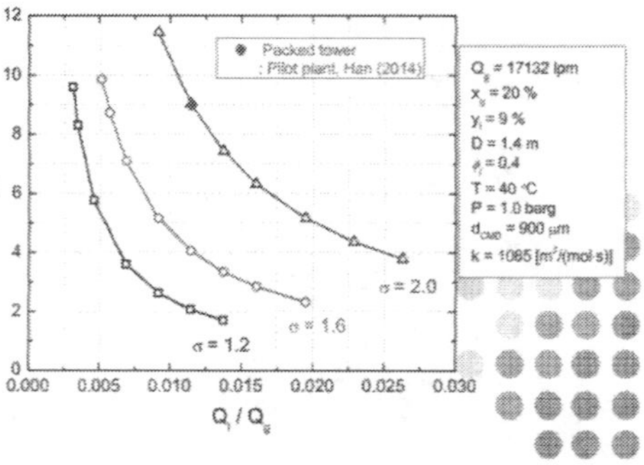


EnvTF Lab.

POSTECH

## Ideal Spray Tower

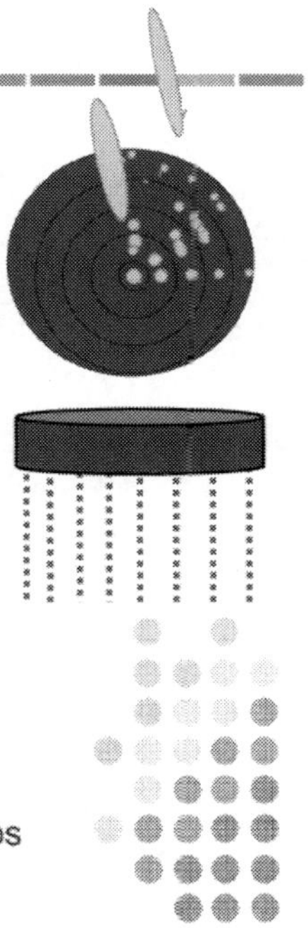

1. target performance
   - maximum capture per solvent mass
   - every droplet follows the same capture/saturation history
2. basic structure
   - uniform vertical injection of mono-disperse droplets
   - spatial distribution of droplets optimized ; radial + axial
3. implementation
   - nozzle plate with multiple micro-nozzles
   - optimized nozzle array pattern and nozzle hole size
4. other characteristics
   - wall loss and fly-back minimized
   - coagulation minimized ; small relative velocity between drops
   - similar conditions along the axial position → scale-up easy

EnvTF Lab.

POSTECH

## Generation of Mono-disperse Droplets

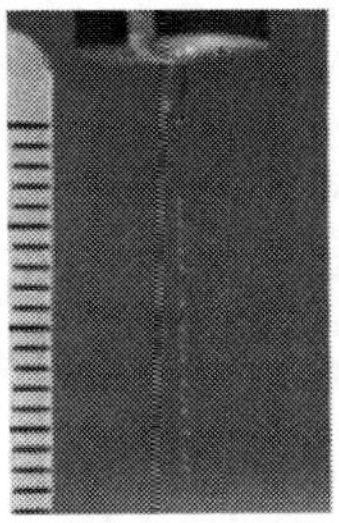

1. single nozzle
   - breakup of liquid column
   - hydrodynamic instability + proper control mechanism
   - f = F(do, $V_j$, liquid properties)
   - $d_p$ = F'(do, $V_j$, liquid properties)
   - $\sigma_g \sim 1.2$

2. nozzle plate
   - micro nozzles of variable shape and size
   - optimum arrangement of nozzle holes
   - vertical injection
   - $\sigma_g \sim 1.2$

EnvTF Lab.

POSTECH

# Experimentation of CO2 Capture

1. spray tower
   - D = 100mm
   - L = 0.5, 1.0 and1.5m → L/D = 5, 10 and 15
2. gas mixture
   - 15% CO2 in N2
   - $V_g$ = 2~4 cm/s
   - gas distributor ; uniform, non-uniform
3. solvent
   - 8% NH3 and 30% MEA
   - $d_{CMD}$ ~ 300μm, $\sigma_g$ = 1.2
4. nozzle plate
   - variety of designs ; do, thickness, contour
   - support
5. temperature ; 25°C

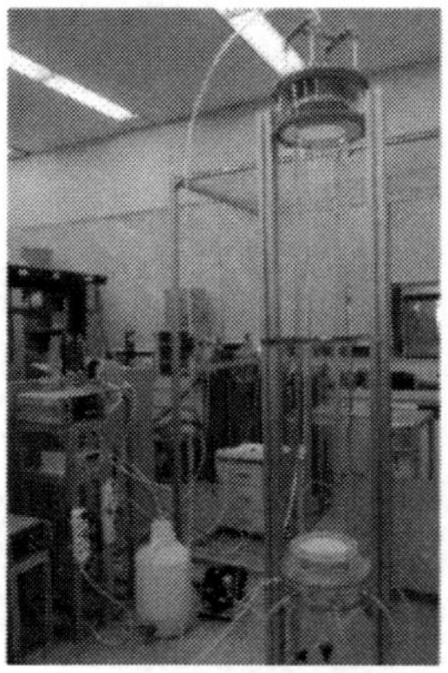

EnvTF Lab.

POSTECH

# Capture Efficiency

1. general shape of the η-curve
   - close to the theoretical or exponential curve up to very high $(Q_L/Q_g)L$ ~ 25
   - not flattened at high η conditions, unlike in most existing systems
2. very high efficiency of 95% attained at 25(mol/mol)m condition
   - mass transfer coefficient ($h_mA$) ; twice as high as in other spray towers
   - similar between NH3 and MEA ; 25% smaller $Q_L$ with MEA

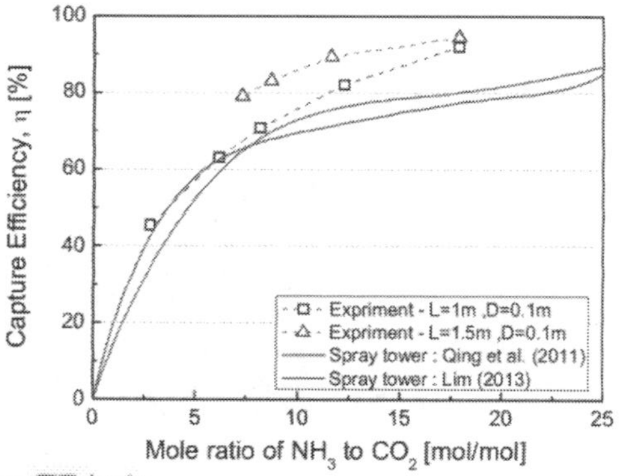


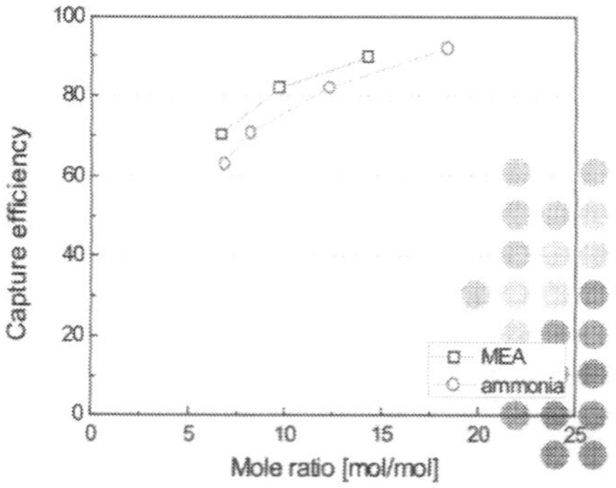


EnvTF Lab.

POSTECH

## Comparison with Numerical Simulation

- numerical results → universal efficiency formula (formula not shown here)
- excellent agreement

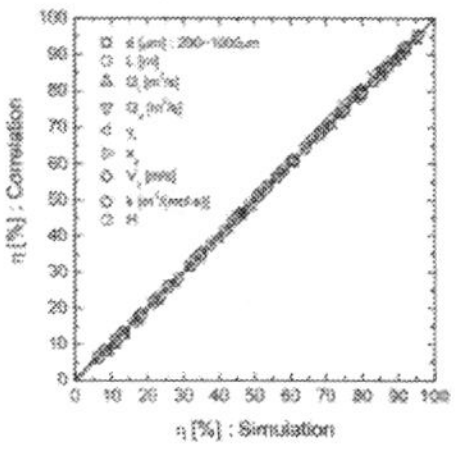


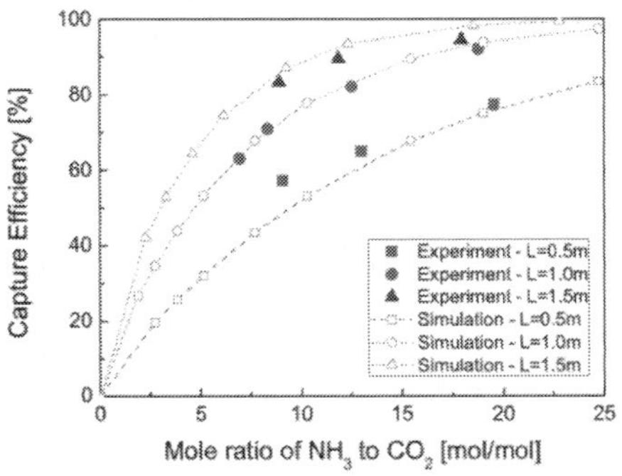


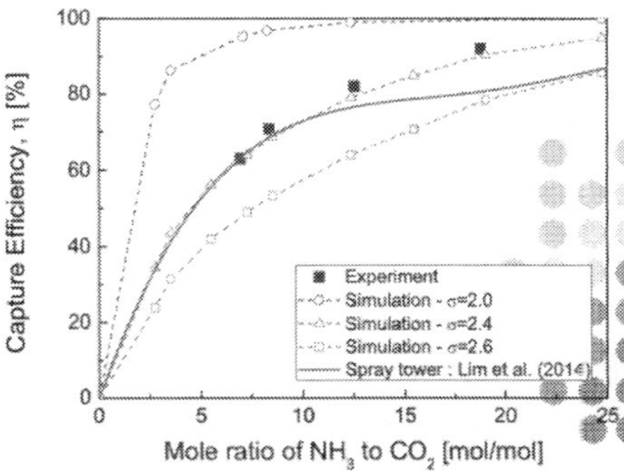


EnvTF Lab.

POSTECH

## Effect of Tower Length

1. potential adverse secondary effects
   - coagulation + wall loss + crossing
2. universal η-curve
   - $\eta = \eta[(Q_L/Q_g)L_{eff}]$ ; $L_{eff} = L_O(L/L_O)^a$
   - experiment ; $\alpha$ = 0.8~0.9
   - simple theory ; $\alpha$ ~ 0.8

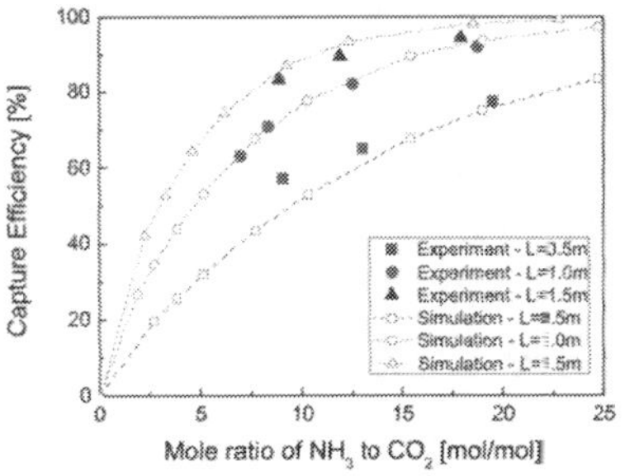


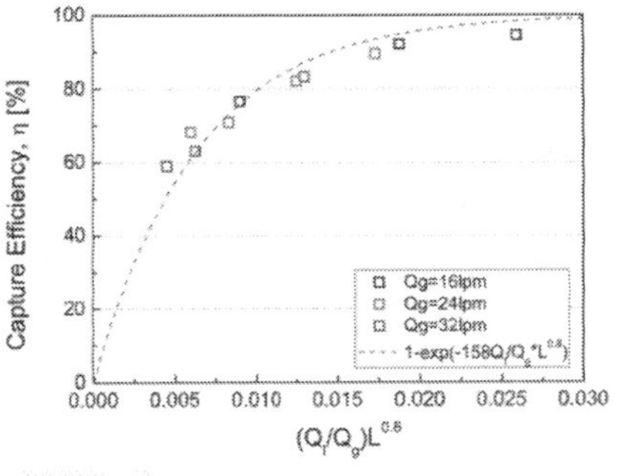


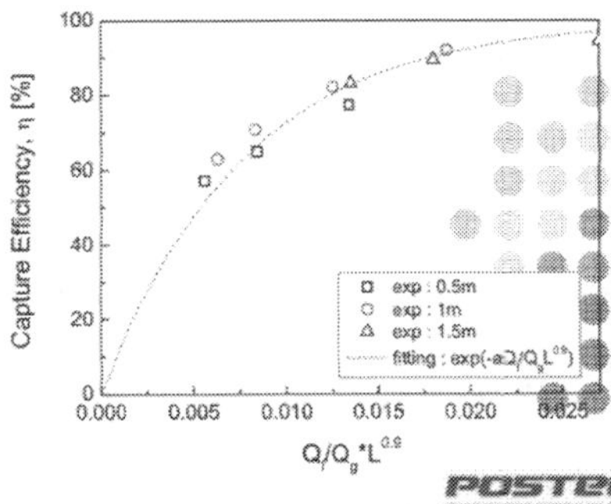


EnvTF Lab.

POSTECH

## Possibility for Further Enhancement

1. longer tower (L)
   - residence time ~ 1 sec → 1/10 ~ 1/5 of other lab-scale towers
   - longer towers provide longer contact time ; $h_m \sim (Q_L/Q_g) L_{eff}/V_f$

2. high gas velocity ($V_g$)
   - $V_f = V_t - V_g$
   - residence/contact time increased

3. drop-size optimized to $V_g$
   - falling velocity optimized independently of $V_g$ → additional d.o.f.
   - $V_t \sim d_P^{1\sim1.5}$

4. optimized nozzle array
   - non-uniform $V_g$
   - gas exit manifold

EnvTF Lab.

POSTECH

## Prediction for Large-Scale Application

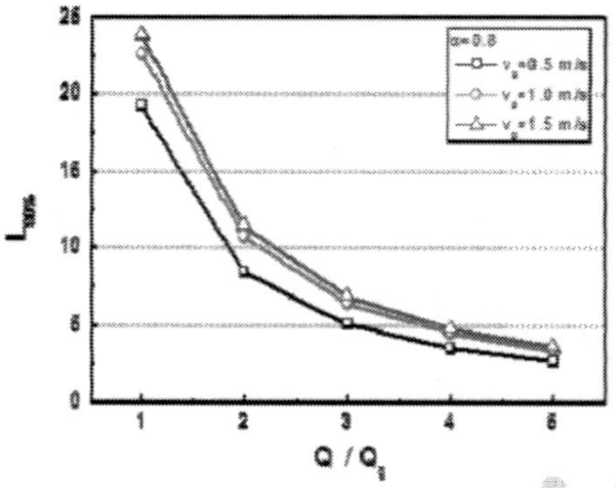


1. demo-scale pilot plant
   - $V_g$ = 0.5~1.5 m/s
   - α = 0.8

2. sample solution for NH3
   1) drop-size optimized to $V_g$
   2) $L_{90}$ ~ 10m with $Q_L/Q_g$ = 2 liter/m³
      → much shorter than any others
   3) $L_{90}$ ~ 20m with $Q_L/Q_g$ = 1 liter/m³
   4) very short tower is sufficient for η = 90% at any gas velocity
      - irrespective of solvent type and gas velocity
      - further reduction in solvent flow possible in taller towers
   5) MEA/DEA ; 25% less absorbent or absorber length

EnvTF Lab.

POSTECH

## Additional Advantages

1. design flexibility – single unit
   - shape of the cross-section flexible
   - (actively) controlled non-uniform size/position distribution possible
   - no viscosity-effect → multi-step liquid injection not needed

2. design flexibility – multi units of small cross-section
   - simultaneous multi-functions ; gas, absorbent, conditions
   - multi-step, recycling, variation of gas load, alternate maintenance

3. overall facility and operation
   - simple structure → low cost for construction and maintenance

4. applicable to regenerators
   - lower $Q_L$ ← (effective mean internal dispersion) → lower $T_r$

EnvTF Lab.

POSTECH

## Summary and Conclusions

1. Basic concept of the ideal spray tower developed
   - mono-disperse droplets + vertical injection
   - scale-up easy and straight-forward with little side effect

2. Capture performance verified ; lab-scale
   - performance curve close to that of an ideal reactor
   - best-ever performance of 95% observed with NH3
   - performance enhanced irrespective of the type of solvent

3. Feasibility of application to full-scale confirmed
   - very short tower is sufficient for $\eta$ = 90% for any solvent gas velocity
     $L_{90}$ ~ 10m with $Q_L/Q_g$ = 2 liter/m$^3$ ; $L_{90}$ ~ 20m with $Q_L/Q_g$ = 1 liter/m$^3$
   - applicable to regenerators with similar benefit
   - substantial (< 1/2) cost reduction expected relative to packed towers thru much shorter tower and no packing material

EnvTF Lab.

POSTECH

## For More Information or Discussion

POSTECH


**Jin W. LEE, Ph.D.**

Professor
Department of Mechanical Engineering

**POHANG UNIVERSITY OF SCIENCE AND TECHNOLOGY**
77 Cheongam-Ro, Nam-Gu, Pohang, Gyeongbuk, 37673, Korea
TEL. +82-54-279-2170 FAX. +82-54-279-3199 CP. +82-10-3234-2170
Email: jwlee @postech.ac.kr


EnvTF Lab.

POSTECH

UNIVERSITY OF
MICHIGAN

# Mobile Carbon Dioxide Removal: A Baseline Cost Estimate and Testing Protocol

Christina Reynolds, Francisco Sotomayor, & Christian Lastoskie

Department of Civil & Environmental Engineering, University of Michigan
Carbon Management Technology Conference, 20 July 2017


## Acknowledgements

DOW®

UNITED STATES ENVIRONMENTAL PROTECTION AGENCY

RACKHAM GRADUATE SCHOOL
UNIVERSITY OF MICHIGAN

# Vehicles are a significant source of emissions

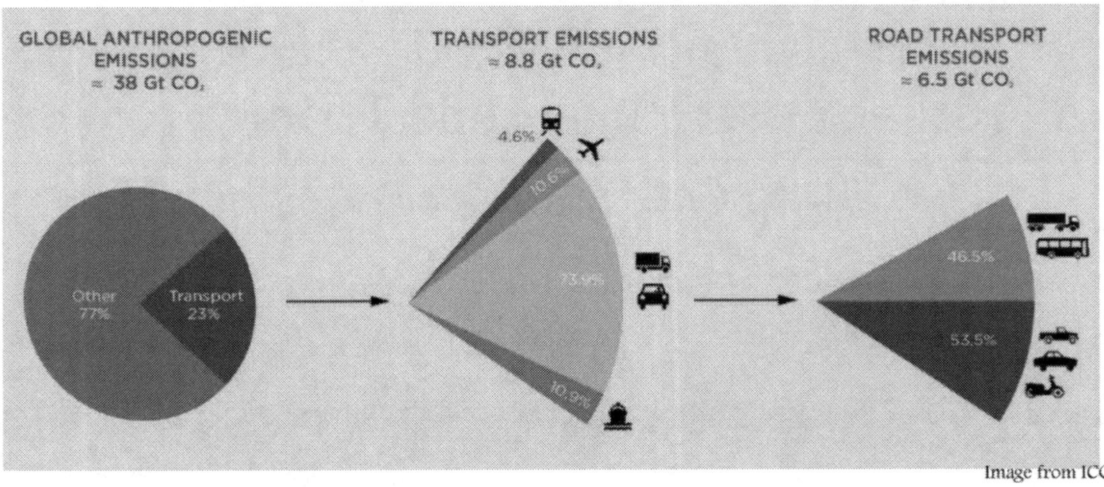


Image from ICCT

# Transportation is the highest emitting sector through 2040

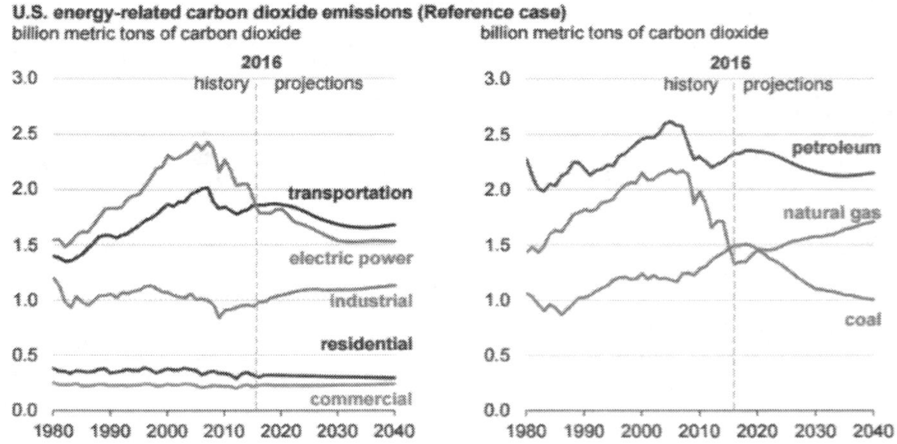


Image from US EIA

# Carbon-fueled vehicles retain the highest market share through 2040

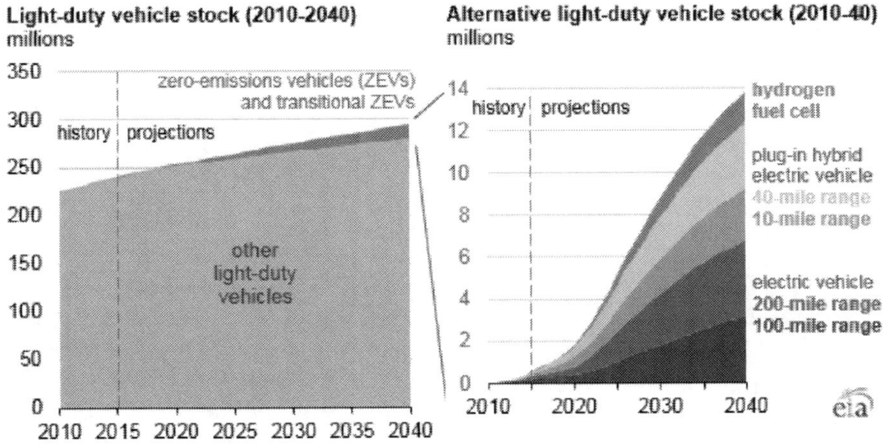


Climate stabilization targets cannot be achieved without carbon dioxide removal from mobile emission sources!

But how do we accomplish this?

# Approaches to carbon capture from stationary sources

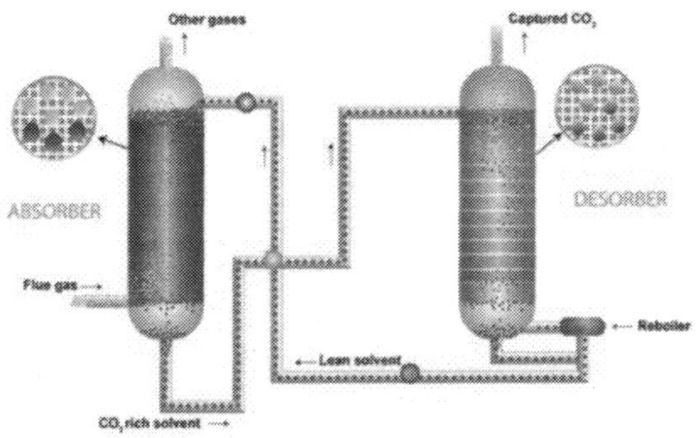


Amine Solvents

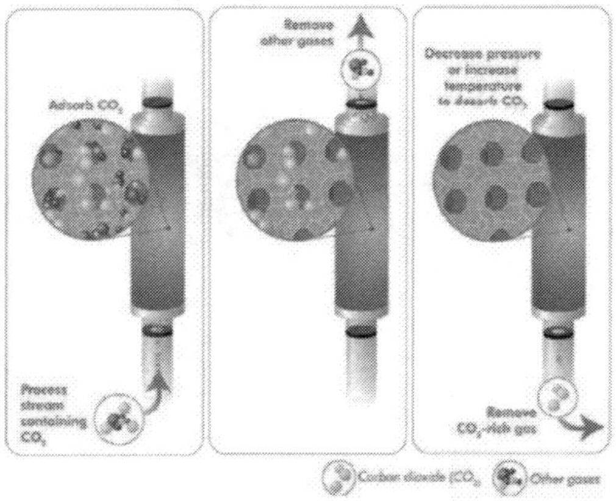


Conventional or Novel Adsorbents (zeolites, activated carbons, MOFs)

Image from CO2CRC

# Direct Air Capture is expensive

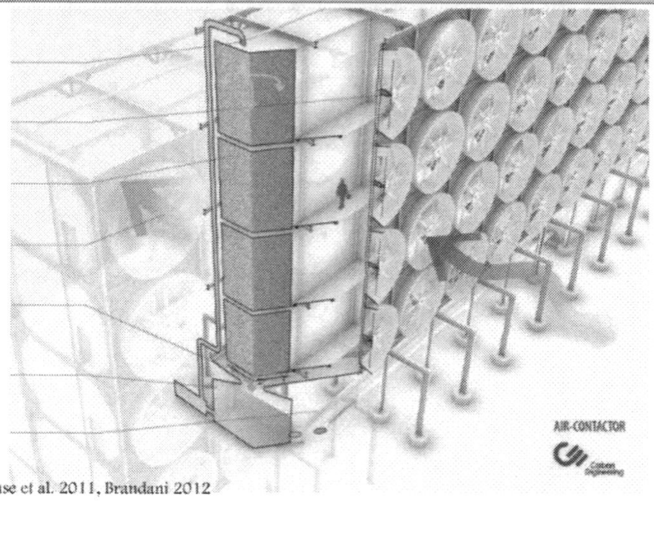


Image from Carbon Engineering; References: Socolow et al. 2011, House et al. 2011, Brandani 2012

## Direct Air Capture is expensive

**Expected cost:**

> $600/tCO_2$

- **10 times** the cost of capture from power plants

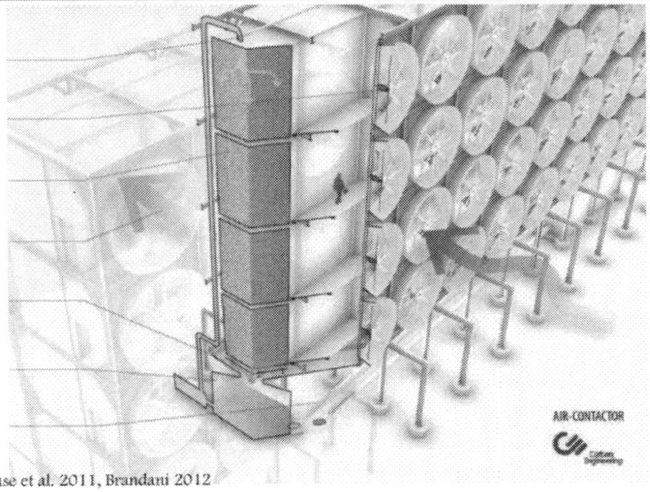


Image from Carbon Engineering; References: Socolow et al. 2011, House et al. 2011, Brandani 2012

## Minimum work to separate increases as $CO_2$ concentration decreases

$$w_{min,100\%} = -\frac{RT}{y_A M_A}\left[y_A \ln(y_A) + (1 - y_A)\ln(1 - y_A)\right]$$

# Minimum work to separate increases as $CO_2$ concentration decreases

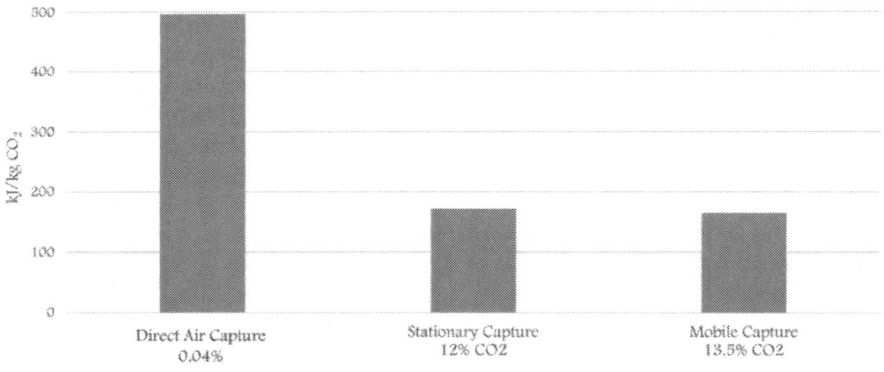


$$w_{min,100\%} = -\frac{RT}{y_A M_A}\left[y_A \ln(y_A) + (1 - y_A)\ln(1 - y_A)\right]$$

So let's consider a mobile carbon capture system modeled along the lines of a stationary capture system…

# Carbon Dioxide Removal from Mobile Emission Sources

# Carbon Dioxide Removal from Mobile Emission Sources

To capture the $CO_2$ emissions from 300 miles of driving in a light-duty vehicle, we need 650 kg of adsorbent with 20 wt% $CO_2$ capacity.

Mass requirements can be significantly decreased by changing several key conditions.

# How do we offload $CO_2$ once it's captured?

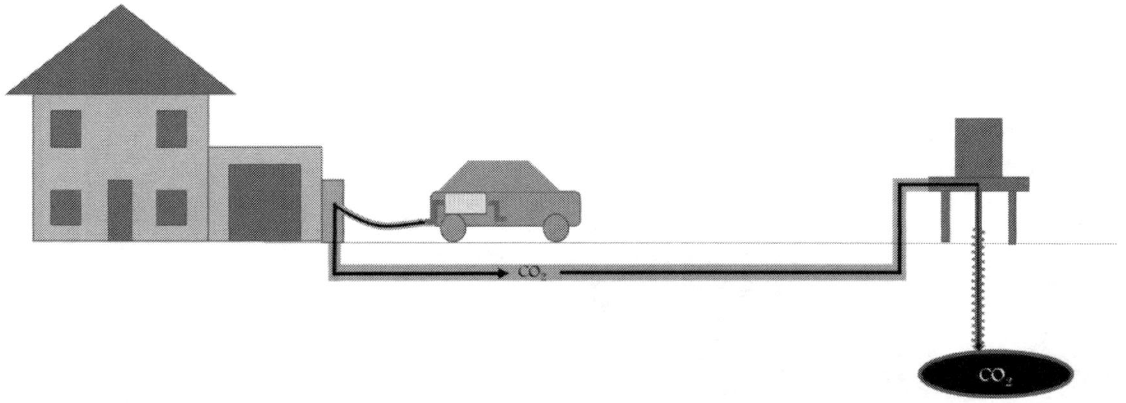


# How do we offload $CO_2$ once it's captured?

Must decide: (1) where and when offloading occurs
(2) what energy source is used

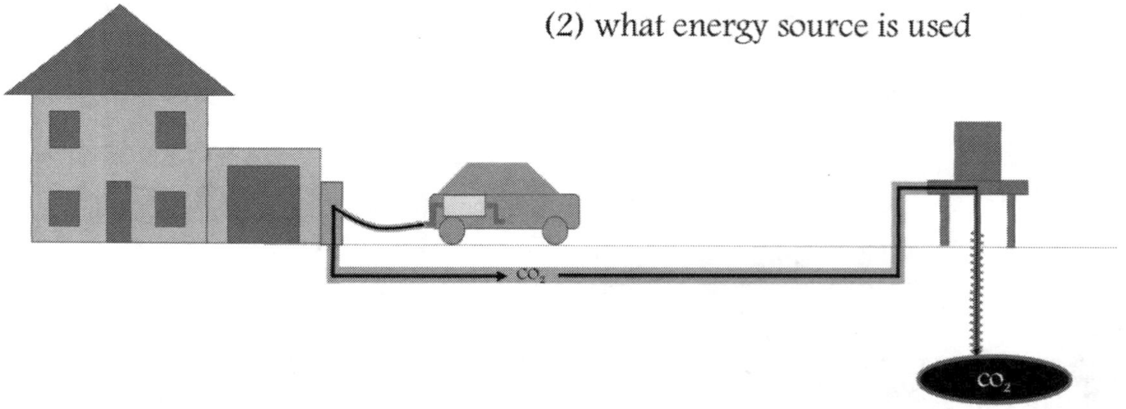

## Where and when to offload $CO_2$?

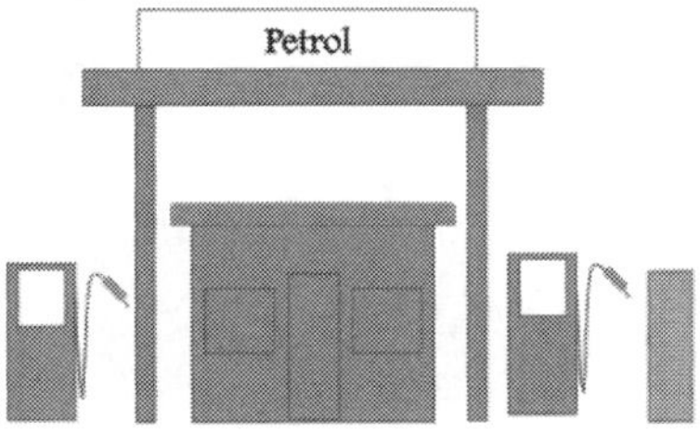


Regeneration at Gas Station
(travel distance = **300 mi**)

Regeneration at Home
(travel distance = **30 mi**)

## What energy source to use?

**Power Plant or Renewables**

$0.10/kWh

**Internal Combustion Engine**

$3/gallon

## Cost categories and relevant assumptions

Weight/fuel penalty

Separation and compression

Capital costs

Transportation and storage

## Cost categories and relevant assumptions

| | | |
|---|---|---|
| Weight/fuel penalty | target fuel economy | 45 mpg |
| | weight → miles per gallon | 7% ↓ in mpg per 10% ↑ in mass |
| Separation and compression | | |
| Capital costs | | |
| Transportation and storage | | |

# Cost categories and relevant assumptions

| | | |
|---|---|---|
| Weight/fuel penalty | target fuel economy | 45 mpg |
| | weight → miles per gallon | 7% ↓ in mpg per 10% ↑ in mass |
| Separation and compression | $\eta_{II}$ separation/compression | 0.40 / 0.85 |
| | carbon intensity of electricity | 0.5 kg/ kWh |
| Capital costs | | |
| Transportation and storage | | |

# Cost categories and relevant assumptions

| | | |
|---|---|---|
| Weight/fuel penalty | target fuel economy | 45 mpg |
| | weight → miles per gallon | 7% ↓ in mpg per 10% ↑ in mass |
| Separation and compression | $\eta_{II}$ separation/compression | 0.40 / 0.85 |
| | carbon intensity of electricity | 0.5 kg/ kWh |
| Capital costs | 100% | separation and compression costs |
| Transportation and storage | | |

# Cost categories and relevant assumptions

| | | |
|---|---|---|
| Weight/fuel penalty | target fuel economy | 45 mpg |
| | weight → miles per gallon | 7% ↓ in mpg per 10% ↑ in mass |
| Separation and compression | $\eta_{II}$ separation/compression | 0.40 / 0.85 |
| | carbon intensity of electricity | 0.5 kg/ kWh |
| Capital costs | 100% | separation and compression costs |
| Transportation and storage | pipeline transport | $2/tonne |
| | distance | 100 km |
| | $CO_2$ emissions | 0.005 kg/tonne-km |
| | storage | $13/tonne |

# Cost estimate for mobile capture scenarios

Offloading Frequency Energy Source

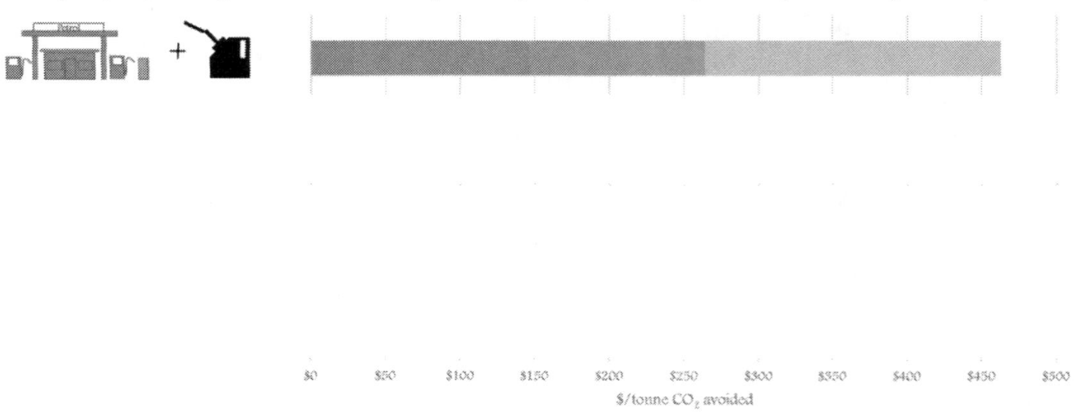

# Cost estimate for mobile capture scenarios

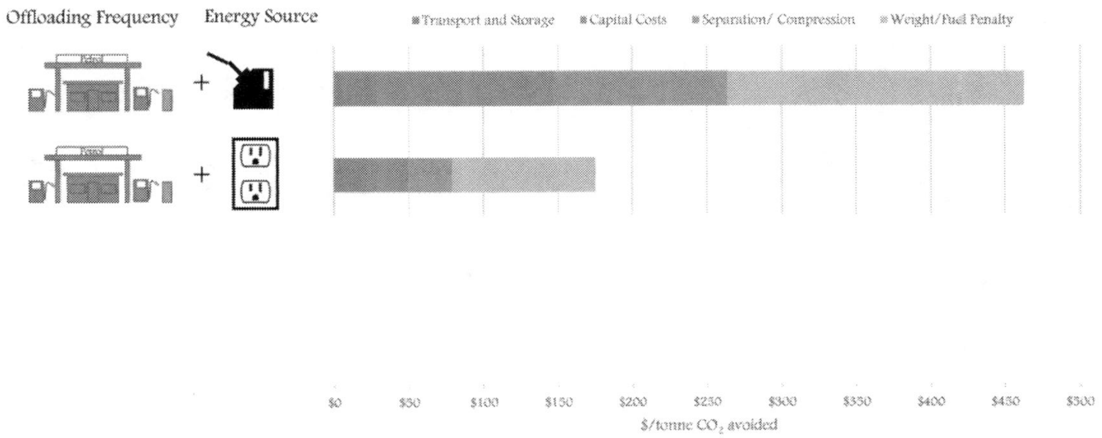


# Cost estimate for mobile capture scenarios

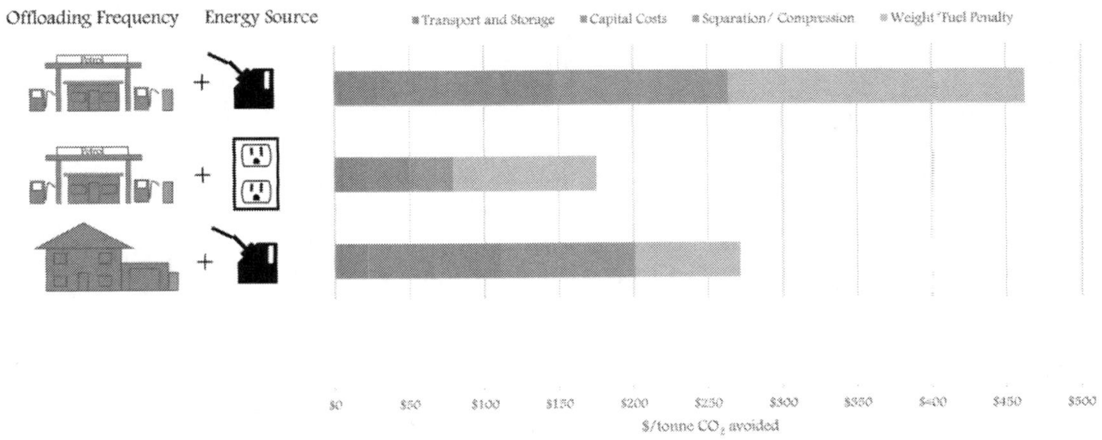

# Cost estimate for mobile capture scenarios

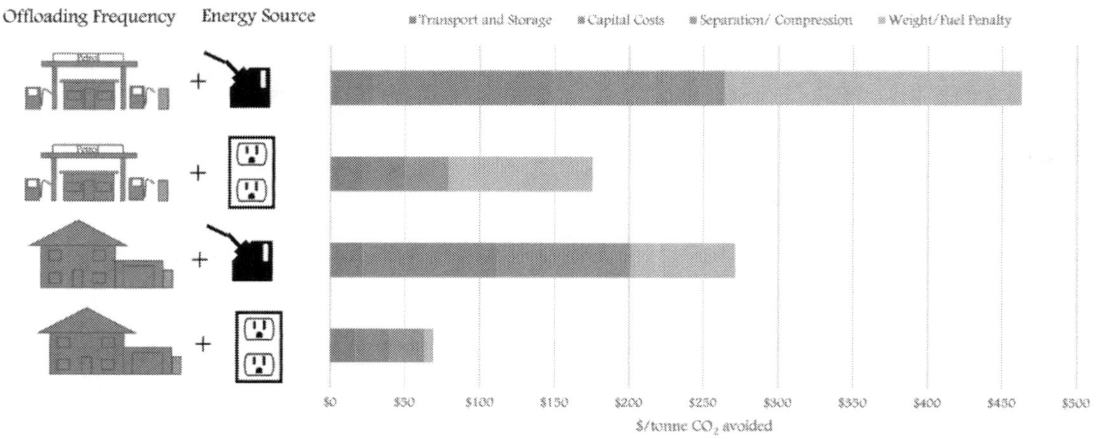


# What does this mean for you?

| | |
|---|---|
| Average car emissions: | 6 tons $CO_2$ per year |
| Price of $CO_2$ abatement: | $70/t |
| Annual cost to capture emissions: | |

# What does this mean for you?

| | |
|---|---|
| Average car emissions: | 6 tons $CO_2$ per year |
| Price of $CO_2$ abatement: | $70/t |
| Annual cost to capture emissions: | **$420** |

# Almost equal share of emissions from heavy duty and light duty vehicles

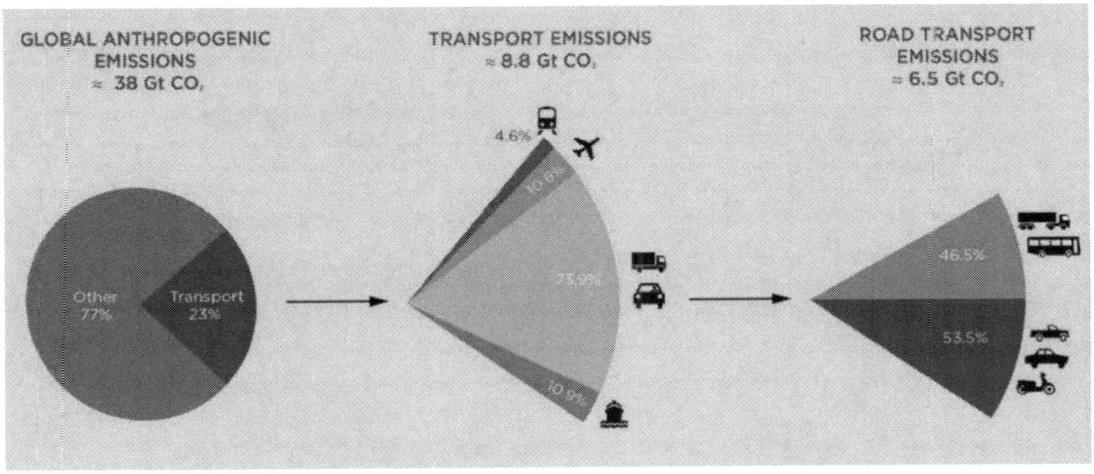

# Almost equal share of emissions from heavy duty and light duty vehicles

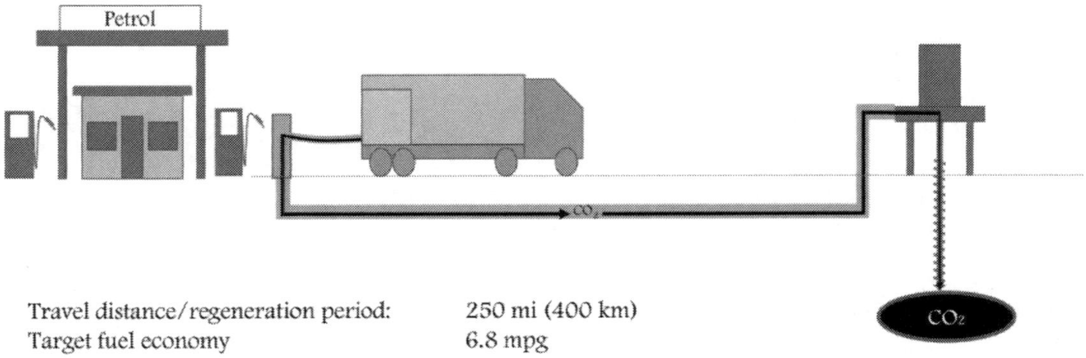


Travel distance/regeneration period: 250 mi (400 km)
Target fuel economy 6.8 mpg

# $CO_2$ capture from heavy duty vehicles is comparable with light duty best case scenario

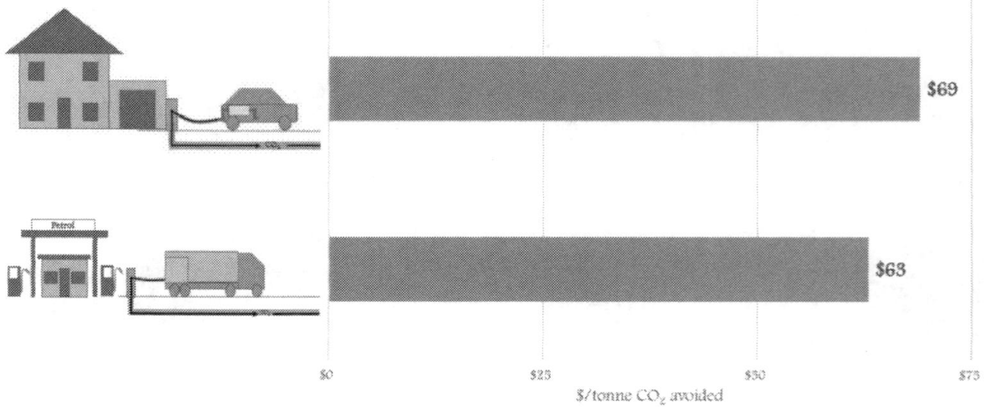

# What does this mean for you?

| | |
|---|---|
| Freight trucking emissions: | 530 M tonnes of $CO_2$ per year |
| Price of $CO_2$ abatement: | \$63/ton |
| Average ton-miles of US freight: | 2.6 B (or 8 ton-miles per person) |
| New annual shipping premium: | |

# What does this mean for you?

| | |
|---|---|
| Freight trucking emissions: | 530 M tonnes of $CO_2$ per year |
| Price of $CO_2$ abatement: | \$63/ton |
| Average ton-miles of US freight: | 2.6 B (or 8 ton-miles per person) |
| New annual shipping premium: | **\$12.60** |

# MCDR costs are on par with other carbon capture methods and significantly less than DAC

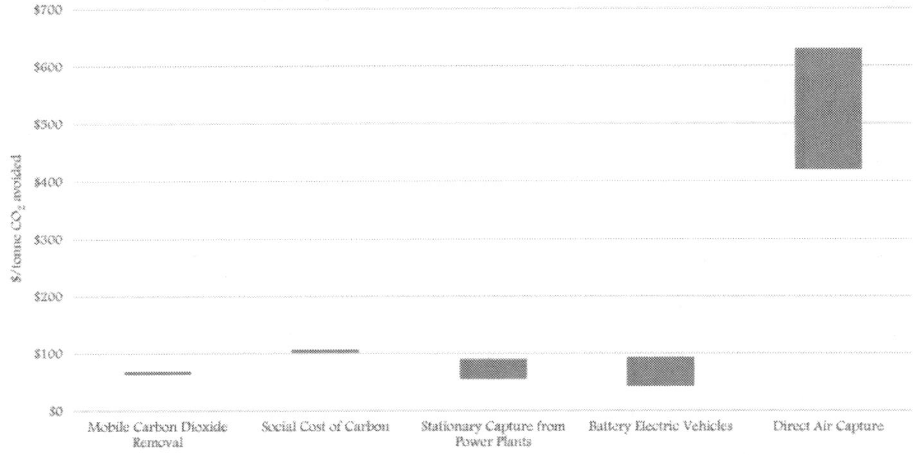


Values from McKinsey, Supekar, Ranjan, Lutsey, & EPA

# Testing protocol permits comprehensive evaluation of material performance

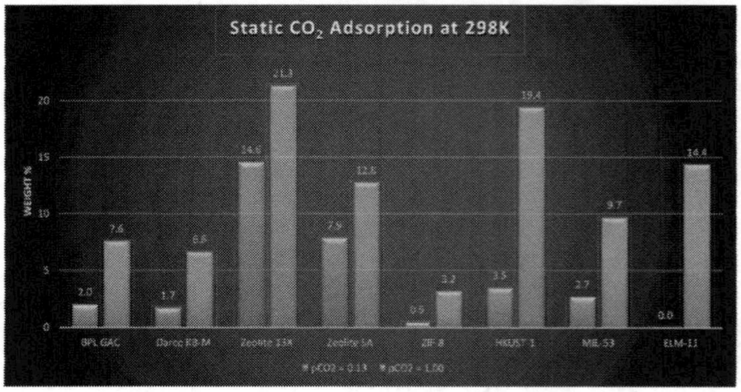


Static (@ 298-343K) →

# Testing protocol permits comprehensive evaluation of material performance

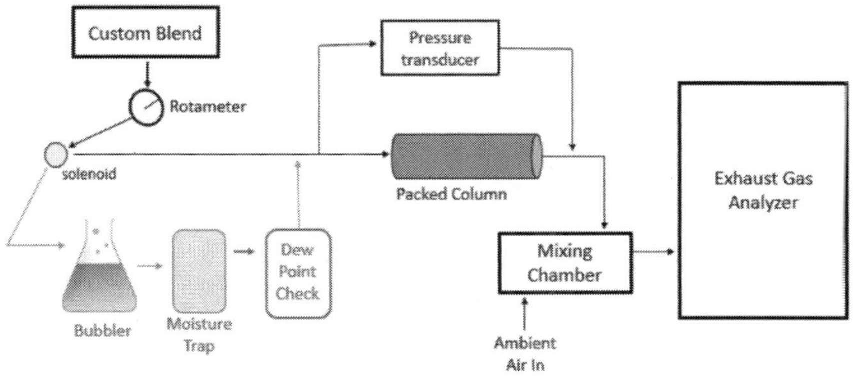


Static (@ 298-343K) → Dynamic, $N_2$-$CO_2$ blend → Dynamic, exhaust blend → Dynamic, wet exhaust blend –

# Testing protocol permits comprehensive evaluation of material performance

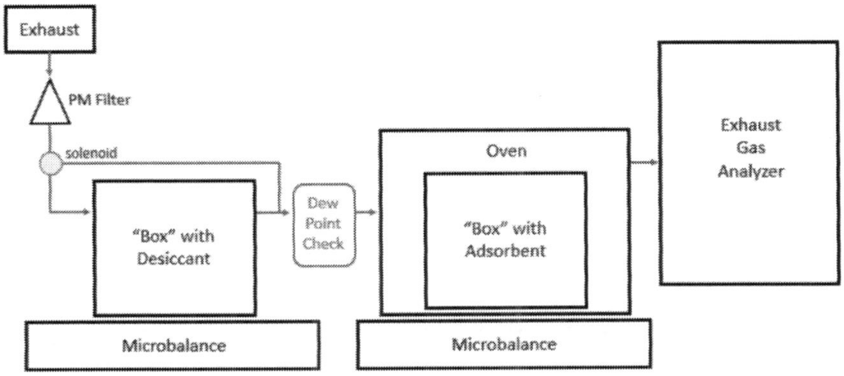


Static (@ 298-343K) → Dynamic, $N_2$-$CO_2$ blend → Dynamic, exhaust blend → Dynamic, wet exhaust blend → Proof-of-concept (1:4 scale) using actual exhaust

## Conclusions and next steps

- Mobile carbon dioxide removal is theoretically feasible and economical
  - *Especially when compared to direct air capture*

## Conclusions and next steps

- Mobile carbon dioxide removal is theoretically feasible and economical
  - *Especially when compared to direct air capture*
- It has the potential to reduce total US GHG emissions by ~20%

## Conclusions and next steps

- Mobile carbon dioxide removal is theoretically feasible and economical
  - *Especially when compared to direct air capture*
- It has the potential to reduce total US GHG emissions by ~20%
- Various adsorbents exist commercially that can capture $CO_2$ at scale

## Conclusions and next steps

- Mobile carbon dioxide removal is theoretically feasible and economical
  - *Especially when compared to direct air capture*
- It has the potential to reduce total US GHG emissions by ~20%
- Various adsorbents exist commercially that can capture $CO_2$ at scale
- Laboratory testing must include actual exhaust

## Conclusions and next steps

- Mobile carbon dioxide removal is theoretically feasible and economical
  - *Especially when compared to direct air capture*
- It has the potential to reduce total US GHG emissions by ~20%
- Various adsorbents exist commercially that can capture $CO_2$ at scale
- Laboratory testing must include actual exhaust
- There is much, much more to learn

## Conclusions and next steps

- Mobile carbon dioxide removal is theoretically feasible and economical
  - *Especially when compared to direct air capture*
- It has the potential to reduce total US GHG emissions by ~20%
- Various adsorbents exist commercially that can capture $CO_2$ at scale
- Laboratory testing must include actual exhaust
- There is much, much more to learn

# CMTC-484390-MS

# A Systematic Reservoir Simulation Study on Assessing the Feasibility of CO2 Sequestration in Shale Gas Reservoirs with Potential Enhanced Gas Recovery

Jie Zhan, University of Calgary; Qingwang Yuan, University of Regina; Allan Fogwill, Canadian Energy Research Institute; Hua Cai, CNOOC Ltd.-Shanghai; Hossein Hejazi, Zhangxin Chen, University of Calgary; Shiqing Cheng, China University of Petroleum-Beijing



This paper was prepared for presentation at the Carbon Management Technology Conference held in Houston, Texas, USA, 17-20 July 2017.



## Abstract

The application of horizontal well drilling coupled with the multistage fracturing technology enables commercial development of shale gas formations, which launches the energy revolution from conventional resources to unconventional resources. With the progress of understanding the nature of shale reservoirs, we find that some shale methane is stored as an adsorbed phase on surfaces of organic carbon. Meanwhile, laboratory and theoretical calculations indicate that organic-rich shale adsorbs CO2 preferentially over CH4. Shale gas reservoirs are recently becoming the promising underground target for CO2 sequestration. In the paper, systematic numerical simulations will be implemented to investigate the feasibility of CO2 sequestration in shale gas reservoirs and quantify the associated uncertainties.

First, a multi-continua porous medium model will be set up to present the matrix, nature fractures and hydraulic fractures in shale gas reservoirs. Based on this model, we will investigate a three-stage flow mechanism which includes convective gas flow mainly in fractures, dispersive gas transport in macro pores and multi-component sorption phenomenon in micro pores. To deal with this complicated three-stage flow mechanism simultaneously, analytical apparent permeability which includes slip flow and Knudsen diffusion will be incorporated into a commercial simulator CMG-GEM. A Langmuir isotherm model is used for CH4 and the multilayer sorption gas model, a BET model, is implemented for CO2. In addition, a mixing rule is introduced to deal with the CH4-CO2 competitive adsorption phenomenon.

In the paper, an integrated methodology is provided to investigate the CO2 sequestration process. Simulation results indicate that a shale gas reservoir is an ideal target for the CO2 sequestration. Even with the reservoir pressure maintenance due to the injection of CO2, the reservoir productivity is not enhanced. Hydraulic fracking which creates freeways for gas flow is the key to improve the reservoir performance. The multicomponent desorption/adsorption is a very important feature in a shale gas reservoir, which should be fully harnessed to benefit the CO2 sequestration process. In addition, we cannot ignore the contribution of slip flow and diffusion to the reservoir performance. Based on the methodology provided in this paper, we can easily deal with the apparent permeability effect using a commercial simulator platform.

## Introduction

As of August 2013, the Geological Survey of Canada estimated that Canada has approximately 4,995 trillion cubic feet (Tcf) of shale gas in place, a large portion of which is located in the Western Sedimentary Basin.[1] Shale gas resources can be found in British Columbia, Alberta, Saskatchewan, Manitoba, Ontario, Quebec, New Brunswick, Nova Scotia, the Northwest Territories, and Yukon.[2] However, gas-in-place estimates are available for certain provinces only, as shown in the following figure.

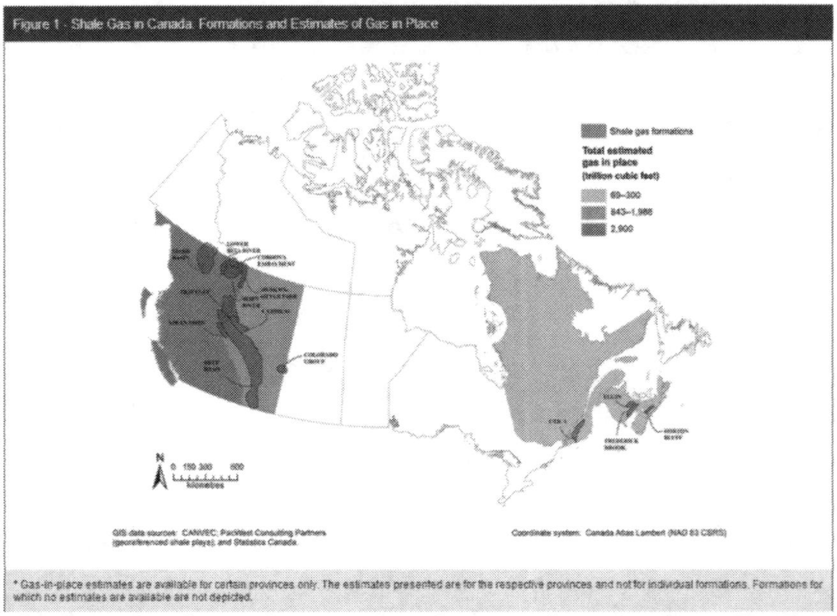


**Fig. 1. Shale gas in Canada: formations and estimates of gas in place (Parliament of Canada)**

The U.S. Energy Information Administration estimates that 573 Tcf of Canadian shale gas is technically recoverable, which represents nearly 8% of the global estimated total. According to this assessment, Canada has the world's fifth largest shale gas deposit, after China (1,115 Tcf), Argentina (802 Tcf), Algeria (707 Tcf) and the U.S. (665 Tcf).[3]

Composed principally of methane (CH4), natural gas produces far fewer potentially hazardous pollutants than coal when combusted. It also produces 43% fewer carbon dioxide ($CO_2$) emissions per unit of produced energy than coal. Used for electricity generation, a combined cycle gas turbine power plant can achieve typical efficiencies greater than 50%.[4] These environmental factors, combined with changes in natural gas markets, have led to the percentage of electricity supplied by natural gas nearly doubling over the last 20 years in the United States, increasing from 11.9% of electricity in 1989 to 23.2% in 2009.[5] Similar trends can be observed globally, where annual natural gas consumption is expected to increase from 108 Tcf in 2007 to 156 Tcf in 2035. Compared with its southern neighbor (U.S.), Canada's shale gas production is still in its early stages and production activities concentrate primarily in western Canada. For Canada, a rise in shale gas production at home and abroad could mean unprecedented economic opportunities and future prosperity.[6]

With the progress of understanding the nature of shale reservoirs, we find that some shale methane is stored as an adsorbed phase on the surfaces of organic carbon. Meanwhile, laboratory and theoretical calculations indicate that organic-rich shale adsorbs $CO_2$ preferentially over $CH_4$. With the help of $CO_2$ tax and the generation of an income stream through potential enhanced recovery of shale gas which recoups some of the cost of capturing and sequestering $CO_2$, shale gas reservoirs become the most economically promising underground target for $CO_2$ sequestration.[7,8] Moreover, with the high well density in shale gas fields, additional wells may not need to be drilled to implement enhanced recovery, further reducing the cost of sequestration. But there are many challenges to prove the viability of sequestration and enhanced recovery in gas shale because the performance of $CO_2$ injection in shale reservoirs is influenced by several engineering parameters.[9] Based on a sophisticated shale gas reservoir model which includes multi-continua, a three-stage flow mechanism and multi-component sorption, a systematic numerical simulation study will be implemented to assess the feasibility of the $CO_2$ sequestration process and quantify its uncertainties.

## Numerical simulation methods

We use CMG-GEM, the advanced general EOS compositional reservoir simulator, to model multiple hydraulic fractures and gas flow in shale reservoirs.[10] In the simulation model, we assumed that gas is flowing into a wellbore through hydraulic fractures with considering a non-Darcy effect. We implement the correlation proposed by Evan and Civan to determine the non-Darcy beta factor which is used in the Forchheimer number.[11,19,20] Local grid refinement with logarithmic spacing is employed to accurately simulate the detailed transient gas flow phenomenon around hydraulic fractures. A dual-permeability model is used to take natural fractures into consideration. The logarithmically spaced, locally refined, dual permeability (LS-LR-DK) methodology has been widely applied to model gas flow in hydraulically fractured shale gas reservoirs.[12, 13, 14] Grid blocks which include hydraulic fractures have a $7 \times 7 \times 1$ local grid refinement. The refined blocks along the center of a parent block have a width of $2\,ft$. The hydraulic fractures are represented by these center blocks. Since a larger width ($2\,ft$) is used for the hydraulic fracture, we need to convert the fracture conductivity in a $2\,ft$ wide grid block. The effective permeability is calculated based on the following equation [10]:

$$K_{eff} = \frac{K_f W_f}{W_{grid}}. \tag{1}$$

where $K_{eff}$ is the effective permeability and $W_{grid}$ is the width of an inner grid block. In the present case, $W_{grid}$ is 2 ft. $K_f$ is the intrinsic permeability of fractures and $W_f$ is the width of fractures.

Because of an increased cross-sectional area from using a larger block width, the actual velocity in the fractures is higher than the velocity in the blocks. Therefore, a non-Darcy correction coefficient is defined so that flow resistance arising from the non-Darcy effect is correctly captured: [10]

$$\beta_{corr} = \left(\frac{K_f}{K_{eff}}\right)^{2-N_{1g}} = \left(\frac{W_{grid}}{W_f}\right)^{2-N_{1g}}. \tag{2}$$

where $\beta_{corr}$ is the non-Darcy correction factor and $N_{1g}$ is the first coefficient in the Forchheimer equation for non-Darcy flow.

In addition, $CH_4$ adsorption on solid surfaces is taken into consideration in the model through implementing the classical Langmuir isotherm.[15] The Langmuir isotherm equation with two fitting parameters is as follows:

$$V(P) = \frac{V_L P}{P + P_L}, \tag{3}$$

where $V(P)$ is the gas volume of adsorption at pressure $P$, $V_L$ is the Langmuir volume, referred to the maximum adsorbed gas volume at the infinite pressure, and $P_L$ is the Langmuir pressure which represents the pressure corresponding to a one-half Langmuir volume.

The multilayer sorption gas model, [16] a general BET model, is implemented for $CO_2$. The general BET model with the four fitting parameters is as follows:

$$V(P) = \frac{V_m C \frac{P}{P_0}}{1 - \frac{P}{P_0}} \left[ \frac{1 - (N+1)\left(\frac{P}{P_0}\right)^N + N\left(\frac{P}{P_0}\right)^{N+1}}{1 + (C-1)\frac{P}{P_0} - C\left(\frac{P}{P_0}\right)^{N+1}} \right]$$

where $P_0$ is the saturation pressure of gas, $V_m$ is the maximum adsorption gas volume when the entire adsorbent surface is covered with a unimolecular layer, $C$ is a constant related to the net heat of adsorption, and $N$ is the maximum number of adsorption layers.

In the model, an extended Langmuir isotherm is implemented to model the competitive multicomponent adsorption and desorption process:[10]

$$wi = \frac{w_{i,max} B_i y_{ig} P}{1 + \sum_j B_j y_{jg}}$$

where $w_i$ is the moles of adsorbed component $i$ per unit mass or rock, $w_{i,max}$ is the maximum moles of adsorbed component $i$ per unit mass or rock, $Bi$ is the parameter for Langmuir isotherm relation, $P$ is the pressure, and $y_{ig}$ is the molar fraction of adsorbed component $i$ in the gas phase.

The transport mechanism of shale gas in matrix pores devidates from the conventional Darcy flow equation, as shown in the following Fig. 1. [17]

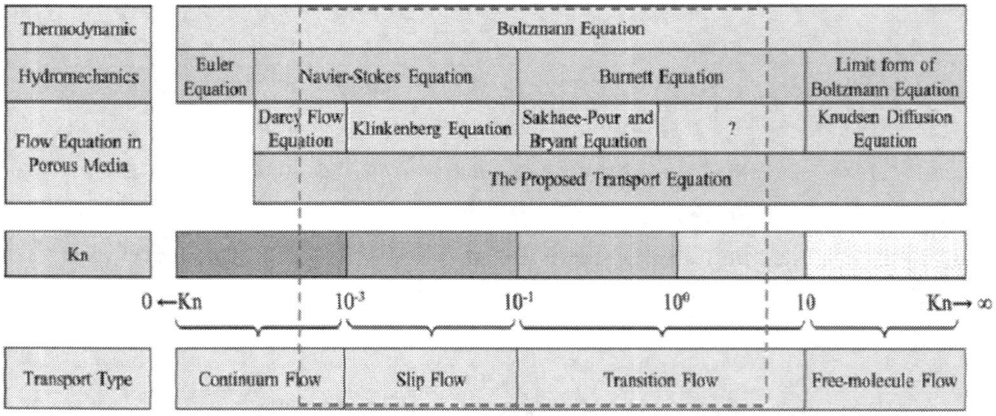


**Fig. 1. Gas flow regimes divided by Kn (Dr. J. Shi)**

Apparent permeability which combines slip flow and Knudsen diffusion will be incorperated into the shale gas reservoir model, [18]

$$k = \left[ \sqrt{\frac{8RT}{\pi M}} \left( \frac{2}{3RT\, p_{avg}} + \frac{\pi}{8 P_{avg}} \left( \frac{2}{\alpha} - 1 \right) \right) r\mu + \frac{r^2}{8} \right]$$

We assume that the average radius of a nano matrix pore is 2 nm. Based on the above equation, we can plot the relationship between the apparent permeability and the reservoir pressure. With a decrease in the reservoir pressure, the apprarent permeability will increase. That is also the reseason that we can see the more-than-expected gas production from a shale gas reservoir.

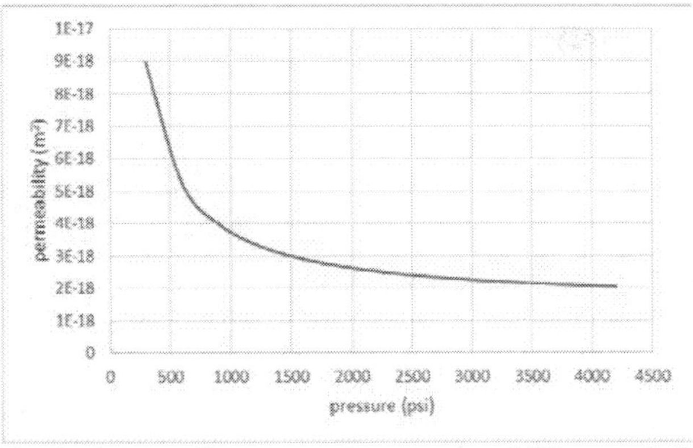


**Fig. 2. The relationship between the apparent permeability and the reservoir pressure**

Here is the ratio of the apparent permeability over the Darcy permeability, which will be integrated into the commercial simulator CMG-GEM to examine the effect of the transport mechanism on the $CO_2$ sequestration process.

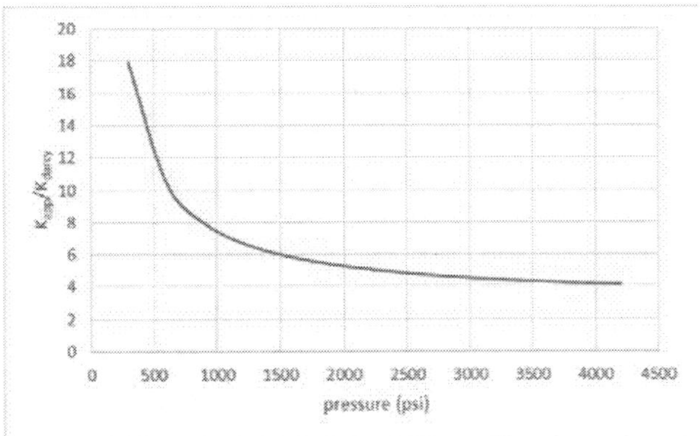


**Fig. 3. The ratio of apparent permeability over the darcy permeability versus the pressure**

## Reservoir model

A shale reservoir is simulated by setting up a homogeneous 3D reservoir model with dimensions of $5000ft \times 3000ft \times 300ft$, which correspond to the length, width and thickness, respectively (Fig. 4). A horizontal well with ten fracturing stages is simulated. The hydraulic fracture spacing and the half-length are 450 ft and 425 ft, respectively. The spacing between the two horizontal wells is 1000 ft. The detailed information of the reservoir and hydraulic fractures is listed in Table 1. The adsorption data and relative permeability data for each domain (matrix, hydraulic fractures and natural fractures) are listed in Table 2. The overall production period is 30 years. First of all, well-1 and well-2 are set to produce for five years. Then well-2 is converted to an injection well (well-3) with the injection rate of $1\times10^6$ $ft^3$/day. The injection time is 5 years.

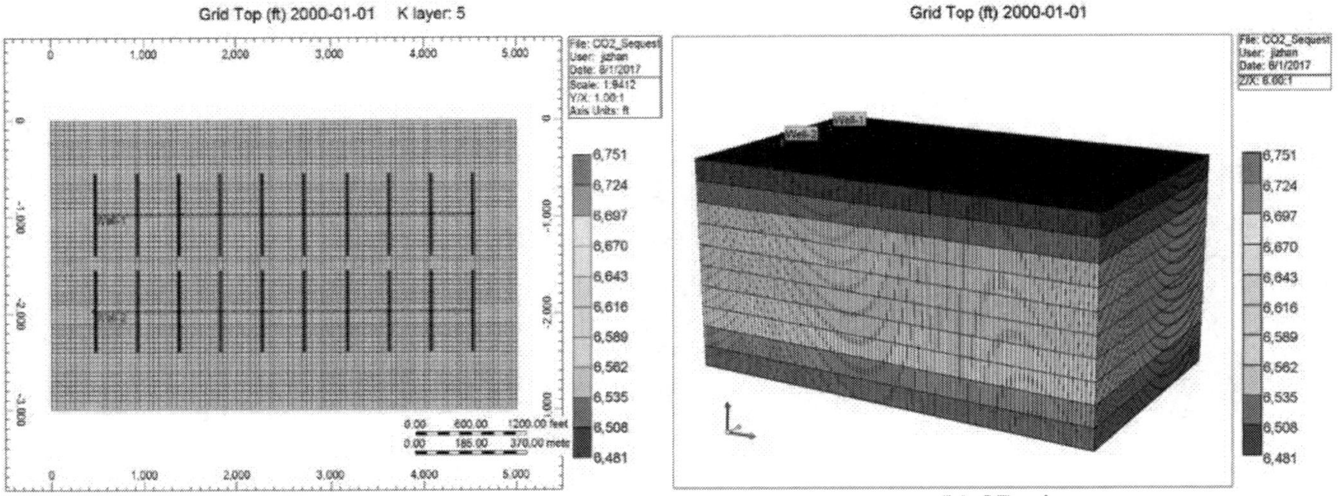


(a) 2D view (b) 3D view

**Fig. 4. Shale gas reservoir model**

**Table 1: Parameters used in the shale gas reservoir model**

| Parameter | Value(s) | Unit |
|---|---|---|
| The model dimensions | 5000×3000×300 | ft |
| Grid Top Depth | 6481 | ft |
| Pore pressure gradient | 0.54 | psi/ft |
| STG | 2000000 | $ft^3$/day |
| BHP | 300 | psi |
| Production time | 30 | year |
| Reservoir Temperature | 150 | F |
| Initial gas saturation | 0.8 | |
| Total compressibility | 1.E-06 | $psi^{-1}$ |
| Matrix permeability | 0.0005 | md |
| Matrix porosity | 0.06 | |
| Fracture permeability | 2.E-05 | md |
| Fracture porosity | 4.E-05 | |
| Fracture Conductivity | 10 | md*ft |

**Table 2: Adsorption data and relative permeability data for different continuum**

| | Matrix | HF | NF |
|---|---|---|---|
| **Adsorption Data** | 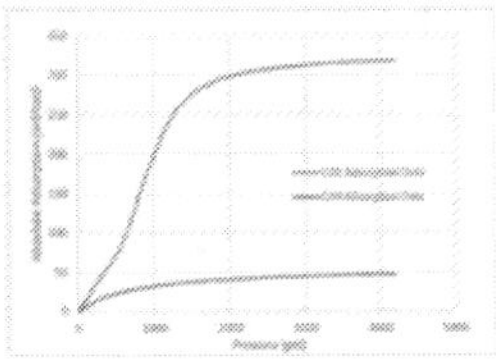 | 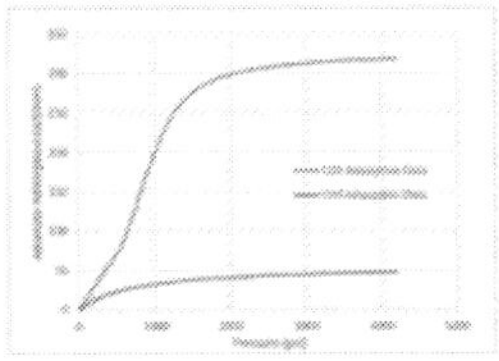 | **No Adsorption** |
| **Relative Perm** | 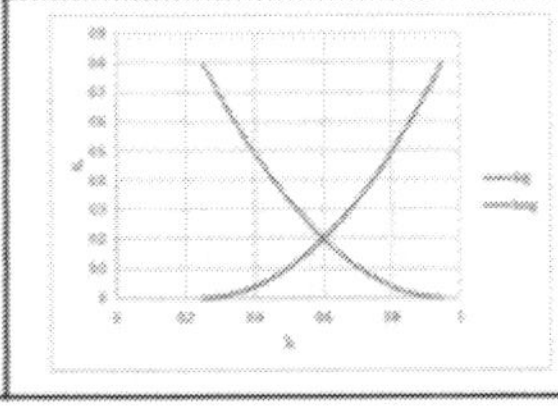 | 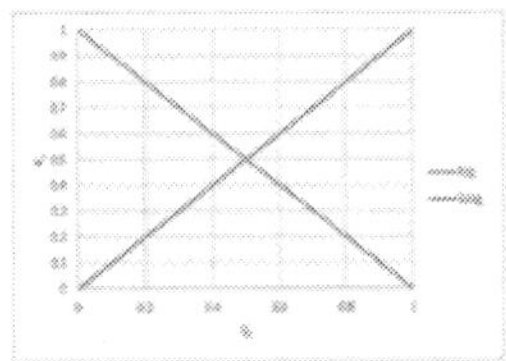 | 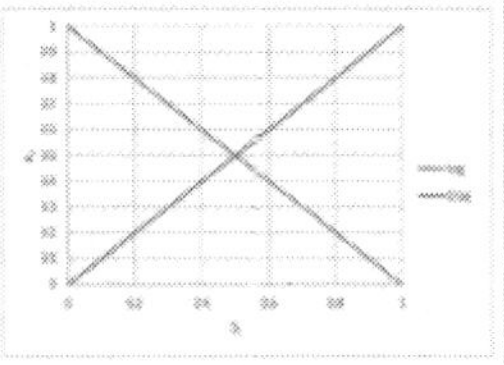 |

As shown in the following figures, 99% of the injected $CO_2$ has been sequestered successfully in the Barnett shale formation for one horizontal well in a 30-year period.

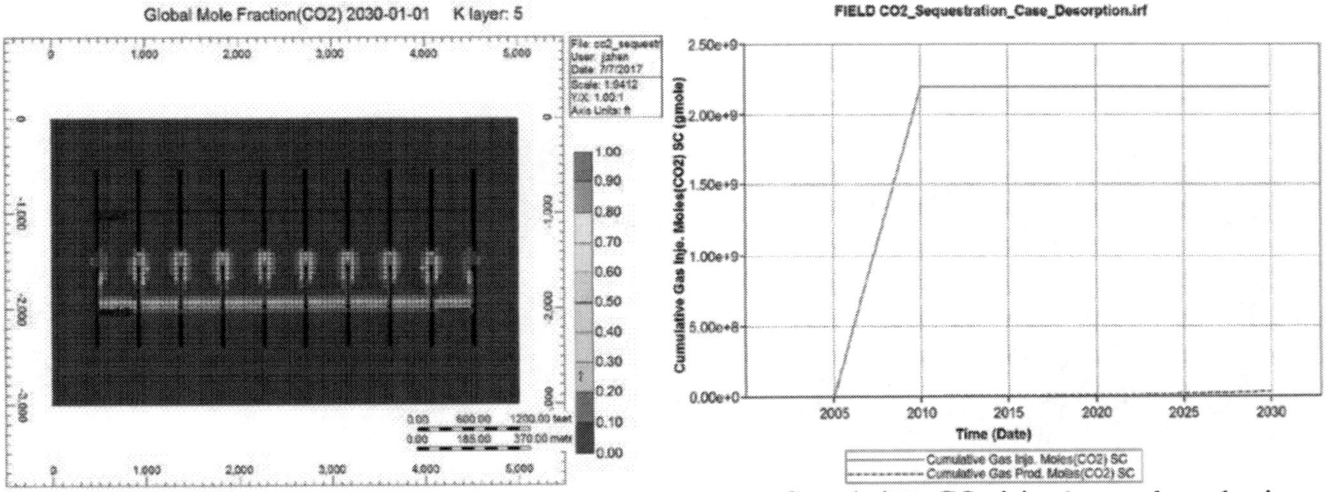


(a) $CO_2$ mole fraction distribution (b) Cumulative $CO_2$ injection and production

**Fig. 5. Simulation results on the CO2 sequestration for the Darcy permeability case**

Due to the $CO_2$ injection, the reservoir pressure is maintained. But the final $CH_4$ recovery is the same for the two cases because of the limitations of the tight formation. Meanwhile, we can observe that the adsorption/desorption process makes contribution to the reservoir pressure maintenance and reservoir productivity, which is a very important feature of a shale gas reservoir. A shale gas reservoir is an ideal target for the $CO_2$ sequestration. Even with the reservoir pressure maintenance due to the injection of the $CO_2$, the reservoir productivity is not enhanced. Hydraulic fracking which creates freeways for gas flow is the key to improve the reservoir performance.

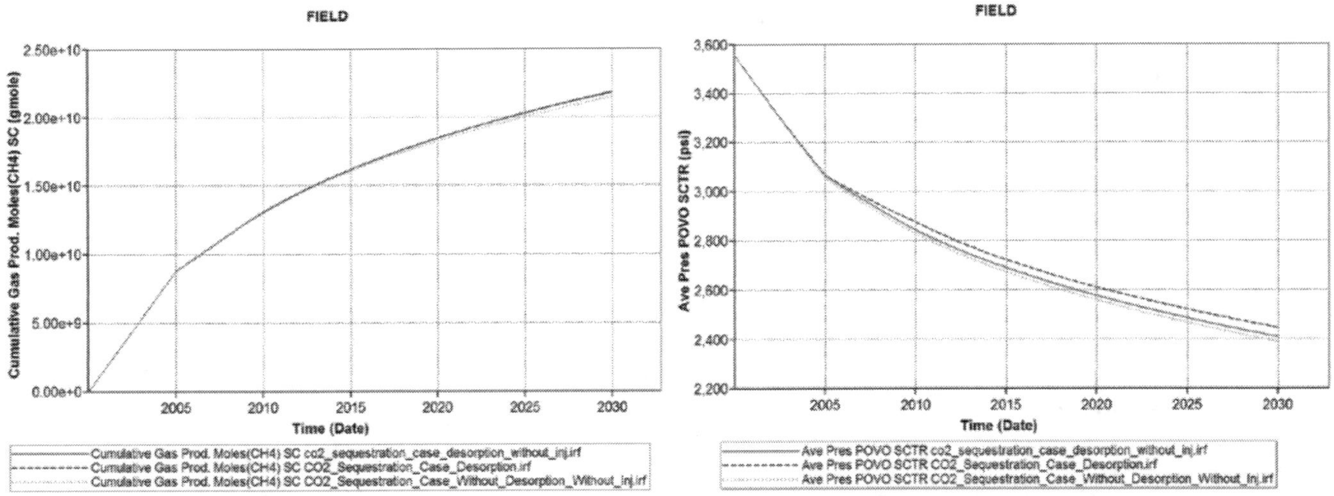


(a) Cumulative $CH_4$ production (b) Average reservoir pressure

**Fig. 6. Simulation results on reservoir performance**

Here, we will examine the effect of the three-stage flow mechanism via integrating the analytical apparent permeability model into the shale gas numerical model.

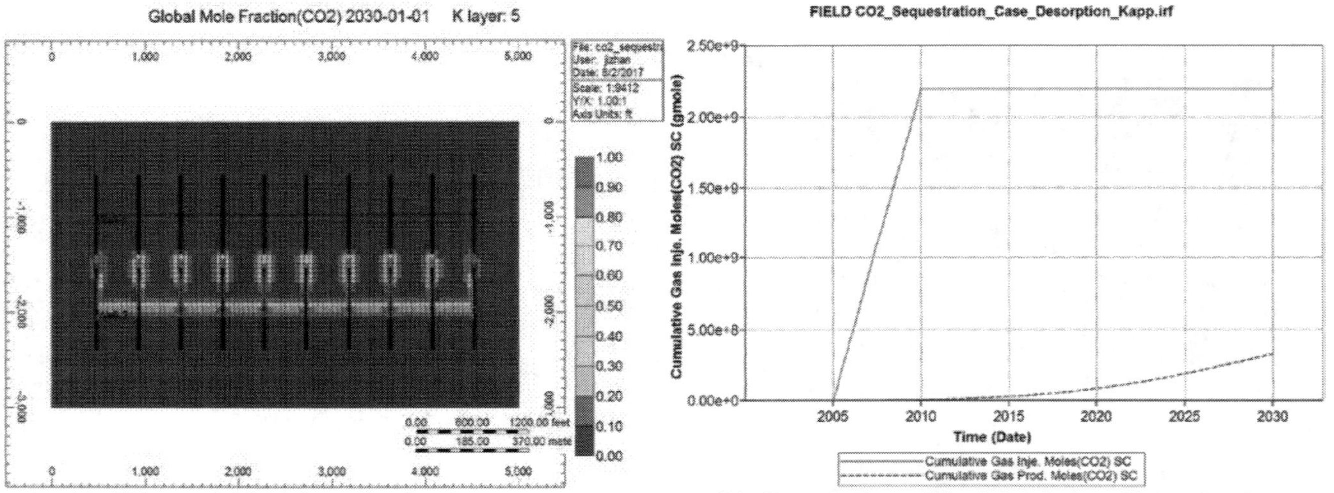


(a) $CO_2$ mole fraction distribution (b) Cumulative $CO_2$ injection and production

**Fig. 7. Simulation results on the CO2 sequestration for the apparent permeability case**

Just as in the following figure, we can observe that the apparent permeability is about 4 times bigger than the Darcy permeability (0.0005 md) during the 30-years production period. Due to the integration of the apparent permeability, the reservoir is more permeable. That is why more $CO_2$ will be produced from the reservoir. Moreover, the apparent permeability is the function of the reservoir pressure. The pressure heterogeneity will lead to the apparent permeability heterogeneity so that we can observe that the CO2 mole fraction distribution is not as uniform as in the Darcy permeability case.

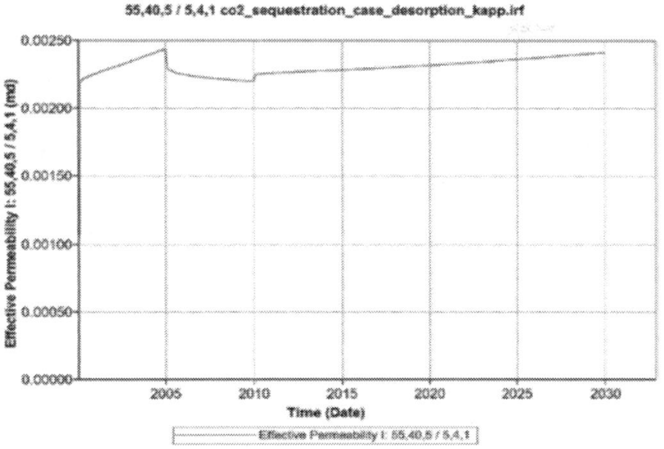


**Fig. 8. Apparent permeability evolution for specific reservoir location with $CO_2$ injection**

There is also a big difference on the average reservoir pressure and the $CH_4$ productivity. Due to the bigger apparent permeability, the pressure is easier to trasnsimit which leads to the bigger productivity. Based on the above analysis, the apparent permeability effect which includes slip flow and diffusion in addtion to the Darcy permeability is not negelible when we deal with a shale reservoir. The slip flow and diffusion make lots of contribution to the reservoir performace as we can observe from the cumulative $CH_4$ gas production.

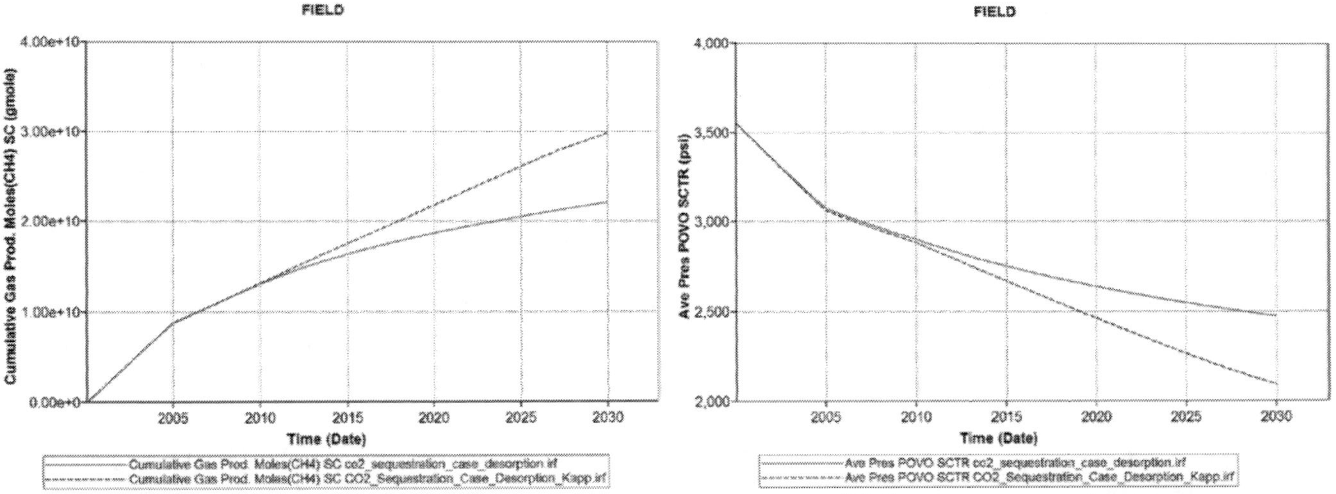


(a) Cumulative $CH_4$ production (b) Average reservoir pressure

**Fig. 9. Reservoir performance comparison between the apparent and darcy permeability case**

We can find one more interesting phenomenon with the integration of the apparent permeability model into the commerical simulator. Due to the injection process which leads to an increase in the reservoir pressure, the apparent permeability decreases with an increase in the reservoir pressure. That is why we can observe slightly bigger cumulative $CH_4$ production for the case without injection. It further validates the point that the permeable pathway for gas flow is the key to improve the reservoir performance at this point compared with the reservoir pressure maintenance.

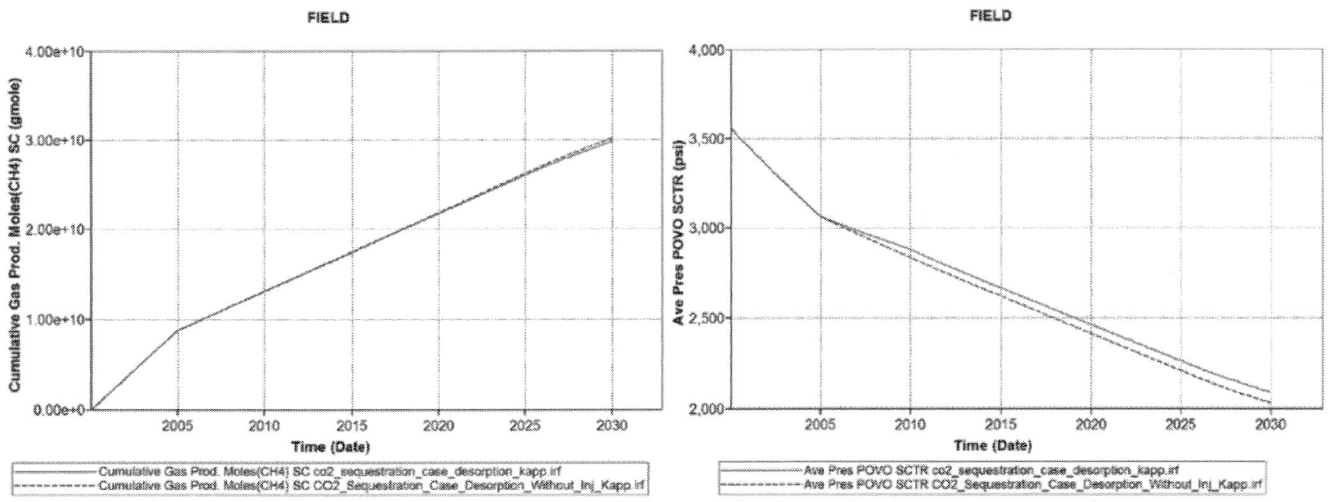


(a) Cumulative $CH_4$ production (b) Average reservoir pressure

**Fig. 10. Simulation results on reservoir performance for the apparent permeability case**

For the apparent permeability evolution, we can observe that there will be a five-year decrease for the case with injection. For the case without injection, after an immediate drop due to the well-2 shut-in, the apparent permeability will increase till the end of the simulation. That is why we can observe the above interesting phenomenon.

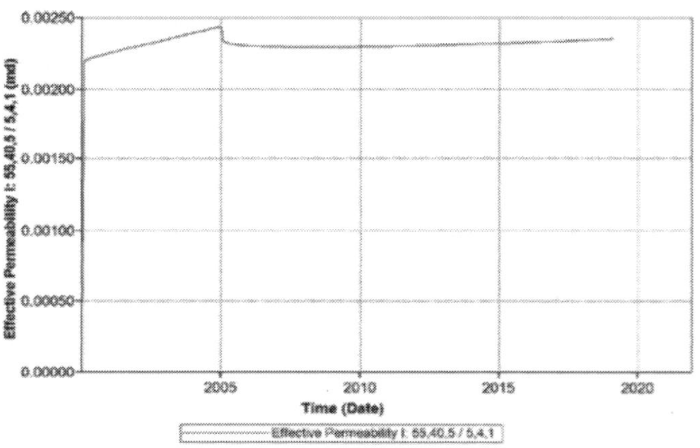


**Fig. 11. Apparent permeability evolution for specific reservoir location without $CO_2$ injection**

## Conclusion and Future Work

An integrated methodology is provided to investigate the $CO_2$ sequestration process. A shale gas reservoir is an ideal target for the $CO_2$ sequestration. Even with the reservoir pressure maintenance due to the injection of $CO_2$, the reservoir productivity is not enhanced. Hydraulic fracking which creates freeways for gas flow is the key to improve the reservoir performance. The multicomponent desorption/adsorption is a very important feature in a shale gas reservoir, which should be fully harnessed to benefit the $CO_2$ sequestration process. In addition, we cannot ignore the contribution of slip flow and diffusion to the reservoir performance. Based on the methodology provided in this paper, we can easily deal with the apparent permeability effect based on a commercial simulator platform. In the near future, we will examine the stress-sensitivity effect on natural fractures in shale reservoirs via implementing the elastic geomechanics model and the Barton-Bandis model. Reservoir heterogeneity will also be taken into account in the future study. Moreover, different well operation schemes will be tested based on the integrated shale gas reservoir model. Based on the improved reservoir understanding, some economic and environmental assessments associated with production efficiency and carbon sequestration will be conducted.

## Acknowledgment

The authors would like to acknowledge the support of Department of Chemical and Petroleum Engineering, University of Calgary and Canadian Energy Research Institute. The Reservoir Simulation Group is gratefully acknowledged. The research is partly supported by NSERC/AIEES/Foundation CMG, AITF iCore, IBM Thomas J. Watson Research Center, and the Frank and Sarah Meyer FCMG Collaboration Centre for Visualization and Simulation.

## Reference

1. Energy and Mines Ministers' Conference (2013).

2. Understanding Shale Gas in Canada. Canadian Society for Unconventional Gas.

3. U.S. Energy Information Administration (2013).

4. NETL, 2010, Fossil Energy Cost and Performance Baseline Studies.

5. EIA, 2010b, Annual Energy Review 2009.

6. U.S. Energy Information Administration, North America leads the world in production of shale gas, Today in Energy, 23 October 2013.

7. McKinsey, 2008, Carbon Capture & Storage: Assessing the Economics.

8. McKinsey, 2009, Pathways to a Low-Carbon Economy: Version 2 of the Global Greenhouse Gas Abatement Cost Curve.

9. Yu, W., Al-Shalabi, E. W., & Sepehrnoori, K. (2014, April 1). A Sensitivity Study of Potential $CO_2$ Injection for Enhanced Gas Recovery in Barnett Shale Reservoirs. Society of Petroleum Engineers. doi: 10.2118/169012-MS

10. CMG, 2015.GEM User's Guide. Computer Modeling Group Ltd.

11. Evan, R.D., and Civan, F.,1994. Characterization of non-Darcy multiphase flow in petroleum bearing formation. Final report, U.S. DOE Contract No. DE-AC22-90BC14659, School of Petroleum and Geological Engineering, University of Oklahoma.

12. Rubin, B. (2010, January 1). Accurate Simulation of Non Darcy Flow in Stimulated Fractured Shale Reservoirs. Society of Petroleum Engineers. doi:10.2118/132093-MS

13. Tavassoli, S., Yu, W., Javadpour, F., & Sepehrnoori, K. (2013, November 5). Selection of Candidate Horizontal Wells and Determination of the Optimal Time of Refracturing in Barnett Shale (Johnson County). Society of Petroleum Engineers. doi:10.2118/167137-MS

14. Liu F, Ellett K, Xiao Y, et al. Assessing the feasibility of CO 2 storage in the New Albany Shale (Devonian–Mississippian) with potential enhanced gas recovery using reservoir simulation[J]. International Journal of Greenhouse Gas Control, 2013, 17: 111-126.

15. Langmuir I. The adsorption of gases on plane surfaces of glass, mica and platinum [J]. Journal of the American Chemical society, 1918, 40(9): 1361-1403.

16. Brunauer S, Emmett P H, Teller E. Adsorption of gases in multimolecular layers [J]. Journal of the American chemical society, 1938, 60(2): 309-319.

17. Shi, J., Zhang, L., Li, Y., Yu, W., He, X., Liu, et al, (2013, November 5). Diffusion and Flow Mechanisms of Shale Gas through Matrix Pores and Gas Production Forecasting. Society of Petroleum

Engineers. doi:10.2118/167226-MS
18. Shabro, V., Torres-Verdin, C., & Javadpour, F. (2011, January 1). Numerical Simulation of Shale-Gas Production: From Pore-Scale Modeling of Slip-Flow, Knudsen Diffusion, and Langmuir Desorption to Reservoir Modeling of Compressible Fluid. Society of Petroleum Engineers. doi:10.2118/144355-MS.
19. Chen, Z., G. Huan & Y. Ma (2006). Computational Methods for Multiphase Flows in Porous Media. Computational Science and Engineering Series, Vol. 2, SIAM, Philadelphia.
20. Chen, Z. (2007). Reservoir Simulation: Mathematical Techniques in Oil Recovery. CBMS-NSF Regional Conference Series in Applied Mathematics, Vol. 77, SIAM, Philadelphia.

## Nomenclature

$K_{eff}$: Effective permeability, *mD*
$K_f$: Fracture permeability, *mD*
$W_f$: Fracture width, *ft*
$W_{grid}$: Block width, *ft*
$\beta_{corr}$: Non-Darcy correction factor
$N_{1g}$: First coefficient in the Forchheimer equation for non-darcy flow
$V(P)$: Gas volume of adsorption at pressure P, *scf/ton*
$V_L$: Langmuir volume, *scf/ton*
$P$: Pressure, *psi*
$P_L$: Langmuir pressure, *psi*
$V_m$ : Maximum gas adsorption volume, *scf/ton*
$C$ : Constant related to the net heat of adsorption
$N$ : Maximum number of adsorption layers
$W_i$ : Moles of adsorbed component I per unit mass or rock, *mole*
$W_{i,\,max}$ : Maximum moles of adsorbed component $i$ per unit mass or rock, *mole*
$B_i$ : Parameter for Langmuir isotherm relation
$y_{ig}$ : Molar fraction of adsorbed component $i$ in the gas phase
$r$ : Pore radius, *ft*
$R$ : Universal gas constant
$M$ : Molar mass, *lbm/mole*
$P_{avg}$ : Average pressure, *psi*
$\rho_{avg}$ : Average gas density, $lbm/ft^3$
$\mu$ : Viscosity, *cp*
$\alpha$: Tangential momentum accommodation coefficient
$T$ : Temperature, *F*

CARBON MANAGEMENT TECHNOLOGY CONFERENCE

# CMTC-484985-MS

# Synergetic $CO_2$ Huff-n-Puff for Edge-Water Fault-Block Reservoir: Laboratory Experiment and Numerical Simulation

Hongda Hao, Jirui Hou, Fenglan Zhao, Zhixing Wang, Zhongfeng Fu, Wengfeng Li, Peng Wang, Meng Zhang, Guoyong Lu and Jian Zhou, China University of Petroleum (Beijing)



This paper was prepared for presentation at the Carbon Management Technology Conference held in Houston, Texas, USA, 17-20 July 2017.



## Abstract

As an effective method for resource utilization, $CO_2$ huff-n-puff can be utilized to reduce $CO_2$ emissions and enhance oil recovery in edge-water flock-block reservoir, which was implemented in Jidong Oil Field, China since 2008 with oil production of $6.5\times10^4$ bbls by 2015. During operation period, synergetic effect was observed in adjacent wells with water cut drops and oil increments in a horizontal well group. Experimental and numerical simulations were conducted to investigate synergetic mechanisms of $CO_2$ huff-n-puff. 3D physical models with a horizontal well group and edge-water-driving system were established in laboratory to simulate the edge-water fault-block reservoir. The formation mechanisms and influence factors of synergetic $CO_2$ huff-n-puff were studied through laboratory experiments. Base reservoir model was also built to further discuss the synergetic types and injection allocations for $CO_2$ huff-n-puff in horizontal well group.

Synergetic $CO_2$ huff-n-puff is a smart gas cycling strategy for the horiontal well group to balance the formation pressure and replace the interwell oil. Experimental and numerical results showed that after $CO_2$ injected into low tectonic position of the reservoir, synergetic effect could be observed in high position well with water cut drops and oil increments. The mechanisms of synergetic effect can be recognized as formation energy supplement, gas sweeping, gravity segregation and $CO_2$-assisted edge-water driving. The stratigraphic dip and heterogeneity are advantages for the formation of synergetic effect. The synergetic types of $CO_2$ huff-n-puff can be summarized as single-well synergy and multi-well synergy. For single-well synergy, edge-water invasion can be effectively controlled by energy supplement after $CO_2$ injected into relatively low position well. For multi-well synergy, better synergetic effect and remaing oil replacement can be achieved after gas injected through different positions of the well group. The development efficiency of synergetic $CO_2$ huff-n-puff can be enlarged with 700t $CO_2$ injected into low position well + 100t $CO_2$ into high position well, and about 5767.9 bbls oil of the well group could be recovered with the soaking time of 50d.

**Key words:** $CO_2$ huff-n-puff; fault-block reservoir; horizontal well; edge water; synergetic effect

# 1 Introduction

As the highly water cut and less potential of enhancing oil recovery in conventional oil reservoirs, the complex oil reservoirs including fault-block reservoirs become the new exploratory development targets which are widely distributed both in China and abroad (Denham R., 1965; Ma C., et al, 1982; Nitzberg K. and Broman W., 1992; Wu Y., et al, 2012). Complex fault-block reservoirs are usually with the properties of broken structures, multiple reservoir types, complicated oil-water systems, etc., and are usually with lower producing levels compared with conventional oil reservoirs (Yu S., et al, 1995; Dai S., et al, 2008). In order to improve the productivity and enhance oil recovery of fault-block reservoirs, various EOR technologies have been operated in the oil field including gas injection, miscible and immiscible $CO_2$ flooding, steamflooding, and chemical combination flooding (Laidlaw W., et al, 1986; Spivak A., et al, 1990; Lim F., et al, 1993; Lu L. and Liu B., 1998; Zhu Y., et al, 2015).

Horizontal wells are usually applied in fault block reservoirs to enlarge the drainage area and improve the productivity. Horizontal wells can overcome the compartments, sealing faults and other lateral discontinuities, and obtain economic and sustained productions of complex reservoirs (Zhou H., et al, 2006; Naderi M., et al, 2015). However, if the reservoirs were adjacent to edge or bottom aquifers, the water invasion problem may occur seriously during the development period of oil field. Water coning and oil bypassing caused early water production and low productivity of crude oil with plenty of oil unrecovered in the edge-water or bottom water driving reservoirs (Hernandez J. and Wojtanowicz A., 2006; Han Z., et al, 2009). Different methods were proposed to restrain the water invasion including production rate control, production pattern management, barrier fluids injection and formation damage inducement (Karp J., et al, 1962; Seright R., et al, 2001; Zeinijahromi A., et al, 2015). Treating strategy should be designed according to actual geology and development of specific oil field to obtain preferable oil recovery in fault block reservoirs with edge-water or bottom-water acquifer.

$CO_2$ huff-n-puff is an effective method to enhance oil recovery in tertiary process. The mechanisms of $CO_2$ huff-n-puff to enhance oil recovery are oil viscosity reduction, oil swelling, well-bore damage removal, solution gas driving, vaporization of light components, water relative permeability reduction and interfacial tensions reduction (Mohammed-Singh L., et al, 2006). Towler B. and Wagle Y. (1992) pointed out that the relative-permeability hysteresis and reservoir pressure increase are the significant mechanisms for the success of $CO_2$ huff-n-puff process. The miscibility between oil and gas also has significant influence on enhancing oil recovery, and Torabi F., et al (2012) revealed that the oil recovery of near-miscible $CO_2$ huff-n-puff was greatly improved compared with immiscible $CO_2$ huff-n-puff. $CO_2$ huff-n-puff technology was firstly utilized to enhance the oil recovery of heavy oil reservoirs (Sayegh S. G., et al, 1984). Then, Monger T. and Coma J. (1988) proposed that $CO_2$ huff-n-puff process can be utilized to enhance oil recovery of light oil with larger slug volume, longer soak period, thicker interval and lower prior water cut. Recent years, the application of $CO_2$ huff-n-puff has been extensive from conventional oil reservoirs to unconventional oil reservoirs, from simple oil reservoirs to complex oil reservoirs (Sorensen J., et al, 2011; Vinassa M., et al, 2015; Ma J., et al, 2015; Sun J., et al, 2016). Moreover, novel applications of $CO_2$ huff-n-puff has been proposed to enhance the oil recovery. For example, Simpson M. (1988) proposed that the $CO_2$ huff-n-puff can be utilized to restrain the bottom water invasion of the oil reservoir, and the oil production of each well can increase to $32m^3/d$ after gas injection. The decreasing in fractional flow of water was caused by the oil swelling and viscosity reduction, which shows an enlightenment of water control with $CO_2$ huff-n-puff in the edge-water flock-block reservoir. Li Z., et al (2011) proposed $CO_2$ and viscosity breaker assisted steam huff and puff technology for horizontal wells was effective in reducing viscosity and improving production of super-heavy oil reservoirs with deep and thin layers, deep and heavy layers, shallow and thin layers. In-situ $CO_2$ huff-n-puff was designed by Wang Y., et al (2015) to enhance oil recovery of fault block reservoirs. Large amount of $CO_2$ was generated near the wellbore formation by adding reagents into the formation, which can decrease the reservoir temperature near the wellbore by 5°C and increase the oil production by 56%.

# 2 Background

One fault block reservoir with sufficient edge-water acquifer is located in Jidong Oil Field, China. The oil-bearing area of the block is 2.26 $ft^2$ with geological reserves of 328.6×$10^4$ bbls and recoverable reserves of 65.75×$10^4$ bbls. Stratigraphic dip of the block is 15-22°, the reservoir depth is 1732.1 m, the water/oil contact is 1735 m, and the reservoir thickness is 26.8 m. The formation temperature is 333.15 K (60℃) and initial formation pressure is 17.03 MPa. The average porosity of the reservoir is 25.9%, and the average permeability is 667×$10^{-3}$ $\mu m^2$. The density of crude oil is 0.87 g/$cm^3$ and the viscosity is 39.5 mPa·s under the formation temperature of 333.15 K.

A horizontal well group including 8 wells was located in the block, and the block was initially developed by edge-water driving. Although the oil production rate was high in the early stage of the development, it declined quickly with serious edge-water invasion. The water cut of several wells which were adjacent to the aquifer reached 99%. Because of the serious edge-water invasion, the oil recovery was less than 10% at the end of natural depletion period.

$CO_2$ huff-n-puff process was then conducted after natural depletion period with an initial intention to restrain the water invasion and improve the productivity of individual well. $CO_2$ was firstly injected into horizontal well of P-3, which was located in low tectonic position of the block and was adjacent to the edge-water aquifer. After 425 t of $CO_2$ injected into the formation, the water cut of P-3 decreased from 99% to 77.9%, and the oil was recovered by 428.2 bbls. More interestingly, a responding well of P-2 was observed with the response of water cut decreasing from 99% to 80.5% and oil increment by 267.1 bbls (As shown in Table 1). Since the responding well of P-2 was adjacent to P-3 and located in relatively high position of the block, synergetic effect of $CO_2$ huff-n-puff can be speculated based on the pilot test.

**Table 1 Result of $CO_2$ huff-n-puff in pilot test**

| Well | | Tectonic position | Water cut/% | | Incremental oil production/bbls |
|---|---|---|---|---|---|
| | | | Before $CO_2$ huff-n-puff | After $CO_2$ huff-n-puff | |
| Injection well | P-3 | Low | 99 | 77.9 | 428.2 |
| Responding well | P-2 | high | 99 | 20.5 | 267.1 |

In order to prove the existence of synergetic effect caused by $CO_2$ injection and study the mechanisms and influence factors of synergetic $CO_2$ huff-n-puff in horizontal well group, a large experimental device including 3D physical models, horizontal well simulation system, and edge-water-driving system was established in the laboratory to simulate the edge-water fault-block reservoir. A base reservoir model was also built to further discuss the synergetic types and injection allocations for $CO_2$ huff-n-puff in horizontal well group.

# 3 Experiments

## 3.1 Materials

The oil sample and water were collected from the block of Jidong Oil Field, China. The density of formation oil is 0.87 g/$cm^3$, the viscosity is 39.5 mPa·s and the gas/oil ratio is 42.42$m^3/^3$ under the formation temperature of 333.15 K (60C°). The composition of formation oil is detailed in Table 2. The minimum miscible pressure (MMP) measured by slim tube test is 24.39MPa, and the initial formation pressure is 17.03MPa. The salinity of the formation water is 937 mg/L. The injected $CO_2$ with purity of 99.99 mol% was from Beijing, China.

**Table 2 Compositions of the formation oil**

| Component | mol% | Component | mol% |
|---|---|---|---|
| $CO_2$ | 0.000 | $nC_4$ | 0.042 |
| $N_2$ | 1.859 | $iC_5$ | 0.026 |
| $C_1$ | 28.641 | $nC_5$ | 0.033 |
| $C_2$ | 1.080 | $C_6$ | 0.066 |
| $C_3$ | 0.189 | $C_{7+}$ | 67.997 |
| $iC_4$ | 0.067 | | |
| Total | 100 | | |

Homogeneous core with a permeability of $500\times10^{-3}$ $\mu m^2$, and heterogeneous cores with permeability of $1000/500\times10^{-3}$ $\mu m^2$ (The permeability contrast is 2) were utilized in the laboratory experiments. The diameter of 3D physical models is 400 mm and the thickness is 45 mm. The 3D artificial cores were fabricated by quartz sand, clay, resin, solidifying agent and ethyl with certain proportion. The physical parameters of the 3D models are tabulated in Table 3.

**Table 3 Physical parameters of 3D models for synergetic $CO_2$ huff-n-puff**

| Scenario. | Model | Permeability contrast | Permeability /$10^{-3}\mu m^2$ | stratigraphic dip | Apparent volume /mL | Pore volume /mL | Porosity /% | Initial oil saturation/% |
|---|---|---|---|---|---|---|---|---|
| 1 | Homogeneous core | 1 | 500 | 15° | 4652 | 792 | 17.02 | 59.32 |
| 2 | Heterogeneous core | 2 | 1000/500 | 0° | 4753 | 839 | 17.65 | 62.69 |
| 3 | | | | 15° | 4505 | 810 | 17.98 | 64.57 |

Fig.1 is the scheme of 3D physical model. In order to simulate horizontal well group of the model, Well-2, Well-3 and Well-4 were pre-located in the middle of the core during the process of fabricating as shown in Fig.1(a). The horizontal wells were parallel with each other as shown in Fig.1(b), which was similar to the well pattern of actual oil field. The horizon section of Well-2 and Well-3 were 80 mm, and the horizon section of Well-4 was 160 mm, which were isometic with the horizon sections of the wells in the oil field. Well-2, Well-3 and Well-4 were perforated to obtain preferable ability of the whole horizon sections.

Well-5 was a vertical well with a constant water injection pressure to simulate the sufficient edge-water supplement, and Well-1 was a vertical well located in the middle of the model to monitor the formation pressure.

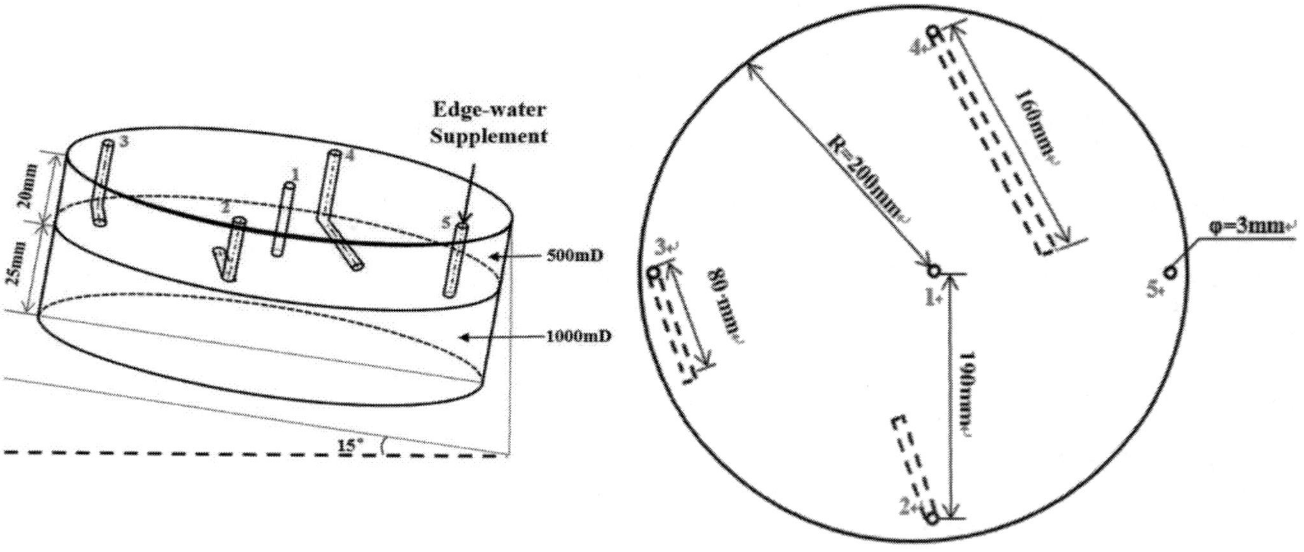


**(a) Side view** **(b) Top view**

**Fig.1. Schematic of 3D physical model**

Fig.2 is the picture of 3D physical model with horizontal well group pre-located in the model. Fig.3 is the 3D coreholder (Jiangsu, China) with maximum operation pressure of 15 MPa, and the coreholder can be rotated from 0°to 90°. After the core model was prepared, it was placed in the coreholder to form different stratigraphic dips of the formation. With the existence of stratigraphic dips, three horizontal wells can be located in different tectonic positions: Well-4 located in low tectonic position, Well-2 located in middle tectonic position, and Well-3 located in high tectonic position (As shown in Fig.1(a)). With the horizontal wells located in different tectonic positions of the model, synergetic effect of $CO_2$ huff-n-puff in edge-water flock-block reservoir can be studied in the laboratory.

**Fig.2. Picture of 3D physical model**

**Fig.3 Coreholder of 3D physical model**

## 3.2 Experimental setup

The experimental setup consisted of six sub-systems: injection system, edge-water injection system, displacement system, production system, temperature control system and data acquisition system (As shown in Fig.4). In the injection system, the formation water, formation oil and $CO_2$ were stored in transfer cylinders and injected into the 3D physical model by a constant pressure and rate pump (HAS-100HSB, Jiangsu, China). In the edge-water injection system, formation water was stored in the transfer cylinder and injected into the 3D physical model through Well-5 by another constant pressure and rate pump. The 3D physical model was placed in the coreholder (As shown in Fig.3), and water was used to provide confining pressure to the coreholder which was usually 2 MPa higher than the injection pressure. In the production system, backpressure regulars (BPRs) were connected with the horizontal wells to control the production pressure. The produced oil and water was recorded by test tubes, while the gas was measured by a gas flow meter (CS200, Beijing, China). The thermostant was used to set the experimental temperature as the reservoir temperature. The formation pressure monitored by Well-1 and the displacement pressure of edge-water were recorded by the data acquisition system.

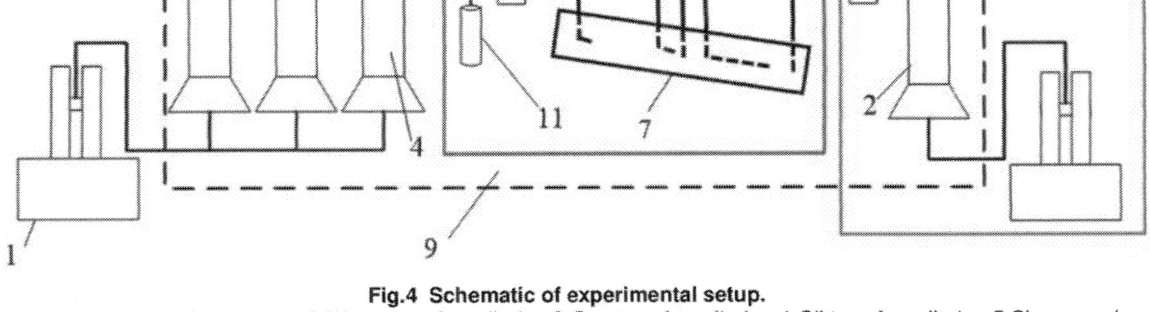


**Fig.4 Schematic of experimental setup.**

1-Constant pressure and rate pump; 2-Water transfer cylinder; 3-Gas transfer cylinder; 4-Oil transfer cylinder; 5-Six-way valve; 6-Backpressure regulator (BPR); 7-3D coreholder; 8-One-way valve; 9-Thermostat; 10-Edge-water injection system; 11-Test tubes; 12-Data acquisition system

### 3.3 Experimental procedure

3D displacement experiments were designed to study the synergetic $CO_2$ huff-n-puff with sufficient edge-water supplement in the laboratory. First, homogeneous and heterogenous models were utilized to investigate the factors of heterogeneity on the formation of synergetic effect, then the heterogenous models were placed with dips of 0° and 15° to discuss the influence of stratigraphic dip on the formation of synergetic $CO_2$ huff-n-puff. The experimental procedures are detailed as follows:

First, 3D physical model and horizontal wells were fabricated and pre-located according to the designing. Then, epoxy resins were coated on the surface of the core to avoid corruption of $CO_2$. The bulk volume of the core was measured before displacement experiment. The core was firstly evacuated and then saturated with formation water. The porosity was determined as the ratio of brine saturation volume to the bulk volume. Finally, the core was displaced by formation oil to reach the residual water saturation condition. The initial oil saturation was calculated as the ratio of injected oil volume to the pore volume.

The detailed sequence of the experiment is explained as follows. ① The 3D physical model was homogeneous with a stratigraphic dip of 15° in Scenario 1, the model was heterogeneous and horizontally arranged in Scenario 2, and the model was heterogeneous with a stratigraphic dip of 15° in Scenario 3. ② The formation temperature was set as 333.15 K (60℃). The backpressure of horizontal wells (Well-2, Well-3 and Well-4) was set as 10 MPa to maintain the formation pressure and the supercritical condition of $CO_2$. Edge-water system was connected with 3D model through Well-5 under constant injection pressure of 10.05 MPa to simulate the sufficient edge-water supplement of the flock-block reservoir. ③ Edge-water driving was conducted firstly in the 3D physical model. Well-2, Well-3 and Well-4 were opened when the edge water was injected into the core. The edge-water driving process was terminated when the water cut of Well-4 reached 98%. ④ $CO_2$ huff-n-puff process was conducted in the 3D model subsequently. First, $CO_2$ was injected into the core through Well-4 with an injection volume of 0.2PV (Subsurface). Then, Well-2, Well-3 and Well-4 were shut-in for 12 h to reach the dissolving equilibrium of $CO_2$ and formation oil. ⑤Well-2, Well-3 and Well-4 were opened simultaneously during the $CO_2$ puff process, and the experiment was terminated when the water cut of Well-4 reached 98% again. During the whole $CO_2$ huff-n-puff process, Well-5 was maintained open to provide sufficient edge-water supplement. Well-1 was used to monitor the change of formation pressure, and oil production, water production and gas production were measured during the experiments.

## 4. Results and discussion

### 4.1 Oil increment by synergetic $CO_2$ huff-n-puff

The experimental results of $CO_2$ huff-n-puff process conducted after natural depletion period are shown in Table 4. During the natural depletion period, the oil recovery of edge-water driving in Scenario 1, Scenario 2 and Scenario 3 were 21.33%, 12.01% and 15.39%, respectively, and the oil recovery of homogeneous model with a stratigraphic dip of 15° was the highest. Because of the better formation condition, larger swept area can be achieved with the gravitational differentiation between oil and edge-water. Similarly, the oil recovery of edge-water driving in Scenario 3 reached 15.39% with the gravitational differentiation. While, the oil recovery of edge-water driving in Scenario 2 was the lowest due to serious edge water invasion. For the production of individual well, the oil recovery of Well-4 was the highest, Well-2 was second, and Well-3 was the least. Because Well-4 was nearest to the aquifer, plenty of formation oil can be displaced to the wellbore of Well-4 by the edge-water. One the other hand, the water cut of Well-4 increasing sharply and burst water flooding firstly occurred in Well-4 which seriously affect the development of the block. When the water cut of Well-4 reached 98%, the water cut of Well-2 also reached 90-95%, and the water cut of Well-3 reached 80-85%.

Fig.5 is the incremental oil recovery of $CO_2$ huff-n-puff process after natural depletion period. After 0.2PV (Subsurface) of $CO_2$ was injected into the 3D model, 10~15% OOIP can be recovered with the horizontal well group in each scenario under the mechanisms of oil sweeping and viscosity reduction. For the individual well, 6.5-8% OOIP can be recovered through Well-4. And for all the scenarios, the

**Table 4 Experimental results of $CO_2$ huff-n-puff in 3D models**

| Scenario. | Model | | | | Horizontal well | | Oil recovery/% | | |
|---|---|---|---|---|---|---|---|---|---|
| | Heterogeneity | Permeability ratio | Permeability /$10^{-3}\mu m^2$ | stratigraphic dip | No. | Tectonic position | Edge-water driving | $CO_2$ huff-n-puff | Total |
| 1 | Homogeneous core | 1 | 500 | 15° | Well-4 | / | 9.15 | 6.69 | 15.84 |
| | | | | | Well-2 | / | 6.35 | 3.20 | 9.55 |
| | | | | | Well-3 | / | 5.83 | 2.73 | 8.56 |
| | | | | | Total | / | 21.33 | 12.62 | 33.95 |
| 2 | Heterogeneous core | 2 | 1000/500 | 0° | Well-4 | Low | 5.65 | 8.01 | 12.66 |
| | | | | | Well-2 | Middle | 3.30 | 2.20 | 5.50 |
| | | | | | Well-3 | High | 3.06 | 1.23 | 4.29 |
| | | | | | Total | / | 12.01 | 11.44 | 23.45 |
| 3 | | | | 15° | Well-4 | Low | 6.78 | 7.86 | 14.64 |
| | | | | | Well-2 | Middle | 4.43 | 4.42 | 8.85 |
| | | | | | Well-3 | High | 4.18 | 3.54 | 7.72 |
| | | | | | Total | / | 15.39 | 15.82 | 31.21 |

formation oil can also be produced through Well-2 and Well-3 unless the gas was only injected through Well-4. This result was consistent with the result of pilot test, which indicated that the synergetic effect indeed existed during $CO_2$ huff-n-puff in horizontal well group.

The incremental oil recovery of $CO_2$ huff-n-puff was 15.82% in the heterogeneous model with a stratigraphic dip of 15°, which achieved the best synergetic effect of $CO_2$ huff-n-puff. The incremental oil recovery of Well-2 and Well-3 in Scenario 3 was 4.42% and 3.54%, respectively, which was also the highest among the three laboratory experiments. After $CO_2$ was injected into the model through low position well, the remaining oil around middle and high position wells can be effectively displaced by the synergetic effect. And the heterogeneity and stratigraphic dip are advantage for synergetic effect of $CO_2$ huff-n-puff.

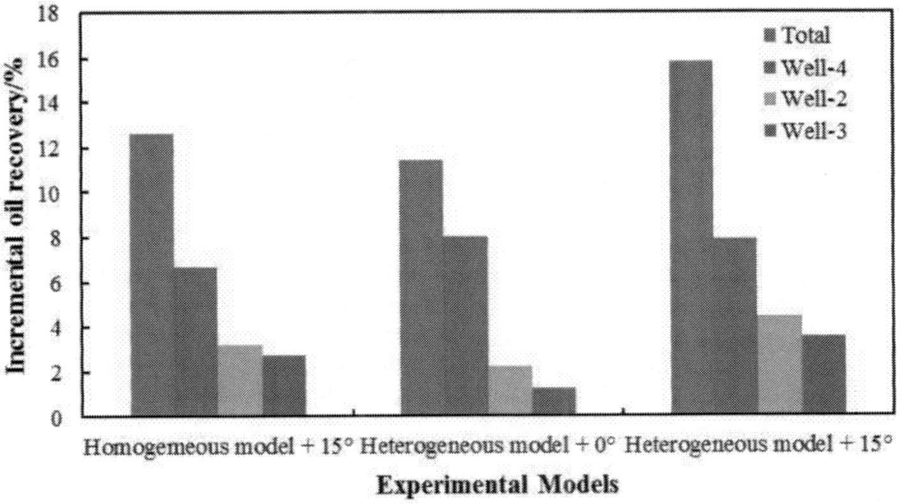


Fig.5 Incremental oil recovery of $CO_2$ huff-n-puff process

## 4.2 Synergetic Mechanisms by $CO_2$ huff-n-puff in horizontal well group

In order to discuss the synergetic mechanisms by $CO_2$ huff-n-puff in horizontal well group, the water cut and the producing gas rate were further analyzed as shown in Fig.6. For illustration purpose, Well-4 was defined as Operation well, and Well-2 and Well-3 were defined as Responding well. For operation well of Well-4 (As shown in Fig.6(a)), water cut dropped and plenty of gas was produced immediately when the well was opened during $CO_2$ puff process. The water cut of operation well can be dropped to 60-70%, which indicated that the edge-water was effectively controlled after $CO_2$ was injected into the model. The water control mechanisms can be explained as follows: (1) As $CO_2$ was injected into the model, the formation pressure increased from 10.65MPa to 13.75MPa. The formation energy supplement restrained the invading front of edge-water. (2) $CO_2$ can be partly dissolved into the invasion water near wellbore and reduce the water cut of the operation well.

For responding well of Well-2 (As shown in Fig.6(b)), Synergetic effect can be observed in all the scenarios with the water cut dropped from 90-95% to 50-60% during $CO_2$ puff process. Plenty of gas was produced immediately when responding well of Well-2 was opened, which indicated that the synergetic effect of $CO_2$ huff-n-puff was mainly caused by gas sweeping. After $CO_2$ dissolved with formation oil and edge-water, the injection gas can be swept to the near-wellbore area of Well-2 with enough injection volume of $CO_2$, and then caused the synergetic effect during $CO_2$ huff-n-puff process.

For Responding of Well-3 (Fig.6(c)), synergetic effect was also observed with the water cut dropped and gas produced during $CO_2$ huff-n-puff process. Unlike the immediate gas production of Well-2, the gas production performance was differential in Responding well of Well-3. In Scenario 1 and Scenario 3

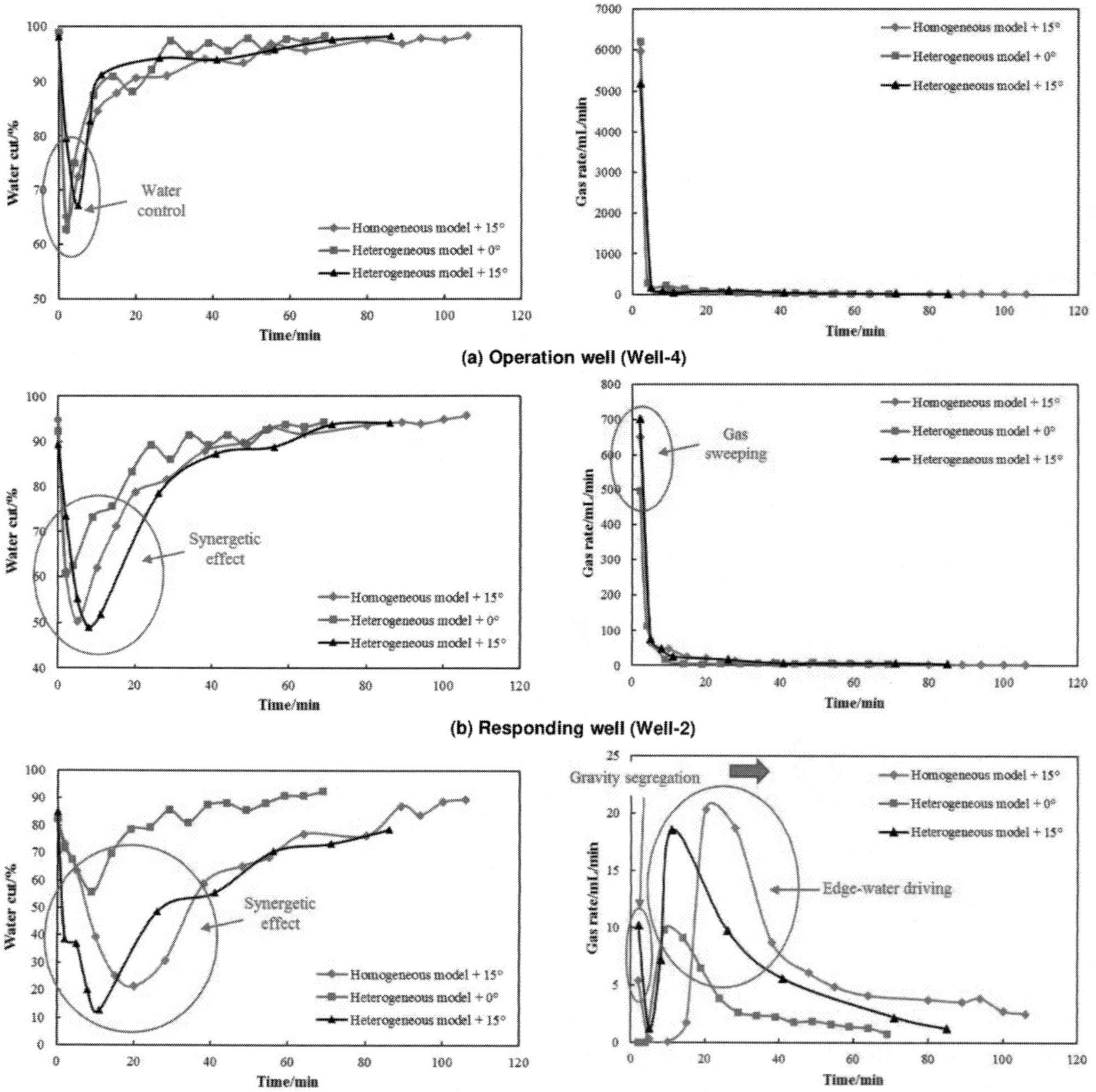


(a) Operation well (Well-4)

(b) Responding well (Well-2)

(b) Responding well (Well-3)

**Fig.6 Water cut and producing gas rate of $CO_2$ huff-n-puff process**

with the existence of stratigraphic dip, some gas was produced immediately when Well-3 was opened, while little gas was produced in Scenario 2 with the heterogenous model horizontally placed. This phenomenon can be explained that as the existence of stratigraphic dip, some gas migrated upward with the gravity segregation between formation oil and gas after $CO_2$ injected through low position well (Well-2). While the producing gas rate of homogeneous model was less than the producing gas rate of heterogeneous model, which means that the heterogeneity is an advantage to accelerate the gas migrating upward.

Unlike the immediate dropping of water cut in Well-2, the water cut of Well-3 reached its minimum after displacement for 10-20 min. Meanwhile, producing gas was also observed as the displacement process, and the gas rate reached its maximum at the same time. After plenty of gas was produced through responding wells at the early stage of $CO_2$ puff process, about 5-10% of free gas can still be remained inside the 3D model. This part of gas was driven upward by the edge-water, and then formed a $CO_2$-assisted edge-water driving to displace the interwell oil in-depth reservoir. Compared Scenario 1, Scenario 2 and Scenario 3, the synergetic effect caused by $CO_2$-assisted edge-water drving was better with a stratigraphic dip. In Scenario 1 and Scenario 3, the injecting $CO_2$ can migrate upward after enough time for soaking and leave some gas remained inside the model. However, the injection $CO_2$ was only remained around the near-wellbore areas of Well-2 and Well-3 in Scenario 2 with no gas migration, and most of the gas was produced at the early stage of $CO_2$ puff process.

With the analysis of production performance of $CO_2$ huff-n-puff process in horizontal well group, the synergetic effect by $CO_2$ huff-n-puff can be summarized as follows.

(1) Formation energy supplement for edge water control: The formation energy supplement can restrain the invading front of edge-water and reduce the water cuts of horizontal well group.

(2) Gas sweeping: Gas can be swept to the near-wellbore area of responding well with enough injection volume of $CO_2$.

(3) Gravity segregation: Gravity segregation can be generated with the existence of stratigraphic dip and heterogeneity.

(4) $CO_2$-assisted edge-water driving: $CO_2$-assisted edge-water driving can be formed with the remaining gas and sufficient edge-water supplement to displace the interwell oil in-depth reservoir.

## 4.3 Numerical simulation of synergetic $CO_2$ huff-n-puff

In order to further further discuss the synergetic types and injection allocations for $CO_2$ huff-n-puff in horizontal well group, base reservoir model was built using Eclipse Compositional Simulator. The model was 350m, 200m, 6.6m in the x, y and z dimensions, respectively. The sand body was normal rhythmic with a stratigraphic dip of 15° as shown in Fig.7. Two horizontal wells of P1 and P2 was performed in the model. P1 was located in the low tectonic position of the model, while P2 was located in the high position of the model. The horizontal sections of P1 and P2 were 70 m, and the distance between wells was 70m. The oil viscosity was 39.5 mPa·s, and the initial oil saturation was 0.55. Other parameters of the base reservoir model were summarized in Table 5.

Several numerical simulations of synergetic $CO_2$ huff-n-puff were conducted of after edge-water driving in the base reservoir model. The synergetic type can be subdivided into single-well synergy and multi-well synergy according to the allocation of $CO_2$ injection volume. In single-well synergy with Scenario 1 and Scenario 2, 800 t of $CO_2$ was injected into the model through low position well (P1), and high position well (P2), respectively. In multi-well synergy with Scenario 3, Scenario 4 and Scenario 5, 800 t of $CO_2$ was injected into the model through low position well (P1) and high position well (P2)

**Table 5 Parameters of the base reservoir model**

| Parameter | Value | Parameter | Value |
|---|---|---|---|
| Reservoir depth (m) | 1760 | Average porosity | 0.26 |
| WOC (m) | 1768 | Average horizontal permeability ($10^{-3}\mu m^2$) | 553 |
| Reservoir temperature (K) | 333.15 | Permeability variation coefficient | 0.5 |
| Initial formation pressure (psi) | 2552 | | |

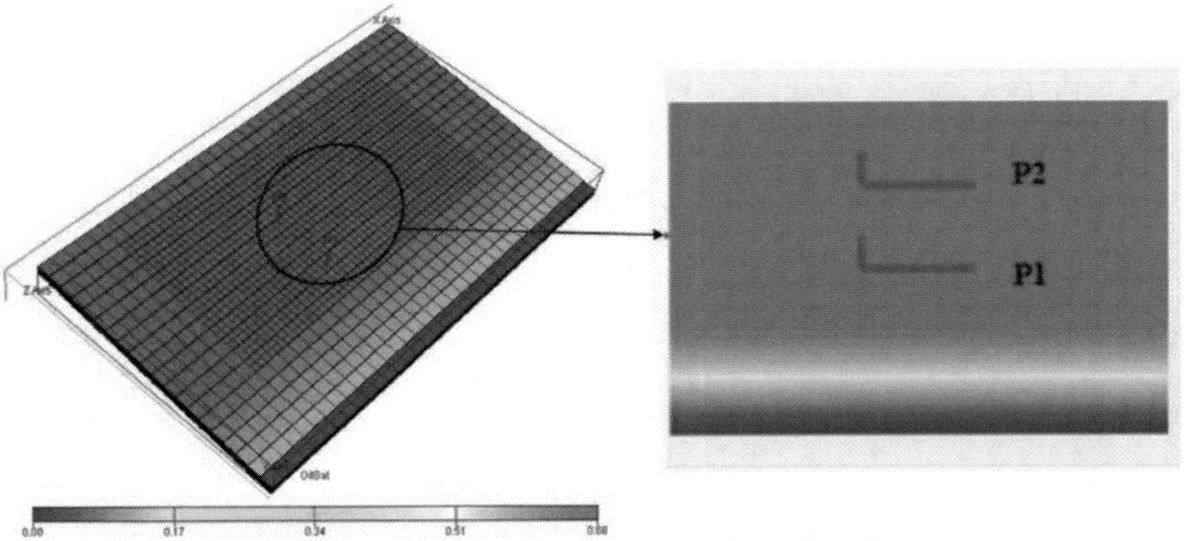


Fig.7 Base reservoir model for synergetic $CO_2$ huff-n-puff

simultaneously with different gas injection allocation. The soaking time in single-well synergy and multi-well synergy was 50 d.

The numerical results of synergetic $CO_2$ huff-n-puff are shown in Table 6. In single-well synergy, the incremental oil production of Scenario 1 was higher than that of Scenario 2. In Scenario 1, the edge water can be effectively controlled by energy supplement with $CO_2$ injected into low position well, and then synergetic effect can be observed with the water cut and oil increment of high position well. The numerical result of Scenario 1 was also in accordance with the 3D experimental results above. Comparatively, less incremental oil was produced with $CO_2$ injected into high position well in Scenario 2. The injecting gas was mostly utilized to dig the remaining oil in high tectonic position of the reservoir, and less synergetic effect was formed during $CO_2$ huff-n-puff process. Thus, it is better to inject $CO_2$ into low tectonic position of the model to strengthen the synergetic effect for enhancing oil recovery in single-well synergy.

Table 6 Numerical results of synergetic $CO_2$ huff-n-puff in base reservoir model

| Scenario | Synergetic Type | Injection Well | Injection Volume (t) | Production of Horizontal Well Group | |
|---|---|---|---|---|---|
| | | | | Incremental Oil Production (bbls) | Water Cut Drops (%) |
| 1 | Single-well Synergy | Low Position Well | 800 | 4780.4 | 47.47 |
| 2 | | High Position Well | 800 | 3333.7 | 20.1 |
| 3 | Multi-well Synergy | Low Position Well + High Position Well | 400 (L) + 400 (H) | 5245.9 | 36.54 |
| 4 | | Low Position Well + High Position Well | 100 (L) + 700 (H) | 5271.0 | 29.69 |
| 5 | | Low Position Well + High Position Well | 700 (L) + 100 (H) | 5767.9 | 47.47 |

The numerical results showed that the incremental oil production of multi-well synergy was higher than that of single-well synergy, and the incremental oil production of Scenario 5 was the highest among the multi-well synergy. The injection allcolation can somehow influence the synergetic effect of $CO_2$ huff-n-puff. In Scenario 5, with 700 t of $CO_2$ injected into low position well, the edge-water invasion can be controlled, and the synergetic effect can be formed. While with 100 t of $CO_2$ injected into high position well, the remaining oil in high tectonic position of the reservoir can also be effectively digged out. Thus, the synergetic effect of $CO_2$ huff-n-puff can be maximumed with proper injection allocation. Scenario 5 illistrated that about 5767.9 bbls oil of the horizontal well group could be recovered with 700 t $CO_2$ injected into low position well + 100 t $CO_2$ into high position well, which provided a feasible injection plan for the application of synergetic $CO_2$ huff-n-puff in the oil field.

## 5. Conclusion

3D experimental device and base reservoir model were utilized to simulate synergetic $CO_2$ huff-n-puff in horizontal well group. Through laboratory and numerical simulations of synergetic $CO_2$ huff-n-puff, some conclusions can be summarized as follows.

(1) Synergetic $CO_2$ huff-n-puff is a smart cyclic gas injection strategy for the well group, and usually occurres with the gas injected into low tectonic position of the edge-water flock-block reservoir. The drop of water cut and the increment of crude oil can be observed both in the operation well and responding well with synergetic effect of $CO_2$ huff-n-puff.

(2) The mechanisms of synergetic effect can be concluded as formation energy supplement, gas sweeping, gravity segregation and $CO_2$-assisted edge-water flooding, and the heterogeneity and stratigraphic dip are advantage for the formation of synergetic $CO_2$ huff-n-puff.

(3) The synergetic types of $CO_2$ huff-n-puff can be summarized as single-well synergy and multi-well synergy. For single-well synergy, edge-water invasion can be effectively controlled by energy supplement with $CO_2$ injected into low position well. For multi-well synergy, better synergetic effect and remaing oil displacement can be achieved after gas injected through different positions of the well group.

(4) About 5767.9 bbls oil of the horizontal well group could be recovered with 700 t $CO_2$ injected into low position well + 100 t $CO_2$ into high position well, which can provide a feasible injection plan for the application of synergetic $CO_2$ huff-n-puff in the oil field.

## Acknowledgements

The project is supported by *China National Major Technology Project (2017ZX05009-004)*. The authors want to acknowledge all the colleagues of *Institute of Enhanced Oil Recovery, China University of Petroleum (Beijing), China.*

## References

Dai S., Zhang C., Yin T., et al, 2008. Development techniques for deep complex fault block reservoirs with low permeability in Zhongyuan Oilfield. Petroleum Exploration and Development, 35 (4), 462-466.

Denham R., 1965. Peripheral pattern waterflood performance, Sholem Alechem Fault Block "E" unit. Journal of Petroleum Technology, 17 (05), 537-541.

Han Z., Yang R., Fan A., et al, 2009. Remaining oil distribution in $Ng_3^3$ bottom water reservoir of Lin 2-6 fault-block in Huimin depression and potential tapping in horizontal well. Mining Science and Technology, 19 (1), 102-107.

Hernandez J., Wojtanowicz A., 2006. Prediction of bypassed oil with correlation in side-water driving reservoirs. Journal of Canadian Petroleum Technology.

Karp J., Lowe D., Marusov N., 1962. Horizontal barriers for controlling water coning. Journal of Petroleum Technology, 14 (07), 783-790.

Laidlaw W., Bayat M., Tehrani A., 1986. Gas injection in the eastern fault block of the Thistle Field. In: European Petroleum Conference, London, United Kingdom.

Li Z., Lu T., Li B., et al, 2011. $CO_2$ and viscosity breaker assisted steam huff and puff technology for horizontal wells in a super-heavy oil reservoir. Petroleum Exploration and Development, 38 (5), 600-605.

Lim F., Saner W., Stilwell W., et al, 1993. Steamflood pilot test in waterflooded, 2500-ft Tar Zone Reservoir, Fault Block II Unit, Wilmington Field, California. In: SPE Annual Technical Conference and Exhibition, Houston, Texas, USA.

Lu L., Liu B., 1998. A feasivility research method and project design on $CO_2$ miscible flooding for a small complex fault block field. In: SPE International Oil and Gas Conference and Exhibition in China, Beijing, China.

Ma C., Shuai S., Yu J., 1982. Development of a complicated fault-block oil field by waterflooding. In: International Petroleum Exhibition and Technical Symposium, Beijing, China.

Ma J., Wang X., Gao R., et al, 2015. Enhanced light oil recovery form tight formations through $CO_2$ huff 'n' puff processes. Fuel, 154, 35-44.

Mohammed-Singh L., Singhal A., Sim S., 2006. Screening criteria for $CO_2$ huff 'n' puff operations. In: SPE/DOE Symposium on Improved Oil Recovery, Tulsa, Oklahoma, USA.

Monger T., Coma J., 1988. A laboratory and field evaluation of the $CO_2$ huff 'n' puff process for light-oil recovery. SPE Reservoir Engineering, 3 (04), 1168-1176.

Naderi M., Rostami B., Khosravi M., 2015. Effect of heterogeneity on the productivity of vertical, deviated and horizontal wells in water drive gas reservoirs. Journal of Natural Gas Science and Engineering, 23, 481-491.

Nitzberg K., Broman W., 1992. Improved reservoir characterization from waterflood tracer movement, Northwest Fault Block,

Prudhoe Bay, Alaska. SPE formation evaluation, 7 (03), 228-234.
Sayegh S., Maini B., 1984. Laboratory evaluation of the $CO_2$ huff-n-puff process for heavy oil reservoirs. Journal of Canadian Petroleum Technology.
Seright R., Lane R., Sydansk R., 2001. A strategy for attacking excess water production. In: SPE Permian Basin Oil and Gas Recovery Conference, Midland, Texas, USA.
Simpson M., 1988. The $CO_2$ huff 'n' puff in a bottomwater-drive reservoir. Journal of Petroleum Technology, 40 (07), 887-893.
Sorensen J., Schmidt D., Knudsen D., et al, 2011. Northwest McGregor field $CO_2$ huff 'n' puff: A case study of the application of field monitoring and modeling techniques for $CO_2$ prediction and accounting. Energy Procedia, 4, 3386-3393.
Spivak A., Garrison W., Nguyen J., 1990. Review of an immiscible $CO_2$ project, Tar Zone, Fault Block V, Wilmington Field, California. SPE Reservoir Engineering, 5 (02), 155-162.
Sun J., Zou A., Sotelo E., et al, 2016. Numerical simulation of $CO_2$ huff-n-puff in complex fracture networks of unconventional liquid reservoirs. Journal of Natural Gas Science and Engineering, 31, 481-492.
Torabi F., Firouz A., Kavousi A., et al, 2012. Comparative evaluation of immiscible, near miscible and miscible $CO_2$ huff-n-puff to enhance oil recovery from a single matrix-farcture system (experimental and simulation studies). Fuel, 93, 443-453.
Tower B., Wagle Y., 1992. Modelling the $CO_2$ huff 'n' puff process in solution-gas drive reservoirs using a black-oil simulator. Journal of Petroleum Science and Engineering, 8 (3), 167-179.
Vinassa M., Cudjoe S., Gomes H., et al, 2015. A comprehensive approach to sweet-spot mapping for hydraulic fracturing and $CO_2$ huff-n-puff injection in Chattanooga Shale Formation. In: SPE/CSUR Unconventional Resource Conference, Calgary, Alberta, Canada.
Wang Y., Hou J., Song Z., et al. Simulation study of in-situ $CO_2$ huff-n-puff in low permeability fault-block reservoirs. In: SPE/IATMI Asia Pacific Oil & Gas Conference and Exhibition, Nusa Dua, Bali, Indonesia.
Wu Y., Weng X., Xu M., et al, 2012. Study of early dynamic evaluation methods in complex small fault-block reservoirs. Energy Procedia, 14, 689-694.
Yu S., Li B., Zhang Q., et al, 1995. Development of Jidong Oilfields with complicated fault-blocks. In: International meeting on Petroleum Engineering, Beijing, China.
Zeinijahromi A., Al-Jassasi H., Begg S., et al, 2015. Improving sweep efficiency of edge-water drive reservoirs using induced formation damage. Journal of Petroleum Science and Engineering, 130, 123-129.
Zhou H., Chang X., Hao J., 2006. Development technique and practice of horizontal wells for complex fault-block reservoirs in Jidong Oilfield. In: SPE International Oil & Gas Conference and Exhibition in China, Beijing, China.
Zhu Y., Liu X., Fan J., et al, 2015. Development of ASP/SP flooding formulations for Huabei Fault Block Reservoir. In: SPE/IATMI Asia Pacific Oil & Gas Conference and Exhibition, Nusa Dua, Bali, Indonesia.

# CMTC-485111-MS

# Feasibility of $CO_2$-EOR in Shale-Oil Reservoirs: Numerical Simulation Study and Pilot Tests

Dheiaa Alfarge, Iraqi Ministry of Oil, Missouri University of Science and Technology; Mingzhen Wei and Baojun Bai, Missouri University of Science and Technology



This paper was prepared for presentation at the Carbon Management Technology Conference held in Houston, Texas, USA, 17-20 July 2017.



## Abstract

Shale oil reservoirs such as Bakken, Niobrara, and Eagle Ford have become the main target for oil and gas investors as conventional formations started to be depleted and diminished in number. These unconventional plays have a huge oil potential; however, the predicted primary oil recovery is still low as an average of 7.5 %. Injecting carbon dioxide ($CO_2$) to enhance oil recovery in these poor-quality formations is still a debatable issue among investigators. In this study, three steps of research have been integrated to investigate the parameters which control the success of $CO_2$ huff-n-puff process in the field scale of shale oil reservoirs. Firstly, a numerical simulation study was conducted to upscale the reported experimental studies outcomes to the field conditions. The second step was to validate these numerical models with the field data from some of $CO_2$-EOR pilots which were performed in Bakken formation, in North Dakota and Montana regions. Finally, statistical methods for Design of Experiments (DOE) have been used to rank the most important parameters affecting $CO_2$-EOR performance in these unconventional reservoirs.

The Design of Experiments approved that the intensity of natural fractures (the number of natural fractures per length unit in each direction, I-direction, J direction, and K direction) and the conductivity of oil pathways (the average conductivity for the entire oil molecules path, from its storage (matrix) to the wellbore) are the two main factors controlling $CO_2$-EOR success in shale oil reservoirs. However, the fracture intensity has a positive effect on $CO_2$-EOR while the later has a negative effect. Furthermore, this study found that the porosity and the permeability of natural fractures in shale reservoirs are clearly changeable with the production time, which in turn, led to a clear gap between $CO_2$ performances in the lab conditions versus to what happened in the field pilots. This work reported that the molecular diffusion mechanism is the key mechanism for $CO_2$ to enhance oil recovery in shale oil reservoirs. However, the conditions of the candidate field and the production well criteria can enhance or downgrade this mechanism in the field scale. Accordingly, the operating parameters for managing $CO_2$-EOR huff-n-puff process should be tuned according to the candidate reservoir and well conditions. Moreover, general guidelines have been provided from this work to perform successful $CO_2$ projects in these complex plays. Finally, this paper provides a thorough idea about how $CO_2$ performance is different in the field scale of shale oil reservoirs as in the lab-scale conditions.

## Introduction

Unconventional liquid-rich reservoirs have different aspects such as shale reservoirs, very tight reservoirs, and source rock reservoirs. Generally, these types of oil and gas reservoirs have two main criteria in common which are very small pore throats, Micro to Nano millimeters, and an ultralow permeability of Micro to Nano millidarcy as shown in **Fig. 1.** According to the recent reports, the oil production from tight formations including shale plays has shared for more than 50% of the total oil production in the US[1]. Hoffman et al., (2016) reported that 4 million barrels per day as an increment in the oil daily production in the US coming from these unconventional oil reservoirs[12]. From 2011 to 2014, Unconventional Liquid Rich (ULR) reservoirs contributed to the all natural gas growth and nearly to 92% of the oil production growth in the US[3]. This revolution in the oil and gas production has mainly happened because shale oil reservoirs have been just increasingly developed due to the advancements in horizontal wells and hydraulic fracturing techniques over the last decade. Several studies have been conducted to estimate the recoverable oil in place in these complex formations indicating huge volumes of oil. The available information refers to about 100-900 Billion barrels the oil in place in Bakken only. However, the predicted oil recovery from the primary depletion stage could lead to 7% only of the original oil in place[5]. Furthermore, some investigators argued that the primary recovery factor is still in a range of 1-2 % in some of these plays[98]. For example, the North Dakota Council reported that "With today's best technology, it is predicted that 1-2% of the reserves can be recovered"[77]. The low oil recovery happens due to the problems in the production sustainability which are the main problems in these unconventional reservoirs. The producing wells usually start with a high production rate. Then, they show a steep decline rate until they get leveled off at a low production rate. According to Yu et al. (2014), the main reason beyond the quick decline in the production rate is due to the fast depletion happening in the natural fractures combined with a slow recharge from the rock matrix (the storage)[105]. Therefore, the oil recovery factor from the primary depletion has been typically predicted to be less than 10%[3,4,7,16,88]. Infill drilling is the current practice to develop these unconventional reservoirs and to get a short-term increment in the oil production; however, the high oil rate from the new wells would not last for a long time as like as the previous wells. In addition, the cost of drilling new horizontal wells with a long lateral length is so expensive. Therefore, the infill drilling strategy might not be the economic practice in these types of reservoirs. Seeking for different options is mandatory. It is known that the main drive mechanism in the most of the shale reservoirs is the depletion drive. This drive mechanism could recover up to 8-12% of OOIP which is the main motivation to apply one of the IOR methods in these reservoirs[60]. Since these reservoirs have a huge original oil in place, any improvement in the oil recovery factor would result in enormous produced-oil volumes. Therefore, IOR methods have a huge potential to be the major starrier in these huge reserves. Although IOR methods are well understood in conventional reservoirs, they are a new concept in the unconventional formations. All basic logic steps such as experimental investigations, simulation studies, and pilot tests for examining the applicability of different IOR methods have just started over the last decade.

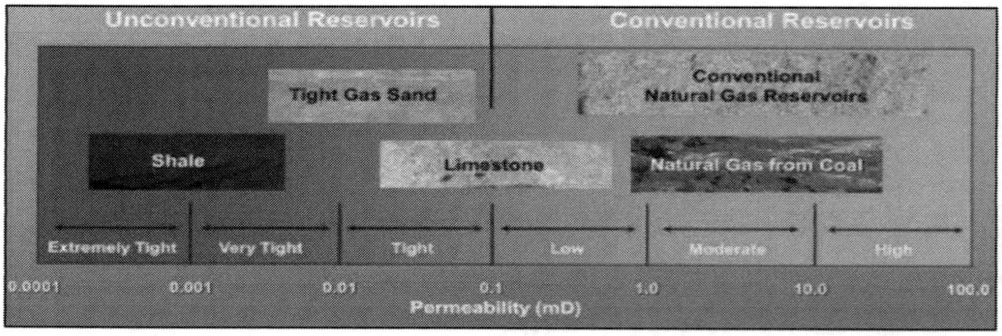


**Fig. 1: Types of oil and gas reservoirs according to the permeability cut offs[5]**

Classically, applying one of the feasible IOR methods in most of the oil and gas reservoirs should be mandatory to increase the oil recovery factor. However, the applications and mechanisms for IOR methods in unconventional reservoirs would not necessarily be the same as in the conventional reservoirs due to the complex and poor-quality properties of these plays. The public understanding of the main critical properties in unconventional reservoirs which might impair any IOR project is the low porosity and the ultralow permeability. Therefore, seeking for the IOR methods which are insensitive to these very small pore throats was the priority. Alfarge et al., (2017a) reviewed more than 70 reports and studies which have been conducted to investigate the applicability of different IOR methods in different unconventional formations of North America as shown in **Fig. 2A**. Different tools have been used in the reviewed studies such as experimental investigations, numerical simulation methods, pilot tests, and mathematical approaches as shown in **Fig. 2B**. Their review reported that the most feasible IOR techniques for these unconventional reservoirs are miscible gases, surfactant, and low-salinity water flooding. However, most of the previous studies recommended that miscible-gases EOR is the best technique for these types of reservoirs. The gases which have been investigated are $CO_2$, $N_2$, and natural gases. $CO_2$-EOR is in the top list of the miscible-gases EOR category to be applied in shale reservoirs. Furthermore, some of the IOR pilot-tests, which have been conducted to investigate the feasibility of natural gases EOR in unconventional reservoirs, showed good results in terms of enhancing oil recovery in these plays. Unfortunately, the results of the pilot-tests for $CO_2$-EOR, huff-n-puff process, were disappointing despite the excellent performance for $CO_2$ in the lab scale. Therefore, this study combined three approaches which are the reported EOR pilot-tests, the reported experimental investigations, and a new numerical simulation study to diagnose the critical parameters which control $CO_2$-EOR success in shale-oil reservoirs.

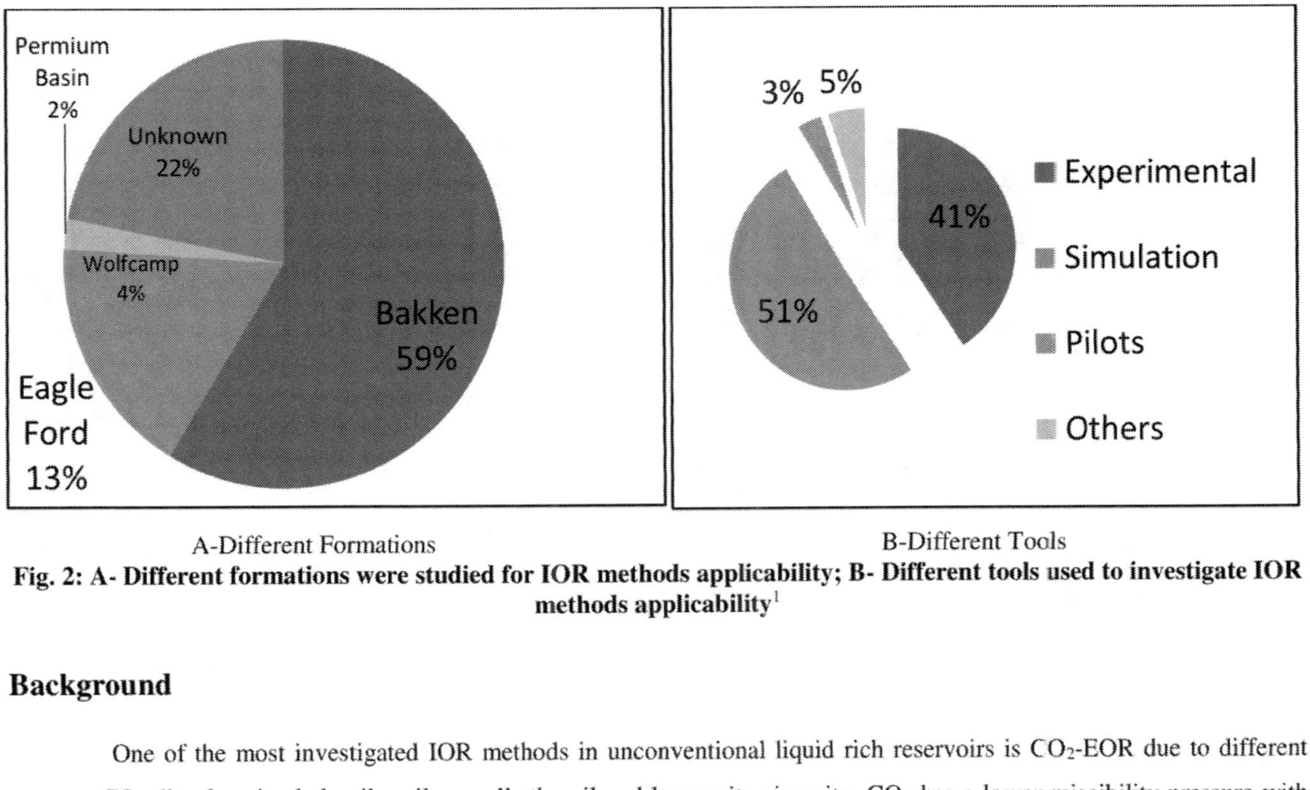


A-Different Formations B-Different Tools

**Fig. 2: A- Different formations were studied for IOR methods applicability; B- Different tools used to investigate IOR methods applicability**[1]

## Background

One of the most investigated IOR methods in unconventional liquid rich reservoirs is $CO_2$-EOR due to different reasons. $CO_2$ dissolves in shale oil easily, swells the oil and lowers its viscosity. $CO_2$ has a lower miscibility pressure with shale oil rather than other gases such as $N_2$ and $CH_4$ [36]. However, the minimum miscible pressure of $CO_2$ in shale oils has a debatable range 2500-3300 psi. The reported low value for the acid number in shale oils might increase the hope to apply $CO_2$ EOR successfully since there would not be much danger of asphaltenes precipitation[60].

The early-published studies investigating $CO_2$-EOR in shale reservoirs started by using modeling methods[79, 97]. The reported models showed that 10-20% of the incremental oil could be recovered by the continuous gas flooding while 5-10% could be recovered by the huff-n-puff gas process[12]. Dong et al., (2013) reported a numerical simulation study evaluating $CO_2$ EOR performance in an interval of Bakken formation in the Sanish field sector[8]. They came up with a scenario to increase $CO_2$ injectivity in that field by drilling more horizontal injection wells. Their scenario predicted the possibility to inject 5000 Mscf/day at a maximum injection pressure of 8000 psi. From their simulation study, they found that using $CO_2$ injection method might increase the oil recovery factor from 5% to 24% in that field. Xu et al., (2014) evaluated the reservoir performance of Elm Coulee field in Eastern Montana under $CO_2$ flooding with different hydraulic fracture orientations[100]. They concluded that transverse fractures would have a higher oil recovery factor, but these transverse fractures would have a lower utilization value than the longitudinal fractures due to the breakthrough problems. Zhu et al., (2015) constructed a model in which the EOR gases could be injected into a hydraulic fracture orienated along a horizontal well and the production process could occur from an adjacent fracture which has an intersection with the same well[108]. They found a substantial improvement in the oil recovery happens by injecting $CO_2$ in the reservoirs which have a fluid flow from fracture to fracture. Pu et al., (2016) introduced a new model which considers capillarity and adsorption effect for the small pores of shale reservoirs[85]. They found that their model would properly simulate $CO_2$ EOR in unconventional reservoirs. Furthermore, the capillarity consideration in the modeling process would predict a higher oil recovery by $CO_2$ injection rather than the cases which would not include a capillarity property.

Regarding lab reports, Song et al. (2013) conducted experimental investigations to compare results from injecting $CO_2$ and water in cores from Canadian-Bakken[81]. They found that the water flooding would enhance oil recovery better than the immiscible $CO_2$ in the huff-n-puff process. However, miscible and near-miscible $CO_2$ huff-n-puff would overcome water performance in enhancing oil recovery. Hawthorne et al., (2013) investigated the mechanism beyond increasing oil recovery by injecting $CO_2$ in Bakken cores[54]. They proved that the diffusion mechanism is the main mechanism for $CO_2$ to increase oil recovery in these complex plays. However, to extract oils from the shale matrix by $CO_2$, long times of exposure combined with large contact areas are required. Gamadi et al., (2014) conducted an experimental work on shale cores from Mancos and Eagle Ford to investigate the EOR potential of $CO_2$ injection in these reservoirs[52]. Their laboratory results indicated that cyclic $CO_2$ injection could improve oil recovery in shale oil cores from 33% to 85% depending on types of shale cores and other operating parameters. Alharthy et al., (2015) compared the performance of injecting different types of gases such $CO_2$, $C_1$-$C_2$ mixtures, and $N_2$ on enhancing oil recovery from Bakken cores[3]. They concluded that injecting gas composed of $C_1$, $C_2$, $C_3$, and $C_4$ could produce nearly as much oil as $CO_2$ injection could produce which was 90% from several Middle Bakken cores and nearly of 40% from the Lower Bakken cores. Also, they found that the counter-current mechanism is the main mechanism for these gases to recover more oils from shale cores. Finally, Yu et al., (2016) investigated $N_2$ flooding process experimentally on Eagle Ford core plugs saturated with dead oil[103]. They examined the effect of different flooding times and different injection pressures on $N_2$ flooding performance. They found that more oil could be produced with a longer flooding time and higher injection pressure. **Table 1** gives a clear summary of the most significant studies which have been conducted to investigate the feasibility of miscible-gases EOR techniques in these unconventional reservoirs.

It is clear from the previous studies that $CO_2$ would have a great potential to enhance oil recovery in these poor-quality reservoirs. However, whether using $CO_2$ in Huff-n-Puff process or injecting $CO_2$ in flooding scenario is still debatable. Due to the low permeability, conformance problems in these reservoirs, and the significant molecular-diffusion rate for $CO_2$ reported in lab conditions, most of the researchers prefer the $CO_2$ Huff-n-Puff process on $CO_2$ flooding. Unfortunately, the results of the pilot tests for $CO_2$-EOR in the cyclic process were disappointing[12]. One of the main reasons for the poor performance for $CO_2$-EOR in the field scale might be due to the wrong prediction for $CO_2$ diffusion mechanism. A detailed study for determining the level of $CO_2$ diffusivity in the real field conditions have been conducted in this work. Identifying the $CO_2$ diffusivity level is the key to the success or failure of the $CO_2$-EOR technique in shale oil reservoirs.

**Table 1: The reported studies for miscible-gases EOR in unconventional reservoirs**

| SN | Authors, Year | Paper n. | Approach | Formation | IOR Method | IOR Mechanism |
|---|---|---|---|---|---|---|
| 1 | Kovscek et al. 2008 | SPE-115679-MS | Experimental | Siliceous shale reservoir core | $CO_2$ | Diffusion |
| 2 | Shoaib et al. 2009 | SPE 123176 | Simulation | Bakken | $CO_2$ | pressure maintenance |
| 3 | Vega et al. 2010 | SPE -135627-MS | Experimental/Simulation | siliceous shale Core | $CO_2$ | Diffusion |
| 4 | Hoteit et al. 2011 | SPE 141937-MS | Mathematical Approach | X | $CO_2$ | diffusion |
| 5 | Hoffman et al. 2012 | SPE 154329 | Simulation | Bakken | $CO_2$/ Natural Gas | X |
| 6 | Dong et al. 2013 | SPE-168827-MS | Simulation | Bakken | $CO_2$ | X |
| 7 | Hawthorne et al. 2013 | SPE-167200 -MS | Experimental | Bakken | $CO_2$ | Extraction |
| 8 | Tao Wan et al. 2013 | SPE 168880 | Simulation | Eagle Ford | $CO_2$ | Oil Viscosity reduction and Pressure m. |
| 9 | Xu et al. 2013 | SPE 168774-MS | Simulation | Bakken | $CO_2$ | pressure maintenance |
| 10 | Kurtoglu et al. 2013 | SPE-168915-Ms | overview/ Simulation | Bakken | $CO_2$ | Oil Viscosity reduction and swelling |
| 11 | Chen et al. 2013 | SPE-164553-MS | simulation | Bakken | $CO_2$ | X |
| 12 | Tovar et al. 2014 | SPE-169022-MS | Experimental | preserved side-wall core X | $CO_2$ | Diffusion/Reduction in Capillary forces |
| 13 | Chen et al. 2014 | SPE-164553-PA | Simulation | Bakken | $CO_2$ | Diffusion |
| 14 | Gamadi et al. 2014 | SPE-169142-MS | Experimental | Mancos and Eagle Ford. | $CO_2$ | Repressurization |
| 15 | Schmidt et al. 2014 | 21-1921 WPC | Pilots | Bakken | Natural gas | Displacement oil in matrix |
| 16 | Tao Wan et al. 2014 | SPE-169069-MS | Simulation | Eagle Ford | $CO_2$ | Oil viscosity reduction and Pressure m. |
| 17 | Adekunle, O. 2014 | PhD dissertation/CSM | Experimental/Simulation | Bakken | $CO_2$/NGL | X |
| 18 | Fai-Yengo et al. 2014 | URTeC:1922932 | Simulation | Bakken | $CO_2$ | Combination |
| 19 | Sheng et al. 2014 | JNGSVolume 22, January 2015, Pages 252–259 | Simulation | X | $CO_2$ | X |
| 20 | Alharthy et al. 2015 | SPE-175034-MS | Experimental/Simulation | Bakken | $CO_2$ | Diffusion |
| 21 | Tao Wan et al. 2015 | SPE 1891403-PA | Simulation | Eagle Ford | $CO_2$ | Diffusion mechanism |
| 22 | Alharthy et al. 2015 | PhD dissertation/CSM | Experimental/Simulation | Bakken | $CO_2$/NGL | Swelling, Repressurization, Diffusion |
| 23 | Sheng et al. 2015 | 2015-438 ARMA Conference Paper - 2015 | Simulation | Wolfcamp shale | Gas | X |
| 24 | Hoffman et al. 2016 | SPE-180270-MS | Pilots | Bakken | CO2/Water flooding | X |
| 25 | Pu et al. 2016 | SPE-179533-MS | Simulation | Bakken | $CO_2$ | Capillarity and Adsorption |
| 26 | Yang et al., 2016 | SPE-180208-MS | Simulation | Eagle Ford | $CO_2$ | CO2 Adsorption |
| 27 | Yu et al., 2016 | SPE-180378-MS | Experimental | Eagle Ford | $N_2$ | Repressurization and fracturing |
| 28 | Yu et al., 2016 | SPE-179547-MS | Experimental | Eagle Ford | $N_2$ | Repressurization |

## Miscible-Gases Pilot Projects

Although there are a few pilots conducted to investigate the applicability of miscible-gases EOR in shale oil reservoirs, this section provides the published results for the pilots which have been mainly conducted in US and Canada. The start point is with the IOR projects which have been conducted in Canadian Bakken. The interesting point is that the pilot tests which have been conducted in Canadian Bakken have approximately the same well pattern, Toe-Heel pattern. Furthermore, the most interesting criteria in these pilots, rather than the pilots which have been conducted in US Bakken, is that the spacing between the injection wells and production wells is very short as 200 ft although the porosity and permeability of Canadian Bakken are much higher than those for US Bakken. This spacing between injectors and producers is much shorter than the spacing between injectors and producers in the pilot tests which have been performed in US Bakken (Alfarge et al., 2017a). This short spacing might be one of the main reasons beyond the encouraging results of the pilot tests in Canadian Bakken. The lateral length for the production and injection wells which were drilled horizontally in Canadian Bakken is approximately equal to one mile. Although the injection process in those pilots was sporadical, any injectivity problems had not been reported. Schmidt et al., (2014) reported a successful project in the Canadian Bakken[22]. Their pilot project covered 1280 acres which were developed by a combination of 80-acre and 160-acre spacing. The fluid and rock properties for their project are shown in **Table 2**. They designed their project by a one-mile horizontal injector and nine perpendicular horizontal producers. The wells pattern was Toe-Heel pattern. Natural gas (primary methane) was used as injectant due to its availability in these reservoirs, its high compressibility, and its low viscosity. They injected a lean gas (with $C_2$-$C_7$ content in the range of 138 bbl/MMCF to 145 bbl/MMcf) at an injection rate of 350-1000 Mscf/day without any reported problems in the injectivity. The reported results of their pilot were encouraging in all nine offset producers where the oil production rate increased from 135 bbl/day to 295 bbl/day as shown in **Fig. 3**. However, there were some problems related to conformance control where some early injected gases got a breakthrough in some of the producers. The gas utilization value had been improved form 10 MCF/bbl to 6.5 MCF/bbl which is very well consistent with the model prediction provided by Alfarge et al. (2017b)[2]. The results from their pilot are motivating. However, the main reasons for the success of their project might be because that Canadian Bakken has a permeability with 1-2 orders of magnitude higher than the permeability for US Bakken and a porosity as a twice larger than that for US Bakken[12]. Furthermore, the short spacing between the injectors and producers could be considered another reason for the success of these pilots.

**Table 2: Summary of fluid and rock properties of project area in Canadian Bakken[22]**

| Parameter | Value | Unit |
|---|---|---|
| Pilot Area | 1280 | Acres |
| Net Pay | 23-26 | ft |
| Porosity | 9-10 | % |
| Permeability | 0.01-0.1 | md |
| Water Saturation | 55-59 | % |
| Original Formation Volume Factor | 1.328 | Rb/STB |
| Bubble Point Pressure | 990 | psi |
| Oil Viscosity | 2-3 | cP |
| Oil Gravity (Stock Tank) | 42 | API |
| OOIP (Pilot Area) | 8000 | MSTB |

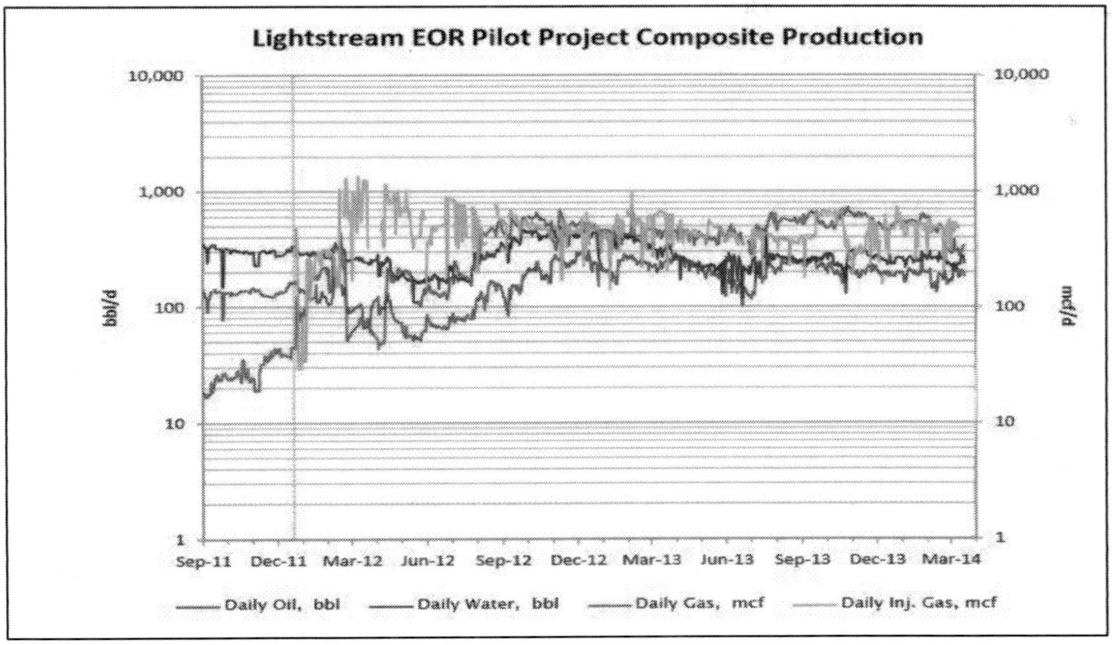


**Fig. 3: Performance of natural gas EOR in Canadian-Bakken conditions[22]**

Hoffman et al., (2016) reported seven IOR pilot-tests conducted in US Bakken, performed in North Dakota and Montana. Four of these seven pilots injected gases[12]. Three of those four pilots injected $CO_2$ while the fourth one injected enriched natural-gas. Some of those pilots were performed as a huff-n-puff process while others were designed in the continuous injection process. **Table 3** shows the pilots distribution and the fluid type injected. The start point is from the projects which were performed in huff-n-puff tests. Pilot test#1 and pilot test#2 were conducted in different parts of US Bakken by two different operators. They injected $CO_2$ as a huff-n-puff process. Both of them did not show problems related to the injectivity where they injected 1000 Mscf/day and 1500-2000 Mscf/day at 2000-3000 psi respectively. However, a clear production increment for any of them had not been well recognized as shown in **Fig. 4**. Pilot test#5 was conducted in a vertical well with 60 ft of middle Bakken pay-thickness to perform a $CO_2$ cyclic process. They injected 300-500 Mscf/day of $CO_2$ for 20-30days. After that, they did shut in the well for 20 days, then the production process was resumed. They observed the injected $CO_2$ produced in an offset well which was 900 ft away from the injection well. It is clear that the operators fractured the vertical well at that high flowrate, so they stopped the operations. The continuous gas injection process had been performed in the pilot test#7. The pilot test#7 has one injector in the center surrounded by four offset wells. Two of the producers which were to the east and the west were located at 2300 ft away from the injector while the other two which were to the north and south were located at 900 ft and 1200 ft respectively away from the injector. They injected an enriched natural gas with approximately 55% methane, 10% nitrogen, and 35% of $C_{2+}$ fractions. The injection rate was 1600 Mscf/day for 55 days at a target surface injection pressure equals to 3500 psi. As a result, all four offset wells had an increment in the production oil rate. However, some people argued whether that oil increment from the injection process or from the frac hits which were going on in the neighboring wells. Once again, the natural gas EOR like what happened in the Canadian Bakken approved to be a promising technique in these reservoirs. To sum up, the reported pilot-tests which used natural gas as injectant were successful. However, $CO_2$-EOR did not show a clear success in the huff-n-puff process which might give a clear indication that the proposed $CO_2$ diffusion mechanism in lab conditions is not the same as in the field conditions.

**Table 3: Summary of pilot tests in the Bakken-North America[12]**

| Name<br>Type | State | Year | Fluid | |
|---|---|---|---|---|
| Pilot Test #1 | ND | 2008 | CO2 | Huff-n-puff |
| Pilot Test #2 | MT | 2009 | CO2 | Huff-n-puff |
| Pilot Test #3 | ND | 2012 | Water | Huff-n-puff |
| Pilot Test #4 | ND | 2012-2013 | Water | Flood |
| Pilot Test #5 | ND | 2014 | CO2 | Vertical inj. |
| Pilot Test #6 | MT | 2014 | Water | Flood |
| Pilot Pilot#7 | ND | 2014 | Nat. gas | Flood |

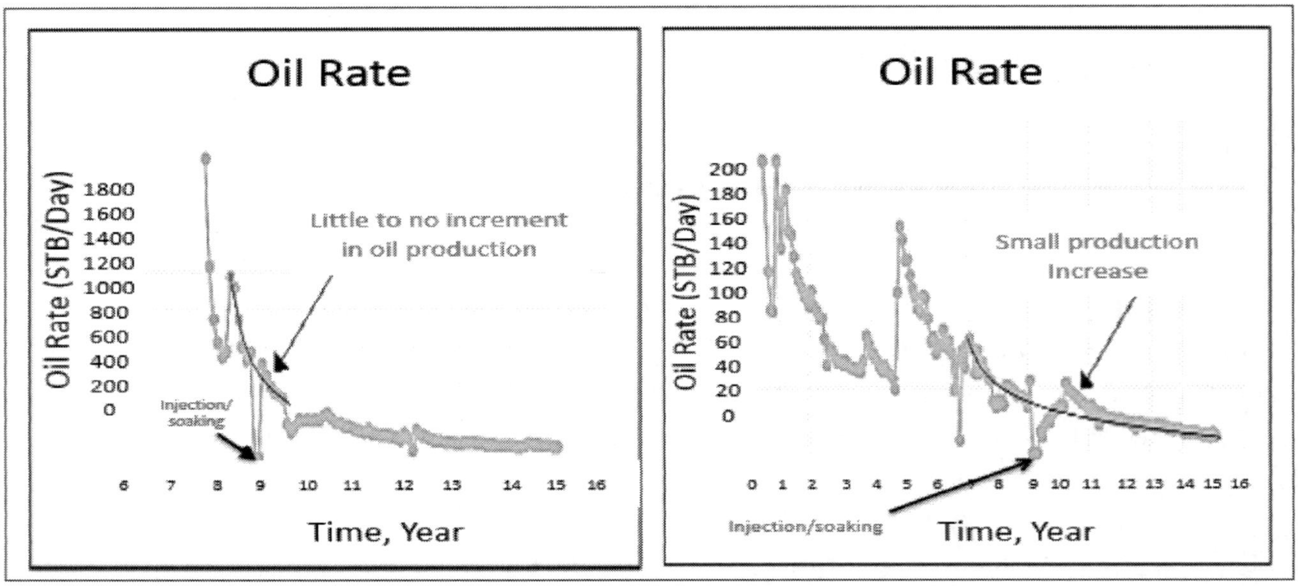


**Fig. 4: Oil production from two Bakken wells performed $CO_2$-EOR in huff-n-puff process[12]**

## Molecular Diffusion

Gravity drainage, physical diffusion, viscous flow, and capillary forces are the common forces which control the fluids flow in the porous media. However, one force might eliminate the contributions of other forces depending on the reservoir properties and operating conditions. Molecular diffusion is defined as the movement of molecules caused by Brownian motion or composition gradient in a mixture of fluids[18]. This type of flow would be the most dominated flow in the fractured reservoirs with a low-permeability matrix when gravitational drainage is inefficient[18, 19]. It has been noticed and approved that gas injection is the most common EOR process affected by the molecular-diffusion considerations. Ignoring or specifying incorrect diffusion rate during the simulation process can lead to overestimate or underestimate the oil recovery caused by the injected gas. This happens not only due to the variance in the miscibility-process between the injected-gas and the formation-oil but also due to the path change for the injected gas species from the fractures to the formation-matrix.

The Péclet number (Pe) is a class of dimensionless numbers which has been used to measure the relative importance of molecular diffusion flow to the convection flow. This number can be calculated as shown in **Eq. 1**. If Pe number is less than 1, the molecular diffusion is the dominant flow. However, if Pe is greater than 50, convection is the dominant flow. The dispersion flow is dominant when Pe in a range of 1 to 50 (Hoteit and Firoozabadi, 2009).

$$\mathbf{Pe} = \frac{diffusion\ time}{convection\ time} = (L2/D)/(L/v) = Lv/D \tag{1}$$

Where v is the bulk velocity, L is a characteristic length, and D is the molecular diffusion coefficient.

## $CO_2$ Molecular-Diffusion Mechanism

Different mechanisms have been proposed for the ability of the injected $CO_2$ to improve oil recovery in unconventional reservoirs as shown in **Table 4**. However, since the matrix permeability in these unconventional reservoirs is in a range of (0.1 –0.00001 mD), $CO_2$ would not be transported by convection flux from fracture to matrix[105]. The main transportation method for the injected $CO_2$ is depending on the difference in the concentration gradient between the concentration of $CO_2$ in the injected gases and the concentration of $CO_2$ in the target-oil. This process of transportation is subjected to Fick's law. The mechanism which is responsible for this process is called the molecular diffusion mechanism. The molecular diffusion process would be more dominated in the tight reservoirs with a significant heterogeneity. Hawthorne et al., (2013) extensively investigated the $CO_2$ diffusion-mechanism in Bakken cores and they proposed five conceptual-steps to explain it[54]. Those conceptual steps include: (1) $CO_2$ flows into and through the fractures, (2) an unfractured rock matrix is exposed to $CO_2$ at fracture surfaces, (3) $CO_2$ permeates the rock driven by pressure, carrying some hydrocarbon inward; however, the oil is also swelling and extruding some oil out of the pores, (4) oil migrates to the bulk $CO_2$ in the fractures via swelling and reduced viscosity, and (5) as the $CO_2$ pressure gradient gets smaller, oil production is slowly driven by concentration-gradient diffusion from pores into the bulk $CO_2$ in the fractures. The importance of considering this mechanism is also depending on the type of injected gases. For example, the shale oil has a high concentration of light components for natural gases such as methane. In the same time, the shale oil has a low concentration of $CO_2$. Therefore, considering this mechanism in the simulation process for the injected $CO_2$ has a significant effect on the obtained oil recovery. However, considering this mechanism in the simulation process for the injected methane has a miner effect on the obtained oil recovery[111]. The effect of the binary molecular diffusion between the injected $CO_2$ and the formation oil was simulated in this work by using the experimental correlation conducted by Sigmund (1976a; 1976b)[112-113]. The following polynomial equation was fitted with their observed experimental values.

$$D_{ij} = \frac{\rho_k^0 D_{ij}^0}{\rho_k} (0.99589 + 0.096016\rho_{kr} - 0.22035\rho_{kr}^2 + 0.032874\rho_{kr}^3 \tag{2}$$

Where $D_{ij}$ is the binary diffusion coefficient in unit of $cm^2/s$ between component i and j in the mixture, $\rho^o D_{ij}^o$ is the zero-pressure limit of the density-diffusivity product, $\rho_k$ is the density of the diffusion mixture in $kg/m^3$, $\rho_{kr}$ is the reduced density which can be calculated by **Eq. 3**, and the subscript k denotes the phase which could be water, oil, or gas. In the simulator, the product of mixture density and diffusion coefficient can be calculated by **Eq. 4**. The diffusion coefficient of component i in the mixture can be calculated by **Eq. 5.**

$$\rho_{kr} = \rho_k \frac{\sum_i^{nc} y_{ik} v_{ci}^{5/3}}{\sum_i^{nc} y_{ik} v_{ci}^{2/3}} \quad (3)$$

Where $y_{ik}$ is the mole fraction of i species in phase k; and $v_{ci}$ is the critical volume of i species.

$$\rho_k^0 D_{ij}^0 = \frac{0.18583\, T^{0.5}}{\sigma_{ij}^2 \mho_{ij} R} \cdot \{\frac{1}{M_i} + \frac{1}{M_j}\}^{0.5} \quad (4)$$

Where R is the universal gas constant, T is the temperature; M is the molecular weight, $\sigma_{ij}^2$ is the collision diameter between i and j, and $\mho_{ij}$ is the collision integral of the Lenard-Jones potential.

$$D_i = \frac{1-y_i}{\sum_{i \neq j} y_i D_{ij}^{-1}} \quad (5)$$

Where $D_i$ is the diffusion coefficient of component i in the mixture and $y_i$ is the mole fraction of component i.

**Table 4: The proposed CO2 EOR mechanisms for improving oil recovery in unconventional reservoirs**

| $CO_2$ mechanism | Approach tool |
|---|---|
| 1-Diffusion | Lab |
| 2-Reduction in Capillary forces | Lab and simulation |
| 3-Repressurization | Lab |
| 4-Extraction | Lab |
| 5-Oil swelling and pressure maintenance | Lab and simulation |
| 6-Oil Viscosity reduction | Lab and simulation |
| 7-Combination of more than one mechanism from above | - |

Most of the previous experimental studies reported that $CO_2$ diffusion mechanism is beyond the increment in the oil recovery obtained in the lab conditions. Then, the observed increment in the oil-recovery and/or the $CO_2$ diffusion-rate obtained in the lab conditions were upscaled directly to the field scale by using numerical simulation methods. This direct upscaling methodology might be so optimistic because the lab-cores have a higher contact area and longer exposure time to the injected $CO_2$ than what might happen in the real-conditions in the field scale. As a result, both of the previous simulation studies and the experimental reports might be too optimistic to predict a quick improvement in the oil recovery from injecting $CO_2$ in these tight formations.

## Numerical Simulation

Most of the reported simulation studies in this area simulated these naturally fractured shale reservoirs by a combination of discrete fractures with a tight formation matrix. They used the refinement process for the grids around the discrete fractures to make the convergence in the numerical calculations happening. We think that their combination, discrete fractures with tight formation matrix, would not capture the real physics for these fractured shale reservoirs. In this simulation study, the LS-LR-DK (logarithmically spaced, locally refined, and dual permeability) model is built to simulate the $CO_2$-EOR in shale reservoirs. The LS-LR-DK method can accurately simulate the fluid flow in fractured shale-oil reservoirs[88]. Furthermore, the representation of the molecular-diffusion mechanism in the previously reported simulation methods would also be misleading because most of the previous studies used the direct upscaling for the lab observations, diffusion coefficients and/or oil increment resulted from $CO_2$ injection in the lab cores. In this paper, an advanced general equation-of-state compositional simulator has been used to build the formation fluid model. Then, both of the models, LS-LR-DK model and fluid model, have been combined to simulate compositional interactions of the reservoir fluid and the injected $CO_2$ during enhanced oil recovery processes. Furthermore, the implementation of the diffusion model in the LS-LR-DK model and fluid model has been conducted. In this study, we tried to build a numerical model which has the typical fluid and rock properties of Bakken formation, one of the most productive unconventional formations in the US. All of the simulation processes have been carried out by using CMG-GEM simulator. In the models of this study, we injected $CO_2$ in different scenarios as Huff-n-Puff process through hydraulically fractured well in Bakken formation. All the mechanisms which were proposed in **Table 4** have been considered in this model. In this field case study, the production well was stimulated with 5 hydraulic fractures. The spacing between the hydraulic fractures is 200 ft. The simulated model includes two regions which are stimulated reservoir volume (SRV) and un-stimulated reservoir volume (USRV) as shown in **Fig. 5**. The dimensions of the reservoir model are 2000 ft x 2000ft x42 ft, which corresponds to length, width, and thickness respectively. The dimensions of the fractured region are 5 fractures with half-length of 350 ft in J direction, width 0.001 ft in I direction, and fracture height of 42 ft in K direction. Fracture conductivity is 15 mD.ft. The other model input parameters are shown in **Table 5**.

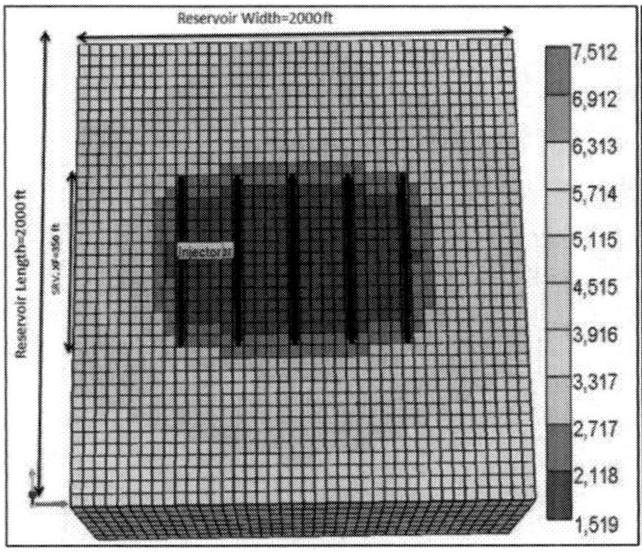


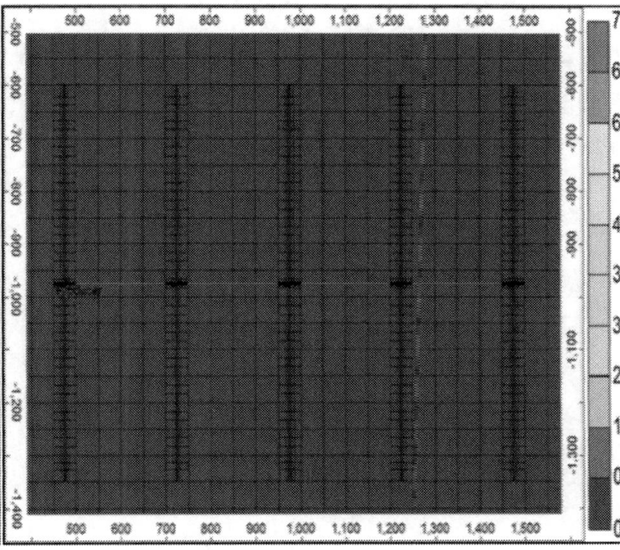


**Figure 5a- Average Pressure in a depleted well in Bakken** **Figure 5b- A closed view for SRV of Production well**

**Table 5: Model input parameters for the base case**

| Parameter | value | Unit |
|---|---|---|
| The model dimensions | 2000x2000x42 | ft |
| Production time | 20 | year |
| Top of reservoir | 8000 | ft |
| Reservoir temperature | 240 | $^oF$ |
| Reservoir pressure | 7500 | psi |
| Initial water saturation | 0.3 | value |
| Total compressibility | $1x10^{-6}$ | $psi^{-1}$ |
| Matrix permeability | 0.005 | mD |
| Matrix porosity | 0.085 | value |
| Horizontal well length | 1000 | ft |
| Total number of fractures | 5 | value |
| Fracture conductivity | 15 | mD-ft |
| Fracture half-length | 250 | ft |
| Fracture Height | 42 | ft |

## Compositional Model for the Formation Fluids

The typical Bakken oil has been simulated in this study. The oil which was used in this model has 42 API°, 725 SCF/STB, and 1850 psi as oil gravity, gas oil ratio, and bubble point pressure respectively. It is known that compositional models are the most time-consumed models' due to the large number of components in the typical formation oil. In our model, we have 34 components so that would take a long time for the simulator to complete running one scenario. The common practice in the numerical simulations for such situation is the careful lump for the reservoir oil components into a short representative list of pseudo-components. These pseudo components could be acceptable if they have matched with the laboratory–measured phase behavior data. The supplied data for such compositional models need to have a description of associated single carbon numbers and their fractions, saturation pressure test results, separator results, constant composition expansion test results, differential liberation test results, and swelling test results[7]. These steps can be used for tuning the EOS to match the fluid behavior. WinProp-CMG has been used to lump the original 34 components into 7 pseudo components as shown in **Table 6**. WinProp is an Equation-of-State (EOS)-based fluid behavior and PVT modeling package. In WinProp, the laboratory data for fluids can be imported and an EOS can be tuned to match the physical behavior for the lab data. Fluid interactions can be predicted and a fluid model can be then created for the use in CMG software[7]. **Table 7** presents the Peng-Robinson EOS fluid description and binary interaction coefficients of the Bakken crude oil with different injected gases. **Fig. 6** represents the two-phase envelope for the Bakken oil which was generated by using WinProp-CMG.

**Table 6: Compositional data for the Peng-Robinson EOS in the model oil**

| Component | Mole fraction | Critical pressure (atm) | Crititical Temp. (K) | Acentric Factor | Molar Weight (g/gmole) |
|---|---|---|---|---|---|
| CO2 | 0 | 7.28E+01 | 3.04E+02 | 0.225 | 4.40E+01 |
| N2-CH4 | 0.2704 | 4.52E+01 | 1.90E+02 | 0.0084 | 1.62E+01 |
| C2H-NC4 | 0.2563 | 4.35E+01 | 4.12E+02 | 0.1481 | 4.48E+01 |
| IC5-CO7 | 0.127 | 3.77E+01 | 5.57E+02 | 0.2486 | 8.35E+01 |
| CO8-C12 | 0.2215 | 3.10E+01 | 6.68E+02 | 0.3279 | 1.21E+02 |
| C13-C19 | 0.074 | 1.93E+01 | 6.74E+02 | 0.5672 | 2.20E+02 |
| C20-C30 | 0.0508 | 1.54E+01 | 7.92E+02 | 0.9422 | 3.22E+02 |

**Table 7: Binary interaction coeficients for Bakken oil**

| Component | CO2 | N2-CH4 | C2H-NC4 | IC5-CO7 | CO8-C12 | C13-C19 | C20-C30 |
|---|---|---|---|---|---|---|---|
| CO2 | | | | | | | |
| N2-CH4 | 1.01E-01 | | | | | | |
| C2H-NC4 | 1.32E-01 | 1.30E-02 | | | | | |
| IC5-CO7 | 1.42E-01 | 3.58E-02 | 5.90E-03 | | | | |
| CO8-C12 | 1.50E-01 | 5.61E-02 | 1.60E-02 | 2.50E-03 | | | |
| C13-C19 | 1.50E-01 | 9.76E-02 | 4.24E-02 | 1.72E-02 | 6.70E-03 | | |
| C20-C30 | 1.50E-01 | 1.45E-01 | 7.79E-02 | 4.27E-02 | 2.51E-02 | 6.00E-03 | |

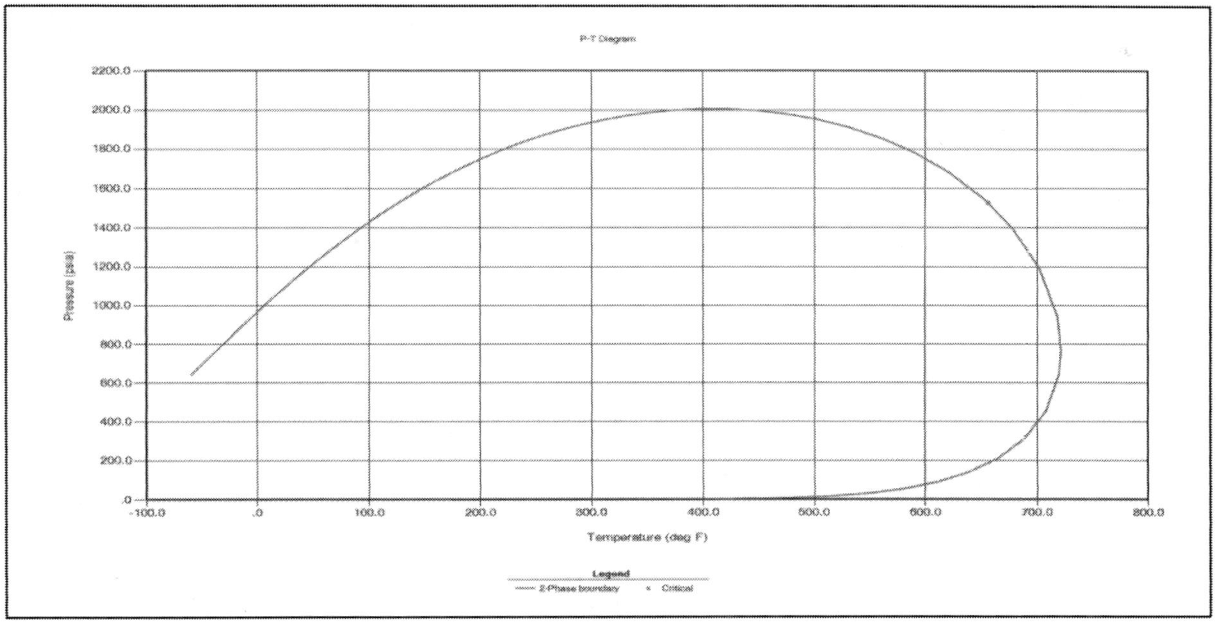


**Fig. 6: The two-phase envelope for Bakken oil which was generated by WinProp-CMG**

## Results and Discussion

***Natural Depletion for Bakken Model.*** The reservoir model was initially run in natural depletion for 7300 days (20 years). The production well, which was hydraulically fractured, was subjected to the minimum bottom-hole pressure of 1500 psi. The simulated Bakken well performance in the natural depletion is shown in **Fig. 7**. In the natural depletion scenario, it has been clear that the production well started with a high production rate initially as shown in **Fig. 7**. Then, it showed a steep decline rate until it got leveled off at a low rate. This is the typical trend to what is happening in the most if not all unconventional reservoirs of North America. If we investigate the pressure distribution in the reservoir model as shown in **Fig. 5**, it is clear that the main reason to that fast reduction in the production rate is due to the pressure depletion in the areas which are close to the production well. However, the reservoir pressure is still high in the areas which are far away from the production well. This explains the poor feeding from neighboring areas in these types of reservoirs due to the what is called the permeability gel (tight formations).

***Flow-Type Determination in the Natural-Depletion Stage.*** We calculated the Péclet number locally in each grid of the model. In the formation-matrix areas, the results indicated that Péclet number is away below 1 for both of gas phase and oil phase which means that the diffusion flow is the most dominant flow in the formation matrix as shown in **Fig. 8.** However, in the hydraulic fractures areas, the viscous flow is clearly dominated where Pe is away above 100. In the natural fractures areas, the results indicated that Péclet number is significantly changeable where it is away below 1 in the areas which are far away from hydraulic fractures; however, it is away above 100 in the areas which are close to the hydraulic fractures as shown in **Fig. 9**. According to the average value of Péclet number in the natural fractures areas, the dispersion flow could be the most dominant flow.

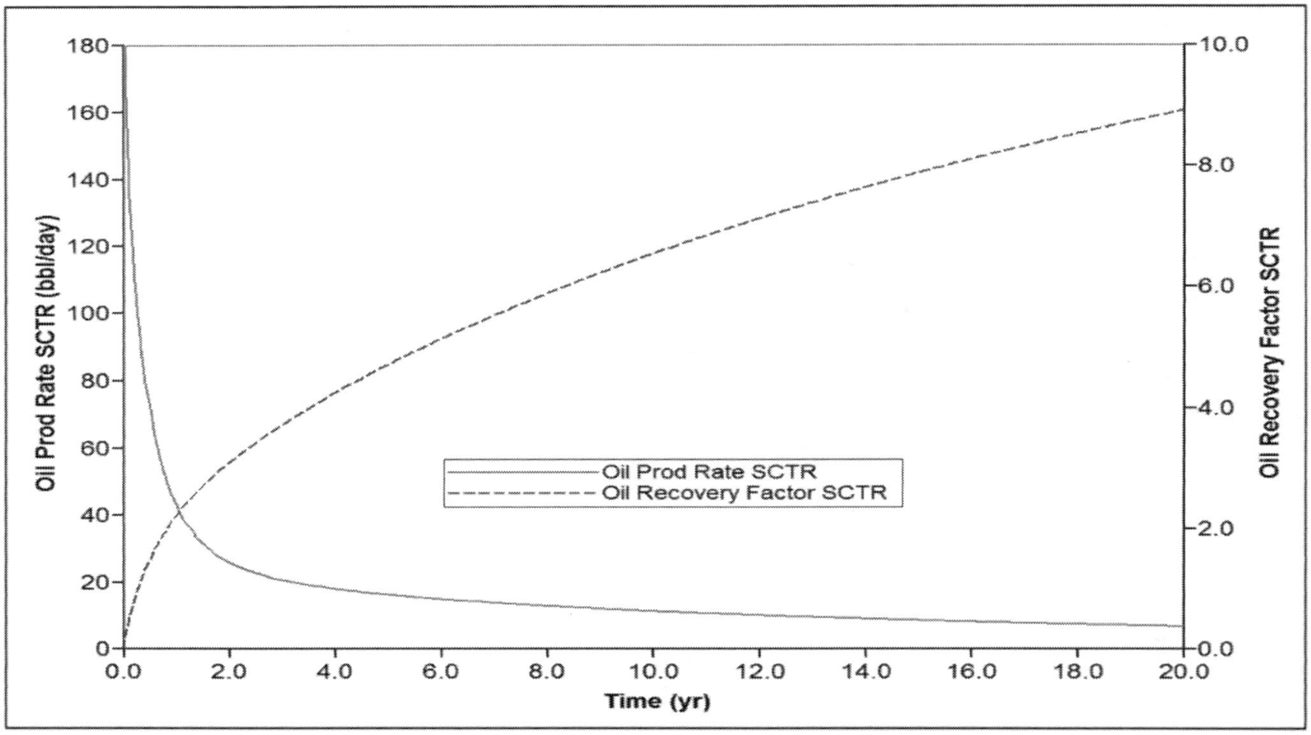


**Fig. 7: The reservoir performance in natural depletion conditions**

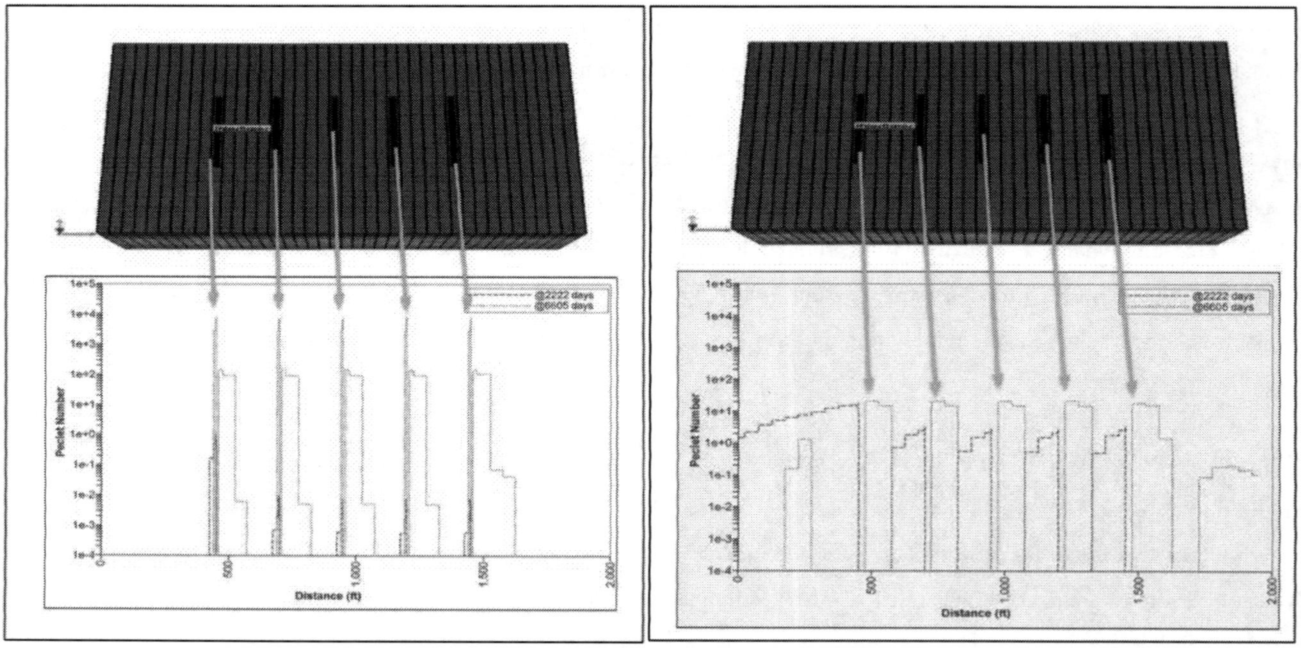

A-Gas Phase B-Oil Phase

**Figure 8: Péclet number distribution a long cross section in the matrix-model**

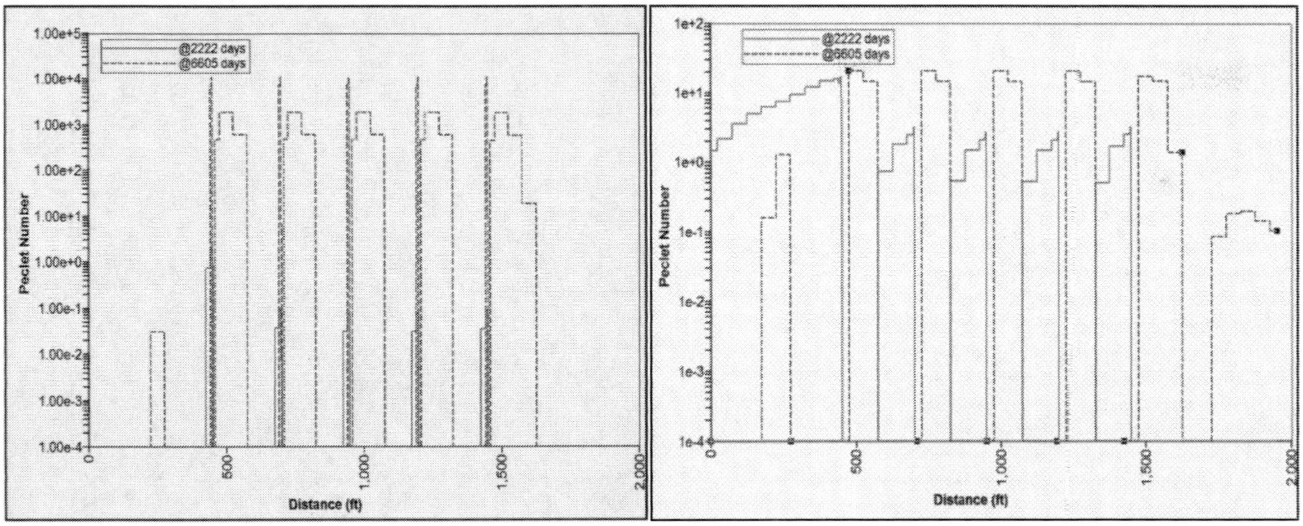


A-Gas Phase B-Oil Phase

**Figure 9: Péclet number distribution a long cross section in the natural fracture-model**

***Effect of Huff-n-Puff Cycles-Number on $CO_2$ Performance.*** $CO_2$ was injected in the production well as a huff-n-puff process in two different scenarios as shown in **Table 8**. Each scenario has two cases: (1) The first case is injecting $CO_2$ assuming there is no molecular-diffusion mechanism for the injected $CO_2$ into formation-oil, (2) The second case is injecting $CO_2$ with molecular diffusion mechanism enabled. The results indicated that the $CO_2$ performance for without-molecular diffusion case did not provide any improvement in the oil recovery from what was obtained at natural depletion production; it is even worse than the base case for both scenarios as shown in **Fig. 10**. If we look closely, we found that the enhancement in the oil production rate from $CO_2$ injection did not offset the loss in the oil production, which was happening during the soaking and injection period. This can be noticed by observing the difference in the slope of oil recovery curves, before and after injecting $CO_2$. However, $CO_2$-EOR, in with the molecular-diffusion cases, has improved the oil recovery and oil production in a significant way as shown in **Fig. 10**. However, the results indicated that the $CO_2$ performance is independent of huff-n-puff cycles-number for the with-diffusion cases. We can notice that the oil recovery obtained for both scenarios, for the with-diffusion cases, is almost the same for both scenarios as shown in **Fig. 10**. However, for without-diffusion cases, the more cycles of $CO_2$ huff-n-puff process is the worst. This can be explained by the soaking period. The soaking period for the scenario which has 2 cycles is longer than that for the 10-cycle scenario. The injected $CO_2$ needs longer soaking periods to perform well in such tight reservoirs.

**Table 8: The agenda and time breakdown for both $CO_2$ huff-n-puff scenarios**

| Scenario 1 | Scenario 2 |
|---|---|
| 2 cycles injected | 10 cycles injected |
| The injection time for each cycle=6 months | The injection time for each cycle=2 months |
| Injection rate= 500 Mscf/day | Injection rate=500 Mscf/day |
| Soaking period=3 months | Soaking period=1 months |
| The production for each cycle=4 years and 3 months | The production for each cycle=9 months |

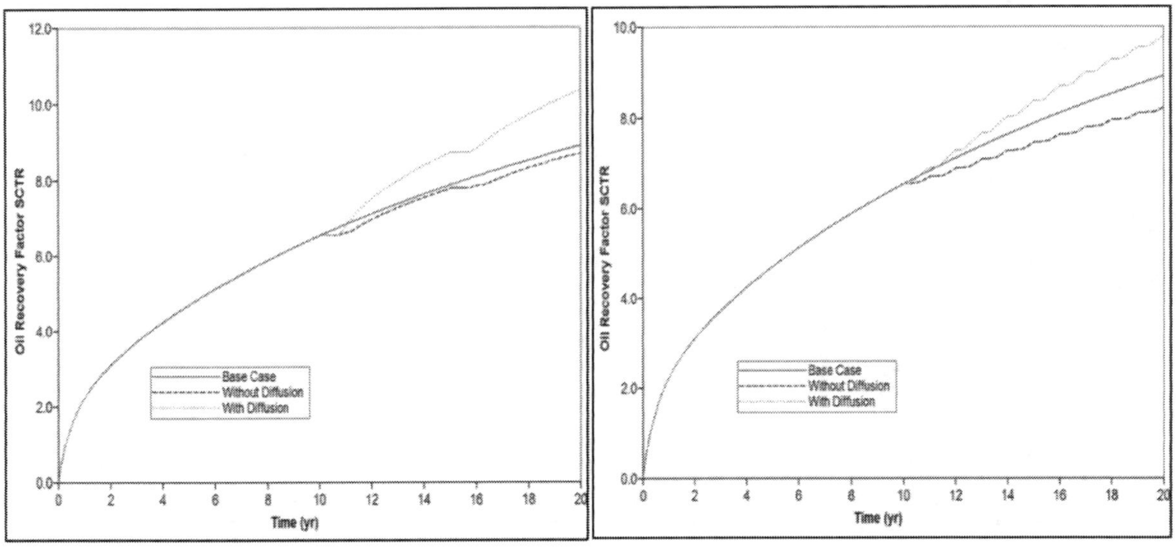


**A-Scenario 1** **B-Scenario 2**

**Fig. 10: Oil recovery factor in natural depletion conditions versus with $CO_2$ Huff-n-Puff**

If we investigate the reasons beyond the role of molecular-diffusion mechanism on $CO_2$ performance to enhance oil recovery in these tight formations, we found this mechanism makes $CO_2$ penetrate deeper into the tight matrix, far way from the hydraulic fractures as shown in **Fig. 11**. However, the case of $CO_2$ injection which does not have a diffusion capacity makes the $CO_2$ penetrate just in the limited areas around the hydraulic fractures. Therefore, for the cases in which $CO_2$ penetrate deeper in the tight matrix, $CO_2$ would swell more volumes of oil, reduce oil viscosity, and finally produce larger quantities of oil by the counter-current mechanism. On the other hand, in the cases which have $CO_2$ with a low molecular-diffusion rate, the injected $CO_2$ would be produced back in the same well very soon. Therefore, producing the injected $CO_2$ back would put another hold on the oil production due to the slippage effect making the enhancement in the oil production from these types of reservoirs even worse.

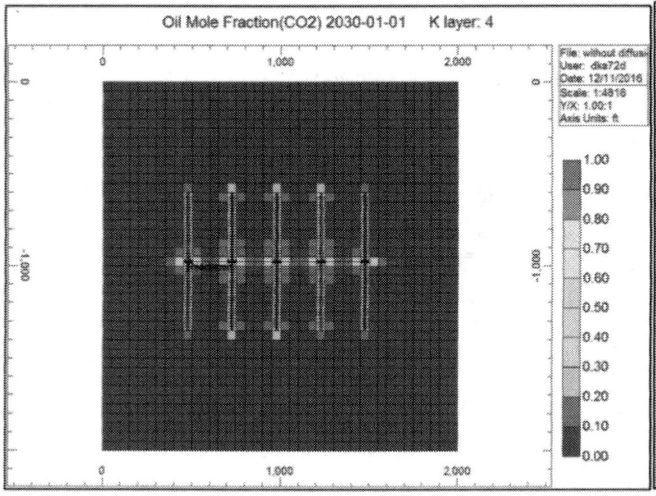


**Fig. 11a- $CO_2$ Injection without-molecular diffusion**

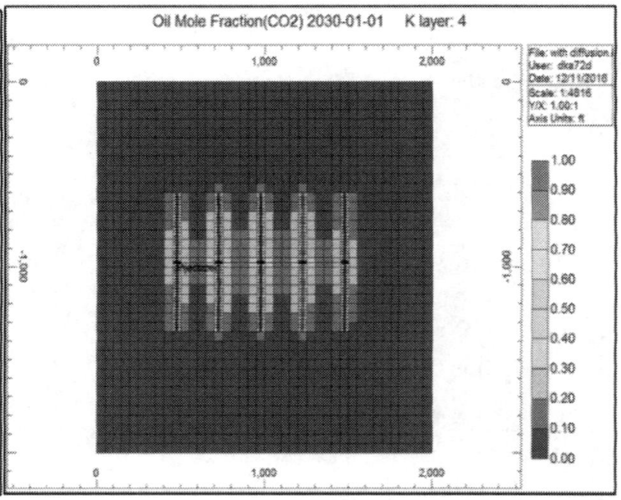


**Fig. 11b- $CO_2$ Injection with Molecular Diffusion**

## Parameters Affecting the Molecular-Diffusion Mechanism for $CO_2$-EOR

***The Exposure Time between the injected $CO_2$ and the Formation-Fluid.*** To investigate the effect of the exposure time between the formation oil and the injected-$CO_2$ on the $CO_2$- molecular diffusion mechanism, different soaking periods have been used for the same scenario. The results confirmed the prediction which is that $CO_2$ would perform better in the cases which have a longer soaking period rather than the cases which have a short soaking period as shown in **Fig. 12**. Another verification has been conducted to verify the effect of the exposure time on the $CO_2$-molecular diffusion. This verification has been done by injecting $CO_2$ in the low-conductivity hydraulic fractures versus injecting $CO_2$ in the high-conductivity hydraulic fractures. The results indicated that the injected $CO_2$ would enhance oil recovery in the reservoirs with low-conductivity fractures more than the reservoirs with high-conductivity fractures. To sum up, as far as the kinetics of the oil recovery process in the productive areas do not exceed the $CO_2$-diffusion rate, the injected $CO_2$ would experience more exposure time with the formation oil before its being produced back.

***The Contact Area between the Injected $CO_2$ and the Formation-Fluid.*** If we need to enhance the $CO_2$-molecuar diffusion in these formations, we need to have a large contact area between the injected $CO_2$ and the formation oil. This can be verified by running some of the model scenarios which have a different contact area between the formation oil and the injected $CO_2$. We did that investigation by running two models which have exactly the same rock and fluid properties. However, one of them injected $CO_2$ in hydraulically fractured well (large contact area) while the other one injected $CO_2$ in non-hydraulically fractured well (small contact area). The results confirmed the prediction which is that $CO_2$ would perform better in the hydraulically fractured well rather than the non-hydraulically fractured well as shown in **Fig. 13** and **Fig. 14**. Another verification has been done by injecting $CO_2$ into an open-hole well versus injecting $CO_2$ into a cased hole. Also, the results confirmed the prediction which is that $CO_2$ would perform better in an open- hole horizontal well rather than a cased-horizontal well.

***Performing Time.*** To investigate the effect of the performing time on $CO_2$-EOR performance, we injected $CO_2$ at a different time from the production well life. In the first scenario, we injected $CO_2$ after 5 years of the production life. However, in the second scenario, we injected $CO_2$ after 10 years from the production life. The results confirmed the prediction which is that $CO_2$ would perform better in the cases which have early $CO_2$-EOR rather than the cases which have late $CO_2$-EOR as shown in **Fig. 15**. This could be explained by the effective-stress principle which might be significantly important to control the permeability and porosity of natural fractures in shale oil reservoirs. As far as the $CO_2$-EOR performed earlier, its performance would be better because the injected $CO_2$ would find a good intensity of natural fracture which helps in enhancing its diffusivity into formation-oil.

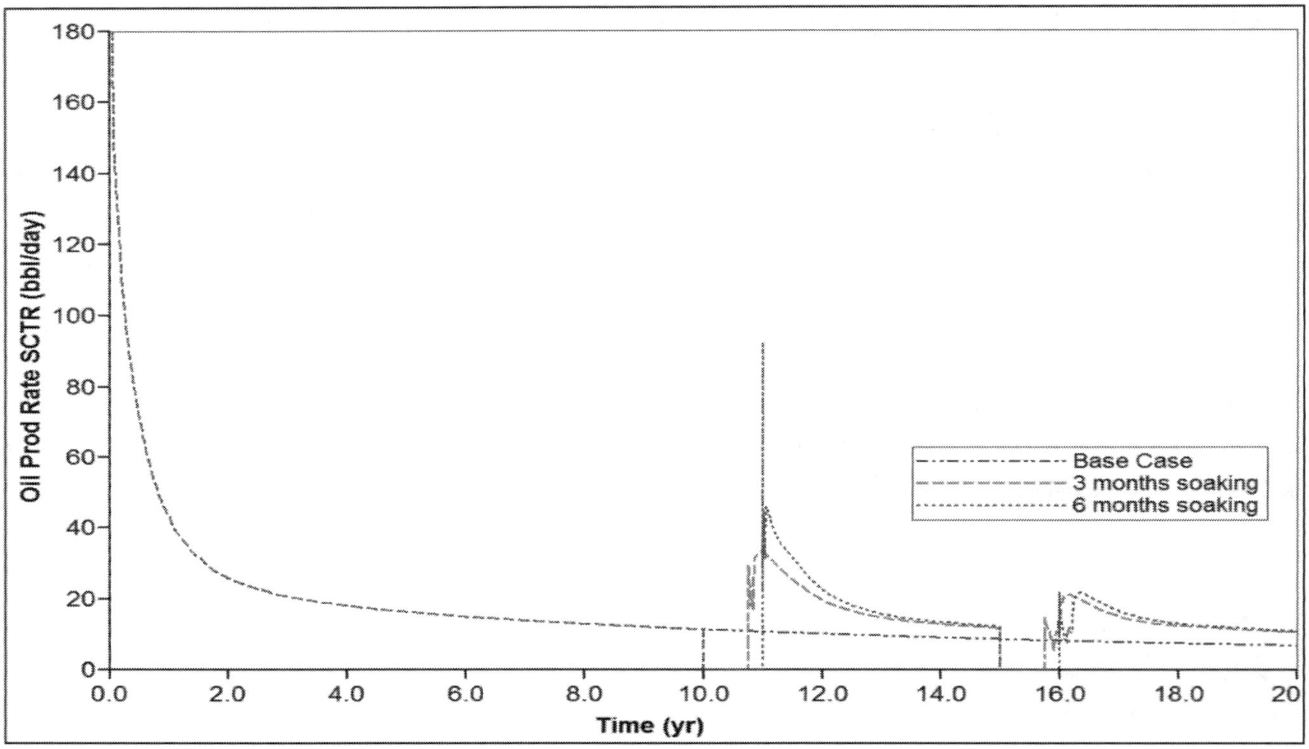


**Fig. 12: Effect of soaking period on $CO_2$ huff-n-puff performance**

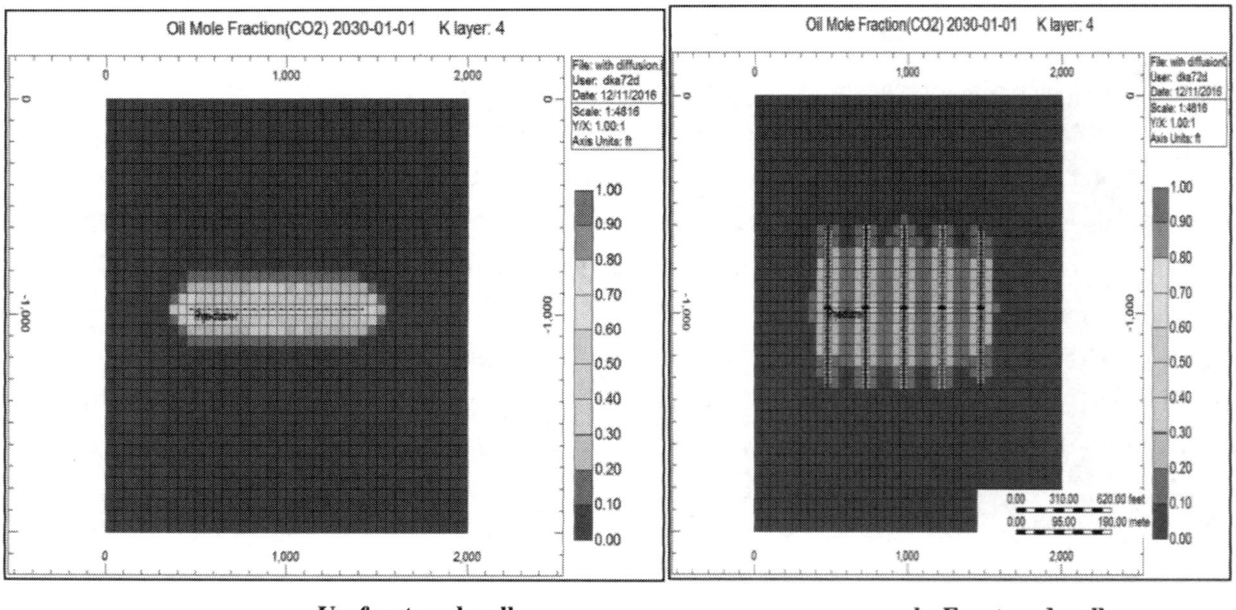


**a- Un-fractured well** **b- Fractured well**

**Fig. 13: Comparison between the penetration of the injected $CO_2$ in the fractured well versus the un-fractured well in Bakken model**

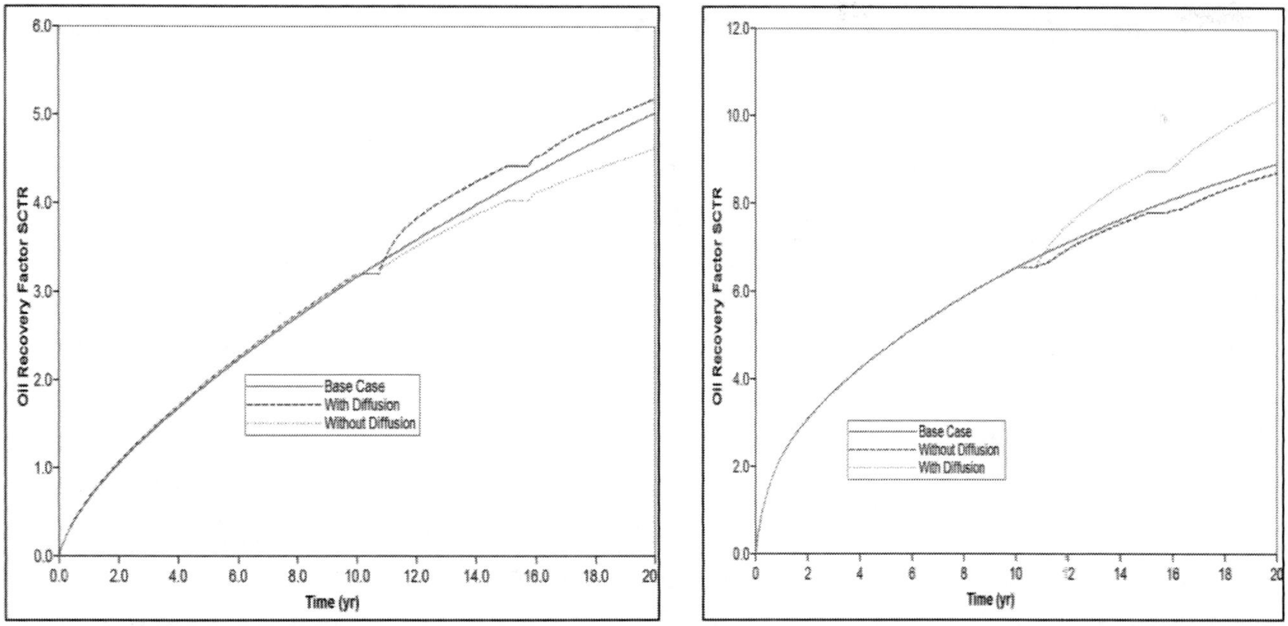


**a-Un-fractured well** **b- Fractured well**

**Fig. 14: Comparison between $CO_2$ performance in fractured well versus un-fractured well in Bakken**

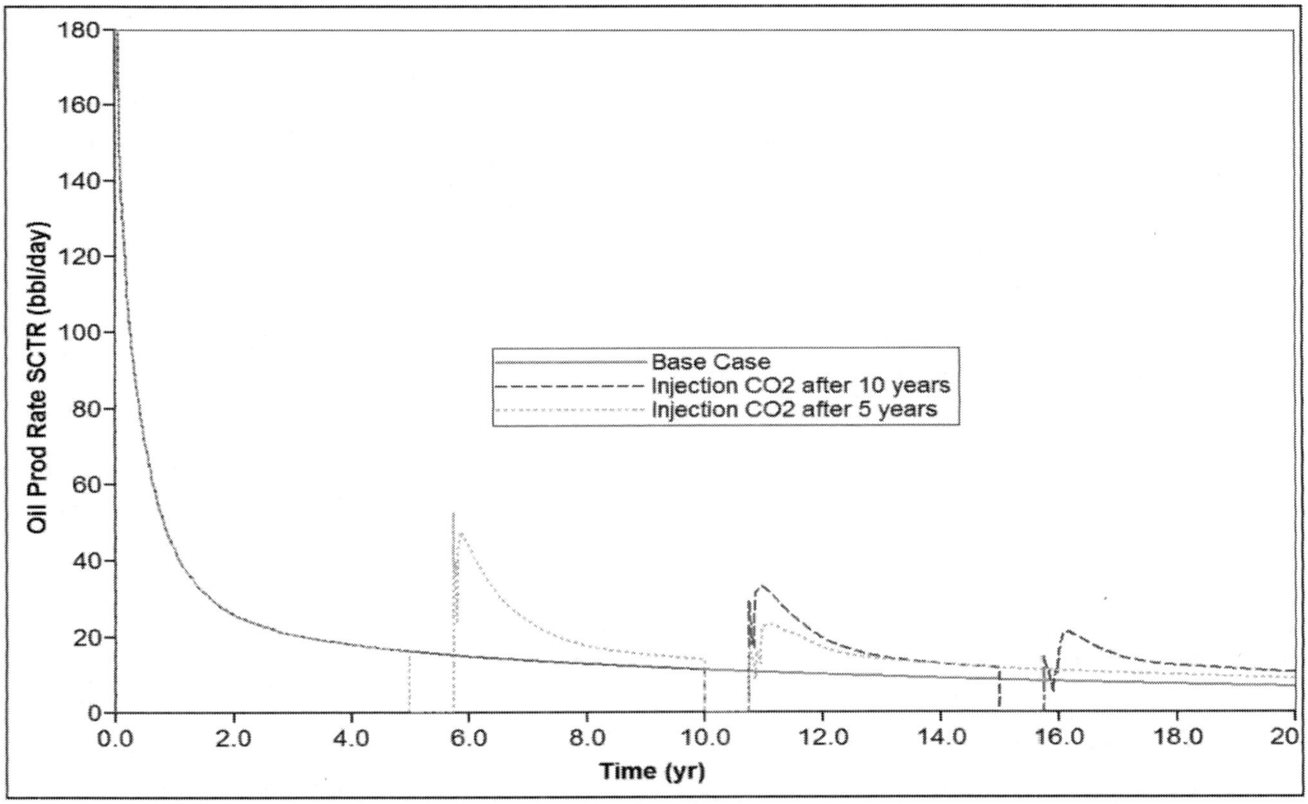


**Fig. 15: Effect of the performing time on $CO_2$-EOR in shale oil reservoirs**

## Design of Experiments for the Factors Affecting the performance of $CO_2$-EOR in the huff-n-puff process

***Sensitivity Analysis.*** The purpose of Sensitivity Analysis is for determining how sensitive an objective function to different parameters qualitatively and quantitively. Identifying the parameters which have a high impact on $CO_2$-EOR huff-n-puff performance would give a good prediction for the $CO_2$-EOR success or failure depending on the reservoir properties prior to the field application. In this part, the objective function which was used is the oil recovery factor at 10 years from $CO_2$-EOR huff-n-puff process. The parameters which were investigated and their range values are listed in **Table 9**. The statistical methods which were used for ranking these parameters are as the following:

- **Sobol Method**: The Sobol method is one of the variance-based sensitivity analysis methods to quantify the amount of variance that each input factor Xi contributes to the unconditional variance of output V(Y) (CMG)[7]. For example, a given case with 3 inputs and one output, if 50% of the output change would happen by changing the first input, 30% by changing the second input, 10% by changing the third one, and 10% due to interactions between the first two input parameters, these percentages are clearly reflected in measures of sensitivity. For more information about the basics and principles of this method, the reference of Sobol, (1992) can be reviewed [110].
- **Morris Method:** The Morris method (also named the Elementary Effects (EE) method) is one of the screening methods which is used to determine the effect of the input parameters on the model outputs (CMG)[7]. Morris approach has two measures, the Mean and the Standard Deviation, which are used together. The Mean reflects the linear influence of an input factor on the output function while the Standard Deviation reflects the nonlinear or interaction functionality. For more information about the basics and principles of this method, the reference of Morris, (1991) can be reviewed[109].
- **Tornado Plot:** a visual tool provides a qualitative and quantitative effect for the input parameters on the output ones, with a higher value meaning more sensitive to that parameter and vice versa (CMG)[7]. For more information about the basics and principles of this method, CMG reference number can be reviewed[7].

***Rank of the High-Impact Parameters which control the performance of $CO_2$-EOR huff-n-puff process.***

- **Formation total porosity (including natural fracture porosity):** Both of Sobol approach and Morris method indicated that the most important factor which affects the obtained oil recovery by $CO_2$-EOR is the total porosity of shale formation as shown in **Fig. 17A** and **Fig. 17B**. We concluded that as formation porosity and fracture intensity increase, oil recovery obtained by $CO_2$-EOR increases, which means that total porosity of the shale formation has a positive effect on $CO_2$-EOR performance as shown in **Fig. 16**. The interpretation which we think behind this behavior is that increasing the total porosity of formation would increase the contact area between the injected $CO_2$ and the formation oil, so $CO_2$-EOR performance would be enhanced.
- **Formation Average Permeability (counting for HF, NF, and matrix permeability):** Both of Sobol approach and Morris method indicated that the second parameter which controls the success of $CO_2$-EOR is the average conductivity of shale formation as shown **Fig. 17A** and **Fig. 17B**. We found that as the conductivity of oil pathways increased, oil recovery obtained by $CO_2$-EOR decreased, which means that the conductivity of oil-pathways has a negative effect on the $CO_2$-EOR performance as shown in **Fig. 17D**. The interpretation which we think behind this behavior is that increasing the conductivity of oil pathways in shale formations would result in increasing the kinetics of oil recovery process in the productive areas. As a result, for a limited effective diffusion rate for the injected-$CO_2$ into formation

oil, $CO_2$ would experience less exposure time with the formation oil before its being produced back. Therefore, $CO_2$-EOR performance would be downgraded with the increasing in oil-pathways conductivity.

- **Molecular Diffusion Rate:** Both of Sobol approach and Morris method indicated that the third parameter which controls the success of $CO_2$-EOR is the molecular diffusion rate between the injected $CO_2$ and the formation oil as shown in **Fig. 17A** and **Fig. 17B**. It is clear that as far as the molecular diffusion rate increased, the oil recovery obtained by the $CO_2$-EOR increased, which means that this parameter has a positive effect on the $CO_2$-EOR performance as shown in **Fig. 17D**. As far as the molecular diffusion rate increased, it would make $CO_2$ penetrate deeper into the tight matrix, far away from the hydraulic fractures. However, the case of $CO_2$ injection which has a low diffusion capacity makes the $CO_2$ penetrate just in the limited areas around the hydraulic fractures. Therefore, for the cases in which $CO_2$ penetrates deeper in the tight matrix, $CO_2$ would swell more volumes of oil, reduce oil viscosity, and finally produce larger quantities of oil by the counter-current mechanism. On the other hand, the cases in which $CO_2$ has a low molecular-diffusion rate would produce the injected-$CO_2$ back very soon.

**Table 9: Parameters with their range which were used in the CMOST analysis**

| Parameters | Range |
|---|---|
| Total Porosity (%) | 0.05-0.11 |
| K in I-direction (mD) | 0.005-0.011 |
| K in J-direction (mD) | 0.005-0.011 |
| K in K-direction(mD) | 0.005-0.011 |
| Diffusion Rate ($cm^2/sec$) | 0.0006-0.01 |
| HF K in I-direction (mD) | 1-10000 |
| HF K in J-direction (mD) | 1-10000 |
| HF K in K-direction (mD) | 1-10000 |

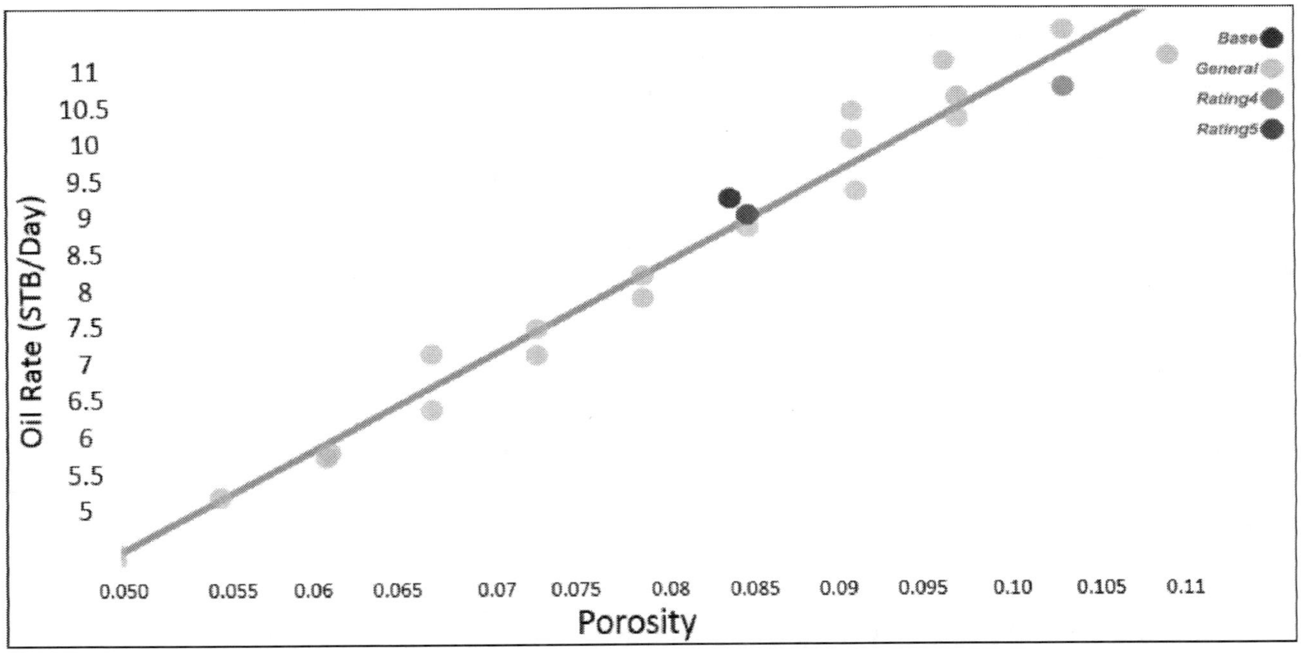


**Fig.16: Effect of the porosity (Natural Fracture intensity) on oil production enhancement by $CO_2$-EOR huff-n-puff**

A-Morris Method B-Sobol Method

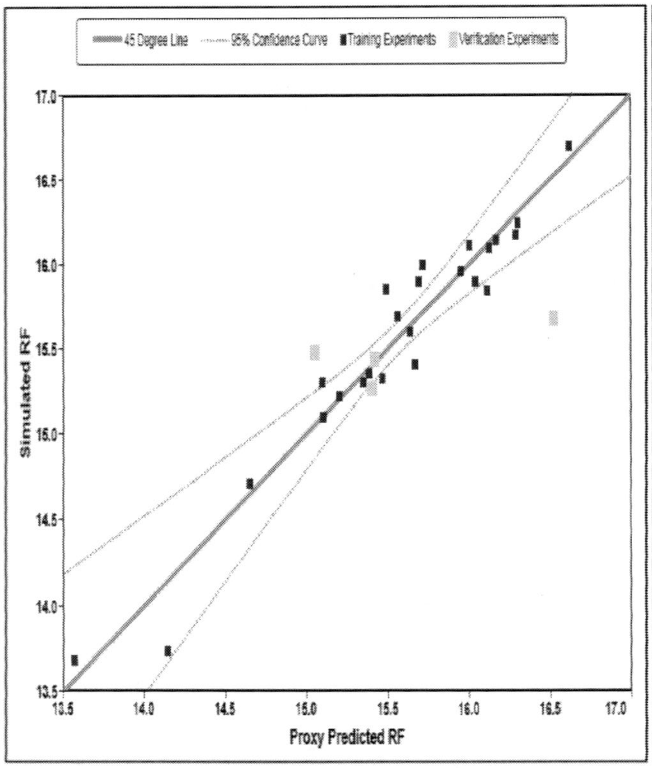


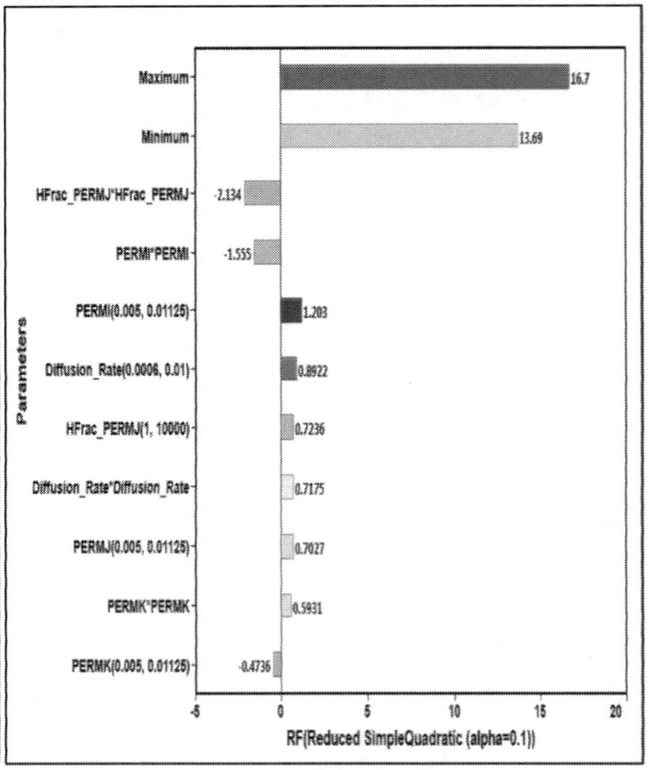


C-Proxy Vs simulated model D-Tornado Plot

**Fig.17: Design of Experiments for factors impacting $CO_2$-EOR huff-n-puff process**

## $CO_2$-Diffusivity Level in the Real Conditions for Shale Oil Reservoirs

We used the typical fluid and rock properties of Bakken to build a model for the Pilot test#2 which have been reported in Hoffman et al., (2016) paper and it was previously explained in this paper. Different scenarios have been run until the best match obtained between the well model and the pilot test as shown in **Fig. 18.** Everything was identical between the model results and pilot tests results which are shown in **Fig. 18**. However, there is just one difference. This difference is that the oil production came quickly after the soaking period in the pilot test; however, it takes longer time in the model case. We believe this is happening due to the reported conformance problems in these pilots where $CO_2$ produced in the offset wells. Therefore, the produced-back $CO_2$ volumes during puff process were small which resulted in less hold up effect on the produced-oil. However, in our model, we did not induce injection fractures. Therefore, $CO_2$ in large volumes produced back during the puff process of our model.

Among different scenario which we investigate, we found that this match can be obtained in a dual permeability model with a low $CO_2$ diffusivity. This means that either of the diffusion rates for the injected $CO_2$, in the reservoir conditions, is too low or the kinetics of the oil recovery process in the production areas are fast. The first possibility which is the low-diffusivity for the injected $CO_2$ in shale reservoirs conditions can be explained by two ways: (1) The contact area between the injected $CO_2$ and the formation oil is small, (2) The exposure time between the injected $CO_2$ and the formation oil is short. The contact area between $CO_2$ and the formation oil is a function of the natural-fractures intensity in shale oil reservoirs. Although it has been reported that these types of reservoirs have a high intensity of natural fractures, the dual permeability model can match the conducted pilot test results with a low intensity of natural fractures. This indicated that either of these natural fractures is not active or they are not connected in good pathways with the hydraulic fractures.

Most of the previous experimental studies reported that $CO_2$ diffusion mechanism is beyond the increment in the oil recovery obtained in the lab conditions. This increment in the oil recovery and/or the diffusion rate which was observed in the lab conditions was directly upscaled by most of the previous researchers to the field scale via numerical simulation methods. This direct upscaling methodology is so optimistic because the lab-cores have a higher contact area and a longer exposure time to the injected $CO_2$ than what happens in these reservoirs conditions. Therefore, both of the previous simulation studies and the experimental reports were optimistic to predict a quick improvement in the oil recovery from injecting $CO_2$ in these unconventional reservoirs. This might explain why the results from the $CO_2$ pilot tests are disappointing. To sum up, the diffusion mechanism for the injected $CO_2$ in the pilot tests had not been well recognized because either of the kinetics for the oil recovery process in the productive areas of these reservoirs are too fast or the $CO_2$ diffusion rate in the field conditions is too slow. According to this study, what happened in the field scale and what should be done is summarized in **Fig. 19**.

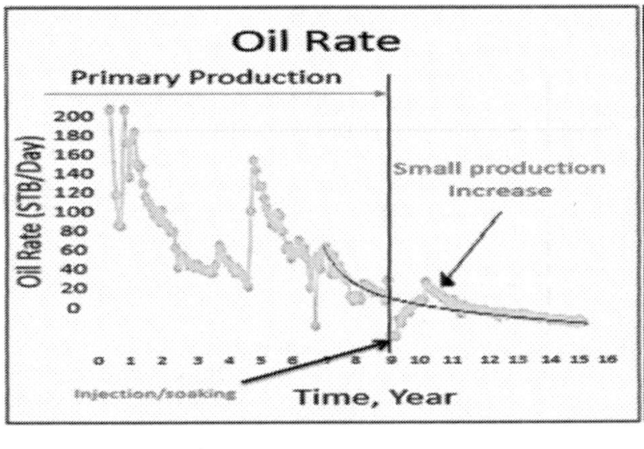


**Fig. 18A: $CO_2$ Pilot test#2 (Hoffman et al., 2016)**

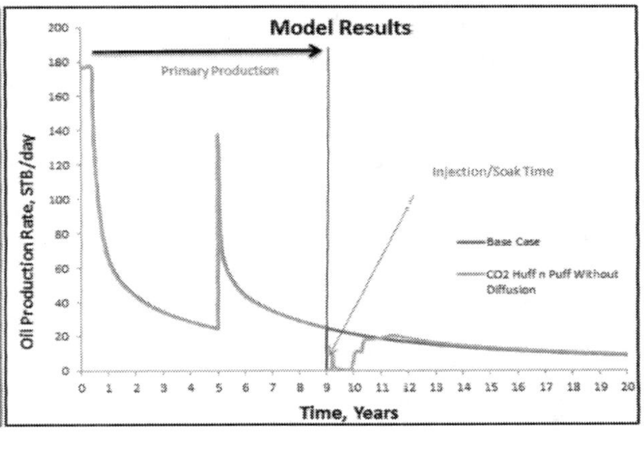


**Fig. 18B: History match from the simulated model**

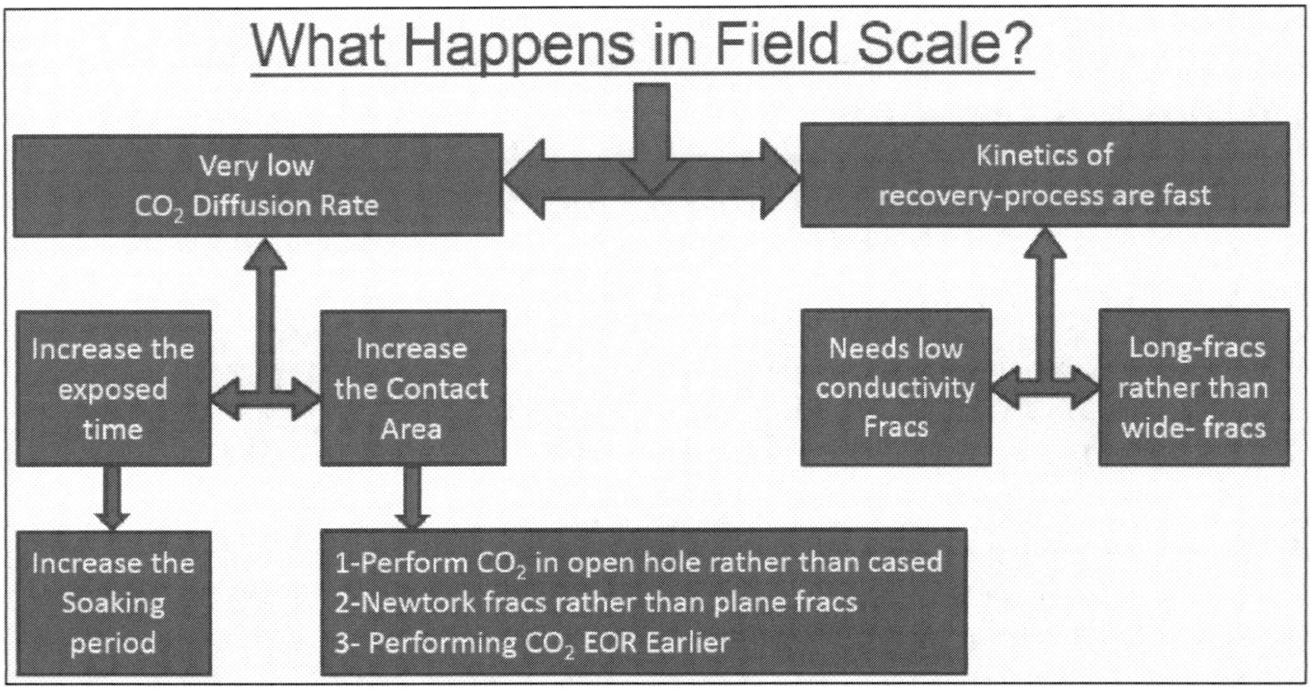


**Fig. 19: What happens at the field scale and what should be done according to this study**

## Conclusions

- From this study, some general guidelines have been provided to understand the $CO_2$-EOR performance in the field scale of unconventional reservoirs in North America.
- Three different approaches which are lab reports, numerical simulations, and pilot tests have been combined and compared in this study for getting an integrated picture about $CO_2$ mechanisms in shale oil reservoirs.
- Design of Experiments approved that the natural fracture intensity and oil-pathways conductivity are the two main factors which control $CO_2$-EOR success in shale oil reservoirs. However, the fractures intensity has a positive effect on $CO_2$-EOR while the later has a negative effect.
- The performing time for $CO_2$-EOR has a significant effect on the $CO_2$ huff-n-puff success.
- Molecular diffusion mechanism is the critical key for $CO_2$-EOR success in shale oil reservoirs. However, the direct upscaling for this mechanism to the field scale via conventional simulation methods by using the same lab-obtained $CO_2$-diffusion rates is misleading.
- To be significant in the field scale, this mechanism requires having either of kinetics for the oil recovery process in the productive areas of these reservoirs are too slow or $CO_2$ diffusion rates in field conditions are too fast.
- The history match with some of the reported pilot-tests indicated that the kinetics of oil recovery process in the productive areas are faster than the diffusion rates for the injected $CO_2$ in those poor-quality reservoirs.

## References


(1) Alfarge, D., Wei, M., and Bai, B., (2017a). IOR Methods in Unconventional Reservoirs of North America Comprehensive Review. SPE-185640-MS prepared for presentation at the SPE Western Regional Meeting held in Bakersfield, California, USA, 23-27 April

(2) Alfarge, D., Wei, M., Bai, B., & Almansour, A. (2017b). Optimizing Injector-Producer Spacing for CO2 Injection in Unconventional Reservoirs of North America. Society of Petroleum Engineers. doi:10.2118/188002-MS2017.

(3) Alharthy, N., Teklu, T., Kazemi, H. et al. 2015. Enhanced Oil Recovery in LiquidRich Shale Reservoirs: Laboratory to Field. Society of Petroleum Engineers. DOI: 10.2118/175034MS.

(4) Alvarez, J.O. and Schechter, D.S. 2016. Altering Wettability in Bakken Shale by Surfactant Additives and Potential of Improving Oil Recovery During Injection of Completion Fluids. Society of Petroleum Engineers. http:10.2118/SPE-179688-MS.

(5) Canadian Society of Unconventional Resources. http://www.csur.com/

(6) Clark, A. J., (2009). Determination of Recovery Factor in the Bakken Formation, Mountrail County, ND. Society of Petroleum Engineers. doi:10.2118/133719-STU.

(7) Computer Modeling Group, GEM Manual. Https: www.CMG.Ca/ accessed 2017.

(8) Dong, C., & Hoffman, B. T. (2013). Modeling Gas Injection into Shale Oil Reservoirs in the Sanish Field, North Dakota. Society of Petroleum Engineers. doi:10.1190/URTEC2013-185Energy Information Administration (EIA), (2015).

(9) Gamadi, T. D., Sheng, J. J., Soliman, M. Y., Menouar, H., Watson, M. C., & Emadibaladehi, H. (2014). An ExperimentalStudy of Cyclic CO2 Injection to Improve Shale Oil Recovery. Society of Petroleum Engineers. doi:10.2118/169142-MS

(10) Hawthorne et al., (2017). Measured Crude Oil MMPs with Pure and Mixed CO2, Methane, and Ethane, and Their Relevance to Enhanced Oil Recovery from Middle Bakken and Bakken Shales. SPE-185072-MS paper was prepared for presentation at the SPE Unconventional Resources Conference held in Calgary, Alberta, Canada, 15-16 February 2017.

(11) Hawthorne, S. B., Gorecki, C. D., Sorensen, J. A., Steadman, E. N., Harju, J. A., & Melzer, S. (2013). Hydrocarbon Mobilization Mechanisms from Upper, Middle, and Lower Bakken Reservoir Rocks Exposed to CO. Society of Petroleum Engineers.

(12) Hoffman, B. T., and Evans J., (2016). Improved Oil Recovery IOR Pilot Projects in the Bakken Formation. SPE-180270-MS paper presented at the SPE Low Perm Symposium held in Denver, Colorado, USA, 5–6 May 2016.

(13) Hoteit, H. and Firoozabadi, A. 2009. Numerical Modeling of Diffusion in Fractured Media for Gas-Injection and Recycling Schemes. SPE J. 14(02): 323–337. SPE-103292-PA. http://dx.doi.org/10.2118/103292-PA.

(14) Jin, L., Sorensen, J.A., Hawthorne, S.B., Smith, S.A., Bosshart, N.W., Burton Kelly, M.E., Miller, D.J., Grabanski, C.B., and Harju, J.A., (2016). Improving oil transportability using CO2 in the Bakken System—a laboratorial investigation: Proceedings of the SPE International Conference & Exhibition on Formation Damage Control, SPE 178948, Lafayette, Louisiana, February 24–26.

(15) Kathel, P. and Mohanty, K.K. 2013. Eor in Tight Oil Reservoirs through Wettability Alteration. Society of Petroleum Engineers. DOI: 10.2118/166281MS.

(16) Kurtoglu, B., Kazemi, H., Rosen, R., Mickelson, W., & Kosanke, T. (2014). A Rock and Fluid Study of Middle Bakken Formation: Key to Enhanced Oil Recovery.Society of Petroleum Engineers. doi:10.2118/171668-MS.

(17) LeFever, J., & Helms, L., 2008, Bakken Formation Reserve Estimates, North Dakota Geological Survey.

(18) Mohebbinia et al., (2017). Molecular Diffusion Calculations in Simulation of Gasfloods in Fractured Reservoirs. SPE-182594-MS Paper presented at the SPE Reservoir Simulation Conference held in Montgomery, TX, USA, 20–22 February 2017.

(19) Moortgat, J. and Firoozabadi, A. 2013. Fickian Diffusion in Discrete-Fractured Media from Chemical Potential Gradients and Comparison to Experiment. Energ Fuel27 (10): 5,793–5,805. http://dx.doi.org/10.1021/ef401141q.

(20) Morris, M.D. Factorial Sampling Plans for Preliminary Computational Experiments. Technometrics (American Statistical Association) 33, no. 2 (May 1991).

(21) Pu, H., and Li, Y., (2016). Novel Capillarity Quantification Method in IOR Process in Bakken Shale Oil Reservoirs. SPE-179533-MS presentated at the SPE Improved Oil Recovery Conference held in Tulsa, Oklahoma, USA, 11–13 April 2016

(22) Schmidt, M., & Sekar, B. K. (2014). Innovative Unconventional 2EOR-A Light EOR an Unconventional TertiaryRecovery Approach to an Unconventional Bakken Reservoir in Southeast Saskatchewan.World Petroleum Congress.

(23) Sheng, J.J. 2015. Enhanced oil recovery in shale reservoirs by gas injection. Journal of Natural Gas Science and Engineering 22: 252–259. http://dx.doi.org/10.1016/j.jngse.2014.12.002

(24) Shoaib, S., & Hoffman, B. T. (2009). CO2 Flooding the Elm Coulee Field. Society of Petroleum Engineers. doi:10.2118/123176-MS.

(25) Sobol, I. Sensitivity Estimates for Nonlinear Mathematical Models. Matematicheskoe Modelirovanie 2, 1993: 112-118.

(26) Song, C., & Yang, D. (2013). Performance Evaluation of CO2 Huff-n-Puff Processes in Tight Oil Formations. Society of Petroleum Engineers. doi:10.2118/167217-MS.

(27) Wan, T., & Sheng, J. (2015). Compositional Modelling of the Diffusion Effect on EOR Process in Fractured Shale-Oil Reservoirs by Gasflooding. Society of Petroleum Engineers. doi:10.2118/2014-1891403-PA.

(28) Wang, D., Zhang, J., Butler, R., and Olatunji, K., (2016). Scaling Laboratory-Data Surfactant-Imbibition Rates to the Field in Fractured-Shale Formations. Socienty of Petroleum Engineers. http://dx.doi.org/10.2118/178489-PA.

(29) Wang, X., Luo, P., Er, V., & Huang, S.-S. S. (2010). Assessment of CO2 Flooding Potential for Bakken Formation,Saskatchewan. Society of Petroleum Engineers. doi: 10.2118/137728-MS.

(30) Wood, T., & Milne, B., (2011). Waterflood potential could unlock billions of barrels: Crescent Point Energy. http://www.investorvillage.com/uploads/44821/files/CPGdundee.pdf

(31) Xu, T., & Hoffman, T. (2013). Hydraulic Fracture Orientation for Miscible Gas Injection EOR in Unconventional OilReservoirs. Society of Petroleum Engineers. doi:10.1190/URTEC2013-189

(32) Yu et al., (2016b). Experimental Evaluation of Shale Oil Recovery from Eagle Ford Core Samples by Nitrogen Gas Flooding. SPE-179547-MS Paper presented at the SPE Improved Oil Recovery Conference held in Tulsa, Oklahoma, USA, 11–13 April 2016.

(33) Yu, W., Lashgari, H., & Sepehrnoori, K. (2014). Simulation Study of CO2 Huff-n-Puff Process in Bakken Tight OilReservoirs. Society of Petroleum Engineers. doi:10.2118/169575-MS
(34) Yu, Y., Li, L. and Sheng J., (2016a). Further Discuss the Roles of Soaking Time and Pressure Depletion Rate in Gas Huff-n-Puff Process in Fractured Liquid-Rich Shale Reservoirs. SPE-181471-MS paper presented in at the SPE Annual Technical Conference and Exhibition held in Dubai, UAE, 26-28 September 2016.
(35) Zhang et al., (2015). Intergrated Method to Screen Tight Oil Reservoirs for CO2 Flooding. SPE-175969-MS paper presented at the SPE/CSUR Unconventional Resources Conference held in Calgary, Alberta, Canada, 20-22 October 2015.
(36) Zhang, K., (2016). Experimental and Numerical Investigation of Oil Recovery from Bakken Formation by Miscible CO2 Injection. Paper SPE 184486 presented at the SPE international Student Paper Contest at the SPE Annual Technical Conference and Exhibition held in Dubai, UAE, 26-28 September 2016.
(37) Zhu, P., Balhoff, M. T., & Mohanty, K. K. (2015). Simulation of Fracture-to-Fracture Gas Injection in an Oil-Rich Shale. Society of Petroleum Engineers. doi:10.2118/175131-MSSPE-180270-MS17
(38) Alharthy, N., Teklu, T., Kazemi, H. et al. 2015. Enhanced Oil Recovery in LiquidRich Shale Reservoirs: Laboratory to Field. Society of Petroleum Engineers. DOI: 10.2118/175034MS.
(39) Alvarez, J.O. and Schechter, D.S. 2015. Wettability Alteration and Spontaneous Imbibition in Unconventional Liquid Reservoirs by Surfactant Additives. Society of Petroleum Engineers. DOI: 10.2118/177057MS.
(40) Alvarez, J. O., Neog, A., Jais, A., & Schechter, D. S. (2014). Impact of Surfactants for Wettability Alteration in StimulationFluids and the Potential for Surfactant EOR in Unconventional Liquid Reservoirs.Society of Petroleum Engineers.doi:10.2118/169001-MS SPE-180270-MS15
(41) Alvarez, J.O. and Schechter, D.S. 2016. Altering Wettability in Bakken Shale by Surfactant Additives and Potential of Improving Oil Recovery During Injection of Completion Fluids. Society of Petroleum Engineers. http:10.2118/SPE-179688-MS.
(42) Baker, et al.,(2016). The Myths of Waterfloods, EOR Floods and How to Optimize Real Injection Schemes. SPE-179536-MS paper was prepared for presentation at the SPE Improved Oil Recovery Conference held in Tulsa, Oklahoma, USA, 11–13 April 2016.
(43) Chen, C., Balhoff, M. T., & Mohanty, K. K. (2014). Effect of Reservoir Heterogeneity on Primary Recovery and CO2 Huff-n-Puff Recovery in Shale-Oil Reservoirs. Society of Petroleum Engineers. doi:10.2118/164553-PA
(44) Clark, A. J., (2009). Determination of Recovery Factor in the Bakken Formation, Mountrail County, ND. Society of Petroleum Engineers. doi:10.2118/133719-STU.
(45) Dawson, M., Nguyen, D., Champion, N., & Li, H. (2015). Designing an Optimized Surfactant Flood in the Bakken. Societyof Petroleum Engineers. doi:10.2118/175937-MS
(46) Dong, C., & Hoffman, B. T. (2013). Modeling Gas Injection into Shale Oil Reservoirs in the Sanish Field, North Dakota. Society of Petroleum Engineers. doi:10.1190/URTEC2013-185Energy Information Administration (EIA), (2015).
(47) Evans, J., Heath, L., Reichhardt, D., & Todd, B. (2015). Elm Coulee EOR Study: Interim Report,Montana Board of Oiland Gas.Ferris, F. G., Stehmeier, L.G., Kantzas, A., Mourits, F. M., (1996). Bacteriogenic mineral plugging,Journal of CanadianPetroleum Technology,13,57–67.
(48) Fitzel, S., Sekar, B., Alvarez, D., & Gulewicz, D. (2015). Gas Injection EOR Optimization Using Fiber-Optic Logging with DTS and DAS for Remedial Work. Society of Petroleum Engineers. doi:10.2118/175891-MS
(49) Flannery, J., and Kraus, J., (2006). Integrated analysis of the Bakken petroleum system, U.S. Williston Basin: American Association of Petroleum Geologists Search and Discovery Article No. 10105.
(50) Foo, D. B., Krislock, J., Meador, T. J., & Cheng, T., (2014). Horizontal Well Injection Profiling Using Distributed Temperature Sensing. Society of Petroleum Engineers. doi:10.2118/171586
(51) Fragoso, A., Wang, Y., Jing, G.et al. 2015. Improving Recovery of Liquids from Shales through Gas Recycling and Dry Gas Injection. Presented at the SPE Latin American and CaribbeanPetroleum Engineering Conference, 18–20 November, Ecuador, Quito. SPE-177278-MS. http://dx.doi.org/10.2118/177278-MS.
(52) Gamadi, T. D., Sheng, J. J., Soliman, M. Y., Menouar, H., Watson, M. C., & Emadibaladehi, H. (2014). An ExperimentalStudy of Cyclic CO2 Injection to Improve Shale Oil Recovery. Society of Petroleum Engineers. doi:10.2118/169142-MS
(53) Gaswirth, S.B., Marra, K.R., Cook, T.A., Charpentier, R.R., Gautier, D.L., Higley, D.K., Klett, T.R., Lewan, M.D., Lillis,P.G., Schenk, C.J., Tennyson, M.E., and Whidden, K.J.2013. Assessment of Undiscovered Oil Resources in the Bakken and Three Forks Formations, Williston Basin Province, Montana, North Dakota, and South Dakota, 2013 U.S. Geological Survey Fact Sheet 2013-3013. U.S. Geological Survey: Denver, CO.
(54) Hawthorne, S. B., Gorecki, C. D., Sorensen, J. A., Steadman, E. N., Harju, J. A., & Melzer, S. (2013). Hydrocarbon Mobilization Mechanisms from Upper, Middle, and Lower Bakken Reservoir Rocks Exposed to CO. Society of Petroleum Engineers. doi:10.2118/167200-MS
(55) Hoffman, B. T. (2012). Comparison of Various Gases for Enhanced Recovery from Shale Oil Reservoirs. Society of Petroleum Engineers. doi:10.2118/154329-MS
(56) Hoffman, B. T., and Evans J., (2016). Improved Oil Recovery IOR Pilot Projects in the Bakken Formation. SPE-180270-MS paper presented at the SPE Low Perm Symposium held in Denver, Colorado, USA, 5–6 May 2016.
(57) Kathel, P. and Mohanty, K.K. 2013. Eor in Tight Oil Reservoirs through Wettability Alteration. Society of Petroleum Engineers. DOI: 10.2118/166281MS.
(58) King, G. E. (2014). 60 Years of Multi-Fractured Vertical, Deviated and Horizontal Wells: What Have We Learned?Societyof Petroleum Engineers. doi:10.2118/170952-MS
(59) Kurtoglu, B., & Salman, A. (2015). How to Utilize Hydraulic Fracture Interference to Improve Unconventional Development. Society of Petroleum Engineers. doi:10.2118/177953-MS
(60) Kurtoglu, B., Kazemi, H., Boratko, E. C., Tucker, J., & Daniels, R. (2013). Mini-Drillstem Tests to Characterize Formation Deliverability in the Bakken. Society of Petroleum Engineers. doi:10.2118/159597-PA
(61) Kurtoglu, B., Kazemi, H., Rosen, R., Mickelson, W., & Kosanke, T. (2014). A Rock and Fluid Study of Middle Bakken

Formation: Key to Enhanced Oil Recovery.Society of Petroleum Engineers. doi:10.2118/171668-MS.
(62) Kurtoglu, B., Sorensen, J. A., Braunberger, J., Smith, S., & Kazemi, H. (2013). Geologic Characterization of a BakkenReservoir for Potential CO2 EOR. Society of Petroleum Engineers. doi:10.1190/URTEC2013-186.
(63) Kurtoglu, B. 2013. Integrated Reservoir Characterization and Modeling in Support of Enhanced Oil Recovery for Bakken.Ph.D. Dissertation, Colorado School of Mines.
(64) Lane, C., Laun, L. E., & Schlosser, D. (2013). Reducing Water Volume in Multistage Fracturing Using Sliding Sleeves and Coiled-Tubing-Deployed Resettable Frac Isolation. Society of Petroleum Engineers. doi:10.2118/163935-MS
(65) Lawal, H., Jackson, G., Abolo, N., & Flores, C. (2013). A Novel Approach To Modeling and Forecasting Frac Hits InShale Gas Wells. Society of Petroleum Engineers. doi:10.2118/164898-MS
(66) LeFever, J., & Helms, L., 2008, Bakken Formation Reserve Estimates, North Dakota Geological Survey.
(67) Li, X., Teklu, W., Abass, A., and Cui, Qi. 2016. The Impact Of Water Salinity /Surfactant On Spontaneous Imbibition Through Capillarity And Osmosis For Unconventional IOR. URTec#2461736 presented at the Unconventional Resources Technology conference held in san Antonio, Texas, USA, 1-3 August.
(68) Liu, G., Sorensen, J. A., Braunberger, J. R., Klenner, R., Ge, J., Gorecki, C. D., Harju, J. A. (2014). CO2-Based Enhanced Oil Recovery from Unconventional Reservoirs: A Case Study of the Bakken Formation. Society of Petroleum Engineers. doi:10.2118/168979-MSMontana Board of Oil and Gas (MBOG), 2015. Live Data Access,http://bogc.dnrc.mt.gov/WebApps/DataMiner/
(69) Morsy, S and Sheng, J.J., 2014. Effect of Water Salinity on Shale reservoir Productivity. Advances in Petroleum Exploration and Development 8(1): 9–14. DOI:10.3968/5604
(70) Morsy, S., Sheng, J. J., & Soliman, M. Y. (2013). Waterflooding in the Eagle Ford Shale Formation: Experimental and Simulation Study. Society of Petroleum Engineers. doi:10.2118/167056-MS.
(71) Nguyen, D., Wang, D., Oladapo, A., Zhang, J., Sickorez, J., Butler, R., Mueller, B., 2014. Evaluation of surfactants for oil recovery potential in shale reservoirs. Paper SPE 169085. Presented at the SPE Improved Oil Recovery Symposium. Tulsa, Oklahoma. April 12–16.
(72) North Dakota Industrial Commission (NDIC), 2015.Oil and Gas Division, https://www.dmr.nd.gov/oilgas/
(73) Patterson, R., (2015). Bakken Decline Rates Worrying For Drillers OilPrice.com, February 25, http://oilprice.com/ Energy/Crude-Oil/Bakken-Decline-Rates-Worrying-For-Drillers.html Prats, M. (1982). Thermal Recovery, SPE Monograph SeriesVol.7,283 pp.
(74) Price, L.C., (1999). Origins and characteristics of the basin-centered continuous-reservoir unconventional oil resource base of the Bakken source system, Williston Basin: unpublished manuscript: www.undeerc.org/Price16SPE-180270-MS
(75) Pu, W., & Hoffman, B. T. (2014). EOS Modeling and Reservoir Simulation Study of Bakken Gas Injection Improved Oil Recovery in the Elm Coulee Field, Montana.Society of Petroleum Engineers. doi:10.15530/urtec-2014-1922538
(76) Pu, H., and Li, Y., (2016). Novel Capillarity Quantification Method in IOR Process in Bakken Shale Oil Reservoirs. SPE-179533-MS presentated at the SPE Improved Oil Recovery Conference held in Tulsa, Oklahoma, USA. 11–13 April 2016
(77) Schmidt, M., & Sekar, B. K. (2014). Innovative Unconventional 2EOR-A Light EOR an Unconventional TertiaryRecovery Approach to an Unconventional Bakken Reservoir in Southeast Saskatchewan.World Petroleum Congress.
(78) Sheng, J.J. 2015. Enhanced oil recovery in shale reservoirs by gas injection. Journal of Natural Gas Science and Engineering 22: 252–259. http://dx.doi.org/10.1016/j.jngse.2014.12.002
(79) Shoaib, S., & Hoffman, B. T. (2009). CO2 Flooding the Elm Coulee Field. Society of Petroleum Engineers. doi:10.2118/123176-MS.
(80) Shuler, P., Tang, H., Zayne, Lu., Tang, Y., 2001. Chemical processes for improved oil recovery from Bakken shale Paper SPE 147531. Presented at the Canadian Unc. Res. Conf. Calgary, Alberta. November 15–17.
(81) Song, C., & Yang, D. (2013). Performance Evaluation of CO2 Huff-n-Puff Processes in Tight Oil Formations. Society of Petroleum Engineers. doi:10.2118/167217-MS.
(82) Sorensen, J. A., Braunberger, J. R., Liu, G., Smith, S. A., Hawthorne, S. A., Steadman, E. N., & Harju, J. A. (2015). Characterization and Evaluation of the Bakken Petroleum System for CO Enhanced Oil Recovery.Society ofPetroleum Engineers.doi:10.2118/178659MS.
(83) Teletzke, G. F., Wattenbarger, R. C., & Wilkinson, J. R. (2010). Enhanced Oil Recovery Pilot Testing Best Practices. Society of Petroleum Engineers. doi:10.2118/118055-PA
(84) Todd, B., (2016). An Evaluation of EOR Potential in the Elm Coulee Bakken Formation, Richland County, Montana, SPERocky Mountain Student Symposium, Butte, MT, Feb. 26.
(85) Tovar, F. D., Eide, O., Graue, A., & Schechter, D. S. (2014). Experimental Investigation of Enhanced Recovery in Unconventional Liquid Reservoirs using CO2: A Look Ahead to the Future of Unconventional EOR. Society ofPetroleum Engineers. doi:10.2118/169022-MS.
(86) Tran, T., Sinurat, P., and Wattenbarger, R.A.2011. Production Characteristics of the Bakken Shale Oil. Presented at the SPE Annual Technical Conference and Exhibition, Denver, CO, USA, 30 October-2 November. SPE-145684.
(87) Valluri et al., (2016). Study of the Rock/Fluid Interactions of Sodium and Calcium Brines with Ultra-Tight Rock Surfaces and their Impact on Improving Oil Recovery by Spontaneous Imbibitio SPE-180274-MS confrernce paper presented at the SPE Low Perm Symposium held in Denver, Colorado, USA, 5–6 May 2016.
(88) Wan, T., & Sheng, J. (2015). Compositional Modelling of the Diffusion Effect on EOR Process in Fractured Shale-OilReservoirs by Gasflooding. Society of Petroleum Engineers. doi:10.2118/2014-1891403-PA.
(89) Wan, T., Meng, X., Sheng, J. J., & Watson, M. (2014). Compositional Modeling of EOR Process in Stimulated Shale Oil Reservoirs by Cyclic Gas Injection. Society of Petroleum Engineers. doi: 10.2118/169069-MS.
(90) Wan, T., Sheng, J. J., & Soliman, M. Y. (2013). Evaluate EOR Potential in Fractured Shale Oil Reservoirs by Cyclic Gas Injection. Society of Petroleum Engineers. doi:10.1190/URTEC2013-187
(91) Wang, D., Butler, R., Liu, H. et al. 2011. Flow Rate Behavior and Imbibition in Shale.SPE Res Eval & Eng14(4): 505–512. SPE-

138521-PA. http://dx.doi.org/10.2118/138521-PA.
(92) Wang, D., Butler, R., Zhang, J. et al. 2012. Wettability Survey in Bakken Shale With Surfactant-Formulation Imbibition. SPE Res Eval & Eng15(6): 695–705. SPE-153853-PA. http://dx.doi.org/10.2118/153853-PA.
(93) Wang, D., Butler, R., Zhang, J., & Seright, R. (2012). Wettability Survey in Bakken Shale With Surfactant-FormulationImbibition.Society of Petroleum Engineers. doi:10.2118/153853-PA.
(94) Wang, D., Butler, R., Zhang, J., 2012.Wettability survey in Bakken shale with surfactant formulation imbibition, Paper SPE 153853 presented at the 18th SPE Improved Oil Recovery Symposium. Tulsa, OK. April 14–18
(95) Wang, D., Butler, R., Zhang, J., 2012.Wettability survey in Bakken shale with surfactant formulation imbibition, Paper SPE 153853 presented at the 18th SPE Improved Oil Recovery Symposium. Tulsa, OK. April 14–18.
(96) Wang, D., Zhang, J. and Butler, R. 2014. Flow Rate Behavior and Imbibition Comparison between Bakken and Niobrara Formations. Presentedat SPE/AAPG/SEG Unconventional Resources Technology Conference, Denver, 25–27 August. SPE-2014-1920887-MS. http://dx.doi.org/10.15530/urtec-2014-1920887.
(97) Wang, X., Luo, P., Er, V., & Huang, S.-S. S. (2010). Assessment of CO2 Flooding Potential for Bakken Formation,Saskatchewan. Society of Petroleum Engineers. doi: 10.2118/137728-MS.
(98) Wang, D., Zhang, J., Butler, R., and Olatunji, K., (2016). Scaling Laboratory-Data Surfactant-Imbibition Rates to the Field in Fractured-Shale Formations. Socienty of Petroleum Engineers. http://dx.doi.org/10.2118/178489-PA.
(99) Wood, T., & Milne, B., (2011). Waterflood potential could unlock billions of barrels: Crescent Point Energy. http://www.investorvillage.com/uploads/44821/files/CPGdundee.pdf
(100) Xu, T., & Hoffman, T. (2013). Hydraulic Fracture Orientation for Miscible Gas Injection EOR in Unconventional OilReservoirs. Society of Petroleum Engineers. doi:10.1190/URTEC2013-189
(101) Yang et al., (2016). Effects of Multicomponent Adsorption on Enhanced Shale Reservoir Recovery by CO2 Injection Coupled with Reservoir Geomechanics. SPE-180208-MS paper was prepared for presentation at the SPE Low Perm Symposium held in Denver, Colorado, USA, 5–6 May 2016.
(102) Yu, Y., Li, L. and Sheng J., (2016a). Further Discuss the Roles of Soaking Time and Pressure Depletion Rate in Gas Huff-n-Puff Process in Fractured Liquid-Rich Shale Reservoirs. SPE-181471-MS paper presented in at the SPE Annual Technical Conference and Exhibition held in Dubai, UAE, 26-28 September 2016.
(103) Yu et al., (2016b). Experimental Evaluation of Shale Oil Recovery from Eagle Ford Core Samples by Nitrogen Gas Flooding. SPE-179547-MS Paper presented at the SPE Improved Oil Recovery Conference held in Tulsa, Oklahoma, USA, 11–13 April 2016.
(104) Yu et al., (2016c). Experimental Investigation of Light Oil Recovery from Fractured Shale Reservoirs by Cyclic Water Injection. SPE-180378-MS paper presented at the SPE Western Regional Meeting held in Anchorage, Alaska, USA, 23–26 May 2016.
(105) Yu, W., Lashgari, H., & Sepehrnoori, K. (2014). Simulation Study of CO2 Huff-n-Puff Process in Bakken Tight OilReservoirs. Society of Petroleum Engineers. doi:10.2118/169575-MS
(106) Zhang, J., Wang, D. and Butler, R. 2013. Optimal Salinity Study to Support Surfactant Imbibition into the Bakken Shale. Presented at SPE Unconventional Resource Conference, Canada, Calgary, 5–7 November. SPE-167142-MS. http://dx.doi.org/10.2118/167142-MS.
(107) Zhang, K., 2016. Experimental and Numerical Investigation of Oil Recovery from Bakken Formation by Miscible CO2 Injection. Paper SPE 184486 presented at the SPE international Student Paper Contest at the SPE Annual Technical Conference and Exhibition held in Dubai, UAE, 26-28 September 2016.
(108) Zhu, P., Balhoff, M. T., & Mohanty, K. K. (2015). Simulation of Fracture-to-Fracture Gas Injection in an Oil-Rich Shale. Society of Petroleum Engineers. doi:10.2118/175131-MS SPE-180270-MS.
(109) Morris, M.D. Factorial Sampling Plans for Preliminary Computational Experiments. Technometrics (American Statistical Association) 33, no. 2 (May 1991).
(110) Sobol, I. Sensitivity Estimates for Nonlinear Mathematical Models. Matematicheskoe Modelirovanie 2, 1993: 112-118.
(111) Hoteit, H. and Firoozabadi, A. 2009. Numerical Modeling of Diffusion in Fractured Media for Gas-Injection and -Recycling Schemes.SPE J14(2): 323–337. SPE-103292-PA. http://dx.doi.org/10.2118/103292-PA.
(112) Sigmund, P. M. 1976. Prediction of Molecular Diffusion at the Reservoir Conditions. Part 1- Measurement and Prediction of Binary Dense Gas Diffusion Coefficients. J Can Pet Technol15(2): 48–57. PETSOC-76-02-05. http://dx.doi.org/10.2118/76-02-05.
(113) Sigmund, P. M. 1976. Prediction of Molecular Diffusion at the Reservoir Conditions. Part II – Estimating the Effectsof Molecular Diffusion and Convective Mixing in Multicomponent Systems. J Can Pet Technol15 (3): 53–62.PETSOC-76-03-07. http://dx.doi.org/10.2118/76-03-07.

# New Amine-Containing Membranes for $CO_2$ Capture

Kai Chen, Yang Han, Witopo Salim, Zi Tong, Dongzhu Wu and W.S. Winston Ho

William G. Lowrie Department of Chemical & Biomolecular Engineering
Department of Materials Science and Engineering
The Ohio State University, Columbus, Ohio, USA

The Carbon Management Technology Conference 2017
Houston, TX, July 17 – 20, 2017

1

# Outline

- $CO_2$ Capture from Flue Gas in Coal- and/or Natural Gas-fired Power Plants
- $CO_2$ Capture from <1% $CO_2$ Concentration Sources, e.g.,
  - Residual flue gas after the primary $CO_2$ capture system
  - Coal-mine gas streams

2

# Introduction

- **Coal-fired power plants**
  - **40% of global $CO_2$ emission**
  - **Remain as major energy supply**

- **Membranes for $CO_2$ capture from flue gas**
  - **System compactness**
  - **Energy efficiency**
  - **Operational simplicity**
  - **Kinetic ability to overcome thermodynamic solubility limitation**

3

# Challenges and Goals

- **High $CO_2$ Permeance** $$P_{CO_2} = \frac{J_{CO_2}}{p_{CO_2,f} - p_{CO_2,p}}$$
  - **Reduce Membrane Area**

  **Thin Membrane Thickness Needed**

- **High $CO_2/N_2$ Selectivity** $$\alpha_{CO_2/N_2} = \frac{P_{CO_2}}{P_{N_2}}$$
  - **Provide High $CO_2$ Purity**

  **High Selective Membrane Material Needed**

**>700 GPU $CO_2$ permeance and >140 $CO_2/N_2$ selectivity at 57°C**
**To meet the DOE target < $40/tonne $CO_2$ (2007 $)**

4

# Amine Polymer Layer Contains Mobile and Fixed Carriers: Facilitated Transport

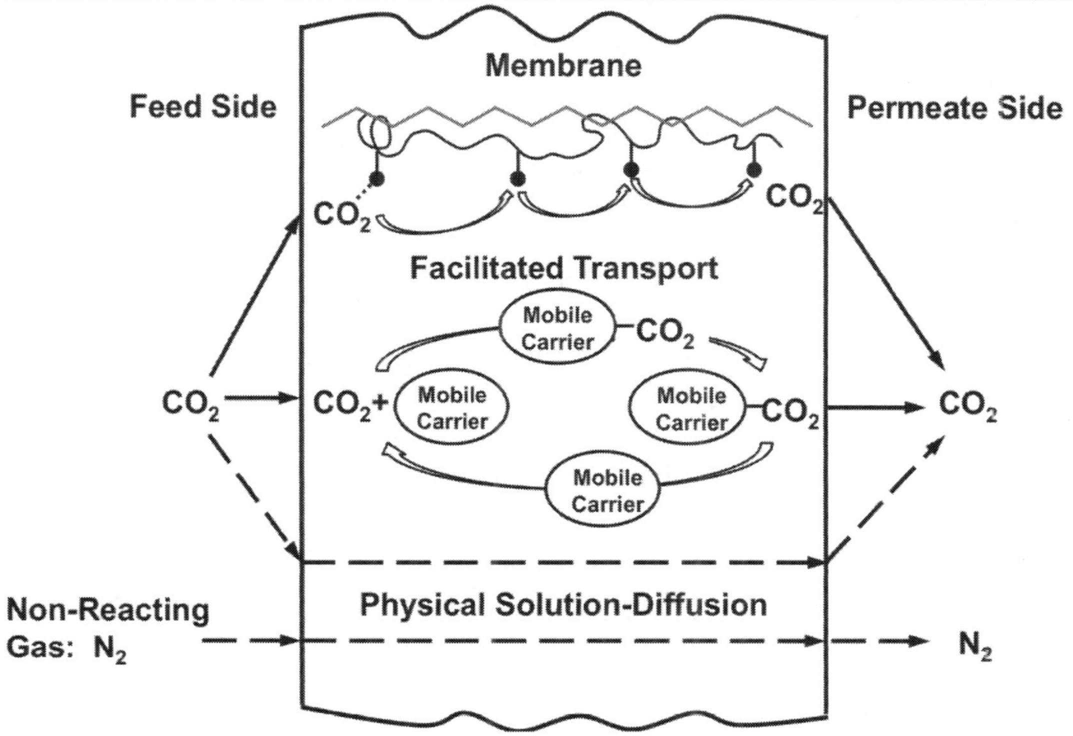


5

# Amine-Containing Polymer Membrane Structure

## Simplicity of Membrane for Low Cost

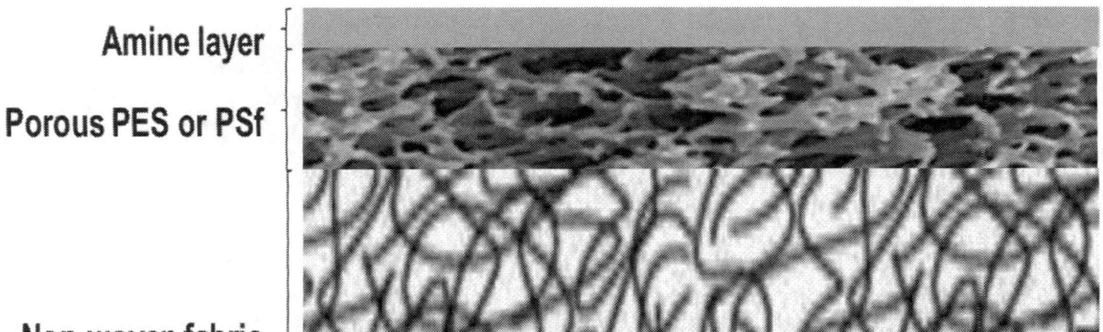


6

# Amine Layer Composition

**Fixed-site carrier:**

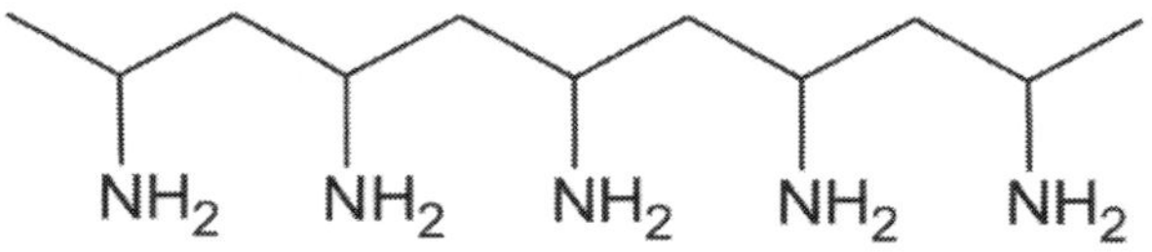

Polyvinylamine (PVAm)

**Mobile carrier:**

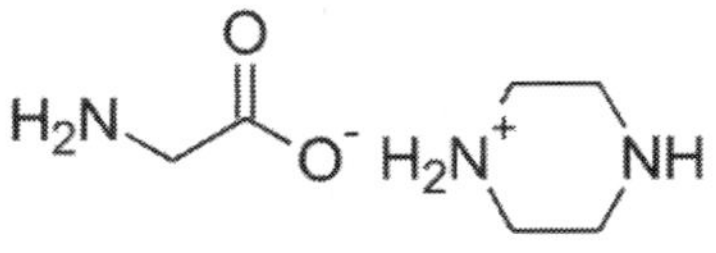

Piperazine glycinate (PG)

**Surfactant:**

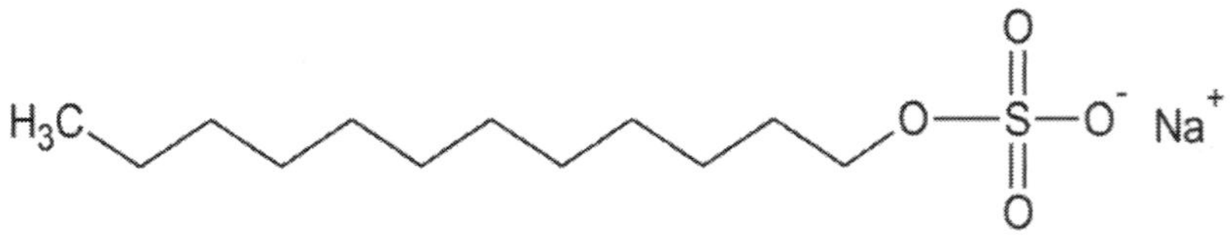

Sodium dodecyl sulfate (SDS)

7

# High-Molecular-Weight PVAm Synthesis

| Polymer | Monomer Conc. (wt.%) | Initiator/Monomer Weight Ratio | 3 wt.% Polymer Solution viscosity (cp) | Weight Average MW |
|---|---|---|---|---|
| PVAm (a) | 30 | 0.5/100 | 486 | 719,000 |
| PVAm (b) | 40 | 0.14/100 | 1,400 | 1,200,000 |
| Lupamin® | N/A | N/A | 50 | 340,000 |

- **MW measured by Dynamic Light Scattering (DLS)**
- **Lupamin® contains ~66% sodium formate**

8

# High-Molecular-Weight PVAm Synthesized

Viscosity:
3 wt.% commercial PVAm: 50 cp
3 wt.% high MW PVAm: 450 – 1950 cp

a) High viscosity
- Decrease coating solution concentration
- Allow for preparation of thinner membrane
- Improve adhesion to support – less defects

b) Low salt amount
- Improve membrane stability

9

# Composite Membrane Synthesized

## Selective Amine Polymer Layer on PES Support

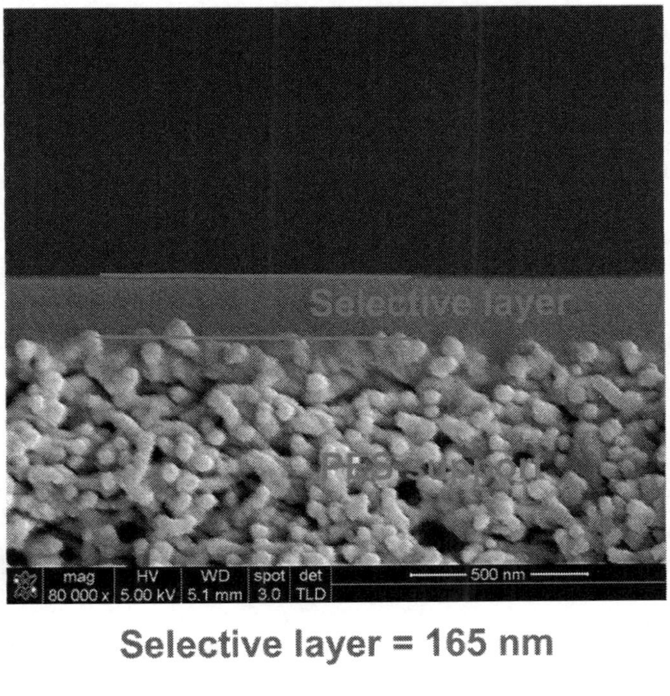


Selective layer = 165 nm

10

## Membrane Scale-up: Continuous Roll-to-Roll Fabrication Machine at OSU

11

## Tuning Membrane Thickness

$$l \times \rho_{dry} = 0.5 \times c \times \rho_{sol} \times l_{gap}$$

$l$: dry membrane thickness (µm)

$\rho_{dry}$: density of dry membrane ($g/cm^3$)

$\rho_{sol}$: density of coating solution ($g/cm^3$)

$c$: total solid concentration of coating solution (by weight)

$l_{gap}$: gap setting of coating knife (µm)

12

# Coating Technique Modification: Insert

| Before Modification | After Modification |
|---|---|

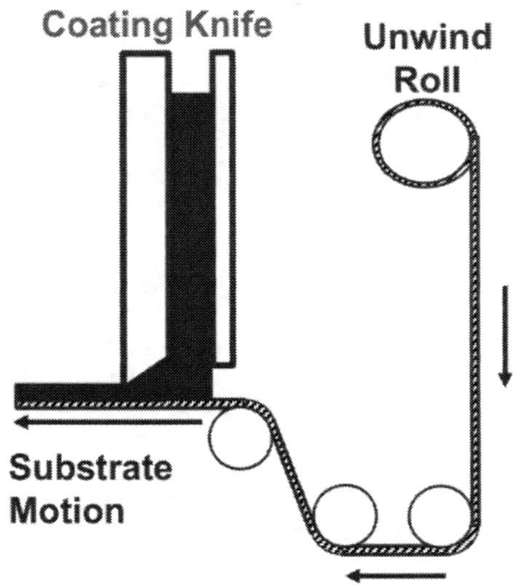


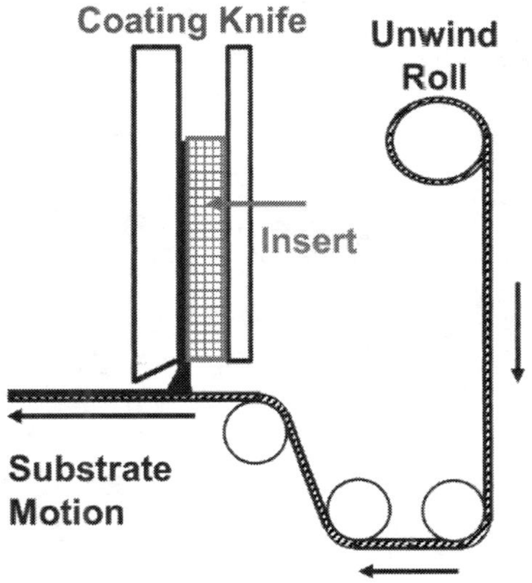


- **A thinner thickness with a higher web speed for a given coating solution supply rate**

13

# Thin 14”-Wide Amine Coatings on Substrates

| Coating knife assembly configuration | Polymer concentration (wt.%) | Viscosity of Coating Solution (cp) | Web speed (ft/min) | Selective amine layer thickness (nm) |
|---|---|---|---|---|
| Insert with 1-inch flow channel height [a] | 1.8 | 870 | 3 | 175 |
| | | | 4 | 135 |
| | | | 5 | 115 |
| | 2.4 | 940 | 3.25 | 210 |
| | | | 3.5 | 195 |
| | 3 | 1180 | 3.5 | 185 |
| Insert with 2-inch flow channel height [a] | 1.5 | 690 | 1 | 230 |
| | | | 2 | 165 |
| | | | 4 | 135 |
| | 1.8 | 870 | 2 | 220 |
| | | | 3 | 155 |
| | 3.4 | 1430 | 1.6 | 210 |
| | | | 2 | 178 |

[a] 2.5 mil of flow channel opening

14

# Membrane Element Fabrication

**Element Rolling Machine**

**Spiral-Wound Membrane Element**

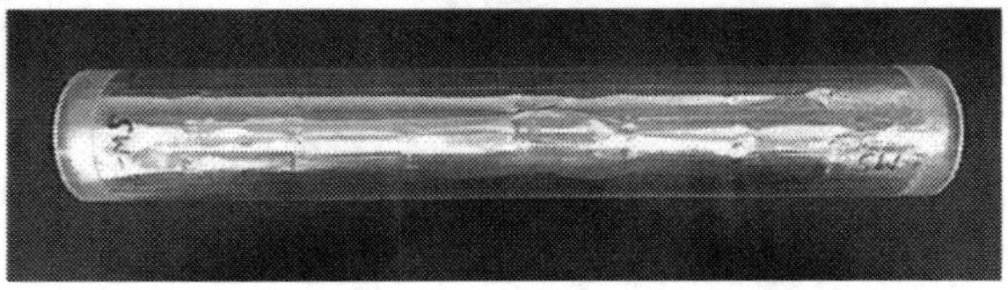

**Membrane Module**

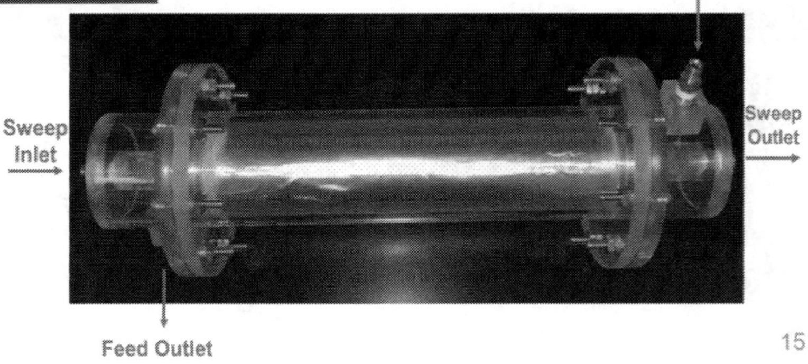


15

# Good Membrane Module Stability Obtained

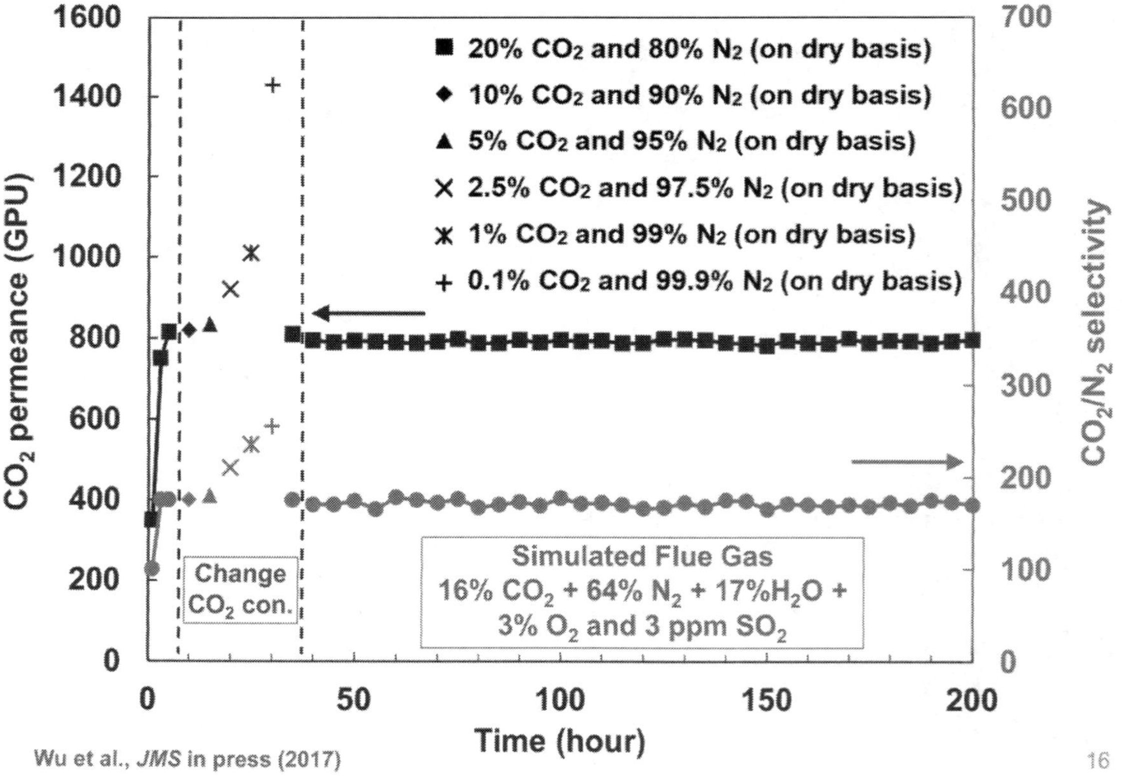


Wu et al., *JMS* in press (2017)

16

# Process Proposed for $CO_2$ Capture from Flue Gas in Coal-Fired Power Plants

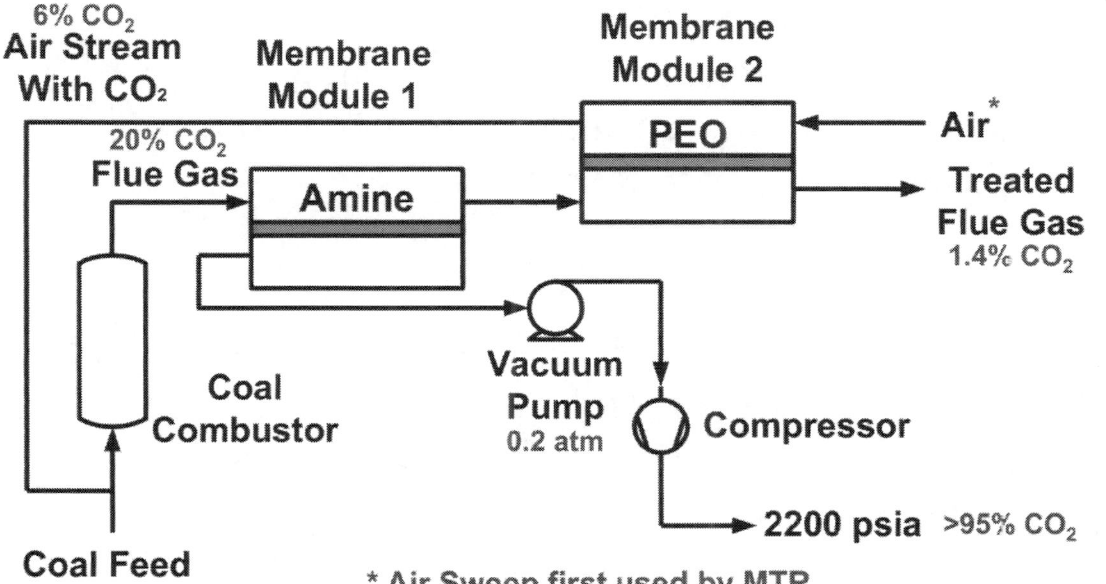


* Air Sweep first used by MTR

- Proposed membrane process does not require cryogenic distillation (compared to competition)

Ramasubramanian et al., *JMS*, *421-422*, 299 (2012) 17

# $SO_2$ Polishing & Membrane Process

- Absorption into 20 wt% NaOH Solution
  - Polishing step based on NETL baseline document
    - Estimated to be ~ \$4.3/tonne $CO_2$ (in 2007 \$, 6.5% COE increase)
  - Non-plugging, low-differential-pressure, spray baffle scrubber
  - High efficiencies (>95%)

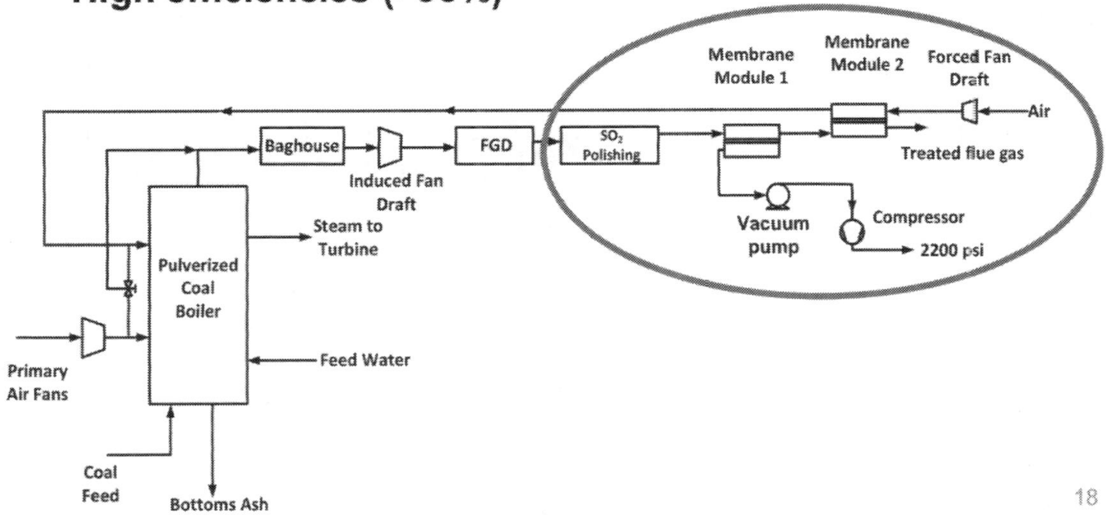


18

# Techno-Economic Analysis

- **Basis: In 2007 dollar**
  - **DOE target: \$40/tonne $CO_2$ set for 2025**
  - **Include membrane module installation cost**
  - **Also include 20% process contingency**

- **Membrane Results at 57°C**
  - **Amine-containing membrane with >140 selectivity key for stand-alone membrane process**
  - **Membrane process does not require cryogenic distillation**
  - **PEO-containing membrane with 2000 GPU $CO_2$ permeance and 20 selectivity**

19

## Composite Membranes Containing Amine Cover Layer: Simulated Flue Gas at 57°C

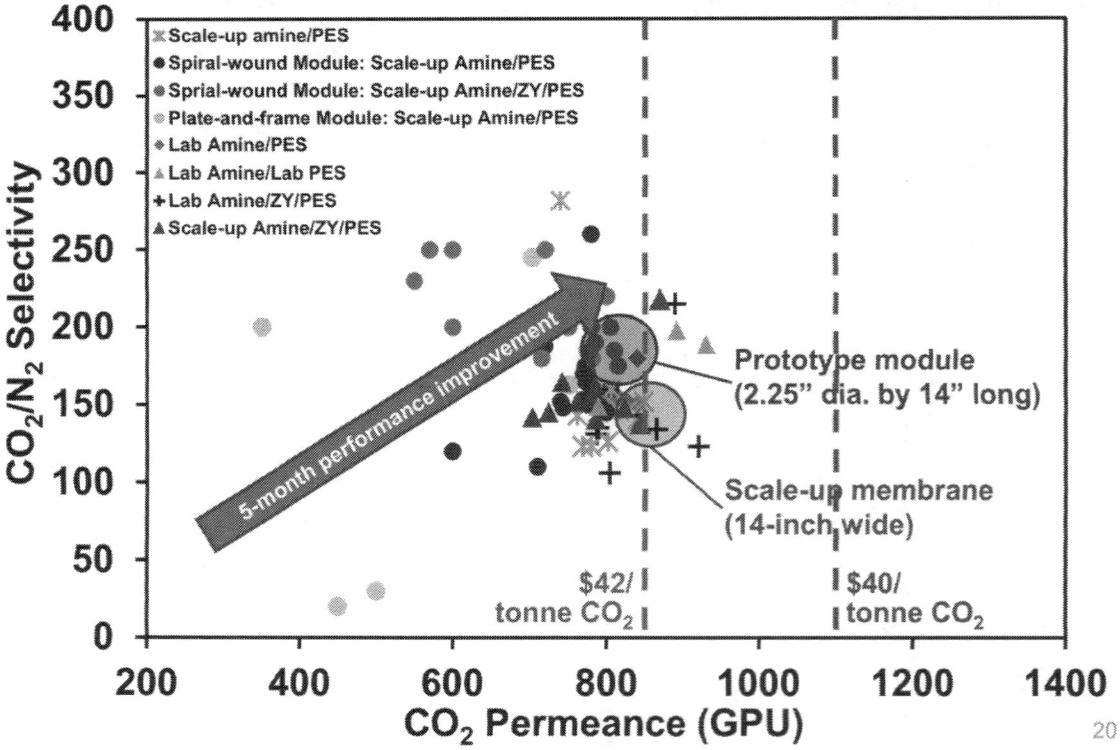


20

# Process Proposed for $CO_2$ Capture from <1% $CO_2$ Sources

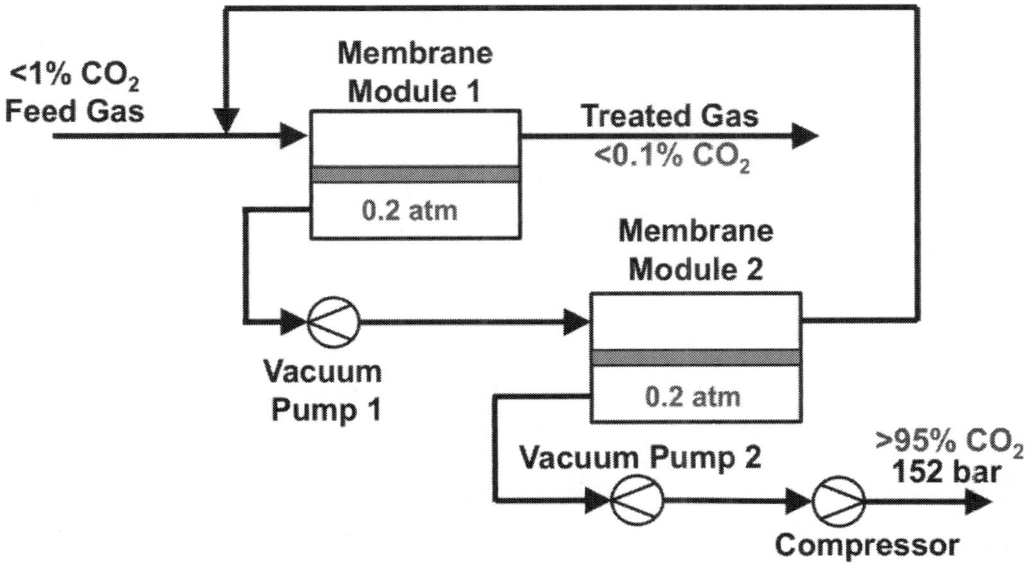


- Proposed membrane process does not require cryogenic distillation (compared to competition)

21

# Location of Proposed Technology in Coal-fired Power Plant

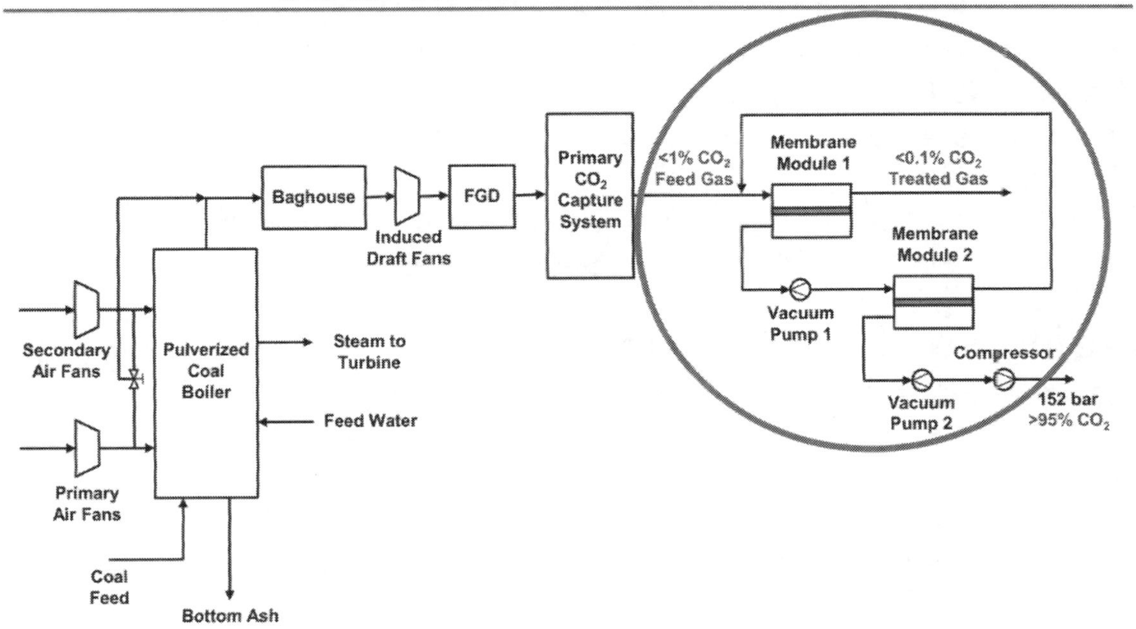


22

## Performance Improves as $CO_2$ Conc. Reduces

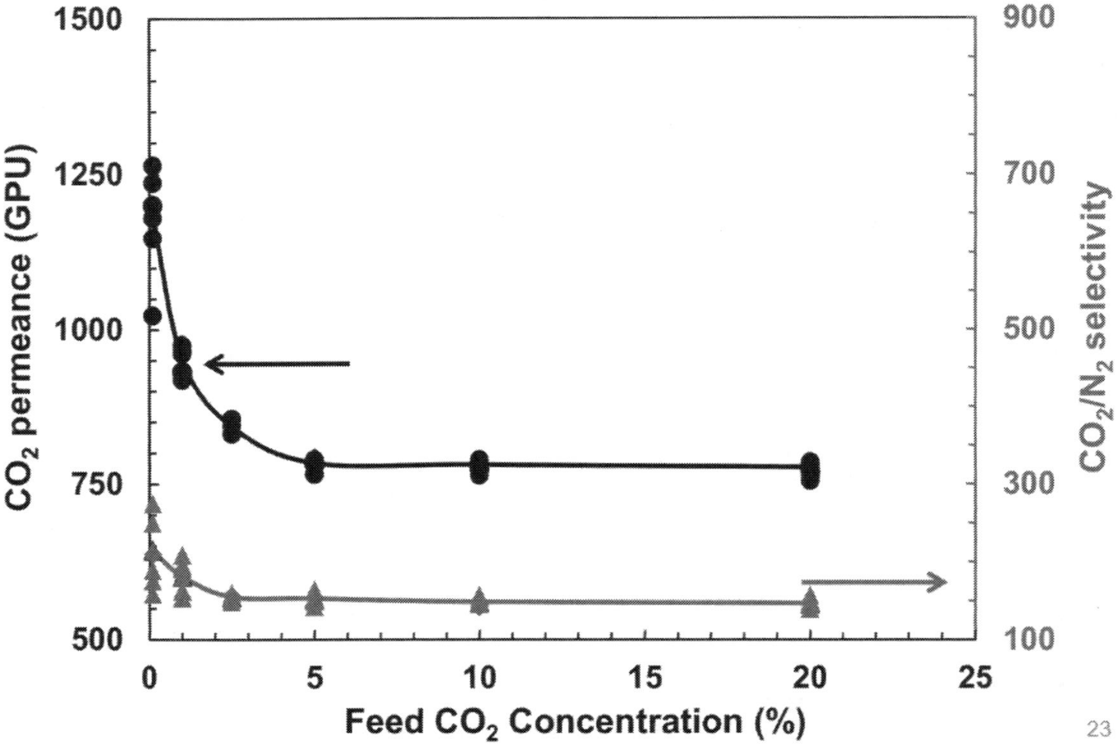


23

## High-Level Techno-Economic Calculations

- **Basis: Membrane Results at 57°C**
  - **982 GPU & 211 Selectivity for 1% $CO_2$ concentration feed gas**
  - **806 GPU & 173 Selectivity for 20% $CO_2$ concentration feed gas**
  - **Include Membrane Module Installation Cost and 20% Process Contingency**
  - **In 2011 dollar: NETL Case 12 of *Updated Costs (June 2011 Basis) for Selected Bituminous Baseline Cases***
- **Calculated Cost Results**
  - **32.1 tonne/h of $CO_2$ captured from 1% $CO_2$ source**
  - **$107.8 million bare equipment cost**
    - Membrane 34%, blowers and vacuum pumps 62%, others 4%
  - **1.76 ¢/kWh (1.24 ¢/kWh capital cost, 0.22 ¢/kWh fixed cost, 0.26 ¢/kWh variable cost, and 0.04 ¢/kWh T&S cost)**
    - COE = 8.09 ¢/kWh for 550 MW supercritical pulverized coal power plant
  - **$302/tonne capture cost ($17.6/MWh × 550 MW/(32.1 tonne/h))**
  - **21.8% Increase in COE (1.76/8.09 = 21.8%)**

24

# Effect of $CO_2$ Permeance on Techno-economic Analysis Results

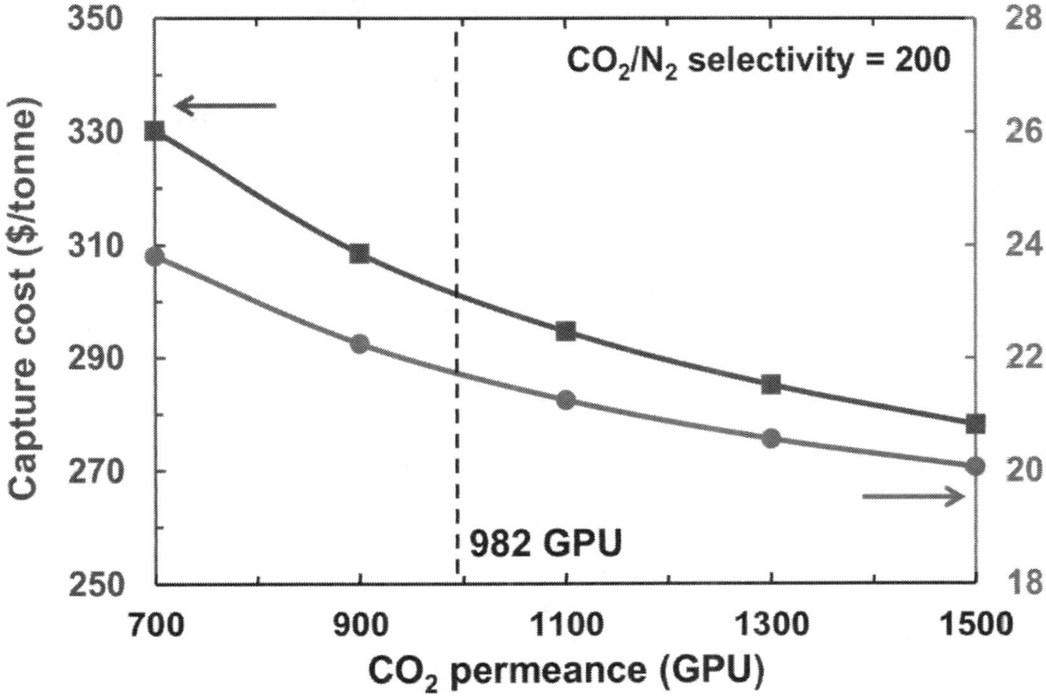


25

# Summary

- **$CO_2$ Capture from Flue Gas**
  - **High-Molecular-Weight PVAm Membrane Synthesized**
  - **Composite Membranes Synthesized in Lab**
    - + ~800 GPU with >200 Selectivity at 57°C
  - **Membrane Scaled up Successfully**
  - **Scale-up Membrane Promising for Meeting DOE Cost Target of $40/tonne $CO_2$ (in 2007 dollar) for 2025**
- **$CO_2$ Capture from <1% $CO_2$ Conc. Sources**
  - **Membrane showed 982 GPU with 211 $CO_2/N_2$ selectivity obtained at 57°C using 1% $CO_2$ concentration feed gas**
    - + 806 GPU with 173 selectivity obtained using 20% $CO_2$ conc. feed gas due to carrier saturation phenomenon
  - **Performance improves as $CO_2$ conc. reduces**
  - **High-level techno-economic analysis conducted**
    - + Capture cost of ~$302/tonne $CO_2$ (in 2011 $)
    - + 21.8% increase in COE

26

# Acknowledgments

José Figueroa
– Strong Efforts & Helpful Inputs for $CO_2$-Sel. Membranes

Yuanxin Chen
Varun Vakharia
Kartik Ramasubramanian
Lin Zhao
– $CO_2$-Selective Membranes

Norman N. Li
Richard Song
– Nanoporous Membrane Support

BASF in Wyandotte, MI
– Free PES Samples

Purolite Corporation, Bala Cynwyd, PA
– Free Ion-Exchange Resin Samples

27

# Acknowledgments for Financial Support

- U.S. Department of Energy
  - DE-FE0007632 and DE-FE0026919

- Ohio Development Services Agency
  - OER-CDO-D-15-09 for Continuous Membrane Fabrication Machine

28

# Decreasing Emissions Preserves Environment

29

ExxonMobil

July 19, 2017

# The potential role of CCS in electricity and liquid fuel production sectors

Energy lives here

Haroon Kheshgi and Bryan Mignone (ExxonMobil Corporate Strategic Research)
Matteo Muratori (National Renewable Energy Lab)
Haewon McJeon, Leon Clarke and Jae Edmonds (Pacific Northwest National Lab)

## Mitigation Cost Increases in Scenarios with Limited Availability of Technologies … Especially CCS

**Table SPM.2 |** Increase in global mitigation costs due to either limited availability of specific technologies or delays in additional mitigation [a] relative to cost-effective scenarios [b]. The increase in costs is given for the median estimate and the 16th to 84th percentile range of the scenarios (in parentheses) [c]. In addition, the sample size of each scenario set is provided in the coloured symbols. The colours of the symbols indicate the fraction of models from systematic model comparison exercises that could successfully reach the targeted concentration level. *{Table 3.2}*

| | **Mitigation cost increases in scenarios with limited availability of technologies** [d] *[% increase in total discounted* [e] *mitigation costs (2015–2100) relative to default technology assumptions]* | | | | **Mitigation cost increases due to delayed additional mitigation until 2030** *[% increase in mitigation costs relative to immediate mitigation]* | |
|---|---|---|---|---|---|---|
| 2100 concentrations (ppm $CO_2$-eq) | no CCS | nuclear phase out | limited solar/wind | limited bioenergy | medium term costs (2030–2050) | long term costs (2050–2100) |
| 450 (430 to 480) | 138% (29 to 297%) 4 | 7% (4 to 18%) 8 | 6% (2 to 29%) 8 | 64% (44 to 78%) 8 | 44% (2 to 78%) 29 | 37% (16 to 82%) 29 |
| 500 (480 to 530) | not available (n.a.) | n.a. | n.a. | n.a. | | |
| 550 (530 to 580) | 39% (18 to 78%) 11 | 13% (2 to 23%) 10 | 8% (5 to 15%) 10 | 18% (4 to 66%) 12 | 15% (3 to 32%) | 16% (5 to 24%) |
| 580 to 650 | n.a. | n.a. | n.a. | n.a. | | |

**Symbol legend—fraction of models successful in producing scenarios (numbers indicate the number of successful models)**

- : all models successful
- : between 80 and 100% of models successful
- : between 50 and 80% of models successful
- : less than 50% of models successful

Source: IPCC AR5 Synthesis Report, Table SPM.2 (2014)

ExxonMobil

2

# Role of CCS Varies Across Integrated Assessment Models (IAMs)

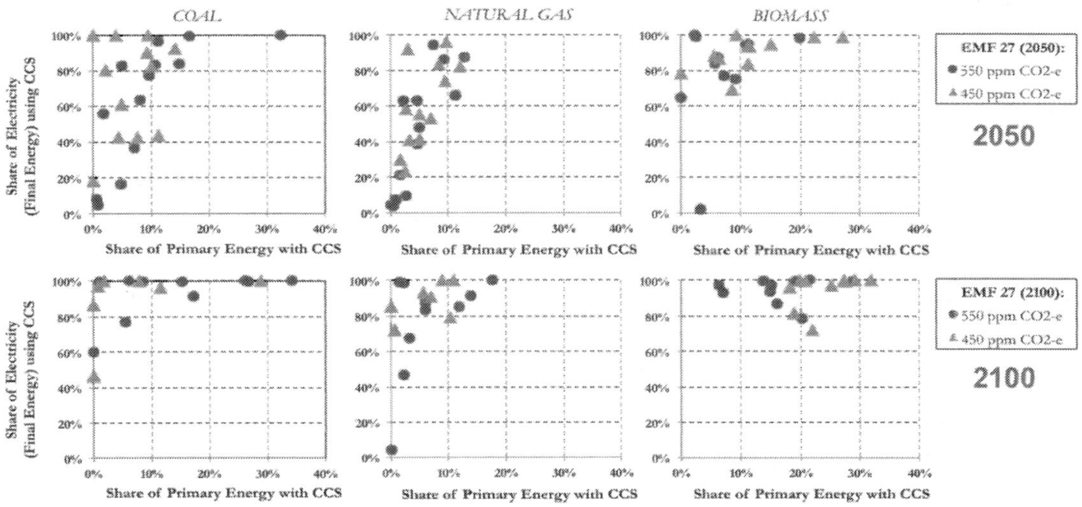


ExxonMobil 3

# A Collaborative Study on CCS Assessment

**A collaboration with researchers at:**

- ExxonMobil Corporate Strategic Research: Haroon Kheshgi and Bryan Mignone
- Pacific Northwest National Lab: Matteo Muratori (now at NREL), Haewon McJeon, Leon Clarke and Jae Edmonds

**Publications:**

- M Muratori, H. Kheshgi, B. Mignone, L. Clarke, H. McJeon, J. Edmonds, "**Carbon Capture and Storage across Fuels and Sectors in Energy System Transformation Pathways**", *International Journal of Greenhouse Gas Control*, 56 (2017) 1–8.
- M Muratori, H. Kheshgi, B. Mignone, H. McJeon, L. Clarke, "**The future role of CCS in electricity and liquid fuel supply**", *Energy Procedia*, Forthcoming.

ExxonMobil 4

## Research Question

What determines the role of CCS in climate change mitigation scenarios?

- Extent of CCS deployment across fuels and sectors?
- Sensitivity to assumptions?

## Method

Examine role of CCS in the integrated assessment model GCAM (Global Change Assessment Model).

ExxonMobil 5

# The Global Change Assessment Model (GCAM)

GCAM is a global long-term integrated assessment model

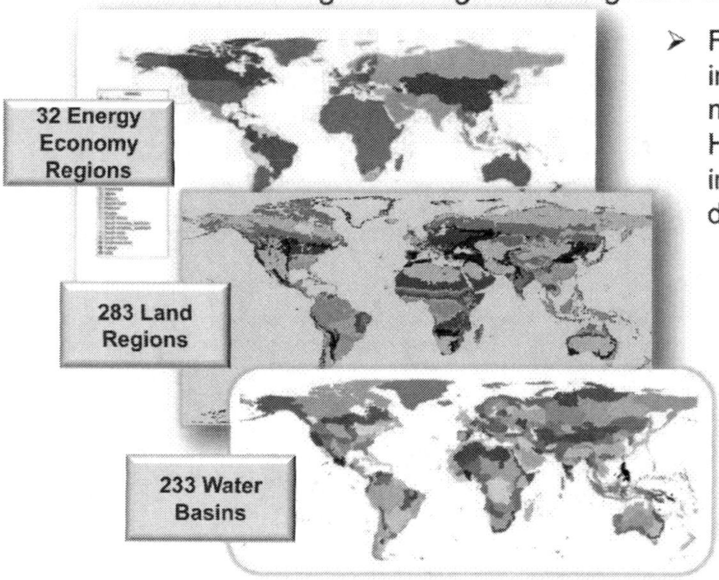


- ➢ For integrated, interdisciplinary research, modeling and analysis of Human-Earth systems to inform policy, strategy and decisions.

Includes:

- Many technology options
- 16 greenhouse gases and aerosols
- Extends through this century

ExxonMobil 6

## The Global Change Assessment Model (GCAM)

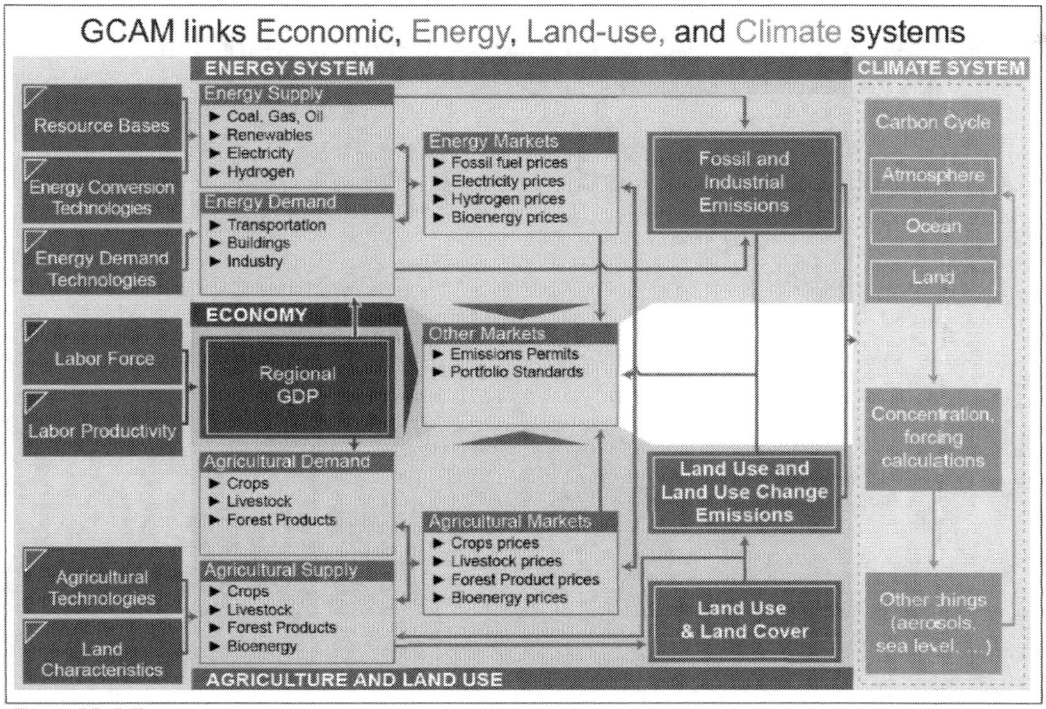


ExxonMobil

7

## GCAM Mitigation Scenarios: Policy Assumptions

- Two emission pathways are prescribed to drive GCAM scenarios that we use to explore the roles of CCS:
  - The IPCC (2013) found that RCP 4.5 and 2.6 were modeled to lead to about 2.4°C and 1.6°C (*likely* below 2°C) temperature rise above pre-industrial by the end of the century.

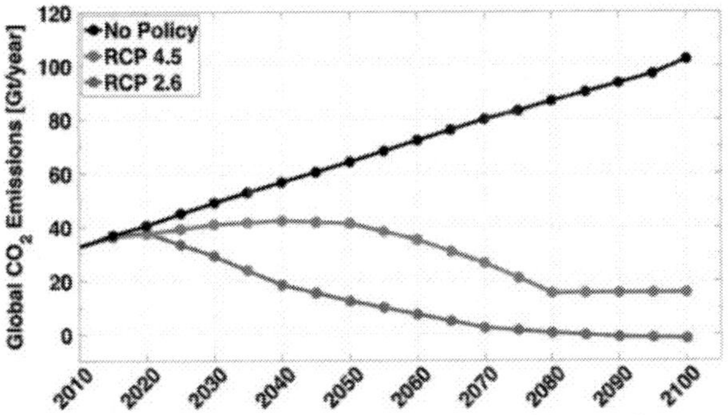


- GCAM applies a price on GHG emissions to achieve the prescribed emission reductions from the No Policy case.
  - Approximates a model least-cost scenario to reach emission pathway.
- Technologies deployed in GCAM scenarios are influenced by their relative economics.

ExxonMobil

8

## GCAM Mitigation Scenarios: $CO_2$ Stored

- The scale of CCS deployment in GCAM depends on the stringency of the climate change mitigation policy.
- The deployment of CCS technologies is <u>not limited to fossil fuels, nor to power plants</u>, as suggested by some studies.

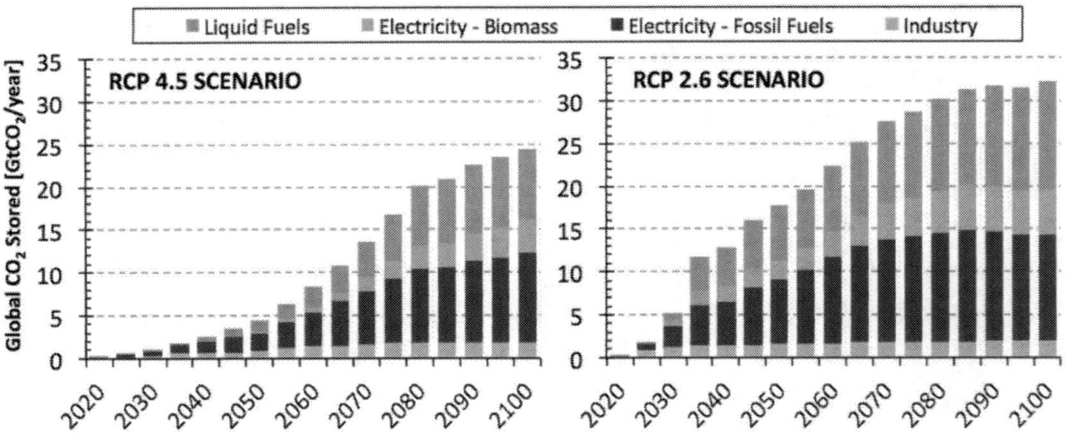


ExxonMobil 9

## GCAM Mitigation Scenarios: Electricity

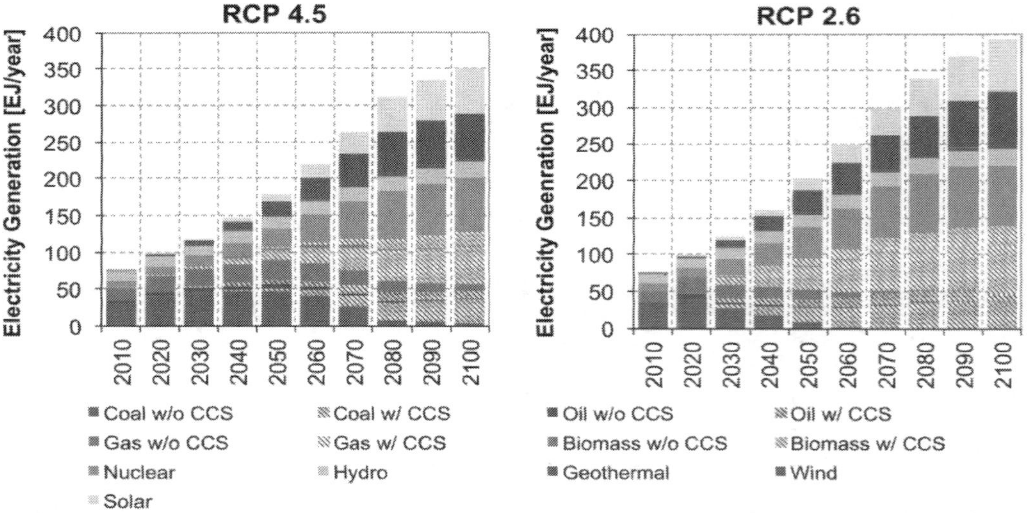


- Both CCS- and biomass-for-electricity are limited at present.
- Both gas & coal with CCS emerge over time; gas demand is robust.
- Nearly all biomass used for electricity includes CCS.

ExxonMobil 10

# GCAM Mitigation Scenarios: Liquid Fuels

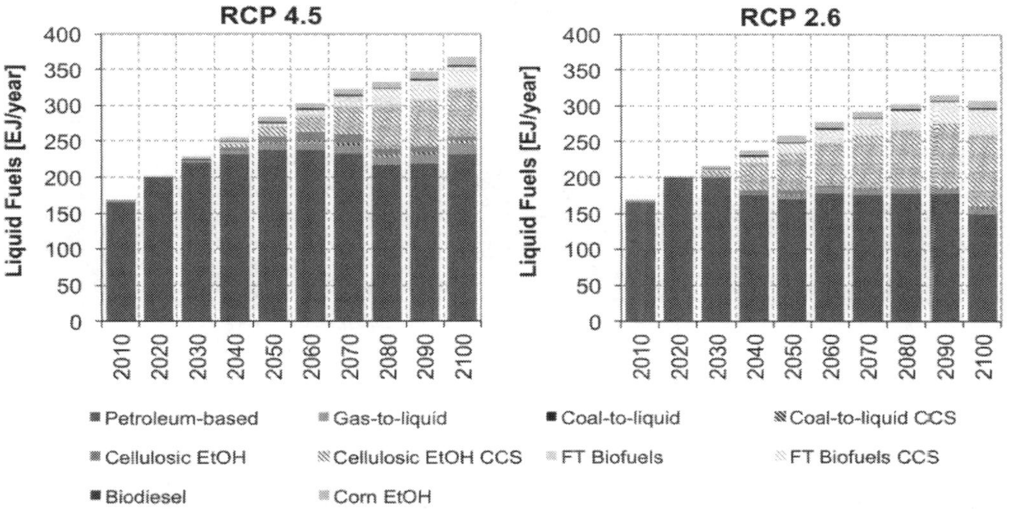


- Petroleum-based fuels currently dominate liquid fuels and continue throughout the century.
- Most bio-ethanol is produced with CCS, not without.

ExxonMobil

11

# CCS in the Electricity Generation Sector

- The fuel choice for CCS applications in the electricity sector in GCAM is driven by the **levelized costs of electricity** (LCOE)
- CCS technologies become competitive at a **sufficiently high carbon prices**.

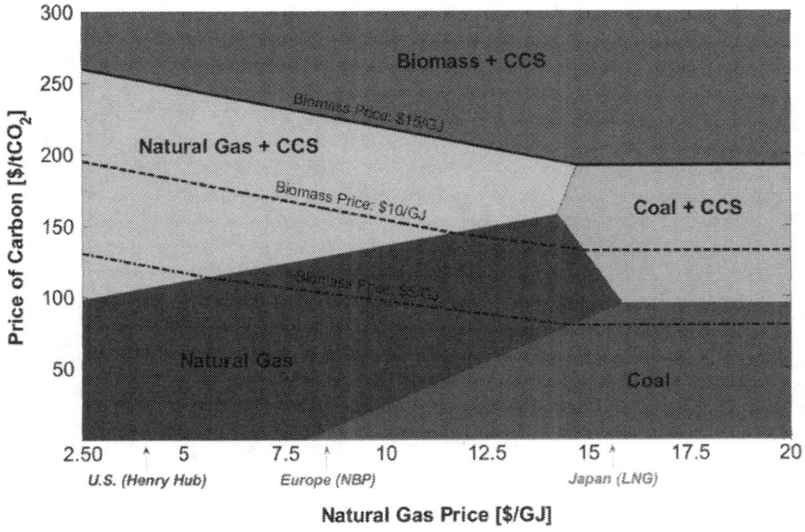


ExxonMobil

12

# Illustrative Economics Shows Why BECCS is Favored when Carbon Price is High

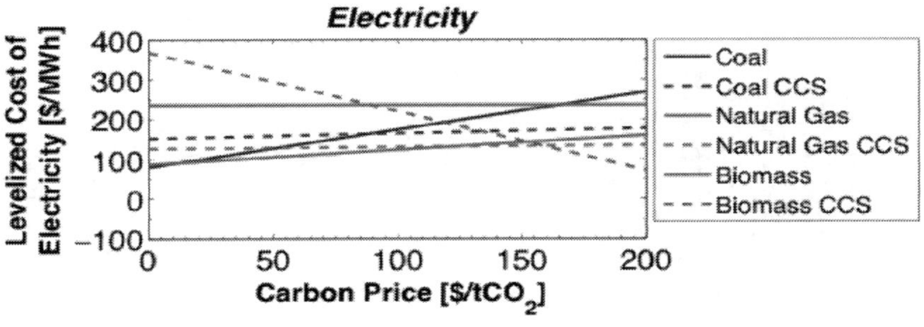


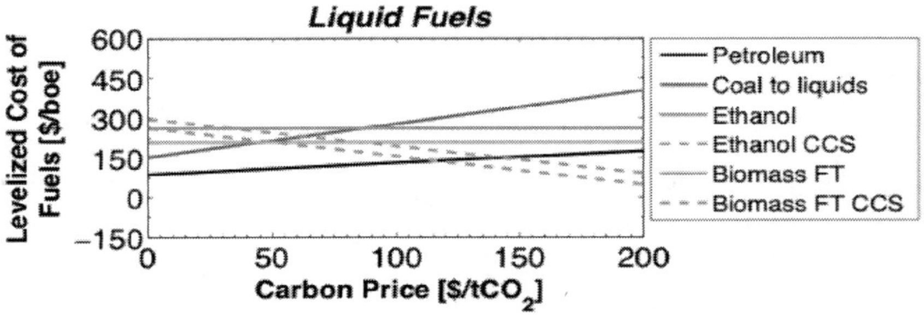


ExxonMobil

13

# CCS across Sectors

Sensitivity of Cumulative $CO_2$ Stored 2020-2100 to Technology Improvement Rates

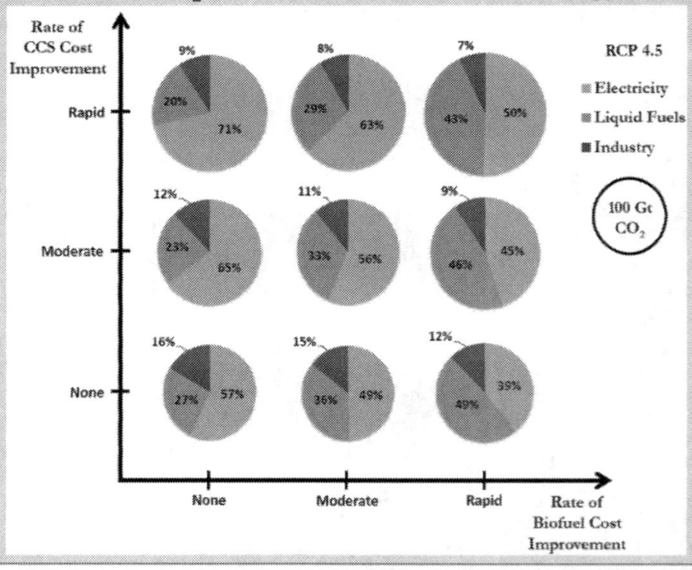


- CCS cost improvement increases CCS use, particularly in Electricity Sector.
- Biofuel cost improvement increases CCS in Biofuels.

ExxonMobil

14

# CCS across Fuels

Sensitivity of Cumulative Primary Energy 2020-2100 in CCS Applications to Technology Improvement Rates

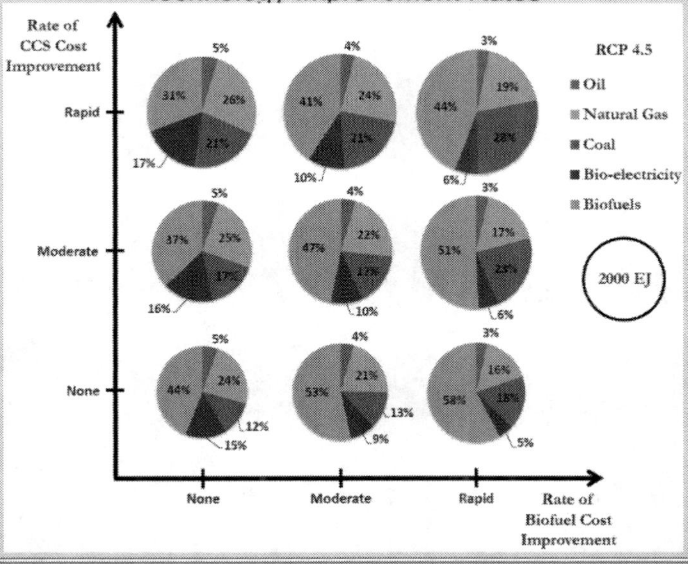


- Biofuels with CCS an important option in the future, but not at this time.
- Natural gas with CCS important fossil source of electricity.

ExxonMobil

15

# Conclusions

- Role of CCS differs across mitigation scenarios of IAMs.
  - IAMs do not agree on overall CCS deployment and on which fuels or sectors CCS would be applied.
- GCAM scenarios can be explained based on underlying cost competition between technology options.
  - Little CCS when carbon price low.
  - Fossil fuel electricity and BECCS (for both electricity and biofuels) when carbon price high.
- Greater scrutiny of CCS cost assumptions and broader assumptions about practical barriers to CCS deployment (and BECCS in particular) is important to further refine scenarios of energy system transformation pathways.

ExxonMobil

16

Thank You

CARBON MANAGEMENT TECHNOLOGY CONFERENCE

## CMTC-485175-MS

# Comparative Study for $CO_2$-EOR and Natural Gases Injection-Techniques for Improving Oil Recovery in Unconventional Oil Reservoirs

Dheiaa Alfarge, Iraqi Ministry of Oil, Missouri University of Science and Technology; Mingzhen Wei, and Baojun Bai, Missouri University of Science and Technology



This paper was prepared for presentation at the Carbon Management Technology Conference held in Houston, Texas, USA, 17-20 July 2017.



## Abstract

Shale formations in North America such as Bakken, Niobrara, and Eagle Ford have huge oil in place, 100-900 Billion barrels of oil in Bakken only. However, the predicted primary recovery is still below 10%. Therefore, seeking for techniques to enhance oil recovery in these complex plays is inevitable. Although most of the previous studies in this area recommended that $CO_2$ would be the best EOR-technique to improve oil recovery in these formations, pilot tests showed that natural gases performance clearly exceeds $CO_2$ performance in the field scale. In this paper, two different approaches have been integrated to investigate the feasibility of three different miscible-gases which are $CO_2$, lean gases, and rich gases. Firstly, numerical simulation methods of compositional models have been incorporated with Local Grid Refinement (LGR) of hydraulic fractures to mimic the performance of these miscible gases in shale-reservoirs conditions. Implementation of a molecular diffusion model in the LS-LR-DK (logarithmically spaced, locally refined, and dual permeability) model has been also conducted. Secondly, different molar-diffusivity rates for miscible gases have been simulated to find the diffusivity level in the field scale by matching the performance for some EOR pilot-tests which were conducted in Bakken formation of North Dakota, Montana, and South Saskatchewan.

The simulated shale-reservoirs scenarios confirmed that diffusion is the dominated flow among all flow regimes in these unconventional formations. Furthermore, the incremental oil recovery due to lean gases, rich gases, and $CO_2$ gas injection confirms the predicted flow-regime. The effect of diffusion-implementation has been verified with both of single porosity and dual-permeability model cases. However, some of $CO_2$ pilot-tests showed a good match with the simulated cases which have low molar-diffusivity between the injected $CO_2$ and the formation-oil. Accordingly, the rich and lean gases have shown a better performance to enhance oil recovery in these tight formations. However, rich gases need long soaking periods, and lean gases need large volumes to be injected for more successful results. Furthermore, the number of huff-n-puff cycles has a little effect on the all injected-gases performance; however, the soaking period has a significant effect. This research project demonstrated how to select the best type of miscible gases to enhance oil recovery in unconventional reservoirs according to the field-candidate conditions and operating parameters. Finally, the reasons beyond the success of natural gases and failure of $CO_2$ in the pilot tests have been physically and numerically discussed.

## Introduction

The Energy Information Administration (EIA) reported that US tight oil production including shale formations will grow to more than 6 million bbl/day in the coming decade, making up most of the total U.S. oil production as shown in **Fig.1**. Oil production from tight formations including shale plays has just shared for more than 50% of total oil production in US (Alfarge et al., 2017). Hoffman et al., (2016) reported that 4 million barrels per day as an increment in US-oil daily production comes from these unconventional oil reservoirs. From 2011 to 2014, Unconventional Liquid Rich (ULR) reservoirs contributed to all natural gas growth and nearly 92% of oil production growth in the US (Alfarge et al., 2017). Specifically, Bakken and Eagle Ford contributed for more than 80% of total US oil production from these tight formations (Yu et al., 2016a). This revolution in oil and gas production happened mainly because shale oil reservoirs have been just increasingly developed due to the advancements in horizontal wells and hydraulic fracturing in last decade. Several studies have been conducted to estimate the recoverable oil in place in these complex formations indicating large quantities of oil in place. The available information refers to 100-900 Billion barrels in Bakken only. However, the predicted recovery from primary depletion could lead to 7% only of original oil in place (Clark, 2009). Furthermore, some investigators argued that the primary recovery factor is still in a range of 1-2 % in some of these plays in North America (Wang et al., 2016). For example, the North Dakota Council reported that "With today's best technology, it is predicted that 1-2% of the reserves can be recovered" (Sheng, 2015). The main problem during the development of unconventional reservoirs is how to sustain the hydrocarbon production rate, which also leads to low oil recovery factor. The producing wells usually start with high production rate initially; however, they show steep decline rate in the first 3-5 years until they get leveled off at very low rate. According to Yu et al., (2014), the main reason beyond the quick decline in production rate is due to the fast depletion of natural fractures networks combined with slow recharging from matrix system, which is the major source of hydrocarbon. Therefore, oil recovery factor from primary depletion has been predicted typically to be less than 10% (LeFever et al, 2008; Clark, 2009; Alharthy et al., 2015; Kathel and Mohanty 2013, Wan et al., 2015; Alvarez et al, 2016).

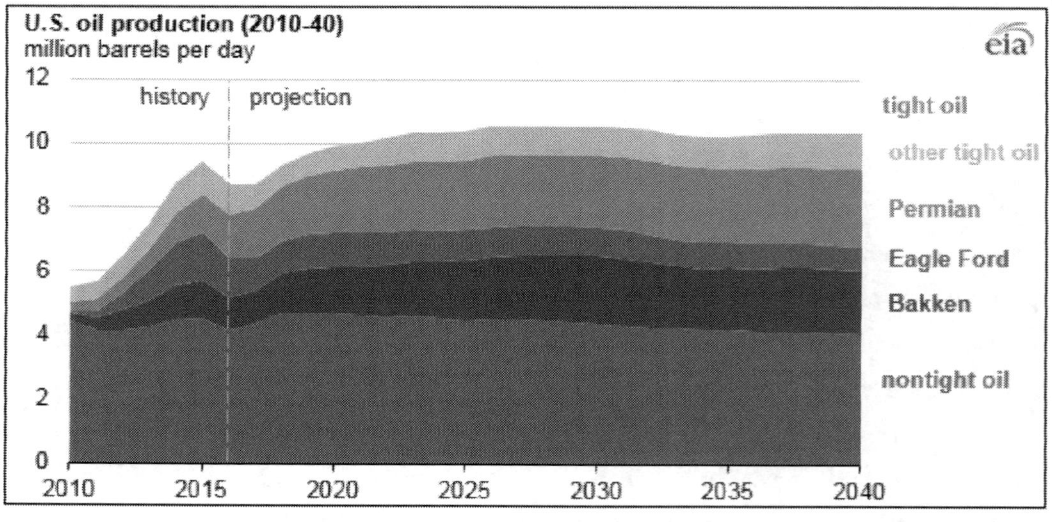


**Fig. 1: Shale and tight oil production in North America from U.S. EIA (Feb-2017)**

Since these reservoirs have huge original oil in place, any improvement in oil recovery factor would result in enormous produced oil volumes. Therefore, IOR methods have huge potential to be the major stirrer in these huge reserves. Although IOR methods are well understood in conventional reservoirs, they are a new concept in unconventional ones. All the basic logic-steps for investigating the applicability of different IOR methods such as experimental works, simulation studies, and pilot tests have just started over the last decade (Alfarge et al., 2017). Miscible-gas injection has shown excellent results in conventional reservoirs with low permeability and light oils. Extending this approach to unconventional reservoirs including shale oil reservoirs in North America has been extensively investigated over the last decade. The gases which have been investigated are $CO_2$, $N_2$, and natural gases. Some of IOR pilot-tests which have been conducted to investigate the feasibility of natural gases in unconventional reservoirs showed good results in terms of enhancing oil recovery as shown in **Fig. 2**. However, most of the studies in this area focused on $CO_2$ due to different reasons. $CO_2$ can dissolve in shale oil easily, swells the oil and lowers its viscosity. Also, $CO_2$ has a lower miscibility pressure with shale oil rather than other gases such as $N_2$ and $CH_4$ (Zhang et al., 2016). Furthermore, experimental studies reported an excellent oil recovery factor could be obtained by injection $CO_2$ in small chips of tight-natural cores (Hawthorne et al., 2017). Unfortunately, the results of pilot-tests for $CO_2$-EOR, huff-n-puff protocol, which have been conducted in unconventional reservoirs of North America were disappointing as shown in **Fig. 3**. This gap in $CO_2$ performance between lab-conditions versus to what happened in field-scale suggests that there is something missing between microscopic-level and macroscopic-level in these plays. Most of the experimental studies reported that the molecular-diffusion mechanism for $CO_2$ is beyond the increment in oil recovery obtained in lab scale (Alfarge et al., 2017). Furthermore, most of the previous simulation studies relied on the lab-diffusivity level for these miscible gases to predict the expected oil increment on field scale (Alfarge et al., 2017). One of the main reasons for the poor-performance for $CO_2$ in the pilot tests might be due to the wrong prediction for $CO_2$ diffusion-mechanism in these types of reservoirs. A detailed study for determining the level of $CO_2$ diffusivity in the real-field conditions have been conducted in this work. Also, comparing $CO_2$ performance with lean gas and rich gas according to different levels of diffusivity has been investigated to clarify the flow and recovery mechanisms for different gases in shale-reservoirs.

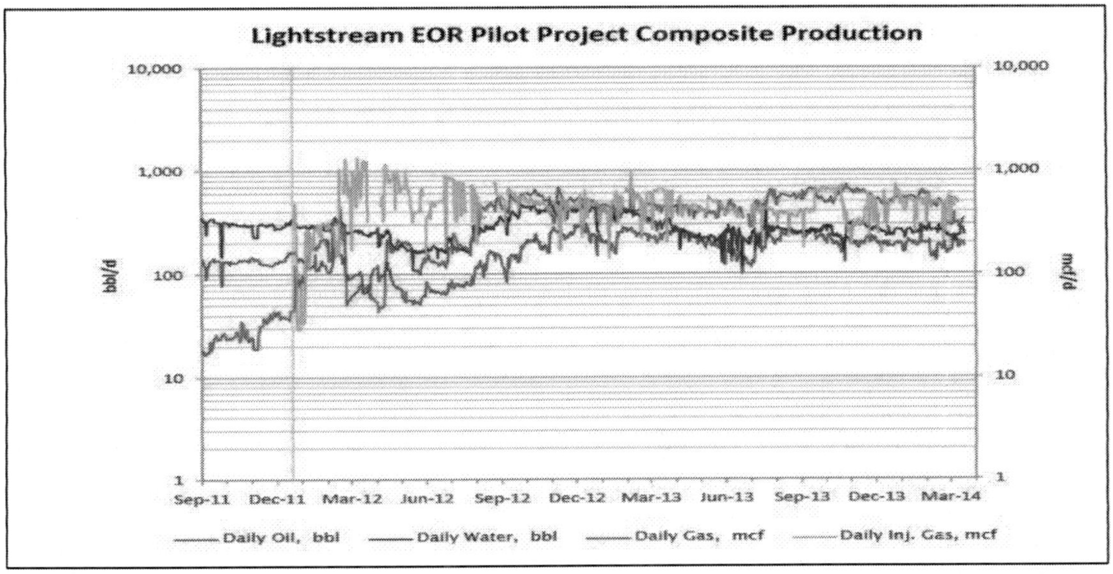


**Fig. 2: Performance of natural gas EOR in Canadian-Bakken conditions (Schmidt et al., 2014)**

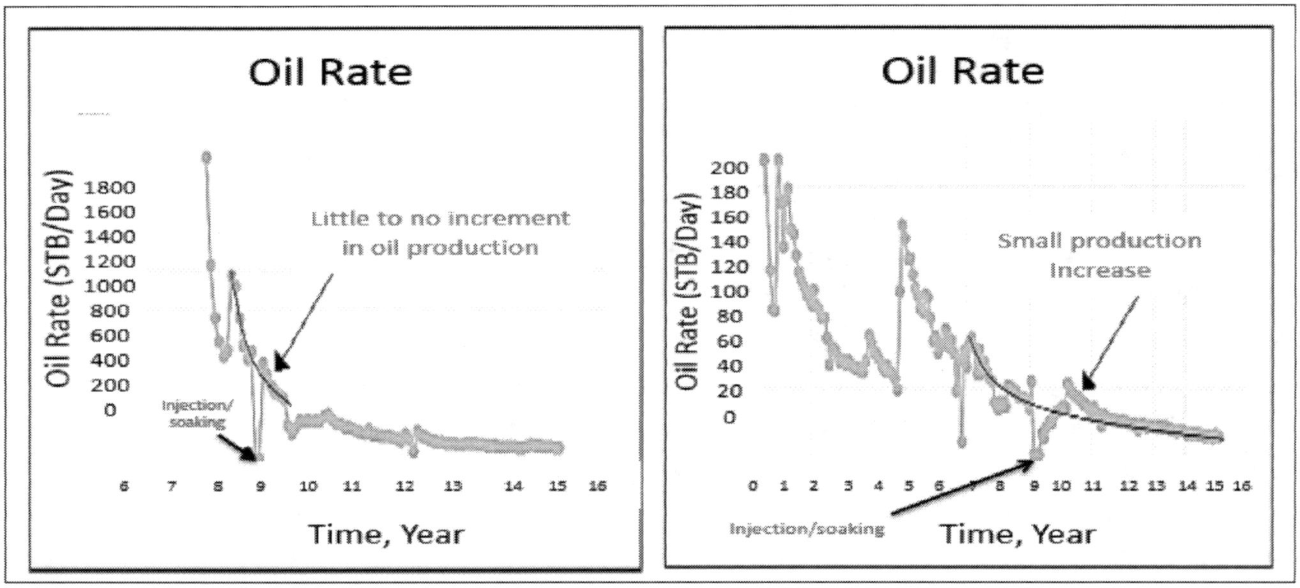


**Pilot test#1** **Pilot test#2**

**Fig. 3: CO2 Pilot Tests Performance in Bakken (Modified from Hoffman and Evans, 2016)**

## Molecular Diffusion

Gravity drainage, physical diffusion, viscous flow, and capillary forces are the common forces which control the fluids flow in porous media. However, one force might eliminate the contributions of other forces depending on the reservoir properties and operating conditions. Molecular diffusion is defined as the movement of molecules caused by Brownian motion or composition gradient in a mixture of fluids (Mohebbinia et al., 2017). This type of flow would be the most dominated flow in fractured reservoirs with a low-permeability matrix when gravitational drainage is inefficient (Moorgate and Firoozabadi, 2013; Mohebbinia et al., 2017). The role of molecular-diffusion flow increases as far as the formation permeability decreases. It has been noticed and approved that gas injection is the most common EOR-process affected by calculations of molecular-diffusion considerations. Ignoring or specifying incorrect diffusion-rate during simulation process can lead to overestimate or underestimate the oil recovery caused by the injected gas. This happens not only due to the variance in miscibility-process between the injected-gas and formation-oil but also due to the path change for the injected-gas species from fractures to the formation-matrix. Da Silva and Belery (1989) reported that molecular diffusion process is happened by three mechanisms:

- Bulk diffusion where fluid-fluid interactions dominate.
- Knudsen diffusion where fluid molecule collides with pore wall (happens when molecular mean free pathway closed to pore size).
- Surface diffusion where fluid molecules transported along adsorbed film (miner unless thick adsorbed layer).

The Péclet number (Pe) is a class of dimensionless numbers which have been used to measure the relative importance of molecular diffusion flow to the convection flow. This number can be calculated as shown in **Eq. 1**. If Pe number is less than 1, diffusion is the dominant flow. However, if Pe is greater than 50, convection is the dominant flow. The dispersion flow is dominant when Pe is in the range of 1 to 50 (Hoteit and Firoozabadi, 2009). **Fig. 4** explains the flow regimes according to Péclet number cutoffs. The movement of fluid components in field conditions is equal to the integration of diffusion, dispersion, and viscous forces. Therefore, the total average velocity of any component is equal to the sum of dispersive velocity and bulk phase velocity (Da Silva and Belery,1989).

$$\mathbf{Pe} = \frac{diffusion\ time}{convection\ time} = (L2/D)/(L/v) = Lv/D \qquad (1)$$

Where v is the bulk velocity, L is a characteristic length, and D is the diffusion coefficient.

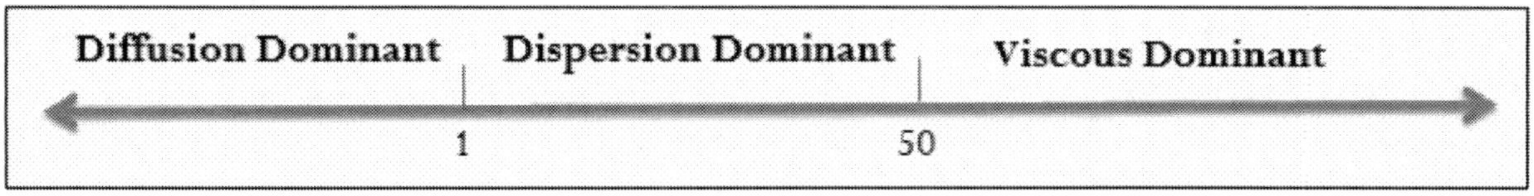


**Fig. 4: Flow Regimes According to Péclet Number Cut offs**

## $CO_2$ Molecular-Diffusion Mechanism

Different mechanisms have been proposed for the injected $CO_2$ to improve oil recovery in unconventional reservoirs as shown in **Table 1**. However, since the matrix permeability in these unconventional reservoirs is in range (0.1 –0.00001 md), $CO_2$ would not be transported by convection flux from fracture to matrix (Yu et al., 2014). The main transportation method for $CO_2$ is happening by the difference in concentration gradient between $CO_2$ concentration in injected gases and the target-oil. This process of transportation is subjected to Fick's law. Hawthorne et al., (2013) extensively investigated the $CO_2$ diffusion-mechanism in Bakken cores and proposed five conceptual-steps to explain it. These conceptual steps include: (1) $CO_2$ flows into and through the fractures, (2) unfractured rock matrix is exposed to $CO_2$ at fracture surfaces, (3) $CO_2$ permeates the rock driven by pressure, carrying some hydrocarbon inward; however, the oil is also swelling and extruding some oil out of the pores, (4) oil migrates to the bulk $CO_2$ in the fractures via swelling and reduced viscosity, and (5) as the $CO_2$ pressure gradient gets smaller, oil production is slowly driven by concentration-gradient diffusion from pores into the bulk $CO_2$ in the fractures.

**Table 1: The proposed CO2 EOR mechanisms for improving oil recovery in unconventional reservoirs**

| $CO_2$ mechanism | Approach tool |
|---|---|
| 1-Diffusion | Lab |
| 2-Reduction in Capillary forces | Lab and simulation |
| 3-Repressurization | Lab |
| 4-Extraction | Lab |
| 5-Oil swelling and pressure maintenance | Lab and simulation |
| 6-Oil Viscosity reduction | Lab and simulation |
| 7-Combination of more than one mechanism from above | - |

Most of the previous experimental studies reported that $CO_2$ diffusion mechanism is beyond the increment in oil recovery obtained in lab conditions. Then, the observed increment in oil-recovery and/or the $CO_2$ diffusion-rate obtained in lab conditions were upscaled directly to field scale by using numerical simulation methods. Although modeling of the diffusion effect on ultimate oil recovery in shale reservoirs is very important to develop these marginal shale-oil projects, evaluation of the recovery contribution from diffusion will help in understanding the recovery mechanisms (Wan and Sheng, 2015). We think that this direct upscaling methodology is so optimistic due to that the lab-cores have higher contact area and longer exposure time to $CO_2$ than what might happen in the real-conditions of unconventional reservoirs. As a result, both of previous simulation studies and experimental works might be too optimistic to predict a quick improvement in oil recovery from injecting $CO_2$ in these tight formations. And, this explains why the previous simulation studies have a clear gap with $CO_2$ pilot-tests performance. It is true that the molecular mechanism is more dominated in naturally fractured reservoirs due to two main reasons (Da Silva and Belery,1989): (1) Dispersive flux through fractures rapidly increase the contact area for diffusion, (2) This mechanism needs small spacing for natural fractures which is so possible to exist in naturally fractured reservoirs. However, the effective diffusion rate in the reported lab conditions is much faster than in the field scale conditions due to the difference in the contact area and the exposure time. This gap in the effective diffusion rates would clearly happen between the lab scale and the field scale of shale-oil reservoirs.

## Numerical Simulation

The majority of the previous diffusion models were developed based on the single-porosity model which requires a tremendous grid refinement to represent an intensely fractured shale-oil reservoir (Wan and Sheng, 2015). In this simulation study, the LS-LR-DK (logarithmically spaced, locally refined, and dual permeability) model is used. It has been reported that the LS-LR-DK method can accurately capture the physics of the fluid flow in fractured tight reservoirs. Also, an advanced general equation-of-state compositional simulator has been used to build an equation-of-state model for Bakken-oil. Then, both models have been combined to simulate compositional effects of reservoir fluid during primary and enhanced oil recovery processes. Furthermore, implementation of a diffusion model in the LS-LR-DK (logarithmically spaced, locally refined, and dual permeability) model has been conducted. Moreover, the counter-current mechanism of molecular diffusion for $CO_2$-EOR, which have been reported by the experimental work for Hawthorne et al., (2013), is simulated in this work. In this study, we tried to build a numerical model which has the typical fluid and rock properties of Bakken formation, one of the most productive unconventional formations in the US. In this model, we injected three different EOR-miscible gases including $CO_2$, lean gas, and rich gas in separated scenarios as Huff-n-Puff protocol through hydraulically fractured well. All the mechanisms which were proposed in **Table 1** have been also incorporated in this model. In this field case study, the production well was stimulated with 5 hydraulic fractures. The spacing between the hydraulic fractures is 200 ft. The simulation model includes two regions which are stimulated reservoir volume (SRV) and un-stimulated reservoir volume (USRV) as shown in **Fig. 5**. The dimensions of the reservoir model are 2000 ft x 2000ft x42 ft, which corresponds to length, width, and thickness respectively. The dimensions of the hydraulically-fractured region are 5 fractures with half-length of 350 ft in J direction, width 0.001 ft in I direction, and fracture height of 42 ft in K direction. Fracture conductivity is 15 md.ft. The other model-input parameters are shown in **Table 2**.

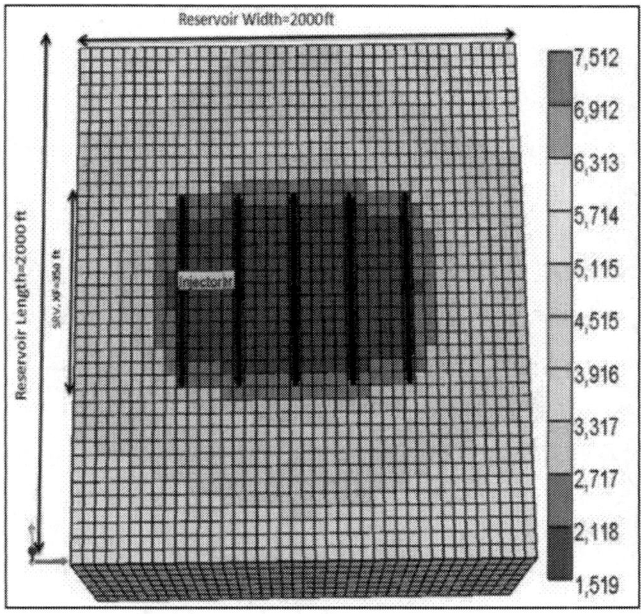


Figure 5a- Average Pressure in a depleted well in Bakken

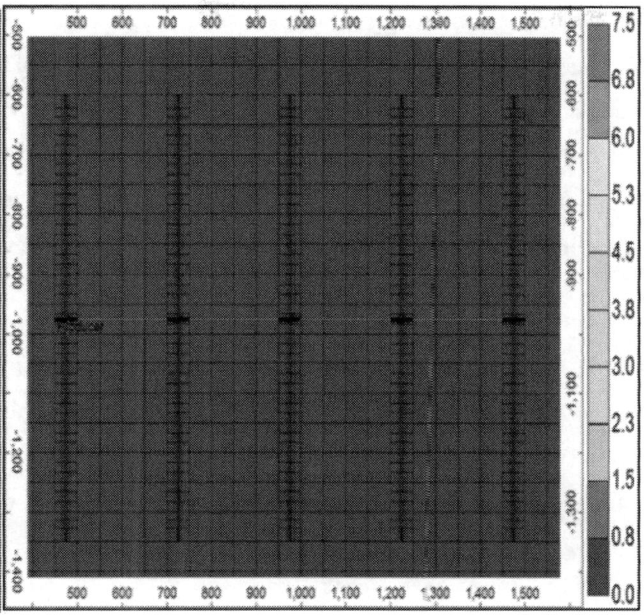


Figure 5b- A closed view for SRV of Production well

**Table 2: Model input parameters for the base case**

| Parameter | value | Unit |
|---|---|---|
| The model dimensions | 2000x2000x42 | ft |
| Production Time | 20 | year |
| Top of Reservoir | 8000 | ft |
| Reservoir Temperature | 240 | °F |
| Reservoir pressure | 7500 | psi |
| Initial Water saturation | 0.3 | value |
| Total compressibility | $1x10^{-6}$ | $psi^{-1}$ |
| Matrix permeability | 0.005 | mD |
| Matrix porosity | 0.085 | value |
| Horizontal Well length | 1000 | ft |
| Total number of fractures | 5 | value |
| Fracture conductivity | 15 | mD-ft |
| Fracture half-length | 250 | ft |
| Fracture Height | 42 | ft |

## Compositional Model for the Formation Fluids

The typical Bakken oil has been simulated in this study. The oil which was used in this model has 42 API°, 725 SCF/STB, and 1850 psi as oil gravity, gas oil ratio, and bubble point pressure respectively. It is known that compositional models are the most time-consumed models due to the number of components in the typical-reservoir oil. In our model, we have 34 components so that would take a long time for the simulator to complete running one scenario. The common practice in numerical simulation for such situation is the careful lump of reservoir oil components into a short representative list of pseudo-components. These pseudo-components would be acceptable if they match the laboratory–measured phase behavior data. The supplied-data for reservoir-oil needs to have a description of associated single carbon numbers and their fractions, saturation pressure test results, separator results, constant-composition expansion-test results, differential liberation test results, and swelling test results. All the available data can be used for tuning the EOS to match the actual fluid behavior.

In our simulation study, we lumped the original 34 components into 7 pseudo components as shown in **Table 3** by using WinProp-CMG. WinProp is an Equation-of-State (EOS)-based fluid behavior and PVT modeling package. In WinProp, laboratory data for fluids can be imported and an EOS can be tuned to match its physical behavior. Fluid interactions can then be predicted and a fluid model can be created. **Table 4** presents the Peng-Robinson EOS fluid description and binary interaction coefficients of the Bakken crude oil with different gases. **Fig. 6** represents the two-phase envelope for Bakken oil which was generated by WinProp-CMG.

**Table 3: Compositional data for the Peng-Robinson EOS in the model oil**

| Component | Mole fraction | Critical pressure (atm) | Critical Temp. (K) | Acentric Factor | Molar Weight (g/gmole) |
|---|---|---|---|---|---|
| CO2 | 0 | 7.28E+01 | 3.04E+02 | 0.225 | 4.40E+01 |
| N2-CH4 | 0.2704 | 4.52E+01 | 1.90E+02 | 0.0084 | 1.62E+01 |
| C2H-NC4 | 0.2563 | 4.35E+01 | 4.12E+02 | 0.1481 | 4.48E+01 |
| IC5-CO7 | 0.127 | 3.77E+01 | 5.57E+02 | 0.2486 | 8.35E+01 |
| CO8-C12 | 0.2215 | 3.10E+01 | 6.68E+02 | 0.3279 | 1.21E+02 |
| C13-C19 | 0.074 | 1.93E+01 | 6.74E+02 | 0.5672 | 2.20E+02 |
| C20-C30 | 0.0508 | 1.54E+01 | 7.92E+02 | 0.9422 | 3.22E+02 |

**Table 4: Binary interaction coefficients for Bakken oil**

| Component | CO2 | N2-CH4 | C2H-NC4 | IC5-CO7 | CO8-C12 | C13-C19 | C20-C30 |
|---|---|---|---|---|---|---|---|
| CO2 | | | | | | | |
| N2-CH4 | 1.01E-01 | | | | | | |
| C2H-NC4 | 1.32E-01 | 1.30E-02 | | | | | |
| IC5-CO7 | 1.42E-01 | 3.58E-02 | 5.90E-03 | | | | |
| CO8-C12 | 1.50E-01 | 5.61E-02 | 1.60E-02 | 2.50E-03 | | | |
| C13-C19 | 1.50E-01 | 9.76E-02 | 4.24E-02 | 1.72E-02 | 6.70E-03 | | |
| C20-C30 | 1.50E-01 | 1.45E-01 | 7.79E-02 | 4.27E-02 | 2.51E-02 | 6.00E-03 | |

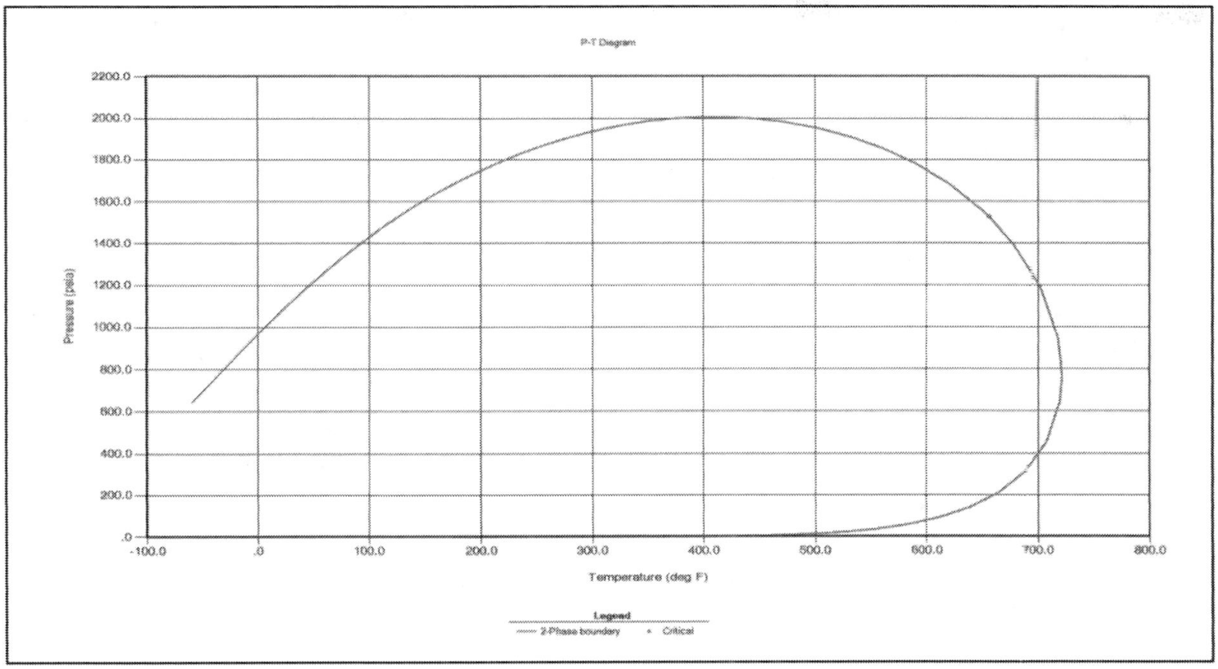


**Fig. 6: The two-phase envelope for Bakken oil which was generated by WinProp-CMG**

## Results and Discussion

***Natural Depletion for Bakken Model.*** The reservoir model was initially run in natural depletion for 7300 days (20 years). The production well, which was hydraulically fractured, was subjected to the minimum bottom-hole pressure of 1500 psi. The simulated Bakken well performance in natural depletion is shown in **Fig. 7**. In the natural depletion scenario, it has been clear that the production well initially started with a high production rate. Then, it showed steep decline rate until it got leveled off at a low rate. This is the typical trend to what happens in the most, if not all, unconventional reservoirs of North America. If we investigate the pressure distribution in the reservoir model as shown in **Fig. 5**, it can be seen that the main reason to that fast reduction in production rate is due to the pressure depletion in the areas which are closed to the production well. However, the reservoir pressure is still high in the areas which are far away from the production well. This explains the poor feeding from neighboring areas in these types of reservoirs due to the tight formation matrix.

***Flow-Type Determination in Natural-Depletion Stage and EOR Stage.*** We calculated the Péclet number locally in each grid in both of natural-depletion stage and EOR-stage. In the formation-matrix areas, the results indicated that Péclet number is away below 1 for both of gas phase and oil phase which means that the diffusion flow is the most dominant flow in formation matrix as shown in **Fig. 8.** However, in the hydraulic fractures parts, the viscous flow is clearly dominated where Pe is away above 100. If we examine how Péclet number changes with time at 10 ft from the hydraulic fracture, we found that Pe number is not changing too much during natural depletion stage; however, it is changeable in EOR stage as shown in **Fig. 9** (EOR stage started after 10 years of production life). Furthermore, we notice that there are two different behaviors for gas phase versus oil phase in EOR stage. Pe number is increasing with time for gas phase; however, Pe number is decreasing with time for oil phase as shown in **Fig. 9**. In the natural fractures areas, the results indicated that Péclet number is significantly changeable where it is away below 1 in the areas which are far away from hydraulic fractures; however, it is away above 100 in the areas which are closed to hydraulic fractures. According to the average value of Péclet number in the natural fractures areas, the dispersion flow could be the most dominant flow. Flow-types regimes for both natural-depletion stage and for EOR-stage have been summarized in **Table 5**.

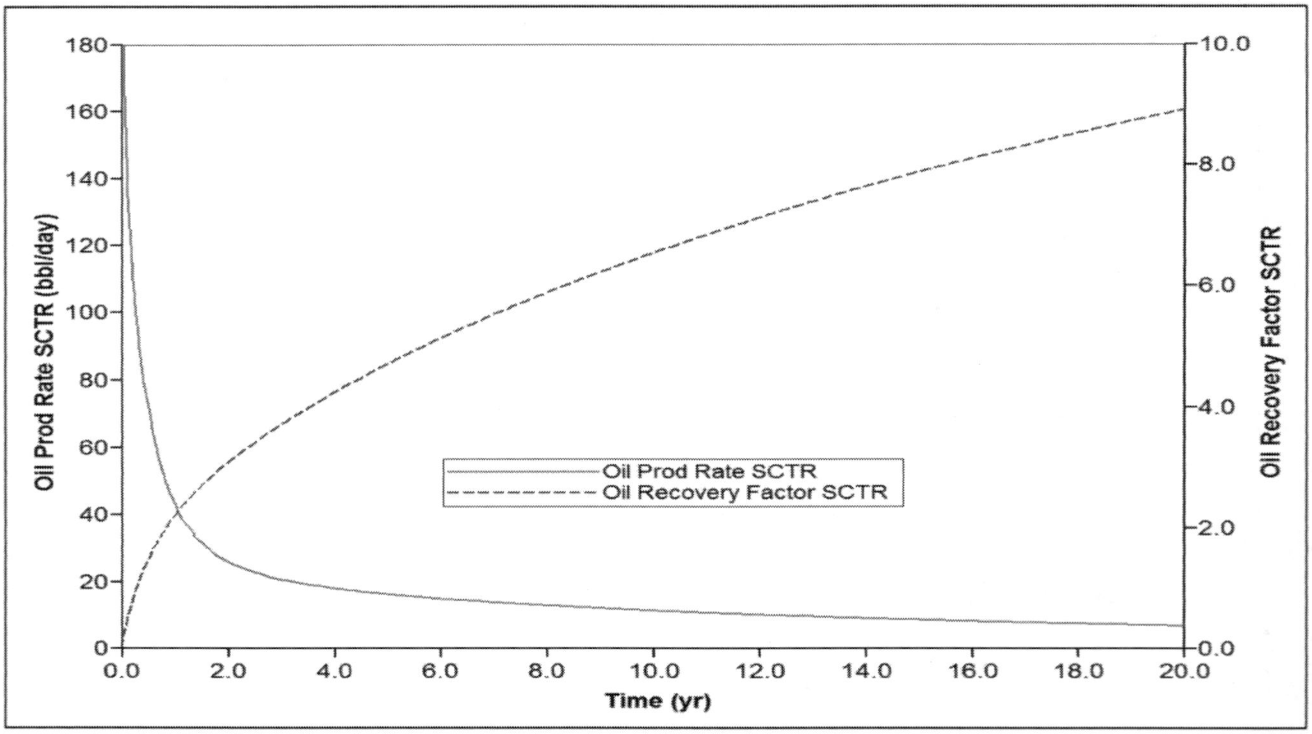


**Fig. 7: The reservoir performance in natural depletion conditions**

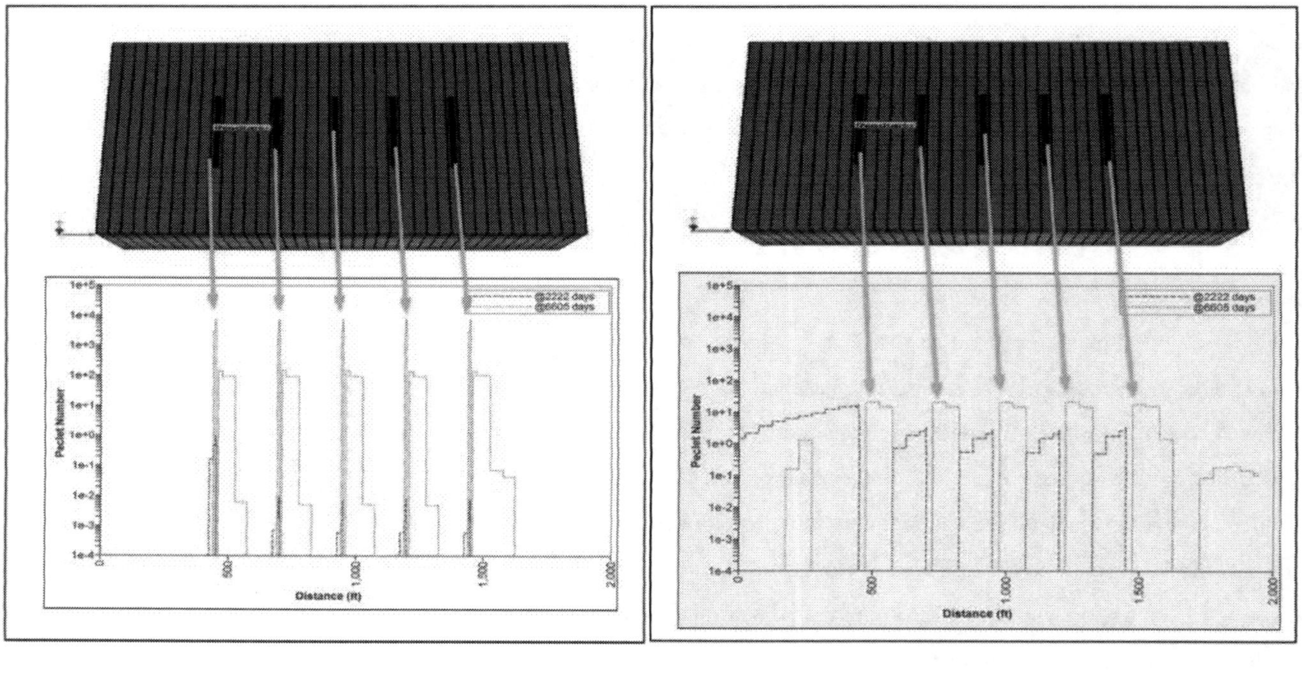


A-Gas Phase

B -Oil Phase

**Fig. 8: Péclet number distribution a long cross section in the matrix-model**

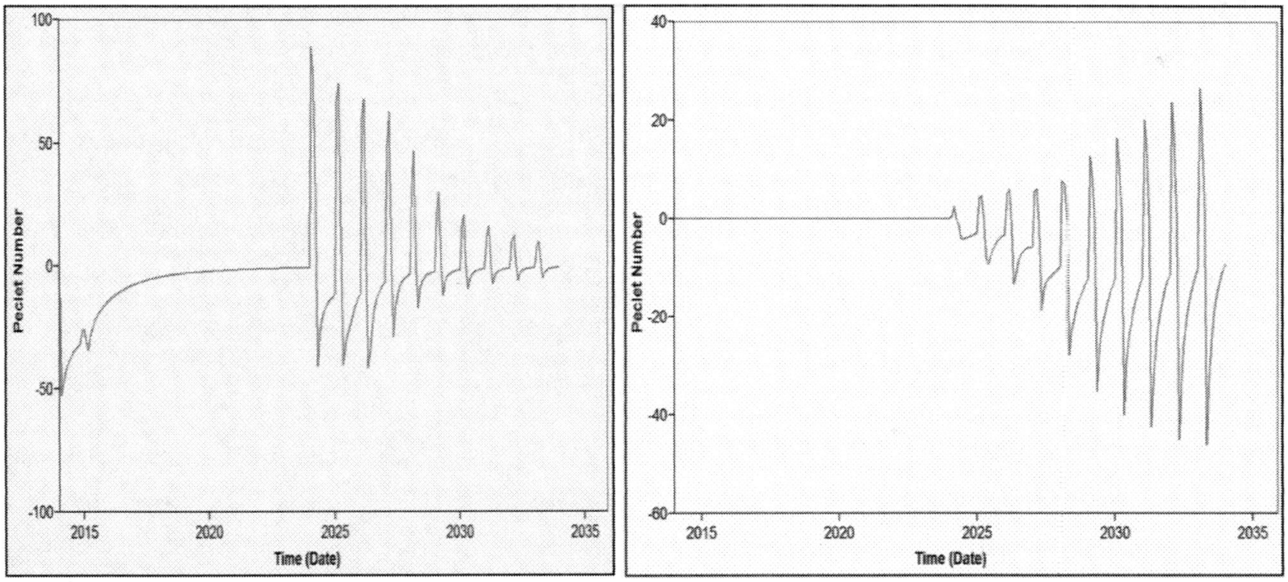


A-Oil Phase B-Gas Phase

**Fig. 9: Péclet number change with time (At 10 ft from the hydraulic fracture)**

**Table 5: Summary for flow-type regimes for natural depletion and EOR stage**

| Primary Production | | | EOR Production | | |
|---|---|---|---|---|---|
| Areas | Oil Phase | Gas Phase | Areas | Oil Phase | Gas Phase |
| HF | Convection | Convection | HF | Convection | Convection |
| NF | Diffusion | Diffusion | NF | Dispersion | Convection |
| Matrix | Diffusion | Diffusion | Matrix | Diffusion | Diffusion |

***EOR Stage for Bakken Model.*** In EOR stage, we injected $CO_2$, lean Gas, and rich gas in the Bakken production-well as Huff-n-Puff protocol in three scenarios. Each of the first scenario and second scenario has two cases. The third scenario has four cases. The EOR stage started after 10 years of natural depletion for all scenarios. In this study, the lean gas contains 90% of $C_1$ and 10% of $C_{2+}$ while the rich gas contains 65% of $C_1$ and 35% of $C_{2+}$.

- **1st Scenario (10 cycles)**: This scenario has two cases. In the first case, the molecular diffusion mechanism is switched on and three different miscible gases which are rich gas, lean gas, and $CO_2$ have been injected. Ten cycles have been injected during 10 years. Each cycle injected 500 Mscf/day for 2 months with a soaking period of 1 month. The results indicated that as far as the molar-diffusion mechanism is switched on, $CO_2$ performance exceeds the performance for both of lean gas and rich gas as shown in **Fig. 10a**. We notice that the performance of miscible gases from the best to the worst as $CO_2$, lean gas, and rich gas respectively. This happens due to the difference in the concentration gradient for the injected gas in the injected fluid and the formation-fluid according to the **Eq. 2**. The concentration gradient is so significant for $CO_2$; however, it is low for both of lean gas and rich gas. The reason causing that lean gas performance exceeds rich gas's is due to the difference in both of molecular-weight and concentration-gradient between lean gas and rich gas. It is known that rich gas has higher molecular weight than that for lean gas, so it needs longer soaking-period to invade the formation-oil. This is particularly true for shale-oil because the composition of shale-oil is usually containing high concentrations of light components (i.e. $C_1$ and $C_2$).

The second case of this scenario is exactly the same of the first case for this scenario. However, the molar diffusion mechanism is switched off in the second case while switched on in the first case. The three different miscible gases have been also injected. All the gases have approximately shown the same performance. Their performance is worse than the base case, natural depletion case, because the obtained increment in oil recovery due to the EOR process does not fully upset the loss in oil production during the injection and soaking periods as shown in **Fig. 10b**.

The results of this scenario are very well consistent with the results which have been reported by Hoteit and Firoozabadi (2009) as shown in **Fig. 11** and **Fig. 12**. In their model, which has been conducted in conventional fractured reservoirs, they observed that methane would perform better than $CO_2$ in the cases which have not considered the molecular-diffusion mechanism. However, the injected $CO_2$ would result in higher increment for oil recovery in the cases which have considered the molecular-diffusion mechanism.

$$\textbf{Rate of Diffusion} = \frac{CD * (C1 - C2) * Ac}{tc} \qquad (2)$$

Where: $C_D$ is the molecular diffusion rate (0.0008-0.0004 $cm^2/s$ was specified in this model), (C1-C2) is the component-concentration difference between the injected fluid and the target fluid, Ac is the contact area between the injected fluid and the target fluid, and $t_C$ is the separation distance between the injected fluid and the target fluid.

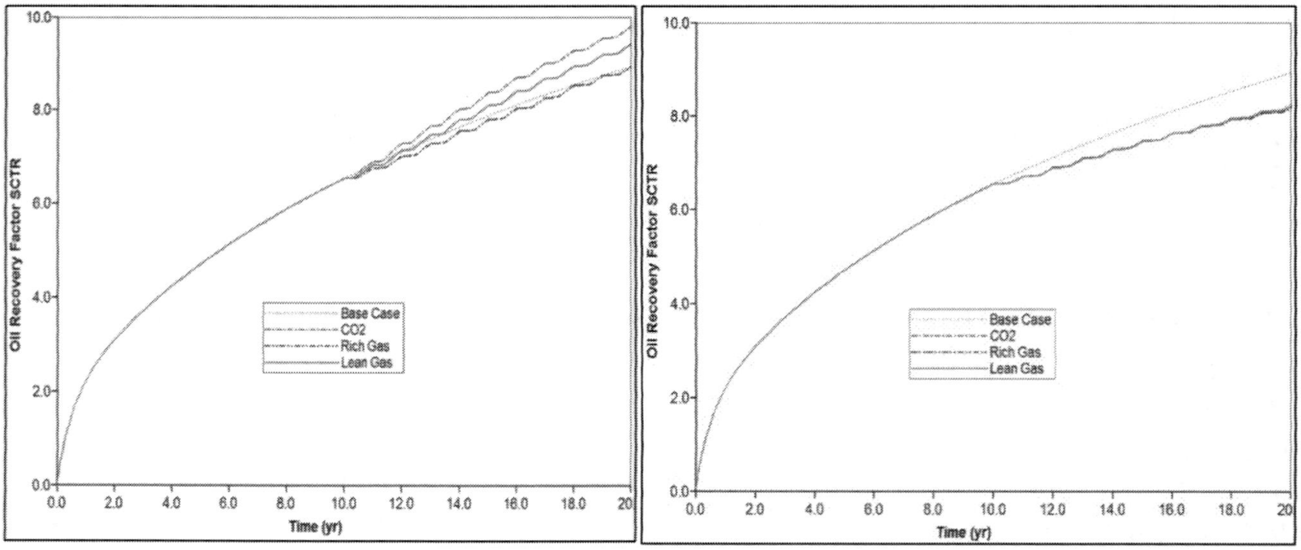


a-(Molar-diffusion mechanism is ON) b-(Molar-diffusion mechanism is OFF)

**Fig. 10: Miscible-Gases Gases performance in Bakken model for the 1st scenario**

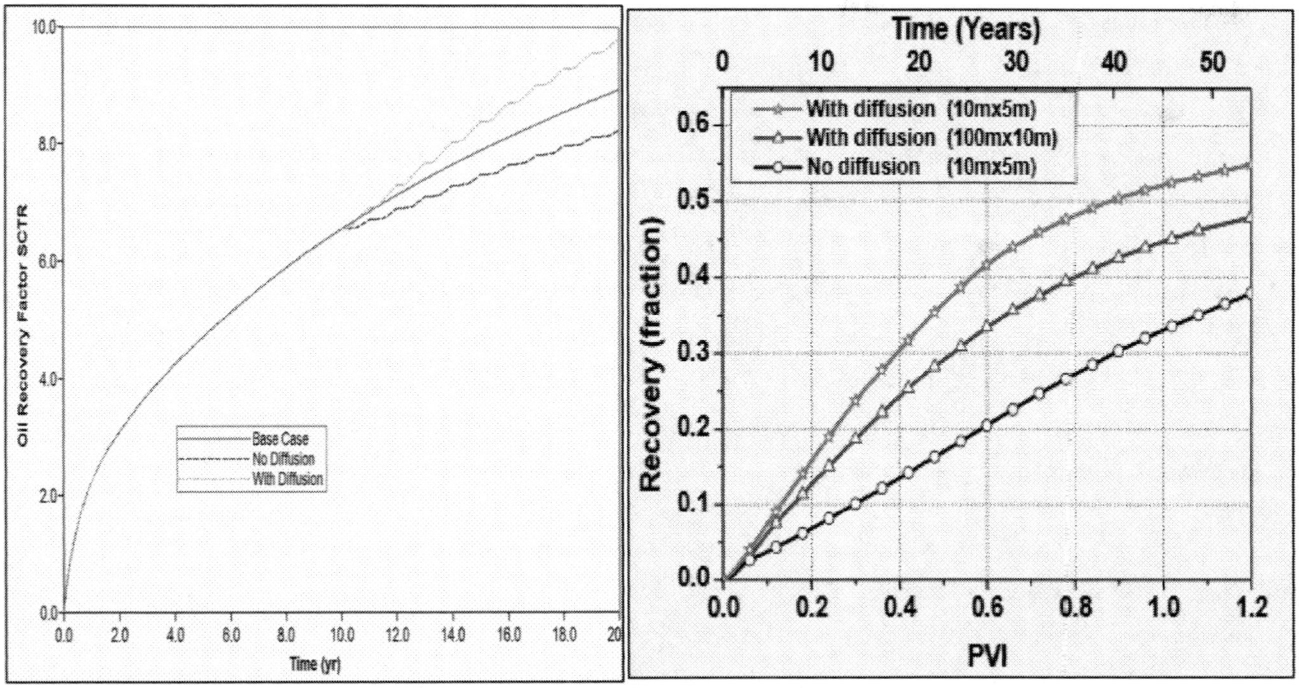


a b

**Fig. 11: Effect of Molecular diffusion on $CO_2$-EOR Performance: (a) $CO_2$ Performance in shale oil reservoirs - Bakken model, b- $CO_2$ performance in fractured conventional reservoirs (Hoteit and Firoozabadi, 2009)**

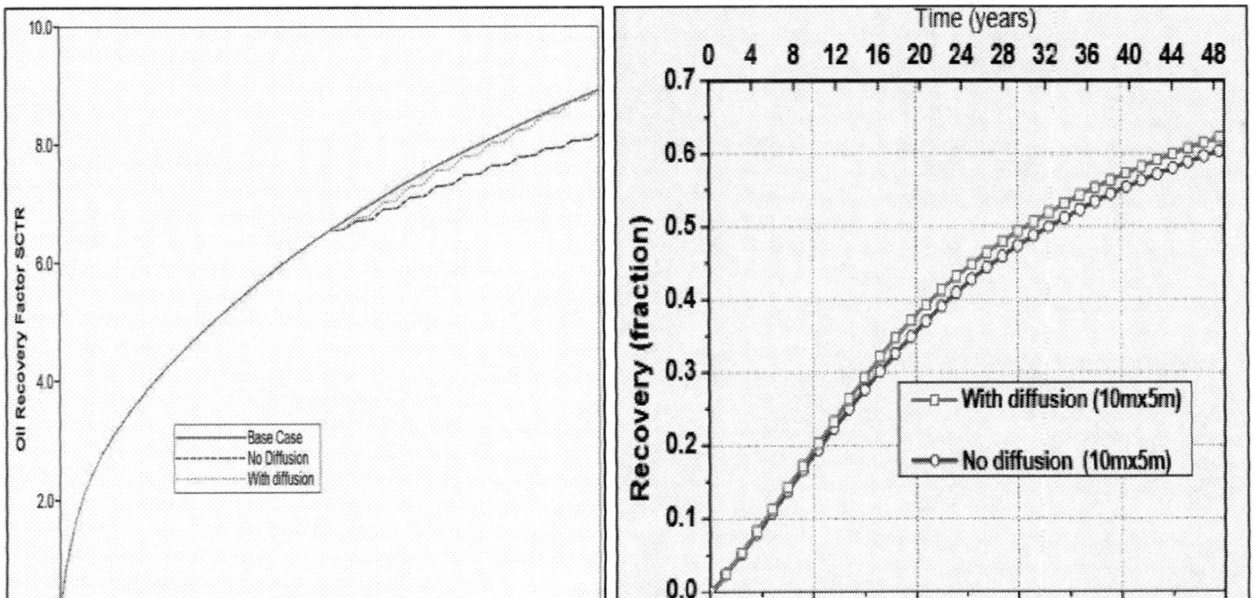


A B

**Fig. 12: Effect of Molecular diffusion on natural gas-EOR Performance: A-Natural gas performance in shale oil reservoirs-Bakken Model, B- NG performance in conventional fractured reservoirs (Hoteit and Firoozabadi, 2009)**

• **2nd Scenario (2 cycles):** This scenario has two cases. In the first case, the molecular diffusion mechanism is switched on and three different miscible gases which are rich gas, lean gas, and $CO_2$ have been injected. Two cycles have been injected during 10 years. Each cycle injected 500 Mscf/day for 6 months with a soaking period of 3 months. The results indicated that $CO_2$ performance is the best as compared with lean gas and rich gas. We notice that the performance of miscible gases from the best to the worst as $CO_2$, rich gas, and lean gas respectively as shown in **Fig. 13a**. Interestingly, the rich-gas performance exceeds the performance of lean gas in the 2nd scenario while the lean-gas performance exceeds rich-gas performance in the 1st scenario. The first case of both scenario is the same; however, the soaking period in this scenario is longer than the soaking period in the previous scenario which favors rich gas on lean gas. The rich gas is showing a good functionality for soaking period as compared with lean gas. This happens due to the difference in the concentration gradient of the miscible-gas between the injected fluid and the formation-fluid according to the **Eq. 2** which favors rich gas on lean gas, but rich gas needs a longer soaking period due to its larger molecular weight.

The second case of this scenario is exactly the same of the first case for this scenario. However, the molar diffusion mechanism is switched off in the second case while switched on in the first case. The three different miscible gases have been also injected. The results indicated that rich-gas performance is the best as compared to $CO_2$ and lean gas as shown in **Fig. 13b**. We notice that the performance of miscible gases from the best to the worst as rich gas, lean gas, and $CO_2$ respectively. The $CO_2$ has the worse performance in this case. This happens mainly due to the large molecules for $CO_2$ as compared with lean gas and rich gas. $CO_2$ would not penetrate into matrix far away from hydraulic fractures if the molecular-diffusion rate is low according to **Eq. 2** and as shown in **Fig. 14** and **Fig. 15**. However, the lean gas and rich gas penetrate deeper into the matrix as compared to what happens in $CO_2$ injection. This happens in all cases in which the molecular-diffusion mechanism switched off.

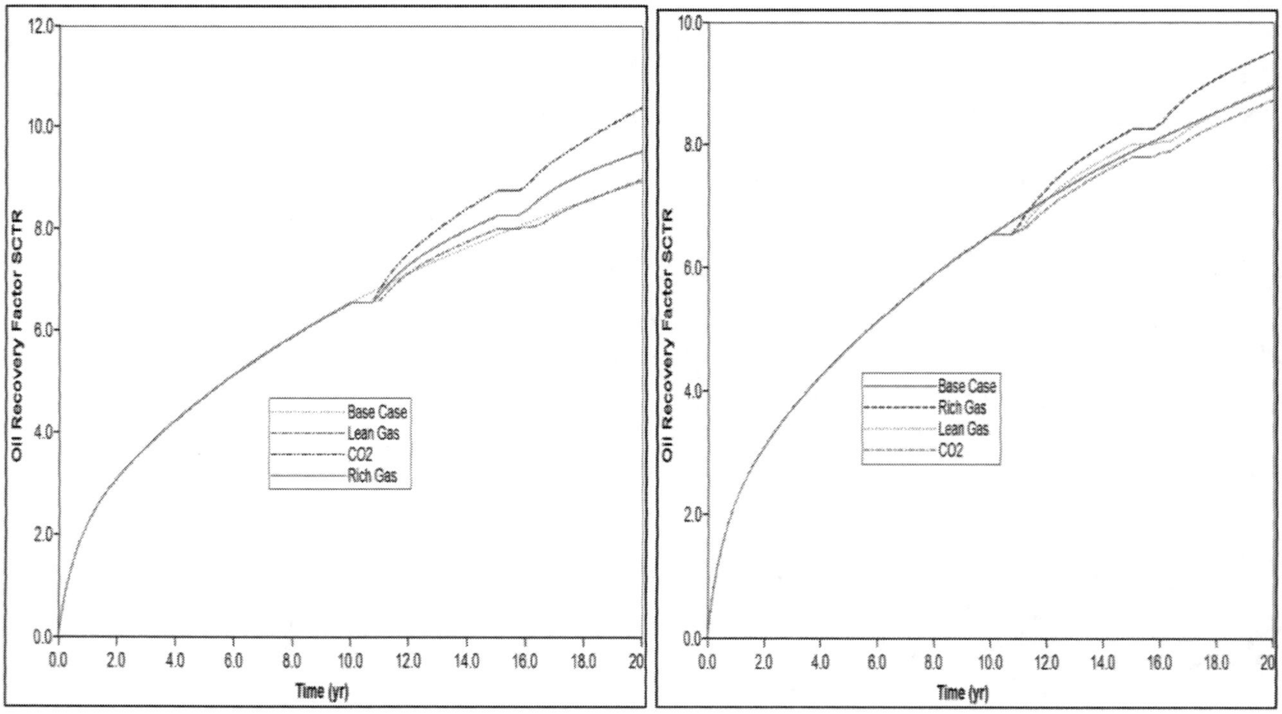


a-(Molar-diffusion mechanism is ON) b-(Molar-diffusion mechanism is OFF)

**Figure 13: Miscible-Gases Gases performance in Bakken model for the 2nd scenario**

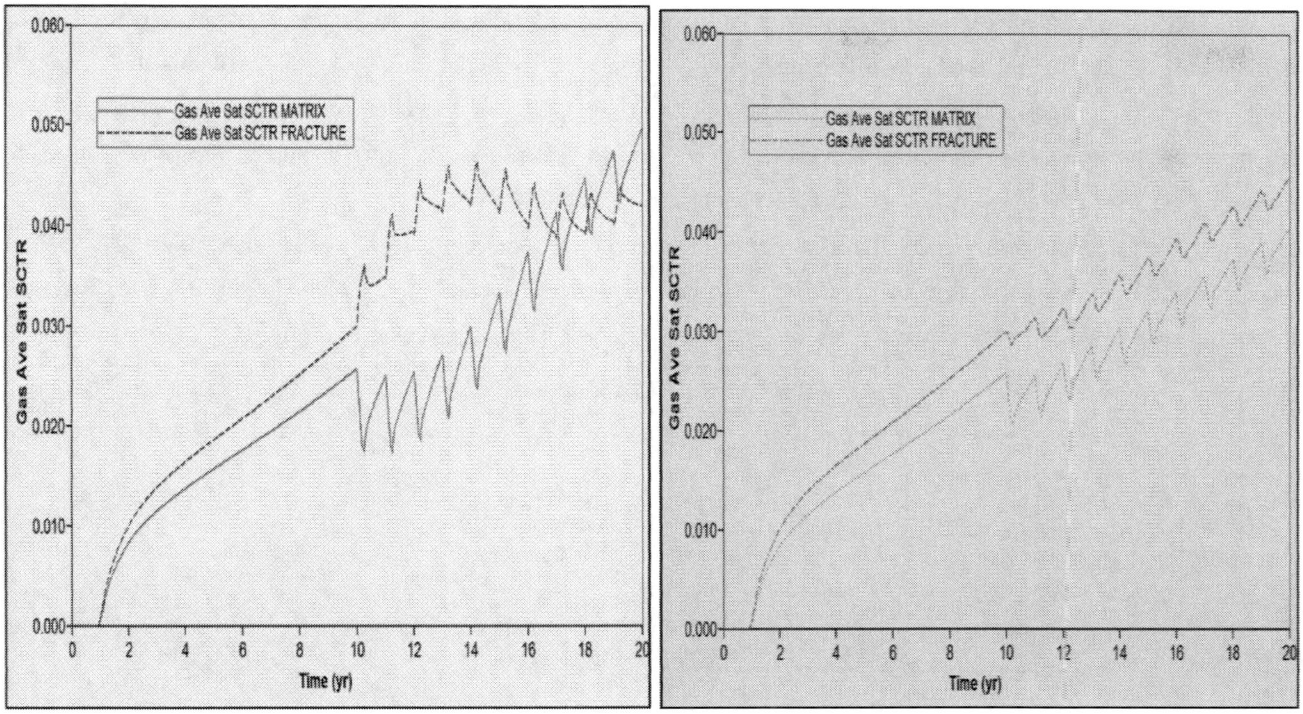


a-Lean Gas b-$CO_2$

**Figure 14: Gas saturation distribution in matrix Vs fracture (Molar-diffusion mechanism is OFF)**

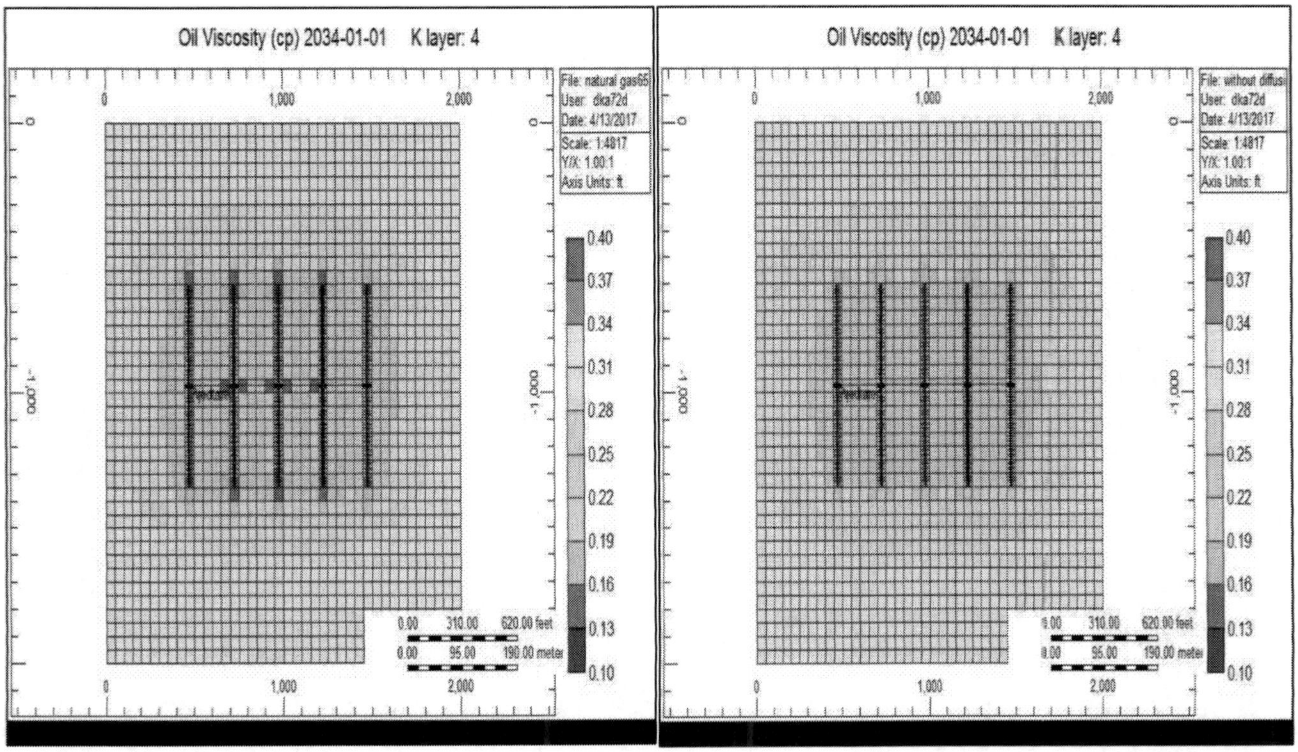


a-natural gas b-$CO_2$

**Figure 15: Effect of the injected gas on oil viscosity (Molar-diffusion mechanism is OFF)**

- **3rd Scenario (Large-Volumes Injected):** This scenario has four cases. In the first case, the molecular diffusion mechanism is switched on and the three-different miscible-gases have been injected. Two cycles have been injected during 10 years. Each cycle injected 1500 Mscf/day for 6 months with a soaking period of 3 months. The results indicated that rich gas performance is the best as compared with lean gas and $CO_2$ as shown in **Fig. 16a**. We notice that the performance of miscible gases from the best to the worst as rich gas, $CO_2$, and lean gas respectively.

The second case is exactly the same of the first case, but the molecular diffusion mechanism is switched off. The results indicated that rich gas performance is the best as compared with lean gas and $CO_2$ as shown in **Fig. 16b**. We notice that the performance of miscible gases from the best to the worst as rich gas, lean gas, and $CO_2$ respectively.

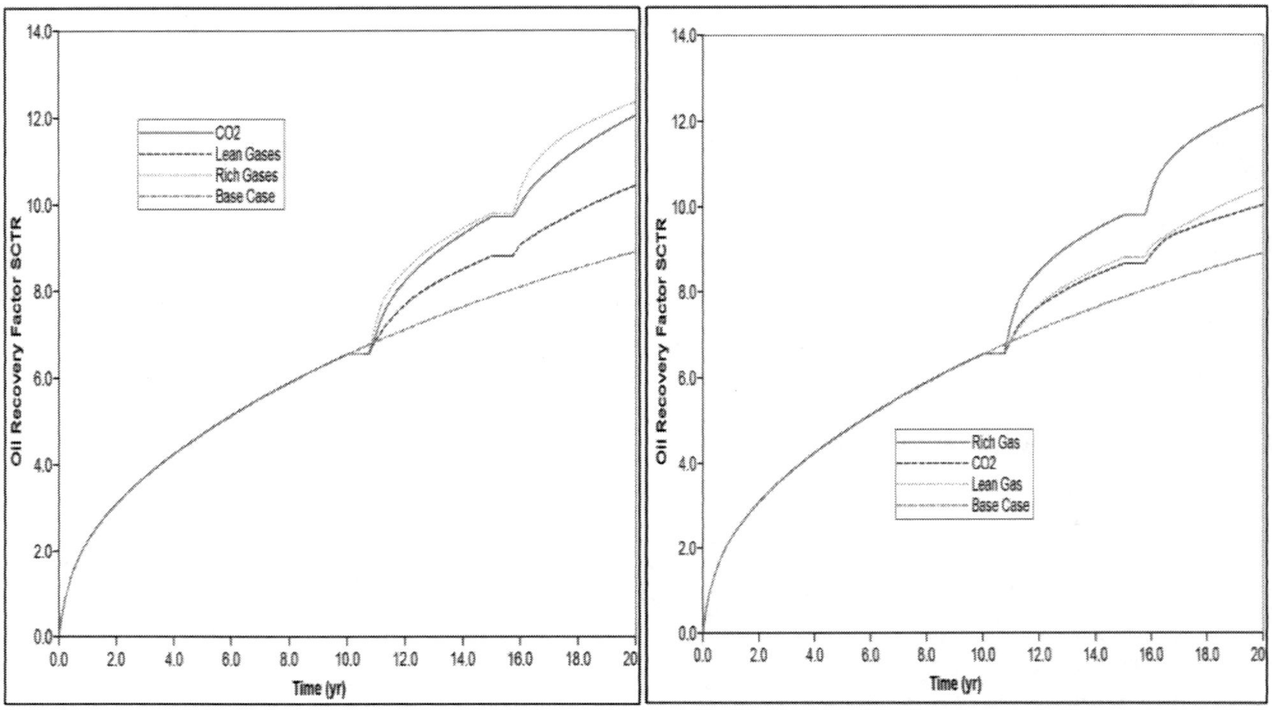


a-Case 1 b-Case2

**Figure 16: Miscible-Gases Gases performance in Bakken model for the 3rd scenario**

In the third case, the molecular diffusion mechanism is switched back on again and the three different miscible gases have been injected. However, in this case, ten cycles have been injected during 10 years. Each cycle injected 1500 Mscf/day for 2 months with a soaking period of 1 month. The results indicated that as far as the molar-diffusion mechanism is switched on, $CO_2$ performance exceeds the performance for both of lean gas and rich gas as shown in **Fig. 17a**. We notice that the performance of miscible gases from the best to the worst as CO2, lean gas, and rich gas respectively.

The fourth case is exactly the same of the third case. However, the molecular diffusion mechanism is switched off. The results indicated that as far as the molar-diffusion mechanism is switched off and large volumes injected, lean gas is the best as shown in **Fig. 17b**. We notice that the performance of miscible gases from the best to the worst as lean gas, rich gas, and $CO_2$ respectively. **Fig. 18** summarizes the Applicability of the three-different miscible gases to enhance oil recovery in Bakken Model.

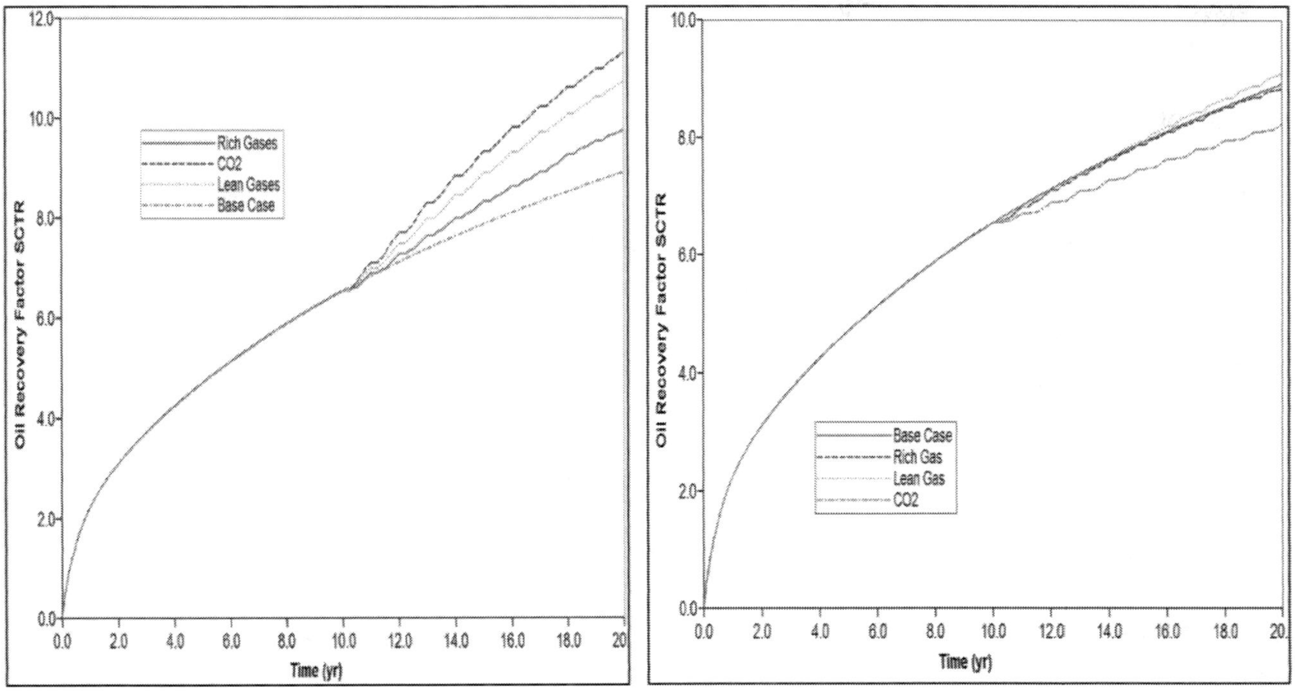


a-Case 3 b-Case 4

**Figure 17: Miscible-Gases Gases performance in Bakken model for the 3rd scenario**

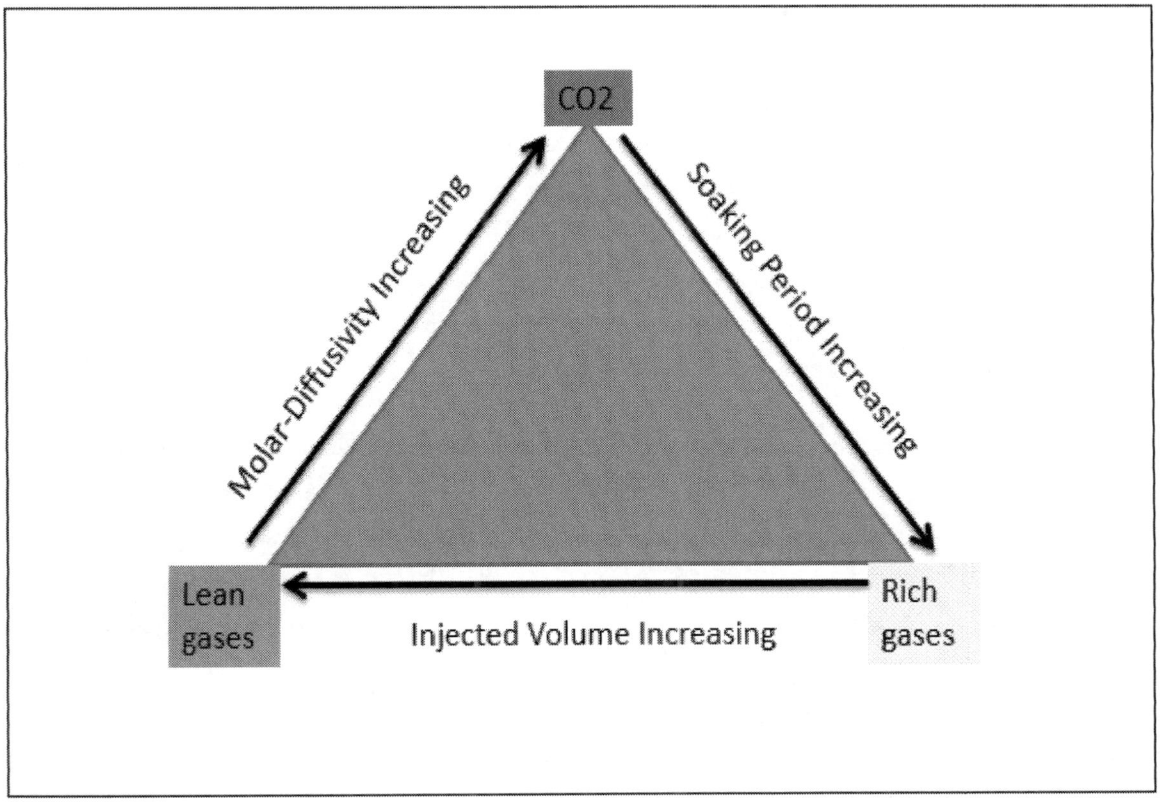


**Figure 18: Applicability of miscible-gases EOR in Bakken Model**

## Molar-Diffusivity Level in the Real Conditions for Shale Oil Reservoirs

Hoffman et al., (2016) reported seven pilot-tests in Bakken formation which was conducted in North Dakota and Montana. We are presenting only one pilot of them in this section. This pilot was mentioned in his paper as pilot test#2. This pilot-test injected $CO_2$ as Huff-n-Puff process in Bakken formation, in Montana portion. They injected 1500-2000 Mscf/day of $CO_2$ for 45 days at 2000-3000 psi. The soaking period was proposed to be 2 weeks. Then, the well was put back in the production process. In puff process, the oil rate had increased slightly above the value which was observed before $CO_2$ injection, but this increment in oil production rate does not reimburse the oil-production lost during the injection and soaking times as shown in **Fig. 19.**

We used the typical fluid and rock properties of Bakken to build a model for that well. Different scenarios have been run until the best match obtained between the well model and the pilot test as shown in **Fig. 19.** Everything was identical between the model results and pilot-test results which are shown in **Fig. 19**. However, there is only one difference. This difference is that the oil production came quickly after the soaking period in the pilot test; however, it takes longer time in the model case. We believe this is happening due to the reported conformance problems in these pilots where $CO_2$ produced in the offset wells. Furthermore, we believe that the conformance problems which were happening in those pilots are due to injection induced fractures (Alfarge et al., 2017). Therefore, the produced-back $CO_2$ volumes in the production well were small which resulted in less hold up effect on produced oil. However, in our model, we have not induced injection fractures. Therefore, $CO_2$ in large volumes produced back during the puff process of our model.

Among different scenarios which we investigate, we found that this match can be obtained in a dual permeability model with low $CO_2$-molecualr diffusivity. This means due to that either of diffusion rate for $CO_2$ in reservoir conditions is too low or kinetics of oil recovery process in the production areas overcomes the $CO_2$ diffusivity. The first possibility which is the low-diffusivity for $CO_2$ in shale reservoirs conditions can be explained by two ways: (1) The contact area between the injected $CO_2$ and formation-oil is small (2) The exposure time between the injected $CO_2$ and the formation-oil is short. The contact area between $CO_2$ and formation oil is mainly a function of natural-fractures intensity in shale oil reservoirs. Although it has been reported these types of reservoirs has high intensity of natural fractures, the dual permeability model can match the conducted pilot test results even with low intensity of natural fractures. This indicated that either of these natural fractures are not active or they are not connected in good pathways with hydraulic fractures.

***Closing Remarks***. Most of the experimental studies reported that $CO_2$ diffusion mechanism is beyond the increment in oil recovery which was obtained in lab conditions. This increment in oil recovery and/or the diffusion rate which were observed in lab conditions were upscaled directly by most of the previous researchers to the field scale by using numerical simulation methods. This direct upscaling methodology is so optimistic due to that the lab-cores have higher contact area and longer exposure time to $CO_2$ than what happened in the real reservoirs conditions. Therefore, both of simulation studies and experimental works were optimistic to predict a quick improvement in oil recovery from injecting $CO_2$ in these unconventional reservoirs. This might explain why the results from pilot tests which were using $CO_2$ as injectant are disappointing and the results from pilot tests which were using natural gases are encouraging (Alfarge et al., 2017). To sum up, diffusion mechanism for $CO_2$ in pilot tests had not been well recognized, which in turn, did not enhance oil production rate in those wells. The reason beyond the low-diffusion rate for $CO_2$ in pilot tests is due to either of the kinetics of oil recovery process in productive areas of these reservoirs are too fast or $CO_2$ diffusion rate in field conditions is too slow (Alfarge et al., 2017). To sum up, the success of $CO_2$ in shale reservoirs is mainly depending on understanding its main mechanisms which are totally different from its mechanisms in conventional reservoirs. Although most of unconventional IOR studies investigated the applicability of $CO_2$, they did not properly investigate its soul mechanism in field scale.

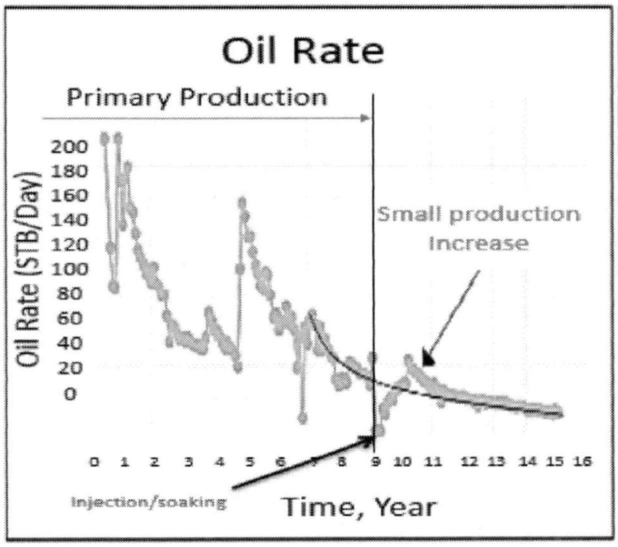


Fig. 19a: CO2 Pilot test#2 (Hoffman et al. 2016)

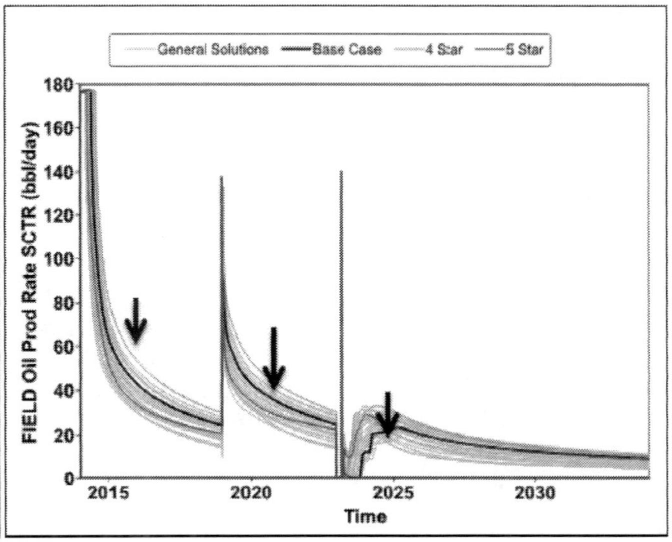


Fig. 19b: History match from the simulated model

## Conclusions

- Although most of the previous studies in this area recommended that $CO_2$ would be the best EOR-technique to improve oil recovery in these formation, pilot tests showed that natural gases performance clearly exceeds the $CO_2$ results in field scale.
- In this study, Péclet number calculations reports a significant flow-type heterogeneity in shale-reservoirs. However, Diffusion flow is the most dominant.
- The simulation results approved that the molecular-diffusion has a significant role on EOR by gas injection in Bakken shale-reservoir. However, $CO_2$ needs a good molar-diffusivity into formation-oil, so it can enhance oil production in these shale reservoirs. Lean gas and rich gas success requires less molar-diffusivity as compared with $CO_2$.
- Some of $CO_2$ Pilot tests showed a good match with the simulated cases which have low-diffusivity between formation-oil and the injected-$CO_2$.
- If the well or field conditions predict a low molar-diffusivity for the injected gases, the rich and lean gases would have a better feasibility than $CO_2$. However, rich gases need long soaking periods and lean gases need large volumes to be injected for more successful results.

## References

Alfarge, D., Wei, M., and Bai, B., (2017). IOR Methods in Unconventional Reservoirs of North America: Comprehensive Review. SPE-185640-MS prepared for presentation at the SPE Western Regional Meeting held in Bakersfield, California, USA, 23-27 April 2017.

Alharthy, N., Teklu, T., Kazemi, H. et al. 2015. Enhanced Oil Recovery in LiquidRich Shale Reservoirs: Laboratory to Field. Society of Petroleum Engineers. DOI: 10.2118/175034MS.

Alvarez, J.O. and Schechter, D.S. 2016. Altering Wettability in Bakken Shale by Surfactant Additives and Potential of Improving Oil Recovery During Injection of Completion Fluids. Society of Petroleum Engineers. http:10.2118/SPE-179688-MS.

Clark, A. J., (2009). Determination of Recovery Factor in the Bakken Formation, Mountrail County, ND. Society of Petroleum Engineers. doi:10.2118/133719-STU.

Computer Modeling Group, GEM Manual. Https: www.CMG.Ca/ accessed 2017.

da Silva, F.V. and Belery, P. 1989. Molecular Diffusion in Naturally Fractured Reservoirs: A Decisive Recovery Mechanism. Presented at the SPE Annual Technical Conference and Exhibition, San Antonio, Texas, USA, 8–11 October. SPE-19672-MS. http://dx.doi.org/10.2118/19672-MS.

Dong, C., & Hoffman, B. T. (2013). Modeling Gas Injection into Shale Oil Reservoirs in the Sanish Field, North Dakota. Society of Petroleum Engineers. doi:10.1190/URTEC2013-185Energy Information Administration (EIA), (2015).

Gamadi, T. D., Sheng, J. J., Soliman, M. Y., Menouar, H., Watson, M. C., & Emadibaladehi, H. (2014). An ExperimentalStudy of Cyclic CO2 Injection to Improve Shale Oil Recovery. Society of Petroleum Engineers. doi:10.2118/169142-MS

Hawthorne et al., (2017). Measured Crude Oil MMPs with Pure and Mixed CO2, Methane, and Ethane, and Their Relevance to Enhanced Oil Recovery from Middle Bakken and Bakken Shales. SPE-185072-MS paper was prepared for presentation at the SPE Unconventional Resources Conference held in Calgary, Alberta, Canada, 15-16 February 2017.

Hawthorne, S. B., Gorecki, C. D., Sorensen, J. A., Steadman, E. N., Harju, J. A., & Melzer, S. (2013). Hydrocarbon Mobilization Mechanisms from Upper, Middle, and Lower Bakken Reservoir Rocks Exposed to CO. Society of Petroleum Engineers.

Hoffman, B. T., and Evans J., (2016). Improved Oil Recovery IOR Pilot Projects in the Bakken Formation. SPE-180270-MS paper presented at the SPE Low Perm Symposium held in Denver, Colorado, USA, 5–6 May 2016.

Hoteit, H. and Firoozabadi, A. 2009. Numerical Modeling of Diffusion in Fractured Media for Gas-Injection and Recycling Schemes. SPE J. 14(02): 323–337. SPE-103292-PA. http://dx.doi.org/10.2118/103292-PA.

Jin, L., Sorensen, J.A., Hawthorne, S.B., Smith, S.A., Bosshart, N.W., Burton Kelly, M.E., Miller, D.J., Grabanski, C.B., and Harju, J.A., (2016). Improving oil transportability using CO2 in the Bakken System—a laboratorial investigation: Proceedings of the SPE International Conference & Exhibition on Formation Damage Control, SPE 178948, Lafayette, Louisiana, February 24–26.

Kathel, P. and Mohanty, K.K. 2013. Eor in Tight Oil Reservoirs through Wettability Alteration. Society of Petroleum Engineers. DOI: 10.2118/166281MS.

Kurtoglu, B., Kazemi, H., Rosen, R., Mickelson, W., & Kosanke, T. (2014). A Rock and Fluid Study of Middle Bakken Formation: Key to Enhanced Oil Recovery.Society of Petroleum Engineers. doi:10.2118/171668-MS.

LeFever, J., & Helms, L., 2008, Bakken Formation Reserve Estimates, North Dakota Geological Survey.

Mohebbinia et al., (2017). Molecular Diffusion Calculations in Simulation of Gasfloods in Fractured Reservoirs. SPE-182594-MS Paper presented at the SPE Reservoir Simulation Conference held in Montgomery, TX, USA, 20–22 February 2017.

Moortgat, J. and Firoozabadi, A. 2013. Fickian Diffusion in Discrete-Fractured Media from Chemical Potential Gradients and Comparison to Experiment. Energ Fuel27 (10): 5,793–5,805. http://dx.doi.org/10.1021/ef401141q.

Pu, H., and Li, Y., (2016). Novel Capillarity Quantification Method in IOR Process in Bakken Shale Oil Reservoirs. SPE-179533-MS presentated at the SPE Improved Oil Recovery Conference held in Tulsa, Oklahoma, USA, 11–13 April 2016

Schmidt, M., & Sekar, B. K. (2014). Innovative Unconventional 2EOR-A Light EOR an Unconventional Tertiary Recovery Approach to an Unconventional Bakken Reservoir in Southeast Saskatchewan. World Petroleum Congress.

Sheng, J.J. 2015. Enhanced oil recovery in shale reservoirs by gas injection. Journal of Natural Gas Science and Engineering 22: 252–259. http://dx.doi.org/10.1016/j.jngse.2014.12.002

Shoaib, S., & Hoffman, B. T. (2009). CO2 Flooding the Elm Coulee Field. Society of Petroleum Engineers. doi:10.2118/123176-MS.

Song, C., & Yang, D. (2013). Performance Evaluation of CO2 Huff-n-Puff Processes in Tight Oil Formations. Society of Petroleum Engineers. doi:10.2118/167217-MS.

Wan, T., & Sheng, J. (2015). Compositional Modelling of the Diffusion Effect on EOR Process in Fractured Shale-Oil Reservoirs by Gasflooding. Society of Petroleum Engineers. doi:10.2118/2014-1891403-PA.

Wang, D., Zhang, J., Butler, R., and Olatunji, K., (2016). Scaling Laboratory-Data Surfactant-Imbibition Rates to the Field in Fractured-Shale Formations. Socienty of Petroleum Engineers. http://dx.doi.org/10.2118/178489-PA.

Wang, X., Luo, P., Er, V., & Huang, S.-S. S. (2010). Assessment of CO2 Flooding Potential for Bakken Formation,Saskatchewan. Society of Petroleum Engineers. doi: 10.2118/137728-MS.

Wood, T., & Milne, B., (2011). Waterflood potential could unlock billions of barrels: Crescent Point Energy. http://www.investorvillage.com/uploads/44821/files/CPGdundee.pdf

Xu, T., & Hoffman, T. (2013). Hydraulic Fracture Orientation for Miscible Gas Injection EOR in Unconventional OilReservoirs. Society of Petroleum Engineers. doi:10.1190/URTEC2013-189

Yu et al., (2016b). Experimental Evaluation of Shale Oil Recovery from Eagle Ford Core Samples by Nitrogen Gas Flooding. SPE-179547-MS Paper presented at the SPE Improved Oil Recovery Conference held in Tulsa, Oklahoma, USA, 11–13 April 2016.

Yu, W., Lashgari, H., & Sepehrnoori, K. (2014). Simulation Study of CO2 Huff-n-Puff Process in Bakken Tight OilReservoirs. Society of Petroleum Engineers. doi:10.2118/169575-MS

Yu, Y., Li, L. and Sheng J., (2016a). Further Discuss the Roles of Soaking Time and Pressure Depletion Rate in Gas Huff-n-Puff Process in Fractured Liquid-Rich Shale Reservoirs. SPE-181471-MS paper presented in at the SPE Annual Technical Conference and Exhibition held in Dubai, UAE, 26-28 September 2016.

Zhang et al., (2015). Integrated Method to Screen Tight Oil Reservoirs for CO2 Flooding. SPE-175969-MS paper presented at the SPE/CSUR Unconventional Resources Conference held in Calgary, Alberta, Canada, 20-22 October 2015.

Zhang, K., (2016). Experimental and Numerical Investigation of Oil Recovery from Bakken Formation by Miscible CO2 Injection. Paper SPE 184486 presented at the SPE international Student Paper Contest at the SPE Annual Technical Conference and Exhibition held in Dubai, UAE, 26-28 September 2016.

Zhu, P., Balhoff, M. T., & Mohanty, K. K. (2015). Simulation of Fracture-to-Fracture Gas Injection in an Oil-Rich Shale. Society of Petroleum Engineers. doi:10.2118/175131-MSSPE-180270-MS17

**CMTC-485460-MS**

# A Guidebook of Carbonate Laws in China and Kazakhstan: Review, Comparison and Case Studies

Yu Liang, University of Texas at Austin
Bin Yuan, University of Oklahoma



This paper was prepared for presentation at the Carbon Management Technology Conference held in Houston, Texas, USA, 17-20 July 2017.



## Abstract

As two of the most important oil and gas players in the world, China and Kazakhstan are experiencing similar CO2 issues related to oil and gas industry. Both countries have been benefiting from oil and gas industry, and striking to reduce the impacts of CO2 emission at the same time. The problem is how to increase the profits while decreasing the CO2 emissions, not only from technical perspective, but also from policy / legislation / permitting perspectives. However, currently there are few studies focusing on the oil and gas laws for these two countries. The objective of this study is to systematically review and compare oil and gas law frameworks, carbonate management, fiscal structures, and taxation systems between China and Kazakhstan.

Therefore, this paper first discusses in details about oil and gas laws in China and oil and gas laws in Kazakhstan, respectively; then a comparison study is conducted between these two countries. In addition, field cases with their fiscal statements are introduced and analyzed to explore the profitability of an oil company in those two counties under CO2 regulation. Finally, advantages and disadvantages of their law frames and fiscal systems are analyzed and discussed.

The royalty rate, taxation and bonuses in China are set in reasonable range subject to applicable laws. The disadvantages of Chinese oil and gas laws system lie in that (1) a windfall tax ranging from 20% to 40% with a threshold of $55/bbl is set in the fiscal system, which can prevent further investment when oil price is high; (2) the clauses and terms in contract system are not flexible enough; (3) an R factor (Columbian model) should be introduced to protect investment when oil price is low and government can receive more revenue when oil price is high; (4) NOC has the right to take over the production operations at any time once the foreign companies have recovered all costs with reasonable reimbursement based on the book values of tangible costs; (5) remittance policy is strict; (6) lack of the incentive for oil and gas industry to decrease CO2 emissions. By contrast, the royalty rate and taxation in Kazakhstan are set within a reasonable range subject to applicable laws. However, according to available and reliable sources, the royalty and bonuses are all negotiable. There is a trend that Kazakhstan government will decrease the relevant taxes rates and extend the tax-exempt range in order to encourage more carbonate investments. The disadvantages Kazakhstan oil and gas laws system include (1) inconsistent policy system and uncertainty in contract types, (2) uncertainty in investment, and (3) uncertainty in disputes settlement. It can be observed from field cases that Kazakhstan government is working on optimizing the law frame and significant progress has been made.

In previous literature, there is no systematic review comparison study of oil and gas laws between China and Kazakhstan. This paper provides insights into oil and gas law systems in these countries for the first time and the results of studies can work as a guidebook of the carbonate management, investment, and legislation for oil and gas companies, potential investors and policy makers.

## Introduction

With strong economic growth, China has become one of the largest oil and gas producing countries in the world. In 2012, China produced 206 Mt oil, accounting for 5% of total oil production in that year, as shown in the Figure1. At same time, China has also become the second largest oil importing country due to large domestic demand, importing 5 MMbbl/day, accounting for about 50% of total demand. In recent year, the main nation- owned oil companies have increased investment in domestic and international oil and gas exploration, development and production in order to maintain oil and gas supplies. It is predicted that in 2020 China's oil import will reach more than 10 MMbbl/day [1].

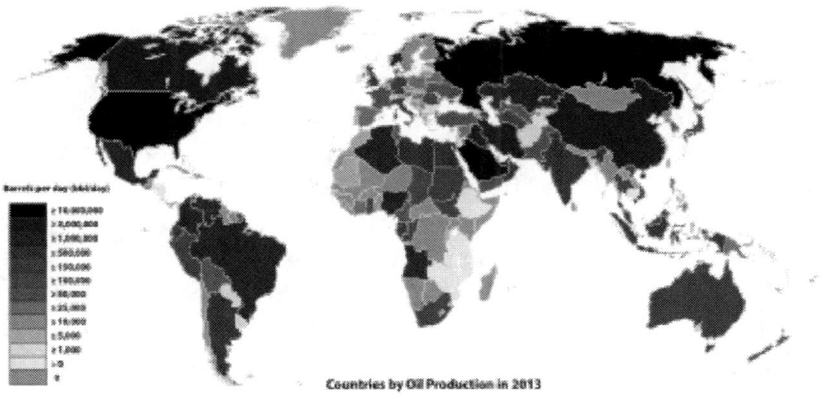


**Figure 1— Countries by oil production in 2013[2].**

There are four major NOCs dominating Chinese oil and gas industry, from upstream to downstream, including China National Petroleum Corporation (CNPC); China Petroleum and Chemical Corporation (Sinopec); China National Offshore Oil Corporation (CNOOC) and China national Chemicals Import and Export Corporation (Sinochem). General speaking, CNPC and Sinopec are the top two NOCs in China, and they have their main business in onshore oil and gas exploration and production while CNOOC focuses on offshore oil and gas resources. Sinochem mainly refines oil, produces fertilizer and chemical products while recently years they also extend their activities into overseas oil exploration, production and refining[3].

Since independence in 1991, Kazakhstan has made of significance progress on economic reform. As a traditional oil producer, Kazakhstan's oil production increased rapidly since 2000, and exceeds 1 MMbbl/day for the first time in 2003. Kazakhstan produces 1.7 million barrels per day in 2014, ranking 18th in top oil producer in the world, as shown in Figure 3. Two large projects, Tengiz field and Karachaganak field, account for 50% of the country's production. The production is predicted to increase in the future due to the third large project, Kashagan, will start to operate during 2016 to 2017. As of 2014, Kazakhstan has proved 30 billion barrels crude oil reserves, raking the twelfth largest oil reserves in the world[4].

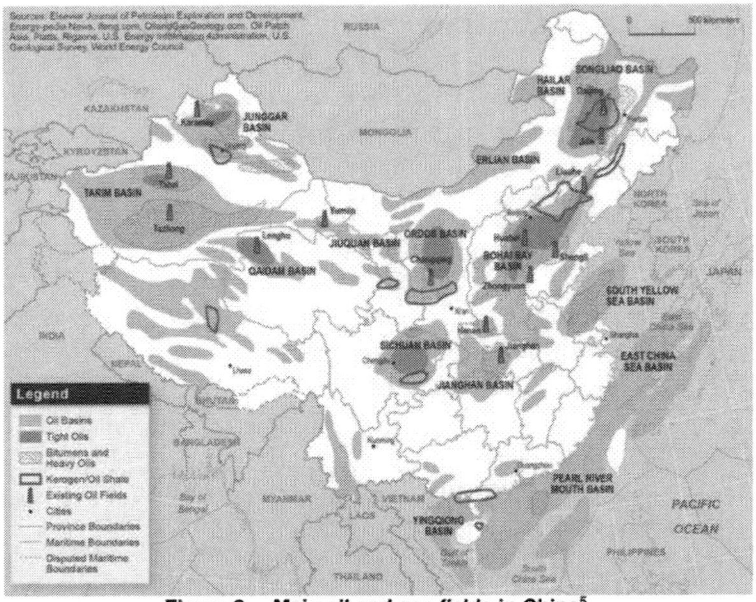


**Figure 2— Main oil and gas fields in China[5].**

Therefore, as two of the most important oil players in the world, China and Kazakhstan have many things in common: both of them used to have similar political system; both of them have large oil and gas reserves with lack of effluent capital and advanced technology; both of them make significant progress on economic development, as well as the oil and gas production. However, there are some differences between two countries, including currently political system, economic system, financial system, and goal of nation interests. Especially for the goal of nation interests, China is determined to become new energy supreme power in the world, while Kazakhstan "will come a factor of energy security in Asia and Europe"[6], which works as an important bridge connecting even controlling the oil and gas from between Asia and Europe. It would be very challenging and rewarding to study and compare the oil and gas laws in China and oil and gas laws in Kazakhstan. In the following sections, oil and gas laws in China and Kazakhstan, in term of ownership and title, oil and gas law framework, operating requirement, roles of national oil companies, foreign investment, private investment, private investment, oil price, fiscal structure, tax regime, contracts, currency regulations and disputes settlement, are studied and compared in detail.

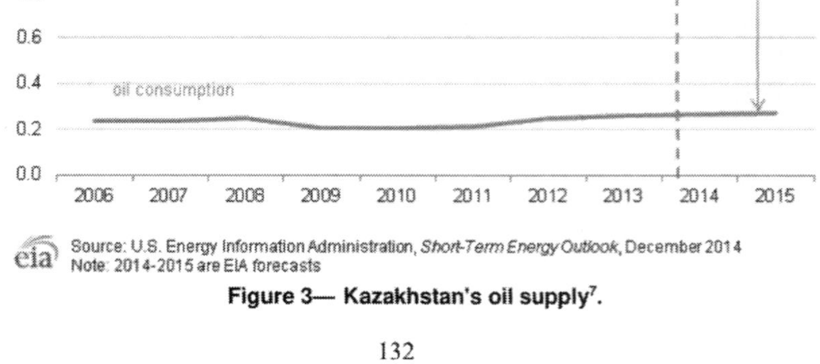


**Figure 3— Kazakhstan's oil supply[7].**

## Oil and Gas Law in China

With strong economic growth, China has become one of the largest oil and gas producing countries in the world recently. In 1979, China government started its first cooperation with foreign investment to explore offshore oil and gas fields in South Sea. Since then, China accelerated to open its onshore and offshore oil and gas resources to international investment. In 1993, China first allowed exploration onshore activities, in the western China Tarim Basin's natural gas reserves. On the other hand, during the cooperation with foreign companies, three Chinese national oil companies, CNPC, Sinopec and CNOOC, have grown rapidly and became top oil companies in the world within 30 years. This section discusses Chinese oil and nature gas law framework, fiscal term, taxation and so on, which can work as a guideline to potential investment.

## Ownership and Title

According to the Constitution of China, the ownership of all minerals, including oil and gas, belong to the state, and the state distributed the rational use of all nature resources. NOCs, as illustrated above, were only, and still are the major operators responsible for onshore and offshore oil and gas exploration, development and production. Because of capital and technology-intensive nature of the exploration, China has signed hybrid contracts, which include the element from concessions, PSA, joint venture and service contracts, with foreign companies for offshore oil field development in South China Sea since 1978, and opened its onshore exploration in Northwest basins to foreign companies since 1993. Local governments and State Ministry of Land And Resource (MOLAR) are responsible for administration of exploration, development and production[8]. Figure 2 shows the onshore and offshore oil and gas fields distributions in China.

## Oil and Gas Law Framework

According to PRC Constitution, National People's Congress, State Council, and local government administrate and regulated exploration and exploitation of oil and gas resources, as a tired structure (arts. 62 89 and 100). The National People's Congress enacts national legislations while the State Council implements national legislations by enactment of regulations and measures. Local governments enforce and administrate the implement of national legislations, regulations and measures. Local regulations are enacted by the people's congress of provinces. Similar to the U.S. legislation system, in term of law enforcement, PRC Constitution is at the top of the legislation structure, followed by national legislations, regulations and measures, and local regulations. In general, the main China legislations directly related to oil and gas exploration, development and production are summarized as following[9]:

- Mineral Resources Law
- Rules for Implementation of the Mineral Resources Law
- Measures for Registration Administration of Exploration Blocks of Mineral Resources
- Measures for Registration Administration of Mineral Resources Exploration
- Measures for the Administration of Transfer of Exploration Right and Exploitation Right
- Regulations on the Exploration of Continental Petroleum Resources in Cooperation with Foreign Enterprises
- Regulations on the Exploration of Offshore Petroleum Resources in Cooperation with foreign Enterprises
- Provisional Regulations on Resources Taxes

In particularly, the Mineral Resources Law, Rules for Implementation of the Mineral Resources Law, Regulations on the Exploration of Continental Petroleum Resources in Cooperation with Foreign Enterprises and Regulations on the Exploration of Offshore Petroleum Resources in Cooperation with Foreign Enterprises are very important therefore discussed below in detail.

In 1986, the Mineral Resources Law was enacted and amended in 1996, stating that the State Council implements the mineral ownership on behalf of the state. The significance of Mineral Resources Law is that it is the first law related to mineral resources, including oil and gas, and is applicable to foreign investment. The purpose of the Law is to promote exploration, development and production of mineral resources. The key points related to oil and gas in the Law include: It is applicable to all mineral resources in China (art.2); Anyone who wish to exploit the resource must make a registration and meet the qualifications (art.3); NOCs are the primary forces in oil and gas development (art. 4); Resources in developed with royalty and government take to the state (art. 5); A unified registration system for mineral exploration and exploration is established (art. 10); Comprehensive assessments are required for the survey of mineral resource (art 21); Original data and sample must be protected and preserved ( art. 24) . Especially, it should be noted that only central government are authorized to review and approve offshore exploration and development proposal, while onshore exploration and development proposal can be reviewed and approved by local government[10].

Rules for Implementation of the Mineral Resources Law were enacted by the State Council in 1994, and are divided into sections to explain the Mineral Resources Law. The purpose of The Rules is to secure lawful rights and interests of exploration licensee and boost development of resources industry. Chapter 1 introduces general information, definition of mineral resources and classifications of mineral resources. The key points in the Rules include: The state should establish a united block register system, and an exploration permit for oil and gas is 2,500 basic unit blocks (art. 3). The State Council designate an organization to examine the approve oil and gas activities (art. 4). The state should establish a licensing system of mineral exploration, development and production. Anyone who wants to exploit mineral resources must apply for registration in accordance with the law (art.5). Mineral resources exploration and exploitation mush by supervised and administrated by the State Council while local governments are responsible for supervision and administration in their respective areas (art. 8). An oil and gas exploration and exploitation license is valid up to seven years, and can be extended to fifteen years (art. 10). After a license is issued, the license is required to begin work within six months (art 18.). With an economic deposit, the exploration rights can be reserved for up to four years for the licensee (art.21). It should be noted that all provisions are applicable to foreign investors[11].

Enacted in 1982 and revised in 2001, Regulations on the Exploration of Offshore Petroleum Resources in Cooperation with Foreign Enterprises is the first petroleum law of PRC, setting up the legal framework in the offshore petroleum development for cooperation between foreign investors and PRC, as well as to ensure nation interest. The key in the law include Chinese government shall protect profits, legitimate rights, and interests of foreign investors (art.3). All offshores petroleum activities shall subject to Chinese lads and jurisdictions (art. 4). CNOOC shall have the overall responsibility and exclusive rights for offshore oil and gas exploration and exploitation in cooperation with foreign enterprises (art. 6) All contracts signed by CNOOC and foreign investors by bidding shall subject to approval by the Ministry of Foreign Trade and Economic Cooperation ( art. 7). All enterprises shall relevant taxes and fees, and all employees shall pay individual income taxes (art. 10). Foreign contractors must report production related activities to CNOOC (art. 14). CNOOC shall have the ownership of all records, samples and other original data (art. 21). Operator and subcontractors must comply with laws and regulations of environment and safety (art 22.). Details about tax, contracts and fiscal structure are discussed in the following sections in this paper. Regulations on the Exploration of Continental Petroleum Resources in Cooperation with Foreign Enterprises is adopted in 1992 and revised in 2001. They are identical to offshore regulations in terms of objective, framework and contents The only difference is that onshore regulations provide CNPC and Sinopec with the exclusion right to cooperate with foreign companies in the exploration, development and production of onshore petroleum resources, as well as new mineral resources, shale oil and shale gas for example[11].

## Operating Requirement

Before exploration the project should be examined, approved, registered and licensed by the State Council. According to the size of the project, the license term run from ten to thirty years. With thirty days before the expiration of a term, the concessioner can extend the license if the certain reporting requirements are met and rents are paid to the state on time. Based on related applicable laws, both exploration and exploitation rights are transferable if certain requirements are satisfied. Once the rights are transferred, the term will continue based on whatever time remains[9].

According to the Regulations on the Payment of Royalty for the Exploitation of Offshore Petroleum Resources, which is approved by the State Council on 2 December 1988, royalty is calculated and paid on the basis of the Annul Gross Production of Crude Oil or Natural gas of each oil or gas field at various rates. The signature bonuses and payment methods vary cases by cases, and is determined and calculated by NOCs. General speaking, based on the general Chinese law, CNOOC is paid a signature fee of $1,000,000 by the contractor with three installments. The contractors pay the first $250,000 within thirty days from the date of the implements of the contract. The second $ 250,000 is paid within the thirty days of the contractor's election exercised. The remains $ 500,000 is paid with the thirty days of the date when the overall development programs for the first oil/gas field in the contracted is approved by the contractors. Again, currently the bonuses, including signature bonuses, production bonuses and reserve based bonuses, are calculated and determines according to negotiations[12].

## Roles of National Oil Companies

In China, oil and gas resources must be explored, developed and produced by National oil companies or Chinese foreign joint ventures. There are three national oil companies, CNPC, Sinopec and CNOOC, are authored to purchase exploration rights, development rights, operation rights, overseas lease rights and overseas oil and gas exploitation rights. In addition, if foreign companies want to explore, develop and produce onshore and/ or offshore oil and gas resource, only CNPC, Sinopec and CNOOC are authorized to negotiate and execute under cooperative on petroleum agreement. Foreign companies can operate with these three NOCs to explore and exploit oil and gas resources under the Mineral Resources Law, and subject to all other laws. The objectives of introduction of foreign capitals and advanced technology include promoting domestically mineral resources; transferring advanced technology; increasing mineral and raw materials supplies, increasing government revenues and improving employment opportunity in China[13].

Due to nation energy security, China government imposes new tariffs to discourage the export of crude oil, and decreases relevant tariffs to encourage import of crude oil products from overseas. There is not such restrictions on the natural gas. The sales of crude oil and natural gas in Chinese markets should be licensed by the MOFCOM with fulfill of certain requirements, such as nosiness status, faculties, capitals and certificates. In particular, foreign companies to a PSA usually sell their share of profit oil to the Chinese NOCs at the price, which is agreed in the PSA, instead of selling and exporting the oil and natural gas outside of China. Figure 4 shows performance of NOCs.

## Private Investment

China government also introduces private capitals to domestic oil and gas exploration and exploitation. The government has approved to establish a 10 billion yuan petrochemical venture, including Shanghai Pengdong Petrochemical and other ten private firms, to compete three NOCs, including CNPC, Sinopec and CNOOC, in order to achieve the goal of nature gas and oil diversity and privatization. The venture, called China Great Wall Petroleum United, was funded by China Chamber of Commerce for Petroleum Industry, with the goal to protect small and medium sized private oil firms in domestic and international nature gas and oil market. However, it should be noted that in China three major NOCs still dominate domestic natural gas and oil market while private oil companies obtain a small share of oil and gas market.

## Tax Regime

In addition to the Tax laws and regulations which are applicable to all companies in China, the operations of natural resource exploration, development and production between NOCs and foreign companies are also subject to several specific regulations related to onshore and offshore oil and gas exploitation. The laws and regulations include: The 1991 Income Tax Law on Enterprises with Foreign Investment and Foreign Enterprises; the 1982 Regulations; the Provisions on the Payment of Royalties for the Exploitation of Offshore Petroleum Resources (1989). In order to encourage investment in onshore and offshore nature gas and oil exploration, development and production, the foreign companies are exempted from the some specific tax, like consolidated industrial commercial tax ( CICT) for the imported tools, equipment, and materials which are directly applied to explore, develop and produce onshore and offshore oil and gas natural resources. Subject to the signed agreement, export tax for profit crude oil can also be exempted. Specifically, as a requirement by WTO and a united tax regime for both of domestic companies and foreign companies will be established, announced by China government recently. The taxes related to offshore and onshore exploration, development and production are summarized in below[14].

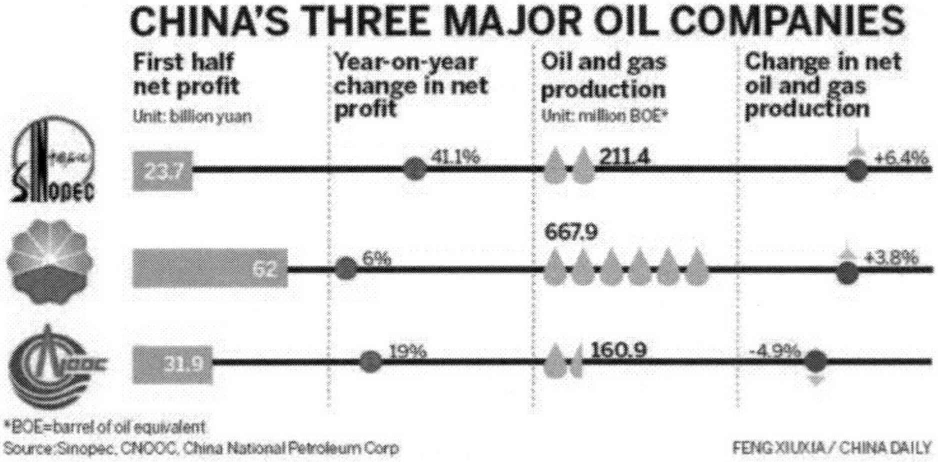


**Figure 4— China's NOCs and related performance[15].**

## Contracts

According to Chinses law," Foreign investors enjoy mining rights as granted by the Chinese authorities, and such rights are to be realized through the execution of petroleum rights". The mining rights include exploration rights and exploitation rights. Exploration rights mean the rights to explore mineral resource with the approval of authorities. Exploitation rights are the rights to develop the mineral resources with the approval of authorities. Both rights are defined in detail by the Law of Mineral Resources[16].

In order to attract foreign investments and avoid exploration risks, a hybrid contract is used as the form of petroleum contracts in China to cooperate with foreign oil companies. It can be seen that China chose the hybrid contract model after a careful study. During 1978 and 1979, in order to design the contract model which best meet national interest, Chinese delegations visit all main oil producing countries in the world, and invited 23 foreign oil companies to China to discuss the pros and cons in each contract model[17].

During exploration phase, foreign companies provide all capitals and technology. If the exploration fails, the investment is unrecoverable. During production phase, both of NOC and foreign companies provide the exploitation capital to develop and produce oil from target oilfield. The hybrid contract contains different elements from concession, PSA, joint venture and service contract, in order to meet the specific needs and interest in different producing stages. Production Sharing Contracts is the official term Chinese government use to describe this contract model, it is flexible and addition terms and clauses can be added to the contract depending on the actual situations.

In term of tax and royalty, first a 12.5 % royalty and 5% Consolidated Industrial Commercial Tax are charged based on the total revenue. It should be noted that after 2011, a resource tax, which is usually 5%, is used to replace the royalty. Then the foreign companies are allowed to use the net revenue to cover the tangible cost first, which is called the cost oil and untaxable. After that, what remaining is called profit oil, and divided to NOC share and company share based on their net revenue interest. Based on the profit oil, a 5% VAT, a 25% Enterprise Income Tax and a 1% Mineral Resources Compensation Fee are charged to the companies. If the oil price exceeds $55 per barrel, an additional tax is added, varying from 20% to 40%. Under some certain circumstances, companies may be exempted from specific types of tax[9].

Given the fact that China is short of capital, management experience and advanced technology, the advantages of hybrid contract include: China government takes no risk during exploration phase and foreign capitals are used to explore potential fields; China government take royalty (or resource tax) first without any risk; foreign companies can recover their cost first, which can encourage further cooperation; during production phase, NOCs can learn advanced technology from foreign companies[11].

The exploration period is usually 5 years, which can be extended to 7 years, and the production period is usually 25 years, which can be extended to 30 years, as agreed by both parties. Most term in the contract is non-negotiable except the work program, X factor and other contributions. In the production phase, the foreign companies must cooperate with a NOC (CNOOC for offshore, and CNPC/Sinopec for onshore) as a joint venture. Usually the NOC owns 51% share of working interest with the option to decrease its WI, while the foreign company is usually appointed as the operator. The operator is also required by the contract: to use advanced technology to develop and produce oil; to design and establish a work program and budget; to share all information with NOC; to establish insurance programs; to establish a training program; to minimize the potential damage to the environment and to ensure employee safety. Especially, the NOC has the right to take over the production operations at any time once the foreign companies has recover all costs with reasonable reimbursement based on the book values of tangible costs[18].

Currently there are some new trends regarding to the cooperation of onshore and offshore oil and gas exploration, development and production between China government and foreign companies, including: NOCs starts their own offshore operations without cooperation with foreign companies due to increasing capitals, technology and experience; CNOOC starts to act as an operator during production phase; and a reverse PSCs in developed, in which foreign companies cooperate with NOC to develop the offshore oil and gas field CNOOC explored. It should be noted that Chinese NOCs have indeed grow to oil giants since 1978 when it first cooperated foreign investment to explore, develop and produce onshore and offshore oil and gas fields, in term of capitals, management, experience and technology.

## Oil and Gas Law in Kazakhstan

Since independence in 1991, Kazakhstan has made of significance progress on economic reform. Currently Kazakhstan owns large reserves of energy, including Tengiz field, Karachaganak field and Kashagan field. The large oil reserves attract international investment from major oil players in the world. On one hand, the state –owned company, KazmunayGas, perform some significant project in those fields. On the other hand, giant international oil players, such as Chevron, ExxonMobil, Lukoil, CNPC, and so on, increase their investment in cooperation to explore, develop and produce gas and oil field in Kazakhstan. President Nazarbayev has claimed "Kazakhstan will come a factor of energy security in Asia and Europe", it means Kazakhstan intend to works as an important bridge connecting even controlling the oil and gas from between Asia and Europe. This section discusses Kazakhstan oil and natural gas law framework, fiscal term, taxation and so on, which can work as a guideline to potential investment[11].

### Ownership and Title

According to the Constitution of Kazakhstan, the ownership of all minerals and subsurface, including oil and gas, are properties of the state. Upon the acceptance of oil and gas agreement approved by government, the investment can be granted by the mineral rights, including exploration rights and exploitation rights.

KazMunayGas is the Kazakh National Oil Companies, holding approximately 40% of total oil reserves. The objectives of government policy include: to ensure and promote energy independence and security; to export oil to west and east; to reform and stimulate domestic economic.

## Oil and Gas Law Framework

The Kazakh code on Subsoil Resources and the Processing of Mineral was adopt in 1992, which was the first law containing ownership of mineral resources, licenses for the exploration and exploitation and roles of central government and local government. The cooperation with foreign investment to explore, develop, and produce onshore oil and gas is subject to regulation under the generally applicable legislation, including the Foreign Investment Law of 17 January 1991, the related Concessions Law of December 1991, the Enterprise Law of 15 January 1991, the Tax Law of 14 February 1991 and foreign exchange legislation. Especially, The Gas Law was adopted in 2011, which establishing a unified system for gas supply and setting measures to fulfil the domestic gas requirements and ensure the implementation of state policy in order to provide provisions and regulations to the gas industry. In addition, the foreign investments are also subject to domestic laws and regulations, including: The 2003 Law on Investment; the Civil Code; the Customs Code and Custom Code of the Customs Union; The Law on Currency Regulation and Currency Control; and The Tax Code, which demonstrates the rights and responsibilities of foreign investment in Kazakhstan.

Due to the Dissolution of Soviet Union, on 1992, President Nazarbaev issued a Decree "On the Implementation of Organization and Activity of the Administrative State Bodies of the Republic of Kazakhstan under Conditions of Economic Reform" to create ministries, state committees and the Bank of Foreign Economic Activity. Right now, there are two ministries responsible for petroleum operations, including the Ministry of Geology and Protection of the Subsoil and the Ministry of Energy and Fuel Resources. The specific duties of each ministry are not well defined, so it is important for foreign companies to identify the appropriate negotiating partner. The Ministry of Geology and Protection of the Subsoil is responsible for exploration phase and the Ministry of Energy and Fuel Resources is responsible for production phase for cooperation with foreign companies. In addition, the Ministry of Industry and the Ministry of Ecology are also involved into oil field exploration and exploitation, which the foreign companies should take into consideration.

## Operating Requirement

In order to obtain the license for the petroleum operations, the company must meet all financial and constituent requirement and register with the following government authorities: the Ministry of Finance; the Ministry of Foreign Economic Relations; and other organization authorized by legislation of Kazakhstan. For the oil and gas operation with foreign investment, an agreement is required by the foreign company according to the Law 'On Enterprises' and the Law 'On Economic Activity and Entrepreneurship'. Based on the result of competitive bidding process and financial and experience check, the local branch of the State Committee of State Property is authorized to grand the agreement. The following documents are required for application: agreement draft between foreign investment and local branch of the State Committee of State Property, technical feasibility study and financial evidence. The surface usage should be licensed which is subject to the land legislation. In addition, the National Foreign Investment Agency of Kazakhstan also works as an authority to grant license for oil and gas exploration and exploitation operations. General speaking, the applicant should obtain a document of mining operation after expert assessment by four government authorities corresponding to mining, environment, safety and health care. Figure 5 shows new developed oil fields in Kazakhstan.

As it can be concluded that the license system in Kazakhstan is quite complicated, and there is not a single authority to grant the license for oil and gas exploration and exploitation operation, therefore it is recommended to approach to the production entity, which holds the operation right, then contact central authority and local branches of the Ministry of Geology.

## Roles of National Oil Companies

There are no national oil companies in Kazakhstan before 2002. KazMunayGas is founded in 2002 by merging Kazakhoil and Oil and Gas Transportation. Before that, foreign companies need to negotiate agreements with regional organizations.

## Private Investment

There are some domestic oil companies, like Temir Oil and on so, in Kazakhstan. However, in term of capital, technology, management exercise, and oil reserves obtained, they cannot compete with KazMunayGas and international oil companies.

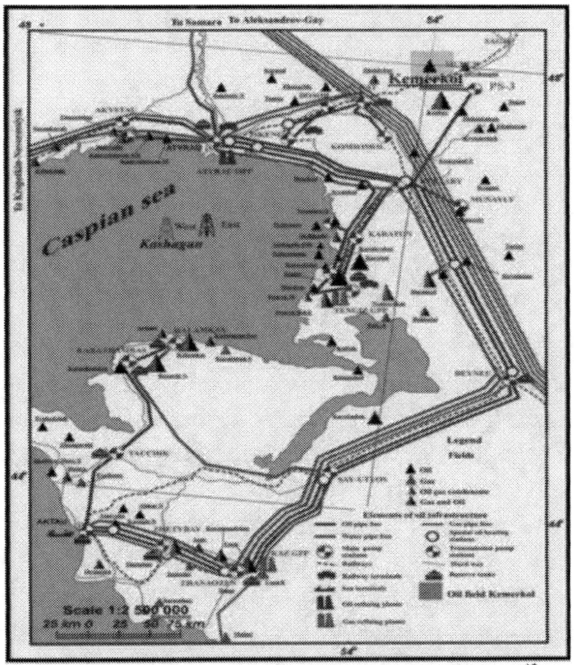


Figure 5— New developed oil fields in Kazakhstan[19].

## Tax Regime

According to the law, there are four types of payments liability for foreign companies to cooperate with Kazakhstan to explore, develop and produce oil and gas fields, including taxes, rentals, and productions based payments (royalties) and income based payments (taxes). Royalties are calculated at a negotiable rate, sometimes as high as 20%. A profit tax with a general tax rate of 35% was charged to the foreign companies. In the joint ventures, if the foreign share exceeds 30%, the profit tax rate is 30% or at the general profit tax rate of 35%. The export tax is levied at a tax rate of 40% on the gross revenues. Rentals, dividends, the amount of income allocated to the construction, maintenance and repair of the highway are deductible from the balance profit. According to the signed agreement, the Kazakhstan can receive a profit share from the profit oil. In addition, in order to meet domestic oil requirement, Kazakhstan government has the priority to purchase all or part of the production from the projects, which is subject to applicable law. Before the commencement of investment activities, bonuses are required to pay to the Kazakhstan government, including signature bonus, production bonus and reserve based bonus. The amounts of bonuses are agreed in the agreement by both parties. According to the concession law, social insurance for employee is mandatory to purchase by foreign companies based on the contract. According to 2009 Kazakh tax code, the Valued Added Taxes was reduced to 12% while the corporate income tax was decreased to 20%. Based on revenues, Kazakhstan applies a flat 11% social tax to companies and a personal income tax rate of 10%.

## Contracts

In Kazakhstan, there are mainly three types of petroleum agreements available when foreign companies negotiated with KazMunayGas or regional organizations: Concession, joint venture and production sharing agreements. The joint venture used to more polpular in Kazakhstan because within the framework of joint ventures Kazakh side can obtain higher level of position in management and administration. A production sharing contracts have been signed with Chevron and Kazakhstan to explore, develop and produce Tengiz field and with Elf for various projects in the Aktyubinsk Oblast. It should be noted that the joint venture and production sharing contracts can be negotiated individually with the regional organizations depending on the situation. In 2005, Kazakhstan government switch to production sharing contracts, and the National Oil Company KazMunayGas is authorized as the operator and take at least 51% share of profit. Currently almost all petroleum agreements are signed based on PSA. 4 shows new developed oil fields near Caspian Sea in Kazakhstan.

## Comparisons and Discussion

The comparisons of oil and gas laws between China and Kazakhstan are summarized in Table 1, from the perspectives of license authority, applicable laws, NOCs, contracts, Royalties, taxation, bonuses, remittance and disputes settlements.

As we can observe from the Table 1, Chinese central government has the absolute power to examine, approve, register and license projects of exploration, development and production of onshore oil fields, either cooperating with foreign investment or investing from NOCs and private companies, while local governments are only responsible for administration of the projects. The law frame is hierarchically structured, like a pyramid. The National People's Congress enacts national legislations while the State Council implements national legislations by enactment of regulations and measures. Local governments enforce and administrate the implement of national legislations, regulations and measures. Local regulations are enacted by the people's congress of provinces. When domestic and/or international companies intend to invest in oil and gas projects, they should directly negotiate with department of central government, and get license from central government by fulfill all requirements, as discussed in previous section. If there is confliction between state legislations and local legislations, the disputes are settled based on state legislations. This hierarchically structured system can ensure the consistency of policy implement.

In order to encourage foreign investment in onshore and offshore nature gas and oil exploration, development and production, the China government started to optimize its law frame and fiscal system since 1978. Currently the law frame in China is relatively complete, and can work efficiently to protect the lawful right of foreign investments. The hybrid contract model is used as the form of petroleum contracts in China to cooperate with foreign oil companies while three NOCs, CNPC, Sinopec and CNOOC, must participate in the all projects with at least 51% share of the project. On one hand, foreign companies can use cost oil to recover cost, which can work as an incentive to stimulate investment and avoid risk in exploration phase. On the other hand, by working with major oil players in the world, Chinese NOCs have made significant progress from perspectives of capitals, technology and management. Currently, CNPC, Sinopec and CNOOC have ranked as the top oil players in the world and start to international oil and gas fields to perform exploration, development and production.

The royalty rate, taxation and bonuses in China are set in reasonable range subject to applicable law. Especially, the 12.5 % royalty is replaced by a 5% resources tax, and there is a trend that China government will decrease the relevant taxes rates and extend the tax-exempt range in order to encourage more investments. When disputes arise, consultations and international arbitrations are expected instead of lawsuit.

The disadvantages in China oil and gas laws system include: a windfall tax ranging from 20% to 40% with a threshold of $55/bbl is set in the fiscal system, which can prevent further investment when oil price is high; the clauses and terms in contract system are not flexible enough, a R factor (Columbian model) should be introduced to protect investment when oil price is low and government can receive more revenue when oil price is high; NOC has the right to take over the production operations at any time once the foreign companies has recover all costs with reasonable reimbursement based on the book values of tangible costs, even though it never happened; Remittance policy is strict, which means government wants to hold money within China.

**Table 1— Comparison Between China and Kazakhstan.**

| | China | Kazakhstan |
|---|---|---|
| License Authority | State Council | Local Council of People's Deputies<br>Ministry of Finance<br>Ministry of Foreign Economic Relations<br>Other authorized agencies |
| Applicable Laws | Mineral Resources Law<br>Rules for Implementation of the Mineral Resources Law<br>Measures for Registration Administration of Exploration Blocks of Mineral Resources<br>Measures for Registration Administration of Mineral Resources Exploration<br>Measures for the Administration of Transfer of Exploration Right and Exploitation Right<br>Regulations on the Exploration of Continental Petroleum Resources in Cooperation with Foreign Enterprises<br>Regulations on the Exploration of Offshore Petroleum Resources in Cooperation with foreign Enterprises<br>Provisional Regulations on Resources Taxes | The Kazakh code on Subsoil Resources and the Processing of Mineral<br>1991 Foreign Investment Law 1991 Concessions Law<br>1991 Enterprise Law<br>1991 Tax Law<br>Foreign exchange legislation<br>2011 The Gas Law<br>The Law on gas and gas supply<br>The 2003 Law on Investment<br>The Civil Code<br>The Customs Code and Custom Code of the Customs Union<br>The Law on Currency Regulation and Currency Control<br>The Tax Code |
| NOCs | CNPC<br>Sinopec<br>CNOOC | KazMunayGas |
| Contracts | Hybrid contract | Concession<br>PSA<br>Joint Venture |
| Royalty | Royalty 12.5% ( Before 2011)<br>Resource Tax 5% ( After 2011) | Negotiable, sometimes as high as 20% |
| Taxation | Valued Added Tax 5%<br>Enterprise Income Tax 25%<br>Mineral Resources Compensation Fee 1%<br>Windfall Tax 20% ~ 40%<br>Production Sharing > 51% | Valued Added Tax 12%<br>Profit Tax 35%<br>Social Tax 11%<br>Personal Income Tax 10%<br>Corporate Income Tax 20%<br>Production Sharing > 51% |
| Bonuses | $1,000,000 | Negotiable |
| Remittance | Withholding taxes 10% | Permitted subject to currency legislation |
| Disputes Settlement | Consultation<br>International Arbitration | International Arbitration<br>Courts of Law<br>Arbitrazh Courts |

Kazakhstan central government distribute its power to examine, approve, register and license projects of exploration, development and production of onshore oil fields, either cooperating with foreign investment or investing from NOC and private companies to local governments. The law frame is not very complete due to relatively short time of independence. When domestic and/or international companies intend to invest in oil and gas projects, sometimes they get confused because they do not know which departments or organization to negotiate.

In order to encourage foreign investment in onshore and offshore nature gas and oil exploration, development and production, the Kazakhstan government started to optimize its law frame and fiscal system since 2000. Even though the law frame and fiscal system is not perfect, it can be seen that the government is working on optimizing the systems to protect and encourage investments. The PSA contract model is used as the form of petroleum contracts in Kazakhstan to cooperate with foreign oil companies while NOC, KazMunayGas, must participate in the project with at least 51% share of the project, which KazMunayGas owns the mineral right. It should be noted that KazMunayGas owns approximately 40% of national oil reserves while the foreign investments owns the rest of reserves.

The royalty rate and taxation in Kazakhstan are set in reasonable range subject to applicable law. However, according to available materials, the royalty and bonuses are all negotiable. Sometimes it is good for foreign companies, but sometimes it also contributes to uncertainty to the cooperation. There is a trend that Kazakhstan government will decrease the relevant taxes rates and extend the tax-exempt range in order to encourage more investments. When disputes arise, the Courts of Law and Arbitrazh Courts determine the settlement of disputes. It should be noted that Courts of Law and Arbitrazh Courts are part of court system, which unequal to the concept of arbitration in western legal systems. There are concerns that investors are limited to settle the disputes by international arbitration. Recently Kazakhstan government starts to allow negotiation and international attributions to settle the disputes.

The disadvantages in Kazakhstan oil and gas laws system include: inconsistent policy system and uncertainty in contract types, uncertainty in investment, and uncertainty in disputes settlement. It can be seen that Kazakhstan government is working on optimizing the law frame and significant progresses have been made.

## Conclusions

It can be seen that both of China and Kazakhstan are working on optimizing their law frame and fiscal systems in order to encourage more investment and advanced technology in onshore and offshore oil and gas exploration, development and production. Both of countries have made significant progress while it would still be a long journey for both countries to achieve their national energy objectives. It is believed that China and Kazakhstan will play more and more important roles in international oil and gas market. In the future, the results of this paper can work as a guidebook for potential oil and gas international or/and domestic investment in both countries.

## Reference

[1] IEA. Oil & Gas Security Emergency Response of IEA Countries.
[2] Wikipedia.
[3] IEA. Developing China's natural gas market: the energy policy challenges.
[4] Elliot, Gordon. "Kazakhstan, Nazarbayev, foreign investment and oil." *La. L. Rev.* 53 (1992): 1243.
[5] EJPE
[6] Nichol, Jim. "Kazakhstan: Recent Developments and US Interests." *Current Politics and Economics of Russia, Eastern and Central Europe* 28.2 (2013): 241.
[7] EIA
[8] Jiang, Julie, and Jonathan Sinton. "Overseas investments by Chinese national oil companies." (2011)
[9] MacBride Jr, William L., and Wang Bei. "CHINESE MINING LAW OVERVIEW (1)." *Mineral Resources Engineering* 10.03 (2001): 365-393.
[10] Lim, Tai-Wei. *Oil and gas in China: the new energy superpower's relations with its region.* Vol. 21. World Scientific, 2010.
[11] Smith, Ernest E., et al. "International petroleum transactions." (1993).
[12] Zweig, David, and Bi Jianhai. "China's global hunt for energy." *Foreign affairs*(2005): 25-38.
[13] Andrews-Speed, Philip, and Roland Dannreuther. *China, oil and global politics.* Routledge, 2011
[14] Keyuan, Zou. "China's governance over offshore oil and gas development and management." *Ocean Development & International Law,* 35.4 (2004): 339-364.
[15] China Daily
[16] Falola, Toyin, and Ann Genova. *The politics of the global oil industry: an introduction.* Greenwood Publishing Group, 2005
[17] Hayton, Bill. *The South China Sea: the struggle for power in Asia.* Yale University Press, 2014.
[18] C. Li, Jennifer. "China's rising demand for minerals and emerging global norms and practices in the mining industry." *Minerals & Energy-Raw Materials Report* 22.3-4 (2007): 105-126
[19] Wikipedia

CARBON MANAGEMENT TECHNOLOGY CONFERENCE

## CMTC-485491-MS

# The Mechanism and Simulation Research of "Foamy Oil " during $CO_2$ Flooding

Ying Jia, E&P Research Institute, SINOPEC, Beijing, China; Lei Huang, Research Institute of Petroleum Exploration and Development, PetroChina, Beijing, China; and Lei Sun, SouthWest Petroleum University, Chengdu, China



## Abstract

Noticeable progress in understanding the high efficiency of "foamy oil " in solution gas drive in heavy oil reservoirs has been made in recent years. However, "foamy oil " during $CO_2$ flooding was still a special and novel phenomenon when $CO_2$ flood asphatic oil. The basic mechanism and corresponding numerical equation of seepage flow of 'foamy oil', in connection with $CO_2$ flooding, have not been reported. This paper presented the mechanism analysis and simulation study that addressed this issue.

A new viewpoint of foamy oil during $CO_2$ flooding was proposed with contrasting the similarity of ''foamy oil"during $CO_2$ flooding front withthat in dissolved gas flooding. During $CO_2$ floodingthe''foamy oil" could form when $CO_2$ contacted with oil because precipitated asphaltene could facilitate bubble nucleation, decrease the critical super-saturation and help in maintaining the dispersed gas flow by suppressing bubble coalescence, whichsimilar to 'foamy oil ' in depletion drive.Two main mechanisms were proposed. The first was enhancing oil recovery obviously by decreasing viscosity of crude oil, reducing interfacial tension, and swelling oil. The second mechanism wassteady gas/oil mobility helping in maintaining high pressure in the reservoirs.

Then, the compositional model considering bubble nucleation, growth and coalescenceand $CO_2$ flowing characteristic was established, combining the advantages of "equilibrium model" with that of "dynamic model". With the establishment of the model, the flowing characteristic of ''foamy oil"during $CO_2$ floodingwas regarded as the function of time and flow condition. The relative permeability curve and critical saturation of bubble were modified to determine the influence of "foamy oil" on oil recovery.

Taken one oil field in north China as example, the compositional model calculation results indicated that bubble was mainly located at the front of carbon dioxide flooding. High efficient flooding potentialwas achievedwith "foamy oil" existed in carbon dioxide displacing front which were in accordance with the anomalously good performance in oil production.This method of calculating foamy oil properties would provide the basics for developing numerical simulation models of foamy oil flowduring $CO_2$ flooding.

## 0 Introduction

Foamy oil flow is the name commonly used to describe a form of two-phase oil-gas flow in porous media which the gas phase remains partially or completely dispersed in the oil[1-4]. there were better expected production behavoir of primary depletion wells in may Canadian and Venzuela heavy oil reservoirs. "foamy oil " during $CO_2$ flooding was still a special and novel phenomenon when $CO_2$ flood asphatic oil.However, no literatures about the formation mechanism of the above phenomena, and corresponding numerical equations to describe the flow law and the effect on the oil development of $CO_2$ flooding, have been reported. This paper presents a theory and model study that address these issues.

Inspired from a kind of phenomena in dissolved gas flooding in cold production of heavy oil, the concept of "foamy oil flow" was introduced into the miscible displacement process, the formation mechanism of the "foamy oil" in the process of $CO_2$ injection was explained. Then, the mathematical model of the "foamy oil flow" in the $CO_2$ process was proposed; finally, the distribution and dynamic characteristics of the "foamy oil flow" in the process of gas injection were numerically simulated.

## 1 "foamy oil" in the process of $CO_2$ displacement

### 1.1 The similarity between the "foamy oil" and the "foamy oil" in the $CO_2$ flooding

In China, special phenomena like "foamy oil flow" during $CO_2$ displacing asphaltic oil were reported. In Zheng 408 block of China, there were small bubbles dispersed in the oil phase when $CO_2$ huff and puff were carried.They mentioned that due to high viscosity, the precipitation of $CO_2$ in crude oil was formed as the formation of "foamy oil", which improved the oil flow capacity, increased elastic energy of crude oil and maintained the formation pressure [5].In Lukeqin oil reservoir of west China, the appearance of unnatural "foamy oil" during gas floodingwere also reported [6]. Experiment proved that natural gas reinjection in a heavy oil field increases foamy oil recovery when the reservoir pressure is below the pseudo-bubble pressure[7,8]. Experiments showed that gas could dissolve in the oil, which would cause oil swelling, viscosity reduction and artificial foamy oil formation.

Above experiment observation and field reports showed that during $CO_2$ flooding the''foamy oil" could form when $CO_2$ contacted with oil which similar to 'foamy oil' in depletion drive.When $CO_2$ contacted with crude oil constantly, pseudo-bubble pressure increased. When the displacement front pressure was lower than the corresponding pseudo-bubble pressure, dissolved gas might flow; however, because of some surfactants existence, gas dispersed as bubbles was formed by the surfactant and could not immediately formed as continuous gas phase. The "foamy oil" in the $CO_2$ flooding is very similar to 'foamy oil ' in depletion drive.

### 1.2 Research status of"foamy oil"flow in dissolved gas drive

Foamy oil flow is the form of two-phase oil-gas flow which the gas phase remains partially or completely dispersed in the oil in dissolved gas flooding in cold production of heavy oil. Many scholars researched on "foamy oil" flow mechanism. Smith[9] was the first scholar who gave detailed accountsabout the rare phenomena. He described the "foamy oil flow" as a fluid characteristic of micro bubbles in the heavy oil. He believed that the mobility of heavy oil containing micro bubbles was several times higher than that of the ordinary single-phase oil. Maini[1] et al. observed this dispersed gas phase and call it "foamy oil" with experiment.He believed that "bubble oil" can exist with conditions as follows, ①viscous force that determine bubble growth should exceed the capillary force;②gravity cannot cause rapid separation in two-phase flow phase;③ interfacial chemistry effect is essential to prevent bubble coalescence. Kraus et al. [10] found that the effect of "foamy oil flow" on dissolve drive includes: ①increasing the compressibility of the fluid;② maintaining the reservoir pressure; ③ and delaying the gas production time.Thesescholarsreached a consensus that the "foamy oil" in dissolved gas flooding has an unusual effect to improve oil recovery.

### 1.3 Analysis of the stability of asphaltene to "foamy oil"

Other authors also proofed that precipitated asphaltene could facilitate bubble nucleation. Smith[9] believed a large number of asphaltene were located at the nucleation site of micro bubble in the bubble formation. However, he had no further research into the physical state of asphaltene and bubble after bubble was nucleated. He also had no further research into the effect of the asphaltene condensed around bubbles on the viscosity of crude oil.The research of Ward and Levart[11] indicated that small bubbles might appear on the asphaltic surface in a certain size under thermodynamic equilibrium, and the crude oil viscosity was reduced due to asphaltene precipitate from crude oil, which increased crude oil production. However, they did not provide a better explanation that the association between asphaltene and bubble was stronger than the asphaltene self-association.Claridge and Prats [12] suggested that asphaltenewas usually micelle like, which affected the viscosity of crude oil. When the pressure declined slowly to the bubble point pressure, the asphaltene and resin migrated to the bubble surface and formed a semi-solid film after the film formation hindered the further growth and merge of the bubble. At any time, the sum of the gas in the bubble and gas dissolved in the crude oil were equal to that of the original gas. Similarly, the quality of asphaltene adsorbed on the bubble and the asphalt with colloidal distribution were equal to the initial asphalt quality. It is also believed that the migration of asphaltene to the surface of the bubble decreased the viscosity of the crude oil. A large number of studies have shown that the bubble formation and the stability of the interfacial film increased with the increase of asphaltene concentration.F.Bauget et al. [13] presented results showing significant shift in the crude properties at asphaltenes concentrations of around 10%. His experimental suggested a significant increase in foamability, bubble stability and viscosity of crude at asphaltenes concentration of 10% and over. I.Adil and B.B.Maini [14] believed that althoughthe viscosity reduction effect was not obvious, asphaltene contributed bubble nucleation and decreased the critical saturation because of asphaltene foaming, which helped keep the dispersion of gas flow.

However, some researchers believed that it was not correct that the adsorption of asphaltene on the bubble surface reduced the viscosity of crude oil. In 1996, Huerta [15] et al. did not observe viscosity decrease when the pressure dropped to the bubble point pressure when they tested the critical gas saturation during heavy oil flowed in porous media. In a word, a large number of studies confirmed that the presence of asphaltenes has promoted the formation of bubbles in crude oil. However, there is no consensus on the effect of asphatene on the viscosity decrease of crude oil.

### 1.4 Numerical simulation of "foamy oil flow"

Numerical simulation of primary depletion in foamy oil reservoirs was still based on the empirical adjustments of conventional solution-gas-drivewith black oil model. Models can be classified into two categories: Equilibrium model and kinetic model [16-19].

① Equilibrium model

These models assume complete local equilibrium between different phases.One model is Pseudo-bubblepoint Model proposed by Kraus et al., another is Modified Fractional-Flow Models proposed by Lebel.Krausdescribed a “pseudo-bubblepoint” model, where the pseudo-bubblepoint pressure is an adjustable parameter in the fluid property description. Modified Fractional-Flow Models attempted to match production behavior by modifying the fractional-flow curve or the gas/oil relativepermeability curves.

②Kinetic Model

Kineticmodel believe that dispersed gas in oil phase is not a thermodynamically stable. Given sufficient timeand environment, the dispersed gas would separateinto free gas and oil phases. In the same time, bubble nucleation would also happen.

Coombe and Maini described a kinetic model that accounted for the physical change of the bubble in oil. They defined three nonvolatile components in the oil phase: dead oil, dissolved gas, and gas dispersed in the form of microbubbles. Sheng simulates the rate of bubble transformation to free gas by using the exponent decay method of super-saturation. The kinetic model was better than the equilibrium model method. However, the rate process of dissolved gas flooding was controlled by the rock / fluid characteristics and capillary number.

# 2 The establishment of mathematical model of "foamy oil flow" in $CO_2$ process

## 2.1 The influence of asphalt on the formation of foamy oil and its conditions

In this paper, precipitated asphaltene can facilitate bubble nucleation, decrease the critical super-saturation and help in maintaining the dispersed gas flow by suppressing bubble coalescence. Both asphaltenes and resins are surfactants in crude oil, and their appearance can form the physical barriers between bubbles and bubbles. In the process of $CO_2$ flooding, $CO_2$ and crude oil contacted which induced bubble pressure rise. At the front of $CO_2$ flooding, when the pressure is lower than the bubble point pressure, bubbles appear in crude oil. At this time, resin and asphaltenedistributed as colloidal state, migrate to the surface of the bubble and form a semi-solid membrane, which prevent the bubble further growing into a continuous gas phase and facilitate the bubble nucleation.

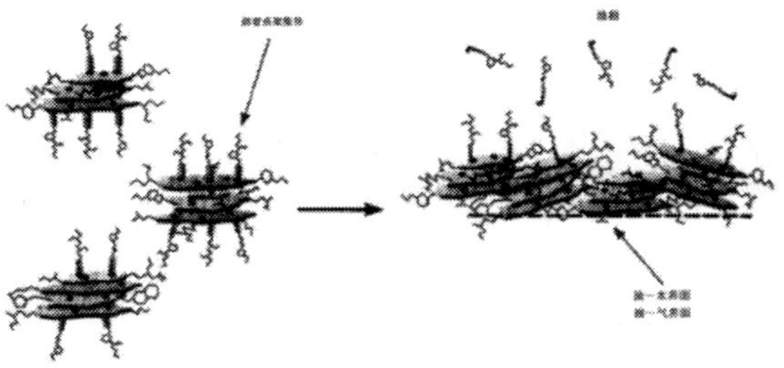

Figure1 depiction of stabilization mechanism of asphaltene

## 2.2 Establishment of numerical model of "foamy oil flow"at the $CO_2$ flooding front

Based on the above analysis, this paper believed that the key to simulate the process of gas injection during "foamy oil flow" is to establish dynamic model to describe the flow characteristics of "bubble oil" with time, at the same time, the bubble flow in"foamy oil flow"is different from continuous gas phase flow, whichis not consistent with the flow rate of crude oil and need to be calculated by interpolation. It is believed that appropriate critical gas saturation value is the key to simulate the bubble nucleation in the simulation of the bubble flow.

In this paper, the multicomponent model considering bubble nucleation and $CO_2$ flowing characteristic was established, combining the advantages of "equilibrium model" with that of "kinetic model".

### *2.2.1 assumptions*

Besides basic hypotheses of conventional compositional model, many other assumptions are considered. During $CO_2$ injection, "foamy oil flow" exist in oil and gas system; there are $n_c$ fixed hydrocarbon components in the oil and gas system, the $n_c$ component presented as bubble which dispersed in the oil phase with the thermodynamic properties of gas phase; the thermodynamic parameters of bubble were calculated bymultiphase equilibrium model of asphatic oil system during $CO_2$ injection near pseudo-bubble pressure, where the properties of the micro bubble is the same as the gas flashed by equations; the flow of bubbles is different from that of the continuous gas phase, which is not consistent with the flow rate of crude oil.

***2.2.2 Compositional Model of "foamy oil" at the $CO_2$ front***

The compositional model of multiphase multicomponent flow considering "foamy oil" under $CO_2$ injection is as follows:

$$\nabla\left[\frac{kk_{ro}}{\mu_o}\rho_o y_{io}\nabla\Phi_o\right]+\nabla\left[\frac{kk_{rg}}{\mu_g}\rho_g y_{ig}\nabla\Phi_g\right]+\sum_{k=1}^{n_f}\sum_j q_{ik}y_{ij}-\frac{\partial\left[\sum_j \phi\rho_j y_{ij}S_j\right]}{\partial t}=0 \tag{1}$$

$$\nabla\left[\frac{kk_{rw}}{\mu_w}\rho_w\nabla\Phi_w\right]+\sum_{k=1}^{n_f}q_{wk}-\frac{\partial(\phi\rho_w S_w)}{\partial t}=0 \tag{2}$$

$$\nabla\left[\frac{kk_{ro}}{\mu_o}\rho_o\nabla\Phi_o\right]+\nabla\left[\frac{kk_{rg}}{\mu_g}\rho_g\nabla\Phi_g\right]+q_t-\frac{\partial(N_o+N_g)}{\partial t}=0 \tag{3}$$

***2.2.3 Phase Equilibrium Equations***

The phase equilibrium equations[20,21] for the components are

$$\ln f_{ig}=\ln f_{io};\quad i=1,\dots,n_c-1 \tag{4}$$

***2.2.4 kinetic equation***

This paper argues that the nucleation of bubbles is a non-equilibrium process.Precipitated asphaltene could facilitate bubble nucleation, decrease the critical super-saturation and help in maintaining the dispersed gas flow by suppressing bubble coalescence,The unbalanced process among $CO_2$, dissolved gas and free gas can cause oil dissolved in gas phase, thereby delay the emission and the formation of transparent bubble than real thermal bubble. This process is affected by the nucleation kinetics. The nucleation of bubbles is a stochastic process controlled by supersaturation, so the supersaturation required for nucleation depends on the length of nucleation time. Therefore, it is not enough to solve the thermodynamic properties of bubbles only with the state of equation. The formation of bubbles is also affected by the adhesion of asphaltene, which is similar to the process of chemical reaction:

In this paper, gas is mainly composed of three parts: light hydrocarbon components, bubbles and $CO_2$. The reactions among them can be described by two reaction equations:

Reaction 1:

$CO_2$+ light hydrocarbon components → bubble (5)

Reaction speed: $x_1=F_1\times[CO_2]+F_2\times$ [light hydrocarbon components]

Reaction 2 :

bubble → $CO_2$+ light hydrocarbon components (6)

Reaction speed: $x_1=F_1\times[CO_2]+F_2\times$ [light hydrocarbon components]

Where: X1 is the reaction rate, Fi frequency coefficient, [] is the mole fraction.

In the calculation, The thermodynamic parameters of gas,bubble and crude oil, molecular weight, compression coefficient, critical pressure and critical temperature, are calculated at the pressurewhenbubble existed in the expansion experiment with EOS.

# 3 Model solution

### 3.1 Interpolation treatment of phase permeability curve

As mentioned above, one imbalance process of "foamy oil flow" during $CO_2$ flooding existsbetween dispersedbubble and free gas. Another imbalance process is related with fluid distribution in reservoir. The capillary force controls the distribution of fluid.Therefore, it is assumed that the fluid is self - distribution, that is, there is lowest free energy distribution at the surface of the fluid / fluid and solid / fluid interface. As a result, according to this assumption, gas begins to flow must be continuous, and isolated bubbles are capillary force trap. This imbalance process is affected by the surface tension, absolute permeability and the maximum flow gradient near the bubble. Therefore, adjusting the relative permeability curve of the bubble is the key to the simulation.

At the same time, the critical gas saturation is an important parameter in the "foamy oil flow". In the critical gas saturation, the continuous gas phase forms and begins to flow. However, the calculation of the critical gas saturation depends on different experimental techniques and data processing methods. Different critical gas saturation was defined. One method of obtaining the critical gas saturation is extrapolating gas phase relative permeability. Another method is to measure the degree of saturation of the gas flow at the beginning of the gas expansion. These two methods are not the same as the measured critical gas saturation. Because of the differences in the value of the critical gas saturation, the determination of the critical gas saturation is another key to the simulation of the gas flow.

**（1）The calculation of relative permeability curve of pseudo"bubble"component**

In this paper, firstly the bubble is treated as a pseudo component, the critical saturation of the relative permeability curve of bubble and gas are defined separately, and then the relative permeability curve of the gas phase is normalized by the original measurement. Next, the definition of relative permeability components are carried out:the critical saturation critical saturation of bubbles is defined lower than other gas components due to the bubble flow before the formation of free gas.

First of all, for gas-fluid systems, the conventional liquid and gas phase permeability curves are shown in figure 2.

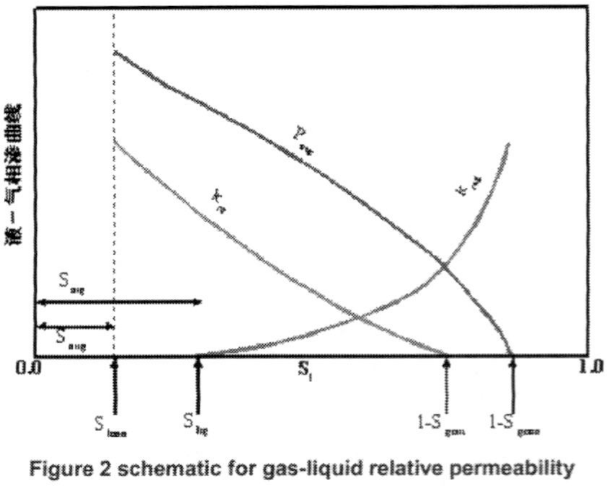


**Figure 2 schematic for gas-liquid relative permeability**

Then, the critical saturation and permeability of the bubble are user defined with experiment data. Because the rate of bubble is slower than that of the free gas, the relative permeability is smaller when the water saturation is relatively small. Therefore, through defining connate water saturation, the relative permeability curve of the bubble is determined in the range of $\left[S_{lcon}, 1-S_{gcrit}\right]$ and $\left[S_{lrg}, 1-S_{gcon}\right]$.For example: the gas phase relative permeability curve data are shown in table 1. The critical saturation of the bubble is 0.25, the relative permeability of the bubble is about 0.05, and the relative permeability of oil and water are 1.0. According to the above method, the relative permeability data are obtained in table 2. Mapping two sets of curves, the bubble flow at small rate is earlier than gas in graph 3.

**Table 1 the relative permeability curve of gas phase**

| $S_l$ | $k_{rg}$ | $k_{rog}$ |
|---|---|---|
| 0.25 | 0.5 | 0 |
| 0.3634 | 0.456 | 0.016 |
| 0.4432 | 0.351 | 0.05 |
| 0.5163 | 0.2543 | 0.1 |
| 0.5923 | 0.1843 | 0.15 |
| 0.6697 | 0.1227 | 0.22 |
| 0.7457 | 0.0817 | 0.3 |
| 0.8216 | 0.0467 | 0.44 |
| 0.8975 | 0.029 | 0.6 |
| 0.9428 | 0.0113 | 0.84 |
| 0.9589 | 0.0043 | 0.93 |
| 0.965 | 0.001 | 0.97 |
| 0.97 | 0.0001 | 0.99 |
| 0.975 | 0 | 1 |

**Table 2 the relative permeability curve of bubble**

| $S_l$ | $k_{rg}$ | $k_{rog}$ |
|---|---|---|
| 0.25 | 0.05 | 0 |
| 0.328207 | 0.0456 | 0.011035 |
| 0.3634 | 0.038886 | 0.016 |
| 0.383241 | 0.0351 | 0.024454 |
| 0.433655 | 0.02543 | 0.045933 |
| 0.4432 | 0.024155 | 0.05 |
| 0.486069 | 0.01843 | 0.079322 |
| 0.5163 | 0.014941 | 0.1 |
| 0.539448 | 0.01227 | 0.115229 |
| 0.591862 | 0.00817 | 0.149712 |
| 0.5923 | 0.008141 | 0.15 |
| 0.644207 | 0.00467 | 0.196944 |
| 0.6697 | 0.003808 | 0.22 |
| 0.696552 | 0.0029 | 0.248265 |
| 0.727793 | 0.00113 | 0.281151 |

| | | |
|---|---|---|
| 0.738897 | 0.00043 | 0.292838 |
| 0.743103 | 0.0001 | 0.297267 |
| 0.7457 | 3.22E-05 | 0.3 |
| 0.746552 | 1.00E-05 | 0.301571 |
| 0.75 | 0 | 0.307931 |
| 0.8216 | 0 | 0.44 |
| 0.8975 | 0 | 0.6 |
| 0.9428 | 0 | 0.84 |
| 0.9589 | 0 | 0.93 |
| 0.965 | 0 | 0.97 |
| 0.97 | 0 | 0.99 |
| 0.975 | 0 | 1 |

**(2)Determination of relative permeability of oil phase**

In the three-phase system, the relative permeability of water phase is determined by the two-phase oil water system. The relative permeability of oil phase in three-phase system is calculated by Stone formula II. Stoneformula II was:

$$k_{ro} = k_{rocw}\left[\left(\frac{k_{row}}{k_{rocw}} + k_{rw}\right)\left(\frac{k_{rog}}{k_{rocw}} + k_{rg}\right) - k_{rw} - k_{rg}\right] \tag{7}$$

In the three-phase system, the relative permeability of water phase is determined by the two-phase oil water system. In this paper, the gas phase is composed of two parts. In order to solve the relative permeability of oil phase, it is the key to determine the average permeability of gas phase. We assume that the average relative permeability of gas bubbles as the weighted average in the gas phase mole fraction of bubble's relative permeability and the relative permeability of gas flow.As shown in Figure 4, the average relative permeability of gas is obtained, and then the relative permeability of oil phaseis determined by formula (7) .

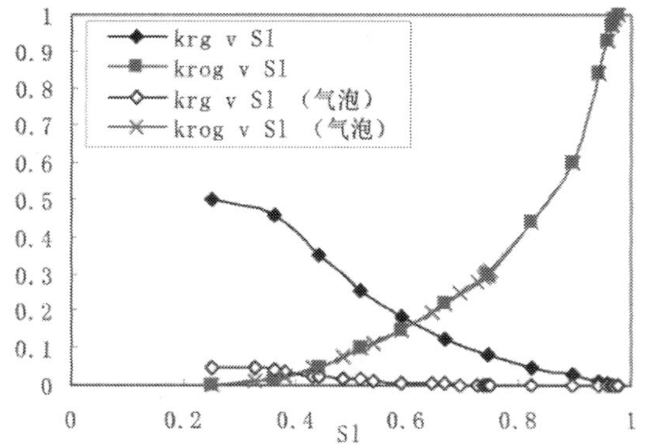


**Figure 3comparsion of the relative permeability between gas and bubble**

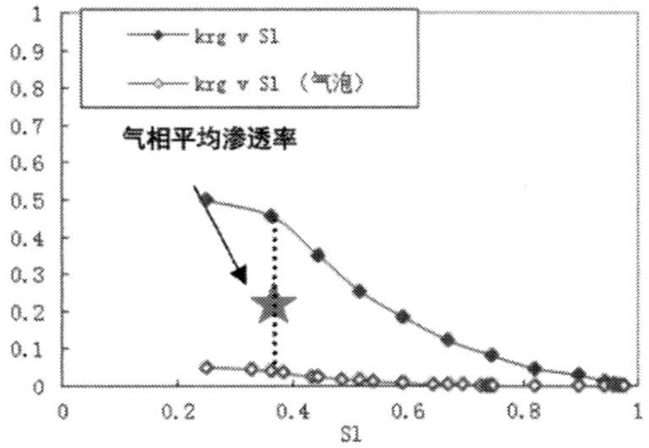


Figure 4 the determination of average relative permeability of gas phase

In the whole system, the relative permeability of oil phase and gas phase are not only a function of saturation, but also the percentage of bubble function.Therefore, the calculation is not easy to converge, which required to adjust the numerical algorithm making the calculation convergence.

### 3.2 Adaptive implicit method for solving component model

The 2n + 3 equations in above are discretized with finite-difference techniques. The resulting discretized equations for the flow of components in the oil and gas phases are:

$$\Delta T_o^m y_{oi}^m \Delta\Phi_o^{n+1} + \Delta T_g^m y_{gi}^m \Delta\Phi_g^{n+1} + Vq_i^m = \frac{V}{\Delta t}\left[N_i^{n+1} - N_i^n\right] \quad (8)$$

Similarly the discretized equation for the flow of water is:

$$\Delta T_w^m \Delta\Phi_w^{n+1} + Vq_w^m = \frac{V}{\Delta t}\left[N_w^{n+1} - N_w^n\right] \quad (9)$$

The discretized equation forthe flow of the total hydrocarbon equation is

$$\Delta T_o^m \Delta\Phi_o^{n+1} + \Delta T_g^m \Delta\Phi_g^{n+1} + Vq_i^m = \frac{V}{\Delta t}\left[N_o^{n+1} + N_g^{n+1} - N_o^n - N_g^n\right] \quad (10)$$

The phase equilibrium equations for the components are:

$$\ln f_{ig} = \ln f_{io}\text{；}\quad i = 1,\ldots,n_c \quad (11)$$

Volume constraint equations are:

$$\sum_j \frac{N_j^{n+1}}{\rho_j} = \phi^{n+1}; j = 0, g, w \quad (12)$$

The superscript n and n+1 represent previous time step and new time step, respectively. The superscript m is equal to n (IMPES method) or n+1 (full implicit method).

The above differential equationsare expanded, and the iterationsare written in the form of iteration allowance

**(1)fugacity equation**（$i = 1,\ldots,n_c - 1$， $n_c$-lequations）

$$R_{n_c - 1} = \ln f_{ig} - \ln f_{io} = 0 \quad (13)$$

**(2) kinetic equation（one equation）**

$$\mathbf{R_{nc} = Rea(CO_2 + light\ hydrocarbon\ components \leftrightarrow bubble)} \quad (14)$$

**(3) Component mobility equation** （$i=1,...,n_c$， $n_c$-1equations）

$$R_{2n_c-1}=\Delta T_o^m y_{oi}^m \Delta\Phi_o^{n+1}+\Delta T_g^m y_{gi}^m \Delta\Phi_g^{n+1}+Vq_i^m-\frac{V}{\Delta t}\left[N_i^{n+1}-N_i^n\right] \quad (15)$$

**(4) Total hydrocarbon equation**

$$R_{2n_c}=\Delta T_o^m \Delta\Phi_o^{n+1}+\Delta T_g^m \Delta\Phi_g^{n+1}+Vq_i^m-\frac{V}{\Delta t}\left[N_o^{n+1}+N_g^{n+1}-N_o^n-N_g^n\right] \quad (16)$$

**(5) Water flow equation**

$$R_{2n_c+1}=\Delta T_w^m y_{wi}^m \Delta\Phi_o^{n+1}+\Delta T_g^m y_{gi}^m \Delta\Phi_g^{n+1}+Vq_i^m-\frac{V}{\Delta t}\left[N_i^{n+1}-N_i^n\right] \quad (17)$$

**(6) Volume constraint equation**

$$R_{2n_c+2}=\sum_j \frac{N_j}{\rho_j}-\phi=0; j=0,g,w \quad (18)$$

The 2n$_c$+2 primary unknowns per grid block are: $p, N_1, N_2,...N_{n_c}, N_w, N_{1g}, N_{2g},..., N_{n_c g}$.This can be obtained by solving simultaneoulsy the system of Equations (18) with Newton's method. As the adaptive method is used. Some gridblocks will be Implicit Pressure.

Therefore, the presented model is based on the compositional model, combining the advantages of "balance model" and "dynamic model", the flow characteristics of “foamy oil flow” as the function of time and the flow conditions, and adjusting the permeability and the critical saturation of bubble to determine the" bubble flow "effect on production performance.

## 4 Case study

### 4.1 Reservoir

XL reservoir oil field locates in the Northeast of China. The field is sandstone reservoir with no gas cap and no active water.The structure is dome anticline close to East-west trending cut by north-south fault.

The reservoir is characterized by low porosities and low permeability.The original formation pressure of reservoir is 15MPa. The field oil-bearing area is 2.66km$^2$, which has an OGIP of 118.2 * 10$^4$t.

There are 25 wells with 4 injectorsand 21 producers (Figure 5), which are divided into four oil production platforms. The well spacing is 300m. Original well pattern is nine point area water injection with N 67.5° E well array direction. October 1994 water was begun to be injected. Up to the present, the oil recovery is 11.7%.

$CO_2$ injection area of the XL oilfield was located in the northern part of the XL oilfield, which is characterized by 2.5 degrees dip angle, 1350-1600m burial depth, no fault and no active aquifer. The characteristics of this area are beneficial for $CO_2$ flooding.

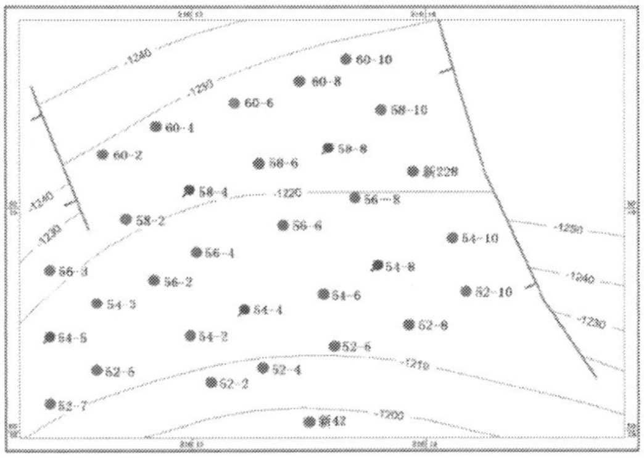


Figure 5 the well location map of JL reservoir

Therefore, for a clear understanding of effect of "foamy oil flow" on oil production and asphaltene deposition, we have done field-scale history matching and prediction.

**Field Grid.**Fig. 6 shows an area view of the field model grid.Locations of wells are displayed on the grid. The model consists of 11 layers. The total simulation grid number is 33440(80 *38* 11).

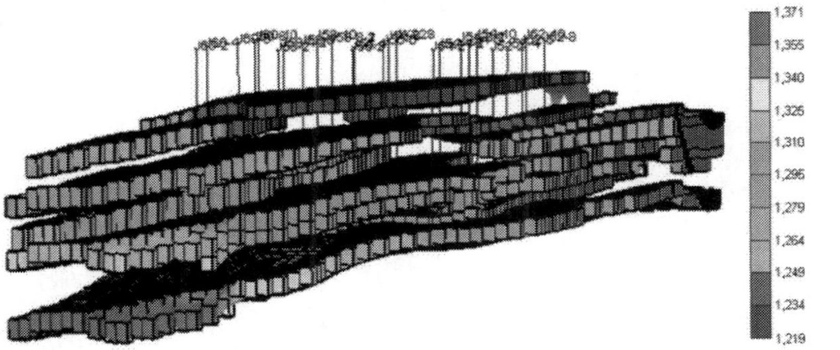


Figure 6 the structural diagram of numerical simulation of $CO_2$ flooding areaof Xinli oil field

## 4.2 $CO_2$ displacement dynamics considering "foamy oil flow"

### *4.2.1 Effect of "foamy oil flow" on oil production and asphaltene deposition*

Simulation research showed that the displacement efficiency considering "foamy oil flow" improved significantly compared with normal numerical simulation without considering “foamy oil”(figure 7(a)) , and gas oil ratio was higher than that without considering the "foamy oil flow"(Figure 7 (b)), which showed"bubble" in the oil phase is beneficial in the production during $CO_2$ flooding, mainly because gas expands to occupy the space occupied by the original liquid, increase the compressibility of the fluid and delay the gas production time. The dispersed bubbles can reduce the relative permeability of the gas phase, thus increase the oil production rate, reduce the production gas oil ratio, and ultimately improve oil recovery.

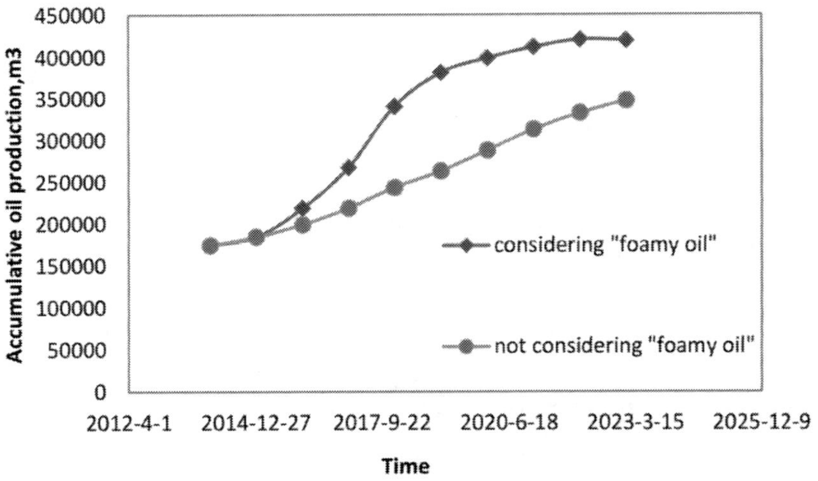


(a) cumulative oil rate

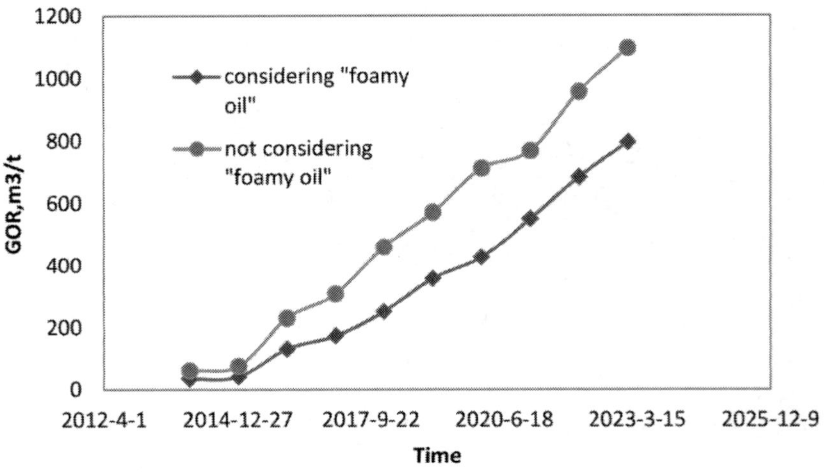


(b) gasoil ratio

Figure 7 the effect of foamy oil on production

### *4.2.2 Distribution law of "foamy oil flow"*

Simulation research showed the relationship between the percentage of $CO_2$ and the percentage of bubbles in different $CO_2$PV. Seen from figure 8 (a) (b),in the early injection stage, bubble is located at the frontier of $CO_2$ flooding, which showed similar flow mechanism of "foamy oil flow" in dissolved gas flooding; at $CO_2$ flooding period, the bubble was still mainly appeared in the flooding front and tend to flow to big porosity area. Therefore, it can be inferred that, when the gas breakthrough, the role of "foamy oil flow" will be reduced. However, before gas breakthrough, the characteristics of "foamy oil flow" can significantly improve the displacement efficiency.

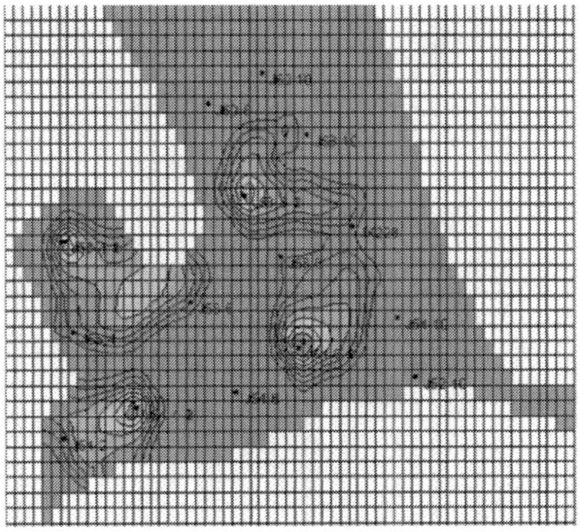

(a)$CO_2$ mole percentage of injection of 0.05 PV

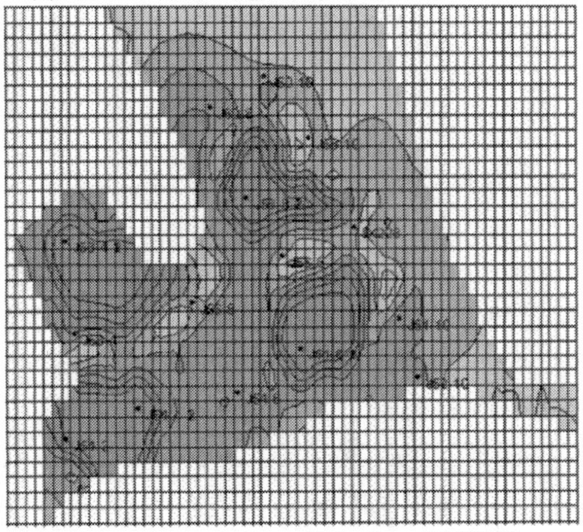

(b)bubble mole percentage of injection of 0.05 PV

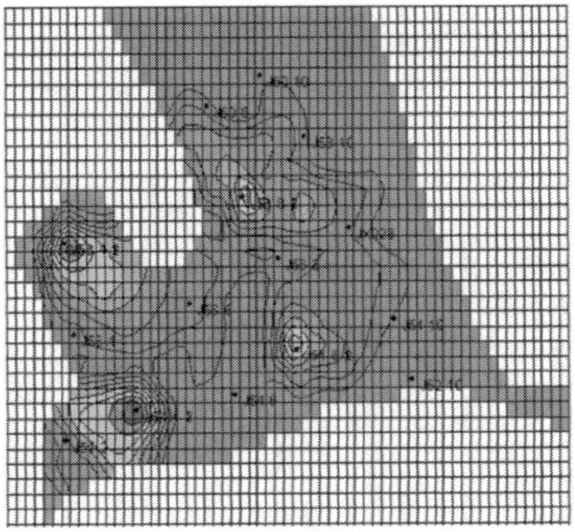

(c)$CO_2$ mole percentage of injection of 0.15 PV

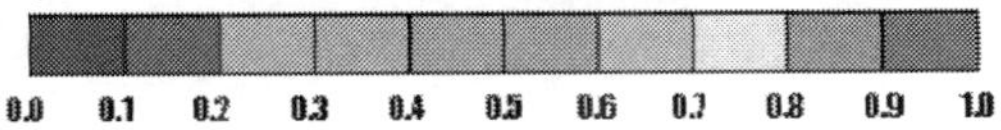

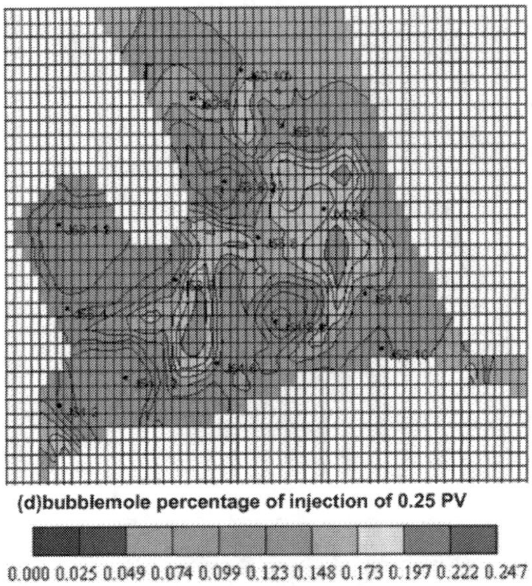


**Figure 8 the distribution graph of bubble content of layer I**

## 5 Conclusions and recommendations

$CO_2$ flooding "foamy oil flow" is a kind of special phenomenon of $CO_2$ displacement of asphaltene crude oil. In this paper, by comparing the similarity between the "foamy oil" in $CO_2$ displacement and the "foamy oil" in the dissolved gas drive, the bubble nuleation mechanism is explained.

Through the research on the surface activity of asphaltene, the possibility of the "bubble" of asphaltene stabilization was confirmed. Considering the influence of asphalt on the formation of "foamy oil" and the dynamic change process of "foamy oil", a mathematical model considering the "foamy oil flow" during $CO_2$ flooding was established.

The calculation results showed that the bubble mainly located at $CO_2$ flooding front, which showed bubble material similar to "foamy oil flow" to stabilize gas injection frontier; when the gas breakthrough, mole percentage of bubble" significantly reduced. Compared with normal numerical simulation without considering “foamy oil”, high efficient flooding potential was achieved with “foamy oil” existed in $CO_2$ displacing front, which deserves further research.

## 6 reference:

1.Maini, B.B., Sarma, H.K., and George, A.E.: Significance of Foamy Oil Behavoir in Primary Production of Heavy Oils, J. Cdn. Pet. Tech., 32, No. 9,50.

2、Li Shilun, Zhang Zhengqing and Ran Xinquan etc. The technology of Enhanced oil recovery with gas injection. Chengdu: Sichuan Science and Technology Press2001.

3、ShenPingping,JiangHuaiyou and ChenYongwu etc. EOR study of $CO_2$ injection[J]. Special Oil & Gas Reservoir,2007,14(3):1-4.

4、GuoWankui, Liao Guangzhi and Shao Zhengbo etc. EOR technology with gas injection. Beijing: Petroleum Industry Press,2003

5、Lu teng, Li Zhaomin and Liuwei etc. Study on $CO_2$ Huff and Puff Technology for Strong Water Sensitive Heavy Oil Reservoir[J]. Journal of Southwest Petroleum University(Science & Technology Edition) 2013,35(1):122-128

6、Chen Jiang, Study on WINPROP phase of foamy oil [J]. Petrochemical Industry Application, 2014,33(11):106-109

7.X.Fei Sun, Y. Zhang and X.Duan etc.: Natural Gas Reinjection in a Heavy Oil Field Increases Foamy Oil Recovery. Paper SPE 169252 presented at the SPE Latin American and Caribbean Petroleum Engineering Conference; Maracaibo, Venezuela, 21-23 May 2014.

8. XinfengJia, Tao Jiang and YonganGu: Enhanced Vapor Extraction through Foamy-oil flow. Paper SPE 165526 presented at the SPE heavy oil conference Canada, Calgary, Alberta, Canada, 11-13 June 2013.

9. Smith, G.E.: Fluid Flow and Sand Production in Heavy-oil Reservoirs Under Solution-Gas Drive, SPE Production Engineering (May 1988) 169-180

10. Kraus W P, McCaffrey W J, Byod G W. Pseudo-bubble Point Model for Foamy oils, paper CIM 93-45 presented at the CIM 1993 Annual Technical Conference, Calgary, Alberta (May 9-12, 1993)

9. Kimbler, O.K., R.L.Reed, and L.H.Silberberg: Physical Characteristics of Natural Films Formed at Crude Oil-water Interfaces. SPEJ, 1966(6): p153-165

10.Eley, D.D., M.J. Hey, J.D.Symonds, and J.H.M. Willison: Electron Micrography of Emulsions of Water in Crude Petroleum, Journal of Colliod and interface science, 1976. 54(3): p. 462-466

11. Ward,C.A. and Levart, Eugene: Conditions for Stability of Bubble Nuclei in Solid Surfaces Contacting a Liquid-gas Solution, J. Appl. Phys. 56(2), 15 July 1984, pp. 491-500

12. Claridge, E.L. and Michael Prats: A Proposed Model for Anomalous Foamy Heavy Oil Behavior. SPE 29243, 1995

13. F.Bauget, D. Langevin, R.Lenormand.: Effects of Asphaltenes and Resins on Foamability of Heavy Oil. SPE 71504

14. I.Adil and B.B. Maini: Role of Asphaltene in Foamy Oil Flow. SPE 94786, 2005

15. M.Huerta, C. Otero, A. Rico, I. Jimenez, etc.: Understanding Foamy Oil Mechanisms for Heavy Oil Reservoirs during Primary Production. SPE 36749, 1996

16.Maini, B.B.:Foamy oil Flow, paper SPE 68885

17. Smith, G.E.: Fluid Flow and Sand Production in Heavy-oil Reservoirs under Solution-Gas Drive, SPE Production Engineering (May 1988) 169-180

18. Maini, B.B., Sarma, H.K., and George, A.E.: Significance of Foamy Oil Behavoir in Primary Production of Heavy Oils, J. Cdn. Pet. Tech., 32, No. 9,50.

19. Coombe, D. and Maini, B.: "Modeling Foamy Oil Flow," presented at the Workshop on Foamy Oil Flow held at the Petroleum Recovery Inst., Calgary, 27 April 1994

20. Jia Ying, Sun Lei and Sun Liangtian etc.The Research on Asphaltene Deposition Mechanism and Its Influence on Development during $CO_2$ Injection. SPE 104417, 2006.

21. Huang Lei, Shen Pingping and Jia Ying etc. Prediction of asphaltene precipitation during $CO_2$ injection. Petroleum Exploration and Development No.03, 2010.

CARBON MANAGEMENT TECHNOLOGY CONFERENCE

# CMTC-485492-MS

# Coalbed Methane Recovery by Injection of Hot Carbon Dioxide

Hon Chung Lau, National University of Singapore, Chaobin Zhao, University of Calgary, Samuel W. Lau, University of Houston



This paper was prepared for presentation at the Carbon Management Technology Conference held in Houston, Texas, USA, 17-20 July 2017.



## Abstract

A new method of coalbed methane (CBM) recovery is proposed wherein hot carbon dioxide at or near supercritical condition is injected into a CBM reservoir to take advantage of enhanced desorption of methane at elevated temperatures and the preferential adsorption of $CO_2$ on coal surfaces compared to methane. The feasibility of this concept was studied by reservoir simulations using one quarter of an inverted five spot pattern using $CO_2$ and $CH_4$ adsorption isotherms published in the literature. The study compared CBM recovery by pressure depletion, $CO_2$ injection and injection of $CO_2$ that is 10°C hotter than the reservoir temperature. Results show that hot $CO_2$ injection can significantly increase the production rate and recovery factor over and above that achievable by reservoir heating or $CO_2$ injection alone. This new method holds promise for enhanced CBM recovery and also $CO_2$ sequestration, especially for high-rank coals where swelling of coal by $CO_2$ injection is minimized. Preliminary considerations suggest that this method can be economic over a range of $CO_2$ and natural gas price if the $CO_2$ comes from natural sources and the CBM field is located near an existing $CO_2$ pipeline. Alternatively, if the source of $CO_2$ is industrial, this method can be profitable if the cost of $CO_2$ capture is offset by the trading price of $CO_2$ and if the CBM project is located near the $CO_2$ source.

## Introduction

Coalbed methane, also known as coal seam gas, is a significant source of natural gas worldwide, with sizable reserves in the United States, Australia, Russia, Canada and China. In 2015, CBM production in the US was 1.27 Tcf (35.97 Bcm), accounting for 3.9% of all US natural gas production (EIA, 2017). Commercial production of CBM occurs in ten US basins with major production coming from the San Juan, Black Warrior and the Central Appalachian. In 2015, CBM production in Australia was 270 Bcf (7.65 Bcm) accounting for 18% of Australia's total natural gas production (Australian Government, 2016).

## CBM Recovery Mechanisms

There are several known methods for CBM recovery. They are summarized below.
**Depletion Drive.** Unlike conventional natural gas deposits where natural gas resides within the pore space of a reservoir, the methane gas in a CBM reservoir is not stored as free gas but rather as sorbed gas on the surface of coal. Figure 1 is a methane sorption isotherm of Pittsburgh coal in the North Appalachian Basin (Rogers *et al.*, 2007). At a reservoir pressure of 800 psi, and temperature of 30°C, the coal is undersaturated with respect to methane. A drop of reservoir pressure to 200 psi will result in desorption of methane from the coal surface, giving a recovery factor of about 27%. This method of CBM recovery is called pressure depletion and is the predominant method used for CBM recovery.

Hitherto, most commonly used CBM production method is pressure depletion wherein a large amount of formation water is produced to lower the reservoir pressure to allow methane desorption from the coal surface. Recovery factor is estimated to be between 20-60% of original gas-in-place, depending on coal permeability, gas saturation, well spacing and other reservoir properties and operational practices.

**Enhanced CBM Recovery by $CO_2$ Injection ($CO_2$-ECBM).** This method utilizes the concept of displacement desorption due to the fact that $CO_2$ has a higher affinity than methane for adsorption on coal surfaces (Stevens *et al.,* 1998). In this process, injected $CO_2$ is preferentially adsorbed at the expense of methane. The displaced methane diffuses into the cleat system and is produced as free gas. Laboratory studies have shown that one to several volumes of injected $CO_2$ are required per volume of incremental produced methane, depending on the coal rank. In general, for high rank coals, the $CO_2/CH_4$ replacement ratio approaches one (Gonzales *et al.,* 2008). However, $CO_2$-ECBM suffers from coal swelling and permeability reduction that occur when $CO_2$ displaces methane in low to medium rank coal (Anggara *et al.,* 2013; Gonzales *et al.,* 2008; Mazumder and Wolf, 2008) This phenomenon, however, is less pronounced in high rank coals, where perhaps the largest potential of $CO_2$-ECBM resides (Gonzales *et al.,* 2008). The effect of $CO_2$ adsorption on coal swelling has been a subject of considerable experimental and numerical research (Law, *et al.,* 2002; van Hemert *et al.,* 2007; Chen *et al.,* 2009).

In the US, there have been several $CO_2$-ECBM pilot projects with various degree of success (Gonzales *et al.*, 2008). This method has also been piloted in China, Canada, Japan and Poland (Pagnier *et al.*, 2005).

**Enhanced CBM Recovery by Nitrogen Injection ($N_2$-ECBM).** This method utilizes the concept of inert gas stripping. Due to its low sorption characteristics, nitrogen acts more like an inert gas in the presence of coal. Nitrogen injected into a CBM reservoir reduces the methane concentration in the gas phase while maintaining the total system pressure. Consequently, the partial pressure of methane is reduced. This allows methane to desorb from the coal surface and diffuse to the cleat network. Simulation and pilot results suggest that $N_2$-ECBM is capable of recovering significant amount of gas-in-place (Puri and Yee, 1990). Where there is no significant $CO_2$ source, $N_2$-ECBM may be an alternative.

**Enhanced CBM Recovery by Underground Coal Gasification.** This method utilizes the concept of thermal desorption. Thermal desorption uses heating to increase the temperature of the coal so as to reduce its adsorption capacity. Thermal treatment can therefore increase gas desorption and diffusion in coal seam gas reservoirs resulting in higher ultimate recovery, lower water production and higher wellhead pressure. The use of underground coal gasification as a thermal recovery process for a CBM reservoir has been proposed (Salmachi *et al.*, 2011) but not been applied in the field.

## Enhanced CBM Recovery by Hot $CO_2$ Injection

Our proposal is to combine $CO_2$ exchange with thermal desorption of CBM by injecting hot $CO_2$ at or near supercritical conditions into a CBM reservoir. This method encompasses the benefits of $CO_2$-ECBM and thermal recovery. It has the following advantages.

Firstly, sorption of gas on coal is governed by the Langmuir isotherm. At high temperatures, the ability of coal to adsorb gas is significantly reduced. Consequently, heating up coal by injecting hot $CO_2$ can significantly enhance desorption of methane from coal surfaces.

Secondly, for an undersaturated coal, an increase in temperature increases the critical desorption pressure at which methane desorption initiates. Consequently, methane production can start at a higher reservoir pressure and less dewatering is required. Furthermore, the total gas recovery for a fixed initial and abandonment pressure increases at higher temperatures as shown in Fig. 1.

Thirdly, transport of methane from the pores to the cleats in the coal is controlled by diffusion. A higher temperature increases the diffusion coefficient and therefore the rate of methane diffusion into the cleats.

Fourthly, isotherms measured in the laboratory have shown that coal can adsorb up to three times as much $CO_2$ by volume than methane, depending on coal rank. Since $CO_2$ preferentially adsorbs on the coal surface, it displaces methane from the coal surface.

Fifthly, hot $CO_2$ can strip water from cleat network, which enhances $CO_2$-methane mass transfer.

Sixthly, this method has the added advantage of sequestering large quantities of $CO_2$ in CBM reservoirs which is beneficial for mitigating greenhouse effects of $CO_2$ emission to the atmosphere.

In summary, the heat supplied by the hot $CO_2$ enhances desorption of methane from the coal surfaces and speeds up its diffusion into the cleats. It also allows desorption to occur at a higher reservoir pressure and therefore reduces the need for dewatering. In addition, $CO_2$ enhances desorption of methane gas from the coal surfaces by nature of its higher affinity on coal.

However, the injection pressure of $CO_2$ should not be too high to fracture the overburden. This process works best in a high rank coal which is usually not susceptible to coal swelling and permeability reduction due to injection of $CO_2$ (Anggara *et al.,* 2013).

## Reservoir Simulation

**Approach.** A dual-porosity compositional reservoir model for a typical high rank CBM reservoir was constructed in a commercial reservoir simulator that included one injection well and one producer well arranged in an one quarter of an inverted five-spot pattern (Fig. 3). This model employed temperature-dependent Langmuir isotherms for methane and $CO_2$. The simulations compared gas rate, cumulative gas production and gas recovery factor between simple pressure depletion versus $CO_2$ injection at reservoir temperature and at 10$^{o}$C higher than reservoir temperature.

**Assumption.** Current commercial reservoir simulators are inadequate to fully simulate the non-isothermal process of hot $CO_2$ injection into CBM which involves the heat exchange between injected $CO_2$ with the reservoir fluid and rock and the temperature dependence of $CO_2$ and methane adsorption. Consequently the following assumptions were made in our simulations.

1. The methane and $CO_2$ adsorption isotherms on coal at 45°C published in the literature (Zhang *et al.*, 2011; Law *et al.*, 2002) are used (Fig. 2).
2. At elevated reservoir temperatures, adsorption capability of $CO_2$ and methane should be reduced, the amount of this reduction with temperature is assumed to form different cases.
3. When the temperature of injected $CO_2$ is different from the reservoir temperature, there is heat exchange between injected $CO_2$ and the reservoir. This process depends on the temperature difference between injected $CO_2$ and the reservoir, $CO_2$ injection rate and volume, and the thermal conductivity of the reservoir fluid and rock. This means that the reservoir temperature during hot $CO_2$ fluid injection is location specific. Therefore, the desorption parameters of $CO_2$ and methane are also location specific. To simplify the problem, it is assumed that the temperature throughout the whole reservoir is instantly increased to the injected $CO_2$ temperature. This will cause a higher gas production rate, but should not affect the ultimate recovery.
4. There is no coal swelling due to $CO_2$ injection.

**Model Description.** A quarter of inverted five-spot model was built for our study (Fig. 3). The dimension of the model was 11x11x2 and grid size in x and y directions was about 25 m x 25 m. Reservoir depth is 750 m. Initial reservoir pressure and temperature are 7.65 MPa and 45°C, respectively, which is inside the supercritical region of the $CO_2$ phase diagram (Fig 4). Coal density is 1,434 kg/m$^3$. At 45°C, dry, ash-free Langmuir pressures for methane and $CO_2$ are 4.6885 MPa and 1.903 MPa, respectively, and Langmuir volumes for methane and $CO_2$ are 0.0118 m$^3$/kg and 0.024 m$^3$/kg, respectively (Zhang *et al.*, 2011). The isotherms are given in Fig. 2. Initial gas content is 6.0 m$^3$/t and gas is 19.1% undersaturated with respect to the isotherm. Porosity is 0.5% and permeability is 2 md. The PVT properties of methane and $CO_2$ under various temperatures and pressures are calculated using the Peng-Robinson Equation of State. Gas and water relative permeability curves are shown in Fig. 5. The model takes into account the competitive adsorption of $CO_2$ in the presence of $CH_4$ by using the partial pressures of the two gases. The cleat is 100% water saturated initially.

**Operational constraints**. The operational constraints for injection and production wells are as following:

- $CO_2$ injection rate (1/4 well) = 1,000 m$^3$/d
- Maximum injection well bottom hole pressure = 10 MPa
- Maximum gas production rate (1/4 well) = 1,000 m$^3$/d
- Minimum production well bottomhole pressure = 0.275 MPa

**Case Design.** To compare CBM recovery by hot $CO_2$ injection to that by pressure depletion, the following cases were designed (Table 1 and Figs. 6-7).

- Case 1: Depletion drive at reservoir temperature of 45°C
- Case 2: $CO_2$ injection at 45°C

- Case 3: $CO_2$ injection at 55°C assuming reservoir temperature equilibrated to 55°C instantly. Langmuir volumes for $CO_2$ and methane are reduced by 10% and 20%, respectively. Langmuir pressure for $CO_2$ is increased by 10% and unchanged for methane.
- Case 4: Same as Case 3 except it is assumed that Langmuir pressure for $CO_2$ remains the same as that at 45°C.
- Case 5: Same as Case 4 except it is assumed that Langmuir volumes for $CO_2$ and methane are reduced by 20% and 10%, respectively, from their values at 45°C.
- Case 6: Depletion development at 55°C. Langmuir volume of methane is unchanged from that at 45°C. Langmuir pressure of methane is reduced by 20% from that at 45°C.
- Case 7: Depletion drive at 55°C. Langmuir volume of methane is unchanged from that at 45°C. Langmuir pressure of methane is reduced by 10% from that at 45°C.

Case 1 is the base case for depletion drive at 45°C. Cases 6 and 7 are the depletion drives at 55°C with different impact of temperature on the Langmuir isotherm. Cases 2 represents the case of $CO_2$ injection without raising reservoir temperature. Cases 3 through 5 are cases of hot $CO_2$ injection at 55°C assuming different impact of temperature on the Langmuir isotherms. We believe the aforementioned assumptions on the impact of temperature on the Langmuir volume and pressure are reasonable and generally agrees with experimental data published by Zhang *et al.* (2011).

## Simulation Results

Simulations results are given in Table 1 and can be summarized below where recovery factor (RF) refers to that at 20 years.

1. Depletion drive at 45°C gives a gas RF of about 44% (Case 1).
2. $CO_2$ injection at 45°C increases the gas RF to 73% (Case 2), a roughly 60% increase.
3. Heating up the reservoir by 10°C increases the gas RF to 50-58% (Cases 6 and 7), a 15-32% increase.
4. However, injection of $CO_2$ at 55°C increases the gas RF to 83% (Cases 3, 4 and 5), a 90% increase.

The gas rate and cumulative gas rate from our simulations are given in Figs. 8 and 9, respectively. It can be seen from these figures that heating up the reservoir by 10°C has a primary impact on production acceleration and a secondary impact on ultimate recovery, although both impacts are significant. This is demonstrated by a higher production peak that occurs sooner for Case 6 compared to Case 1 (Fig. 8). Note that at later times, the production rate for Cases 1 and 6 are the same.

However, $CO_2$ injection has a primary impact on ultimate recovery and a secondary impact on production acceleration. This is demonstrated in Fig. 9 showing that for the first few years, gas cumulative gas production is roughly the same for Cases 1 and 2, but at later times, Case 2 has a much higher cumulative gas production than Case 1.

Consequently, hot $CO_2$ injection has the dual benefits of both accelerating production and increasing ultimate recovery. The effects of heating the reservoir and injecting $CO_2$ are roughly additive. Heating increases ultimate recovery by approximately 30% whereas $CO_2$ injection by approximately 60%. Hot $CO_2$ injection increases ultimate recovery by approximately 90%.

**Model Limitations.** There are, however, limitations to our simulations. Firstly, our simulations are isothermal where we assumed the reservoir to heat up to the injected $CO_2$ temperature instantaneously. This will lead to a faster gas production rate, but should not affect the ultimate recovery. Secondly, $CO_2$ injection is not optimized but assumed to be at a constant rate. In field applications, both $CO_2$ injection rate and duration of injection should be optimized using adsorption isotherms measured from the coal samples. Then economics can be run to determine which combinations will give the best NPV and ultimate recovery. Thirdly, we have only investigated increasing the reservoir temperature by 10°C by injecting hot $CO_2$. In field applications, the $CO_2$ injection temperature should also be optimized by running non-isothermal simulations to take into account the time it takes to heat up the reservoir, based on thermal properties of the coal. Fourthly, the isotherms used in our simulations are taken from the literature. We make reasonable assumptions on the impact of temperature increase on the isotherms. For field applications, experimental measured isotherms from the target coal seams should be used in simulations. Fifthly, the assumption that $CO_2$ has no impact on coal swelling needs to be verified with coal samples from the field. These limitations notwithstanding, our simulation study demonstrated the potential of hot $CO_2$ injection as a CBM recovery mechanism.

## Conceptual Design

The following is a conceptual design of the proposed method.

1. Carbon dioxide from either a natural or industrial source is captured, filtered and transported by pipeline to a CBM field under supercritical conditions.
2. The $CO_2$ is heated to a temperature that is 10°C or higher than the reservoir temperature by heat exchangers using produced natural gas from the field.
3. Hot $CO_2$, at or near supercritical condition, is injected into a CBM formation through an injection well. Natural gas and water is produced from producers with injectors and producers arranged in a usual well pattern such as inverted five-spot or inverted nine-spot.
4. An alternative design is to inject hot $CO_2$ into a CBM well. The well is shut in for a soaking period after which it is opened for production. The process can be repeated. The injection, soaking and production times can be optimized by reservoir simulations.

## $CO_2$ Sources and CBM fields in US

The source of $CO_2$ is a major consideration in the proposed process. The $CO_2$ source can be either natural or industrial. Figure 10 shows the natural and industrial $CO_2$ sources, existing $CO_2$ pipelines and CBM fields in the US. Natural sources of $CO_2$ exist in Mississippi (Jackson Dome), New Mexico (Bravo Dome) and Colorado (Sheep Mountain and McElmo Dome). Industrial $CO_2$ sources exist in Texas, Kansas, Oklahoma, Wyoming, North Dakota and Michigan. Extensive $CO_2$ pipeline exists in the Gulf coast, Texas, New Mexico, Colorado, and Wyoming. In the US, CBM fields are located in New Mexico (San Juan Basin), Colorado (Piceance Basin), Wyoming (Powder River Basin), Kansas (Arkoma Basin), Oklahoma (Cherokee Platform), Alabama (Black Warrior Basin), Virginia (Central Appalachian Basin) and Pennsylvania (Northern Appalachian Basin). Many of these are located close to existing $CO_2$ pipelines making them especially attractive for enhanced CBM recovery by hot $CO_2$ injection. Outside of the US, $CO_2$ pipeline infrastructure exists in UK and Canada, but not elsewhere (IEA, 2013).

## The Economic Case

Where a CBM reservoir is located near an existing $CO_2$ pipeline and the $CO_2$ source is natural, e.g. in Texas or Wyoming, the cost of $CO_2$ can be around $0.50-1.0/Mcf. In existing pipelines, $CO_2$ is usually transported in dense liquid phase near supercritical conditions (31°C, 74 bar). Therefore the line pressure of $CO_2$ is usually high enough for $CO_2$ injection into the CBM without the need for compression. The other cost is the heating up of $CO_2$. Since produced methane can be used as fuel to heat up the $CO_2$, the heating cost can be assumed to be a fraction of the produced methane. In this case, an economic case can be constructed for various price of natural gas using the results of reservoir simulations. Since our simulations were not optimized for $CO_2$ injection, no detailed economic case is given here. However, assuming that no additional well and facility cost is incurred and heating of $CO_2$ consumes up to 20% of produced natural gas, our rough estimate suggests that hot $CO_2$ injection can be economic under a low $CO_2$ price ($0.5/Mscf) and a low natural gas price (~ $3.0/Mscf) scenario or a high $CO_2$ price ($1.0/Mscf) and moderate to high natural gas price scenario (> $4.0/Mscf).

When $CO_2$ has be to captured from industrial sources such as flue gas from power generation plants, then its capturing cost can fall within the range of $0.8-1.0/Mscf (Iwasaki *et al.*, 2004). Furthermore, when the distance between the power plant and the CBM reservoir is 100 km or less, the cost of pipelines used to transport $CO_2$ is relatively small. The same is true if the power plant is located near an existing $CO_2$ pipeline. In this case, it is conceivable that an enhanced CBM project by hot $CO_2$ injection can also be economical under a moderate to high natural gas price scenario (>$4.0/Mscf).

If, on the other hand, a trading price of $CO_2$ emission is applicable, the economic case can be considerably improved. For $CO_2$, $1.0/Mscf is equivalent to $19/ton. Therefore, when the trading price of $CO_2$ emission exceeds $20/ton, the cost of $CO_2$ capture from flue gas can be completely offset by that price, and an enhanced CBM recovery project by hot $CO_2$ injection can result in a large profit. Current estimate is that the price of $CO_2$ will be close to $20/ton in year 2020 (Luckow *et. al.*, 2015).

The economic case will be worsened if the source of industrial $CO_2$ is located far from CBM fields or existing $CO_2$ pipelines. In this case, the cost of constructing a long $CO_2$ pipeline needs to be included.

## Future Research

Coal swelling experiments using field samples in the presence of $CO_2$ should be used to screen potential candidate CBM reservoirs for hot $CO_2$ injection. Only those samples that do not show significant swelling should be considered as candidates. Next, the adsorption isotherms of both methane and $CO_2$ in candidate CBM reservoirs should be measured using coal samples from target coal seams over a range of temperature at and above reservoir temperature. The thermal conductivities of coal should also be measured. The experimental results can then be input into a reservoir simulator to investigate the potential of CBM recovery by hot $CO_2$ injection. To fully simulate the process, a non-isothermal CBM reservoir simulator will be needed.

## Conclusions

The following conclusions can be made from the present study.

1. A new method of CBM recovery by injection of hot $CO_2$ is proposed. This method takes advantage of the competitive adsorption of $CO_2$ over methane on coal surfaces and enhanced desorption of methane at elevated temperatures. In addition, it holds promise to sequester large

amount of industrial $CO_2$ in a CBM reservoir to mitigate the adverse effect of $CO_2$ on the environment.

2. Simulation studies using a quarter of an inverted five spot pattern were conducted to determine the effectiveness of this process by using literature derived adsorption isotherms for $CO_2$ and $CH_4$ and by making various assumptions on the effect of temperature on these isotherms.
3. Simulation results show that heating has a primary impact on gas production acceleration and a secondary impact on ultimate recovery, whereas the opposite is true for $CO_2$ injection. The combined result is that hot $CO_2$ injection can lead to significant production acceleration and increase in ultimate recovery.
4. A hot $CO_2$ injection project can be economically attractive if $CO_2$ comes from a natural source and the CBM field is located near an existing $CO_2$ pipeline, or where the source of $CO_2$ is industrial and the trading price for $CO_2$ emission offsets the cost of $CO_2$ capture and transportation.

## References


Australian Energy Update 2016. Office of the Chief Economist, Department of Industry, Innovation and Science. www.inudstry.gov.au/oc

Anggara, F., Sasaki, K. and Sugai, Y. 2013. Swelling Measurements of a Low Rank Coal in Supercritical CO2. *International J. of Geoscience*, 4, 863-870.

Chen, Z., Liu, J., Elsworth, D., Connell, L., and Pan, Z. 2009. Investigation of $CO_2$ Injection Induced Coal-Gas Interactions. Paper ARMA 09-099 presented at the 43rd US Rock Mechanics Symposium and 4th US-Canada Rock Mechanics Symposium, Asheville, NC, 26 June to 1 July.

Energy Information Administration. 2017. United Statutes Government Publication. https://www.eia.gov/dnav/ng/ng_prod_coalbed_s1_a.htm

Gonzalez, R., Schepers, K., Riestenberg, D., Koperna, G., Oudinot, A. 2008. Assessment of the Potential and Economic Performance of ECBM Recovery and $CO_2$ Sequestration. Presented at the SPE Latin American and Caribbean Petroleum Engineering Conference, Cartagena, Colombia, 31 May-3 June. SPE 121157.

International Energy Agency (IEA). 2013. $CO_2$ Pipeline Infrastructure. Report 2013/18. December.

Iwasaki, S., Kamijo, T., Takashina, T. and Tanaka, H. 2004. Large Scale Flue Gas $CO_2$ Recovering /$CO_2$ Cost and Total System for $CO_2$ Enhanced Oil Recovery. Mitsubishi Heavy Industries, Ltd. Technical Review, Vol. 41, No. 4, August.

Law, D. H.-S., van der Meer, L. G. H. and Gunter, W. D. 2002. Numerical Simulator Comparison Study for Enhanced Coalbed Methane Recovery Processes, Part I: Pure Carbon Dioxide Injection. Presented at the SPE Gas Technology Symposium, Calgary, Alberta, Canada, 30 April – 2 May. SPE 75669.

Luckow, P., Stanton, E. A., Fields, S., Biewald, Jackson, S., Fisher, J. and Wilson, R. 2015. 2015 Carbon Dioxide Price Forecast. Synapse Energy Economics, Inc., Cambridge, Massachusetts, USA.

Mazumder, S. and Wolf, K. H. 2008. Differential Swelling and Permeability Change of Coal in Response to $CO_2$ Injection for ECBM. Presented at the SPE Asia Pacific Oil & Gas Conference and Exhibition, Perth, Australia, 20-22 October. SPE 114791.

Pagnier, H. J. M., van Bergen, F., Kreft, E., van der Meer, L. G. H. and Simmelink, H. J. 2005, Field Experiment of ECBM-$CO_2$ in the Upper Silesian Basin of Poland. Presented at the SPE Europec/EAGE Annual Conference, Madrid, Spain, 13-16 June. SPE 94079.

Puri, R. and Yee, D. 1990. Enhanced Coalbed Methane Recovery. Presented at the 65th Annual Conference and Exhibition of SPE, New Orleans, LA, 23-26 September. SPE 20732.

Rogers, R., Ramarthy, M., Rodvelt, G. and Mullen, M. 2007. *Coalbed Methane: Principles and Practices*. Third Edition, Oktibbeha Publishing Co.
Salmachi, A., Haghighi, M., Bedrikovetsky, P., and Xu, C. 2011. Thermal Gas Recovery from Coal Seam Gas Reservoirs Using Underground Coal Gasification. Presented at the SPE Europec/EAGE Annual Conference and Exhibition, Vienna, Austria, 23-26 May. SPE 140772.
Stevens, S. H., Spector, D., Riemer, P. 1998, Enhanced Coalbed Methane Recovery Using $CO_2$ Injection. Presented at the SPE International Conference and Exhibition, Beijing, China, 2-5 November. SPE 4881.
Van Hemert, P., Wolf, K-H. A. A., and Bruining, J. 2007. The Intrinsic Reliability of Manometric Sorption Apparatus Using Supercritical Carbon Dioxide. Presented at the SPE Annual Technical Conference and Exhibition, Anaheim, California, USA, 11-14 November. SPE 110497.
Zhang, L., Aziz. N., Ren, T. and Wang, Z. 2011. Influence of Temperature on the Gas Content of Coal and Sorption Modelling. *Procedia Engineering*. 26, 1430-1439.

Table 1 – Simulation results

| Case | 1 | 2 | 3 | 4 | 5 | 6 | 7 |
|---|---|---|---|---|---|---|---|
| Recovery mechanism | Depletion drive at 45°C | $CO_2$ injection at 45°C | $CO_2$ injection at 55°C | $CO_2$ injection at 55°C | $CO_2$ injection at 55°C | Depletion drive at 55°C | Depletion drive at 55°C |
| Initial reservoir temperature, °C | 45 | 45 | 45 | 45 | 45 | 55 | 55 |
| Initial reservoir pressure, MPa | 7.64 | 7.64 | 7.64 | 7.64 | 7.64 | 7.64 | 7.64 |
| Injected $CO_2$ temperature, °C | NA | 45 | 55 | 55 | 55 | NA | NA |
| $CO_2$ Langmuir Pressure, MPa | NA | 1.903 | 2.100 | 1.903 | 1.903 | NA | NA |
| $CO_2$ Langmuir Volume, $m^3/t$ | NA | 24.0 | 21.6 | 21.6 | 19.2 | NA | NA |
| $CH_4$ Langmuir Pressure, MPa | 4.689 | 4.689 | 4.689 | 4.689 | 4.689 | 4.689 | 4.689 |
| $CH_4$ Langmuir Volume, $m^3/t$ | 11.8 | 11.8 | 9.44 | 9.44 | 10.62 | 9.44 | 10.62 |
| Cummulative gas production, MMscf | 2.11 | 3.50 | 4.03 | 3.97 | 4.01 | 2.79 | 2.42 |
| Gas recovery factor, % | 43.8 | 72.6 | 83.5 | 82.4 | 83.2 | 57.9 | 50.2 |

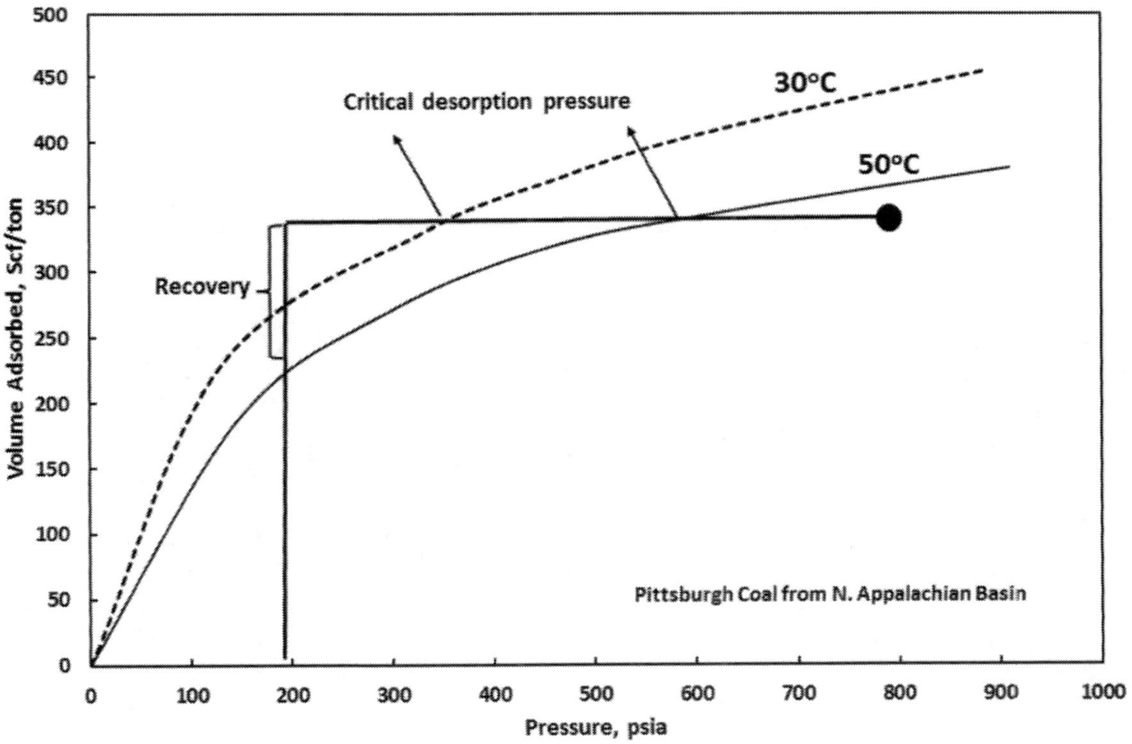


Figure 1 – Effect of temperature on methane sorption isotherms (data from Rogers *et al.* 2007)

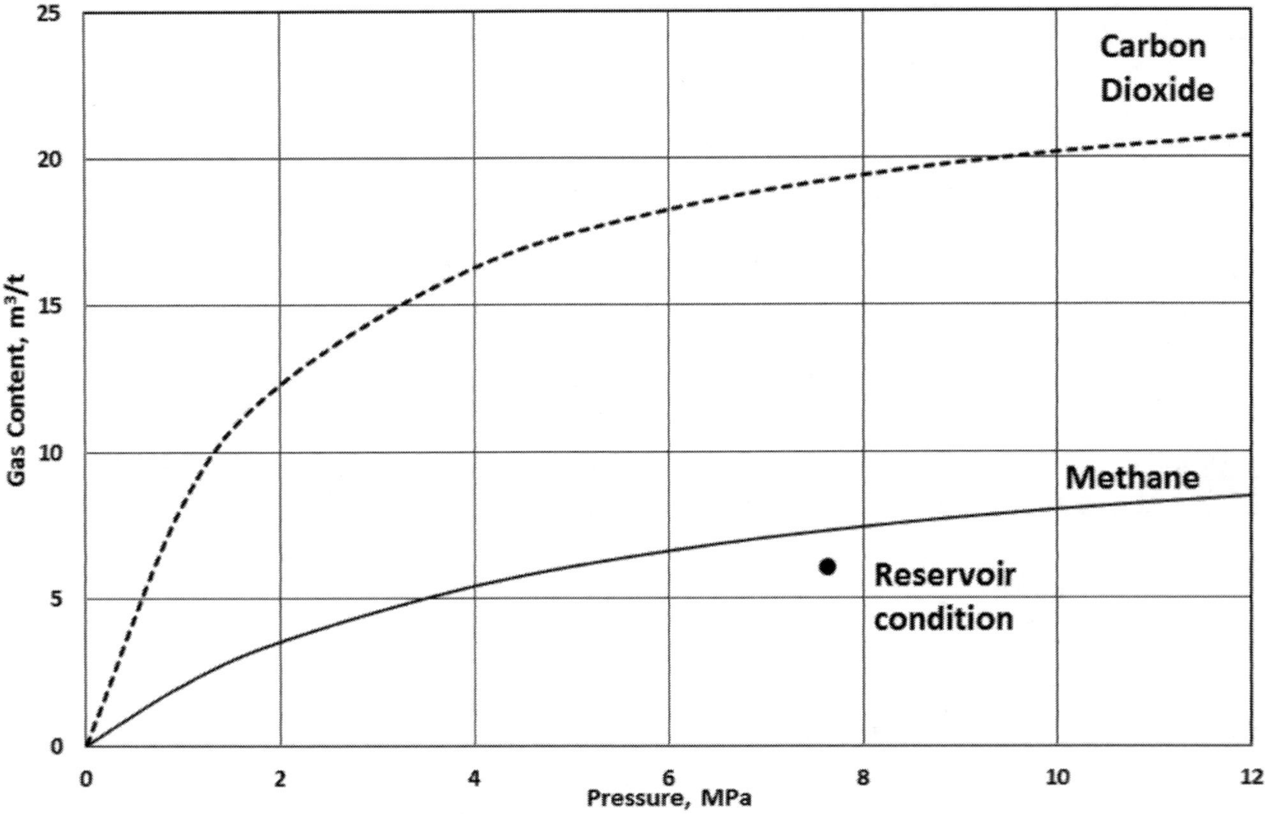


Figure 2 – Langmuir isotherm for methane and carbon dioxide at 45°C and initial reservoir condition

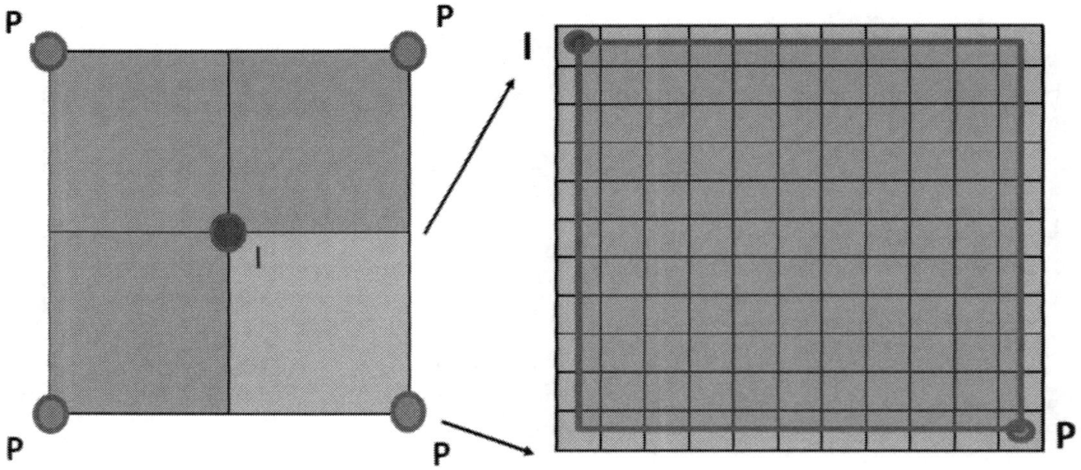


Figure 3 – Schematic of an inverted five spot well pattern

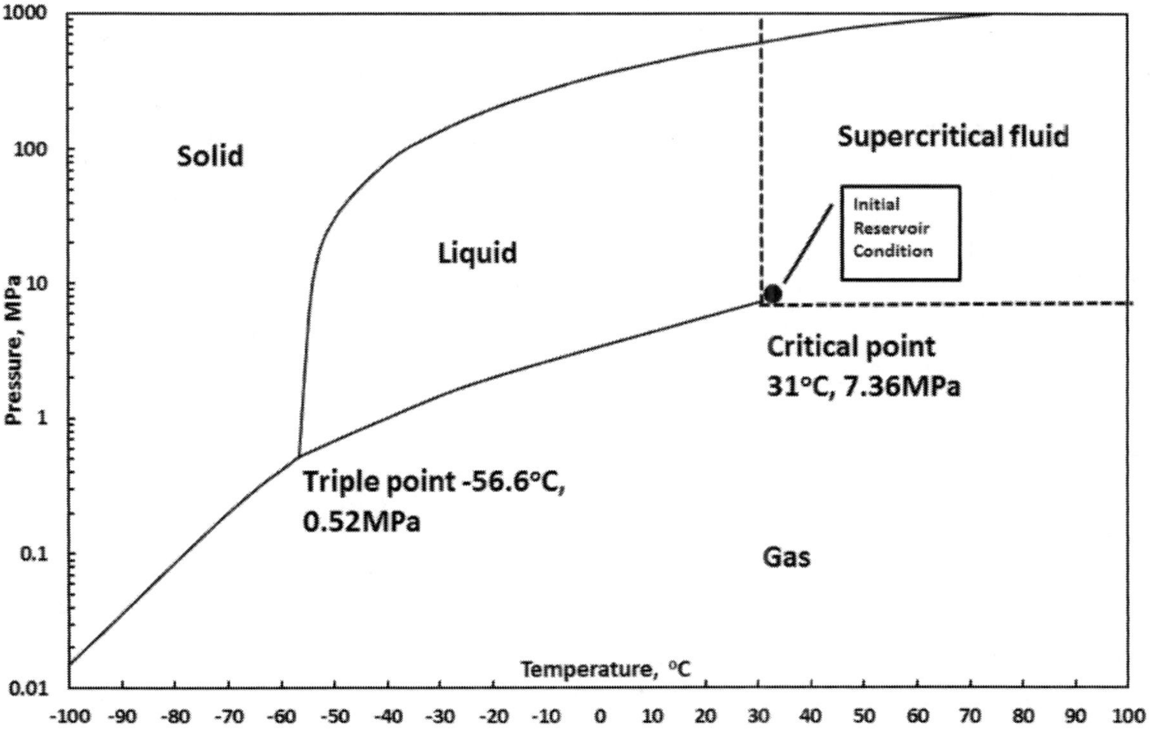


Figure 4 – $CO_2$ phase diagram and initial reservoir condition

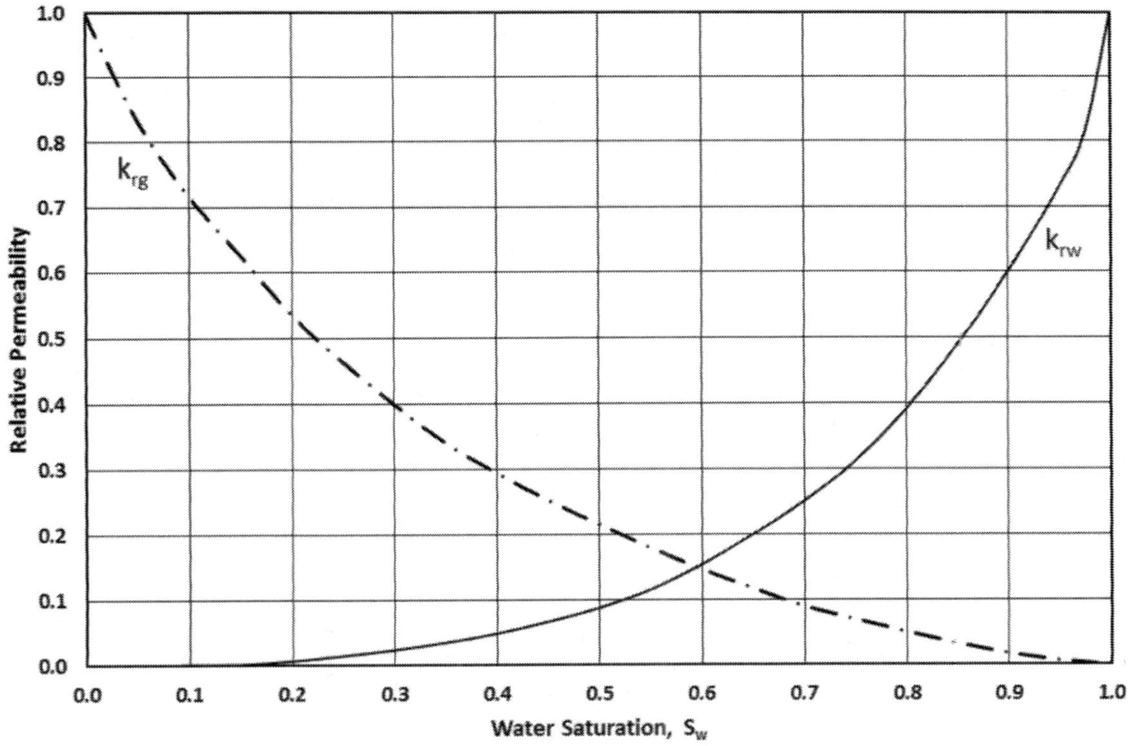


Figure 5 – Relative permeability curves

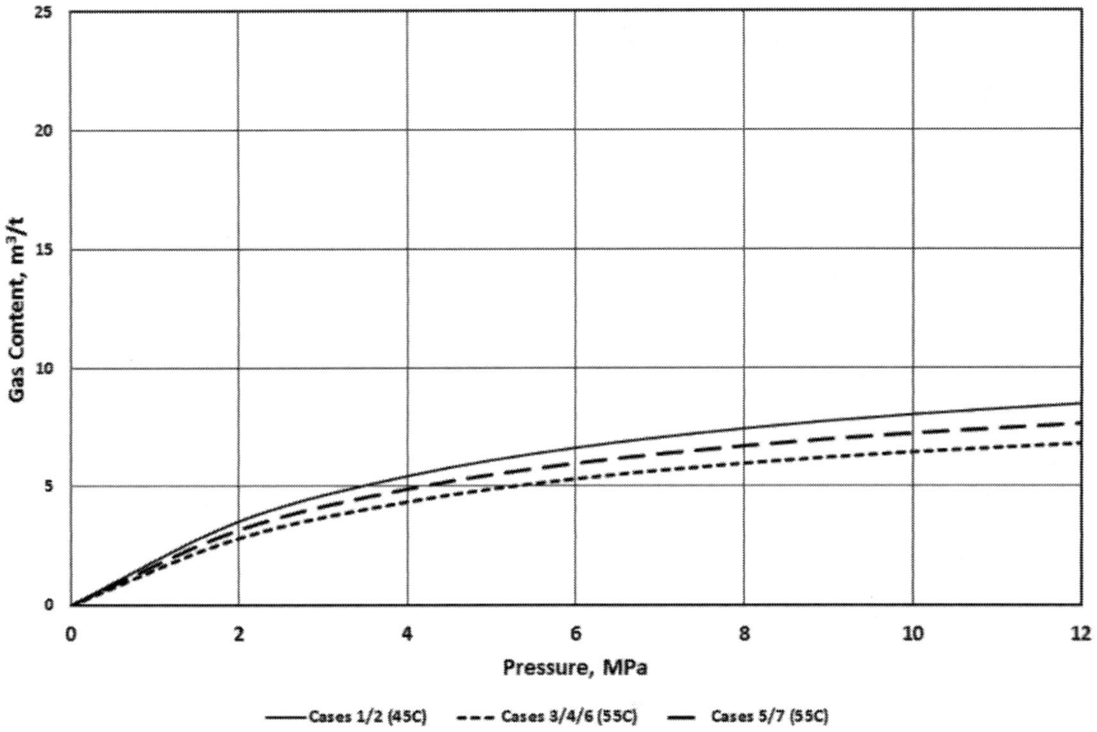


Figure 6 – $CH_4$ Langmuir isotherms for different simulation cases

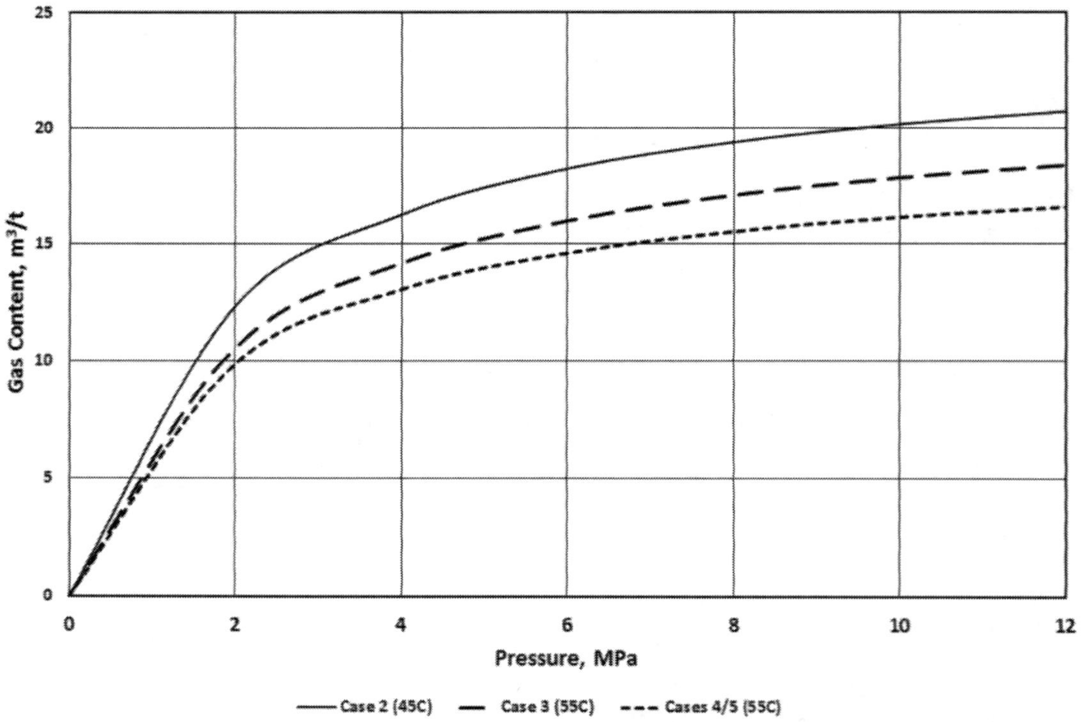


Figure 7 – $CO_2$ Langmuir isotherms for different simulation cases

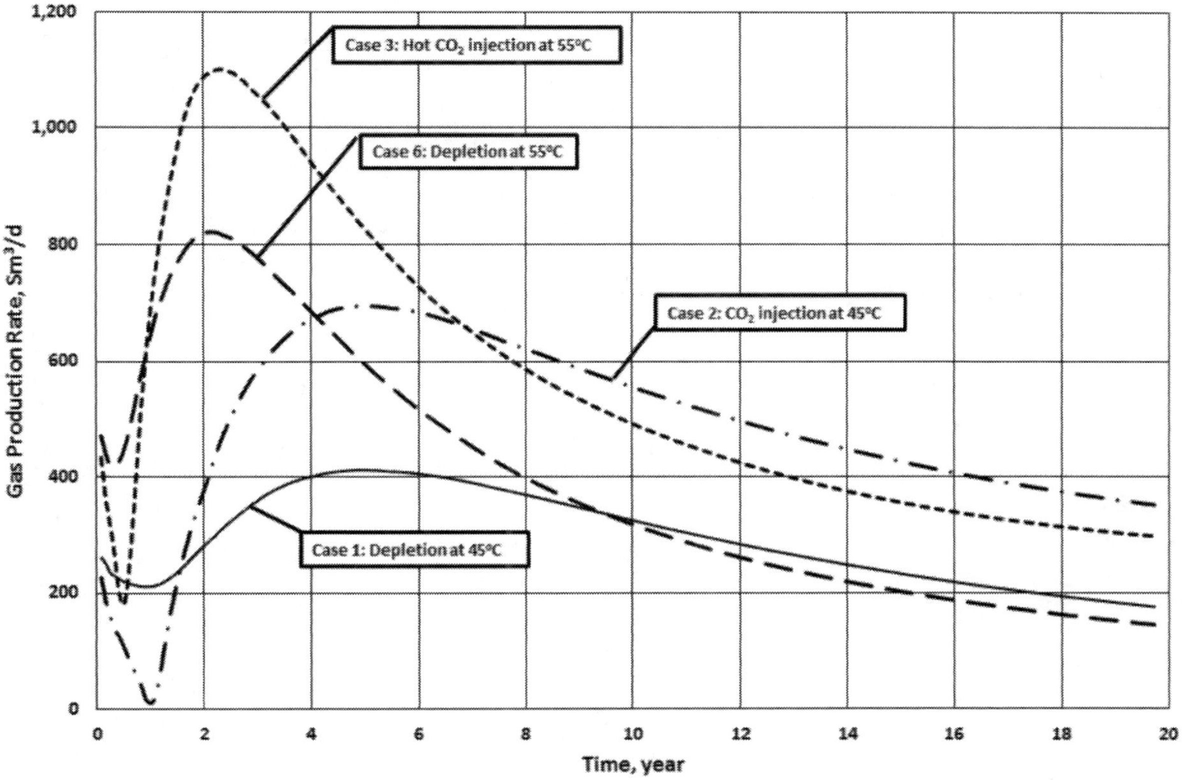


Figure 8 – Gas production rate from simulations

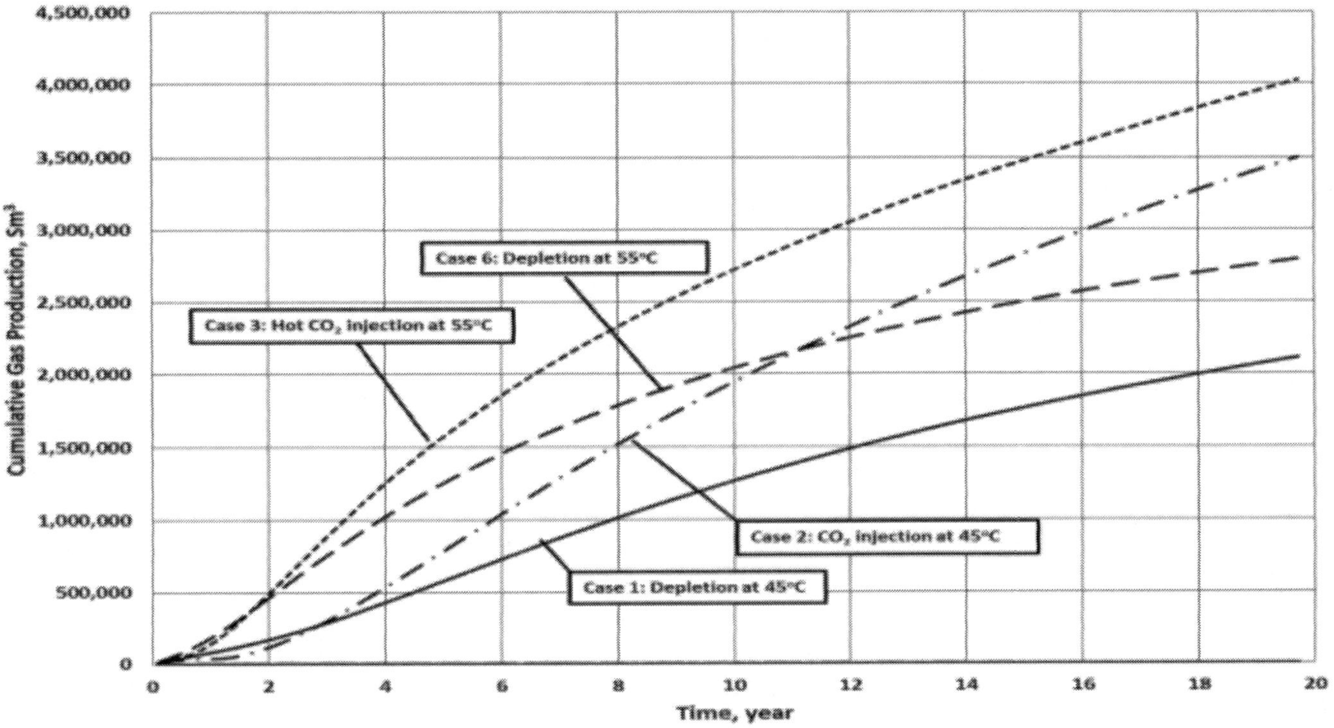


Figure 9 – Cumulative gas production from simulations

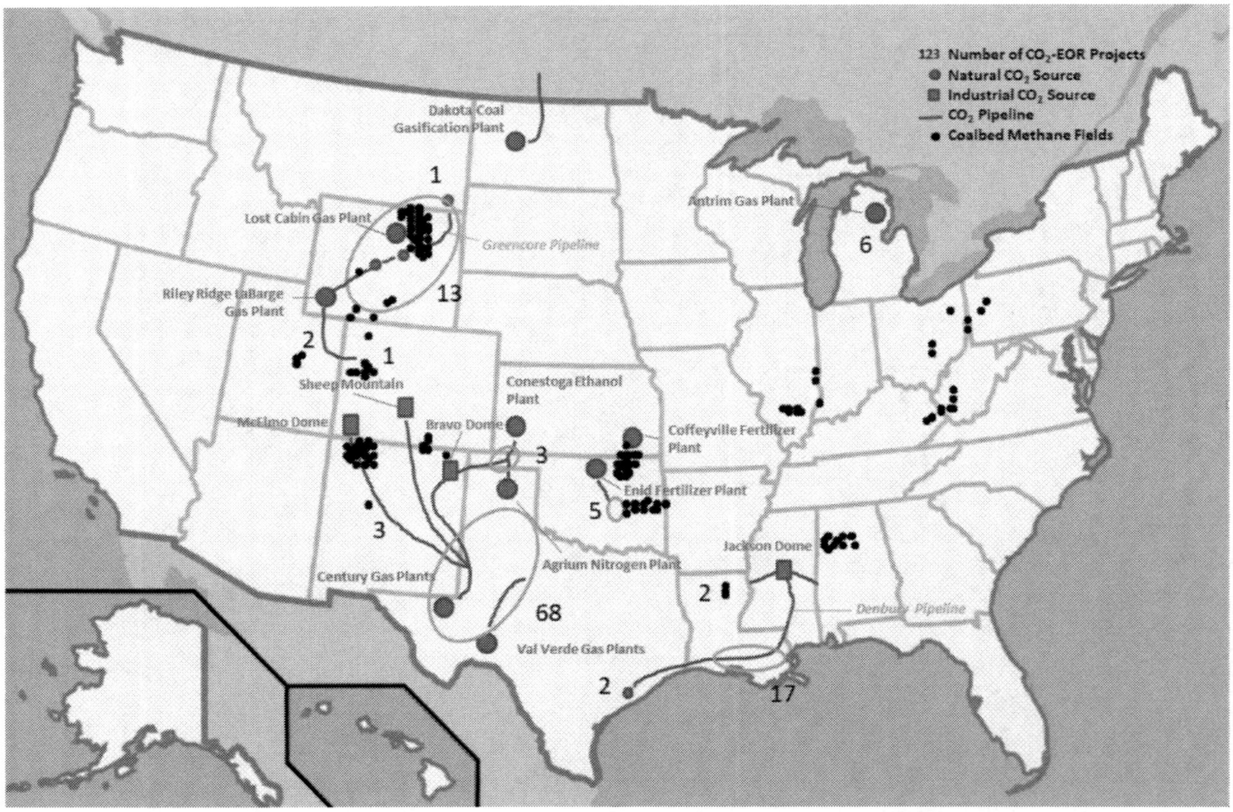


Figure 10 – Location of CBM fields, $CO_2$ sources and $CO_2$ pipelines in the US

OIL AND GAS CLIMATE INITIATIVE

# More energy, lower emissions

Catalyzing practical action on climate change

July 19, 2017

CMTC 2017 Houston, Texas

OIL AND GAS CLIMATE INITIATIVE

## OGCI members

**Current members** of the initiative are BP, CNPC, Eni, PEMEX, Reliance Industries, Repsol, Saudi Aramco, Shell, Statoil and TOTAL.

**OGCI members represent** diverse national and international companies.

**OGCI members produce** over one-fifth of global oil and gas production and over 10% of energy supply.

bp CNPC eni PEMEX Reliance Industries Limited REPSOL أرامكو السعودية saudi aramco Statoil TOTAL





## OGCI Communications

2015 OGCI Report

2016 Web site

2016 OGCI Report

bp CNPC eni PEMEX Reliance Industries Limited REPSOL أرامكو السعودية saudi aramco Statoil TOTAL

OIL AND GAS CLIMATE INITIATIVE

OIL AND GAS CLIMATE INITIATIVE

## OGCI work streams

OGCI has been focusing on three key areas:

**Natural gas / methane**

**Low emission opportunities**

**Long term solutions**

June 2017

6

OIL AND GAS CLIMATE INITIATIVE

## Long Term Solution - CCS work programs

**Pathways to CCS market commercialization in a low carbon economy:**

Aiming at developing a common position on the market mechanisms for CCS commercialization and to determine how Oil and Gas companies could act to facilitate CCS development at scale.

**Availability of $CO_2$ storage capacity in key markets:**

Aiming at having a storage capacity methodology mutually agreed between OGCI companies and externally, and then collaborate with local institutes of targeted countries to trigger adequate and robust determination of underground $CO_2$ storage capacity.

June 2017

OIL AND GAS CLIMATE INITIATIVE

## Pathways to CCS market commercialization in a low carbon economy

Objectives:

**Identify** key qualifiers and milestones which will facilitate commercial CCS deployment.

**Define** oil and gas company contributions towards pathways to CCS commercialization that will enable CCS on gas and oil facilities.

**Develop** a clear understanding of the different elements of CCS cost reduction and whether/how oil and gas company collaboration can deliver against these elements.

June 2017

OIL AND GAS CLIMATE INITIATIVE

## Availability of $CO_2$ storage capacity in key markets

Objectives:

**To define a methodology** for reporting of $CO_2$ storage reserves/resources.

**Identify and collaborate with regional parties** in key markets such as China and India to undertake storage capacity assessment.

**Update storage capacity estimates** based on our agreed methodology for key regions of the world.

June 2017

## Path forward

June 2017

OIL AND GAS CLIMATE INITIATIVE

## OGCI collaborations

| International Energy Association (IEA) | Global CCS Institute (GCCSI) | Carbon Capture and Storage Association (CCSA) | Society of Petroleum Engineers (SPE) |
|---|---|---|---|

June 2017

11

# The Oil and Gas Climate Initiative Climate Investments

OGCI CLIMATE INVESTMENTS

OIL AND GAS CLIMATE INITIATIVE

# Reducing greenhouse gas emissions possible actions

# Focus Areas

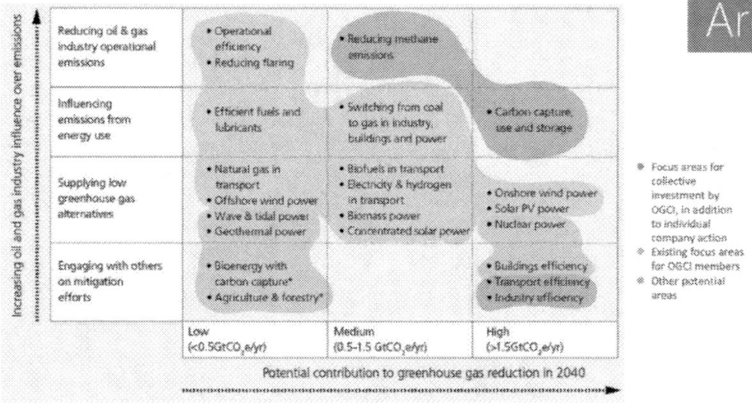


Our strategic assessment explored the actions identified by the IEA and others as having the potential to reduce greenhouse gas emissions sufficiently by 2040 to remain on track for a 2°C scenario. We mapped these actions on a matrix showing the emissions reduction potential (horizontal) and the oil and industry's ability to influence the activity (vertical). Actions located in the upper right-hand half of the most relevant, with some more appropriate for individual companies and some for collaborative OGCI initiatives.

**Source: OGCI** * These actions could have a more significant impact even before 2040, but will both be crucial elements of emissions reduction in the second half of the century.



# OGCI Climate investments The Multiplier Effect

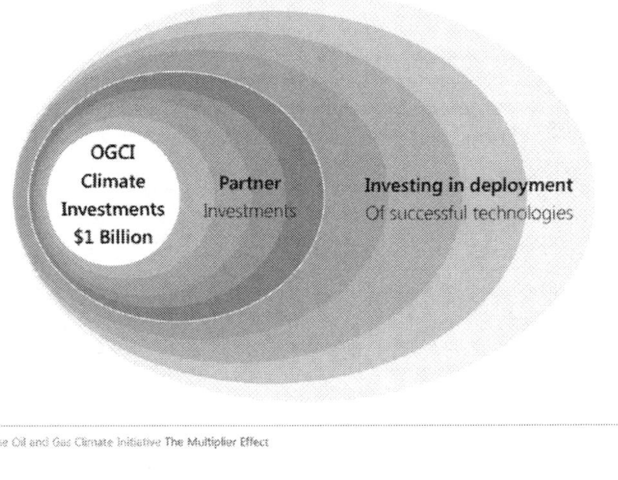


We expect our investments to have a considerable multiplier effect.

This will come from partners investing along OGCI Climate Investments, as well as our own and other companies' investments in deploying commercialized technologies.

The overall impact on reducing emissions will be on the scale of a gigatonne or more over the decade.

OIL AND GAS CLIMATE INITIATIVE

June 2017

CARBON MANAGEMENT TECHNOLOGY CONFERENCE

CMTC-485740-MS

# Performance Evaluation for Carbon Sequestration in Shale Gas Reservoir Systems

Langtao Zhu, China University of Petroleum (Beijing), Xinwei Liao, China University of Petroleum (Beijing), Zhiming Chen, China University of Petroleum (Beijing), Xuyang Cheng, China University of Petroleum (Beijing)



This paper was prepared for presentation at the Carbon Management Technology Conference held in Houston, Texas, USA, 17-20 July 2017.



## Abstract

Storage $CO_2$ in shale formation is considered as a promising option to reduce $CO_2$ emissions and enhance shale gas recovery. Many simple analytical and semi-analytical techniques have been proposed to support screening analysis and performance assessment for potential carbon sequestration sites. However, these analytical techniques have ignored the effect of fracture occurrences, which are important to the carbon sequestration in shale formation.

In this paper, numerical simulation technology is applied to model different complex hydraulic fracture occurrences and evaluate the $CO_2$ seepage rule during carbon sequestration. First, multi-component Langmuir isotherm is applied to simulate the adsorption desorption phenomenon of $CO_2$. Combining Langmuir isotherm with the reservoir parameters of Barnett shale formation, shale reservoir simulation model is constructed to simulate the $CO_2$ seepage law during the $CO_2$ sequestration. Then, based on the shale reservoir simulation model, local grid refinement technology is used to characterize the five typical fracture occurrences caused by hydraulic fracturing. Subsequently, the numerical simulation cases with five different fracture occurrences are run to evaluate the performance during $CO_2$ storage process. Finally, some critical important parameters, including engineering and geologic parameters, are evaluated through the sensitivity study. In order to identify and illustrate the performance which progressively occur over time, the log-log plot of $CO_2$ injection rate versus time is adopted.

The simulation results indicate that the flow regimes of $CO_2$ injection can be classified into eight flow regions. Among these flow regions, the inner boundary dominated flow regime is a unique one which can occur only when realizing strongly stimulated reservoir volume. Further, the sensitivity study indicates that the fracture topologies and the primary fracture conductivity are the two key factors dominating the early-time flow behavior during the carbon sequestration. And the final-time flow behavior is dominated by the shale reservoir parameters.

This work systematically analyzes the effect of fracture topologies on carbon sequestration and enlarges our knowledge of $CO_2$ storage in shale gas systems.

## 1. Introduction

Shale gas has become an increasingly important source of natural gas ($CH_4$) in the United States over the last decade. Due to its unconventional characteristics, injecting carbon dioxide ($CO_2$) to enhance shale gas recovery ($CO_2$-EGR) is a potentially feasible method to increase gas-yield while realizing $CO_2$ sequestration (CS). Therefore, it is more and more important to quantitatively demonstrate the influence of various reservoir and completion parameters on performance of multiply-hydraulic fractured horizontal wells (MFHW) in the process of $CO_2$ injection.

Ultra-tight shale reservoirs present numerous challenges to modeling and understanding. These reservoirs typically require hydraulic fracturing, which will increase the productivity and create complex fracture profiles. Additionally, $CO_2$ has a much higher sorption affinity to kerogen than $CH_4$, which will induce strongly multi-component competitive adsorption phenomenon. This phenomenon will remarkably affect the flow event and result in an unintuitive (difficult-to-model) pressure profile behavior.

According to single-component and multi-component competitive adsorption isotherm based on the Langmuir isotherm, various analytical (Bumb and Mckee, 1988), semi-analytical, and numerical models have been proposed to characterize rate and pressure behavior versus time in shale gas systems featuring horizontal wells with multi-cluster staged hydraulic fractures:

1) Analytical and semi-analytical models. At present, the analytical and semi-analytical models can only describe the adsorption phenomenon of single-component based on the Langmuir isotherm. And the fundamental methods are Stehfest function and the point source method proposed by Ozkan (Brown et al., 2011; Stehfest, 1970). Applying these methods, a lot of analytical and semi-analytical empirical models were proposed to evaluate the production data. In these models, MFHWs with symmetric or asymmetric bi-wing fractures (Fig.4-a) were initiatively proposed to reveal the seepage rule based on the production performance(Liu et al., 2015; Ren and Guo, 2015; Sureshjani and Clarkson, 2015; Wang, 2014). Meanwhile, much fracturing imaging strongly shows that complex fracture networks can easily develop along bi-wing fractures for the natural fractures (Fig.4-d)(Stalgorova, 2012; Stalgorova and Mattar, 2013). With the development of the multiply-hydraulic fracturing technology in recent years, complex fracture networks can easily develop along horizontally fractured well, as shown in Fig. 1&Fig.4-e (Dongyan et al., 2015; Ali et al., 2013; Chen et al., 2015). Although a lot have been achieved in the performance predicts of MFHWs with fracture networks by means of analytical and semi-analytical methods, these methods can't accurately and effectively describe the performance of MFHW for the complexity of fracture network and multi-component competitive adsorption phenomenon.

2) Numerical simulation models. Compared with the analytical and semi-analytical methods, the numerical simulation method is becoming the main method to study the shale gas reservoir because it can model complex fracture networks and simulate multi-component competitive adsorption phenomenons. The local grid refinement method (LGR) is adopted to model the hydraulic fracture to accurately simulate nonlinear flow. Additionally, the extended Langmuir isotherm (EL) is applied to systematically evaluate the potential of $CO_2$-EGR and CS in the shale reservoir (Ruthven, 1984; Ritter and Yang, 1987). Many scholars adopt numerical simulation method to quantitatively demonstrate the influence of fracture topologies (as shown in Fig.4) on performance of fractured vertical or horizontal wells in tight gas and shale gas reservoir systems (Godec et al., 2013; Bacon et al., 2015; Freeman et al., 2013). And they further study the influence of multiply-hydraulic fracturing (Fig. 1) on performance of MFHW (Jones et al., 2013). Despite a lot of published numerical studies about the effect of fracture topologies, there are few people synthetically study the flow behavior of injecting $CO_2$ through MFHW.

In this paper, the EL isotherm is modeled to simulate the competitive adsorption phenomenon between $CO_2$ and $CH_4$. And we further systematically evaluate the influence of various reservoir and completion parameters on the $CO_2$ injection procedure:(1) A dual-porosity model is incorporated in the reservoir simulation model. (2) Local grid refinement (LGR) is adopted in the simulation runs to improve the simulation accuracy of hydraulic fractures. (3) Base on the topology of SRV (Fig.1 & Fig.4-e), $CO_2$ injection rate behavior analysis is conducted to confirm the seepage rule of $CO_2$ injection in shale gas reservoir systems. (4) Sensitivity analysis is conducted to confirm the effect of various geologic, engineering, and operating parameters on the performance of MFHW.

This work presents a novel perspective on investigating the flow behavior of $CO_2$ injection in shale gas system.

## 2. Description of numerical model

To assess the flow event of the $CO_2$ injection process in shale gas reservoir, a set of numerical simulation studies run to simulate the process of $CO_2$ injection. Barnett shale is selected as the representative shale gas reservoir and the correlation parameters are presented in Table.1 (seen in Appendices). And a model for multi-component transport in dual-porosity sorbing media is constructed. The following introduces key models describing porosity and permeability evolution in both matrix and fracture networks including the coupling between these two media.

2.1. Field and constitutive equation

Field and constitutive equations for gas flow and transport in shale are defined (Eshkalak et al., 2014). These equations are coupled through porosity and permeability evolution driven by Langmuir sorption. Generally, the multi-component fluid flow of gas in a dual-porosity model is considered in constructing the geologic model of shale gas reservoirs based on a modified Warren and Root's model. In this model, the natural and hydraulic fractures are the flow channel. And the matrix is the main reservoir space. The intercommunication between the inter-granular void spaces is controlled by diffusion law driven by Langmuir sorption. The coupled flow and mass transport processes involving fluid-rock interactions can be written as: (given by Eq.(1)).

$$\begin{aligned}\frac{V}{\Delta t}(N_i^{n+1}-N_i^n) &= \Delta T_g \rho_g y_i \phi (convection)\\ &+ \Delta A D_{ig} \Delta(\rho_g y_{ig})(diffusion + dispersion)\\ &+ q_i (injeciton\ /\ production)\end{aligned} \tag{1}$$

where $N^n$ denotes the moles per unit of grid block volume at the n moment, $V$ is the grid block volume, $\Delta t$ is time step, The first term to the right of the equation is the convection term, $T$ is the transmissibility of gas phase, $\rho$ is the molar density of gas phase, y is the mole fraction in gas phase, $\phi$ is the porosity. The second term to the right is the diffusion and dispersion term, where A is an empirical correction factor of $D$, $D$ is the diffusion and dispersion coefficient. The last term denotes injection or production rate. Subscripts $i$ and $g$ respectively represent the number of component and the gas phase.

The following assumptions apply:

1) The shale reservoir is a homogenous, isotropic and elastic continuum. The system is isothermal.

2) Gas present within the system is ideal.

3) Gas flow through the fractures in shale conforms to Darcy's Law (the water phase is not considered in this study and the original gas component is 100% $CH_4$). Gas transport in the shale matrix is assumed to obey Fick's Law.

4) Gas sorption only occurs within the matrix.

2.2 Competitive adsorption and desorption between $CH_4$ and $CO_2$

Organic matter of shale formation has shown stronger adsorption potential of carbon dioxide ($CO_2$) compared to methane ($CH_4$) according to experimental studies. The gas desorption from or adsorption onto kerogenic media has been studied extensively in coal-bed methane reservoirs. However, the sorption, adsorption, and transport properties of shale are not necessarily analogous to coal (Schettler and Parmely, 1991). Currently the commonly used empirical model for single-component surface sorption is the Langmuir isotherm (given by Eq. (1)) and the Freundlich isotherm. For the multi-component competitive adsorption phenomenon, the extended Langmuir isotherm is proposed (given by Eq. (2))(Ruthven, 1984; Ritter and Yang, 1987b).

$$V_{ads,i}(P) = V_{Li}\frac{y_i P}{P_{Li}(1+\sum_{j=1}^{m} y_j \frac{P}{P_{Lj}})}, i = 1, \ldots\ldots, m, \ldots\ldots \quad (2)$$

Where the $V_L$ term represents the total storage at infinite pressure and the pressure at which half of this volume is stored ($P_L$), $V_{ads}$ is the potential releasable-gas content in scf/ton, $P$ is pore pressure (assumed as the average reservoir pressure), y is hydro carbon mole fraction in gas phase, m is the total number of component in gas phase. Subscripts $i$ and $j$ represent the number of component. Additionally, the extended Langmuir model assumes instantaneous equilibrium of the sorptive surface and the storage in the pore space i.e., there is no transient lag between pressure drop and desorption or adsorption response.

2.3. Permeability model for hydraulic fracture

The multiply-hydraulic fracturing can create complex fracture network which obviously improves the conductivity near horizontal well. In order to improve the simulation accuracy for regions around hydraulic fractures, local grid refinement (LGR) is widely applied to model different fracture occurrences in the simulation runs (Eshkalak et al., 2014b; Yu et al., 2014). The transmissibility of the fracture blocks was adjusted to obey the following relationship (given in Eq.3). In addition, when the fractures intersect, the staggered mesh is equal to the primary fractures not to the induced secondary fractures.

$$k_f^{'} \times \Delta f = k_f \times w_f \quad (3)$$

Where $k_f^{'} \times \Delta f$ is the product of fracture permeability and width (fracture conductivity), and $k_f \times w_f$ is the corresponding product in the simulation model.

2.4 Model implementation and numeric simulation scheme

A series of numerical models have been conducted to quantitatively determine the influence of various reservoir and completion parameters on performance of MFHW in the process of $CO_2$ injection(Kalantari-Dahaghi et al., 2015; Liu et al., 2013). These models follow the transport of binary components in a fractured porous shale reservoir pierced by a MFHW. The horizontal well is an injection well located at the center of a circular section that cuts the reservoir vertically as shown in Fig. 1. The numerical simulation model adopts corner-point grid and the simulation area approximates circle. And the pressure of bottom hole is 4MPa higher than the original reservoir pressure.

General, the shale gas reservoir easily realize strongly SRV because of natural fractures and the fragile shale. Therefore, we selected the fracture topology of SRV as the base case, given as Case A1 in Table 2, to figure out the flow regimes of $CO_2$ injection process. The specific parameters of base case are presented in Table 1. Further, based on the base case, sensitivity studies were conducted to synthetically evaluate the effect of various reservoir and completion parameters on performance of $CO_2$ injection process. The numerical simulation schemes of sensitivity studies run as follows:

1) In the first set of simulation models, the LGR method is employed to model five kinds of fracture topologies (as shown in Fig.1 and Fig.4). We conduct reservoir numerical simulation to analyze the effect of the fracture topologies on the behavior of $CO_2$ injection process.

2) Multiply-hydraulic fracturing can easily obtain stimulated reservoir volume(SRV)(Mayerhofer et al., 2008). Based on the topology of SRV, the second set of grids evaluates the effects of primary fractures and secondary fractures (as shown in Fig.1&Fig4.e) on performance of $CO_2$ injection process. Additionally, we further evaluate the effect of the width of fractured region.

3) The third set examines the effects of various reservoir parameters, including the pressure and permeability of shale reservoir, on performance of multiply-fractured horizontal wells in $CO_2$ injection process.

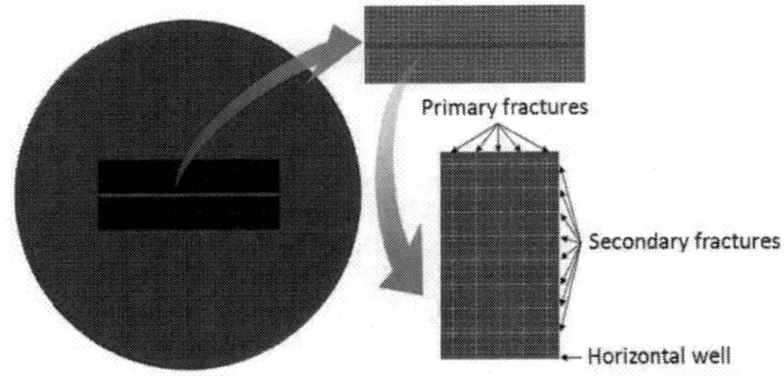


Fig. 1 Top view of SRV fracturing simulation model—Base case

## 3. Results and analysis

In order to intuitively analyze the effect of various parameters on the performance of $CO_2$ injection processes, all results are presented in dimensional form. The log-log plot of $CO_2$ injection rate versus time (LOG-LOG) curve is adopted to identify and illustrate the various flow regimes. This kind of plot can intuitively reveal the different flow regimes and clearly demonstrate the effect of completion and reservoir parameters on the $CO_2$ injection process.

### 3.1 Base case results

The Fig.2 presents the LOG-LOG curve obtained from the simulation results of the base case corresponding to case A1 in Table 2. Compared with the result of the triple-porosity model which interprets the flow event of natural gas (100%CH4) in shale reservoir (Al-Ahmadi and Wattenbarger, 2011), the flow regimes are largely consistent. This indicates that the numerical simulation result is reliable. According to the numerical simulation results of the base case A1, eight flow regions can be identified while the $CO_2$ seepages in the fractured system (as shown in Fig.2-b).

1) Region 1 (IAF): The initial $CO_2$ adsorption period. During this period, the initial $CO_2$ injection behaviors are mainly affected by the permeability of hydraulic fractures and the adsorption capability of different component. The $CO_2$ injection rate remains almost unchanged in a special short time.

2) Region 2 (FLF): The linear flow period in the hydraulic primary fractures, during which the LOG-LOG curve appears with a constant slope (as shown in Fig.2-b). In this period, the gas flow in the reservoirs mainly appears as a linear flow parallel to the surfaces of the hydraulic primary fractures (as shown in Fig.3-a), and hydraulic primary fractures have no effect on each other. Due to the competitive adsorption between $CO_2$ and $CH_4$, the LOG-LOG curve did not appear an apparent slope of -1/2 which is obvious in triple-porosity model (as shown in Fig.2-a).

3) Region 3 (FBF): First bilinear flow period, during which the LOG-LOG curve keeps a constant slope of -1/4. During this period, a linear flow occurs in each primary fracture, and the gas flow in the secondary fracture appears as a linear flow perpendicular to the surfaces of the hydraulic primary fractures at the same time (as shown in Fig.3-b). This region corresponds to the region 2 proposed by Al-Ahmadi (as shown in Fig.2-a)(Al-Ahmadi and Wattenbarger, 2011).

4) Region 4 (SLF): The linear flow in the secondary fractures, it will occur once the transient flow in the primary fractures and secondary fractures ends indicating the end of bilinear flow. For the competitive adsorption between $CO_2$ and $CH_4$, this flow region also does not exhibit an obvious negative half-slope on the LOG-LOG curve which is different from the region 3 of the triple-porosity model (as shown in Fig.2-a).

5) Region 5 (SBF): The second bilinear flow in both secondary fractures and natural fractures, as shown in Fig.3-c, it is caused by the linear flow in the natural fractures while the secondary fractures are still in transient flow. This flow region exhibits a negative quarter-slope on the LOG-LOG curve, as shown in Fig.3-c, which is similar to the transient flow in region 4 (as shown in Fig.2-a).

6) Region 6 (IBDF): The inner boundary dominated flow. This transient flow regime will only occur when the conductivity ratio between fractured region and shale reservoir is high. The high conductivity ratio will lead to most of the injected $CO_2$ lagging in fractured region, as shown in Fig.3-d. The $CO_2$ injection rate will exhibit an exponential decline due to constant bottom-hole pressure and the significant increase of inner fractured region. This flow regime has been proposed by many scholars (Fikri Kuchuk., 2014; Zhou et al., 2012).

7) Region 7 (TLF): The linear flow from fractured region to the un-stimulated shale gas reservoir. However, this transient flow regime only can occur when the length of fractured region is much bigger than the width of that. Most of injected $CO_2$ will flow to shale gas reservoir perpendicular to the horizontal well, as shown in Fig.3-e. This region exhibits a negative half-slope on the LOG-LOG curve, and it is similar to the region 5(as shown in Fig.2-a). This flow region also has been proposed by Ali (Chen et al., 2015).

8) Region 8 (OBDF): The outer boundary dominated flow, it starts when the pressure transmits to the outer boundary. This flow is governed by exponential decline due to constant bottom-hole pressure, as shown in Fig.3-f.

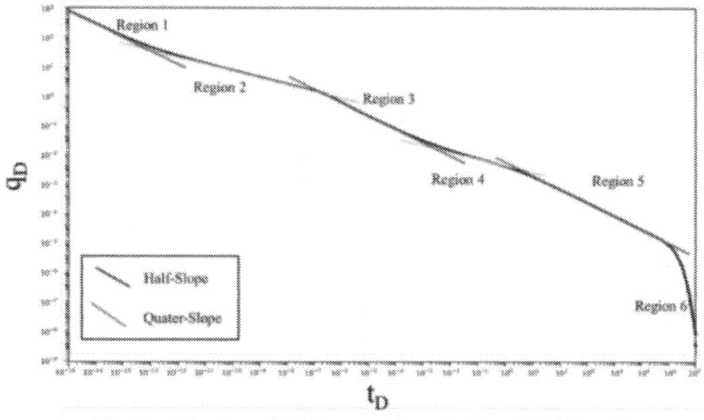


a. The log-log plot of triple-porosity solution (Al-Ahmadi).

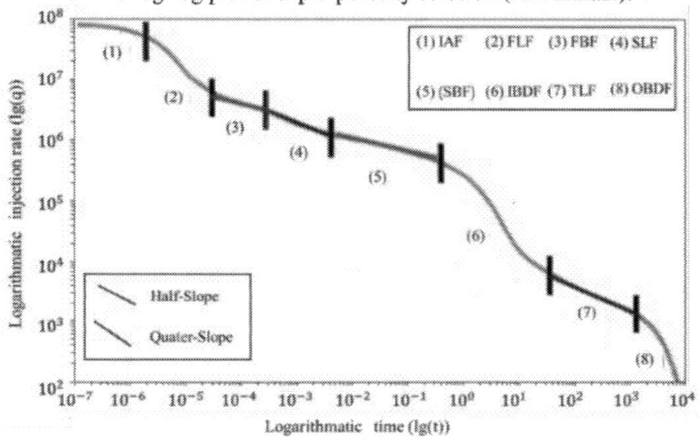


b. The log-log plot of numerical simulation results.

Fig. 2 The comparison between analytical and numerical simulation results

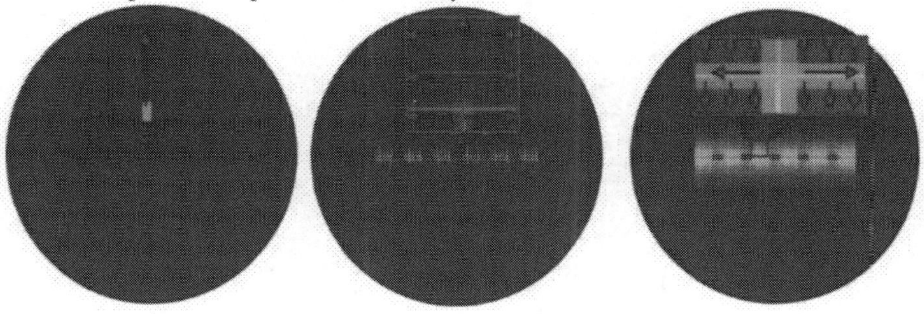

a. Region 2 b. Region 3 c. Region 5

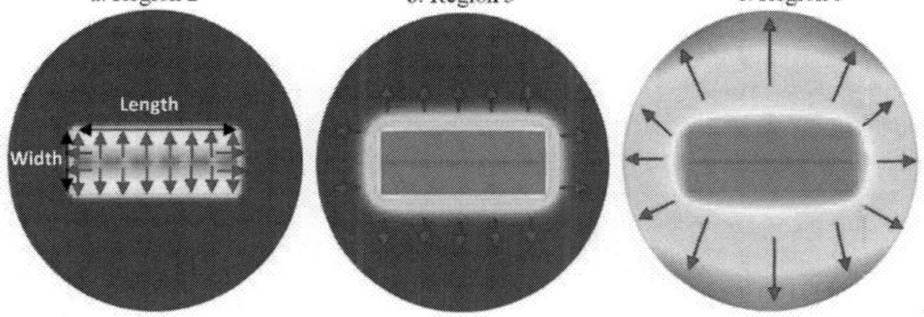


d. Region 6 e. Region 7 f. Region 8

Fig. 3 Flow regimes of a HWFN in numerical simulator—Base case

### 3.2 Effect of fractured region

#### 3.2 .1 Effect of fracture topology

In order to further analyze the influence of various fracture topologies on performance of MFHW in the process of $CO_2$ injection, the method of LGR is applied to model the hydraulic fractures to obtain another four kinds of fracture topologies, as shown in Fig.4. The numerical simulation results are shown in Fig.5, and it indicates that fracture topologies mainly affect the flow regimes at early time.

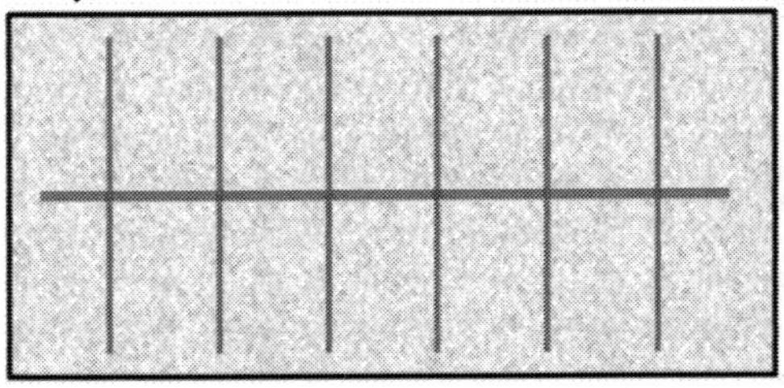
a. Model 1: Bi-wing planar

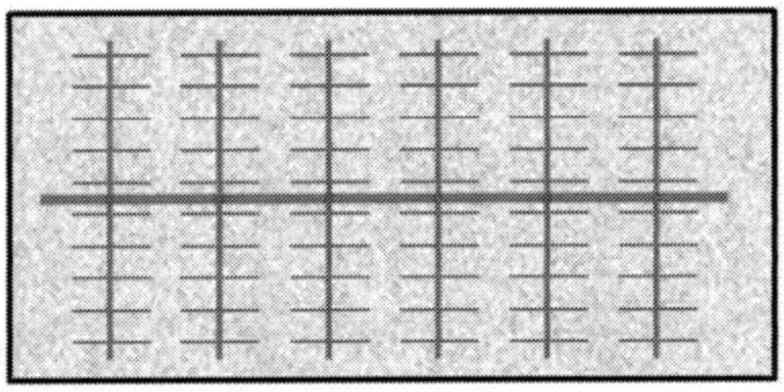
b. Model 2: Planar-decoupled

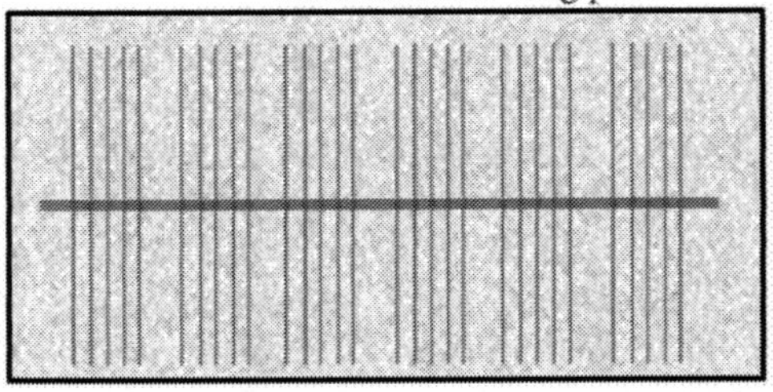
c. Model 3: Complex Planar (NO communication)

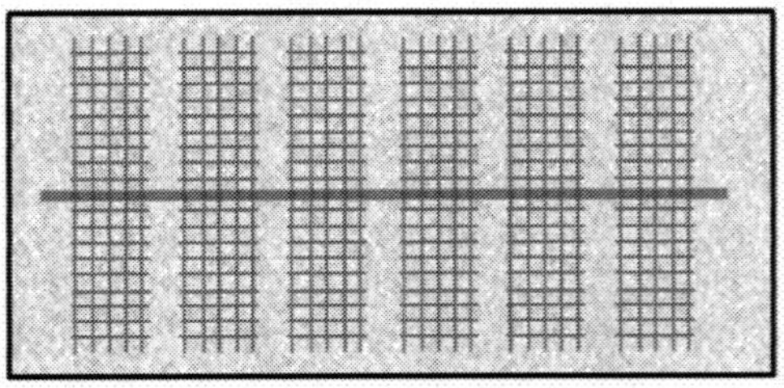
d. Model 4: Complex Planar (Communication)

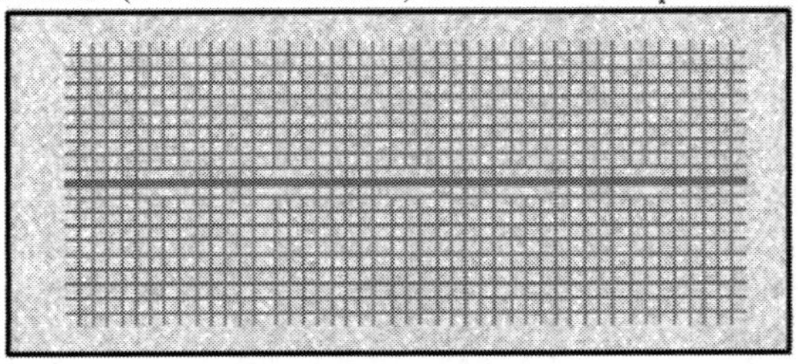
e. Model 5: Network (Base case)

Fig.4 Top view of fracture growth and complexity scenarios

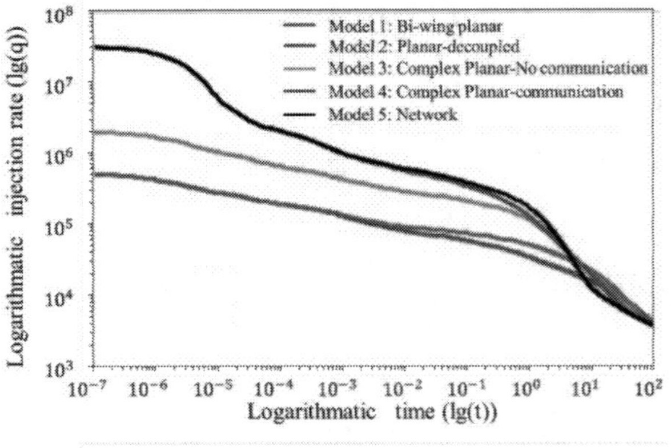


Fig.5 The effect of fracture topologies.

Comparing the LOG-LOG curves of different fracture topologies, the obvious inner boundary dominated region only occur when numerical simulation model is Model 4 or Model 5, as shown in Fig.5, which is notably different from the other three models. For the reason of failing to form complex fracture network (volume fracturing), the other three kinds of fracture topologies can't occur region 4, region 5, and region 6 (as shown in Fig 2-b). We also can figure out that the inner boundary dominated regime reacts more obviously with the increase of stimulated volume. The five LOG-LOG plots will eventually merge in the region 7 at the 100$^{th}$ day (the linear flow from fractured region to shale gas reservoir), such phenomenon indicates that the hydraulic fracturing can only affect the initial $CO_2$ injection capacity and reservoir physical properties are the main factor controlling injection capacity.

### 3.2.2 Effect of fracture conductivity

Both natural fractures and fragile shale are unique features of shale reservoirs, and multi-cluster staged hydraulic fracturing can easily create complex fracture network (as shown in Fig.1). Unfortunately, proppant transport cannot be reliably modeled when fracture growth is complex. At present, the distribution of proppant can be classified into three scenarios (Fig.6): 1) If the proppant is confined within a dominant primary fracture (Fig.6-a), this would result in higher conductivity in primary fractures. 2) When the proppant is excessively concentrated in a dominant primary fracture and evenly distributed throughout the secondary fractures (Fig.6-b), the permeability of primary fractures is lower than that of the secondary fractures. 3) Recent studies have illustrated that if the proppant is evenly distributed throughout the whole complex fracture network (Fog.6-c), this would make the permeability of primary fracture equal to that of the secondary fractures. Numerical simulation models were constructed to extensively study the conductivity of propped fractures. In all numerical simulation cases involving primary and secondary fractures, we treat the fracture as possessing a fixed dimensionless conductivity as described in Eq. (4) and Eq. (5) in order to compare the conductivity with values obtained using other models, defined as follows:

$$F_{cd1} = (k_{f1} \times w_{f1}) / (k_f \times x_f) \quad (4)$$

$$F_{cd2} = (k_{f2} \times w_{f2}) / (k_f \times x_f) \quad (5)$$

Where $F_{cd}$ is the dimensionless conductivity between hydraulic fractures and that of natural fractures, $k_f$ is the permeability of the hydraulic fractures, $w_f$ is the width of hydraulic fractures, $k_f$ is the permeability of the natural fractures, and $x_f$ is the width of the natural fractures. Subscripts 1 and 2 respectively represent the primary fractures and the secondary fractures.

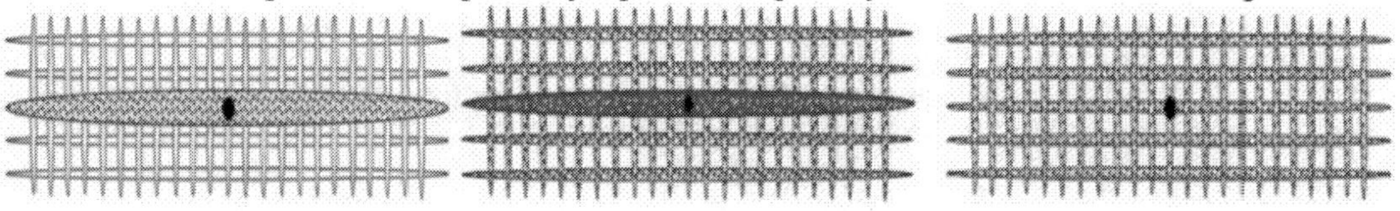

a. Concentrated in a dominanted fracture b. Concentrated &evenly Distributed c. Evenly Distributed

Fig. 6 Proppant transport scenarios-top view (Cipolla et al., 2008)

#### 3.2.2.1 Effect of secondary fractures conductivity

The LOG-LOG curve presented in Figure 7 illustrates the effect of the secondary fracture conductivity on performance of $CO_2$ injection. The five cases shown in Fig.7 correspond to different secondary fracture conductivities, ranging from a low value of 0.5 to a high value of 100. The sensitivity parameters are given as Cases B1, B2, B3, B4 and B5 in Table 3.

With the increase of the secondary fracture conductivity, the second bilinear regime (Fig.3-c) will be more obvious and the slope of inner boundary dominated flow regime becomes steeper. When the dimensionless conductivity of secondary fractures is 0.5, the second bilinear flow regime will change seriously as shown in Fig.8-a. This pressure map is quite different from that flow regime shown in Fig.3-c. At the same time, the inner boundary dominated flow regime gradually disappears, and the all LOG-LOG curves will finally merge at the linear flow time (from fractured region to shale reservoir), at about 100$^{th}$ day, as shown in Fig.8-b. This illustrates that the variation of secondary fractures conductivity mainly affects the second bilinear and inner boundary dominated flow regimes. Further, we can also find that the other flow regimes almost stay unchanged with the increase of the secondary fractures permeability. This indirectly indicates primary fractures mainly affect the performance of the initial injection flow regime and the physical properties are the principal influence factors in the later injection process.

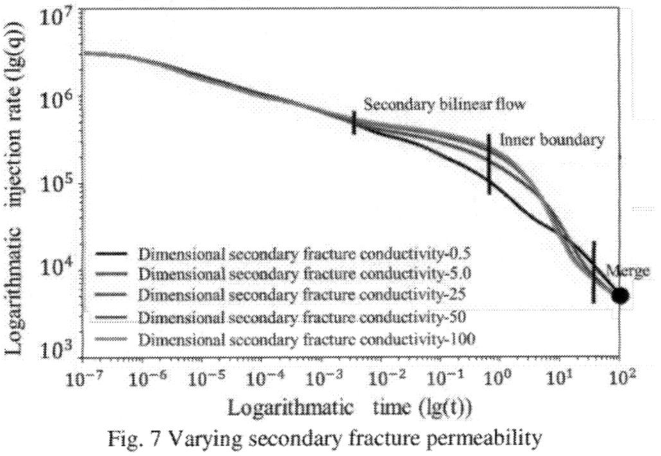


Fig. 7 Varying secondary fracture permeability

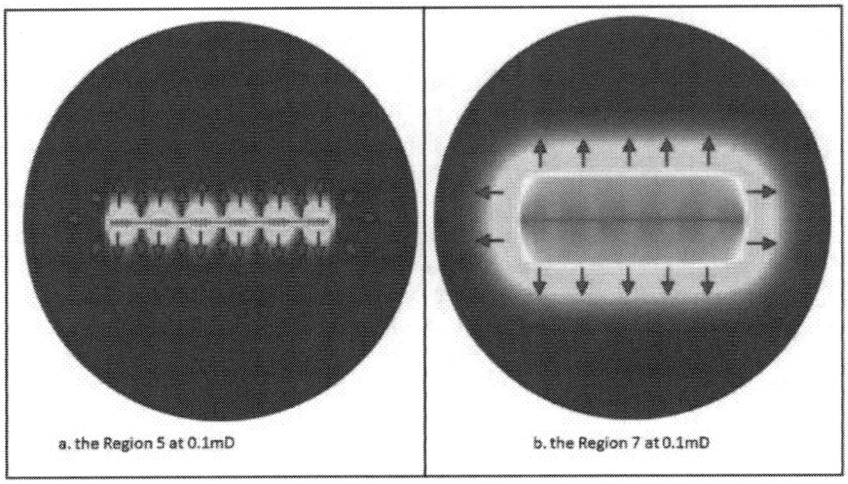


Fig. 8 Varying the conductivity of secondary fracture

### 3.2.2.2 Effect of primary fractures conductivity

When proppant appears as Fig.6-b shown, the primary fractures permeability will be lower than the secondary fractures permeability. The five cases shown in Fig.9 correspond to different primary fracture conductivities, ranging from a low value of 0.5 to a high value of 100. The sensitivity parameters are given as Cases C1, C2, C3, C4 and C5 in Table 3.

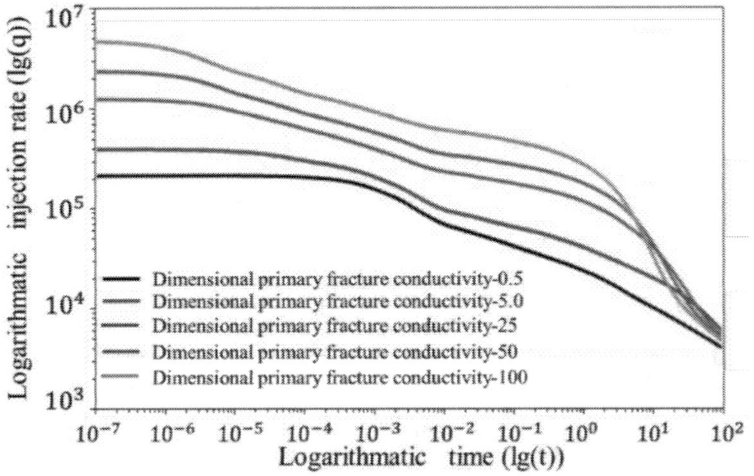


Fig. 9 Varying primary fracture permeability

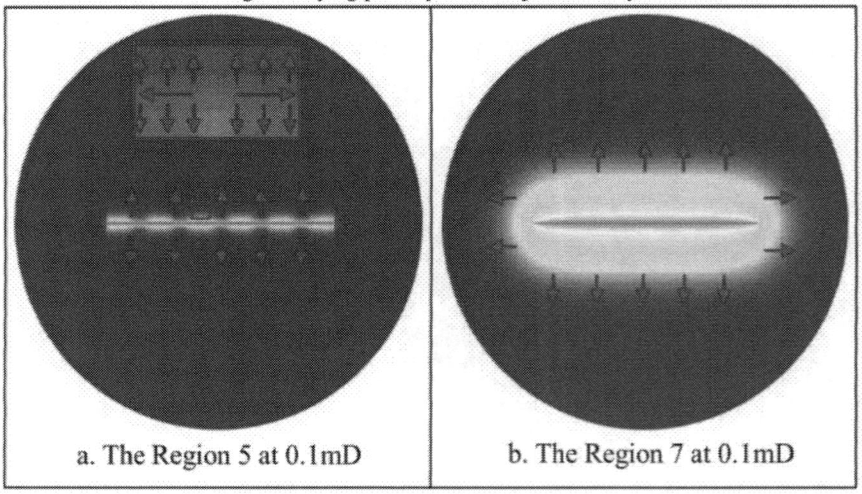


Fig. 10 Varying the conductivity of the primary fracture

Fig.9 and Fig.10 present the effect of the primary fracture conductivity on performance of $CO_2$ injection. As can be seen, with the decrease of the primary fracture conductivity, the onrushing phenomenon of injected $CO_2$ along the primary fracture severely weakens leading to much $CO_2$ lagging near horizontal well (Fig.10-b). And the time of the initial $CO_2$ adsorption period obviously extending. Meanwhile, we also find that the linear flow in primary fractures and inner boundary dominated flow fade away with the decrease of the primary fracture conductivity, the second bilinear flow regime (as shown in Fig.9) almost stays unchanged. When the $F_{cd1}$ is 0.5, most initial injected $CO_2$ would largely lag near the horizontal well. So the second bilinear flow will just happen near horizontal well (as shown in Fig.10-a) which is quite different from that of the base case (as shown in Fig.5-c). All the LOG-LOG curves will finally merge at the linear flow, from fractured region to shale reservoir, for the properties of shale formation.

### 3.2.2.3 Effect of fracture-network conductivity

When proppant distributes as Fig.6-c shown, the primary fractures permeability is identical with the secondary fractures permeability. The five cases shown in Fig.11 correspond to different hydraulic fracture conductivities, ranging from a low value of 0.5 to a high value of 100. The sensitivity parameters are given as Cases D1, D2, D3, D4 and D5 in Table 3.

Fig.11 presents the effect of the primary hydraulic permeability on performance of $CO_2$ injection. As can be seen, when the permeability of the primary and secondary fractures decrease at the same time, the variation of the LOG-LOG curves is similar with that of Fig.9. This indicates that the permeability of primary fractures is the major factor controlling the $CO_2$ injection process.

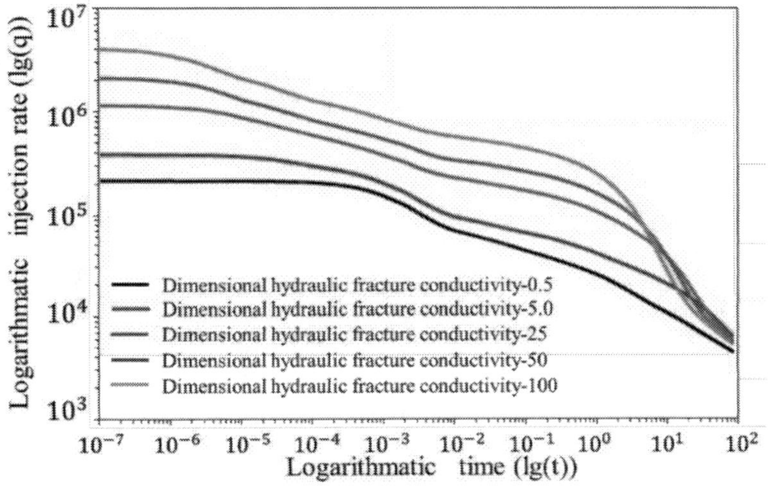


Fig. 11 Varying the permeability of primary and secondary fracture

### 3.2.3 Effect of SRV size

Fig.12 presents the effect of SRV size on performance of $CO_2$ injection process. The three cases shown in Fig.12 correspond to different primary fracture conductivities, ranging from a low value of 85m to a high value of 255m. The sensitivity parameters are given as Cases E1, E2, and E3 in Table 4.

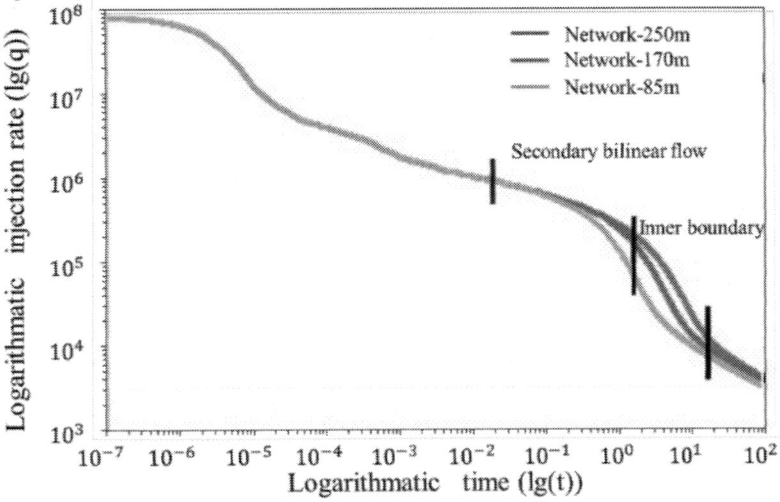


Fig. 12 Varying the width of the fractured region

We can find that the width can only affect the second bilinear and inner boundary dominated flow regimes. With the increase of the fractured region width, the emerging time of the second bilinear flow regime gradually extends and the inner boundary dominated flow regime occurs later. The descending slop of the LOG-LOG curves are identical, which indicates that the width cannot affect the inner boundary dominated flow regime but just postpones the emerging time of inner boundary flow. In general, the width of fractured region has a little influence on the seepage flow process of the $CO_2$ injection process.

### 3.3 Effect of reservoir parameters

#### 3.3.1 Effect of permeability of natural fractures

Fig.13 presents the effect of natural fracture permeability on performance of $CO_2$ injection process. As we can see, the natural fracture permeability can obviously influence the inner boundary dominated flow regime. With the decrease of the dimensionless flow conductivity between hydraulic fracture and natural fracture, the less $CO_2$ will lag in the fractured region during the inner boundary dominated flow time. This phenomenon makes the inner boundary dominated flow regime fade away, and it manifests as the slop of the LOG-LOG curve gradually decreases. All the LOG-LOG curves can't merge, which further indicates that the reservoir parameter is the main factor affecting the flow event of $CO_2$.

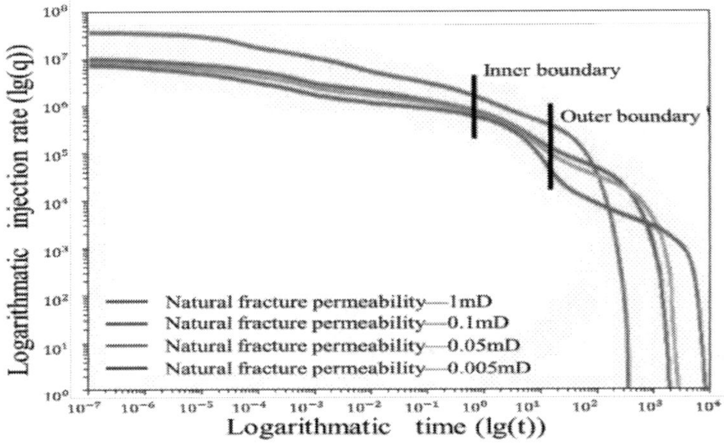


Fig. 13 Varying the permeability of natural fractures

#### 3.3.2 Effect of reservoir pressure

Fig.14 describes the effect of reservoir pressure on performance of $CO_2$ injection process. By comparison with the type curves getting from the base case, the reservoir pressure has little effect on the performance on the $CO_2$ injection process. With the reservoir pressure decreasing, the compression properties of $CO_2$ and $CH_4$ gradually enhance and the early $CO_2$ injection rate gradually decreases. And the flow regimes after the inner boundary dominated regime occur later. Finally, all the LOG-LOG curves will merge at the 100th day, which indicates that the main control factor is not reservoir pressure but the reservoir permeability.

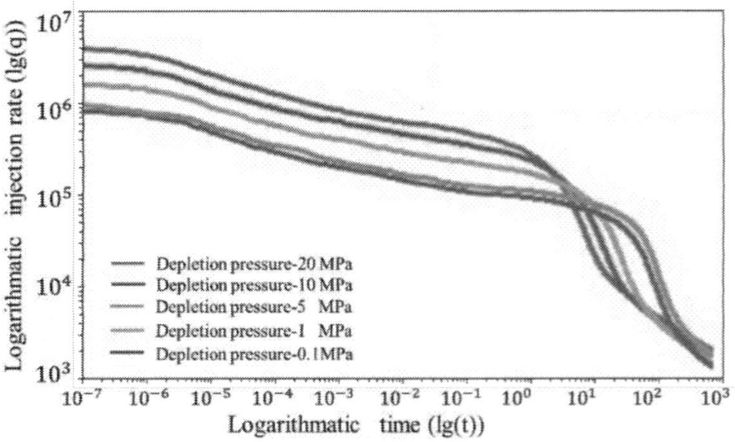


Fig. 14 varying the reservoir pressure

## 4. Conclusion

To characterize the influence of various reservoir and completion parameters on $CO_2$ injection process in shale gas systems, this work employs numerical method to analyze the pressure performance of MFHW. Local grid refinement method is applied to model complex fracture networks. After series of numerical simulation studies, we draw some meaningful conclusions:

- Simulation results of base case illustrate that eight flow regimes can be identified: (1) IAF. (2) FLF. (3) FBF. (4) SLF. (5) SBF. (6) IBDF (7) TLF. (8) OBDF. The occurrence of IBDF and OBDF is related to permeability contrast between fracture networks and shale reservoir.
- Fracture geometries with complex fracture network have three unique flow regimes: (1) SLF; (2) SBF (3) IBDF. The occurrence of these three flow regimes is related to forming a huge stimulated reservoir volume.
- Flow regimes affected by permeability of secondary fractures are SBF and IBDF. As the conductivity of secondary fracture increases, the SBF will extend longer and the IBDF occurs later.
- Flow regimes affected by conductivity of primary fractures are FLF, FBF, IBDF, and SBF. As the conductivity of primary fracture decreases, the FLF, FBF, and IBDF gradually disappear and the SBF hardly changes.
- Reservoir parameters can dominate the IBDF. Decreasing the permeability of natural fractures, the IBDF will extend longer. Moreover, the effect of reservoir pressure on the pressure performance is little.

This work employs numerical method for synthetically forecasting the pressure performance of MFHW in $CO_2$ injection process. Given that complex fracture networks and complex $CO_2$ adsorption phenomenon exist in real cases, the numerical approach can not accurately simulate all cases. However, this work can enlarge our knowledge of CO2 injection in shale gas systems.

## Nomenclatures

| | |
|---|---|
| $A$ | empirical correction factor for D |
| $D$ | diffusion and dispersion coefficient |
| $k$ | permeability |
| $N$ | the moles per unit of grid block |
| $P$ | pressure |
| $P_L$ | Langmuir pressure |
| $t$ | time |
| $T$ | transmissibility factor |
| $V$ | gird block volume |
| $V_{ads}$ | the potential releasable-gas content |
| $w$ | the width of fracture |
| $y$ | the mole fraction in gas phase |
| $\rho$ | molar concentration |
| $\emptyset$ | porosity |

## Acknowledgement

We are grateful to any reviewers for their insightful and constructive comments for improving the manuscript. This work was funded by National Basic Research 973 Program of China (2015CB250900) and The Important Project of science and technology in developing great oil&gas field and coal bed gas (2016ZX05047-004).

# References:


Al-Ahmadi, H.A. and Wattenbarger, R.A., 2011. Triple-porosity Models: One Further Step Towards Capturing Fractured Reservoirs Heterogeneity. Presented at the SPE/DGS Saudi Arabia Section Technical Symposium and Exhibition, 15-18 May, Al-Khobar, Saudi Arabia.

Ali, A.J., Siddiqui, S. and Dehghanpour, H., 2013. Analyzing the production data of fractured horizontal wells by a linear triple porosity model: Development of analysis equations. Journal of Petroleum Science and Engineering, 112: 117-128.

Bacon, D.H., Yonkofski, C.M.R., Schaef, H.T., White, M.D. and McGrail, B.P., 2015. CO2 storage by sorption on organic matter and clay in gas shale. Journal of Unconventional Oil and Gas Resources, 12: 123-133.

Brown, M., Ozkan, E., Raghavan, R. and Kazemi, H., 2011. Practical Solutions for Pressure-Transient Responses of Fractured Horizontal Wells in Unconventional Shale Reservoirs. SPE Reservoir Evaluation & Engineering, 14(06): 663 - 676.

Bumb, A.C. and Mckee, C.R., 1988. Gas-Well Testing in the Presence of Desorption for Coalbed Methane and Devonian Shale. Spe Formation Evaluation, 3(1): 179-185.

Chen, Z., Liao, X., Zhao, X., Dou, X. and Zhu, L., 2015. Performance of horizontal wells with fracture networks in shale gas formation. Journal of Petroleum Science and Engineering, 133: 646-664.

Cipolla, C.L., Warpinski, N.R., Mayerhofer, M.J., Lolon, E. and Vincent, M.C., 2008. The Relationship Between Fracture Complexity, Reservoir Properties, and Fracture Treatment Design. Presented at the SPE Annual Technical Conference and Exhibition, 21-24 September, Denver, Colorado, USA. SPE-115769-MS.

Dongyan, F., Jun, Y., Hai, S., Hui, Z. and Wei, W., 2015. A composite model of hydraulic fractured horizontal well with stimulated reservoir volume in tight oil & gas reservoir. Journal of Natural Gas Science and Engineering, 24: 115-123.

Eshkalak, M.O., Al-Shalabi, E.W., Sanaei, A., Aybar, U. and Sepehrnoori, K., 2014a. Simulation study on the CO2-driven enhanced gas recovery with sequestration versus the re-fracturing treatment of horizontal wells in the U.S. unconventional shale reservoirs. Journal of Natural Gas Science and Engineering, 21: 1015-1024.

Eshkalak, M.O., Al-Shalabi, E.W., Sanaei, A., Aybar, U. and Sepehrnoori, K., 2014b. Simulation study on the CO2-driven enhanced gas recovery with sequestration versus the re-fracturing treatment of horizontal wells in the U.S. unconventional shale reservoirs. Journal of Natural Gas Science and Engineering, 21: 1015-1024.

Fikri Kuchuk., D.B., 2014. Pressure-Transient Behavior of Continuously and Discretely Fractured Reservoirs.

Freeman, C.M., Moridis, G., Ilk, D. and Blasingame, T.A., 2013. A numerical study of performance for tight gas and shale gas reservoir systems. Journal of Petroleum Science and Engineering, 108: 22-39.

Godec, M., Koperna, G., Petrusak, R. and Oudinot, A., 2013. Potential for enhanced gas recovery and CO2 storage in the Marcellus Shale in the Eastern United States. International Journal of Coal Geology, 118: 95-104.

Jones, J.R., Volz, R. and Djasmari, W., 2013. Fracture Complexity Impacts on Pressure Transient Responses From Horizontal Wells Completed With Multiple Hydraulic Fracture Stages.

Kalantari-Dahaghi, A., Mohaghegh, S. and Esmaili, S., 2015. Data-driven proxy at hydraulic fracture cluster level: A technique for efficient CO2- enhanced gas recovery and storage assessment in shale reservoir. Journal of Natural Gas Science and Engineering, 27: 515-530.

Liu, F., Ellett, K., Xiao, Y. and Rupp, J.A., 2013. Assessing the feasibility of CO2 storage in the New Albany Shale (Devonian - Mississippian) with potential enhanced gas recovery using reservoir simulation. International Journal of Greenhouse Gas Control, 17: 111-126.

Liu, M. et al., 2015. Sensitivity analysis of geometry for multi-stage fractured horizontal wells with consideration of finite-conductivity fractures in shale gas reservoirs. Journal of Natural Gas Science & Engineering, 22: 182-195.

Mayerhofer, M.J., Lolon, E., Warpinski, N.R. and Cipolla, C.L., 2008. What is Stimulated Reservoir Volume (SRV). Presented at the SPE Shale Gas Production Conference, 16-18 November, Fort Worth, Texas, USA.

Ren, J. and Guo, P., 2015. A novel semi-analytical model for finite-conductivity multiple fractured horizontal wells in shale gas reservoirs. Journal of Natural Gas Science & Engineering, 24: 35-51.

Ritter, J.A. and Yang, R.T., 1987a. Equilibrium adsorption of multicomponent gas mixtures at elevated pressures. Industrial & Engineering Chemistry Research, 26(8): 1679-1686.

Ritter, J.A. and Yang, R.T., 1987b. Equilibrium adsorption of multicomponent gas mixtures at elevated pressures. Industrial & Engineering Chemistry Research, 26(8): 1679-1686.

Ruthven, D.M., 1984. Principles of adsorption and adsorption processes. Wiley, 168-179 pp.

Schettler, P.D. and Parmely, C.R., 1991. Contributions to Total Storage Capacity in Devonian Shales. Paper SPE-23422-MS presented at the SPE Eastern Regional Meeting, Lexington, Kentucky， 22-25 October.

Stalgorova, E.M.L., 2012. Practical Analytical Model to Simulate Pro duction of Horizontal Wells with Branch Fractures.

Stalgorova, K. and Mattar, L., 2013. Analytical Model for Unconventional Multifractured Composite Systems. SPE Reservoir Evaluation & Engineering, 16(03): 246-256.

Stehfest, H., 1970. Algorithms 368： Numerical Inversion of Laplace Transforms. Commun. of the ACM, 13(1): 47-49.

Sureshjani, M.H. and Clarkson, C.R., 2015. Transient linear flow analysis of constant-pressure wells with finite conductivity hydraulic fractures in tight/shale reservoirs. Journal of Petroleum Science & Engineering, 133(3): 455-466.

Wang, H.T., 2014. Performance of multiple fractured horizontal wells in shale gas reservoirs with consideration of multiple

mechanisms. Journal of Hydrology, 510(3): 299-312.
Yu, W., Al-Shalabi, E.W. and Sepehrnoori, K., 2014. A Sensitivity Study of Potential CO2 Injection for Enhanced Gas Recovery in Barnett Shale Reservoirs. Paper SPE-169012-MS presented at the SPE Unconventional Resources Conference, The Woodlands, Texas, USA, 1-3 April.
Zhou, W., Banerjee, R., Poe, B.D., Spath, J. and Thambynayagam, M., 2012. Semi-Analytical Production Simulation of Complex Hydraulic Fracture Network.

# Appendices

Table 1 Average parameters of Barnet shale in Case A1 from Table 2.

| Types | Parameters | Value | Unit | Source |
|---|---|---|---|---|
| Reservoir | Grid spacing | 20*20*2.0 | m | \ |
| | Circular model diameter | 1620 | m | (Heller and Zoback, 2014) |
| | Reservoir thickness | 10 | m | \ |
| | Porosity of matrix | 2 | % | (Heller and Zoback, 2014) |
| | Porosity of fracture | 0.1 | % | (Heller and Zoback, 2014) |
| | Natural fracture permeability | 0.01 | mD | (Heller and Zoback, 2014) |
| | $k_v/k_h$ | 0.1 | \ | (Heller and Zoback, 2014) |
| | Initial reservoir pressure, $P_i$ | 24 | Mpa | (Heller and Zoback, 2014) |
| | Density of rock, | 2500 | $kgm^{-3}$ | (Heller and Zoback, 2014) |
| | Reservoir temperature | 115 | ℃ | (Heller and Zoback, 2014) |
| Shale gas | Langmuir volume of $CH_4$ | 40 | atm | (Heller and Zoback, 2014) |
| | Langmuir pressure of $CH_4$ | 2.1 | $cm^3/g$ | (Heller and Zoback, 2014) |
| | Langmuir volume of $CO_2$ | 32.8 | atm | (Heller and Zoback, 2014) |
| | Langmuir pressure of $CO_2$ | 4.2 | $cm^3/g$ | (Heller and Zoback, 2014) |
| Fractured well | Horizontal well length, $L_H$ | 1000 | m | (Heller and Zoback, 2014) |
| | Fracture half-length | 170 | m | (Heller and Zoback, 2014) |
| | Primary fracture permeability, $K_{pf}$ | 30 | mD | (Heller and Zoback, 2014) |
| | Secondary fracture permeability, $K_{sf}$ | 30 | mD | (Heller and Zoback, 2014) |
| | Injection bottom-hole pressure | 28 | MPa | \ |

Table 2 Description of the simulation runs and sensitivity analyses varying fracture topologies.

| Case | Permeability | | Fracture topology |
|---|---|---|---|
| | Hydraulic fractures/mD | Natural fracture/mD | |
| A1 | 30 | 0.01 | Base case |
| A2 | 30 | 0.01 | Model 1 |
| A3 | 30 | 0.01 | Model 2 |
| A4 | 30 | 0.01 | Model3 |
| A5 | 30 | 0.01 | Model 4 |

Table 3 Description of the simulation runs and sensitivity analyses varying fracture conductivity.

| Case | Permeability | | | | | Fracture topology |
|---|---|---|---|---|---|---|
| | Primary fracture/mD | Secondary fracture/mD | Natural fracture/mD | $F_{cd1}$ | $F_{cd2}$ | |
| B1 | 30 | 0.1 | 0.01 | 150 | 0.5 | Base case |
| B2 | 30 | 1 | 0.01 | 150 | 5.0 | Base case |
| B3 | 30 | 5 | 0.01 | 150 | 25 | Base case |
| B4 | 30 | 10 | 0.01 | 150 | 50 | Base case |
| B5 | 30 | 20 | 0.01 | 150 | 100 | Base case |
| C1 | 0.1 | 30 | 0.01 | 0.5 | 150 | Base case |
| C2 | 1 | 30 | 0.01 | 5.0 | 150 | Base case |
| C3 | 5 | 30 | 0.01 | 25 | 150 | Base case |
| C4 | 10 | 30 | 0.01 | 50 | 150 | Base case |
| C5 | 20 | 30 | 0.01 | 100 | 150 | Base case |
| D1 | 0.1 | 0.1 | 0.01 | 0.5 | 0.5 | Base case |
| D2 | 1 | 1 | 0.01 | 5.0 | 5.0 | Base case |
| D3 | 5 | 5 | 0.01 | 25 | 25 | Base case |
| D4 | 10 | 10 | 0.01 | 50 | 50 | Base case |
| D5 | 20 | 20 | 0.01 | 100 | 100 | Base case |

Table 4 Description of the simulation runs and sensitivity analyses varying fractured region width and reservoir parameters.

| Case | Permeability | | | Reservoir pressure/MPa | Fractured region width/m | Fracture topology |
|---|---|---|---|---|---|---|
| | Hydraulic fractures/mD | Natural fracture/mD | $F_{cd1}= F_{cd2}$ | | | |
| E1 | 30 | 0.01 | 150 | 22 | 85 | Base case |
| E2 | 30 | 0.01 | 150 | 22 | 170 | Base case |
| E3 | 30 | 0.01 | 150 | 22 | 255 | Base case |
| F1 | 30 | 0.005 | 300 | 22 | 170 | Base case |
| F2 | 30 | 0.05 | 30 | 22 | 170 | Base case |
| F3 | 30 | 0.1 | 3.0 | 22 | 170 | Base case |
| G1 | 30 | 1 | 0.3 | 22 | 170 | Base case |
| G2 | 30 | 0.01 | 150 | 20 | 170 | Base case |
| G3 | 30 | 0.01 | 150 | 10 | 170 | Base case |
| H1 | 30 | 0.01 | 150 | 5 | 170 | Base case |
| H2 | 30 | 0.01 | 150 | 1 | 170 | Base case |
| H3 | 30 | 0.01 | 150 | 0.1 | 170 | Base case |

# Bench Scale Testing of Next Generation Hollow Fiber Membrane Modules (DE-FE0026422)

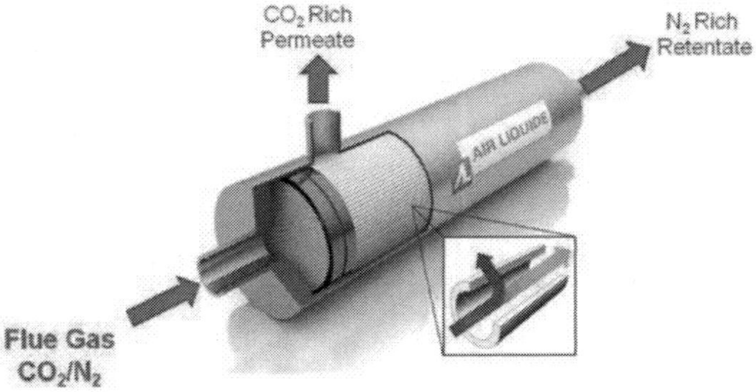



**Shilu Fu**
Research Scientist
shilu.fu@airliquide.com


**Carbon Management Technology Conference 2017**

**Houston, TX**

1

## Agenda

1. Air Liquide & $CO_2$

2. Project Overview

3. Current Technology Status

2

# 1

## Air Liquide & $CO_2$

Air Liquide

## Air Liquide & $CO_2$

Market position

- $CO_2$ as a product – 24 $CO_2$ plants in US
- Huge potential market in EOR application

Air Liquide **Cryocap™:** new range of products for this market

- Cryocap™ $H_2$: $CO_2$ capture during Hydrogen production
- Cryocap™ Oxy: $CO_2$ capture with Oxycombustion
- Cryocap™ NG: $CO_2$ capture from Natural Gas
- Cryocap™ Steel: $CO_2$ removal from Steel Production

provider

Air Liquide creative oxygen

4

## Air Liquide -- $CO_2$ separation solution provider

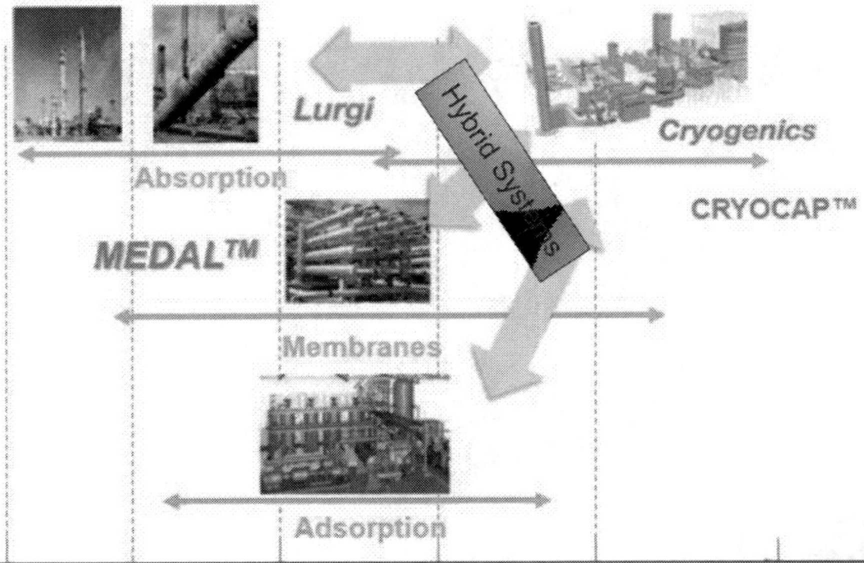


Utilizing all existing techniques for the whole range of $CO_2$ feed and product purities from any $CO_2$ source

Air Liquide
creative oxygen

5

## Large Air Liquide $CO_2$ Projects

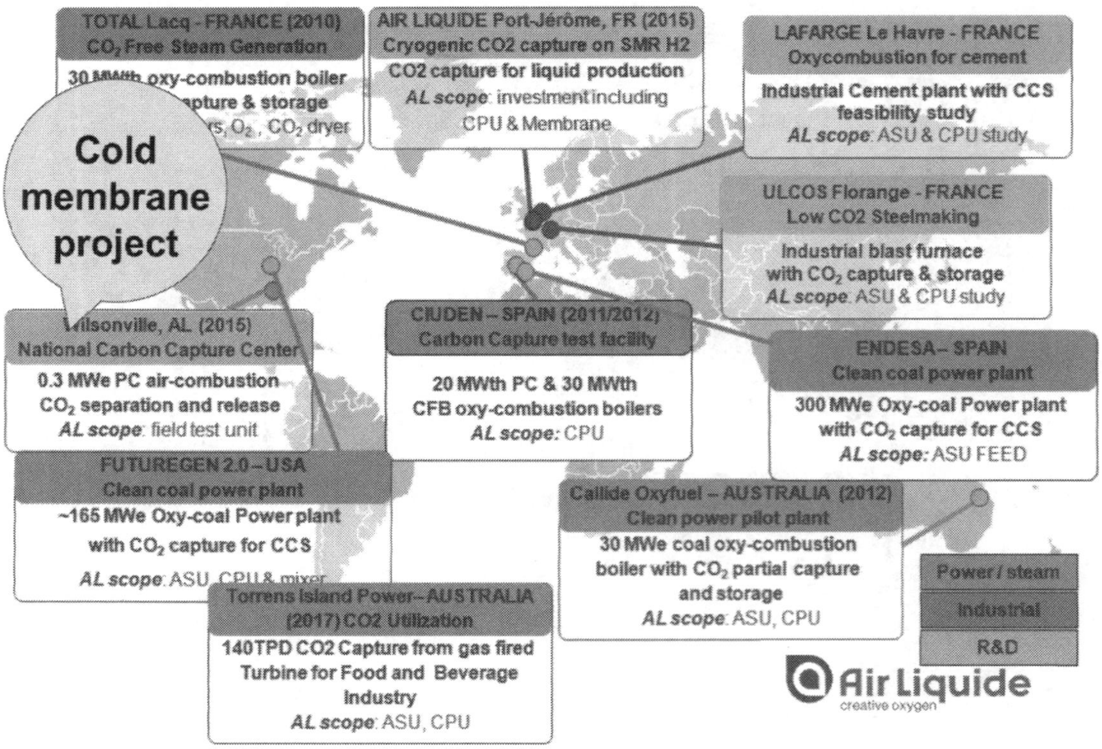


Air Liquide
creative oxygen

## Cold Membrane Technology Roadmap

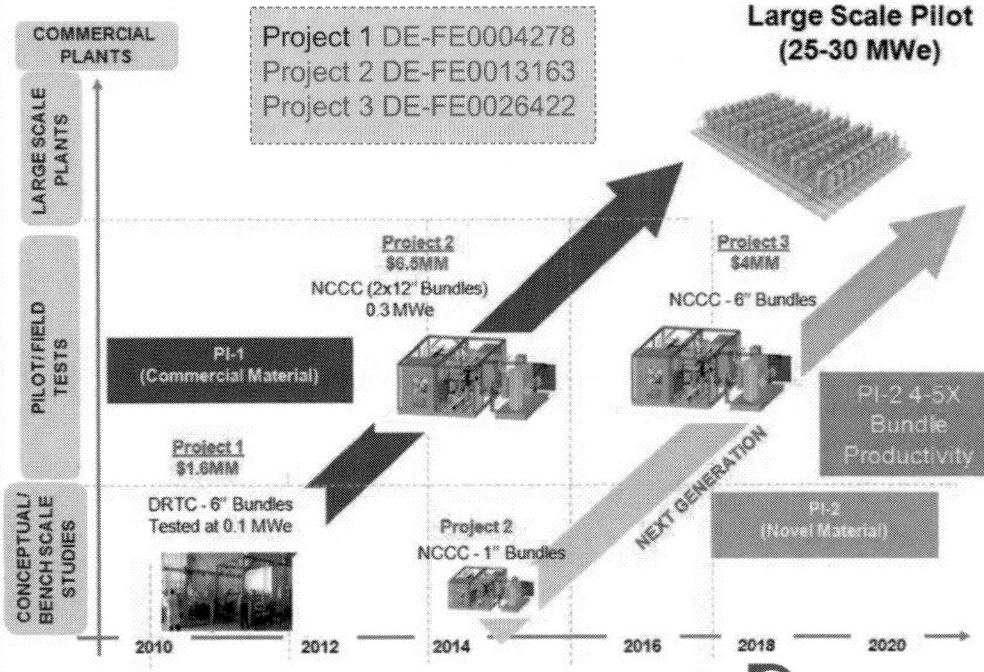


Air Liquide creative oxygen 7

# 2

## Project Overview

Air Liquide

## Project Overview: $CO_2$ capture by cold membrane

***Project target:*** $CO_2$ capture from coal fired power plant flue gas with AL cold membrane technology at $40/tonne:
***Total Budget:*** $4.0MM, DOE Funding - $3.0MM, AL Cost Shares - $1.0MM
***Period of Performance:*** 10/01/2015 through 12/31/2018 over 2 budget periods

| | Expenditures |
|---|---|
| Budget Period 1 (Oct 2015- June 2017) | $1,600,000 |
| Budget Period 2 (Jul 2017- Dec 2018) | $2,370,000 |

***NETL Project Manager:*** José Figueroa
***Project Partners:*** ALAS (membrane scale-up and manufacturing), NCCC (field testing), E&C (engineering cost analysis), Parsons (TEA review)

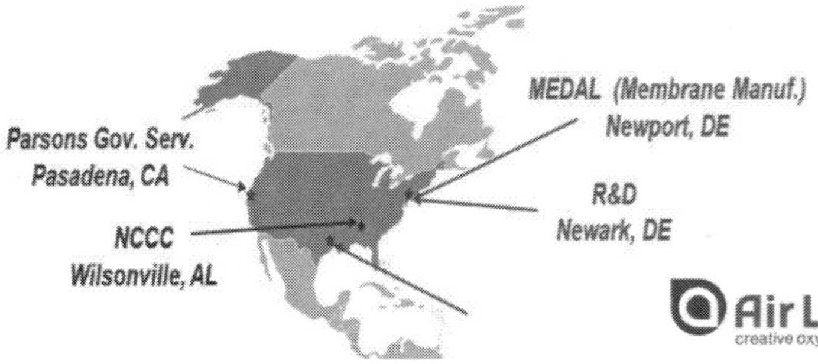


Air Liquide
creative oxygen



## Motivation: Membrane? Why cold? Why PI-2

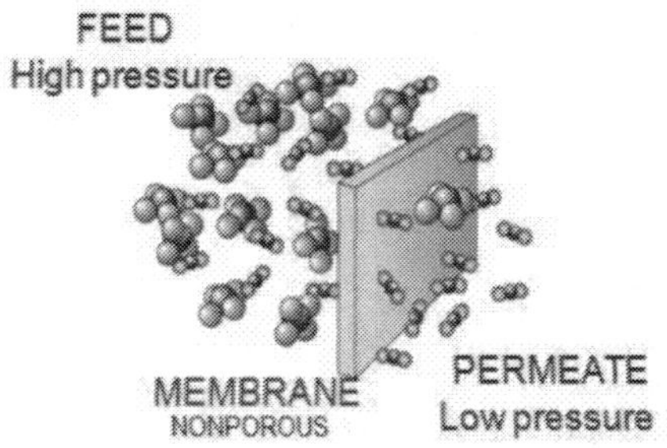


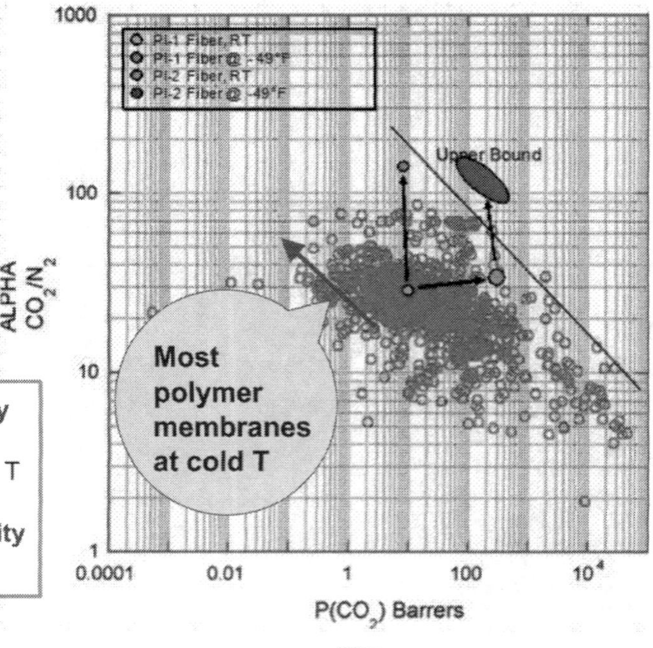


- ❖ **Why cold? -- improve selectivity**
- ★ Discovered by Air Liquide R&D in 2008: Improved $CO_2$ selectivity at T below ambient
- ❖ **Why PI-2? -- improve productivity**
- ★ 4-5 X higher $CO_2$ productivity

Air Liquide
creative oxygen

## Air Liquide Capture Technology Summary

❖ **Air liquide hybrid cold membrane - liquefaction process**
- Novel cold membrane technology
- Well established liquefaction

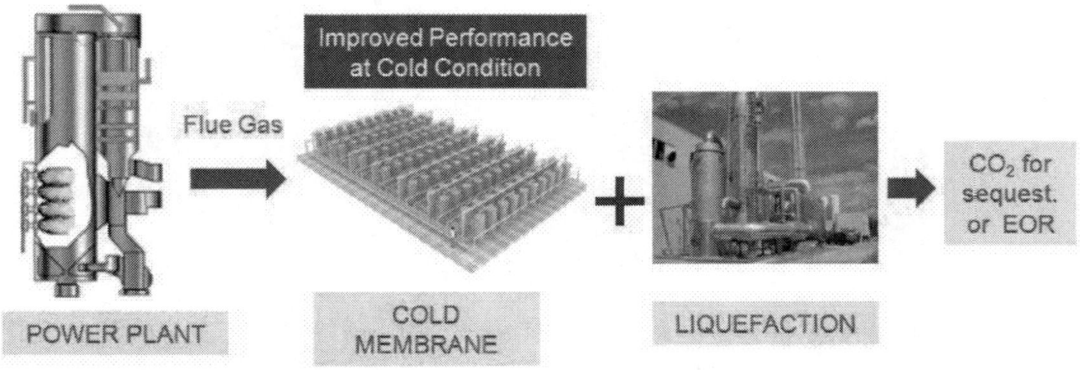


Synthetic flue gas at 0.1 MWe (TRL4) bench scale skid → Actual flue gas from coal fired power plant at 0.3 MWe (TRL5) field test unit at NCCC

## Synthetic flue gas testing

**0.1MWe (TRL 4) bench scale skid**

Air Liquide
creative oxygen

12

## Actual flue gas testing at NCCC

### 0.3MWe (TRL 5) field test unit

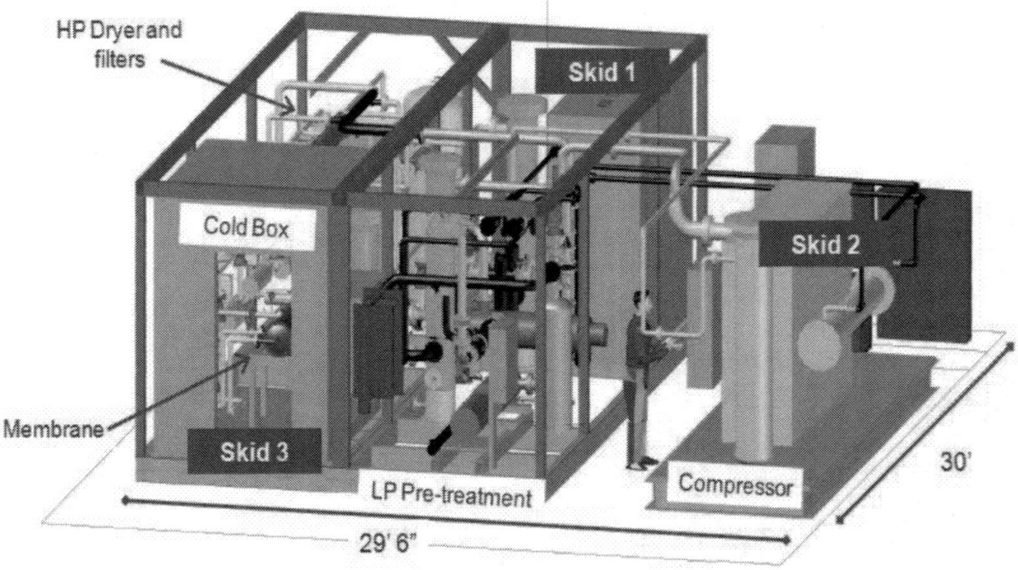


Air Liquide
creative oxygen

13

## Actual flue gas testing at NCCC

### 0.3MWe (TRL 5) field test unit

de

14

# 3

## Current Technology Status

Air Liquide

## Membrane bundles - lab scale to commercial scale

| | Diameter in. (mm) | Length ft. (m) | Fiber Count |
|---|---|---|---|
| **Mini permeator** | 0.25 - 0.5" (6.4 – 12.7) | 0.5-2.8 (0.15-.085) | <1000 |
| **Permeator** | 1" (25.4) | 1.6 (0.5) | 2 – 20x |
| **Prototype Permeator** | 2.5" (63.5) | | 10 – 100x |
| **R&D prototype bundle** | 4" (101.6) | 2.8 (0.85) | 100 – 400x |
| **6" bundle (commercial)** | 6" (152.4) | | 200 - 1000x |
| **12" bundle (commercial)** | 12" (304.8) | | 500 - 4000x |

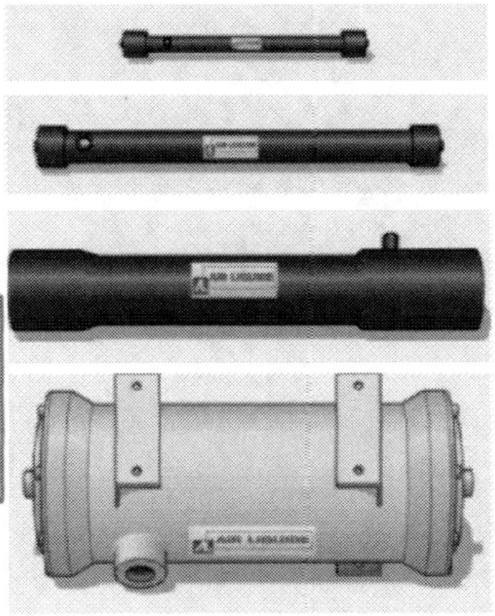

Air Liquide creative oxygen

16

## PI-2 1" mini permeator tested at NCCC (2016)

★ Testing for viability of PI-2 membrane in treated flue gas service
★ Tested during the previous DOE-AL project (DE-FE0013163)

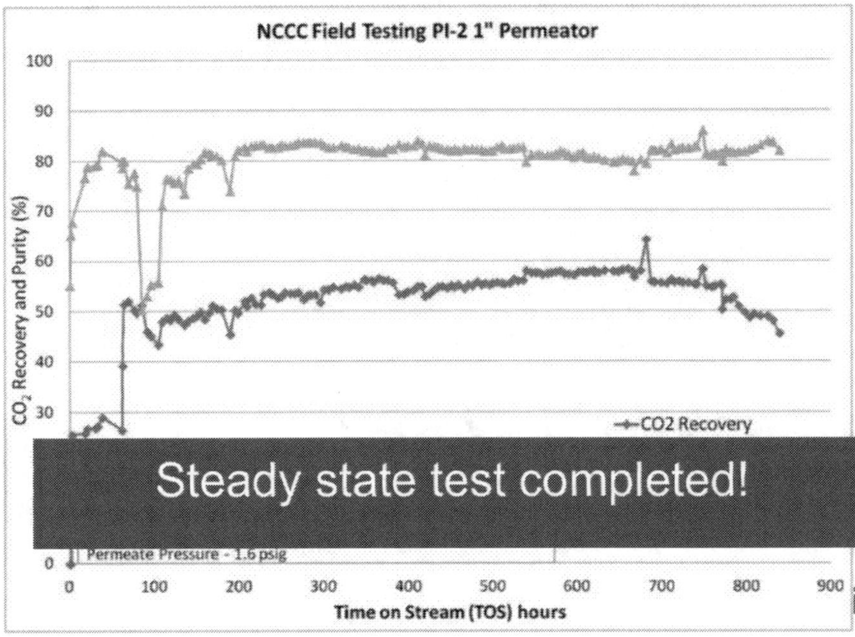


17

## Design/Manufacture 4" PI-2 Bundles (2017)

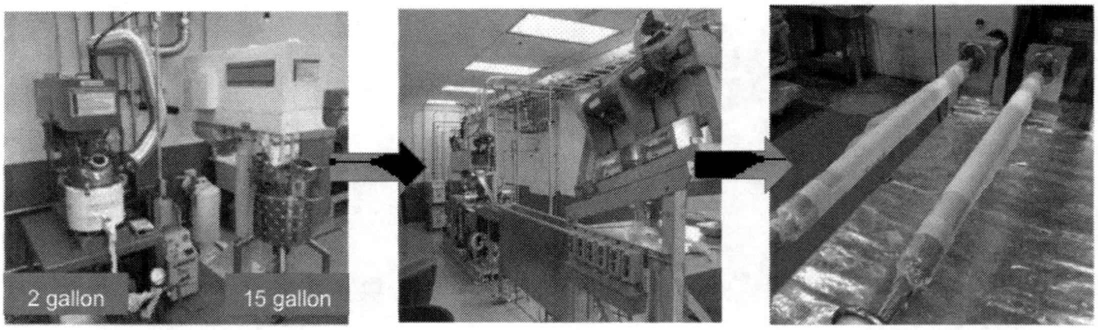


**Pilot mixers** **DSU unit Multi-filament spin line** **Two prototype bundles**

**Development spin unit (DSU) - representative of commercial spin line**

18

## 4" Bundles tested with synthetic flue gas

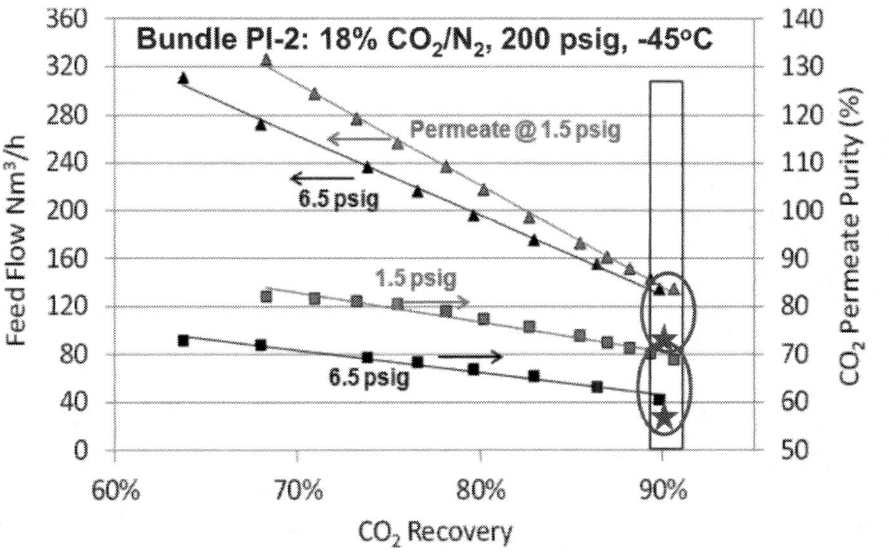


★ ***Exceeding the project target:*** > 90 $Nm^3/h$ feed at 90% $CO_2$ recovery, >58% $CO_2$ purity

creative oxygen

19

## Ongoing & future work

- **Reducing $CO_2$ capture cost with PI-2:**

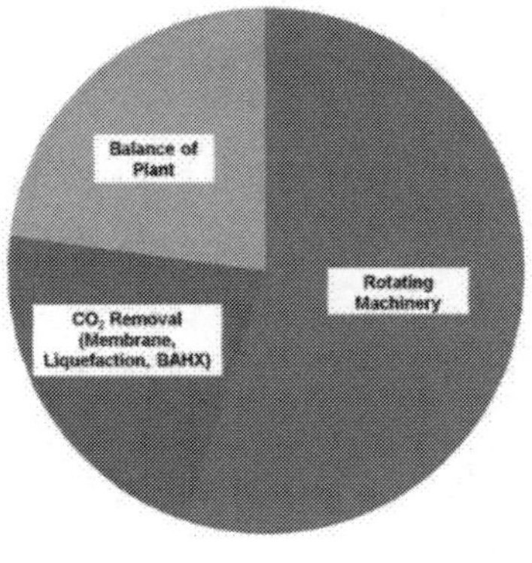


**Projected 12" PI-2 bundle cost breakdown**
(TEA study from Project 2)

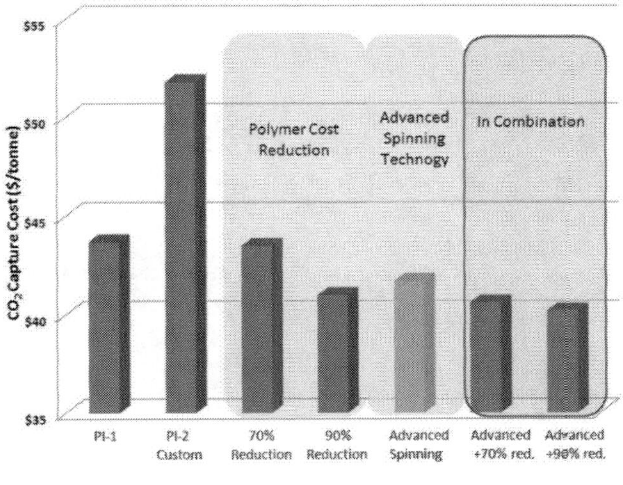


★ **Reduced costs moving from lab production to bulk production**
★ **Advanced vs. traditional membrane formation method**

20

## Ongoing & future work

- **Field test of 6" PI-2 bundle at NCCC**

21

## Summary

★ Air Liquide is utilizing all existing techniques for the whole range of $CO_2$ feed and product purities from any $CO_2$ source

★ ***Exceeding the project target:*** > 90 $Nm^3/h$ feed at 90% $CO_2$ recovery, >58% $CO_2$ purity

★ Getting ready for the 0.3 MWe field test of 6" PI-2 bundle

Air Liquide
creative oxygen

22

## Acknowledgement / Disclaimer

- US DOE: José Figueroa, Sheldon Funk
- NCCC: Frank Morton, Tony Wu
- Air Liquide: Rob Gagliano, Alex Augustine, Trapti Chaubey, Sudhir Kulkarni, Dave Hasse, Mike Bennett, Dean Kratzer, Jean-Marie Gauthier, Ted Li, Judy Huss, Randy Hutchison, Mike Turney

- Some material in this presentation is based on work supported by the Department of Energy National Energy Technology Laboratory under Award Number DE-FE0004278 (completed), DE-FE0013163 (completed), and DE-FE0026422.

"This report was prepared as an account of work sponsored by an agency of the United States Government. Neither the United States Government nor any agency thereof, nor any of their employees, makes any warranty, express or implied, or assumes any legal liability or responsibility for the accuracy, completeness, or usefulness of any information, apparatus, product, or process disclosed, or represents that its use would not infringe privately owned rights. Reference herein to any specific commercial product, process, or service by trade name, trademark, manufacturer, or otherwise does not necessarily constitute or imply its endorsement, recommendation, or favoring by the United States Government or any agency thereof. The views and opinions of authors expressed herein do not necessarily state or reflect those of the United States Government or any agency thereof."


Air Liquide
creative oxygen

23

Research & Development

# Opening new ways

**THANK YOU FOR YOUR ATTENTION**

Membrane Technology & Research

# Pilot Test Results from a Polaris™ Membrane 1 $MW_e$ $CO_2$ Capture System

Richard Baker, Carlos Casillas, Ken Chan, Brice Freeman,
Don Fulton, Pingjiao Hao, Jenny He, Jurgen Kaschemekat,
Jay Kniep, Jennifer Ly, Tim Merkel, Vincent Nguyen, Zhen Sun,
Xuezhen Wang, Xiaotong Wei, and Steve White

**Carbon Management Technology Conference**
Houston, TX
Wednesday, July 19, 2017

# MTR Introduction

- Privately-held, 60 employees mostly located in Newark, California
- Started in 1982 funded by U.S. SBIR grants
- Commercial products in petrochemical, natural gas and refinery industries
- Provides complete turn-key solutions with over 200 membrane systems installed worldwide
- 15 person R&D group with expertise in membrane materials/formation and process design

# Background: MTR $CO_2$ Capture Process

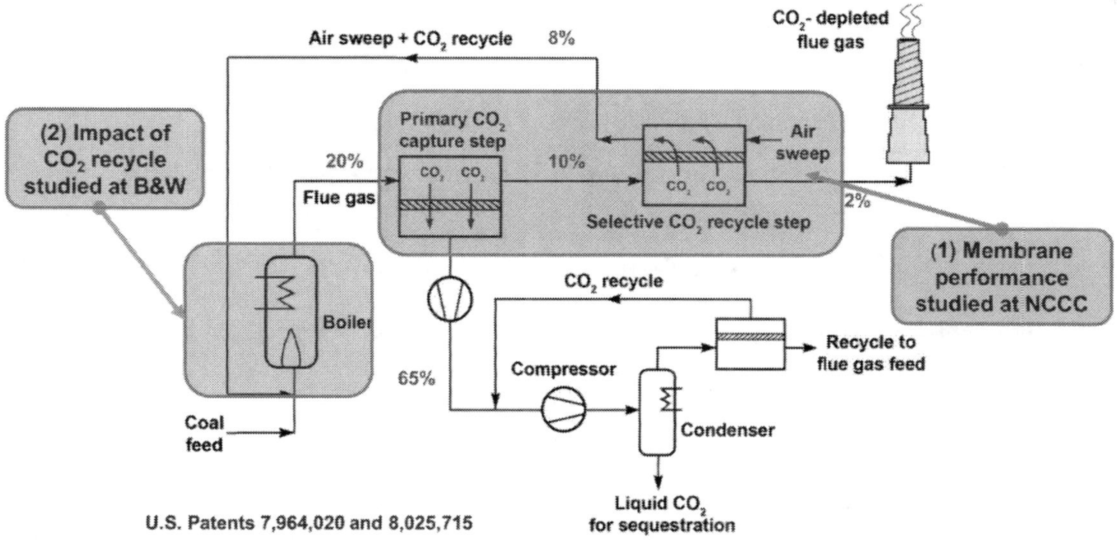


U.S. Patents 7,964,020 and 8,025,715

**Benefits of selective recycle:**

- Increases $CO_2$ concentration going to the capture step, and
- Reduces the fractional $CO_2$ removal required by the capture step

3

MTR

# MTR $CO_2$ Capture Development Timeline

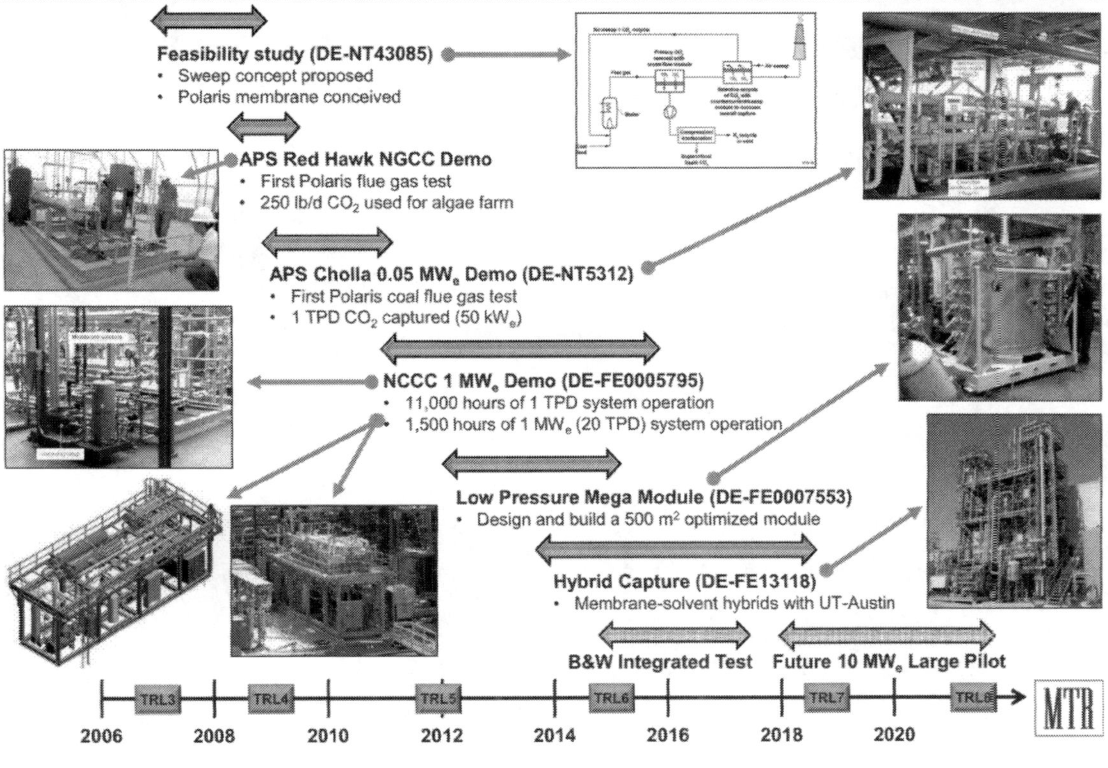


MTR

# Pros and Cons of a Membrane Post-Combustion Capture Process

**Benefits:**
- No hazardous chemical handling, emissions, or disposal issues
- Not affected by oxygen, $SO_x$ or $NO_x$
- Water use lower than other technologies (recovers $H_2O$ from flue gas)
- No steam use → no modifications to existing boiler/turbines
- Near instantaneous response; high turndown possible
- Modular and compact; relatively simple installation/operation
- Particularly well-suited for partial capture (40-60%)

**Challenges:**
- Very low partial pressure driving force favors high permeance membranes
- Unknown impact of real flue gas on membrane-module lifetime
- Module pressure drop and flow distribution issues

5

MTR

# Polaris™ Membranes Continue to Improve

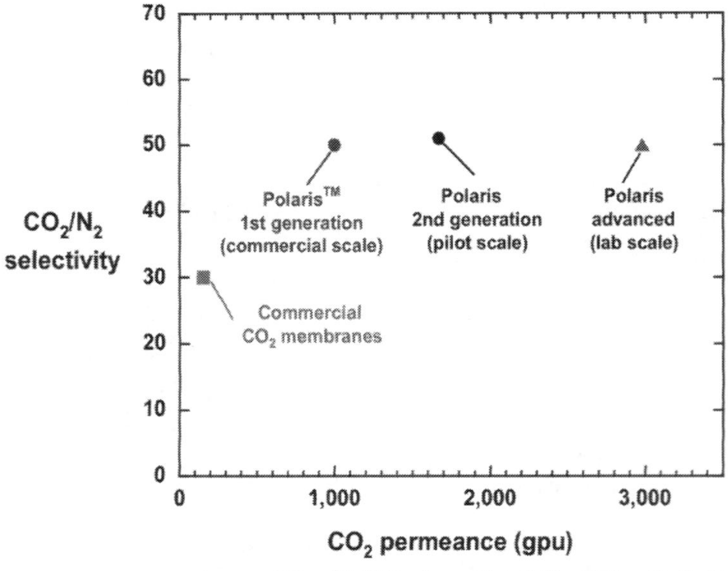


1 gpu = $10^{-6}$ cm³(STP)/(cm² s cmHg)= 3.35 x $10^{-10}$ mol/(m² s Pa)

- In addition to lowering costs, these improvements are important to shrink the size of the capture system

6

MTR

# 0.05 $MW_e$ (1 TPD) Testing at NCCC

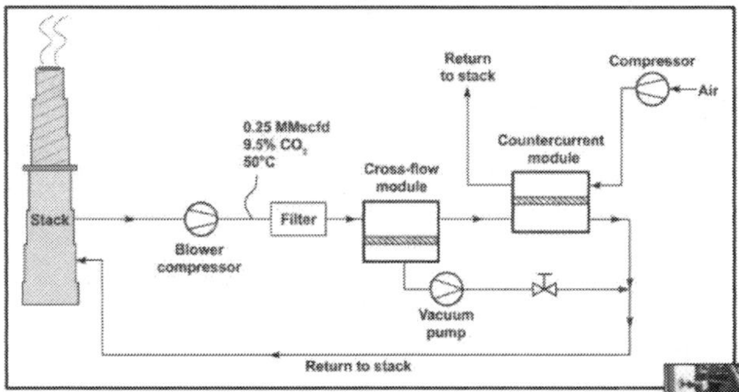


- Field laboratory that allows lifetime evaluation (>11,000 hours cumulative), and validation testing of new membranes
- Many lessons learned (i.e., ammonium sulfate deposition) applied to scale up

- System is testing vacuum and air sweep membrane steps
- Sized to capture 1 ton $CO_2$/day using commercial sized module
- System was in operation from Spring 2012 through Summer 2015

7

# 1 $MW_e$ (20 TPD) System at NCCC

- System installation completed Aug 2014
- Operation during January – June 2015
- Decommissioned in July 2015

- System showed stable performance over 1,500 hours of operation, including demonstration of prototype low-pressure drop plate-and-frame sweep modules

8

# 1 MW$_e$ System Shows Stable Performance

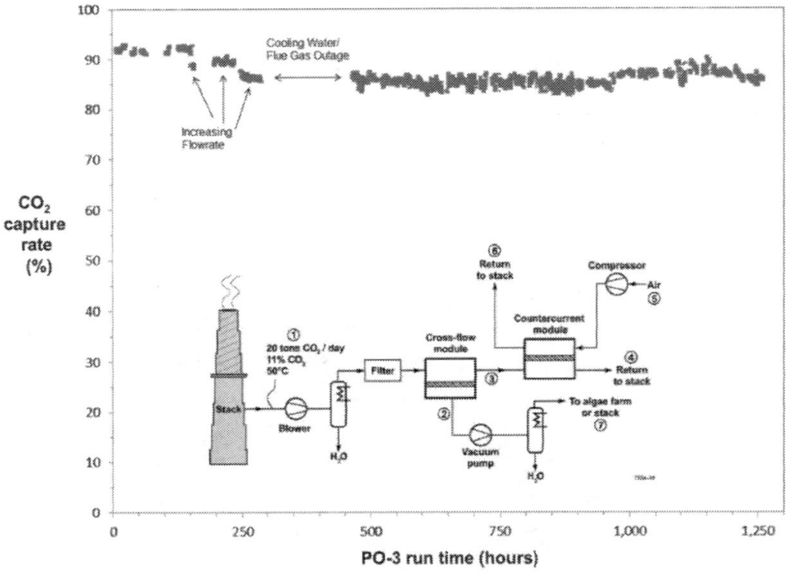


- System operated in slipstream mode (no recycle to boiler) with varying ambient conditions (sub-freezing in January to >95°F in June)
- Stable performance, reaching up to 90% capture
- System goes from cold start to steady state in ~15 minutes

**Figure data from NCCC campaign PO3 (May to July 2015)**

MTR

9

# New Modules Demonstrate Improved Pressure Drop Performance

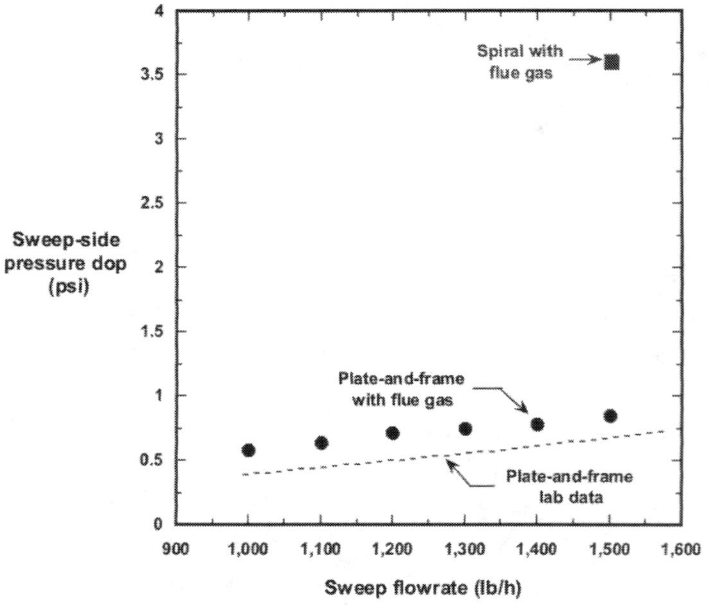


- Field data is consistent with lab results, and confirms much lower air sweep pressure drop in new modules
- At full scale, the difference in pressure drop amounts to savings of about 10 MW$_e$

MTR

10

# B&W Studies of $CO_2$ Recycle Impact on Boiler Performance

**Phase I – CFD modeling**

- B&W modeled 2 boiler configurations (radiant boiler firing bituminous coal and SWUP firing PRB coal) and 2 sweep recycle cases (constant secondary air flow and constant stoichiometry)
- Main conclusion of modeling study: secondary air laden with $CO_2$ appears feasible as a retrofit in either of the boiler configurations examined if oxygen mass flow to boiler is fixed

**Phase II – Pilot testing**

- B&W's SBS-II 1.8 $MW_{th}$ pilot boiler operated with $CO_2$-laden combustion air
- Two coals evaluated: a western sub-bituminous coal and a highly volatile bituminous coal
- $O_2$ content of windbox air varied from 21% to 16% through $CO_2$ dilution
- Monitored flame stability, length, and shape; unburned combustibles in fly ash, and furnace exit gas temperature
- Radiant furnace and convective pass heat absorptions were measured
- Boiler efficiencies for air and sweep firing were determined

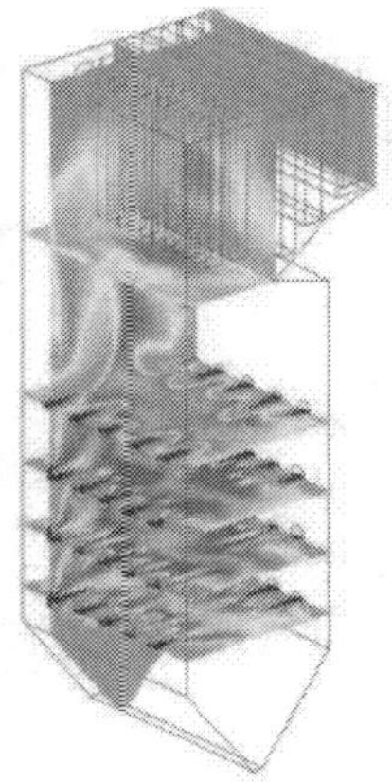

MTR

11

# B&W Boiler Study Highlights

- Stable and attached flames with air (21% $O_2$) and $CO_2$-enriched air (16-18% $O_2$)
- $CO_2$-enriched flame was less luminous than air-fired case
- Lower furnace heat absorption but higher convection pass/air heater heat transfer for $CO_2$-enriched operation relative to air
- For bituminous coal, 30% lower $NO_x$ emissions with $CO_2$-enriched air
- No burner modifications necessary
- Net reduction in plant efficiency of ~0.75% at 18% $O_2$

**Flame image from combustion of PRB coal with air (21% $O_2$)**

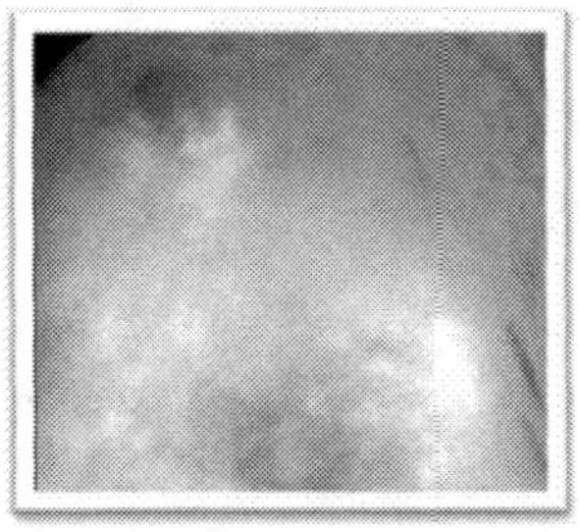

**Flame image from combustion of PRB coal with $CO_2$–enriched (18% $O_2$)**

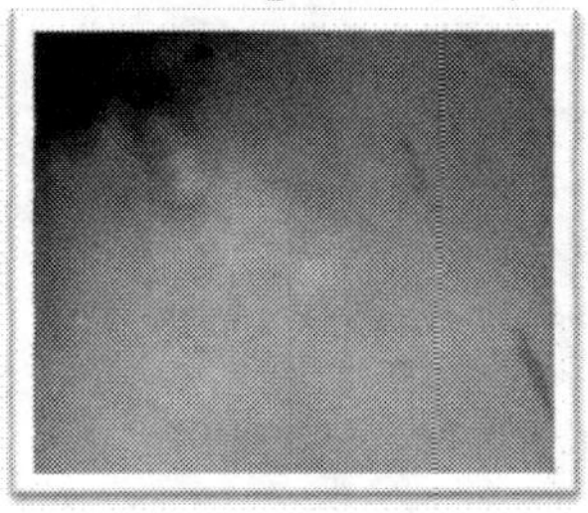

MTR

12

# B&W Pilot Boiler / Membrane Schematic

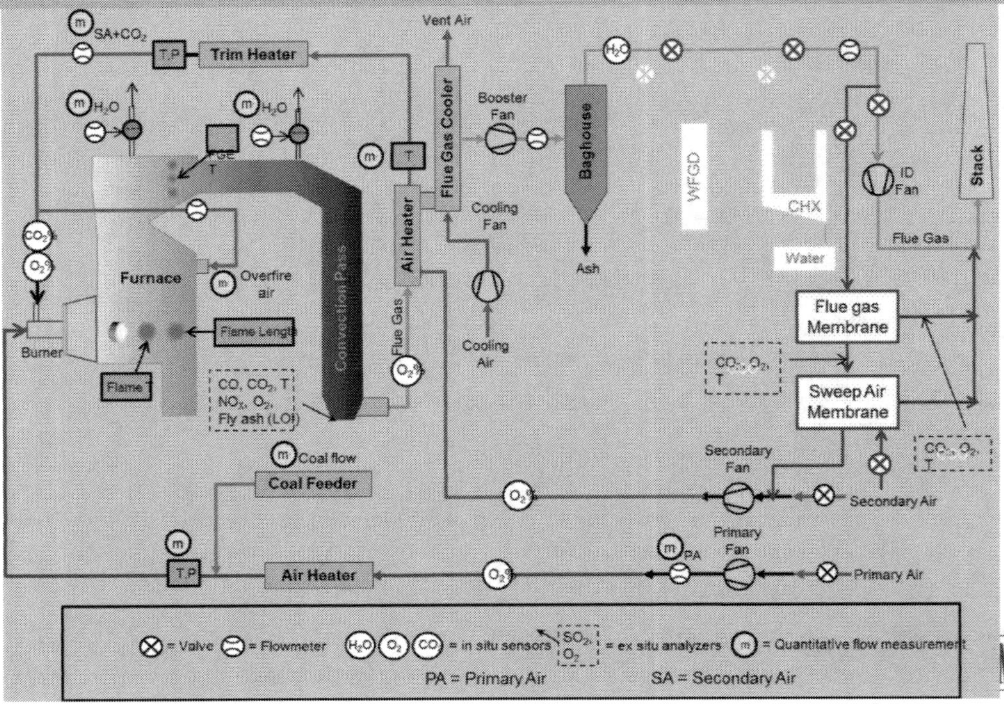


13

MTR

# MTR Skid During Transport and Installation at B&W – May 2016

Skid arriving at B&W ⟶

⟵ Installation of 2nd floor

MTR

# MTR Skids at B&W's SBS-II Research Facility

Main skid and smaller low-pressure drop sweep module anchored to foundation

15

MTR

# Sample Results from B&W Integrated Tests

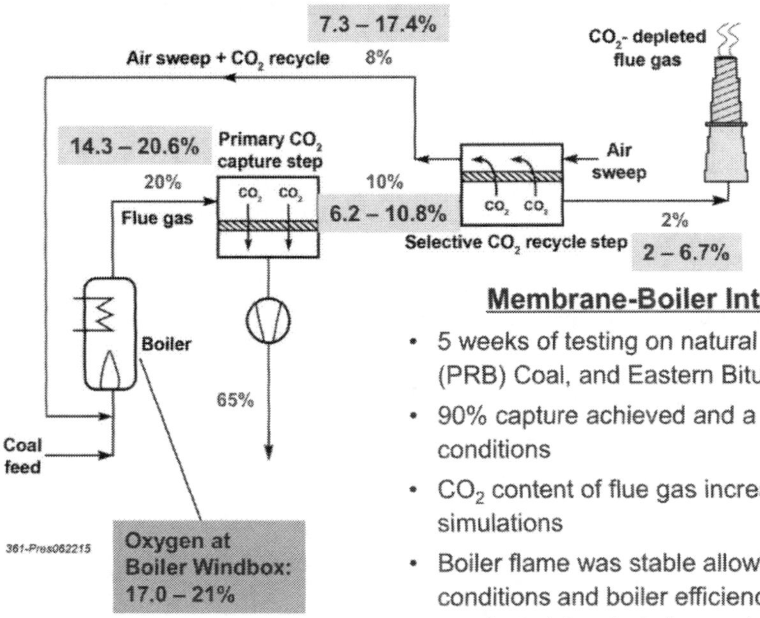


**Membrane-Boiler Integrated Test Plan**

- 5 weeks of testing on natural gas, Powder River Basin (PRB) Coal, and Eastern Bituminius Coal
- 90% capture achieved and a variety of partial capture conditions
- $CO_2$ content of flue gas increased as expected in simulations
- Boiler flame was stable allowing a full battery of stream conditions and boiler efficiency measurements to be conducted (analysis is ongoing)

MTR

# 1 $MW_e$ System Sweep Flow Rate Parametric Results from Integrated Tests

Influence on the Overall $CO_2$ Capture Rate

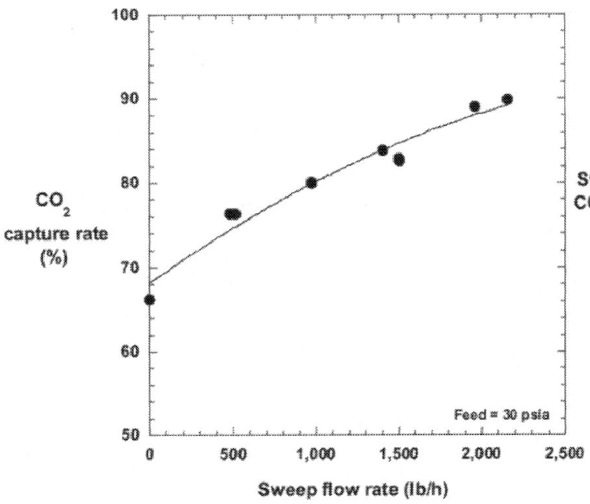


Influence on Sweep Step $CO_2$ Capture Rate Efficiency

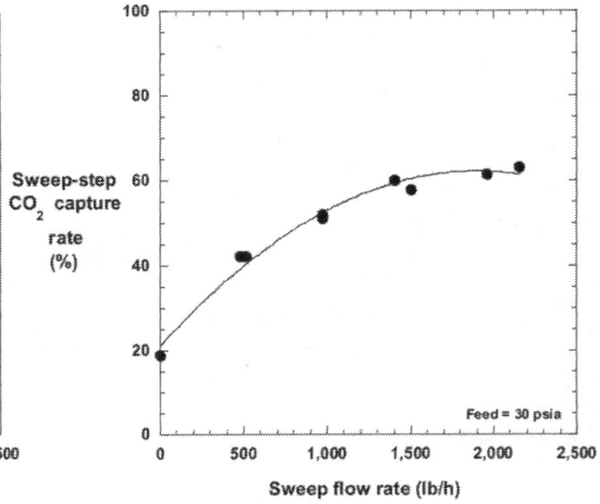


MTR

# Integrated Boiler/Membrane Systems Transition Response

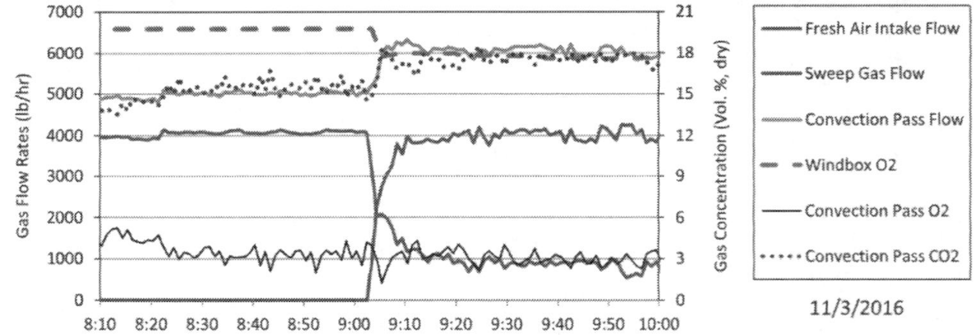


11/3/2016

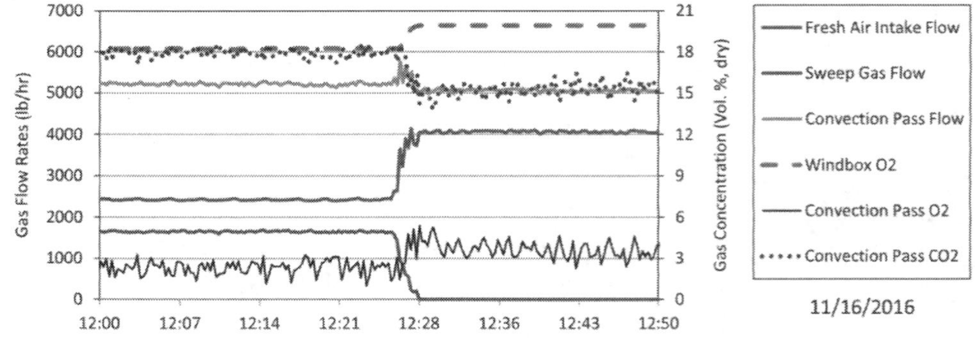


11/16/2016

MTR

# Integrated Boiler/Membrane Systems Transition Response to E-Stop

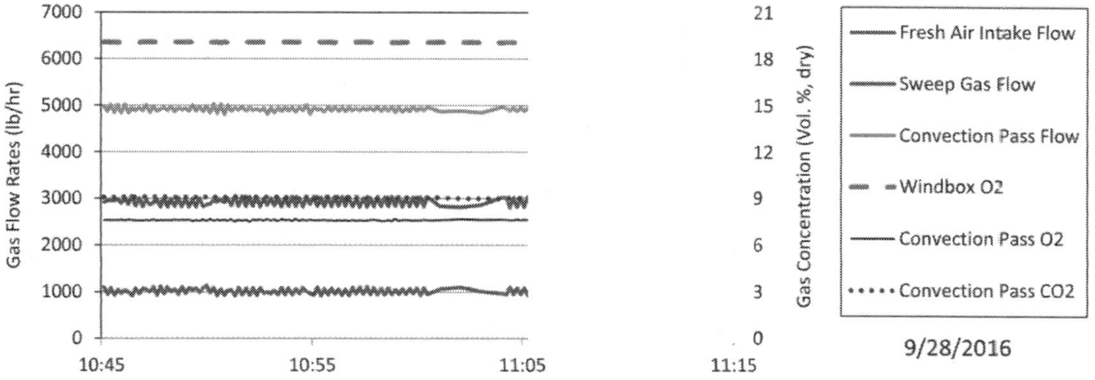


MTR

# B&W's Analysis of $CO_2$ Recycle Impact on Boiler Operation

- Furnace heat absorption is lower resulting
- "Furnace" refers to the radiant heat transfer section of the boiler upstream of the tube banks in the convection pass.
- Convection pass heat absorption is higher
- Convection pass outlet heat flux is higher
- Air heater heat absorption is higher
- Air heater flue gas outlet heat flux is higher
- Total heat absorption is slightly reduced

| Test Name | | Coal 30P M1 & M2 | Coal 27P M2 Only |
|---|---|---|---|
| Date | | 20-Oct-16 | 18-Oct-16 |
| Test Duration | (h:mm) | 7:00 | 7:15 |
| Fuel | | PRB | PRB |
| Load | (MW) | 1.5 | 1.4 |
| FEGT | (°C) | 1,179 | 1,259 |
| Convection Pass Exit Temperature | (°C) | 397 | 380 |
| Air Heater Exit Temperature (Flue Gas) | (°C) | 217 | 210 |
| Membrane Secondary Air Ratio | | 53% | 0% |
| Furnace Absorption | (MW) | 0.52 | 0.66 |
| Convection Pass Absorption | (MW) | 0.96 | 0.91 |
| Convection Pass Outlet Heat Flux | (MW) | 0.50 | 0.43 |
| Total Heat Absorption | (MW) | 1.62 | 1.68 |
| Air Heater Absorption | (MW) | 0.19 | 0.16 |
| Air Heater Outlet Heat Flux (Flue Gas) | (MW) | 0.31 | 0.27 |

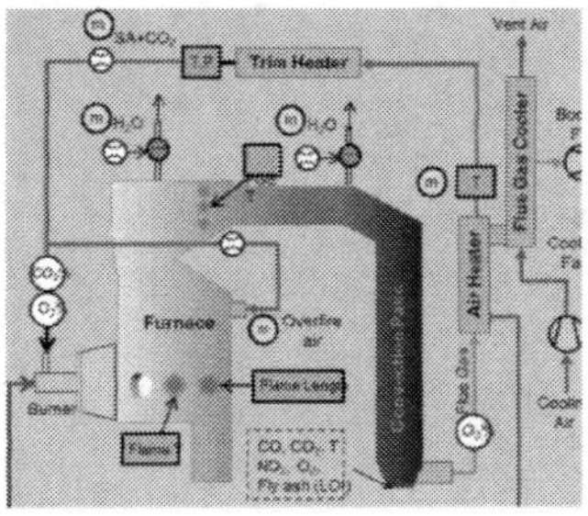


MTR

# Summary

- $CO_2$ capture membrane performance continues to improve and has been validated on the 0.05 $MW_e$ slipstream system with over 11,000 hours of runtime at NCCC
- 1 $MW_e$ small pilot operation at NCCC was completed in 2015. Testing successfully demonstrated optimized modules (low $\Delta p$, low cost) with over 1,500 hours of runtime
- 1 $MW_e$ small pilot was successfully integrated with the B&W research boiler for five weeks of integrated testing with $CO_2$ recycle to the boiler in late 2016
- The integrated membrane-boiler field test experimentally validated simulated system performance
- Boiler flame was stable throughout parametric testing allowing a full battery of stream conditions and boiler efficiency measurements

21 MTR

# Acknowledgements


- **U.S. Department of Energy, National Energy Technology Laboratory**
  - José Figueroa
  - Mike Mosser
- **Southern Company Services (NCCC)**
  - Tony Wu
- **Babcock & Wilcox**
  - Hamid Farzan
  - Jennifer Sivy
  - Andrew Mackrory


22 MTR

CARBON MANAGEMENT TECHNOLOGY CONFERENCE

# CMTC-486032-MS

# Risk Based Approach to Identify the Leakage Potential of Wells in Depleted Oil and Gas Fields for $CO_2$ Geological Sequestration

Muhammad Zulqarnain, Mehdi Zeidouni, Richard G. Hughes, Louisiana State University



This paper was prepared for presentation at the Carbon Management Technology Conference held in Houston, Texas, USA, 17-20 July 2017.



## Abstract

The selection of depleted oil and gas fields as potential $CO_2$ geological storage sites has both positive and negative aspects that need to be considered. The positives are that the storage capacity or pore volume can be reliably estimated from field's production history, and reservoir characterization can be performed with more readily available well, log or seismic data without additional expenses. The main drawback is the presence of wells in the field, as each well may provide a leakage pathway for injected $CO_2$. The leakage potential of a well is a function of its proximity to injection wells, cement coverage in the potential storage zone, well abandonment conditions including cementing of the annular space, and the nature of any barriers to prevent $CO_2$ leakage to the surface. Qualitative and quantitative risk-based approaches can be used to identify the wells that have comparatively higher leakage probabilities in comparison to other wells. The objective of this study is to use a risk-based approach to identify and categorize wells based on their leakage potential in depleted oil and gas fields. This will not only help in planning injection strategies but may also help in selection of remediation strategies. The model may be presented well by using the Fault Tree Analysis (FTA) technique. It implements screening criteria and a tier-based approach in which wells are screened and categorized into different tiers based on different well characteristics. The well characteristics include the physical distance from injection wells, the quality and portion of cement coverage of wells in the target zone, the regulations at the time of well completion, the leakage potential of sealing barriers for the targeted zone, the number of overlying shale and sand intervals and leakage of either $CO_2$ or brine to shallower wells, the nature and quality of permanent or temporary well abandonment procedures, and the quality and length of annular space covered with cement for shallower well casings or sections. Existing models for well leakage are used to quantitatively estimate the leakage rate. The risk of leakage is presented qualitatively and quantitatively in the form of leaked $CO_2$ volume to shallow aquifers or to the atmosphere. The approach is used for a representative depleted oil and gas field in southern Louisiana to show an example application of the process. The developed model provides a means to systemically identify the wells that are more likely to leak and have high consequences. Due to the reduced order nature of the tool, it should prove to be a useful tool in the planning and execution phase of the $CO_2$ sequestration process.

## Introduction

A fundamental step in the selection of a storage site for $CO_2$ sequestration is to make sure that the selected site not only meets project economics but also has good storage and long terms retention features. The safety of the storage site becomes the prime importance and dictates the decision of selecting a particular site. The selection of depleted oil and gas fields as a potential $CO_2$ geological storage site has both positive and negative aspects that need to be considered. The positives are that the storage capacity or pore volume can be reliably estimated from field's production history, and reservoir characterization can be performed with more readily available well, log or seismic data without additional expenses. The main drawback is the presence of wells in the field, as each well may provide a leakage pathway for injected $CO_2$. In addition to wells, $CO_2$ may also leak from failure of cap rock and from faults or fractures. Possible $CO_2$ leakage pathways are depicted in Figure 1.

Depending on the flow area available in a well, the pressure differential, proximity of leaky well to the $CO_2$ injection well and the nature of the spread of $CO_2$ (sweep) in the field due to heterogeneity, wells coming into contact with the $CO_2$ plume may act as possible leakage pathways. For brine leakage only the pressure differential and a wellbore offering the least resistant path to flow may be the only necessary conditions for leakage to occur. The main emphasis in this study is placed on the leakage of $CO_2$. The information of wellbore flow path available may be estimated from publically available data sources. This work utilizes the Louisiana Department of Natural Resources SONRIS system (SONRIS, 2017). Based on the available data, the wells in a depleted field may be categorized by different features, including drilling date and the nature of cementing regulations at that time, plug and abandonment data and corresponding regulations or procedures adopted to plug a well (Watson and Bachu, 2009).

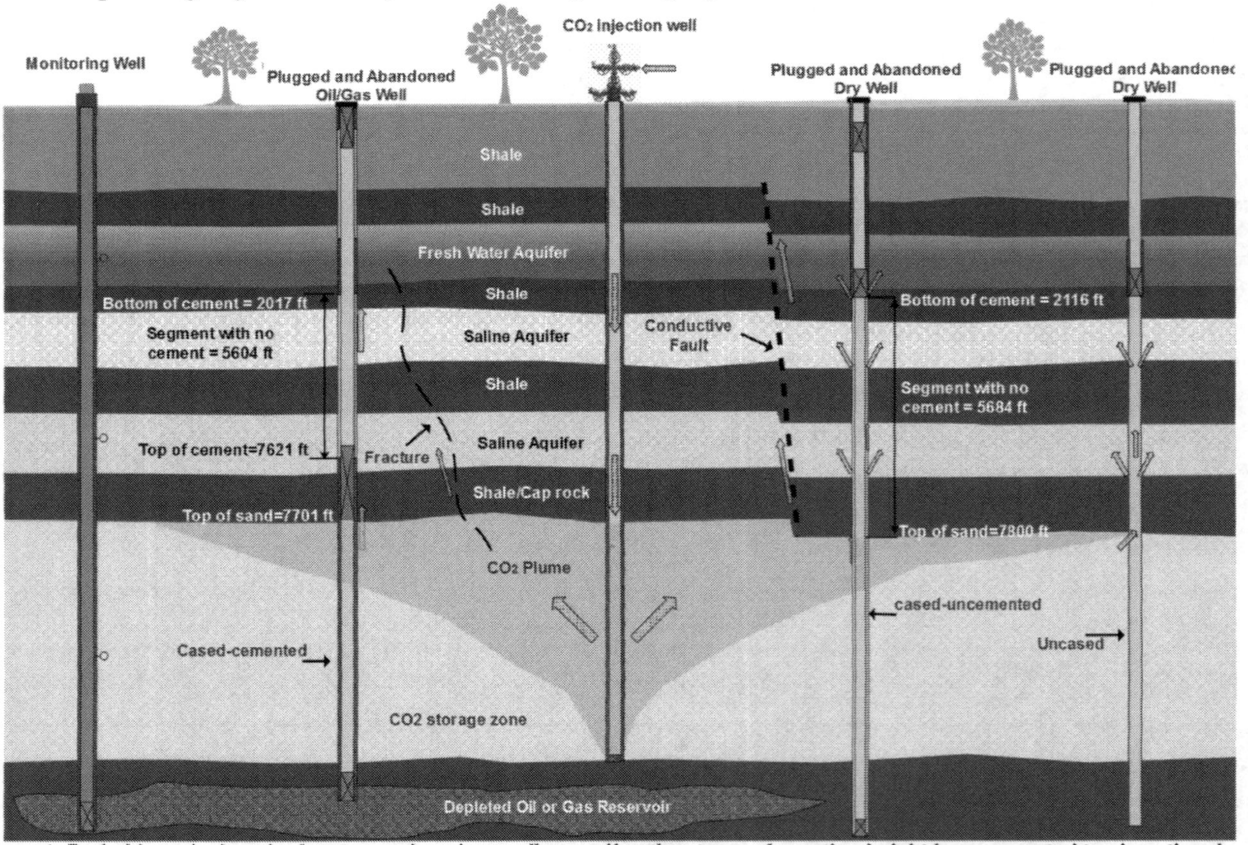


**Figure 1: Probable co leak paths for storage in a deep saline aquifer, the storage formation height is exaggerated to show the plume spread, three well types based on the cement coverage in the storage zone are shown**

Risk based approaches have been used in the past by researchers to estimate $CO_2$ leakage risk from storage zones via wellbores. A brief review of some of the most relevant work is presented here. Watson and Bachu (2007) developed a model to identify the wells with higher leakage potential by using the regulatory data. They used a score-based approach to evaluate the deep and shallow leakage potential of a wellbore. They used the spud date, abandonment data and other wellbore information to form the basis of their score system. They pointed out that elastomer bridge material used during the well plugging procedure may be damaged when it comes into contact with $CO_2$. In another study, Watson and Bachu (2008) used regulatory and wellbore data to evaluate the wells for gas or $CO_2$ leakage. They noted that majority of the leakage factors depends on the processes adopted during drilling, completion and abandonment phases of a well. Stauffer et al., (2009) used wellbore permeability as a key quantitative measure of a well's leakage potential. They used the amount of $CO_2$ or brine that could leak along the degraded cement intervals as the basis of their criteria. Celia et al. (2009) studied the effects of depth on the injection rates and showed that the leakage risk decreases for deeper storage zones and for zones with smaller number of wells penetrating the storage zone.

Nogues et al. (2012) proposed a simplified formulation to estimate the leakage along old wells and pointed out that there could be high uncertainties associated with different parameters in estimation of maximum probable leakage. Duguid et al. (2013) analyzed the cementing data of some old wells and concluded that cement for most of the wells was largely intact and was not degraded due to brine exposure. Syed et al. (2014) studied the interaction of $CO_2$ with well cement and presented a relationship showing the reduction in cement permeability as a result of this interaction. Duguid et al., (2014) studied the cement integrity of an old well for $CO_2$ injection project and found mixed results for annular cement. They found that, in some well sections the cement had very poor quality, and it will not work as a barrier against leaking fluids; however, in some sections the cement retained its properties and can act as a barrier. Gaurina and Mavar (2017) investigated the $CO_2$ leakage risk from a storage zone and looked at different leakage sources and categorized the leaks according to their severity. Results of these findings along with additional parameters of prime importance forms the basis of well leakage risk criteria proposed in this study.

The main objective of the current study is to form a criteria to categorize the leakage potential of plugged and abandoned wells, and identify the wells that are most likely to leak, and suggest the strategies to reduce the leakage risk. According to cement coverage of a well in the storage zone, the wellbore may be categorized as either

- Cased-cemented
- Cased-uncemented
- Uncased

We briefly go over each well category, describing in detail the probable leakage pathways and other important parameters necessary for leakage risk categorization.

### Category-1: Wells with complete cement coverage in the storage zone

In these wells, casing in the entire storage zone is cemented. Based on the historical data (Ozyurtkan et al., 2011) and references therein, the fluids from the storage zone may migrate to shallower permeable formations through either cement sheath, casing-cement or cement-formation micro annuli, or may flow inside the casing if casing integrity has been compromised over the years. These possible flow paths have been depicted in Figure 2 (a). These flow path assumptions are valid only if formations have not collapsed in the wellbore. It is highly probable in south Louisiana, that some of wellbores may have collapsed over time. In that case some of the above mentioned flow paths may be unavailable for leaky fluids.

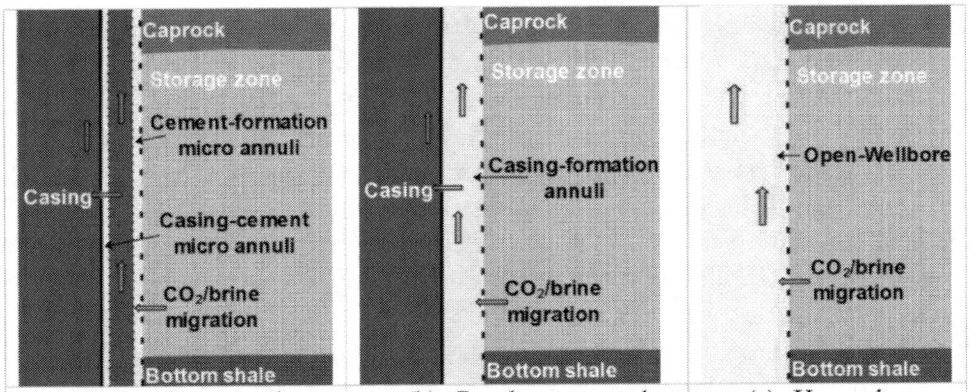

(a) Cased-cemented (b) Cased-uncemented (c) Uncased

**Figure 2: Possible fluid migration paths along wellbore inside the storage zone, for (a) a cemented casing, (b) non-cemented casing, (c) no casing, size of the micro annuli has been exaggerated to show details**

### Category-2: Wells with no cement coverage in storage zone

In these wells the casing in the storage zone is not cemented. Casing was set deeper than the storage zone and cement top is deeper than the storage zone's top. For these wells the casing-formation annular gap is open to flow, unless formation collapses has occurred in the cap rock that might hinder the fluid migration from storage zone. The possible leak flow paths for this wellbore type are depicted Figure 2 (b).

### Category-3: Wells with no casing

It is possible that some of the dry wells may only have surface casing for the protection of fresh water aquifer and the rest of the deeper well section may not be cased at all. In a worst case scenario, the entire wellbore area may provide a path for leakage fluids to escape from the storage zone, shown in Figure 2 (c).

In the next section the rational adopted for estimating the leakage risk potential of these three type of wells is briefly explained.

## Leakage risk classification criteria

A well's leakage potential from a storage zone can be attributed to the following factors

- Wellbore type: Higher leakage rate and high probability of leakage is expected for uncased wellbore sections in comparison to cased-uncemented and cased-cemented wellbores;
- Injector-Leaky well distance: The distance a potential leaky well is from the injection well is another important parameter in $CO_2$ leakage risk classification. Due to operational conditions and dynamic storage zone variable constraints, the $CO_2$ plume may not reach every well in the field. Therefore wells in the immediate vicinity of an injector well are more likely to have higher $CO_2$ leakage than the ones at greater distance;
- Storage zone boundaries: Depending on the storage zone extent, the boundary may behave as a closed, semi-closed or open boundary system. Pressure buildup rate is much higher in bounded storage zones as compared to semi-closed or open boundary zones. Higher pressure buildup may translate to higher leakage rates;
- Overlaying buffer layers (segments): In unconsolidated sands, it is likely that over time some portions of the wellbores may have been blocked by formation collapse. This may greatly influence the leakage rates, especially for uncased wells, as these may greatly alter the permeability of the buffer or collapsed zone.

The proposed leakage risk criteria in the form of a flow chart is shown in Figure 3(a). It starts with site specific data collection, sand and cement top calculations and finally calculating the leakage risk. The corresponding fault tree is shown in Figure 3(b) as well. Fault tree analysis (FTA) is a top-down approach and is a logical representation of the many events and component failures that may combine to cause the system or top event failure (Stamatelatos, 2002 and Zulqarnain, 2015). It uses 'logic gates' (mainly AND or OR gates) to show how 'basic events' may combine to cause the critical 'top event'. The top event in the present study is the leakage from the storage zone. One important aspect that is highlighted by the fault tree of a cased-cemented wellbore (Figure 3(b)) is the fact that leakage potential is a combination of flow potential and flow area available. A weakly cemented wellbore segment may not necessarily imply $CO_2$ leakage, unless other conditions also exist. Therefore flow potential and flow paths are connected by AND gate. An AND gate is activated when all the inputs are available, while an OR gate becomes active if any one of the input is available. For example, when flow potential exists, the fluid may leak through either of the three potential leak paths. Similar well specific fault trees may be constructed for cased-uncemented and uncased wells. Brine leakage FTA will be slightly different than the fault tree for $CO_2$ leakage as plume extent or sweep are not involved. As pressure builds up the brine may leak, through available flow paths.

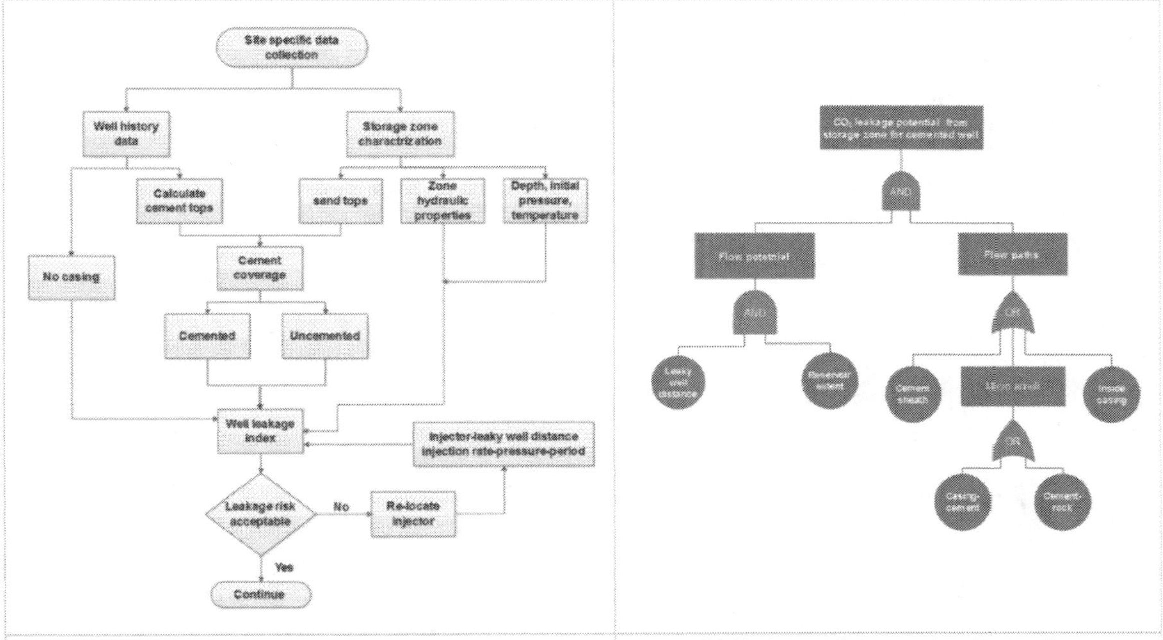


(a) Process flow chart

(b) Fault tree of cemented casing section

**Figure 3: (a) Flow chart of a well's leakage risk classification, (b) Fault tree of category-1 well section**

Next, field specific data to be used for quantitative measurement of the effect of different parameters in defining a well's leakage potential is examined.

## Field Specific data analysis

Bayou Sorrel is a nearly abandoned oil and gas field in southern Louisiana, and is selected as an example case to show the steps necessary to evaluate the leakage potential through wells. In this field, the majority of the wells were drilled in the 1950's and 60's, as shown in Figure 4 (a). Here the well permit date is used as approximate proxy for the actual drilling dates, as for some of the wells the actual start of drilling (spud date) was not available. A representative set of 14 wells were randomly selected from the total of 176 wells in the area to assess the leakage risk of each well. The approximate locations of these 14 wells are shown in Figure 4 (b). A majority of the abandoned oil and gas wells are located in the center of the field where the productive sands were located, while wells towards the outer boundary are mostly dry wells and are sparsely located.

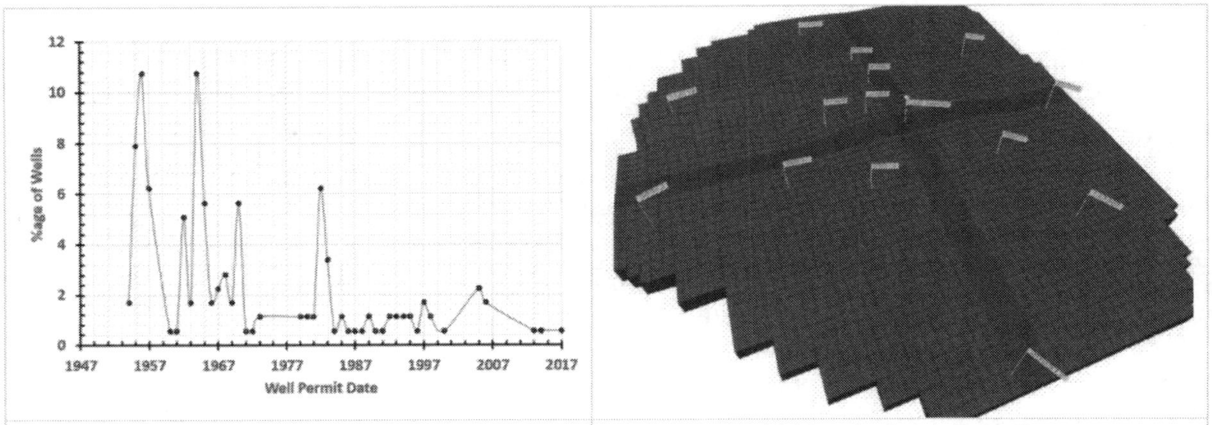


| (a) well permit date | (a) 14 randomly selected wells |
|---|---|

**Figure 4: (a) well permit dates and (b) 14 selected wells for cement coverage calculation**

Well log data is used to identify the top and bottom of the storage zone, with an average depth of 7,900 ft. The injector is initially located in the center of the field to maximize the distance from the zone boundaries. Cement tops are determined by using the well cementing data and by using the following formulation, Bourgoyne et al., (1991).

$$L = \frac{\left(Sacks \times \frac{cement\ volume}{sack} - cement\ volume\ left\ in\ casing\right)}{annular\ capacity \times cement\ access\ factor}$$

For wells in which cement access factor was not calculable, a conservative value of 2 was used, as usually an access factor of 1.5 - 1.75 is used, Bourgoyne et al., (1991). The storage zone, cemented intervals, nature of the well (dry plugged or abandoned production well), cement plugs and year in which either the well was plugged or the last workover are given in Figure 5.

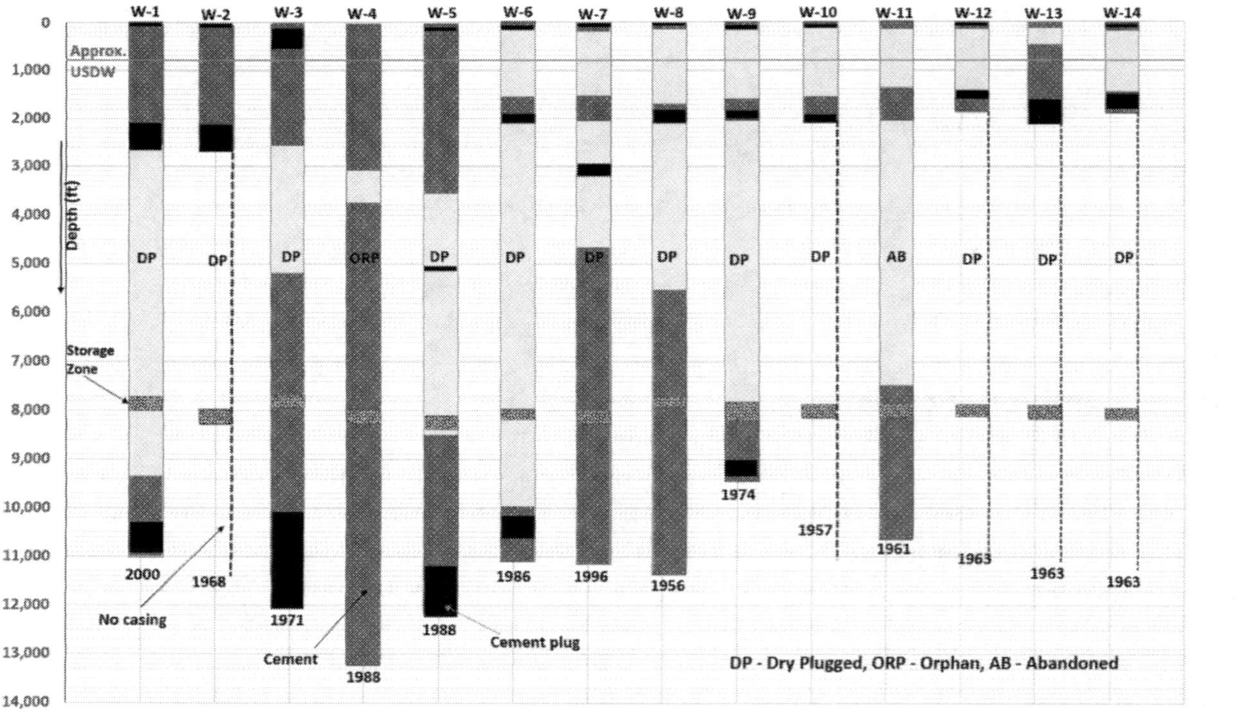


**Figure 5: Data of 14 selected wells annuli is shown with storage zone, cemented intervals, nature of well (dry plugged or abandoned oil and gas well), cement plugs and year in which last work was performed on well**

The USDW lower depth for this area is in the range of 300-500 ft. A conservative value of 800 ft is selected to allow some margin for depth variations. There are approximately 27 aquifer layers in between the selected storage zone at an average depth 7900 ft and the USDW bottom which is at approximately 800 ft. The average sand and shale thickness of these buffer layers are 131 and 115 ft respectively.

A closer examination of the data shows that dry and plugged wells drilled in 1950's and 60's may only have surface casing installed to protect the fresh water aquifer, with an average casing setting depth of 2,000 ft. These wells may provide the largest leak threat, provided that the wellbore has not collapsed in these wells. These wells need special attention due to the large flow area available to leaking fluids and very high permeabilities in the region. Leakage models used in this study are briefly explained in the next section.

## Leakage rate modeling

The amount of $CO_2$ leakage is the basic element of well leakage risk assessment, consequences can be presented in the form of leaked volume to shallow fresh water aquifers or to the atmosphere. The multi-segment Wellbore Model (MWM) and Cemented Wellbore Model (CWM), available in the NRAP-Well Leakage Assessment (WLA) toolset are used (Huerta and Vasylkivska, 2016). A brief description of the models are provided below.

### Cemented wellbore model (CWM)

This model is based on the results of 3-D numerical simulations of injection into a storage zone with an abandoned wellbore (Jordan et al., 2015). Leakage is treated as a flow through porous media by using Darcy's law, (Huerta and Vasylkivska, 2016). In its simplest form the flow rate is estimated from

$$Q = k_{eff} A \frac{\psi_L - \psi_T}{L}$$

where $Q$ is the volumetric flow rate, $k_{eff}$ is the effective permeability, A is the cross sectional area of flow, $\psi_L$ and $\psi_T$ are leakage potential at the leakage source and sink respectively and $L$ is the leak path length. This model can be used to calculate leakage to an overlying shallow aquifer and a thief zone. The thief zone is of fixed thickness and should not be located at a depth less than one third the depth of the storage zone (Huerta and Vasylkivska, 2016). Time varying pressure and $CO_2$ saturation data from reservoir simulation are used as model inputs. In this study this model is only used to study the effect of storage boundary type on $CO_2$ leakage rates.

### Multi segment wellbore model (MWM)

This model can calculate leakage to multiple overlying aquifers or thief zones and was developed by Nordbotten et al. (2009). This model focuses on modeling flow across large distances and does not take into account the flow in cement fractures and cracks. Flow inside the annulus is modeled and wellbore permeability along each overlying shale zone is prescribed along with the aquifer permeabilities. The model assumes constant density of $CO_2$ and does not incorporate any geochemical or geomechanical processes taking place inside the wellbore. This model is used to carry out the sensitivity analysis of wellbore type (cased-cemented, cased-uncemented or uncased), distance between the injector and leaky well and number of buffer layers (or barrier zones) between storage zone and leak outlet.

## Wellbore permeability calculations

For pure cement, permeability is reported in the range of micro to millidarcy (Ozyurtkan et al., 2013), while the permeability of a cemented wellbore is reported to be in the range of 1.7 mD to 170mD (Gasda et al., 2013). Therefore in this study for a cemented wellbore an average value of this reported range, 86 mD is used. For cased-uncemented annuli and uncased wellbore, we use the following two equations to find the permeability in these cases. Flow rate for a circular porous medium is expressed as by Darcy law

$$q = k\pi r^2 \frac{\Delta P}{\mu L}$$

while the Hagen–Poiseuille's equation is used for flow inside circular pipe

$$q = \frac{\pi r^4}{8} \frac{\Delta P}{\mu L}$$

Comparing these two equations, the equivalent permeability for a circular pipe can be expressed as

$$k = \frac{r^2}{8}$$

The calculated values of the permeabilities for a wellbore with ID = 9.875″, and casing OD = 7″, are shown in Table 1.

**Table 1: Permeability of cemented, cased-uncemented and uncased wellbore**

| Wellbore Type | Cased-cemented (mD) | Cased-uncemented (mD) | Uncased (mD) |
|---|---|---|---|
| Segment Permeability | 8.48E+01 | 7.91E+12 | 1.59E+13 |

Now we select the three wells representing the cased-cemented, cased-uncemented and uncased wellbore from 14 well data set and calculate the average wellbore permeabilities for these categories. Well No.4 is selected as a representative example of a cemented wellbore as it has longer segments of cemented annulus. This wellbore has three segments, two cemented and one cased-uncemented segment. We used an average permeability in series to find the average permeability of the wellbore, which is the used in the wellbore leakage model. The average permeability is given by the following expression

$$k_{avg} = \frac{\sum_{i=1}^{n} L_i}{\sum_{i=1}^{n} \left(\frac{L}{k}\right)_i}$$

To represent cased-uncemented and uncased wellbore, well # 6 and 12 are selected, based on their cement coverage. The value of average wellbore permeabilities of these three representative example wells are shown in Table 2.

**Table 2: Permeability of the representative three wellbore type**

| Wellbore | Cased-cemented (mD) | Cased-uncemented (mD) | Uncased (mD) |
|---|---|---|---|
| Average wellbore permeability | 9.23E+01 | 8.81E+02 | 1.02E+03 |

In the next section we form the basis of the well leakage risk assessment criteria.

## Well leakage index (WLI)

Now we form the criteria to define the well leakage index, which can be used to identify the wells that have relatively large leakage risk in comparison to other wells. We assign a range of values from 0 to 1 for each of the four factors of wellbore type, distance of leaky well to injection well, storage zone boundary types and number of buffer layers. Where the value 0 shows no influence and a value of 1 shows maximum influence. These four factors can be used to estimate the leakage potential of the well by calculating the well leakage index

$$WLI = CI \times DI \times LI \times BI$$

where *WLI* is the Well leakage index, *CI* is the cement index, *DI* is the distance index, *LI* is the layer index and *BI* is the boundary type index. Results of the $CO_2$ leakage models are used to assign values to these indices. Therefore the computed value of the well leakage index will provide a quantitative measure of a well's leakage potential.

Based on the WLI a tier-based approach can be developed, bounded by the limits of WLI from 0 to 1. The assumed well tiers based on the WLI are shown in Table 3.

**Table 3: Well tiers for a specific filed based on the WLI**

| Well Tiers | WLI range (fraction of field's maximum WLI) | Remarks |
|---|---|---|
| 1 | WLI <= 0.25 | Wells with minor leakage risk |
| 2 | 0.25 <= WLI <=0.50 | Wells with moderate leakage risk |
| 3 | 0.50< WLI <=0.75 | Wells with high leakage risk |
| 4 | 0.75 < WLI | Wells with severe leakage risk |

Now the results of a sensitivity study of wellbore types, injector to leaky well distance, storage zone boundary type and effect of buffer layers are presented in next section. MWM and CWM leakage models are used to carry out this sensitivity analysis.

## Results and discussions

The results of the sensitivity analysis of the different indices are presented in this section. The $CO_2$ leakage volume to a fresh water aquifer or to the atmosphere is computed and is normalized by the highest leaked volume. For example for wellbore type sensitivity analysis, the leaked volume for uncased or open wellbore is used to normalize the leaked volume of all three well types, as this represents the highest leaked rate. In this way the highest leaked volume has a fraction of 1 and other categories have values that are a fraction of 1. These fractions are than used to define corresponding indices, which are used to calculate the well leakage index.

### Sensitivity of wellbore type

The multisegment wellbore model (MWM) is used to carry out this sensitivity analysis. The model allows a maximum of 29 aquifers with 30 shale layers in-between the storage zone and the atmosphere. The Bayou Sorrel field sand data was used to define 29 gross sand intervals between the storage zone at an average depth of 7,900 ft and the atmosphere. A single value of permeability of 500 mD is used for shallow aquifers.

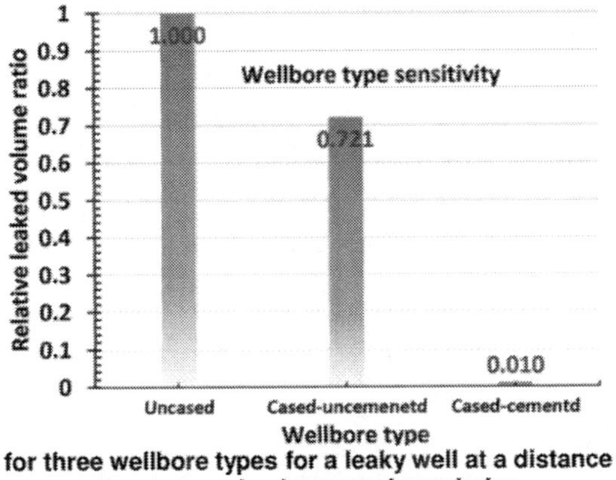


**Figure 6: Leakage normalized volume, for three wellbore types for a leaky well at a distance of 328 ft from injector location with a storage zone having open boundaries**

The average permeabilities of uncased, cased-uncemented and cased-cemented wellbores shown in Table 2 are used. The normalized leakage volume for the three well types for an active injection period of 30 years is shown in Figure 6. A 30% reduction in the total leaked $CO_2$ volume is observed when the category shifts from uncased open wellbore to cased-uncemented wellbore and an even more significant reduction is observed for a cased-cemented wellbore. We use this information to define the cement index specified in Table 1.

### Sensitivity of injector-leaky well distance

The multisegment wellbore model (MWM) is used to carry out this sensitivity analysis. Due to the nature of the $CO_2$ displacement in the storage zone, the distance between the injection and the leaky well becomes one of the important parameter for $CO_2$ leakage risk classification. In this section we examine the results of sensitivity of leaked volume to injector-leaky well distance.

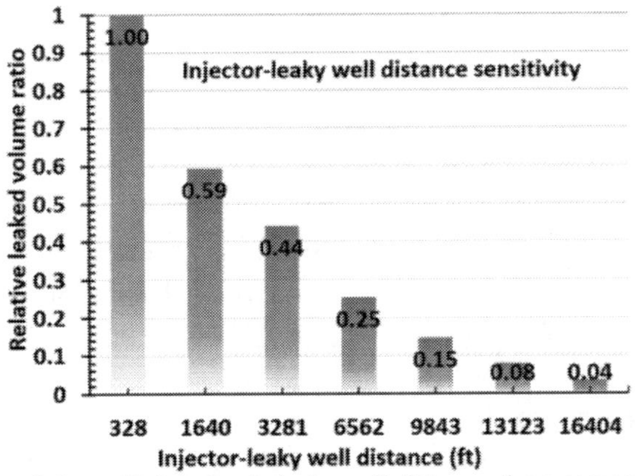


**Figure 7: Normalized leaked volume variation with injector to leaky well distance, for a cemented wellbore and with a storage zone having open boundaries**

A leaky well at a distance of 328 ft is used to normalize the leaked volume, as it has the highest leaked volume. The results of the leakage model for a cemented wellbore configuration with open zone boundaries scenario for an active injection period of 30 years are shown in Figure 7.A nearly 66% drop can be noticed for the first 3,281 ft and then the cumulative leakage volume decreases slowly. Maximizing the injector-leaky well distance especially the distance from uncased wells will results in substantially reducing the $CO_2$ leakage volume through wellbore over the course of a projects' life. This parameter can be optimized during the planning phases of a project.

### Sensitivity of storage zone boundary type

The storage zone boundary condition dictates the pressure buildup in the zone. Pressure builds rapidly for a closed boundary system, as compared to a semi-closed or an open boundary system. Reservoir simulations are performed to obtain pressure and $CO_2$ saturation profiles for a cemented wellbore model set of inputs. Three simulation runs were carried out for a closed, semi-closed and open boundary storage zone, at a constant injection rate of 2.46 Mton/year (0.1 $m^3/s$). A constant rate scenario is assumed to be consistent, as for other wellbore leakage models a constant rate of 0.1 $m^3/s$ was used. For a closed system the injection rate is reduced as the well bottom hole pressures reaches 80% (6267 psi) of the fracture pressure, and injection is continued for a total 52 years, Figure 8 (a). This 52 year time frame is selected due to the fact that in nearly this time, the $CO_2$ front reaches the storage zone boundary for the more open boundary scenarios, detailed information about problem setup can be found in (Zulqarnain, Zeidouni, & Hughes, 2017). Therefore the same injection period is used for all three boundary types. In order to capture the sensitivity of pressure buildup only, a well located in close proximity (328 ft m) to the injection well is selected, so that the plume extent does not effects the results. The $CO_2$ plume extent for the three boundary types is shown in Figure 8. Commercially available software Petrel (2014) and CMG-GEM (2017) are used to create 3D geological model and perform reservoir simulations respectively. Plume extent is taken care by the injector-leaky well distance sensitivity analysis.

The smallest spread of the $CO_2$ plume is seen in the closed system for the studied constant injection rate Figure 8 (b). For this case the bottom hole pressure limit is reached within 5.41 years, with an increase of 2491 psi from initial zone pressure. The $CO_2$ plume is mainly concentrated around the injection well and a very large portion of the storage zone remain upswept by $CO_2$. This will have implications for $CO_2$ and brine leakage rates. For a majority of the wells, $CO_2$ leakage will not occur during the injection period, but brine may leak excessively to overlaying zones due to high pressure buildup. Since the $CO_2$ plume is of limited extent, the injection well should be located such that it is away from the leaky well locations towards the edges of the storage zone to further reduce the $CO_2$ leakage risk. The majority of wells are concentrated in the middle of the field, where the injector is currently located in the simulations.

In the semi-closed system Figure 8 (c), it was assumed that the pressure in the storage zone is supported by a limited extent neighboring aquifer having a size 3 times the size of the storage zone size. Pressure increases with time and an increase of 2184 psi was noted in the well bottom-hole pressure after a period of 52 years. The plume during the injection period does not reach the zone boundary. Also the spread of $CO_2$ is not homogeneous, therefore some of the wells towards the outer north-west and south-west edges of storage zone do not encounter the $CO_2$ plume.

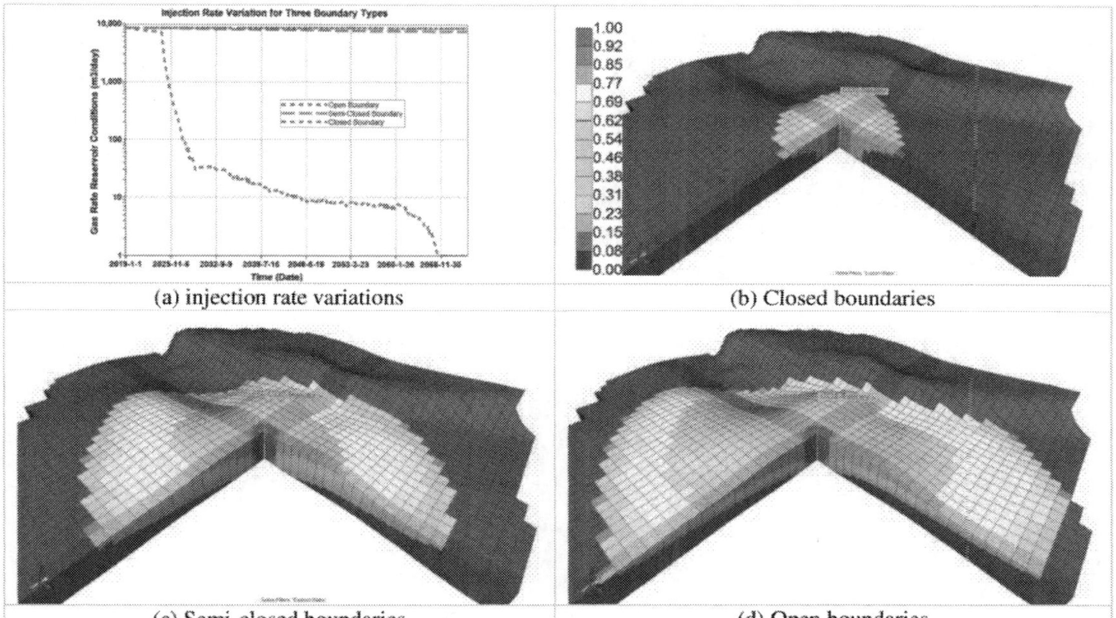


(a) injection rate variations

(b) Closed boundaries

(c) Semi-closed boundaries

(d) Open boundaries

**Figure 8: $CO_2$ plume extent for (a) closed boundary, (b) semi-infinite and (c) open boundary storage zone**

For the open boundary system, pressure of the storage zone does not increase substantially, with an increase of only 201 psi noted in the injection well's bottomhole pressure over the injection period of 52 years. The extent of the $CO_2$ plume is largest in open boundary system as can be seen in Figure 8 (d). However, there are still some portions of the storage zone in which the $CO_2$ plume has not yet reached. The $CO_2$ plume may keep spreading due to buoyancy even after $CO_2$ injection has stopped, but at a substantially reduced rate. The results presented here are only for the time interval of active $CO_2$ injection period.

When we collectively look at the results, we can observe the following trends. In the case of the closed boundary system, the pressure increase is the highest and plume spread is lowest. While in the case of the open boundary system, the pressure increase is the smallest and the plume extent is largest. The storage zone with semi-closed boundary behaves in-between these open and closed systems.

The temporal profile of pressure and $CO_2$ saturation for leaky wells are extracted from the reservoir simulation data, and are used as inputs to the Cemented Wellbore Model (CWM). The cumulative $CO_2$ leaked volume is used as an indicator to see the relative difference of different boundary types. The leaked $CO_2$ volume is normalized using the closed boundary scenario data and results are shown in Figure 9.

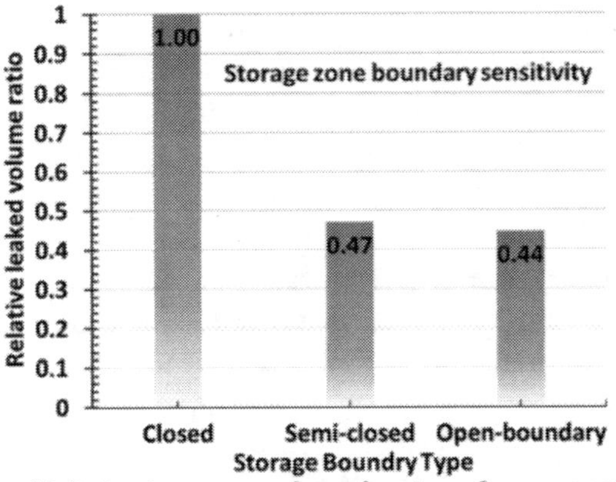


**Figure 9: Normalized leaked volume sensitivity to storage zone boundary type, for a cemented wellbore type having a distance of 3oo ft from the injector location**

The difference in the leaked volume ratio between the closed and the semi-closed boundary type is substantial and there is only a small change between the semi-closed and the open boundary scenario.

## Sensitivity of buffer layers

In south Louisiana with unconsolidated formations, it is possible that over time some of the wellbore sections may have been collapsed. A majority of the drilled formations are made up of shale and therefore shale is the cause of most of the wellbore problems. A collapsed wellbore section may act as a barrier and the wellbore permeability may be altered substantially. This is especially true in the case of an uncased wellbore. This in turn may prevent the migration of leaky fluids through the wellbore. In this section we study the hypothetical but plausible scenario that some wellbore sections may have been collapsed over time and the permeability of these sections is reduced substantially as compared to an open wellbore. A good section of cemented wellbore may also fall under this category.

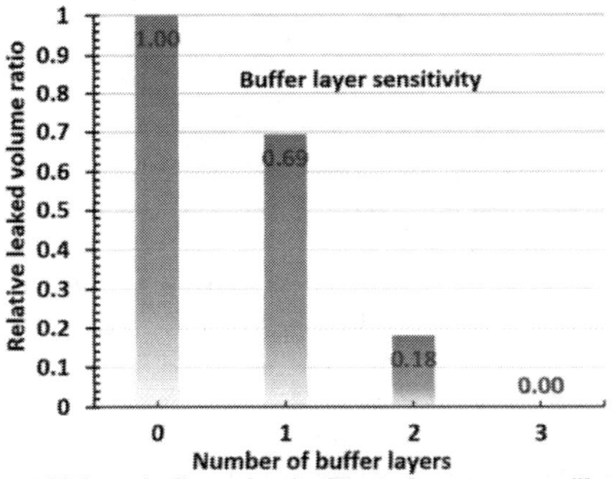


**Figure 10: Normalized leaked volume sensitivity to buffer or barrier layers, for an open wellbore configuration at a distance of 300 ft from injector location for a storage zone with open boundaries**

An uncased wellbore configuration is used to study the buffer layer effects. We assume the thickness of each buffer layer is approximately 100 ft (30 m) and select shale layers having nearly the same thickness in the upper portions of the well with less than 4,000 ft depth and carry out the sensitivity analysis. A permeability value of 0.01 mD is assumed for these buffer layers. The results of the leakage rate to fresh water aquifers are reported in Figure 10. A nearly 30% reduction in leakage volume is noted if one such buffer layer or segment exists and a higher reduction of nearly 82% is noticed if two buffer or collapsed zones are present each of which has 100 ft length and 0.01 mD permeability.

## Calculation of the wellbore leakage index

Based on the results of the sensitivity of four wellbore parameters, we now formulate the criteria to assign the indices to different parameters. The suggested values based on the modeling results are shown in Table 4.

**Table 4: Assign indices to different variable categories**

| Variables | Category-1 | Category-2 | Category-3 |
|---|---|---|---|
| **Wellbore Type** | **Cased-cemented** | **Cased-uncemented** | **Uncased** |
| Cement Index (CI) | 0.01 | 0.72 | 1 |
| | | | |
| **Injector-leaky well distance (m)** | **5000** | **1000** | **100** |
| Distance index (DI) | 0.04 | 0.44 | 1 |
| | | | |
| **Boundary Type** | **Open boundary** | **Semi-closed** | **Closed** |
| Boundary Index | 0.44 | 0.47 | 1 |
| | | | |
| **No. of Buffer Layers** | **2** | **1** | **0** |
| Layer Index (LI) | 0.18 | 0.69 | 1 |

Now we apply this criteria to the sample of 14 selected wells. The field is bounded by a fault to the north, so most probably it will behave like a semi-closed system. The well information and assigned indices are shown in Table 5. After calculation of the well leakage index for each of the wells, we use the criteria specified in Table 3, to assign the well tiers, in order to identify the wells that have relatively higher leakage potential. Initially we use the WLI value for no buffer layers as a conservative approach to assign the well tiers.

**Table 5: Assigned variable indices for the 14 selected wells for well types 1-cased-cemenetd, 2-cased-uncemented, 3-uncased**

| Well Sr. No. | Wellbore Type | Cement Index | Injector-leaky well distance (ft) | Distance Index | Storage zone boundary type | Well leakage index when number of buffer layers | | |
|---|---|---|---|---|---|---|---|---|
| | | | | | Semi-Closed | 0 | 1 | 2 |
| **1** | 2 | 0.72 | 984 | 0.73 | 0.47 | 2.4602E-01 | 1.6975E-01 | 2.3180E-02 |
| **2** | 3 | 1.00 | 2297 | 0.52 | 0.47 | 2.4272E-01 | 1.6748E-01 | 2.2563E-02 |
| **3** | 1 | 0.01 | 4921 | 0.33 | 0.47 | 1.5370E-03 | 1.0605E-03 | 9.0477E-07 |
| **4** | 1 | 0.01 | 1640 | 0.60 | 0.47 | 2.8203E-03 | 1.9460E-03 | 3.0462E-06 |
| **5** | 2 | 0.72 | 9843 | 0.15 | 0.47 | 5.2372E-02 | 3.6137E-02 | 1.0505E-03 |
| **6** | 2 | 0.72 | 5906 | 0.28 | 0.47 | 9.5333E-02 | 6.5780E-02 | 3.4806E-03 |
| **7** | 1 | 0.01 | 6890 | 0.24 | 0.47 | 1.1440E-03 | 7.8937E-04 | 5.0123E-07 |
| **8** | 1 | 0.01 | 984 | 0.73 | 0.47 | 3.4169E-03 | 2.3577E-03 | 4.4715E-06 |
| **9** | 2 | 0.72 | 16404 | 0.03 | 0.47 | 9.4121E-03 | 6.4943E-03 | 3.3927E-05 |
| **10** | 3 | 1.00 | 11483 | 0.12 | 0.47 | 5.4734E-02 | 3.7766E-02 | 1.1473E-03 |
| **11** | 1 | 0.01 | 328 | 1.00 | 0.47 | 4.7002E-03 | 3.2431E-03 | 8.4606E-06 |
| **12** | 3 | 1.00 | 4921 | 0.33 | 0.47 | 1.5370E-01 | 1.0605E-01 | 9.0477E-03 |
| **13** | 3 | 1.00 | 2625 | 0.48 | 0.47 | 2.2713E-01 | 1.5672E-01 | 1.9757E-02 |
| **14** | 3 | 1.00 | 9843 | 0.15 | 0.47 | 7.2739E-02 | 5.0190E-02 | 2.0264E-03 |

The results of the assigned well tiers are shown in Figure 11. From this distribution it is easy to identify the wells with the highest leakage potential. Wells assigned to tier-4 need special attention, as these have the highest leakage potential.

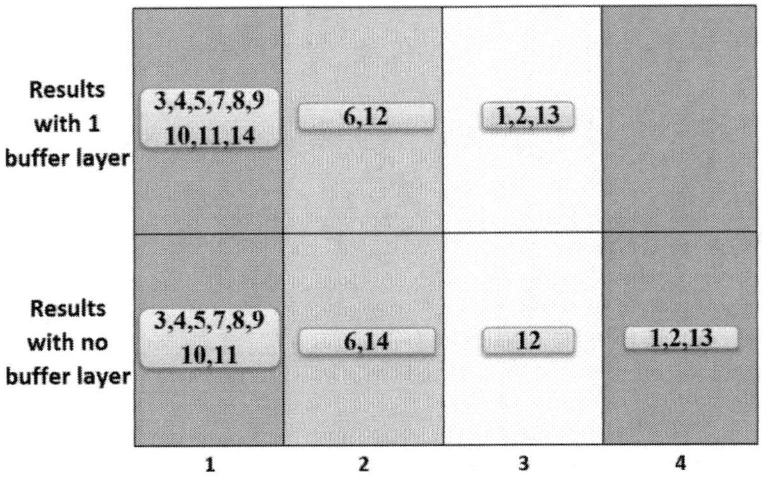


**Figure 11: Well tiers assigned to 14 selected wells, based on their well leakage index value**

Well 1 falls in tier-4 primarily due to its close proximity to the injection well and its wellbore category of cased-uncemented, which have relatively higher permeability values than cased-cemented wellbores. Similar arguments are true for other wells as well. Now we examine the effect of buffer layers. As can be seen in Figure 11, the presence of only one buffer segment or layer reduces the leakage risk of some of the wells, visible by their well tier shift to lower values. The presence of more buffer layers or well segments that may act as barriers will further reduce the leakage risk of wells.

The criteria used to define indices have a range from 0 to 1. These ranges may be revised and may be calibrated to account for the relative effect of different indices. In that case the ranges may not necessary be restricted between 0 and 1 and relatively higher values of such an index would be the indicators for concern.

## Conclusions

A risk based approach is presented that can be used to identify wells that have relatively higher leakage potential as compared to other wells penetrating a storage zone. The method uses qualitative and quantitative measures of assessing a well's leakage potential. The approach uses well leakage index as the primary variable to categorize the wells into four tiers, with tier-1 having lowest and tier-4 having highest leakage potential. The cement data of a representative sample of 14 wells from a Louisiana oil field is used to categorize the wellbore types into cased-cemented, cased-uncemented or uncased wellbores to calculate their respective leak permeabilities. The results of well cement data show that dry and plugged wells drilled in the 1950's and 60's need special attention. These wells may only have surface casing installed to protect fresh water aquifers and well segments passing through the deeper storage zones may not be cased. These wells may provide the largest flow area to leaking fluids provided that the wellbore has not collapsed. It was also observed that all wells had some sort of protection for the fresh water aquifers, either cemented surface casing or cement plugs installed at an average depth of 2,000 ft. Well leakage index is based on the cement coverage of the well section passing through the storage zone, the well's proximity to the injection well, the nature of the storage zone boundaries and the number of buffer barrier layers or zones between the storage zone and the base of the USDW. The $CO_2$ leakage to the USDW is calculated for a period of 30 years and average leaked volume is estimated for a constant storage zone injection rate of 2.46 Mt/year. The leaked volume is normalized for each variable category and a well leakage index is calculated based on these normalized values. The well leakage index provides a quantitative measure of a well's leakage risk. It is also noted that optimization of injector location is of prime importance for well leakage risk assessment. If possible it should be located in the field where wells are sparsely located. The proposed risk based model to categorize the wells based on the well leakage index should facilitate in the planning and execution stages of a project.

## Nomenclature

| | |
|---|---|
| FTA | Fault tree analysis |
| *CMW* | Cemented wellbore model |
| *MWM* | Multisegment wellbore model |
| *WLI* | Wellbore leakage index |
| *CI* | Cement-index |
| *DI* | Distance index |
| *BI* | Boundary index |
| *LI* | layer index |
| $k_{eff}$ | Effective permeability (mD) |
| $k_{avg}$ | Average permeability (mD) |
| $\psi$ | Leakage potential |

## Acknowledgements

The work is financially supported by U.S. Department of Energy for Carbon Storage Assurance and Facility Enterprise project (Grant # FE0029274).

## References

Bourgoyne Jr, A.T., Millheim, K.K., Chenevert, M.E., Young Jr, F.S., (1991). Applied Drilling Engineering, Society of Petroleum Engineers, SPE Textbook Series Vol. 2, ISBN: 978-1-55563-001-0

Celia, M.A., Nordbotten, J.M., Bachu, S., Dobossy, M., Court, B., 2009. Risk of Leakage versus Depth of Injection in Geological Storage. Energy Procedia 1, 2573–2580. doi:10.1016/j.egypro.2009.02.022

CGM-GEM, 2017. Compuertr Modeling Group. Calgary, Canada.

Duguid, A., Butsch, R., Carey, J.W.,., Celia, M,A.,, Chugunov, N., Gasda, S., Ramakrishnan, T.S., Stamp, V., Wang, J., 2013. Pre-injection baseline data collection to establish existing wellbore leakage properties. Energy Procedia 37, 5661–5672. doi:10.1016/j.egypro.2013.06.488

Duguid, A., Carey, J.W., Butsch, R., 2014. Well integrity assessment of a 68 year old well at a $CO_2$ injection Project. Energy Procedia 63, 5691–5706. doi:10.1016/j.egypro.2014.11.602

Gasda, S.E., Celia, M. a., Wang, J.Z., Duguid, A., 2013. Wellbore Permeability Estimates from Vertical Interference Testing of Existing Wells. Energy Procedia 37, 5673–5680. doi:10.1016/j.egypro.2013.06.489

Gaurina, M. N., Mavar, K.N., 2017. Depleted hydrocarbon reservoirs and $CO_2$ injection wells – $CO_2$ leakage assessment. The Mining-Geology-Petroleum Engineering Bulletin, 15–27. doi:10.17794/rgn.2017.2.3

Huerta, N. J.; Vasylkivska, V.S., 2016. Well Leakage Analysis Tool (WLAT) User's Manual. ; NRAP-TRS-III-011-2016; NRAP Technical Report Series; U.S. Department of Energy, National Energy Technology Laboratory: Albany, OR, 2016; p 44.

Jordan, A.B., Stauffer, P.H., Harp, D., Carey, J.W., Pawar, R.J., 2015. A response surface model to predict $CO_2$ and brine leakage along cemented wellbores. Int. J. Greenh. Gas Control 33, 27–39. doi:10.1016/j.ijggc.2014.12.002

Nogues, J.P., Court, B., Dobossy, M., Nordbotten, J.M., Celia, M.A., 2012. A methodology to estimate maximum probable leakage along old wells in a geological sequestration operation. Int. J. Greenh. Gas Control 7, 39–47. doi:10.1016/j.ijggc.2011.12.003

Nordbotten, J.A.N.M., Kavetski, D., Celia, M.A., 2009. Model for $CO_2$ Leakage Including Multiple Geological Layers and Multiple Leaky Wells 43, 743–749. doi:10.1021/es801135v

Ozyurtkan, M.H., Altun, G., Mihcakan, I.M., Serpen, U., 2013. An Experimental Study on Mitigation of Oil Well Cement Gas Permeability. International Petroleum Technology Conference, 16577, pp. 26–28.

Petrel, 2014. Petrel Exploration & Production software platform, Schlumberger, Houston, USA.

SONRIS, 2017. Strategic Online Natural Resources Information System. Louisiana State, USA.

Stamatelatos Michael, 2002. Fault tree Handbook with Aerospace Applications, NASA Office of Safety and Mission Assurance, NASA Headquarters, Washington, DC 20546.

Stauffer, P.H., Viswanathan, H.S., Pawar, R.J., Guthrie, G.D., 2009. A System Model for Geologic Sequestration of Carbon Dioxide. Environ. Sci. Technol. 43, 565–570. doi:10.1021/es800403w

Syed, A., Shi, J.Q., Durucan, S., Korre, A., Nash, G., 2014. Experimental and numerical investigations into $CO_2$ interactions with well infrastructure and its impact on long term well integrity. Energy Procedia 63, 5707–5714. doi:10.1016/j.egypro.2014.11.603

Watson, T., Bachu, S., 2009. Evaluation of the Potential for Gas and $CO_2$ Leakage Along Wellbores. SPE Drill. Complet. 24, 115–126. doi:10.2118/106817-PA

Watson, T.L., Bachu, S., 2008. Identification of Wells with High $CO_2$ -Leakage Potential in Mature Oil Fields Developed for $CO_2$ -Enhanced Oil Recovery. 2008 SPE Improv. Oil Recover. Symp. 10. doi:10.2118/112924-MS

Watson, T.L., Bachu, S., 2007. Evaluation of the Potential for Gas and $CO_2$ Leakage Along Wellbores, in: E&P Environmental and Safety Conference. Society of Petroleum Engineers. doi:10.2118/106817-MS

Zulqarnain, M., 2015. Deepwater Gulf of Mexico Oil Spill Scenarios Development and Their Associated Risk Assessment. (PhD), Louisiana State University, Baton Rouge, LA, USA. (etd-07022015-181918)

Zulqarnain, M., Zeidouni, M., & Hughes, R. G., 2017. Static and Dynamic CO2 Storage Capacity Estimates of a Potential $CO_2$ Geological Sequestration Site in Louisiana Chemical Corridor. CMTC-486020-MS, Paper presented at the Carbon Management Technology Conference, Houston, Texas, USA.

CARBON MANAGEMENT TECHNOLOGY CONFERENCE

## CMTC-486093-MS

# Research on Enhancing Heavy Oil Recovery Mechanism of Flue Gas Assisted Steam Flooding

Zhuangzhuang Wang, Zhaomin Li, Teng Lu, China University of Petroleum (East China); Qingwang Yuan, University of Regina; Jianping Yang, Hongyuan Wang, Shizhong Wang, PetroChina Liaohe Oilfield company



This paper was prepared for presentation at the Carbon Management Technology Conference held in Houston, Texas, USA, 17-20 July 2017.



## Abstract

Flue gas is industrial waste gas produced by the burning of fossil fuels. Its main compositions are 10% - 15% of carbon dioxide and 80% - 85% of the nitrogen, two key components needed for gas flooding. Adding a certain amount of flue gas into steam in displacement could decrease steam partial pressure, improve the steam quality and reduce heat loss, resulting in reduction of steam injection amount and improvement of development performance. The objective of the research is to investigate the mechanism of enhanced oil recovery (EOR) of flue gas assisted steam flooding which has a dual significance of reducing greenhouse gas emissions and improving oil displacement efficiency.

In this paper, PVT measurements at high temperature and high pressure (HTHP) were firstly conducted to analyze the effect of flue gas on property of heavy oil. Then sandpack displacement experiments containing 5 sub-experiments: steam flooding, flue gas assisted steam flooding, first steam flooding then flue gas assisted steam flooding, water flooding, flue gas assisted water flooding, were operated to compare the contribution of heat and gas on recovery. In the experiment of flue gas assisted steam flooding, the production rate and composition of gas were measured and analyzed, and the form of the produced oil was also compared with steam flooding.

The PVT measurements results show that with flue gas dissolved, viscosity of heavy oil declines and volume expands. The solubility of flue gas in heavy oil decreases with temperature and increases with pressure. The greater the solubility of flue gas, the lower viscosity of heavy oil, the larger volume. At the same solubility, as the temperature increases, the viscosity reduction effect of flue gas weakens and the volume expansion effect enhances.

The displacement experiments results indicate that the addition of flue gas to steam can significantly improve oil displacement efficiency compared with steam injected alone. In the test of flue gas assisted steam flooding, the heavy oil was produced in the form of foamy oil because of dissolution of flue gas, especially $CO_2$, which could expand oil volume and reduce the flow resistance of heavy oil to a certain extent. Besides, the cumulative volume of the produced gas was smaller than the injected gas, and in the produced gas the proportion of $CO_2$ was less than the injected proportion. Furthermore, the contribution of flue gas to recovery when injected with steam is greater than with water because of the synergistic effect of heat and gas.

This research reveals the interactions between flue gas and oil in displacement and the synergistic effect of steam and flue gas on EOR, which provides the theoretical support for the application of flue gas assisted steam flooding.

## Introduction

In recent years, with greenhouse effect and global warming becoming increasingly serious, energy-saving and emission reduction has attached people's focus. Flue gas generated by the burning of fossil fuels is a major source of greenhouse gas. CCUS ($CO_2$ capture, utilization and storage) is an effective means to reduce greenhouse gas emissions, which has good application prospect in oilfield[1-6].

Steam flooding is a conventional exploiting method at present, but there are some issues including huge consumption of steam, high water cut and low oil recovery[7-8]. In order to improve the development performance of steam flooding, flue gas assisted steam flooding was proposed that flue gas is injected with steam together. The main components of flue gas are 80-85% of $N_2$ and 15-20% of $CO_2$. $N_2$ has the effect of increasing reservoir energy and heat insulation, and $CO_2$ has the effect of dissolution, viscosity reduction and volume expansion[9-10]. To inject flue gas with steam into reservoir not only improves the oil recovery of steam flooding, but also reduces the greenhouse gas emissions.

Now around flue gas assisted steam flooding, a lot of experiments and simulations have been carried out. Although many mechanisms of flue gas has been put forward to account for EOR, such as decreasing steam partial pressure, improving the steam quality and reducing heat loss and so on[11-14], no one is confirmed. So in order to investigate the effect of flue gas in displacement, PVT measurements at HTHP and sandpack flooding experiments were carried out in this paper. In PVT measurements, the dissolution law of flue gas into heavy oil was measured first and then the effect of flue gas on property of heavy oil was analyzed. Five tests of sandpack flooding were conducted to compare the contribution of heat and gas on recovery. The composition of the produced gas and the form of the produced oil in the experiments of flue gas assisted steam flooding were analyzed to study the interactions between flue gas and heavy oil in displacement.

## Experiments

### Materials

The heavy oil used in the experiments was sampled from Liaohe oilfield, China. Its basic properties are shown in the Table 1.

The flue gas used in the experiments was prepared in laboratory with $N_4$ and $CO_2$ with mole fraction of 80% and 20% respectively.

In sandpack floodings, the sandpack models were packed by refined silica sand, the parameters of which are shown in Table 2.

**Table 1 Properties of the heavy oil used in the experiments**

| Property | Value |
|---|---|
| Density at 20℃(kg/m$^3$) | 936.8 |
| Viscosity at 50℃(mPa·s) | 8860 |
| Saturate content (wt%) | 44.23 |
| Aromatic content (wt%) | 30.76 |
| Resin content (wt%) | 21.60 |
| Asphaltene content (wt%) | 3.41 |

### Apparatus

In the PVT test, a rotatable PVT device and a falling-ball viscometer were used, as is shown in Fig.1. The rotatable PVT device mainly consists of a heated piston cylinder, a hand-cranked pump and a temperature and rotation controller. Gas is mixed with oil and dissolved in the piston cylinder. The hand-cranked pump with scale is used to provide pressure and measure volume. The accuracy of the scale is ±0.05mL and the accuracy of the pressure gauge is ±0.01MPa. The falling-ball viscometer is connected with the piston cylinder, in which the viscosity of saturated oil is measured. Its measurement range is 100-100000 mPa·s. The BPR is used to provide a pressure bigger than the saturation pressure to prevent solution gas from releasing.

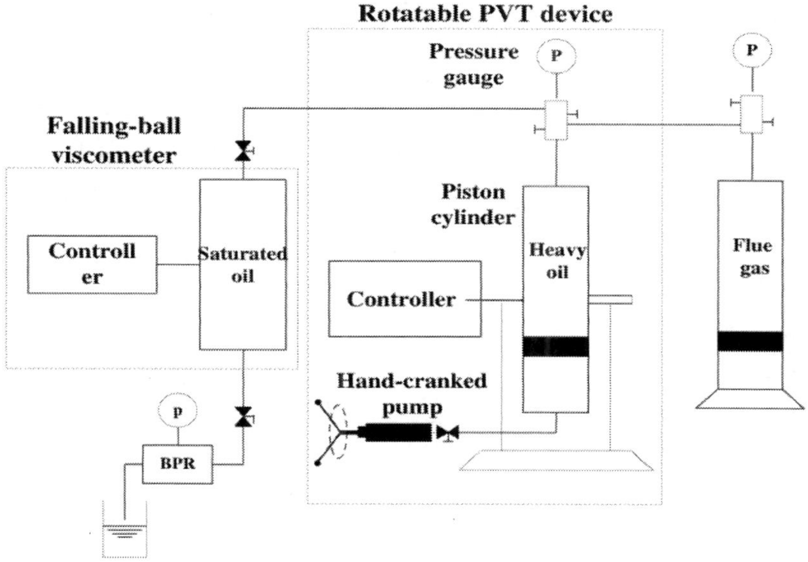


**Fig.1 Schematic diagram of the PVT measurement apparatus**

The schematic of the apparatus used in the experiments of sandpack flooding is shown in Fig.2. It mainly consists of injection system, multifunction displacement system, data acquisition system and production system. Injection system mainly includes an ISCO pump, a steam generator, a gas flowmeter , a gas cylinder and an oil cylinder. The injection rate of steam and oil are controlled by ISCO pump with accuracy of ±0.001mL. The gas injection rate is controlled by gas flowmeter with accuracy of 0.01Sccm. Multifunction displacement system mainly includes sandpack model, incubator and back-pressure regulator (BPR). Sandpack model has a length of 60.00cm and an inner diameter of 2.54cm, on which two pressure detecting points are distributed evenly. Data acquisition system records pressure by pressure transducers that are connected to a computer. Production system consists of liquid-gas separator, oil-water separator, gas mass-flow meter and balances. The produced gas is measured by gas mass-flow meter with accuracy of 0.01Sccm. The produced oil and water are measured separately by balance with accuracy of 0.01g.

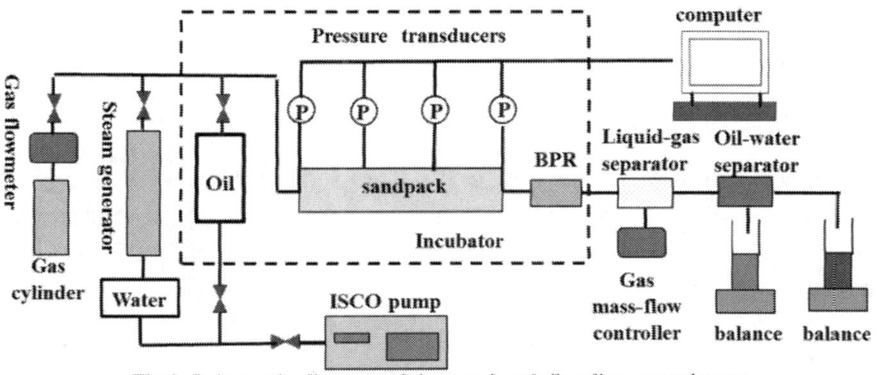


**Fig.2 Schematic diagram of the sandpack flooding experiment**

## Experimental procedures

### *Experiment steps for PVT test*

(1) A certain volume of gas was injected into the piston cylinder containing oil according to the solution gas-oil ratio.
(2) The mixture of oil and gas was provided a higher pressure by the hand-crank pump with the cylinder rotating to make sure the gas could be dissolved into the oil completely and quickly.
(3) Part of the saturated oil was injected into the falling-ball viscometer to measure the viscosity of the live oil first.
(4) For the remaining live oil, its volume and corresponding pressure were recorded with the piston withdrawing gradually.
(5) On the pressure curve against volume, the pressure at the inflection point could be regarded as saturation pressure. Meanwhile, the volume factor could be calculated according to known saturation volume and dead oil volume.

### *Experiment steps for sandpack flooding*

(1) Sandpack model was prepared with proper permeability and porosities. Then it was saturated with water after evacuated for 4 hours. The pore volume and permeability could be measured and calculated.
(2) After the temperature of sandpack model reached stable in the incubator of 70℃, it was saturated with oil at the rate of 0.1mL/min for more than 2PV. Then the irreducible water saturation and initial oil saturation were calculated.
(3) For steam flooding, steam of 250℃ was injected at the rate of 2mL/min until the water cut of the produced liquid reached 98%. The produced gas, water and oil were measured respectively.
(4) For flue gas assisted steam flooding, flue gas was injected with steam together at the gas-water ratio of 1 with other parameters unchanged.
(5) For water flooding and flue gas assisted water flooding, the temperature of injection fluid was same as the temperature of the incubator. The detailed parameters are shown in Table 2.

**Table 2 Parameters of the sandpack flooding**

| Test No. | Flooding pattern | Permeability (mD) | Porosity (%) | Initial oil saturation (%) | Injection rate (mL/min) | |
|---|---|---|---|---|---|---|
| | | | | | Water | Flue gas |
| 1# | Steam flooding | 3033 | 37.26 | 89.25 | 2 | 0 |
| 2# | Flue gas assisted steam flooding | 3150 | 37.78 | 89.86 | 1 | 1 |
| 3# | First steam flooding then flue gas assisted steam flooding | 3090 | 37.56 | 88.78 | First 2 then 1 | First 0 then 1 |
| 4# | Water flooding | 3120 | 38.03 | 89.26 | 2 | 0 |
| 5# | Flue gas assisted water flooding | 3065 | 37.73 | 89.52 | 1 | 1 |

# Results and discussions

## PVT measurements

The PVT tests were conducted at 60℃, 90℃ and 120℃ to study the effect of dissolution of flue gas on property of heavy oil. The dissolution law of flue gas at different temperature and pressure were analyzed first. Then the viscosity and volume of saturated oil were measured. On this basis, viscosity reduction factor and volume factor under different conditions were calculated.

### *Dissolution law of flue gas*

Fig.3 shows the relationship of saturation pressure and solution gas-oil ratio at different temperatures. From the picture we can see, at the same temperature, as the saturation pressure increases, solution gas-oil ratio increases gradually. At the same saturation pressure, solution gas-oil ratio decreases with the increase of temperature. This is because at higher temperature, gas molecular thermal motion accelerates, which inhibits the dissolution of flue gas into oil.

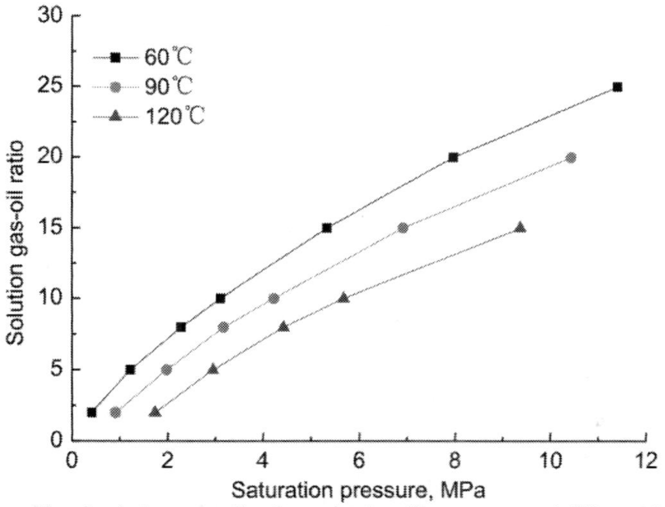


**Fig.3 Relationship of solution gas-oil ratio and saturation pressure at different temperatures**

***Viscosity reduction property of flue gas on heavy oil***

To investigate the viscosity reduction effect of flue gas on heavy oil, the viscosity reduction rate was used and defined as the ratio of reduction of viscosity to original viscosity, as is shown in equation (1):

$$Y = \frac{X_1 - X_2}{X_1} \times 100\% \tag{1}$$

Where Y is viscosity reduction rate, %; $X_1$ is the viscosity of original oil, mPa·s; $X_2$ is the viscosity of saturated oil, mPa·s.

The relationship of viscosity reduction rate and solution gas-oil ratio at different temperatures is shown in Fig.4. As solution gas-oil ratio increases, viscosity reduction rate increases, meaning that the viscosity of saturated oil decreases gradually with more flue gas dissolved. Under the same solution gas-oil ratio, the higher the temperature, the lower the viscosity reduction rate, which indicates that the effect of viscosity reduction of flue gas is weakened at higher temperature.

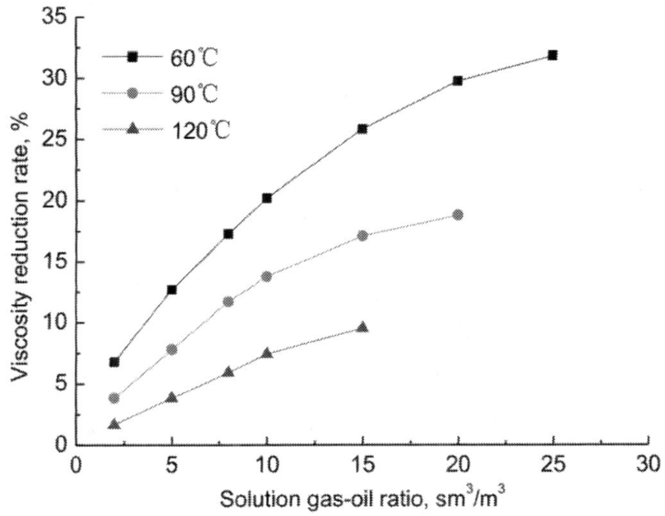

Fig.4 Relationship of solution gas-oil ratio and viscosity reduction rate at different temperatures

### *Volume expansion property of flue gas on heavy oil*

In order to analyze the expansion effect of solution gas on heavy oil, the volume factor is used and defined by the equation (2),

$$B = \frac{V_2}{V_1} \times 100\% \tag{2}$$

Where B is volume factor; $V_1$ is the volume of oil on the ground, mL; $V_2$ is the volume of saturated oil at saturation condition, mL.

The relationship of volume factor and solution gas-oil ratio is shown in Fig.5. It can be seen that at the same temperature, as solution gas-oil ratio increases, the volume factor increases gradually. Although the pressure increases with the increase of solution gas-oil ratio, the expansion effect of solution gas on oil is greater than the compression of pressure. Under the same solution gas-oil ratio, the volume factor increases with the temperature, for the thermal-expansion effect becomes bigger.

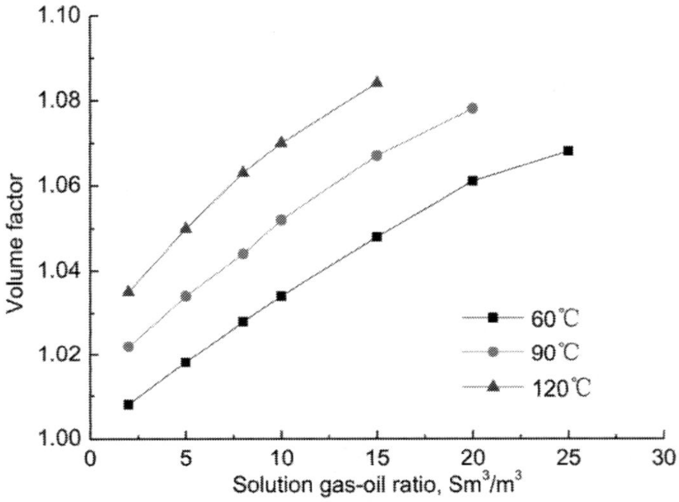


Fig.5 Relationship of solution gas-oil ratio and volume factor at different temperatures

## Sandpack flooding

In order to investigate the effect of flue gas on the displacement law, three experiments of steam flooding, flue gas assisted steam flooding and first steam flooding then flue gas assisted steam flooding, were conducted, as shown in Table 2 (No.1, No.2 and No.3). The production law of oil, water and gas were compared in the three experiments. In order to investigate the synthetic effect of flue gas and heat on EOR, two more experiments of water flooding (No.4) and flue gas assisted flooding (No.5) were conducted to compare the contribution of different factors on recovery.

### *Oil production rate and form of produced oil*

Fig.6 shows the change curves of oil production rate along with injection volume in different displacements. As we can see, with fluid injected, oil production rate first rises quickly then declines rapidly in all three experiments. But in the experiment of flue gas assisted steam flooding, the oil production rate is greater than in steam flooding. And in the experiment of No.3, when the flue gas is injected with steam subsequently, the oil production rate increases significantly. This shows the addition of flue gas is really beneficial to the production of oil no matter when the gas is injected.

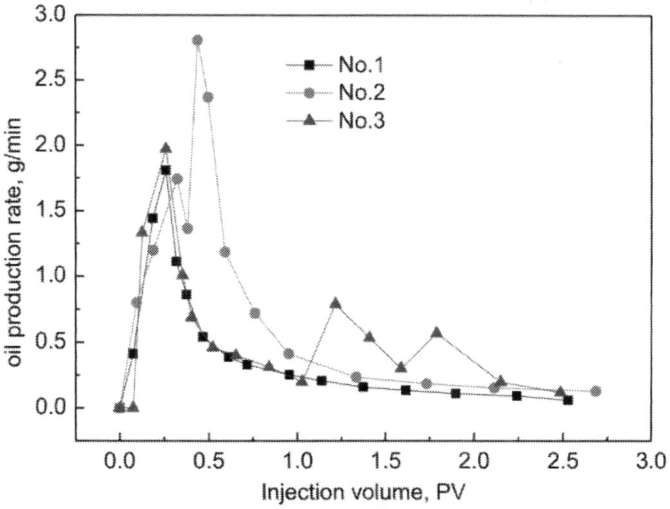


**Fig.6 Oil production rate of different flooding patterns**

Fig.7 compares the form of produced oil in test No.1 and test No.2. Fig.7(a) presents the form of the produced oil in test No.1, and Fig.6(b) shows the form of the produced oil in test No.2. We can see that the produced oil in the two tests is quite different in form. A large number of bubbles exist in the produced oil of test No.2, which forms a special state of mixture of heavy oil and solution gas, namely foamy oil. Flue gas is dissolved into oil under high pressure in displacement and then releases when produced. But due to high viscosity of heavy oil, it is difficult for the released solution gas to coalesce and form continuous free gas phase. As a result, it exists in heavy oil in the form of micro-bubble. The formation of foamy oil can expand the volume of original heavy oil and reduce its flow resistance, which can greatly promote the production of heavy oil.

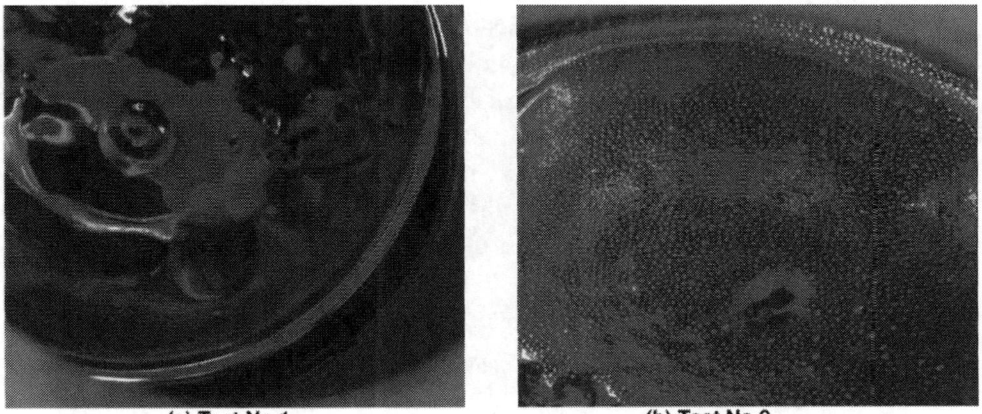

**(a) Test No.1** **(b) Test No.2**

**Fig.7 Form of produced oil**

***Water cut***

Fig.8 shows the change curves of water cut along with injection volume in different displacement tests. From the picture we can see, there exists a brief period in which oil is produced without water. Once the injected water breaks through, water cut in produced liquid will rise sharply. And when the water cut exceeds 80%, corresponding to the injection volume about 0.5PV, water cut will increase slowly. Compared with steam flooding, the water cut in the experiment of flue gas assisted steam flooding is lower at the same injection volume, especially early in displacement. In the test of No.3, when flue gas is injected, the water cut first declines significantly then increases gradually. It can be concluded that when steam and flue gas flow together, the flue gas occupies the main flow channel and inhibits water channeling to some extent, resulting in lower water cut in the tests of No.2 and No.3.

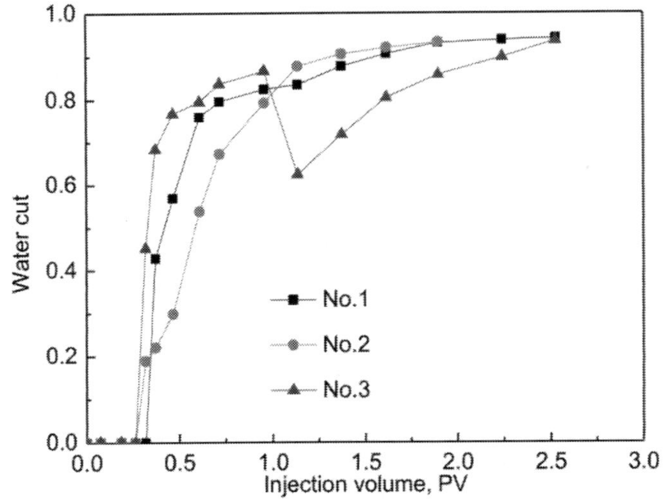


**Fig.8 Water cut of different flooding patterns**

***Gas production and composition change***

In flue gas assisted steam flooding (Test No.2), some complicated interactions happen between flue gas and oil, such as multiphase flow, gas dissolution and diffusion, gas retention, which has an important influence on EOR. So the gas production law and composition change of produced gas was investigated to analyze the effect and mechanism of gas on EOR.

Fig.9 shows cumulative gas injection volume, cumulative gas production volume and cumulative gas retention volume against injection volume. Cumulative gas retention volume can be defined as the difference between cumulative injection volume and cumulative production volume. As can be seen from the picture, cumulative gas injection volume begins to increase gradually, while cumulative gas production volume has little change. This indicates that injected gas is captured in the model with dissolution and diffusion occurring. As the gas injected volume increases, gas breaks through and gas production rises sharply with quantities of oil carried out in the form of foamy oil. After breakthrough, the injected gas was almost all produced in the steady displacement stage. when the displacement was over, the cumulative gas production volume was less than the cumulative gas injection volume, with about 200mL gas, one-fourth of the cumulative gas injection volume, remaining in the sandpack model or captured in the produced oil in the form of micro-bubble.

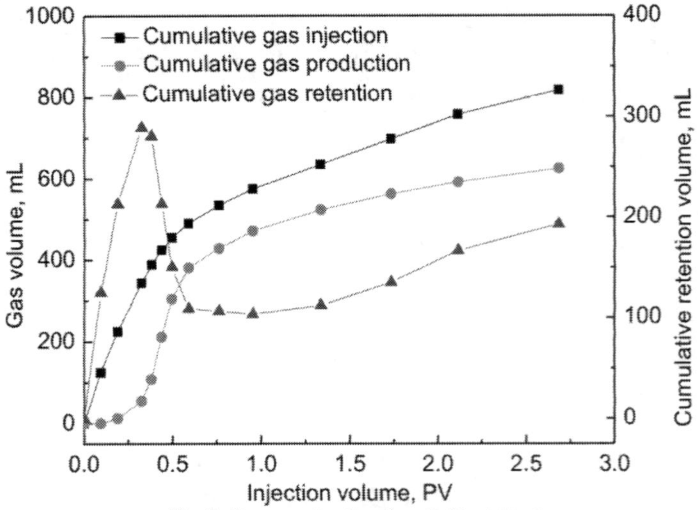


**Fig.9 Gas production law in Test No.2**

Table 3 and Fig.10 show the composition of produced gas at different displacement stage of Test No.2. We can see that the proportion of flue gas is always less than 1 and decreases with time. This is because some hydrocarbon gas generated at high temperature condition. As the displacement progresses, the proportion of $N_2$ in the produced gas declines and the proportion of $CO_2$ increases gradually. Before gas channeling the proportion of $N_2$ is greater than the injected proportion, while the proportion of $CO_2$ is less than the injected proportion. It might be because the solubility of $CO_2$ in heavy oil is greater than $N_2$, more $CO_2$ is dissolved into heavy oil. But after gas channeling, the previous dissolved $CO_2$ releases and is produced, leading to an opposite proportion of $N_2$ and $CO_2$ to before gas channeling. The change of the produced gas composition reflects the interactions of dissolution, diffusion and release between flue gas and heavy oil in displacement.

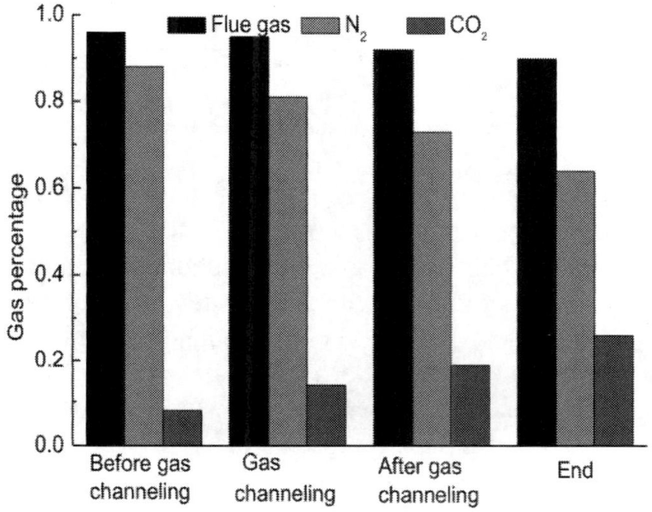


**Fig.10 Composition Change of produced gas of Test No.2**

**Table 3 Proportion of different gas in produced gas at different stages**

| Stage | Proportion of flue gas | Proportion of $N_2$ | Proportion of $CO_2$ |
|---|---|---|---|
| Before gas channeling | 0.96 | 0.88 | 0.08 |
| Gas channeling | 0.97 | 0.81 | 0.16 |
| After gas channeling | 0.95 | 0.75 | 0.2 |
| End of displacement | 0.91 | 0.64 | 0.27 |

### *Oil recovery of different flooding patterns*

Fig.11 shows the change curves of oil recovery with injection volume in different displacement patterns. As can be seen from Fig.11, it can improve oil recovery significantly to inject flue gas both for steam flooding and for water flooding. Especially in the test No.3, the oil recovery rose dramatically when flue gas was injected with steam together, considering the recovery no longer increased clearly when steam injected alone. This fully demonstrates it is effective for EOR to add flue gas in steam flooding.

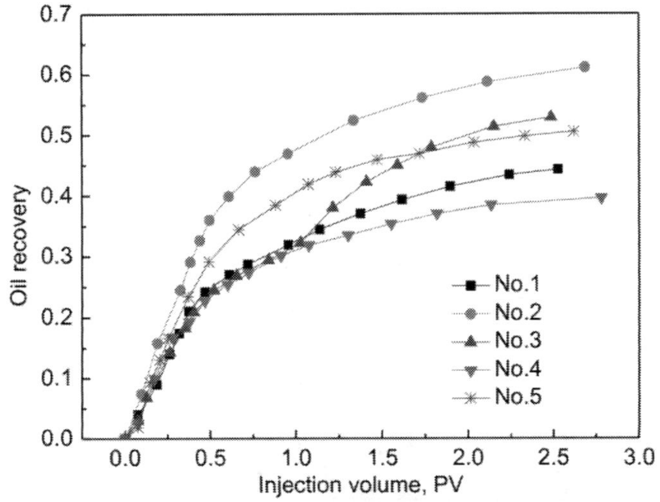


**Fig.11 Oil recovery of different flooding patterns**

In order to analyze the EOR mechanism of flue gas, the contribution of different factors on oil recovery was compared in Table 4. By comparing the recovery of steam flooding and water flooding, the contribution of heat on recovery without flue gas can be obtained, which is 4.7%. By comparing the recovery of flue gas assisted steam flooding and flue gas assisted water flooding, the contribution of heat with flue gas can be obtained, which is 10.6%. It is obvious that the contribution of heat with flue gas is larger than without flue gas, which indicates heat can play a greater role in displacement in the presence of flue gas. Similarly, the contributions of flue gas with and without heat were obtained by comparison, 16.8% and 10.9% respectively. The difference represents the synergistic effect of flue gas and heat that flue gas can carry steam into deep core, decrease heat loss along the model and expand heat-swept zone, and the heat in turn can facilitate gas drive.

**Table 4 Contribution of different factors on recovery**

| Flooding pattern | Oil recovery /% | Contribution of different factors/% | | |
|---|---|---|---|---|
| | | Flue gas | Heat | Synergistic effect |
| Steam | 44.3 | 16.8 | 4.7 | 5.9 |
| Steam+flue gas | 61.1 | | 10.6 | |
| Water | 39.6 | 10.9 | / | |
| Water+flue gas | 50.5 | | / | |

## Conclusion

(1) The solubility of flue gas in heavy oil increases with pressure, and decreases with temperature. With the flue gas dissolved, the viscosity of heavy oil declines and the volume expands, which is a significant EOR mechanism.

(2) When flue gas injected in displacement, the heavy oil was produced in the form of foamy oil. The formation of foamy oil can expand the volume of original heavy oil and reduce its flow resistance, which can greatly promote the production of heavy oil.

(3) The composition of the produced gas changes with time. Before gas channeling, the proportion of $CO_2$ in the produced gas was less than the injected proportion because of its dissolution. The proportion of $CO_2$ increased and the proportion of $N_2$ decreased gradually with time. At last, a quarter of cumulative injected gas was captured in the core approximately.

(4) Oil recovery can be improved significantly in the test of flue gas assisted steam flooding. A synergistic effect of heat and gas is generated when steam and flue gas is injected together.

## References

[1] Wilson M, Monea M. IEA GHG Weyburn $CO_2$ monitoring and storage project summary report 2000-2004. The 7th international conference on greenhouse gas control technologies, 5-9 September, 2004.

[2] Paul Z, Mike H. Permitting issues for $CO_2$ capture, transport and geological storage: a review of Europe, USA, Canada and Australia. Int J Greenhouse Gas Control 2007;1:94-100.

[3] Karaoguz O K, Topguder N N S, Lane R H, Kalfa U, Celebioglu D. Improved sweep in Bati Raman heavy-oil $CO_2$ flood: bullhead flowing gel treatments plug natural fractures. SPE Reserv Eval Eng 2007;10:1–12.

[4] Secaeddin S, Ulker K, Demet C. Bati Raman filed immiscible $CO_2$ application-atatus quo and future plans. In: Latin American & Caribbean petroleum engineering conference, 15–18 April, 2008, 2008, SPE 106575.

[5] Kleinsteiber S W. The wertz tensleep $CO_2$ flood: design and initial performance. J Petrol Technol 1988;42:630–6.

[6] Agustssen H, Grinestafr G H. A study of IOR by $CO_2$ injection in the Gullfaks field. In: SPE/DOE symposium on improved oil recovery, 17–21 April, 2004 Tulsa, Oklahoma, 2004, SPE 89338.

[7] Mohammadpoor.M. An extensive review on the effective sequence of heavy oil recovery[C]. SPE 157964, 2012.

[8] CHAI Li-wen. Main problems and countermeasures of pilot steam drive in mid-deep massive heavy oil reservoir[J]. Special Oil and Gas Reservoirs, 2005, 12(6): 44-47.

[9] WANG Jia-huai, LI Yun. Improve steam soaking performance by injecting nitrogen[J]. Journal of Southwest Petroleum Institute, 2002, 24(3):46-49.

[10] Chung F T H, Burchfield T E. Research aimed at immiscible $CO_2$ flooding[J]. Oil and Gas Journal, 1987, 04:76-82

[11] Li S, Li Z, Sun X. Effect of flue gas and n-hexane on heavy oil properties in steam flooding process[J]. Fuel, 2017, 187:84-93.

[12] Zhang C, Li ZM, Wang HY, et al. Performance of flue gas and foam assisted steam huff and puff with horizontal well in enhanced recovery of Fengcheng Oil field. SPE 170105, 2014.

[13] Fan Y, Liu Y F, Ting R U, et al. A Feasibility Study on Exploitation of Heavy Oil Reservoirs by High Temperature Compound Gas Drive[J]. Xinjiang Petroleum Geology, 2010, 31(5):530-532.

[14] Ma C, Duan Y. Experimental study on compound steam-flue gas drive in block Wa 38[J]. Special Oil & Gas Reservoirs, 2001.

Carbon Clean Solutions

# Carbon Capture Technology for Flue Gas Applications

CMTC Conference in Houston, TX July 17-20, 2017

Dr. Avi Patkar, PE – VP US Projects

Prateek Bumb – Chief Technology Officer

CCS-Houston-18Jul17.pptx

## Agenda

- Introduction
- Technology Overview
- Case Studies

2

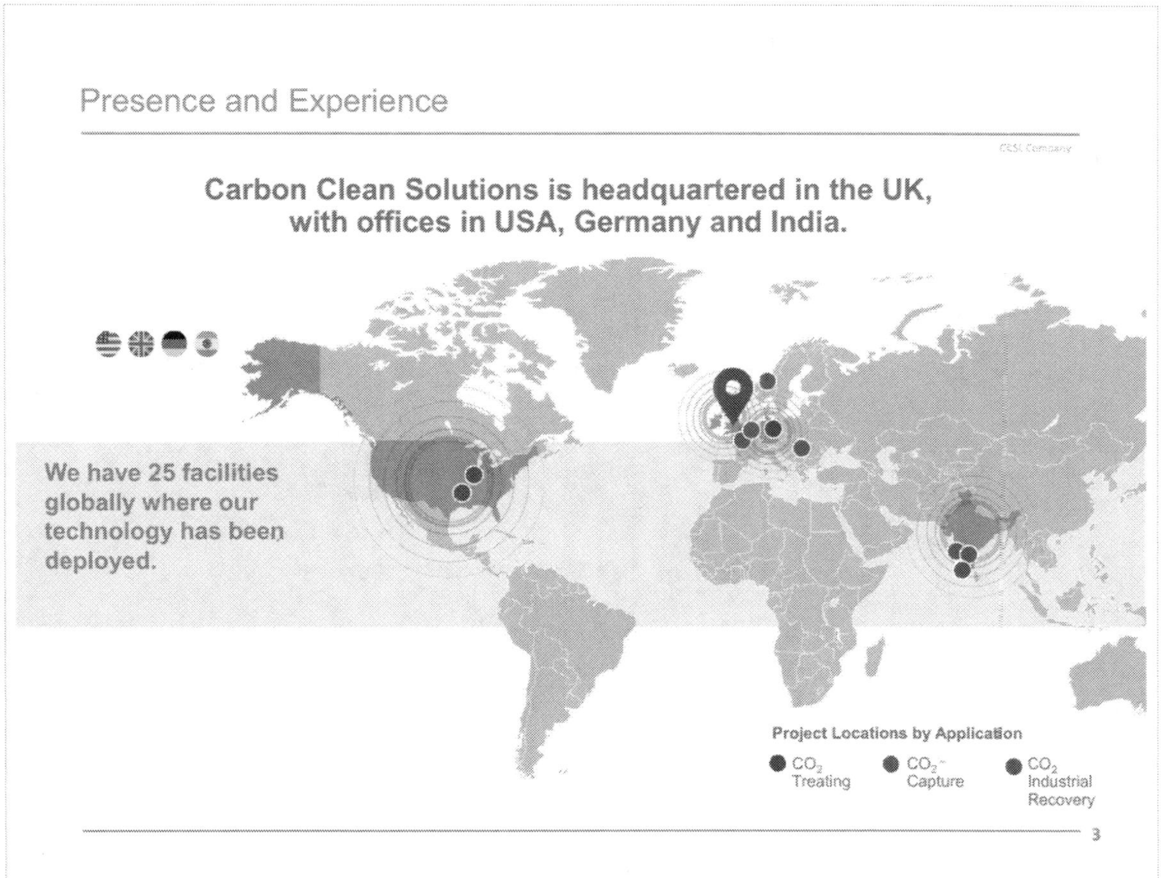
Presence and Experience
Carbon Clean Solutions is headquartered in the UK, with offices in USA, Germany and India.
We have 25 facilities globally where our technology has been deployed.
Project Locations by Application
$CO_2$ Treating
$CO_2$ Capture
$CO_2$ Industrial Recovery
3

Introduction to Carbon Clean Solutions
We've made industrial decarbonisation a commercial reality
Carbon Clean Solutions is an established global leader in providing $CO_2$ capture technology.
Our advanced technology is proven to dramatically reduce to the cost of carbon capture – enabling industrial emitters to reduce $CO_2$ emissions.
TK1252

## Applications in Industry

CCSL Company

### Our breakthrough technologies make carbon capture a commercial reality in a variety of industrial settings

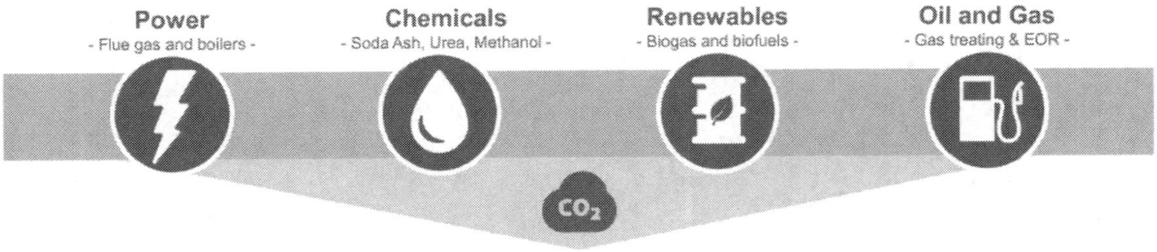


### We can recover $CO_2$ for industrial reuse

**Our systems are proven at 25+ sites around the world:** affordable carbon capture solutions can improve the bottom line while reducing greenhouse gas emissions.

5

## $CO_2$ Capture Economics

CCSL Company

### Our technologies can achieve carbon capture at a cost of USD $40 per metric ton, a figure that continues to improve.

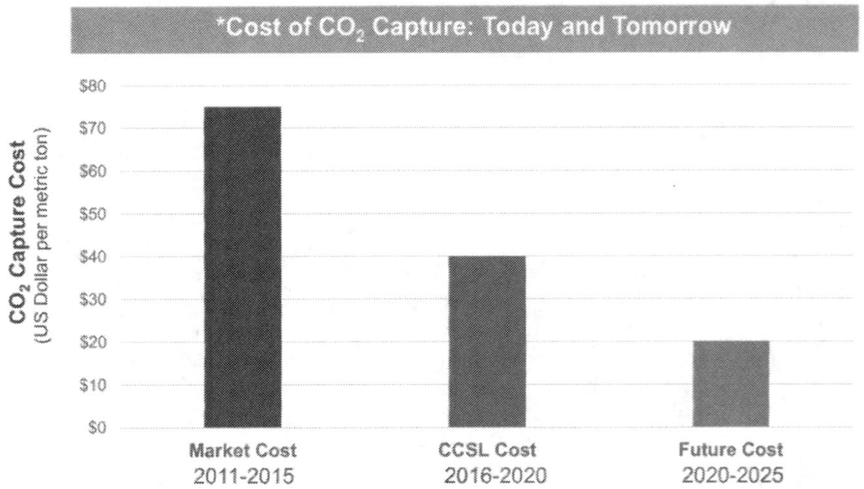


**Lifecycle cost: includes capital and operating costs, W. European location basis, excludes downstream $CO_2$ compression*

6

## Partner benefits

CCSL Company

Carbon Clean Solutions

- Our focus is on providing the most cost-effective $CO_2$ capture and separation technology – basic engineering design and solvent
- Flexible business models help to drive projects at a local level with engineering, procurement, construction (EPC) partners
- Our engineering know-how and patented solvent chemistry makes it easier to design and execute $CO_2$ capture projects
- Our technology creates significant savings by driving down the operating and capital costs for $CO_2$ capture plants to $40/ton and a target of less than $40/ton

VEOLIA

TNO

APPLY SØRCO

U.S. DEPARTMENT OF ENERGY

Department of Energy & Climate Change

TECHNOLOGY CENTRE MONGSTAD

WORLD ECONOMIC FORUM
Technology Pioneer

7

## Performance Chemistry

CCSL Company

### Drop-In solvent for improved performance

Our advanced solvents reduce operating costs, emissions, and waste for new and existing Co2 capture systems.

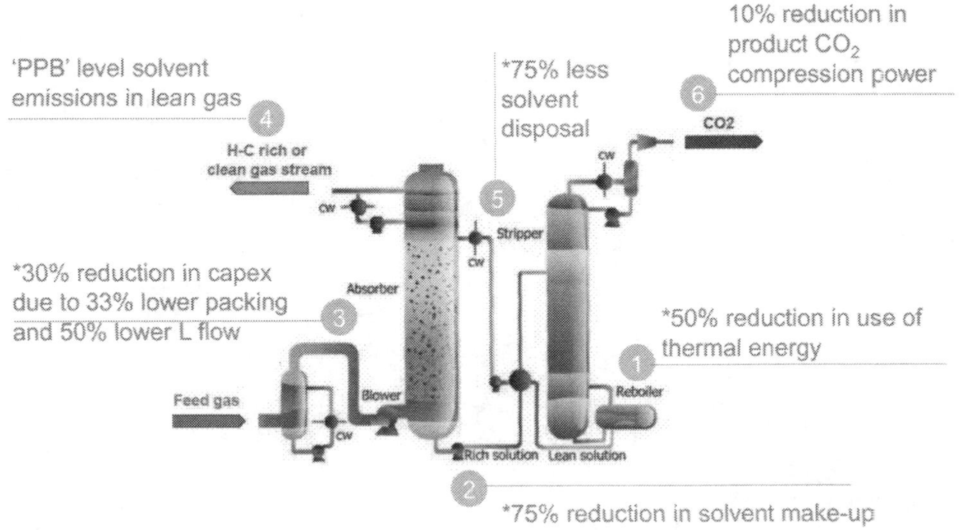


*Conventional CO2 recovery benchmark is a monoethanolamine (MEA) based chemical absorption process

8

CCSL Company

## Technology

9

## Global Technology Case Studies

CCSL Company

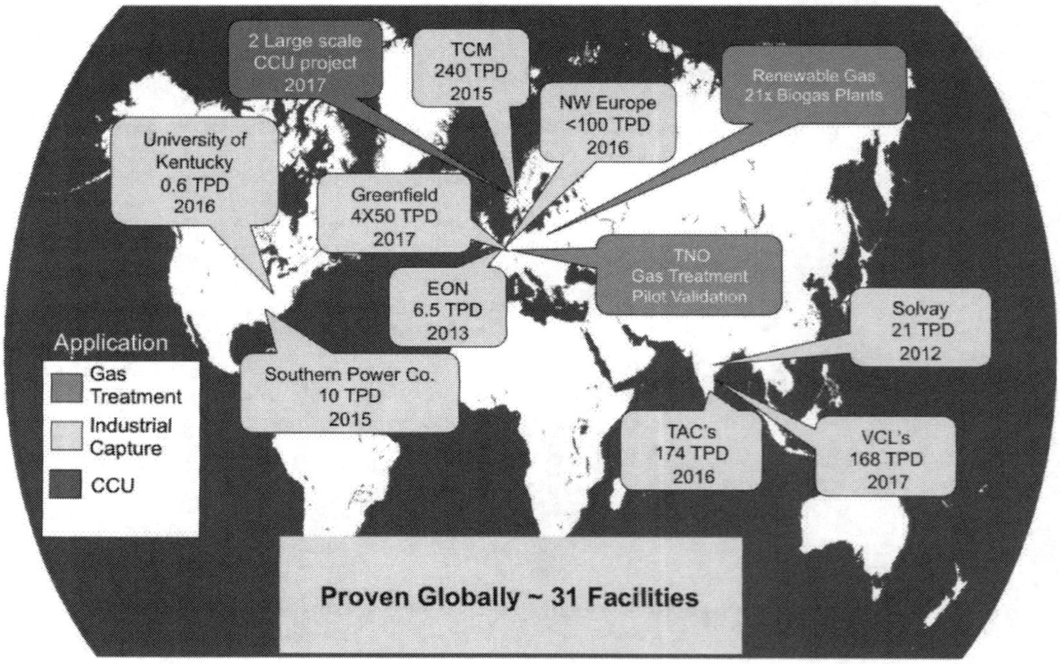


10

## Validation Case Studies - CDRMax

CCSL Company

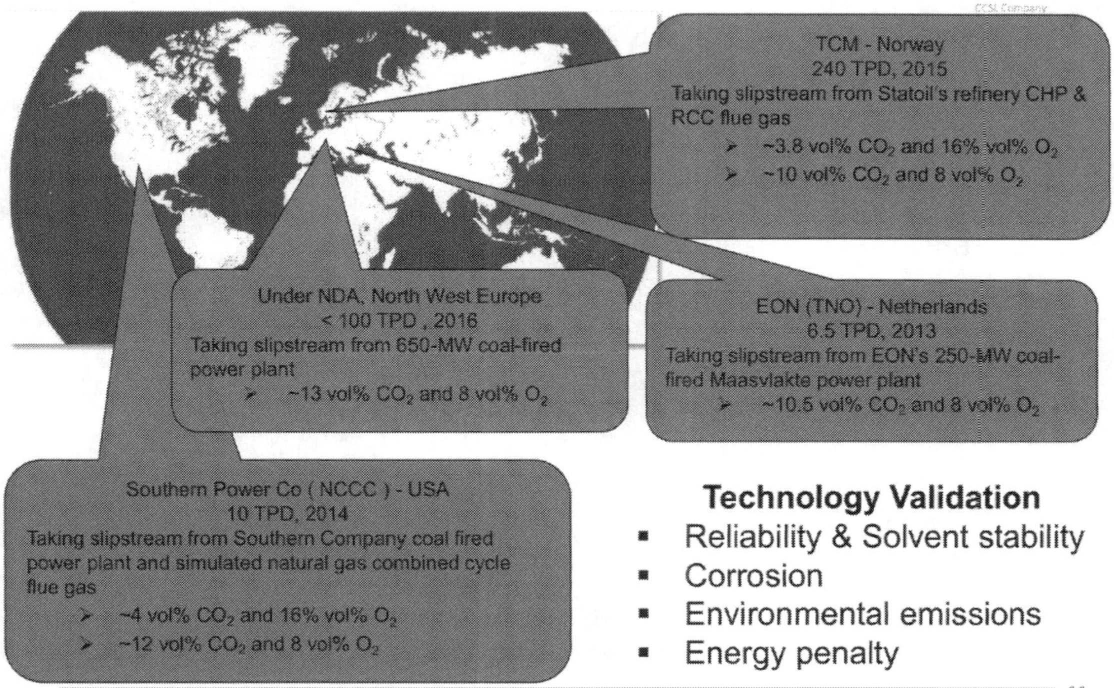


**Technology Validation**

- Reliability & Solvent stability
- Corrosion
- Environmental emissions
- Energy penalty

11

## 100% Plant Reliability

CCSL Company

- Successfully demonstrated $CO_2$ capture of more than 25,000 tons $CO_2$
- No loss of run time due to CDRMax solvent issues

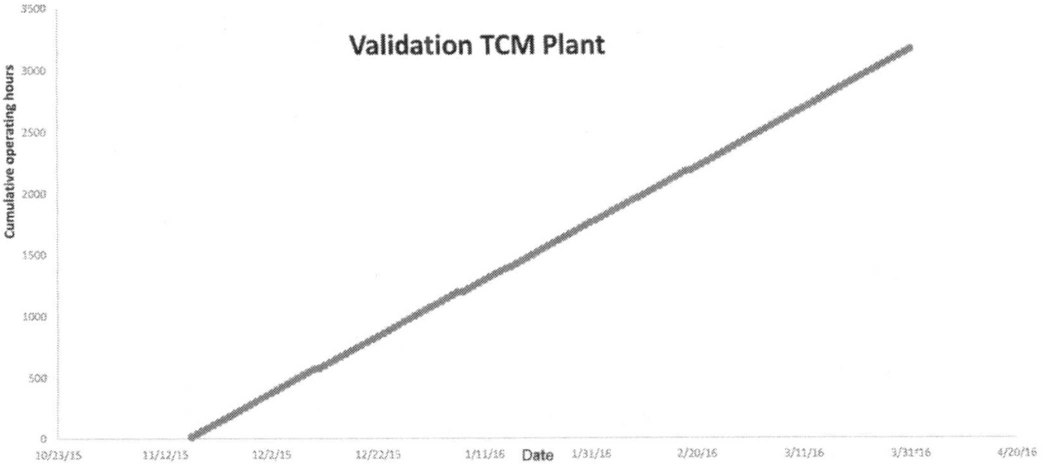


**Validation of 100% Plant Reliability at NCCC, E.ON, TCM & NWE plant**

12

## Superior Solvent Stability

CCSL Company

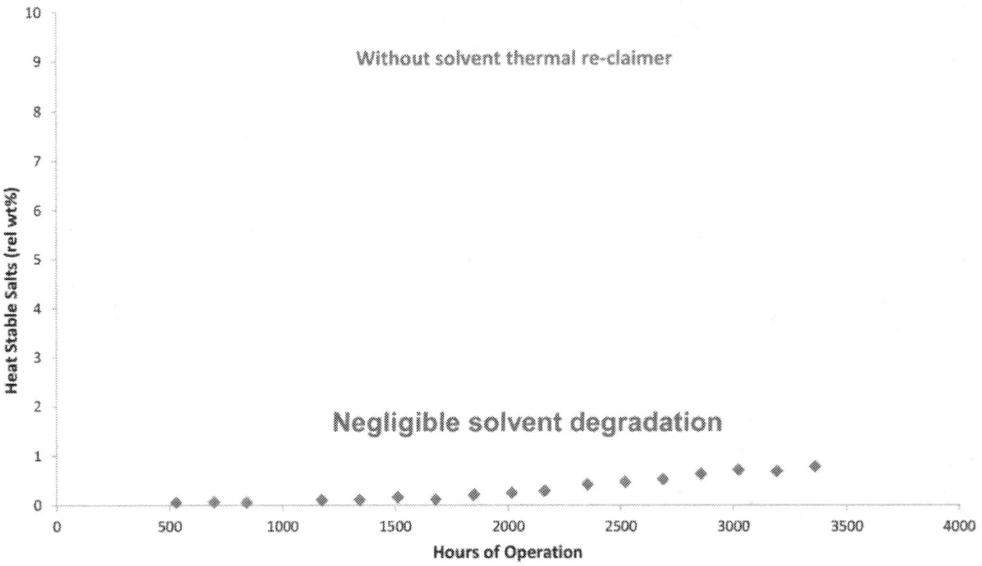


**Validation TCM plant**

13

## Superior Solvent Stability

**MEA visual change in 1000 operating hours**

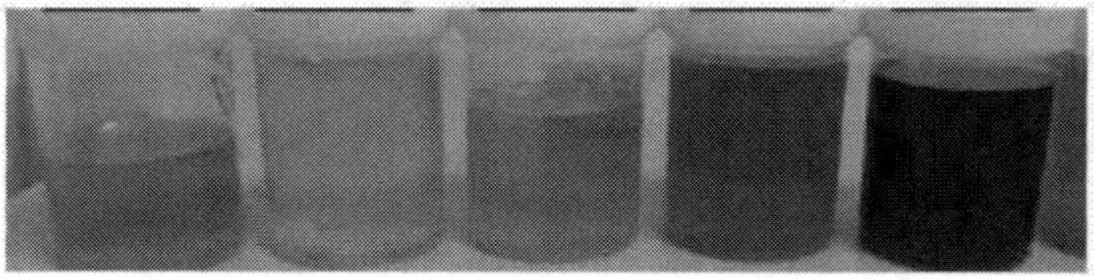

**APBS visual change in 1000 operating hours**

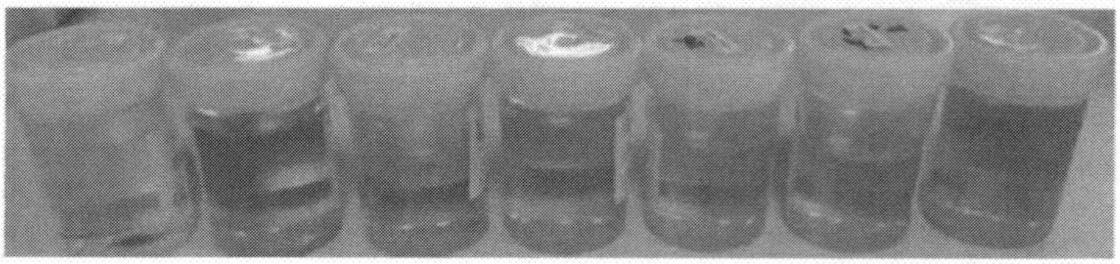

**Validation E.ON plant (without solvent re-claimer)**

14

## Corrosion Data - E.ON Plant

CCSL Company

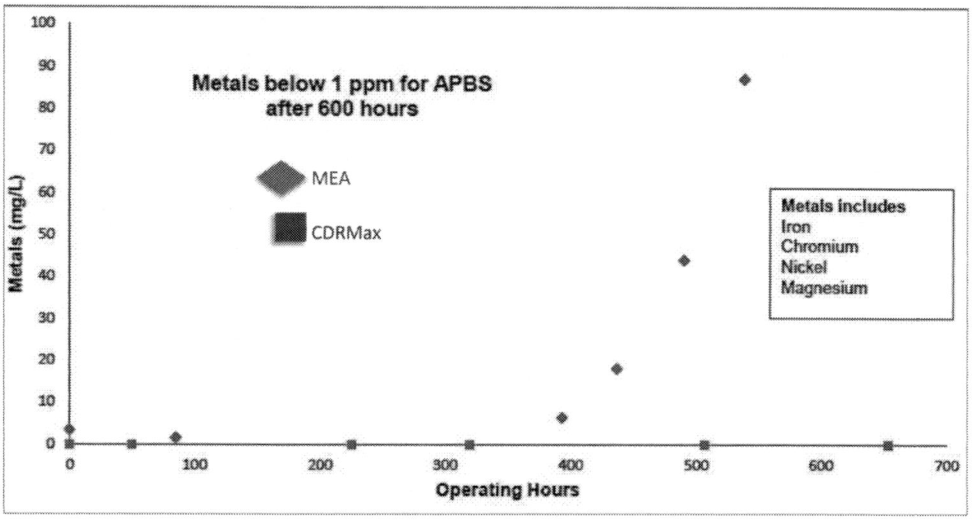


Dissolution of metals from SS304 with CDRMax solvent is 90 times lower than MEA

15

## Corrosion Data - NCCC

CCSL Company

Dissolution of metals from with CDRMax solvent MEA solvent comparison

| Dissolved Metals in Fresh and Rich Solvents (all ppb wt) [1, 2] | | | | |
|---|---|---|---|---|
| **Metal** | **Fresh MEA** | **Fresh CDRMax** | **Aged MEA** | **Aged CDRMax** |
| Arsenic | < 12 | 53.2 | 219 | 114 |
| Barium | < 12 | <10 | 265 | 11.8 |
| Cadmium | < 12 | < 5 | < 10 | < 5 |
| Chromium | < 12 | 42.2 | 45,090 | **2,120** |
| Selenium | 44.1 | 41.8 | 1,950 | **660** |

1) MEA testing in 2012 (Wheeldon, June 2012). No corrosion inhibitor. 316L SS. 300 hrs operations
2) CDRMax testing in 2014. No corrosion inhibitor. 316L SS. ~ 500 hrs operations .

16

## Corrosion Data - TCM

CCSL Company

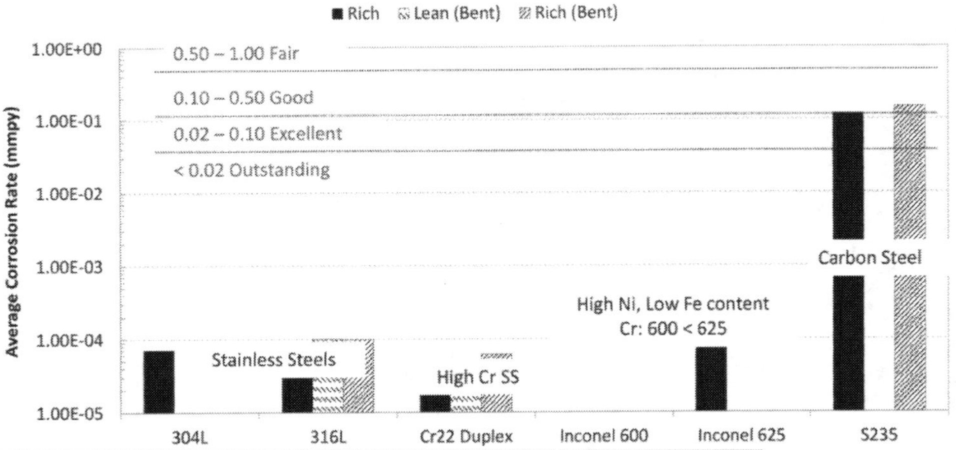


| Parameter | Loading | T | P | pH | [CO2] | Linear Velocity | Duration |
|---|---|---|---|---|---|---|---|
| UoM | Lean/Rich | °C | bar(g) | | mol/mol | m/s | Hours |
| Value | Lean | 119 | 5.5 | 10.4 | 0.13 | 1.1 | 3500 |
| | Rich | 105 | 6.0 | 9.2 | 0.46 | 1.1 | 3500 |

Average conditions of corrosion coupon exposure environments at the TCM campaign

17

## Corrosion Data – NWE

CCSL Company

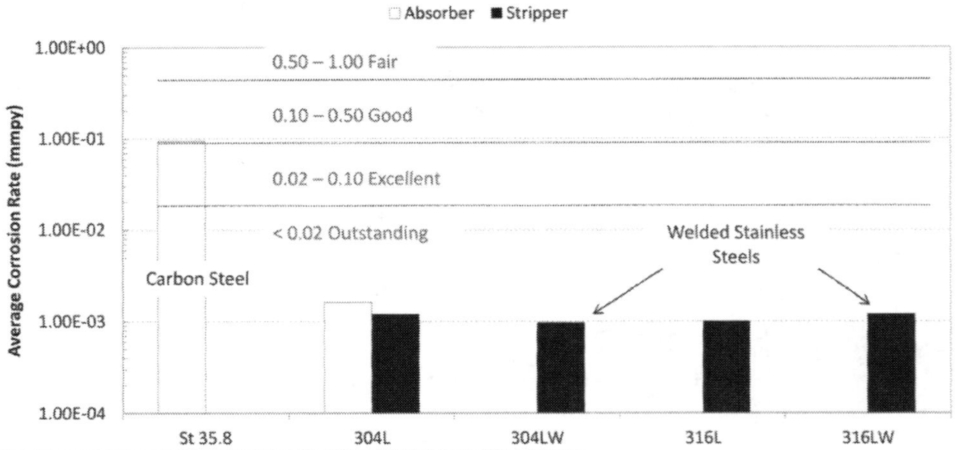


| Parameter | Location | Temperature | Pressure | [CO2] | Duration |
|---|---|---|---|---|---|
| UoM | N/A | °C | bar(g) | mol/mol | Hours |
| Value | Absorber | 23 – 46 | 0 | 0.41 | 1000 |
| | Stripper | 107 – 123 | 0.3 – 1.3 | 0.12 | 1000 |

Average conditions of corrosion coupon exposure environments at the N.W.E. campaign

18

## Oxidative Degradation Data - E.ON

CCSL Company

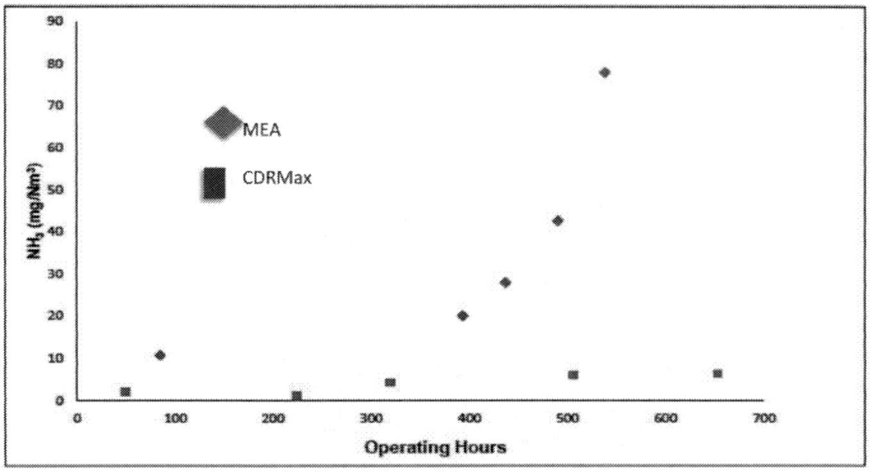


- Ammonia emissions indicator of the oxidative degradation
- CDRMax solvent top-up 8-10 times less than that for MEA

19

## Far Below Current Emissions Permissible Limit

CCSL Company

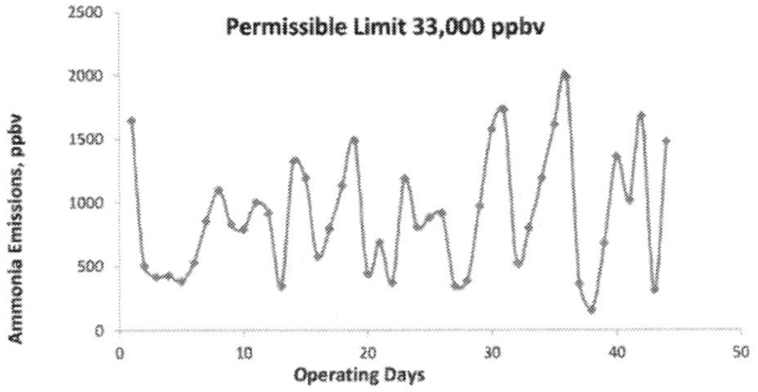


With CDRMax solvent 15-30 times less ammonia emissions than permissible limit

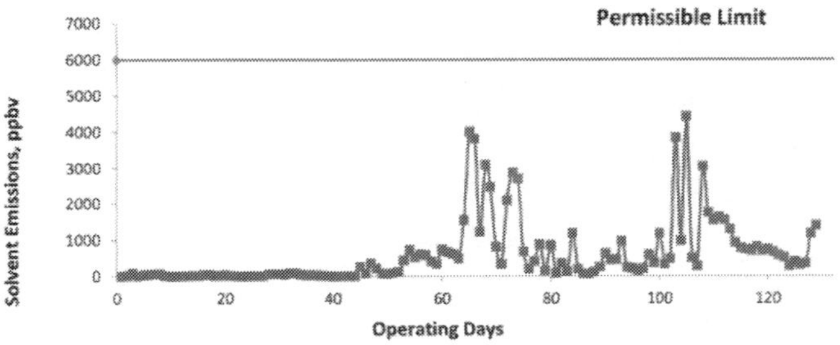


Successfully operated consistently within the permissible limit of solvent emissions.

20

## 55x times less aerosols emissions

CCSL Company

CDRMax solvent exhibit inherent capability to limit aerosols solvent losses & meet stringent emission limits

Significant improvement in health & safety aspects

CDRMax aerosols emissions
28 mg/$Nm^3$

MEA aerosols emissions
1587 mg/$Nm^3$

21

## Continuous Improvement in Energy Consumption

CCSL Company

Energy comparison to MEA
(Coal fired post combustion $CO_2$ capture plant)

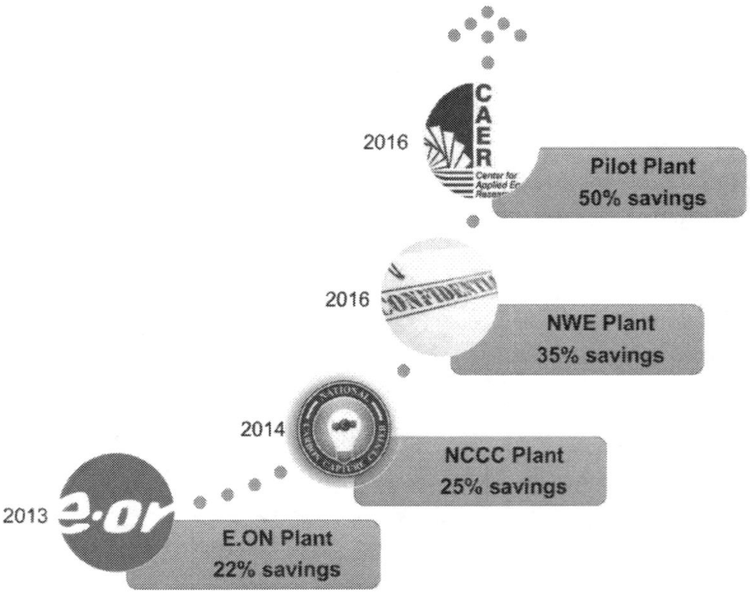


22

## Key Unique Selling Points (USP) of CDRMax Technology

CCSL Company

**Reductions in capital and operating costs**

- 40 % lower OPEX and 30 % Lower CAPEX than conventional $CO_2$ capture technology

**Low solvent emissions**

- Atmospheric solvent emission at the 'parts per billion' (ppb) scale

**Solvent Technology Readiness Level (TRL) = 8/9**

- Technology proven and deployed at full scale
- Commercially available, globally

**High availability and reliable performance**

- No operational issues resulting in 100 % plant availability and efficient performance

**Validated scale up models**

- Developed simulations and models for technology and process scale-up, which have been validated at all scales

**Extensive experience on any type of flue gases**

- Versatile solvent technology can be applied to any type of flue gas for efficient and economical $CO_2$ capture

23

CCSL Company

## Case Studies

24

## 10MW CO2 Capture to Chemicals Project in India

10MW Coal power plant CO2 capture:

- CO2 converted to chemicals
- FEED completed 2015
- Commissioning October, 2016
- CCSL's engineering design

41

## Pilot: 6.5 TPD at EON Benelux, Netherlands

**1000 hour test campaign:**

- Coal / biomass fired power plant campaign ($CO_2$ concentration 12 vol%)

**Result Vs MEA campaign:**

- Minimal degradation of solvent
- 15x reduction in corrosion
- 10x reduction in ammonia emissions
- 50x reduction in aerosol emission

MEA visual change in 1000 operating hours

APBS visual change in 1000 operating hours

42

## Demo: 240 TPD at Technology Centre Mongstad, Norway

World Largest Most Advanced CO2 Capture Facility

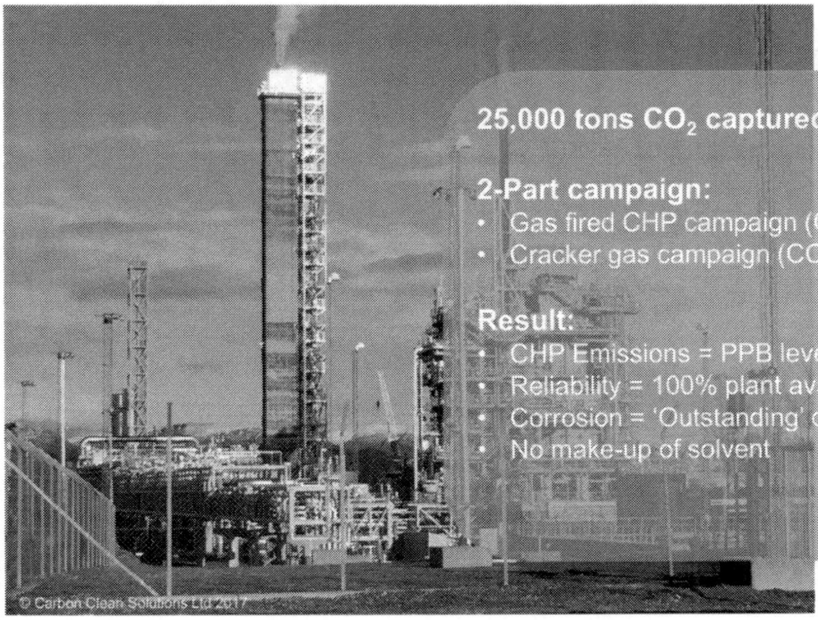

**25,000 tons $CO_2$ captured over 3500 hours**

**2-Part campaign:**
- Gas fired CHP campaign ($CO_2$ 3.6 vol%)
- Cracker gas campaign ($CO_2$ 13 vol%)

**Result:**
- CHP Emissions = PPB level solvent emissions
- Reliability = 100% plant availability
- Corrosion = 'Outstanding' quality for SS 304L
- No make-up of solvent





## Specialty chemicals: greenfield project

**Client**: Vishnu Chemicals

**Location**: India

**CO2 capture**: 168 TPD

**Project description**: Produce own CO2 for use in production of Sodium Dichromate, Strontium Carbonate and Barium Carbonate, used in ceramic glazes.

**Timing**: Currently in construction, delivering in 2017

## Large Scale CCUS Project – greenfield project

**Client**: KEA (Waste to Energy Plant Oslo city)

**Location**: EU

**CO2 capture**: 1500 TPD (75 MW) flue gases

**Project description**: Complete Decarbonisation of Waste to Energy Plant of Oslo city

**Timing**: Pre-FEED / FEED Stage





## Large Scale CCUS Project – greenfield project

**Client**: Yara

**Location**: EU

**CO2 capture**: 1000 TPD (50 MW) flue gases

**Project description**: CO2 capture from reformer flue gases of Urea plant.

**Timing**: Pre-FEED / FEED Stage

Carbon Clean Solutions

## Contact Us


Prateek Bumb, Co-founder & CTO
prateek@carboncleansolutions.com
Mobile: +44 7795460345

47 Castle Street, Reading RG1 7SR,
Berkshire, United Kingdom

Dr. Avi Patkar, PE, VP US Projects
avinash@carboncleansolutions.com
Mobile: 513-802-1849
8090 Clearmeadow Drive,
West Chester, OH 45069


48

CMTC-486215-MS

# Stabilized $CO_2$ Foam for EOR Applications

Ahmed Farid Ibrahim, Arezoo Emrani, Hisham Nasraldin, Texas A&M University



## Abstract

The practice of injecting $CO_2$ for oil production was initiated in the 1950's. Today, $CO_2$ flooding is an established technique to enhance oil recovery (EOR), and $CO_2$ capture and storage in deep geologic formations is being studied for mitigating carbon emissions. $CO_2$-foam has been used to improve the sweep efficiency as a replacement for polymers to avoid formation damage. Although it is common to use surfactants to generate and stabilize foams, they tend to degrade at high temperatures (212°F), high-salinity environments, and in contact with crude oil. Adding nanoparticles is a new technique to stabilize $CO_2$ foams. The present work evaluates new foaming solutions that incorporate nanoparticles and viscosifiers to investigate the mobility-control performance when these foams are used as EOR fluids.

This study investigates the stability of alpha olefin sulfonate (AOS) foam and the corresponding mobility-reduction factor (MRF) for different foam solutions in the presence of nanoparticles and viscosifiers. To achieve this objective, foam stability was studied for various solutions to find the optimal solution at which higher foam stability in the $CO_2$ foam system can be reached. Coreflood tests were also conducted on different Buff Berea sandstone cores at 150°F saturated initially with a dead crude oil. The $CO_2$ foam was injected with 80% quality as tertiary recovery mode. The oil recovery and the pressure drop across the core were then measured for different foam solutions.

Adding silica nanoparticles (0.1 wt%) and viscoelastic surfactant (VES) (0.4 wt%) to the AOS (0.5 wt%) solution improves both foam stability and MRF. In contact with crude oil, unstable oil-in-water emulsion formed inside the foam lamella that decreased the foam stability. A weak foam was formed for AOS solutions, but the foam stability increased by adding nanoparticles and VES. The oil recovery from the conventional water flooding (as a secondary recovery before foam injection) ranged from 40 to 48% of the original oil-in-place. AOS was not able to enhance the oil recovery with an apparent viscosity similar to that for the water/gas system (with no AOS in the solution), and no more oil was recovered by AOS foam. The addition of nanoparticles and VES to the solution improved the foam MRF and allowed extra oil production (8% in presence of nanoparticles, 15% by adding nanoparticles and VES).

Surfactants may not be able to stabilize foam at high temperatures and in contact with crude oil. Protecting surfactants from being degraded and improving the foam stability is of great importance and can be achieved by adding nanoparticles and viscosifiers to the solution. Adding nanoparticles is highly recommended for EOR applications, particularly at high temperatures.

## Introduction

Carbon dioxide ($CO_2$) flooding has been used as an enhanced oil recovery (EOR) method. However, there are several challenges such as poor sweep efficiency and low oil recovery because of gravity segregation and viscous fingering. Moreover, the viscosity of $CO_2$-water/brine mixture without adding surfactant decreases as temperature increases (Chen et al. 2012). $CO_2$-foam has been investigated since 1960's to improve the sweep efficiency and oil recovery (Bian et al. 2012).

The first idea of using foam to displace oil was suggested and patented by Bond and Holbrook in 1958 (Wang and Li 2014). Fred (1961) conducted extensive laboratory experiments to further confirm that the foam can perform as a blocking agent to reduce the mobility of the displacing solvent (Wang and Li 2014). Based on Fred's work, the first idea to dissolve a surfactant into $CO_2$ during an EOR method for the purpose of mobility control foam was suggested by Bernard and Holm (1967) (Xing et al. 2012).

One of the interests of $CO_2$-EOR is to find surfactants soluble in $CO_2$ phase which can be injected with $CO_2$ to generate foam in reservoirs (Chen et al. 2012). The first surfactant families selected for EOR method was petroleum and synthetic aromatic sulfonates because of their availability and relatively low cost (Bavlere et al. 1988). For EOR process, the main requirement for using surfactants is to provide low interfacial tension (IFT) in multiphase systems formed in porous media.

$CO_2$-foam, consisting of low amounts of water and high amounts of compressed gas, minimize formation damage to unconventional reservoirs through fast cleanup and recovered permeability, as most of the gas flows back after depressurization (Xiao et al. 2016). However, $CO_2$-foams have some potential weaknesses. Foam stability improvement, especially in severe reservoir conditions, is a key factor to increase sweep efficiency. $CO_2$-foam can increase the viscosity of $CO_2$ relative to oil and water, which can cause improved sweep efficiency (Chen et al. 2012).

Foams are thermodynamically and kinetically unstable because the surface energy of gas/liquid interfaces tends to decrease as they degenerate into separate gas and liquid phases. At high-temperature reservoir conditions, surfactants generally tend to degrade. In addition, surfactant loss in a reservoir due to adsorption in the porous medium results in a large consumption of chemicals in $CO_2$-foam flooding. Thus, high reservoir temperature can be a major feature governing the economic viability of $CO_2$-foam flooding, and can also be another factor of foam instability (Chen et al. 2012).

The addition of thickeners such as polymers or VES for the $CO_2$-foam has been addressed to improve the foam stability (Zhao et al. 2015). Mixing surfactant and polymer may reduce fluid viscosity at high temperatures. However, adding nanoparticles to the mixture may enhance the liquid phase viscosity and stabilize $CO_2$-foam at high pressures and high temperatures (HP/HT). It is known that nanoparticles can adsorb at the gas/liquid interfaces to stabilize bubbles in foams by creating a rigid protective barrier around dispersed bubbles, which can reduce liquid film drainage.

Nanoparticles have been researched extensively as a means to stabilize foams used in oil production operations (Emrani and Nasr-El-Din 2017; Kalyanaraman et al. 2017; Singh et al. 2016; Mohd et al. 2014; and Mo et al. 2012). Nanoparticles are easily dispersed in water and able to flow through porous media (Mohd et al. 2014). Various other aspects of the nanoparticle-stabilized foams, such as the effect of salinity on the nanoparticle concentrations, have been established by Espinoza et al. (2010). They showed foam stability was improved as nanoparticle concentration increased under high salinity conditions. Although longer-lasting foams have been generated by various systems, the stability of these foams in the presence of crude oil and corefloods at high temperature has not been explored extensively. Yu et al. (2012) showed that the equilibrium adsorption of nanoparticles in different porous media is very low. Also, nanoparticles cannot change the core permeability based on their coreflood tests. They suggested that foam mobility increases with an increase of volumetric $CO_2$/nanoparticle dispersion phase ratio and injection flow rates. San et al. (2017) studied the effect of different ions on nanosilica-stabilized $CO_2$-foam generation. The results showed that stable $CO_2$-foam was generated as the synthetic produced water and nanosilica dispersion/$CO_2$ flowed through a porous medium (San et al. 2017).

Mixing surfactants can also improve the stability of foams especially in contact with crude oil (Andrianov et al. 2012). Mixing non-ionic surfactant to an anionic surfactant enhances the foam stability because of the formation of a viscous surface layer which raises the interfacial viscosity of the foam thereby reducing the rate of film thinning. Betaines are known for their foam enhancing properties generated a stronger and more stable foam in the presence of oil. Betaine increases the critical capillary pressure required for droplet entry which is related to the disjoining pressure, thus the foam stability increases (Basheva et al. 2000).

The capacity to reduce the gas-liquid mobility during the injection of foam into a porous media is known as foam blocking ability. The mobility reduction factor (MRF) is used as an index to indicate foam blocking ability and stability in this research. Therefore, large blocking ability represents a strong and stable foam (Wang et al. 2015).

The mobility reduction factor (MRF), defined as the ratio of total mobility of $CO_2$/brine to foam mobility, increases with the foam life. When foams become more stable, more resistance to flow results in a higher mobility reduction. Using $CO_2$-foam for mobility control was first proposed by Bond and Holbrook in 1958 (Yin et al. 2009). It has been proven that the addition of surfactants aid in the generation of foam wherever the $CO_2$ flows, especially if they are partially $CO_2$ soluble. The MRF can be calculated by comparing the pressure drop across the core during foam injection to the pressure drop measured at gas only (**Eq. 1**) (Bian et al. 2012):

$$\text{Mobility Reduction Factor} = \text{MRF} = \frac{\left[\frac{kA\Delta p}{QL}\right]_f}{\left[\frac{kA\Delta p}{QL}\right]_g} = \frac{\Delta p_f}{\Delta p_g}, \quad \text{......} \quad (1)$$

where Q is the flow rate, k is the absolute core permeability, A is the cross-section area of the core, L is the core length, $\mu$ is the viscosity, $\Delta p$ is the pressure drop across the core. The subscripts "f" and "g" represent the experiment with and without foam, respectively.

The objective of this study is to investigate using nanoparticles and VES to improve foam properties and increase MRF in EOR applications. The experimental studies included: (1) foamability for different solutions combined with optical microscope analysis to examine the effect of the nanoparticles and VES on foam stability, and (2) coreflood tests to understand the effect of nanoparticles, VES in presence of oil on MRF.

## Experimental Studies

**Material.** An anionic alpha-olefin sulfonate surfactant (AOS) was used to prepare the aqueous phase for a series of coreflood experiments. The general structure of this class of surfactants is $R\text{-}SO_3^-Na^+$, where (R) represents the hydrophobic group. The number of carbon atoms in the surfactant was 14-16 with an average molecular weight of 315 g/mol. The surfactant was obtained from a local chemical company as a solution containing 40 wt% active material. Foam solutions were prepared using brine containing 5 wt% NaCl in deionized water. $CO_2$ gas with a purity of 99.99 mol% was used to pressurize the system and form the foams under investigation. A chemical company supplied silica nanoparticles ($SiO_2$) of size 140 nm. Viscoelastic surfactant cocamidopropyl betaine (cocobetaine) (VES) was utilized to stabilize foam. The additives concentrations of the foam solutions were 0.5, 0.4, and 0.1 wt% for AOS, VES, and $SiO_2$, respectively. The solutions were mixed using a magnetic stirrer then homogenized with an ultrasonic homogenizer.

Berea sandstone cores were used with permeability ranging from 106 to 115 md. **Table 1** shows the mineralogy of the Berea sandstone cores. A dead crude oil was used in this study, and its properties are given in **Table 2.**

| Mineral | Quartz | Kaolinite | Microline | Muscovite | Smectite |
|---|---|---|---|---|---|
| Concentration, wt% | 91 | 3 | 4 | 1 | 1 |

Table 1 – Mineralogy of Berea sandstone cores.

| Temperature, °C | Viscosity, cp | Density, g/cm$^3$ |
|---|---|---|
| 25 | 3.03 | 0.82 |
| 30 | 2.44 | 0.81 |
| 40 | 2.04 | 0.81 |
| 50 | 1.71 | 0.80 |
| 60 | 1.88 | 0.79 |
| 70 | 1.61 | 0.78 |

Table 2 – The viscosity and density of dead crude oil.

**Methods**

***Faomability Study.*** Foams are stabilized by surfactant-based solutions in the presence of a viscosifier or nanoparticles to prevent bubble coalescence. Foams are generally described in terms of their foamability, which is the ability of a foaming solution (surfactant only or surfactant with VES or nanoparticles) to form a foam. Foamability of the solutions was studied by performing a shake test.

***Coreflood Study.*** The capacity to reduce the gas/liquid mobility during the injection of foam into a porous medium is known as foam blocking ability; hence the larger the blocking ability is, the more stable the foam would be. Mobility reduction factor (MRF) is used as an index to indicate foam blocking ability and stability in the present work.

**Fig. 1** shows a schematic diagram of the coreflood setup used. Two stainless-steel piston accumulators with a capacity of two liters each were used to store the synthetic brine and the solutions (2). A one-liter accumulator was used to store the $CO_2$ (1) and was attached to a $CO_2$ cylinder (8). A syringe pump (7) was used to displace the solutions from the piston accumulators. Valves (v1, v3) were installed at the accumulator's outlet to control the fluids alternating during the injection. To monitor the pressure at the core inlet, a pressure gauge was installed at the coreholder inlet (G1). A hassler type core holder (3) was used to hold the core during the coreflood test. The coreholder was installed in an oven (11) that can be used to increase the system temperature. A backpressure regulator (10) was installed at the core outlet to maintain the outlet pressure. It was adjusted by a nitrogen cylinder (9). A hand pump (6) was used to apply overburden pressure around the core. A pressure transducer (4) was used to measure the pressure drop across the core and send the measurements to a data acquisition system (5) to a computer that records the data through a software.

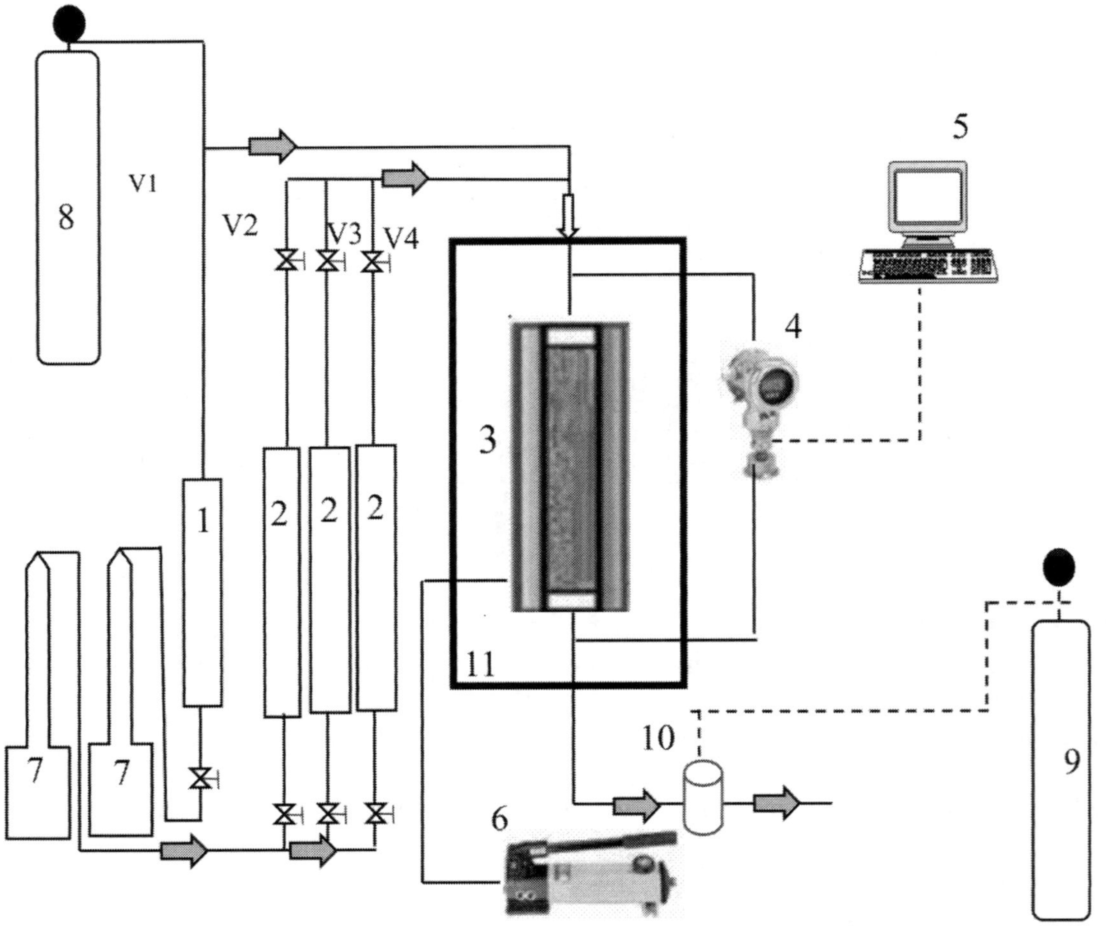


**Fig. 1 – Schematic for coreflood setup, where 1 = $CO_2$ accumulator, 2 = brine and solution accumulators, 3 = core holder, 4 = pressure transducer, 5 = PC recorder, 6 = hand pump for overburden pressure, 7 = syringe pump, 8 = $CO_2$ cylinder, 9 = $N_2$ cylinder, and 10 = back pressure regulator, 11 = oven.**

***CT-Scan Analysis.*** X-ray computed tomography (CT) has emerged as a powerful tool for nondestructive imaging due to its simplicity and high-resolution images (Taud et al. 2005). When a CT scan is operated, the X-ray source rotates around the object and the transmitted X-ray intensity is recorded by a series of detectors. The recorded data is converted into numerical CT numbers. CT numbers are represented in Hounsfield units. A Hounsfield unit represents a relative change in the attenuation density. The scale is set so that air has a value of -1000, DI water is zero, and compacted bone is 1000.

Two different sample tubes were filled with the crude oil, and 5 wt% NaCl brine and scanned to obtain the CT number. CT for the NaCl brine and crude oil were estimated to be 66 and -170, respectively.

***Core Preparation.*** Four cylindrical Buff Berea sandstone cores were used (6 in. length and 1.5 in. diameter). Each core was first dried in an oven at 150°F for 24 hrs. The weight of the dried core was measured ($W_{dry}$) and the core was scanned with computerized tomography (CT) scan ($CT_{dry}$). The core was then saturated with 5 wt% NaCl under vacuum for four hours. The core was then inserted on the coreholder and 5 wt% NaCl was injected at a constant flow rate at room temperature. Once the pressure drop across the core stabilized, the initial core permeability was measured. The weight of the saturated core was then measured ($W_{W\text{-}sat}$) and the core scanned again with CT scan ($CT_{W\text{-}sat}$). The core pore volume (PV) was calculated from the brine density ($\rho_{brine}$ = 1.035 g/cm$^3$ at 70°F) and weight difference in both dry and saturated cases (**Eq. 2**). The cores were stored on the brine until it was time to run the experiment.

$$PV = \frac{W_{W-sat} - W_{dry}}{\rho_{brine}} \quad (2)$$

Where PV is the core pore volume, $W_{W-sat}$ and $W_{dry}$ are the core weight in brine saturated and dry cases, respectively. $\rho_{brine}$ is the density of the brine, respectively.

Moreover, the porosity distribution along the core was estimated from the CT scan. The difference in the recorded CT number between the dry and saturated cases arises from the difference between both the brine and air density. This difference is proportional to the volume occupied by either fluid (pore volume), and estimation of the porosity can be calculated using **Eq. 3** (Akin and Kovscek 2003).

$$\emptyset = \frac{CT_{w-sat} - CT_{dry}}{CT_{brine} - CT_{air}} \quad (3)$$

Where $\emptyset$ is the core porosity, $CT_{w-sat}$ and $CT_{dry}$ are the CT numbers of the core in brine saturated and dry cases, respectively. $CT_{brine}$ and $CT_{air}$ are the CT numbers of the brine and air, respectively.

***Experiment procedures.*** **Fig. 2** shows schematic for the procedures of coreflood experiments. After pore volume and permeability measurements, the core saturated with the crude oil. The core inserted vertically on the core holder and the back pressure was set at 700 psi and the overburden pressure was set at 1500 psi. The oven temperature was set at 150°F. The crude oil was then injected at 0.2 $cm^3$/min for 5 PV till no water is coming out then this step repeated at 0.5, 1 and 2 $cm^3$/min for 2 PV to ensure reaching the maximum oil saturation. The initial oil inplace on the core is equal to the displaced water was collected. The weight of the oil-saturated core was measured ($W_{O-sat}$) and the core was scanned with CT scan ($CT_{O-sat}$). The initial oil inplace also confirmed from the weight difference between the water saturated and oil saturated cases (**Eq. 4**).

$$V_{oi} = \frac{W_{W-sat} - W_{O-sat}}{\rho_{brine} - \rho_{oil}} \quad (4)$$

Where $V_{oi}$ is the initial oil-in-place on the core, $W_{W-sat}$ and $W_{O-sat}$ are the core weight in brine saturated and oil saturated cases, respectively. $\rho_{brine}$ and $\rho_{oil}$ are the density of the brine and crude oil, respectively.

The initial oil saturation distribution along the core was estimated from the CT scan (**Eq. 5**).

$$S_{oi} = \left(\frac{CT_{w-sat} - CT_{O-sat}}{CT_{brine} - CT_{O}}\right) / \emptyset \quad (5)$$

Where $S_{oi}$ and $\emptyset$ are the initial oil saturation and the core porosity, respectively. $CT_{W-sat}$ and $CT_{O-sat}$ are the CT numbers of the core in brine saturated and oil saturated cases, respectively. $CT_{brine}$ and $CT_{O}$ are the CT numbers of the brine and crude oil, respectively.

Water flooding stage was then followed as a secondary recovery mode. 5 wt% NaCl was injected at 0.5 $cm^3$/min, and the produced oil collected with time for 5 PV until no oil observed on the production, then the flow rate increased to 1 and 2 $cm^3$/min to ensure reaching the residual oil saturation. The weight of the water-flooded core was measured ($W_{W-flood}$) and the core was scanned with CT scan ($CT_{w-flood}$).

The total oil recovery from the water flooding was confirmed from the weight difference between the water flooded and oil saturated cases (**Eq. 6**).

$$V_{op-wf} = \frac{W_{W-flood} - W_{O-sat}}{\rho_{brine} - \rho_{oil}} \quad (6)$$

Where $V_{op-wf}$ is the total oil recovery from the water flooding, $W_{W-flood}$ and $W_{O-sat}$ are the core weight in water flooded and oil saturated cases, respectively. $\rho_{brine}$ and $\rho_{oil}$ are the density of the brine and crude oil, respectively.

The residual oil saturation distribution along the core after water flooding was estimated from the CT scan (**Eq. 7**).

$$S_{or-wf} = \left(\frac{CT_{w-sat} - CT_{w-flood}}{CT_{brine} - CT_{O}}\right) / \emptyset \quad \text{.......} (7)$$

Where $S_{or-wf}$ and $\emptyset$ are the residual oil saturation and the core porosity, respectively. $CT_{w-sat}$ and $CT_{w-flood}$ are the CT numbers of the core in water-saturated and water-flooded cases, respectively. $CT_{brine}$ and $CT_{O}$ are the CT numbers of the brine and crude oil, respectively.

The last stage was the foam injection. On this stage, 5 wt% NaCl brine was injected first to pressurize the system and no more oil produced as it was a continuation of the water flooding stage. The foam solution was injected for two pore volumes to satisfy the adsorption requirements for the rock. $CO_2$ was then coinjected with the prepared foam solutions simultaneously to generate foam with 80% quality (gas volume / total volume = 80%). The foam was injected at 2 $cm^3$/min, and the produced oil collected with time for 5 PV where no more oil is producing. The weight of the foam-flooded core was measured ($W_{F-flood}$) and the core was scanned with CT scan ($CT_{F-flood}$).

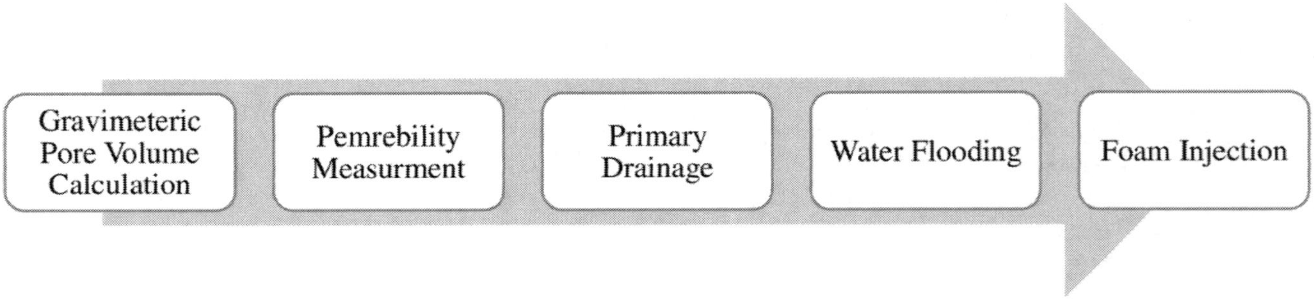


**Fig. 2 – Schematic for the procedures of coreflood experiments.**

**Table 3** shows the experimental design and physical properties of the cores that used on the core flood experiments.

| Core # | Initial Core Permeability, md | PV, $cm^3$ | Initial oil saturation vol% | Solution |
|---|---|---|---|---|
| 1 | 113 | 32.4 | 57 | 5 wt% NaCl Brine |
| 2 | 106 | 32.9 | 59.6 | (0.5 wt%) AOS |
| 3 | 110 | 33.3 | 57.6 | (0.5 wt%) AOS + (0.1 wt%) $SiO_2$ |
| 4 | 115 | 33.4 | 61.8 | (0.5 wt%) AOS + (0.4 wt%) VES+ (0.1 wt%) $SiO_2$ |

Table 3– Sandstone cores properties and the experimental conditions.

## Results and Discussion

**Shake test results.** To generate foam, 5 $cm^3$ of the different foam solutions, (a) AOS solution, (b) a mixture of the AOS solution and $SiO_2$ nanoparticles, and (c) a mixture of the AOS solution, VES, and $SiO_2$ nanoparticles, were shaken at ambient conditions for one minute. **Fig. 3** shows the results of the foam shake test at 77°F and subsequent foam degradation over a time interval of 5 hours. In the absence of crude oil, a fine texture foam was generated, but the bubbles coalescence with time and the bubble size increased. The foam height remains unchanged with time for 5 hrs. on the three solutions. In presence of crude oil, the foam becomes unstable and decay very fast. AOS foam decayed in 30 min. The addition of nanoparticles to the AOS solution slightly increased the foam stability and the foam decayed completely in 1 hr. By adding VES to the foam solutions, the foam stability increased, where the foam remains for more than 5 hrs.

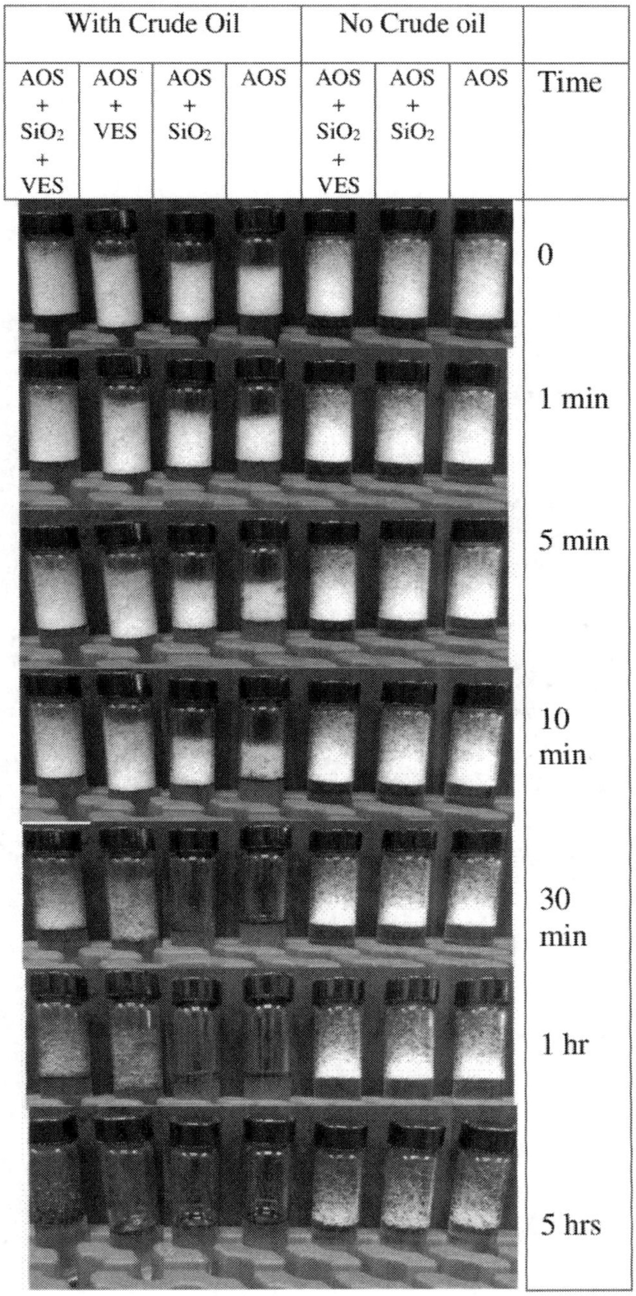


Fig. 3 – Foam shake test at 77°F for the different foam solutions in presence and absence of crude oil.

Two reasons for the deleterious effect of oil on the foam stability. First, the oil droplets tend to spread on the gas/liquid interface. As a result, the stable gas/liquid interface changes to unstable gas/oil interface, which accelerates the rupture of the foam lamella. Second, oil forms an emulsion on the foam lamella.

The oil droplets on the unstable emulsion agglomerate together and accelerate the drainage of the foam lamella, hence, the foam decays faster (Koczo et al. 1992).

**Fig. 4** shows microscope picture for AOS foam in contact with crude oil. The crude oil forms layers on the interface between the gas and liquid phases. Moreover, an emulsion was formed inside the foam lamella. **Fig. 5** shows microscope pictures for the foam lamella in the case of AOS foam in the presence of oil. The oil droplets on the unstable emulsion agglomerate together and accelerate the drainage, where the lamella thickness decreases with time.

By adding nanoparticles and VES to the solution the foam becomes more stable. **Figs. 6** and **7** show the microscopic analysis for AOS, VES, and nanoparticles foam in contact with oil. The generated foam was stable, where the formed oil in water emulsion was very stable and remains dispersed. As a result, the foam lamella thickness didn't change with time and the foam lasts for a longer time as appears on the shaking tests.

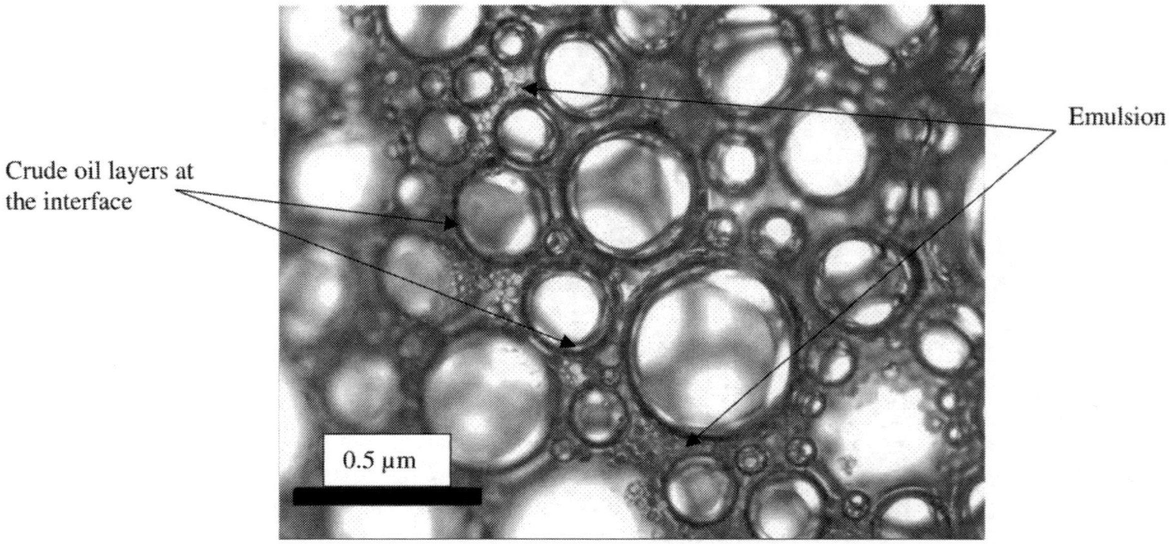


Fig. 4 – Microscope analysis for AOS foam in contact with oil (5x).

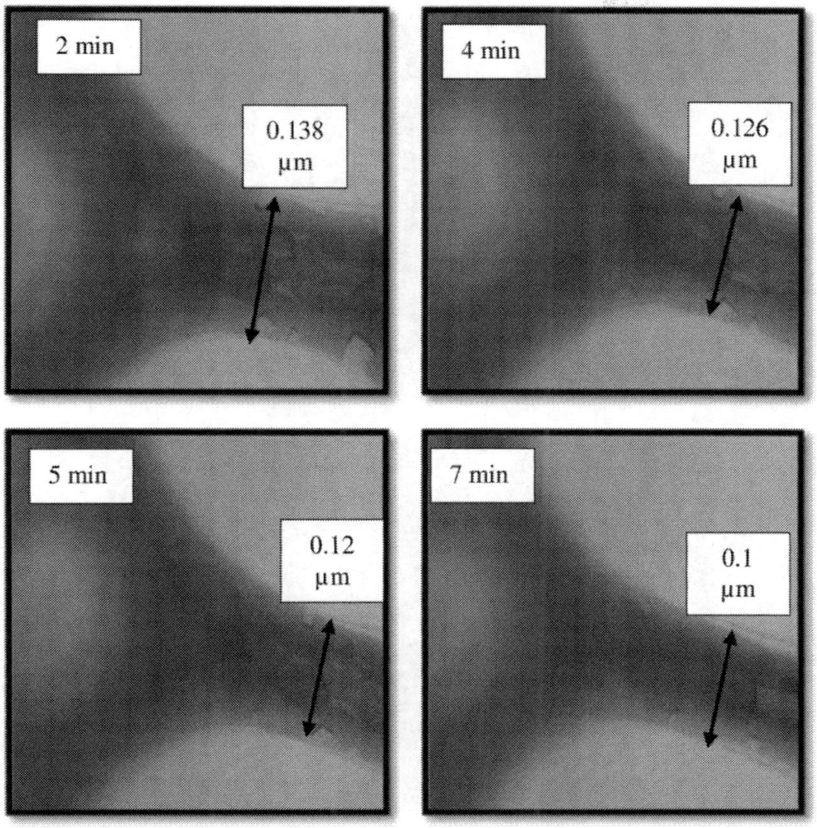


Fig. 5 – Microscope analysis for AOS foam to track the unstable emulsion and draining of the foam lamella with time in contact with oil (20x).

Fig. 6 – Microscope analysis for AOS, VES, and nanoparticles foam in contact with oil (5x).

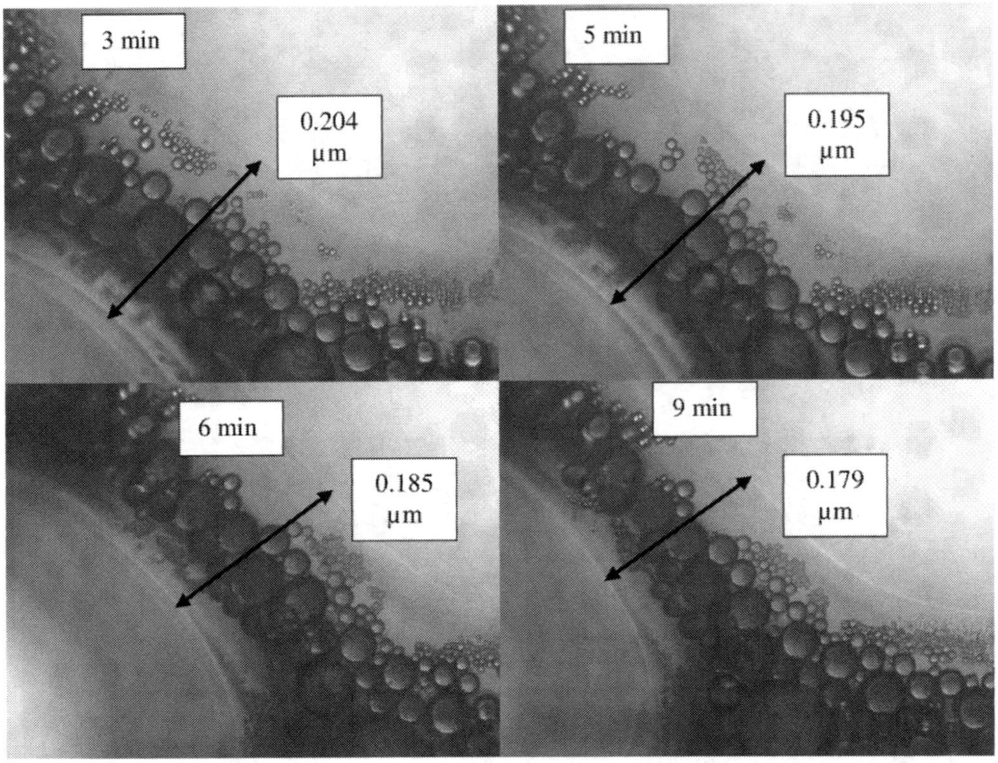


Fig. 7 – Microscope analysis for AOS, VES, and nanoparticles foam to track the emulsion and draining of the foam lamella with time in contact with oil (20x).

**Coreflood results.** Four coreflood experiments were conducted to emphasize the effect of adding nanoparticles and VES on the foam stability and the EOR performance. The initial oil saturation in the four cores varied between 57 to 61.8 vol% with initial oil-in-place between 18.5 to 20.6 $cm^3$. The oil recovery from the water recovery as a secondary mode was between 40 to 48 % of the initial oil-in-place and residual oil saturation 30 to 35 vol%. **Fig. 8** shows the CT scan along core #3 for different cases including dry, water-saturated, oil-saturated, and water-flooded cases. The CT number values increase the density of the fluids on the pores increases. Lowest values appeared on the dry case where the pores filled with air, and highest values in the case of the water-saturated core where the pores filled with brine. As the water replaced with oil in oil-saturated case, the CT number values decreased due to low oil density (0.82 $g/cm^3$) comparing to brine density (1.035 $g/cm^3$). Finally, by water flooding and displacing the oil from the pores, the CT number values increased but did not reach the water-saturated case due to the presence or residual oil in the pores. Eqs. 3, 5, and 7 was used to calculate the porosity, initial, residual oil saturations along the core. **Fig. 9** shows the porosity, initial, and residual oil saturations distribution along the core. The average core porosity was 18.9 vol%. For the oil saturation stage (primary drainage), the crude oil was injected at different flow rates until no water was observed on the outlet flow to ensure reaching the residual water saturation. The oil saturation trend shows higher oil saturation at the inlet than that at the outlet due to viscous fingering. The same behavior was found in the case of the residual oil saturation after water flooding. **Fig. 10** shows the pressure drop across the core at different injection rates (0.5 to 2 $cm^3$/min) and the corresponding recovery factor during the water flooding stage. At 0.5 cm3/min, most of the oil was produced (recovery factor = 49%) and then the recovery factor slightly increased (50%) after the flow rate increased to 1 cm3/min.

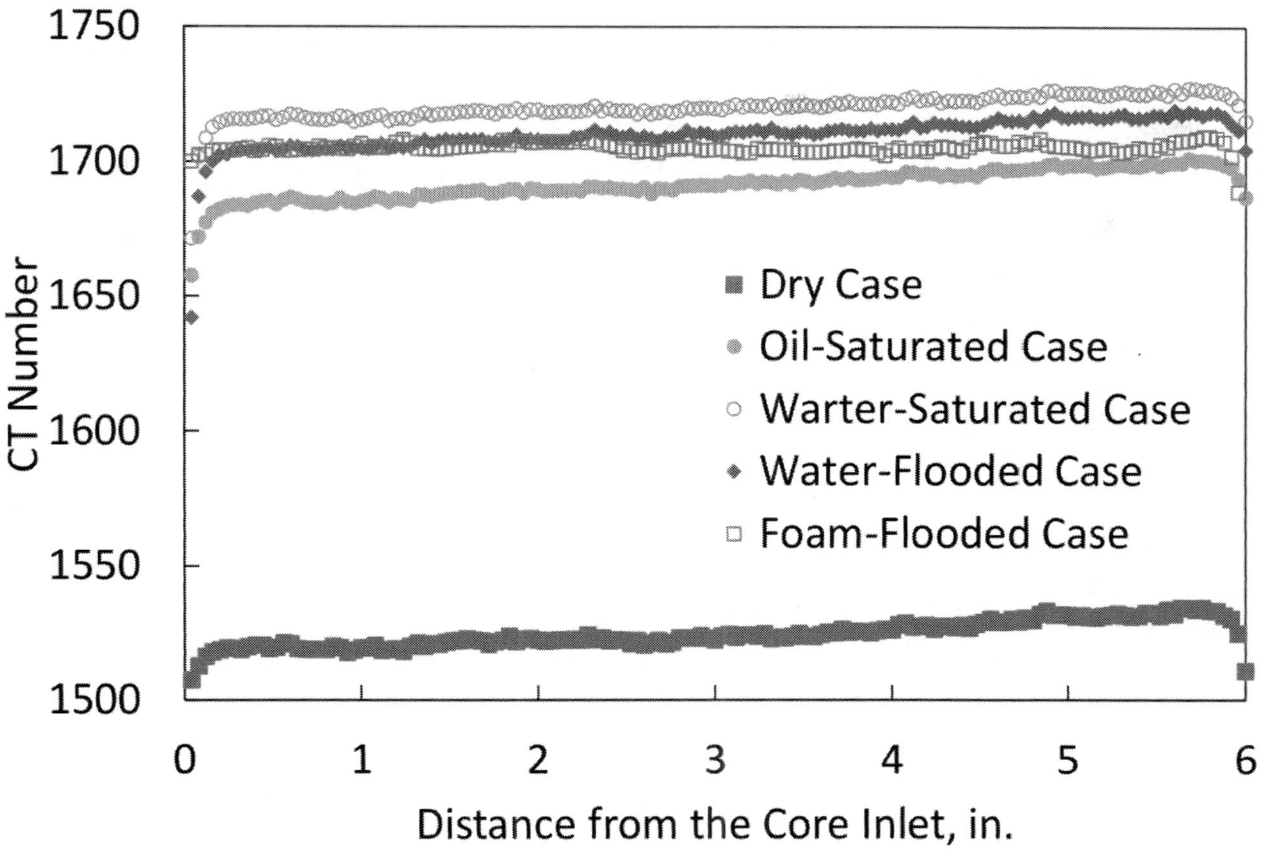


Fig. 8 - CT scan trend along core #3 for different cases.

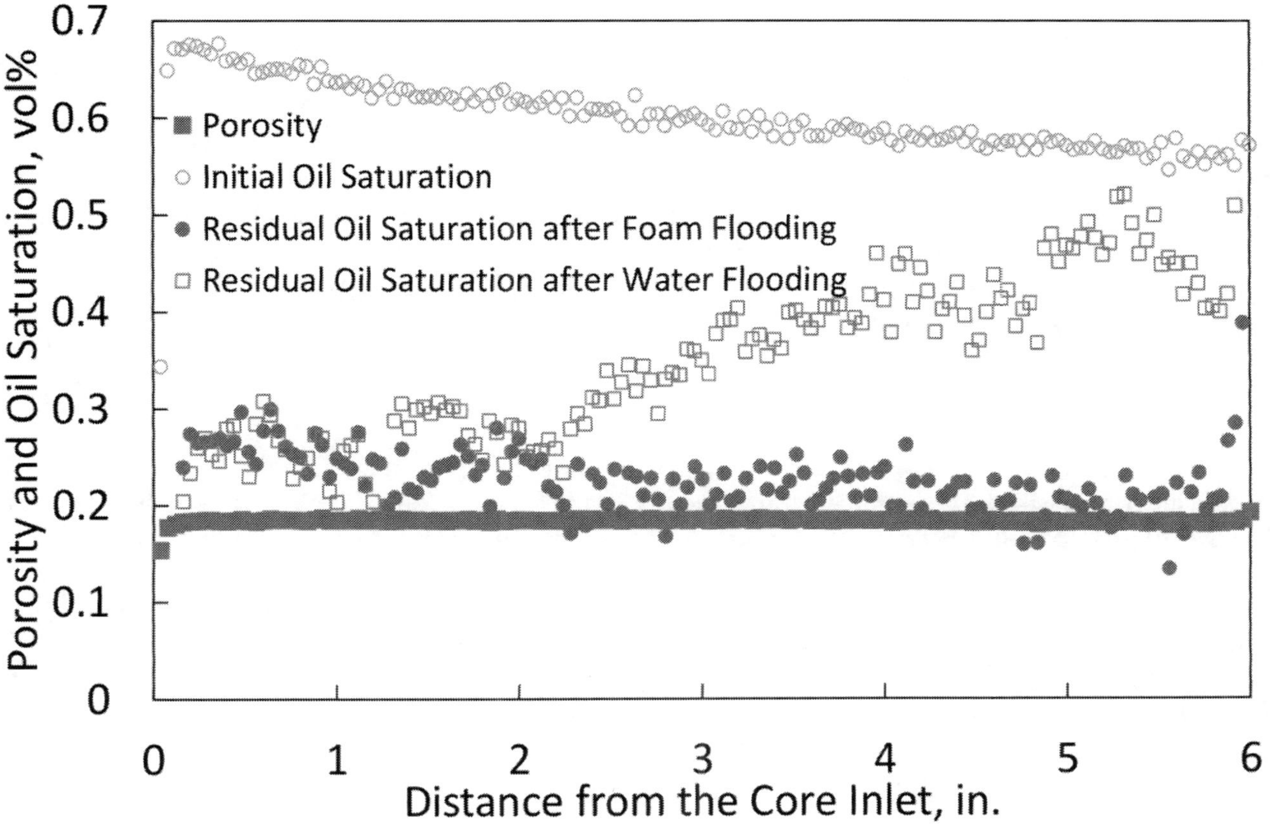

Fig. 9 – Porosity, initial, and residual oil saturations distribution along the core.

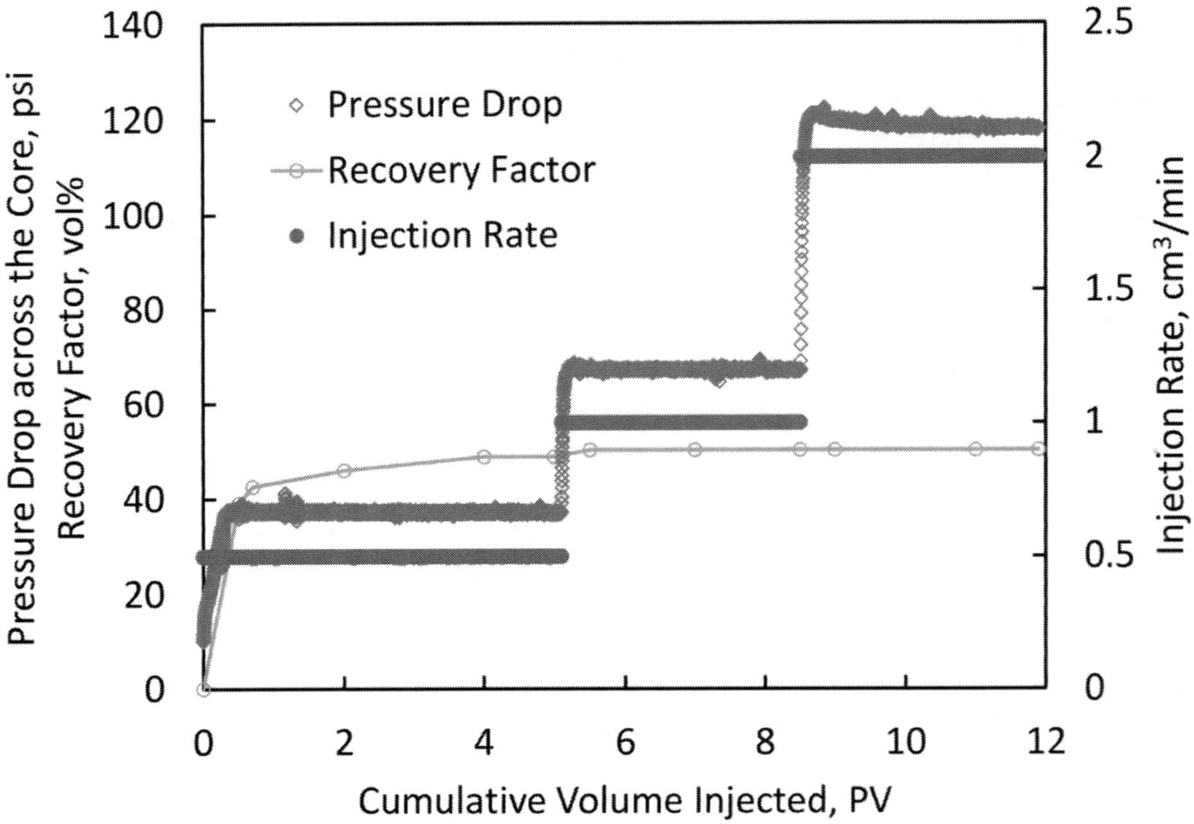


Fig. 10 – Pressure drop across the core at different injection rates and the corresponding recovery factor during the water flooding stage.

**Fig. 11** shows the pressure drop across the core for the coreflood experiments during the tertiary recovery mode. **Fig. 12** shows the recovery factor from the tertiary mode (oil recovery/oil-in-place before foam injection). The pressure drop is an indication for the mobility control factor as defined by Eq. 1. Lower pressure drop was observed on the first core, where $CO_2$ was coinjected with NaCl brine with no foam generation and only 29.6% of the residual oil after water flooding was produced. The reason for that is the viscosity fingering and lower displacement efficiency. With adding 0.5 wt% AOS, a weak foam was generated and the pressure drop slightly increased. Hence, the AOS did not control the gas mobility. As a result, AOS was not able to enhance the oil recovery with an apparent viscosity similar to that for the water-alternative-gas system (with no AOS in the solution). Oil recovery from the AOS foam increased by 2% compared to the water/gas system.

The pressure drop across the core increased in the presence of nanoparticles in the solution, which indicates increasing the mobility control factor. Emrani et al. (2017) explained that the nanoparticles adsorb to the interface between the liquid/gas phases and minimizes the contact area between them; as a consequence, it can build a strict barrier that prevents droplets coalescence and improves the foam stability. Increasing the mobility reduction factor reflects improving the displacement efficiency. As a result, the oil recovery increased (38% of the residual oil after water flooding).

Adding VES to the system stabilizes the emulsion and delay the foam decay that improves the foam stability as shown from the foamability tests. Therefore, mobility reduction factor improved and the oil recovery increased to 45.7% as well. Moreover, adding VES to the foam solutions increases the liquid viscosity, and the viscosity slows down the drainage of liquid initially. This keeps the liquid film thicker and causes the foam to stay in a stable state for a longer time.

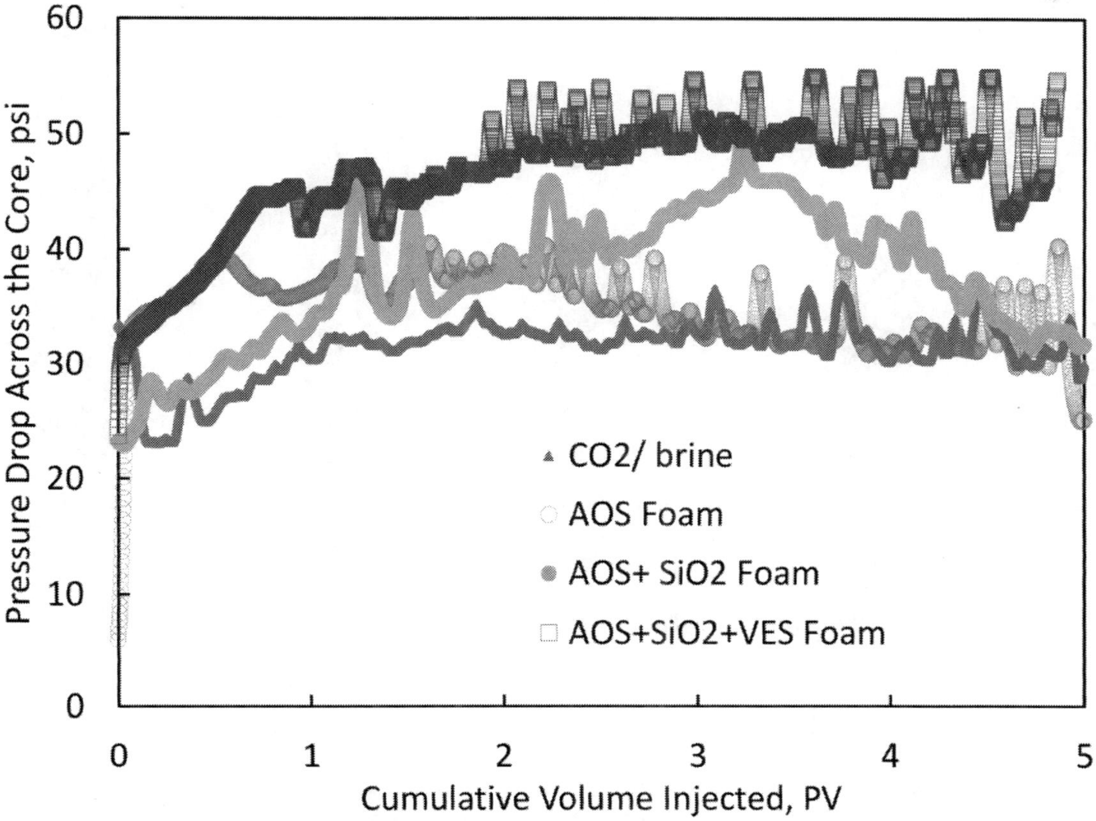

Fig. 11 – Pressure drop across the core during the tertiary recovery mode.

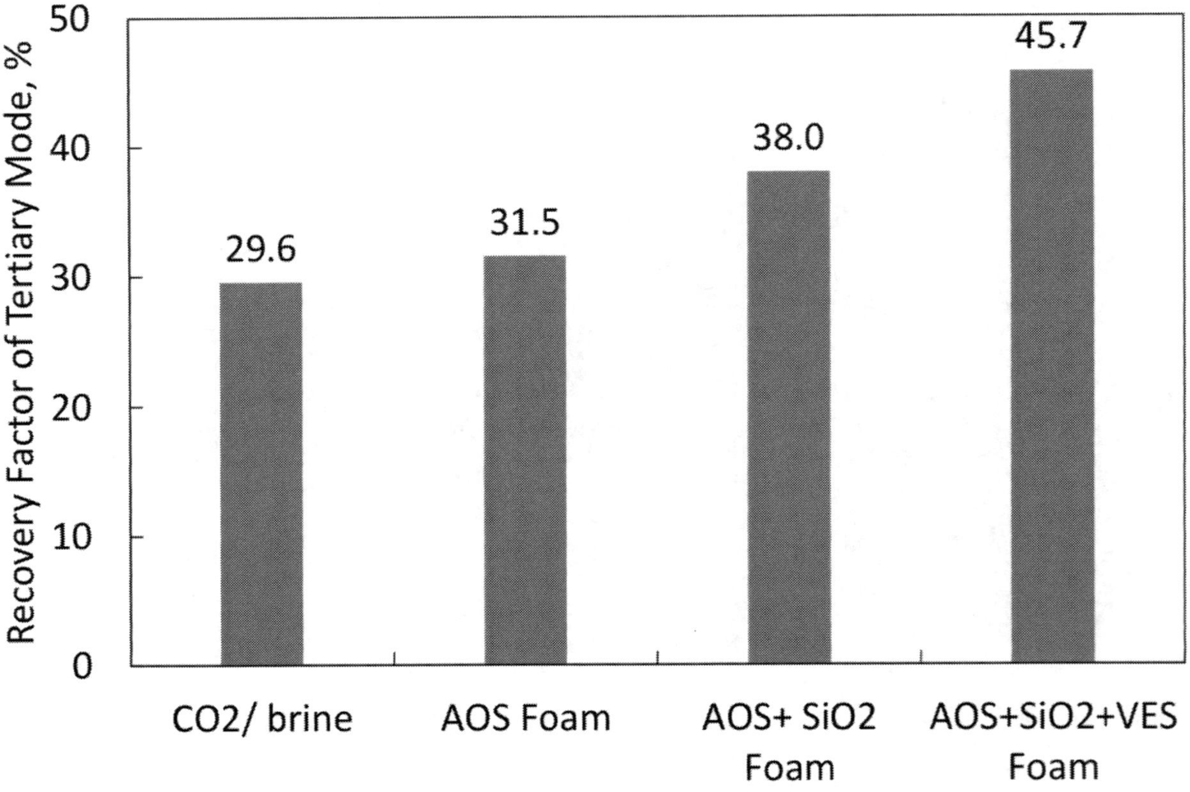


Fig. 12 – Recovery factor from the tertiary recovery mode for the different foam solutions.

**Fig. 13** summarizes the CT scan results for the four cores on the different injection stages. CT number values varies from 1400 to 1950. The lower CT values indicated by blue color and as the CT number increase, it changes to red then white. In the dry case, the cores had the lowest CT number where the pores were filled with air with CT number of (-1000). However, the water saturated cores had the highest CT number, where the pores were filled with brine with CT number of (66). The other three cases (oil-saturated, water-flooded, and foam flooded cases) the pores were filled with oil and brine, as a result, the CT number will be between water-saturated and dried cases (oil CT number is -170). As the oil saturation increases, the core CT number decreases and the color will tend to blue. CT scan images in the case of core 1 and 2 after foam flooding indicate lower displacement efficiency and more crude oil left behind. As the foam quality increases in the case of nanoparticles and VES, the displacement efficiency increases and the residual oil decreases.

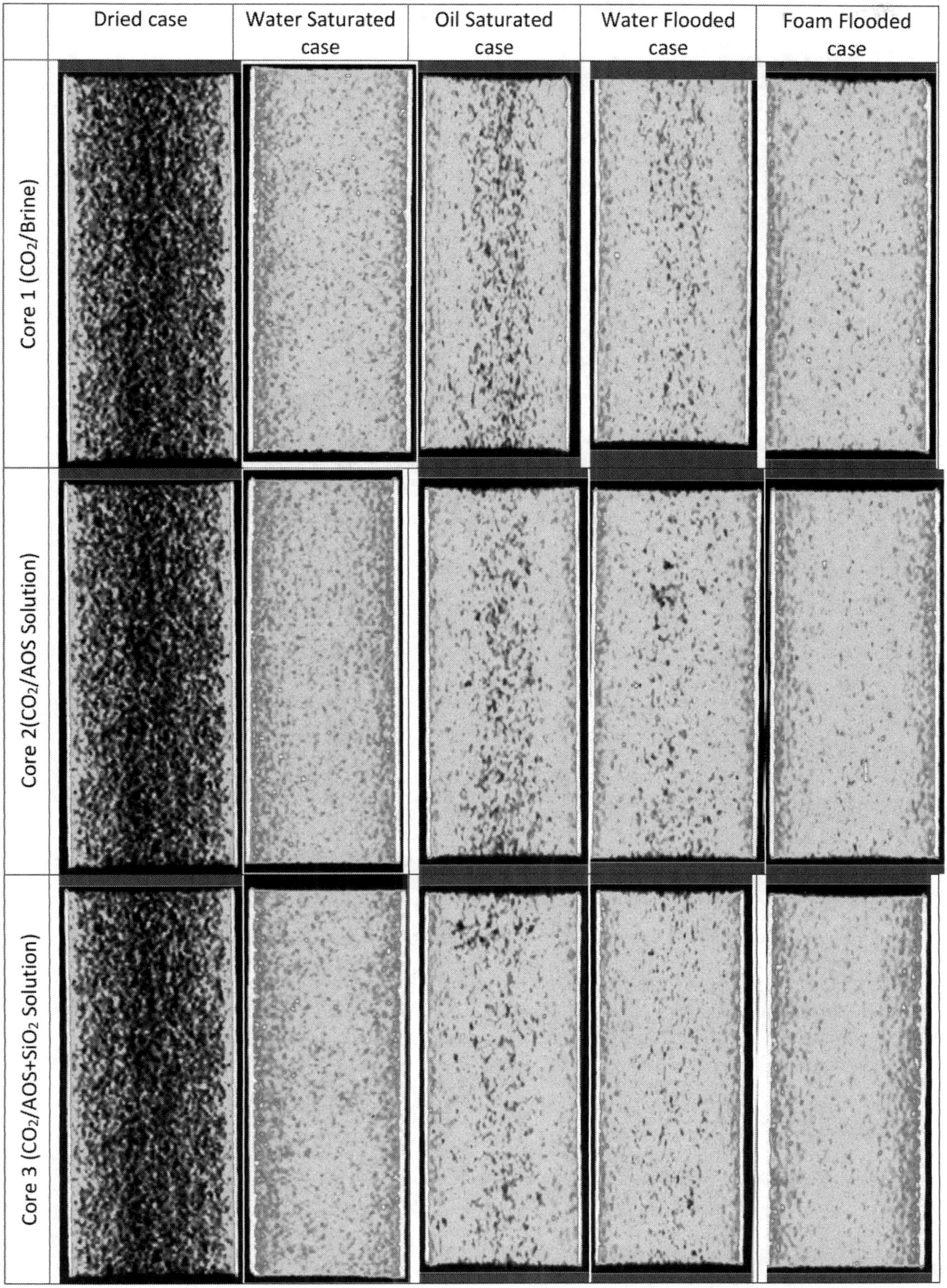
Dried case
Water Saturated case
Oil Saturated case
Water Flooded case
Foam Flooded case
Core 1 ($CO_2$/Brine)
Core 2($CO_2$/AOS Solution)
Core 3 ($CO_2$/AOS+$SiO_2$ Solution)

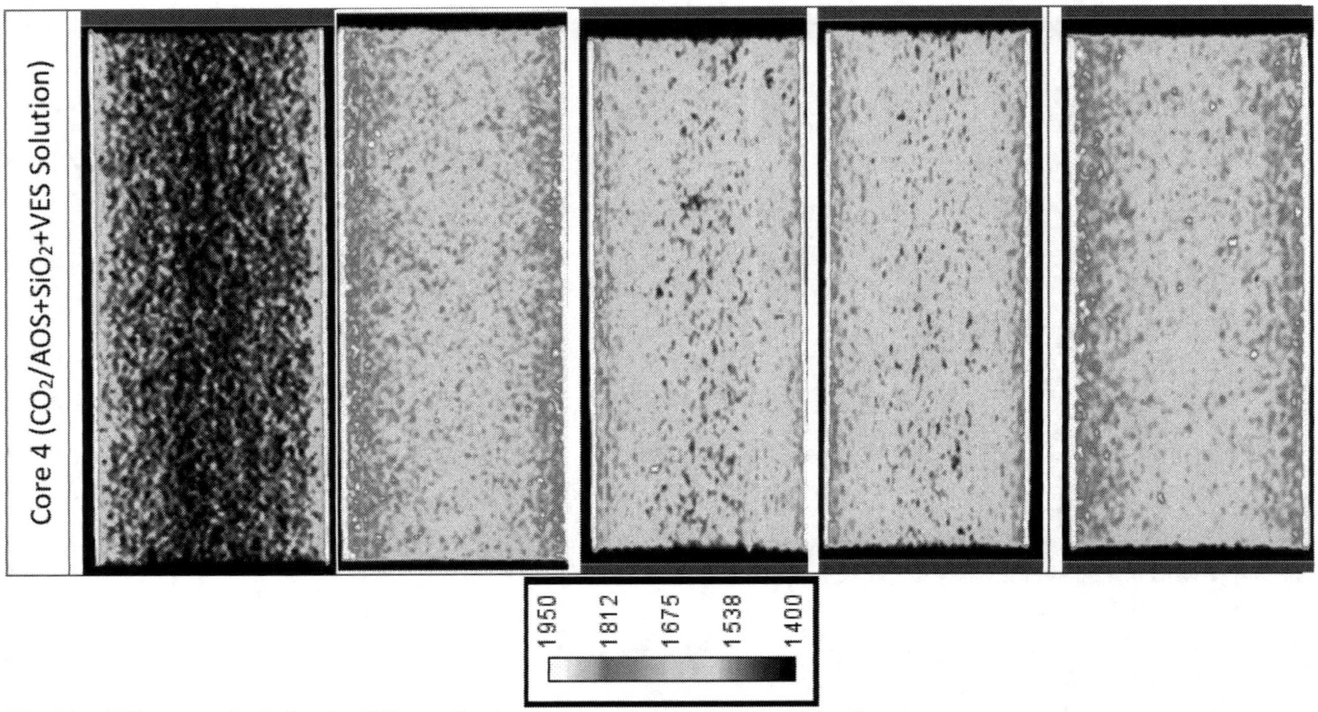


Fig. 13 – CT scan analysis for the different foam solutions after the different injection stages.

## Conclusions

This study investigates the effect of VES and silica nanoparticles on foam stability and MRF in porous media. The main conclusions are summarized as follows:

1. Adding silica nanoparticles and viscoelastic surfactant to the AOS solution improves both foam stability and MRF.
2. In contact with crude oil, unstable oil-in-water emulsion formed inside the foam lamella and oil layer formed on the gas/liquid interface that decreased the foam stability and accelerates the lamella drainage.
3. A weak foam was formed for AOS solutions, but the foam stability increased by adding nanoparticles and VES.
4. AOS was not able to enhance oil recovery with an apparent viscosity similar to that for the water-alternative-gas system (with no AOS in the solution), and almost no more oil was recovered by AOS foam.
5. The addition of nanoparticles and VES to the solution improved the foam MRF and allowed extra oil production (8% in the presence of nanoparticles, 15% by adding nanoparticles and VES).

## Acknowledgements

The authors thank the Crisman Institute and Texas A&M University for its support to conduct this work.

## Abbreviations

| | |
|---|---|
| AOS | alpha olefin sulfonate |
| EOR | enhanced oil recovery |
| HT | high temperature |
| HP | high pressure |
| MRF | mobility reduction factor |
| VES | viscoelastic surfactant |

| | |
|---|---|
| CT | computed tomography |
| PV | core pore volume |

## Nomenclature

| | |
|---|---|
| A | cross-section area of the core, $cm^2$ |
| K | absolute core permeability, md |
| L | core length, in. |
| Q | total flow rate, $cm^3$/min |
| S | fluid saturation, volume fraction |
| V | oil volume, $cm^3$ |
| W | core weight, gm |
| $\rho_{brine}$ | brine density, g/$cm^3$ |
| $\emptyset$ | core porosity, volume fraction |
| $\Delta p$ | pressure drop across the core, psi |
| $\mu$ | fluid viscosity, cp |

## Subscripts

| | |
|---|---|
| w | water |
| o | oil |
| $w_{-flood}$ | water flooding |
| r | residual |

## References


Akin, S., and Kovscek, A. R. 2003. Computed tomography in petroleum engineering research. Geological Society, London, Special Publications. 215 (01): 23-38. https://doi.org/10.1144/GSL.SP.2003.215.01.03.

Andrianov, A., Farajzadeh, R., Mahmoodi Nick, M. et al. 2012. Immiscible Foam for Enhancing Oil Recovery: Bulk and Porous Media Experiments. *Industrial & Engineering Chemistry Research* **51** (05): 2214-2226. http://dx.doi.org/10.1021/ie201872v.

Basheva, E. S., Ganchev, D., Denkov, N. D. et al. 2000. Role of Betaine as Foam Booster in the Presence of Silicone Oil Drops. *Langmuir* **16** (03): 1000-1013. http://dx.doi.org/10.1021/la990777.

Bian, Y., Penny, G. S. and Sheppard, N. C. 2012. Surfactant Formulation Evaluation for Carbon Dioxide Foam Flooding in Heterogeneous Sandstone Reservoir. Presented at the SPE Improved Oil Recovery Symposium, Tulsa, Oklahoma, 14-18 April. SPE 154018. http://dx.doi.org/10.2118/ 154018-MS.

Chen, Y., Elhag, A. S., Poon, B. M., et al 2012. Ethoxylated Cationic Surfactants for $CO_2$ EOR in High Temperature, High Salinity Reservoirs. Presented at the SPE Improved Oil Recovery Symposium, Tulsa, Oklahoma, 14-18 April. SPE 154222. http://dx.doi.org/10.2118/154222-MS.

Emrani, A. S., Ibrahim, A. F., and Nasr-El-Din, H. A. 2017. Mobility Control using Nanoparticle-Stabilized $CO_2$ Foam as a Hydraulic Fracturing Fluid. Presented at the SPE Europec featured at 79$^{th}$ EAGE Conference and Exhibition, Paris, France, 12-15 June. SPE-185863-MS. https://doi.org/10.2118/185863-MS.

Emrani, A. S., and Nasr-El-Din, H. A. 2017. An Experimental Study of Nanoparticle-Polymer-Stabilized $CO_2$ Foam. *Colloids and Surfaces, A: Physicochemical and Engineering Aspects*, **524** (01): 17-27.

Espinoza, D. A., Caldelas, F. M., Johnston, K. P., et al. 2010. Nanoparticle-Stabilized Supercritical $CO_2$ Foams for Potential Mobility Control Applications. Presented at the SPE Improved Oil Recovery Symposium, Tulsa, Oklahoma, 24-28 April. SPE 129925. http://dx.doi.org/10.2118/ 129925-MS.

Kalyanaraman, N., Arnold, C., Gupta, A., et al. 2016. Stability Improvement of $CO_2$ Foam for Enhanced Oil-Recovery Applications Using Polyelectrolytes and Polyelectrolyte Complex Nanoparticles. *Journal of Applied Polymer Science,* **134** (01):1-15. http://dx.doi.org/10.1002/app.44491.

Koczo, K., Lobo, L. A., and Wasan, D. T. 1992. Effect of Oil on Foam Stability: Aqueous Foams Stabilized by Emulsions. *Journal of Colloid and Interface Science* **150** (02): 492-506. DOI: http://dx.doi.org/10.1016/0021-9797(92)90218-B.

Mo, D., Yu, J., Liu, N., et al. 2012. Study of the Effect of Different Factors on Nanoparticle-Stablized $CO_2$ Foam for Mobility Control. Presented at the SPE Annual Technical Conference and Exhibition, San Antonio, Texas, 8-10 October. SPE 159282. http://dx.doi.org/10.2118/159282- MS.

Mohd, T. A. T., Shukor, M., Ghazali, N. et al. 2014. Relationship between Foamability and Nanoparticle Concentration of Carbon Dioxide ($CO_2$) Foam for Enhanced Oil Recovery (EOR). *Applied Mechanics and Materials*. **548** (01): 67-71. Trans Tech Publications.

San, J., Wang, S., Yu, J. et al., 2017. Nanoparticle-Stabilized Carbon Dioxide Foam Used in Enhanced Oil Recovery: Effect of Different Ions and Temperatures. *SPEJ*. Preprint. http://dx.doi.org/10.2118/179628-PA.

Singh, R., and Mohanty, K. K. 2014. Synergistic Stabilization of Foams by a Mixture of Nanoparticles and Surfactants. Presented at the Improved Oil Recovery Symposium, Tulsa, 12–16 April. SPE-169126-MS. https://doi.org/10.2118/169126-MS.

Xiao, C., Balasubramanian, S. N., and Clapp, L. W. 2016. Rheology of Supercritical $CO_2$ Foam Stabilized by Nanoparticles. Presented at the SPE Improved Oil Recovery Conference, Tulsa, Oklahoma, USA, 11-13 April. SPE-179621-MS. http://dx.doi.org/10.2118/179621-MS.

Xing, D., Wei, B., McLendon, W. J. et al. 2012. $CO_2$-Soluble, Nonionic, Water-Soluble Surfactants That Stabilize $CO_2$-in-Brine Foams. *SPE J.* **17** (04). http://dx.doi.org/10.2118/129907-PA.

Wang, C., and Li, H. Z. 2014. Foam Stability of Solvent/Surfactant/Heavy-Oil System under Reservior Conditions. Presented at the SPE International Heavy Oil Conference and Exhibition, 8-10 December, Mangaf, Kuwait. SPE-172888-MS. http://dx.doi:10.2118/172888-MS.

Wang, D., Hou, Q., Luo, Y., et al. 2015. Stability comparison between particles-stabilized foams and polymer-stabilized foams. *Journal of Dispersion Science and Technology*, **36** (2): 268-273.

Yin, G., Grigg, R. B., and Svec, Y. 2009. Oil Recovery and Surfactant Adsorption during $CO_2$-Foam Flooding. Presented at the Offshore Technology Conference, Houston, Texas, USA, 4-7 May. OTC-19787-MS. http://dx.doi:10.4043/19787-MS.

Yu, J., Liu, N., Li, L., et al. 2012. Generation of Nanoparticle-Stabilized Supercritical $CO_2$ Foams. Presented at the Carbon Management Technology Conference, Orlando, Florida, 7-9 February. CMTC 150849. http://dx.doi.org/10.7122/150849-MS.

Zhao, G., Fang, J., Dai, C. et al. 2015. Enhanced Foam Stability by Adding Dispersed Particle Gel: A New 3-Phase Foam Study. Presented at the SPE Asia Pacific Enhanced Oil Recovery Conference, Kuala Lumpur, Malaysia, 11-13 August. SPE-174597-MS. http://dx.doi:10.2118/174597-MS.

CMTC-486239-MS

# Feasibility of Near-Miscible Flooding By Production Gas Rejection with Varying $CO_2$ Content in Qinhuangdao Oilfield

Hao Chen, China University of Petroleum, Beijing; Xiansong Zhang, Baozhen Li, CNOOC; Yuan Mei, He Tang, China University of Petroleum, Beijing, and Shuai Wang, Changqing Oilfield Company, PetroChina, Xifeng, China



This paper was prepared for presentation at the Carbon Management Technology Conference held in Houston, Texas, USA, 17-20 July 2017.



## Abstract

A newly discovered offshore oilfield in QHD 29-2 block is facing the problem of selecting the appropriate developing method. It covers approximately 2300 km$^2$ and with an average water depth of 27.6m. The depth of the exploration well is over 3600 m with thick sand layer and oil zone. Conventional waterflood cannot be implemented due to the reservoir characteristics of small pores and throats, complex lithology, strong heterogeneity and low water injectivity. Near-miscible flooding is proposed considering the wide range of $CO_2$ content (24-90 mol%) in the production gas.

Slim tube test and slim tube simulation are conducted successively to determine the minimum miscible pressure (MMP) of the production gas and oil samples from the targeted reservoir. The relationship of displacement efficiency (DE), interfacial tension (IFT) and displacement pressure are provided and chosen as the basis for the division of the pressure interval of near-miscible flooding. The lower limit of the $CO_2$ content in the production gas to achieve near-miscible flooding are determined for the well 29-2E-5 with the well depth of 3308-3330 m. On this basis, an adjustment measures of adding intermediate components of ($C_2$-$C_6$) is proposed and assessed. The amount of the adding components is calculated and provided correspondingly.

Boundaries of the pressure region of near-miscible flooding are obtained for different $CO_2$ contents. Considering the reservoir conditions (112.1 °C, 31.96 MPa), the lower $CO_2$ content of 64% is estimated to be able to achieve near-miscible flooding for the targeted well. Accordingly, 2.3, 6.5, and 10.3 mol% of ($C_2$-$C_6$) are determined to be the lowest amounts for the adding components to achieve near-miscible injection for the $CO_2$ contents of 55%, 40%, and 24%, respectively.

Thus, the evaluation of the feasibility and optimization measures of near-miscible flooding by production gas re-injection with varying $CO_2$ content in a newly discovered offshore reservoir was conducted. Specific regions in the vicinity of MMP for impure $CO_2$ near-miscible flooding on the basis of comprehensive analysis of displacement efficiency and IFT from the views of both engineering and physicochemical were determined. Also, an ideal adjustment measures of adding intermediate hydrocarbons ($C_2$-$C_6$) were proposed and assessed. Moreover, the tuned compositional model of the paper could be beneficial for providing a further pilot and field case study of production gas re-injection in this reservoir.

## 1. Introduction

Carbon dioxide miscible flooding is a proven oil recovery process, generally one of the most efficient and promising methods for enhancing oil recovery (EOR) methods (Farajzadeh et al., 2010). However, in view of the limitations of gas sources, sedimentary environments and oil characteristics, the application and promotion of miscible flooding is very difficult in China (Chen et al., 2013). Even though some reservoirs are experiencing miscible flooding, maintaining high pressures required for miscible injection is very difficult (Dong et al., 2001). However, laboratory tests and some field tests have fully verified that near-miscible flooding can also obtain very satisfactory recovery (Shyeh-Yung et al, 1991; Schechter et al., 1998; Crigg et al., 1997). Thomas et al. (1994) claims that many successfully miscible flooding would be better described as near-miscible, depending on the evaluation techniques. Theoretically, near-miscible flooding refers to the gas injection of close but not complete miscibility with the oil (Sohrabi et al., 2007). It sounds very attractive from both economic and operational standpoints. For one thing, lower pressure reduces the gas mobility, a certain degree of mobility control effect can be obtained. For another, injection gas volume and compression cost decrease with the pressure decrease (Shyeh-Yung and Stadler, 1995).

The first step in determining whether a field is a good $CO_2$ flooding candidate is to conduct a screening study to assess the injection performance. The minimum miscibility pressure (MMP) is a key parameter used to distinguish the flooding type, either immiscible or miscible. However, near miscible flooding is much more complicated. Both of the upper and lower boundaries of the near-miscible region in the vicinity of MMP has to be determined. There are several methods to determine MMP or miscibility degree for gas flooding, including both empirical, experimental and numerical simulation studies. Empirical correlations are easy to use and always serve as a tool for quick MMP prediction (Alomair and Iqbal, 2014). However, the accuracy is not very satisfactory because most of the formulas are established based on very limited data within certain reservoir conditions (Teklu et al, 2012). Comparatively, laboratory MMP measurements are more accurate, but time consuming and very expensive. The most widely used experimental method is slim tube test (Yellig and Metcalfe, 1980). Due to the drawbacks of the experimental methods and empirical formulas, computational methods for MMP determination are proposed over the years and are successfully applied to reproduce the MMP between the injected gas and crude oil using computational methods based on the principles of fluid flow and phase equilibrium, which is often modeled based on cubic equation of State (EOS) and flash calculations (Teklu et al, 2012).

It has been recognized that the $CO_2$ MMP for a reservoir oil depends on the reservoir temperature, oil composition, and the purity of injected $CO_2$. MMP increases with increasing reservoir temperature. The effect of oil composition has been observed by several researchers and reached consensus to substantially affect MMP. Pure $CO_2$ is not always available as an injection gas. Impure $CO_2$ streams, however, are available from a variety of sources, including natural reservoirs and process plant waste streams. Typically, the $CO_2$ -rich produced gas streams contain a wide variety of components from methane and nitrogen to intermediate hydrocarbons-such as ethane, propane, butane, and $H_2S$. Metcalfe (1982) has shown that the presence of impurities can affect the pressure required to achieve miscible displacement. In general, the presence of $CH_4$ or $N_2$ in $CO_2$ can substantially increase the $CO_2$ MMP, while the presence of $H_2S$, $C_2H_6$, or intermediate hydrocarbons (such as $C_3$, $C_4$) can reduce the $CO_2$ MMP.

Use of a $CO_2$ source diluted by impurities can substantially improve the economic prospects of an EOR project. Also, $CO_2$ flood candidates that cannot become miscible with pure $CO_2$ may be viable with enriched $CO_2$ streams. For gravity-stable $CO_2$ floods, density adjustment while maintaining miscibility can be achieved by diluting the $CO_2$ stream with selected components. Contamination of $CO_2$ by $C_1$ or $N_2$ has been shown to adversely affect the MMP. Conversely, the addition of $C_2$, $C_3$, $C_4$, or $H_2S$ to $CO_2$ has been shown to have the effect of lowering the MMP. Rutherford (1962) empirically found pseudocritical temperature to show a correlation with the MMP required for hydrocarbon miscible floods. Jacobson (1972) applied the same concept with the acid gases of $CO_2$ and $H_2S$. Shang et al (2014) studied the effect of adding $CH_4$ and $N_2$ in the injection $CO_2$ on MMP. It was found that with the increase of the contaminants, MMP increases significantly. Comparatively, the impact of $N_2$ is more obvious. Hou et al (2013) summarized the main factors of MMP. It was concluded that for a certain gas composition and reservoir temperature, the more the intermediate components ($C_2$-$C_6$) in the oil and the smaller the relative molecular mass, the smaller the MMP is. On the contrary, the heavier the oil, the more difficult to achieve miscibility. For a certain reservoir, the reservoir temperature and oil composition are mostly stable, thus, the intermediate components ($C_2$-$C_6$) in the injection gas is one of the key factors influencing the miscibility of oil and gas system. Actually, the detrimental effect of nitrogen and methane can be offset by the beneficial effects of intermediate hydrocarbons. Thus, the produced gas from a $CO_2$ flood may be miscible with the oil at reservoir conditions even though it contains a high level of impurities.

QHD 29-2 block is an offshore oilfield of China with very thick oil zone, complex lithology, and strong heterogeneity. Conventional waterflood cannot be implemented due to the small pores and throats, and low water injectivity. The discovery of a big natural gas sources with varying $CO_2$ content just on top of the oil layer, provides big potential for the production gas re-injection to develop the oilfield effectively. Thus, in view of the wide range of $CO_2$ contents and huge costs of gas separation in an offshore oilfield, near-miscible flooding, is presented for the rational development of the reservoir.

So far, researches focusing on near miscible flooding is still very limited. Firstly, there is no reliable way to determine the pressure interval of near-miscible flooding, especially for impure $CO_2$ near-miscible flooding. In addition, the effect of contaminants in the injection gas on miscibility degree is still unclear. Thus, in this paper, the well 29-2E-5 with the well depth of 3308-3330 m in QHD offshore oilfield is taken as an example, pressure interval of near-miscible flooding under different $CO_2$ contents are determined. Also, the lower limit of the $CO_2$ content to implement near-miscible flooding under reservoir conditions is predicted. For the $CO_2$ content blow the lower limit, the effect and feasibility of adding intermediate components ($C_2$-$C_6$) in the injection gas are studied. On this basis, the amount of ($C_2$-$C_6$) needed for different $CO_2$ content are forecasted.

## 2. Methodology

Both experimental and simulation studies were conducted. The experimental part includes chromatographic analysis, PVT test, and slim tube test. The simulation part includes phase matching of reservoir fluids, compositional modeling and slim tube simulation.

### 2.1 Materials

Physical properties of the crude oil samples collected from the well 29-2E-5 with the depth of 3308-3330 m and 3475-3500 m, were shown in Table 1. Fig.1 showed the composition comparison of the oil-solvent system for these two layers. Obviously, for the upper layer, the content of light components ($C_1$+$N_2$) is higher while the content of intermediate components ($C_2$-$C_6$) is lower, which is more difficult to achieve miscibility with the injection gas. Table 2 is the compositions of the gas samples used in this paper, determined by gas chromatography (GC) analysis.

**Table 1 Physical properties of crude oil in well 29-2E-5**

| | Parameters | well 29-2E-5 | | Units |
|---|---|---|---|---|
| | | 3308-3330 | 3475-3500 | m |
| Reservoir condition (*) | Pressure | 31.96 | 34.04 | MPa |
| | Temperature | 112.1 | 116.7 | °C |
| Fluid properties | Saturation pressure | 17.92 | 17.49 | MPa |
| | Gas/oil ratio | 107.50 | 84.30 | $m^3/m^3$ |
| | Volume coefficient of the oil | 1.3360 | 1.2650 | $m^3/m^3$ |
| | Oil density* | 0.6986 | 0.7354 | $g/cm^3$ |
| | Oil viscosity* | 0.29 | 1.39 | mPa.s |
| | Died oil density | 0.8391 | 0.8502 | $g/cm^3$ |
| Composition of Well-flow content | $C_1+N_2$ | 42.92 | 38.40 | % |
| | $CO_2+C_2\sim C_{10}$ | 27.33 | 29.11 | % |
| | $C_{11}^+$ | 29.75 | 32.49 | % |

**Table 2 compositions of the injection gases in the test**

| Component | $CO_2$ | $N_2$ | $C_1$ | $C_2$ | $C_3$ | $iC_4$ | $nC_4$ | $iC_5$ | $nC_5$ | $C_6$ |
|---|---|---|---|---|---|---|---|---|---|---|
| Fraction, % | 100 | 0.00 | 0.00 | 0.00 | 0.00 | 0.00 | 0.00 | 0.00 | 0.00 | 0.00 |
| | 85 | 0.16 | 13.67 | 0.72 | 0.27 | 0.04 | 0.08 | 0.02 | 0.02 | 0.03 |
| | 70 | 0.33 | 27.35 | 1.44 | 0.53 | 0.08 | 0.15 | 0.04 | 0.04 | 0.05 |
| | 55 | 0.49 | 41.02 | 2.16 | 0.80 | 0.12 | 0.23 | 0.05 | 0.06 | 0.08 |
| | 40 | 0.65 | 54.69 | 2.87 | 1.07 | 0.16 | 0.31 | 0.07 | 0.08 | 0.11 |

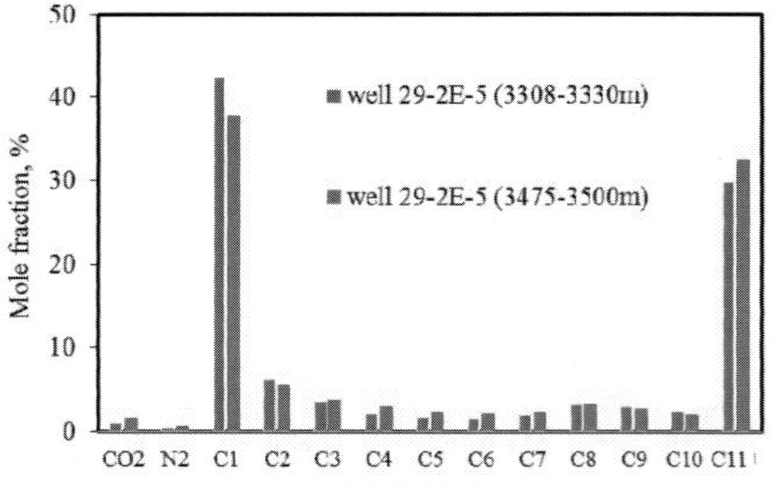


**Fig.1 Composition of oil-solvent system (well 29-2E-5)**

## 2.2 Slim tube test

RUSKA-2730 high pressure high temperature visual PVT apparatus produced by Ruska Company (Houston, Texas, USA) and CFS-200 $CO_2$ core flooding system produced by Corelab Company (Houston, Texas, USA) were used together to recombine the live oil and conduct the slim tube test, respectively. Experiments were conducted according to the measurement method for MMP by slim tube test (SY/T

6573-2016) of oil and gas industry standard of People's Republic of China. Physical parameters of the slim tube was shown in Table 3. More details of the experiments can be found in previous publications (Chen et al., 2017).

**Table 3 Physical parameters of the slim tube**

| Parameter | Length, m | Diameter, mm | Porosity, % | Pore volume, $cm^3$ | Gas permeability, $10^{-3}um^2$ |
|---|---|---|---|---|---|
| Value | 12.51 | 4.58 | 37.17 | 76.60 | 5275 |

### 2.3 slim tube simulation

PVTi of the Eclipse 300 was used to construct the compositional model by component lumping technique and tuning Peng-Robinson Equation of State (PREOS) based on the experimental data. More parameters such as displacement efficiency, IFT, GOR, viscosity and density of both liquid and gas phases, can be obtained.

## 3. Results and discussion

Based on the oil samples collected from the upper and lower depths of the well 29-2E-5, MMP and MNMP under different $CO_2$ contents were obtained from both slim tube test and slim tube simulation. On this basis, the effect of impurities of the injection gas on the miscibility was discussed. Moreover, the feasibility of adding intermediate components ($C_2$-$C_6$) to enhance the miscibility degree was studied.

### 3.1 Determination of pressure interval of near-miscible flooding

IFT and displacement efficiency are two parameters for the determination of MMP from physicalochemical and engineering point of views, respectively. Different from miscible flooding, a pressure interval with two endpoints, instead of a single point, is needed for determining near-miscible flooding. Fig.2 is the relationship of IFT, displacement efficiency and injection pressure for the oil sample from the well 29-2E-2 (3308-3330 m) with pure $CO_2$. Obviously, with the pressure increase, IFT of oil and gas system decrease sharply and then slows down. While the displacement efficiency increase linearly first then slows down and finally leveled off. Thus, it is not easy to distinguish a region from the smooth curves.

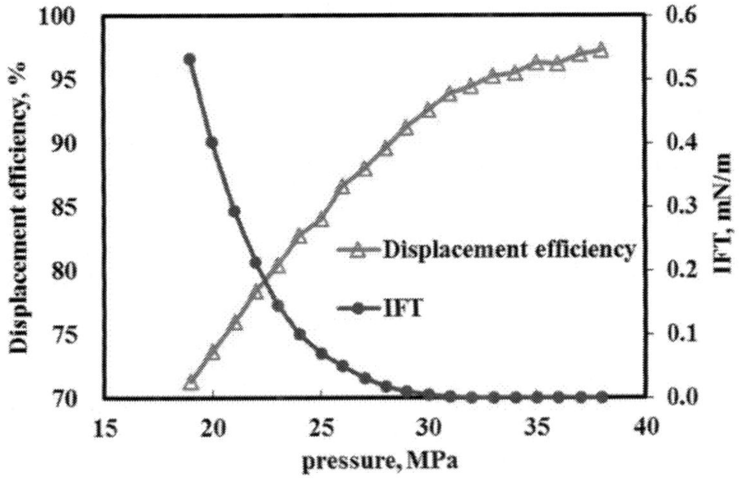


**Fig.2 IFT, displacement efficiency and injection pressure (100% $CO_2$)**

In this paper, the IFT data obtained from the slim tube simulation was transformed to semilog coordinate with the pressure. A much more obvious breakpoint can be identified, which is regarded as the lower boundary of the pressure region of near-miscible flooding. From the engineering point of view, 0.001 mN/m is generally chosen as the ultra-low IFT which can be used as the sign of complete miscibility. Thus, a pressure region of near-miscible flooding can be determined. Similarly, the upper and lower boundaries can also be obtained according to the relationship of displacement efficiency and pressure. More test points are needed to represent the trend of the curve clearly. As is shown in Fig.3, in the vicinity of MMP determined by conventional slim tube test, a new region can be distinguished which is significantly different from the immiscible region and miscible region. Thus, pressure interval of near-miscible flooding can be determined and compared by both IFT and displacement efficiency. And what we find is, statistically, the displacement efficiency and IFT of this transition interval are about 88%-96% and 0.001-0.05 mN/m, respectively.

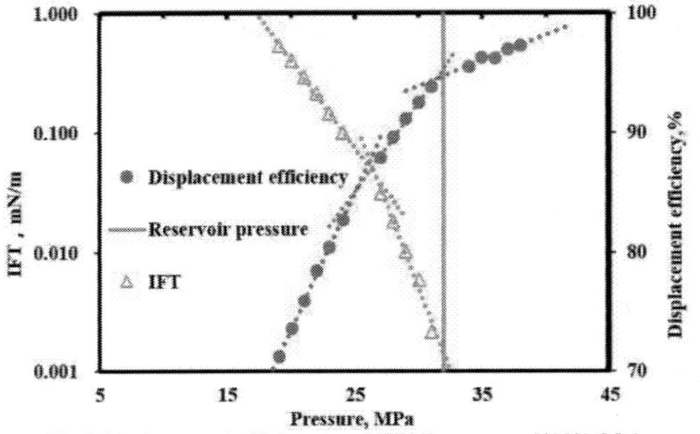

**Fig.3 Displacement efficiency and IFT VS pressure (100% $CO_2$)**

Slim tube tests with more test points under three different $CO_2$ contents (100%, 80%, and 55%) were conducted for oil samples from the well depth of 3308-3330 m of the well 29-2E-5. On this basis, slim tube simulation under 5 different $CO_2$ contents (100%, 85%, 70%, 55%, 40%) were conducted using the software Eclipse based on the compositional model matched with the data of PVT tests. Similarly, the upper and lower boundaries of the pressure region were calculated for different $CO_2$ contents. Three regions includes immiscible flooding, near-miscible flooding, and miscible flooding, were divided by these two boundaries. Table 4 is the summary of the MMP and MNMP determined by slim tube test and slim tube simulation. Obviously, the simulation results obtained by displacement efficiency and IFT are very close. Thus, the average values were taken as the MMP or MNMP. Compared with the slim tube results, the relative error is all less than 8%.

**Table 4 Summary of the determined MMP and MNMP of the well 29-2E-5 (3308-3330 m)**

| $CO_2$ content, mol% | MNMP, MPa | | | | MMP, MPa | | | |
|---|---|---|---|---|---|---|---|---|
| | simulation | | | test | simulation | | | test |
| | $E_D$ | IFT | AVE. | | $E_D$ | IFT | AVE. | |
| 100 | 26.85 | 26.18 | 26.52 | 27.26 | 33.21 | 32.53 | 32.90 | 30.49 |
| 85 | 28.47 | 29.46 | 28.97 | --- | 35.43 | 35.22 | 35.19 | --- |
| 70 | 31.80 | 30.79 | 31.30 | --- | 36.87 | 37.10 | 37.09 | --- |
| 55 | 33.38 | 33.68 | 33.53 | 34.14 | 41.08 | 39.09 | 40.11 | 38.56 |
| 40 | 34.95 | 35.01 | 34.98 | --- | 42.72 | 41.51 | 42.52 | --- |

Note: ED: Results obtained by displacement efficiency; IFT: Results obtained by IFT;

### 3.2 The effect of mixing gas on miscibility

Fig.4 is the relationship of the forecasted average upper and lower boundaries of near-miscible flooding and the $CO_2$ content for the well 29-2E-5. Considering the mixing gases of the injection gas, $CH_4$ is the main components. Thus, with the decrease of $CO_2$ content, the miscibility becomes more and more difficult. Both MMP and MNMP increase greatly. It is found that the trends basically obey liner relationship. Specifically, for the oil samples of the well 29-2E-5, for each 10% of impurities increase in the injection gas, MMP and MNMP increase 1.6 MPa and 1.4 MPa, respectively. It can also be concluded that with the decrease of $CO_2$ content, the potential scope, which belongs to near-miscible flooding, enlarges. In addition, compared with the reservoir pressure of 31.96 MPa, $CO_2$ content below 64% cannot achieve near-miscible flooding for the well depth of 3308-3330 m of the well 29-2E-5. Thus, for the production gases of the reservoir, the lower limit of the $CO_2$ content can be re-injected to implement near-miscible flooding is 64%.

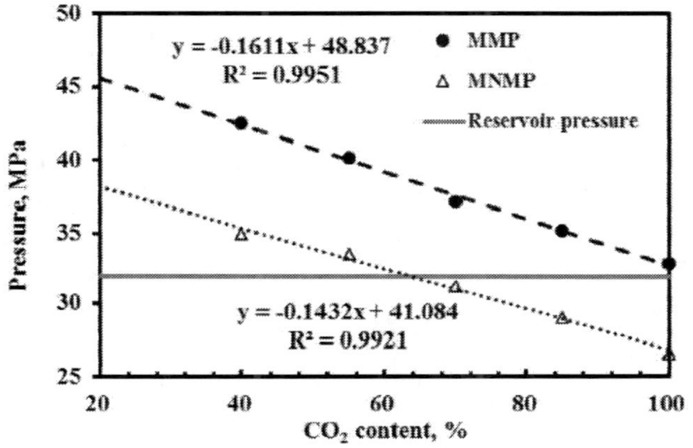


Fig.4 MNMP and MMP VS $CO_2$ content

### 3.3 Miscibility control by adding intermediate components

Generally, the mixing of $N_2$, $O_2$, and $CH_4$ can increase the difficulty of miscibility, while the mixing of $H_2S$, $SO_2$, and intermediate hydrocarbons ($C_2$-$C_6$) are beneficial for the miscibility. For near-miscible flooding, the whole process is the dual function of combined condensing/vaporizing mechanism. Taken the oil samples from the well depth of 3308-3330 m of the well 29-2E-5 as an example, the feasibility of adding a certain amount of intermediate components ($C_2$-$C_6$) in the injection gas to lower the pressure interval of near-miscibility was studied. This will definitely be helpful for implementing and enlarging the potential of near-miscible flooding in QHD offshore oilfield.

Firstly, for a certain $CO_2$ content, different contents of intermediate hydrocarbons ($C_2$-$C_6$) were added to three groups of gas samples (55%, 40%, and 24% $CO_2$) to test the effect on the lower and upper boundaries of pressure interval. Obviously, the adding of 5 mol% ($C_2$-$C_6$) greatly reduce the MMP and MNMP of the targeted oil samples, the reducing ranges are 8% and 11%, respectively. As is shown in Fig.5, the corresponding $CO_2$ content for the pressure range of near-miscible flooding expands to (34.5%-76.6%). The lower limit of $CO_2$ content decreased by nearly 30%, which effectively enlarge the $CO_2$ content limit for implementing near-miscible flooding. Two of the three gas samples can achieve near-miscible flooding due to the adding of 5 mol% intermediate hydrocarbons. It should also be noted that with the adding of light components, the appropriate range of $CO_2$ contents for implementing near-miscible flooding also expands, nearly 1/3 for the targeted well depth.

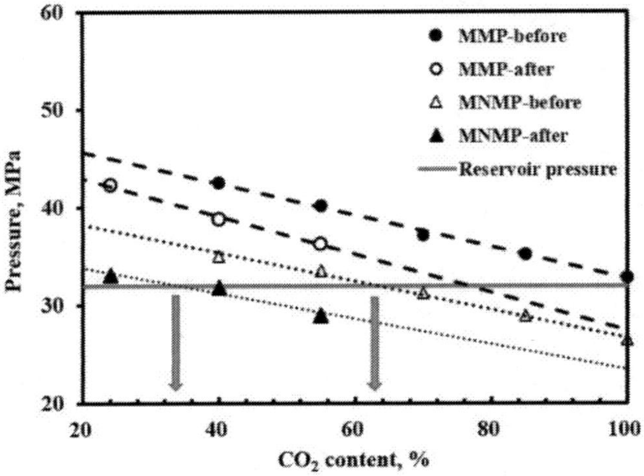


Fig.5 the effect of 5% ($C_2$-$C_6$) on MNMP

It is well known that the adding of intermediate components can effectively reduce the MNMP and MMP, and enlarge the range of near-miscible flooding. However, the amount of light components needed to add for achieving near-miscible flooding for different $CO_2$ contents was still unclear. Thus, different contents (1%, 3%, 5%, 8%, 10%, and 15%) of intermediate components were added to the injection gases with the $CO_2$ contents of 55%, 48%, 40%, 32% and 24%. The relationship of the amount of ($C_2$-$C_6$) needed for each $CO_2$ content are obtained, as is shown in Fig.6. It can be concluded that for each 10% contaminants content increase in the well depth of 3308-3330 m in well 29-2E-5, about 2.6% intermediate hydrocarbons ($C_2$-$C_6$) is needed to keep near-miscible flooding.

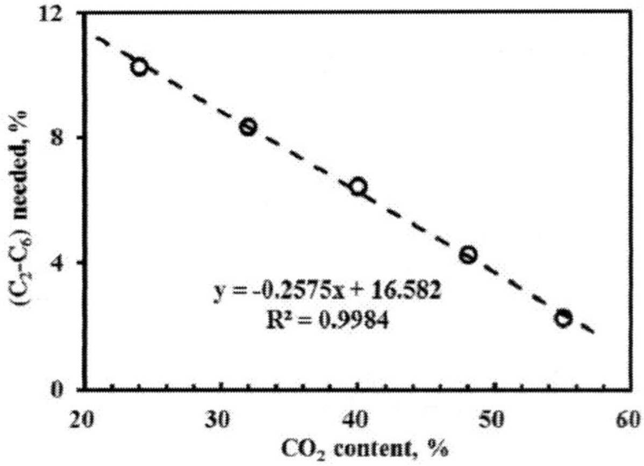


Fig.6 $CO_2$ content VS intermediate components ($C_2$-$C_6$) needed to achieve near-miscible flooding

From Fig.6, it can be concluded that for the $CO_2$ content of 55%, 40%, and 24%, about 2.29%, 6.46% and 10.28% of the intermediate components ($C_2$-$C_6$) are needed for achieving near-miscible flooding in the well QHD 29-2E-5 with the depth of 3308-3330 m, as is shown in Table 5.

Table 5 The amount of ($C_2$-$C_6$) needed for different $CO_2$ contents

| $CO_2$, mol% | 55 | 48 | 40 | 32 | 24 |
|---|---|---|---|---|---|
| ($C_2$-$C_6$) needed, mol % | 2.29 | 4.26 | 6.46 | 8.38 | 10.28 |

## 4. Conclusions

(1) Based on the data obtained from slim tube test and slim tube simulation with denser test points, pressure regions of near-miscible flooding with different $CO_2$ content were determined for the well depths of 3308-3330 m of the well 29-2E-5.

(2) For the reservoir conditions, the lower limit of $CO_2$ content for achieving near-miscible flooding in well depth of 3308-3330 m of the well 29-2E-5 is 64%.

(3) The mixing of the contaminants (mainly methane) can greatly enhance the difficulty to achieve miscibility. For the targeted layer of the well 29-2E-5, for each 10% increase of contaminant contents, MNMP and MMP increase 1.4 MPa and 1.6 MPa, respectively.

(4) By adding 5 mol% light components of ($C_2$-$C_6$), the lower limit of $CO_2$ content for achieving near-miscible flooding can be decreased effectively. The corresponding $CO_2$ content range for near-miscible flooding can be decreased to (34.5-76.6%) with 1/3 interval enlarge.

(5) The ($C_2$-$C_6$) contents needed for achieving near-miscible flooding in the upper layer of the well QHD 29-2E-5 under different $CO_2$ contents was obtained. For each 10% decrease of $CO_2$ content, 2.6% of the intermediate components is needed.

## Acknowledgments

This study was supported by Beijing Natural Science Foundation (Grant 3173044), Science Foundation of China University of Petroleum, Beijing (No. 2462013YXBS003), Open Foundation of State Key Laboratory of Offshore Oil Exploitation (2015-YXKJ-001).

## References

1. Alomair, O., Iqbal, M., 2014. $CO_2$ Minimum miscible pressure (MMP) estimation using multiple linear regression (MLR) technique. In: Proceedings of SPE Saudi Arabia Section Technical Symposium and Exhibition. Al-Khobar, Saudi Arabia, 21-24 April.
2. Chen H., Yang, S., Ren, S., Yu, D. H., Lu, H., Li, F., & Zhang, X., 2013. Crude oil displacement efficiency of produced gas re-injection. International Journal of Green Energy. 10 (6): 566-573.
3. Chen, H., Zhang, X., Chen Y., Tang, H., Mei Y., Li, B., and Shen, X., 2017. Study on pressure interval of near-miscible flooding by production gas Re-injection in QHD offshore oilfield, Journal of Petroleum Science and Engineering, https://doi.org/10.1016/j.petrol.2017.07.045.
4. Dong, M., Huang, S., Srivastava, R., 2000. Effect of solution gas in oil on $CO_2$ minimum miscibility pressure. Journal of Canadian Petroleum Technology. 39(11), 53-61.
5. Farajzadeh, R., Andrianov, A., & Zitha, P. L., 2010. Investigation of Immiscible and Miscible Foam for Enhancing Oil Recovery. Industrial & Engineering Chemistry Research, 49(4), 1910-1919.
6. Grigg, R.B. and Schechter, D.S., 1997. State of the industry in $CO_2$ floods; In: Proceedings of the SPE Annual Technical Conference and Exhibition, San Antonio, TX, October 5-8.
7. Hou, D., Luo P., Sun, L., Tang, Y., Sun, Y., and Pan, Y., 2013. An improved model for predicting minimum miscibility pressure of hydrocarbon gas crude oil system. Xinjiang Petroleum Geology. 34 (6): 684-688.
8. Jacobson H A. Acid Gases and Their Contribution to Miscibility. Journal of Canadian Petroleum Technology, 1972, 11(2):56-59.
9. Metcalfe R S. Effects of Impurities on Minimum Miscibility Pressures and Minimum Enrichment Levels for $CO_2$ and Rich-Gas Displacements. Society of Petroleum Engineers Journal, 1982, 22(2): 219-225.
10. Rutherford W M. Miscibility Relationships in the Displacement of Oil By Light Hydrocarbons[J]. Society of Petroleum Engineers Journal, 1962, 2(04): 340-346.
11. Schechter, D.S., Grigg, R., and Guo, B., 1998. Wellman unit $CO_2$ flood: reservoir pressure reduction and flooding the water/oil transition zone. In: Proceedings of SPE Annual Technical Conference and Exhibition. New Orleans, LA,

September 27-30.
12. Shakiba, M., Ayatollahi, S., & Riazi, M., 2016. Investigation of oil recovery and $CO_2$ storage during secondary and tertiary injection of carbonated water in an Iranian carbonate oil reservoir. Journal of Petroleum Science and Engineering, 137: 134-143.
13. Shang, B., Liao, X., Zhao, X., Cai, Z., Li, J., He, Y., and Su, X., 2014. Research about the influence of impurities on MMP and crude oil properties for $CO_2$ flooding. Petroleum Geology and Recovery Efficiency, 21(6):92-94.
14. Shyeh-Yung, J.J., 1991. Mechanisms of miscible oil recovery: effects of pressure on miscible and near-miscible displacements of oil by carbon dioxide. In: Proceedings of SPE Annual Technical Conference and Exhibition. Dallas, Texas. October 6-9.
15. Shyeh-Yung, J.J., Stadler, M.P., 1995. Effect of injectant composition and pressure on displacement of oil by enriched hydrocarbon gases. SPE Reservoir Engineering, 10(2): 109-115.
16. Sohrabi, M., Danesh, A., Tehrani, D. H., Jamiolahmady, M., 2007. Microscopic mechanisms of oil recovery by near-miscible gas injection. Transport in Porous Media, 72(3), 351-367.
17. Thomas, F.B, Holowach, N., Zhou, X., Bennion, D.B., Benion, D.W., 1994. Miscible or near-miscible gas injection, which is better? In: Proceedings of SPE/DOE Symposium on Improved Oil Recovery. Tulsa, April 17–20.
18. Teklu, T. W., Ghedan, S. G., Graves, R. M., & Yin, X., 2012. Minimum Miscibility Pressure Determination: Modified Multiple Mixing Cell Method. In: Proceedings of SPE EOR Conference at Oil and Gas West Asia, Muscat, Oman, April 16-18.
19. Yellig,W.F., Metcalf., R.S., 1980. Determination and prediction of $CO_2$ minimum miscibility pressures. J. Pet. Tech. 32(1):160-168.

# CMTC-486243-MS

# Investigation of the Oil Recovery Mechanism during Laboratory $CO_2$ EOR Experiment with Unconventional Shale Cores through Compositional Simulation

Amy Zou, and David Schechter, Texas A&M University



This paper was prepared for presentation at the Carbon Management Technology Conference held in Houston, Texas, USA, 17-20 July 2017.



## Abstract

After implications of hydraulic fracturing operations, the commercial production of tight formations and shale plays were successfully achieved in past decades. Due to the rapid decline rate after primary depletion of fractured reservoirs, extracting the remaining liquid hydrocarbon from the nano-Darcy permeability matrix becomes the next step.

Previously conducted laboratory experiments demonstrated promising results by successfully recovering liquid hydrocarbon from preserved and unfractured sidewall unconventional core plugs. However, what are the driving forces behind this observed result was not well understood. In other words, is the hydrocarbon recovery associated with commonly known recovery mechanisms during $CO_2$ EOR processes, such as viscous displacement, oil volume expansion, viscosity reduction and vaporization of lighter hydrocarbon components? Or, is it driven by other mechanisms that are frequently considered insignificant during conventional $CO_2$ EOR processes?

This study utilizes a commercial compositional simulator to investigate the oil production mechanisms from the matrix into the fractures. The process includes constructing a fine grid 3D model to simulate the previously conducted laboratory experiment, performing systematic sensitivity analysis, and evaluating the mechanisms that could potentially contribute to the oil recovery observed during the experiments. With laboratory scale modeling, the dominating mass transfer mechanism between the matrix and fractures, which in turn translates into oil recovery mechanism, is concluded to be diffusion. The work provided in this study can be used to enhance the accuracy for upscaled field simulations. However, whether $CO_2$ EOR will unlock the unconventional liquid reservoir potential and make significant economic impacts at field scale needs to be carefully evaluated on a case by case basis.

## Introduction

For the past few decades, due to the increasing demand of energy and the advancements in horizontal drilling and hydraulic fracturing technologies, the industry reallocated its resources into exploring ways to produce oil from the previously unprofitable shale plays. The current technique to produce from the shale plays is through primary depletion. However, studies have shown that the recovery factor remains low, and large sums of the reserves are left behind in the reservoirs (Liu *et al.* 2014). To look ahead into the future production of the remaining reserves because of the declination of primary recovery to an uneconomic level, operators are shifting their attentions into research that focuses on EOR methods in shale reservoirs.

Enhanced oil recovery (EOR), also known as tertiary recovery, serves the purpose of producing the remaining hydrocarbons after the depletion of the reservoir energy or after pressurizing the reservoir and displacing the hydrocarbon with gas or water. EOR processes are necessary when pressure maintenance becomes insufficient or in the cases where secondary recovery processes are not applicable due to the low recovery potential. When investigating methods to recover from unconventional reservoirs with low injectivity, EOR is considered as a suitable candidate. Because unlike primary and secondary recovery, EOR processes increase production mainly by altering the rock and fluid properties.

However, due to the many differences in properties between conventional and unconventional reservoirs, the EOR methods that are applicable in conventional reservoirs need to be re-evaluated when applied in unconventional reservoirs. For instance, during a conventional EOR scheme, the overall recovery is determined by both macroscopic and microscopic displacement efficiency. The macroscopic displacement efficiency is affected by the reservoir heterogeneities and anisotropy, characteristics of the matrix, fluid mobility, flood patterns, etc. The microscopic displacement efficiency is affected by interfacial tension, wettability, capillary pressure and relative permeability (Terry 2001). Because of the differences in mass transfer mechanisms, some of the above factors remain significant in unconventional cases, whereas some factors do not affect recovery. Laboratory experiments conducted by various researchers have shown success in oil recovery when applying $CO_2$ as the EOR agent to unconventional cores. In these studies, considering diffusion as a critical oil recovery mechanism was brought to attention (Hawthorne *et al.* 2013; Kovscek *et al.* 2008; Vega *et al.* 2010).

To continue on the journey of understanding the underlying oil recovery mechanism of oil recovery from unconventional rocks when $CO_2$ is applied as the EOR agent to aid in the future evaluation of the economic potential of this method, work shown in this paper is composed of simulation studies conducted using data collected from the first $CO_2$ experiment described in the paper by Tovar *et al.* (2014). During the experiment, oil was successfully extracted from shale cores that are in preserved and un-fractured condition. To Summarize, this study is designed to address the following two questions: During the $CO_2$ experiment described by Tovar *et al.* (2014),

1) Which microscopic characteristics of the rock have predominant effects on recovery (Sensitivity Analysis)?
2) What is the dominating oil recovery mechanisms?

## Laboratory Experiment

The core scale simulation in this study is based on the experimental data obtained from the first set of the experiment described in the work by Tovar *et al.* (2014). During the experiment, two preserved sidewall cores from the same well in an unconventional play were used. The petrophysical properties and the saturations of the cores were not measured before the experiments to preserve its original conditions. However, since the cores are not stored in a pressurized environment, the fluids saturating the cores can be assumed as dead oil and water. The dimensions of the cores are summarized in **Table 1**.

| Core Number | 1-1 | 1-2 |
|---|---|---|
| Core Diameter, cm | 2.53 | 2.53 |
| Core Length, cm | 3.97 | 3.48 |
| Core Bulk Volume, $cm^3$ | 19.94 | 17.50 |

**Table 1 – The dimensions of the cores used in the experiment (Modified after Tovar et al. 2014).**

Glass beads were packed outside the cores, to allow $CO_2$ to have direct contact with the cores and to simulate the presence of a highly permeable fracture around the cores. Two Berea sandstone cores were placed on each end of the set up to block the glass beads from entering the production tube. The core holder was then placed in a water bath which circulates hot water to simulate the reservoir temperature. And, the assembly was mounted in a CT (computed tomography) scanner. The cores were kept horizontally during the experiment. The schematic of the experimental setup can be seen in **Fig. 1**.

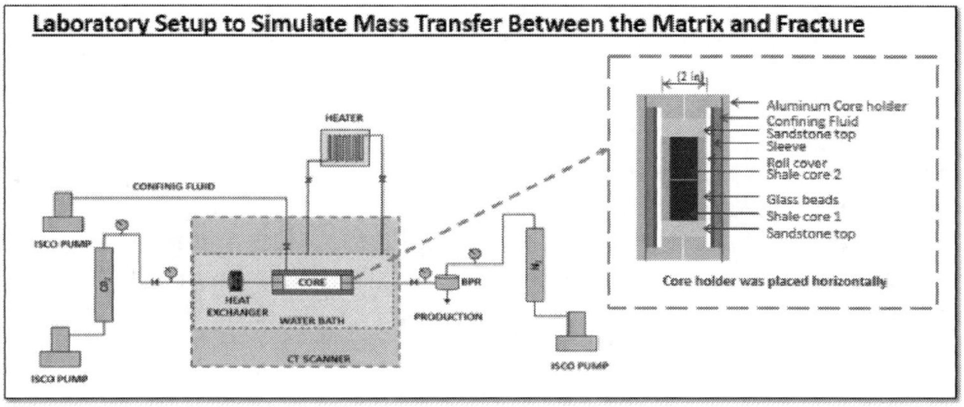


**Fig. 1 – Schematic of the experiment setup.**

The experiment was performed under constant pressure conditions (where viscous displacement was absent) at 3000 psi at 150 °F. Periodic scanning of the cores with the CT scanner was performed throughout the experiment. The experiment lasted approximately 100 hrs. Production was allowed twice a day on average by lowering the pressure at the outlet, which is controlled by a back pressure regulator. The CT images of the core revealed constant density/saturation/composition changes of the resident fluid during both experiments, which indicated $CO_2$ was constantly penetrating into the core throughout the process. A final oil production volume of 0.4 ml was recorded after the experiment.

## Build Core-Scale Simulation Models

The purpose of the numerical simulation is to provide a mathematical model that validates the observed physical phenomena during laboratory experiments or field production and to be used for forecasting and optimization. Since one of the objectives is to investigate the mass transfer mechanisms behind the observed oil recovery during experiments, the following section provides theoretic background on material transfer and the calculations behind the simulator used in this study.

## Mass Transfer Theory

### Dispersion

Dispersion is the mixing of fluids within a porous medium. Dispersion of the fluid is controlled by two main mechanisms: convection and molecular diffusion (Bird *et al.* 2002; Deen 1998). The convective transport is accompanied by bulk motion. It is the mechanism caused by the heterogeneities of the porous medium, which caused local fluid velocity variations (Lie 2013). On the other hand, molecular diffusion is caused by the compositional difference or chemical potential of the species. It results from the motion of the molecules, and it is related to small pore-scale molecular displacement (Deen 1998). Longitudinal dispersion in a porous medium is represented by **Eq. 1** (Jarrell *et al.* 2002).

$$K_l = D_e + \alpha_1 v \quad \text{(Eq. 1)}$$

$K_l$ is the longitudinal dispersion coefficient, $D_e$ is the effective molecular diffusion coefficient, $\alpha_1$ is the longitudinal dispersivity and $v$ is the superficial velocity which is the product of porosity and interstitial velocity (Jarrell *et al.* 2002). Based on Eq. 1, the rate of dispersion is a sum of the convective and diffusive spreading.

To understand the dominant mechanism for dispersion, Lie (2013) used the characterization of the flow described by Sahimi (2011), which explains the five different regimes using the Péclet number. The Péclet number is defined as the ratio between convection and diffusion transport as shown in **Eq. 2**.

$$P_e = \frac{d_g v}{D_m} = \frac{D_L}{D_m} \quad \text{.................................................................}(Eq.\ 2)$$

Where $d_g$ is the average diameter of the particle in meter, v is the interstitial velocity in m/s, and $D_m$ is the molecular diffusion coefficient in $m^2/s$. $D_L$ is the longitudinal dispersion coefficient, which is proportional to the average fluid velocity (Perkins and Johnston, 1963), and describes the transport in the direction of the bulk flow. **Fig. 2** below shows the characterization of the five dispersion regimes in terms of the Péclet number.

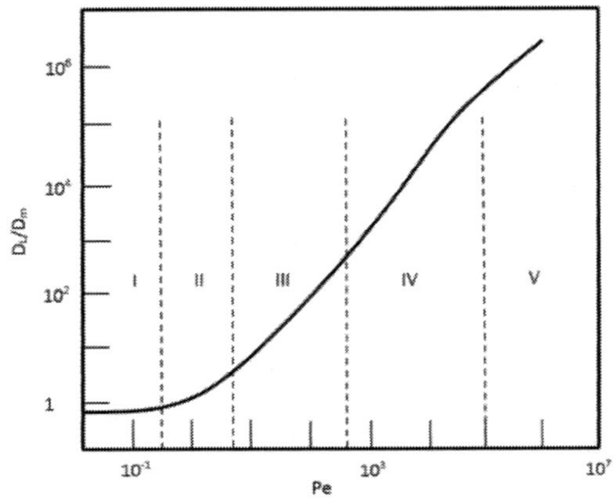


**Fig. 2 – Five dispersion regime characterization. (Adapted from Lie 2013).**

In **Region I**, where the Péclet number is smaller than 0.3, the dispersion is dominated by diffusion. In this region, the fluid velocity is too low, and it gives enough time for molecular diffusion to take place.

In **Region II**, where the Péclet number is greater than 0.3 but smaller than 5, diffusion is still dominating. However, convection will also aid in the process of dispersion.

In **Region III**, where the Péclet number is greater than 5 but smaller than 300, the velocity is large enough for convection to become the dominant mechanism. However, the effect of diffusion on dispersion cannot be neglected in this flow regime. In this region, the flux caused by diffusion is inversely proportional to the diffusion coefficient of the species. This is because when diffusion influx is coming from the transverse direction, the higher diffusion coefficient reduces the amount of time for the molecules to move forward along with the longitudinal convective flow.

In **Region IV**, where the Péclet number is greater than 300 but smaller than $10^5$ the dispersion is completely dominated by convection due to the considerably higher interstitial velocity. This is also referred to as mechanical dispersion.

**Region V**, which is also called the turbulent regime, is not of interest of the dispersion mechanism between shale matrix and the fractures.

During a $CO_2$ injection in conventional reservoirs, either under miscible or immiscible conditions, $CO_2$ is able to flow through the permeable rock matrix. The oil recovery mechanism is a combination of gravity drainage, light component stripping, oil swelling, viscosity reduction, and viscous displacement. Perkins and Johnston (1963) stated that if the fluid is able to flow through the porous media, the flow mechanism is dominated by convective dispersion, meaning the velocity of the bulk fluid is too high for diffusion to have any significant effect on dispersion. However, to consider the scenario when $CO_2$ flows through the fractures in an unconventional reservoir, the flow will not significantly go through the rock matrix. Therefore, convection could be eliminated as the main mechanism responsible for the mass transfer between the matrix and fracture (Hawthorne *et* al. 2013).

In the scope of this study, which is to investigate the mechanism of oil recovery from the shale matrix into the fractures, the dispersion process is characterized into dispersion Region I because of the nano scale permeability values. Therefore, when modeling $CO_2$ EOR in unconventional reservoirs, molecular diffusion is an important factor to be included. Many studies have reached similar conclusions (Hawthorne *et al.* 2103; Wan and Sheng 2015; Lie 2013; Grogan and Pinczewski 1987; Darvish *et al.* 2006; Hoteit and Firoozabadi 2009).

### Diffusion

Molecular diffusion describes the random movement of molecules inside a system. Diffusion differs from convection because it happens with or without external forces, such as a pressure gradient and gravity. In an isobaric system, when there is a temperature or concentration gradient, the gradient tends to disappear as a function of time due to diffusion. In an isothermal and isobaric system in a porous medium, the rate of mass influxes due to diffusion is a function of pressure, temperature, composition, particle sizes, and the characteristics of the porous medium. The modeling of mass transfer across a phase boundary has been extensively studied in the field of chemical engineering. The common approach is based on the film model, which assumes thermodynamic equilibrium at the gas-oil interface and the continuity of component fluxes across the interface (Hoteit 2013).

The rate and direction of the diffusion process are the diffusion flux. There are different models describing the calculation of the diffusion influx. The next two sections discuss the two models available in the simulator used in this study.

### The Classical Fick's Law

The classical Fick's law is a very simple conceptual model to describe the movement of chemical components from one location to another. Analogous to Newton's Law of motion, Ohm's Law of electrical conduction and Fourier's Law of heat conduction, the Fick's Law describes the relationship between mass flux quantity and concentration gradient as shown in **Eq. 3**.

$$J_i = -D_i \frac{\partial c_i}{\partial d} \quad \text{.....(Eq.3)}$$

Where

$J_i$ is the diffusion flux,

$D_i$ is the diffusion coefficient,

$\frac{\partial c_i}{\partial d}$ is the concentration gradient.

The classical Fick's law is ideal for diffusion calculation of a binary ideal fluid system. It simply assumes the diffusion process of each component is based on its concentration gradient and is independent of other components, which does not hold true in a realistic case with multicomponent mixture. Even for a binary system, one cannot predict when and where it may fail. Hoteit (2013) provided examples of both binary and multi-component system diffusion studies. The examples showed the use of different diffusion coefficients for different components violated the molar flux balance and caused pressure variations in an isothermal and isobaric system. The solution to eliminating this pressure variance is to set the diffusion coefficients of all components to be equal or modify the diffusion coefficient of the last component so that the overall molar balance is honored. The variation in pressure is more significant when the system is under non-ideal conditions. Therefore, this study uses the irreversible thermal dynamic model, which uses the chemical potential to determine the diffusion flux instead of using the concentration gradient.

### The Irreversible Thermodynamic Model

The irreversible thermodynamic model for diffusion flux calculation considers the chemical potential gradient to be the driving force for diffusion. The calculation takes the form as shown in **Eq. 4**.

$$J_i = -cD_i^a x_i \frac{1}{RT}\frac{\partial}{\partial d}\left[\mu_i - M_i g(h-h_o) + M_i D_i^T \ln(T)\right] \text{.........................} (Eq.\ 4)$$

The above equation takes gravity and temperature gradients in addition to the chemical potential gradients into consideration. For an isothermal system that ignores gravity, Eq. 4 can be written as **Eq. 5**.

$$J_i = -cD_i^a x_i \frac{\partial}{\partial d}\ln(f_i) \text{.........................................................} (Eq.\ 5)$$

$D_i^a$ is the activity-corrected diffusion coefficient. Combining Eq. 3 and Eq. 5, the activity-corrected diffusion coefficient can be expressed by **Eq. 6**.

$$D_i^a = \frac{D_i}{\partial \ln(f_i)/\partial \ln(x_i)} \text{.........................................................} (Eq.\ 6)$$

#### *Chemical Potential and Fugacity*

According to Eq. 5, the contributing parameters involved in the diffusion flux calculation in an oil and gas system are porosity, oil and gas saturation, molar density of the oil and gas, mole fractions of the components, diffusion coefficients, and fugacity. Out of the above parameters, the molar density, diffusion coefficient, and fugacity are functions of pressure and temperature. Therefore, although the initiation of the diffusion process is not governed by the system pressure, the rate of diffusion is controlled by the system pressure.

To understand the effect of physical conditions on diffusion flux, the term chemical potential needs to be briefly reviewed. This concept was introduced by Gibbs in 1957. This section will briefly explain the definition of chemical potential and fugacity. The detailed definition can be found in the work published by Firoozabadi (1999).

For an open system in which there is no restriction of material and energy transfer, the differential form of the fundamental equation is expressed in **Eq. 7**.

$$dU = \left(\frac{\partial U}{\partial S}\right)_{V,\underline{n}} dS + \left(\frac{\partial U}{\partial V}\right)_{S,\underline{n}} dV + \sum_{i=1}^{c}\left(\frac{\partial U}{\partial n_i}\right)_{S,V,n_i} dn_i \text{ ...........................} (Eq.\ 7)$$

Where

$\underline{n} \equiv (n_1, n_2 ... n_c)$,

$U$ is the internal energy of the system,

$S$ is the entropy of the system,

$V$ is the volume.

The first two partial derivatives on the right side of Eq. 7 are given as:

$$\left(\frac{\partial U}{\partial S}\right)_{V,\underline{n}} = T \text{ ..................................................................} (Eq.\ 8)$$

$$\left(\frac{\partial U}{\partial V}\right)_{S,\underline{n}} = -P \text{ ................................................................} (Eq.\ 9)$$

Where

$T$ is the thermodynamic temperature,

$P$ is the pressure.
And the coefficient of the last term is defined as the chemical potential, denoted by $\mu$ :

$$\mu_i = \left(\frac{\partial U}{\partial n_i}\right)_{S,V,n_j} \quad \text{.......................................(Eq. 10)}$$

The function of the chemical potential is similar to that of the pressure and temperature of a system. When a chemical potential gradient exists within a system, diffusion occurs from regions with high chemical potential to regions with low chemical potential despite gravity or other external forces.

Combining Eq. 7 to Eq. 10, the expression for internal energy change can be written as **Eq. 11**.

$$dU = TdS - PdV + \sum_{i=1}^{c} \mu_i dn_i \quad \text{.......................................(Eq. 11)}$$

In general, for two cells separated by a permeable wall, which permits free transport of all components, the direction of the flow is solely determined by the chemical potential difference between the two cells. It is a form of work done by an isothermal and isobaric system due to composition changes to achieve thermal equilibrium.

The entropy of the two systems are:

$$dS^1 = \frac{dU^1}{T^1} + \frac{P^1}{T^1}dV^1 - \sum_{i=1}^{c}\left(\frac{\mu_i^1}{T^1}\right)dn_i^1 \quad \text{.......................................(Eq. 12)}$$

$$dS^2 = \frac{dU^2}{T^2} + \frac{P^2}{T^2}dV^2 - \sum_{i=1}^{c}\left(\frac{\mu_i^2}{T^2}\right)dn_i^2 \quad \text{.......................................(Eq. 13)}$$

Use the maximum entropy principle and rearranging Eq. 11 to Eq. 13:

$$dS = d(S^1 + S^2) = (\frac{1}{T^1} - \frac{1}{T^2})dU^1 + (\frac{P^1}{T^1} - \frac{P^2}{T^2})dV^1 - \sum_{i=1}^{c}(\frac{\mu_i^1}{T^1} - \frac{\mu_i^2}{T^2})dn_i^1 = 0 \quad \text{........(Eq. 14)}$$

The conditions for the two cells to achieve thermal equilibrium are $T^1 = T^2$, $P^1 = P^2$ and $\mu_i^1 = \mu_i^2$
Therefore, for an isothermal and isobaric system, the two cells need to have equal chemical potential instead of having equal mole fraction as defined in the classical Fick's law shown in Eq. 5.
Chemical potential is the partial molar Gibbs free energy, $G$. The Gibbs free energy is a function of temperature, pressure, and composition.

$$\mu_i = \left(\frac{\partial G}{\partial n_i}\right)_{T,P,n_j} \quad \text{.......................................(Eq. 15)}$$

Where

$$dG = VdP - SdT \quad \text{.......................................(Eq. 16)}$$

For an isothermal system,

$$dG = VdP \quad \text{.......................................(Eq. 17)}$$

For an ideal gas

$V = \dfrac{RT}{P}$, therefore combining Eq. 16 and Eq. 17

$$dG_{Ideal} = \frac{RT}{P} dP = RTd\ln P \text{ ....................................} (Eq.\ 18)$$

Combining Eq. 15 and Eq. 18

$$d\mu_i = RTd\ln(f)_{T,\underline{n}} \text{ and } \lim_{P \to 0}\left(\frac{f_i}{x_i P}\right) = 1 \text{ ..........................} (Eq.\ 19)$$

Where

$x_i$ is the mole fraction,

$f$ is the fugacity.

The term $\frac{f_i}{x_i P}$ is the fugacity coefficient. At low pressure, the fugacity coefficient equals one, which indicates the fluid behaves like an ideal fluid. In other words, fugacity is defined to represent the correct chemical potential for a real fluid with pressure $P$. The unit for fugacity is the same as the unit for pressure.

In an isothermal system, with diffusion as the only driving mechanism for mass transfer, the rate of mass transferring depends on the chemical potential (fugacity) of each components existing in the different phases.

For ideal gases, the fugacity is simply the pressure. For non-ideal gases, the fugacity does not equal the true pressure and needs to be corrected in terms of the compressibility factor and the pressure because of the change in Gibbs free energy.

$$\ln\left(\frac{f}{P}\right) = \int_0^P (z-1)\frac{dP}{P} \text{ ..............................................} (Eq.\ 20)$$

For liquid, the fugacity at low-pressure ideal conditions equals to the saturation pressure. At higher pressure, since the fluid properties do not change significantly, although there is a correction term for non-ideal liquid fugacity, the Gibbs free energy of a non-ideal liquid does not change much compared with ideal liquid.

## Diffusion Coefficient

Numerous studies demonstrated or proposed methods to either physically measure the diffusion coefficient in the laboratory or estimate it using correlations.

Physically measuring the diffusion coefficient for a multi-component mixture can be challenging and sometimes inaccurate due to the fact that other factors such as gravity, capillary forces, convective mixing could interfere with the results. Correlations such as the ones proposed by Sigmund and Wilke-Chang are widely used in literature to predict the diffusion coefficient of mixtures. New models for diffusion coefficient estimation, such as the one developed by Leahy-Dios and Firoozabadi (2007), are not available in commercial software and could be considered for future studies. The correlation used in this study is the Sigmund correlation (Sigmund 1976a; Sigmund 1976b) shown in **Eq. 21** to **Eq. 30**.

$$D_{ij} = \frac{\rho_k^0 D_{ij}^0}{\rho_k} \cdot \left(0.99589 + 0.096016\rho_{kr} - 0.22035\rho_{kr}^2 + 0.032874\rho_{kr}^3\right) \text{ ..............} (Eq.\ 21)$$

$$\rho_{kr} = \rho_k \left( \frac{\sum_{i=1}^{n_c} y_{ik} v_{ci}^{5/3}}{\sum_{i=1}^{n_c} y_{ik} v_{ci}^{2/3}} \right) \text{ ..............................................................} (Eq.\ 22)$$

$$\rho_k^0 D_{ij}^0 = \frac{0.0018583 T^{1/2}}{\sigma_{ij}^2 \Omega_{ij} R} \cdot \left( \frac{1}{M_i} + \frac{1}{M_j} \right)^{1/2} \quad \text{(Eq. 23)}$$

$$\sigma_i = (2.3551 - 0.087\omega_i) \cdot \left( \frac{T_{ci}}{P_{ci}} \right)^{1/3} \quad \text{(Eq.24)}$$

$$\varepsilon_i = k_B (0.7915 + 0.1963\omega_i) T_{ci} \quad \text{(Eq. 25)}$$

$$\sigma_{ij} = 0.5(\sigma_i + \sigma_j) \quad \text{(Eq. 26)}$$

$$\varepsilon_{ij} = \sqrt{\varepsilon_i \varepsilon_j} \quad \text{(Eq. 27)}$$

$$T_{ij}^* = \frac{k_B}{\varepsilon_{ij}} \quad \text{(Eq. 28)}$$

$$\Omega_{ij} = 1.06306(T_{ij}^*)^{-0.1561} + 0.193\exp(-0.47635 T_{ij}^*) + 1.03587\exp(-1.52996 T_{ij}^*) + 1.76474\exp(-3.89411 T_{ij}^*) \quad \text{(Eq. 29)}$$

Lastly, the diffusion coefficient for component I in the mixture is calculated using an equation based on Wilke formula:

$$D_{ik} = \frac{1 - y_{ik}}{\sum_{j \neq i} y_{ik} D_{ij}^{-1}} \quad \text{(Eq. 30)}$$

As mentioned in numerous studies, the diffusion coefficients for the components are parameters subjected to adjustment during history matching. Therefore, the estimated diffusion coefficients only serve as baseline values for future adjustments.

## Commercial Simulators

Since one of the critical oil recovery mechanisms, diffusion, is being studied in this study, Commercial simulator A and Commercial simulator B were both investigated because the two simulators offer different capabilities when modeling diffusion.

### Commercial Simulator A

Attempts to use Simulator A was made at first for the laboratory modeling. When using single porosity model, Simulator A has the capability of activating the diffusion coefficients calculated using either Sigmund or Wilke-Chang correlation or enabling the user to define the value of the diffusion coefficients explicitly. However, it was discovered that Simulator A is only capable of calculating the intra-phase diffusion for single porosity models, which means the cross phase diffusion is being neglected. When there is phase discontinuity a t the boundary between the gas bearing fracture and the oil bearing matrix, small amounts of $CO_2$ are mixed into the oil phase due to the initial pressure difference. However, as the pressures in the fracture and the matrix regions become equal, no more $CO_2$ will enter the oil phase. The amount of $CO_2$ that is already mixed with the oil phase will diffuse within the oil phase and evenly distributed inside the homogeneous core given enough time.

Lie (2013) mentioned in his work that Simulator A is unable to model the diffusion of supercritical $CO_2$, which is treated as a gas-phase in the simulation, into the oil phase. Also, the simulation results do not agree with the observation from experimental work as shown in the simulation performed by Vega *et al.* (2010). With the above reason, the single porosity model with Simulator A was not select for this study and is not recommended to use for simulations where diffusion between two phases is considered.

### Commercial Simulator B

Simulator B has the capability of modeling both intra phase and cross phase diffusion fluxes with single porosity models. There are two models available for the calculation of diffusion fluxes. The model selected for this study is based on the irreversible thermodynamics model discussed earlier, in which the diffusion is driven by the chemical potential gradient. To model the cross phase diffusion process, the cross phase diffusion coefficient is calculated with **Eq. 31.**

$$Dog_i^a = \sqrt{Do_i^a \times Dg_i^a} \qquad \text{(Eq. 31)}$$

## Fluid Model

Workflow of modeling the dead oil used in the experiments is shown in the **Fig. 3**.

**Fig. 3 – Process diagram of building the fluid model.**

After the regression, a dead oil model is obtained to be used in the simulation. **Table 2** summarizes the properties of the fluid mode.

| Component | Mole Fraction, % | Critical Pressure, atm | Critical Temperature, K | Acentric Factor | Molecular Weight, g/mol | Critical Volume, $m^3$/kg-mol |
|---|---|---|---|---|---|---|
| $CO_2$ | 1E-6 | 72.80 | 304.20 | 0.225 | 44.01 | 0.09 |
| COMP1 | 33.18 | 34.30 | 559.74 | 0.254 | 95.33 | 0.36 |
| COMP2 | 30.51 | 23.97 | 593.36 | 0.399 | 124.12 | 0.54 |
| COMP3 | 22.90 | 17.17 | 673.34 | 0.617 | 181.06 | 0.77 |
| COMP4 | 13.41 | 14.74 | 634.14 | 0.719 | 259.99 | 1.07 |

**Table 2 – Properties of the fluid model after regression.**

The MMP and FCM were calculated with PVTsim. At 150 °F, the MMP was determined at 1727 psi, and the FCM was achieved at 1934 psi. The MMP value is further confirmed with a simulated slim-tube experiment. A 1D model with 200 cells was used.

## Slab Core Model

Before building a 3D core model to simulate the actual experiment, a 2D model was used for property sensitivity study. The slab model is shown in **Fig. 4**. It is more discretized both in the matrix and the fracture regions compared with the actual 3D model.

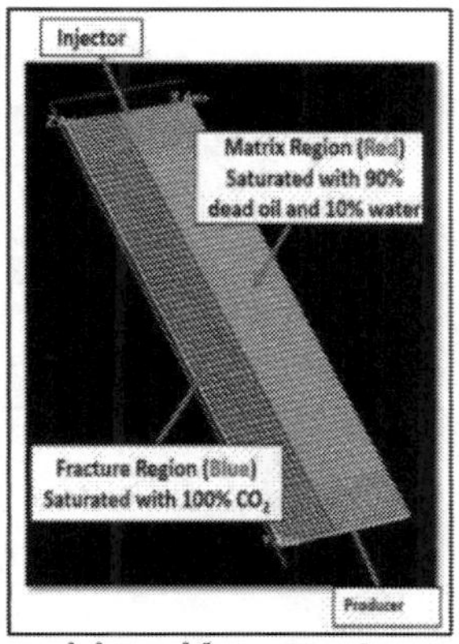


**Fig. 4– The slab model used for property sensitivity analysis.**

The sensitivity analysis was conducted using the 2D slab model for the following properties: matrix porosity, matrix permeability, fracture porosity, fracture permeability, water saturation, capillary pressure, diffusion coefficient, and system pressure. Results will be shown in the discussion section.

## Heterogeneous 3D Core Model

### Grid Sensitivity

Both radial and Cartesian grids of different degrees of refinement were investigated during this study. It was considered more reasonable to use a radial grid model to represent the flow behavior during the experiment, however, the Cartesian grid was also considered for the purpose of capturing the heterogeneous properties of the cores. Grid sensitivity analysis was performed to ensure a Cartesian grid model yields the same results as a radial grid model. **Fig. 5** below shows the 5 models built for grid sensitivity analysis. **Table 3** shows the dimensions of each model.

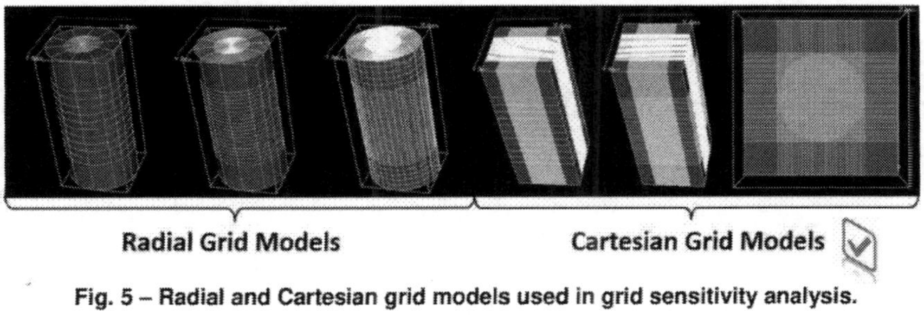


**Fig. 5 – Radial and Cartesian grid models used in grid sensitivity analysis.**

| Model | Grid Type | DR(DX) | Dθ(DY) | DZ | Single Matrix Block Dimension |
|---|---|---|---|---|---|
| Radial 1 | Radial | 11 | 10 | 12 | 0.127 cm x 36º x 0.691 cm |
| Radial 2 | Radial | 30 | 10 | 22 | 0.0635 cm x 36º x 0.3455 cm |
| Radial 3 | Radial | 60 | 36 | 50 | 0.03175 cm x 10º x 0.17275 cm |
| Cartesian 1 | Cartesian | 52 | 52 | 12 | 0.0508 cm x 0.0508 cm x 0.691 cm |
| Cartesian 2 | Cartesian | 52 | 52 | 32 | 0.0508 cm x 0.0508 cm x 0.23033 cm |

**Table 3 – Grid sensitivity on both radial and Cartesian grids with different degrees of refinement.**

As seen in **Fig. 6** below, although the models did not agree very well in the early times, the ultimate recovery yielded were fairly close. In addition, the Cartesian models gave fairly close results to the radial models. To accomplish the ultimate goal of this study, Cartesian 1 grid, which agrees well with the radial models in both early and late time, is selected for the 3D modeling.

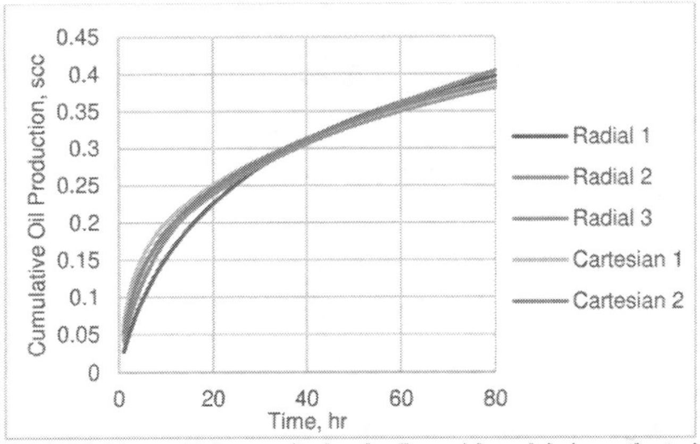


**Fig. 6 – Grid sensitivity results for the five grid models investigated.**

## Incorporating Heterogeneous Property Maps

To determine the porosity and permeability maps of the core model, the CT images from the experiment was carefully examed. As seen in **Fig. 7** below, the core displayed very distinguished bedding plan features. Since CT number corresponds to density, the change in density can be clearly observed in the images. Fig. 7 also shows the changes in density occurred mostly along the bedding plane, and the rest of the core experienced very minimal to no change in density.

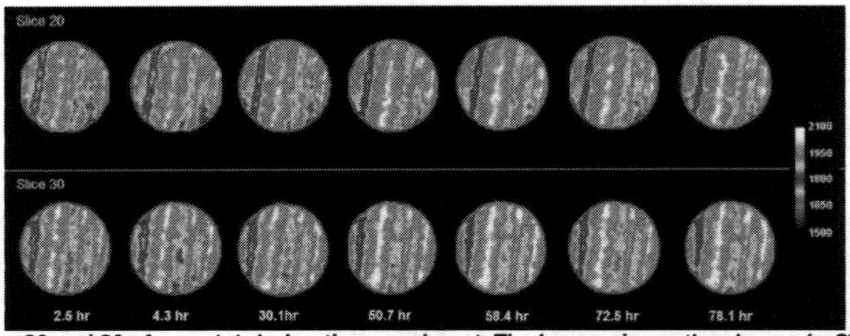


**Fig. 7 – CT images for slice 20 and 30 of core 1-1 during the experiment. The image shows the change in CT number as a function of time.**

In this case, Changes in CT number could be regarded as the only indication for the presence of $CO_2$, because either $CO_2$ adoption or mixing with the oil phase would cause potential density changes. On the other hand, there are three reasons for no density change. One is there is no oil in the regions, which means the pores are filled with water, and the density change of water when $CO_2$ is dissolved in it is negligible. The second reason is the regions are isolated, which means $CO_2$ cannot reach those areas. And, the third reason is the porosity is zero. The first and second reasons are most likely to be the case in experiment I. However, since $CO_2$ solubility in water is not included in this study, and defining a saturation map purely based on the CT images is not practical and does not add any values, the voxels with zero delta CT values are defined with very small porosity values in the range of 0 to 0.01. In future studies where the process of $CO_2$ diffuses through the water phase and recovers the trapped oil is being modeled, defining a saturation map becomes necessary.

After processing the CT images, each image was constructed into a gridded structure. The grid of the image corresponds to the dimension of the Cartesian grid model described in the earlier section with a $\Delta$CT value associated with each block. Co-kriging was used to aid in the process of defining the property maps. Various realizations were made. The goal of picking the appropriate porosity distribution is to match the ultimate 4 ml production with the least need of changing the diffusion coefficients. **Fig. 8** shows a sample slice of the porosity map used in the simulation. The diffusion coefficients used to match the cumulative production volume are listed in **Table 4**.

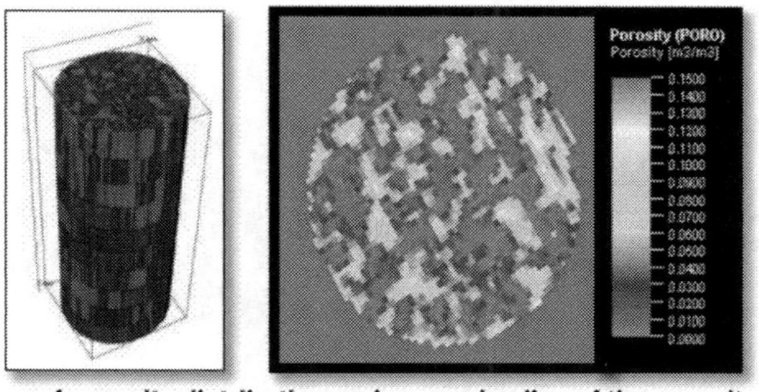


**Fig. 8 – Overview of porosity distribution and a sample slice of the porosity map.**

| Component | $CO_2$ | COMP1 | COMP2 | COMP3 | COMP4 |
|---|---|---|---|---|---|
| Gas, $cm^2$/hr | 1.04E-03 | 5.14E-05 | 4.63E-05 | 3.26E-05 | 2.45E-05 |
| Oil, $cm^2$/hr | 3.55E-04 | 4.22E-05 | 3.91E-05 | 3.14E-05 | 2.46E-05 |

**Table 4 – Diffusion coefficients used to match the ultimate recovery volume.**

After obtaining a porosity value for each grid block, the permeability value was determined based on porosity using the correlation proposed by Sigal (2002) shown in **Eq. 32**.

$$k = 60 r_{eff}^2 \phi^m \quad \text{(Eq. 32)}$$

where

$r_{eff}$ is the effective pore throat size in micron,

$\varphi$ is porosity,

$m$ is the cementation factor, which is chosen to be 2 in this study.

Based on Sigal (2013), the effective pore throat size for the particular type of rock that is similar to the one in this study measured with mercury injection was approximately 0.001 micron. Water saturation of the core was explicitly assigned at an immobile value 0.1.

After obtaining the above properties, a fine micro-scaled core model was built to mimic experiment I. Property maps were input for the matrix grid cells. The model contains two saturation regions with individually assigned relative permeability and capillary pressure data. Injectors and producers were placed at each end of the grid. The condition of the simulation is at 3000 psi and 150 °F.

## Results of Experiment Simulation

The simulation with the heterogeneous core model described above follows the same $CO_2$ injection and oil production schedule as performed during the experiment. A cumulative oil production of 0.4 mL was achieved at the end of the simulation as shown in **Fig. 9**.

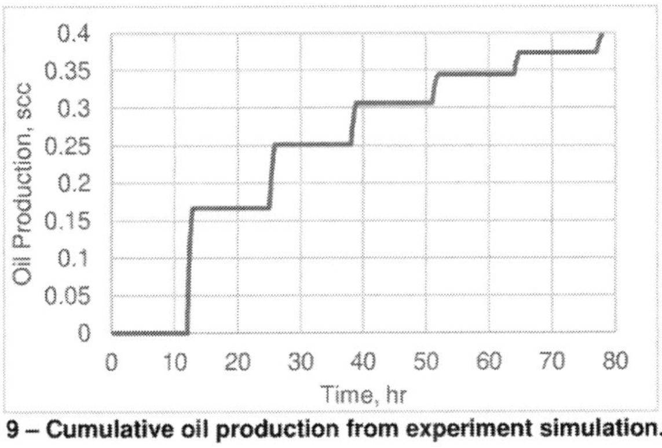


**Fig. 9 – Cumulative oil production from experiment simulation.**

**Fig. 10** below shows the oil density change inside the core over time. The slice is the same slice as shown in Fig. 8.

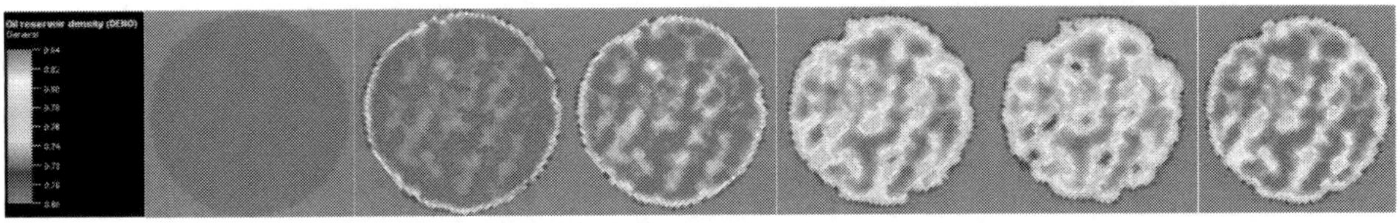

**Fig. 10 – Oil density change for the experiment I 3D heterogeneous model at time 0hr, 5hr, 10hr, 24hr, 48hr, and 72hr.**

In addition, as shown in Fig. 10, the rate at which oil density changes differs from cell to cell because of the different porosity and diffusivity values. And regions that have larger changes in density correspond to the streaks with high porosity values defining the bedding plane existed in the core.

## Discussion on Property Sensitivity

**Table 5** shows the results of the sensitivity analysis using the 2D slab model. For each property, several cases were run and oil recovery results were compared with a baseline value (10.59%). **Table 6** shows the baselined diffusion coefficient calculated using the Sigmund correlation.

| Property | Property Value | Recovery Factor (RF), % | ΔRF, % | Property Baseline Value |
|---|---|---|---|---|
| Matrix Porosity, % | 12 | 9.76 | -0.83 | 0.08 |
| | 4 | 11.22 | 0.63 | |
| Matrix Permeability, mD | 10 | 10.59 | 0 | 0.0001 |
| | 0.001 | 10.59 | 0 | |
| | 0.00001 | 10.65 | 0.06 | |
| Fracture Porosity, % | 20 | 7.79 | -2.8 | 0.45 |
| Fracture Permeability, mD | 10000 | 10.59 | 0 | 1000 |
| | 10 | 10.59 | 0 | |
| Water Saturation, % | 20 | 10.62 | 0.03 | 10 |
| | 30 | 10.65 | 0.06 | |
| | 40 | 10.66 | 0.07 | |
| Capillary Pressure @ 1600 psi | Low | 8.79 | -1.8 | Zero |
| | Mid | 8.79 | -1.8 | |
| | High | 8.82 | -1.77 | |
| Diffusion Coefficient | Baseline X 10 | 21.81 | 11.22 | *Diffusion Coefficient Table |
| | Baseline X 0.1 | 3.43 | -7.16 | |
| System Pressure, psi | 1600 | 8.79 | -1.8 | 3000 |
| | 1800 | 9.66 | -0.93 | |

**Table 5 – Summary of property sensitivity analysis.**

| Component | $CO_2$ | COMP1 | COMP2 | COMP3 | COMP4 |
|---|---|---|---|---|---|
| Gas, $cm^2/hr$ | 8.69E-04 | 4.28E-05 | 3.86E-05 | 2.72E-05 | 2.04E-05 |
| Oil, $cm^2/hr$ | 2.96E-04 | 3.52E-05 | 3.26E-05 | 2.62E-05 | 2.05E-05 |

**Table 6 – Diffusion Coefficient calculated using the Sigmund correlation.**

An interesting phenomenon was observed during the sensitivity study of matrix permeability. As seen in Table above, high permeability cases are associated with very slightly lower oil recovery. To look further into this phenomenon, few more cases were run.

**Table 7** shows the permeabilities and corresponding recovery factors. **Fig. 11** below shows the cumulative oil production for each case.

| | Permeability, mD | Recovery Factor, % |
|---|---|---|
| Case 1 | 10 | 10.59 |
| Case 2 | 1.00E-03 | 10.59 |
| Case 3 | 1.00E-04 | 10.59 |
| Case 4 | 1.00E-05 | 10.65 |
| Case 5 | 1.00E-06 | 11.12 |

**Table 7 – Permeability values and the corresponding recovery factors for cases with different matrix permeability values.**

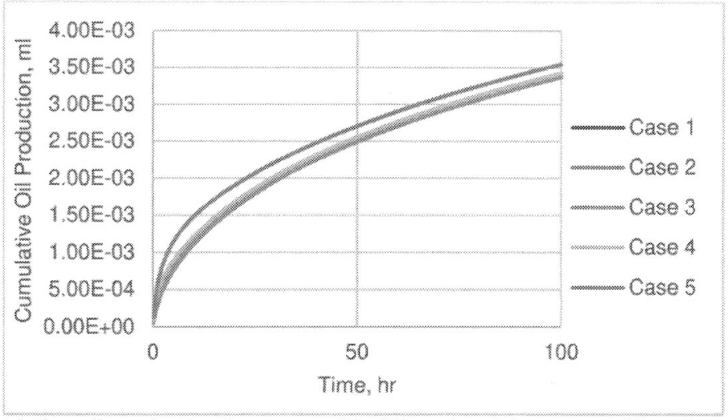


**Fig. 11 – Cumulative oil production for cases with different matrix permeability values (Scenario 2).**

The results still turned out to be counterintuitive. Case 5, which has the smallest permeability, yielded the highest recovery than the rest of the cases. Whereas the rest of the cases have similar production values. Pressure profiles of the core for each case were plotted and compared as shown in **Fig. 12**. It was discovered that in this case, the pressure inside the core played a role in determining recovery because it directly affects the chemical potential for each of the components. Since the permeability was different for each case, according to Darcy's law, the rate at which pressure builds up inside the core is directly proportional to the permeability value. As one can observe from Fig, 12, Case 5 has significantly lower pressure than the rest of the cases for approximately 40 hours inside the core region. Although the fugacity of a non-ideal liquid does not change much with pressure, it still has some effect. For Case 5, the fugacity of the fluid inside the core is less than that of the rest of the cases, whereas the fugacity in the fracture is the same. This fugacity gradient resulted in the higher mass transfer rate for Case 5 compared with the other cases. Again, as mentioned in the earlier section, since the fugacity of fluid does not change significantly with pressure, the increase in production is only minor.

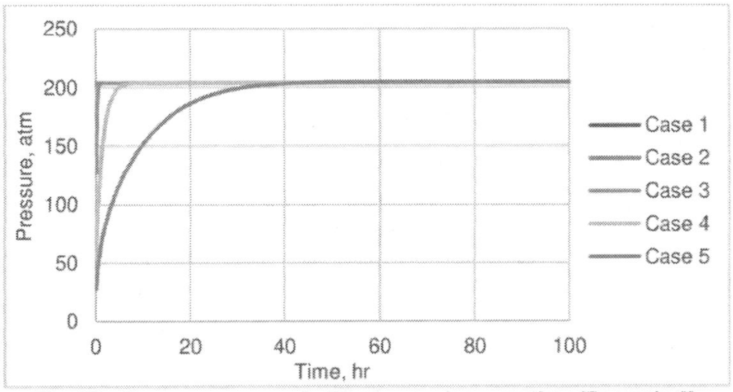


**Fig. 12 – Average pressure in the core region over time (Scenario 2).**

In addition, as mentioned in Eide *et al.* (2014), the concentration of a component is a linear function of the square root of time if the flow is dominated by diffusion. **Fig. 13** shows the oil production versus the square root of time. The early time shows a nonlinear relationship between the two due to the initial convection caused by the pressure gradient between the fracture and matrix region. After that, a linear relationship can be observed, indicating diffusion is the dominant mechanism.

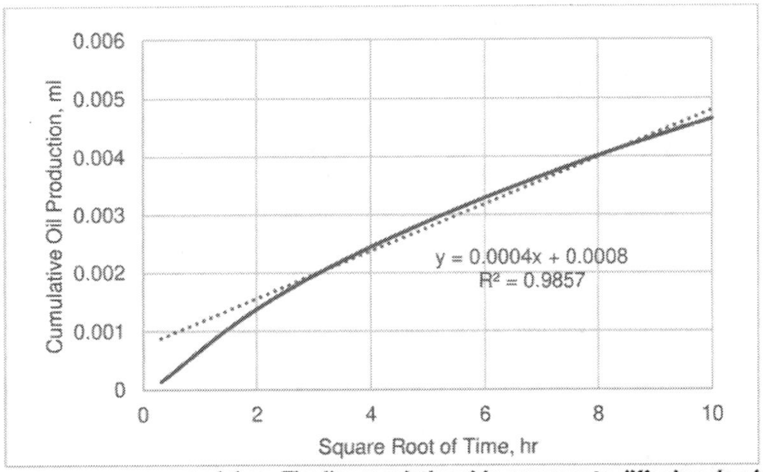


**Fig. 13 – Production versus square root of time. The linear relationship represents diffusion dominated flow.**

Overall, the permeability of the matrix has only minor effects on oil recovery when diffusion is the dominating recovery mechanism. The study conducted by Vega *et al.* (2010) reached a similar conclusion that heterogeneity has limited effects on cumulative recovery.

## Discussion on Recovery Mechanism

The recovery mechanisms being investigated in this paper include:

1) Diffusion
2) Viscous displacement
3) Gravity effect
4) Oil viscosity reduction
5) Vaporization of light hydrocarbon components

### Diffusion

The importance of diffusion was already highlighted in the sensitivity analysis section. To further study the effects of diffusion, diffusion coefficients were deactivated in this part of the study as shown in **Table 8**. And, the effective diffusion coefficient (served as the baseline) used in case 3 are shown in Table 6.

| | Diffusion Coefficient, $cm^2/hr$ | Recovery Factor, % |
|---|---|---|
| Case 1 | No Diffusion Coefficient | 0 |
| Case 2 | Diffusion Coefficient X 0.1 | 3.43 |
| Case 3 | Diffusion Coefficient | 10.59 |
| Case 4 | Diffusion Coefficient X 10 | 21.81 |

**Table 8 – Diffusion coefficient values and the corresponding recovery factors.**

For all cases shown in Table 8, a porosity of 8%, permeability of 1E-4 mD and water saturation of 0.1 were used. In case 1, only shallow $CO_2$ penetration into the core was observed in early time, and no hydrocarbon was produced from the matrix. The observation from case 1 not only proves that the mass transfer mechanism is diffusion, it also shows that oil swelling has little contribution in the oil recovery process, which agrees with the conclusion made by Hawthorn *et al.* (2013). The second case and fourth case used the baseline diffusion coefficients multiplied by a factor. As observed in the recovery factors listed in Table 8 and **Fig. 14**, increasing the diffusion coefficients significantly increases the recovery factor.

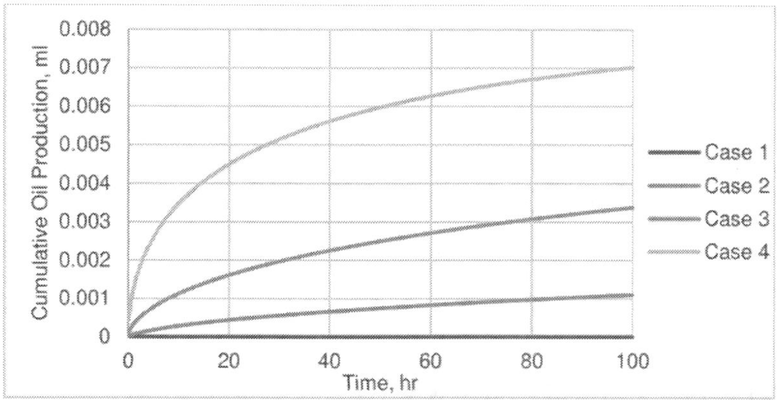


**Fig. 14 – Cumulative oil production for cases with different diffusion coefficient values.**

## Viscous Displacement

In a scenario where $CO_2$ is injected into the system where the core matrix has high permeability, viscous displacement is considered as a main source of oil recovery. In the scope of this study, the simulation runs are designed to mimic the actual experimental processes and conditions mentioned in the previous section. During the experiment, $CO_2$ is slowly injected into the porous glass beads (simulating fractures) through the inlet, while the outlet of the system is being controlled by a back pressure regulator and remained shut until the pressure of the system reaches the desired targets. The process itself differs from a conventional core flood process in a sense that the displacement of material should happen mainly within the glass beads regions, not the core regions. To look further into the role of viscous displacement during the experiment, the possibility of viscous displacement taking place during the process was eliminated by eliminating the initial pressurizing process of the system using $CO_2$. The initial pressure was set at 3000psi for all the cases shown in **Table 9.** Different permeability scenarios were run for the purpose of comparison, and the resulting recovery factors are listed. The cumulative oil production and production rate from the matrix into the fracture was plotted versus time as shown in **Fig. 15**.

| | Permeability, mD | Recovery Factor, % |
|---|---|---|
| Case 1 | 10 | 14.46 |
| Case 2 | 1.00E-03 | 14.46 |
| Case 3 | 1.00E-04 | 14.46 |
| Case 4 | 1.00E-05 | 14.46 |

**Table 9 – Permeability values for different cases and the corresponding yielded recovery factors.**

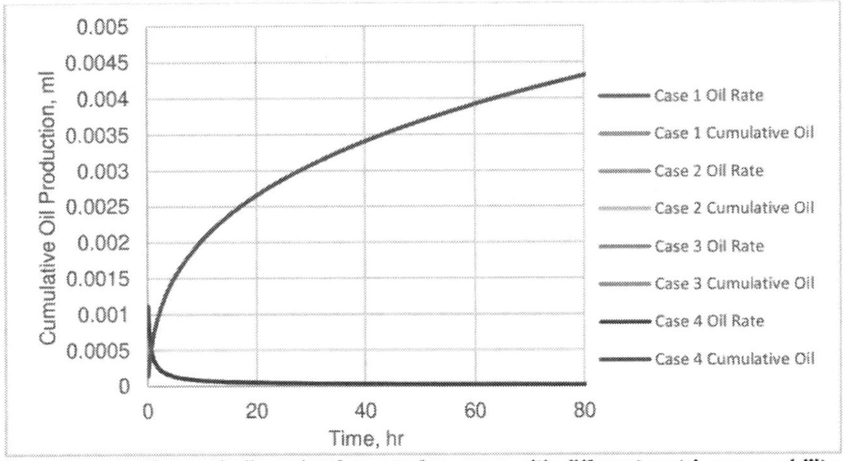


**Fig. 15 – Cumulative oil production and oil production rate for cases with different matrix permeability values.**

As seen in Fig. 15, the recovery factors for all cases are equal despite the large permeability contrast. The result further confirmed that the production is only controlled by molecular diffusion, which is independent of pressure or bulk fluid movement. The production rate in this case is only influenced by the chemical potential and the diffusion coefficient of each component, which are the same in all cases.

## Gravity Effect

The model was built vertically instead of horizontally as in the experiment. Therefore, gravity effect was investigated. By inspecting the simulation result images from the 2D slab model, gravity only affected the oil distribution in the fracture region, the amount of $CO_2$ penetration and oil saturation were symmetrical in the top and bottom slices inside the matrix region. **Fig. 16** below shows the mole fraction profile for COMP2 at 10 hours. Since the mole fraction inside the fracture region for COMP2 is very small in comparison to $CO_2$, the figure on the right shows the mole fraction of COMP2 at 10 hours in a log scale, which clearly shows the gravity effect in the fracture region. In addition, the slices closer to the two ends of the core observed lower COMP2 mole fraction than the slices in the middle section of the core because these slices were affected by both radial and axial flows. No gravity effect was observed from the results. It is only reasonable due to the high capillary pressure that exists inside the matrix region.

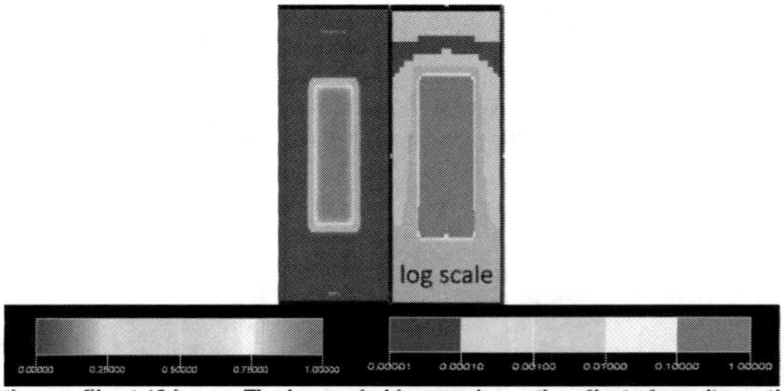


**Fig. 16 – COMP 2 mole fraction profile at 10 hours. The log scaled image shows the effect of gravity on the oil distribution in the fracture region.**

## Oil Viscosity Reduction

As mentioned in many studies, $CO_2$ reduces the viscosity of the oil to enhance the oil mobility, which can also be observed from the simulation as shown in **Fig. 17** below.

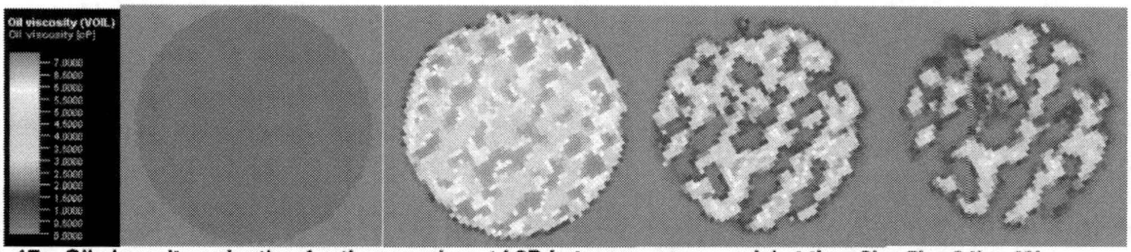

Fig. 17 – Oil viscosity reduction for the experiment I 3D heterogeneous model at time 0hr, 5hr, 24hr, 48hr.

However, since it has been discussed earlier that diffusion governs the mass transfer process, the reduced viscosity will not be observed if the diffusion of $CO_2$ into the oil phase is deactivated. In other words, oil viscosity reduction occurs after the initiation of dispersion processes and may have played a role in enhancing the magnitude of produced oil volume.

## Vaporization of Light Hydrocarbon Component

**Fig. 18** shows the molar fraction in the exported oil for each individual oil component, namely COMP1, COMP2, COMP3, and COMP4. Similar to a miscible $CO_2$ flooding scheme, the "lighter" components have the highest total production volume. This is the result of the lighter components having higher diffusion coefficients, and also easier to be vaporized into gas phase by $CO_2$. However, similar to oil viscosity reduction, light component stripping occurs after diffusion initiated dispersion and may have played a role in enhancing the magnitude of produced oil volume.

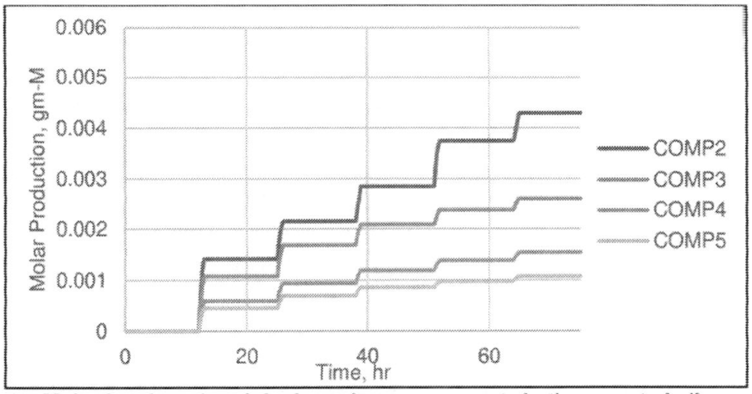


Fig. 18 – Molar fraction of each hydrocarbon components in the exported oil.

# Conclusion

The following five conclusions can be drawn from this study

1. Cross-phase diffusion must be included to properly model the diffusion process.
2. During the experiment, diffusion was the dominant mass transfer mechanism. Deactivating diffusion resulted in zero oil production despite the fact that certain amounts of $CO_2$ were able to enter the core due to the initial pressure difference between the fracture and the matrix (convection).
3. Oil viscosity reduction and light component stripping served the purpose of enhancing the magnitude of oil recovery.
4. The oil recovery factor is sensitive to the following parameters: matrix porosity, saturation, region pressure, and diffusion coefficient.
5. The oil recovery factor is not sensitive to the following parameters: matrix permeability, relative permeability, and capillary pressure.

## References


Bird, R.B., Stewart, W.R., and Lightfoot, E.N. 2002. Transport Phenomena, second edition. New York, New York: John Wiley & Sons, Inc.

Darvish, G.R. Lindeberg, E.G., Holt, T., and et al. 2006. Reservoir Conditions Laboratory Experiment of CO2 Injection into Fractured Cores. SPE paper presented at the SPE Eueopec/EAGE Annual Conference and Exhibition, Vienna, Austria. SPE-99650-MS.

Deen, W.M. 1998. Analysis of Transport Phenomena.

Eide, O., Ferno, M.A., and Graue, A. 2014. Visualization of CO2 EOR by Diffusion in Fractured Chalk. SPE paper presented at the SPE Annual Technical Conference and Exhibition, Amsterdam, The Netherlands, 27-29 October. SPE-170920-MS. http://dx.doi.org/10.2118/170920-MS.

Firoozabadi, A. 1999. Thermodynamics of Hydrocarbon Reservoirs. New York, New York. McGraw-Hill Companies, Inc.

Grogan, A.T. and Pinczewski, W.V. 1987. The Role of Molecular Diffusion Processes in Tertiary CO2 Flooding. Journal of Petroleum Technology. 39(5):5991-602. SPE-12706-PA.

Hawthorne, S.B., Gorecki, C.D., Sorensen, J.A. and et al. 2013. Hydrocarbon Mobilization Mechanisms from Upper, Middle, and Lower Bakken Reservoir Rocks Exposed to CO2. SPE paper presented at the SPE Unconventional Resources Conference, Calgary, Alberta, Canada. SPE-167200-MS.

Hoteit, H. Proper Modeling of Diffusion in Fractured Reservoirs. 2013. SPE paper presented at the SPE Reservoir Simulation Symposium, The Woodlands, TX, USA. SPE-141937-MS.

Hoteit H. and Firoozabadi, A. 2009. Numerical Modeling of Diffusion in Fractured Media for Gas-Injection and Recycling Schemes. SPE J. 14(2): 323-337.

Jarrell, P.M., Fox, C.E., Stein, M.H., et al. 2002. Practical Aspects of CO2 Flooding. Society of Petroleum Engineers.

Kovscek, A.R., Tang, G.Q., and Vega, B. 2008. Experimental Investigation of Oil Recovery from Siliceous Shale by CO2 Injection. SPE paper presented at the SPE Annual Technical Conference and Exhibition, Denver, CO, USA. SPE-115679-MS.

Lie, S.H. 2013. Diffusion as an Oil Recovery Mechanism during CO2 Injection in Fractured Reservoirs. Master Thesis in Reservoir Physics, University of Bergen.

Liu, G., Sorensen, J.A., Braunberger, J.R. and et al. 2014. CO2-Base Enhanced Oil Recovery from Unconventional Reservoirs: A Case Study of the Bakken Formation. SPE paper presented at the SPE Unconventional Resources Conference, The Woodlands, TX, USA. SPE-168979-MS.

Perkins, T.K. and Johnston, O.C. 1963. A Review of Diffusion and Dispersion in Porous Media. Society of Petroleum Engineers Journal.

Sahimi, M. 2011. Flow and Transport in Porous Media and Fractured Rock. New York, New York: Wiley.Vch Verlag Gmbh & Co.

Sigal, R.F. 2002. The Pressure Dependence of Permeability. Petrophysics J. 43(2): 92-102.

Sigal, R.F. 2013. Mercury Capillary Pressure Measurements on Barnett Core. SPE Reservoir Evaluation & Engineering J.

Sigmund, P. M. 1976a. Prediction of Molecular Diffusion at Reservoir Conditions. Part 1 – Measurement and Prediction of Binary Dense Gas Diffusion Coefficients. Petroleum Society of Canada.

Sigmund, P. M. 1976b. Prediction of Molecular Diffusion at Reservoir Conditions. Part II – Estimating the Effects of Molecular Diffusion and Convective Mixing in Multicomponent Systems. Petroleum Society of Canada.

Terry, R.E. 2001. Enhanced Oil Recovery. Encyclopedia of Physical Science and Technology, third edition, 18, 503-518.

Tovar, F.D., Eide, O., Graue, A., and Schechter, D. Experimental Investigation of Enhanced Recovery in Unconventional Liquid Reservoirs Using CO2: A Look Ahead to the Future of Unconventional EOR. SPE paper presented at SPE Unconventional Resources Conference, The Woodlands, TX, USA. SPE-169022-MS.

Vega, B., O'Brien, W.J., and Kovscek, A.R. 2010. Experimental Investigation of Oil Recovery From Siliceous Shale by Miscible CO2 Injection. SPE paper presented at the SPE Annual Technical Conference and Exhibition, Florence, Italy, 19-22 September. SPE-135627-MS. http://dx.doi.org/10.2118/135627-MS.

Wan, T. and Sheng, J. 2015. Compositional Modelling of the Diffusion Effect on EOR Process in Fractured Shale-Oil Reservoirs by Gasflooding. SPE paper presented at SPE/AAPG/SEG Unconventional Resources Technology Conference, Denver, CO, USA. Journal of Canadian Petroleum Technology. SPE-1891403.

AIChE ASME AIME IEEE ASCE
Carbon Management

# Optimizing Reservoir Oil Displacement and $CO_2$ storage on the Texas Farnsworth Unit CMTC-486252-MS


Authors: Edward Ennin, SPE, Reid Grigg, SPE

**New Mexico Petroleum Recovery Research Center**
**New Mexico Institute of Mining and Technology**



AIChE ASME AIME IEEE ASCE
Carbon Management


## Presentation Outline

- Introduction
- Background
- Experimental Set-up
- Theories and Principles [ $CO_2$ storage ]
- Results [$CO_2$ storage]
- Discussion [ $CO_2$ storage ]

CMTC - 486251 - MS

AIChE ASME AIME IEEE ASCE
Carbon Management

# Presentation Outline

- Theories and Principles [ WAG & SWAG ]
- Results [WAG & SWAG]
- Discussion [ WAG & SWAG]
- Conclusions
- Recommendations

CMTC - 486251 - MS

AIChE ASME AIME IEEE ASCE
Carbon Management

# INTRODUCTION

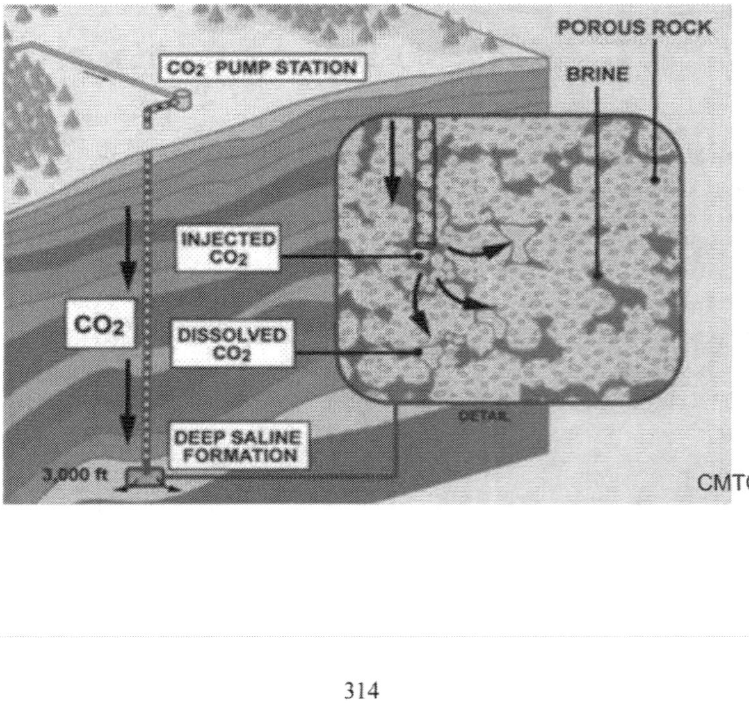


CMTC - 486251 - MS

AIChE ASME AIME IEEE ASCE
Carbon Management

# INTRODUCTION

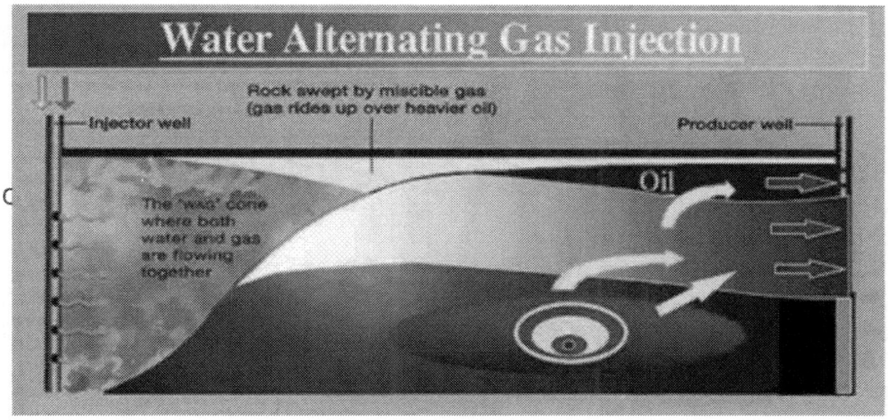


CMTC - 486251 - MS

AIChE ASME AIME IEEE ASCE
Carbon Management

# BACKGROUND

**Why measure displacement of brine and $CO_2$?**

*$CO_2$ injection into Saline Aquifers is the future of carbon capture and storage (CCS)

*It has been shown that brine salinity change can affect the recovery rate in a waterflood, thus does it have an effect in $CO_2$ displacement?

*FWU salinity has had a significant decrease over time with the waterflooding using low salinity water as injectant.

CMTC - 486251 - MS

## BACKGROUND

*WAG is widely used in field applications.

*SWAG has not been used in field applications.

*Worth knowing the effect of injection fluid dynamics of recovery factor.

*PH difference between produced water and injection fluid.

*Observe the risk of simultaneous injection of fluid.

CMTC - 486251 - MS

AIChE ASME AIME IEEE ASCE
Carbon Management

## BACKGROUND

### Regional Carbon Sequestration Partnerships

*Development Phase $CO_2$ Injection Volumes*

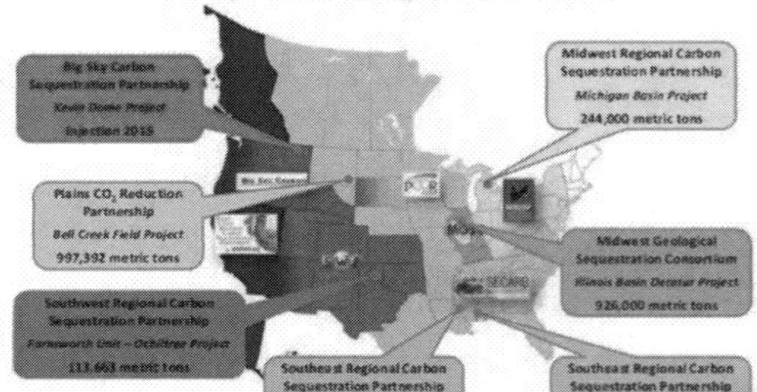


CMTC - 486251 - MS

AIChE ASME AIME IEEE ASCE
Carbon Management

## BACKGROUND

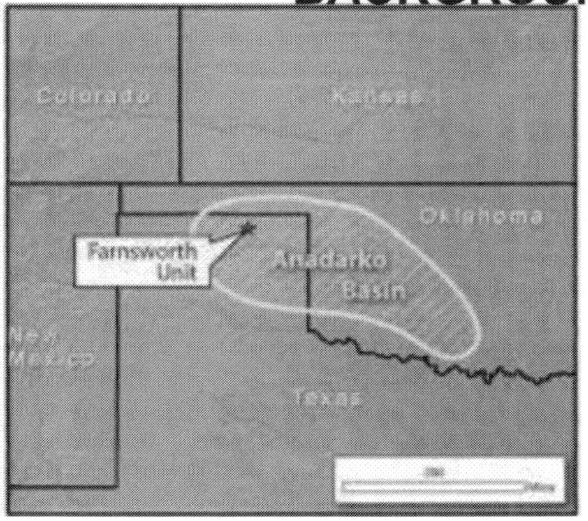


Farnsworth Unit (FWU)

Texas Panhandle

Discovered: 1950's

Waterflood started: 1964

$CO_2$ flood started: December 2010

CMTC - 486251 - MS

AIChE ASME AIME IEEE ASCE
Carbon Management

## Theories and Principles [CO2 Storage ]

*Absolute permeability in rock with single fluid, Darcy's law:

$$q = \frac{k\,A\,\Delta p}{\mu\,L}$$

*Relative permeability determine with two or more fluid:

$$k_r = \frac{(q * \mu)_j * L}{A * (dp)_j * k_{abs}}$$

*Brine Saturation ratio of bri $S_W = \frac{V_{bp}}{PV}$ in rock:

CMTC - 486251 - MS

AIChE ASME AIME IEEE ASCE
Carbon Management

# Theories and Principles [$CO_2$ Storage]

•Drainage > non-wetting fluid displaces wetting fluid.

*Imbibition > wetting fluid displaces non-wetting fluid.

*Fluids present are brine and $CO_2$

*Brine wetting phase and $CO_2$ non-wetting phase.

CMTC - 486251 - MS

AIChE ASME AIME IEEE ASCE
Carbon Management

# Experiemental Setup

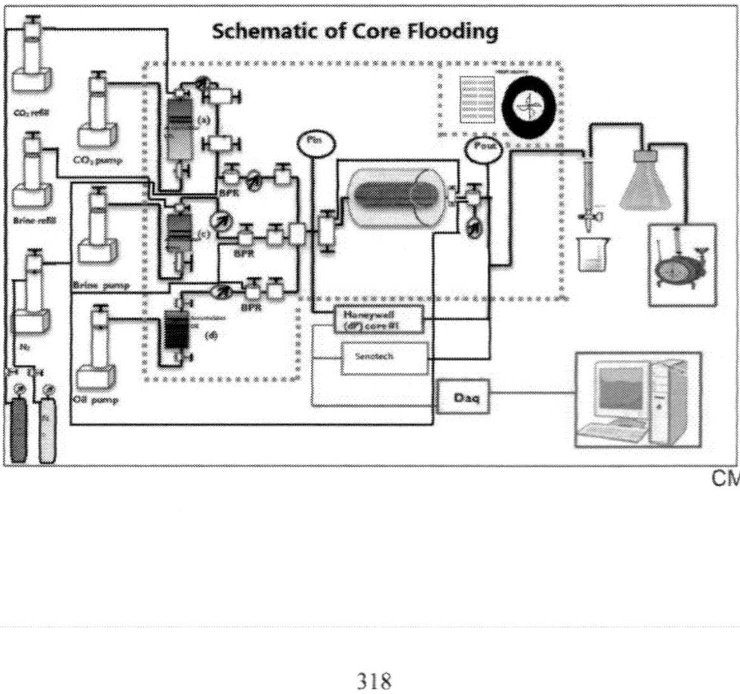


CMTC - 486251 - MS

# EXPERIMENTAL SET-UP

* Physical properties and dimensions of core used in this study.

| CORE (7- 13-15) | | Units |
|---|---|---|
| Length | 17.52 | Cm |
| Diameter | 3.77 | Cm |
| Pore Volume | 24.48 | Cm^3 |
| Porosity | 12.48 | % |
| Peameabilty | 48.39 | md |

CMTC - 486251 - MS

# EXPERIMENTAL SET-UP(CO2 Storage)

**Injections Fluids Used :**

*Supercritical $CO_2$

*Brine(4000ppm)Farnsworth unit

*Brine(35000ppm)Sea Water

**All tests performed at reservoir conditions of 168°F and 4400 psi**

CMTC - 486251 - MS

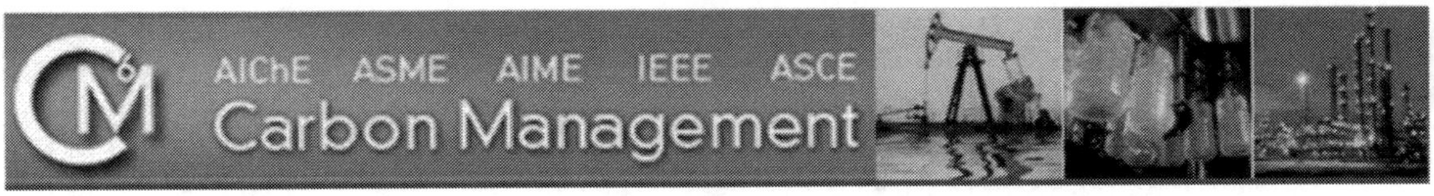


# Results [CO2 storage]

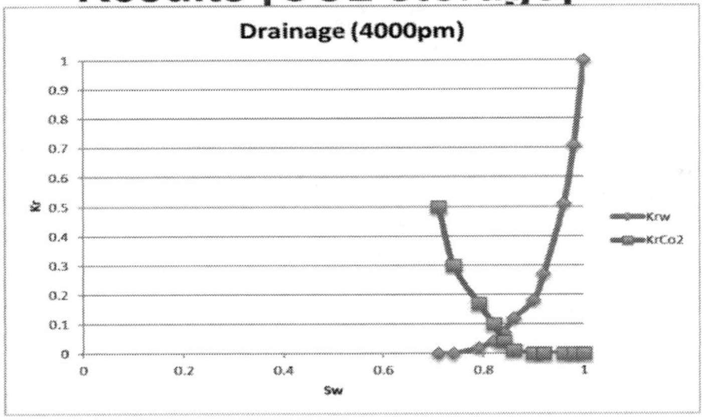


CMTC - 486251 - MS

# Results [CO2 storage]

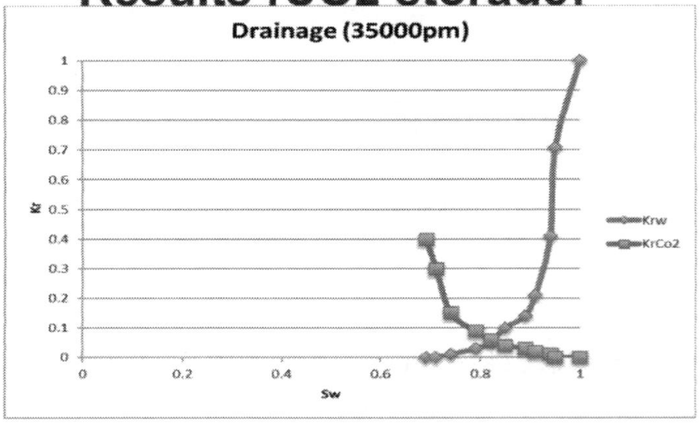


CMTC - 486251 - MS

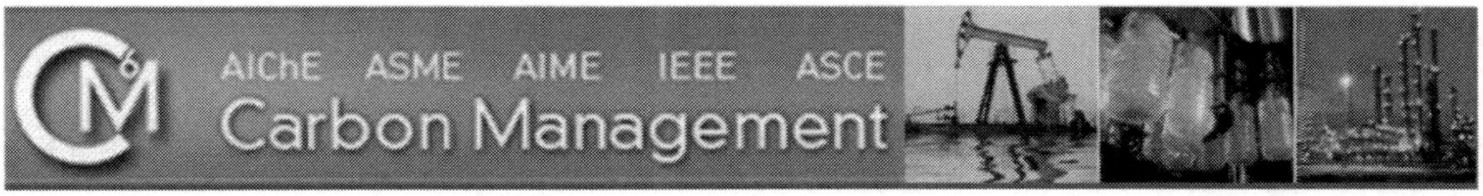


# Results [CO2 storage]

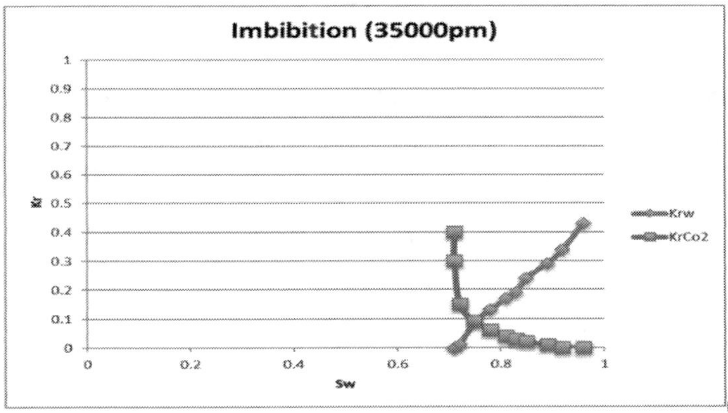


CMTC - 486251 - MS

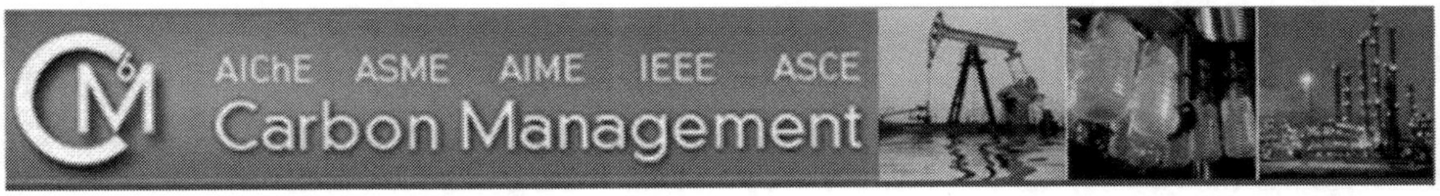


# Results [CO2 storage]

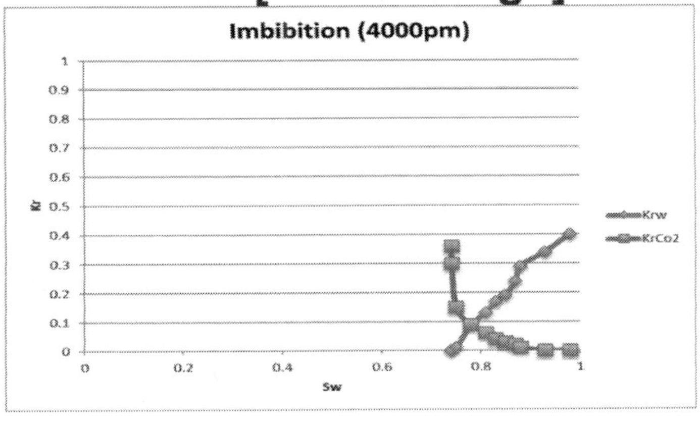


CMTC - 486251 - MS

AIChE ASME AIME IEEE ASCE
Carbon Management

## Results [$CO_2$ storage]

**End point values for gas and water saturations**

| Relative Permeability | Sw | Sg | Vol. of CO2 Stored(mL) |
|---|---|---|---|
| Drainage with 4000ppm | 0.68 | 0.32 | 7.8336 |
| Drainage with 35000ppm | 0.69 | 0.31 | 7.5888 |
| Imbibition with 4000ppm | 0.74 | 0.26 | 6.3648 |
| Imbibition with 35000ppm | 0.73 | 0.27 | 6.6096 |

CMTC - 486251 - MS

AIChE ASME AIME IEEE ASCE
Carbon Management

## Discussion [ $CO_2$ storage ]

* Brine salinity of the aquifer had little or no effect on $CO_2$ storage in the cores tested under the conditions studied.

*Higher $CO_2$ storage was achieved during drainage, though significant storage remained after brine injection that followed the $CO_2$ injection.

CMTC - 486251 - MS

AIChE ASME AIME IEEE ASCE
Carbon Management

## Theories and Principles [ WAG & SWAG ]

*WAG is when water and gas are injected alternately for better sweep efficiency

*SWAG is when water and gas are injected simultaneously for better sweep efficiency

*Recovery Factor is ratio of recoverable oil to estimated oil in place.

CMTC - 486251 - MS

AIChE ASME AIME IEEE ASCE
Carbon Management

## Theories and Principles [ WAG & SWAG ]

*Gas Cut is the ratio of gas produced to the volume of total fluids produced.

*Residual oil saturation is fraction of pore volume occupied by oil at the end of oil displacement.

.

CMTC - 486251 - MS

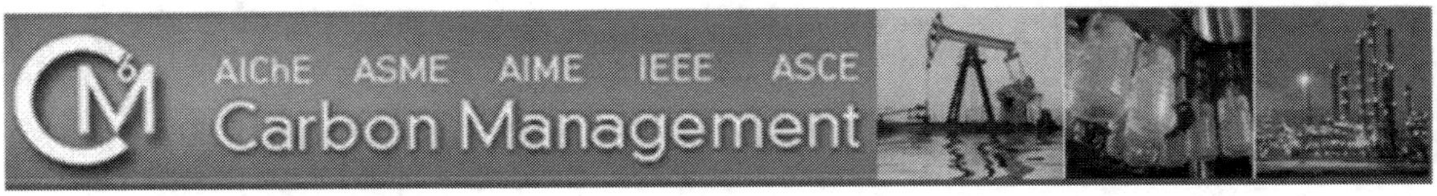


# Results [WAG & SWAG]

| WATER AND ALTERNATING GAS | RF | FW | Gas Cut |
|---|---|---|---|
| SWAG 50%/50% | 0.80 | 0.56 | 0.31 |
| SWAG 75%$CO_2$/25%WATER | 0.87 | 0.14 | 0.63 |
| SWAG 25%$CO_2$/75%WATER | 0.75 | 0.75 | 0.10 |
| WAG WATER FF $CO_2$ | 0.73 | 0.42 | 0.45 |
| WAG $CO_2$ FF WATER | 0.73 | 0.42 | 0.43 |

CMTC - 486251 - MS

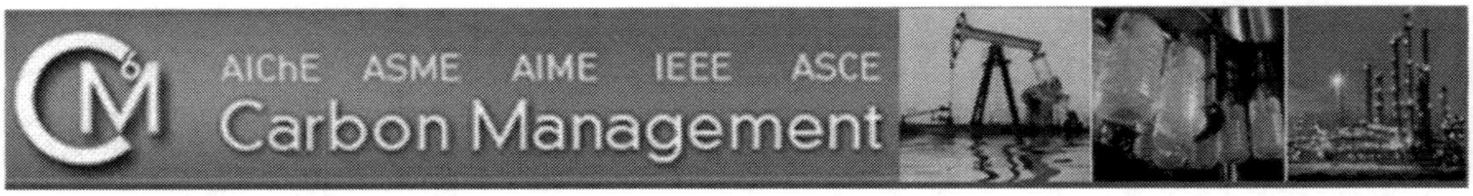



# Results [WAG & SWAG]


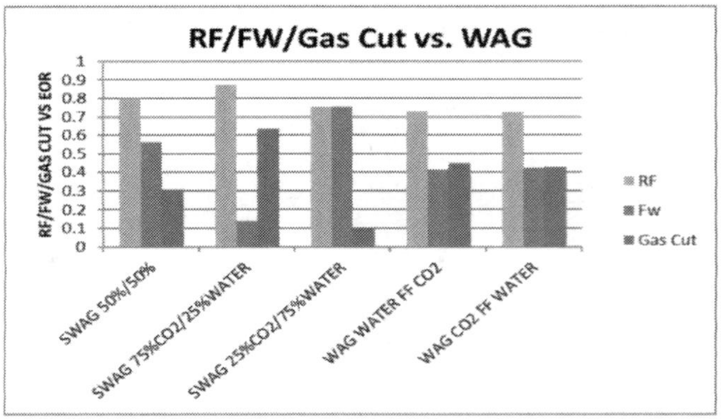



CMTC - 486251 - MS

AIChE ASME AIME IEEE ASCE
Carbon Management

## Discussion [ WAG & SWAG]

*WAG with 75% $CO_2$ and 25% recorded highest oil recovery

*WAG with 25% $CO_2$ and 75% recorded the least oil recovery

*WAG with 50%$CO_2$ and 50% water recorded a sizable recovery factor.

CMTC - 486251 - MS

AIChE ASME AIME IEEE ASCE
Carbon Management

## Conclusions

*The salinity of aquifer does not have an effect on its ability to store $CO_2$.

*Drainage which the practical method of storing records the best results.

*SWAG is an effective EOR method.

*pH of injection fluids were lower the pH of produced water.

CMTC - 486251 - MS

AIChE ASME AIME IEEE ASCE
Carbon Management

## Recommendations

*Studies should be conducted on field application of SWAG

*Research into making $CO_2$ a stronger wetting agent should done to Aid sequestration.

CMTC - 486251 - MS

CARBON MANAGEMENT TECHNOLOGY CONFERENCE

# Carbon Capture Utilization and Storage (CCUS) as a Regional Development Tool: Planning and Design Considerations

***Presenter : Kevin C OBrien, PhD***

***CMTC 2017***
***Houston, Texas***
***July 17-20, 2017***

PRAIRIE RESEARCH INSTITUTE · ACS · Linde · BASF · NATIONAL ENERGY TECHNOLOGY LABORATORY · AEI Affiliated Engineers · U.S. DEPARTMENT OF ENERGY · I

1

---

**Disclaimer**



PRAIRIE RESEARCH INSTITUTE · ACS · Linde · BASF · NATIONAL ENERGY TECHNOLOGY LABORATORY · AEI Affiliated Engineers · U.S. DEPARTMENT OF ENERGY · I

2

## Prairie Research Institute: Illinois-focused Resource Research and Service

*Addressing societal challenges that impact Illinois and the global community*

I 1867

↓

PRAIRIE RESEARCH INSTITUTE

↓

- ILLINOIS NATURAL HISTORY SURVEY — PRAIRIE RESEARCH INSTITUTE
- ILLINOIS SUSTAINABLE TECHNOLOGY CENTER — PRAIRIE RESEARCH INSTITUTE
- ILLINOIS STATE GEOLOGICAL SURVEY — PRAIRIE RESEARCH INSTITUTE
- ILLINOIS STATE ARCHAEOLOGICAL SURVEY — PRAIRIE RESEARCH INSTITUTE
- ILLINOIS STATE WATER SURVEY — PRAIRIE RESEARCH INSTITUTE

3

## Overview

*Examining from a state-wide perspective*

- Status of large scale $CO_2$ capture pilot
  - Phase 2 proposal submitted to DOE for 15 MW large scale capture pilot
- Select utilization options synergistic with regional economy
  - Preferred options can vary throughout one state or region
- Identify relevant workforce development partners
  - Related to utilization options and part of $CO_2$ value chain
- Develop tools to connect $CO_2$ utilization with existing economy
  - Tools to examine dispatching of $CO_2$

4

Phase 2 proposal (Design, Build, Operate) submitted

# STATUS OF LARGE SCALE PILOT



# Host Site: Abbott Power Plant

*Ideal site for large scale pilot testing of coal and natural gas*

- Seven boilers total: three are coal based (Chain-grate stoker design) others natural gas
- **Coal side has completely separate treatment system from natural gas side**
- For testing will run two coal boilers
- Illinois high sulfur coal is burned
- Electrostatic precipitators and a wet Flue Gas Desulfurizer (FGD) in place
- **Tradition of evaluating new emission technologies**
- **Tradition of showcasing technologies to other power plants and education groups**

Major advantage that University owns and operates Host Site

## Overview of Capture System for Large Pilot Plant

*Technology features*

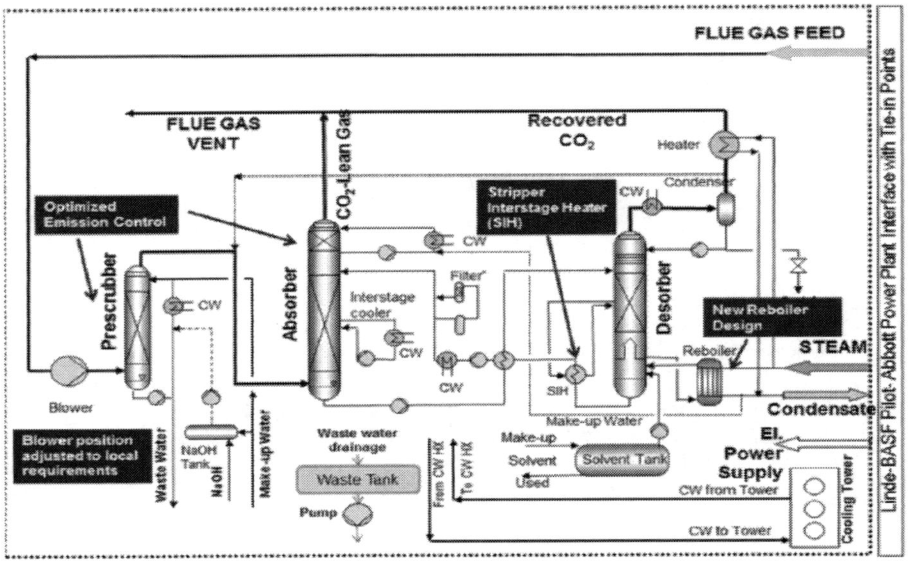


7

## Overview of Phase 2 Project Schedule

*More than just a design, build, operate project*

**BP2 (13 months)**
- Basic Design/Engineering
- Detailed Design/Engineering
- Permitting Application
- EH&S Analysis
- Stakeholder Engagement

**BP3 (17 months)**
- Equipment & Materials Procurement
- Construction & Installation
- Permit Compliance
- Stakeholder Engagement

**BP4 (18 months)**
- Commissioning & Startup
- Pilot Plant Operation
- Testing Data Analysis & TEA
- Permit Compliance
- Stakeholder Engagement and Education Operations
- Commercialization Plan

- ***Stakeholder Engagement helps educate , understand market needs, and propagate technology***
- ***Education: workforce development for existing and future operators and engineers***
- ***Demonstrating not only the technology but how to create jobs and drive regional economies***

8

## Phase 2: Project Organization Chart

*Added expertise in aerosols, OSBL procurement / construction, and dry-bed emissions reduction*

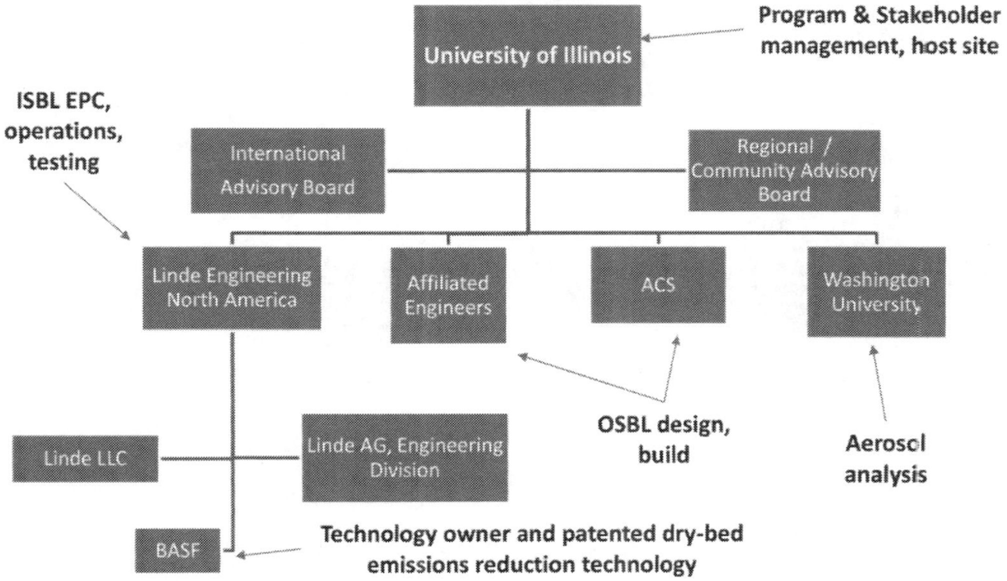




## Site for Carbon Capture Plant Established and Evaluated

*Located close to Abbott Power Plant*

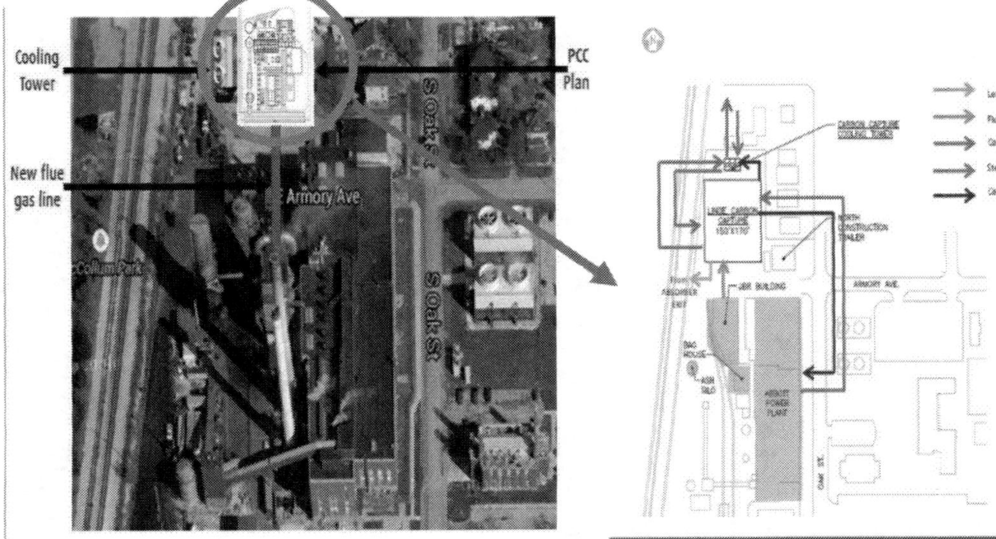


*Extract flue gas POST CEMS Unit*

## Regional & Global Test Bed for CCUS

*Concentration of natural resources and intellectual capital*

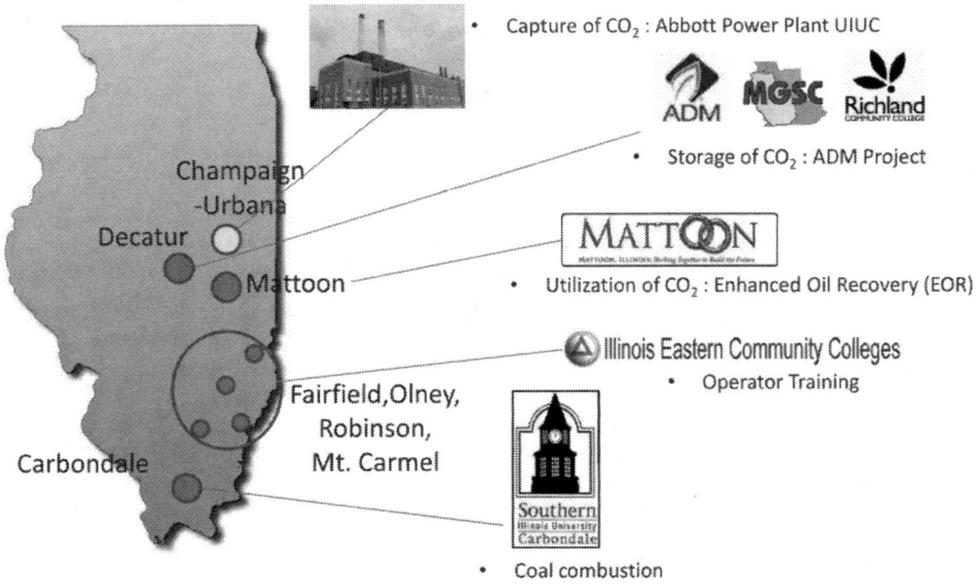


11

Important to consider regional economy

# SYNERGISTIC UTILIZATION OPTIONS

12

# CarbonSAFE ILLINOIS

*Funded to match carbon "sources" with carbon "sinks"*

**$12M funding - Commercial-scale CCS opportunities for 50+ million tonnes $CO_2$ capture and storage in the Illinois Basin**

- **Geological characterization and utilization options such as EOR**
  - **drilling, core, modeling**
- **Source suitability, options, and proximity to storage**
- **Transportation needs**
- **Business case scenarios**
- **Pre-Feasibility and Feasibility studies**

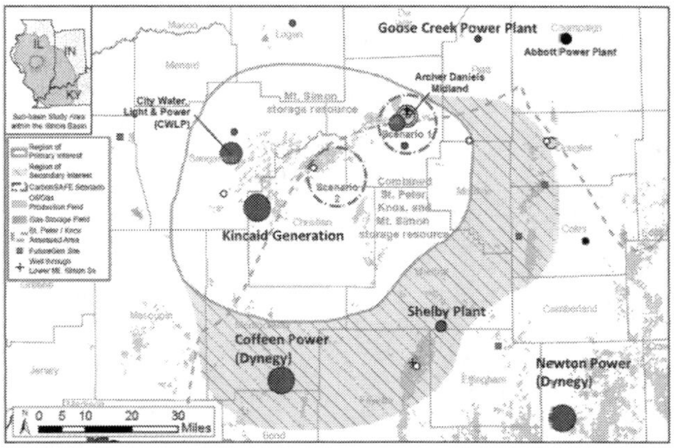


ILLINOIS STATE GEOLOGICAL SURVEY
PRAIRIE RESEARCH INSTITUTE

ILLINOIS SUSTAINABLE TECHNOLOGY CENTER
PRAIRIE RESEARCH INSTITUTE

PRAIRIE RESEARCH INSTITUTE | ACS | Linde | BASF | NATIONAL ENERGY TECHNOLOGY LABORATORY | AEI Affiliated Engineers | ENERGY

# Connection Between Coal-Fired Plants and Agriculture

*Long standing relationship in Illinois*

Coal fired Power Plants

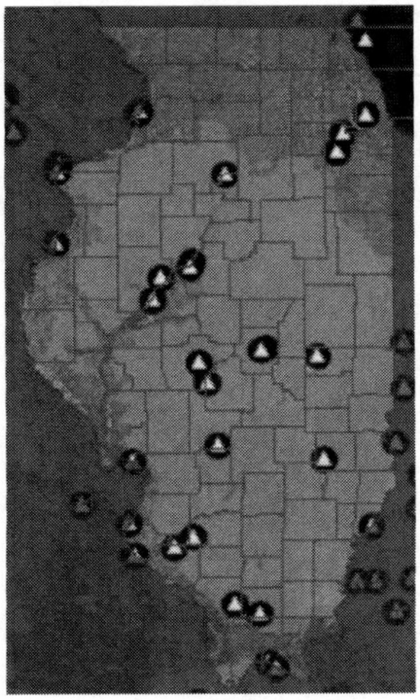

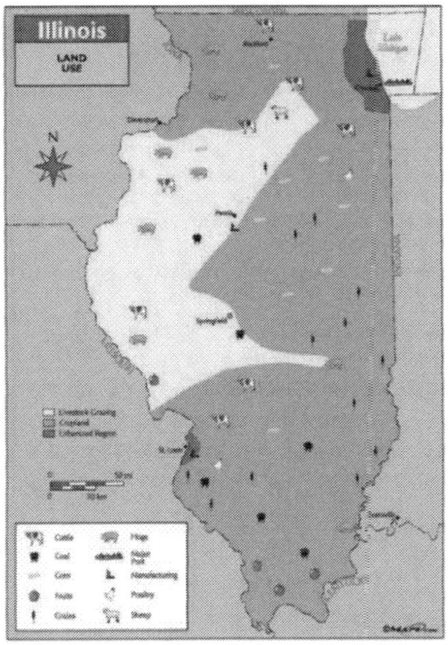


PRAIRIE RESEARCH INSTITUTE | ACS | Linde | BASF | NATIONAL ENERGY TECHNOLOGY LABORATORY | AEI Affiliated Engineers | ENERGY

14

# $CO_2$ Utilization with Algae

*Synergistic with agricultural economy in Illinois*

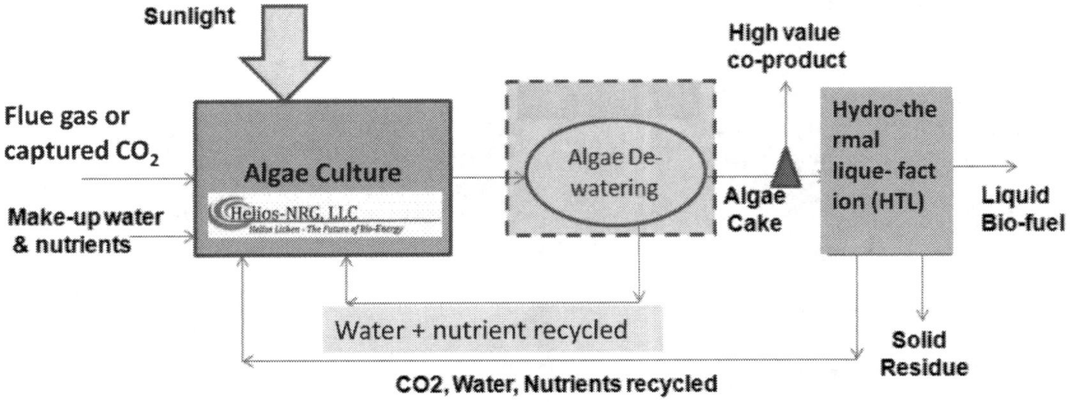


# Nutrient Loss Reduction: Critical to Illinois Agriculture

*Plan required to reduce nutrients (N and P) carried in rivers and to the Gulf of Mexico*

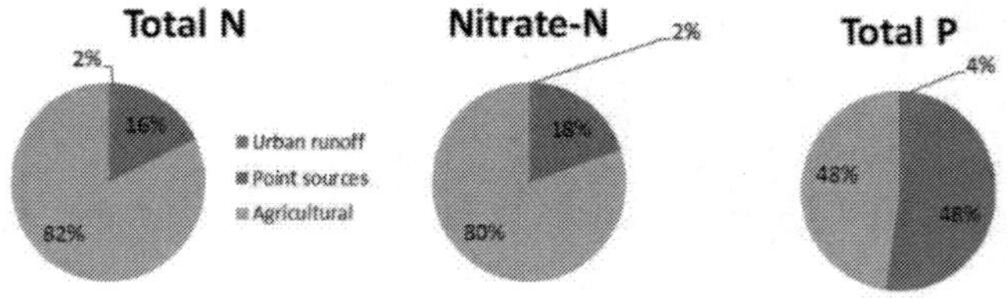


Nutrient sources in Illinois contributing to riverine nutrient export from the state

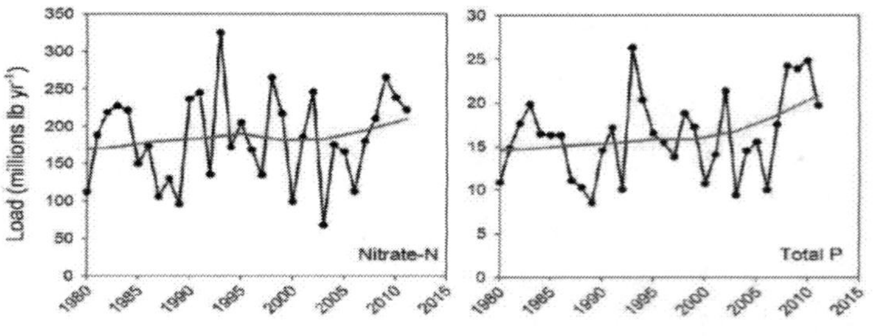


Annual nitrate-nitrogen and total phosphorus loads from Illinois

16

## SoyFACE: Evaluating Elevated $CO_2$ Levels on Crop Growth

*Free Air Concentration Enrichment (FACE) approach requires no enclosure*

**FACE ring**. Wind Direction and velocity are measured in the center, then a computer controls the release of gases to simulate future possible conditions.

Fumigation ring is 30 m in diameter. At the center of the ring, wind speed and direction is monitored in real time

*http://soyface.illinois.edu/*

17

Partners connected into relevant supply chains

# WORKFORCE DEVELOPMENT

18

## Training Operators and Engineers

*Partners already connected into existing supply chains*

ILLINOIS EASTERN COMMUNITY COLLEGES

Southern Illinois University Carbondale

Association of Illinois Electric Cooperatives

Eastern Illinois University

Coles-Moultrie Electric Cooperative

COLES Together

19

## ACKNOWLEDGEMENTS


| Name | Organization |
|---|---|
| Bruce W Lani | National Energy Technology Laboratory / US Department of Energy |
| Yongqi Lu, Vinod Patel, Hong Lu, Sallie Greenberg, Randy Locke, Deb Jacobson, Gary Miller | Prairie Research Institute / University of Illinois |
| Michael Larson, David Wilcoxen, Richard Rundus, Tracy Malvestuto | Abbott Power Plant / University of Illinois |
| Krish Krishnamurthy, Makini Byron, Joseph Naumovitz, Torsten Stoffregen, Ali Jangi, Andy Poplin | Linde |
| Sean Rigby | BASF |
| David Guth, Steve Bennett | Affiliated Engineers Inc. |
| Steve Davis, John Lowrey | Association of Illinois Electric Cooperatives |

**This project is supported by the U.S. Department of Energy / National Energy Technology Laboratory (DOE/NETL) through Cooperative Agreement No. DE-FE0026588**


20

CMTC-486336-MS

# Minimizing Solvent Oxidation with $NO_2$ Prescrubbing

Andrew Sexton, Darshan Sachde, Austyn Vance, Kevin Fisher, Trimeric Corporation
Joseph Selinger, Kent Fischer, Gary Rochelle, The University of Texas at Austin



## Abstract

A novel method to remove nitrogen dioxide ($NO_2$) from the flue gas of coal-fired power plants with $CO_2$ capture has been advanced for commercial implementation. The technology leverages the equipment and chemistry in an existing (sulfur dioxide) $SO_2$ polishing scrubber upstream of the main $CO_2$ capture unit to remove the $NO_2$, preventing degradation of the $CO_2$ capture solvent and formation of nitrosamines. The report focuses on further evaluation of the chemical additives and operating conditions associated with the $NO_2$ removal process to define conditions for commercial scale testing and deployment. Experimental work systematically evaluated a series of potential additives to minimize the oxidation of sulfite in a representative $SO_2$ pre-scrubber solution. The additive combinations and concentrations were varied alongside important process conditions such as temperature, oxygen concentration, and metals present in solution to mimic the conditions expected in a commercial system. Important results of the parametric experimental work include identifying a new, potent sulfite oxidation inhibitor, revealing the importance of combining inhibitors with metal chelating agents, validation of a low-cost additive process, and development of a new semi-empirical model to represent mechanisms associated with sulfite oxidation. In addition, the experimental work revealed the impact of operating at higher temperatures (representative of a field test unit), which will guide the selection and concertation of additives as well. Techno-economic analysis identified potential net savings as large as \$1.30/tonne $CO_2$ captured and quantified the potential benefit of low cost additive options actively being pursued by the development team. Finally, the experimental results and engineering analysis supported the development of a detailed field testing plan and protocol to evaluate the technology at near-commercial scale. The field test preparation included development of procedures to introduce chemical additives to an existing $SO_2$ polishing unit and identification of representative flue gas conditions based on a review of existing plants. These activities will have direct bearing on operation and design of commercial units.

# 1 Background

Coal is used to generate approximately 40% of the electricity in the United States, with historical capacity numbers approaching 60% of total electricity generation. A variety of market, environmental, and regulatory factors play a role in determining the mix of fuels used in electricity generation. In the case of coal, environmental considerations have been an important part of competing with other fossil fuels (e.g., natural gas) and alternative sources of energy (wind, nuclear, hydropower, etc.). Maximizing the value of coal as a fuel requires balancing the need for reliable, low-cost electricity with management of the associated environmental impact of coal use.

Coal-fired power plants have made significant progress in reducing emissions of sulfur dioxide ($SO_2$), nitrogen oxides ($NO_X$), particulate matter (PM), and recently mercury (Hg), since the passage of the Clean Air Act. Future regulations may require carbon capture on any new coal-fired power plant. To prepare for upcoming regulations, significant research and development is currently being pursued for new technologies to capture carbon from flue gas streams produced by coal-fired electric generating power plants.

Solvent-based technologies appear to be a viable option for large-scale carbon capture from coal-fired utilities. The United States Department of Energy (DOE) is currently supporting multiple small- and large-scale R&D projects to demonstrate the technical and economic feasibility of carbon capture using a variety of solvent technologies, but many of these solvent systems have similar challenges external to advances made to reduce both the capital cost and the regeneration energy penalty. These challenges include oxidation and degradation of the solvents, which increase the costs and environmental impact of the capture process. The focus of this paper and associated research is to address some of these solvent management challenges for amine-based $CO_2$ capture solvents.

## 1.1 Introduction

A significant operating cost of post-combustion $CO_2$ capture with amine scrubbing is solvent degradation and loss of amine scrubbing efficiency, including solvent oxidation. Solvent oxidation of amine solvents by $SO_X$ in flue gas is well-documented. Most modern coal-fired power plants are equipped with flue gas desulfurization (FGD) systems, which reduce the concentration of $SO_X$ to approximately 40 ppmv (DOE/NETL 2013). This concentration of $SO_X$ can still generate a significant amount of sulfate heat-stable salts in the amine circulation loop and contribute to aerosol formation in the $CO_2$ absorber, so it is recommended to use an alkaline pre-scrubber to reduce $SO_X$ in the amine capture system to 1 ppmv or less.

Typical coal-fired boiler flue gas contains nitrogen oxides ($NO_X$) composed of NO and $NO_2$ that ultimately reach the $SO_2$ polishing unit. While the bulk of the $NO_X$ is in the form of NO, NO does not readily absorb in amine solvents (Fine 2015). $NO_2$ does absorb readily into amine solutions and poses the larger risk to the $CO_2$ capture process. Research conducted at The University of Texas (UT) has indicated that $NO_2$ absorbs into amine solvents to ultimately oxidize the amine and generate nitrosamine (Fine, 2015). In practice, very little $NO_2$ is removed in the $SO_2$ polisher and eventually reaches the $CO_2$ absorber. Effectively all of the $NO_2$ is absorbed into the amine solution as nitrite, forming an amine radical in the process.

$$R_2NH + NO_2 \rightarrow HNO_2 + R_2N^{\cdot}$$

The amine free radical may then react with itself or oxygen to stoichiometrically oxidize the amine.

$R_2N^{\cdot} + R2N^{\cdot} \rightarrow$ Oxidation products

In addition, the amine free radical can react with nitric oxide (NO) in the high-temperature portion of the capture process to produce nitrosamine or other oxidation products (Fine 2015):

$R_2N^{\cdot} + NO^{\cdot} \rightarrow R_2NNO$

The nitrosamine accumulates in the solvent until it decomposes to oxidation products and $N_2O$ at the highest operating temperatures within the stripper. It is hypothesized that 1 mole of $NO_2$ in the inlet flue gas will oxidize 2 to 4 moles of amine within the $CO_2$ capture system (Fine 2015).

In addition to the economic challenges regarding the oxidation of the solvent, nitrosamines also pose a significant environmental concern. Many nitrosamine compounds are considered carcinogenic to humans at low concentrations. Additionally, nitrosamine accumulation in the solvent will be a problem for spills and for waste solvent handling and disposal and may be present in cleaned flue gas emissions from the absorber (Maree, Nepstad and de Koeijer 2013).

As $CO_2$ capture is implemented, advances need to be made in order to increase the economic feasibility of the amine capture process by reducing reclaiming and solvent makeup costs, and to mitigate additional environmental risks associated with the solvent capture process through the formation of nitrosamines and other solvent oxidation products. The University of Texas (UT) has determined that it is economically attractive to remove $NO_2$ in the sulfur dioxide ($SO_2$) pre-scrubber by adding sodium thiosulfate ($Na_2S_2O_3$)/ethylenediaminetetracetic acid (EDTA) or a tertiary amine to the caustic (NaOH) scrubbing liquor.

In the case of $NO_2$ absorption by sulfite solutions, oxidation of the sulfite can significantly limit the viability of the process. $NO_2$ absorption produces sulfite radical species, which can react with oxygen to generate additional free radicals. In this way, each mole of $NO_2$ absorbed can generate many moles of sulfite radical. These radicals eventually are oxidized to sulfate or terminated in a reaction with another free radical. This reaction mechanism is outlined by Nash (1979) and Huie and Neta (1984).

$NO_2 + SO_3^{2-} \rightarrow NO_2^- + SO_3^{\cdot-}$

$SO_3^{\cdot-} + O_2 \rightarrow SO_5^{\cdot-}$

$SO_5^{\cdot-} + SO_3^{2-} \rightarrow SO_4^{\cdot-} + SO_4^{2-}$

$SO_4^{\cdot-} + SO_3^{2-} \rightarrow SO_3^{\cdot-} + SO_4^{2-}$

$2SO_3^{\cdot-} \rightarrow S_2O_6^{2-}$

Thiosulfate provides an additional termination step, which can slow the rate of sulfite oxidation. After the same absorption and free radical formation steps, thiosulfate can create a new termination pathway. Equations below are outlined by Owens (1984).

$NO_2 + SO_3^{2-} \rightarrow NO_2^- + SO_3^{\cdot-}$

$SO_3^{\cdot-} + O_2 \rightarrow SO_5^{\cdot-}$

$SO_5^{\cdot-} + S_2O_3^{2-} \rightarrow SO_5^{2-} + S_2O_3^{\cdot-}$

$SO_5^{2-} + SO_3^{2-} \rightarrow 2SO_4^{2-}$

$2S_2O_3^{\cdot-} \rightarrow S_4O_6^{2-}$

This chemistry provided the basis for initial experimental work at UT that demonstrated the viability of the $NO_2$ absorption process (Fine 2015). Minimal equipment modification is required for an existing pre-scrubber to accomplish $NO_X$ scrubbing and therefore near-zero additional capital costs are required to implement this solution. While the use of sodium thiosulfate could potentially be cost prohibitive at certain operating conditions, particularly at low $SO_2$ concentrations in the flue gas (e.g., low sulfite production rates), alternate inhibitors and additive pathways are evaluated in this work.

Absorption rates of $NO_2$ in tertiary amines are also very fast (Fine 2015). Only 10 to 20 mM tertiary amine is required to provide 90% $NO_2$ removal at typical packing heights in an $SO_2$ pre-scrubber. Therefore, an inexpensive amine such as triethanolamine (TEA) could be sacrificed in the $SO_2$ pre-scrubber, thus avoiding nitrosamine formation and the more expensive loss of the main amine in the capture system.

With funding from the U.S. Department of Energy (DOE), UT conducted additional bench-scale experiments to further quantify the use of thiosulfate, EDTA, tertiary amines, and other proprietary additives in an $SO_2$ pre-scrubber to remove $NO_2$. The experimental and analytical methods (including a batch reactor) were based off of the methods used by Fine at UT.

Trimeric then used the data from the UT experiments, along with a thorough literature review, to develop a pilot field test and sampling plan in conjunction with the National Carbon Capture Center. An initial phase of pilot-scale field testing was completed by UT in the summer of 2017.

Finally, Trimeric used the bench-scale data collected by UT to conduct a preliminary concept engineering design and cost analysis. This engineering work demonstrates that such an approach could be implemented at a full-scale using commercially available chemicals and equipment, in a cost effective way, while managing any safety and environmental issues. Furthermore, no fatal flaws were identified during the course of this engineering analysis.

# 2 Experimental Section

The primary focus of the bench-scale research in this work was to perform experiments that simultaneously quantified the effect of pre-scrubber additives on sulfite oxidation and $NO_2$ absorption in a bench scale reactor apparatus that is described in detail in subsequent sections.

## 2.1 Methods/Test Plan

Operating conditions and additive concentrations were selected to prepare for the conditions at the pilot test site. The National Carbon Capture Center (NCCC), operating at the Alabama Power E.C. Gaston power generation facility, has an $SO_2$ polisher that treats a slipstream of flue gas from a full-scale coal-fired boiler and flue gas desulfurization (FGD) unit. This host site provides representative conditions to test the commercial application of the technology.

**Table 1: Experimental Parameters for Oxidation Reactor Tests**

| Parameter | Value | Comments |
|---|---|---|
| Inlet $NO_2$ Concentration (ppmv) | 1.5 - 5 | 2 ppm reflects target concentration for field testing |
| Temperature (°C) | 25 - 55 | Higher T represents process conditions expected at pilot and commercial scale |
| Sulfite Concentration (mmol/kg) | 4 - 50 | 15 mmol/kg reflects target concentration for 90% $NO_2$ removal in field test unit |
| Thiosulfate Concentration (mmol/kg) | 0 - 200 | 200 mmol/kg identified as point of diminishing returns on thiosulfate. Operate field test at this "maximum" value to test effectiveness of thiosulfate-based oxidation |

| | | |
|---|---|---|
| | | inhibition system. |
| Metals Concentration (mmol/kg) | 0.1 - 0.5 | Guided by separate metal solubility experiments. See Section 3. |
| EDTA Concentration (mmol/kg) | 0.02 - 1 | Low level of EDTA for both sets of experiments confirmed as important to chelate background levels of dissolved metal in solution. |
| Tertiary Amine Concentration (mmol/kg) | N/A | Range of conditions omitted to protect proprietary information |
| pH | 9.5 | The field $SO_2$ Polisher is expected to operate in a range between 8.5 and 9. See discussion below. |
| Total Gas Rate (SLPM) | 2.1 | Experimental parameter only. |

Additional proprietary additives and inhibitors were developed during the bench-scale testing. General results for proprietary additives will be discussed in this report.

### *2.1.1 Laboratory Apparatus*

A High Gas Flow apparatus (HGF, Figure 1) is used to conduct experiments demonstrating $NO_2$-catalyzed sulfite oxidation. Dry air exits a mass flow controller and passes through a temperature controlled water saturator to maintain water balance in the HGF. $CO_2$ and dilute $NO_2$ is then directly mixed with the hydrated air stream. The resulting gas stream is sparged directly into the bottom of the HGF through 350 mL of the experimental absorbing solution. A constant operating temperature within the HGF apparatus is maintained via an oil bath circulating fluid through the jacket of the reactor.

The gas stream is sparged through the solution, and the effluent gas stream from the HGF is analyzed using the Thermo Scientific model 42i trace level $NO_X$ analyzer before being exhausted back to the fume hood. A vacuum pump pulls the exhaust gas through the analyzer. The gas stream can also be bypassed around the HGF and diluted with air before entering the NOX analyzer to evaluate the feed $NO_2$ concentrations.

Liquid samples are taken from a sample port at the bottom of the HGF at regular intervals and analyzed for sulfite and sulfate concentration using anion chromatography.

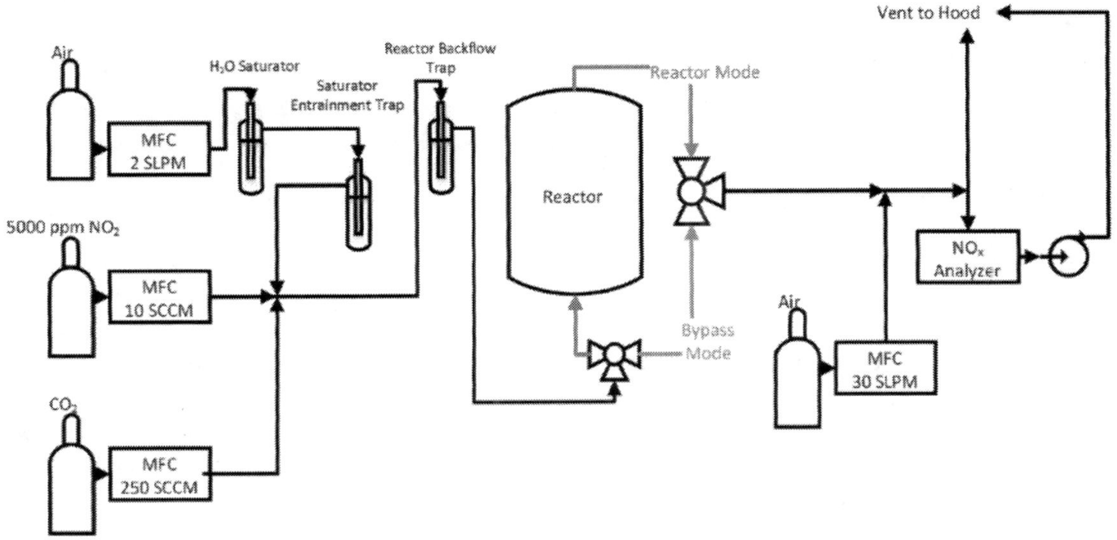


**Figure 1: High Gas Flow Apparatus**

### *2.1.2 Test Procedure*

Each experiment starts with the preparation of 400 mL of synthetic absorbing solution (350 mL for the reactor, 50 mL for pre-rinse). The following additives may be included in the solution preparation:

- Sodium Sulfite
- Sodium Thiosulfate
- Sodium Carbonate
- Sodium Bicarbonate
- Ferrous Sulfate (Iron source to simulate the effect of catalytic oxidation)
- EDTA
- Tertiary Amine
- Deionized Water

The sodium carbonate and bicarbonate are added in roughly a 3:1 ratio to simulate a solution in equilibrium with the $CO_2$ in the gas stream as estimated from literature VLE data (Hilliard 2008) and a solution pH of approximately 9.5.

The solution is stirred for 60 to 90 minutes before the experiment. The high gas flow apparatus is then rinsed with deionized water and 50 mL of sulfite solution prior to every experiment. The apparatus is then charged with the remaining 350 mL of solution.

### *2.1.3 Experiment and Sampling*

The experiment begins with the gas flow in bypass mode (see Figure 1) to measure the initial $NO_2$ concentration in the gas stream. The valve is then turned to "reactor" mode to initiate the absorption-oxidation experiment. The effluent gas from the reactor is measured continuously during the experiment to quantify outlet $NO_2$ concentration. Liquid samples are collected every 6-10 minutes. The sample collection starts by switching the apparatus to bypass mode so a liquid sample can be removed by pipette from the sampling port near the bottom of the reactor. The sample is immediately preserved via the addition of formaldehyde to convert the sulfite to methylsulfonic acid (MSA). Two dilution steps are performed to prepare the sample for ion chromatography.

Experiment time can range from 45 minutes to 2 hours depending on the rate of oxidation. Longer experiments are required when oxidation is strongly inhibited to allow sufficient accumulation of oxidation products.

### *2.1.4 Analytical Methods*

The gas leaving the HGF apparatus (or in bypass mode to measure the inlet concentration) is sent continuously to the Thermo Scientific model 42i trace level $NO_X$ analyzer. The liquid samples from the HGF are analyzed using ion chromatography. Anion chromotography is used to quantify formate, nitrite, sulfate, sulfite, and thiosulfate concentrations with a Dionex ICS-3000 anion chromatograph with an IonPac AS15 column. In the case of sulfate, sulfite, and thiosulfate, the sample must be preserved because sulfite is oxidatively unstable at room temperatures.

## 2.2 Experimental Results

The experimental test results are summarized in Table 2. Please note that in the first three experiments in the table, the $NO_X$ analyzer was not functioning properly, so $NO_2$ absorption could not be measured.

**Table 2: High Gas Flow $NO_2$-Catalyzed Sulfite Oxidation Experimental Results**

| | Test | Sulfite Conc. | Thiosulfate Conc. | Amine Conc. | EDTA Conc. | $Fe^{2+}$ Conc. | T | $O_2$ | $NO_2$ inlet conc. | $NO_2$ outlet conc. | $k^{ox} * 10^3$ | $NO_2$ absorption rate * $10^6$ | F-value ($SO_3$ ox/ $NO_2$ abs) |
|---|---|---|---|---|---|---|---|---|---|---|---|---|---|
| | | mmol/kg | mmol/kg | mmol/kg | mmol/kg | mmol/kg | °C | Vol% | ppm | ppm | $min^{-1}$ | mol/kg/min | mol/mol |
| Low T Vary Sulfite | 1 (Base) | 47 | 100 | 0 | 0.02 | 0 | 25 | 21 | 5 | unk | 3.2 | 1.07[1] | 108 |
| | 2 | 23 | 100 | 0 | 0.02 | 0 | 25 | 21 | 5 | unk | 2 | 1.05[1] | 55 |
| | 3 | 7.4 | 100 | 0 | 0.02 | 0 | 25 | 21 | 5 | unk | 3.8 | 0.96[1] | 19 |
| | 4 | 6 | 100 | 0 | 0.02 | 0 | 25 | 21 | 4.87 | 1.11 | 1.5 | 0.9 | 10 |
| | 5 (Base) | 43.6 | 100 | 0 | 0.02 | 0 | 25 | 21 | 5.67 | 1.19 | 2.4 | 1 | 115 |
| | *FINE (Base)* | *40* | *100* | *0* | *0.02* | *0* | *20* | *21* | *5* | *0* | *3.4* | *3.6* | *37.9* |
| High T Vary $SO_3^{2-}$, $NO_2$, $S_2O_3^{2-}$ | 6 | 39.3 | 100 | 0 | 0.02 | 0 | 53 | 21 | 5.35 | 2.8 | 5.5 | 0.6 | 296 |
| | 7 | 4.2 | 100 | 0 | 0.02 | 0 | 55 | 21 | 5.01 | 3.01 | 5.1 | 0.5 | 38 |
| | 8 | 10.4 | 100 | 0 | 0.02 | 0 | 55 | 21 | 5.8 | 2.88 | 7.6 | 0.7 | 89 |
| | 9 | 13.8 | 200 | 0 | 0.02 | 0 | 55 | 21 | 1.71 | 1.21 | 3.3 | 0.1 | 328 |
| | 10 | 16.5 | 200 | 0 | 0.02 | 0 | 55 | 21 | 5.86 | 3.4 | 5.2 | 0.6 | 126 |
| Metals | 11 | 12.1 | 200 | 0 | 0.02 | 0.5 | 53 | 21 | 3.95 | 1.72 | 5.2 | 0.5 | 102 |
| | 12 | 14.1 | 200 | 0 | 0.02 | 0.1 | 54 | 21 | 2.12 | 1.53 | 3.9 | 0.1 | 402 |
| | 13 | 16.4 | 200 | 0 | 1 | 0.5 | 52 | 21 | 2.21 | 0.92 | 3.7 | 0.3 | 200 |
| | 14 | 14.6 | 200 | 0 | 0.2 | 0.1 | 52 | 21 | 1.67 | 0.78 | 3 | 0.2 | 177 |

[1] For the first three experiments, $NO_2$ absorption measurements were not obtained due to problems with the $NO_X$ analyzer. The absorption rates were calculated from a model that will discussed later in this section. The F-value is derived from this calculated absorption rate.

**Table 2: High Gas Flow $NO_2$-Catalyzed Sulfite Oxidation Experimental Results (Continued)**

| | Test | Sulfite Conc. | Thiosulfate Conc. | Amine Conc. | EDTA Conc. | $Fe^{2+}$ Conc . | T | $O_2$ | $NO_2$ inlet conc. | $NO_2$ outlet conc. | $k^{ox}$ * $10^3$ | $NO_2$ absorption rate * $10^6$ | F-value ($SO_3$ ox/ $NO_2$ abs) |
|---|---|---|---|---|---|---|---|---|---|---|---|---|---|
| | | mmol/kg | mmol/kg | mmol/kg | mmol/kg | mmol/kg | °C | vol % | ppm | ppm | $min^{-1}$ | mol/kg/min | mol/mol |
| Oxygen | 15 | 13.6 | 200 | 0 | 0.02 | 0 | 52 | 8 | 2.02 | 0.52 | 2.2 | 0.3 | 88 |
| Oxygen | 16 | 16.6 | 200 | 0 | 0.02 | 0 | 52 | 4 | 2 | 1.46 | 1.1 | 0.1 | 121 |
| Oxygen | 17 | 15.4 | 200 | 0 | 0 | 0 | 52 | 4 | 2 | 1.15 | 7 | 0.2 | 530 |
| Tertiary Amine | 18 | 17.3 | 0 | 20 | 0.02 | 0 | 52 | 21 | 2 | 0.1 | 0.4 | 0.5 | 17 |
| Tertiary Amine | 19 | 15.2 | 0 | 20 | 0.02 | 0 | 25 | 21 | 2 | 1.6 | 0.3 | 0.1 | 65 |
| No Inhibitor | 24 | 16.2 | 0 | 0 | 0.02 | 0 | 53 | 21 | 2 | 1.4 | 22.7 | 0.2 | 980 |

The series of parametric tests in the experimental results in Table 2 will be discussed below, followed the regression of an empirical model to explain the effects of experimental parameters over the entire dataset.

### *2.2.1 Explanation of Measured Results Variables*

Three measured results are quantified and presented in Table 2: first-order sulfite oxidation rate constant ($k_{ox}$), $NO_2$ absorption rate, and the F-value (defined by Fine as the ratio of moles of sulfite oxidized per mole of $NO_2$ absorbed). Absolute absorption or oxidation rates can be misleading due to underlying experimental conditions, so the F-value was defined as a convenient metric to indicate the effectiveness of an additive in limiting the oxidation for every mole of $NO_2$ absorbed. Therefore, lower F-values are indicative of more effective additives/inhibitors. A full theoretical background for these processes is not within the scope of this paper – details are available in the literature (Fine 2015) (Shen 1997).

### *2.2.2 Comparison to Previous Work (Fine 2015) (Tests 1 and 5)*

The baseline experiments associated with tests 1 through 5 attempted to duplicate the operating conditions of Fine (with the exception of slightly higher temperature, 25°C vs. 20°C). When comparing the results to Fine, the measured oxidation rates ($k_{ox}$) are lower in the most recent experiments, though Case 1 is similar to the results of Fine.

Despite the lower oxidation rate, the F-value in Table 2 is significantly higher for Case 5 than the comparable Fine baseline experiment. One explanation is that the $NO_2$ absorption rate is lower in the current experiments due to the lower overall gas rate to the HGF (2.1 vs. 7.5 SLPM). However, even accounting for this difference, the relative catalytic effect of $NO_2$ appears to be higher in the current experiments than in Fine's previous work at these base case conditions.

One potential source of error in the experiments is the measurement of $NO_2$ absorption. As noted by Fine, and experienced again throughout this experimental work, the $NO_X$ analyzer is very sensitive to the presence of water. The gas in this experiment is saturated to water, and in the high temperature experiments in particular, this presents a risk for condensation in the reactor outlet. Removing the water (via desiccant or other method) risks removing $NO_2$ in the gas phase as well via absorption, providing an alternate path for error in the absorption measurements. Therefore, the F-values calculated in this work are subject to the limitations of the $NO_2$ absorption measurements, and the oxidation rates will provide a better representation of the effect of changing experimental conditions. The absorption measurements will be discussed in a subsequent section. An alternate drying method has been identified (membrane-based dryer) that will selectively remove water from the gas-phase, limiting the risk of condensation and absorption. This approach will be used in future work.

### *2.2.3 Effect of Thiosulfate (Test 8 vs. Test 10)*

Fine previously established that thiosulfate at low concentrations (25 mmol/kg) significantly inhibited oxidation, and concentrations of 100 mmol/kg provided the expected additional benefit (~2x reduction in the sulfite oxidation rate). As noted in the discussion of the test matrix development, one bench-scale test objective was to establish if there is a point of diminishing returns associated with the addition of thiosulfate, and apply that condition to field testing to provide a margin for potentially higher rates of oxidation in field systems (presence of metals or other contaminants, loss of additives via blowdown, etc.).

Experiments 8 and 10 in Table 2 provide nearly identical experimental conditions with thiosulfate concentration increasing from 100 to 200 mmol/kg. As the oxidation rate indicates ($k_{ox}$), doubling the thiosulfate concentration leads to a 1.4x reduction in the oxidation rate. This is consistent with the square root dependence of oxidation on thiosulfate concentration. Thus, even at the higher value of thiosulfate concentration, the inhibitor effect appears to still be following the expected mechanism.

### *2.2.4 Effect of Metals and EDTA (Tests 11-14)*

Tests 11 and 12 directly examined the effect of metals in the presence of low baseline levels of EDTA. As the results indicate, the addition of 5 times as much iron (0.5 vs. 0.1 mmol/kg) did increase the oxidation rate by 33%. This result indicates that the additional iron must be soluble to some degree in solution to catalyze oxidation – it has been hypothesized that EDTA may be providing the mechanism to solubilize additional iron by chelating iron present in solution. Furthermore, while the oxidation rate increases, it is not as large as the increase in metals concentration might indicate. This may indicate that a solubility limit is reached or that EDTA is effectively limiting the impact of metals.

### *2.2.5 EDTA vs. Oxidation Inhibitors (Test 16 vs. 17, Test 22 vs. 23, Test 24)*

A series of experiments were run to isolate the effect of EDTA (reducing oxidation by chelating metals) from the effect of the oxidation inhibitors (terminating oxidation chain reactions).

*Test 16 vs. Test 17: Thiosulfate vs. EDTA*

Experiments 16 and 17 are identical with the exception that one test uses a baseline concentration of EDTA, while the other experiment is performed in the absence of EDTA. Experiment 16 pairs a high thiosulfate concentration known to be effective in limiting oxidation (200 mmol/kg) with a baseline level of EDTA (0.02 mmol/kg). Test 17 removes the EDTA. Results indicate that the sulfite oxidation rate increases by a factor of 7 when the EDTA is removed, and even small amounts of EDTA are effective at limiting catalytic effects of trace levels of metal which can lead to significant oxidation regardless of the presence of oxidation inhibitors.

*Test 22 vs. Test 23: Proprietary Inhibitor vs. EDTA*

Tests 22 and 23 are similar to Tests 16 and 17, with the addition of a proprietary inhibitor. The detailed experimental conditions and results are not presented here, but general results will be addressed. Removing EDTA causes the oxidation rate to increase by a factor of 18 over the test conducted with the proprietary inhibitor and EDTA, indicating the mechanism of EDTA is not dependent on the type of inhibitor present. These measured oxidation rates reinforce that the presence of EDTA is important to manage oxidation.

*Test 24: EDTA without Inhibitor*

The final experiment tested EDTA without the presence of an oxidation inhibitor. The oxidation rate in this case was the highest measured in this work. This result indicates that EDTA and oxidation inhibitors are acting on two largely independent mechanisms that catalyze oxidation of sulfite, with both mechanisms being relevant in the conditions of the $SO_2$ pre-scrubber. The oxidation inhibitors are likely following the thiosulfate free-radical termination mechanism associated with the absorption of $NO_2$ (see Equations 1.6 – 1.8). The EDTA is likely acting on the metal catalysis of the sulfite –oxygen reaction in solution to oxidize the solvent. Therefore, in order for the $NO_2$ removal technology to be effective, both EDTA and an oxidation inhibitor are required. Additional experiments may evaluate the final composition of the solution, including byproducts and the inhibitors, to try to develop a deeper understanding of these mechanisms.

### *2.2.6 Effect of Oxygen Concentration (Test 9 vs. Tests 15 and 16)*

Tests 9, 15, and 16 were all operated at similar conditions with decreasing concentrations of oxygen in the inlet synthetic flue gas (21 vol%, 8 vol%, and 4 vol%, respectively). The high oxygen case (21%) represents a "worst-case" or accelerated oxidation scenario, while 4-5% oxygen is more representative of a flue gas stream from a coal-fired boiler. Previous work by Fine hypothesized that oxygen concentrations would only be relevant at low concentrations (< 5%) due to depletion of oxygen in the boundary layer impacting the bulk solution oxidation. However, results indicate that the oxidation rate drops proportionally with the oxygen concentration over the entire range of the experiments. Therefore, it is important to account for the effect of oxygen when modeling or predicting the performance of the process.

## 2.3 Modeling Experimental Results

Two sets of modeling activties were developed as part of the experimental analysis. First, an $NO_2$ absorption model that Fine validated at low temperatures (20°C) was modified with an Arrhenius temperature term to see if the absorption measurements at higher temperatures in this work could be effectively predicted by accounting for temperature dependence.

The second activity developed a general empirical model for oxidation rates using the theoretical basis for oxidation mechanisms (Shen 1997) to guide selection of parameters for the model. This oxidation model served as the basis for the subsequent techno-economic analysis in Section 4 and generalized the effect of the different parameters tested in the experimental work.

### 2.3.1 $NO_2$ Absorption Rate Temperature Dependence

The $NO_2$ absorption rate data in these experiments is limited by the concerns regarding water entering the $NO_X$ analyzer at higher temperatures as outlined above. However, the early experimental data after the initial repair of the analyzer (following Test 3) is the most reliable before the analyzer performance began to degrade towards the end of the test matrix. This dataset includes a transition from 25°C to 55°C, allowing an opportunity to test the effect of temperature. Fine previously developed a model for ambient temperature and anticipated that the $NO_2$ absorption rate would not show a temperature dependence based on analysis by Shen (Shen 1997). However, he was unable to independently corroborate this, and simply used the model at higher temperatures where he could not take measurements. The equation below adds a temperature dependent term to his original model:

$$N_{NO2} = \frac{(1-e^{-Nog})[NO_2]_{FEED}V}{m_{Solution}} * \exp\left[\frac{E_A}{R} * \left(\frac{1}{T_{REF}} - \frac{1}{T_{ACTUAL}}\right)\right]$$

Where:

- $N_{OG}$ = Number of gas phase transfer units;
- V = Molar flow of feed gas to apparatus;
- $m_{solution}$ = Mass of solution in HGF;
- $T_{REF}$ = 298.15 K.

The term for the number of gas phase transfer units ($N_{OG}$) contains a regression parameter to fit absorption rate data (Fine 2015). This term was not manipulated in this work – only the temperature dependence ($E_A$) was regressed. Experiments 4 through 10 were included in the regression in an effort to avoid the later data that faced analyzer issues. The results are depicted in Figure 2.

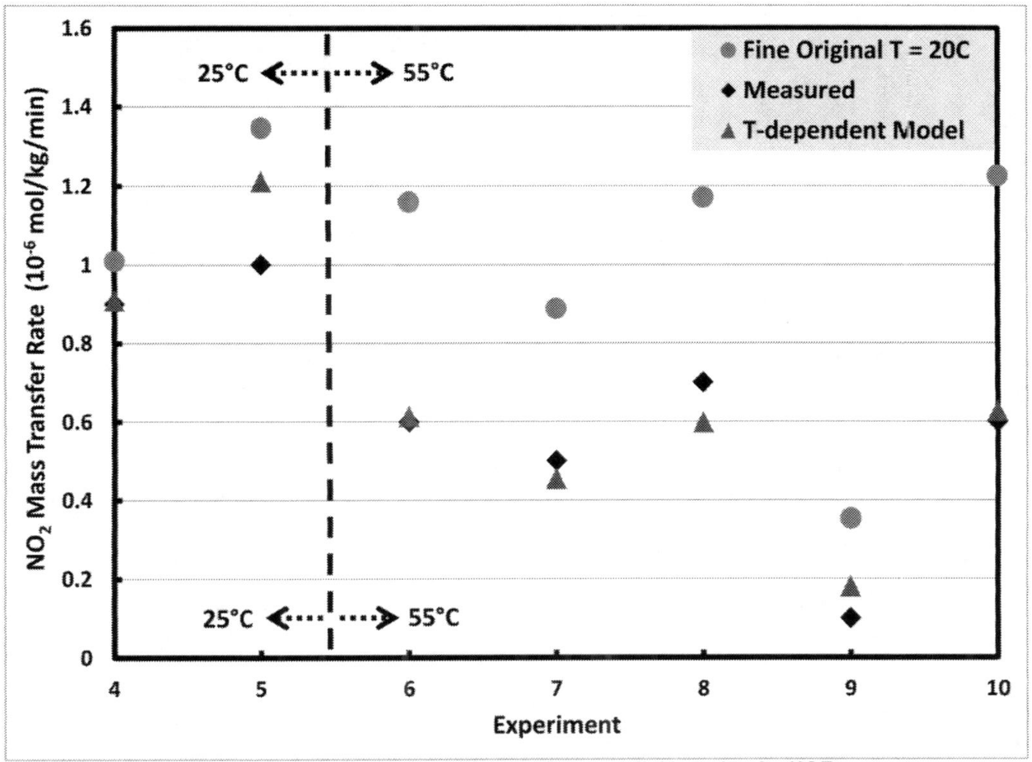


Figure 2: Temperature-Corrected $NO_2$ Absorption Rate predicitions for the HGF apparatus

An activation energy of -15.3 kJ/mol was used to fit the data, indicating that increases in temperature reduce the $NO_2$ absorption following Equation 3.3. The figure illustrates this effect, as Fine's original model generally over-predicts the mass transfer rate for these cases, particularly the 55°C cases. The new temperature dependent model shows good agreement with measured data.

### *2.3.2 Generalized Model for Sulfite Oxidation in the Presence of Inhibitors*

A power-law empirical model was regressed using the sulfite oxidation rate data ($k_{ox}$) collected in this work in addition to data from Fine's work to create a database of 25 experiments for the regression. Experiments conducted with metals added, without EDTA or an inhibitor, and with tertiary amine were all excluded. The metals experiments were excluded because the mechanism is unlikely to be captured by the generic model. The general form of the model is as follows:

$$k_{OX} = C_1 \left( \frac{y_{NO_2,Feed}\left[1-\exp(-N_{OG})\right]}{y_{NO_2,REF}} \right)^{X1} \left( \frac{[SO_3^{2-}]}{[SO_3^{2-}]_{REF}} \right)^{X2} \left( \frac{[S_2O_3^{2-}]}{[S_2O_3^{2-}]_{REF}} \right)^{X3} \left( \frac{[O_2]}{[O_2]_{REF}} \right)^{X4} \exp\left[ \frac{E_A}{R} * \left( \frac{1}{T_{REF}} - \frac{1}{T_{ACTUAL}} \right) \right]$$

Where:

$k_{OX}$ = Rate of sulfite oxidation ($min^{-1}$)
C1, x1, x2, x3, x4, $E_A$ = regressed parameters;
$y_{NO2}$ = $NO_2$ in feed (ppm);

The equation represents the different contributors to the oxidation rate: nitrite concentration, sulfite concentration, thiosulfate concentration, oxygen concentration and a term to account for temperature dependence embedded in rate constants, diffusion coefficients, and Henry's constants. The results of the regression for thiosulfate are summarized in Table 3.

**Table 3: Regression parameters for oxidation rate model**

| | Regressed Model | Theoretical (Shen) |
|---|---|---|
| **C1** | 7.30 | N/A |
| **Ea (kJ/mol)** | 23.8 | N/A |
| **X1 (Nitrite)** | 0.55 | 0.5 |
| **X2 (Sulfite)** | -0.05 | 0 |
| **X3 (Thiosulfate)** | -0.39 | -0.5 |
| **X4 ($O_2$)** | 0.18 | 0.5 |

The regression results for the full data set are compared to the expected theoretical dependence. Interestingly, the general power law form captures the theoretical dependence of the model well. Only thiosulfate and oxygen dependence deviate significantly from theory. For each of these parameters, the sensitivity to the parameter may vary depending on the range over which the variable is tested. For example, Fine indicated that most of his data above 5% oxygen was insensitive to oxygen concentration, but was expected to be much more sensitive below that range (Fine 2015).

The predictions of the full model regression for thiosulfate are compared to the measured oxidation rates in Figure 3.

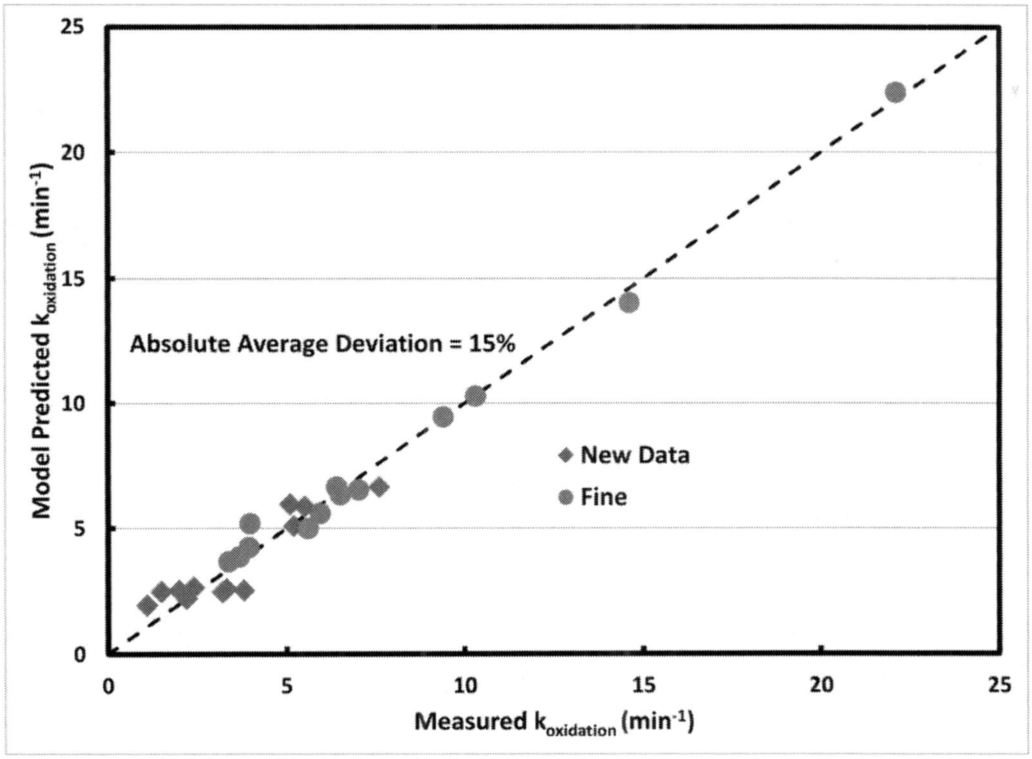


**Figure 3: Model-Predicted vs. Experimental Oxidation Rates**

Figure 3 illustrates that the model does a good job of predicting the oxidation rates for both data sets over the wide range of conditions and rates tested.

While the model appears to provide a strong fit of the data and the interpretation relative to theory provides some insight, the limited data and large number of parameters necessitates caution if the model is used for significant extrapolation. The model will be used for conditions that fall largely within the range of values in the experimental analysis as the experiments were designed to represent conditions applied in the field. Therefore, the model should be adequate for use in the techno-economic analysis to derive concentrations of additives such as thiosulfate.

# 3 Field Test

The pre-scrubbing system at the NCCC consists of a packed column, buffer tank, and a pump that recirculates the sodium hydroxide solution from the buffer tank to the packed column. A slipstream of flue gas exiting the FGD from the main power generation facility flows through the pre-scrubber, which uses dilute caustic solution to remove a majority of the remaining $SO_2$ from the flue gas. The dilute caustic solution, stored in a buffer tank, is pumped through the pre-scrubber using a recirculation pump. Concentrated sodium sulfate solution from the $SO_2$ pre-scrubber is blown down to wastewater neutralization, while the cooled flue gas exits the pre-scrubber to a flue gas blower. The pre-scrubber equipment specifications and operating conditions are summarized in Table 4.

**Table 4: National Carbon Capture Center Pilot Scale $SO_2$ Pre-Scrubber**

| | | |
|---|---|---|
| Packing Type | MP202Y | |
| Packing Height | 20 | ft |
| Diameter | 29.25 | in. |
| Specific Area | 200 | $m^2/m^3$ |
| Operating Temperature | 55 | °C |
| Operating Pressure | 101 | kPa |
| Buffer Tank Volume | 1420 | gallons |
| Liquid Recirculation Rate | 15 | gpm |
| Gas Rate | 8,000 – 12,000 | lb/hr |
| $SO_2$ in Feed Gas | 30 - 40 | ppmv |
| $NO_2$ in Feed Gas | 0.5 - 1 | ppmv |

The pre-scrubber is operated in a semi-batch mode: 50 wt% caustic is added in batches as the pH approaches ~8 to keep the operating pH in the 8-9 range and to maintain a concentration of approximately 10 wt% NaOH in the pre-scrubber.

The UT-Trimeric project team also modified the pilot unit for direct injection of $NO_2$ into the flue gas stream upstream of the $SO_2$ pre-scrubber. This provided the ability to vary the $NO_2$ over a wide range of conditions.

The field test plan objectives are broadly summarized as follows:

- Vary $NO_2$ feed concentration
    - Installed $NO_2$ injection system to raise inlet $NO_2$ up to 5 ppmv
- Vary additive combinations and concentrations
    - Semi-batch operation = additive concentrations vary with time (reaction, dilution)
- Analyze liquid samples
    - Quantify oxidation rates
    - Waste stream characterization/disposition

## 3.1 Results

At the time of the development of this paper, the field testing at NCCC was in progress. The results in this paper should be treated as preliminary results with additional data and analysis expected after completion of the field testing.

The experiments at NCCC thus far have demonstrated the effectiveness of the thiosulfate inhibitor system. After initial addition of thiosulfate, the sulfite concentration in the pre-scrubbing system accumulated to approach a steady state value, validating the performance of the thiosulfate as an inhibitor. Figure 4 indicates that $NO_2$ removal above 95% was achieved with 85 to 90% removal consistently attained. The field testing also validated liquid sampling methods, the injection of $NO_2$ into the flue gas, gas-phase $NO_2$ measurements, and batch operation and control of the pre-scrubbing system.

Additional testing will confirm the consumption of the additives (primarily thiosulfate) with time and evaluate the performance of the system as the concentrations of the key components vary (including the effect of batch addition of caustic solution to change the pH of the solution).

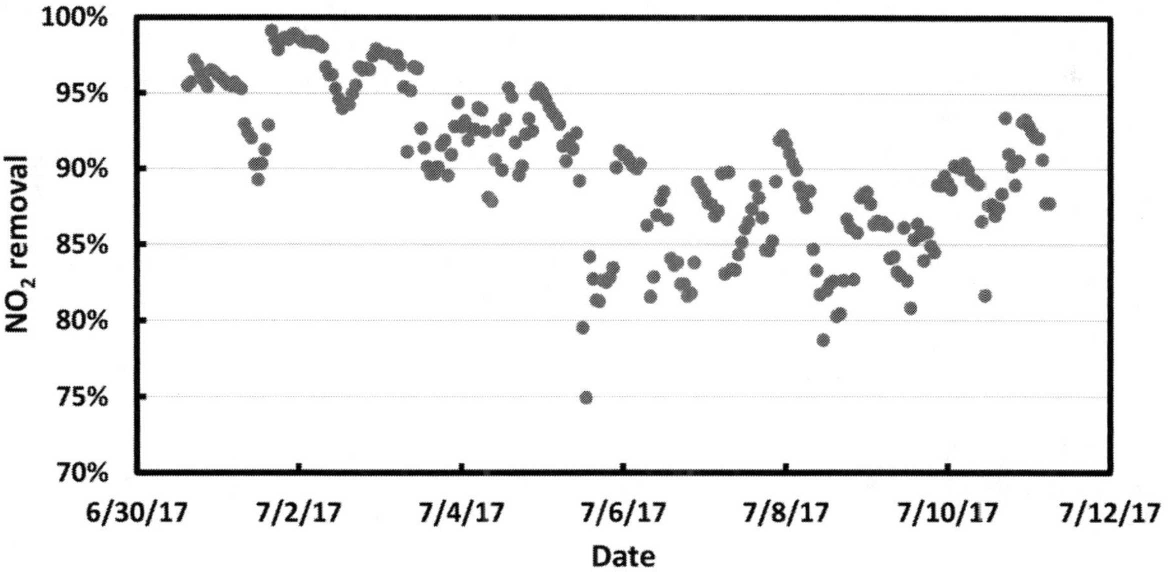


Figure 4: Experimental Field Tests $NO_2$ Removal Results

# 4 Techno-Economic Analysis

This portion of the paper will focus on the conceptual design and evaluation of a commercial scale system based on the data and findings of experiments, literature review, and field test preparation.

## 4.1 Conceptual Process Design

Figure 5 depicts the $SO_2$ polisher, where the additives are used to facilitate the concurrent absorption of $NO_2$, in relation to the direct contact cooler (DCC) and the absorber unit in the main $CO_2$ capture process.

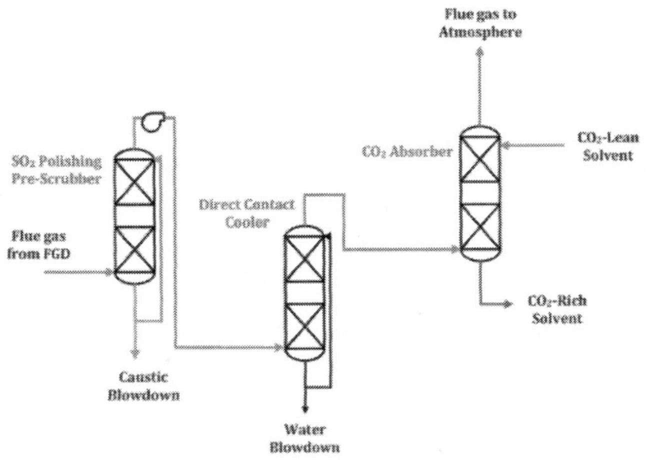


Figure 5: Process Flow Diagram

The process flow diagram illustrates an important design choice for the process. While integrating the $SO_2$ polisher and DCC might appear desirable from a capital cost standpoint (elimination of a column), the dual-column design is important for the chemical consumption of the $SO_2$ polisher. The blowdown rate (or equivalently, chemical make-up rate) in the individual $SO_2$ polisher is nominally dictated by the solubility limit of sodium sulfate that accumulates as $SO_2$ is absorbed. If the DCC is integrated, the purge rate will be set at a much higher level to remove the water that is condensed from gas in the DCC to maintain the water balance in the unit (and prevent dilution). This effect is even more important when $NO_2$ removal is integrated into the polishing scrubber. The primary cost of the $NO_2$ removal process is the chemical additives that are used. If purge rates are artificially high to remove water, the economics of the $NO_2$ removal will suffer.

Furthermore, combining the unit operations effectively increases the volume of waste solution that must be handled from the process. In the independent columns, a comparatively smaller, concentrated waste stream leaves the $SO_2$ polisher, while a large volume of process quality water leaves the direct contact cooler for potential re-use elsewhere in the $CO_2$ capture process or at the utility host site.

The sequence of the columns is also important. The experimental results in Section 3 indicate that higher operating temperatures reduce $NO_2$ absorption and increase sulfite oxidation. This would seem to incentivize moving the DCC upstream of the polishing scrubber to reduce the temperature of the feed gas. However, the issue of isolating waste streams arises again. The direct contact cooler now sees flue gas with higher levels of contaminants - in addition to the $NO_2$ and $SO_2$, impurities such as metals that may be removed in the $SO_2$ polisher instead accumulate in the water leaving the DCC. In addition, the introduction of $SO_2$ into the DCC creates a corrosive environment, necessitating the use of more expensive materials of construction. Therefore, the specific arrangement and function of the unit operations upstream of the capture unit are important to maximizing the viability of the $NO_2$ removal process and reducing the volume of waste for the capture process as a whole.

The basic equipment requirements for the $NO_2$ removal process are satisfied by an existing $SO_2$ polisher unit – the addition of specific chemical additives to the existing polisher is the basis of the technology. At most, a small storage tank or chemical tote might be required if chemicals cannot be introduced directly into the $SO_2$ polisher buffer tank. The assumption in the subsequent economic analysis will be that no additional capital costs are incurred as part of the process implementation. However, the economics of the process can be improved at specific operating conditions (e.g., low $SO_2$ content flue gas, which results in lower steady-state sulfite concentration in solution) by adding incremental packing to the column over what is nominally required for $SO_2$ removal (Fine 2015). In those specific cases, the initial design of the column must include the costs of additional packing.

Finally, the inlet $NO_2$ concentration to the $SO_2$ polisher is an important aspect in determining performance and cost savings potential of the $NO_2$ scrubbing technology. Coal-fired power plants may have a variety of emissions technologies upstream of the $SO_2$ polisher associated with the $CO_2$ capture unit that impact the $NO_2$ concentration. As a rule of thumb, $NO_2$ is up to ~5% of the total $NO_X$ leaving a boiler. In some large international markets, where coal-fired plants do not have $NO_X$ restrictions (e.g., Australia, China), the total $NO_X$ concentrations may range from 100 – 700 ppm depending on the specific conditions at the site (type of coal, boiler, age, etc.) [e.g., (Cousins, et al. 2012)]. This corresponds to $NO_2$ concentrations as high as 35 ppm.

For the U.S. market, which is the initial focus of this work, there are restrictions on $NO_X$ emissions via the Clean Air Act and Clean Air Act Amendments. A detailed review and analysis conducted in 2005 evaluated the technology response of coal-fired power facilities to tightening $NO_X$ regulations (Srivastava, et al. 2005). The results of their evaluation indicated a wide range of $NO_2$ concentrations reflecting, in part, underlying variations such as the type of coal or age of the facility as well as the technology approach used to control $NO_X$. However, their results indicated $NO_2$ levels generally below 15 ppmv with plants with SCR installed providing the lowest $NO_2$ levels.

## 4.2 Methods

The approach to evaluate the economic viability of the $NO_2$ removal technology in this work relies on a baseline $CO_2$ capture plant cost assessment for a process utilizing piperazine (PZ) and a custom reclaiming model/solvent degradation model developed for amine solvents such as PZ monoethanolamine (MEA), and a blend of methyldiethanolamine (MDEA) and PZ. While this analysis focuses on PZ, the technology will have value for amine solvents in general – previous research has shown that all classes of amines (primary, secondary and tertiary) will absorb $NO_2$ and form degradation products as a result (Fine 2015) . The benefit of the technology will scale to some level with the cost of the amine itself (more expensive solvents will yield a larger benefit). In addition, the level of nitrosamine formation (which varies greatly across amines) may provide additional incentive. In general, applying the $NO_2$ removal technology enhances the ability to select amines based on performance in the main capture system rather than being constrained by degradation or environmental/safety considerations.

The $CO_2$ capture techno-economic evaluation of PZ included a bottom up cost assessment of all of the equipment, chemicals, and operating costs associated with a capture process, including a baseline value for the cost of a thermal reclaimer. The approach mirrored the one used in the Cost and Performance Baseline Document produced by National Energy Technology Laboratory within the U.S. Department of Energy and provided the cost scaling factors to roll up the installed equipment cost into the impact on total plant costs (National Energy Technology Laboratory 2010). The $SO_2$ polisher, which is relevant for the modeling of the $NO_2$ absorption process, was included but was not specified in detail. Therefore, the NCCC pre-scrubber test unit was used to define equipment specifications such as packing height and type, while the diameter was scaled up to the commercial flow rates in the PZ cost analysis. The relevant baseline conditions and assumptions from the PZ analysis and $SO_2$ polisher are summarized in Table 5 and Table 6.

**Table 5: Baseline Assumptions for Techno-Economic Analysis**

| | | |
|---|---|---|
| Total equivalent electrical capacity | 723 | MWe |
| Flue Gas Flow Rate | 25.7 | kmol/s |
| $NO_2$ Concentration (ppmv) | 1 - 5 | ppmv |
| $SO_2$ Concentration (ppmv) | 10 - 300 | ppmv |
| $CO_2$ Capture Rate (@ 90% capture, 100% Capacity Factor) | 4,339,178 | tonne/yr |
| Capacity Factor | 85% | |

**Table 6: $SO_2$ Polisher Information**

| | | |
|---|---|---|
| Packing Type | Sulzer MP202Y | |
| Height of Packing | 6.1 | m |
| Diameter of Column | 17 | m |
| Specific Area of Packing (ap) | 200 | $m^2/m^3$ |
| Operating Temperature | 55 | °C |
| Operating Pressure | 101.325 | kPa |

The cost savings from removing the $NO_2$ upstream of the capture unit are directly associated with the amine solvent reclaimer and solvent losses. Removing $NO_2$ upstream results in a lower reclaimer flow rate, which should reduce the size of the reclaimer and associated capital costs and reduce the amine losses associated with reclaiming. In addition, the reduced nitrosation/degradation of the solvent further reduces amine losses. The combination of these savings is the gross economic benefit of removing the $NO_2$ upstream of the capture unit before accounting for costs of the process.

### *4.2.1 Cost of $NO_2$ removal process*

The cost associated with the removal of $NO_2$ is the consumption of the additives associated with the process. In this analysis, it was assumed that the baseline $SO_2$ polisher installed for the capture process (Table 6) was not modified to implement the $NO_2$ scrubbing process. This is the most desirable scenario as the process can leverage existing capital investment.

The $NO_2$ removed was defined by the packing characteristics in Table 6 and by the concentration of sulfite or amine (when the amine is used as a scavenger). With the packing in the column fixed by the existing $SO_2$ polisher design, the only way to reach 90% removal is to vary the sulfite or tertiary amine concentration. Once this concentration is defined, in the case of the sulfite system, the balance between oxidation of sulfite and absorption of $SO_2$ defines the concentration of inhibitor required. At steady state, the oxidation (sulfite destruction) rate will be equal to the $SO_2$ absorption (sulfite production) rate. This allows direct calculation of the inhibitor concentration at steady state.

## 4.3 Results of Techno-Economic Analysis

The methods to calculate the savings generated by $NO_2$ removal and costs associated with additives were used to evaluate the economic viability of the $NO_2$ removal process. Figure 6 presents the net cost savings of applying the $NO_2$ removal technology as a function of the $NO_2$ and $SO_2$ concentration in the flue gas the thiosulfate inhibitor system.

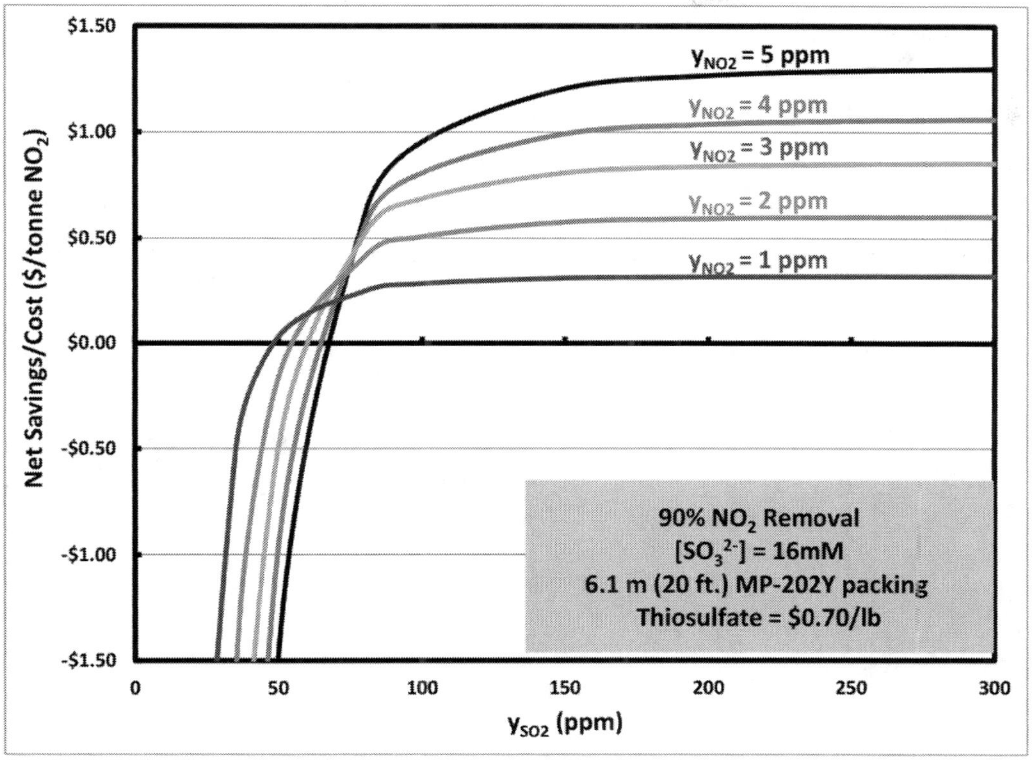


Figure 6: Net Savings of $NO_2$ removal using thiosulfate as an inhibitor

The figure illustrates several important points:

- As expected, increasing $SO_2$ concentrations in the flue gas improve the economics significantly by replenishing any sulfite that is oxidized. While the purge rate also increases, the reduced concentration of inhibitors in the purge more than offsets the increased flow.
- Increasing $NO_2$ has two effects. Low $NO_2$ concentrations reduce the rate of oxidation – this is apparent in the preceding figure as lower $NO_2$ concentrations "break-even" at lower levels of $SO_2$ – less sulfite needs to be made up due to lower oxidation rates. However, the lower $NO_2$ limits the maximum cost savings that can be achieved as the impact of $NO_2$ on the main capture process is reduced at lower $NO_2$ concentrations.
- The figure appears to reach an asymptote as a function of $SO_2$ concentration for each $NO_2$ concentration. This reflects an approach to the maximum cost savings.
- The approach to this asymptote, or maximum savings, represents the effectiveness of the specific inhibitor.

To evaluate the impact of alternative additive formulations, the results for the proprietary inhibitor identified in this work is compared to thiosulfate for specific operating conditions in Figure 7.

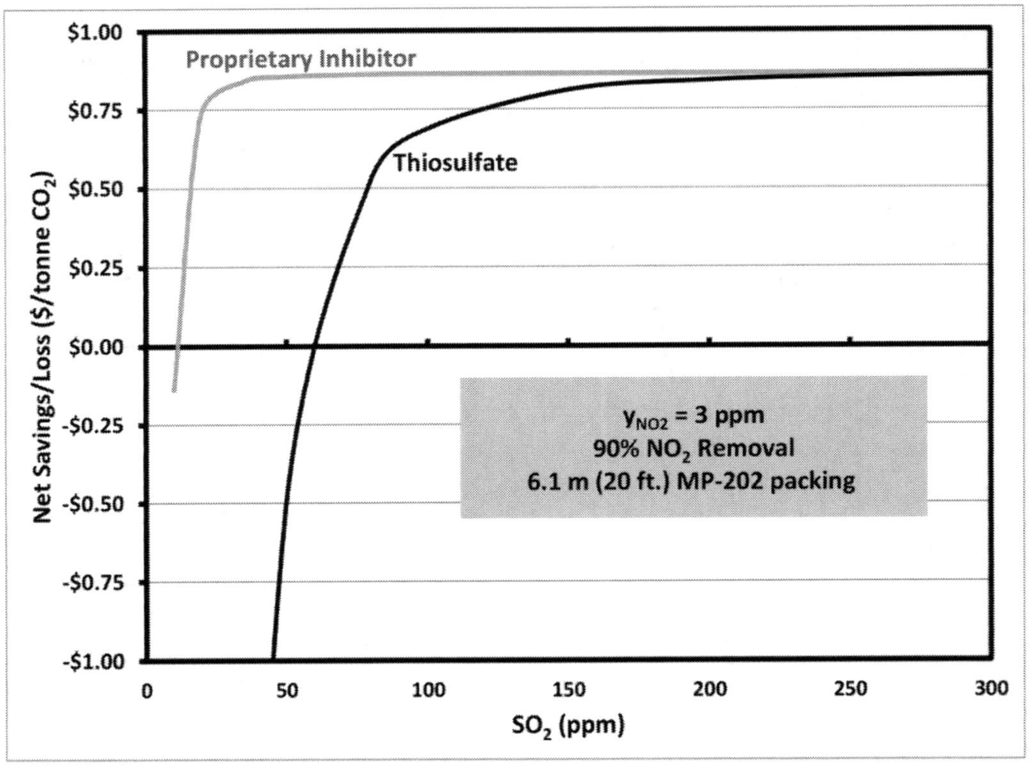


**Figure 7: Net Savings of $NO_2$ removal: Thiosulfate vs. proporitary inhibitor**

As the figure indicates, the proprietary inhibitor expands the economically attractive operating window (i.e., lower $SO_2$ concentrations) and generates larger costs savings compared to the baseline thiosulfate technology across the range of conditions in the figure. The result sin the figure highlight one of several options developed by the team to improve the economics and performance of the $NO_2$ removal technology compared to the thiosulfate system.

# 5 Conclusions

The experimental evaluation, field testing, and techno-economic evaluation of the novel $NO_2$ removal technology accomplished several important goals of the research:

- Experimental data were obtained which provide the necessary proof-of-concept required to demonstrate the effectiveness of the technology. Specifically, the team identified concentrations and combinations of additives required to minimize sulfite oxidation across a wide range of representative operating conditions.
- A greater understanding of the complex relationships between the operating conditions and the effectiveness of the technology has been obtained.
- Optimal additives and potential solutions to manage secondary impacts of the technology were identified (e.g., sending pre-scrubber blowdown to the FGD unit).
- Specific operating conditions were analyzed to guide field testing and to develop knowledge for commercial-scale operation (e.g., transient behavior of the pre-scrubbing system during start-up and for semi-batch operation). Expected operating conditions at commercial sites were analyzed to guide further testing and optimization (e.g., review of expected $NO_2$ concentration reaching the pre-scrubber).

- A techno-economic evaluation of the entire $CO_2$ capture process with the PZ solvent and a reclaiming model to predict the effects of degradation were leveraged to evaluate the total cost savings for 90% removal of $NO_2$ at feed concentrations of 1 – 5 ppm $NO_2$. The range of potential savings \$0.30 to \$1.30.

In summary, the absorption of $NO_2$ with sulfite plus oxidation inhibitors has proven to be viable at bench scale and field scale. Multiple routes to low-cost chemical additives, including novel inhibitors, have been identified to enhance the probability of adoption at commercial scale. The technology has garnered interest across the value chain of solvent suppliers, $CO_2$ capture technology suppliers, and utility owners and operators and there are plans for the field and laboratory work to continue with industry support moving forward.

## References


Chang, S.G., D. Littlejohn, and N.H. Lin. "Kinetics of Reaction in a Wet Flue Gas Simultaneous Desulfurization and Denitrification System." *ACS Symposium Series*, 1982: 127-152.

Cousins, Ashleigh, Aaron Cottrell, Anthony Lawson, Sanger Huang, and Paul H.M. Feron. "Model verification and evaluation of the rich-split process modification at an Australian-based post combustion $CO_2$ capture pilot plant." *Greenhouse Gases Science and Technology*, 2012: 329-345.

Cullinane, John Timothy. *Thermodynamics and Kinetics of Aqueous Piperazine with Potassium Carbonate for Carbon Dioxide Absorption.* PhD Dissertation, Austin, TX: The University of Texas at Austin, 2005.

Dai, Ning, and William A. Mitch. "Influence of Amine Structural Characteristics on N-Nitrosamine Formation Potential Relevant to Postcombustion $CO_2$ Capture Systems." *Environmental Science and Technology*, 2013.

DOE/NETL. *Cost and Performance Baseline for Fossil Energy Plants Volume 1: Bituminous Coal and Natural Gas to Electricity* . Washington D.C.: Depatment of Energy, 2013.

Fine, Nathan Andrew. *Nitrosamine Management in Aqueous Amines for Post-Combustion Carbon Capture.* PhD Dissertation, Austin: The University of Texas at Austin, 2015.

Frailie, Peter. *Modeling of Carbon Dioxide Absorption/Stripping by Aqueous Methyldiethanolamine/Piperazine.* Dissertation, Austin, TX: The University of Texas at Austin, 2014.

Hilliard, Marcus D. *A Predictive Thermodynamic Model for an Aqueous Blend of Potassium Carbonate, Piperazine, and Monoethanolamine for Carbon Dioxide Capture from Flue Gas.* Dissertation, Austin, TX: The Univeristy of Texas at Austin, 2008.

Huie, Robert E., and P. Neta. "Chemical behavior of sulfur trioxide(1-) (SO3-) and sulfur pentoxide(1-) (SO5-) radicals in aqueous solutions." *The Journal of Physical Chemistry*, 1984: 5665-5669.

IEAGHG. *Evaluation of Reclaimer Sludge Disposal from Post-Combustion $CO_2$ Capture.* Stoke Orchard, UK: International Energy Agency, 2014.

Lee, Yungli J., and Lewis B. Benson. United States of America Patent 4,976,937. 1990.

Maree, Yolandi, Sissel Nepstad, and Gelein de Koeijer. "Establishment of Knowledge base for Emission Regulation for the $CO_2$ Technology Centre Mongstad." *GHGT-11.* Kyoto, Japan, 2013.

Nash, T. "The Effect of Nitrogen Dioxide and of some Transition Metals on the Oxidation of Dilute Bisulphite Solutions." *Atmospheric Environment*, 1979: 1149-1154.

National Energy Technology Laboratory. *Cost and Performance Baseline for Fossil Energy Plants Volume 1: Bituminous Coal and Natural Gas to Electricity.* Report, Washington D.C.: U.S. Department of Energy, 2010.

Oblath, S.B., S.S. Markowitz, T. Novakov, and S.G. Chang. "Kinetics of the Formation of Hydroxylamine Disulfonate by Reaction of Nitrite with Sulfites." *The Journal of Physical Chemistry*, 1981: 1017-1021.

Owens, DR. *Sulfite Oxidation Inhibited by Thiosulfate.* Master's Thesis, Austin: The University of Texas at Austin, 1984.

Shen, Chen H. *Nitrogen Dioxide Absorption in Aqueous Sodium Sulfite.* PhD Dissertation, Austin: The University of Texas at Austin, 1997.

Srivastava, Ravi K., Robert E. Hall, Sikander Khan, Kevin Culligan, and Bruce W. Lani. "Nitrogen Oxides Emission Control Options for Coal-Fired Electric Utility Boilers." *Journal of the Air & Waste Management Association*, 2005: 1367-1388.

Susianto, Petrissans M., and A. Zoulalian. "Influence of the pH on the Interactions between Nitrite and Sulfite Ions. Kinetics of the Reaction at pH 4 and 5." *Industrial Enginerring and Chemistry Research* 40, No. 26 (2001): 6068-6072.

CARBON MANAGEMENT TECHNOLOGY CONFERENCE

# CMTC-486335-MS

# Miscible $CO_2$ Flooding Using Horizontal Multi-Fractured Wells in San Andres Formation, TX – a Feasibility Study

J. Yang, Y. Oruganti, and P. Karam, Baker Hughes, a GE Company; D. Doherty, J. Doherty, and J. Chrisman, Riley Exploration



This paper was prepared for presentation at the Carbon Management Technology Conference held in Houston, Texas, USA, 17-20 July 2017.



## Abstract

The San Andres is a well-known dolomitic enhanced oil recovery target with low matrix permeabilityin the area of interest (Yoakum County, TX). A reservoir simulation study was undertaken to investigate the feasibility of using horizontal multi-fractured wells in low permeability miscible floods. A reservoir model was developed for the area of interest and was history-matched with the primary production data from the field. The model was then used to illustrate the $CO_2$ miscible flood potential by quantifying the incremental recovery over the primary production scenario.
Compositional modeling was used in the study to evaluate $CO_2$ flooding feasibility and efficiency. A holistic workflow including PVT modeling, petrophysical analysis, geomodeling, and hydraulic fracture modeling, provided integrated input into the reservoir model. Continuous $CO_2$ flooding was explored as an operating strategy. Furthermore, water alternating gas (WAG) cases were designed and run as a more realistic and cost-effective method of implementing miscible flooding. Based on the history-matched model, sensitivity analyses were conducted on hydraulic fracture geometry, well spacing, injection patterns and operating conditions for the primary production scenario, continuous $CO_2$ flooding and WAG scenarios.

Field surveillance and observations during the history-matching process showed that the wells had undergone damage from scaling. Sensitivity analysis showed that 300ft to 400ft cluster spacing resulted in the highest oil production during the first 10 years. Interdependent parameters such as well spacing and fracture half-length were studied together; this sensitivity review showed that the differential oil recovery from 128 acres to 160 acres was larger than that from 160 acres to 213 acres, leading to the recommendation that 160 acres could be the optimized well spacing. In the optimized design, the continuous $CO_2$ injection case showed an incremental oil recovery of 22% (compared to primary production). The $CO_2$ utilization factor was between 7 and 8, which was consistent with the reported value from literature. WAG sensitivity analysis showed that longer hydraulic fractures did not necessarily improve WAG efficiency, but led to earlier $CO_2$ breakthrough. This observation confirmed our early suspicion that smaller hydraulic fracturing treatment could be a more cost-effective design for miscible flooding in this reservoir. In addition, sweep efficiency and recovery were sensitive to WAG ratio, but not to injection slug size in each cycle.

The current study sheds light on the feasibility of conducting a $CO_2$ miscible flood using horizontal multi-fractured wells in low permeability reservoirs – a topic that is yet to be explored widely in petroleum engineering literature and in the industry. Incremental production that can be expected from a miscible $CO_2$ flood is estimated and recommendations are provided for optimal well spacing, WAG ratio and operating constraints to help determine a viable field development plan.

## Introduction

San Andres formation is a carbonate reservoir in west Texas, and is well known as a $CO_2$ flooding target. Aside from the highly heterogenous dolomitic reservoir quality, other challenges associated with the formation include high residual oil saturation, and high water cut from primary production. Multiple publications defined this formation as a naturally water flooded residual oil zone (ROZ) (Melzer, et al., 2006; Koperna, et al., 2006; Honarpour, et al., 2010; Harouaka, et al., 2013). San Andres in the area of study, however, is believed to be the main pay zone (MPZ), and the reason is two-fold. First, due to the waterflood-like nature of the ROZ, the produced water from the San Andres ROZ typically has low TDS (10,000-50,000ppm) as it is diluted by the meteoric water recharge (Trentham, 2011), while our field data shows much higher TDS (180,000-200,000ppm). Second, oil saturation in both Chambliss and Brahaney formations is at least 50% for most of the pay zone, which is higher than what is typically observed in ROZ. (Fig 1). Rather than being a natural water flooded ROZ, it is possible that the area of interest was originally wet that was most likely partially filled from oil spillover from Wasson and Brahaney fields when the Laramide Uplift to the west/northwest caused tilting of these fields, resulting in spilling and trapping of oil due to the stratigraphic pinching out of the San Andres to the west/northwest of Wasson. Petrophysical logs in the area of interest typically show significant oil saturation, but primary production often yields higher than normal water cut. One explanation could be the extremely heterogeneous porosity distribution, with oil being trapped in the poorly connected pores (Cannon and Rossmiller, 1984). Another theory is the mixed wettability (Patel, et al., 1987; Honarpour, et al., 2010). At early time, wells produce from water-wet fractures and vuggy porosity, and later on, oil-wet matrix porosity starts contributing to the production. This is consistent with what is observed in the field, in that wells produce higher water cut initiallythat gradually decreases with time.

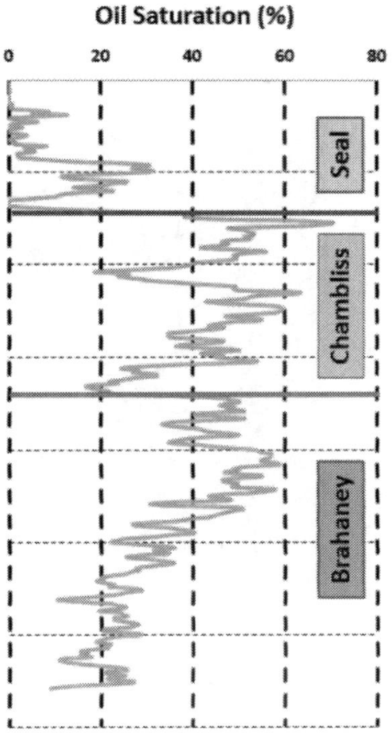


**Figure 1. Oil saturation profile in San Andres Formation. The main pay zone consists of Chambliss and Brahaney formations**

$CO_2$ miscible flooding is recognized as a possible strategy to effectively produce from the San Andres formation. The application of horizontal wells with hydraulic fractures in miscible flooding is yet to be fully understood. Numerous researchers reported results from their $CO_2$ miscible flooding simulation studies in Slaughter field dolomite (Guillot, 1995), west Texas carbonate (Lim et al., 1992; Lim et al.,

1996), Prodhoe Bay sandstone (McGuire, et al., 1998), Bakken (Xu and Hoffman, 2013), but few have considered miscible flooding using multi-fractured horizontal wells. The objective of this study is to investigate the feasibility of $CO_2$ flooding using multi-fractured horizontal wells, and to optimize fracture design, well spacing, and estimate the hydrocarbon recovery from various field development strategies, such as continuous $CO_2$ flooding and Water Alternating Gas (WAG) processes.

The area of study is 1x1 sq. mile acerage in Yoakum County, Texas, with two producing horizontal wells at the time of the study, 1H and 4H. Due to the low permeability, all wells were hydraulically fractured with 120 ft cluster spacing to improve the productivity. About one year of historical production data is available from each well. In vertical direction, the main reservoir formations are the Chambliss and Brahaney dolomite, with an anhydrite sealing layer at the top and a water bearing layer below.

## Model Development

A dynamic reservoir model was built from the upscaled geomodel. The workflow of the geomodeling was reported earlier (An et. al. 2017) indicating the variation of facies within the reservoir. The model for history matching covers an area of interest of about 1.5x1.5 $mi^2$ with cell size of 50ft x100ft in the horizontal planes to accommodate the hydraulic fractues (Fig 2). In the vertical direction, the model incorporated 40 layers including one for the sealing layer on top, 11 for Chambliss, 18 for Brahaney, and 10 for the bottom water bearing zone (Fig 3). Hydraulic fracture geometry was determined by numerical modeling with history matched treatment pressure, details of which are not included in this paper. Local grid refinement (LGR) was applied in the near-fracture zone to capture accurately the pressure transient and fluid flow (Fig 4). As the level of refinement could numerically affect the simulation result, LGR used in all simulation runs remain the same regardless of the HF geometry.

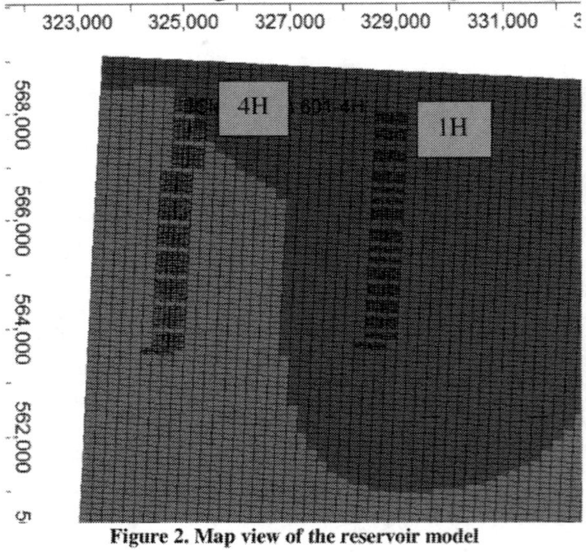


**Figure 2. Map view of the reservoir model**

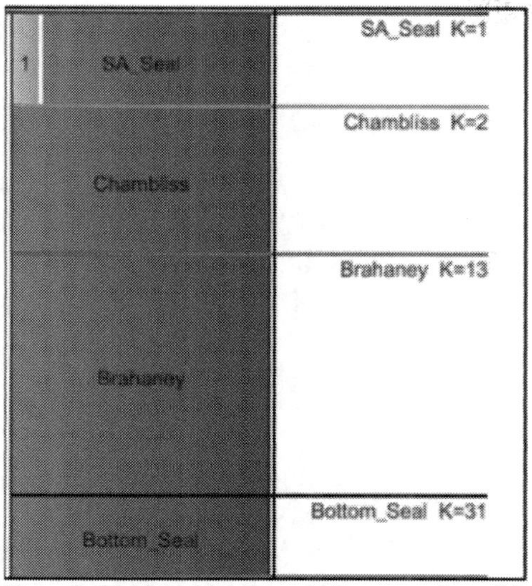

**Figure 3. Stratigraphic column in the reservoir model**

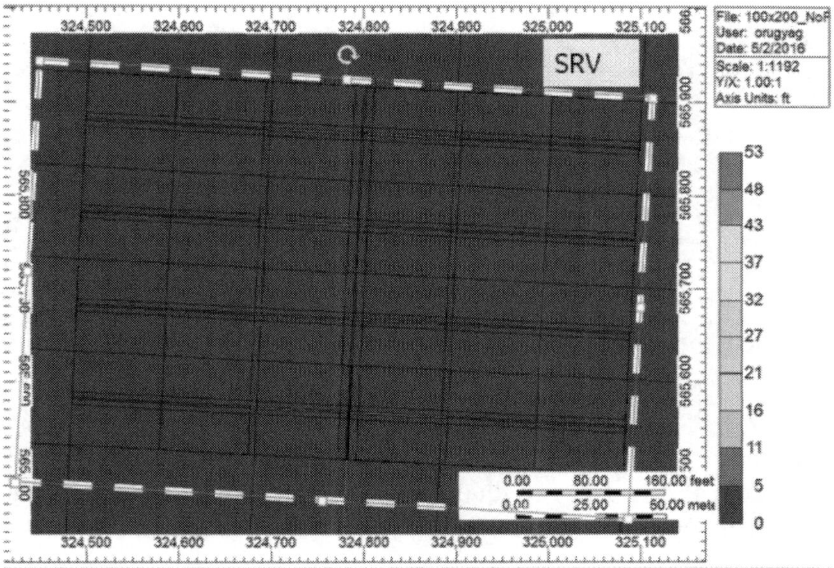

**Figure 4. Zoomed-in map view to illustrate LGR and SRV. Legend displays permeability in mD**

PVT lab test including Constant Composition Expansion (CCE), Differential Liberation (DL), and separator test was available from San Andres formation in an anolog field. PVT modeling was conducted by matching the lab test data. The outcome of this exercise is the compositional PVT model with eight lumped components. $CO_2$ and the last pseudocomponent were kept as separate components to enhance the modeling accuracy. DL data was adjusted to the separator condition before being fed into the model (Al-Marhoun, 2003).

In general, dolomite/dolostone in the Permian Basin has been widely recognized as a highly heterogenous reservoir with multi-porosity system. Additionally, natural fractures could also contribute to fluid flow in addition to matrix (Mathis and Sears 1984; Quijada, 2005; Mohamed, et al., 2012). In our study, however, neither the borehole imaging nor core data indicated pronounced evidences of natural fracturing (An et al., 2017). Therefore, a single porosity model was used in the simulation. However, to account for the possibility of other types of secondary porosity, the model consists of a near-fracture zone referred to as stimulated reservoir volume (SRV), represented by an enhanced permeability that was fine-tuned during the history matching process. The permeability in the fracture and SRV is believed to be affected by multiple mechanisms such as in-situ stress change due to depletion, proppant crashing, proppant embedment, and clay swelling (Han, el al., 2015). The model was simulated using GEM, which is CMG's finite difference compositional simulator.

Reservoir and fluid properties listed in Table 1 show a summary of the reservoir input parameters. According to historical GOR data, the reservoir was believed to be undersaturated at the time of the study, and the minimal miscible pressure (MMP) is assumed to be the same as saturation pressure. Based on the water saturation distribution from the geomodel, capillery pressure profile was computed to account for the initial equillirium in the reservoir.

**Table 1. Reservoir and fluid properties**

| Pressure (psi) | Temperature (°F) | Permeability (mD) | Porosity |
|---|---|---|---|
| 1,800-2,000 | 130 | 0.2 | 0.08 |
| Depth (ft) | MMP = $P_{sat}$ (psi) | API (°) | Initial GOR (SCF/STB) |
| 5,200+ | 1,500 | 31 | 800 -1,000 |

## History Matching and Forecasting

At the time of the study, the area of interest was under primary production with two producing horizontal wells. Based on the upscaled geomodel, some reservoir and fluid properties, including matrix permeability, SRV permeability, fracture conductivity, relative permeability (curvature and end points), and skin are considered as parameters with high uncertainty and modified to match the production data. The operator experienced inorganic scaling issue from both wells, and that is why skin is considered responsible to productivity reduction.

With the consideration of all the mechanisms mentioned above, the simulation yielded a very good history match (Fig. 5 and 6). With liquid rate serving as the constraint, GOR, BHP, oil rate, water rate, and water cut matches were evaluated. Both GOR and water cut show a good match on the overall trend but missed some details in the early stages. Once a satisfying history match was achieved, the model was run under a constant BHP for 30 years to forecast the primary recovery. (Fig 7 and 8)

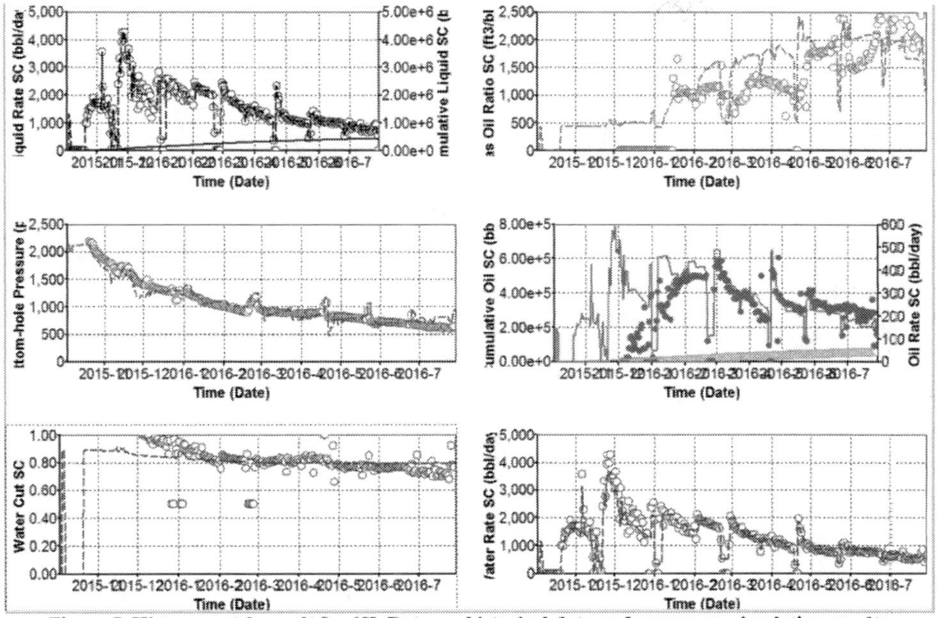

**Figure 5. History match result for 1H. Dots are historical data and curves are simulation results. From left to right and top to bottom, the plots are liquid rate, GOR, bottom-hole pressure, oil rate, water cut, and water rate. The same is applicable for Fig 6-8**

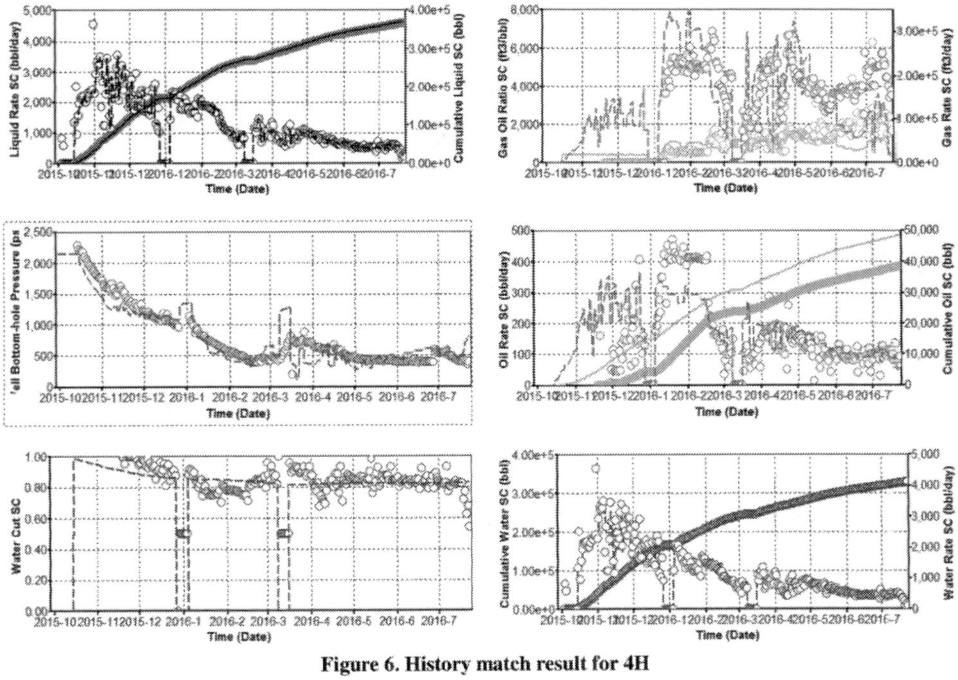

**Figure 6. History match result for 4H**

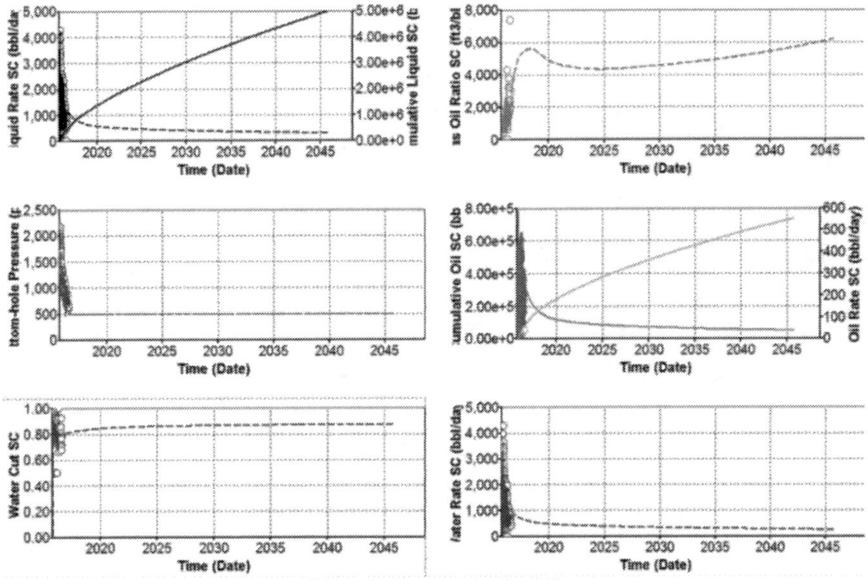


**Figure 7. Forecasting for 1H**

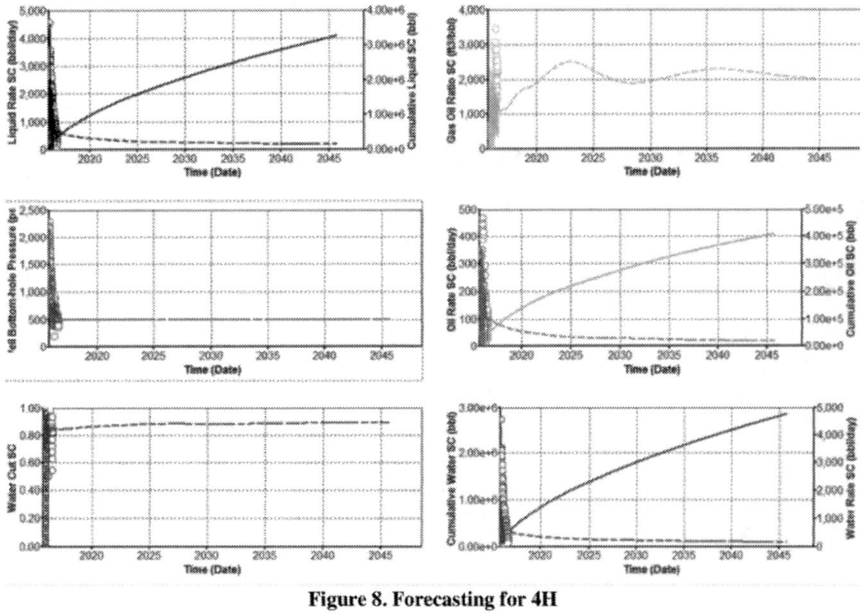


**Figure 8. Forecasting for 4H**

## Sensitivity Analysis

One of the objectives of the simulation study is to investigate the field development strategy. Optimized well spacing is desired to effectively drain the reservoir with good sweep efficiency and without inter-well interference. Additionally, cluster spacing or hydraulic fracture spacing optimization is considered to balance the productivity and operational investment. For the current $CO_2$ flooding model, traditional massive fracturing treatments for primary production purpose may not be feasible, as fractures act as highly permeable conduits which could result in early $CO_2$ breakthrough and low sweep efficiency. Therefore, analyzing these parameters is important for a better understanding of the optimized strategy.

### Cluster Spacing

Seven scenarios of cluster spacing, varying from 100 ft to 700 ft with 100 ft increments, were tested in the study. The following assumptions were made to simplify the study.

- Green field reservoir
- Single well model with 160-acre spacing
- BHP follows the decline behavior of 1H for the first year and remains constant at 620 psi for the rest of the simulation
- Identical local grid refinement for the seven cases
- Surface fluid rate is capped to 3000 bbl/day due to facility constraints

As illustrated in Fig. 9, where the field cumulative oil production is plotted against cluster spacing, curves represent simulation results from different time steps. In the early stages, particularly for the first 2 years, tighter cluster spacing yielded higher cumulative oil, indicating negligible inter-fracture interference. However, at later stages, wider cluster spacings start outperforming the closer spacing. But too wide a cluster spacing leads to lower cumulative production (500ft, 600ft, and 700ft cases). This suggests that the optimal cluster spacing occurs at ~300ft-400ft with higher production by the 10$^{th}$ year. Therefore, from a longer term perspective, 300 or 400 ft cluster spacing could be a better design, as the interference between fractures is not significant enough to hinder production.

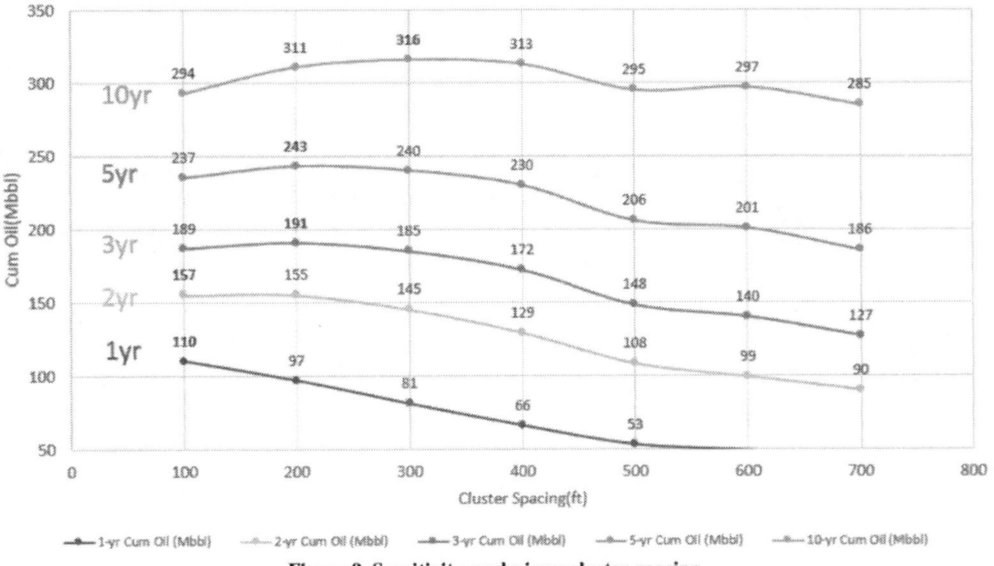


**Figure 9. Sensitivity analysis on cluster spacing**

In Fig 10, cumulative production per fracture, or fracture production efficiency is being considered. Ideally, the optimized case should have both high overall cumulative production and fracture efficiency.

Economic modeling (not shown here) suggests that 400 ft cluster spacing is the optimized design for this particular study.

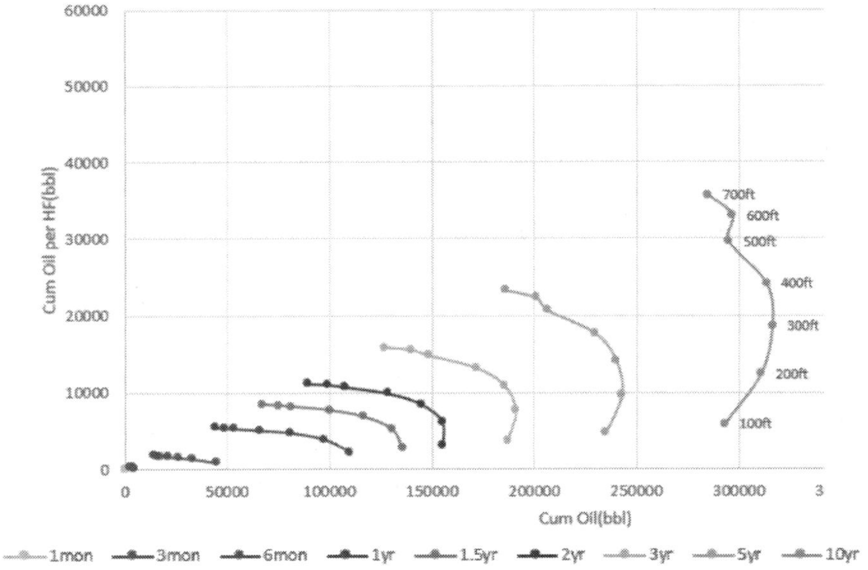


**Figure 10.Sensitivity analysis on cluster spacing, considering both cum production and HF efficiency**

## Well Spacing and Fracture Half Length

Three well spacing cases, 213 acre (3 wells), 160 acre (4 wells), and 128 acre (5 wells), are considered in the study (Fig 11). It is noted that even after fracture modeling, fracture half length ($x_f$) still has high uncertainty among the fracture geometry parameters, due to the limited data available away from the wellbore. In reality, engineers have little control on the $x_f$ from an operational point of view. As inter-dependent parameters, well spacing is coupled with half length in the sensitivity analysis. The assumptions are listed below.

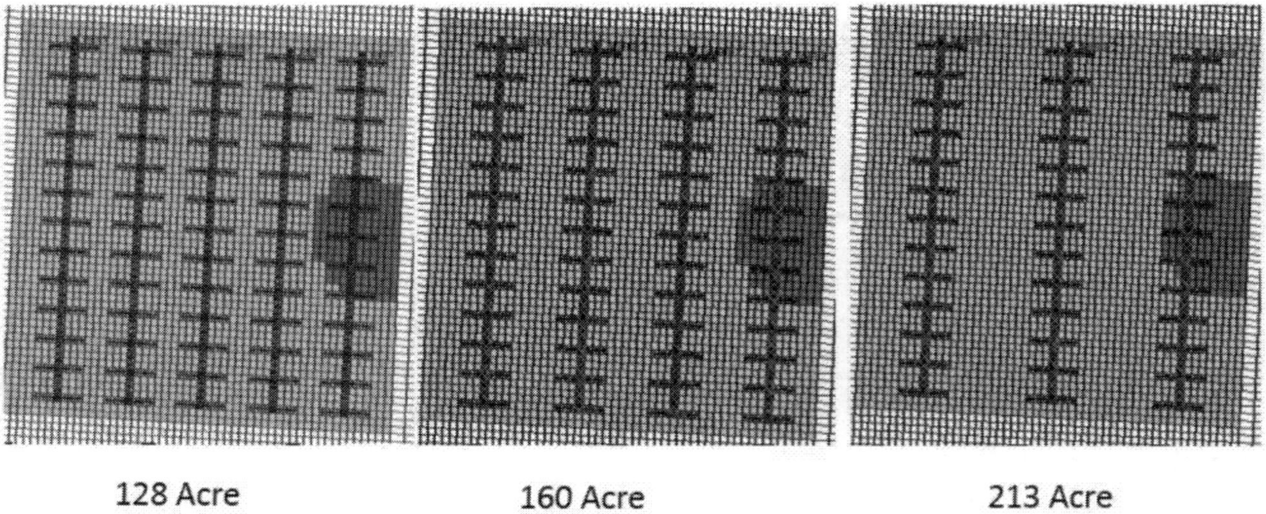


**Figure 11. Model setup for well spacing and half length sensitivity analyses**

- Green field reservoir
- 400ft cluster spacing

- BHP follows the decline behavior of 1H for the first year and remains constant at 620 psi for the rest of the simulation
- Fluid rate is capped to 3,000 bbl/day
- Three half length values are considered, 150ft, 250ft, 350ft. In total, there are nine simulation cases.

The simulation results for different well-spacing and hydraulic fracture half length are shown in Fig 12, 13 and 14. Focusing on later times, i.e., at the end of 10, 20, and 30 years, incremental production from 3-well spacing to 4-well spacing is higher than that from 4-well to 5-well scenario, regardless of the fracture half length. In other words, the production benefit from the 4th well is higher than that from the 5th well, which leads to the conclusion that solely based on the production, 4 wells per section, or the 160 acre spacing case is a better design. The observation was later confirmed by the economic modeling.

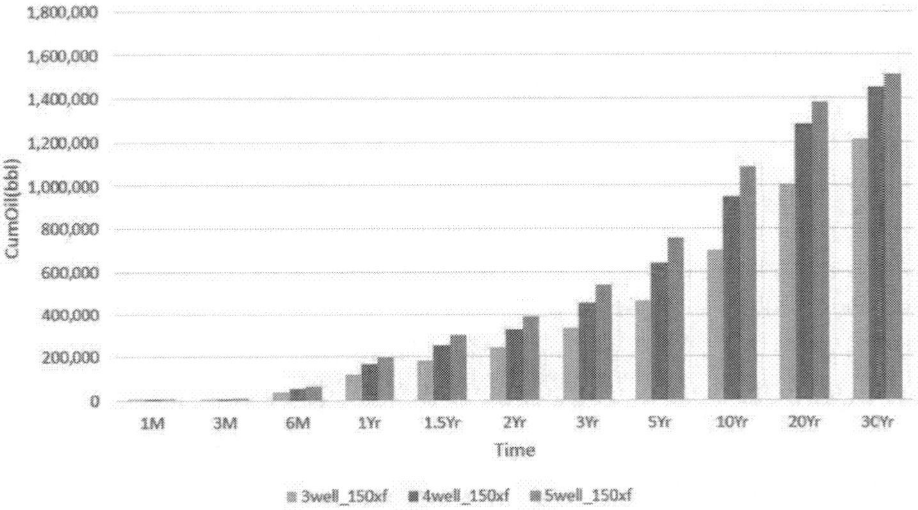


**Figure 12. Sensivitity analysis on well spacing for $x_f$ = 150 ft**

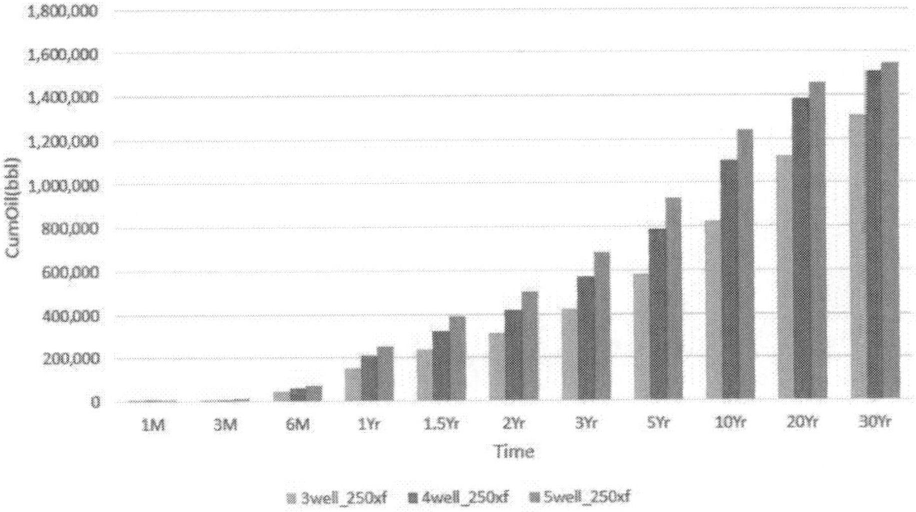


**Figure 13. Sensivitity analysis on well spcacing for $x_f$ = 250 ft**

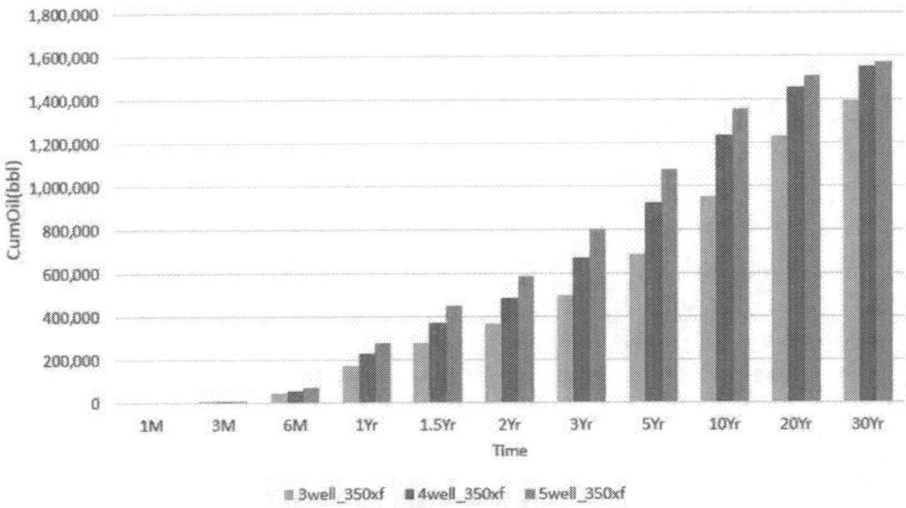


**Figure 14. Sensivity analysis on well spacing for $x_f$ = 350 ft**

## Continuous $CO_2$ Injection

Continuous $CO_2$ flooding feasibility analysis was conducted after a good understanding of primary production. A series of vertical injectors is placed in between horizontal producers as an analog of line drive flooding pattern. Because of symmetry, the model is further simplified as a single well model as shown in Fig 15. The simulation model consists of multiple vertical injectors, and two horizontal wells to the sides with one wing hydraulic fractures. To ensure that the $1/3^{rd}$ partial model is representative of the full model, a quick validation run to compare both models was made, as shown in Fig 16. The simulation result from the $1/3^{rd}$ model is multiplied by three for both injected gas and produced oil to obtain the equivalent results from the full model. As the results from both models agree with each other, the partial model is used in subsequent studies.

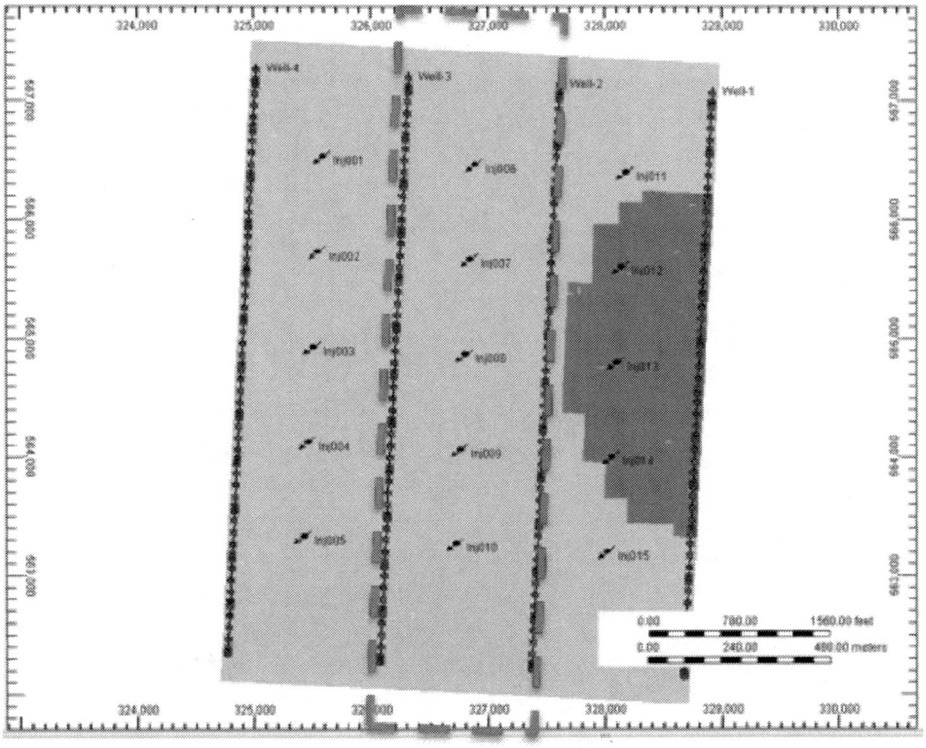

**Figure 15. Map view of the model with injectors and producers. Red box indicates the partial model**

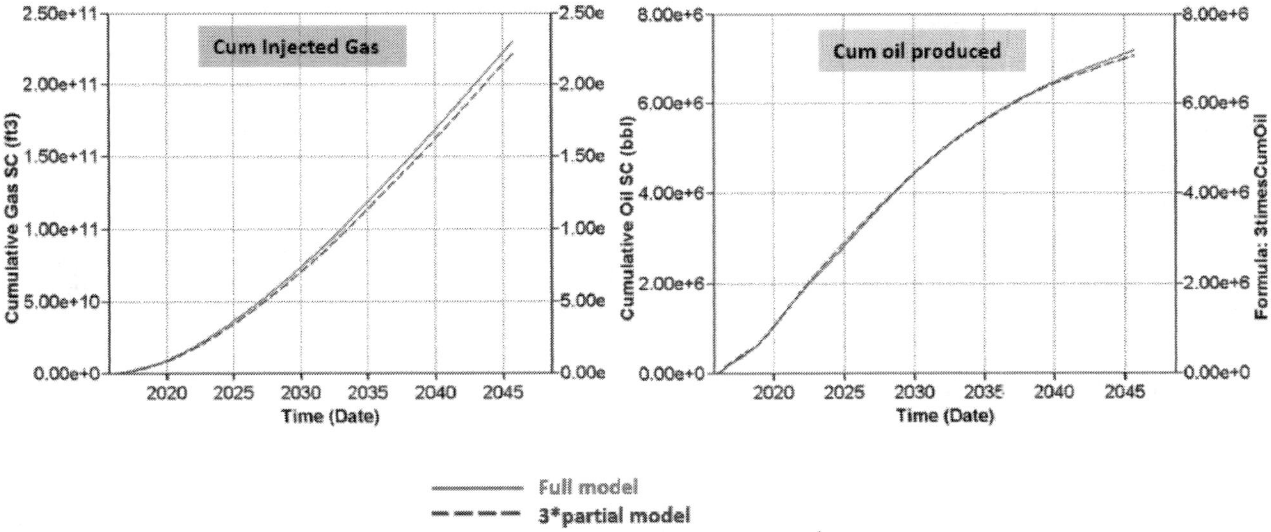


**Figure 16. Validation runs to compare the full model and the 1/3rd partial model**

In such tight reservoirs, a major challenge to $CO_2$ flooding is the injectivity. Feasibility test shows that injectors with no stimulation could barely have reasonable amount of injection. A set of sensitivity analysis were used to figure out the stimulation strategy for the injectors, as shown in Table 2. Original perforation design only covers the Brahaney interval to minimize $CO_2$ gravity override. Additionally, the study considers a longer perforation interval scenario that extends to the top of the Chambliss. Negative skin factor represents acidizing treatment near the wellbore. Single stage hydraulic fracturing of the vertical injectors is another alternative to increase the injectivity.

**Table 2. Sensitivities to Completion Design**

| Case | #injector | Skin | Perf'd K layers | HF (Y/N)? | Recovery (%) | HCPVI | Cum $CO_2$ injected (BCF) | $CO_2$ utilization factor (MCF/BBL) |
|---|---|---|---|---|---|---|---|---|
| Primary depletion | 0 | - | - | N | 2.8 | -- | -- | -- |
| NoSkin_noHF_ShorterPerfInt | 5 | 0 | 15-30 (Brahaney) | N | 11 | 0.81 | 28.9 | 7.38 |
| NoSkin_noHF_LongerPerfInt | 5 | 0 | 2-33 (Chambliss, Brahaney) | N | 13 | 0.91 | 34.1 | 7.77 |
| wSkin_noHF_Shorter PerfInt | 5 | -3 | 15-30 (Brahaney) | N | 14 | 1.18 | 46.5 | 7.51 |
| NoSkin_wHF_ShorterPerfInt | 5 | 0 | 15-30 (Brahaney) | Y | 17 | 1.76 | 61.6 | 7.63 |
| NoSkin_wHF_Longer PerfInt | 5 | 0 | 2-33 (Chambliss, Brahaney) | Y | 21 | 1.92 | 81.3 | 7.36 |

The simulations were run for 30 years and the result is shown in Fig 17. While the primary depletion scenario yields less than 3% recovery, the cases with $CO_2$ injection yield 11% to 21% oil recovery. The best case scenario in terms of the highest recovery and injectivity, is the case with hydraulically fractured injectors with long perforation intervals. $CO_2$ utilization factor is an indicator of $CO_2$ flooding efficiency, and is defined as follows:

$$CO_2\ utilization\ factor = \frac{Cum\ CO_2\ injected - Cum\ CO_2\ produced}{Incremental\ oil\ production\ over\ primary\ depletion}$$

where all the volumes in the equation are at reservoir conditions. The lower the $CO_2$ utilization factor, the higher the flooding efficiency.

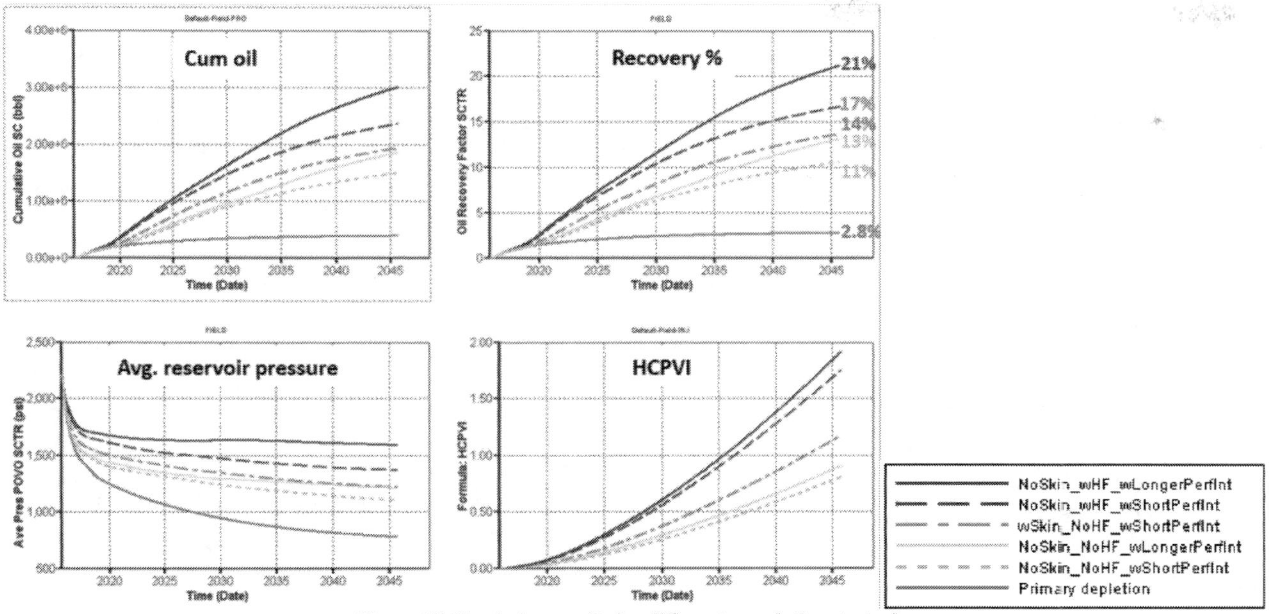


**Figure 17. Simulation results for different completion strategies**

Additionally, as illustrated in the 2D side view (Fig 18), significant $CO_2$ override is not observed until 10 years due to the low permeability. To maximize the injectivity, it is beneficial to have injectors with extended perforation length that are hydraulically fractured.

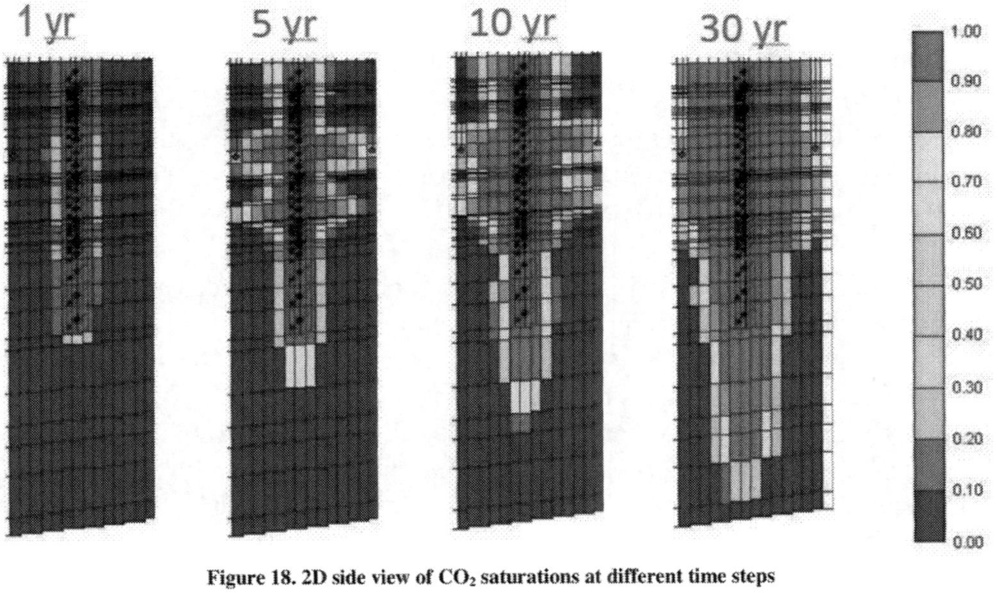


**Figure 18. 2D side view of $CO_2$ saturations at different time steps**

In addition, the number of injectors in the center of the model was also investigated. Five scenarios from 3 to 7 injectors were tested to evaluate the recovery and efficiency of the flooding. The results are in Table 3. The 7-injector scenario has both the highest recovery and efficiency, and this case was subsequently used as the base case in the following studies.

Table 3. Sensitivity to the number of injectors

| Case | Cum Oil (MMSTB) | Recovery (%) | Cum $CO_2$ inj (BCF) | HCPVI | $CO_2$ utilization factor (MCF/BBL) |
|---|---|---|---|---|---|
| 3 injectors | 2.26 | 15.93 | 57 | 1.39 | 7.89 |
| 4 injectors | 2.79 | 19.66 | 71 | 1.73 | 7.56 |
| 5 injectors | 3.01 | 21.22 | 81 | 1.91 | 7.36 |
| 6 injectors | 3.47 | 24.46 | 92 | 2.25 | 7.21 |
| 7 injectors | 3.81 | 26.85 | 99 | 2.41 | 7.07 |

## WAG Design

Despite the high recovery, continuous $CO_2$ injection has limited field application due to the constraints on $CO_2$ resource and high cost. Alternatively, WAG is a more cost-effective technique. To address the uncertainty in the fracture geometry, two cases with different fracture half lengths were run. The design is shown as follows.

- $x_f$ = 150ft and 250ft
- 7 vertical injectors
- WAG cycle: 180 days of $CO_2$ injection followed by 180 days of water injection
- Injection constraint: max. BHP = 4,000 psi (fracture pressure = 4,500 psi)
- Production constraint: min. BHP = 620 psi

As shown in the set up of the model (Fig. 19), the injectors are staggered with the cluster locations to minimize early breakthrough. While the primary production scenario yields less than 3% recovery, both WAG scenarios reach more than 25% recovery in 30 years (Fig. 20). The Recovery vs HCPVI plot illustrates the $CO_2$ flooding efficiency. The curves of $x_f$ = 250ft and $x_f$ = 150ft on the Recovery vs. HCPVI plot are fairly close to each other. It is obvious that the $CO_2$ flooding efficiency is not very sensivite to the fracture half length. However, longer fracturesresulted in early $CO_2$ breakthrough. For these particular cases, $CO_2$ mole fraction from producers starts ramping up in early 2017 for xf = 250ft, while the ramp-up for the xf = 150ft case happens roughly one year later than the xf = 250ft case (Fig. 21). To avoid early breakthrough, small fracturing treatment volume, in terms of proppant and frac fluid, is recommended.

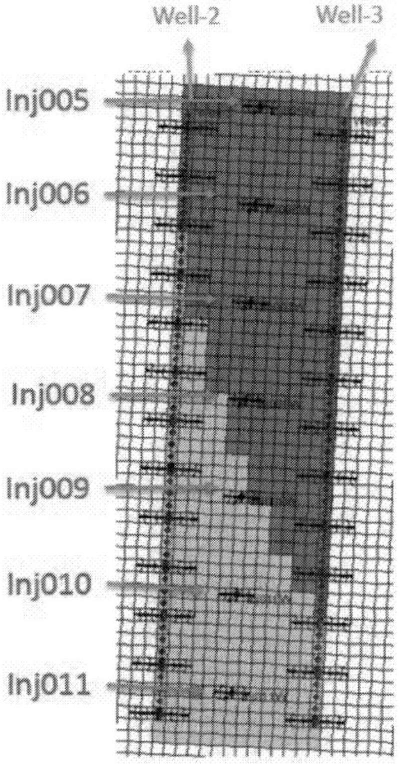


**Figure 19. Model setup for WAG**

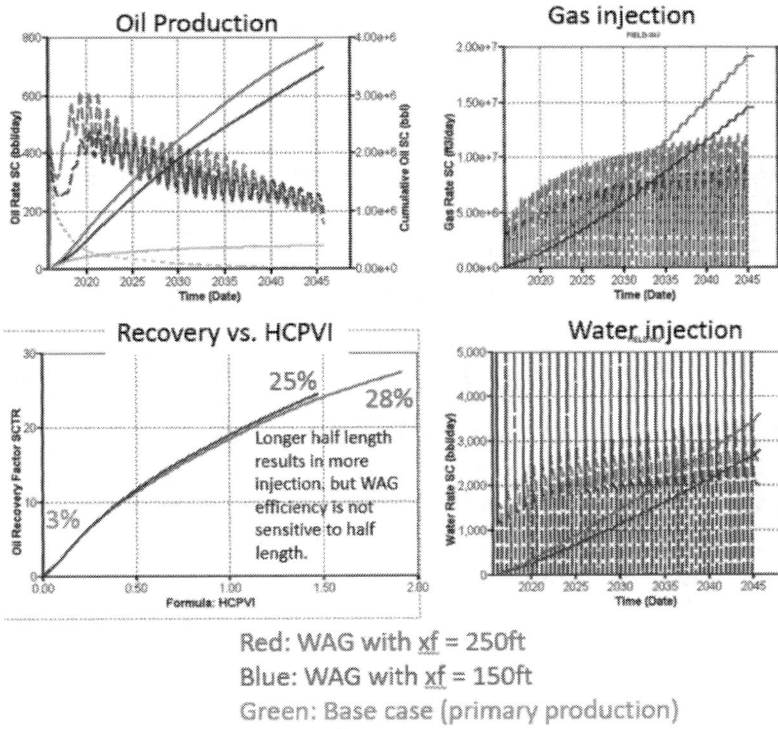


**Figure 20. 1/3$^{rd}$ model WAG simulation results**

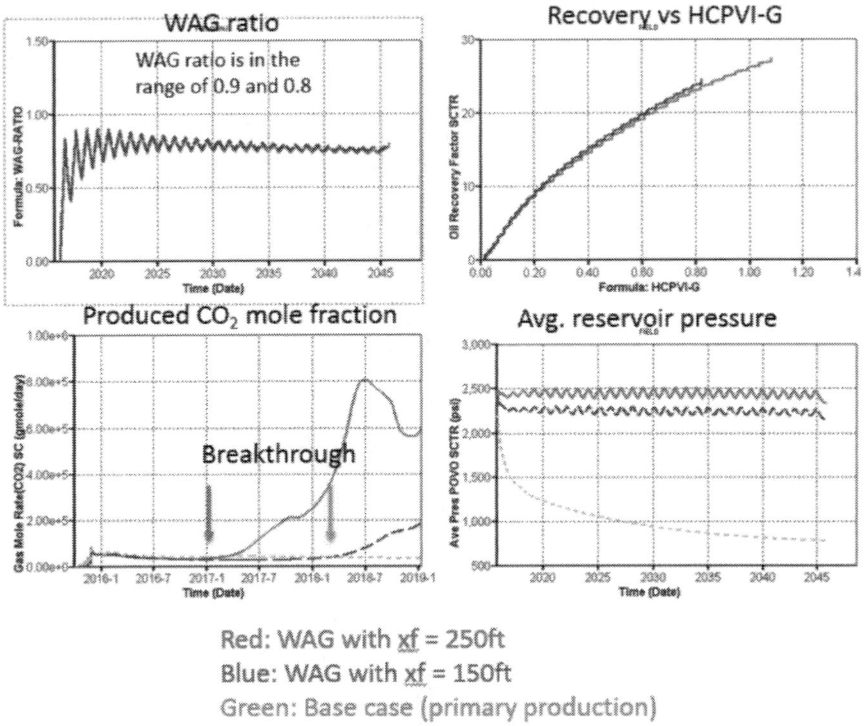


**Figure 21. 1/3rd model WAG simulation results showing $CO_2$ breakthrough**

## Conclusion

The San Andres formation is a tight dolomitic reservoir with mixed wet matrix. High residual oil saturation is responsible for the high water cut from primary production, despite the large amount of immobile oil trapped underground. $CO_2$ miscible flooding is proved to be an effective method to enhance the oil recovery ecomically. Due to the low permeability of the matrix, horizontal wells with multi-fractures show high potential to significantly increase the oil recovery. As one of the pioneering investigations on this topic, this simulation study shows that smaller hydraulic fracture treatment volumes with larger cluster spacing than is traditionally observed in the field, and shorter half lengths could be beneficial in avoiding early $CO_2$ breakthrough while maintaining high flooding efficiency. The $CO_2$ flooding efficiency in the cases examined is not very sensitive to fracture half length. The study also demonstrates that gravity override is not pronounced in a tight reservoir such as the San Andres formation. It is encouraging that miscible $CO_2$ flooding is able to produce more than 20% incremental oil recovery, with $CO_2$ utilization factors between 7 and 8 MCF/BBL. This integrated study demonstrates that not only is it feasible to implement miscible $CO_2$ flooding in tight oil reservoirs using horizontal multi-fractured wells, but it can also improve recovery significantly compared to primary depletion. There are few simulation studies in literature that discuss miscible $CO_2$ flooding in tight reservoirs, and therefore, the findings from the current study have enormous implications in the industry for EOR applications in low permeability reservoirs.

## Acknowledgments

We thank Baker Hughes, a GE Company and Riley Exploration management teams for permission to publish this paper.

## References


Al-Marhoun, M.A. *Adjustment of Differential Liberation Data to Separator Conditions*, SPE 68234 SPE Reservoir Evaluation & Engineering, Vol 6, pp142-146, 2003

An, X., Karam, P., Morris, S., Doherty, J. Doherty, D., *Facies Modeling of Complex Carbonate Platform Architecture Identified from Borehole Image Log and Other Logging Technologies*, SPWLA Annual Logging Symposium Jun 17-21, 2017

Cannon, D.E. and Rossmiller J. W. *Oil Saturation Evaluation for EOR in a Carbonate*, SPE 13288 Annual Technical Conference and Exhibition, Sep 16-19, 1984

Guillot, S.N. *Horizontal Well Applications in a Miscible $CO_2$ Flooding, Sundown Slaughter Unit, Hockley County, Texas.* SPE 30742 Annual Technical Conference and Exhibition, Oct 22-26, 1996

Han, J., Hurt, R., and Sookprasong, A. *Stress Field Change Due to Reservior Depletion and Its Impact on Refrac Treatment Design and SRV in Unconventional Reservoirs*, SPE 178496 Unconventional Resources Technology Conference, Jul 20-22, 2015

Harouaka, A., Trentham, B., and Melzer, S. *Long Overlooked Residual Oil Zones (ROZ's) Are Brought to the Limelight*, SPE 167209 Unconventional Resources Conference-Canada. Nov 5-7, 2013

Honarpour, M.M., Nagarajan, N.R., Grijalba, A.C., Valle, M., and Adesoye, K. *Rock-Fluid Characterization for Miscible $CO_2$ Injection: Residual Oil Zone, Seminole Field, Permian Basin*, SPE 133089 Annual Technical Conference and Exhibition, Sep 19-22, 2010

Koperna, G. J., Melzer, L. S., and Kuuskraa, V. A. *Recovery of Oil Resourcs From the Residual and Transitional Oil Zones of the Permian Basin*, SPE 102972 Annual Technical Conference and Exhibition, Sep 24-27, 2006

Lim, M.T., Khan, S.A., Sepehrnoori, K., and Pope, G.A. *Simulation of Carbon Dioxide Flooding Using Horizontal Wells*, SPE 24929 Annual Technical Conference and Exhibition, Oct 4-7, 1992

Lim, M.T., Pope, G.A., and Sepehrnoori, K. *Mechanistic Study of Carbon Dioxide Flooding Using Horizontal Wells*, The Journal of Canadian Petroleum Technology, Vol 35, pp16-24, 1996

Mathis, R.L., and Sears, S.O. *Effect of $CO_2$ Flooding on Dolomite Reservoir Rock, Denvor Unit, Wasson (San Andres) Field*, TX, SPE 13132 Annual Technical Conference and Exhibition, Sep 16-19, 1984

McGuire, P.L., Redman, R.S., Mathews, W.L., and Carhart, S.R. *Unconventional Miscible EOR Experience at Prudhoe Bay*, SPE 39628 Improved Oil Recovery Symposium, Apr 19-22, 1998

Melzer, L. S., Koperna, G. J., and Kuuskraa, V. A. *The Origin and Resource Potential of Residual Oil Zones*, SPE 102964 Annual Technical Conference and Exhibition, Sep 24-27, 2006

Mohamed, I.M., He, J., and Nasr-El-Din, H.A. *Carbon Dioxide Sequestration in Dolomite Rock*, IPTC 14924 International Petroleum Technology Conference, Feb 7-9 2012

Patel, P. D., Chrstman, P.G., and Gardner J.W. *Investigation of Unexpectedly Low Field-Observed Fluid Mobilities During Some $CO_2$ Tertiary Floods*, SPE Reservoir Engineering, Vol 2, pp 507-513, 1987

Quijada, M.G. *Optimization of A $CO_2$ Flooding Design Wasson Field – West Texas*, M.S. Thesis, Texas A&M University, 2005

Trentham, B. *Residual Oil Zone: The Long term Future of Enhanced Oil Recovery in the Permian Basin and Elsewhere*, AAPG Southwest Section meeting, Jun5-7, 2011

Xu, T., and Hoffman, T. *Hydraulic Fracture rientation for Miscible Gas Injection EOR in Unconventional Oil Reservoirs*, SPE 168774, Unconventional Resources Technology Conference, Aug 12-14, 2013

CARBON MANAGEMENT TECHNOLOGY CONFERENCE

# CMTC-486365-MS

# Carbon Dioxide in Situ Generation for Enhanced Oil Recovery

Shuoshi Wang, University of Oklahoma; Mohannad Kadhum, Cargill; Qingwang Yuan, University of Regina; Bor-Jier (Ben) Shiau, University of Oklahoma; Jeffrey H. Harwell, University of Oklahoma



This paper was prepared for presentation at the Carbon Management Technology Conference held in Houston, Texas, USA, 17-20 July 2017.



## Abstract

Carbon dioxide flooding of oil fields around the world is proven as a successfully adopted practice in increasing oil production particularly in marginal wells with low production rates. However, the limitations of this technology lie in the limited supply of carbon dioxide, high capital cost, and infrastructure corrosion. In this work, we present an alternative CO2 flooding method which generates CO2 inside the reservoir to increase oil recovery. The process involves the injection of a concentrated CO2 producing solution of ammonium carbamate (AC). Chemical solvent CO2 capture technology was widely used for years. Carbamates were formed when aqueous amines absorbed CO2. The new proposed in situ CO2 generation EOR technique provides a way to directly apply the product of the CO2 capture technology for outstanding tertiary recovery.

Ammonium carbamate (CH6N2O2), highly water-soluble chemicals, can dissociate at reservoir temperature producing carbon dioxide and ammonia. The carbon dioxide migrates to the oil phase, causing oil phase swelling and reducing oil viscosity, and therefore increasing oil production. The ammonia dissolves in the water, and the ammonia-water solution increases the water wettability of the rock.

Flow experiments were conducted using 6" Ottawa sand packs. The experiments demonstrated that the decomposition of a 35% AC solution injected to the sand packs resulted in further lowering of the residual oil saturation following a standard water flood. The tertiary recovery in the high-pressure sand pack experiments was found to average 27%.

In the proposed process, AC can be dissolved in produced reservoir fluids or seawater and injected into the reservoir to generate CO2 in situ and increase oil production as it decomposes. The benefits of this process compared to CO2 flooding lie in the simplicity of adapting this technology to an existing waterflood, and the lack of the complicated infrastructure needed in a typical CO2 project, such as compression and gas handling facilities. An additional advantage lies in the ability to deliver the CO2 in the form of a room temperature solid, alleviating the need for a pipeline. In situ CO2 generation by injecting CO2 producers has the potential to bring many inshore and offshore oil fields into the economic production landscape by making them amenable to CO2 flooding without needing to be near a traditional source of CO2.

## Introduction

To address excessive conformance control issues of Water Alternating CO2(Han and Gu 2014, Lei et al. 2016) and CO2 solubility limitation of Carbonated Water Injection (CWI)(Dong et al. 2011, Mosavat and Torabi 2014, Fathollahi and Rostami 2015), i.e., injecting water with dissolved CO2, In situ CO2 generation EOR method(Gumersky et al. 2000, Bakhtiyarov 2008, Li et al. 2013, Wang, Hou, and Tang 2016, Abdelgawad and Mahmoud 2015, Chen et al. 2016) was proposed as an alternative approach in the past few years.

Among these efforts, to achieve reasonable oil recovery, quite a few rather complex formulations were used, involving a combination of elevated concentrations of surfactant, polymer, plus additives like alkali or acids. Our target is to develop a new in situ CO2 generation approach. The potential benefits of this improved formulation may include:

I. Not relying on the natural CO2 sources and installation of CO2 transportation pipeline.
II. Better sweep efficiency than CO2 WAG process.
III. Several folds increase of GWR (Gas Water Ratio) comparing to CWI.
IV. Simple and cheap (directly link CO2 capture and CO2 EOR).
V. Desirable tertiary recovery performance at both above and below minimum miscibility pressure conditions.

CO2 capture applications by using Amine scrubbing technique were studied for years. (Bonenfant, Mimeault, and Hausler 2003, McCann, Maeder, and Attalla 2008, Rochelle 2009). Aqueous amines solution could absorb CO2 from flue gas and form carbamates by the following reaction(Khatri et al. 2006):

$$2R_1R_2 + CO_2 \leftrightarrow R_1R_2NH_2^+ + R_1R_2NCOO^- \cdots\cdots\cdots\cdots (1)$$

At elevated temperature condition, the carbamates could decompose and release the absorbed CO2 along with the ammonia. In the meantime, the benefits of CO2 and ammonia(van den Pol et al. 2014) on enhanced oil recovery were extensively proved. Therefore, delivering of these EOR chemicals could be done by injecting carbamate solution. To be able to do lab scale experiments, ammonium carbamate was chosen to be the simplest representative of the solute compounds in typical CO2 captured carbamate solution. It could decompose in aqueous solution at elevated temperature.

$$NH_2CO_2NH_4 \rightarrow 2NH_3 + CO_2 \cdots\cdots\cdots\cdots (2)$$

Batch experiments conducted in our labs have shown that AC in aqueous solution dissociates to release CO2 and ammonia either with elevated temperature (above 95℃) or by the titration with acids such as hydrochloric acid or citric acid(Shiau et al. 2010). The acid lowers the decomposition temperature down to room temperature. In the previously proposed technique, the in situ CO2 generation was not the main EOR mechanism. The gas generating agent was used in low concentration with low generated gas volume. The outstanding tertiary recovery was the synergistic effect from low inter facial tension flooding and in situ CO2 generation.

In the current work, the mechanism of in situ CO2 generation EOR was isolated and studied closely without surfactants or polymers. The involved EOR mechanisms were oil swelling and viscosity reduction caused by CO2. Moreover, the generated ammonia could provide the benefits form alkali flooding. Other than an EOR method, it also had the potential for CO2 sequestration(Shen, Moghanloo, and Tian 2015).

## Material & Apparatus

Ammonium carbamate (99 wt.%) NaCl (99.5 wt.%) and Dodecane (99.5 wt.%) were purchased from Sigma-Aldrich. Besides pure normal alkane, two more crude oil were tested. Crude oil with 40 API and 27 API had a viscosity at 4.6 cP and 22 cP respectively. F-95 Ottawa sand was purchased from U.S. silica. F-95 Ottawa sand size distribution was between 0.075mm and 0.3 mm with d50 at 0.145mm. The F-95 Ottawa sand packed stainless steel had a porosity at 34.5% and permeability at 4 D. The length and the diameter of the porous media in the sand pack test were 6'' and 0.834'' respectively.

To test the in situ CO2 generation system as a tertiary recovery method, all the tests were conducted in high-pressure sand pack flooding system.

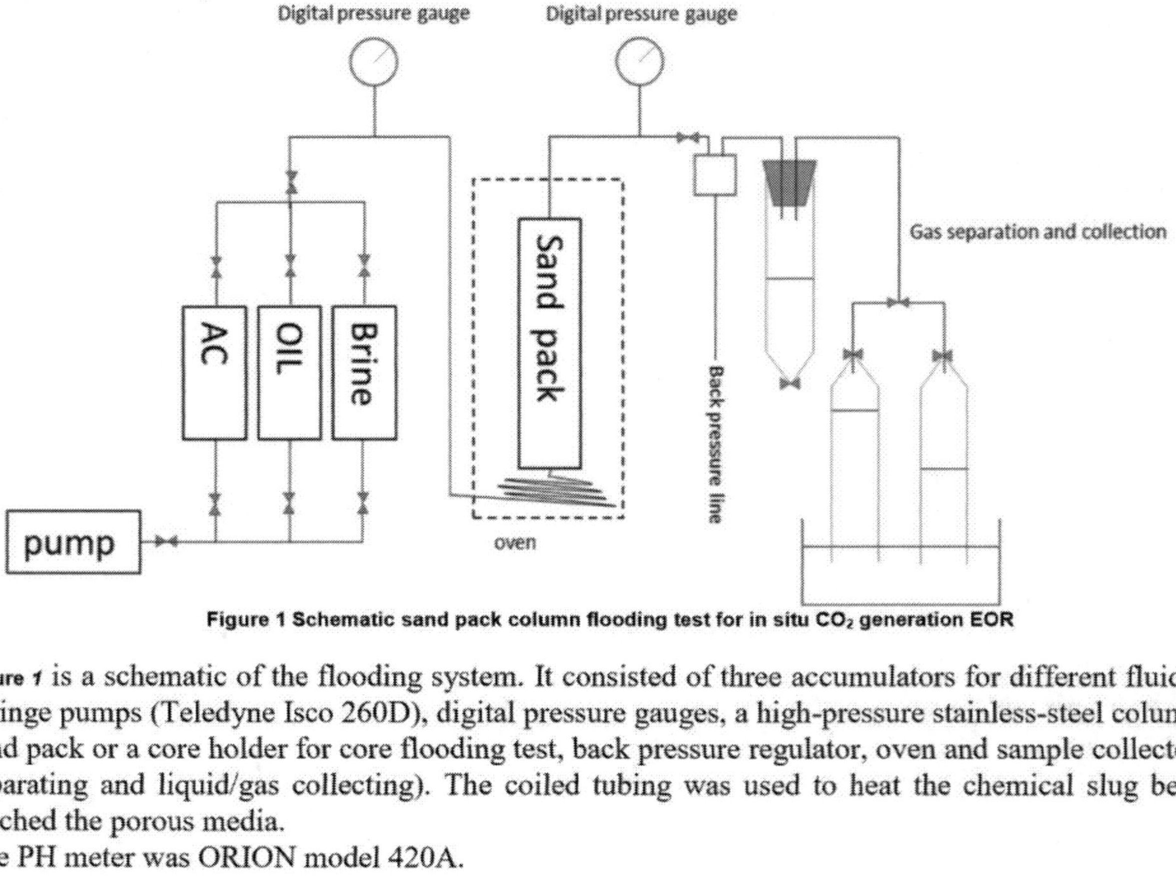


**Figure 1 Schematic sand pack column flooding test for in situ $CO_2$ generation EOR**

Figure 1 is a schematic of the flooding system. It consisted of three accumulators for different fluids, two syringe pumps (Teledyne Isco 260D), digital pressure gauges, a high-pressure stainless-steel column as a sand pack or a core holder for core flooding test, back pressure regulator, oven and sample collector (gas separating and liquid/gas collecting). The coiled tubing was used to heat the chemical slug before it reached the porous media.
The PH meter was ORION model 420A.

## Experiment Procedure

Sand packs were pre-saturated by injecting a known amount of crude oil. The sand pack and core were aged at 80℃ for two months to reach a higher residual oil saturation value. The temperature and back pressure of the system were set to 120℃ and 1500 psi through the whole sand pack flooding experiment. Injection rate of all the slugs were 0.03 mL/min. Then all the sand packs were flooded in the same sequence. In the beginning, water flooding was done till oil cut reached zero. Then it was followed by two pore volumes injection of gas generating agent slug. After the gas generating agent injection, water flooding was resumed till no more oil produced. Once the oil cut reached zero again, two more pore volumes of the gas generating agent slug were injected. Following the injection, the flow stopped for 48 hrs (shut-in cycle). After the shut-in, the water flooding resumed till oil cut reached zero again. Pressure and temperature were automatically recorded. The recovered liquid and gas were recorded manually at certain injected pore volume. The oil produced by water flooding and in situ CO2 generation was collected separately. Then Gas chromatography analyzed the compositional differences of these crude oils.

The Brine used in this work was 5 wt. % NaCl solution. The gas generating agent slug was 35 wt.% ammonium carbamate and 5 wt. % NaCl solution.
For the PH measurement solution, the concentration of the NaCl was fixed at 5 wt. %. The concentration of AC was from 0.5 wt. % to 30 wt. %.

# Experiment Result

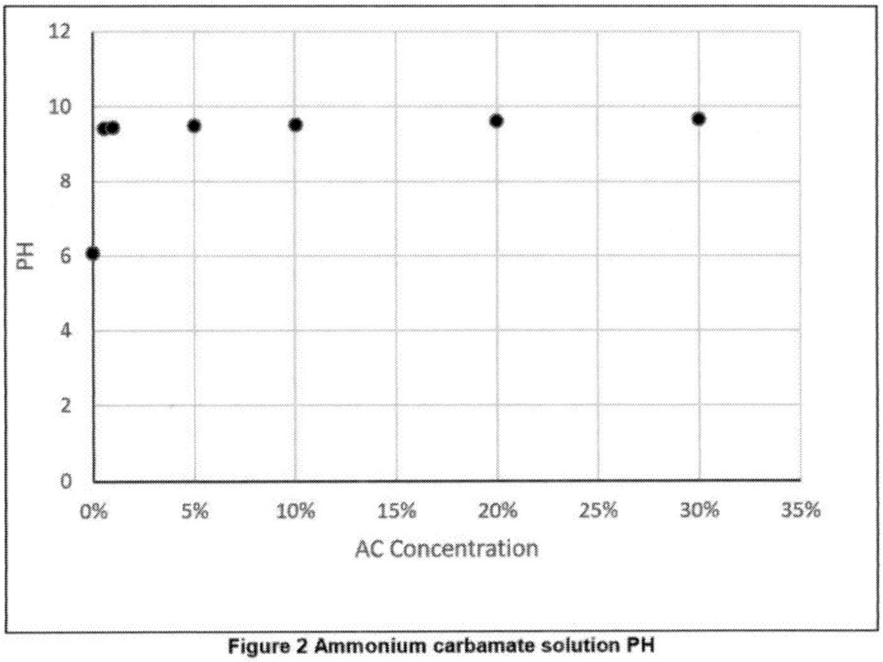


**Figure 2 Ammonium carbamate solution PH**

Figure 2 showed the PH measurement of ammonium carbamate brine solutions. The PH of 5 wt. % NaCl solution was around 6. After adding a small amount of ammonium carbamate, the 0.5 wt. % ammonium carbamate solution showed a high PH at 9.4. Then the PH of the solution reached its plateau (PH=9.5) when the concentration of ammonium carbamate was higher than 1 wt. %. The gas generating solution in the in situ CO2 generation EOR test was acting as a PH buffer. An alkaline solution containing ammonium could help produce crude oil in tertiary recovery stages(van den Pol et al. 2014).

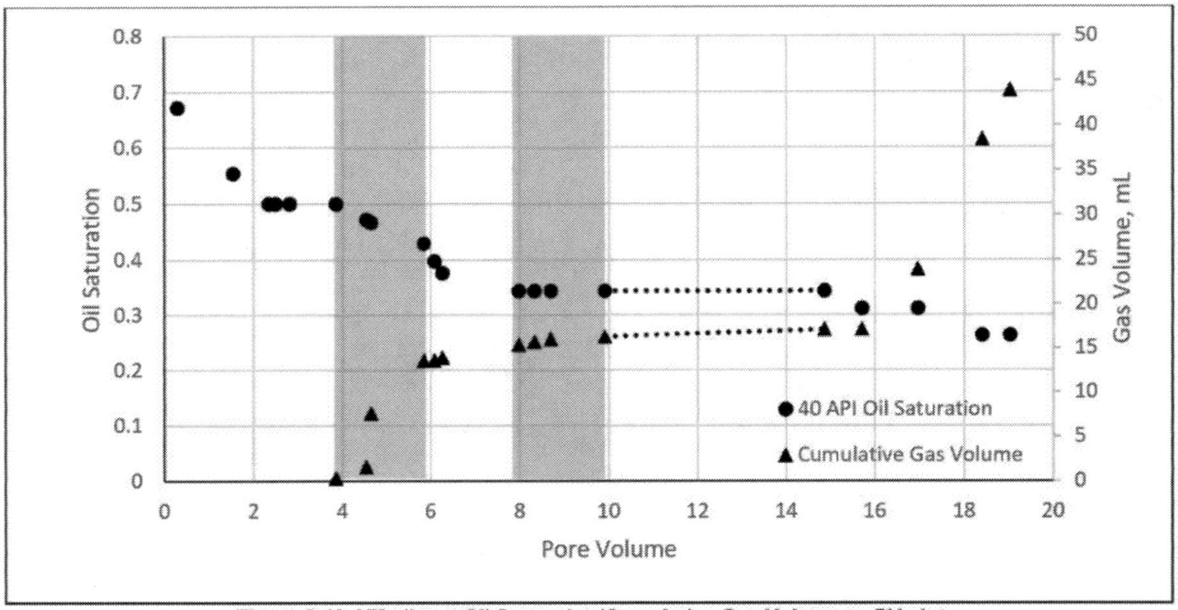


**Figure 3 40 API oil test Oil Saturation/Cumulative Gas Volume vs PV plot**

Figure 3 showed the oil saturation, and cumulative gas volume results from the 40 API crude oil in-situ CO2 generation test. The shaded area indicated the injection of the gas generating agent. The dashed line indicated the flow stoppage. At the first water flooding stage, the oil cut reached zero within 4 pore volumes brine injection. The oil saturation stopped changing at 50% after water flooding. The oil breakthrough happened at about half pore volume gas generating agent injection. CO2 gas breakthrough time was the same as the oil breakthrough. Oil was continuously produced during gas generating agent injection stage. Then the oil cut reached zero within 2 pore volume brine injection. In the second gas generating agent slug injection, oil collector did not receive more crude oil. However, gas production did not stop after the gas generating agent injection. After 2 days shut-in, water flooding produced more crude oil. Moreover, the gas production was at the similar rate as stages before shut-in.

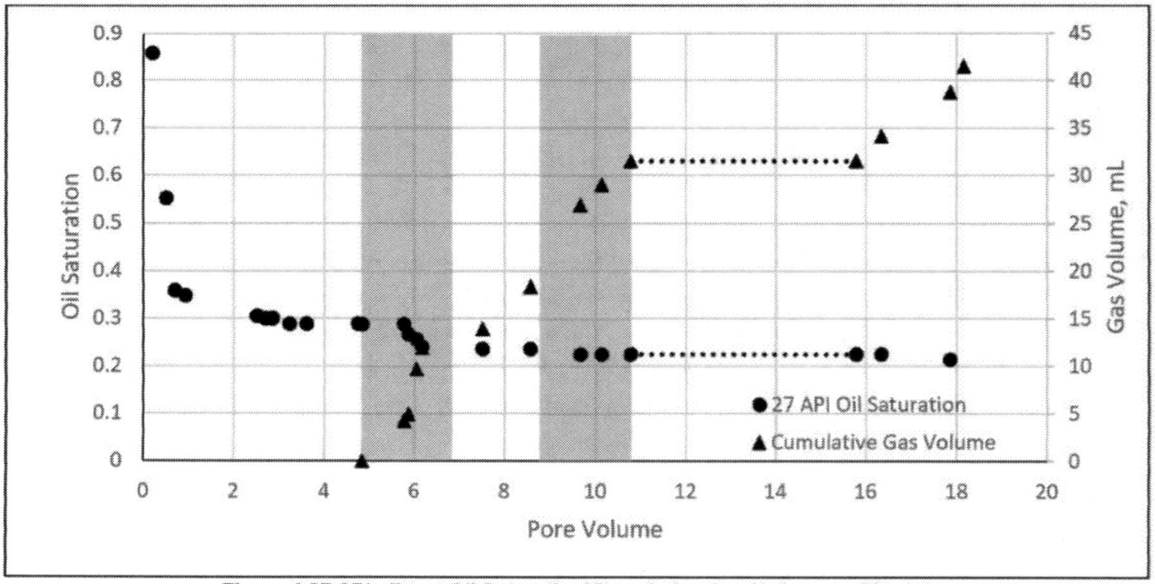


**Figure 4 27 API oil test Oil Saturation/Cumulative Gas Volume vs PV plot**

Figure 4 showed the test result of a heavier oil. The viscosity of this oil was 20 cP. Because of the difference between crude oils. The aging process was not helping on increasing the residual oil saturation of this sand pack. After the water flooding, the oil saturation dropped to 30%. At the first gas generating agent injection stage, no oil breakthrough was observed after 1 pore volume injection. However, gas collector observed CO2 breakthrough at 1 pore volume injection. The oil breakthrough happened right after the gas breakthrough. The high crude oil viscosity did affect the mobility ratio of the flowing system. And the oil cut reached zero during the injection of the gas generating agent. Part of the reason would be that the residual oil was too low to see an extended oil production stage. The gas production rate was constant after the breakthrough till the system shut-in. The produced oil volume by 2 days shut-in was small. The gas production rate was the same as the rate before the shut-in.

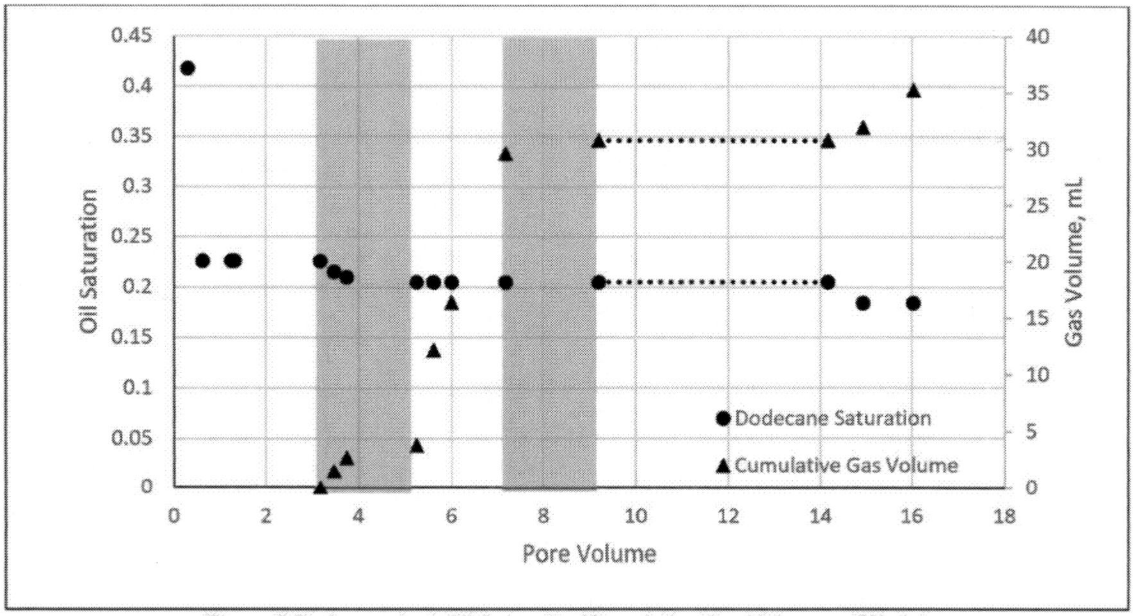


**Figure 5 Dodecane test, Oil Saturation/Cumulative Gas Volume vs PV plot**

Figure 5 showed the Dodecane flooding case. Dodecane was a pure normal alkane. Therefore, the aging process was designed to keep consistency between this test and previous tests. The aging would not help on the residual oil saturation. Because the viscosity of Dodecane was lower than the brine in the water flooding, the sand pack reached its residual oil saturation at the first 0.5 pore volume brine injection. In the gas generating agent flooding stages, the oil breakthrough was at 0.5 pore volume along with the gas breakthrough. The shut-in treatment produced a small amount of oil. However, considering the low volume of residual oil, it is still a significant production. The gas rate is still the same constant before and after the shut-in reaction.

From all the tests, the oil breakthrough was dependent on the oil viscosity. The collected CO2 volume was corresponding to the volume of collected liquid. All the tests showed the tertiary recovery ability of the in situ CO2 generation technique.

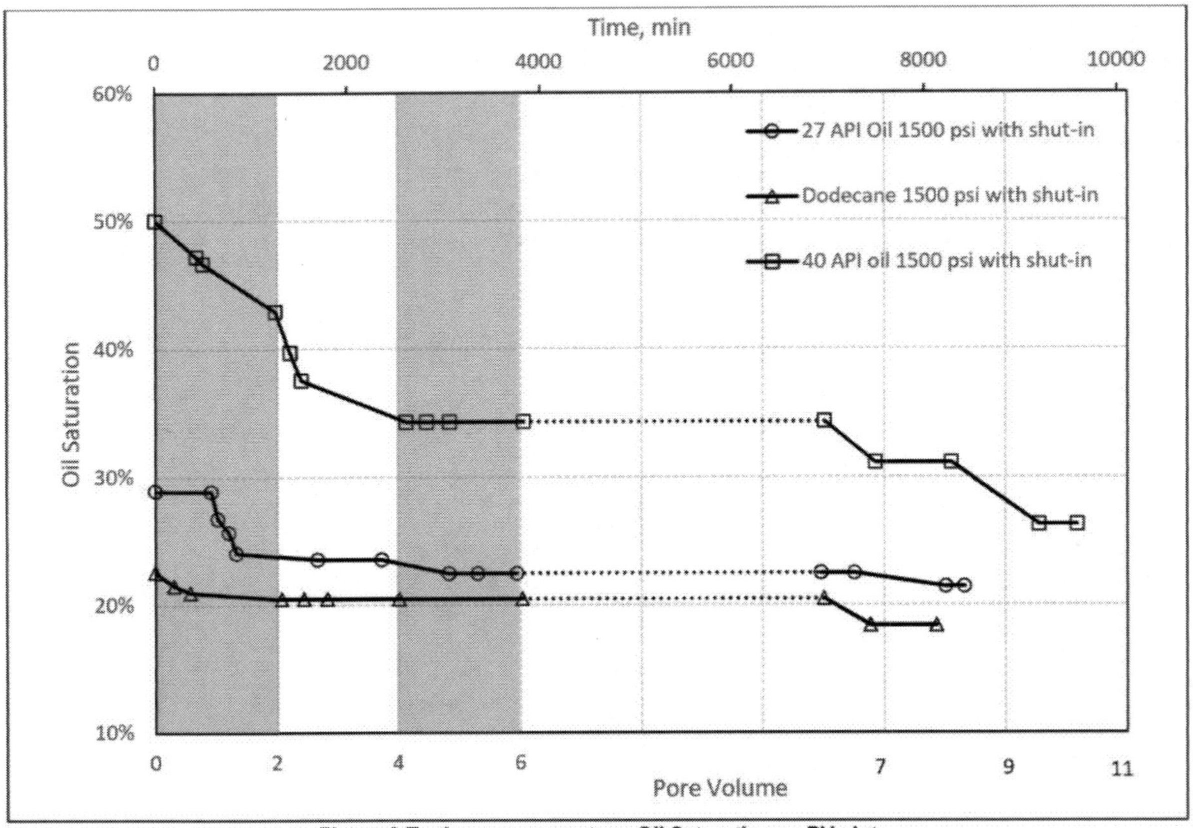


**Figure 6 Tertiary recovery stage Oil Saturation vs PV plot**

Figure 6 compared the oil saturation changing of all the tests during the tertiary recovery stage.

For 40 API oil test, the residual oil saturation after water flooding was 50%. After the tertiary recovery by in situ CO2 generation, the final oil saturations was 26.41%. The tertiary recovery was 47.19%.

For 27 API oil test, the residual oil saturation after water flooding was 28.93%. After the tertiary recovery by in situ CO2 generation, the final oil saturations was 20.66%. The tertiary recovery was 28.6%.

For 40 API oil test, the residual oil saturation after water flooding was 22.57%. After the tertiary recovery by in situ CO2 generation, the final oil saturations was 18.37% The tertiary recovery was 18.6%.

Light oil with high residual oil saturation showed best tertiary recovery. Because all the tests were run at below minimum miscibility pressure condition for all the oil, the CO2 oil solubility was the only parameter that controlled the CO2 partition between the aqueous phase and oil phase when the CO2 water solubility was fixed at certain pressure and temperature. CO2 oil solubility was higher in light oil than it in heavy oil. Therefore, more generated CO2 from the gas generating agent slug could transfer to the oil phase. This phenomenon would lead to better swelling factor and viscosity reduction in light oil test. Dodecane was the lightest oil among all the tested oil. However, the low residual oil saturation limited its performance. Therefore, 40 API oil showed the best tertiary recovery.

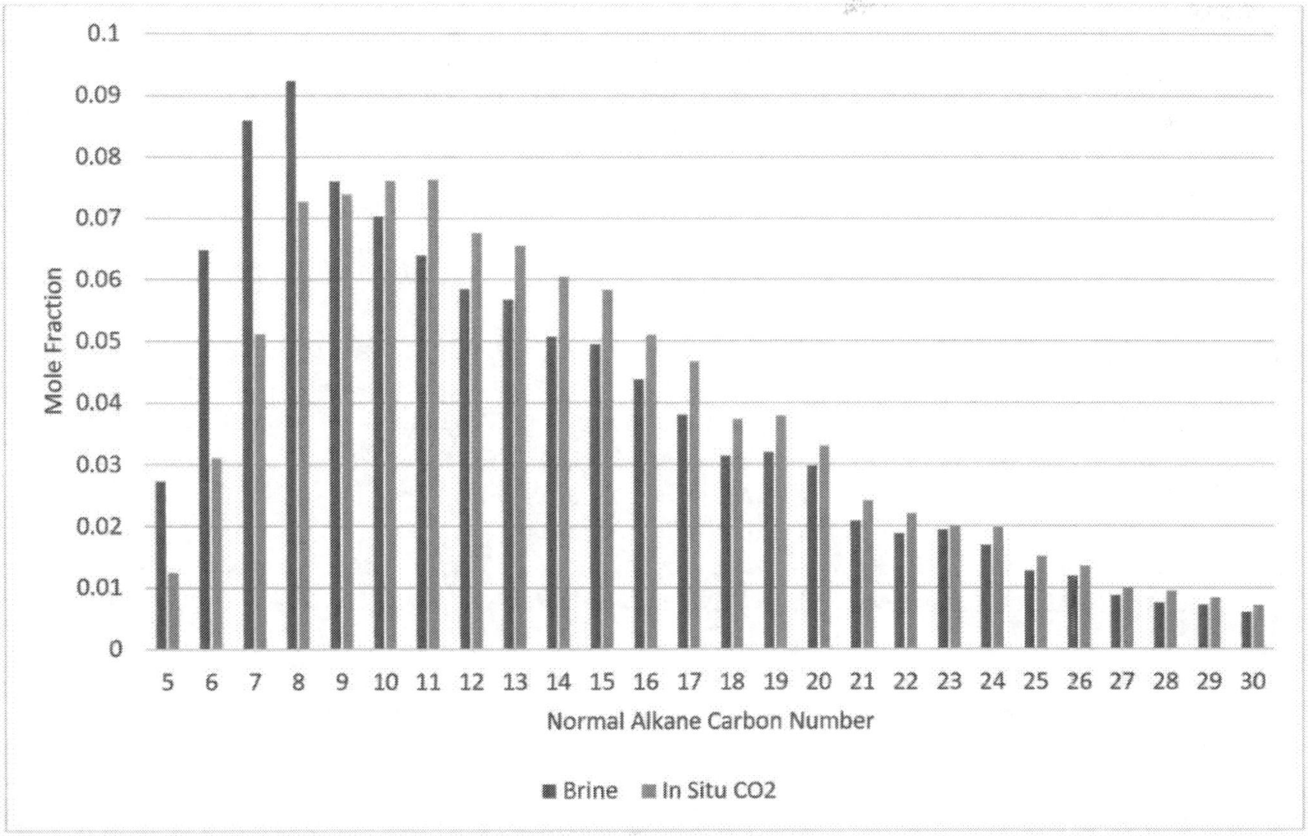


**Figure 7 Oil Composition Change**

Figure 7 was the oil compositional analysis of the 27 API oil test. The recovered oil from secondary recovery (water flooding) and tertiary recovery (in situ CO2 generation) was collected separately. Therefore, the changing of the oil composition caused by in situ CO2 generation EOR could be detected by gas chromatography analysis. In water flooding produced oil sample, the mole fraction of lighter components below C9 was higher than them in in situ CO2 generation produced sample. Comparing to water flooding sample, in situ CO2 generation produced oil sample contained more components heavier than C9.

This measurement indicated that this test did not have a multiple contact miscibility process since it could make the produced oil lighter. There was no separate gas phase in the flooding system, which was consistent with our visual observation. Asphaltene liberation caused by ammonium(Flury et al. 2013) was the reason of more heavy components in the tertiary recovered oil sample.

## Conclusion

1. Single chemical ammonium carbamate system can be used to deliver CO2 to the oil reservoir and generate a significant amount of CO2 to enhance oil recovery.
2. The new proposed in situ CO2 generation system shows comparable tertiary recovery efficiency to previously proposed complex in situ CO2 generation system.

## Reference

Abdelgawad, KZ, MA Mahmoud. In-Situ Generation of CO 2 to Eliminate the Problem of Gravity Override in EOR of Carbonate Reservoirs. Society of Petroleum Engineers.

Bakhtiyarov, Sayavur. 2008. Technology on in-situ gas generation to recover residual oil reserves, New

Mexico Institute Of Mining And Technology.

Bonenfant, Danielle, Murielle Mimeault, Robert Hausler. 2003. Determination of the structural features of distinct amines important for the absorption of CO2 and regeneration in aqueous solution (in *Industrial & engineering chemistry research* **42** (14): 3179-3184.

Chen, Changlong, Mohannad J Kadhum, Marissa C Mercado et al. 2016. Surfactant-Only Stabilized Dispersions of Multiwalled Carbon Nanotubes in High-Electrolyte-Concentration Brines (in *Energy & Fuels* **30** (11): 8952-8961.

Dong, Yannong, Birol Dindoruk, Claudia Ishizawa et al. An experimental investigation of carbonated water flooding. Society of Petroleum Engineers.

Fathollahi, Alireza, Behzad Rostami. 2015. Carbonated water injection: Effects of silica nanoparticles and operating pressure (in *The Canadian Journal of Chemical Engineering* **93** (11): 1949-1956.

Flury, Christopher, Artin Afacan, Marjan Tamiz Bakhtiari et al. 2013. Effect of caustic type on bitumen extraction from Canadian oil sands (in *Energy & Fuels* **28** (1): 431-438.

Gumersky, KH, IS Dzhafarov, A Kh Shakhverdiev et al. In-situ generation of carbon dioxide: New way to increase oil recovery. Society of Petroleum Engineers.

Han, Longyu, Yongan Gu. 2014. Optimization of Miscible CO2 Water-Alternating-Gas Injection in the Bakken Formation (in *Energy & Fuels* **28** (11): 6811-6819.

Khatri, Rajesh A, Steven SC Chuang, Yee Soong et al. 2006. Thermal and chemical stability of regenerable solid amine sorbent for CO2 capture (in *Energy & Fuels* **20** (4): 1514-1520.

Lei, Hao, Shenglai Yang, Lihua Zu et al. 2016. Oil Recovery Performance and CO2 Storage Potential of CO2 Water-Alternating-Gas Injection after Continuous CO2 Injection in a Multilayer Formation (in *Energy & Fuels* **30** (11): 8922-8931.

Li, Yanlai, Kuiqian Ma, Yingxian Liu et al. 2013. Enhance Heavy Oil Recovery by In-Situ Carbon Dioxide Generation and Application in China Offshore Oilfield. Proc.

McCann, Nichola, Marcel Maeder, Moetaz Attalla. 2008. Simulation of enthalpy and capacity of CO2 absorption by aqueous amine systems (in *Industrial & engineering chemistry research* **47** (6): 2002-2009.

Mosavat, Nader, Farshid Torabi. 2014. Application of CO2-saturated water flooding as a prospective safe CO2 storage strategy (in *Energy Procedia* **63**: 5619-5630.

Rochelle, Gary T. 2009. Amine scrubbing for CO2 capture (in *Science* **325** (5948): 1652-1654.

Shen, Tong, Rouzbeh Ghanbarnezhad Moghanloo, Wei Tian. Ultimate CO 2 Storage Capacity of an Over-Pressurized 2D Aquifer Model. Society of Petroleum Engineers.

Shiau, Bo Jier Ben, Tzu-Ping Hsu, Bruce Lynn Roberts et al. Improved Chemical Flood Efficiency by In Situ CO2 Generation. Society of Petroleum Engineers.

van den Pol, Esther, Carl H. T. van Rijn, Diederik W. van Batenburg et al. 2014. Alkali Surfactant Polymer Flooding Using Ammonia for Offshore use. Proc.

Wang, Yong, Jirui Hou, Yong Tang. 2016. In-situ CO 2 generation huff-n-puff for enhanced oil recovery: Laboratory experiments and numerical simulations (in *Journal of Petroleum Science and Engineering* **145**: 183-193.

CMTC-486396-MS

# Development and Evaluation of an Iridium Oxide Based Chemical Sensor for Downhole $CO_2$ Monitoring--Part I

Sai Wang, University of North Dakota;
Kegang Ling, University of North Dakota;
Ning Liu, University of Louisiana at Lafayette;
Hongsheng Wang, University of North Dakota



This paper was prepared for presentation at the Carbon Management Technology Conference held in Houston, Texas, USA, 17-20 July 2017.



## Abstract

Geological carbon sequestration represents a long-term storage of $CO_2$, in which large-scale $CO_2$ is injected into the subsurface geologic formations, such as the deep saline aquifers or depleted oil and gas reservoir. In the $CO_2$ sequestration process, the injected $CO_2$ is expected to remain in the reservoir and not to migrate to the earth surface. To better understand the $CO_2$ movement undersurface and obtain real time information in carbon sequestration, an iridium oxide-based Severinghaus-type $CO_2$ chemical sensor was constructed and tested in this study.

The $CO_2$ sensor was designed and constructed based on the intersection inspiration from electrochemistry idea. The principle of the $CO_2$ sensor design is dramatically rely on the pH detection of the electrolyte solution which generated by the hydrolysis process of $CO_2$. The developed $CO_2$ sensor includes a couple of Iridium-Oxide electrodes. To meet the working purpose, iridium oxide nanoparticles was prepared and electrodeposited for the thin $IrO_2$ film generation on the surface of metal substrate. The other critical parts, such as a thin gas-permeable silicone membrane, a porous metal supporting material, and the bicarbonate-based electrolyte solution are prepared for the sensor's preparation. The assembled sensor was tested in aqueous solution with different $CO_2$ concentrations. Then the sensor was settled in harsh, high-pressure environments, in order to invest the performance of the $CO_2$ sensor under reservoir conditions.

## Introduction

The definition of $CO_2$ sequestration was the whole process of the $CO_2$ capture and the $CO_2$ long-term storage [1]. It had been treated as a potential method to decelerate the accumulation process of greenhouse gas which generated from the fossil fuels burning and other source [2]. While for the geologic sequestration, it means to put the captured $CO_2$ in the geological formation for the aim of long-term storage.

Fig. 1 represents the options for the geological $CO_2$ storage. Vast amount of formations such as the deep saline aquifers, the depleted oil & gas reservoirs and the coal beds can provide the optical space and seals for the $CO_2$ trapping in deep underground. Those chosen formations should be deeper than 2600ft and have the good porosity for the large-scale $CO_2$ storage. Also, the required permeability could permit the large and fast injection rate of $CO_2$.

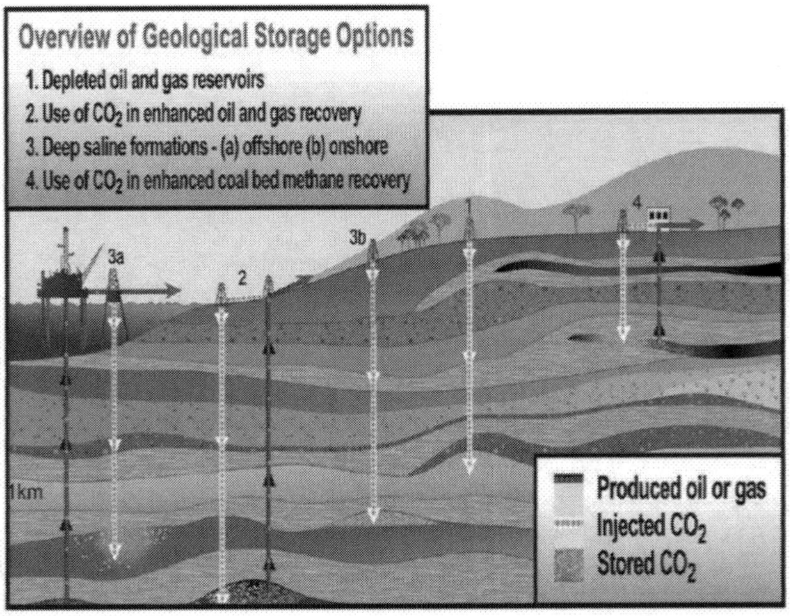


**Fig. 1 Styles of geological formations and reservoirs applied for carbon sequestration [3]**

During the geological $CO_2$ storage process, the injected $CO_2$ is expected to stay in the subsurface formations. However, public concerns of $CO_2$ migration and possible leakage to surface require comprehensive studies and understanding the $CO_2$ migration behavior in subsurface. An in-situ $CO_2$ monitoring approach is highly desirable for this purpose.

Comparing with other technologies, which may had largely time consuming or costly requirement, a preferred method for in-situ underground $CO_2$ migration monitoring is to install down-hole $CO_2$ sensors directly in the underground. This potential approach may realize by the interaction with the electrochemical idea, which is, transforming the $CO_2$ detection to the pH measurement [4]. At present a great deal of research is carrying out on the $CO_2$ chemical sensor. For example, in 1998, Trapp *et al.* disclose a coulometric $CO_2$ gas sensor, which could continuously measure the concentration of $CO_2$ under atmospheric conditions, but not for the harsh environment such as high pressure, high temperature and high salinity conditions [5]. U.S. patent 5554272 revealed a planar bicarbonate sensor combined with a pH sensor that could detect the $CO_2$ concentration in the water environment [6]. However, the sensor did not work efficiently under the high pressures prevalent underground. The $CO_2$ sensor developed by Maier *et al.* was reported to work under high pressure condition while not the high humidity circumstance [7].

The purpose of this study is to develop a downhole $CO_2$ sensor which can be used in carbon sequestration to in situ and continuously monitor $CO_2$ movement. This study primarily presented the development and test of a robust, novel downhole $CO_2$ sensor which can monitor $CO_2$ concentration changes continuously at subsurface condition [8]. Differed from the traditional sensor, this type of the $CO_2$ sensor used a metal-oxide electrode instead of a glass electrode to improve the resistant ability for the harsh downhole environments encountered [9]. Up to date, this Severinghaus-type sensor will be the first downhole, chemical $CO_2$ sensor for monitoring $CO_2$ movement under simulated reservoir conditions. And that will generate an invaluable effect for large-scale storage of $CO_2$. Distinct from all other monitoring methods, this small size chemical sensor is easy to handle and could provide the monitoring data continuously, timely and accurately.

## Experimental

### Sensor preparation

The $CO_2$ sensor was designed and constructed across the intersection with electrochemistry idea. The principle of the $CO_2$ sensor design is dramatically based on the pH detection the electrolyte solution which generated by the hydrolysis process of $CO_2$. The $CO_2$ molecules spread through the thin gas-permeable silicone membrane and then dissolve into the internal solution, after that the $CO_2$ will have a reaction with $H_2O$ to generate the carbonic acid. While the carbonic acid is not stable, it will dissociate into $HCO_3^-$ and $H^+$ (as seen in Equation 1). Consequently, the generated $H^+$ will vary the pH value of the solution, which is measured through the electrodes.

$$\mathbf{CO_{2\,(aq)} + H_2O \leftrightarrow H_2CO_3 \leftrightarrow H^+ + HCO_3^-} \qquad \textbf{[1]}$$

For the sensor applied under high pressure conditions and simulated reservoir conditions, the construction procedure was described as Fig. 2. Firstly, the working electrode was prepared through the electrodeposition method, in which the iridium oxide nanoparticles solution was electrochemically deposited using cyclic voltammetry onto the metal substrates [10]. Then the electrodes were fixed through a stainless steel cap and stabilized by silicone sealant. Following a stain steel porous filter element was settled for the function of supporting the whole frame, which means that it can enable the sensor to work under harsh downhole environment, especially under high pressure condition. This support unit was coated by silicone sealant onto the inner and outer surface in case of contacting with electrodes. What followed is the above stabilized electrodes were inserted into the support unit where inside being full of bicarbonate-based internal electrolyte solution All the joints were sealed by the silicone sealant in order to preventing the internal solution leakage Finally, as above mentioned sealing method, both outlets of the gas-permeable membrane were blocked off by the silicone sealant. Meanwhile, some other tiny materials were applied in the sensor manufacture for the purpose of support and potential transfer.

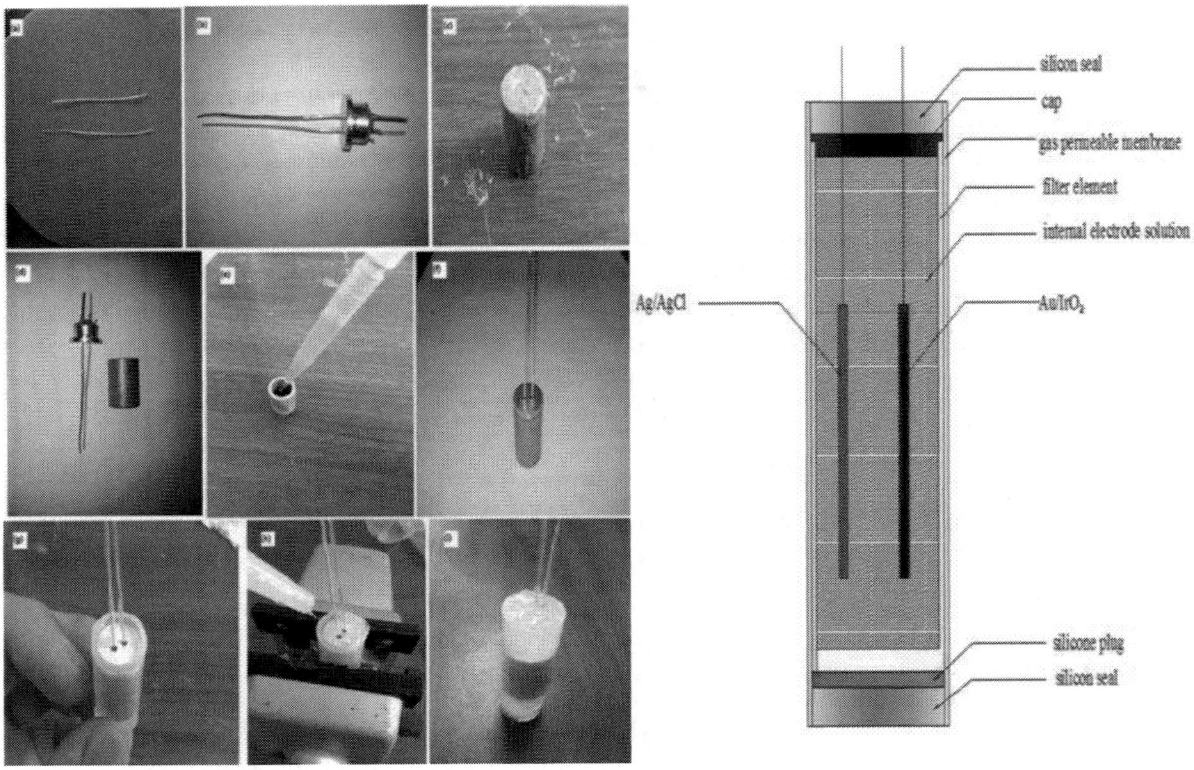


**Fig. 2 Construction of $CO_2$ sensor and schemetic of the $CO_2$ sensor.**

## Test of the performance of the $CO_2$ sensor under different $CO_2$ concentrations.

The following steps were designed to investigate the $CO_2$ sensor's performance. All potentiometric measurements were carried out at room temperature of 25°C±2°C.

Fig. 3 shows the schematic figure for setting up the experiment. Firstly, the $CO_2$ sensor was settled in the stainless steel tank. Then the tank was injected with 0.2M citric acid and tightly sealed. The stainless steel tubing and the connecting wires from the sensor came out from the top of the tank. The ISCO pump was connected with this stainless steel tubing (0.375 in.) and used for controlling the pump injection and the tank pressure. The connecting wires of the sensor were connected with a 34401A 61/2 digit multi-meter. Before carrying out the tests, the base potential of the $CO_2$ sensor was measured by maintaining the $CO_2$ sensor into the citric acid solution under a certain pressure. After the potential reached a stable value, the certain amount of the $NaHCO_3$ solution (0.1M) was injected in the tank and then the tank was sealed immediately. Finally the ESCO pump was began to increase the tank pressure. The sensor output signal was measured by the digit multi-meter. The whole process was conducted under the pressure of 3000psi.

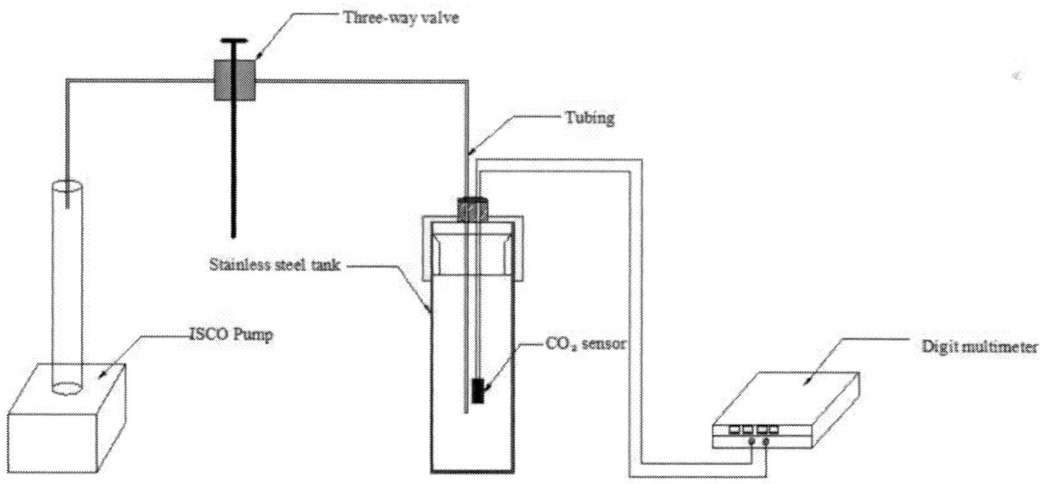


**Fig. 3 Schematic structure of the high pressure plant.**

## CO2/brine core flooding test

Core flooding is a technique used for conducting experiments on core samples in conditions simulated to be as close to the natural environment as possible. A coreflooding system was constructed to simulate the $CO_2$ storage process and evaluate the performance of the $CO_2$ sensor in the process. Fig. 4 shows the schematic diagram of the core flooding apparatus. Shown in Fig.4, a coreholder mounts the core sample under high overburden pressure, similar to conditions in formations used for underground $CO_2$ storage. The downhole $CO_2$ sensor is placed into the core and monitors the $CO_2$ movement. After $CO_2$ and brine are filled in two different accumulators and connected with two different backpressure regulators (BPR), the two pumps start to inject distilled water into the accumulator and push $CO_2$ and brine out from the accumulator. The mixture flows through two filters with pore size of 0.5 μm. Then the $CO_2$/brine mixture flows into the core sample where $CO_2$/brine contacts with the downhole $CO_2$ sensor. A change on the output signal of the sensor is used for monitoring the movement of $CO_2$ and concentration change in the core. In the experiments, the core sample is first flooded with brine to saturate the core. Then liquid $CO_2$ and brine are injected into the core. After $CO_2$ and brine emerge from the core, the mixture flows through another BPR, which controls the operation pressure in the core. Then the mixture separates as liquid phase and gas phase after emerging from the BPR. The liquid phase (brine) is collected in a liquid trap and the gas phase ($CO_2$) emits to the outside of the building through 0.125 in. tubing. The entire apparatus, except for the pumps, is contained in an air bath box to maintain a constant temperature throughout the experiment.

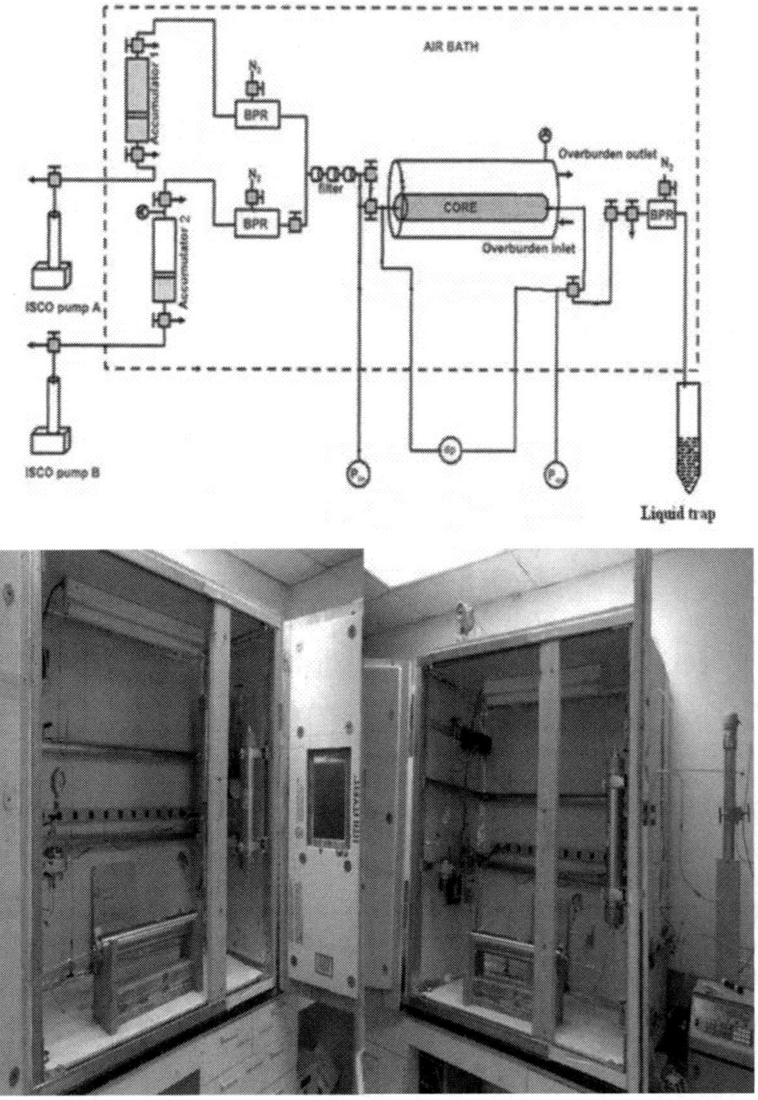


**Fig. 4 The schematic diagram and pictures of the core flooding apparatus.**

# Results and Discussion

## $CO_2$ sensor performance test under high pressure (3000 psi)

Response time t and potential differential ΔE are two important parameters to evaluate the $CO_2$ sensor sensitivity. The potential differential was defined as: ΔE=E ($CO_2$)-E (base), the response potential caused by $CO_2$. Therefore, the $CO_2$ responses for each sensor were evaluated by using ΔE and response time t in this experiment.

Figure 5 illustrates a typical potential response of the $CO_2$ sensor. The measurement was divided into three stages: 1) the potential stabilized at 80 mV in the bicarbonate-based internal electrolyte solution; 2) the $CO_2$ diffused through the gas permeable silicone membrane to the internal electrolyte solution. As a result, the potential dramatically increased from 80 mV to 126 mV; 3) the reaction reached the equilibrium and the potential stabilized at 126 mV.

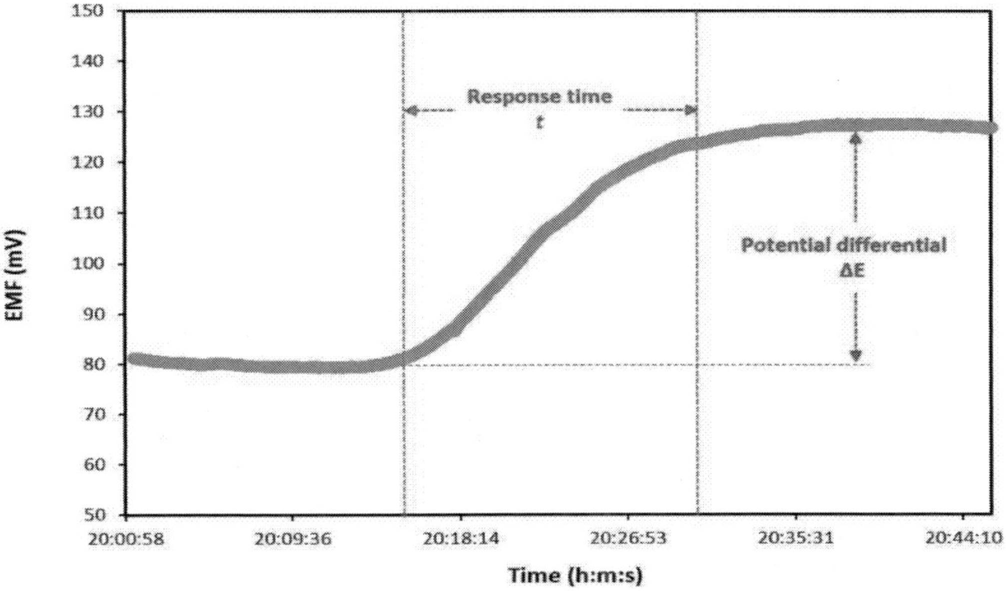


**Fig. 5 A typical potential response of the $CO_2$ sensor.**

Generally, an excess of citric acid was present in the system solution, All the $NaHCO_3$ reacted with the acid and transferred to the $CO_2$ following equation [2]. Thus, a certain quantity of $CO_2$ can be generated and employed automatically.

$$2HCO_3^- + 2H^+ = 2CO_2 + 2H_2O \qquad [2]$$

Fig. 6 illustrates the potential responses of the $CO_2$ sensor under pressure of 3000 psi. As shown in Fig.6, the base potential stabilized at 99 mV in 4101s. Then, $NaHCO_3$ was rapidly injected to the solution. The sensor output potential immediately increased and leveled off at 120 mV. The response time was 3461s as the $CO_2$ concentration was 0.001M. With the increase $CO_2$ concentration to 0.002 M, a similar behavior was observed, as depicted in Fig. 6. It can be found that the potential change is proportional to the $CO_2$ concentration directly. The potential change increased as the $CO_2$ concentration increased. The response time of the $CO_2$ sensor was found to be relied on the $CO_2$ concentration. Also the response time for the $CO_2$ sensor is on the order of $10^3$ seconds at the operation pressure of 3000psi.

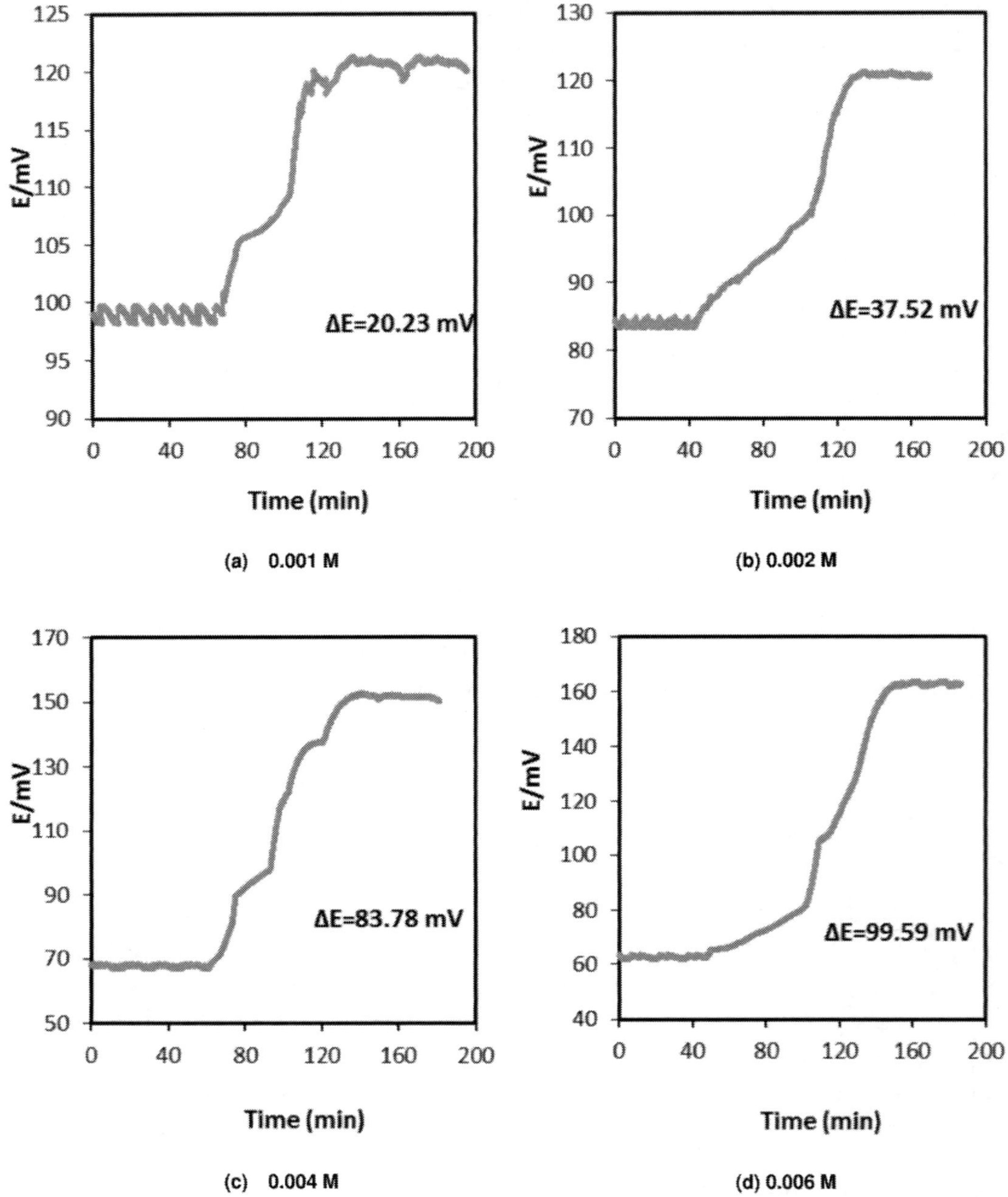


**Fig. 6 Potential response of $CO_2$ sensor with the addition of $NaHCO_3$ in the 0.2 M citric acid-HCl solution at 3,000 psi, [$NaHCO_3$] is (a) 0.001M, (b) 0.002 M, (c) 0.004 M and (d) 0.006 M.**

## Monitor $CO_2$ migration in simulated $CO_2$ storage process

Fig. 7 presented the potential reponsse of the $CO_2$ sensor for the coreflooding process. The results demonstrate that, when 2.0% brine was injected into the core, the output signal of the sensor was around 43.6 mV. The output signal stayed constant as 1 PV of brine flooded the core, indicating that the sensor was equilibrated with ambient conditions. Then, when $CO_2$ began injection into the core, the output potential was the same as that in the brine flooding test for almost 3000s and then it suddenly increased to 243.1 mV. Apparently, as $CO_2$ injection was initiated, $CO_2$ moved forward along the core and took 30 min to reach the sensor. The output potential of the sensor stayed unchanged before the $CO_2$ plume arrived. When the injected $CO_2$ reached the sensor, the $CO_2$ molecule diffused into the sensor and changed the pH of the internal solution, resulting in an immediate potential increase to 243.1 mV. The response time of the sensor, due to the high pressure and concentration of the $CO_2$, was around 20 min. After that, the output potential leveled off, implying new equilibrium was reached in the sensor's internal solution. The change of the output potential, based on the results of Fig. 7, is 199.5 mV.

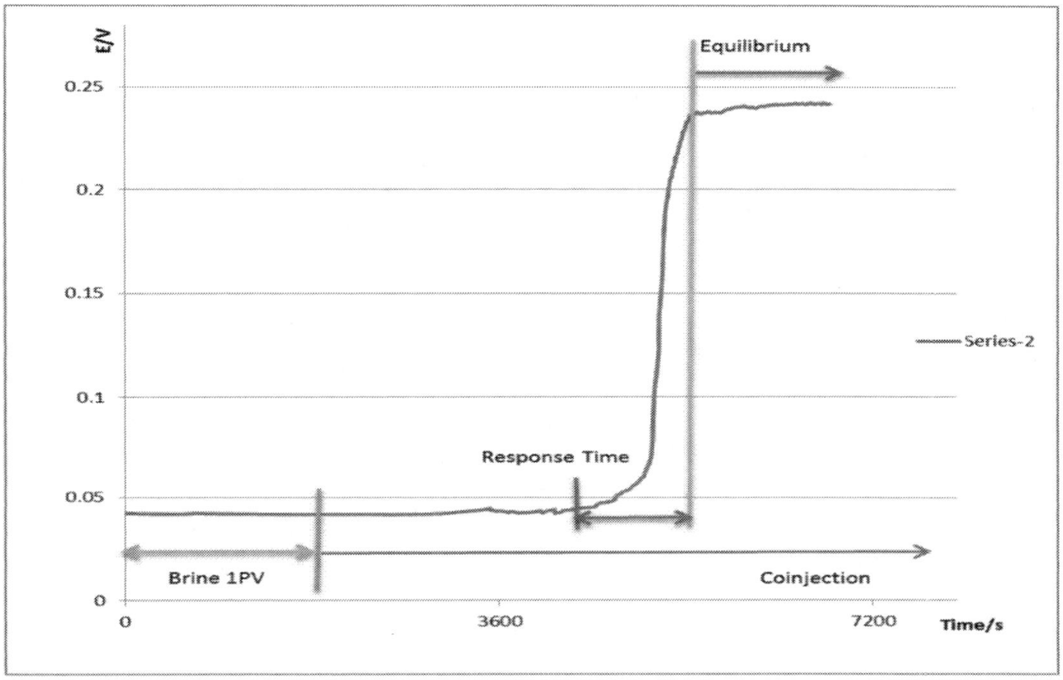


**Fig. 7 Potential response of $CO_2$ sensor vs. Time---Series 2. Brine injection rate: 100mL/h; Co-injection rate: 100 mL/h**

A second $CO_2$/brine core flooding test was performed with the $CO_2$ sensor. This procedure was similar to the first test, except that the sensor was flooded with brine after the $CO_2$/brine core flooding. Fig. 8 shows the result of the sensor output potential change during the test. The sensor output potential was around 61.3 mV during the brine flooding. Then the output potential started to slowly increase after 0.5 PV of $CO_2$/brine was injected into the core, indicating some dissolved $CO_2$ diffused into the sensor. After that, as 1 PV of $CO_2$/brine was injected into the core, an abrupt potential increased was observed. This high potential increase indicated the $CO_2$ plume reached the sensor, as shown in Fig. 8. Then after the sensor was equilibrated with the $CO_2$ plume, the core switches to brine flooding. The result indicates that, after 7 PV brine flooded the core, the sensor recovered almost to its initial state (the output potential was 70.1 mV and initial output potential was ~61.3 mV). This implies that our downhole $CO_2$ sensor can be recovered by brine flooding in $CO_2$ storage process. This is promising for the sensor application in WAG $CO_2$ storage operation. However, further tests need to be conducted to understand the sensor performance and lifetime in WAG operation.

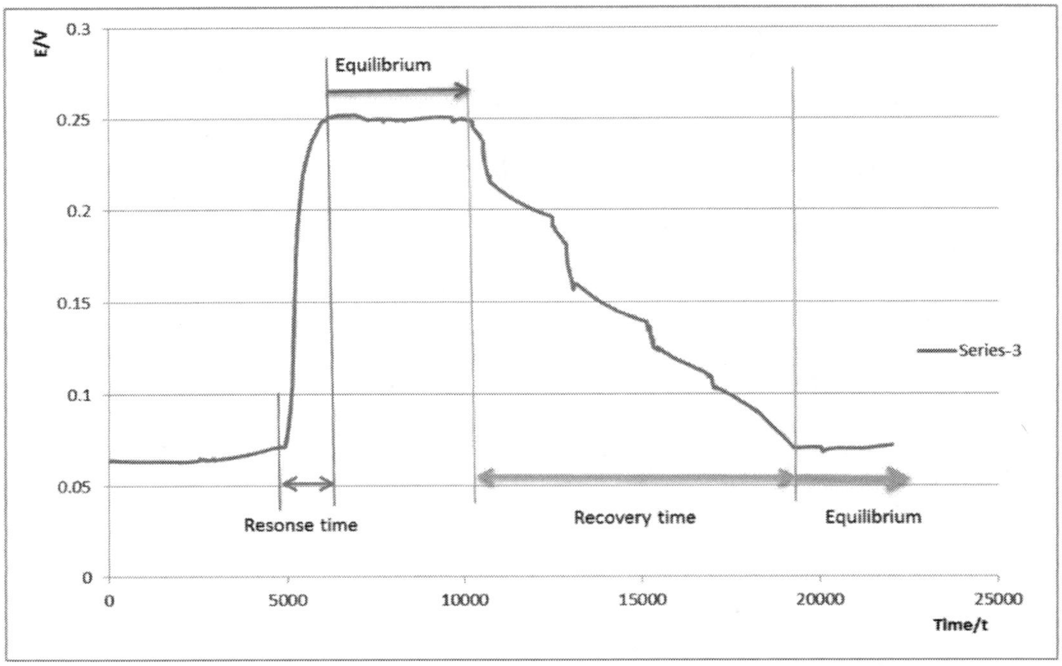


**Fig. 8 Potential response of $CO_2$ sensor vs. Time---Series 3. Brine injection rate: 100mL/h; Co-injection rate: 100 mL/h; Secondary brine injection rate: 100mL/h**

## Conclusion

An electrochemical downhole $CO_2$ sensor was developed and tested in the solution with different concentrations of $CO_2$ and displayed very good $CO_2$ sensing performance under high pressure. $CO_2$/brine coreflooding tests were carried out to evaluate the performance of the $CO_2$ sensor in simulate $CO_2$ storage process. The results indicated that the sensor could detect $CO_2$ movement under the simulated reservoir conditions. Further studies showed that the sensor could be reused by brine flooding after $CO_2$/brine flushed the core. The results of the core flooding tests demonstrated that the sensor had potential application for $CO_2$ monitoring in carbon storage.

## Acknowledgement

The authors are grateful to The Petroleum Engineering Department of University of North Dakota. Grateful acknowledgement delivered to the support of the Department of Energy through the National Energy Technology Laboratory under contract number DE-FE0009878.

## Reference

[1] Sedjo R., Sohngen B., Carbon Sequestration in Forests and Soils. Annual Review of Resource Economics, 4: 127–144, 2012.

[2] Hodrien C., Squaring the Circle on Coal - Carbon Capture and Storage. Claverton Energy Group Conference, May 9, 2010.

[3] Sally M. Benson, David R. C., $CO_2$ Sequestration in Deep Sedimentary Formations, Elements, Vol. 4, PP: 325-331, 2008.

[4] ChengGang X., Bhavani R.C., Andrew J., Goiran, P.O., Berrim, A., Downhole pH As a Novel Measurement Tool in Carbonate Formation Evaluation and Reservoir Monitoring. Petrophysics, 29-2:159-171, 2008.

[5] Trapp T., Ross B., Cammann K., Schirmer E., Berthold C., Development of A Coulometric $CO_2$ Gas Sensor. Sensors and Actuators B: Chemical, 50: 97-103, 1998.

[6] John S., Joseph S., "Planar Bicarbonate Sensor". United States Patent, 5,554,272, 1996.

[7] Maier, J., Holzinger, M., Sitte, W., Fast Potentiometric $CO_2$ Sensors with Open Reference Electrodes." Solid State Ionics, 74: 5-9, 1994.

[8] Sai Wang, Jingshan San, Jianjia Yu, Robert Lee, Ning Liu, A Downhole $CO_2$ Sensor to Monitor $CO_2$ Movement In-situ for Geologic Carbon Storage, Internal Journal of Greenhouse Gas Control.

[9] Sai Wang, Jingshan San, Ning Liu, A Chemical $CO_2$ Sensor Monitoring $CO_2$ Movement Under Reservoir Conditions, Carbon Management Technology Conference, 17-19 Nov,2015, Sugarland, Texas, USA

[10] Munawar Khalil, Sai Wang, Jianjia Yu, Robert L. Lee, and Ning Liu, Electrodeposition of Iridium Oxide Nanoparticles for pH Sensing Electrodes. Journal of the Electrochemical Society, 163 (9) B485-B490 (2016).

**Ashwin Pasumarti**
Samin Raziperchikolaee
Joel Sminchak
Shuvajit Bhattacharya
Mark Kelley
Neeraj Gupta

# Assessing the Effective $CO_2$ Storage Capacity of a Reservoir Using a Geomechanical Framework

A case study of a site in the Arches Province of the Midwest U.S.

1 BATTELLE

## Outline

- Background
  - $CO_2$ storage in deep saline aquifers
  - Injection-induced stress changes in the reservoir
- Modeling the Arches Province Site
  - Methodology: workflow, model construction, assumptions
  - Sensitivity scenarios
- Analysis
  - Geomechanical impacts
  - Effective capacity estimate

2 BATTELLE

# CONCEPTUAL REVIEW

3

BATTELLE

# $CO_2$ Storage in Deep Saline Aquifers

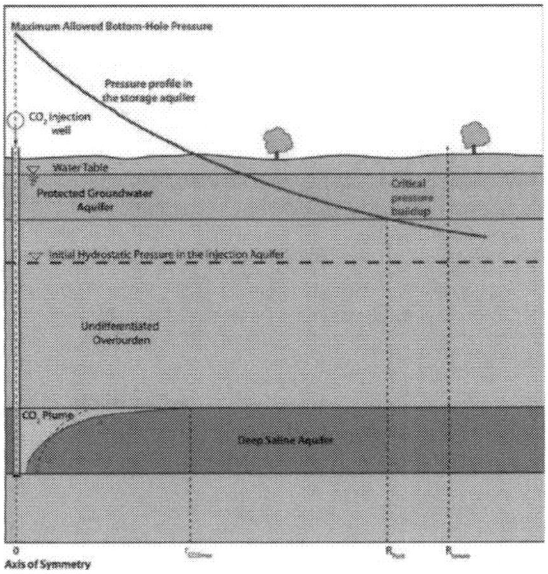


- $CO_2$ injection into an aquifer creates a plume.
- Pressure profile generated.
- All injected $CO_2$ is stored either in supercritical state or by dissolution into the brine.

Bachu, S., 2015, Review of $CO_2$ storage efficiency in deep saline aquifers: International Journal of Greenhouse Gas Control, v. 40, p. 188-202.

BATTELLE

# Coupled Fluid Flow – Reservoir Geomechanics Simulation

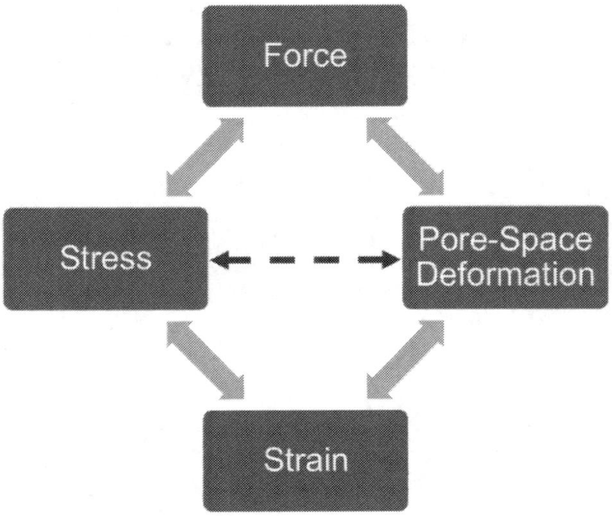


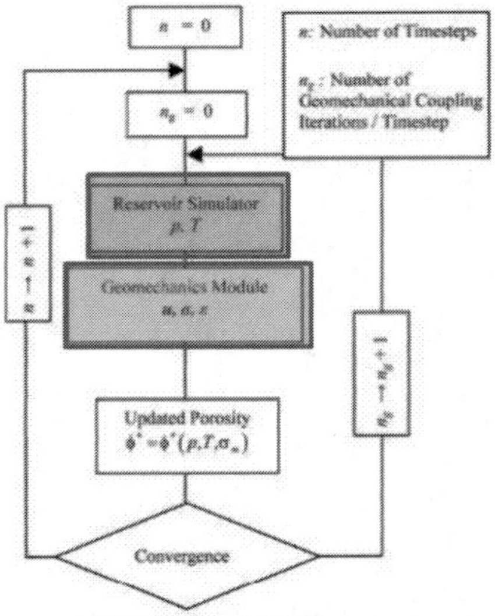


modified from Tran et al, 2005



BATTELLE

# Geomechanical Impacts of Injection

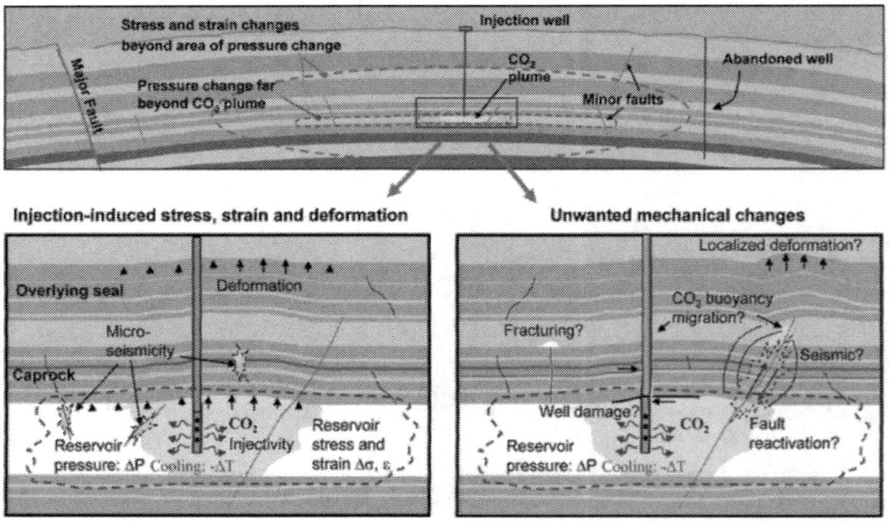


Rutqvist, J. (2012). The geomechanics of $CO_2$ storage in deep sedimentary formations. Geotechnical and Geological Engineering, 30(3), 525-551.



BATTELLE

# Overview

### Geologic/Geomechanical Properties

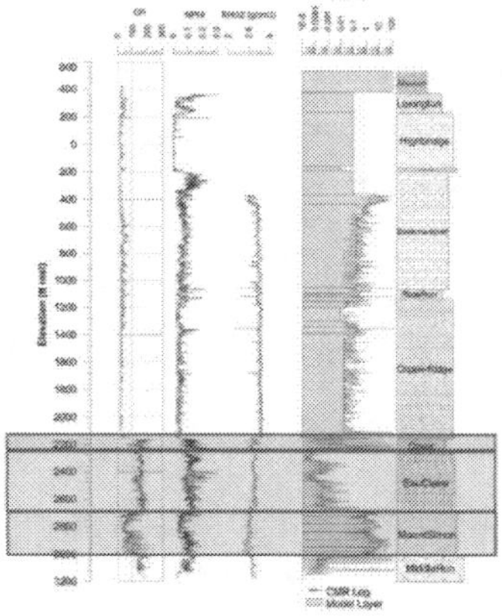

Caprock
Intermediate
Aquifer

Delineated geomechanical units

### Reservoir Model

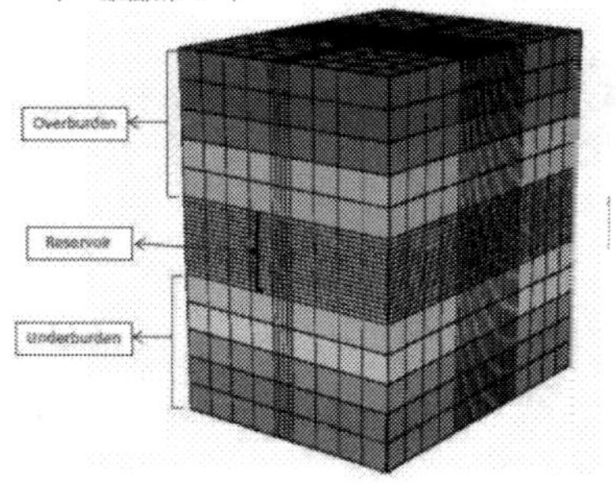


### $CO_2$ Injection

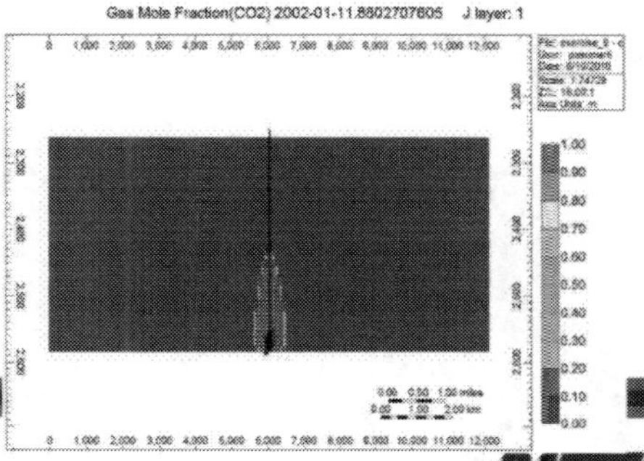




BATTELLE

# Mohr Circle Analysis

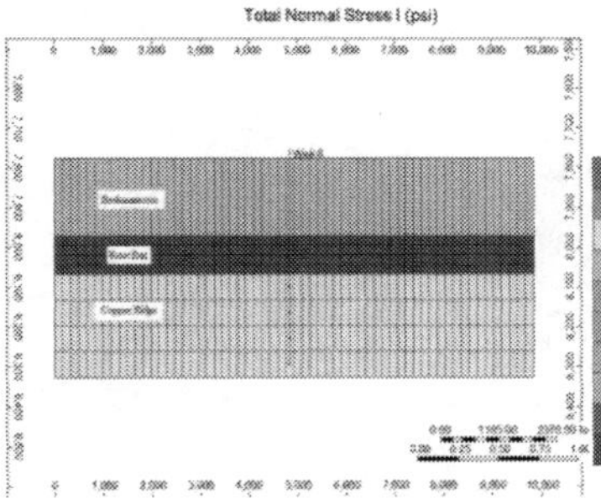


Total stress increases as a result of injection

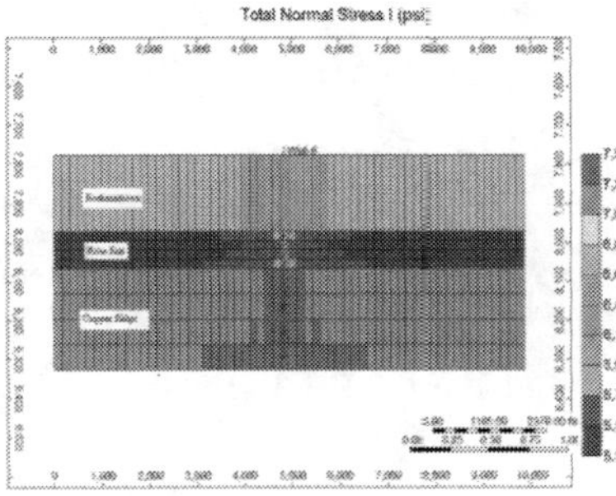


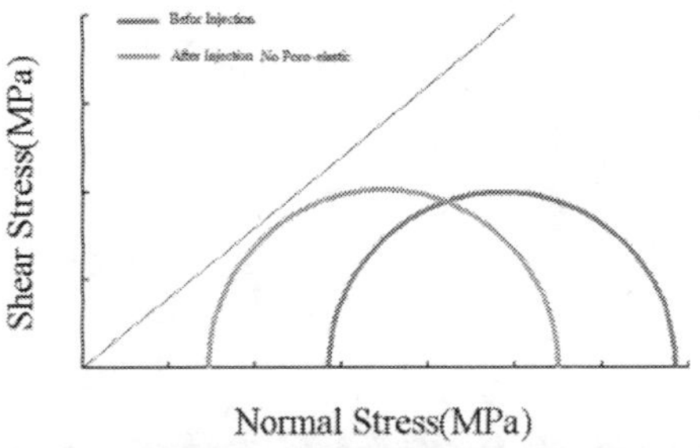


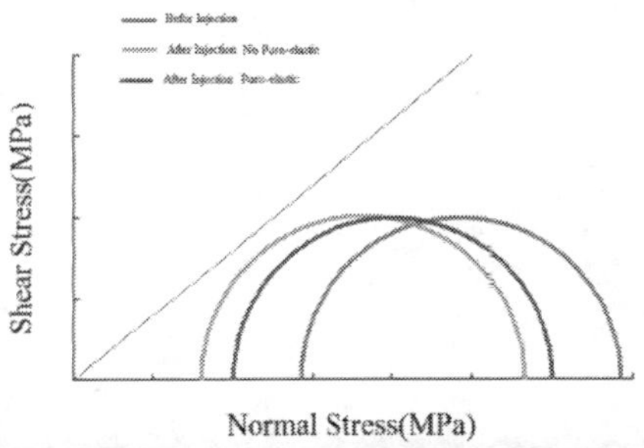




BATTELLE

# Intact & Fractured Rock Failure

(a) *This case is for illustrative purposes only – normally the Mohr circle would not move without changing size in a real injection scenario.*

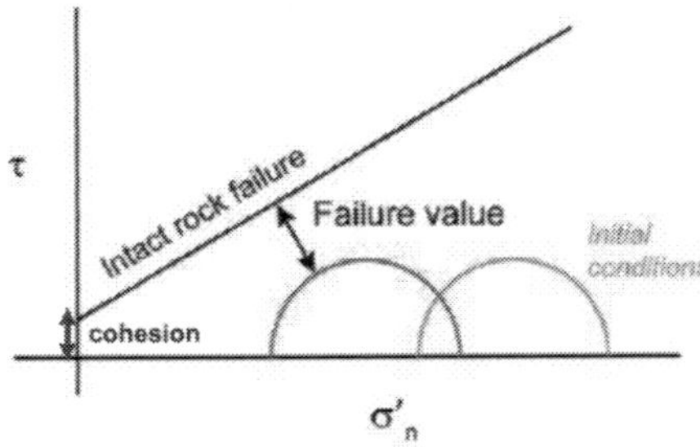


(b) *This case is illustrative of injection in a strike slip stress regime where the Mohr circle enlarges due to the poroelastic effect.*

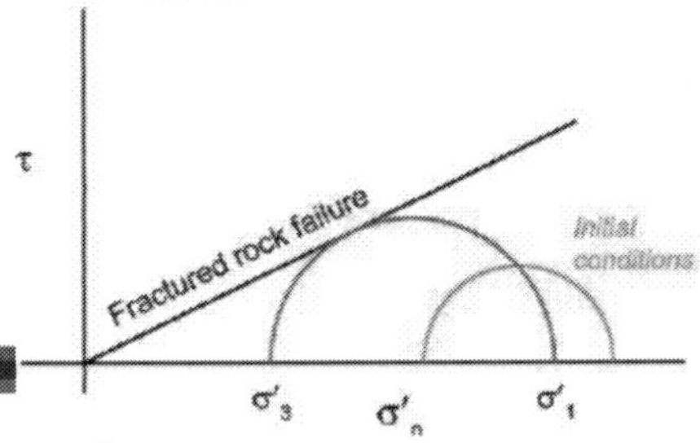


- ❑ Shear failure is said to occur if the Mohr circle plotted after injection hits the failure envelope.
- ❑ Distance from the envelope implies minimal risk of fracture activation.
- ❑ A weak or highly naturally fractured rock has a very low value of rock cohesion.

Olden, P., G. Pickup, M. Jin, E. Mackay, S. Hamilton, J. Somerville, and A. Todd. 2012. Use of rock mechanics laboratory data in geomechanical modelling to increase confidence in CO2 geological storage. International Journal of Greenhouse Gas Control, 11, 304-315.

9

BATTELLE

# MODEL DEVELOPMENT

10

BATTELLE

# Analysis Framework

**Input**

**Model Construction** — Input: Estimates for:
- Permeability/Porosity
- Thickness/Layering
- Geomechanical Parameters
- Other model assumptions

Output: Simulation-ready Reservoir and Geomechanical grids, with *Aquifer*, *Caprock*, and *Overburden*.

**Baseline Simulations** — Input: Numerically tuned shoe-box or single-well radial model.

Output: Pressure and stress-field response in *Aquifer*, *Caprock*, and *Overburden* for a baseline injection scenario.

**Sensitivity Studies** — Input: Uncertainty bandwidths for geomechanical parameters.

Other site-specific sensitivities (e.g. varying injection rate, boundary conditions, etc.)

Delineated injection scenarios and corresponding models.

Output: Assessment of vertical uplift, risk of shear and tensile failure

**Additional Modeling** — Input: Natural fractures

Stress-enhanced permeability

Output: **Quantify effective capacity**

**Output**



BATTELLE

# Arches Province in the Midwest US: East Bend Well Site

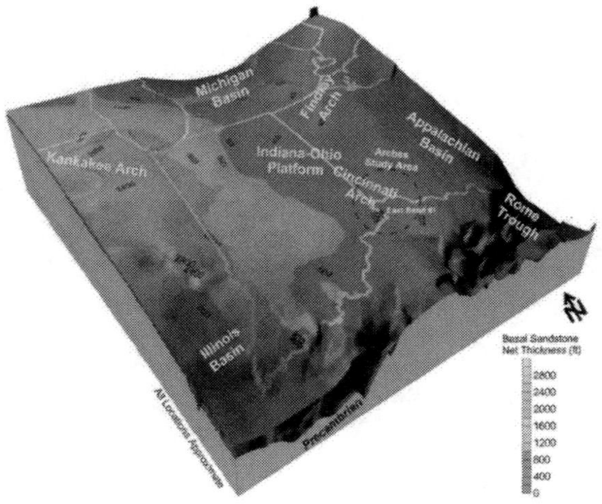


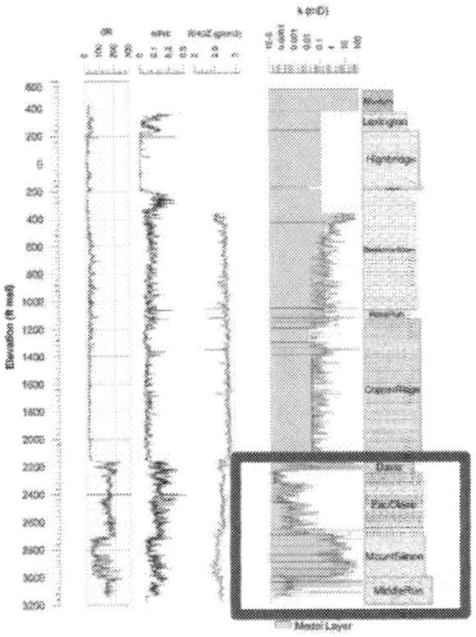



BATTELLE

# Model Construction

- 25,000 Acres
- 1000 x 1000 ft
- Grid refinement around Injector
- 30 years of injection

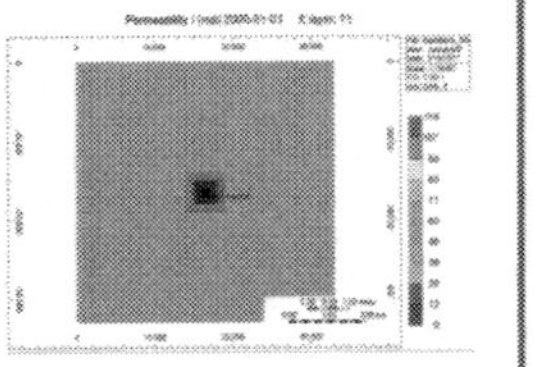


- Collocated Geomechanical and Reservoir grids
- Middle Run (underburden) to Surface.
- Log or literature-based stress gradients, static Young's modulus, Poisson's ratio

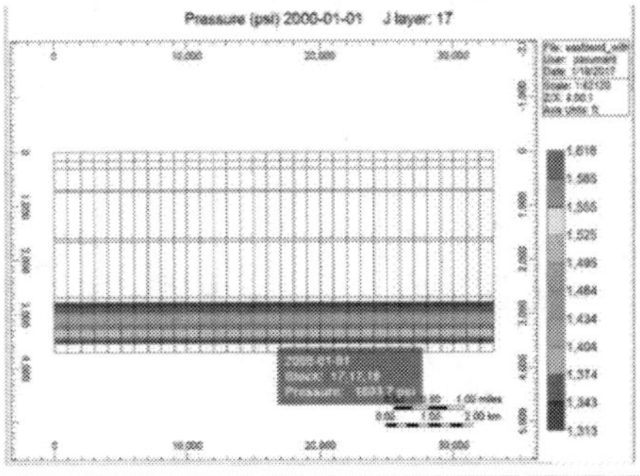


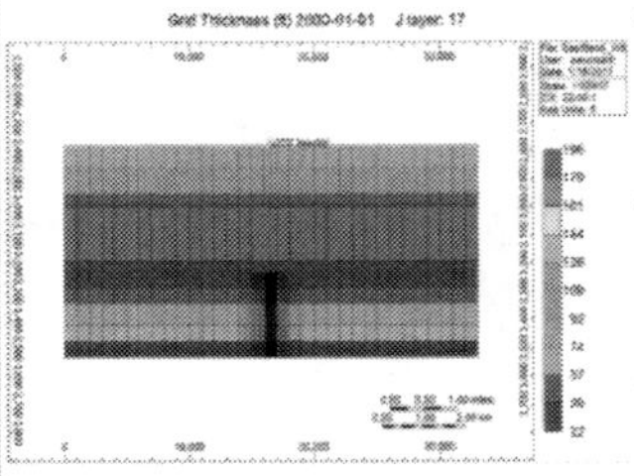


- Davis overlying the Eau Claire overlying the Mt. Simon
- 0.733 psi/ft injection limit or 2500 psi in Mt Simon.
- Assumptions for relative perm, homogeneity, etc.

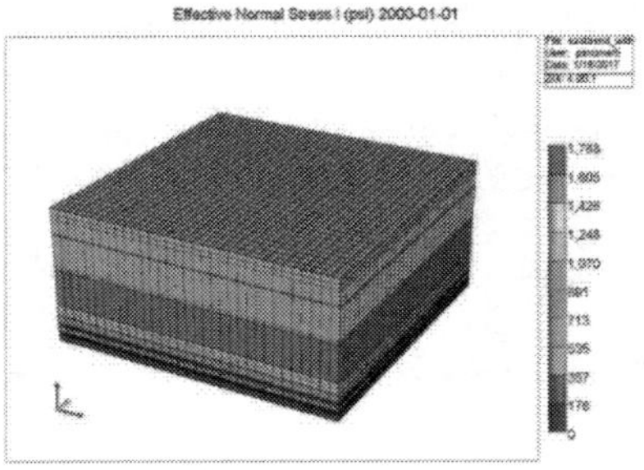


13 **BATTELLE**

# Characterization: Permeability

- ❑ Pressure Transient-Analysis on 3 brine injection fall-off tests
- ❑ ~1000 ft radius of investigation

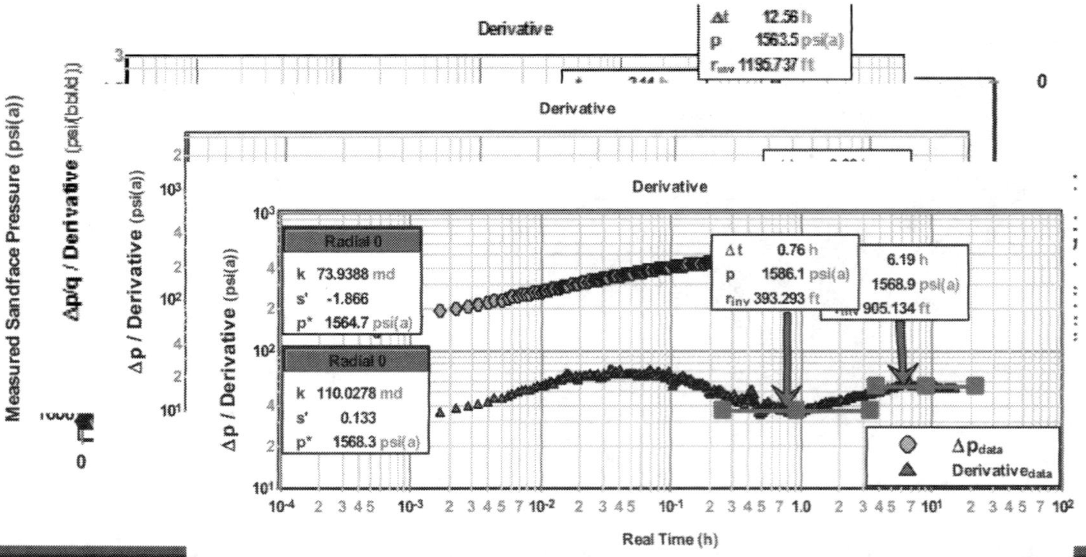


Battelle. 2010. CO2 Injection Test in the Cambrian-Age Mount Simon Formation, Duke Energy East Bend Generating Station, Boone County, Kentucky. Conducted by the Midwest Regional Carbon Sequestration Partnership (MRCSP). Morgantown, WV: U.S. Department of Energy, National Energy Technology Laboratory, DOE-NETL Cooperative

14 **BATTELLE**

# Characterization:
## Minimum Horizontal Stress

Log, mini-frac test data and literature-based regional geomechanical trends enable constraining the stress in the Mt. Simon.

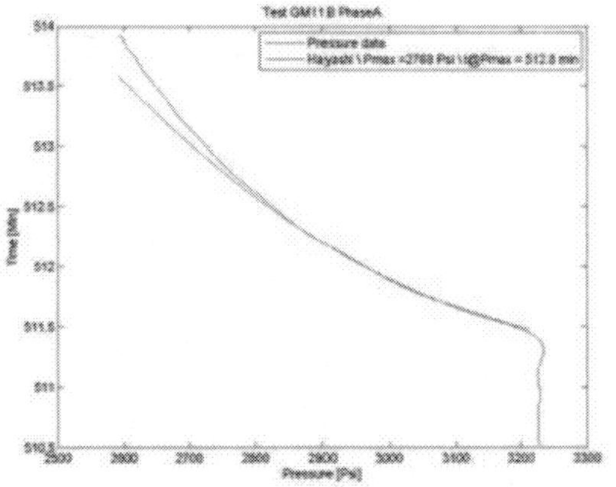

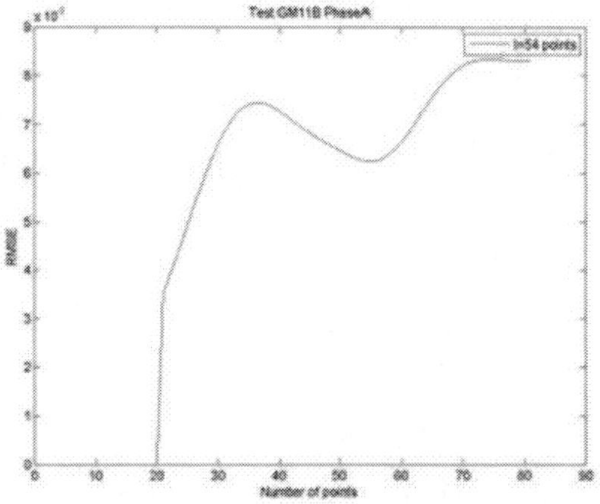

Cornet, F., and Battelle. 2014. Results from the In Situ Stress Characterization Program, Phase 1: Hydraulic Tests Conducted in the FutureGen Stratigraphic Pilot Well. Conducted by The FutureGen Industrial Alliance, Inc., . Washington,DC: U.S. Department of Energy, DOE Award NumberDE-FE0001882

15

BATTELLE

# Sensitivity Scenarios

| Scenario # | Boundary Condition | Biot's Coefficient | Young's Modulus | Max. Horizontal Stress Gradient |
|---|---|---|---|---|
| 1. Base Case (Most Conservative) | Closed | 1 | Low | High |
| 2. | Closed | 1 | Low | Low |
| 3. | Closed | 1 | High | Low |
| 4 | Closed | 0.8 | High | Low |
| 5 (Most Optimistic) | Open | 0.8 | High | Low |

16

BATTELLE

# RESULTS AND ANALYSIS

17

BATTELLE

# Injection into the Mount Simon:
## $CO_2$ Volumes and Pressure Increase

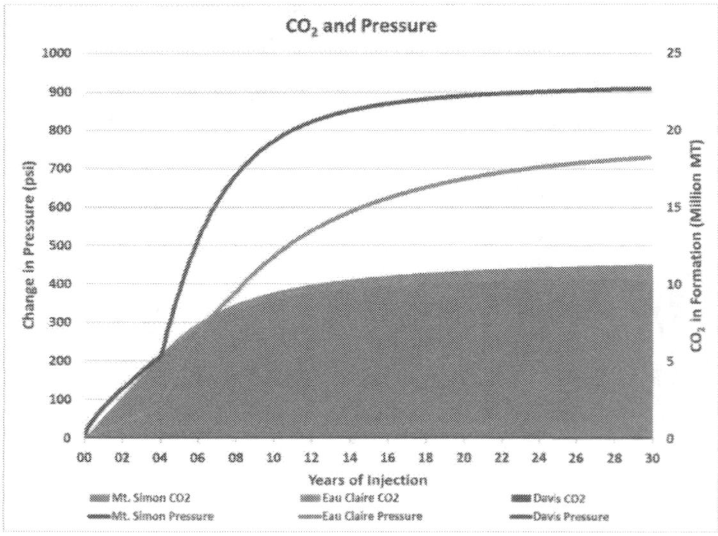


- ❑ Δp of ~900 psi in the Mt. Simon.
- ❑ Δp of ~725 psi in the Eau Claire.
- ❑ Approx. 11.25 million MT of $CO_2$ stored.
- ❑ ~1 million MT of $CO_2$ migrates upward into the Eau Claire.

18

BATTELLE

## Injection into the Mount Simon:
### $CO_2$ Plume and Pressures Attained

**Gas Saturation**

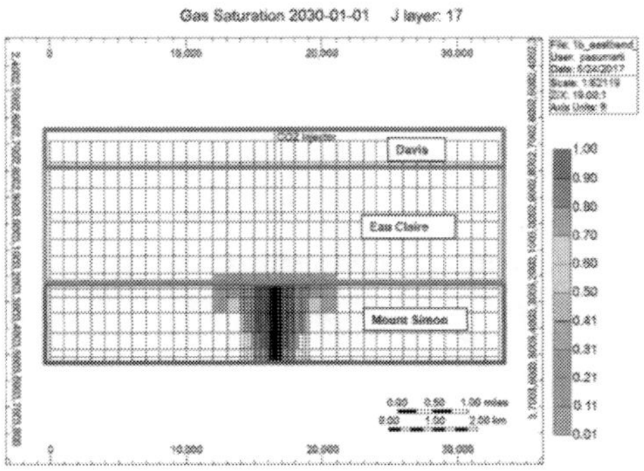


$CO_2$ plume is around ~5000 ft wide and penetrates up to lower Eau Claire.

**Pressure**

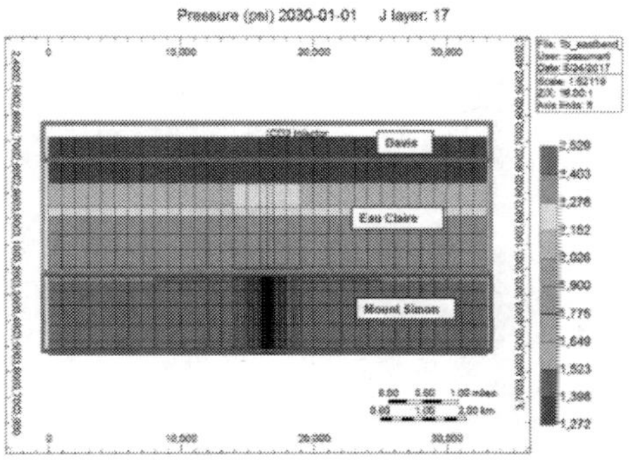


Lower Eau Claire pressure increases to ~2300 psi.

19

BATTELLE

## Injection into the Mount Simon:
### Stress-Strain Impact

**Δ Min. Effective Stress**

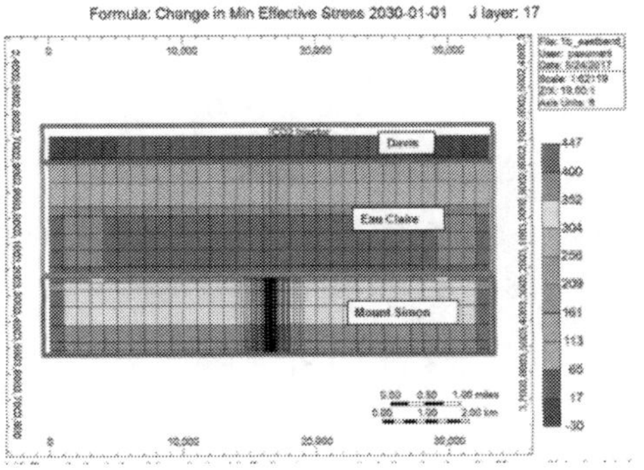


Reduction in effective stress is more pronounced in lower Eau Claire.

**Volumetric Strain**

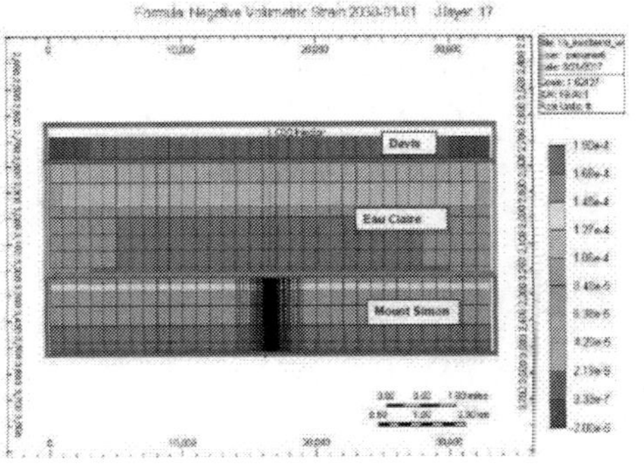


Pore-space deformation occurs mostly in the Eau Claire and Mt. Simon.

20

BATTELLE

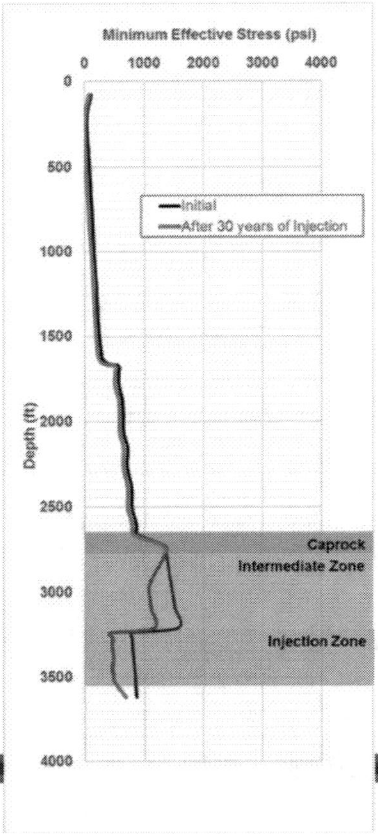


# Injection into the Mount Simon:
## Localized Stress Effects

- The minimum effective stress-profile from the underburden to the surface before and after injection.
- Negligible impact on layers caprock and above.



BATTELLE

# Injection into the Mount Simon:
## Surface Uplift

### Areal Displacement

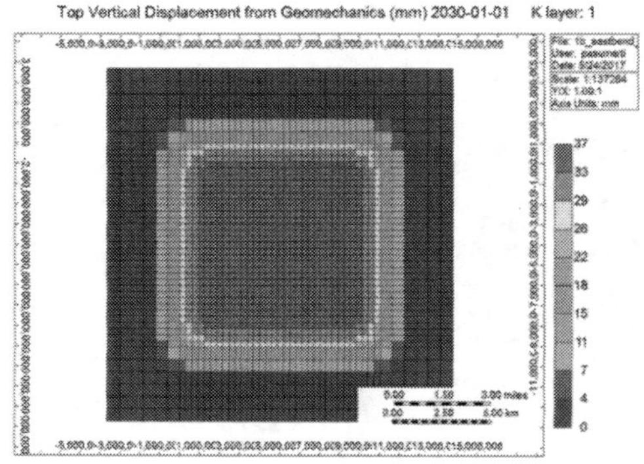


Near uniform uplift across 25,000 acres at the end of injection.

### Areal Displacement w.r.t. $CO_2$ Volume

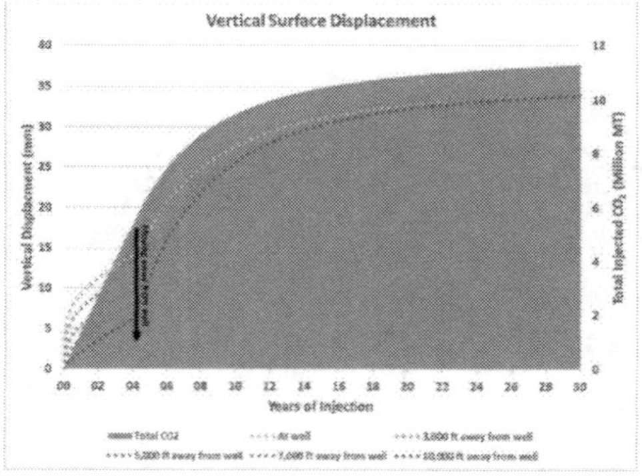


Surface uplift of approx. 32 mm with ~11.25 million MT of injection.



BATTELLE

# Injection into the Mount Simon:
## Caprock Integrity (Davis)

### Principle Effective Stresses

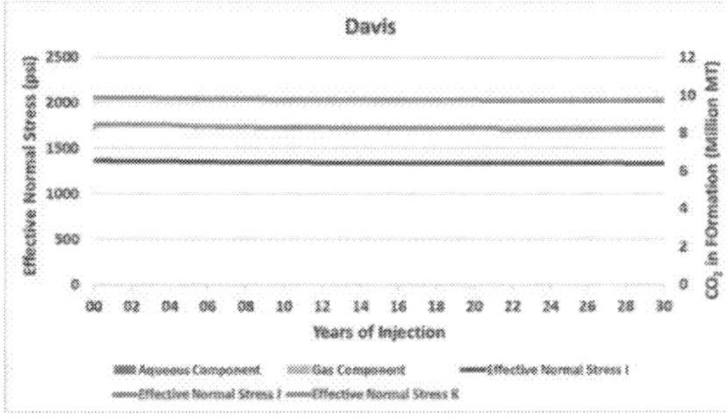


Stresses in caprock are unchanged.

### Mohr Circle

No shear or tensile failure.

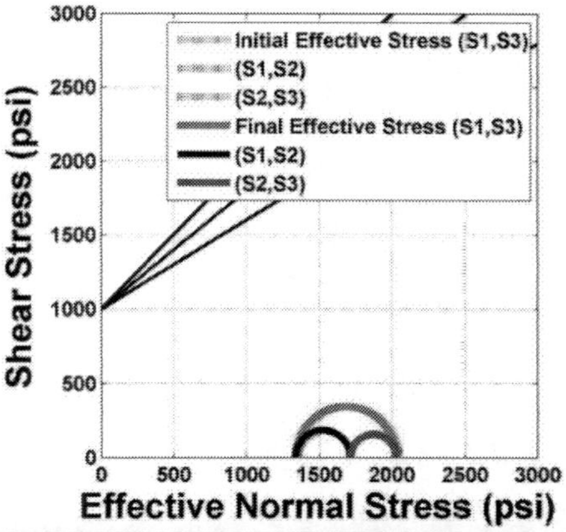




BATTELLE

# Injection into the Mount Simon:
## Intermediate Zone Integrity (Eau Claire)

### Principle Effective Stresses

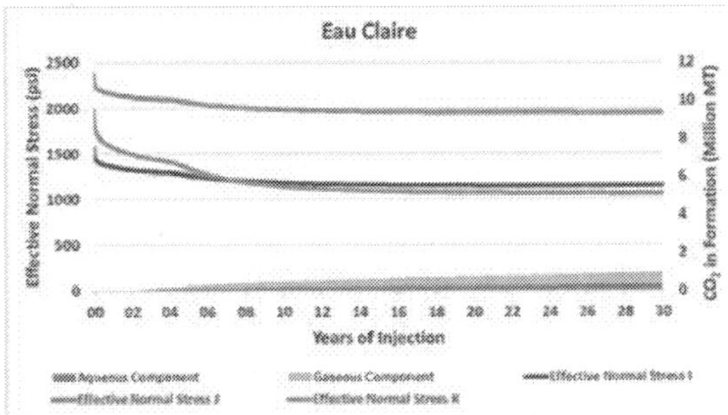


Vertical stress affected more than horizontal.
Regime change after 8 years of injection.

### Mohr Circle

No shear or tensile failure.

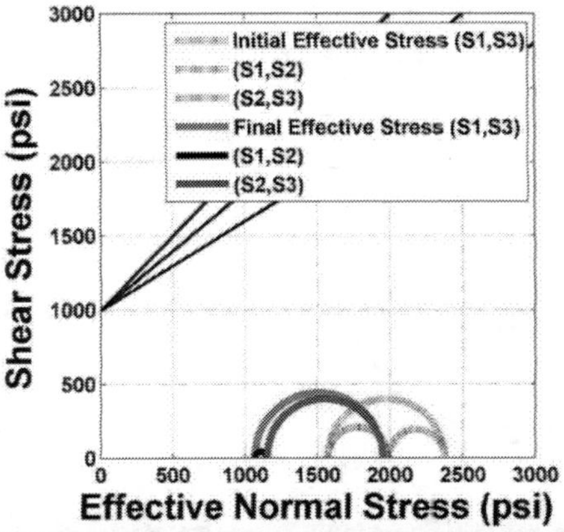




BATTELLE

# Injection into the Mount Simon:
## Reservoir Integrity

### Principle Effective Stresses

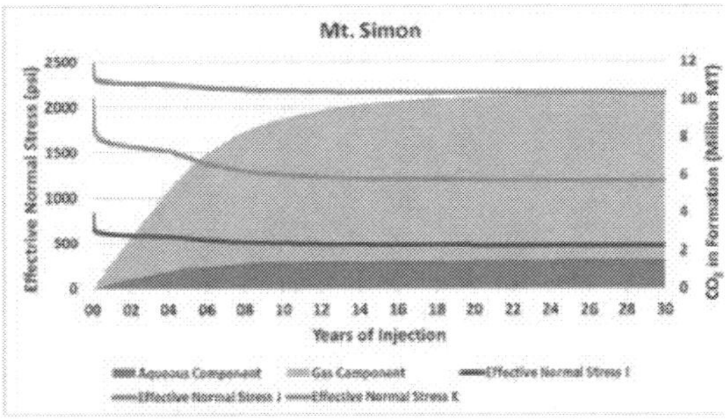


Vertical stress affected more than horizontal.
No regime change.

### Mohr Circle

No shear or tensile failure.

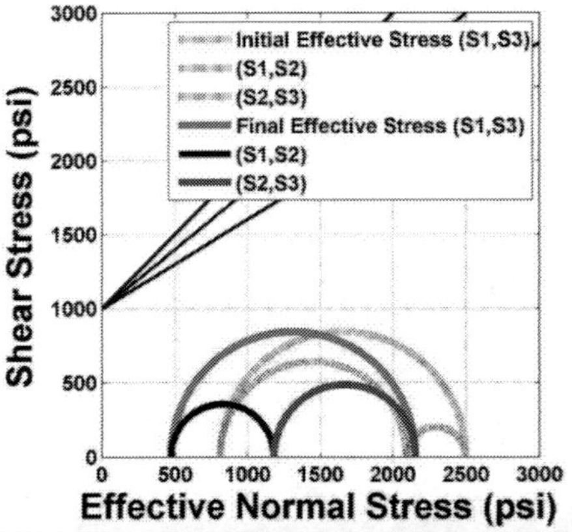




# Sensitivity Scenarios

| Scenario # | Shear Failure? | Tensile Failure? | Surface Uplift (mm) | Storage Capacity (millions of MT) |
|---|---|---|---|---|
| 1 - Base Case (Most Conservative) | No | No | 32 | 11.25 |
| 2 | No | No | 32 | 11.25 |
| 3 | No | No | 27 | 11.25 |
| 4 | No | No | 22 | 12.5 |
| **5 (Most Optimistic)** | **No** | **No** | **1.2** | **38.25** |

## Summary

- ❑ Simulations suggest minimal risk of tensile or shear failure.
  - ❑ Minimum effective stress in all three zones is >500 psi.
  - ❑ Even the most conservative shear envelope is sufficiently distant from Mohr's circles.
- ❑ Up to 32 mm of uplift may be expected.
- ❑ A stress-regime change may occur in the Intermediate Eau Claire
  - ❑ Simulations do any indicate that this poses any threat to the operation.
- ❑ Evidence in literature for optimistic modelling conditions.
- ❑ **Conservative estimate of effective capacity is ~11.25 million MT.**

27

BATTELLE

THIS RESEARCH WAS SUPPORTED BY THE U.S. DOE / NATIONAL ENERGY TECHNOLOGY LABORATORY (CONTRACT DE-FE0023330) AND THE OHIO DEVELOPMENT SERVICES AGENCY OHIO COAL DEV. OFFICE (GRANT CDO-D-14-17).

28

BATTELLE

CARBON MANAGEMENT TECHNOLOGY CONFERENCE

# CMTC-486411-MS

# Analytical Corrections to $CO_2$-Brine Core Relative Permeability for Low Rate Flow Modeling

Avinoam Rabinovich, Tel Aviv University



## Abstract

Relative permeability of $CO_2$ and brine is one of the fundamental parameters controlling flow related to carbon storage in saline aquifers. Core samples recovered from subsurface formations are characterized in laboratory experiments to determine effective core relative permeability curves. Typically, coreflooding experiments are conducted at high injection rates so that the resulting flow is viscous dominated. However, at lower rates, it has been shown that the effective curves may change as capillary heterogeneity effects become significant. Using relative permeability determined by conventional coreflooding in simulations with low flow rates, e.g., to model $CO_2$ migration in aquifers, may incur significant error.

A new method for calculating low flow rate relative permeability curves is presented. The method is based on approximate analytical solutions for effective relative permeability under steady state and capillary limit flow conditions. Derivation is carried out using power law averaging, assuming log normally distributed core permeability. We validate the analytical solution by comparison to numerical solutions for a wide range of cases. An additional correction for the $CO_2$ curves is shown to be necessary and derived by matching analytical and numerical results. Given a core which has been characterized by conventional high rate coreflooding experiments, the current method gives a fast and efficient correction for low flow rate applications. It circumvents the need for additional experiments or computationally expensive coreflooding simulations.

# Introduction

Prediction of $CO_2$ flow behavior in aquifers is essential for successful large scale storage operations. Numerical modeling is the main tool for conducting such predictions and investigations. A first step in flow modeling is to characterize the properties of the storage reservoir, particularly estimation of the relative permeability, $k_{rj}$, where $j$ designates phase ($CO_2$/brine). A common method for obtaining $k_{rj}$ estimations is by $CO_2$ coreflooding experiments. High injection rates are often used in these experiments to minimize capillary end effects and the resulting curves are therefore limited to high flow rate (viscous dominated) simulations, e.g., flow in regions near $CO_2$ injection wells. However, accurate modeling of the $CO_2$ plume migration requires considering the gravity-capillary dominated flow far from the wells (or after wells have been shut off, i.e., post-injection), where flow rates are low. In this case, using the conventional $k_{rj}$ curves may lead to errors in prediction. The goal of this work is to derive analytical formulas which correct the conventional, high flow rate $k_{rj}$ measurements for use in low rate, post-injection $CO_2$ storage modeling.

Aquifer properties are characterized on various scales. Simulations are conducted on reservoir models of dimension on the order of 100 m to 1 km. These are discretized into grid blocks of typical size 1-10 m while relative permeability is measured on rock samples of smaller, core-scale resolution (order of 10 cm). The measurements are upscaled and assigned to grid blocks for simulation. It is common practice to assign a single set of relative permeability curves to the core sample and to assume that the core is homogeneous, ignoring any sub-core heterogeneity. However, a number of investigations have focused on the small scale, sub-core permeability variations and show that these can affect the large scale flow (Ringrose et al., 1993; Perrin and Benson, 2010; Kuo et al., 2010; Kuo and Benson, 2015; Li and Benson, 2015). The sub-core scale is considered here to be the smallest scale in which Darcy's law still holds (order of 1 mm). In this work, we distinguish between sub-core scale relative permeability curves, $k_{rj}$, which are considered to be intrinsic properties (we refer to these as characteristic curves), and the core effective (or equivalent) relative permeability $k_{rj}^{\text{eff}}$.

An important feature of sub-core heterogeneity is spatial variations in capillary pressure curves known as capillary heterogeneity. These have been measured directly by Pini et al., 2012 and Pini and Benson (2013). Capillary heterogeneity is often related to the heterogeneities in porosity and permeability, as suggested by the well-known Leverett J-function scaling relationship (Leverett, 1941). Flow at sufficiently low rates is driven by large capillary forces compared to viscous ones and phenomenon such as capillary trapping occurs. In this regime, capillary heterogeneity will also impact coreflooding measurements of $k_{rj}^{\text{eff}}$ and these will differ from the characteristic curves ($k_{rj}$). As a result, $k_{rj}^{\text{eff}}$ curves are in fact rate dependent. This has been the focus of attention recently and investigated on the core scale by Perrin et al. (2009); Kuo et al. (2010); Krause (2012a); Kuo and Benson (2015); Krause and Benson (2015) and on larger scales by Virnovsky et al. (2004); Lohne et al. (2006); Neuweiler et al. (2011); Rabinovich et al. (2015).

Evidence has shown that a slight inaccuracy in characterizing the multiphase flow properties can accumulate and lead to significant error in modeling long-term $CO_2$ sequestration (Li and Benson, 2015). Particularly, using relative permeability curves which do not incorporate the impact of capillary heterogeneity may incur error in $CO_2$ storage simulations (Rabinovich et al., 2015). It is therefore paramount to accurately characterize core relative permeability. Coreflooding experiments conducted to estimate relative permeability are based on the extended Darcy's law using measurements of overall pressure drop in the core. Results represent core effective properties ($k_{rj}^{\text{eff}}$) and incorporate capillary heterogeneity effects when these are significant. Therefore, it should be possible to use a single set of $k_{rj}^{\text{eff}}$ curves for a given flow rate to characterize the core, thus avoiding sub-core resolution flow modeling. This is demonstrated in Li et al. (2013) Fig. 10 and the discussion therein.

Currently, coreflooding experiments are typically conducted at high flow rates. This is to minimize the capillary end effect, a phenomenon in which the wetting-phase saturation at the core outlet increases due to the abrupt change of capillary pressure ($P_c$) to zero when phases exit the core (Richardson et al., 1952). Therefore, coreflooding experiments usually measure $k_{rj}^{\text{eff}}$ associated with a viscous dominated flow which is equivalent to the characteristic curves, i.e., $k_{rj}^{\text{eff}} = k_{rj}$ is the conventionally used core relative permeability.

For accurate estimation of $k_{rj}^{\text{eff}}$, rate dependent curves departing from $k_{rj}$ should be considered. Ideally, this would entail conducting multiple coreflooding experiments spanning a wide range of injection rates. However, this approach is not likely to be adopted due to the cost and complexity of carrying out many experiments for each core sample. We note that if the core has sufficiently low heterogeneity, $k_{rj}^{\text{eff}}$ would not vary significantly and a single set of curves can be used for all flow rates. A more feasible approach is using corefooding simulations to calculate rate dependent $k_{rj}^{\text{eff}}$ (Krause and Benson, 2015). However, this requires estimating an accurate sub-core permeability field which is a difficult problem in itself. Current methods for sub-core permeability estimation (Krause et al., 2013) entail imaging the core with CT scans and applying numerous coreflooding simulations in an iterative process, creating possible cost and complexity issues. It would be very useful to have a simple method to transition from high flow rate $k_{rj}^{\text{eff}}$ (equivalent to the characteristic curves) to low flow rate curves for modeling $CO_2$ migration.

This work presents a semi-analytical approximate solution to core effective relative permeability for capillary dominated flow. It relates between sub-core properties: $k_{rj}$, $P_c$, $k_G$ (permeability geometric mean) and $\sigma_y^2$ (variance of log-permeability, $y = \ln k$), and core effective properties $k_{rj}^{\text{eff}}$. The solution allows to easily transition from the conventional $k_{rj}^{\text{eff}} = k_{rj}$, measured from viscous dominated (high rate) coreflooding experiments, to new curves which can be used to model low rate $CO_2$ migration more accurately. The effective property problem is simplified by assuming capillary limit conditions, a reasonable assumption for upscaling $CO_2$ storage simulations (Mouche et al., 2010; Behzadi and Alvarado, 2012). Log normally distributed permeability ($k$) is further assumed and an estimation of the core $k$ statistical properties ($k_G$ and $\sigma_y^2$) is required. The solution is based on power law averaging and heuristic formulas to correct the $CO_2$ relative permeability curves. Corrections are shown to be particularly important for cases of $P_c$ curves exhibiting capillary entry pressure. Extensive validation is carried out by comparing the new solution to numerical calculations for a wide range of cases, i,e, varying $\sigma_y^2$, $P_c$ curves and $k_{rj}$ curves.

We note that the current formula considers only drainage coreflooding experiments and deriving $k_{rj}^{\mathrm{eff}}$ for imbibition is not addressed. Furthermore, we consider isotropic $k$ here and anisotropic correlation structures should be considered in the future. This work is an extension to the two-dimensional case of effective relative permeability presented in Rabinovich et al. (2016). The solution is extended here to three-dimensions and also considers additional complications such as capillary entry pressure effects. In fact, the $CO_2$ relative permeability solution here is found to be quite different from the 2D case due to these effects. Moreover, Rabinovich et al. (2016) do not consider the various aspects of applications to core modeling.

# Problem statement

We model coreflooding experiments as immiscible flow of supercritical $CO_2$ and brine assuming incompressible rock and flowing phases. Similar coreflooding models have been considered by Krause (2012b) and Li (2014). The governing equations are obtained from a mass balance and the extended Darcy's law for each phase. They are given by

$$\phi\frac{\partial S_j}{\partial t}-\nabla\cdot\left[\frac{k_{rj}}{\mu_j}\mathbf{k}\cdot\nabla\left(p_j+\rho_j g z\right)\right]=0, \tag{1}$$

where $\phi$ is the porosity of the rock, $S_j$ the saturation of phase $j$ ($j=\mathrm{CO_2}$ or $j=w$ for water/brine phase), $k_{rj}$ the relative permeability to phase $j$, $\mu_j$ the viscosity of phase $j$, $\mathbf{k}$ the absolute permeability tensor, $p_j$ the pressure of phase $j$, $\rho_j$ is phase density, $g$ is gravitational acceleration, and $z$ is the vertical coordinate. The constraint $S_w+S_{\mathrm{CO_2}}=1$ applies and the pressures of the nonwetting phase and the wetting phase are related by

$$p_{\mathrm{CO_2}}-p_w=P_c(S_w), \tag{2}$$

where $P_c$ is capillary pressure.

Given the three sub-core properties: $\mathbf{k}$, $P_c(S_w)$ and $k_{rj}(S_w)$, we seek to find the core effective property $k_{rj}^{\mathrm{eff}}$. Isotropic permeability is assumed so that $k(x,y,z)$ is a scalar function of space, i.e., $k_x=k_y=k_z$ and (x,y,z) is the three-dimensional Cartesian coordinate system, where $x$ is aligned with the direction of flow injection. The $P_c(S_w)$ curve can be obtained using various laboratory measurement techniques (Omoregle, 1988). We adopt the $J$-function approach (Leverett, 1941) to model capillary heterogeneity; i.e.,

$$P_c(S_w,k)=\alpha\sqrt{\frac{\phi}{k}}J(S_w), \tag{3}$$

where $\alpha$ is a scaling coefficient which accounts for interfacial tension, contact angle and unit conversion. Two of the most widely used $J$-functions are the Brooks-Corey (Brooks and Corey, 1966) (denoted BC) and van Genuchten (van Genuchten, 1980) (denoted VG) models given by

$$J_{BC}(S_w)=(\widetilde{S}_w)^{-1/\lambda} \tag{4}$$

and

$$J_{VG}(S_w)=\left[(\widetilde{S}_w)^{-1/m}-1\right]^{1-m}, \tag{5}$$

respectively, where $\lambda$ and $m$ are fitting parameters, $\widetilde{S}_w = (S_w - S_{wi})/(1 - S_{wi})$ and $S_{wi}$ is the irreducible wetting phase saturation. We note that these are drainage curves and therefore hysteresis effects are not considered.

Eq. (3) with a single $J$-function has been used to represent sub-core capillary pressure in numerical and experimental work (e.g., Kuo et al. (2010),Krause et al. (2013) and Wei et al. (2014)). Sometimes, it is further assumed that the global (effective) core capillary pressure has the same structure (Pini et al., 2012; Krause et al., 2013). In this work, only the sub-core scale $P_c$ is assumed to be of the type presented in Eq. (3) with a given $J$-function. However, in experimental settings, $P_c$ should be measured carefully to insure it is representative of the sub-core scale as it may differ from the core effective $P_c$ curves.

Core characteristic relative permeability $k_{rj}$ are typically measured directly via high flow rate coreflooding experiments. The most widely applied techniques are steady state or unsteady methods. Considering the steady state method, a mixture of the two phases is injected into the fully brine saturated core at a given rate and fractional flow until steady state is reached. The curves are measured using extended Darcy's law as follows

$$k_{rj} = \frac{q_j \mu_j L}{k^{\text{eff}} A \Delta p_j}, \tag{6}$$

where $q_j$ is the injection rate (flux) of phase $j$, $\Delta p_j$ is the phase pressure difference between inlet and outlet, $L$ is core length, $A$ is the area perpendicular to flow and $k^{\text{eff}}$ is the core effective absolute permeability. Each experiment consisting of a given injection fractional flow produces a point in the $k_{rj}^{\text{eff}}$ curves and fractional flows are varied between 100% and 0% to obtain the full curves. The experiment with 100% brine injection results in single phase flow which by Darcy's Law yields the absolute permeability, i.e.,

$$k^{\text{eff}} = \frac{q_w \mu_w L}{A \Delta p_w}. \tag{7}$$

We consider here a common model for uniform drainage characteristic curves given by

$$k_{rw} = (\widetilde{S}_w)^{n_w}, \tag{8a}$$

$$k_{rCO_2} = \kappa_r^{\max}(1 - \widetilde{S}_w)^{n_{CO_2}}, \tag{8b}$$

where $\kappa_r^{\max}$, $n_w$ and $n_{CO_2}$ are fitting parameters determined by matching Eqs. (8a,b) to experimental results (Eqs. (6) and (7)). We note that for coreflooding measurements to be equivalent to characteristic curves, $k_{rj}$ must be uniform in the core. This is typically assumed for core samples of a given lithofacies and consistent with the assumption of a single $J$-function in the core.

If total injection rate $q = q_w + q_{CO_2}$ in the coreflooding experiments described above is not large enough and the core is heterogeneous, the measured relative permeability curves will no longer represent characteristic curves, but rather effective curves $k_{rj}^{\text{eff}} \neq k_{rj}$. At sufficiently low $q$, when the capillary forces dominate compared to viscosity (capillary limit), $k_{rj}^{\text{eff}}$ will converge to a given function of saturation, independent of injection rate. These low injection rate curves are the capillary limit (CL) effective curves and our goal is to derive semi-analytical expressions for these $k_{rj}^{\text{eff}}(S_w)$ curves.

**Capillary limit effective relative permeability**

The CL method of upscaling is described in the literature (e.g., Pickup and Sorbie (1996), Rabinovich et al. (2016)) and a short summary is presented in the following. We assume steady state, small viscous pressure gradients compared to capillary pressure gradients and negligible impact of gravity. The flowing phases are therefore in capillary equilibrium in which $P_c$ is constant. Thus, for a given constant value of $P_c(\widetilde{S}_w, k) = \widehat{P}_c$, the saturation field is obtained by inversion as follows

$$\widetilde{S}_w(x,y,z) = P_c^{-1}(\widehat{P}_c, k(x,y,z)), \tag{9}$$

where constant core porosity $\phi$ is assumed. The spatial variation of relative permeability is now given by the relationship $k_{rj}(x,y,z) = k_{rj}(\widetilde{S}_w)$ with $\widetilde{S}_w$ given by Eq. (9). Furthermore, the phase permeability is known and given by

$$K_j(x,y,z) = k_{rj}(\widetilde{S}_w) \cdot k(x,y,z) = k_{rj}(P_c^{-1}(\widehat{P}_c, k(x,y,z))) \cdot k(x,y,z). \tag{10}$$

If we now relax the CL assumption, allowing flow in the system, but still use the CL relationships, the problem reduces to two separate flow problems

$$\nabla \cdot \left[ \frac{K_w(x,y,z)}{\mu_w} \nabla p_w \right] = 0, \tag{11a}$$

$$\nabla \cdot \left[ \frac{K_{CO_2}(x,y,z)}{\mu_{CO_2}} \nabla p_{CO_2} \right] = 0, \tag{11b}$$

where $K_j(x,y,z)$ are known functions given by (10). Typical boundary conditions for an effective property problem are considered - given constant pressure at the inlet ($p_j^{in}$) and outlet ($p_j^{out}$) of the core and no-flow condition on the curved face of the cylinder. The core effective phase permeability is defined similar to a single phase flow $k^{\text{eff}}$ problem as

$$K_j^{\text{eff}} = \frac{\langle q_{j,x} \rangle \mu_j L}{A\left(p_j^{in} - p_j^{out}\right)}, \tag{12}$$

where $\langle q_{j,x} \rangle$ is the average flux of phase $j$ in the direction of flow injection. The core effective relative permeability is then given by

$$k_{rj}^{\text{eff}} = \frac{K_j^{\text{eff}}}{k^{\text{eff}}}. \tag{13}$$

The procedure for CL $k_{rj}^{\text{eff}}$ calculation can be summarized as follows:

1. Calculate $k^{\text{eff}}$ from a coreflood experiment by Eq. (7).

2. Choose a value $\widehat{P}_c$ of capillary pressure.

3. Calculate the saturation field $\widetilde{S}_w(k)$ by Eq. (9) and $\left\langle \widetilde{S}_w \right\rangle$ by arithmetic averaging of $\widetilde{S}_w$.

4. Calculate $K_j^{\text{eff}}$ by Eq. (12) for a given realization of $k$ (or by an analytical approach as done in the following), substituting Eq. (10) in Eqs. (11a) and (11b) and solving numerically to obtain $\langle q_{j,x} \rangle$.

5. Obtain the point $\left(\left\langle \widetilde{S}_w \right\rangle, k_{rj}^{\text{eff}}\right)$, with $k_{rj}^{\text{eff}}$ from Eq. (13).

6. Repeat steps 2-5 for different choices of $\widehat{P}_c$, scanning values of $\left\langle \widetilde{S}_w \right\rangle$ from 0 to 1 to generate the complete $k_{rj}^{\text{eff}}$ curves.

# Semi-analytical solution

We now derive semi-analytical expressions for $k_{rj}^{\text{eff}}$. The CL method is used as described in the previous section, however, the numerical solution of the equations in step 4 will not be necessary and instead, closed form formulas for $K_j^{\text{eff}}$ will be proposed. This will allow carrying out calculations with hardly any computational cost.

Assuming the VG $J$-function of Eq. (5) in capillary pressure (Eq. (3)) we can invert as described in Eq. (9) to obtain

$$\widetilde{S}_w(\widetilde{P}_c, k) = \left[\left(\widetilde{P}_c\sqrt{k}\right)^{\frac{1}{1-m}} + 1\right]^{-m}, \tag{14}$$

where $\widetilde{P}_c = \widehat{P}_c/(\alpha\sqrt{\phi})$ is a constant which will be varied between zero and infinity to cover the full range of saturations $(0 < \left\langle \widetilde{S}_w \right\rangle < 1)$, as noted in step 6 in the previous section. Then, an expression for $K_j$ is derived by substitution of Eq. (14) in Eqs. (8a,b) to arrive at

$$K_w(\widetilde{P}_c, k) = k\left[\left(\widetilde{P}_c\sqrt{k}\right)^{\frac{1}{1-m}} + 1\right]^{-n_w m} \tag{15a}$$

$$K_{CO_2}(\widetilde{P}_c, k) = \kappa_r^{\max} k \left\{1 - \left[\left(\widetilde{P}_c\sqrt{k}\right)^{\frac{1}{1-m}} + 1\right]^{-m}\right\}^{n_{CO_2}}. \tag{15b}$$

A similar derivation is carried out for the BC $J$-function of Eq. (4) to arrive at

$$\widetilde{S}_w(\widetilde{P}_c, k) = \begin{cases} \widetilde{P}_c k^{-\lambda/2}, & \widetilde{P}_c k^{-\lambda/2} < 1 \\ 1 & \widetilde{P}_c k^{-\lambda/2} > 1 \end{cases}, \tag{16}$$

which then by substitution in Eq. (8a,b) gives

$$K_w(\widetilde{P}_c, k) = \begin{cases} \widetilde{P}_c^{n_w} k^{1-\lambda n_w/2}, & \widetilde{P}_c k^{-\lambda/2} < 1 \\ 1 & \widetilde{P}_c k^{-\lambda/2} > 1 \end{cases} \tag{17a}$$

$$K_{CO_2}(\widetilde{P}_c, k) = \begin{cases} \kappa_r^{\max} k \left(1 - \widetilde{P}_c k^{-\lambda/2}\right)^{n_{CO_2}}, & \widetilde{P}_c k^{-\lambda/2} < 1 \\ 0 & \widetilde{P}_c k^{-\lambda/2} > 1 \end{cases}. \tag{17b}$$

The piecewise functions in Eqs. (16) and (17a,b) are used to ensure that unphysical values of $\widetilde{S}_w > 1$ are eliminated. This is an important feature resulting from the inclusion of capillary entry pressure in the BC type $P_c(\widetilde{S}_w)$ model, i.e., for $\widetilde{S}_w = 1$, $P_c$ is nonzero as seen in Eq. (4). Such $K_{CO_2}$ fields are challenging to average because many methods such as the geometric mean will (by definition) average to zero if any point in the field is zero, resulting in possibly meaningless averages.

We first consider layered permeability briefly. It is well known that $k^{\text{eff}} = k_A$, the arithmetic mean, for flow parallel to the layers, and $k^{\text{eff}} = k_H$, the harmonic mean, for flow perpendicular to the layers. This result has been extended to CL effective relative permeability by Corey and Rathjens (1956), Virnovsky et al. (2004) and Rabinovich et al. (2016), where an exact solution is given by

$$k_{rj}^{\text{eff}} = \frac{[K_j]_A}{k^{\text{eff}}}, \tag{18}$$

for flow parallel to the layers, and

$$k_{rj}^{\text{eff}} = \frac{[K_j]_H}{k^{\text{eff}}}, \tag{19}$$

for flow perpendicular to the layers, where $K_j$ are given here by Eqs. (15a,b) or (17a,b). Therefore, if core samples exhibit layered permeability, a transition to low flow rate curves is straightforward using Eqs. (18) and (19).

The core can be further modeled as comprised of numerous layers, each with permeability drawn from a distribution of known probability density function (PDF). Eqs. (18) and (19) are then expressed as the following integrals

$$k_{rj}^{\text{eff}} = \frac{1}{k_A} \int_{-\infty}^{\infty} K_j(\widetilde{P}_c, k) p(k) dk, \tag{20}$$

$$k_{rj}^{\text{eff}} = \frac{1}{k_H} \left[ \int_{-\infty}^{\infty} K_j(\widetilde{P}_c, k)^{-1} p(k) dk \right]^{-1}, \tag{21}$$

where $p(k)$ is the PDF of the distribution. The corresponding arithmetic average of $\widetilde{S}_w$ is given by

$$\langle \widetilde{S}_w \rangle = \int_{-\infty}^{\infty} \widetilde{S}_w(\widetilde{P}_c, k) p(k) dk. \tag{22}$$

Next, log-normal permeability fields are considered such that $y = \ln k$ is normally distributed and characterized by the mean $k_G$ and variance $\sigma_y^2$. Such fields are commonly considered on a range of scales Freeze (1975), and have been found to apply also to sub-core permeability (see Fig. 9 in Rabinovich (2017)). An isotropic correlation structure is assumed, i.e., $l_x = l_y = l_z$, where $l_x$, $l_y$ and $l_z$ are the dimensionless correlation lengths in the $x$, $y$ and $z$ directions (nondimensionalized by the domain length in the corresponding direction). Under these conditions, Rabinovich (2017) found that the approximation

$$k_{rw}^{\text{eff}} = \frac{[K_w]_G \left(1 + \sigma_{\Gamma_w}^2/6 + \sigma_{\Gamma_w}^4/72\right)}{k^{\text{eff}}} \tag{23}$$

applies to cases of sufficiently small variance of log phase permeability $\sigma_{\Gamma_w}^2$, where $\Gamma_w = \ln K_w$. However, it has been shown in Rabinovich et al. (2016) that for $k_{rCO_2}^{\text{eff}}$, variance $\sigma_{\Gamma_{CO_2}}^2$ are much larger and errors for Eq. (23) are generally large. We therefore adopt a different approach, also used in Rabinovich et al. (2016) and Rabinovich (2017), of power law averaging.

Power law averaging is a heuristic method for calculating effective permeability in one-, two- and three-dimensional $k$ fields (Journel et al., 1986; Deutsch, 1989). For a discrete permeability field, e.g., a numerical grid, the averaging is as follows

$$k^{\text{eff}} = \left( \sum_{i=1}^{N} v_i k_i^{\omega} \right)^{1/\omega}, \tag{24}$$

where $v_i$ is the volume fraction associated with permeability $k_i = \{k_1, k_2, k_3 ... k_N\}$ and $\omega$ is the power used for averaging. For continuous $k$ the power average is given by

$$k^{\text{eff}} = \left\{ \int_{-\infty}^{\infty} k^{\omega} p(y) dy \right\}^{1/\omega}, \tag{25}$$

where $p(y)$ is the PDF of $y = \ln k$.

In analogy to Eq. (25) and following Rabinovich et al. (2016); Rabinovich (2017), we propose using power law averaging for the calculation of $k_{rj}^{\text{eff}}$, i.e.,

$$k_{rj}^{\text{eff}} = \frac{1}{k^{\text{eff}}} \left\{ \int_{-\infty}^{\infty} \left[ K_j(\widetilde{P}_c, k) \right]^{\omega} p(y) dy \right\}^{1/\omega}, \tag{26}$$

where $K_j(\widetilde{P}_c, k)$ is given by Eqs. (15a,b) or (17a,b). The value of $\omega$ used in Eq. (26) is determined by solving Eq. (25) where $k^{\text{eff}}$ is known from coreflooding measurements. Values are between $\omega = -1$ (harmonic average), representing flow perpendicular to layered media, and $\omega = 1$ (arithmetic average), representing flow parallel to layers. For 2D isotropic media $\omega$ tends to zero, equivalent to geometric averaging.

The solution for $k_{rw}^{\text{eff}}$ obtained by substituting Eqs. (15a,b) or (17a,b) in Eq. (26) is semi-analytical (although we may refer to this solution as analytical) since an integral remains to be solved numerically. A fully analytical solution is achievable by solving the integrals using series expansions (see Rabinovich et al. (2016) Section 5), however performing the numerical quadrature is favorable for many cases, both in simplicity and calculation time, and in the next section, calculations are carried out with the semi-analytical solution. We note that the solution proposed here is an approximate one and the error will be evaluated in the next section, where corrections to the $CO_2$ curves will also be discussed.

# Results

In this section, the approximate semi-analytical solution for the CL effective relative permeability curves $k_{rw}^{\text{eff}}$ is tested against numerical calculations. We assume the sub-core characteristic curves have been measured in high flow rate coreflooding experiments and fitted by the function shown in Eqs. (8a,b). A few values of $n_w$ and $n_{CO_2}$ will be considered in these equations. A value of $\kappa_r^{\max} = 0.8$, i.e., the maximum $CO_2$ relative permeability (obtained at $\widetilde{S}_w = 0$), is taken consistently. This parameter is a constant in our formulation and changing its value is not expected to have a significant impact on the results presented here. We further assume that $P_c(\widetilde{S}_w)$ curves have been measured experimentally, e.g., using mercury injection capillary pressure, and the $J$-function results have been fitted with a VG (Eq. (4)) or BC (Eq. (5)) model. The impact of varying $\lambda$ and $m$ parameters will be considered.

The numerical solutions are obtained through the following procedure. Three dimensional permeability realizations are generated by sequential Gaussian simulation (Deutsch and Journel, 1992) using the Stanford Geostatistical Modeling Software, SGeMS (Remy et al., 2009). Core samples are taken to be rectangular for simplicity. Then, Eqs. (11a,b), with specified phase pressure boundary conditions, are solved using Stanford's General Purpose Research Simulator, GPRS (Cao, 2002), and $K_j^{\text{eff}}$ is calculated via Eq. (12), which gives $k_{rj}^{\text{eff}}$ by Eq. (13). In all cases the permeability fields are defined on 50×25×25 grids. A separate simulation is required to generate each data point on every effective relative permeability curve.

## Van Genuchten capillary pressure

Results for VG capillary pressure are considered in this section. In Fig. 1, a comparison between the new analytical $k_{rw}^{\text{eff}}$ and numerical calculations is presented for three variances ($\sigma_y^2$). Parameters of $m = 0.75$ and $n_w = n_{CO_2} = 2$ are taken, used in the past for $CO_2$-brine coreflooding simulations (Bielinski, 2007; Rabinovich et al., 2016). The analytical water curves, given by substituting Eq. (15a) in Eq. (26), are shown to accurately match the numerical ones for all 3 cases. The $CO_2$ curves (substituting Eq. (15b) in Eq. (26)) show a slight error for larger $\sigma_y^2$, however, a correction, which will be discussed in the following, is used and the corrected curves (dashed lines) are accurate. The values of $\omega$ used in Eq. (26) are indicated below the figures and it can be seen they are not sensitive to changes in $\sigma_y^2$.

The characteristic curves are also plotted in Fig. 1 for comparison. It can be seen that as heterogeneity increases (larger $\sigma_y^2$), $k_{rj}^{\text{eff}}$ curves are more distant from the characteristic curves, which is expected due to the increasing impact of capillary heterogeneity. Furthermore, the $CO_2$ curves (without correction) show more departure from the numerical curves as $\sigma_y^2$ grows. This is also observed in Fig. 2a, where the $k_{rCO_2}^{\text{eff}}$ error (red line) is plotted as a function of $\sigma_y^2$ and seen to be increasing. The error is taken here as the maximum difference between analytical and numerical solutions, i.e., $\max\left|\left(k_{rj}^{\text{eff}}(\langle\widetilde{S}_w\rangle)\right)_A - \left(k_{rj}^{\text{eff}}(\langle\widetilde{S}_w\rangle)\right)_N\right|$, where $A$ denotes analytical and $N$ denotes numerical. The water curves (blue line), however, do not exhibit a consistent increase in error with $\sigma_y^2$ and furthermore the errors are smaller in comparison to the $CO_2$ curves.

The impact of $\sigma_y^2$ on the accuracy of the analytical method and the different behaviour of the water and $CO_2$ curves was investigated extensively in Rabinovich et al. (2016) for the two-dimensional case with geometric averaging (as opposed to the power law averaging used here). It was shown that larger variance of $\ln K_j$ (denoted $\sigma_{\Gamma_j}^2$) are associated with larger error in the method. This can be explained with effective property theory for geometric averaging in the 2D case, however the same theory does not apply to our case. Nevertheless, we found the behaviour to be similar, as shown in Fig. 2b. Here, it can be seen that increasing $\sigma_y^2$ leads to a larger $\left\langle\sigma_{\Gamma_j}^2\right\rangle$ and this in turn is associated with a larger error as seen in Fig. 2a. Average $\sigma_{\Gamma_j}^2$ is taken here because each point on the curve has a different $\sigma_{\Gamma_j}^2$ corresponding to $\langle\widetilde{S}_w\rangle$ values.

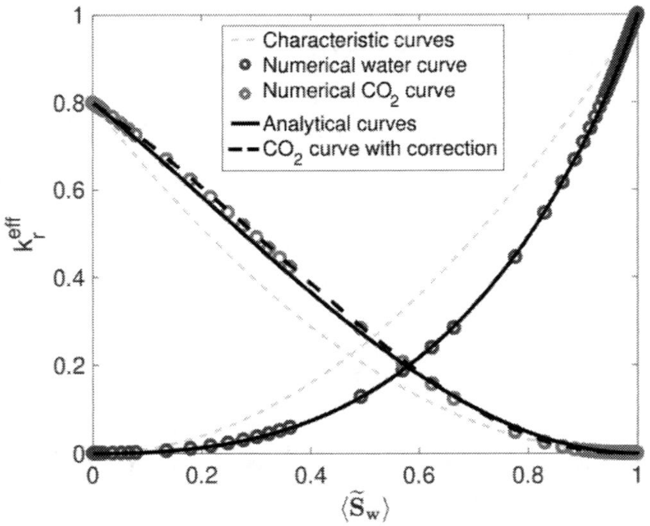


**(a)** $\sigma_y^2 = 1$**,** $\omega = 0.36$

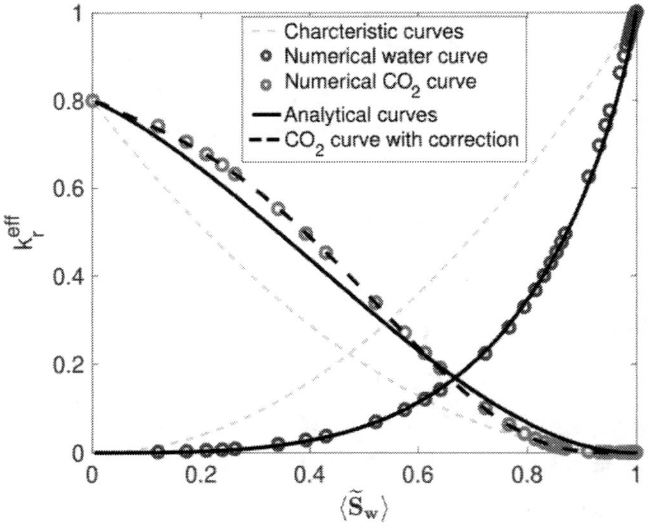


**(b)** $\sigma_y^2 = 3$**,** $\omega = 0.33$

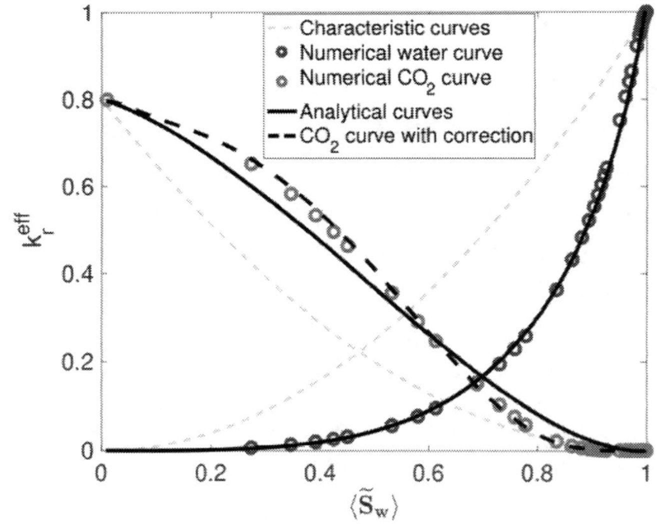


**(c)** $\sigma_y^2 = 4$**,** $\omega = 0.34$

**Figure 1: Core effective relative permeability as a function of water saturation in isotropic media with** $l_x = l_y = l_z = 0.1$**,** $k_G = 100$ **md, assuming VG J-function with** $m = 0.75$**.**

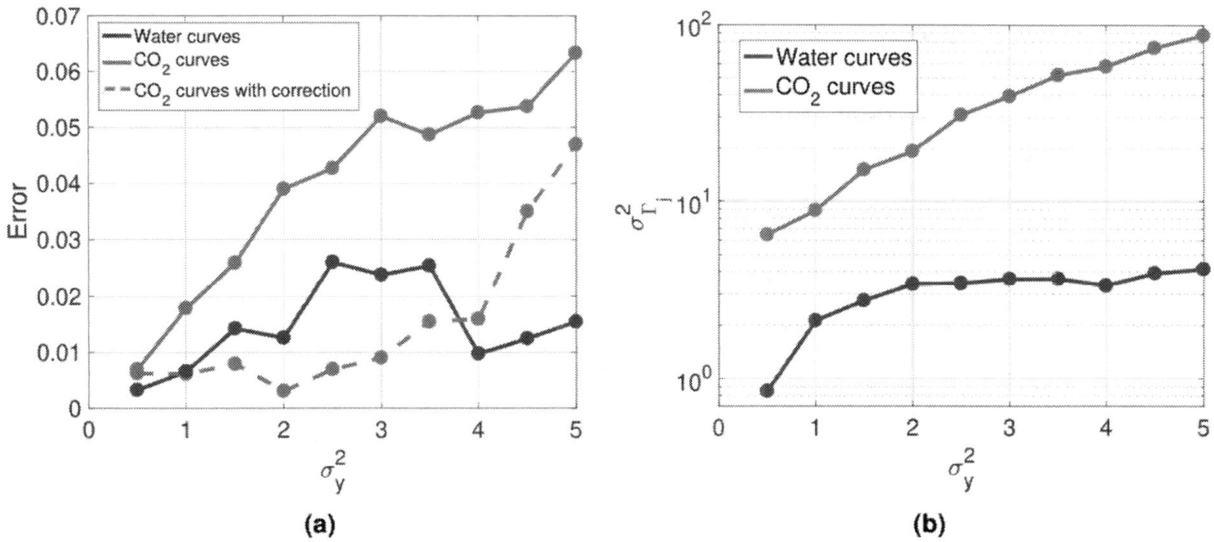


**Figure 2: (a) Analytical method error and (b) variance of log phase permeability ($\sigma^2_{\Gamma_j}$) as a function of variance of log permeability ($\sigma^2_y$).**

The water curves, on the other hand, show much smaller values of $\left\langle \sigma^2_{\Gamma_j} \right\rangle$ which hardly increase with $\sigma^2_y$ and therefore the errors in Fig. 2a are smaller and do not increase consistently. We note that all errors in Fig. 2a are considered to be reasonable (under 0.1), even for $\sigma^2_y$ as large as 5 and the correction improves the error in all cases.

As discussed above, the analytical solution $k^{\text{eff}}_{r\text{CO}_2}(\widetilde{P}_c,k)$ of Eq. (26) exhibits lower accuracy for large $\sigma^2_y$ and therefore a correction is derived. We consider a different power $\omega$ in Eq. (26) for each point on the $k^{\text{eff}}_{r\text{CO}_2}$ curve. The new power is defined by

$$\widetilde{\omega}(\widetilde{P}_c,m) = \omega \cdot \omega_c(\widetilde{P}_c,m) \tag{27}$$

where $\omega_c$ is determined by matching the analytical and numerical solution for each point on the curve. The advantage of expressing corrections as a function of $\widetilde{P}_c$ and $m$ is that these are existing parameters of the solution. The new solution for the $CO_2$ curves, incorporating the correction, is given by

$$k^{\text{eff}}_{rj} = \frac{1}{k_\omega}\left\{ \int_{-\infty}^{\infty} \left[K_j(\widetilde{P}_c,k)\right]^{\widetilde{\omega}(\widetilde{P}_c,m)} p(y)dy \right\}^{1/\widetilde{\omega}(\widetilde{P}_c,m)}. \tag{28}$$

In Fig. 3 the correction $\omega_c(\widetilde{P}_c)$ is presented for different values of $\sigma^2_y$ and $m$. It evident that the correction remains approximately the same for $k$ realizations of different $\sigma^2_y$ while varying $m$ has a significant impact on $\omega_c(\widetilde{P}_c)$. The structure of the curves appears to remain similar for changing

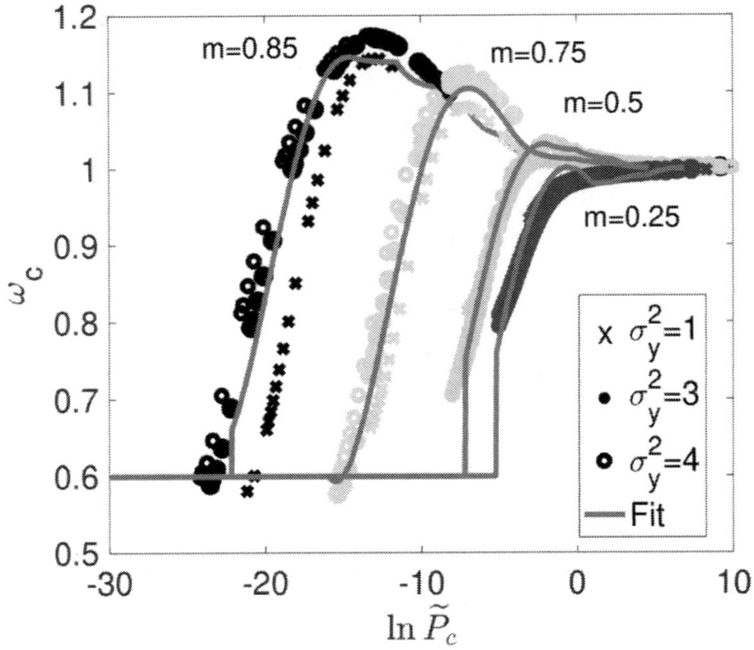


**Figure 3: The correction to power law averaging (Eqs.** (27),(28)) **matching analytical and numerical $CO_2$ curves. VG J-function, varying $m, \sigma_y^2$ and $n_w = n_{CO_2} = 2$ are considered.**

$m$ so we proceed by fitting a polynomial function as follows

$$\omega_c(\widetilde{P}_c, m) = \begin{cases} 0.6 & \ln \widetilde{P}_c < L_{\text{lower}} \\ \widehat{\omega}_c(\widetilde{P}_c, m) & L_{\text{lower}} < \ln \widetilde{P}_c < L_{\text{upper}} \\ \left[\widehat{\omega}_c(e^{L_{\text{upper}}}, m) - 1\right]\left[\ln \widetilde{P}_c - \ln \widetilde{P}_c^{\max}\right] / \left[L_{\text{upper}} - \ln \widetilde{P}_c^{\max}\right] + 1 & \ln \widetilde{P}_c > L_{\text{upper}} \end{cases}, \tag{29}$$

where

$$\widetilde{\omega}(\widetilde{P}_c, m) = A_1\left[\ln \widetilde{P}_c - B_1(m)\right]^4 + A_2\left[\ln \widetilde{P}_c - B_1(m)\right]^3 + A_3\left[\ln \widetilde{P}_c - B_1(m)\right]^2 + A_4 \ln \widetilde{P}_c + B_2(m),$$

$$B_1(m) = C_1 m^3 + C_2 m^2 + C_3 m + C_4 \quad , \quad B_2(m) = D_1 m^2 + D_2 m + D_3 - A_4 B_1(m),$$
$$A_1 = 1.429 \cdot 10^{-4} \ , \ A_2 = 4.776 \cdot 10^{-3} \ , \ A_3 = 4.76 \cdot 10^{-2} \ , \ A_4 = 0.161 \ , \ C_1 = -228.92,$$
$$C_2 = 317.68 \ , \ C_3 = -144.46 \ , \ C_4 = 26.23 \ , \ D_1 = 0.345 \ , \ D_2 = -0.138 \ , \ D_3 = 1.105,$$
$$L_{\text{upper}} = -531.43m^3 + 813.14m^2 - 389.36m + 55.82 \ , \ L_{\text{lower}} = -58.26m^2 - 35.89m - 10.57,$$

and $\widetilde{P}_c^{\max}$ is the maximum chosen value for $\widetilde{P}_c$.

We now test the approximate analytical solution given by Eqs. (15a,b), (22) and (27) - (29) for various values of $m$ and $\sigma_y^2$. The results are presented in Table 1, showing the maximum difference between analytical and numerical solutions. It can be seen that errors are fairly low with a largest error of 0.0628 and values under 0.03 for the vast majority of cases tested. We note that $m = 0.65, 0.8$ considered here were not used in the calibration of the correction (see

| | Water curves ($k_{rw}^{\text{eff}}$) | | | | $CO_2$ curves ($k_{rCO_2}^{\text{eff}}$) | | | |
|---|---|---|---|---|---|---|---|---|
| | $\sigma_y^2 = 1$ | $\sigma_y^2 = 2$ | $\sigma_y^2 = 3$ | $\sigma_y^2 = 4$ | $\sigma_y^2 = 1$ | $\sigma_y^2 = 2$ | $\sigma_y^2 = 3$ | $\sigma_y^2 = 4$ |
| $m = 0.85$ | 0.0119 | 0.0089 | 0.0102 | 0.0169 | 0.0264 | 0.0193 | 0.0381 | 0.0628 |
| $m = 0.8$ | 0.0104 | 0.0168 | 0.0181 | 0.0111 | 0.0166 | 0.0161 | 0.0199 | 0.0293 |
| $m = 0.75$ | 0.0066 | 0.0126 | 0.0237 | 0.0097 | 0.0062 | 0.0031 | 0.009 | 0.0159 |
| $m = 0.65$ | 0.0108 | 0.0245 | 0.0360 | 0.0384 | 0.0068 | 0.0115 | 0.0169 | 0.0194 |
| $m = 0.5$ | 0.0066 | 0.0204 | 0.0294 | 0.0283 | 0.0031 | 0.0058 | 0.0213 | 0.0618 |
| $m = 0.25$ | 0.0034 | 0.0126 | 0.0216 | 0.0313 | 0.0017 | 0.0059 | 0.0056 | 0.01 |

**Table 1: Error** $\max\left|\left(k_{rj}^{\text{eff}}(\langle\widetilde{S}_w\rangle)\right)_A - \left(k_{rj}^{\text{eff}}(\langle\widetilde{S}_w\rangle)\right)_N\right|$ **for VG J-function with cases of different** $m$ **and** $\sigma_y^2$ **(**$n_w = n_{CO_2} = 2$**).**

| | Water curves ($k_{rw}^{\text{eff}}$) | | | $CO_2$ curves ($k_{CO_2g}^{\text{eff}}$) | | |
|---|---|---|---|---|---|---|
| | $\sigma_y^2 = 1$ | $\sigma_y^2 = 3$ | $\sigma_y^2 = 4$ | $\sigma_y^2 = 1$ | $\sigma_y^2 = 3$ | $\sigma_y^2 = 4$ |
| $n_w = n_{CO_2} = 3$ | 0.0071 | 0.0097 | 0.0111 | 0.0106 | 0.0143 | 0.0175 |
| $n_w = n_{CO_2} = 5$ | 0.007 | 0.0242 | 0.0103 | 0.0149 | 0.0299 | 0.0320 |

**Table 2: Error** $\max\left|\left(k_{rj}^{\text{eff}}(\langle\widetilde{S}_w\rangle)\right)_A - \left(k_{rj}^{\text{eff}}(\langle\widetilde{S}_w\rangle)\right)_N\right|$ **for VG J-function with cases of different** $n_w, n_{CO_2}$ **and** $\sigma_y^2$ **(**$m = 0.75$**).**

Fig. 3) and this strengthens the reliability of its application. The current correction is intended for use only in the range $0.25 \geq m \leq 0.85$ and values outside this range are not presented here. Nevertheless, tests conducted indicate that for values of $m < 0.25$ the correction should still apply while for values of $m > 0.85$, error grows rapidly.

The analytical solution is tested further considering different values of the characteristic curve powers, $n_w$ and $n_{CO_2}$ in Eqs. (8a,b). Results are presented in Table 2, where maximum error is calculated as in the previous Table. It can be seen that increasing $n_w$ and $n_{CO_2}$ does not have a significant impact on the water curve error (see Table 1, $m = 0.75$ for comparison), while for the $CO_2$ curves a moderate increase of error is observed. We conclude from these results that the analytical solution remains accurate for larger values of $n_w$ and $n_g$ and an additional correction is not necessary.

## Brooks-Corey capillary pressure

We now consider the BC $J$-function (Eq. (4)) in our formulation. The solution in this case is obtained by substituting Eqs. (17a,b) in Eq. (26). Results are presented in Fig. 4 for the parameters $n_w = n_{CO_2} = 2$ and $\lambda = 2$, used in previous literature for $CO_2$-brine flow modeling (Li, 2014; Rabinovich et al., 2015). It can be seen that the water curve analytical solution is in agreement with the numerical one, showing little to no error. The $CO_2$ curves, on the other hand, deviate substantially from the numerical solution, particularly for large $\sigma_y^2$. A correction is necessary to accurately use the analytical solution for $CO_2$ curve modeling and this is discusses in the following. The corrected curves are plotted in Fig. 4 (dashed lines) and approximately match the numerical curves showing reasonable error.

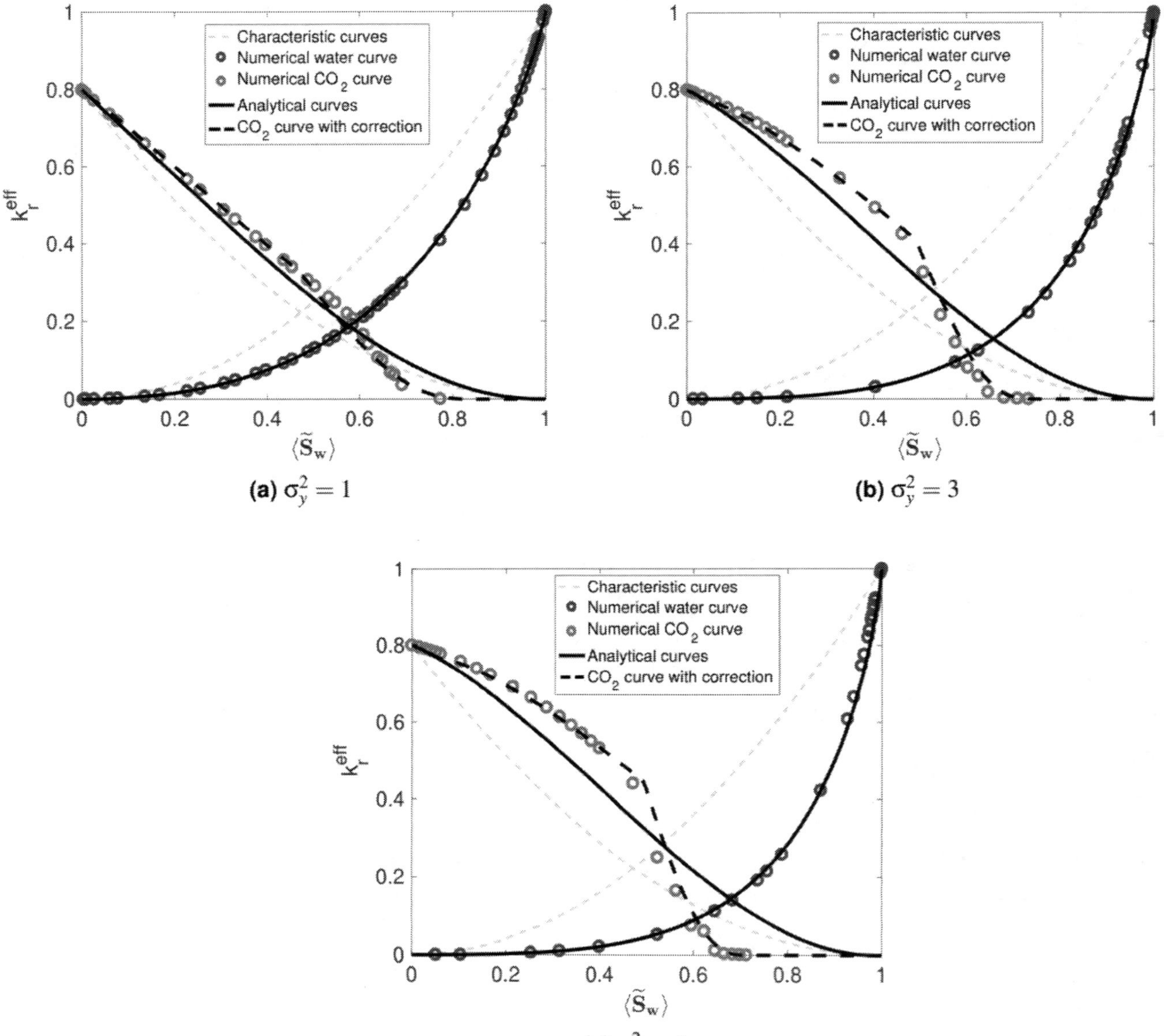


(a) $\sigma_y^2 = 1$

(b) $\sigma_y^2 = 3$

(c) $\sigma_y^2 = 4$

**Figure 4: Core effective relative permeability as a function of water saturation in isotropic media with $l_x = l_y = l_z = 0.1$, $k_G = 100$ md, assuming BC J-function with $\lambda = 2$.**

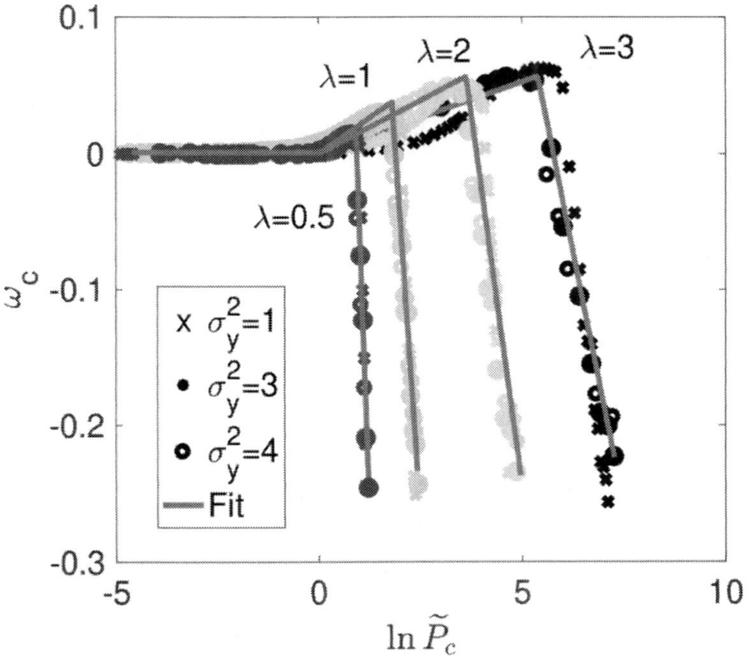


**Figure 5: The correction to power law averaging (Eqs.** (28), (30)**) matching analytical and numerical $CO_2$ curves. BC J-function, varying $\lambda, \sigma_y^2$ and $n_w = n_{CO_2} = 2$ are considered.**

The $k_{rCO_2}^{\text{eff}}$ structure apparent in Fig. 4 is different than in the VG case (see Fig. 1 for comparison) showing a sharp decrease at $\langle \widetilde{S}_w \rangle \approx 0.4$ and reaching zero at smaller saturations. This is a result of capillary entry pressure, which by Eq. (17b) leads to many zero values in the $K_{CO_2}$ field, reducing $k_{rCO_2}^{\text{eff}}$ at large $\langle \widetilde{S}_w \rangle$. This effect is not captured by the analytical solution and therefore the correction is paramount.

The corrections for $k_{rCO_2}^{\text{eff}}$ is similar to that described for VG capillary pressure, i.e., considering different powers $\omega$ in Eq. (26). However, we find higher accuracy is achieved by defining the new power $\widetilde{\omega}(\widetilde{P}_c, \lambda)$ slightly different than in the previous case and as follows:

$$\widetilde{\omega}(\widetilde{P}_c, \lambda) = \omega + \omega_c(\widetilde{P}_c, \lambda). \quad (30)$$

In Fig. 5 the correction $\omega_c(\widetilde{P}_c)$ is presented for different values of $\sigma_y^2$ and $\lambda$. As in Fig. 3, the correction $\omega_c(\widetilde{P}_c)$ shows little variation with $\sigma_y^2$ and a somewhat systematic change with $\lambda$. We fit the results with first degree polynomial (linear) functions. The result is given by

$$\omega_c(\widetilde{P}_c, \lambda) = \begin{cases} 0 & \ln \widetilde{P}_c \leq 0 \\ A_1'(\lambda)\left(0.0155 \ln \widetilde{P}_c + 2.939 \cdot 10^{-4}\right) & 0 < \ln \widetilde{P}_c < \widetilde{P}_c^{\text{intersect}} \\ A_2'(\lambda)\left(0.905 - 0.235 \ln \widetilde{P}_c\right) + A_3'(\lambda) & \ln \widetilde{P}_c \geq \widetilde{P}_c^{\text{intersect}} \end{cases}, \quad (31)$$

where

$$A'_1(\lambda) = B'_1\lambda^3 + B'_2\lambda^2 + B'_3\lambda + B'_4$$

$$A'_2(\lambda) = C'_1\lambda^3 + C'_2\lambda^2 + C'_3\lambda + C'_4 \quad , \quad A'_3(\lambda) = D'_1\lambda^5 + D'_2\lambda^4 + D'_3\lambda^3 + D'_4\lambda^2 + D'_5\lambda + D'_6,$$

$$B'_1 = 0.224, \quad B'_2 = -1.33, \quad B'_3 = 2.078, \quad B'_4 = 0.369,$$

$$C'_1 = -0.588, \quad C'_2 = 3.846, \quad C'_3 = -8.346, \quad C'_4 = 6.947$$

$$D'_1 = 0.114, \quad D'_2 = -1.264, \quad D'_3 = 5.536, \quad D'_4 = -12.145, \quad D'_5 = 13.985, \quad D'_6 = -7.09$$

and $\widetilde{P}_c^{\text{intersect}}$ is the intersection of the two lines in Eq (31).

| | Water curves ($k_{rw}^{\text{eff}}$) | | | | $CO_2$ curves ($k_{rCO_2}^{\text{eff}}$) | | | |
|---|---|---|---|---|---|---|---|---|
| | $\sigma_y^2 = 1$ | $\sigma_y^2 = 2$ | $\sigma_y^2 = 3$ | $\sigma_y^2 = 4$ | $\sigma_y^2 = 1$ | $\sigma_y^2 = 2$ | $\sigma_y^2 = 3$ | $\sigma_y^2 = 4$ |
| $\lambda = 3$ | 0.0161 | 0.0127 | 0.0093 | 0.006 | 0.0749 | 0.0644 | 0.0283 | 0.0883 |
| $\lambda = 2.5$ | 0.0155 | 0.0137 | 0.0095 | 0.0084 | 0.0826 | 0.0543 | 0.0169 | 0.095 |
| $\lambda = 2$ | 0.0104 | 0.0123 | 0.0105 | 0.0083 | 0.0477 | 0.0293 | 0.0446 | 0.0881 |
| $\lambda = 1.5$ | 0.007 | 0.0041 | 0.0053 | 0.0036 | 0.0507 | 0.0524 | 0.0442 | 0.0129 |
| $\lambda = 1$ | 0.001 | 0.0073 | 0.0122 | 0.0136 | 0.0265 | 0.0209 | 0.0127 | 0.046 |
| $\lambda = 0.75$ | 0.0036 | 0.0131 | 0.0168 | 0.0243 | 0.0112 | 0.0073 | 0.0295 | 0.0297 |
| $\lambda = 0.5$ | 0.0063 | 0.0172 | 0.0305 | 0.0394 | 0.0083 | 0.0017 | 0.0055 | 0.0214 |

**Table 3: Error** $\max\left|\left(k_{rj}^{\text{eff}}(\langle\widetilde{S}_w\rangle)\right)_A - \left(k_{rj}^{\text{eff}}(\langle\widetilde{S}_w\rangle)\right)_N\right|$ **for BC J-function with cases of different** $\lambda$ **and** $\sigma_y^2$ **(**$n_w = n_{CO_2} = 2$**).**

Table 3 presents results for the analytical solution given by Eqs. (17a,b), (22), (28), (30) and (31) for various values of $\lambda$ and $\sigma_y^2$. It can be see that the error for the water curves is small in all cases with a maximum of 0.0394. For the $CO_2$ curves, error is larger with a maximum of 0.883. While all of the errors are within what we consider a tolerable range ($< 0.1$), the $CO_2$ curve errors are much larger than in the case of VG capillary pressure (see Table 1 for comparison). We note that $\lambda = 0.75, 1.5, 2.5$ considered here were not used in the calibration of the correction (see Fig. 5).

In Table 4, we consider different values of the characteristic curve powers, $n_w$ and $n_{CO_2}$ in Eqs. (8a,b). Analytical error in comparison with the numerical solution is presented, where maximum error is calculated as in the previous Table. It can be seen that increasing $n_w$ does not have

| | Water curves ($k_{rw}^{\text{eff}}$) | | | $CO_2$ curves ($k_{rCO_2}^{\text{eff}}$) | | |
|---|---|---|---|---|---|---|
| | $\sigma_y^2 = 1$ | $\sigma_y^2 = 3$ | $\sigma_y^2 = 4$ | $\sigma_y^2 = 1$ | $\sigma_y^2 = 3$ | $\sigma_y^2 = 4$ |
| $n_w = n_{CO_2} = 3$ | 0.012 | 0.0099 | 0.0066 | 0.0212 | 0.0493 | 0.0915 |
| $n_w = n_{CO_2} = 5$ | 0.0043 | 0.0033 | 0.003 | 0.0185 | 0.082 | 0.1287 |

**Table 4: Error** $\max\left|\left(k_{rj}^{\text{eff}}(\langle\widetilde{S}_w\rangle)\right)_A - \left(k_{rj}^{\text{eff}}(\langle\widetilde{S}_w\rangle)\right)_N\right|$ **for BC J-function with cases of different** $n_w, n_{CO_2}$ **and** $\sigma_y^2$ **(**$\lambda = 2$**).**

a significant impact on the water curve error (see Table 3, $\lambda = 2$ for comparison). For the $CO_2$ curves at low variance ($\sigma_y^2 = 1$), the errors remain small, however, for larger variance ($\sigma_y^2 = 3,4$) we begin to see a significant departure fom the numerical solution.

# Summary and conclusions

A method to correct core relative permeability for use in low flow rate $CO_2$ storage modeling is presented. Given the following core properties: capillary pressure, characteristic relative permeability (from high injection rate coreflooding experiments) and sub-core permeability geometric mean and log variance, we derive low flow rate effective relative permeability. The correction is given by an approximate semi-analytical solution assuming capillary limit conditions. The solution is based on power law averaging of the fine scale phase permeability field and is an exact solution in the case of one-dimensional (layered) permeability.

The semi-analytical solution is compared to a numerical one to test its accuracy. A wide range of cases are considered by varying the sub-core permeability field ($\sigma_y^2$), capillary pressure $J$-function type (VG or BC), $J$-function parameters ($m$, $\lambda$) and characteristic curve model (power in $k_{rj}(\widetilde{S}_w)$ - $n_w$, $n_{CO_2}$). It is shown that for all cases the analytical water curves approximately match the numerical solution. The $CO_2$ curves for a VG $J$-function type are accurate in cases of sufficiently small log permeability variance ($\sigma_y^2$). For larger $\sigma_y^2$ we derive a correction to the power $\omega$ in the averaging to improve accuracy. With the correction, it is shown that $CO_2$ curves are also estimated accurately for all cases. Considering a BC $J$-function type the analytical $CO_2$ curves show a large error due, in part, to the capillary entry pressure effect. A correction is presented, shown to reduce the error and reasonable agreement with the numerical solution is achieved in almost all cases. Only for the case of large powers in the characteristic curve model ($n_w = n_{CO_2} = 5$) with permeability of large $\sigma_y^2$, errors exceeded the tolerable limit.

While the work presented here offers some insight on a general effective property problem, our main goal is to developed a method for practitioners to correct the core relative permeability measured in standard high rate coreflooding experiments. The analytical solution should allow for simple corrections without having to conduct additional low rate experiments or computationally demanding simulations. We have shown the method is robust, applicable to various types of cores and can be used in cases with significant capillary trapping (i.e., $P_c$ curves with nonzero capillary entry pressure). Currently, the method is limited to drainage processes and an extension of the method for imbibition (in which hysteresis is important) should be considered in the future. Furthermore, it may be possible to extend the method to cores with permeability distributions other than the log-normal ones considered here.

# Acknowledgements

This research was supported by Grant No. 2016182 from the United States-Israel Binational Science Foundation (BSF).

# References


H. Behzadi and V. Alvarado. Upscaling of upward $CO_2$ migration in 2D system. *Advances in Water Resources*, 46:46–54, 2012.

A. Bielinski. *Numerical simulation of $CO_2$ sequestration in geological formations.* PhD thesis, University of Stuttgart, Stuttgart, Germany, 2007.

R.H. Brooks and A.T. Corey. Properties of porous media affecting fluid flow. *Journal of the Irrigation and Drainage Division*, 92(2):61–90, 1966.

H. Cao. *Deveopment of Techniques for General Purpose Simulators.* PhD thesis, Stanford University, Stanford, CA, USA, 2002.

A.T. Corey and C.H. Rathjens. Effect of stratification on relative permeability. *Journal of Petroleum Technology*, 8(12):69–71, 1956.

C.V. Deutsch. Calculating effective absolute permeability in sandstone/shale sequences. *SPE Formation Evaluation*, 4:343–348, 1989.

C.V. Deutsch and A.G. Journel. *GSLIB: Geostatistical Software Library and User's Guide.* Oxford University Press, New York, 1992.

A.R. Freeze. A stochastic-conceptual analysis of one-dimensional groundwater flow in nonuniform homogeneous media. *Water Resources Research*, 11(5):725–741, 1975.

A.G. Journel, C.V. Deutsch, and A.J. Desbarats. Power averaging for block effective permeability. In *SPE California Regional Meeting.* Society of Petroleum Engineers, 1986. SPE 15128.

M. Krause, S.C. Krevor, and S.M. Benson. A procedure for the accurate determination of sub-core scale permeability distributions with error quantification. *Transport in porous media*, 98(3):565–588, 2013.

M.H. Krause. Modeling and investigation of the influence of capillary heterogeneity on relative permeability. In *SPE Annual Technical Conference and Exhibition.* Society of Petroleum Engineers, San Antonio, Texas, SPE-160909-STU, 2012a.

M.H. Krause. *Modeling and Investigation of the Influence of Capillary Heterogeneity on Multiphase Flow of $CO_2$ and Brine.* PhD thesis, Stanford University, 2012b.

M.H. Krause and S.M. Benson. Accurate determination of characteristic relative permeability curves. *Advances in Water Resources*, 83:376–388, 2015.

C-W. Kuo and S.M. Benson. Numerical and analytical study of effects of small scale heterogeneity on $CO_2$/brine multiphase flow system in horizontal corefloods. *Advances in Water Resources*, 79:1–17, 2015.

C-W. Kuo, J.C. Perrin, and S.M. Benson. Effect of gravity, flow rate, and small scale heterogeneity on multiphase flow of $CO_2$ and brine. In *SPE Western Regional Meeting.* Society of Petroleum Engineers, Anaheim, CA, SPE 132607, 2010.

M.C. Leverett. Capillary behavior in porous solids. *Transactions of the AIME*, 142:152–169, 1941.

B. Li. *Modeling geological $CO_2$ sequestration: Translations across spatial scales and advancements in nonlinear Newton solver.* PhD thesis, Stanford University, 2014.

B. Li and S.M. Benson. Influence of small-scale heterogeneity on upward $CO_2$ plume migration in storage aquifers. *Advances in Water Resources*, 83:389–404, 2015.

B. Li, H.A. Tchelepi, and S.M. Benson. Influence of capillary-pressure models on $CO_2$ solubility trapping.

*Advances in Water Resources*, 62:488–498, 2013.

A. Lohne, G.A. Virnovsky, and L.J. Durlofsky. Two-stage upscaling of two-phase flow: From core to simulation scale. *SPE Journal*, 11(3):304–316, 2006.

E. Mouche, M. Hayek, and C. Mügler. Upscaling of $CO_2$ vertical migration through a periodic layered porous medium: The capillary-free and capillary-dominant cases. *Advances in Water Resources*, 33 (9):1164–1175, 2010.

I. Neuweiler, A. Papafotiou, H. Class, and R. Helmig. Estimation of effective parameters for a two-phase flow problem in non-gaussian heterogeneous porous media. *Journal of contaminant hydrology*, 120: 141–156, 2011.

Z.S. Omoregle. Factors affecting the equivalency of different capillary pressure measurement techniques. *SPE formation evaluation*, 3(1):146–155, 1988.

J.C. Perrin and S.M. Benson. An experimental study on the influence of sub-core scale heterogeneities on $CO_2$ distribution in reservoir rocks. *Transport in Porous Media*, 82(1):93–109, 2010.

J.C. Perrin, M. Krause, C.W. Kuo, L. Miljkovic, E. Charoba, and S.M. Benson. Core-scale experimental study of relative permeability properties of $CO_2$ and brine in reservoir rocks. *Energy Procedia*, 1(1): 3515–3522, 2009.

G.E. Pickup and K.S. Sorbie. Scaleup of two-phase flow in porous media using phase permeability tensors. *SPE Journal*, 1(4):369–382, 1996.

R. Pini and S.M. Benson. Characterization and scaling of mesoscale heterogeneities in sandstones. *Geophysical Research Letters*, 40(15):3903–3908, 2013.

R. Pini, S.C.M Krevor, and S.M Benson. Capillary pressure and heterogeneity for the $CO_2$/water system in sandstone rocks at reservoir conditions. *Advances in Water Resources*, 38:48–59, 2012.

A. Rabinovich. Estimation of sub-core permeability statistical properties from coreflooding data. *Advances in Water Resources*, DOI: 10.1016/j.advwatres.2017.07.012, In Press, 2017.

A. Rabinovich, K. Itthisawatpan, and L.J. Durlofsky. Upscaling of $CO_2$ injection into brine with capillary heterogeneity effects. *Journal of Petroleum Science and Engineering*, 134:60–75, 2015.

A. Rabinovich, B. Li, and L.J. Durlofsky. Analytical approximations for effective relative permeability in the capillary limit. *Water Resources Research*, 52(10):7645–7667, 2016.

N. Remy, A. Boucher, and J. Wu. *Applied Geostatistics with SGeMS: A User's Guide*. Cambridge University Press, Cambridge, UK, 2009.

J.G. Richardson, J.K. Kerver, J.A. Hafford, and J.S. Osoba. Laboratory determination of relative permeability. *Journal of Petroleum Technology*, 4(8):187–196, 1952.

P.S. Ringrose, K.S. Sorbie, P.W.M Corbett, and J.L. Jensen. Immiscible flow behaviour in laminated and cross-bedded sandstones. *Journal of Petroleum Science and Engineering*, 9(2):103–124, 1993.

M.T. van Genuchten. A closed-form equation for predicting the hydraulic conductivity of unsaturated soils. *Soil science society of America Journal*, 44(5):892–898, 1980.

G.A. Virnovsky, H.A. Friis, and A. Lohne. A steady-state upscaling approach for immiscible two-phase flow. *Transport in Porous Media*, 54(2):167–192, 2004.

N. Wei, M. Gill, D. Crandall, D. McIntyre, Y. Wang, K. Bruner, X. Li, and G. Bromhal. $CO_2$ flooding properties of liujiagou sandstone: influence of sub-core scale structure heterogeneity. *Greenhouse Gases: Science and Technology*, 4(3):400–418, 2014.

CARBON MANAGEMENT TECHNOLOGY CONFERENCE

# CMTC-486447-MS

# An Optimal Injection Scheme for Maximizing Oil Recovery in $CO_2$ Miscible Flooding

Qingwang Yuan, University of Regina; Jie Zhan, University of Calgary; Jinjie Wang, China University of Geosciences; Fanhua Zeng, University of Regina; Kelvin D. Knorr, and Muhammad Imran, Saskatchewan Research Council



This paper was prepared for presentation at the Carbon Management Technology Conference held in Houston, Texas, USA, 17-20 July 2017.



## Abstract

As an efficient enhanced oil recovery (EOR) technique, carbon dioxide ($CO_2$) miscible flooding can greatly reduce the viscosity of oil and improve its mobility, and has great potential to achieve higher oil recovery. However, the disadvantage of $CO_2$ flooding, when compared with waterflooding, is the relatively larger viscosity ratio between $CO_2$ and oil. Under such unfavorable conditions, frontal instabilities, or viscous fingering, can easily develop. This may affect the performance of $CO_2$ msicible flooding and result in less sweep efficiency and oil recovery.

In the present study, nonlinear numerical simulations were conducted to model the $CO_2$ miscible flooding in subsurface porous media. Both concentration-dependent diffusion and a varying dispersion that is closely related with flow rates were incorporated into the mathematical model. The development of frontal instabilities with time was simulated with highly accurate numerical methods. More importantly, to reduce the unstable displacement and improve sweep efficiency, a time-dependent injection rate involving periodic alternation of injection and extraction was employed. Different from the widely used constant injection rate, this time-dependent displacement rate led to different flow dynamics and sweep efficiency, although the amount of $CO_2$ injected was the same. In particular, the effect of a cycle period on the propagation of $CO_2$ was carefully examined. It was found that a longer period led to earlier breakthrough of $CO_2$ and less sweep efficiency. However, a shorter period with faster alternation of injection and extraction had a stabilizing effect. In particular, a later breakthrough was achieved and higher sweep efficiency at breakthrough was obtained compared with that of a constant injection rate. This indicates that pulsed displacement through fast switching of injection and extraction has the potential to maximize oil recovery in $CO_2$ miscible flooding.

## Introduction

Miscible flooding is proven as an economical EOR process and can be used for a variety of different reservoirs. In $CO_2$ miscible flooding, one of the most promising EOR techniques, the $CO_2$ can become miscible with the oil when the pressure is higher than the minimum miscibility pressure (MMP). It can therefore improve oil recovery and sequester $CO_2$ at the same time in an environmentally friendly way.

Two main factors may affect the performance of $CO_2$ miscible flooding. One is the displacement efficiency at the pore level in which the $CO_2$ dissolves into oil and causes oil swelling. The oil is then extracted and flows with $CO_2$ and is produced (Huang et al., 2017; Yan et al., 2017). Reduced sweep efficiency on the field scale is another major factor that results from $CO_2$ viscous fingering in a relatively homogeneous reservoir. For instance, at reservoir conditions with temperature of 110°F and pressure of 2,500 psi, the $CO_2$ viscosity is 0.06 cp (Stalkup, 1978). Consequently, the viscosity ratio would range from around 17 to 170 for oil viscosity from 1 to 10 cp. The injected $CO_2$ has a great tendency to develop frontal instabilities, or viscous fingering, resulting in low sweep efficiency and oil recovery. For reservoirs with relatively high-viscosity oils, it is even worse (Zhou et al., 2017).

To maximize sweep efficiency in $CO_2$ miscible flooding field projects, the well count, well pattern, $CO_2$ injection bank size and/or water/$CO_2$ ratio in water-alternating-gas (WAG) steps have been optimized (Pipes and Schoelling, 2014). These measures are technically successful but may reduce the economic performance of $CO_2$ miscible flooding. However, optimizing $CO_2$ injection rates is a good option for improving sweep efficiency while reducing costs. Different from the commonly used constant injection rate, a cyclic time-dependent displacement rate involving pulsed injection and extraction of $CO_2$ are employed in the present study. This is very promising since it can improve sweep efficiency with the same amount of $CO_2$ injected into reservoirs as that of a constant injection rate. In other words, a cyclic displacement rate has the potential to enhance oil recovery with zero additional cost. The study of the influence of different cyclic time-dependent displacement scenarios on viscous fingering in $CO_2$ miscible flooding is essential for the optimal injection scheme in order to maximize oil recovery.

Enhancement of oil recovery with this cyclic pulsed time-dependent injection rate has been proven by field pilot tests, mainly in immiscible waterflooding. Several full-scale field trials were conducted in 1998–1999 in Canada. The first field-scale experiment was done in Morgan Field near Lloydminster, Alberta (Spanos et al., 2003). Thirteen out of 19 oil production wells were put into operation. Water pulsing at 6–8 strokes per minute was employed for 10.5 weeks. At the end of the pulsing, 11 out of the 13 oil wells showed clear oil production improvement. Spanos et al. report that the total oil production was increased by 37%, while water cut was reduced by more than 20%. Another two projects with cyclic water injection were performed in heavy oil reservoirs in Alberta and Saskatchewan with oil viscosity values of 1,600 cp and 10,300 cp, respectively. The decline of oil production rate in these two projects was clearly slowed down during and after cyclic waterflooding (Spanos et al., 2003). In the same year, cyclic water injection with 5–6 pulses per minute was applied to six oil wells in a weakly consolidated oil reservoir in Germany with oil viscosity 90–120 cp (Groenenboom et al., 2003). After around 6 months' cyclic water injection, four oil wells showed a clearly increasing or stable oil production rate, while no enhancement of oil production could be observed in the other two wells.

Pilot tests indicate that the cycle period in such time-dependent displacements plays an important role in affecting sweep efficiency and oil recovery. To address this, numerical simulations of field-scale cyclic waterflooding were performed. Surguchev et al. (2002) established a 3D model for the highly heterogeneous Lower Tilje/Åre reservoirs located in the Norwegian Sea. The length of time for a pressurizing/depressurizing half-cycle is around 5–15 days. They reported that, compared with constant water injection, cyclic injection with ratios of frequent pulsing injection to no injection of 1:2 and 1:3 is able to improve oil recovery by 5.6% in a 10-year period, while water cut is reduced by 10–20%. Similar to the findings by Surguchev et al. (2002), Stirpe et al. (2004) reported that the cyclic injection scenario with 1 month of injection and 3 months of shut-in yielded the highest oil recovery among all of the cases studied. They built both 2D and 3D models with cyclic water injection for a 5-spot pattern in the Lagocinco field, Venezuala. Higher oil recovery was achieved for both scenarios, while the enhancement of the 2D model was greater. Meanwhile, they found that the effect of cyclic water injection was more obvious for reservoirs with pressure higher than bubblepoint pressure with shorter well spacing. Later, the numerical simulations by Shchipanov et al. (2008) indicated that the cyclic water injection scenario with a long cycle period could improve oil recovery 5% over that of constant continuous water injection. However, the case with a shorter cycle period could only produce about 3%

incremental oil. They also pointed out that incorporating small-scale reservoir heterogeneities was important for accurate numerical modeling.

In both field-scale pilot tests and numerical simulations, the mechanisms behind the cyclic displacement scenario need to be further investigated for a better understanding and higher increment of oil recovery through optimizing the time-dependent injection rates. To address this, lab-scale experimental studies were conducted to examine the improvement of sweep efficiency by cyclic fluid injection. Davidson et al. (1999) performed cyclic water injection experiments with pressure pulsing in both cylindrical and rectangular sandpacks. Compared with no-pulsing water injection, the sweep efficiency could be improved by suppressing viscous fingering instabilities. A scenario with high amplitude and low frequency led to a more stable displacement front between water and oil. Similar conclusions were obtained by Zschuppe (2001), who used a consistent pulsing source (CPS) in waterflooding in a crude-oil-saturated medium. Results indicated that the pulsed tests could increase the oil recovery rates and final sweep efficiency by 10% over non-pulsed tests. Ivanov and Araujo (2006) carried out an experimental study on immiscible pulsed flow in waterflooding. Several parameters were examined, including bead size, cycle period, injection rate and oil viscosity. Unlike the studies reported by Davidson et al. (1999) and Zschuppe (2001), it was found that the shorter cyclic periods led to the highest oil recovery. While most of these studies were conducted for waterflooding experiments, there is also application of time-dependent displacement rates in solvent-based EOR processes. A vapor extraction (Vapex) process using a cyclic solvent injection was investigated by Muhamad et al. (2012). Important findings were reported in terms of the temporal variation of solvent injection, which is more beneficial for enhancement of heavy oil recovery. Compared with constant solvent injection, the long pulse improved oil recovery by around 19%, while the short pulse was even more promising, as it resulted in 24% additional oil recovery. This time-dependent injection of solvent is favorable for the mixing of solvent and oil and consequently reduces oil viscosity and improves oil recovery. An optimized propane injection pressure versus time was also investigated in heavy oil recovery by the same group (Muhamad et al., 2016). Compared with constant pressure propane injection, this optimal injection scheme was able to improve oil production by 31% on average.

As to the theoretical and mathematical modeling of time-dependent injection rates in miscible and immiscible flooding processes, several researchers have established analytical and numerical models. The suppression of viscous fingering instabilities of two-phase flows with time-dependent flow rates was investigated in 2D and 3D geometry (Dias, 2013; Dias et al., 2010; Dias and Miranda, 2010, 2009). An optimal injection rate with a piecewise-constant injection was also analyzed as to its stabilization effects (Dias et al., 2010). For the miscible displacements considering diffusion of two components, a nonlinear numerical simulation was conducted by Chen et al. (2010) with a similar injection rate to that of Li et al. (2009). The stabilizing effect of the time-dependent injection rate was reported. In addition, our previous studies have also investigated, both analytically and numerically, the time-dependent scenarios in miscible displacements considering diffusion (Yuan and Azaiez, 2014a), inertia (Yuan and Azaiez, 2015a), reaction and heat transfer (Yuan and Azaiez, 2014c; 2015c).

As discussed above, the previous studies of time-dependent displacement on oil recovery are focused either on immiscible cases or miscible cases considering only diffusion. No dispersion has been taken into account in the varying injection scenarios, whereas in typical $CO_2$ miscible flooding, the flow rate is large enough such that the effect of dispersion should be considered. The role of dispersion is therefore non-negligible, especially when it is induced by variations of time-dependent injection rates. In the present study, the effects of time-dependent displacement rates on frontal instabilities are examined in $CO_2$ miscible flooding. The way to control such instabilities and improve sweep efficiency is also discussed. This would allow for long-term $CO_2$ storage and highly efficient $CO_2$-EOR processes.

## Model development

At certain reservoir conditions, the injected $CO_2$ can become fully miscible with oil when the pressure in reservoirs is higher than the MMP (Wang et al., 2015). The mixture of $CO_2$ and crude oil is in the form of a single phase when first-contact miscibility is achieved. In an actual $CO_2$ miscible flooding process, very complex phase behaviors are involved that depend on the reservoir pressure and temperature and on properties of the oil and rock of the formation.

To simplify the physical problem and simulate the detailed development of frontal instabilities in $CO_2$ miscible flooding, we assume that the miscible displacement takes place in a 2D horizontal porous medium with length *Lx* and width *Ly*. The effect of gravity on the displacement process is therefore neglected. Moreover, the porosity and permeability of the porous media are assumed to be constant and homogeneous. The porous medium is initially saturated with crude oil with viscosity $\mu_2$. The $CO_2$ with viscosity $\mu_1$ is then injected from one side of the domain to displace oil. Since the viscosity of $CO_2$ is less than that of oil, under this unfavorable viscosity contrast, the contact area between these two components will become unstable, and finger-like patterns will develop (Tan and Homsy, 1988; Yuan, 2014). The viscosity at the contact area of $CO_2$ and oil is mainly affected by their concentration and calculated by mixing rule. In addition, to focus on the effect of time-dependent injection rate on frontal instabilities, complex mechanisms such as swelling effect of oil, rock wettability alternation, and gas adsorption/desorption in porous media are not considered in the present study (Wang et al., 2016a, 2016b).

The mass transfer between $CO_2$ and oil in this miscible displacement process is governed by the following equations:

$$\nabla \cdot \boldsymbol{u} = 0 \tag{1}$$

$$\nabla P = -\mu \boldsymbol{u} \tag{2}$$

$$\frac{\partial c}{\partial t} + \boldsymbol{u} \cdot \nabla c = \nabla \cdot [D \cdot \nabla c] \tag{3}$$

The above equations are given in the dimensionless form as are those discussed in the study by Ghesmat and Azaiez (2008). The flow rate is denoted by a 2D velocity vector $\boldsymbol{u} = (u, v)$. In Eq. 2, the viscosity $\mu$ is scaled by that of the $CO_2$ and represents the mobility of fluid by incorporating the constant permeability of the porous media. The pressure term can be eliminated by taking the curl of Eq. 2, as in the studies by Zimmerman and Homsy (1991) and Ghesmat and Azaiez (2008). In addition, the concentration is also in dimensionless form in the convection-dispersion equation 3, where *D* is the sum of diffusion coefficient and dispersion coefficient.

In most of the previous studies on nonlinear simulations of miscible displacements, the coefficient *D* is assumed to be constant and only diffusion is considered. However, in $CO_2$ miscible flooding, a concentration-dependent diffusion coefficient is more realistic at reservoir conditions (Grogan et al., 1988). The dimensionless form of concentration dependence of diffusion coefficient is given by Yuan et al. (2017b, 2017c, 2017d). More practically, the dispersion is velocity-induced as defined by Ghesmat and Azaiez (2008). This is very important in the cyclic time-dependent displacement scenarios involving both injection and extraction of $CO_2$ and oil. The role of dispersion may vary quite significantly in the different stages. Especially for a long-term displacement, the repeated injection and extraction will result in very complex flow dynamics when convection, diffusion and dispersion take effect together.

To solve the governing equations 1–3 with proper initial and boundary conditions, the highly accurate spectral method and fully implicit alternating direction implicit (ADI) are used. Readers who are interested in the details of these numerical methods can refer to the studies by Islam (2003) and Yuan et al. (2017a).

## Time-dependent displacement rates

Different time-dependent displacement rates have been adopted to study the influence on frontal instabilities in both immiscible and miscible displacements (Chen et al., 2010; Dias, 2013; Yuan and Azaiez, 2014a, 2014c). To easily analyze the change of displacement rates on frontal instabilities, a cyclic displacement rate consisting of two-stage piecewise-constant rate is used in the present study (Dias et al., 2010; Yuan and Azaiez, 2015a). A dimensionless form of displacement rate is given in Fig. 1.

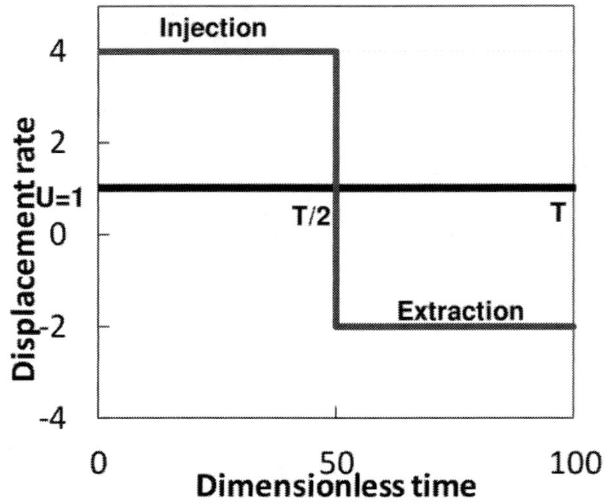


Fig. 1—Time-dependent displacement rate with injection and extraction in dimensionless form.

The time-dependent displacement rate consists of two stages: the injection stage with $U_1$=4 and extraction stage with $U_2$=-2. The period of a complete change of one cycle is $T$. To further simplify the analysis of instability mechanisms, the length for injection or extraction is $T/2$ so that the amount of $CO_2$ injected is the same as that of the constant injection case with $U$=1 in any cycle. This ensures that the frontal instabilities are comparable between constant and time-dependent displacement rates with the same criteria. It should also be mentioned that it is a repeated displacing process. The cycle period is varied from $T$=1 to $T$=200 and its effects are examined in the present research.

In the following discussion on variations of frontal instabilities in $CO_2$ miscible flooding, we fixed the parameters as aspect ratio of domain $A=Lx/Ly=2$, Peclet number $Pe$=500, and viscosity or mobility ratio $MR$=20.

## Results and discussion

In this section, the frontal instabilities in $CO_2$ miscible flooding in a 2D domain are discussed. Both qualitative and quantitative analysis is conducted by plotting the variations of concentration profiles and the $CO_2$ front with time. We will focus on the effects of time-dependent displacement rates and compare them with those of constant injection.

The concentration profiles for the constant injection rate are shown in Fig. 2 at dimensionless times $t$=100 and 300, respectively. Under the unfavorable viscosity ratio $MR$=20, the viscous fingering of injected $CO_2$ forms as the displacement continues and eventually dominates the flows. The development of the unstable front can enhance the mixing between $CO_2$ and oil by increasing the contact area of the two components, while it also results in earlier breakthrough. The longer the viscous fingers are, the more unstable the displacements will be. Similar results have been reported by other researchers for different cases with constant injection (Islam and Azaiez, 2005; Meiburg and Chen, 2000; Tan and Homsy, 1988; Yuan and Azaiez, 2015c). However, the scenario for time-dependent displacement rates considering various diffusion coefficients and dispersion is investigated for the first time in this study.

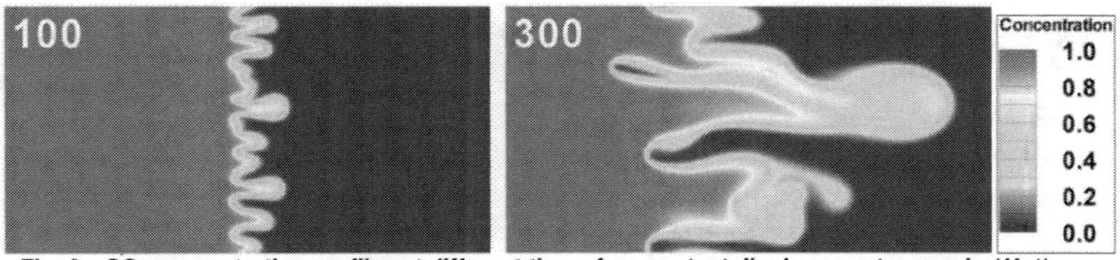

**Fig. 2—$CO_2$ concentration profiles at different times for constant displacement scenario ($U$=1).**

## Cyclic displacements

In the present study, we focus on the effect of cycle period on frontal instabilities in $CO_2$ miscible displacements. The displacement in the dimensionless form starts with an injection at $U_1$=4, followed by an extraction $U_2$=-2 where the injected $CO_2$ is extracted under the pressure gradient. As discussed above, the length of time for injection and extraction is the same and equals $T/2$.

**Fig. 3—$CO_2$ concentration profiles of time-dependent displacement rate at $T$=1, $U_1$=4 and $U_2$=-2.**

Fig. 3 shows the concentration profiles for the time-dependent injection rate at cycle period $T$=1. Such a short period indicates a fast switching between injection and extraction in any period. Very different development of frontal instabilities can be observed. In particular, the viscous fingering seems to develop much later compared with the constant injection. At $t$=100, the contact zone between $CO_2$ and oil is still very stable, as shown in Fig. 3, while the front becomes clearly unstable in Fig. 2 at the same time. However, a careful observation of the width of mixing zone suggests that it seems to be smoother for the time-dependent case. This is mainly because of the stronger injection and extraction in both stages, which are favorable for the $CO_2$ and oil to more uniformly blend into each other. In other words, the time-dependent displacement rate with higher amplitude can enhance mixing through stable diffusion and dispersion, rather than unstable viscous fingering.

The initial uniform front consequently results in the less unstable displacements at $t$=300 in Fig. 3. It can be seen that four fingers develop with similar length and width, although one of them is slightly larger. In comparison, only one dominant viscous finger can develop for the displacement with constant injection, and it is much longer at the same time. There is no doubt that viscous fingers will eventually develop under such an unfavorable viscosity ratio for either constant injection or time-dependent injection, while the variations of displacement rates with time indeed take effect and stabilize the displacements. This also indicates that, to achieve a relatively stable displacement in $CO_2$ miscible flooding, the fast oscillation of displacement rates with short periods should be used. This finding is similar to the experimental results in waterflooding by Ivanov and Araujo (2006) and the Vapex process by Muhamad et al. (2012), in which the shortest cyclic periods lead to the highest oil recovery. However, the mechanisms are different because of the nature of the different displacements. It should also be noted that the conclusion in the present study is clearly distinct from the case considering diffusion and inertia in our previous research (Yuan and Azaiez, 2015b) where the frequent switching between injection and extraction at very short intervals has a negligible effect on frontal instabilities. Comparing these two different physical problems with the same time-dependent displacement rates, one can conclude that it is the dispersion that plays a stabilizing role in the fast cyclic displacement process.

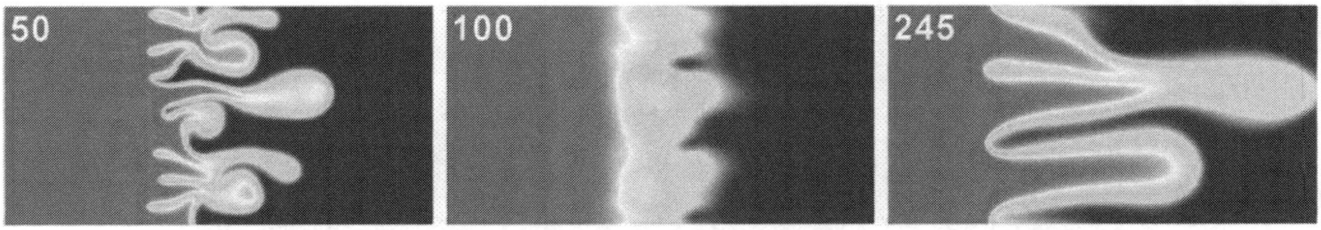


**Fig. 4—$CO_2$ concentration profiles of time-dependent displacement rate at $T$=100, $U_1$=4 and $U_2$=-2.**

Whereas the short period tends to stabilize the displacements, a longer period exhibits different characteristics. The longer the period is, the longer the injection in any complete period would be. Especially if injection in the first half of the period is long enough during $t\in[0, T/2]$, viscous fingering would easily develop and grow to be larger by the stronger injection.

Fig. 4 depicts the concentration profile for the time-dependent case with $T$=100. It is reasonable that at the end of first injection stage at $t$=50, the displacements become very unstable, with complex interactions among the viscous fingers. This is followed by an extraction stage in the next half of the period during $t\in[50, 100]$ in which the oil is displacing $CO_2$. At the end of one complete period $t$=100, a much wider mixing zone can be observed at the contact between $CO_2$ and oil compared with the cases with $T$=1 and constant injection. This indicates that the enhancement of mixing can be much stronger by intentionally triggering frontal instabilities with a longer period but with the same amount of $CO_2$ injected. If the displacement continues, the viscous fingering further develops and at $t$=245, the fingers reach the boundary and breakthrough happens. For constant injection, by contrast, the breakthrough happens later than $t$=300. For $T$=1, the breakthrough is even later. Although the longer period enhances the mixing of $CO_2$ and oil, which is favorable for the reduction of oil viscosity, it also results in earlier breakthrough of $CO_2$.

Having discussed the effect of period ranging from 1 to 100 for time-dependent displacement rates, we can conclude that the shorter period stabilizes the $CO_2$ miscible flooding, while the longer period destabilizes the displacements. The effect of the increase of cycle period on instabilities seems to be monotonic. Such changes as well as instabilities can be measured by quantitative analysis of the front location with time.

## Front location analysis

The front of the injected $CO_2$ in displacements is defined by the relative location at transversely averaged concentration *Cavg*=0.01 in the streamwise direction. The average concentration is defined by Sajjadi and Azaiez (2014). Since the numerical calculation is conducted in the dimensionless form and in a moving frame reference, the relative location of the $CO_2$ front therefore always starts with 0.5. It can be very conveniently converted to real space by considering the practical problems with certain values of diffusion, dispersion and the size of domain.

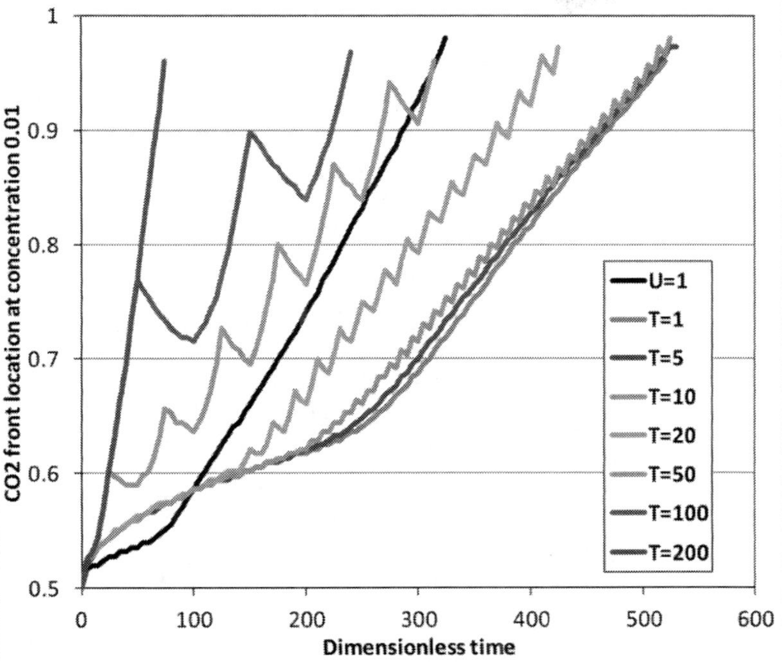


**Fig. 5—$CO_2$ front location at different times for constant injection (black curve) and various periods at $U_1$=4 and $U_2$=-2.**

Fig. 5 depicts the variations of $CO_2$ front with time for constant injection ($U$=1) and time-dependent injection rates at different periods. Clearly, the initial $CO_2$ front for time-dependent cases is larger than that for constant injection. This is the main contributor to the larger amplitude of injection and extraction stages which enhances the uniform mass transfer. However, the fast oscillation of displacement rates would suppress the development of instabilities and result in the slower propagation of the front at the later times. The shortest period $T$=1 consequently leads to the lowest values of the front at a longer time, indicating the least unstable displacement. An increase in period from $T$=1 to 10 does not significantly change the propagation of the front but clearly shows the effect of different stages. Further increase in period to be $T$=20 enhances the movement of the front, but this is still less than that for constant injection at breakthrough until the period is no shorter than $T$=50. Similar with the finding of our previous research on inertia (Yuan and Azaiez, 2015a), this range for $T \in [0, 50]$ is referred to as a period-stabilizing range in which the displacements for time-dependent scenarios are less unstable than for constant injection. However, there is fundamental difference between these two cases in terms of physical problems and mechanisms. In the present study, the velocity-induced dispersion plays a role, while the inertia is the dominating factor in the previous research.

When the period is long enough, the front always moves faster than with constant injection because of the larger amplitude of injection and extraction and also because of the earlier development of viscous fingering.

The front is also analyzed by plotting the dimensionless time at which the $CO_2$ reaches a certain location in the domain, shown in Fig. 6. The first set of bars are for constant injection rate with $U$=1 without any pulsing. It is worth noting that $U$=1 does not mean the period $T$ is infinitely small or large since we have fixed the values of two-stage piecewise-constant rate as $U_1$=4 and $U_2$=-2. For the less unstable displacements at shorter periods, it is reasonable for the $CO_2$ to reach the certain location at a later time. At a longer period, by contrast, the time is much shorter and the $CO_2$ fingers move much faster in the domain. A calculation of swept ratio (SR) at the breakthrough time of constant injection is also conducted and the values are labelled in Fig. 6. The definition of SR is given by Yuan and Azaiez (2012). Clearly, the swept ratio is larger for a relatively less unstable displacement at short periods $T \in [0, 50]$. The increment of swept ratio at the time at which breakthrough happens for constant injection is as large as 21% for the case at $T$=1. For longer periods, on the other hand, the swept ratio would be smaller and breakthrough would happen earlier than for constant injection.

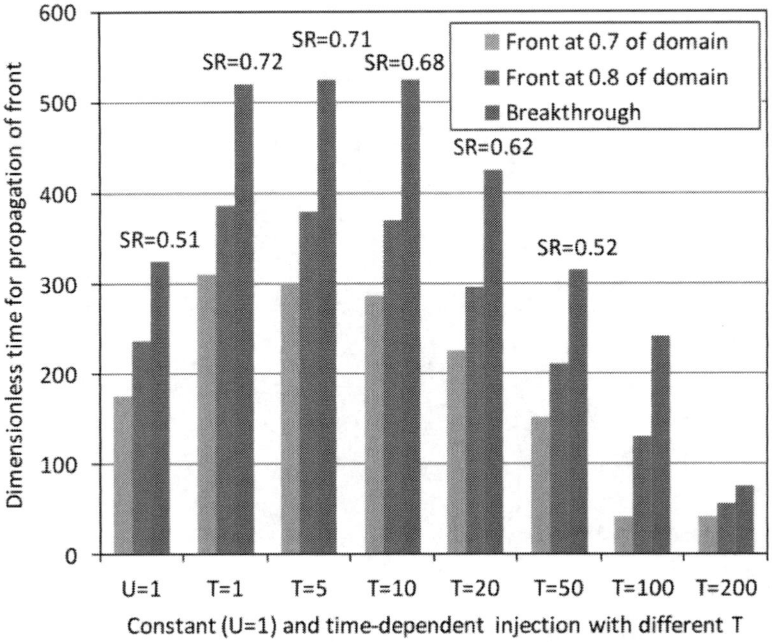


**Fig. 6—The time at which the front of injected $CO_2$ reaches a certain location of domain and swept ratio (SR) at $t$=230.**

In $CO_2$ miscible flooding, viscous fingering is the dominant factor which may result in unstable displacements and earlier breakthrough. Therefore, to improve sweep efficiency and maximize oil recovery by reducing viscous fingering, the fast oscillation of cyclic displacement rates involving stronger injection and extraction is a better option. This is a very promising way to enhance oil recovery by $CO_2$ miscible flooding with nearly zero additional cost.

## Conclusions

The effects of time-dependent displacement rates on $CO_2$ miscible flooding are discussed for the first time in the present study by considering varying diffusion and dispersion, processes, which are functions of concentration and flow rates. Several important conclusions can be obtained through the analysis of development of frontal instabilities.

It is found that both the dispersion and frontal instabilities can enhance the mixing between $CO_2$ and oil although their mechanisms are different. The dispersion induces uniform mass transfer and has a stabilizing effect, whereas the frontal instabilities result in a wider mixing by increasing the contact zone of the two components.

Quite different flow dynamics in $CO_2$ miscible flooding can be observed for time-dependent displacement rates compared with those for constant injection. Depending on the length of cycle period, the variations of displacement rates may either stabilize or destabilize the frontal instabilities. A short period tends to stabilize the displacements and leads to a significant increase of swept ratio when compared with widely used constant injection. An optimal injection scheme would be one with very short periods rather than one with a constant injection rate in $CO_2$ miscible flooding.

The present study also indicates that the time-dependent displacement rates can be further optimized through suppression of the frontal instabilities by considering the amplitude and other types of rates. Even higher sweep efficiency can probably be achieved. It would be a very promising way to enhance oil recovery at nearly no additional cost in $CO_2$ miscible flooding and other solvent injection processes.

## Acknowledgements

Financial support from the Mitacs and the Saskatchewan Research Council (SRC) is gratefully acknowledged. The authors would like to thank WestGrid for providing computational resources. The authors also appreciate SRC for granting permission for publication of this paper.

## Nomenclature

| Symbol | Description |
|---|---|
| $A$ | Aspect ratio of domain |
| $c$ | Dimensionless concentration |
| $Cavg$ | Averaged concentration in transverse direction |
| $D$ | Diffusion and dispersion coefficient ($m^2$/s) |
| $Lx$, $Ly$ | Length and width of domain, respectively (m) |
| $MR$ | Viscosity or mobility ratio |
| $P$ | Dimensionless pressure |
| $Pe$ | Peclet number |
| $SR$ | Swept ratio |
| $t$ | Dimensionless time |
| $T$ | Cycle period |
| $U$ | Dimensionless constant injection rate, $U$=1 |
| $U_1$ | Dimensionless injection rate |
| $U_2$ | Dimensionless extraction rate |
| $\boldsymbol{u}$ | 2D velocity vector, $\boldsymbol{u}=(u,v)$ |
| $u$ | Velocity in streamwise direction |
| $v$ | Velocity in transverse direction |
| $\mu$ | Dimensionless viscosity or mobility |
| $\mu_1$ | Dimensionless viscosity or mobility of $CO_2$ |
| $\mu_2$ | Dimensionless viscosity or mobility of oil |

## Reference

Chen, C.Y., Huang, C.W., Wang, L.C., Miranda, J.A., 2010. Controlling Radial Fingering Patterns in Miscible Confined Flows. Phys. Rev. E - Stat. Nonlinear, Soft Matter Phys. 82, 1–8.

Davidson, B.C., Spanos, T., Dusseault, M.B., Wang, J., 1999. Laboratory Experiments on Pressure Pulse Flow Enhancement in Porous Media. In: The CIM Regina Technical Meeting. Petroleum Society of CIM, Regina, Saskatchewan, Canada.

Dias, E.O., 2013. Viscous-Fingering Minimization in Uniform Three-Dimensional Porous Media. Phys. Rev. E - Stat. Nonlinear, Soft Matter Phys. 88, 1–5.

Dias, E.O., Miranda, J.A., 2009. Interfacial Instabilities in Periodically Driven Hele-Shaw Flows. Phys. Rev. E - Stat. Nonlinear, Soft Matter Phys. 80, 1–8.

Dias, E.O., Miranda, J.A., 2010. Control of Radial Fingering Patterns: A Weakly Nonlinear Approach. Phys. Rev. E - Stat. Nonlinear, Soft Matter Phys. 81, 1–7.

Dias, E.O., Parisio, F., Miranda, J.A., 2010. Suppression of Viscous Fluid Fingering: A Piecewise-Constant Injection Process. Phys. Rev. E - Stat. Nonlinear, Soft Matter Phys. 82, 1–4.

Ghesmat, K., Azaiez, J., 2008. Viscous Fingering Instability in Porous Media: Effect of Anisotropic Velocity-Dependent Dispersion Tensor. Transp. Porous Media 73, 297–318.

Groenenboom, J., Wong, S., Meling, T., Zschuppe, R., Davidson, B., 2003. Pulsed Water Injection during Waterflooding. In: SPE International Improved Oil Recovery Conference. Society of Petroleum Engineers, Kuala Lumpur.

Grogan, A.T., Pinczewski, V.W., Ruskauff, G.J., Orr, F.M.J., 1988. Diffusion of CO2 at Reservoir Conditions: Models and Measurements. SPE Reserv. Eng. 2, 93–102.

Huang, T., Zhou, X., Yang, H., Liao, G., Zeng, F., 2017. CO2 Flooding Strategy to Enhance Heavy Oil Recovery. Petroleum 3, 68–78.

Islam, M.N., 2003. Simulation of Viscous Fingering in High Mobility Ratio Miscible Displacements. University of Calgary.

Islam, M.N., Azaiez, J., 2005. Fully Implicit Finite Difference Pseudo-Spectral Method for Simulating High Mobility-Ratio Miscible Displacements. Int. J. Numer. Methods Fluids 161–183.

Ivanov, D.A., Araujo, M., 2006. Dynamics of Two-Phase Immiscible Pulsed Flow. In: SPE/DOE Symposium on Improved Oil Recovery. Society of Petroleum Engineers, Tulsa, Oklahoma, USA.
Li, S., Lowengrub, J.S., Fontana, J., Palffy-Muhoray, P., 2009. Control of Viscous Fingering Patterns in a Radial Hele-Shaw Cell. Phys. Rev. Lett. 102, 1–4.
Meiburg, E., Chen, C.-Y., 2000. High-Accuracy Implicit Finite-Difference Simulations of Homogeneous and Heterogeneous Miscible-Porous-Medium Flows. SPE J. 5, 129–137.
Muhamad, H., Upreti, S.R., Lohi, A., Doan, H., 2012. Performance Enhancement of VAPEX by Temporal Variation of Solvent Injection Pressure. J. Pet. Sci. Eng. 96–97, 93–101.
Muhamad, H., Upreti, S.R., Lohi, A., Doan, H., 2016. Optimal Control Study to Enhance Oil Production in Labscale Vapex by Varying Solvent Injection Pressure with Time. Optim. Control Appl. Methods 37, 424–440.
Pipes, J.W., Schoelling, L.G., 2014. Performance Review of Gel Polymer Treatments in a Miscible CO2 Enhanced Recovery Project, SACROC Unit Kelly-Snyder Field. In: The SPE Imroved Oil Recovery Symposium. Society of Petroleum Engineers, Tulsa, Oklahoma, USA, pp. 1–14.
Sajjadi, M., Azaiez, J., 2014. Improvement of Sweep Efficiency of Miscible Displacement Processes in. In: SPE Heavy Oil Conference. Society of Petroleum Engineers, Calgary, Alberta, Canada, pp. 1–7.
Shchipanov, A.A., Surguchev, L.M., Jakobsen, S.R., 2008. Improved Oil Recovery by Cyclic Injection and Production. In: SPE Russian Oil & Gas Technical Conference and Exhibition. Society of Petroleum Engineers, Moscow, Russia.
Spanos, T., Davidson, B., Dusseault, M., Shand, D., Samaroo, M., 2003. Pressure Pulsing at the Reservoir Scale: A New IOR Approach. J. Can. Pet. Technol. 42, 16–28.
Stalkup, F.I., 1978. Carbon Dioxide Miscible Flooding: Past, Present, and Outlook for the Future. J. Pet. Technol. 30, 1102–1112.
Stirpe, M.T., Guzman, J., Manrique, E., Alvarado, V., 2004. Cyclic Water Injection Simulations for Evaluations of its Potential in Lagocinco Field. In: SPE/DOE Symposium on Improved Oil Recovery. Society of Petroleum Engineers, Tulsa, Oklahoma, USA.
Surguchev, L., Koundin, A., Melberg, O., Rolfsvag, T.A., Menard, W.P., 2002. Cyclic Water Injection: Improved Oil Recovery at Zero Cost. Pet. Geosci. 8, 89–95.
Tan, C.T., Homsy, G.M., 1988. Simulation of Nonlinear Viscous Fingering in Miscible Displacement. Phys. Fluids 31, 1330–1338.
Wang, J., Dong, M., Li, Y., Gong, H., 2015. Prediction of Nitrogen Diluted CO2 Minimum Miscibility Pressure for EOR and Storage in Depleted Oil Reservoirs. Fuel 162, 55–64.
Wang, J., Wang, B., Li, Y., Yang, Z., Gong, H., Dong, M., 2016a. Measurement of Dynamic Adsorption–Diffusion Process of Methane in Shale. Fuel 172, 37–48.
Wang, J., Yang, Z., Dong, M., Gong, H., Sang, Q., Li, Y., 2016b. Experimental and Numerical Investigation of Dynamic Gas Adsorption/Desorption–Diffusion Process in Shale. Energy & Fuels 30, 10080–10091.
Yan, Y., Li, C., Dong, Z., Fang, T., Sun, B., Zhang, J., 2017. Enhanced Oil Recovery Mechanism of CO2 Water-Alternating-Gas Injection in Silica Nanochannel. Fuel 190, 253–259.
Yuan, Q., 2014. Miscible Cyclic Time-Dependent Displacements in Porous Media. University of Calgary.
Yuan, Q., Azaiez, J., 2012. Simulation of Miscible Cyclic Injection Flows in Porous Media. In: SPE Heavy Oil Conference Canada. Society of Petroleum Engineers, Calgary, Alberta, Canada.
Yuan, Q., Azaiez, J., 2014a. Miscible Displacements in Porous Media with Time-Dependent Injection Velocities. Transp. Porous Media 104, 57–76.
Yuan, Q., Azaiez, J., 2014b. Cyclic Time-Dependent Reactive Flow Displacements in Porous Media. Chem. Eng. Sci. 109, 136–146.
Yuan, Q., Azaiez, J., 2014c. Viscous Fingering Instabilities In Sinusoidal Time-dependent Flows. In: International Conference on Heat Transfer and Fluid Flow. Prague, Czech Republic, pp. 1–7.
Yuan, Q., Azaiez, J., 2015a. Inertial Effects in Cyclic Time-Dependent Displacement Flows in Homogeneous Porous Media. Can. J. Chem. Eng. 93, 1490–1499.
Yuan, Q., Azaiez, J., 2015b. Non-Isothermal Displacements with Step-Profile Time Dependent Injections in Homogeneous Porous Media. In: World Congress on Mechanical, Chemical, and Material Engineering. Barcelona, Spain, pp. 1–9.
Yuan, Q., Azaiez, J., 2015c. Inertial Effects of Miscible Viscous Fingering in a Hele-Shaw Cell. Fluid Dyn. Res. 47, 15506.
Yuan, Q., Wang, J., Yao, S., Zeng, F., Knorr, K.D., Imran, M., 2017a. Highly Accurate Simulation of Frontal Instabilities at Extremely Large Mobility Contrasts in Solvent-Based Oil Recovery. In: SPE Reservoir Characterisation and Simulation Conference and Exhibition. Society of Petroleum Engineers, Abu Dhabi, UAE.
Yuan, Q., Zhou, X., Zeng, F., Knorr, K.D., Imran, M., 2017b. Nonlinear Simulation of Miscible Displacements with Concentration-Dependent Diffusion Coefficient in Homogeneous Porous Media. Chem. Eng. Sci. 172, 528–544.
Yuan, Q., Zhou, X., Zeng, F., Knorr, K.D., Imran, M., 2017c. Effects of Concentration-Dependent Diffusion on Mass Transfer and Frontal Instability in Solvent-Based Processes. In: SPE Canada Heavy Oil Technical Conference. Society of Petroleum Engineers, Calgary, Alberta, Canada.
Yuan, Q., Zhou, X., Zeng, F., Knorr, K.D., Imran, M., 2017d. Investigation of Concentration-Dependent Diffusion on Frontal Instabilities and Mass Transfer in Homogeneous Porous Media. Can. J. Chem. Eng. 9999, 1–16.

Zhou, X., Yuan, Q., Zeng, F., Zhang, L., Jiang, S., 2017. Experimental Study on Foamy Oil Behaviour Using a Heavy Oil-Methane System in the Bulk Phase. J. Pet. Sci. Eng.

Zimmerman, W.B., Homsy, G.M., 1991. Nonlinear Viscous Fingering in Miscible Displacement with Anisotropic Dispersion. Phys. Fluids A 3.

Zschuppe, R.P., 2001. Pulse Flow Enhancement in Two-Phase Media by. University of Waterloo.

AIChE ASME AIME IEEE ASCE
Carbon Management

# Analytical Temperature Modeling for Early Detection and Rate Estimation of $CO_2$ Wellbore Leakage

Yilin Mao and Mehdi Zeidouni, Louisiana State University

CARBON MANAGEMENT

LSU

Slide 2

# Outline

- ➢ Model description and introduction
- ➢ Methodology
- ➢ Result and procedure
- ➢ Conclusion

CARBON MANAGEMENT

# Model description

Slide 3

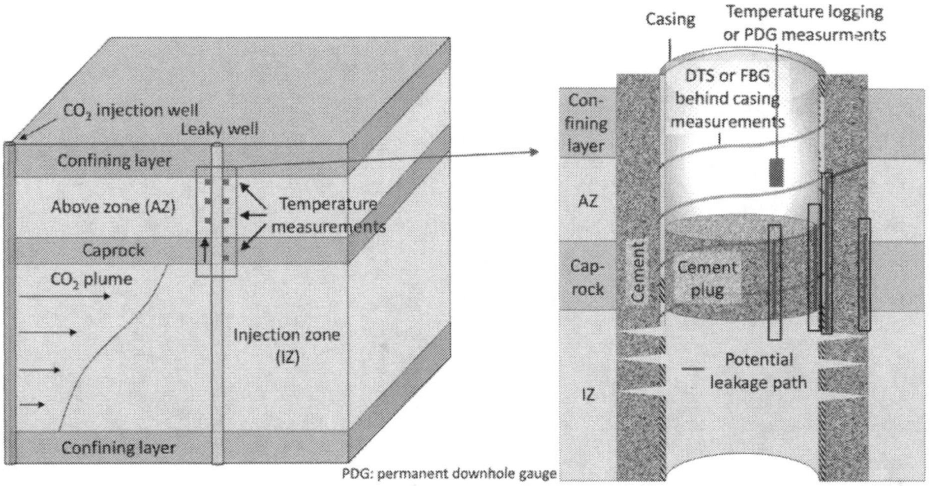


PDG: permanent downhole gauge

DTS: distributed temperature sensing

FBG: Fiber Bragg grating sensor

Wellbore leakage estimation from temperature • Yilin Mao

# Introduction - thermal behavior

Slide 4

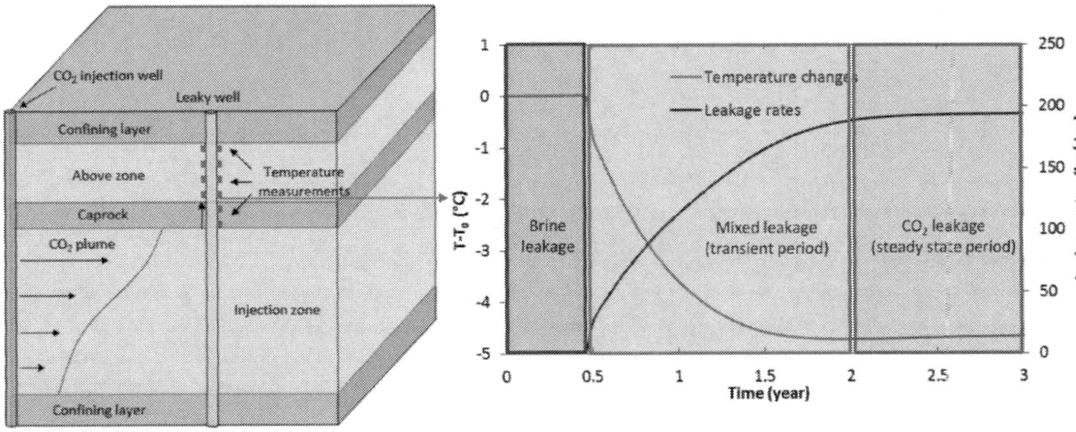


Wellbore leakage estimation from temperature • Yilin Mao

## Introduction - permeability

Slide 5

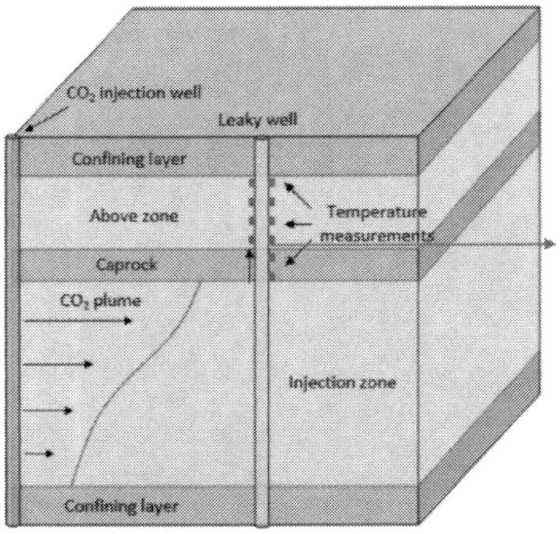


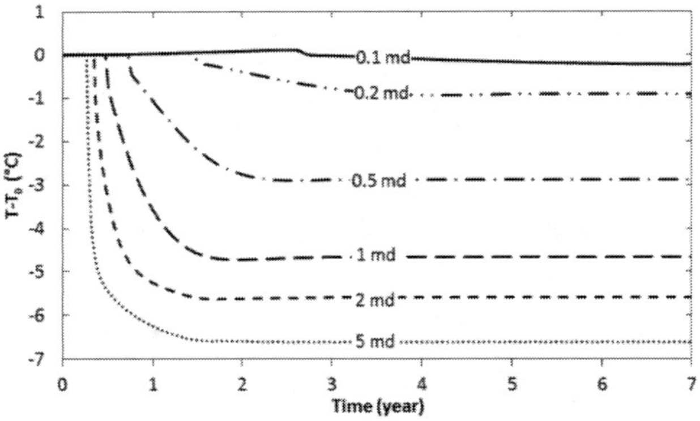


Wellbore leakage estimation from temperature • Yilin Mao

## Outline

Slide 6

- Model description and introduction
- Methodology
- Result and procedure
- Conclusion

Wellbore leakage estimation from temperature • Yilin Mao

# Control volume

Slide 7

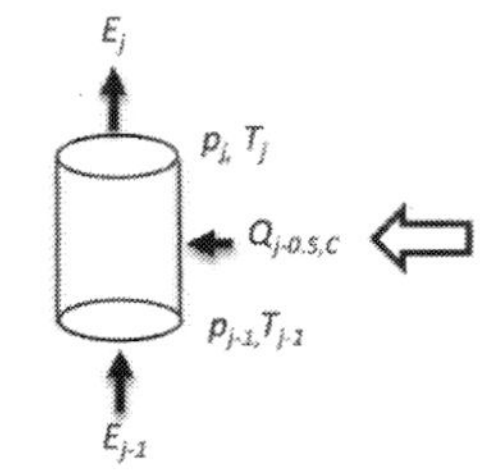


**Discretized**

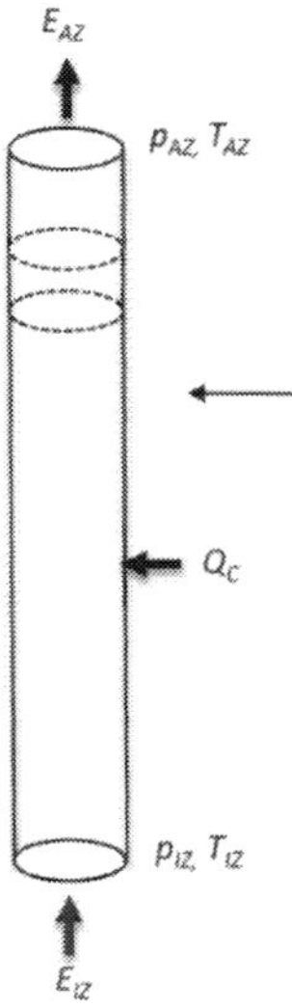


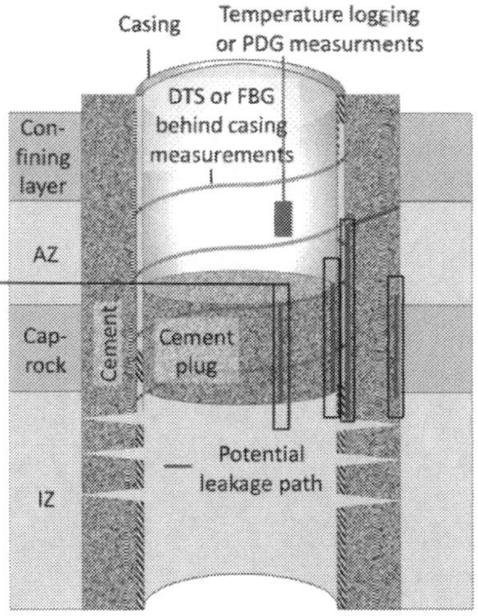


*E*: energy

*Q*: heat

*p*: pressure

*T*: temperature

Wellbore leakage estimation from temperature • Yilin Mao

# Energy balance

Slide 8

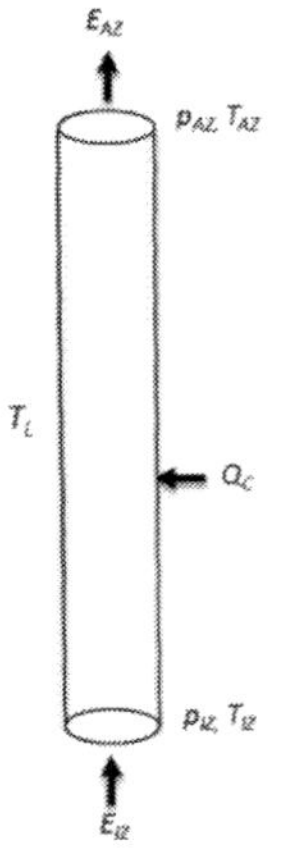


$$E_{AZ} - E_{IZ} = Q_C$$ (Steady state condition)

(Internal and potential energy)

$$m\left(\hat{H} + gz\right)_{AZ} - m\left(\hat{H} + gz\right)_{IZ} = Q_C$$

$$d\hat{H} = \left(\frac{\partial \hat{H}}{\partial p}\right)_T dp + \left(\frac{\partial \hat{H}}{\partial T}\right)_p dT = -\mu_{JT} c_p dp + c_p dT \qquad \dot{Q}_c = \frac{2\pi HK\left(T_{i,L} - T_L\right)}{\ln\left[1 + \left(c - \frac{1}{a + \sqrt{t_D}}\right)\sqrt{t_D}\right]}$$

$$\dot{m} c_p T_{AZ} - \dot{m} c_p T_{IZ} + \dot{m} g H = \dot{m} c_p \mu_{JT}\left(p_{AZ} - p_{IZ}\right) + \frac{2\pi HK\left(T_{i,L} - T_L\right)}{\ln\left[1 + \left(c - \frac{1}{a + \sqrt{t_D}}\right)\sqrt{t_D}\right]}$$

$$\dot{m} = \frac{2\pi HK\left(T_{i,L} - T_L\right)}{\left[c_p \mu_{JT}\left(p_{IZ} - p_{AZ}\right) + c_p\left(T_{AZ} - T_{IZ}\right) + gH\right]\ln\left[1 + \left(c - \frac{1}{a + \sqrt{t_D}}\right)\sqrt{t_D}\right]}$$

$\hat{H}$: enthalpy

*z*: height

$\mu_{JT}$: Joule-Thomson (JT) coefficient

$c_p$: specific heat

*K*: conductivity

*a and c*: contants

$t_D$: dimensionless time for conduction

Wellbore leakage estimation from temperature • Yilin Mao

Slide 9

# Analytical approach

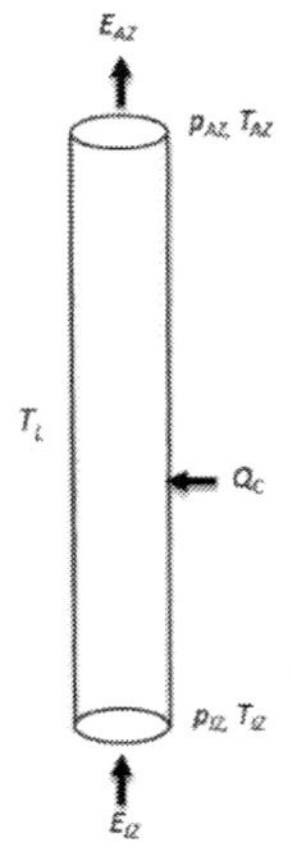


Convection

$$\dot{m} = \frac{1}{c_p \mu_{JT}\left(p_{IZ} - p_{AZ}\right) + c_p\left(T_{AZ} - T_{IZ}\right) + gH} \frac{2\pi HK\left(T_{i,L} - T_L\right)}{\ln\left[1 + \left(c - \frac{1}{a + \sqrt{t_D}}\right)\sqrt{t_D}\right]}$$

Conduction

Joule-Thomson (JT) effect

Gravity effect

$$\dot{m}_j = \frac{2\pi h_j K_{j-0.5}\left(T_{i,j-0.5} - T_{j-0.5}\right)}{\left[c_j \mu_{JT,j-0.5}\left(p_{j-1} - p_j\right) + c_j\left(T_j - T_{j-1}\right) + gh_j\right]\ln\left[1 + \left(c - \frac{1}{a + \sqrt{t_{D,j}}}\right)\sqrt{t_{D,j}}\right]}$$

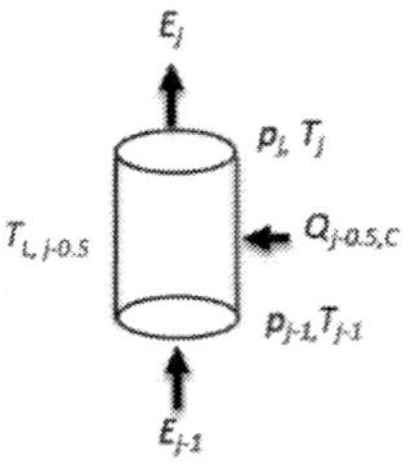


Wellbore leakage estimation from temperature • Yilin Mao

Slide 10

# Outline

- ➢ Model description and introduction
- ➢ Methodology
- ➢ Result and procedure
- ➢ Conclusion

Wellbore leakage estimation from temperature • Yilin Mao

# Model and leakage properties

Slide 11

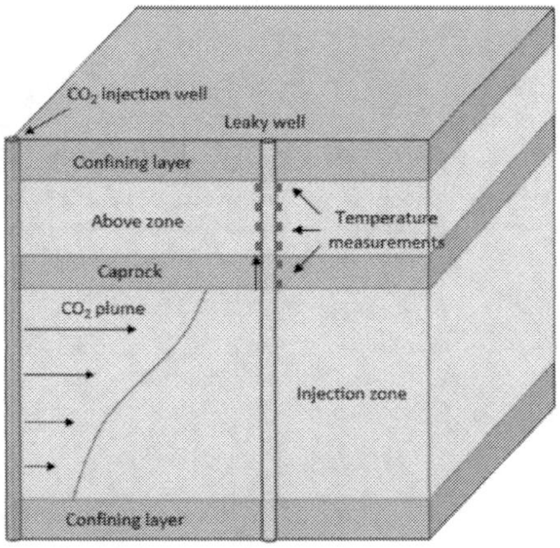


Source: Mao, Y., Zeidouni, M., Askari, R., 2017. Effect of leakage pathway flow properties on thermal signal associated with the leakage from $CO_2$ storage zone. Greenh Gases 7, 512-529.

| Model properties | | Leakage properties | |
|---|---|---|---|
| Porosity (fraction) | 0.3 | Leak-injector distance (m) | 100 |
| Vertical permeability (md) | 10 | Leakage area ($m^2$) | 0.04-25 |
| Lateral permeability (md) | 100 | Leakage vertical permeability (md) | 0.1-5 |
| Initial temperature at IZ (degC) | 45 | Leakage rates (kg/day) | 8-2600 |
| Geothermal gradient (degC/m) | 0.03 | | |
| Injection pressure (MPa) | 13 | | |
| Initial pressure at IZ (MPa) | 8 | | |
| Caprock thickness (m) | 10 | | |
| IZ thickness (m) | 55 | | |
| Depth of IZ top (m) | 1000 | | |

Wellbore leakage estimation from temperature • Yilin Mao

# Leakage rate estimation

Slide 12

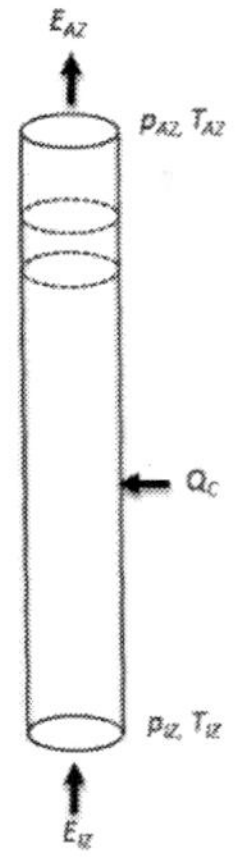


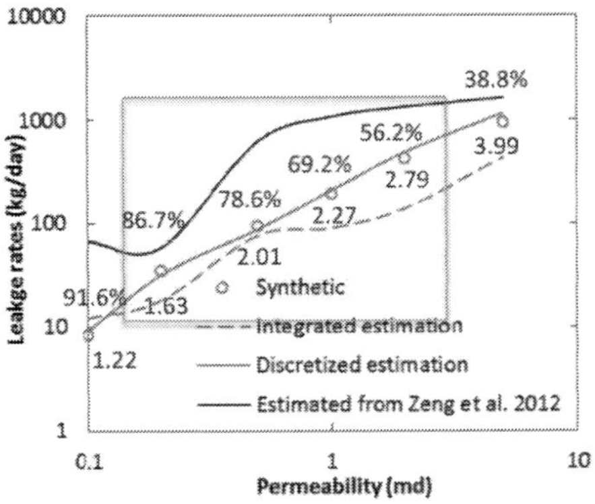


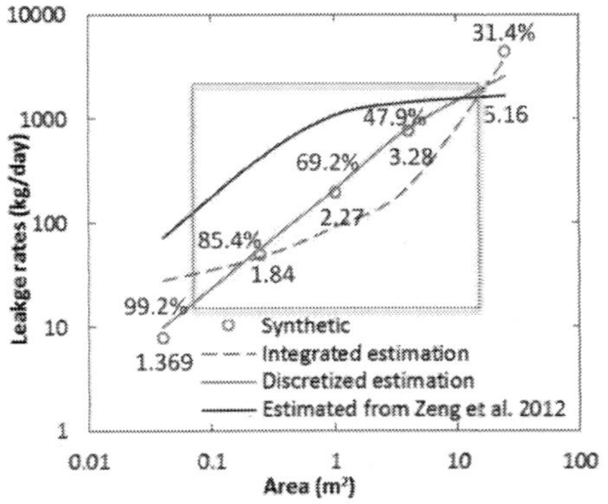


*Scope: 1.5 – 3.5*

Source: Zeng, F.H., Zhao, G., Zhu, L.J., 2012. Detecting $CO_2$ leakage in vertical wellbore through temperature logging. Fuel 94, 374-385.

Wellbore leakage estimation from temperature • Yilin Mao

# Leakage rate estimation

Slide 13

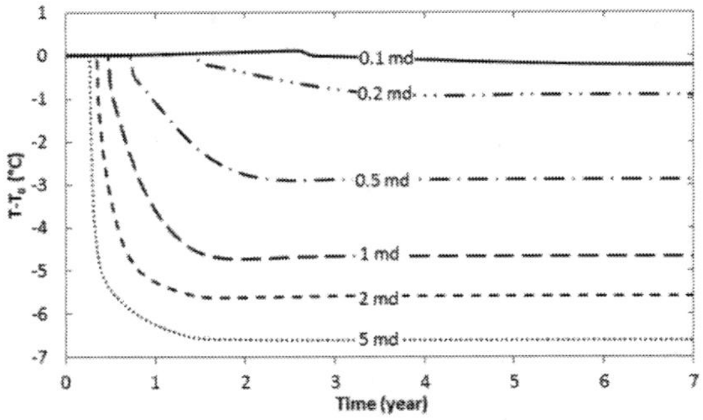


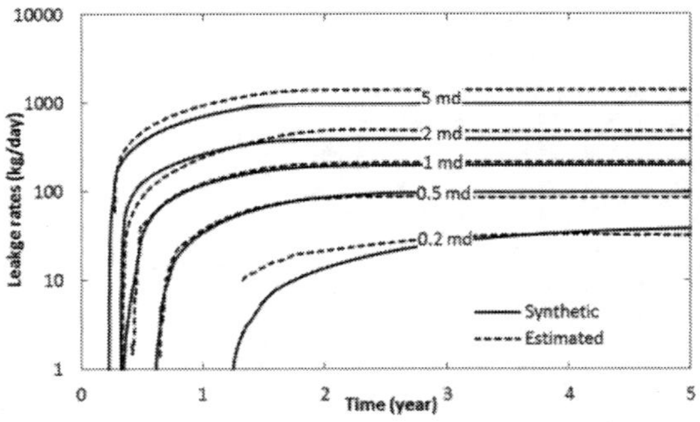




# Scope of this approach

Slide 14

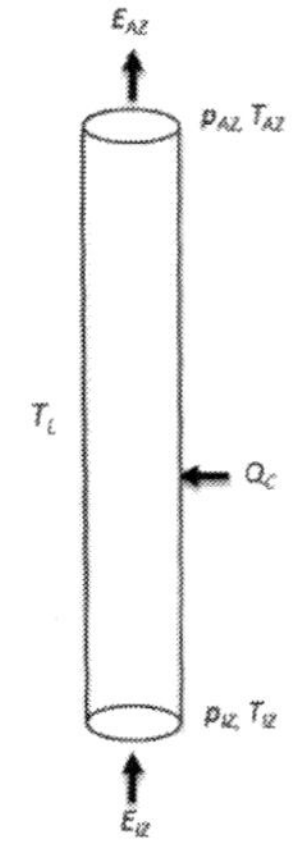


Convection

$$\dot{m} = \frac{1}{c_p\mu_{JT}(p_{IZ}-p_{AZ}) + c_p(T_{AZ}-T_{IZ}) + gH}\,\frac{2\pi HK(T_{i,L}-T_L)}{\ln\left[1+\left(c-\frac{1}{a+\sqrt{t_D}}\right)\sqrt{t_D}\right]}$$

Conduction

Joule-Thomson (JT) effect

~~Gravity effect~~

$$ratio = \frac{1}{\dot{m}c_p\mu_{JT}(p_{AZ}-p_{IZ})}\,\frac{2\pi HK_r(T_L-T_{i,L})}{\ln\left[1+\left(c-\frac{1}{a+\sqrt{t_D}}\right)\sqrt{t_D}\right]}$$

⇩

$$N_{JT/cond} = \frac{\dot{m}c_p\mu_{JT}\Delta p}{KL\Delta T}$$

# Procedures

Slide 15

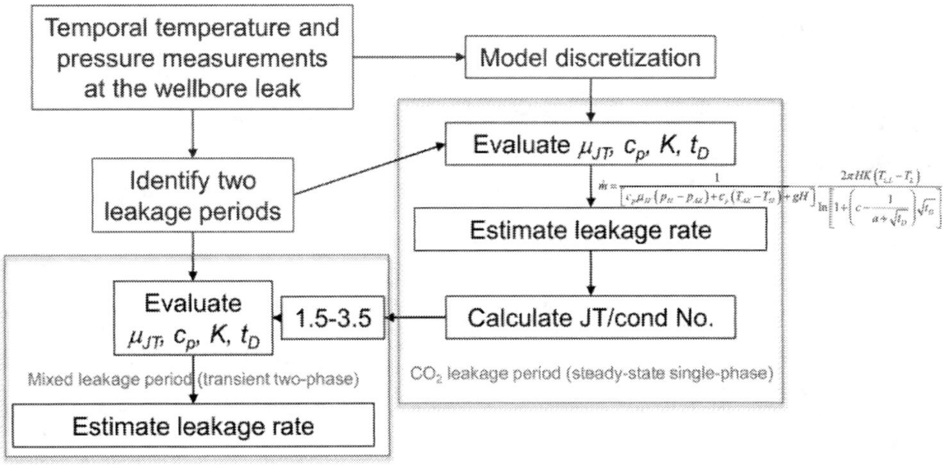


Wellbore leakage estimation from temperature • Yilin Mao

# Conclusions

Slide 16

- ✓ Thermal behaviors of wellbore leakage
- ✓ Analytical modeling of leakage rate
- ✓ Estimation validation for both periods
- ✓ Scope of this approach
- ✓ Practical procedures

Wellbore leakage estimation from temperature • Yilin Mao

# CMTC-486486-MS

# Comparing Two Anionic Surfactants for Mobility Control With Supercritical $CO_2$ in Miscible EOR

Mohammed A. Almobarky, Zuhair Al Yousef, David Schechter, Texas A&M University



This paper was prepared for presentation at the Carbon Management Technology Conference held in Houston, Texas, USA, 17-20 July 2017.



## Abstract

Carbon Dioxide ($CO_2$) flooding is one of the most globally used EOR processes to enhance the oil recovery. However, the low gas viscosity and density result in gas channeling and gravity override which lead to poor sweep efficiency. Foam application for mobility control is a promising technology to increase the gas viscosity which leads to lower mobility and better sweep efficiency inside the reservoir. Foam is generated inside the reservoir by co-injection surfactant and gas. Although there are many surfactants that can be used for such purpose, their performance with Supercritical $CO_2$ ($ScCO_2$) is weak which leads to poor or loss of mobility control. This experimental study evaluates a newly developed surfactant (CNF) that was introduced for $ScCO_2$mobility control in comparison with a common foaming agent, anionic Alpha Olefin Sulfonate (AOS) surfactant. Experimental work was divided into three stages: foam static tests, interfacial tension measurements, and foam dynamic tests. Both surfactants were investigated at different conditions. In general, results showed that both surfactants are good foaming agents to reduce the mobility of $ScCO_2$ with better performance of CNF surfactant. Shaking tests in presence of crude oil showed that foam life for CNF extends to more than 24-hr but less than that for AOS. Moreover, CNF features lower CMC, higher adsorption and smaller area/molecule at the liquid-air interface. Furthermore, entering, spreading, and bridging coefficients interpretations indicated that CNF surfactant produces very stable foam with light crude oil in both DI and saline water, whereas AOS was stable only in DI water. At all conditions for mobility reduction evaluation, CNF exhibited stronger flow resistance, higher foam viscosity, and higher mobility reduction factor than that of AOS surfactant. In addition, CNF and $ScCO_2$ simultaneous injection produced 8.83% higher oil recovery than that of the baseline experiment and 7.87% higher than that of AOS. Pressure drop profiles for foam flooding using CNF was slightly higher than that of AOS indicating that CNF is better in terms of foam-oil tolerance which resulted in higher oil recovery.

## Introduction

Oil reservoir has three recovery stages: primary, secondary and tertiary. After the primary and secondary stages, it is estimated that two-thirds of the OOIP are left underground (Green and Willhite 1998). For tertiary recovery, many Enhanced Oil Recovery (EOR) processes can be employed to extract more oil. Among these EOR processes, $CO_2$ injection is one the most utilized processes globally (Taber et al. 1997). However, gas injection processes face many challenges such as gas channeling and gravity override that lead to poor sweep efficiency (Healy et al. 1994). Many techniques have been applied to enhance the sweep efficiency such as water alternating gas (WAG), polymer, and foam. Foam is a promising technology that can be used to reduce the mobility of the injected gas by increasing its viscosity (Enick et al. 2012) and diverting the flow toward lower permeability zones where the remaining oil exists (Fried 1961).

The surfactant is the main element in foam system. It facilitates the foam generation by reducing the $\sigma_{g/w}$ which reduces the work required to generate foam, and it adsorbs at the interfaces to provide the foam with the required stability by stabilizing the thin-films between bubbles (Schramm, 2000).Thus, the surfactant screening is the first step toward successful surfactant-stabilized-foam project (Boeije et al. 2017). In foam applications, in general, and particularly in $CO_2$ EOR, surfactant structure is a significant factor that affects the efficiency in every aspect of the process: gas viscosity, mobility control, and EOR (Adkins et al. 2010). These effects are related to the different interactions of surfactants and $CO_2$ than that with air (Adkins et al. 2010). Moreover, $ScCO_2$ presence results in low pH acidic environment where some types of surfactants hydrolyze and lose their interfacial activity such as sulfates (Talley 1988).

Alpha olefin sulfonate (AOS) is hydrolytically and thermally stable, and soluble at low to medium hard water (Porter 1994). Farajzadeh et al. (2010) experimentally investigated the use of AOS for mobility control and EOR in miscible and immiscible flooding with the aid of CT scanner for simultaneous monitoring of the flooding process. They reported 19% more oil recovery with $ScCO_2$ than that of the immiscible $CO_2$ flooding. However, no sharp front was observed with the use of $ScCO_2$. They attributed this to the poor foam stability with oil. Haugen et al. (2012) experimentally used AOS for mobility control and EOR in fractured oil wet and water wet cores, and reported that the pre-generated foam is better than in-situ foam generation in terms of mobility reduction and oil recovery. They attributed the results to the poor foam-oil tolerance. Li et al. (2012) used AOS surfactant in Surfactant-Alternating-Gas (SAG) injection mode for foam generation using $N_2$. Their experiments were conducted in a two-dimensional sand pack with 19 to 1 permeability contrast. They attributed the poor sweep efficiency to the weak foam stability in presence of crude oil. They suggested that enhancing the foam-oil tolerance could provide higher oil recovery because this may enhance the sweep efficiency. Indeed, mixing the surfactant with a foam booster CTAB zwitterionic surfactant improved the foam-oil tolerance, provided better displacement efficiency, and resulted in higher oil recovery.

This study is to evaluate a newly developed anionic surfactant (CNF) to control the mobility of $ScCO_2$. Moreover, CNF surfactant results are compared with C14-16 AOS anionic surfactant which is heavily used in literature with $CO_2$ in gaseous and supercritical states. This new surfactant, and the major challenges for the surfactants' utilization with $ScCO_2$, may provide more opportunities for foam applications in foam-assisting miscible $CO_2$ EOR projects.

## Materials

Table 1 shows the general properties of both surfactants used in experimental work. Surfactants were diluted using LabChem DI water ASTM, type II. Moreover, tests were conducted at 0.5-wt% surfactant concentration. Besides DI water, the salinity effect was investigated using brine solutions at 10,000, 20,000, and 30,000 ppm with Sodium Chloride (NaCl) purchased from Cole-Parmer. The crude oil used in this study is brought from North Burbank Unit, OK USA (NBU). It is light crude oil with 33.7°API, 8-cp viscosity at room temperature 23°C, and 39.5°API and 3.27-cp at 50°C which is the reservoir temperature. The properties of the glass-beads used to make the glass-beads pack are 2.5 specific gravity and 100-μm diameter purchased from Potters Industries LLC.

**Table 1: The properties of the surfactants**

| Surfactant | Form | Chemical Family | pH | Density [gm/ml] | Charge | Flash Point [°C] | Carbon Chain Length |
|---|---|---|---|---|---|---|---|
| CNF | Liquid | Complex Nanofluid | 7.73 | 1.07 | Anionic | >93.3 | -- |
| AOS | Liquid | Alpha Olefin Sulfonate | 8.2 | 1.06 | Anionic | >94 | 14 -16 |

## Methodology

The experimental work was divided into three stages: static foam tests, interfacial tension measurements, and foam dynamic tests. The foam dynamic tests are divided into three sections: mobility reduction evaluation in high permeability glass beads pack, mobility reduction evaluation in lower permeability Bentheimer sandstone, and core flooding experiments. As mentioned earlier, the surfactants concentrations were kept constant at 0.5-wt% diluted with DI water and prepared in three NaCl salinities: 10,000, 20,000 and 30,000 ppm.

### Foam Static Tests

Foam was generated by shaking 3-ml of surfactant solutions in 13 X 100-mm (9-ml) Pyrex glass test tubes. Care has been taken to perform 10 to 15 gentle and uniform shakings for all samples. Samples were prepared at 0.5-wt% concentrations in DI water, 10,000, 20,000, and 30,000 NaCl brine solutions. After the foam has been generated with shaking inside the test tube, the foam columns were monitored by taking images at different times. Then, the foam columns lengths were measured from images using ImageJ software. The foaming ability was investigated using the initial foam column length ($h_{fi}$), and the foam stability was measured by the Foam Half-Life (FHL) which is the time at which the foam column loses half of the hfi. The samples were prepared for static tests without oil and stirred for about 12 hr before testing. For static tests with crude oil, the samples were prepared at 0.5-wt% concertation and stirred for 12-hr. Then, the surfactant solution is placed in 9-ml test tubes above which the crude oil was simply poured, then the sample was shaken immediately.

### Interfacial Tension Measurements

Air-water surface tension measurements ($\sigma_{g/a}$) were conducted at different surfactant concentrations in DI water using Dataphysics OCA 15 Pro IFT instrument, pendant drop method. The surface measurements were used for CMC determination and interfacial activity predictions for both surfactants.

The surface tension $\sigma_{a/w}$ measurements vs. the logarithmic values of the concentrations below the CMC is a linear relationship with a straight line. The slope of this straight line can be used to interpret the interfacial activities: adsorption and area/molecule at the interface. According to Gibbs adsorption equation, the higher the slope is the higher the adsorption at the air-water interface. Furthermore, the higher the adsorption at the density also results in smaller area/molecule which indicates a stronger packing at the interface which induces higher foam stability (Rosen and Kunjappu 2004).

The interfacial tension measurements, also, can be also used to investigate the foam-oil tolerance by calculating the entering coefficient (E), spreading coefficient (S), bridging-coefficient (B), and the lamellae number (L) using equations 1, 2, 3, and 4 below where $\sigma_{o/w}$ and $\sigma_{o/g}$ are the oil-water and oil-gas interfacial tension, respectively.

$$E=\sigma_{a/w} + \sigma_{o/w} - \sigma_{o/g} \quad (1)$$

$$S=\sigma_{a/w} - (\sigma_{o/w} + \sigma_{o/g}) \quad (2)$$

$$B=\sigma^2_{a/w} + \sigma^2_{o/w} - \sigma^2_{o/g} \quad (3)$$

$$L = 0.15\ \frac{\sigma_{a/w}}{\sigma_{o/w}} \quad (4)$$

## Foam Dynamic Tests

These experiments were designed for mobility reduction evaluation and oil recovery investigation by conducting core flood experiments.

### Mobility Reduction Evaluation in High Permeability Glass Beads Pack

These experiments were conducted at different shear rates 317-sec$^{-1}$, and one low shear rate 9.51-sec$^{-1}$. Furthermore, three injection qualities were applied 50, 70, and 90%. All experiments were conducted at constant pressure and temperature 1800-psi and 50°C to ensure the supercritical conditions of $CO_2$. The foam was generated by simultaneously injecting surfactant and $ScCO_2$ through the glass beads pack. The pressure drop was measured using two sets of pressure transducers: 500-psi for high range and 50-psi for low range. The pressure drop data were collected using data acquisition system. The onset of strong foam generation was recognized as a rapid increase in pressure drop according to Dicksen et al. (2002). The flow continued with monitoring the pressure drop data until the steady state pressure was attained. Then, the steady state pressure drop collected data were averaged and used to calculate the mobility, foam effective viscosity, and Mobility Reduction Factor (MRF) using equations 5, 6 and 7 below.

$$\lambda = \frac{k}{\mu} = \frac{q\,l}{A\,\Delta P} \quad (5)$$

$$\mu_{eff} = \frac{k}{\lambda} \quad (6)$$

$$MRF = \frac{\Delta P_{foam}}{\Delta P_{baseline}} \quad (7)$$

Where λ for mobility, k is the permeability, μ viscosity, q is the flow rate, *l* is the length of the porous media, A is the cross sectional area of the glass beads pack

Table 2 shows the dimensions and the petrophysical properties.

**Table 2: Glass beads properties**

| | |
|---|---|
| **Glass Beads Size** | 100 μm |
| **Length** | 13 in |
| **Diameter** | 0.18 in |
| **Pore Volume** | 1.625 ml |
| **Porosity** | 30% |
| **Permeability** | 17.1-Darcy |

Figure 1 below show the experimental setup for the mobility reduction evaluation.

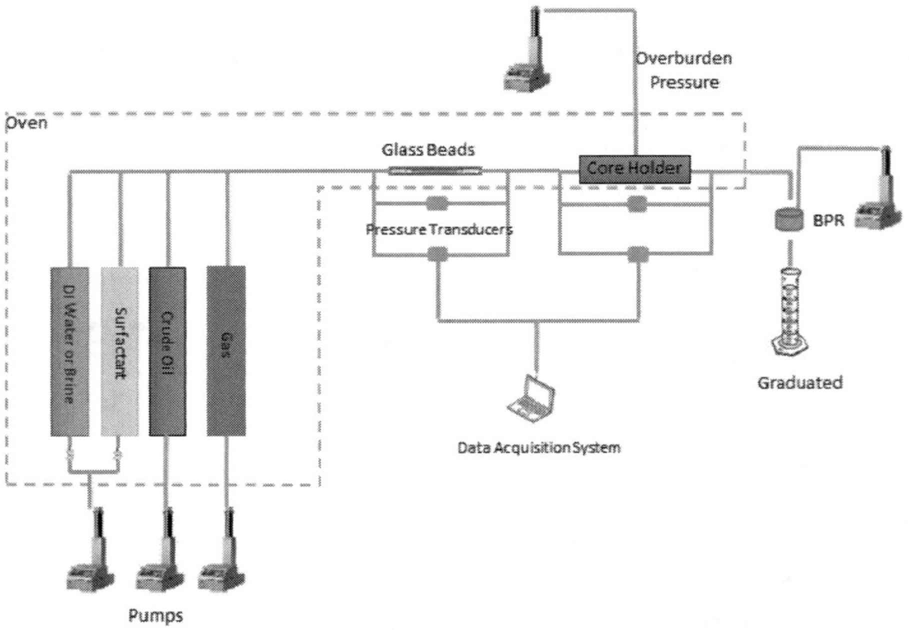


**Figure 1: Mobility reduction evaluation and core flooding experimental setup**

Table 3 shows the experimental conditions for the experiments conducted in the glass beads pack for mobility reduction evaluation at 317-$sec^{-1}$ and 9.51-$sec^{-1}$ shear rates. Run No. 20 is the base line experiment at which ScCO2 were used without surfactant for comparison purposes.

**Table 3: Glass beads pack experimental conditions**

| Run # | Surfac-tant | Salinity NaCl [wt%] | Injection Quality [%] | Q [ml/min] | Shear rate [1/sec] |
|---|---|---|---|---|---|
| 1 | AOS | -- | 90 | 0.5 | 317 |
| 2 | AOS | 1 | 90 | 0.5 | 317 |
| 3 | AOS | 2 | 90 | 0.5 | 317 |
| 4 | AOS | 3 | 90 | 0.5 | 317 |
| 5 | AOS | -- | 70 | 0.5 | 317 |
| 6 | AOS | 1 | 70 | 0.5 | 317 |
| 7 | AOS | 2 | 70 | 0.5 | 317 |
| 8 | AOS | 3 | 70 | 0.5 | 317 |
| 9 | AOS | -- | 50 | 0.5 | 317 |
| 10 | AOS | 1 | 90 | 0.015 | 9.51 |
| 11 | CNF | -- | 90 | 0.5 | 317 |
| 12 | CNF | 3 | 90 | 0.5 | 317 |
| 13 | CNF | 1 | 90 | 0.015 | 9.51 |
| 14 | -- | 1 | -- | 0.5 | 317 |

## Mobility Reduction Evaluation in Low Permeability Bentheimer Sandstone

These experiments were conducted using 1" diameter 12" length homogeneous Bentheimer sandstone. The core was left in an oven overnight for drying. Then, it was mounted in the core holder and 500-psi overburden pressure was applied. After that, the air was removed from the core using vacuum pump followed by saturating the core with 10,000 ppm NaCl brine solution at which the pore volume and porosity can be measured. Then, the overburden pressure was kept at 500-psi more than the test pressure, the experimental setup along with the core was pressurized using the back pressure regulator, as shown in figure 1. Experimental conditions for all runs are listed in Table 4. After the system was pressurized and the temperature was maintained at 50°C, the absolute permeability was measured by obtaining the pressure drop at different flow rate using Darcy law. Then, 5 to 6 pore volumes of brine solution were injected to ensure 100% core saturation. Although the XRD tests for these rocks showed that their composition is 100% quartz, 1 pore volume of surfactant solution was injected into the core at 5 ft/day (~ 9-sec-1 shear rate) to mitigate the effect of surfactant adsorption on rock surfaces. After that, the foam was applied by simultaneously injecting surfactant and $ScCO_2$ or $N_2$ gas at 5 ft/day, too. The foam injection was continued until the steady pressure was attained. The recorded steady state pressure drop data were averaged and used to calculate the mobility, foam effective viscosity, and MRF using equations 5, 6, and 7 above. Figure 1 above shows a schematic diagram of the experimental setup for mobility evaluation in sandstone. Table 4 below shows the properties and the experimental conditions for the experiments conducted in Bentheimer sandstone. Moreover, the last two experiments in Table 4 (runs 9 and 10) are baseline experiments conducted using $N_2$ and $ScCO_2$ injection for comparison purposes, respectively. Table 4: Petrophysical properties for Bnetheimer Sandstone in mobility reduction evaluation and experimental conditions

**Table 4: Petrophysical properties for Bnetheimer Sandstone in mobility reduction evaluation and experimental conditions**

| Run # | Surfac-tant | Concentr-tion [wt%] | NaCl Salinity [wt%] | P [psi] | Velocity [ft/day] | ShearRate [sec$^{-1}$] | Injection Quality [%] | Gas | K [Darcy] |
|---|---|---|---|---|---|---|---|---|---|
| 1 | AOS | 0.5 | 1 | 1800 | 5 | 8.92 | 90 | $ScCO_2$ | 1.7 |
| 2 | AOS | 0.5 | 1 | 1800 | 5 | 8.92 | 70 | $ScCO_2$ | 1.58 |
| 3 | CNF | 0.5 | 1 | 1800 | 5 | 8.92 | 90 | $ScCO_2$ | 1.7 |
| 4 | CNF | 0.5 | 1 | 1800 | 5 | 8.92 | 70 | $ScCO_2$ | 1.7 |
| 5 | AOS | 0.5 | 1 | 850 | 5 | 9 | 90 | $N_2$ | 1.57 |
| 6 | AOS | 0.5 | 1 | 850 | 10 | 18 | 90 | $N_2$ | 1.57 |
| 7 | CNF | 0.5 | 1 | 850 | 5 | 9. | 90 | $N_2$ | 1.7 |
| 8 | CNF | 0.5 | 1 | 850 | 10 | 18 | 90 | $N_2$ | 1.7 |
| 9 | -- | -- | 1 | 850 | 5 | 9 | -- | $N_2$ | 1.62 |
| 10 | -- | -- | 1 | 1800 | 5 | 9 | -- | $ScCO_2$ | 1.62 |

## Core Flooding Experiments

The core flooding experiments were also conducted in 1" diameter 12" length homogeneous Bentheimer sandstone. The core was left in an oven overnight for drying. Then, it was saturated with water following the same procedure conducted in the mobility reduction evaluation. After that, 5 to 6 pore volumes were injected into the core at a low flow rate to ensure 100% water saturated porous media followed by the absolute permeability measurement using Darcy law. The setup was pressurized to 1450-psi keeping the overburden pressure 500-psi higher than the test pressure, and the test temperature was 50°C. Once the pressure and temperature were stable, crude oil was injected at 5 ft/day until no more water was observed in the effluent. The water production was collected in a graduated cylinder for the OOIP estimation. Then, water flooding was applied by injecting 5 pore volumes of brine solution at 5 ft/day until no more oil production was observed. The high amount of water injection was to ensure that water flooding reached the optimum oil recovery, no more oil can be produced by water injection, and to ensure the removal of any end effects might exist. Then, the second stage was to inject 1 to 1.5 pore volumes of the surfactant solution (surfactant pre-flush) to mitigate the surfactant adsorption on the rock. After that, 5 pore volumes of simultaneous injection of surfactant and $ScCO_2$ was applied for 24-hr at 5 ft/day. Pressure drop was recorded for the three oil recovery stages. Figure 1 shows the schematic diagram for the core flooding experimental setup. One baseline experiment was conducted by injecting $ScCO_2$ only for comparison purposes with oil recovery of CNF and AOS foam floods. Table 5 shows the properties for the Bentheimer sandstone used to conduct the core flooding experiments and the experimental conditions.

**Table 5: Petrophysical properties for the Sandstone and the experimental conditions in core flooding experiments.**

| Run # | Surfactant | Length [in] | Diameter [in] | Pore Volume [cc] | Porosity [%] | Permeability [Darcy] | Type |
|---|---|---|---|---|---|---|---|
| 1 | -- | 12 | 1 | 33.52 | 21.71 | 1.87 | Baseline |
| 2 | AOS | 12 | 1 | 34.74 | 22.5 | 1.71 | Foam Flood |
| 3 | CNF | 12 | 1 | 33.74 | 21.85 | 1.91 | Foam Flood |

## Results and Discussion

### Foam Static Tests

Foaming Ability (or foamability) was investigated by measuring the initial foam heights ($h_{fi}$) for both surfactants from the shaking tests. Both surfactants gave almost the same $h_{fi}$. Therefore, both CNF and AOS are good foaming agents in terms of foaming ability. This also indicates the efficiency of both surfactants to reduce the air-liquid surface tension. The surface tension at the air-liquid interface will be discussed shortly.

Figure 2 shows the FHL for both surfactants in DI water, 10,000, 20,000, and 30,000 ppm NaCl salinities. Both surfactants provided good foam stability. As observed, foam stability decreases as the salinity increase which is attributed to the reduction in repulsive forces between the surfactant molecules due to the addition of salts. Moreover, CNF surfactant is slightly better than AOS in DI water and in all salinities. For both surfactants, DI water foam stability is the best. Foam stability decreases with the addition of salts and with increasing the salinity up to 20,000 ppm, but both surfactants exhibited foam stability enhancement at 30,000 ppm NaCl. Liu et al. (2005) reported that foam stability with $CO_2$ decreased with increasing salinity up to 2-wt%. Then, a further increase in salinity enhanced the foam stability. However, foam stability plateaued shortly with further increase in salinity. Zhao et al. (2012) reported that the developed surfactant mixture exhibited better stability as the salinity increased, whereas a further increase in salinity weakens the foam stability considerably.

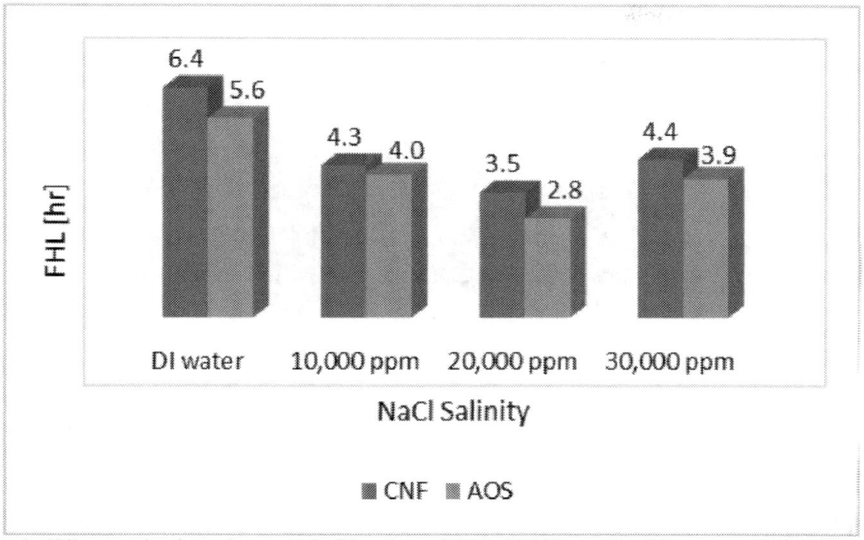


**Figure 2: Foam half-lives or both surfactants in DI water, 10,000, 20,000, and 30,000 ppm at 0.5-wt% concentrations.**

The foam stability with crude oil was impressive for both surfactants with FHL extends to more than a day for CNF and 12 -18 hr for AOS. Figure 2 shows two images for CNF left and AOS right. CNF image was taken after 24-hr, whereas AOS image were taken after 18-hr. Each image shows samples in DI water, 10,000, and 20,000 ppm salinity left to right. Clearly, both surfactants produced stable foam with crude oil, but CNF foam was significantly stronger than that of AOS in presence of crude oil.

However, shaking involves eventually very high shear rates which provide high energy for any surfactant to give its optimum performance as foaming agents regardless of how the shaking was performed. Therefore, it is difficult to recognize the differences in foaming ability and foam stability as well. Therefore, combining the shaking tests observations with the interfacial tension measurements are next.

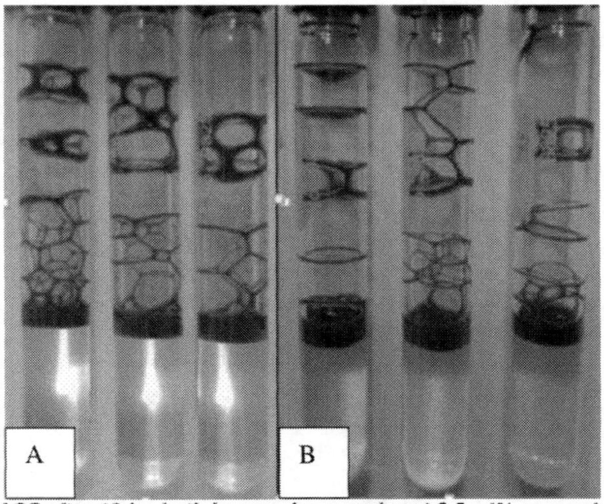


**Figure 3: A) CNF after 24-hr and B) AOS after 18-hr, both images for samples at 0.5-wt% concentration in DI water, 10,000 and 20,000 ppm.**

## Interfacial Tension Measurements.

As mentioned earlier, surface tensions were measured a different concentrations for samples prepared in DI water. The measurements are shown in surface tension vs. log (concentration) plot in Figure 4. The

CMC values for CNF and AOS are 0.011-wt% and 0.028-wt%, respectively. Surfactant concentration for foam application are recommended above the CMC (Nikolov et al. 1986). Using surfactant concentration above the CMC provides the best foam stability, whereas foam has less opportunity to be of good stability when using surfactant concertation below the CMC (Rafati et al. 2012). The CMC defines the foaming efficiency of the surfactants. The lower the CMC is the higher the foaming efficiency (Rosen and Kunjapoo 2004). Moreover, Mannhardt et al. (2000) found experimentally the foaming ability decreases with decreasing the concentrations and no foam generation would be possible for concentrations below the CMC. Therefore, the lower the CMC of the surfactant the better in many ways but the most important is lowering the project costs as the lower CMC enables the use lower surfactant concentrations for good foaming efficiency.

According to Gibbs surface adsorption equation 8, the higher the slope is the higher the adsorption of the surfactant at the liquid-air interface, and consequently, the better the foamability and foam stability (Rosen and Kunjappo 2004). Moreover, the higher the adsorption of a foaming agent at the air-liquid interface is the smaller the area/molecule at the interface. Therefore, such properties in a foaming agent indicate that it provides better foam stability because of the smaller area/molecule which has stronger packing of the foaming agent molecules at the interface.

$$\Gamma = -\frac{1}{RT}\left(\frac{d\gamma}{d\ln C}\right) \tag{8}$$

Where Γ surfactant adsorption at the air-liquid interface, R: gas constant, T temperature, ϒ: surface tension and C concentration.

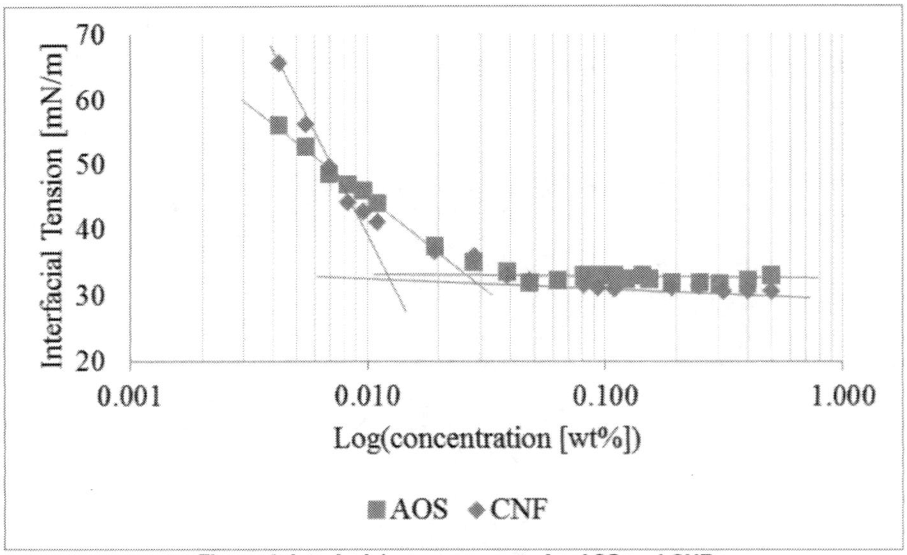


**Figure 4: Interfacial measurements for AOS and CNF**

Furthermore, for almost all concentrations in **figure 4**, CNF is able to reduce the surface tension lower than that of AOS which also indicates that CNF is predicted to perform better than AOS in foam generation. Table 6 and 7 show the $\sigma_{a/w}$ for both surfactants at 0.5-wt% concentration in DI and saline water.

Table 6: O/W and A/W IFT for AOS surfactant concentration at 23°C

| | | NaCl Salinity [ppm] | | |
|---|---|---|---|---|
| | DI Water | 10,000 | 20,000 | 30,000 |
| $\sigma_{a/w}$ | 32.5 | 32.3 | 32.1 | 32.15 |
| $\sigma_{o/w}$ | 1.4 | 0.52 | 0.44 | 0.38 |

*Table 7*: $\sigma_{a/w}$ and $\sigma_{o/w}$ for CNF at 0.5-wt% surfactant concentration at 23°C

| | | NaCl Salinity [ppm] | | |
|---|---|---|---|---|
| | DI Water | 10,000 | 20,000 | 30,000 |
| $\sigma_{a/w}$ | 30.7 | 31 | 31.25 | 31.11 |
| $\sigma_{o/w}$ | 5.88 | 3.94 | 3.51 | 3.11 |

## Foam Stability in Presence of Crude Oil.

Table 6 and 7 show the $\sigma_{o/w}$ measurements for both surfactants in DI and saline water. The addition of salts with CNF showed no much reduction in $\sigma_{o/w}$ as in AOS. This means that CNF has high probability to be more efficient in terms of foam-oil tolerance inside the reservoir.

Tables 8, 9, 10, and 11 show the entering, spreading, bridging coefficients and lamellae number, respectively. Table 8 for entering coefficient gives positive values for all which clarifies that oil will enter the lamellae. However, table 9 for spreading coefficient implies that oil will spread at the A/W interface to destabilize the foam for AOS foam in saline solutions only. The spreading coefficient for CNF is negative for all samples. Moreover, table 10 gives negative values for CNF at all conditions. These observations imply that CNF is going produce very stable foam with oil. However, AOS produces stable foam with oil in DI water. Furthermore, lamellae number values in table 11 confirm the same. CNF foam are in stable region in DI water, and all samples in salinities are in the semi-stable regions. However, AOS in DI water is in the semi-region, and unstable for all samples in saline water.

*Table 8*: Entering coefficients at 0.5-wt% surfactant concentration at 23°C

| | | NaCl Salinity [ppm] | | |
|---|---|---|---|---|
| Surfactant | DI water | 10,000 | 20,000 | 30,000 |
| AOS | 2.4 | 1.32 | 1.04 | 1.03 |
| CNF | 5.08 | 3.44 | 3.26 | 2.72 |

*Table 9*: Spreading coefficients at 0.5-wt% surfactant concentration at 23°C

| | | NaCl Salinity [ppm] | | |
|---|---|---|---|---|
| Surfactant | DI water | 10,000 | 20,000 | 30,000 |
| AOS | -0.4 | 0.28 | 0.16 | 0.27 |
| CNF | -6.68 | -4.44 | -3.76 | -3.5 |

*Table 10*: Bridging coefficients at 0.5-wt% surfactant concentration at 23°C

| | | NaCl Salinity [ppm] | | |
|---|---|---|---|---|
| Surfactant | DI water | 10,000 | 20,000 | 30,000 |
| AOS | 65.96 | 51.31 | 38.35 | 41.52 |
| CNF | -15.19 | -15.73 | -3.37 | -14.75 |

*Table 11*: Lamellae number at 0.5-wt% surfactant concentration at 23°C

| | | NaCl Salinity [ppm] | | |
|---|---|---|---|---|
| **Surfactant** | **DI water** | **10,000** | **20,000** | **30,000** |
| AOS | 3.48 | 9.32 | 10.94 | 12.69 |
| CNF | 0.78 | 1.18 | 1.34 | 1.50 |

## Foam dynamic Tests in Glass Beads Pack (High Permeability)

Fourteen runs were conducted in the high permeability glass beads pack at high and low shear rates for both surfactants with $ScCO_2$ foam. Moreover, one baseline experiment at which the $ScCO_2$ injection was not able to provide measurable pressure data unless at 317-$sec^{-1}$ high shear rate. The comparisons below are made based on the foam effective viscosities measured using the steady state pressure drop across the glass beads pack. Results are listed in table 12 below and the experimental conditions for these tests are listed previously in table 3.

Table 12: Experimental Conditions and results for AOS and CNF with $CO_2$ in Glass beads pack

| Run # | Surfac-tant | Salinity NaCl [wt%] | Injection Quality [%] | Q [ml/min] | Shear rate [1/sec] | $\Delta P_{ss}$ [psi] | Mobility [md/cp] | $\mu_{eff}$ [cp] | MRF |
|---|---|---|---|---|---|---|---|---|---|
| 1 | AOS | -- | 90 | 0.5 | 317 | 208 | 118 | 144 | 5225 |
| 2 | AOS | 1 | 90 | 0.5 | 317 | 164 | 150 | 114 | 4100 |
| 3 | AOS | 2 | 90 | 0.5 | 317 | 161 | 156 | 109 | 4025 |
| 4 | AOS | 3 | 90 | 0.5 | 317 | 216 | 459 | 37 | 5400 |
| 5 | AOS | -- | 70 | 0.5 | 317 | 175 | 142 | 121 | 4375 |
| 6 | AOS | 1 | 70 | 0.5 | 317 | 99 | 249 | 69 | 2475 |
| 7 | AOS | 2 | 70 | 0.5 | 317 | 93 | 265 | 64 | 2325 |
| 8 | AOS | 3 | 70 | 0.5 | 317 | 91 | 273 | 63 | 2275 |
| 9 | AOS | -- | 50 | 0.5 | 317 | 122 | 202 | 85 | 3050 |
| 10 | AOS | 1 | 90 | 0.015 | 9.51 | 9.97 | 74.15 | 230 | 250 |
| 11 | CNF | -- | 90 | 0.5 | 317 | 162 | 153 | 112 | 4050 |
| 12 | CNF | 3 | 90 | 0.5 | 317 | 118 | 209 | 82 | 2950 |
| 13 | CNF | 1 | 90 | 0.015 | 9.51 | 16.09 | 46.00 | 371 | 402 |
| 14 | -- | 1 | -- | 0.5 | 317 | 0.04 | 616,270 | 0.03 | -- |

The effect of injection quality is shown Figure 5. It shows the effect of three injection qualities on both surfactants' foam viscosities at high shear rate in DI water. AOS foam viscosity decreases as the injection quality decreases, whereas CNF foam viscosity increases as the injection quality increases. The opposite behaviors of these foaming agents can be attributed to many reasons that were not covered in this study. In fact, the relationship of the foam viscosity with injection quality is still controversial. In an experimental work conducted by Marsden and Khan (1966), they found that the higher the injection quality is the higher the foam viscosity. Lee and Heller (1990) experimentally found the opposite where the increase in injection quality decreases the viscosity. Foam viscosity depends on factors such flow rate, the permeability of the porous media, and foam texture (Hirasaki and Lawson 1985).

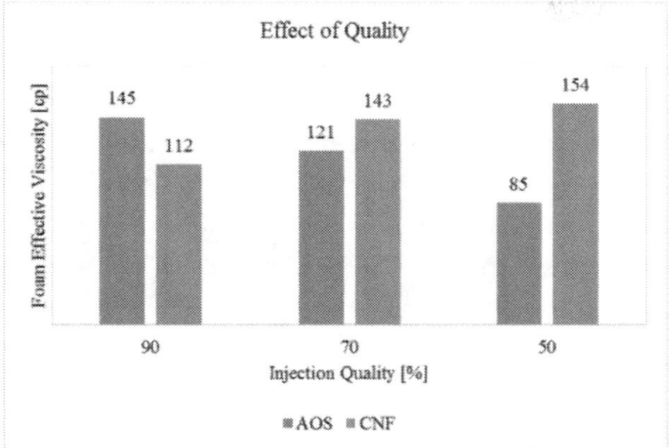


**Figure 5 : Effect of injection quality of AOS and CNF in glass beads pack at 317-sec$^{-1}$**

The effect of shear rate at 90 % injection quality for both AOS and CNF at high shear rate 317-sec$^{-1}$ and low shear rate 9.51-sec$^{-1}$ at 10,000 ppm NaCl salinity are shown in Figure 6. The higher the shear rate is the lower viscosity because of the shear thinning nature of foam. As shown, foam viscosities of CNF at both shear rates are better than that of AOS. Figure 7 also shows all results for CNF and AOS at low shear rate 9.51-sec$^{-1}$ at 90% injection quality in 10,000 ppm NaCl salinity. Again, CNF proves its powerful performance with higher steady state pressure drop, lower mobility, higher foam viscosity, and higher mobility reduction factor (MRF).

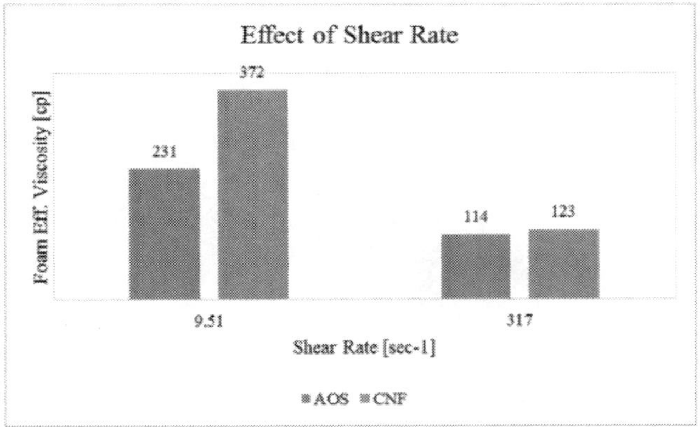


**Figure 6 : High vs. low shear rate foam viscosities for AOS and CNF at 0.5-wt% surfactant concentration at 90% injection quality**

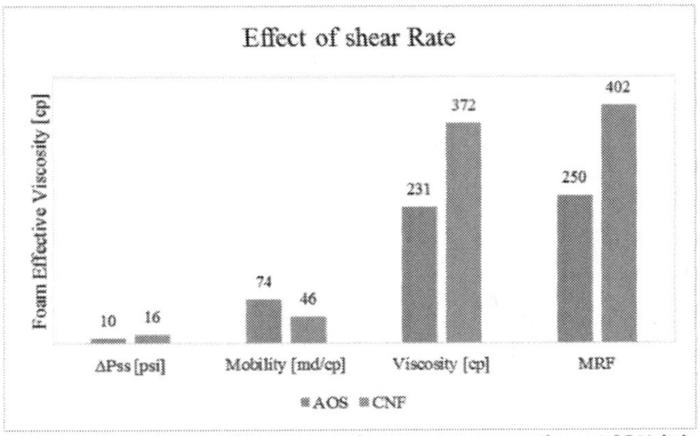


**Figure 7: low shear rate results at 0.5-wt% surfactant concentration at 90% injection quality**

## Foam Dynamic Tests in Bentheimer Sandstone (low Permeability)

Two among 10 experiments were conducted using $N_2$ injection at 850-psi and $ScCO_2$ injection at 1800-psi as baseline experiments for comparison. All experimental conditions are listed in table 4. Results are listed in table 13.

***Table 13*: Experimental Conditions for AOS and CNF in sandstone**

| Run # | Surfactant | Gas | Injection Quality [%] | $\Delta P_{ss}$ [psi] | Mobility [md/cp] | Foam Viscosity [cp] | MRF |
|---|---|---|---|---|---|---|---|
| 1 | AOS | $ScCO_2$ | 90 | 0.4 | 411 | 4.13 | 1.67 |
| 2 | AOS | $ScCO_2$ | 70 | 25.42 | 6.79 | 237.11 | 105.9 |
| 3 | CNF | $ScCO_2$ | 90 | 0.60 | 301.06 | 7.21 | 2.5 |
| 4 | CNF | $ScCO_2$ | 70 | 2.00 | 89.00 | 20.00 | 8.3 |
| 5 | AOS | $N_2$ | 90 | 1.35 | 129.98 | 12.69 | 3.4 |
| 6 | AOS | $N_2$ | 90 | 3.46 | 101.43 | 16.27 | 8.6 |
| 7 | CNF | $N_2$ | 90 | 12.89 | 13.61 | 124.88 | 32.2 |
| 8 | CNF | $N_2$ | 90 | 34.4 | 10.20 | 166.64 | 86 |
| 9 | -- | $N_2$ | -- | 0.4 | | | -- |
| 10 | -- | $ScCO_2$ | -- | 0.24 | 680 | 2.5 | -- |

The effect of permeability is shown in Figure 8 which compares the foam viscosities for AOS and CNF with $ScCO_2$ at 90% injection quality at 9-sec$^{-1}$ shear rate. The permeabilities of the cores and glass beads pack are 1.7 to 1.98 Darcy and 17.1-Darcy, respectively. CNF is repeatedly prove to be better than AOS by generating higher foam viscosity in both cases at high and low permeability porous media. These results in figure 8 are in agreement with the fact that foam favors the higher permeability. The foam viscosities for surfactants in the glass beads pack are extremely higher than that in sandstone.

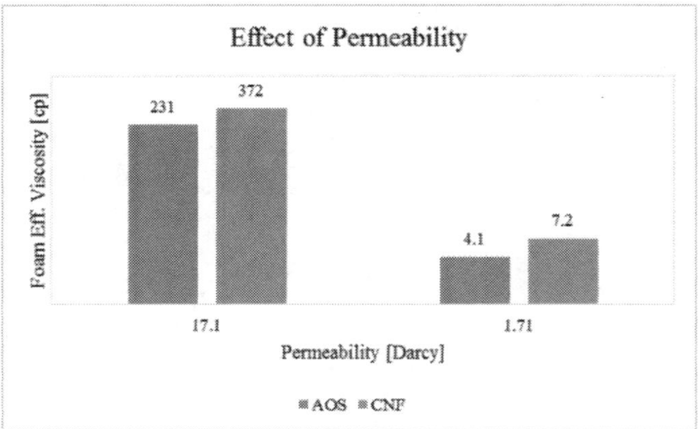

**Figure 8: AOS and CNF at ScCO2 at 90% injection quality**

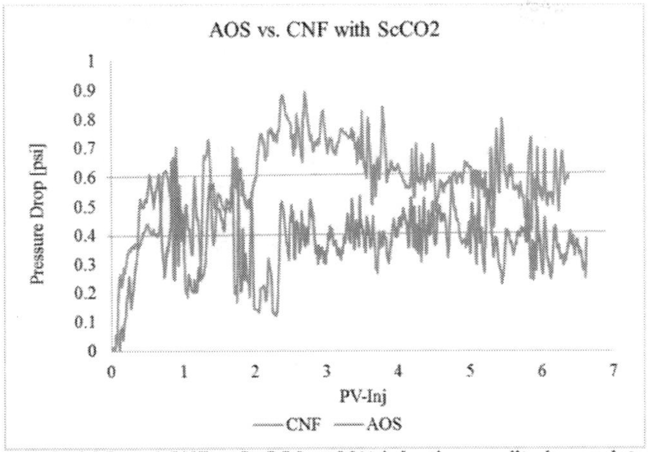

**Figure 9: AOS and CNF at ScCO2 at 90% injection quality in sandstone**

The effect of injection quality at low permeability sandstone were investigated using two injection qualities 90% and 70% for both surfactants with $ScCO_2$ at 9.51 $sec^{-1}$ shear rate. The pressure profiles for both surfactants are shown in Figures 9 for 90% and 10 for 70% injection quality. According to the pressure profiles, CNF foam is stronger at 90%, whereas AOS is stronger at 70% injection quality. Therefore, the results suggest that CNF is better at more realistic conditions (i.e. low shear rate and sandstone reservoir), CNF co-injection with $ScCO_2$ provide higher foam viscosity at higher injection qualities, while AOS requires low injection qualities for better performance. This also indicates that using CNF for mobility control with $ScCO_2$ will eventually reduce the cost as the amount of liquid decreases as the injection quality increases.

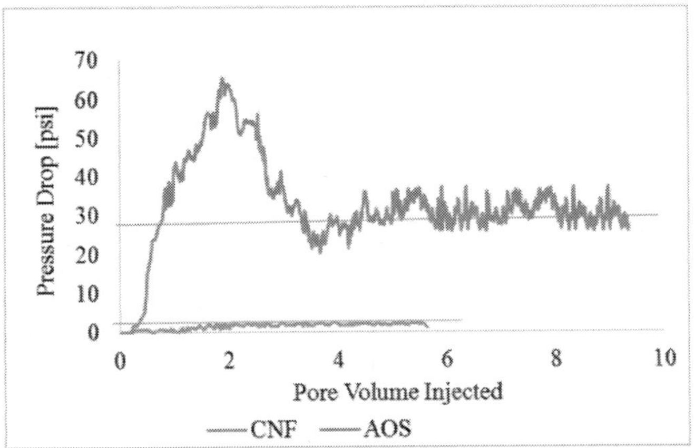

**Figure 10: AOS and CNF at ScCO2 at 70% injection quality in sandstone**

The mobility control with $N_2$ gas was also tested for both surfactants. The pressure profiles for AOS and CNF are shown in Figure 11. These experiments were conducted in sandstone at 0.5-wt% surfactant concentrations prepared in 10,000 ppm NaCl brine solution at two velocities 5 and 10 ft/day (9 and 18-$sec^{-1}$) at 90% injection quality. Each surfactant was injected simultaneously with $N_2$ gas at 5 ft/day until the steady state pressure drop was attained, then the velocity was raised to 10 ft/day until the steady state pressure drop was attained for the new velocity. From Figure 11, CNF appeared always better in terms of flow resistance at both velocities than AOS. Moreover, Figure 12 compares the foam viscosities for AOS and CNF at both velocities. Surprisingly, the viscosity increases as the velocity (i.e shear rate) increases for both surfactants. Although this is shear thickening behavior, foam is known of its non-Newtonian shear thinning nature (Schramm and Wassmuth 1994). This behavior could be related to the procedure of the experimental work. Precisely, the reason could be performing two velocities in one experiment. At 5 ft/day, the surfactant was very efficient to provide high flow resistance due to the gas blockage effect. Therefore, the gas relative permeability is already low, and increasing the velocity to 10 ft/day at such conditions would promote the foam generation because of the high shear rate in a blocked porous media. As a result, foam contradicted its shear thinning nature by providing higher viscosity at higher shear rate, lesson learned.

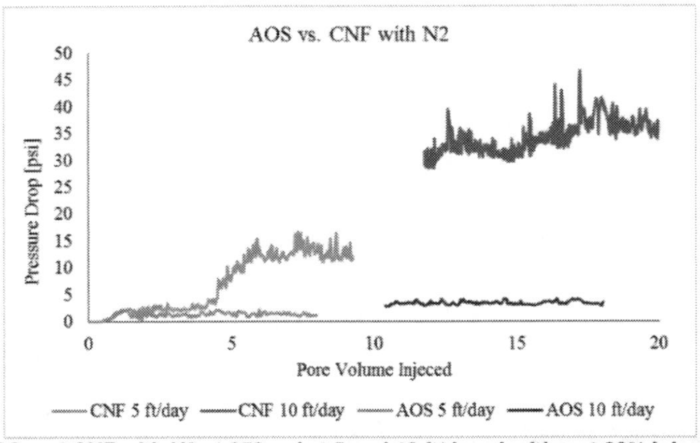


**Figure 11: AOS and CNF with N2 at 850-psi at 5 and 10 ft/da velocities at 90% injection qualities**

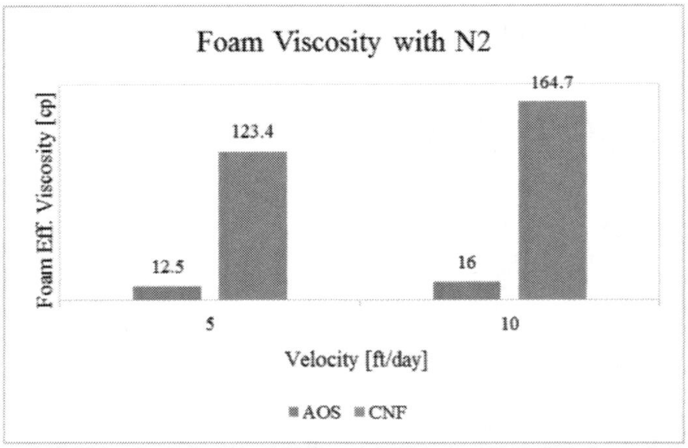


**Figure *12*: Foam viscosity at 5 and 10 ft/day for 0.5-wt% of CNF and AOS in 1-wt% NaCl brine solution at 90% injection quality with $N_2$ gas**

## Core Flood Experiments

The baseline experiment oil recovery is shown as a function of pore volume injected in Figure 13. The ultimate oil recovery of water flooding is 36% of the OOIP. The water flooding was followed by 5 to 6 pore volume of continuous $ScCO_2$ injection which resulted in 27.54% of the OOIP more oil recovery. The total oil recovery from the baseline experiment is 76.2% of the OOIP.

Figure 14 shows the results for AOS as a foaming agent after water flooding. The same procedure were conducted. This run started with injecting 4.56 pore volumes of water, water flooding, which resulted is 35.42% of the OOIP oil recovery. Then, 1.62 pore volume of AOS surfactant solution were injected to reduce the surfactant adsorption on the rock surfaces. The surfactant pre-flush stage resulted in 4.75% of the OOIP oil recovery. The third stage is the foam flood with simultaneous injection of 5 pore volumes of

AOS/Sc$CO_2$. AOS foam flood resulted in 28.5% additional oil recovery, and 68.67% of the OOIP total oil recovery. The additional recovery by AOS foam accounts for 1% higher than that of baseline experiment.

CNF foam core flood results are shown in figure 15. Water flooding stage produced 39.66% of the OOIP. Moreover, no oil production was observed during the injection of around 1.5 pore volumes of CNF solution to reduce the adsorption effect. The final stage, CNF-Sc$CO_2$ simultaneous injection or foam flooding resulted in 36.3% of the OOIP oil recovery after water flooding. This amount of oil produced by CNF foam is 7.87% higher than that of AOS foam, and 8.83% higher than baseline experiment. The total recovery for this core flood is 76.3%.

In addition, the pressure drop for the recovery stages in foam core floods are shown in figure 14 and 15. The CNF foam pressure drop in figure 15 is slightly higher than that of AOS foam in figure 14. Both pressure drops are high enough o CNF foam in figure 16 and AOS foam in figure 15. However, during the last 1 pore volume, the CNF foam pressure drop maintained higher pressure drop than that of AOS foam. This is attributed to the higher CNF foam stability with oil.

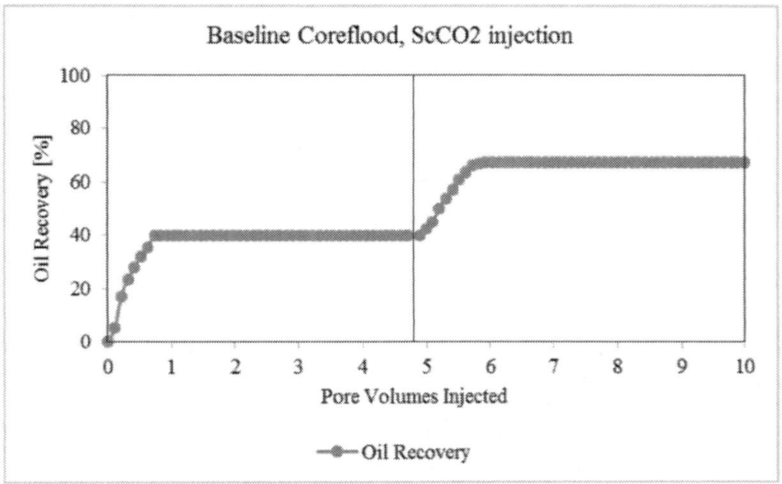


**Figure 13: Baseline experiment**

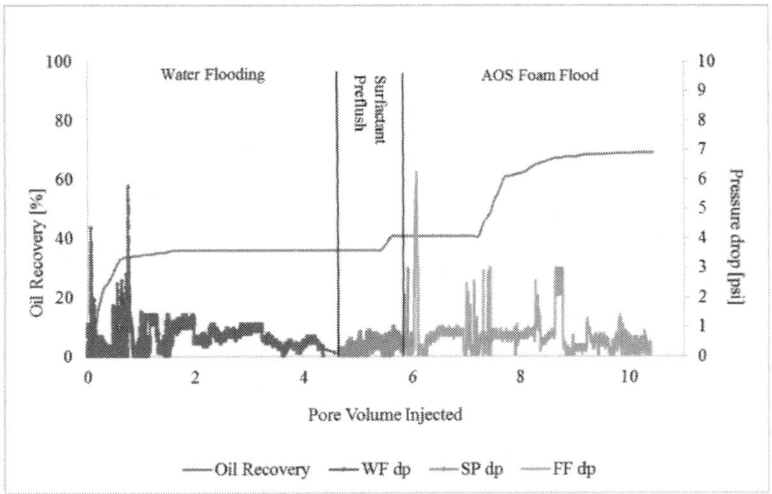


**Figure *14*: AOS foam flood experiment**

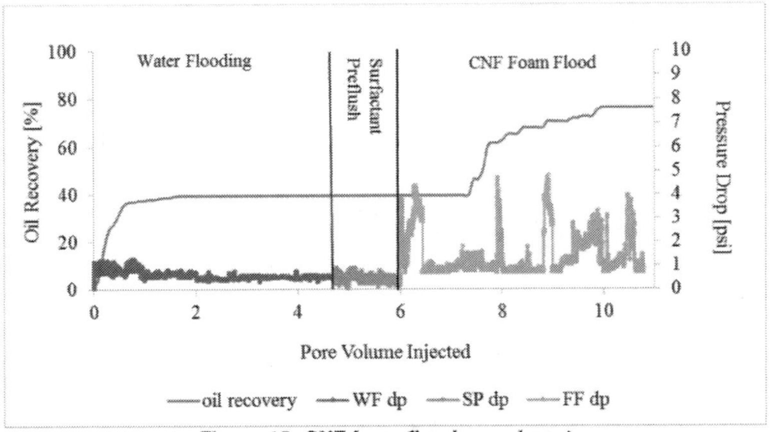

**Figure 15: CNF foam flood experiment**

## Conclusion

1. This study investigated two surfactants for the mobility control with $ScCO_2$: CNF newly developed surfactant and the foaming agent anionic surfactant AOS.
2. Both surfactants showed good foaming ability and foam stability without oil in shaking tests. Such behavior is not representative to the actual foaming ability of both surfactants because the shaking tests are naturally involved high shear rate which enforce the surfactants to perform at their optimum abilities as foaming agents. However, CNF was able to reduce the A/W IFT lower than that of AOS.
3. The newly developed surfactant, CNF, showed impressive foam oil tolerance than AOS. It does not reduce the O/W IFT to low value same as AOS in saline water. This makes it a good surfactant for the applications of mobility control.
4. CNF provided negative values of entering, spreading, and bridging coefficients, whereas AOS provided negative values in DI water only. Moreover, lamellae number indicates that CNF foam is stable in DI water and semi-stable with the addition of salts in terms of foam-oil tolerance. However, the lamellae number for AOS showed semi-stable in DI water, and unstable at all salinities. These results showed that CNF is very stable with oil more than AOS.
5. At all conditions for mobility reduction with $ScCO_2$, and with $N_2$ at lower pressures, CNF showed higher foam viscosity, and better mobility reduction than AOS.
6. For oil recovery with $ScCO_2$ mobility control, AOS produced 1% more than the baseline experiment, whereas CNF produced almost 7.78% more than AOS foam, and 8.38% more than the baseline experiments.
7. It is not recommended to test two shear rates at one experiment. Such procedure would result in dramatic errors in foam viscosity measurements.

## Nomenclature

| | |
|---|---|
| OOIP | original oil in place |
| EOR | enhanced oil recovery |
| CNF | complex nanofluid |

| | |
|---|---|
| AOS | alpha olefin sulfonate |
| $ScCO_2$ | supercritical $CO_2$ |
| MRF | Mobility reduction factor |
| DI | deionized water |
| API | American petroleum institute |
| Ppm | part per million |
| Ml | milliliter |
| $\sigma_{g/w}$ | gas-water interfacial tension |
| $\sigma_{o/w}$ | oil-water interfacial tension |
| $\sigma_{o/g}$ | oil-gas interfacial tension |
| $h_{fi}$ | initial oam height in shaking tests |
| FHL | foam half life |
| E | entering coefficient |
| S | spreading coefficient |
| B | bridging coefficient |
| L | lamellae number |

## References


Adkins, S. S., Chen, X., Nguyen, Q. P., Sanders, A. W., & Johnston, K. P. 2010. Effect of branching on the interfacial properties of nonionic hydrocarbon surfactants at the air–water and carbon dioxide–water interfaces. Journal of Colloid and Interface Science, 346 (2), 455-463. DOI: 10.1016/j.jcis.2009.12.059

Bera, A., Ojha, K., & Mandal, A. 2013. Synergistic Effect of Mixed Surfactant Systems on Foam Behavior and Surface Tension. *Journal of Surfactants and Detergents.* **16** (4): 621-630.

Boeije, C. S., Bennetzen, M., & Rossen, W., 2017. A Methodology for Screening Surfactants for Foam Enhanced Oil Recovery in an Oil-Wet Reservoir. doi: 10.2118/185182-pa.

Dicksen, T., Hirasaki, G. J., & Miller, C. A. 2002. Conditions for Foam Generation in Homogeneous Porous Media. SPE/DOE Improved Oil Recovery Symposium, 13-17 April, Tulsa, Oklahoma, USA. SPE-75176-MA.

Enick, Robert Michael, David Kenneth Olsen, James Robert Ammer et al. 2012. Mobility and Conformance Control for $CO_2$ EOR via Thickeners, Foams, and Gels -- A Literature Review of 40 Years of Research and Pilot Tests. SPE Improved Oil Recovery Symposium, Tulsa, Oklahoma, USA. SPE-154122-MS.

Farajzadeh, R., A. Andrianov, and P. L. J. Zitha. 2010. Investigation of Immiscible and Miscible Foam for Enhancing Oil Recovery. Industrial & Engineering Chemistry **49** (4): 1910-1919.

Fried, A. N. The Foam-Drive Process for Increasing the Recovery of Oil. 1961. U.S. Bureau of Mixes, Rep. Inv., 5866, Washington, D. C.

Hirasaki, G., Lawson J. 1985. Mechanisms of Foam Flow in Porous Media: Apparent Viscosity in Smooth Capillaries. SPE Journal, **25** (2):176-190.

Green, D.W. and Willhite, G.P. 1998. Enhanced Oil Recovery. Volume 6, SPE Textbook Series, Richardson, Texas: SPE.

Haugen, Åsmund, Martin A. Fernø, Arne Graue, and Henri J. Bertin. 2012. Experimental Study of Foam Flow in Fractured Oil-Wet Limestone for Enhanced Oil Recovery. SPE Reservoir Evaluation & Engineering **15** (2). SPE-129763-PA.

Healy, R. N., Holstein, E. D., & Batycky, J. P. 1994. Status of Miscible Flooding Technology. 14th World Petroleum

Congress, 29 May-1 June, Stavanger, Norway. WPC-26169.
Lee, H. O., & Heller, J. P. 1990. Laboratory Measurements of $CO_2$-Foam Mobility. SPE Journal. SPE-17363-PA.

Li, Robert F., George Hirasaki, Clarence A. Miller, and Shehadeh K. Masalmeh. 2012. Wettability Alteration and Foam Mobility Control in a Layered, 2d Heterogeneous Sandpack. SPE Journal **17** (04). SPE-141462-PA.

Liu, Y., Grigg, R. B., & Bai, B. 2005. Salinity, pH, and Surfactant Concentration Effects on $CO_2$-Foam. SPE-93095-MS.

Mannhardt, K., Novosad, J., & Schramm, L. 2000. Comparative evaluation of foam stability to oil. SPE Reservoir Evaluation & Engineering, 3 (01), 23-34.

Marsden, S. S., and Suhail A. Khan. 1966. The Flow of Foam through Short Porous Media and Apparent Viscosity Measurements. SPE Journal **6** (01). SPE-1319-PA.

Nikolov, A., Wasan, D., Huang, D., & Edwards, D. 1986. The Effect of Oil on Foam Stability: Mechanisms and Implications for Oil Displacement by Foam in Porous Media. SPE Annual Technical Conference and Exhibition. New Orleans, Louisiana, USA. SPE-15443-MS.

Porter, M. R. 1994. Handbook of Surfactants, Blackie Academic & Professional, United Kingdom.

Rafati, R., Hamidi, H., Idris, A. K., & Manan, M. A. 2012. Application of sustainable foaming agents to control the mobility of carbon dioxide in enhanced oil recovery. Egyptian Journal of Petroleum, **21** (2): 155-163.

Rosen, M. J., & Kunjappu, J. T. 2004. Foaming and Antifoaming by Aqueous Solutions of Surfactants and Interfacial Phenomena John Wiley & Sons, Inc.: 308-335.

Schramm, L. L., & Wassmuth, F. 1994. Foams: Basic Principles Foams: Fundamentals and Applications in the Petroleum Industry. Volume 242, American Chemical Society: 03-45.

Schramm, L.L. Surfactants: Fundamentals and Applications in the Petroleum Industry. 2000. Cambridge University Press: 03-50.

Taber, J.J., Martin, F.D. and Seright, R.S. 1997. EOR Screening Criteria Revisited - Part 1: Introduction to Screening Criteria and Enhanced Recovery Field Projects. SPERE 12 (3): 189-198. SPE-35385-PA.

Talley, L. D. 1988. Hydrolytic Stability of Alkylethoxy Sulfates. SPE-14912-PA.
Zhao, L., Li, A., Chen, K., Tang, J., & Fu, S. 2011. Development and evaluation of foaming agents for high salinity tolerance. Journal of Petroleum Science and Engineering. **81**: 18-23. DOI:10.1016/j.petrol.2011.11.006.

CARBON MANAGEMENT TECHNOLOGY CONFERENCE


## CMTC-#486524-MS


# Numerical Analyses of the Effect of the Impurity in the Gas on the Solubility Trapping in the $CO_2$ Sequestration


Lingyu Mu, Xinwei Liao, Xiaoliang Zhao, Zhiming Chen, Langtao Zhu and Biao Luo, China University of Petroleum(Beijing), Beijing, China





## Abstract

The $CO_2$ dissolved in the aquifer will increase the density of brine, which can result in the instability of the gravity and prompt the onset of the viscous finger. The viscous finger will lead to the convective mix, accelerating the process of $CO_2$ solution in the brine. However, the gas stream in the $CO_2$ storage usually contains the impurities such as $N_2$, $O_2$, and $SO_2$, which can change the density difference in the process of solution, and affect the solubility trapping in the $CO_2$ sequestration.

In this paper, a numerical simulation method was used to study the effect of different impurities on the solubility trapping in the process of $CO_2$ storage. Firstly, based on the PR-HV model, this paper calculated the solubility of $CO_2$, $N_2$, $O_2$, and $SO_2$ with different temperature and salinity and analysed the variation of the solubility. Then a multi-component numerical simulation model based on a certain aquifer layer was established to compare the $CO_2$ dissolution rate and the onset time of the instability and analyze the influence of impurities in the $CO_2$ stream on the solubility trapping. Finally, this paper clarified the impact on the $CO_2$ storage and suggested that the concentration of the impurities should be controlled in a rational range for the perspective of the economy and efficiency.

The results show that the solubility of $CO_2$ is higher than $N_2$ and $O_2$ in the saline water, and close to that of $SO_2$. We applied the solubility data to the numerical simulation. The results of the numerical simulation shows that with the increase of the concentration of $N_2$ or $O_2$, $CO_2$ dissolution rate has a decrease, and the onset time of the instability has an increase. It meas the longer time $CO_2$ plume keeping in the state of good flowing capability and low density. The onset of viscous finger will be postponed, leading to a negative influence on the solubility trapping and the risk of the $CO_2$ leak through fractures and faults. On the contrary, $SO_2$ can shorten the onset time of the instability, which accelerates the viscous finger and prompts the solubility trapping. A further conclusion is that the effect of $SO_2$ on the viscous fingering is more significantly than $N_2$ and $O_2$.

This paper deepens the understanding about the effect of the impure $CO_2$ on the solubility trapping, and clarifies the effect of different impurities.

## 1 Introduction

Global warming and greenhouse effect caused by the emission of $CO_2$ and other greenhouse gases are attracting more and more serious public concern. $CO_2$ capture and storage (CCS) is one of the most promising methods to reduce the emission of greenhouse gases, which refers to the process consisting of the separation of $CO_2$ from industrial and energy-related sources, transport to a storage location and long-term isolation from the atmosphere, shown in Figure 1. The CCS system consists of capture, transport, geological storage, ocean storage, mineral carbonation, industrial utilization, etc[1-6].

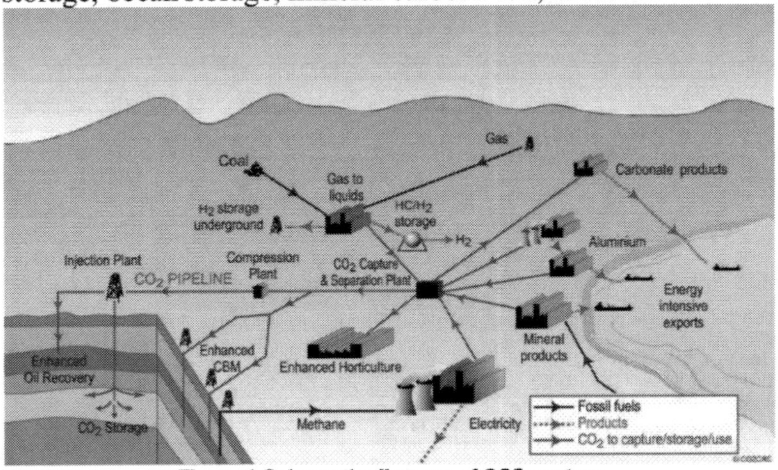

**Figure 1 Schematic diagram of CCS system**

Geological storage is generally regarded as the most mature and promising option for isolating the captured $CO_2$ from the atmosphere[7]. Geological storage means that $CO_2$ is stored in the geological structure, such as oil and gas fields, coal beds or deep saline water[7] ,as shown in Figure 2. $CO_2$ enhanced oil/gas (EOR/EGR) technology has been very mature with the development of the oil industry[9, 10]. Because of characteristics of the oil industry, $CO_2$ is generally used for recycling and a great quantity of gas will be produced to the ground with oil, which cannot meet the storage capacity requirements and long-term isolation. Coal beds stored with $CO_2$ need long-term isolation and cannot be mined for tens to hundreds of millions of years[3]. However, in hundreds of years, we may mine the coal beds due to advances in technology and energy demand, which means that $CO_2$ storage in coal beds has certain limitations. $CO_2$ stored in the deep saline water will not influence the ecosystem, and storage capacity is expected to reach 1000~10000 giga tons[11]. It is believed that deep saline formations have by far the largest capacity for $CO_2$ storage and are much more widespread than other options.

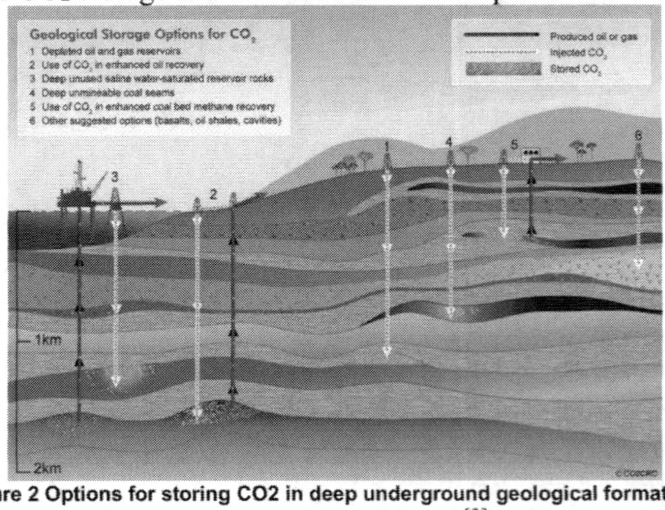

**Figure 2 Options for storing CO2 in deep underground geological formations**

There are four main mechanisms for geological storage of $CO_2$[3]. A low-permeability geological barrier such as mudstone or halite impedes the migration of buoyant $CO_2$ via high capillary pressure, which is how the stratigraphic and structural trapping mechanism acts, shown in Figure 3 (a). However, $CO_2$ will enter the caprock when the pressure of the reservoir is high enough[3]. With the migration of $CO_2$, some $CO_2$ is trapped in the pores of rock due to the capillary pressure (gas-liquid interface tension)[12],shown in shown in Figure 3 (b). Residual trapping plays an important role at the early stage of $CO_2$ storage, for core-scale experiments show that residual trapping occurs in a few days or months[12]. When $CO_2$ dissolves into the saline water, the brine saturated with $CO_2$ is denser and will sink, shown in Figure 3 (c). The content of dissolved $CO_2$ will increase with time elapses. Solubility trapping is a major way for $CO_2$ geological storage, and the low flow rate of the formation water can ensure the stability of long-term storage[13]. $CO_2$ dissolved in the brine will react with rocks to form carbonate mineral precipitation, shown in Figure 3 (d), thus $CO_2$ is storaged with the most stable state[14, 15]. Although this chemical reaction exists at the early stage, mineral trapping is a rather long process that may last tens of thousands of years. The type of the trapping mechanism depends on the specific geological conditions, and in general it contains the mechanisms mentioned above. The occurrence of a specific mechanism depends on the length of time and migration of the $CO_2$.With the increase of time, the dominated trapping mechanism will change, and the security of $CO_2$ storage gradually increases, shown in Figure 4.

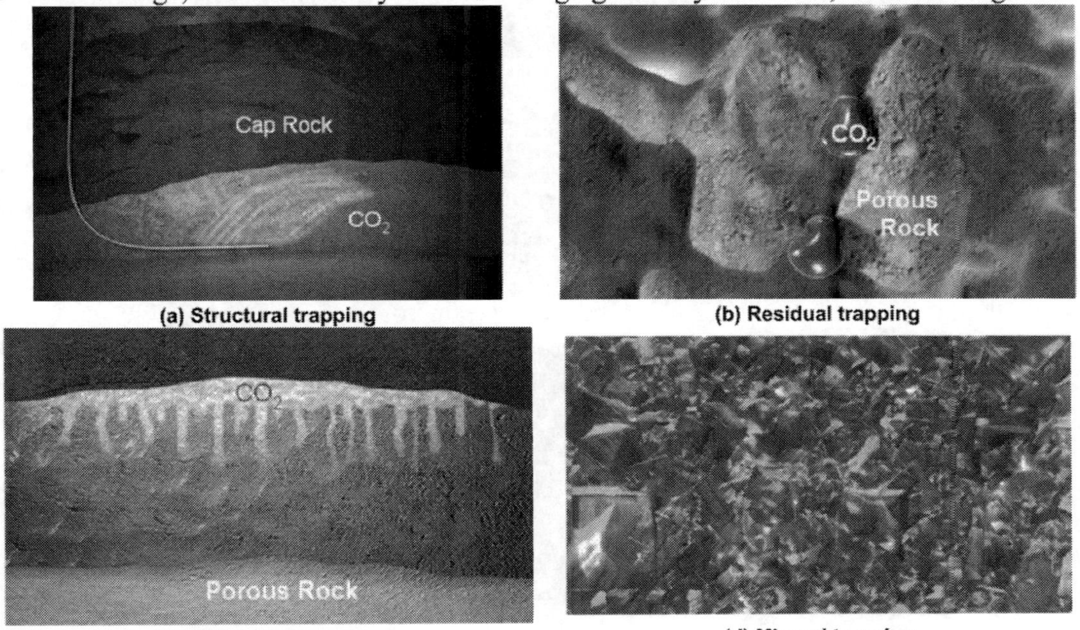


(a) Structural trapping (b) Residual trapping

(c) Solubility trapping (d) Mineral trapping

**Figure 3 Schematic diagram of mechanisms for geological storage of CO2**

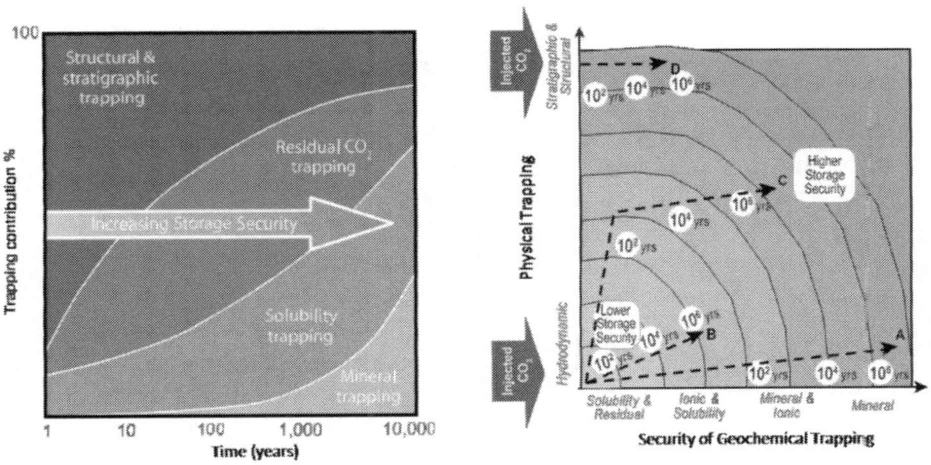


**Figure 4 Trapping contribution and security of physical and geochemical trapping mechanism**

For solubility trapping has a bearing on $CO_2$ geological storage, we mainly study the mechanism of the solubility trapping in this paper. The dissolved $CO_2$ in the aquifer will increase the density of brine[16, 17]. The increase of the density is heterogeneous in the different part of the formation. A local increase of density, coupled with the heterogeneity of the aquifer such as permeability heterogeneity[18], will result in the instability of the gravity and prompt the onset of the viscous finger[19-22]. The viscous finger will lead to the convective mix, accelerating the process of $CO_2$ solution in the brine and reducing the possibility of $CO_2$ leakage. In view of its importance for long-term $CO_2$ storage, the convective mix driven by density difference has aroused many researchers' attention. Linear steady-state analysis and numerical simulation are two commonly used methods[23-27].

**Table 1 The range of contents of the impurities**

| Component | The range of contents, % | |
|---|---|---|
| | Capture after combustion | Capture before combustion |
| $CH_4$ | <0.010 | <0.035 |
| $N_2$ | <0.170 | <0.600 |
| $H_2S$ | trivial | <3.400 |
| CO | <0.001 | <0.400 |
| $O_2$ | <0.010 | trivial |
| $NO_x$ | <0.005 | / |
| $SO_x$ | <0.001 | / |
| $H_2$ | trivial | <3.000 |
| Ar | trivial | <0.050 |

Whichever method of is used to capture $CO_2$, the $CO_2$ stream contains a certain species and concentration of impurities, such as $N_2$, $O_2$, Ar, $SO_x$, $NO_x$, $H_2S$, $H_2$ and so on, which depends on the fuel used, the energy conversion process and the $CO_2$ capture technology[28]. The power industry is the main source of $CO_2$ emissions, whose main methods of $CO_2$ capture are capture after the combustion and capture before the combustion. The contents of the impurities[29] in the two methods are listed in Table 1. Even the smallest amount of the impurities could have a profound effect.

**Table 2 The properties of the gases**

| Component | Molar weight(g/mol) | Critical temperature(°C) | Critical pressure(MPa) |
|---|---|---|---|
| $CO_2$ | 44 | 31.04 | 7.38 |
| $SO_2$ | 64 | 157.49 | 7.88 |
| $O_2$ | 32 | -183.0 | 5.08 |
| $N_2$ | 28 | -147.05 | 3.39 |

The dissolution of $CO_2$ can increase the density of the formation fluid and cause convective mix[25, 26]. However, the physical and chemical properties of the impurities co-injected with $CO_2$ is markedly different with those of $CO_2$, which may have an impact on the process of $CO_2$ storage[30]. According to the content of the impurities, we choose $SO_2$, $O_2$ and $N_2$ as representative to study the effect of the impurities on the solubility trapping. The properties of $CO_2$, $SO_2$, $O_2$ and $N_2$ are listed in Table 2. The molar weight of the $SO_2$ is heavier than that of $CO_2$, while the molar weight of $O_2$ and $N_2$ are lighter. In other words, the increase of the density difference when $SO_2$ dissolves in the brine is bigger than that of the other three gases. The critical temperature and pressure of $O_2$ are lower than that of $CO_2$. Therefore, the mixture of $CO_2$ and $O_2$ will exist as supercritical fluid in the aquifer. Moreover, $N_2$ is in the same condition. $SO_2$ has a similar critical pressure with $CO_2$, while the critical temperature of $SO_2$ is rather higher. However, the mixture of $CO_2$ and $SO_2$ can be regard as supercritical fluid considering the content of $SO_2$ in the mixture.

## 2 The dissolution law of the impurities

### 2.1 Theoretical model of solubility

Peng-Robinson equation of state cannot apply to the activity calculation of nonpolar substances. Besides, the application of Vander Waals mixing rule is limited, for the gas-liquid mixture is a non-ideal and highly asymmetric system[31]. Therefore, fugacity coefficient model established by Huron[32] and Peng-Robinson equation of state are adopted to characterize the thermodynamic properties of the CO2-water system. PR-HV model[33] is as followed:

$$p=\frac{RT}{v-b_m}-\frac{a_m}{v(v+b_m)+b_m(v-b_m)} \tag{1}$$

$$a_m=b_m\left[\sum_{i=1}^{n}x_i\frac{a_i\alpha_i}{b_i}-\frac{G_\infty^E}{c_0}\right] \tag{2}$$

$$b_m=\sum_{i=1}^{n}x_ib_i \tag{3}$$

$a_i$, $b_i$ and $\alpha_i$ are expressed by the following formulas:

$$a_i=0.477235\frac{R^2T_{ci}^2}{p_{ci}}\alpha_i \tag{4}$$

$$b_i=0.077796\frac{RT_{ci}}{p_{ci}} \tag{5}$$

$$\alpha_i=\left[1+m\left(1-T_{ri}^{0.5}\right)\right]^2 \tag{6}$$

Without the defined pressure, the Gibbs free energy is calculated as follows:

$$G_\infty^E=\sum_{i=1}^{n}x_i\frac{\sum_{j=1}^{n}x_jC_{ji}G_{ji}}{\sum_{k=1}^{n}x_kG_{ki}} \tag{7}$$

According to the principle of chemical thermodynamics, the fugacity coefficient of component i with (T, p) as independent variable is:

$$RT\ln\varphi_i=\frac{\partial}{\partial n_i}\left[\int_{v_t}^{\infty}\left(p-\frac{nRT}{v_t}\right)dv_t\right]T,v_t,n_{j\neq i}-RT\ln Z \tag{8}$$

Then, the fugacity coefficient of the mixture calculated by PR-HV model is:

$$\ln\varphi_m=\frac{b_i}{b_m}(Z_m-1)-\ln\left[Z_m\left(1-\frac{b_m}{v_m}\right)\right]-\frac{1}{2\sqrt{2}RT}\left[\frac{a_i}{b_i}-\frac{RT\ln\gamma_i}{c_0}\right]\ln\left[\frac{v_m+(\sqrt{2}+1)b_m}{v_m-(\sqrt{2}-1)b_m}\right] \tag{9}$$

### 2.2 The dissolution of the gases in saline water

PR-VW model is used to calculate the solubility of $CO_2$, $SO_2$, $O_2$ and $N_2$ in different temperature and salinity. The solubility of $CO_2$, $SO_2$, $N_2$ and $O_2$ in pure water is listed in Table 3. The solubility of $N_2$ is the smallest, and is the same order of magnitude as that of $O_2$. $SO_2$ is the most soluble within the impurities, whose solubility is about thousands of times than $N_2$ and scores of times than $CO_2$. With the temperature increases, the solubility of the gases decrease. The decreases in the solubility of $CO_2$, $SO_2$, $N_2$ and $O_2$ are 82.22%, 82.45%, 68.02%, and 67.09% respectively. The solubility of the gases in the saline water is shown in Figure 5, Figure 6, Figure 7 and Figure 8. With the salinity increases the solubility of the gases decrease.

Table 3 The solubility of the gases in pure water

| Temperature(℃) | solubility(g/kg) | | | |
|---|---|---|---|---|
| | $CO_2$ | $SO_2$ | $N_2$ | $O_2$ |
| 0 | 3.3333 | 228 | 0.0295 | 0.0691 |
| 10 | 2.5185 | 171 | 0.0243 | 0.0561 |
| 20 | 1.6889 | 114 | 0.0190 | 0.0435 |
| 30 | 1.2444 | 79 | 0.0160 | 0.0359 |
| 40 | 0.9778 | 56 | 0.0136 | 0.0309 |
| 50 | 0.7704 | 45 | 0.0112 | 0.0268 |
| 60 | 0.5926 | 40 | 0.0094 | 0.0227 |

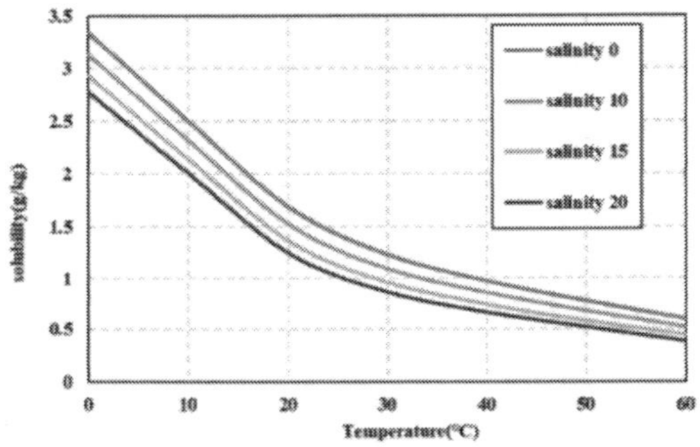


Figure 5 The solubility of $CO_2$ in the saline water

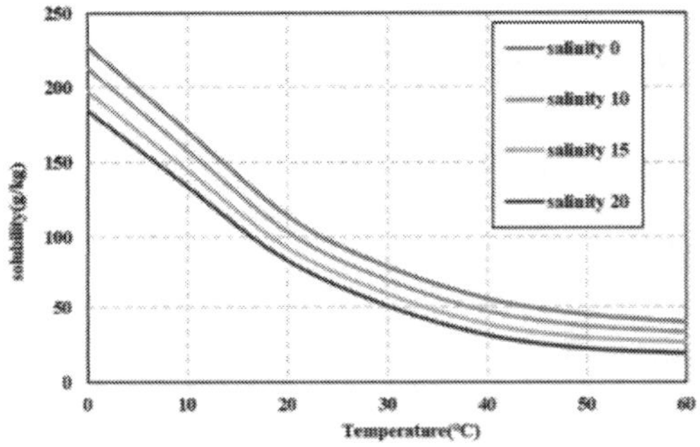


Figure 6 The solubility of $SO_2$ in the saline water

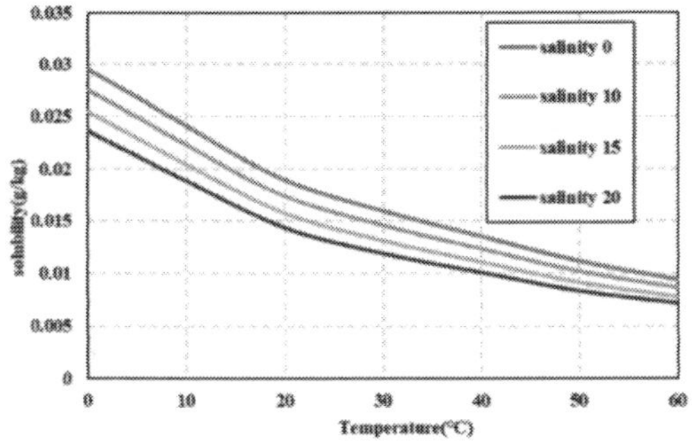


Figure 7 The solubility of $N_2$ in the saline water

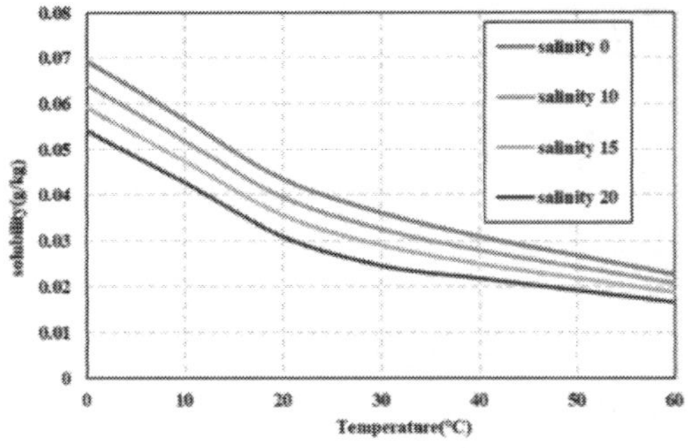


Figure 8 The solubility of $O_2$ in the saline water

## 2 Numerical simulation model

To study the effect of the impurities on the solubility trapping, a numerical simulation model is established to simulate the process of the solubility trapping in the $CO_2$ sequestration. The numerical model is established with the commercial software ECLIPSE 300. As shown in Figure 9, there is an impermeable caprock impeding the migration of buoyant $CO_2$ and the $CO_2$ stream is injected from the injection well to the saline water. The schematic diagram of the numerical simulation model is shown in Figure 10 and the parameters of the model is listed in Table 4.

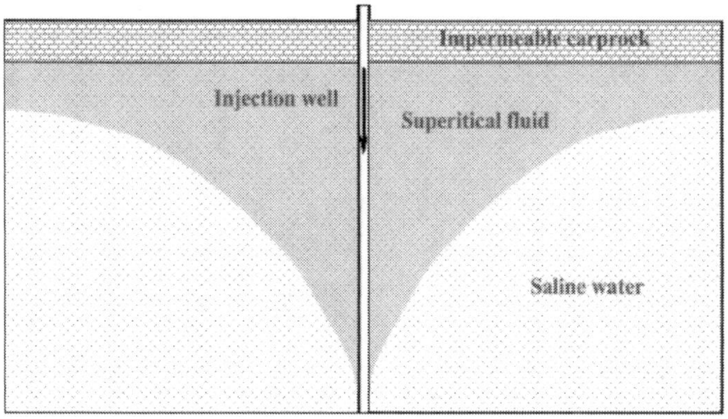

Figure 9 Schematic diagram of the model

Figure 10 Numerical simulation model

Table 4 Reservoir parameters

| Reservoir parameters | Value | Reservoir parameters | Value |
|---|---|---|---|
| Size of the model | 30×10×30 | Rock compressibility | $5.8\times10^{-4}MPa^{-1}$ |
| Top depth | 1400m | Initial pressure | 17.4MPa |
| Porosity | 0.27 | Temperature | 32.2℃ |
| Permeability X | 400mD | irreducible water saturation | 0.3 |
| Permeability Y | 400mD | salinity | 0~20 |
| Permeability Z | 40mD | | |

Ten kinds of the compostions are set to study the impact of the concentration of the impurities, shown in Table 5. The range of the concentration of the impurities is 0% ~ 10% according to the composition of the $CO_2$ streams captured.

Table 5 Injected gas compositions

| Injected gas compositions | Dissolved gas compositions | | | |
|---|---|---|---|---|
| (mol%) | $CO_2$(×$10^{-5}$ mol%) | $SO_2$(×$10^{-5}$ mol%) | $N_2$(×$10^{-5}$ mol%) | $O_2$(×$10^{-5}$ mol%) |
| 100% $CO_2$ | 2095.30 | - | - | - |
| 98% $CO_2$+2%$SO_2$ | 2094.99 | 209.04 | - | - |
| 95% $CO_2$+5% $SO_2$ | 2094.19 | 516.12 | - | - |
| 90% $CO_2$+10% $SO_2$ | 2093.50 | 1020.48 | - | - |
| 98% $CO_2$+2% $N_2$ | 2047.20 | - | 4.52 | - |
| 95% $CO_2$+5% $N_2$ | 1981.70 | - | 8.79 | - |
| 90% $CO_2$+10% $N_2$ | 1876.80 | - | 14.62 | - |
| 98% $CO_2$+2% $O_2$ | 2049.48 | - | - | 7.31 |
| 95% $CO_2$+5% $O_2$ | 1983.98 | - | - | 11.58 |
| 90% $CO_2$+10% $O_2$ | 1879.08 | - | - | 17.41 |

## 3 Results and discussion

The change of the gas compositions is expected to have an impact on the convection process. Different compositions of the $CO_2$ streams will result in different density changes. According to the compositions of injection gases (Table 5), we can get the variations of density difference with different impurity concentrations, as shown in Figure 11.

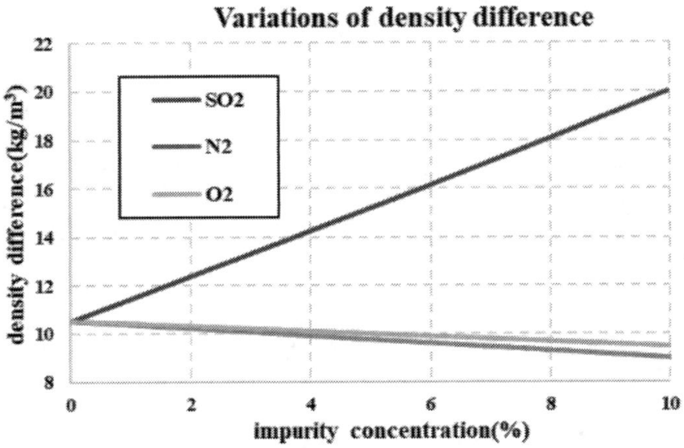

Figure 11 Variations of density difference

For the convective mix can accelerate the process of $CO_2$ solution in the brine and reduce the possibility of $CO_2$ leakage, Onset time of the instability represents the validity of the solubility trapping mechanism to a certain extent. There is no commonly accepted definition of onset time of the instability. According to the method proposed by George S.H. Pau[34], onset time of the instability is defined as the time when dissolution rate, referring to $CO_2$ molar flux from top boundary to the simulated region, is 2% higher than diffusive flux. Although the process of the convection is complicated and it is difficult to quantify, the total inventory of $CO_2$ dissolved into the saline water is rather stable and measured. Eventually, we choose dissolution rate, onset time and total inventory dissolved as the parameters of the impacts of different $SO_2$, $N_2$ and $O_2$ concentrations on solubility trapping.

### 3.1 Dissolution rate and onset time of the instability

The dissolution rate of $CO_2$ is an important parameter of solubility trapping. As shown in Figure 12, Figure 13 and Figure 14, the dissolution rate can be divided into three stages. These are diffusion-dominated stage, convection-dominated stage and convection-decay stage. Before the process of the convection, the downward transport of the dissolved gas is dominated by molecular diffusion and the dissolution rate decays at the rate of $t^{-0.5}$.

At this stage, the concentrations of the impurities have a limited influence on the dissolution rate. The dissolution rate decreases slightly with the increase of the concentrations of $N_2$ and $O_2$ (Figure 13 and Figure 14). At the convection-dominated stage, due to the collision, fusion and merging of the viscous finger, the dissolution rate curve exhibits a significant irregular change, which also denotes the inherent instability of the convection process. At this stage, the effect of the impurities on the dissolution rate is much greater than the diffusion-dominated stage. $N_2$ and $O_2$ will reduce the dissolution rate of $CO_2$ stream and $SO_2$ will increase the dissolution rate on the contrary. With the increase of concentration of the $SO_2$, the dissolution rate will increase. By contrast, the dissolution rate will decrease with the increase of the concentration of $N_2$/ $O_2$. Under the same impurity concentration, the variation of the dissolution rate of $CO_2$-$SO_2$ system is much larger than that of $CO_2$-$N_2$ system or $CO_2$-$O_2$ system. At the convection-decay stage, the dissolution rate decreases and molecular diffusion becomes the dominant fluid transport mechanism again. As a result, the difference among the dissolution rates of the impurities are also quite small.

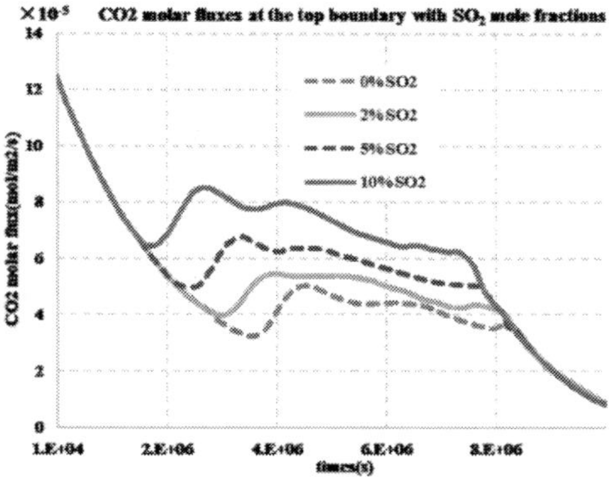


**Figure 12 $CO_2$ molar fluxes with $SO_2$ mole fractions**

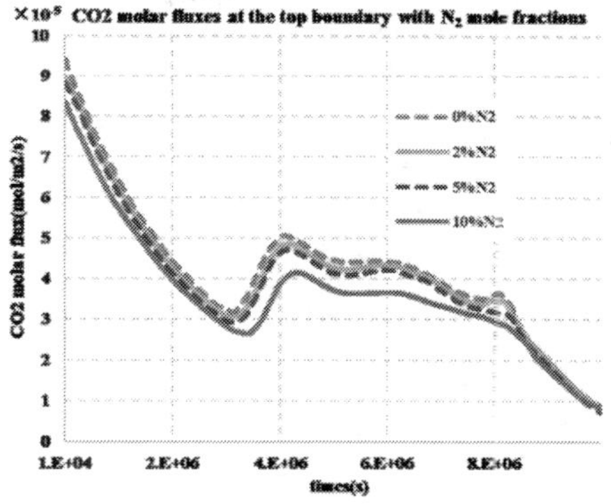


**Figure 13 $CO_2$ molar fluxes with $N_2$ mole fractions**

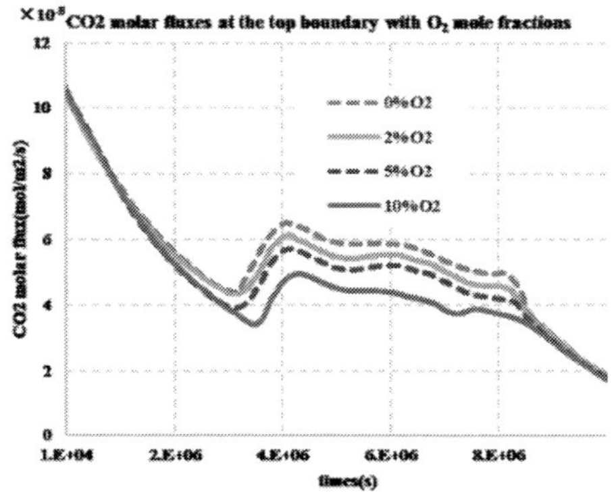


**Figure 14 $CO_2$ molar fluxes with $O_2$ mole fractions**

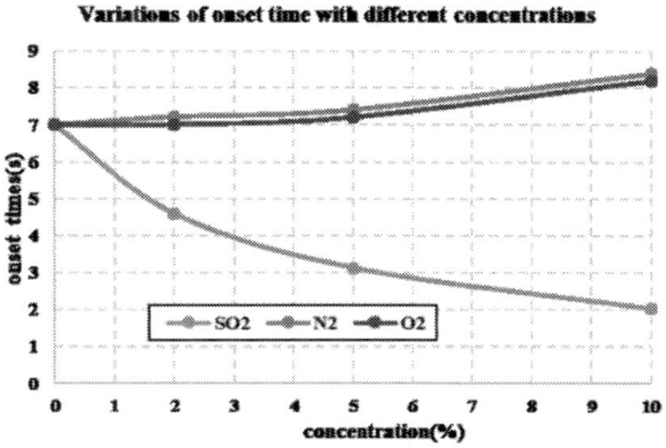


**Figure 15 Onset time with different concentrations**

Figure 15 shows the onset time of the impurities with different concentrations. With the increase of concentration of the impurities, the onset time of $CO_2$-$SO_2$ system decreases and the onset time of $CO_2$-$N_2$ / $CO_2$-$O_2$ system increases. Moreover, the decrease of onset time of $CO_2$-$SO_2$ system is greater than the increase of the $CO_2$-$N_2$ system and $CO_2$-$O_2$ system increases. The increase of onset time of $CO_2$-$N_2$ system is slightly greater than that of $CO_2$-$O_2$ system. The inclusion of $N_2$ and $O_2$ in the $CO_2$ streams delays the onset time, which is negative to the dissolution of $CO_2$ and the solubility trapping mechanism. $SO_2$ will encourage the convection and is conducive to $CO_2$ storage.

### 3.2 Total inventory of $CO_2$ dissolved

The total inventory of $CO_2$ dissolved into the saline water is stable and can be used to measure the efficiency of solubility trapping mechanism. Figure 16, Figure 17 and Figure 18 show the $CO_2$ inventory with different concentrations of the impurities at the early stage. Figure 19, Figure 20 and Figure 21 show the $CO_2$ inventory with different concentrations of the impurities at the end of the simulations. Since the convection can increase the dissolution rate of $CO_2$ largely, the total inventory of $CO_2$ dissolved is much larger than the total inventory of $CO_2$ diffused after the onset of the convection. At the diffusion-dominated stage, the total inventory of $CO_2$ dissolved is equivalent to the total inventory of $CO_2$ diffused. With the concentration of $SO_2$ increasing, the total inventory of $CO_2$ increases significantly. With the concentration of $N_2$ and $O_2$ increasing, the total dissolved CO2 inventory decreases slightly.

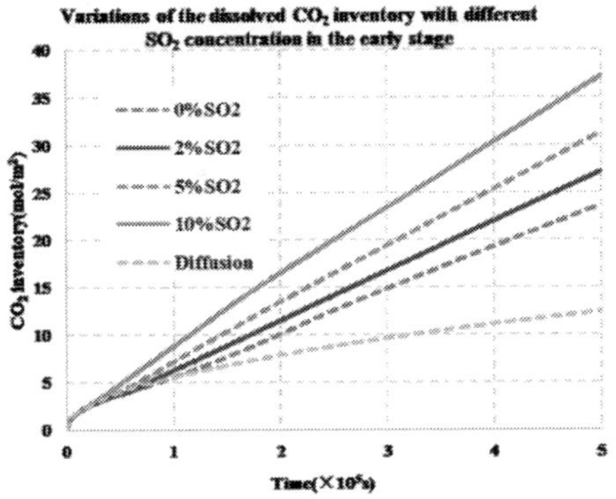


Figure 16 $CO_2$ molar fluxes with $SO_2$ mole fractions

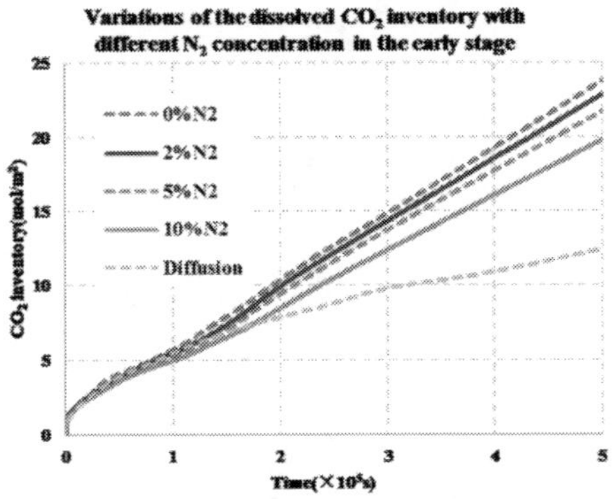


Figure 17 $CO_2$ molar fluxes with $N_2$ mole fractions

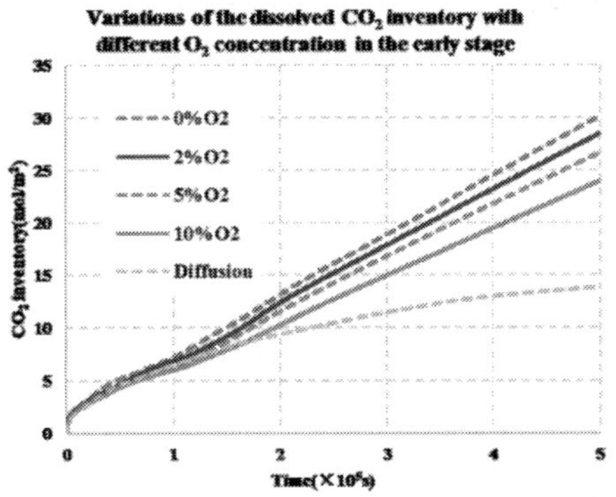


Figure 18 $CO_2$ molar fluxes with $O_2$ mole fractions

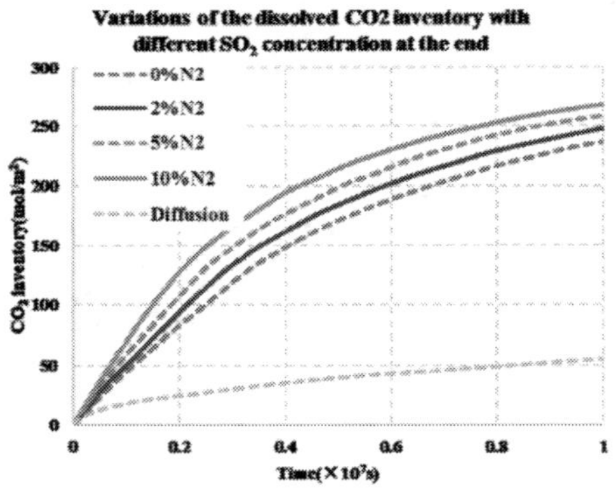


Figure 19 $CO_2$ molar fluxes with $SO_2$ mole fractions

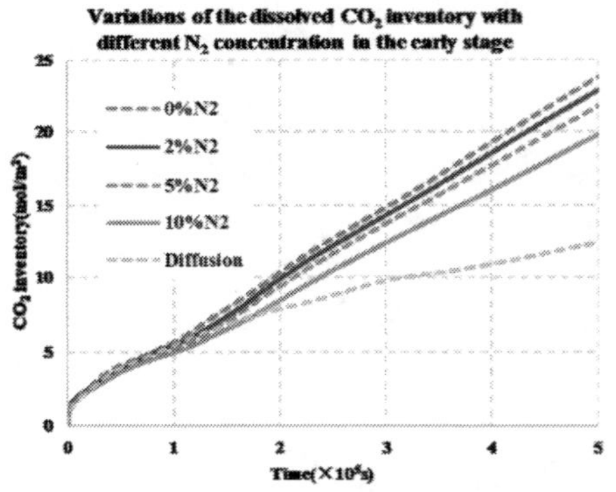


Figure 20 $CO_2$ molar fluxes with $N_2$ mole fractions

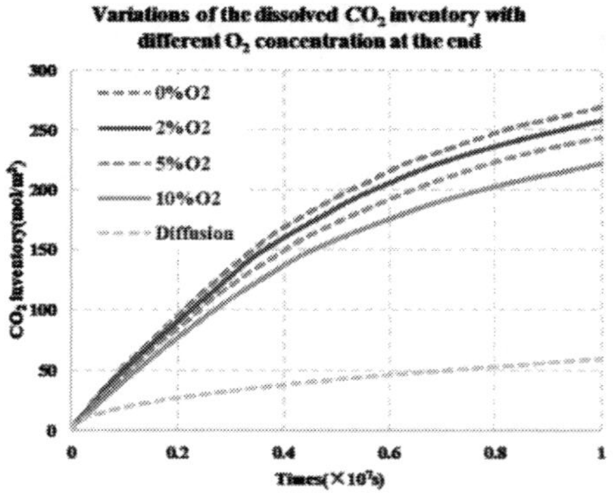


Figure 21 $CO_2$ molar fluxes with $O_2$ mole fractions

## 3.3 The effect of the salinity

According to the dissolution rule of the impurities in the brine mentioned above, with the salinity increases the solubility of the gases decrease, which will decrease the dissolution rate of the $CO_2$. The increase of the salinity will lead to the decrease of onset time of the instability, shown in Figure 22. Furthermore, total inventory of $CO_2$ dissolved will decrease with the increase of the salinity, shown in Figure 23.

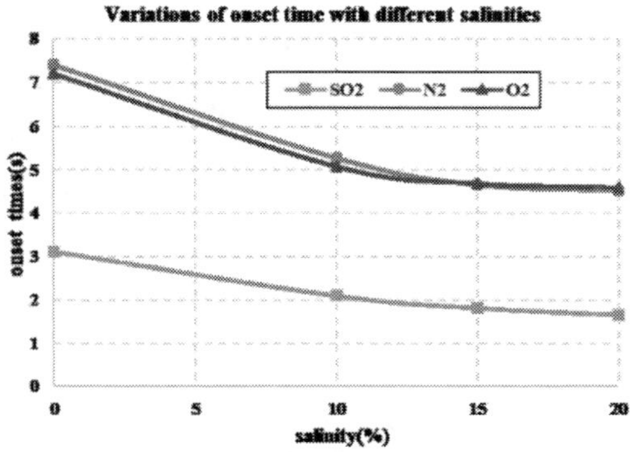

Figure 22 $CO_2$ molar fluxes with $SO_2$ mole fractions

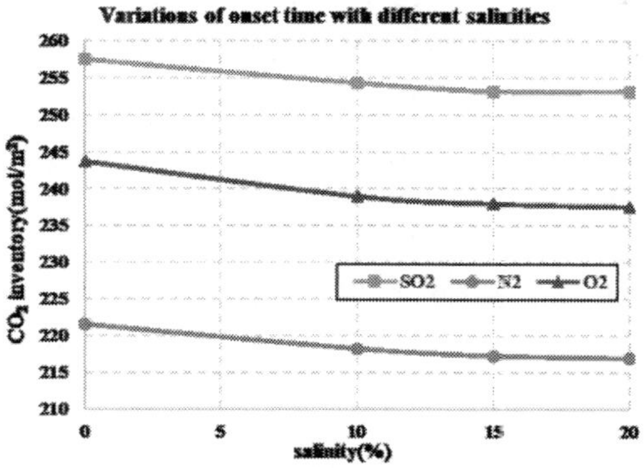

Figure 23 $CO_2$ molar fluxes with $N_2$ mole fractions

## 4 Conclusions and future work

The presence of impurities in the $CO_2$ have a significant impact on the $CO_2$ storage. It will be costly to remove the impurities in the $CO_2$ stream. So allowing the presence of impurities in the $CO_2$ streams will cut down the cost of the $CO_2$ capture. However, the impurities have a complex effect on the process of the $CO_2$ storage. In this paper, a numerical simulation model considering dissolution and diffusion is established by Eclipse to simulate the effect of the impurities on the solubility trapping. The convection driven by the density difference will encourage the dissolution of $CO_2$ and increase the dissolution rate apparently in the solubility trapping mechanism. The dissolution of gas in the aquifer will change the density of brine. $SO_2$ is the most soluble within the impurities, whose solubility is about thousands of times than $N_2$ and scores of times than $CO_2$.moreover, the salinity of the formation water have a significant effect on the solubility. The density difference caused by the gases will have a variation. When the gases is co-injected with $CO_2$ into the formation, the process of solubility trapping will be influenced.With the concentration of $O_2$ or $N_2$ increasing, the onset time of the instability will be postponed. The dissolution rate and the total inventory of $CO_2$ dissolved into the formation water will decrease. On the contrary, $SO_2$ will reduce the onset of the instability and increase the dissolution rate and the total inventory. In a word, $SO_2$ is conducive to $CO_2$ storage, while $N_2$ and $O_2$ are negative to the $CO_2$ storage. Besides, the effect of $SO_2$ is more obvious than that of $O_2$ and $N_2$ under the same circumstance. The concentration of the impurities should be controlled in a rational range for the perspective of the economy and efficiency.

## Nomenclature

$p$ *pressure(MPa)*

$T$ *temperature(℃)*

$R$ *general gas constant(8.3147MPa·cm$^3$/(mol·K))*

$a_m$ *gravitational constant of the mixed systems*

$b_m$ *repulsive constants of the mixed system*

$v$ *molar volume(cm$^3$/mol)*

$n$ *mole of the system*

$x_i$ *mole fraction of i component in the liquid*

$a_i$ ,$b_i$ *parameters of EOS of i component*

$\alpha_i$ *a function about temperature*

$G_\infty^E$ *excess Gibbs Free Energy under Infinite Pressure(J/mol)*

$c_0$ *constant*( $=\frac{1}{2\sqrt{2}_m}\ln\frac{2+\sqrt{2}}{2-\sqrt{2}}$ )

$T_{ci}$ ***critical temperature of the i component(℃)***

$p_{ci}$ ***critical pressure of the i component(MPa)***

$T_{ri}$ ***reduced temperature of the i component(℃)***

$p_{ri}$ ***reduced pressure of the i component(MPa)***

$m$ ***a function about acentric factor***

$C_{ji}$ $C_{ji}=g_{ji}-g_{ii}$

$g_{ji}, g_{ii}$ ***Boltzmann factor between molecules***

$k_{ij}$ ***interaction coefficient between i component and j component***

$G_{ji}, G_{ki}$ ***temperature-dependent parameters that can be adjusted***

$\alpha_{ji}, \alpha_{ki}$ ***non-random parameters between i, j components and between i, k components***

$\varphi_i$ ***fugacity coefficient of i component***

$Z$ ***deviation factor***

$n_i, n_j$ ***mole fraction of i, j component***

$\varphi_m$ ***fugacity coefficient of the mixture***

$Z_m$ ***deviation factor of the mixture***

$\gamma_i$ ***activity coefficient of i component***

## Acknowledgement

This work is supported by Science Foundation of China University of Petroleum, National Basic Research 973 Program of China (2015CB250900), and National Natural Science Foundation of China (U1262101).

## Reference

[1] Holloway S. Carbon dioxide capture and geological storage[J]. Philosophical Transactions of the Royal Society A: Mathematical, Physical and Engineering Sciences. 2007, 365(1853): 1095-1107.

[2] De Silva P N K, Ranjith P G. A study of methodologies for CO2 storage capacity estimation of saline aquifers[J]. Fuel. 2012, 93: 13-27.

[3] Metz B, Davidson O, De Coninck H, et al. IPCC special report on carbon dioxide capture and storage[Z]. Intergovernmental Panel on Climate Change, Geneva (Switzerland). Working Group III, 2005.

[4] Burnside N M, Naylor M. Review and implications of relative permeability of CO2/brine systems and residual trapping of CO2[J]. International Journal of Greenhouse Gas Control. 2014, 23: 1-11.

[5] Bachu S. CO2 storage in geological media: Role, means, status and barriers to deployment[J]. PROGRESS IN ENERGY AND COMBUSTION SCIENCE. 2008, 34(2): 254-273.

[6] Leung D, Caramanna G, Maroto-Valer M M. An overview of current status of carbon dioxide capture and storage technologies[J]. RENEWABLE & SUSTAINABLE ENERGY REVIEWS. 2014, 39: 426-443.

[7] Michael K, Golab A, Shulakova V, et al. Geological storage of CO2 in saline aquifers-A review of the experience from existing storage operations[J]. INTERNATIONAL JOURNAL OF GREENHOUSE GAS CONTROL. 2010, 4(4): 659-667.

[8] Bachu S, Adams J J. Sequestration of CO2 in geological media in response to climate change: capacity of deep saline aquifers to sequester CO2 in solution[J]. 2003, 44(20): 3151-3175.

[9] Shen P P, Liao X W. The Technology of Carbon Dioxide Stored in Geological Media and Enhanced Oil Recovery[J]. 2009.

[10] Bachu S, Shaw J C, Pearson R M. Estimation of Oil Recovery and CO2 Storage Capacity in CO2 EOR Incorporating the Effect of Underlying Aquifers[J]. Society of Petroleum Engineers Journal. 2004.

[11] Birkholzer J T, Zhou Q L, Tsang C F. Large-scale impact of CO2 storage in deep saline aquifers: A sensitivity study on pressure response in stratified systems[J]. INTERNATIONAL JOURNAL OF GREENHOUSE GAS CONTROL. 2009, 3(2): 181-194.

[12] Ghesmat K, Hassanzadeh H, Abedi J. The effect of anisotropic dispersion on the convective mixing in long - term CO2 storage in saline aquifers[J]. 2011, 57(3): 561-570.

[13] Spycher N, Pruess K, Ennis-King J. CO2-H2O mixtures in the geological sequestration of CO2. I. Assessment and calculation of mutual solubilities from 12 to 100°C and up to 600 bar[J]. 2003, 67(16): 3015-3031.

[14] Bachu S, Gunter W D, Perkins E H. Aquifer disposal of CO 2: hydrodynamic and mineral trapping[J]. Energy Conversion and Management. 1994, 35(4): 269-279.
[15] Oelkers E H, Gislason S R, Matter J. Mineral Carbonation of CO2[J]. ELEMENTS. 2008, 4(5): 333-337.
[16] Bachu. CO2 storage in geological media: role, means, status and barriers to deployment [J]. Progress in Energy and Combustion Science. 2008, 2(34): 254-273.
[17] Gilfillan S, Lollar B S, Holland G, et al. Solubility trapping in formation water as dominant CO2 sink in natural gas fields[J]. NATURE. 2009, 458(7238): 614-618.
[18] Lengler U, De Lucia M, K Hn M. The impact of heterogeneity on the distribution of CO2: Numerical simulation of CO2 storage at Ketzin[J]. International Journal of Greenhouse Gas Control. 2010, 4(6): 1016-1025.
[19] Elenius. Convective mixing in geological carbon storage[Z]. 2011.
[20] Sainz-Garcia A, Abarca E, Nardi A, et al. Convective mixing fingers and chemistry interaction in carbon storage[J]. International Journal of Greenhouse Gas Control. 2017, 58: 52-61.
[21] Torabi M, Peterson G P, Torabi M, et al. A thermodynamic analysis of forced convection through porous media using pore scale modeling[J]. International Journal of Heat and Mass Transfer. 2016, 99: 303-316.
[22] Emami-Meybodi H. Stability analysis of dissolution-driven convection in porous media[J]. Physics of Fluids. 2017, 29(1): 14102.
[23] Jiang X. A review of physical modelling and numerical simulation of long-term geological storage of CO2[J]. Applied Energy. 2011, 88(11): 3557-3566.
[24] Ennis-King J P, Paterson L. Role of Convective Mixing in the Long-Term Storage of Carbon Dioxide in Deep Saline Formations[J]. Spe Journal. 2005, 10(3): 349-356.
[25] Ennis-King J, Paterson L. Coupling of geochemical reactions and convective mixing in the long-term geological storage of carbon dioxide[J]. International Journal of Greenhouse Gas Control. 2007, 1(1): 86-93.
[26] Ennisking J, Preston I, Paterson L. Onset of convection in anisotropic porous media subject to a rapid change in boundary conditions[J]. Physics of Fluids. 2005, 17(8): 239-248.
[27] Meng Q, Jiang X. Numerical analyses of the solubility trapping of CO2 storage in geological formations[J]. Applied Energy. 2014, 130: 581-591.
[28] White C M, Strazisar B R, Granite E J, et al. Separation and capture of CO2 from large stationary sources and sequestration in geological formations - Coalbeds and deep saline aquifers[J]. JOURNAL OF THE AIR & WASTE MANAGEMENT ASSOCIATION. 2003, 53(6): 645-715.
[29] Oosterkamp. State-of-the-Art Overview of CO2 Pipeline Transport with relevance to offshore pipelines[Z]. Norway: PolytecR & DFoundation, 2008.
[30] Li D, Jiang X. A numerical study of the impurity effects of nitrogen and sulfur dioxide on the solubility trapping of carbon dioxide geological storage[J]. Applied Energy. 2014, 128: 60-74.
[31] Yan W, Huang S, Stenby E H. Measurement and modeling of CO2 solubility in NaCl brine and CO2 – saturated NaCl brine density[J]. International Journal of Greenhouse Gas Control. 2011, 5(6): 1460-1477.
[32] Marie-José H. New mixing rules in simple equations of state for representing vapour-liquid equilibria of strongly non-ideal mixtures[J]. Fluid Phase Equilibria. 1979.
[33] Dali H. Experimental research and theoretical model for CO2 solubility in water under high temperature and high pressure[J]. 2015.
[34] Pau G S H, Bell J B, Pruess K, et al. High-resolution simulation and characterization of density-driven flow in CO2 storage in saline aquifers[J]. Advances in Water Resources. 2010, 33(4): 443-455.

# CMTC-486556-MS

# Numerical Simulation and Performance Evaluation of $CO_2$ Huff-n-Puff Processes in Unconventional Oil Reservoirs

Mengjing Cao, Xiaodong Wu, Yongsheng An,Yi Zuo and Ruihe Wang, China University of Petroleum -Beijing; and Peng Li, Shengli Oilfield, Sinopec



This paper was prepared for presentation at the Carbon Management Technology Conference held in Houston, Texas, USA, 17-20 July 2017.



## Abstract

Unconventional oil, such as tight oil and shale oil, has become one of the most significant contributors of oil reservoirs and production growth. Due to low porosity and ultra-low permeability, unconventional oil reservoirs require multistage hydraulic fracturing technique to maximize production. However, the primary recovery remains very low to narrow the profit margin heavily. Although $CO_2$ huff-n-puff process holds great potential to increase oil recovery and has a chance to sequester $CO_2$ to reduce environmental footprint, our current knowledge of the performance of this process is very limited.

With numerical simulation, we performed a series of sensitivity work to present the impacts of reservoir properties, fracture properties and operation parameters such as $CO_2$ injection rate, injection time, soaking time, number of cycle of $CO_2$ on enhanced oil recovery in the tight oil formation. What's more, the method of analysis of variance (ANOVA) was used to evaluate the performance of $CO_2$ huff-n-puff process and beneficial result from $CO_2$ EOR technology. Simulation results showed that bottom hold pressure and injection cycles impose more significant impose on oil recovery increment than injection time, injection rate and production time per cycle.Based on the typical reservoir and fracture properties from tight oil reservoir, the numerical models were estabilished to evaluate the performance of four EOR methods: $CO_2$ huff-n-puff, water huff-n-puff, nanofluids huff-n-puff and water alternating gas (WAG). With the camparison of oil recovery and its increment of four EOR methods and depletion method, it is found that $CO_2$ huff-n-puff method would lead to much more incremental oil recovery than other three methods, which reveals its huge potentials of enhancing oil recovery and improving development profit in unconventional reservoirs. The conclusion of this work has the potential to advance our understanding of the role of $CO_2$ in developing unconventional oil reservoirs, which will benefit both energy economy and environment with $CO_2$ geological sequestration.

## Introduction

Recently, uncoventional oil reservoirs have attracted much more attention. Unfortunately, low permeability of the tight and shale oil formations prevents them from being developed effectively and benefitially. Generally speaking, even with long horizontal drilling and multistage hydraulic fracturing techniques, the primary revovery remains low to only 5-10% of original oil in place (Christensen JR et al, 2001). Song Chengyao et al (2012) thought the water alternating miscrible flooding process was the most favourable scheme for tight formations in terms of both recovery efficiency and fluid injectivity. Luky Hendraningrat et al (2013) proposed the optimizing nanofluids concentration to maximize oil recovery in the low-permeability Berea sandstone with the mechanism of wettability alternation and interfacial tension reduction. WAG and Nano-EOR have been proven to not only increase oil procudion, but also improve fluid injectivity, which would be nice way to extract oil from unconventional formations.With experimental researches and field applications in conventional oil reservoirs, $CO_2$ flooding schemes have shown favourable revovery potential of enhancing oil recovery (EOR). S.M. Ghaderi et al (2013) proposed the coupled methodology to evaluate profitability and the risk of failure for multiple recovery meahanism sequences as well as the need for incentives to make $CO_2$ EOR profitable in tight oil reservoir.H. Wang et al (2014) estabilished the compositional numerical model of $CO_2$ flooding in tight oil reservoir to show that the minimum miscrible pressure (MMP) and the total gas injection volume were two key factors of $CO_2$ flooding effect. However, fractured horizontal well requires huag $CO_2$ consumption to dissolve into the crude oil and erious early gas breakthrough occurs in the complex fractures network, which inhibits the performance of $CO_2$ flooding.K. Zhang et al (2015) presented an integrated method for $CO_2$ flooding reservoir criteria with the consideration of asphatene precipitation, oil recovery performance and rish analysis, which would be used as guidance to select suitable candidateds for $CO_2$ flooding. Bing Kong et al (2016) compared the performances of waterflooding and $CO_2$ huff-n-puff and studied the inter-well interference during $CO_2$ huff-n-puff process. Compared with synchronous $CO_2$ huff-n-puff, asynchronous pattern performed much better.Chengyao Song and Daoyong Yang (2017) combined experimental technique with numerical simulation method to evaluate the $CO_2$ huff-n-puff process in a tight oilfield in Bakken formation. Experimentally, oil recovery was significantly enhanced by $CO_2$ huff-n-puff much more than by waterfloodiung. Coreflooding simulation and reservoir simulation were developed to evaluate the revovery performance of $CO_2$ huff-n-puff by using the CMG GEM simulator. They found that injection pressure and production pressure influenced the ultimate oil recovery more significantly. Despite many attempts and efforts already made to eavulate recovery performance of $CO_2$ huff-n-puff, the mechainsms have not been well understood and there is little field trial of $CO_2$ huff-n-puff for tight oil exploitation. Escpecially under the condition of low oil prices, $CO_2$ projects have to take large investment risk. Fortunately, world's largest carbon-capture project, the Petra Nova project, began commercial operation in 2017, which would attract much more interests from researches and field trials.

In this paper, numerical simulation is conducted to evaluate the recovery performance of four EOR methos in tight oil reservoir: $CO_2$ huff-n-puff, water huff-n-puff, nanofluids huff-n-puff and water alternating gas (WAG). The method of analysis of variance (ANOVA) is carried out to present the impacts of operation parameters such as $CO_2$ injection rate, injection time, soaking time, number of cycle of $CO_2$ on enhanced oil recovery.

## Numerical Simulation

A tight oil reservoir has been selected as the targeted reservoir. The reservoir covers an area of 2000 $\times$500 m with a payzone thichness 15 m. The formation matrix permeability is 0.015 mD and its porosity is 10%. The initial pressure is 45 MPa. The length of fractured horizontal well is 1800 m, while half length and numbers of fracture are 150 m and 23, respectively. The reservoir model is created by using the commercial software Eclipse. The model has a grid system of 200$\times$50$\times$3, among which each grid is 10 m in the x and y direction, while it is 5 m in the z direction. With the constant oil production rate, the well produces for 2160 days and then it is converted to a $CO_2$ injection well with injection rates of 100000 $m^3$/d for 20 days. Then the well is shut-in and soaking for 20 days. Finally the well is put back into production for 300 days with constant bottom hole pressure, 6 MPa. This is one cycle of $CO_2$

huff-n-puff process. After that, another cycle of $CO_2$ huff-n-puff process continues. In this paper, the total times of cycles are 6 and the model is modelled totally for 5400 days.

The properties and parameters of the reservoir and the fractured horizontal well are presented in Table 1.

Table 1—Parameters of numerical simulation model

| Parameters | Value | Parameters | Value |
|---|---|---|---|
| Reservoir length, m | 2000 | Length of horizontal well, m | 1800 |
| Reservoir width, m | 500 | Injection rate, $m^3/d$ | 100000 |
| Effective thickness, m | 15 | Injection time per cycles, days | 20 |
| Initial reservoir pressure, MPa | 45 | Soaking time per cycles, days | 20 |
| Porosity, % | 10 | Production time per cycles, days | 300 |
| Permeability, mD | 0.015 | Cycles, times | 6 |
| Initial oil saturation | 0.7 | Bottom hole pressure, MPa | 6 |
| Fracture numbers | 23 | Modelling time, days | 5400 |
| Half length of fracture, m | 150 | $CO_2$ huff-n-puff time, days | 3140 |

Figure 1 is the oil production profile and shows oil production as a function of production time. As it can be seen from the figure 1, the oil production starts to decrease after 1500 days and the well is shut-in at 2160 days to carry out $CO_2$ huff-n-puff injection. The profile indicates that oil production is largest in the first cycle and most oil is produced in the first two cycles. From the third cycle to the last cycle, the oil production during the puff process becomes less significant.

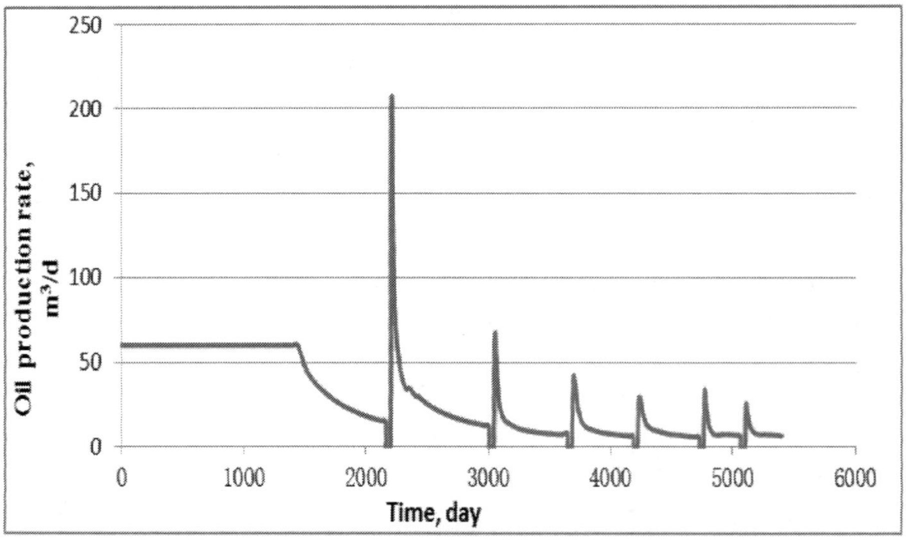


Figure 1—Oil production rate of $CO_2$ huff-n-puff process

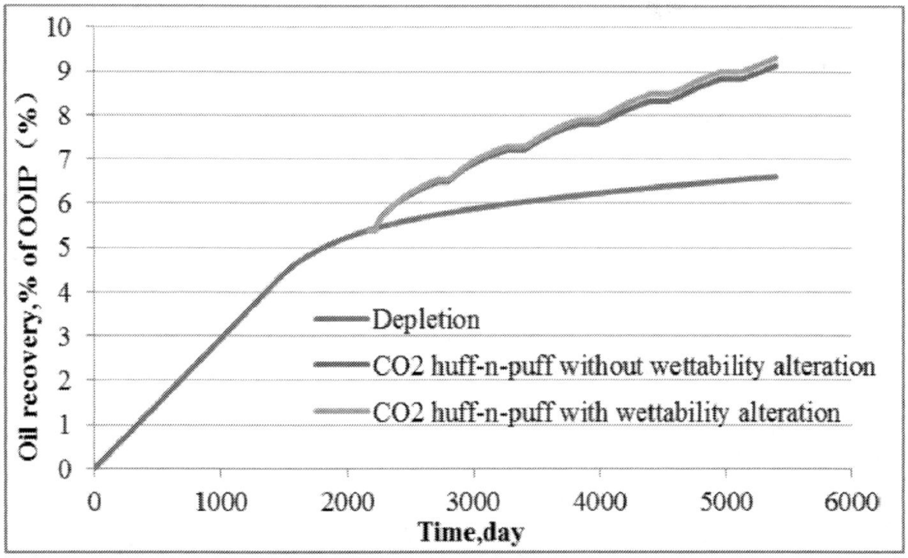

**Figure 2—Comparison of oil recovery for depletion and $CO_2$ huff-n-puff**

Figure 2 is the oil recovery profile of three scenarios: depletion, $CO_2$ huff-n-puff without wettability alternation and $CO_2$ huff-n-puff with wettability alternation. First of all, it is obviously found that $CO_2$ huff-n-puff brings significant recovery increment, about 3%. What's more, with the consideration of the mechanism of wettability alternation, the oil recovery is increased further, which suggests that wettability alternation is an important mechanism of enchancing oil recovery.

## Analysis of Variance

The above numerical model with the consideration of wettability alternation is further applied to evaluate the recovery performance of $CO_2$ huff-n-puff under various operation parameters such as $CO_2$ injection rate, injection time, soaking time, number of cycle of $CO_2$. The simulated results of ANOVA are shown in Figure 3-8, respectively.

As can be seen in Figure 3, the ultimate oil recovery factors under injection rate of 40000 $m^3$/d, 60000 $m^3$/d, 80000 $m^3$/d, 100000 $m^3$/d and 120000 $m^3$/d are found to be 8.1%, 8.3%, 8.4%, 8.9%, 8.4%.

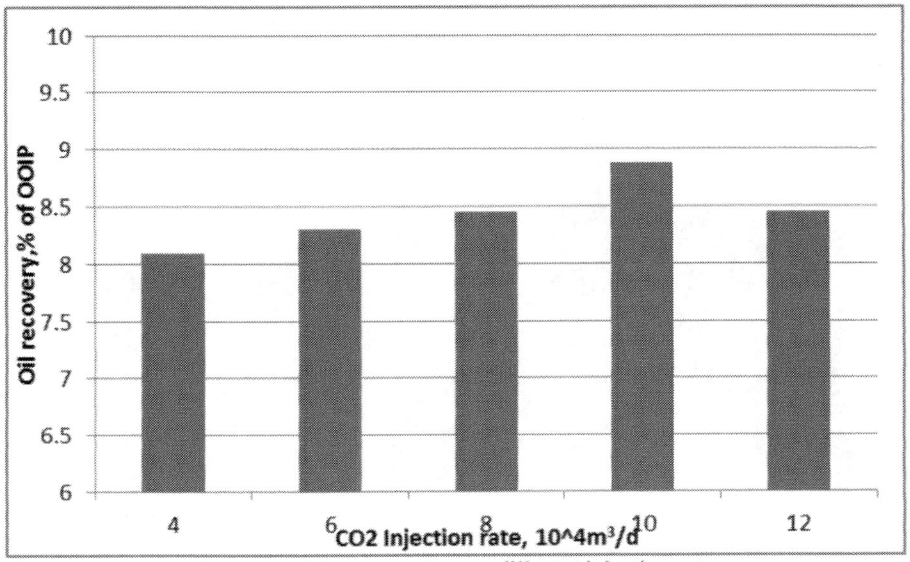

**Figure 3—Oil recovery versus different injection rate**

The oil recovery under various injection time is shown in Figure 4. The ultimate oil recovery factors under injection time of 5 days, 10 days, 15 days, 20 days and 25 days are found to be 7.5%, 7.8%, 8.7%, 8.9%, 8.4%.

Figure 5 shows the oil recovery under various soaking time. When the soaking rate is set to be 10 days, 15 days, 20 days and 25 days, the factor is 8.1%, 8.1%, 8.9%, and 8.5%, respectively.

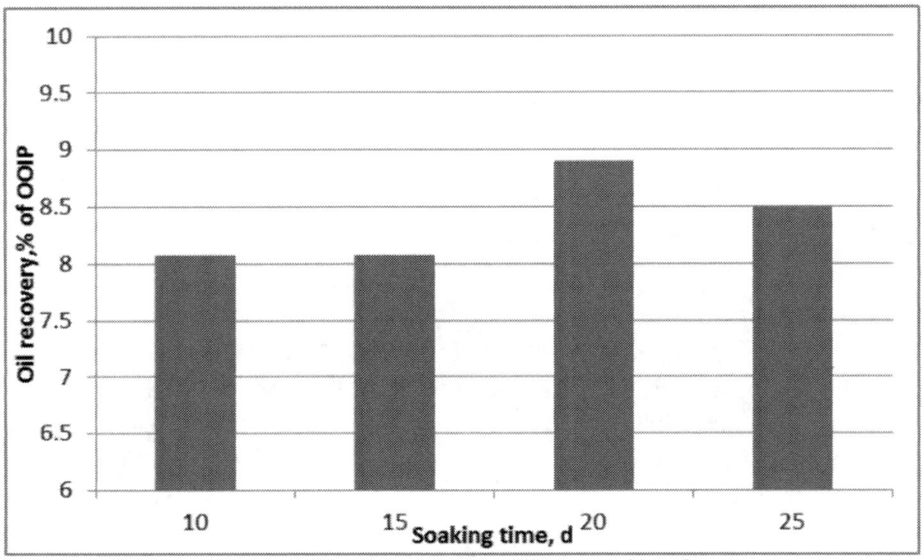


Figure 4—Oil recovery versus different soaking time

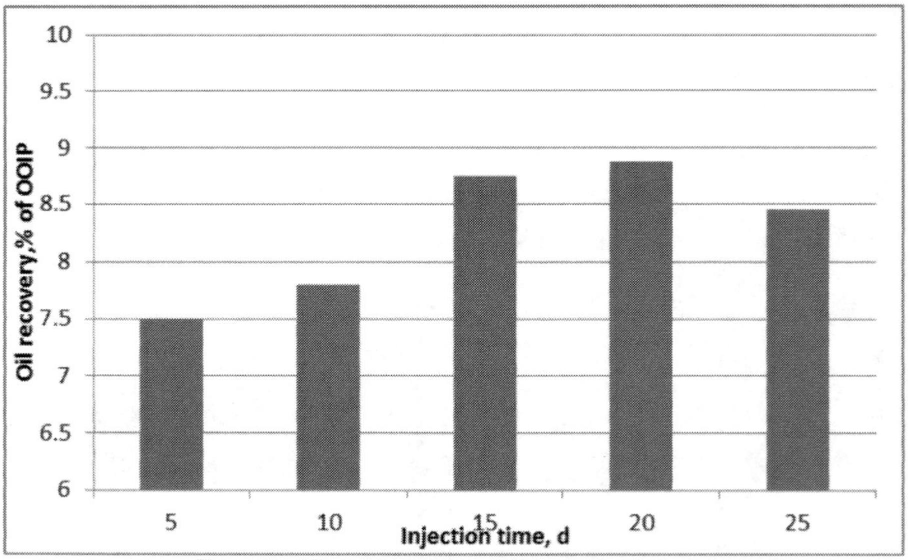


Figure 5—Oil recovery versus different injection time

The oil recovery performance is shown in Figure 6, where oil recovery is increased by a longer production tiem per cycles from 100 days to 500 days.

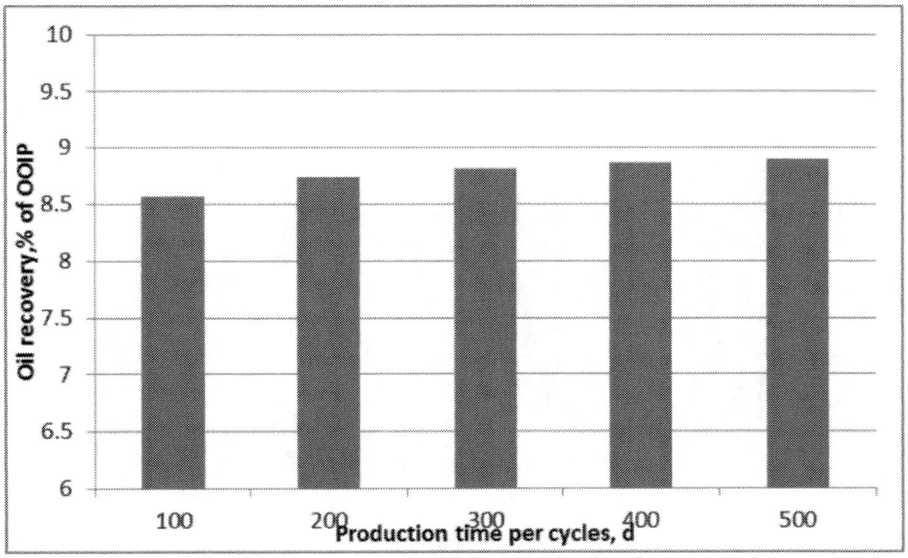


**Figure 6—Oil recovery versus different production time per cycles**

In Figure 7, where oil recovery factor decreases with increasing bottom hole pressure from 6 MPa to 12 MPa, the difference of oil recovery between the highest and the lowest value is about 3%.

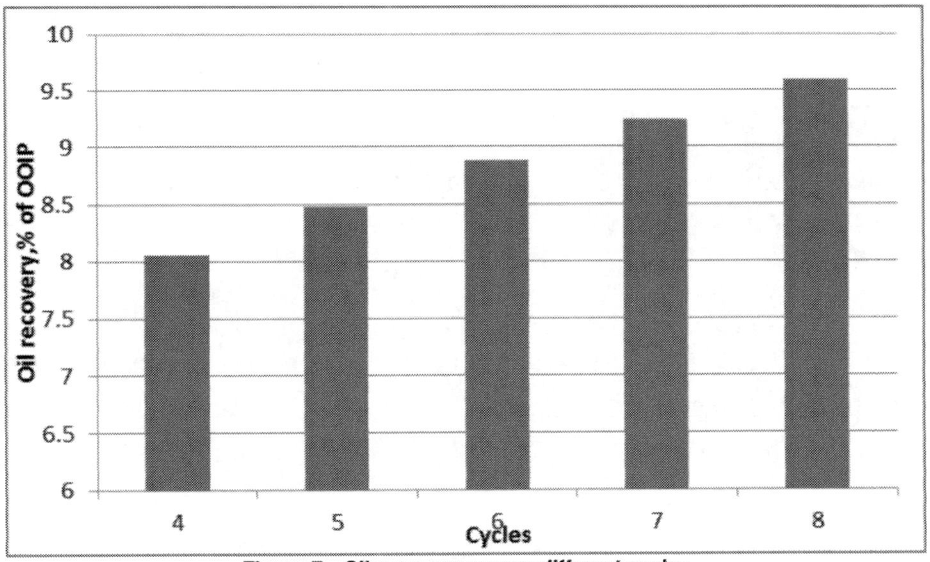


**Figure 7—Oil recovery versus different cycles**

In figure 8, with the increading of cycles, the oil recovery factor increases from 8% to 9.6%.

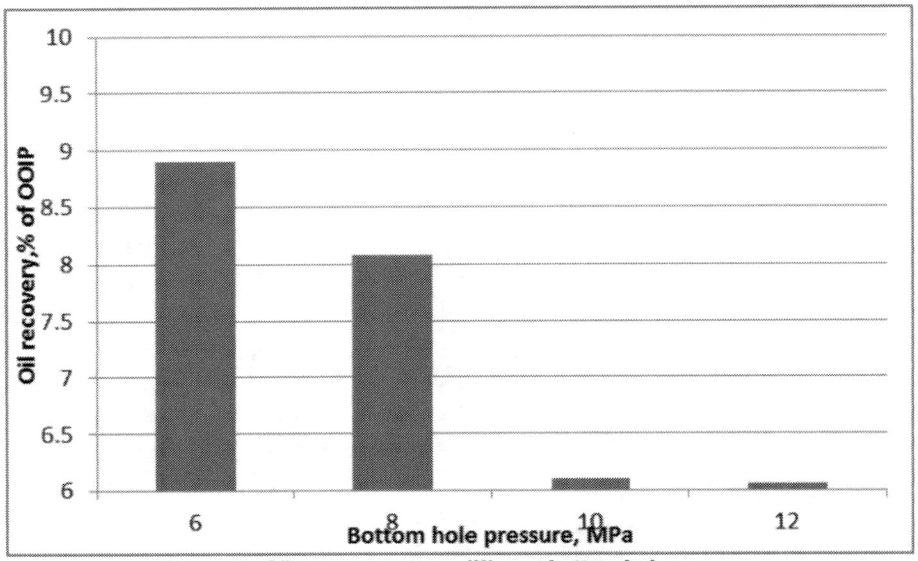


Figure 8—Oil recovery versus different bottom hole pressure

With comparison of oil recovery factor under various parameters, it can be found that the most important parameter is bottom hole pressure, followed by cycles, soaking time, injection time, injection rate, production time per cycle. Besides, the range of oil recovery factor is obtained as 6%-9.6%.

## Performance Evaluation

There have been several EOR methods applied into conventional reservoirs development. According to the analysis of K. Zhang et al (2015), four typical EOR methods are highly possible to become favourable recovery potential for unconventional reservoirs exploitation: $CO_2$ huff-n-puff, water huff-n-puff, nanofluids huff-n-puff and water alternating gas (WAG).

Based on the above numerical model, five cases, including depletion (case #1), CO2 huff-n-puff (case #2), water huff-n-puff (case #3), nanofluids huff-n-puff (case #4) and WAG (case #5), are conducted to evaluate their recovery performance.

The oil recovery of five cases is presented in Figure 9. When compared with depletion (case #1), four methods could indeed enhance oil recovery of tight oil reservoir. In addition, tt can be suggested that CO2 huff-n-puff (case #2) leads to largest oil recovery, followed by WAG (case #5), nanofluids huff-n-puff (case #4) and water huff-n-puff (case #3).

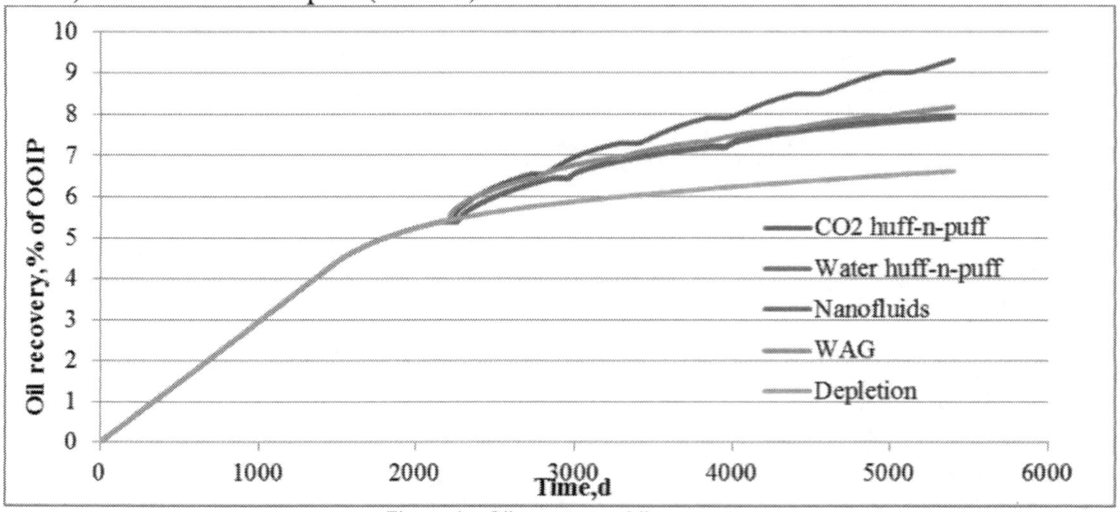


Figure 9—Oil recovery of five cases

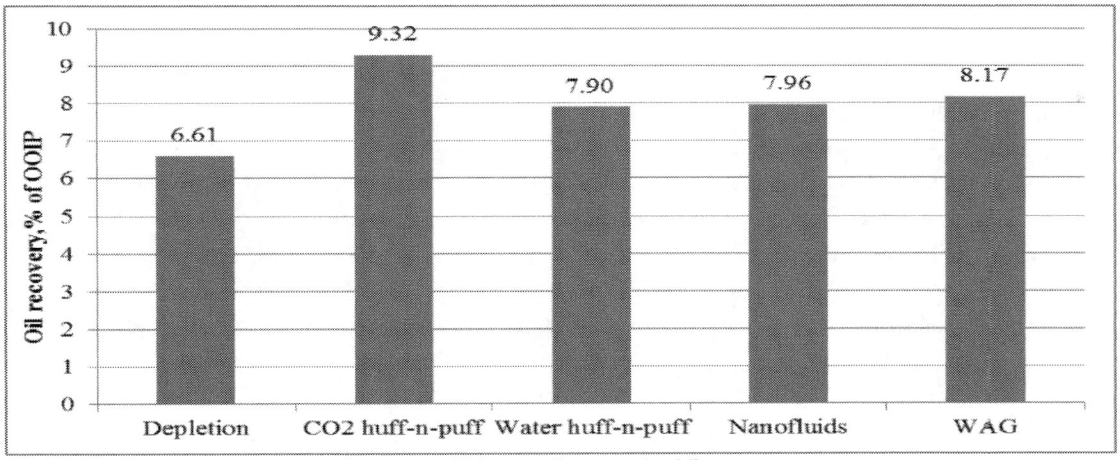


**Figure 10—Ultimate oil recovery of five cases**

In Figure 10, without any EOR method, the oil recovery of depletion at 5400days of production is 6.61%, which is the primary recovery. When water huff-n-puff, nanofluids huff-n-puff and water alternating gas (WAG) is applied respectively, the oil recovery is less than 8.2% and the range of the incremental factor is 1.3%-1.5%. However, $CO_2$ huff-n-puff could enhance oil recovery factor to 9.32% with the incremental facor 2.7%.

## Conclusions

This paper conduted numerical simulation to model $CO_2$ huff-n-puff process and evaluate the recovery performance in targeted tight oil reservoir. With the method of analysis of variance, operation paratmeters, such as CO2 injection rate, injection time, soaking time and number of cycles, are anyalysized to evaluate the impacts on the performance. Five cases are modeled to study the application of four EOR methods: CO2 huff-n-puff, water huff-n-puff, nanofluids huff-n-puff and WAG.

In our work, the primary recovery in tight oil reservoir is 6.61%. It's found that $CO_2$ huff-n-puff brings significant recovery increment, about 3%. With the results of ANOVA, it can be seen that bottom hole pressure is the most important parameter, followed by cycles, soaking time, injection time, injection rate, production time per cycle. Through the five cases calculation and comparsion, $CO_2$ huff-n-puff reveals the greatest advantages of enhancing oil recovery in unconventional reservoirs against other three EOR mehods. As a result, it is suggested that $CO_2$ huff-n-puff process should attract more attention to make greater contribution to the exploitation of unconventional reservoirs and carbon geological sequestration

## Acknowledgments

This research is supported by the National Program onKey Basic Research Project (973 Program, 2015CB250906). We would like to thank the Computer Modeling Group (CMG) Foundation for its support to this work. And we also thank the assistance of Key Laboratory of Petroleum Engineering of China University of Petroleum-Beijing.

## References

1. Christensen JR, Stenby EH, Skauge A. Review of WAG field experience. SPE Reservoir Eval Eng ,2001;4(2):97–106. SPE-71203-PA. https://doi.org/10.2118/71203-PA.

2. Chengyao Song, Daoyong Yang. Experimental and numerical evaluation of $CO_2$ huff-n-puff processes in Bakken formation. Fuel, 2017,190:145–162. http://dx.doi.org/10.1016/j.fuel.2016.11.041

3. Chengyao Song, Daoyong Yang. Experimental and numerical evaluation of $CO_2$ huff-n-puff processes in Bakken formation. Fuel, 2017,190:145–162. http://dx.doi.org/10.1016/j.fuel.2016.11.041
4. Yu W, Lashgari HR, Wu K, Sepehrnoori K. CO2 injection for enhanced oilrecovery in Bakken tight oil reservoirs. Fuel 2015;159(3):354–63.
5. Chen C, Balhoff MT, Mohanty KK. Effect of reservoir heterogeneity on primary recovery and CO2 huff 'n' puff recovery in shale-oil reservoirs. SPE Reservoir Eval Eng 2014;17(3):404–13.
6. Kong B, Wang S, Chen S. Simulation and optimization of CO2 huff-and puff processes in tight oil reservoirs. Paper SPE 179668, presented at the SPE improved oil recovery conference, Tulsa, OK, April 11–13, 2016.
7. Sheng JJ. Increase liquid oil production by huff-n-puff of produced gas in shalegas condensate reservoirs. J Unconventional Oil Gas Resour 2015;11(4):19–26.
8. Computer Modeling Group Ltd. WinProp and GEM Program. Calgary, AB,Canada; 2014.
9. Zhang Y, Yang D, Song C. A damped iterative EnKF method to estimate relative permeability and capillary pressure for tight formations from displacement experiments. Fuel 2016;167(5):306–15.
10. Cudjoe S, Vinassa M, Gomes JHB, Barati RG. A comprehensive approach to sweet-spot mapping for hydraulic fracturing and CO2 huff-n-puff injection in Chattanooga shale formation. J Nat Gas Sci Eng 2016;33(6):1201–18.

CMTC-486571-MS

# The First Integrated Approach for $CO_2$ Capture and Enhanced Oil Recovery in China

Xiangzeng Wang, Shaanxi Yanchang Petroleum Group Co.; Qingwang Yuan, University of Regina; Shuoshi Wang, University of Oklahoma; and Fanhua Zeng, University of Regina



This paper was prepared for presentation at the Carbon Management Technology Conference held in Houston, Texas, USA, 17-20 July 2017.



## Abstract

As the only company which owns the right for coal mines and oil and gas production in China, Yanchang Petroleum Group has the unique advantages to implement integrated $CO_2$ capture and enhanced oil recovery (EOR) in Ordos Basin. This area is also one of the largest $CO_2$ emission areas in China. The coal to chemicals plant was built to efficiently co-utilize coal, oil and natural gas. The energy efficiency is about 8.9% higher than the world average level. Two corresponding $CO_2$ capture plants were built with the capacity of 50,000 and 360,000 tonnes per year. The cost for $CO_2$ capture is as low as $17.5/tonne, much cheaper than most of the other $CO_2$ capture project in the world. By optimizing the distance between $CO_2$ capture plants and EOR sites, the shortest distance, the shortest distance for $CO_2$ transportation is only 10 kilometers. It is estimated that the cost for $CO_2$ transportation is $2.58/tonne. Meanwhile, the $CO_2$ is used for enhanced oil recovery in Yanchang oil fields. Extensive research has been done to investigate the suitable geological conditions for $CO_2$-EOR. Experiments have also been conducted to study the behaviors of $CO_2$-crude oil mixture. Two pilot tests including Qiaojiawa 203 block and Wuqi Yougou are now in operation with well production being doubled or tripled. In addition, more than 87% of reservoirs in Yanchang oil field in Ordos Basin are suitable for $CO_2$-EOR with estimated billions of $CO_2$ storage capacity.

## Introduction

Greenhouse gas emissions are regarded as one of the most important factors resulting in global warming and climate change. In all the greenhouse gases, the proportion of carbon dioxide ($CO_2$) emitted is the greatest and around 76%, according to the statistics from the Intergovernmental Panel on Climate Change (IPCC) in 2013.

The $CO_2$ emissions are mainly from the consumption of fossil fuels such as coal, oil and natural gas which provide 85% of worldwide energy needs for human activities (BP, 2017). While in these $CO_2$ sources from energy consumptions, the burning of coal produces more $CO_2$ than oil or natural gas at the same equivalent electricity generated. Meanwhile, in 2016, the coal consumption consists of 28% of total world energy, while this number is 62% in China where the coal reserves are the third largest in the world. Table 1 and Fig. 1 are the comparisons of energy consumptions for China, USA and the total world in 2016 (BP, 2017).

**Table 1. Comparison of energy consumptions for China, USA and total world in 2016, unit: Million tonnes oil equivalent.**

| | Oil | Natural gas | Coal | Nuclear energy | Hydro-electricity | Renewables | Total |
|---|---|---|---|---|---|---|---|
| China | 578.7 | 189.3 | 1887.6 | 48.2 | 263.1 | 86.1 | 3053 |
| USA | 863.1 | 716.3 | 358.4 | 191.8 | 59.2 | 83.8 | 2272.7 |
| Total world | 4418.2 | 3204.1 | 3732 | 592.1 | 910.3 | 419.6 | 13276.3 |

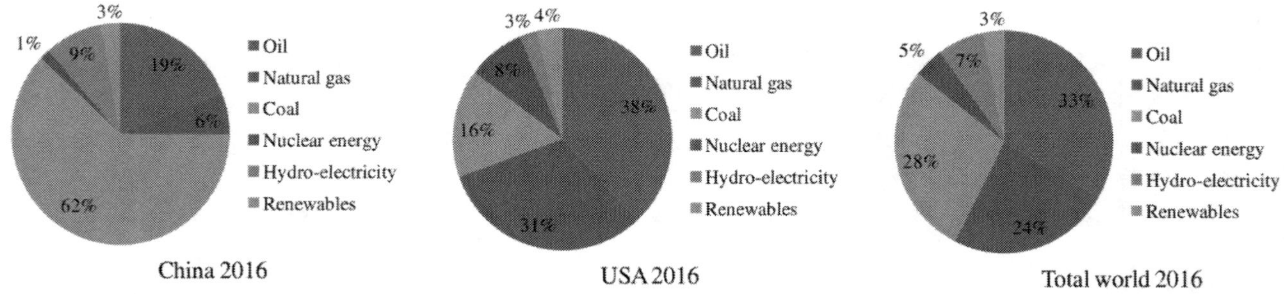


**Figure 1. Percentage of energy consumptions for China, USA and total world in 2016.**

In 2016, China's carbon emissions are estimated to be 9.1 billion tonnes, while the $CO_2$ emissions will further increases with the development of economy. Analyzing the $CO_2$ emission intensity, which is the ratio of $CO_2$ emitted to gross domestic product (GDP), can help us to find the measures to deal with $CO_2$ emissions and meanwhile develop the economy. The spatial distribution of $CO_2$ emission intensity is depicted in Fig. 2 for different provinces in China (Zhao et al., 2014).

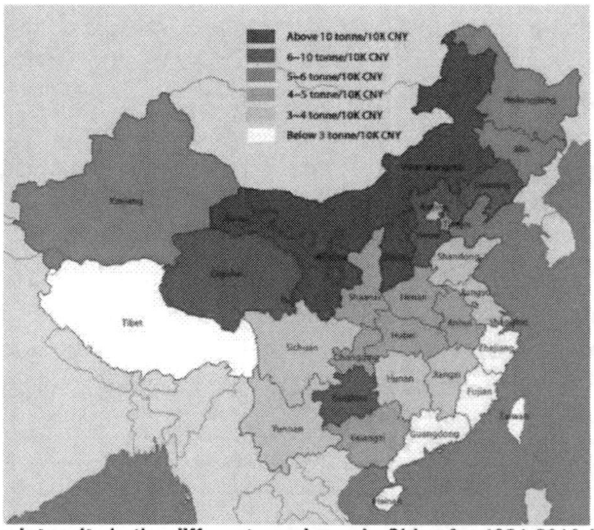

**Figure 2. $CO_2$ emission intensity in the different provinces in China for 1991-2010 (Zhao et al., 2014).**

The dark red color means the more intensive $CO_2$ emitted through human activities as per GDP produces. The areas with the highest $CO_2$ emission intensity are located in the north part in the Ordos Basin which is ranked as the second largest sedimentary basin in China. It is found that the coal resources are as large as 3.98Tt in this region which accounts for around 39% of China's total coal reserves (Wang et al., 2017). Currently, there are 6 coal mines and this basin is therefore the largest $CO_2$ emission area in China.

Efforts towards reduction of coal consumptions and improvement of energy efficiency are therefore the key in our pursuit of sustainability excellence. The carbon capture utilization and storage (CCUS) project is an excellent option to greatly reduce $CO_2$ emissions. Especially for the Ordos Basin, there are great potentials to reduce $CO_2$ emissions while developing the local economy through the integrated $CO_2$ capture and enhanced oil recovery approach which is also the next generation of CCUS and $CO_2$-EOR project.

# Overview of integrated approach for $CO_2$ capture and EOR

## Advantages for integrating CCUS and EOR

Previous CCUS project and most of other international projects for $CO_2$ capture and EOR are usually separated. The distance between $CO_2$ capture and the $CO_2$-EOR sites are very long. For example, in the Weyburn project in Canada, the $CO_2$ is captured in North Dakota in USA and transported through pipelines with the distance around 325 kilometers (Lotz and Brent, 2008). Such long distance between $CO_2$ source and $CO_2$-EOR limits the integrated implementation of CCUS and EOR application.

There is a great advantage to implement this integrated approach for Yanchang Petroleum Group in the Ordos Basin in China. In the recent years, several coal to chemical plants were built in this area and an extremely large amount of $CO_2$ is emitted. Meanwhile, the Ordos Basin not only has China's one-third of coal resources but also is the fourth largest oil and gas basin. The distance between coal to chemical plants and oil fields is very short, ranging from 10 to 140 kilometers, as shown in Fig. 3 (Wang et al., 2017).

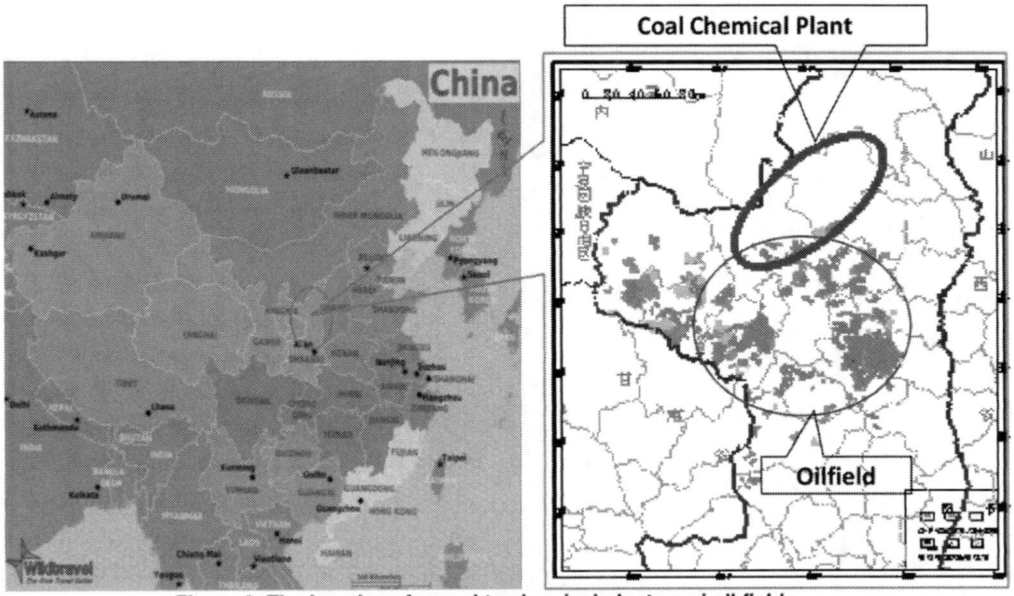


Figure 3. The locations for coal to chemical plants and oil fields.

In addition, because of the low and ultra-low permeability of the oil reservoirs and the lack of water in this area, a large amount of $CO_2$ is needed for maintaining reservoir pressure and enhancing oil recovery. Most importantly, the Yanchang Petroleum Group has the right for both coal and oil resources in Ordos Basin. The design of $CO_2$ capture and CCUS project can take full advantage of the $CO_2$-EOR and therefore implement the integrated approach by sequestrating $CO_2$ and improving oil recovery at the same time.

## Current coal to chemicals plants

The Yanchang Petroleum Group has developed modern coal to chemicals plants by comprehensively co-utilizing coal, oil and gas in Ordos Bain. Through coal gasification and direct liquefaction, the coal is converted to the production such as methanol, gasoline and diesel, acetic acid, olefin, aldehyde, ketone, acid, dimethylether and so on. A diagram of the modern coal to chemicals production line is shown in Fig. 4.

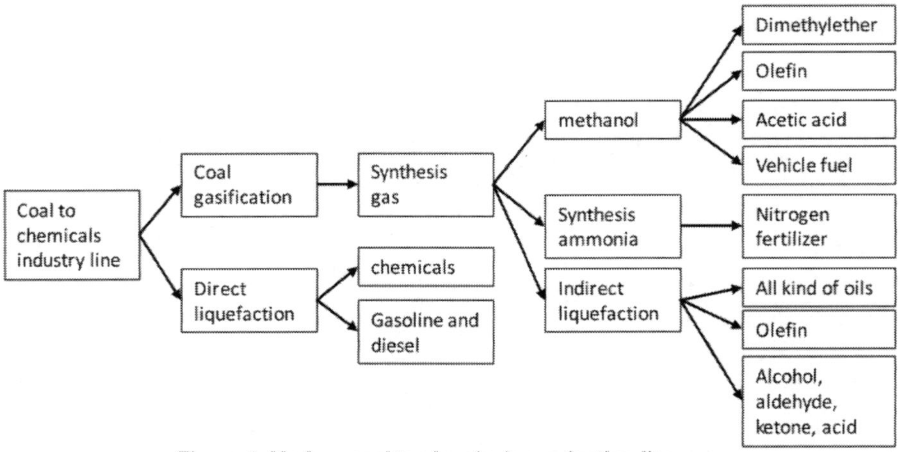


**Figure 4. Modern coal to chemicals production line.**

Two $CO_2$ capture plants were built corresponding to two coal to chemicals plants in Yanchang Petroleum Group. The first one has the capacity of 50,000 tonnes/year $CO_2$ capture. It was built in 2012 in order to reduce the $CO_2$ emissions from the coal to chemicals plant located in Yulin of Shaanxi province which has the capacity of 30,000 tonnes/year production of acetic acid. The $CO_2$ is mainly captured from fuel gas which usually has higher concentration of $CO_2$ in coal to chemicals processes. The 52,000 tonnes of $CO_2$ were captured in 2015 with 80% of mole fraction captured through low-temperature adsorption in methanol wash process.

Another $CO_2$ plant is designed to capture 360,000 tonnes/year. The $CO_2$ source is mainly from the second energy to chemicals plant operated by Yanchang Petroleum Group. The products include 2-ethyl hexanol (0.2 Million tonnes, Mt), polyethylene (0.25 Mt), polypropylene (0.45 Mt), and oils (0.6 Mt).

# $CO_2$-EOR projects

## Screen criteria of pilots

The screen criteria have been widely studied for the $CO_2$-EOR since the first commercial implementation in 1972 (Bachu, 2016). A large number of $CO_2$-EOR pilot tests were reported (Brock and Bryan, 1989; Enick et al., 2012; Manrique et al., 2010). Technically, the recognized criteria for the successful operations of $CO_2$-EOR are given as follows.

**Table 2. Screen criteria for $CO_2$-EOR**

| Factors | $CO_2$ Miscible flooding (Brashear and Kuuskraa, 1978) | $CO_2$ immiscible flooding (Klins and Farouq Ali, 1982) |
|---|---|---|
| Oil viscosity, mPa.s | <12 | 100-1000 |
| Oil density, kg/m$^3$ | <876.2 | 904-1000 |
| Formation depth, m | >900 | >700 |
| Reservoir pressure, MPa | >10 | >7 |
| Reservoir temperature, °C | NC | NC |
| Average permeability, mD | NC | NC |
| Initial oil saturation, Soi, % | >25 | >50 |
| Porosity×Soi, f | >0.04 | >0.08 |
| Positive conditions | Thin layer, high dip angle, no bottom or blank water, small Kv | |
| Negative conditions | High heterogeneity, strong fractures, gas cap, active bottom and blank water | |

Based on the previous screening criteria for the successful operations of $CO_2$-EOR, we established a selection system with 13 major factors. Each factor has a certain range of variations and typical values are used to form a combination of all the possible options. Table 3 lists all the 13 factors and values of them. By conducting sensitive analysis, four major factors are found to be very sensitive to the oil recovery including formation depth, thickness, initial oil saturation and sedimentary rhythm. Several major factors are not sensitive to the oil recovery such as dip angle, reservoir temperature, Ky/Kx and coefficient of variation. Moreover, three large categories are also listed as the first level for geological conditions, reservoir fluid properties and field development. The weight of geological conditions is relatively large. A second level of weight under each category is also given in Table 3. In addition, the main effects of all the factors on final oil recovery are carefully analyzed.

**Table 3. Selection criteria of oil reservoirs for $CO_2$-EOR.**

| No. | Factors | Values | Sensitivity | Weight level I | Weight level II | Conclusions |
|---|---|---|---|---|---|---|
| 1 | Sedimentary rhythm | Positive, negative | Strong | 0.65 | 0.16 | Positive rhythm is more positive effect on oil recovery |
| 2 | Initial oil saturation, Soi, % | 30, 40, 50, 60, 65 | Strong | | 0.16 | Larger Soi has positive effect on oil recovery |
| 3 | Effective thickness of oil formation, m | 5, 7.5, 10, 13, 15, 20, 30 | Strong | | 0.16 | Negative effect on oil recovery |
| 4 | Formation pressure, MPa | 3, 5, 7, 9, 12, 18, 25 | Strong | | 0.16 | Higher pressure has positive effect on oil recovery |
| 5 | Average permeability, mD | 0.1, 0.5, 1.5, 10, 50 | Yes | | 0.10 | Higher permeability has positive effect at an earlier time but negative effect later |
| 6 | Kv/Kh | 0.001, 0.01, 0.1, 0.3, 0.5 | Yes | | 0.10 | More interlayers have positive effect on oil recovery |
| 7 | Coefficient of variation | 0.1, 0.2, 0.35, 0.5, 0.7 | No | | 0.10 | Larger values have negative effect on oil recovery |
| 8 | Dip angle of reservoir, ° | 5, 10, 15, 20, 25, 35 | Minor | | 0.06 | Larger values have positive effect on oil recovery |
| 9 | Ky/Kx | 5, 10, 20, 50, 100 | Minor | | 0.04 | Larger values have slightly positive effect on oil recovery |
| 10 | Temperature, ℃ | 19, 30, 40, 55, 75 | No | | 0.04 | Nearly no effect on oil recovery |
| 11 | Oil viscosity | calculated by 5 components | Yes | 0.25 | 0.50 | Higher values have negative effect on oil recovery |
| 12 | Well pattern | 5-spot, 7-spot, inverse 7-spot, 9-spot, inverse 9-spot | Yes | 0.10 | 0.50 | 5-spot, inverse 7-spot and inverse 9-spot can achieve the higher oil recovery |
| 13 | Well space, m | 100, 200, 300, 400, 500 | Yes | | 0.50 | Larger values have negative effect on oil recovery |

Not all of the field data can be easily obtained and there may be large uncertainties for the data from fields. In all of the factors which may have impact on oil recovery of $CO_2$-EOR, six of them are selected and given weight in Table 4.

**Table 4. Final weight of different factors.**

| Factors | Weight | The worse value | The optimal value |
|---|---|---|---|
| Effective thickness of oil formation | 0.2592 | 30 | 1 |
| Formation depth | 0.2304 | 300 | 2500 |
| Average permeability | 0.1872 | 0.1 | 1.5 |
| Temperature | 0.0432 | 75 | 19 |
| Oil density | 0.14 | 1000 | 700 |
| Oil viscosity | 0.14 | 1000 | 1 |

These six factors are then normalized in terms of their weight. Assume there is $l$ number of reservoirs, each reservoir has m factors. The value of $j$ factor in the $k$ reservoir is $X'_{k,j}$. The normalized value is there,

$$X_{k,j} = \frac{|X'_{k,j} - X^*_{w,j}|}{|X^*_{o,j} - X^*_{w,j}|} \tag{1}$$

Where, $X^*_{o,j}$ is the optimal value and $X^*_{w,j}$ is the worse value. Therefore, the optimal and worse values of all six factors are given in Table 4.

Based on the above analysis and method, we evaluated all the 172 oil reservoirs in Yanchang oil fields for the applicability of $CO_2$-EOR. It is found that 87% of reservoirs can be used for $CO_2$-EOR, which accounts for 80% of all reserves in Yanchang Petroleum Group. There is therefore a great potential for $CO_2$-EOR and $CO_2$ sequestration.

## $CO_2$-crude oil properties

$CO_2$-EOR is either miscible or immiscible depending on the properties of mixture at certain reservoir pressure and temperature conditions. The different scenarios have significant influence on field performance and oil recovery. It is therefore necessary to investigate the $CO_2$-crude oil properties through experiments.

The bubble point pressure is first determined as shown in Fig. 5. With the increase of $CO_2$ concentration, the bubble pressure increases. The initial bubble point pressure of reservoir is 7.45MPa. It increases slowly when the $CO_2$ concentration is low, while it increases very fast once the $CO_2$ concentration is higher than 41%. The bubble point pressure is 25.25 MPa when the $CO_2$ concentration is 67.13%.

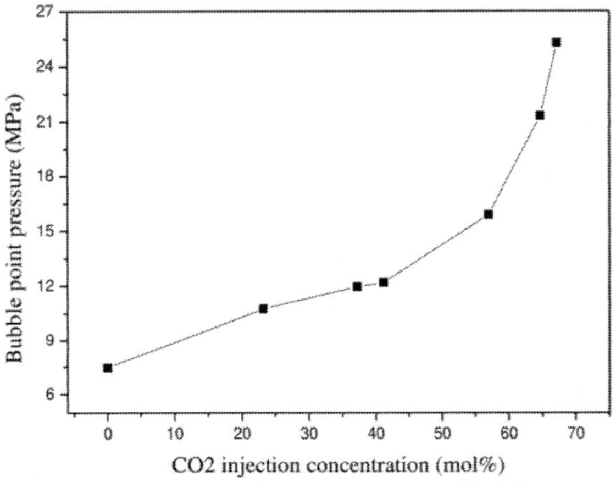


**Figure 5. The variations of bubble point pressure with $CO_2$ concentration.**

The density and volume coefficient of $CO_2$-crude oil mixture are also measured and the results are shown in Fig. 6. At the same $CO_2$ concentration, the density of fluids decreases as the pressure decreases. Once the reservoir pressure is below bubble point pressure, the density decreases much faster. Moreover, with the increase of $CO_2$ concentration, the trend of density of the mixture becomes smoothly. At the same reservoir pressure, the fluid density decreases with the increase of $CO_2$ concentration.

The volume coefficient versus pressure is also tested and shown in Fig. 6. With the increase of $CO_2$ concentration in the mixture, the volume coefficient increases. The value is 1.12 at reservoir pressure 30 MPa without any $CO_2$ injected. While when the $CO_2$ concentration is as high as 67.13%, the volume coefficient is 1.81 at the same pressure. This is therefore beneficial for the improvement of oil recovery in $CO_2$-EOR processes.

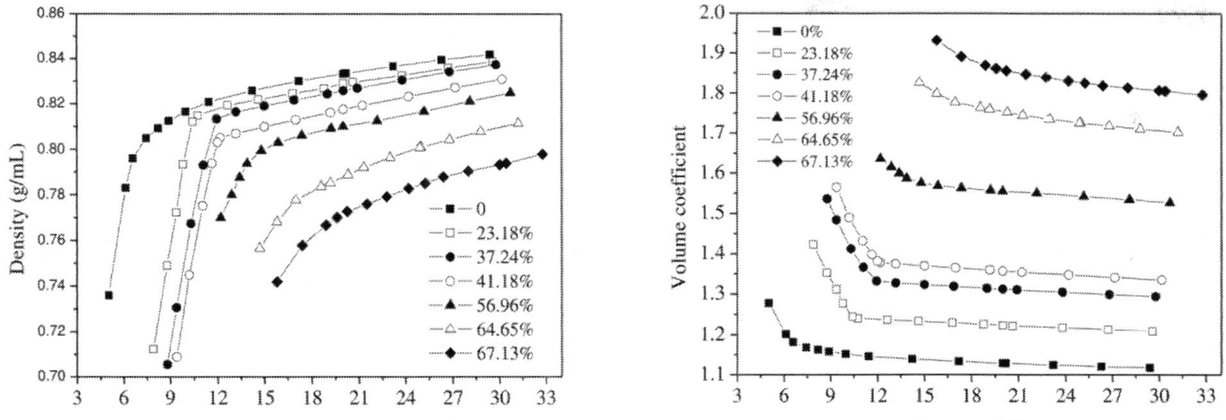


**Figure 6. The variations of mixture density and volume coefficient with pressure.**

The minimum miscibility pressure (MMP) is one of the most important factors affecting the oil recovery. Once the miscibility between $CO_2$ and oil is achieved, the surface tension tends to be zero and more $CO_2$ can dissolve into oil. A miscible displacement process takes place in $CO_2$-EOR (Yuan and Azaiez, 2015, 2014). The reduction of oil viscosity, improvement of its mobility and swelling effects are all beneficial for enhancing oil recovery (Huang et al., 2017; Or et al., 2016; Seyyedsar et al., 2017; Thomas, 2008; Yuan et al., 2017). A variety of methods have been used to determine the MMP (Wang et al., 2015). The interfacial tension vanish method is used in the present study and the $CO_2$-oil visualization before and after miscibility is shown in Fig. 7-a and b. Once the miscibility is achieved, there is no clear interface between them and the oil drop cannot keep its shape. The MMP is determined as 23.8 MPa.

In addition, the slim tube method is also used to determine MMP. A tube with 18 meters long which is packed with 80-100 mesh sands is adapted. This method is regarded as more reliable than other methods (Nobakht et al., 2008). The MMP is 22.15 MPa which is slightly less than that measured by interfacial tension vanish method.

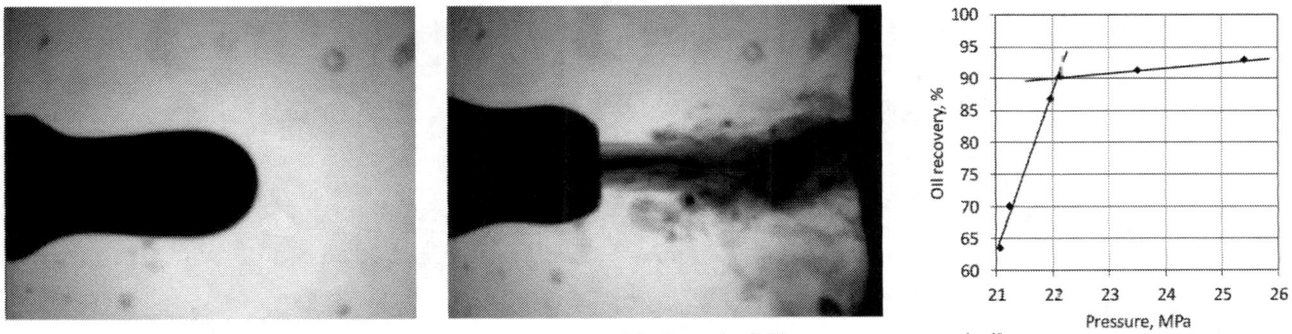


**a) before achieving miscibility** **b) after achieving miscibility** **c) oil recovery versus pressure**

**Figure 7. Experimental determination of minimum miscibility pressure.**

## $CO_2$-EOR field pilot tests

The first $CO_2$-EOR pilot test is conducted in Qiaojiawa 203 block from 2012. The average oil production is around 0.2 tonnes/day/well and the water cut is about 63.6%. However, the oil recovery at that time is only 1.72%. To improve oil recovery, this block is chosen as pilot test. There were 5 $CO_2$ injection wells and 21 oil wells. By the end of 2014, the total $CO_2$ injection is nearly 34,500 tonnes with the average 20 tonnes/day/well. The injection pressure is around 6-8 MPa. As of February 2016, there were 5 $CO_2$ injectors and 34 oil production wells. Compared with that of 2014, the total of 24,900 tonnes additional $CO_2$ were injected, while the average oil production is slightly reduced to about 19 tonnes/day/well. So far, the oil increased by $CO_2$-EOR is totally 2,200 tonnes, while a negligible amount of $CO_2$ was produced.

In December 2014, the first well Wu 38-28 in the second field pilot Wuqi Yougou began to injection $CO_2$ for oil recovery. After the first month $CO_2$ injection, the oil production of 16 oil wells was found to increase. As to the June 2016, the total amount of $CO_2$ injected is 9,221.2 tonnes with the average injection rate 19.3 tonnes and injection pressure 10.3 MPa. Currently, there are 5 $CO_2$ injectors and 18 oil producers in this pilot test. The pilot test also shows that the liquid rate increases by 14.0%, while the oil production rate is improved by 23.32%. Meanwhile, because of the $CO_2$ injection, the average water cut is reduced by 5.5%. Numerical simulation indicates that the oil recovery with $CO_2$-EOR can be improved by 11.8% compared with water flooding.

## Summary

The Yanchang Petroleum Group provides a great advantage to implement the integrated $CO_2$ CCUS and oil recovery approach. This is also the first integrated in China by directly capturing $CO_2$ from coal to chemicals plant and injecting into the nearby reservoirs for enhanced oil recovery.

The development of coal to chemical enables the co-utilization of coal, oil and natural gas. A large amount of useful chemical products have been produced including methanol, acetic acid, olefin, aldehyde, ketone, acid, dimethylether and all kinds of oil. Two coal to chemicals and energy to chemicals plants are now in operation. In addition, two $CO_2$ capture plants with capacity of 50,000 and 360,000 tonnes per year are now providing the $CO_2$ sources needed for enhanced oil recovery in the Yanchang oil fields. A series of processes have been optimized through the $CO_2$ capture project management, location of $CO_2$ capture plants and EOR onsite as well as connections of different projects. For one thing, the efficiency for energy consumptions is improved to be 8.9% higher than world average level. For another, the cost for $CO_2$ capture is around \$17.5/tonne which is much cheaper than most of the $CO_2$ capture projects in the world (Wang et al., 2017). The shortest distance for $CO_2$ transportation is 10 kilometers which is much shorter than the other CCUS projects such as Weyburn project with the distance 325 kilometers. This also greatly reduces the $CO_2$ transportation cost to be only \$2.58/tonne. Meanwhile, it provides very stable $CO_2$ supply for field EOR projects.

Research shows that around 87% of reservoirs in Yanchang oilfields are suitable for the $CO_2$-EOR with the estimated geological reserves $1.74\times10^9$ tonnes. A rough evolution of the cumulative $CO_2$ storage would be $6.57\times10^8$ tonnes. If considering the saline aquifer in this area, the amount of $CO_2$ for sequestration is probably 10 billion tones or even more in the Ordos Basin.

## Acknowledgements

The authors would like to acknowledge the support from Yanchang Petroleum Group Co., Ltd and the University of Regina. The authors also appreciate the Yanchang Petroleum Group for granting permission for publication of this paper.

## Reference

Bachu, S., 2016. Identification of Oil Reservoirs Suitable for $CO_2$-EOR and $CO_2$ Storage (CCUS) Using Reserves Databases, with Application to Alberta, Canada. Int. J. Greenh. Gas Control 44, 152–165.

BP, 2017. BP Statistical Review of World Energy June 2017.

Brashear, J.P., Kuuskraa, V.A., 1978. The Potential and Economics of Enhanced Oil Recovery. J. Pet. Technol. 30, 1231–1239.

Brock, W., Bryan, L., 1989. Summary Results of $CO_2$ EOR Field Tests, 1972-1987. In: SPE Joint Rocky Mountain Region/Low Permeability Reservoirs Symposium and Exhibition. Society of Petroleum Engineers, Denver, Colorado, USA.

Enick, R.M., Olsen, D., Ammer, J., Schuller, W., 2012. Mobility and Conformance Control for $CO_2$ EOR via Thickeners , Foams , and Gels - A Literature Review of 40 Years of Research and Pilot Tests. In: Eighteenth SPE Improved Oil Recovery Symposium. Society of Petroleum Engineers, Tulsa, Oklahoma, USA.

Huang, T., Zhou, X., Yang, H., Liao, G., Zeng, F., 2017. $CO_2$ Flooding Strategy to Enhance Heavy Oil Recovery. Petroleum 3, 68–78.

Klins, M.A., Farouq Ali, S.M., 1982. Heavy Oil Production By Carbon Dioxide Injection. J. Can. Pet. Technol. 21, 64–72.

Lotz, M., Brent, a C., 2008. A review of Carbon Dioxide Capture and Sequestration and the Kyoto Protocol's Clean Development Mechanism and Prospects for Southern Africa. J. Energy South. Africa 19, 13–24.

Manrique, E., Thomas, C., Ravikiran, R., Izadi, M., Lantz, M., Romero, J., Alvarado, V., 2010. EOR: Current Status and Opportunities. In: SPE Improved Oil Recovery Symposium. Society of Petroleum Engineers, Tulsa, Oklahoma, USA.

Nobakht, M., Moghadam, S., Gu, Y., 2008. Determination of $CO_2$ Minimum Miscibility Pressure from Measured and Predicted Equilibrium Interfacial Tensions. Ind. Eng. Chem. 47, 8918–8925.

Or, C., Sasaki, K., Sugai, Y., Nakano, M., Imai, M., 2016. Swelling and Viscosity Reduction of Heavy Oil by $CO_2$-Gas Foaming in Immiscible Condition. SPE Reserv. Eval. Eng. 19, 294–304.

Seyyedsar, S.M., Farzaneh, S.A., Sohrabi, M., 2017. Investigation of Low-Density $CO_2$ Injection for Enhanced Oil Recovery. Ind. Eng. Chem. Res. 56, 5443–5454.

Thomas, S., 2008. Enhanced Oil Recovery-An Overview. Oil Gas Sci. Technol. 63, 9–19.

Wang, J., Dong, M., Li, Y., Gong, H., 2015. Prediction of Nitrogen Diluted $CO_2$ Minimum Miscibility Pressure for EOR and Storage in Depleted Oil Reservoirs. Fuel 162, 55–64.

Wang, X., Zeng, F., Gao, R., Zhao, X., Hao, S., Liang, Q., Jiang, S., 2017. Cleaner Coal and Greener Oil Production: An Integrated CCUS Approach in Yanchang Petroleum Group. Int. J. Greenh. Gas Control 62, 13–22.

Yuan, Q., Azaiez, J., 2014. Miscible Displacements in Porous Media with Time-Dependent Injection Velocities. Transp. Porous Media 104, 57–76.

Yuan, Q., Azaiez, J., 2015. Inertial Effects in Cyclic Time-Dependent Displacement Flows in Homogeneous Porous Media. Can. J. Chem. Eng. 93, 1490–1499.

Yuan, Q., Zhou, X., Zeng, F., Knorr, K.D., Imran, M., 2017. Effects of Concentration-Dependent Diffusion on Mass Transfer and Frontal Instability in Solvent-Based Processes. In: SPE Canada Heavy Oil Technical Conference. Society of Petroleum Engineers, Calgary, Alberta, Canada.

Zhao, X., Burnett, J.W., Fletcher, J.J., 2014. Spatial Analysis of China Province-Level $CO_2$ Emission Intensity. Renew. Sustain. Energy Rev. 33, 1–10.

# $CO_2$ Storage Resource and Reservoir Feasibility Assessment for Deep Saline Cambrian-Ordovician Formations in Eastern Ohio

**Isis Fukai**, Priya Ravi Ganesh, Mackenzie Scharenberg, Joel Main

July 20th, 2017

**Carbon Management Technology Conference**

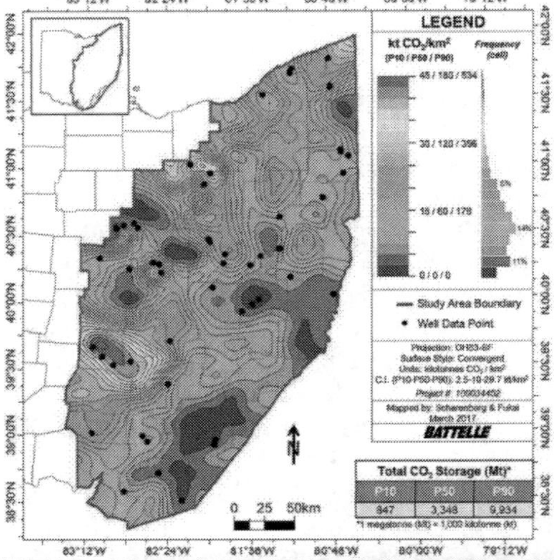


1 **BATTELLE**

# Acknowledgments

*This work is funded by the Ohio Development Services Agency – The Ohio Coal Development Office and the U.S. Department of Energy - National Energy Technology Laboratory.*

*Thanks to colleagues at NETL for guidance with $CO_2$-SCREEN; Coauthors & team members who contributed to the research*

Ohio | Development Services Agency

U.S. DEPARTMENT OF ENERGY | NATIONAL ENERGY TECHNOLOGY LABORATORY

MRCSP MIDWEST REGIONAL CARBON SEQUESTRATION PARTNERSHIP

2 **BATTELLE**

# Project Overview

## Problem

- Ohio responsible for ~25% of total $CO_2$ emissions reported in the United States[1]
- Geologic resources available in-state for $CO_2$ storage & climate change mitigation not well defined

## Objective

Characterize deep saline formations & overlying caprocks to help define geologic storage framework for OH

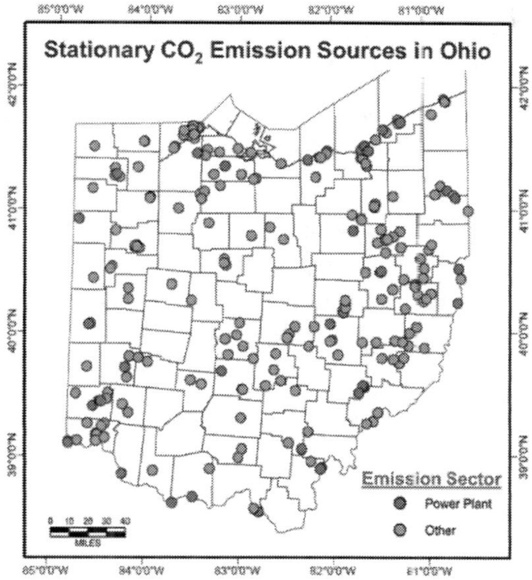


**Map showing the locations of stationary $CO_2$ emission sources in Ohio (EPA GHGRP, 2016; Battelle, 2017). Total reported 2015 emissions: 113 megatonnes $CO_2$**

1. Based on 2015 emissions reported to EPA GHGRP (2016)



BATTELLE

# Project Overview: Organization & Approach

Project divided into five main tasks

**Focus: Static Storage Resource Assessment**

Formation-scale evaluation

Component of Reservoir Feasibility Analysis

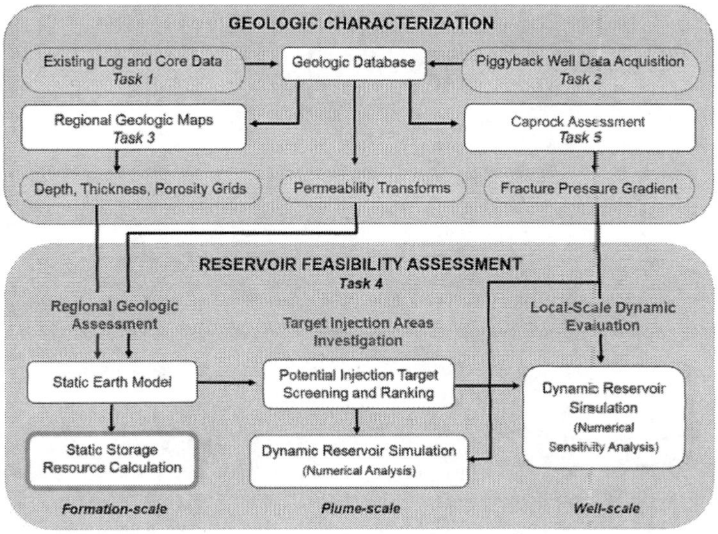


**Schematic showing overall project approach and task organization, with an emphasis on components of the Reservoir Feasibility Assessment (Battelle, 2017).**



BATTELLE

# Study Area & Formations of Interest

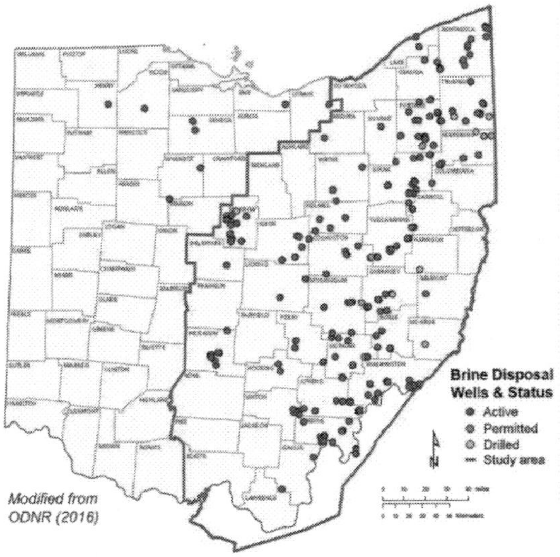


**Study Area:** 23,643 mi$^2$ (61,236 km$^2$) area in eastern Ohio comprising corridor of brine disposal wells

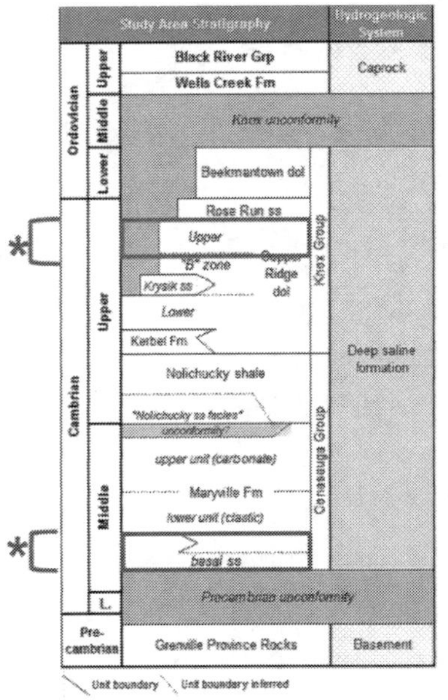


**Saline formations (reservoirs):** brine-saturated sandstones and dolomites

5 **BATTELLE**

# Formations of Interest

## Upper Copper Ridge Dolomite

**Thin-section photomicrograph in plane-polarized light showing an Upper Copper Ridge dolomite sample from the south-central portion of the study area (blue: porosity).**

## Basal Cambrian Sandstone

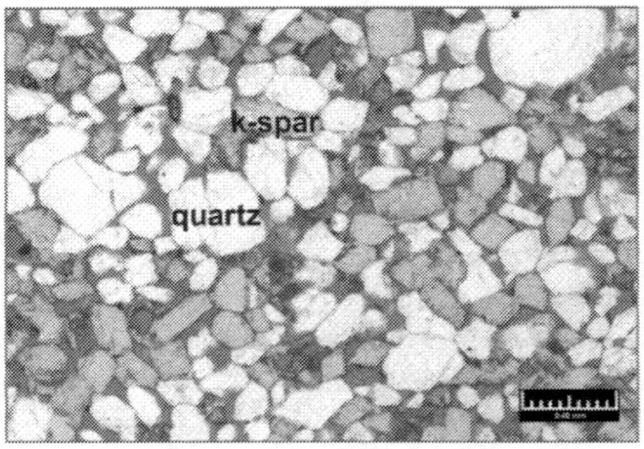


**Thin-section photomicrograph in plane-polarized light showing a basal Cambrian sandstone sample from the south-central portion of the study area (blue: porosity).**

| Formation | Area (mi$^2$)* | Avg. Depth (ft.) | Avg. Gross Thickness (ft.) | Avg. Porosity (%) |
|---|---|---|---|---|
| Upper Copper Ridge dolomite | 23,643 | 5,157 | 184 | 5 |
| basal Cambrian sandstone | 23,643 | 5,834 | 131 | 8 |

* Specific to the eastern Ohio study area

6 **BATTELLE**

# Static Storage Resource Assessment

**Static (volumetric) method:** Uses subsurface pore volumes & in-situ fluid saturation to estimate an equivalent quantity of $CO_2$ that could potentially be stored

Total Pore Volume = **Theoretical Maximum Storage Resource**

Storage Efficiency

Net Pore Volume = **Prospective Storage Resource**[1]

**DOE-NETL Storage Resource Equation**[1]

$$G_{CO2} = \underbrace{A_t\, h_g\, \phi_t\, \rho_{CO2res}}_{\text{Theoretical Maximum Storage Resource}}\, E_{saline}$$

$G_{CO2}$: Prospective Storage Resource

| Parameter[1] | Definition |
|---|---|
| Total Area ($A_t$) | Total area of formation |
| Gross Thickness ($h_g$) | Gross formation thickness |
| Total Porosity ($\phi_t$) | Total formation porosity (interconnected+isolated+clay-bound) |
| $\rho_{CO2res}$ | Density of $CO_2$ at reservoir temperature and pressure |
| Storage Efficiency ($E_{saline}$) | Fraction of pore volume that is technically accessible for $CO_2$ storage |

1. DOE-NETL, 2010; 2012; Goodman et al., 2011; 2016



# Static Storage Resource Calculation

**Two grid-based methods:**

1. Static Earth Model (SEM)
   (3D, high resolution, deterministic)
2. $CO_2$-SCREEN[1]
   (low resolution, stochastic)

Same input data used for both grid calculations

Results compared to assess effects of model resolution and calculation method on Storage Resource outcomes

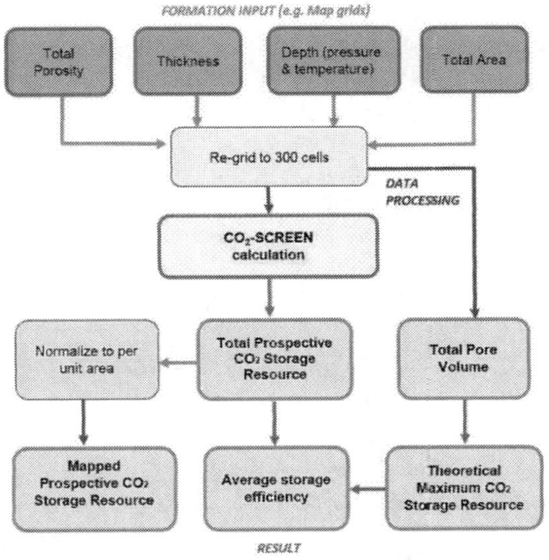


**Schematic showing the workflow for the $CO_2$-SCREEN calculation, with 2D maps grids from the SEM re-gridded to the required resolution and then used to calculate and map the Prospective Storage Resource (Battelle, 2017).**

1. Sanguinito et al., 2016; https://edx.netl.doe.gov/organization/co2-screen.

# Static Storage Resource Calculations: Data Input

New and existing data compiled in previous tasks

- Well logs from 413 wells in the study region: basic and advanced wireline logs
- Core data from ~15 wells

Data used to construct and populate Static Earth Model (SEM) for Storage Resource calculations

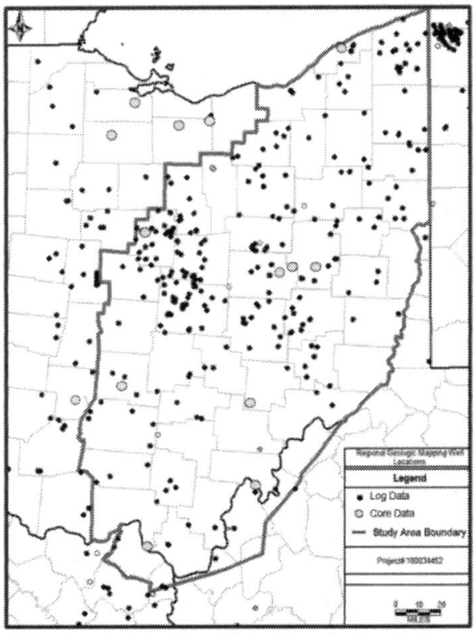


Locations of wells with log data and core data used to construct structure, thickness, and porosity maps for the formations of interest in the study area (Battelle, 2017).

9

BATTELLE

# Static Storage Resource Calculation: Static Earth Model (SEM)

**SEM**: 3D model of subsurface geology and pore volume for Cambrian-Ordovician deep saline formations in the study area

- Formation Structure
  (depth, thickness, lateral continuity)
- Petrophysical Properties
  (porosity, permeability)

2D map grids from SEM coarsened & used as input for $CO_2$-SCREEN

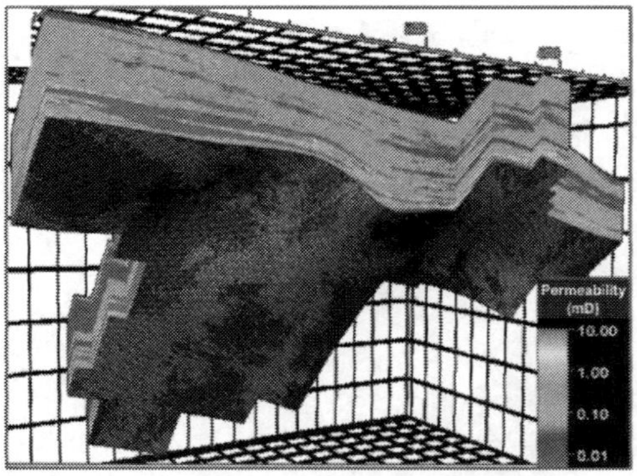


Portion of the SEM developed for the eastern OH study area clipped to show the basal Cambrian sandstone interval at a selected site. The entire SEM has a total of 145,413,098 grid cells with X-Y-Z dimensions of 1,875 x 1,875 x 3ft (Battelle, 2017).

$CO_2$ Storage Resource calculated deterministically directly in model using DOE-NETL equation & $E_{saline}$ from $CO_2$-SCREEN

$$\underbrace{G_{CO2}}_{\text{Prospective Storage Resource}} = \underbrace{A_t\, h_g\, \phi_t\, \rho_{CO2res}}_{\text{Theoretical Maximum Storage Resource}}\, E_{saline}$$

10

BATTELLE

# Static Storage Resource Calculation:
## DOE-NETL $CO_2$-SCREEN Tool[1]

### INPUT

| Storage Efficiency Input | Auto-populated | | User Specified | | |
|---|---|---|---|---|---|
| | $P_{10}$ | $P_{90}$ | $P_{10}$ | $P_{90}$ | $X_{10}$ |
| Net-to-Total Area | 0.20 | 0.80 | 0 | 0 | -1.39 |
| Net-to-Gross Thickness | 0.21 | 0.76 | 0 | 0 | -1.32 |
| Effective-to-Total Porosity | 0.62 | 0.78 | 0 | 0 | 0.49 |
| Volumetric Displacement | 0.18 | 0.63 | 0 | 0 | -1.52 |
| Microscopic Displacement | 0.39 | 0.82 | 0 | 0 | -0.45 |

**Physical Parameters**

Mean and standard deviation values for each grid cell

| Grid cell # | Area* (km²) | Gross Thickness* (m) | | Total Porosity*(%) | |
|---|---|---|---|---|---|
| | Mean | Mean | Std Dev | Mean | Std Dev |
| 1 | 109.2 | 97.1 | 0.0 | 4.4 | 0.0 |
| 2 | 109.2 | 104.9 | 0.0 | 4.5 | 0.0 |
| 3 | 109.2 | 116.6 | 0.0 | 4.1 | 0.0 |
| 4 | 109.2 | 135.1 | 0.0 | 3.8 | 0.0 |
| 5 | 63.8 | 157.2 | 0.0 | 2.9 | 0.0 |
| 6 | 109.3 | 76.4 | 0.0 | 4.3 | 0.0 |
| 7 | 109.2 | 92.6 | 0.0 | 5.2 | 0.0 |
| 8 | 109.2 | 103.4 | 0.0 | 5.6 | 0.0 |
| 9 | 109.2 | 110.1 | 0.0 | 4.8 | 0.0 |
| 10 | 109.2 | 124.0 | 0.0 | 3.7 | 0.0 |

1. Sanguinito et al., 2016; https://edx.netl.doe.gov/organization/co2-screen.
2. IEAGHG, 2009

### RESULTS

| | $GCO_2$ (Mt) | | |
|---|---|---|---|
| Grid Cell # | P10 | P50 | P90 |
| 1 | 2.4 | 8.1 | 19.5 |
| 2 | 2.7 | 9.0 | 21.6 |
| 3 | 2.8 | 9.2 | 22.2 |
| 4 | 2.9 | 9.7 | 23.5 |
| 5 | 1.6 | 5.2 | 12.5 |
| 6 | 1.8 | 6.1 | 14.8 |
| 7 | 2.7 | 9.1 | 21.9 |
| 8 | 3.3 | 10.9 | 26.3 |
| 9 | 3.0 | 10.0 | 24.1 |
| 10 | 2.7 | 8.8 | 21.2 |

| $CO_2$ Storage Statistics | P10 | P50 | P90 | Units |
|---|---|---|---|---|
| Summed $CO_2$ Total | 564 | 1873 | 4517 | Mt |
| | 0.6 | 1.9 | 4.5 | Gt |

- Grid-based input & results: 300 cells
- Storage Efficiency defaults[2]
- Probabilistic Prospective Storage Resource estimates: P10, P50, P90

11

BATTELLE

# Storage Resource Results: Upper Copper Ridge Dolomite

### Theoretical Maximum $CO_2$ Storage Resource

| | |
|---|---|
| Total Pore Volume | 195 km³ |
| Theoretical Maximum Storage Resource | 155 Gt* |

* Gt: gigatonnes: $10^9$ tonnes

### Storage Efficiency

| $E_{saline}$ | | |
|---|---|---|
| P10 | P50 | P90 |
| 0.7% | 2.2% | 5.3% |

### Prospective $CO_2$ Storage Resource

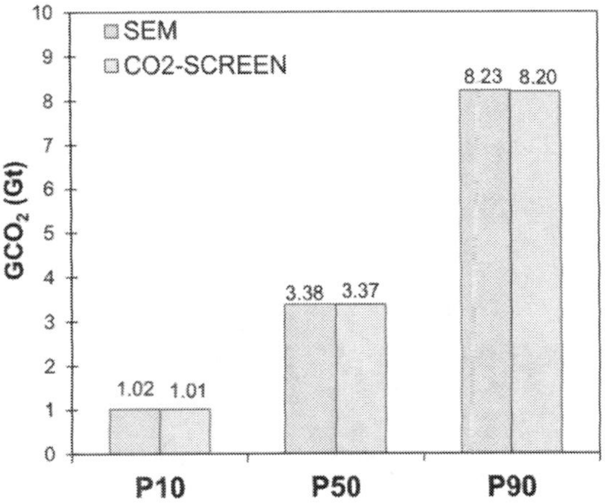


Less than 1% difference between SEM and $CO_2$-SCREEN results

12

BATTELLE

# Storage Resource Results: Upper Copper Ridge Geospatial Analysis

**Local Storage Resource Highs:**

- Northeastern Ohio
  (Lake, Ashtabula counties)
- South-central Ohio
  (Pickaway, Fairfield, Hocking counties)

Areas are constrained by data from three or more wells within < 20 mi (32 km) distance

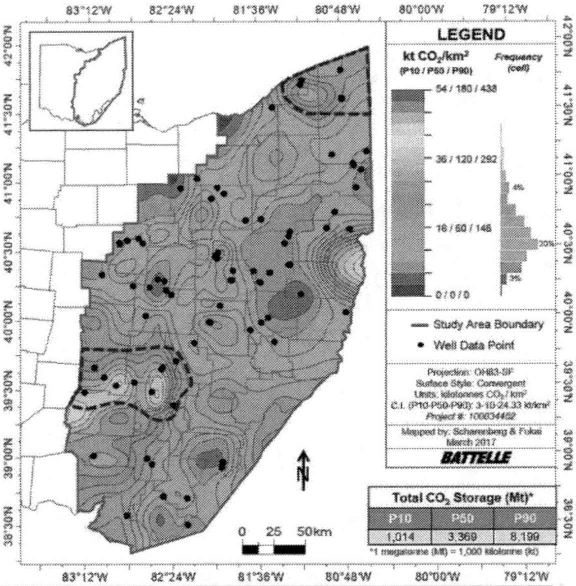


**Right: Map showing the Prospective $CO_2$ Storage Resource (in kt/km$^2$) for the Upper Copper Ridge Dolomite in the eastern Ohio study area (Battelle, 2017).**

13

BATTELLE

# Storage Resource Results: Basal Cambrian Sandstone

### Theoretical Maximum $CO_2$ Storage Resource

| | |
|---|---|
| Total Pore Volume | 138 km$^3$ |
| Theoretical Maximum Storage Resource | 111 Gt* |

* Gt: gigatonnes: $10^9$ tonnes

### Storage Efficiency

| $E_{saline}$ | | |
|---|---|---|
| **P10** | **P50** | **P90** |
| 0.8% | 3.0% | 8.9% |

### Prospective $CO_2$ Storage Resource

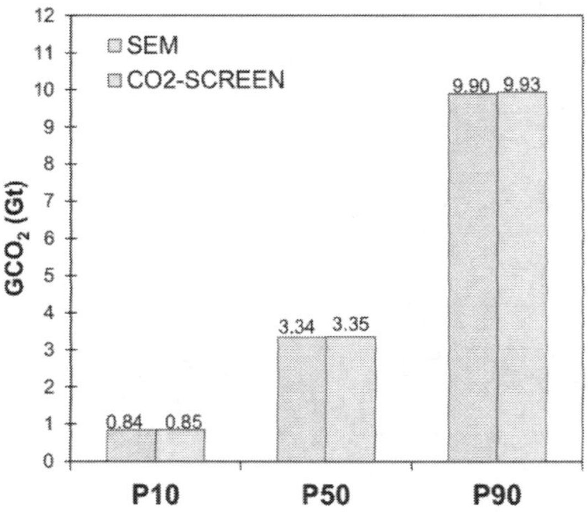


Less than 1% difference between SEM and $CO_2$-SCREEN results

14

BATTELLE

# **Storage Resource Results:** Basal Cambrian Sandstone Geospatial Analysis

**Local Storage Resource Highs:**

- Northeastern Ohio
  (Lake, Ashtabula, Cuyahoga counties)
- East-central Ohio
  (Medina, Wayne, Coshocton counties)
- South-central Ohio
  (Pickaway, Fairfield, Hocking counties)

Storage resource highs in the northeast and south-central areas similar to Upper Copper Ridge

Local highs used to delineate site-scale models for dynamic analysis

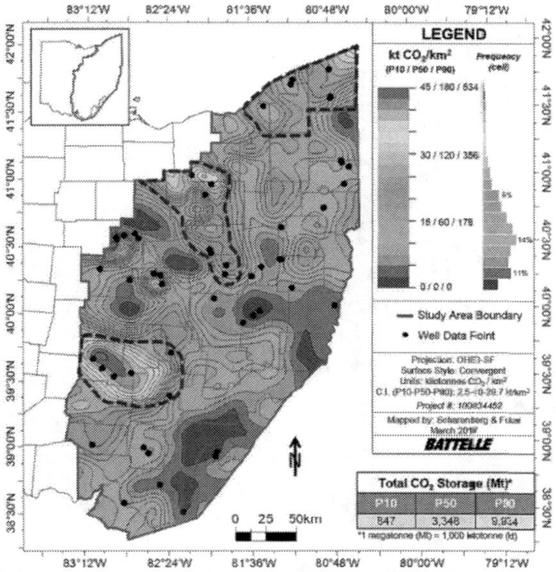


**Right: Map showing the Prospective $CO_2$ Storage Resource of the basal Cambrian sandstone in the eastern Ohio study area (Battelle, 2017).**



BATTELLE

# Summary & Conclusions

A systematic workflow of static modeling exercises is used to help define the regional geologic storage framework of the eastern Ohio study area

**Quantification of $CO_2$ Storage Resource for major deep saline formations**

- Theoretical Storage Resource: 111-155 Gt
- Prospective Storage Resource: 3.4 Gt (P50)
- $CO_2$ storage efficiency: 2.2 – 3.0% (P50)

Less than 1% difference between SEM & $CO_2$-SCREEN results

**Generation of Storage Resource maps**

- Spatial distribution of storage resource
- Help guide site selection for further analysis
- Potential for stacked storage in some areas

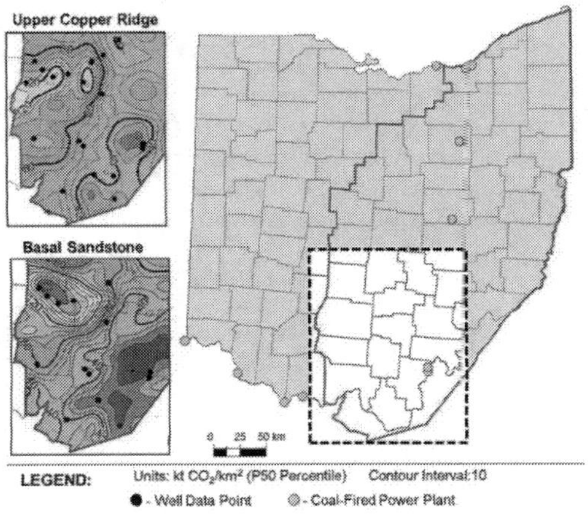




BATTELLE

BATTELLE

It can be done

800.201.2011 | solutions@battelle.org | www.battelle.org

# BACKUP SLIDES

# Storage Efficiency

$$E_{saline} = E_{An/At}\, E_{hn/hg}\, E_{\phi e/\phi t}\, E_v\, E_d$$

| Parameter | Symbol | Definition |
|---|---|---|
| Net-to-Total Area | $E_{An/At}$ | Fraction of the total area (map view) available for $CO_2$ storage |
| Net-to-Gross Thickness | $E_{hn/hg}$ | Fraction of the gross thickness available for $CO_2$ storage |
| Effective-to-Total Porosity | $E_{\phi e/\phi t}$ | Fraction of the total porosity that is interconnected, available for $CO_2$ storage |
| Volumetric Displacement Efficiency | $E_V$ | Combined fraction of net volume surrounding an injection well that can be contacted by $CO_2$ as a consequence of density, buoyancy effects |
| Microscopic Displacement | $E_d$ | The fraction of pore space occupied by immobile in-situ fluids |

# Input & well counts

| Formation | Depth | | Thickness | | Porosity | |
|---|---|---|---|---|---|---|
| | Avg (ft.) | Well Data Count (n) | Avg (ft.) | Well Data Count (n) | Avg (%) | Well Data Count (n) |
| Upper Copper Ridge | 5,157 | 252 | 184 | 202 | 2.1 | 84 |
| basal sandstone | 5,834 | 109 | 151 | 101 | 8.7 | 55 |



Add data classification or delete this footer

BATTELLE

# AUTOMOTIVE CO2 MITIGATION USING AN ONBOARD BOSCH REACTOR: ANALYSIS

MICHAEL BUCKNER, DR. THOMAS BRADLEY

COLORADO STATE UNIVERSITY

CMTC: JULY 20TH, 2017

## OBJECTIVE

- Determine the technical and economic requirements of an automotive $CO_2$ reduction system
  - Determine preferred carbon product and method of expulsion from vehicle
  - Determine cost contributors
- Conduct a techno-economic and lifecycle analysis of implementing an automotive Bosch reactor
  - Determine the optimal operating conditions
  - Determine the optimal design

# AUTOMOTIVE $CO_2$ REDUCTION

- Preferred carbon product
  - Non-corrosive
  - Cheap to dispose
  - Non-gaseous
- Cost Goals
  - Minimize the amount of hydrogen required to store onboard
  - Determine feasibility of regenerative and non-regenerative catalysts
  - Need to mitigate carbon "coking" on catalyst surface

# BOSCH REACTION

- Composed of 4 different reactions

| Reaction | Formula | $\Delta H$ |
|---|---|---|
| Reverse Water Gas Shift | $CO_2 + H_2 \rightarrow CO + H_2O$ | 41 kJ/mol |
| Hydrogenation | $CO + H_2 \rightarrow C(s) + H_2O$ | -131 kJ/mol |
| Boudouard | $2CO \rightarrow C(s) + CO_2$ | -172 kJ/mol |
| Sabatier | $CO_2 + 4H_2 \rightarrow CH_4 + 2H_2O$ | -165.4 kJ/mol |

## REACTION RATES (EQUILIBRIUM CONSTANTS)

- Developed by solving each species' transport equation simultaneously with its energy balance
- Temperature dependent equilibrium constants developed by comparing fugacity ratios to differences in Gibb's free energies at actual and reference conditions (chemical activity)

$$K_i = \exp\left[\frac{\lambda_1}{T^2} + \frac{\lambda_2}{T} + \lambda_3\right]$$

| Description | Reaction | Equation | $\lambda_1$ | $\lambda_2$ | $\lambda_3$ |
|---|---|---|---|---|---|
| 1. $CO_2$ Methanation | $CO_2 + 4H_2 \rightleftharpoons CH_4 + 2H_2O$ | Eq. 1 | -730,726.0 | 24,125.3 | -26.9616 |
| 2. CO Methanation | $CO + 3H_2 \rightleftharpoons CH_4 + H_2O$ | Eq. 2 | -538,798.1 | 28,062.7 | -30.7759 |
| 3. Reverse WGS | $CO_2 + H_2 \rightleftharpoons CO + H_2O$ | Eq. 4 | -191,928.1 | -3,937.4 | 3.8143 |
| 4. Hydrogenation | $CO + H_2 \rightleftharpoons C(s) + H_2O$ | Eq. 5 | -121,003.4 | 16,573.0 | -17.3858 |
| 5. Boudouard | $2CO \rightleftharpoons C(s) + CO_2$ | Eq. 6 | 70,924.7 | 20,510.4 | -21.2000 |

## REACTION RATES (GAS PHASE EQUILIBRIA)

- Methanation

$$r_1 = \frac{k_1 K_{CO_2,1} K_{H_2,1}^4 z_{CO_2} z_{H_2}^4 P^5}{\left(1 + K_{CO_2,1} P z_{CO_2} + K_{H_2,1} P z_{H_2}\right)^5}(1-\eta)$$

- RWGS

$$r_3 = k_3 \rho P \left(z_{CO_2} z_{H_2} K_3 - z_{CO} z_{H_2O}\right)$$

- Hydrogenation

$$r_4 = k_4 \rho \left(z_{CO} z_{H_2} K_4 P - z_{H_2O}\right)$$

- Boudouard

$$r_5 = \frac{\rho_{cat} \epsilon_o k_B^+ K_{CO,5}\left(P z_{CO} - \frac{1}{K_B^*}\frac{z_{CO_2}}{z_{CO}}\right)}{\left(1 + K_{CO,5} P z_{CO} + \frac{1}{K_{O,CO_2} K_{CO,5}}\frac{z_{CO_2}}{z_{CO}}\right)^2}$$

# PROPOSED REACTOR DESIGN

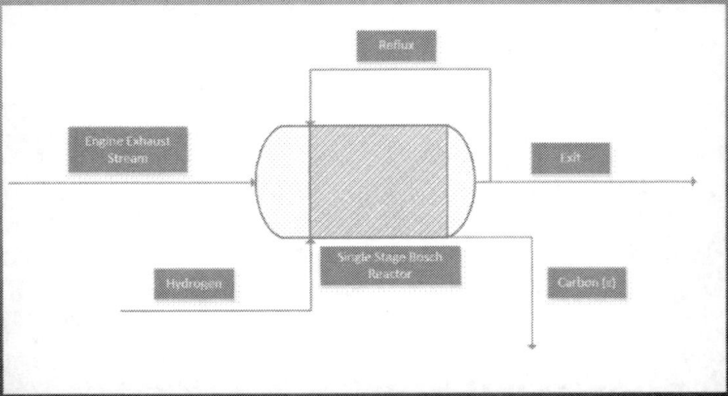


# CATALYST COKING

- Assume nickel catalyst is completely regenerative
  - Needs mechanical or magnetic agitation
- For ~105kg $CO_2$ in 1 tank of fuel
  - 1.38 kg steel wool required (packing density = 0.2g/cm$^3$)
    - Volume steel wool required to avoid coking disadvantage = 0.0069m$^3$

# PROCESS CONSIDERATIONS

- Reaction rate altering characteristics
  - Temperature
  - Fuel Air Equivalence Ratio
  - Reactor Volume
  - Reflux Rate
  - Catalyst

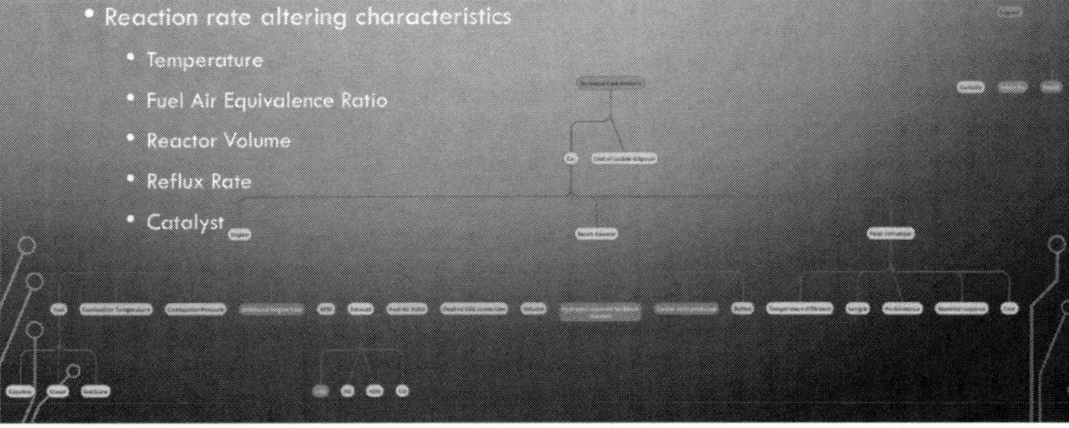

# ENGINE EXHAUST STREAM

- Higher Fuel/Air Equivalence Ratio → More $H_2$ produced in combustion
  - Less $H_2$ required in onboard tanks
  - Higher concentration of CO
  - Decreases fuel economy

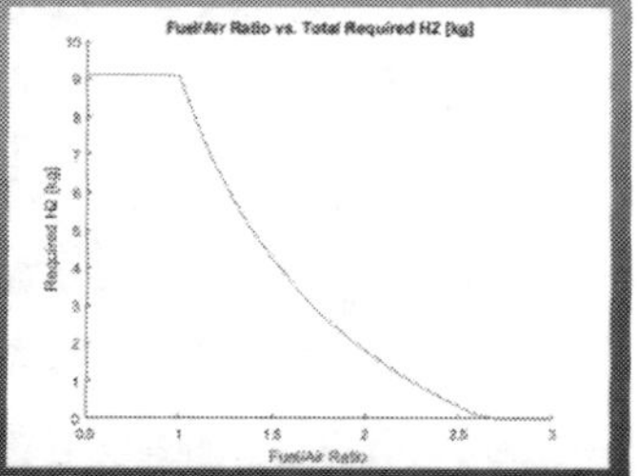


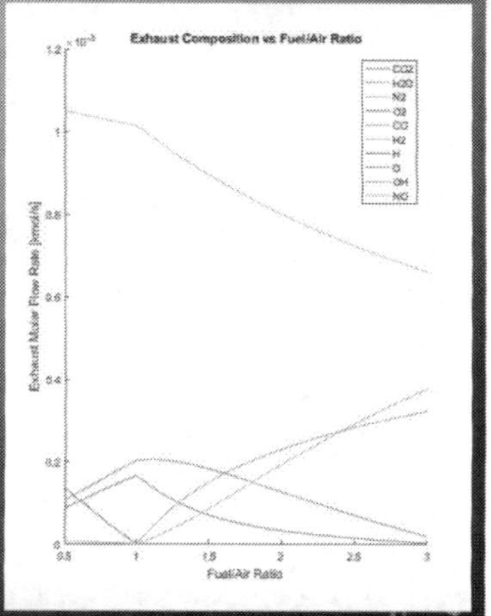

# EFFECT OF REACTOR VOLUME

- Higher Volume → Lower Conversion
  - Lower pressure (diffuse)
  - Lower reaction rate
- Lower Volume → Higher Conversion
  - Higher pressure
  - Higher activity
  - Higher reaction rate
  - Need to minimize exit pressure

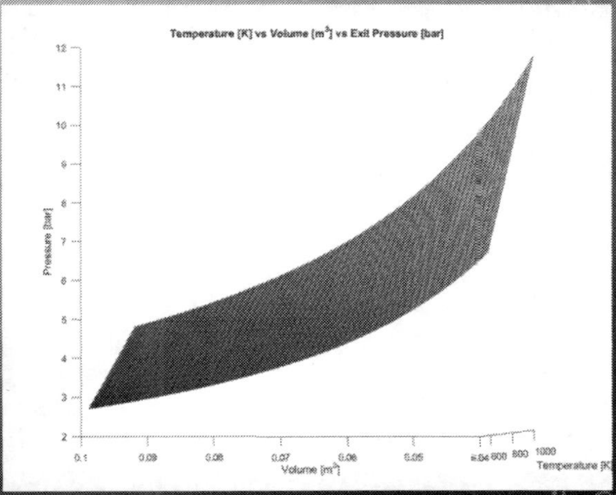


# EFFECT OF TEMPERATURE (RATES)

- Directly affects reaction rate constant
  - Higher T → RWGS
  - Lower T → Methanation

Fuel Air Ratio= 2.3

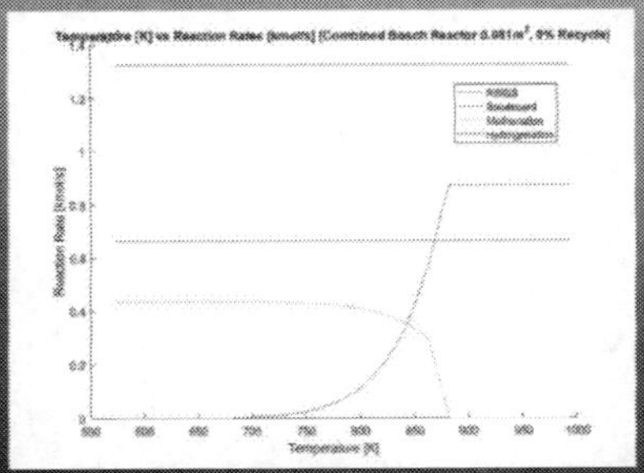


Fuel Air Ratio= 1.2

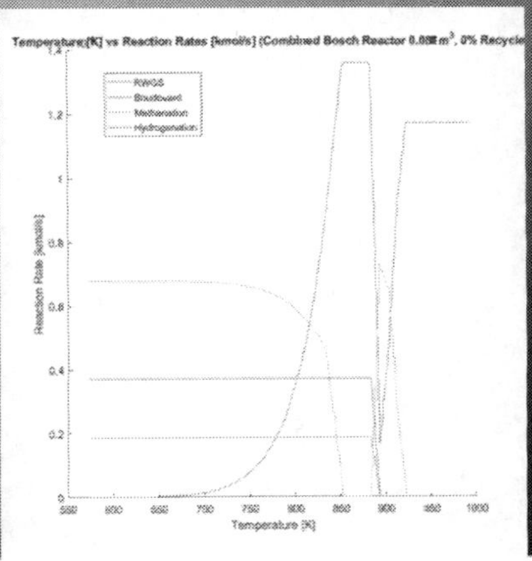

# EFFECT OF TEMPERATURE (CONVERSION)

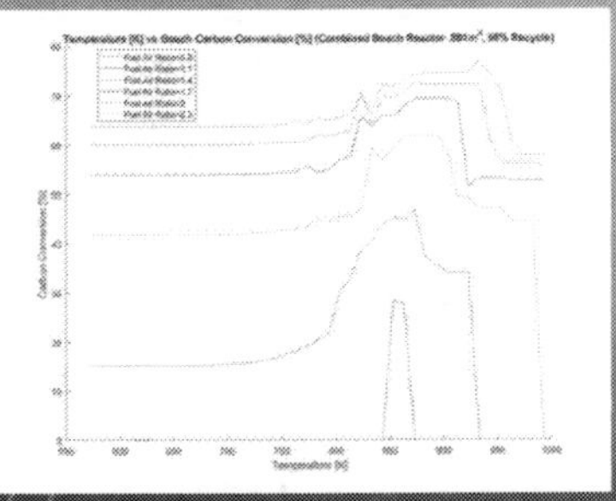

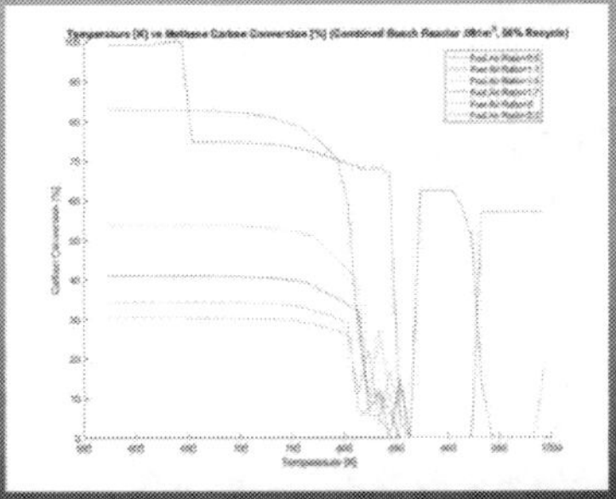

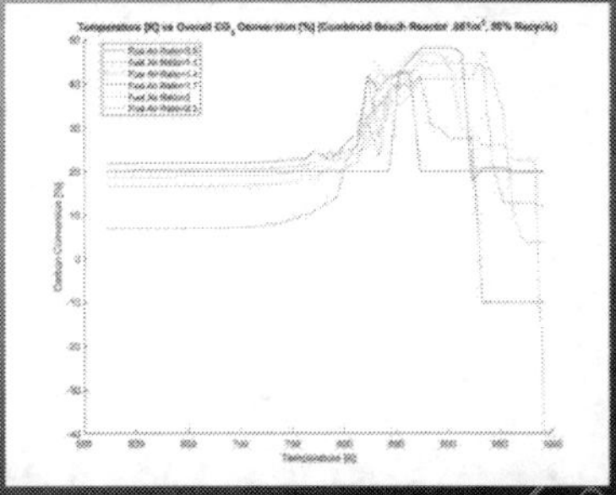

# WHY INCREASE FUEL AIR RATIO?

- Produce more CO and $H_2$ in combustion equilibrium
- Bypass RWGS reduction (endothermic)
- Increased power generation
  - Higher fuel content
  - Improved turbo-diesel operation
- One less methanation stage
- Can mitigate higher fuel use by decreasing engine size
  - Lower Volumetric Flow Rate

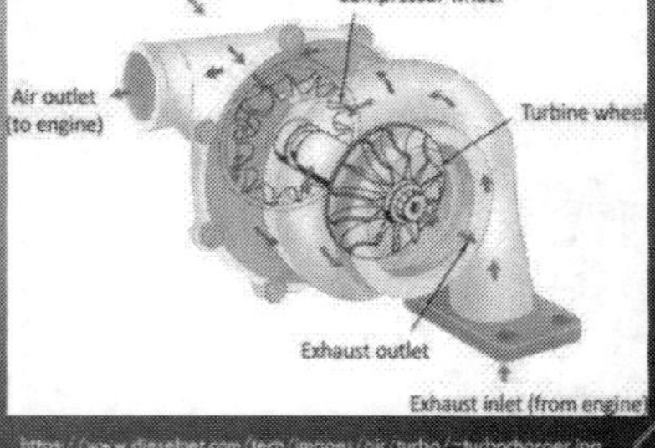


https://www.dieselnet.com/tech/images/air/turbo/~turbocharger.jpg

# ENERGY & CARBON BALANCE

- Carbon expelled per tank ~= 74% * 30.1 kg C / tank = 22.7 kg carbon / tank

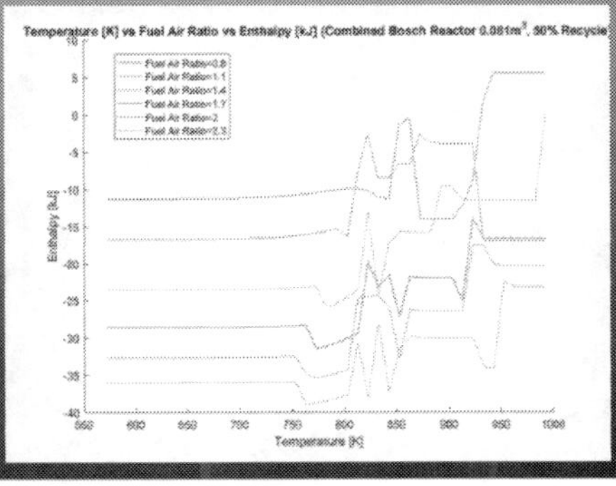


# COST FACTORS

| Cost Factor | Cost |
|---|---|
| $H_2$ Tanks | $250/tank |
| Reactor Cost | $1600 or $50/tank |
| $H_2$ Cost / kg | $1.80/kg |
| Additional Fuel Cost / gal | $2.27/gal |

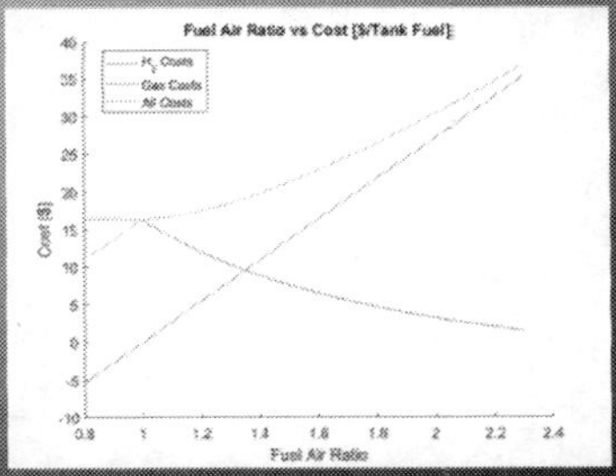

# ACKNOWLEDGEMENTS

- Dr. Thomas Bradley
- Dr. Allan Kirkpatrick
- EcoCAR3 at Colorado State University

# REFERENCES


- Abney, Morgan B., and J. Matthew Mansell. "The Bosch Process - Performance of a Developmental Reactor and Experimental Evaluation of Alternative Catalysts." Ntrs.nasa.gov. 9 Jun. 2016. Web. 22 Mar. 2017. <https://ntrs.nasa.gov/archive/nasa/casi.ntrs.nasa.gov/20100033001.pdf>
- Swickrath, Michael J., and Molly Anderson. " The Development of Models for Carbon Dioxide Reduction Technologies for Spacecraft Air Revitalization." Ntrs.nasa.gov. 10 Apr. 2013. Web. 23 Mar. 2017. <https://ntrs.nasa.gov/archive/nasa/casi.ntrs.nasa.gov/20120004277.pdf>
- N.a. "Cast steel prices per lb, kg and ton in China for price evaluation." *Iron-foundry.com*. 26 Feb. 2017. Web. 4 May 2017. <http://www.iron-foundry.com/cast-steel-prices-lb-kg-ton.html>
- Akse, James. "Regenerative Bosch Reactor" Ntrs.nasa.gov Web. 23 Mar. 2017 <http://techport.nasa.gov/file/13504>
- Ferguson, Colin R., and Allan Kirkpatrick. Internal Combustion Engines: Applied Thermosciences. Chichester, West Sussex, United Kingdom: John Wiley & Sons, 2016. Print.

THE UNIVERSITY OF TEXAS AT AUSTIN — WHAT STARTS HERE CHANGES THE WORLD

# Matching Environmental and Economic Performance of CCUS systems:

## an approach to a decision-making methodology for sustainable development

Focus in CCUS Power Plant for CO2-EOR

Ramón Gil-Egui[1], Vanessa Nuñez-López[1]
CTMC 2017
Houston, July 2017

[1]Bureau of Economic Geology, Jackson School of Geosciences, The University of Texas at Austin

BUREAU OF ECONOMIC GEOLOGY — TEXAS Geosciences, The University of Texas at Austin, Jackson School of Geosciences — GCCC GULF COAST CARBON CENTER

THE UNIVERSITY OF TEXAS AT AUSTIN — WHAT STARTS HERE CHANGES THE WORLD

# Content:

- Brief overview
- System Characterization
  - CCUS in Electric Power generation
  - CO2-EOR
  - Business Models
- Decision Making (methodological Approach)
  - Common practices
  - Integrated Framework
- Methodological proposal
- Conclusions

# Main Objectives:

- Identify and frame critical aspects of the CCUS Electric Power for CO2- EOR.
- A quick review of the integrated assessment methodology for decision-making in complex systems
- Develop a first approach to a broad decision-making framework for CCUS Electric Power for CO2- EOR systems

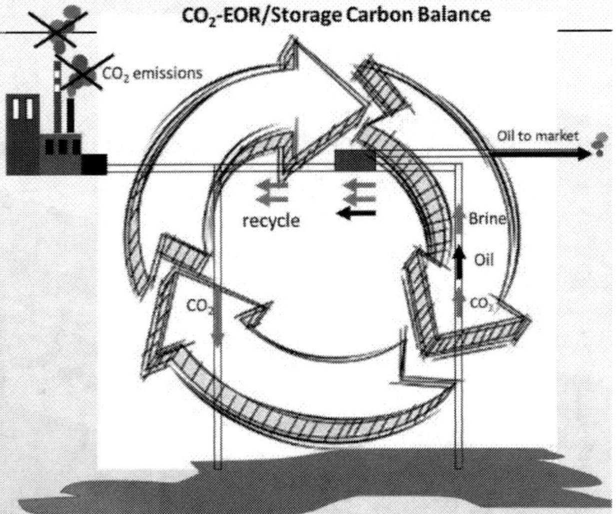


3



# Brief overview

- Climate Change is real, it is occurring NOW!
- Anthropogenic prominent role.
- Paris Agreement
- The Goal was set (UNFCCC & IPCC): -2°C

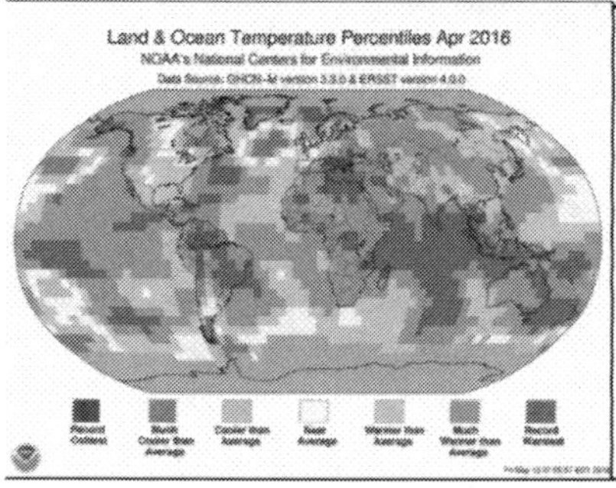


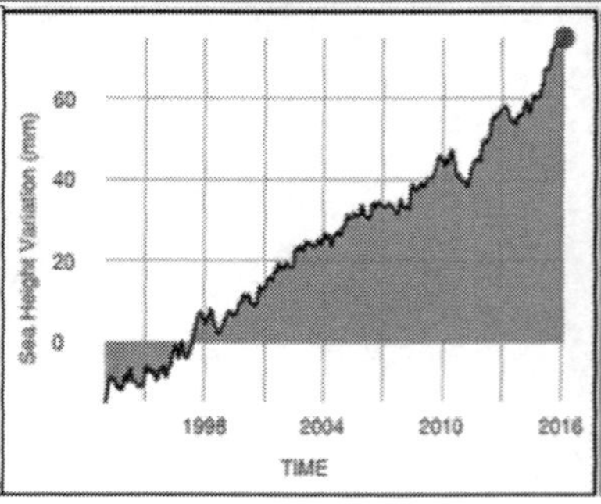


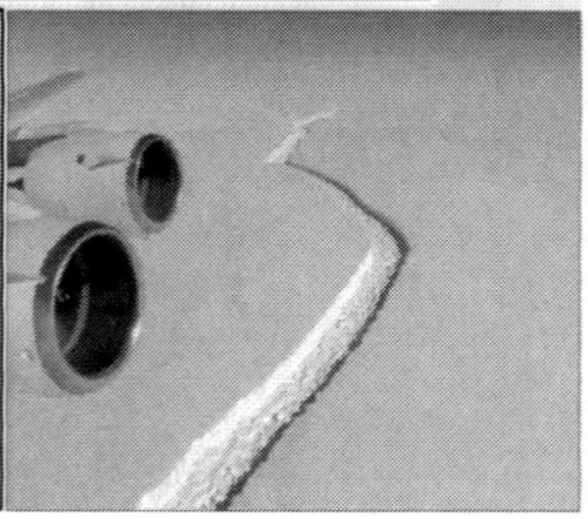

# BROWN TO GREEN: THE G20 TRANSITION TO A LOW-CARBON ECONOMY | 2017 by Climate Transparency

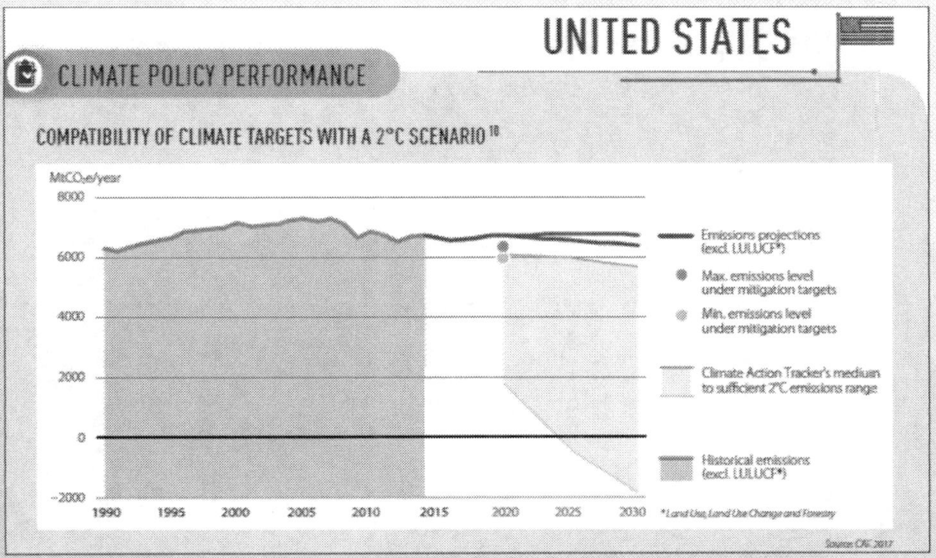


THE UNIVERSITY OF TEXAS AT AUSTIN
WHAT STARTS HERE CHANGES THE WORLD

# Brief overview

- Means Energy de-carbonization:
  - Transportation electrification
  - Phase-out large number of coal-fired Power Plants by 2030
  - Significantly share growth of renewables (+70%)
  - Carbon Capture, Utilization and Sequestration (CCUS)
  - Side demand energy efficiency.
- Simultaneous implementation of these technologies
- CCUS should be a priority [IPCC, 2013]
- All the Global Climate Change models necessarily include CCUS

THE UNIVERSITY OF TEXAS AT AUSTIN

WHAT STARTS HERE CHANGES THE WORLD

# Brief overview

What is CCUS?

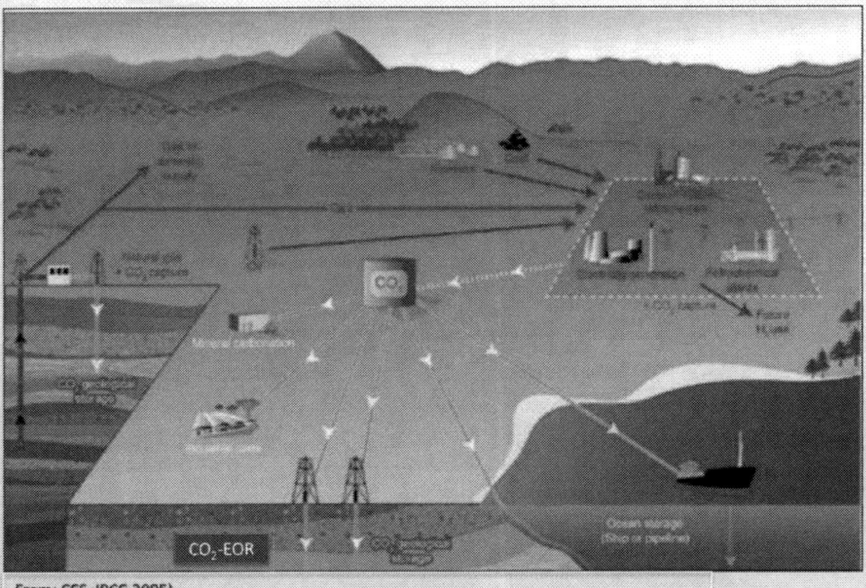


From: CCS, IPCC,2005)

THE UNIVERSITY OF TEXAS AT AUSTIN

WHAT STARTS HERE CHANGES THE WORLD

# System Characterization

Electric Power Sector

- Strategical High Value Sector
- High Capital Intensive
- Integrated and interconnected system
- Major changes in markets pressure the system to its operable limits
- Planning is becoming increasingly complex

THE UNIVERSITY OF TEXAS AT AUSTIN

WHAT STARTS HERE CHANGES THE WORLD

# Figure from: DOE - Quadrennial Energy Review

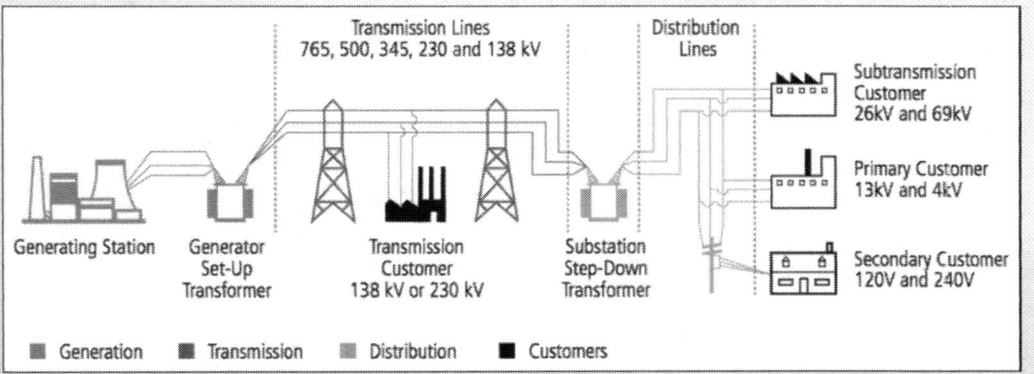


THE UNIVERSITY OF TEXAS AT AUSTIN

WHAT STARTS HERE CHANGES THE WORLD

## System Characterization

CCUS in Electric Power Sector

- CCUS is a key factor in least-cost transitions to a low-carbon electricity system in 2050
- The scale of Power Plants force thinking their integration to a CCUS system
- Capture technology is expensive and energy intense
- Energy penalties reduce competitiveness
- Low energy prices, low demand growth, more renewals share and others limits investments on CO2 capture
- Power Sector CO2 supply require long term demand

# System Characterization

## CO2-EOR

- Promissory CCUS technology
- Mature process. Tertiary Recovery
- Additional 4-15% of OIP (Mezler, 2012)

American Oil & Gas Reporter

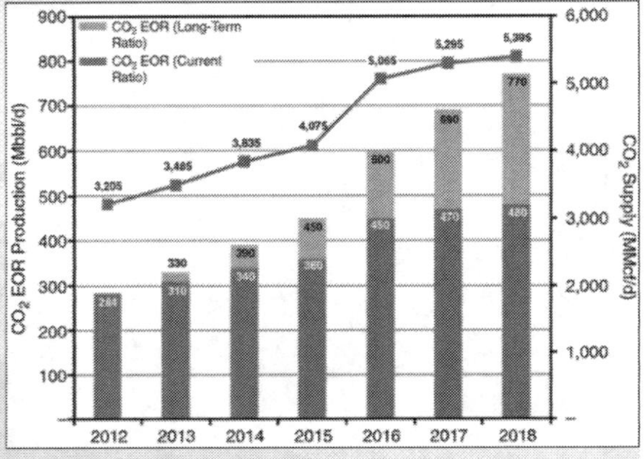


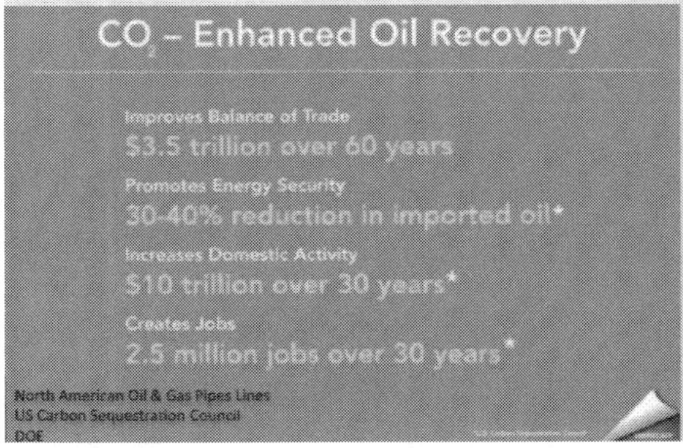


North American Oil & Gas Pipes Lines
US Carbon Sequestration Council
DOE

- CO2-EOR potential requires expanded access to CO2 sources
- Historically Oil Optimization
- Next Generation Tecnologies improves oil production and CO2 storage capacity

THE UNIVERSITY OF TEXAS AT AUSTIN
WHAT STARTS HERE CHANGES THE WORLD

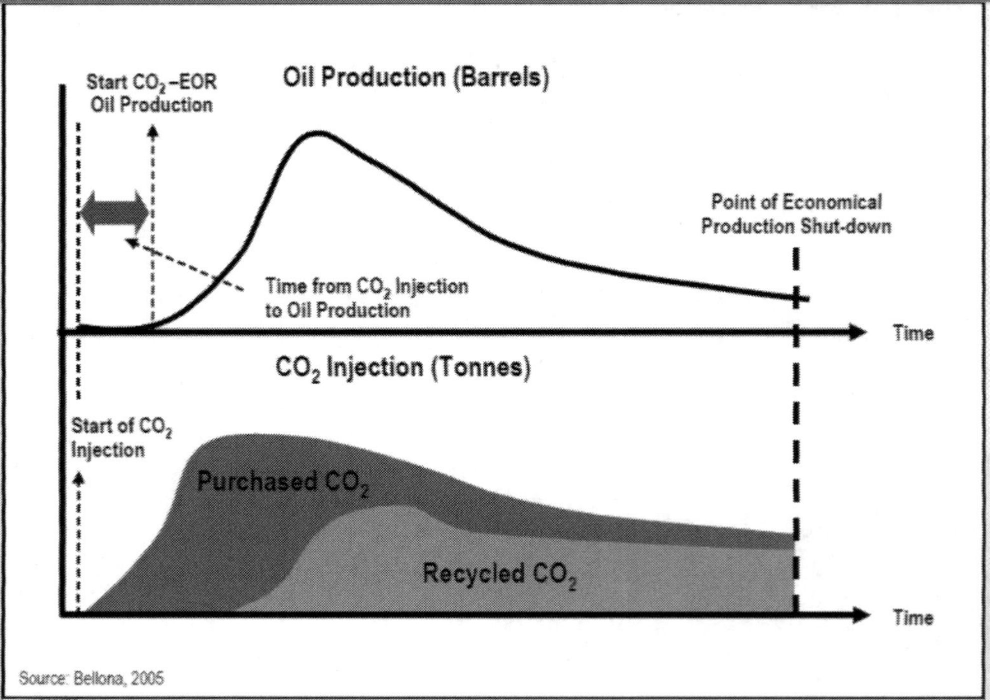


Source: Bellona, 2005

JAF028275.PPT

THE UNIVERSITY OF TEXAS AT AUSTIN — WHAT STARTS HERE CHANGES THE WORLD

# Business Models

- $CO_2$-EOR: activity dominated by independents
- Predominant $CO_2$ source: Gas Processers and Fertilizer plants
- Operators look for Upstream integration building pipelines
- Two large independent companies control half of total $CO_2$-EOR production and expanding
- Petranova, CCUS Coal firered Power Plant. New player build a downstream integration model.

THE UNIVERSITY OF TEXAS AT AUSTIN — WHAT STARTS HERE CHANGES THE WORLD

# MAIN DIFILCULTIES INTEGRATING THIS CCUS SYSTEM 1/2

- **Alignment of the industries sectors that integrate the CCUS system (economic performance)**
- Strategic sectors has high regulatory intervention that has to harmonized
- Product optimization and emission intensity
- $CO_2$ emissions: externality vs commodity

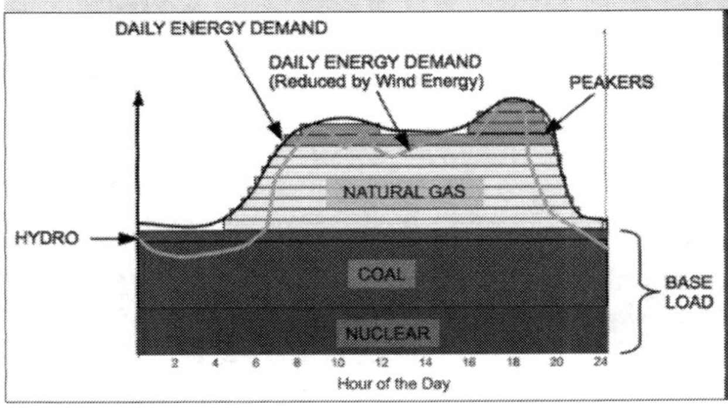


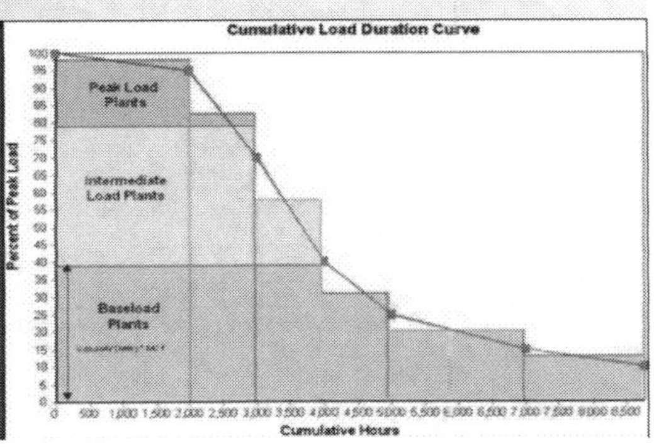

THE UNIVERSITY OF TEXAS AT AUSTIN — WHAT STARTS HERE CHANGES THE WORLD

# MAIN DIFILCULTIES INTEGRATING THIS CCUS SYSTEM 2/2

- Assignment of the environmental responsibilities (the environmental performance -LCA)
- Different methodological approaches to emissions accounting and allocation processes.
- DOE-NETL recommends system expansion with displacement (Skone et al., 2016)
  - which product or process of the systems (up, middle, or downstream),
  - under which criteria (cost or emission efficiency, marginal or average), and
  - All or part of it?

THE UNIVERSITY OF TEXAS AT AUSTIN — WHAT STARTS HERE CHANGES THE WORLD

# Decision Making (methodological Approach)

- Common practices
  - two different methods are commonly known, not integrated for environmental and economic evaluation (LCA and LCC)
  - Different purpose different method and scope
  - From LCA-LCC to ELCC

From Norris, G, 2001

| Tool/Method | LCA | LCC |
|---|---|---|
| Purpose | Compare relative environmental performance of alternative product systems for meeting the same end-use function, from a broad, societal perspective | Determine cost-effectiveness of alternative investments and business decisions, from the perspective of an economic decision maker such as a manufacturing firm or a consumer |
| Activities which are considered part of the 'Life Cycle' | All processes causally connected to the physical life cycle of the product; including the entire pre-usage supply chain; use and the processes supplying use; end-of-life and the processes supplying end-of-life steps | Activities causing direct costs or benefits to the decision maker during the economic life of the investment, as a result of the investment |
| Flows considered | Pollutants, resources, and inter-process flows of materials and energy | Cost and benefit monetary flows directly impacting decision maker |
| Units for tracking flows | Primarily mass and energy; occasionally volume, other physical units | Monetary units (e.g., dollars, euro, etc.) |
| Time treatment and scope | The timing of processes and their release or consumption flows is traditionally ignored; impact assessment may address a fixed time window of impacts (e.g., 100-year time horizon for assessing global warming potentials) but future impacts are generally not discounted | Timing is critical. Present valuing (discounting) of costs and benefits. Specific time horizon scope is adopted, and any costs or benefits occurring outside that scope are ignored |

THE UNIVERSITY OF TEXAS AT AUSTIN — WHAT STARTS HERE CHANGES THE WORLD

# Decision Making (methodological Approach)

- Integrated Framework
  - The CCUS system for EOR integrates productive sectors of significant national relevance in terms of economy, security and environment
  - System optimization and appropriate cost-benefit assessment necessarily goes through an overview both cross-sector and public-private trade-off
  - Integrated Analysis for: Feed Stock, PowerPlant, CO2-EOR site, Refinery and Product combustion

THE UNIVERSITY OF TEXAS AT AUSTIN — WHAT STARTS HERE CHANGES THE WORLD

# Decision Making (methodological Approach)

- ***System Assessment***
- *Risk and Uncertainty Technology*
- *Planning and Projection*
  - *Technology Readiness Levels*
  - *Technology Roadmapping*
  - *Expert Elicitation*
  - *Experience Curve Analysis*
- *Analysis Tools and Metrics*
  - *Life-Cycle Assessment Overview*
    - *Greenhouse Gas Emissions*
    - *Other Emissions*
    - *Water Use*
    - *Land Use*
  - *Materials and Criticality*
  - *Reliability and Resilience*
  - *Other Metrics*

- ***System Assessment (cont..)***
- *Economics metrics*
  - *Levelized Cost of Energy*
  - *Life-Cycle Cost -LCC- (e.g PTLaser, TCAce)*
  - *Environmental Life-Cycle Cost - ELCC-*
- *Evaluation Tools*
  - *Options Space Analysis*
  - *Wedge Analysis*
  - *Integrated Assessment Models*
  - *Science of Human Decision Making*
  - *Real Options Valuation*

# General Methodological Proposal

- ***System Assessment***
- *Risk and Uncertainty Technology*
  - *CO2-EOR site selection-Characterization and technology implementation*
  - *CCUS Power Plant*
  - *Vehicle efficiency*
- *Planning and Projection*
  - *Technology Readiness Levels*
  - *Technology Roadmapping*
  - *Expert Elicitation*
  - *Experience Curve Analysis*

- ***System Assessment (cont..)***
- *Analysis Tools and Metrics*
  - *Life-Cycle Assessment*
    - *GHG*
    - *Water Use*
    - *Land Use*
- *Economics metrics*
  - *Levelized Cost of Energy*
  - *LCC (e.g PTLaser, TCAce)*
  - *ELCC*
- *Evaluation Tools*
  - *Integrated Assessment Models*
  - *Real Options Valuation*

Key aspects of this proposal would be modeling the operation and investments of the Grid (e.g. ERCOT) system by minimizing the cost to meet the emission reduction goals

THE UNIVERSITY OF TEXAS AT AUSTIN
WHAT STARTS HERE CHANGES THE WORLD

# Conclusion

- CCUS in Power Plants for CO-EOR is a very complex cross-sector system that require be develop at the minimum economic and environmental cost
- The real value of CCUS can only be determined by an integrated analysis of economic and ecological performance
- The integrated assessment models require greater diffusion and validation that allows standardization and implementation in different levels of analysis. In particular for the making of private decisions
- Methodological proposal must be reviewed and refined in order to improve the decision making in the CCUS system

THE UNIVERSITY OF TEXAS AT AUSTIN

WHAT STARTS HERE CHANGES THE WORLD

# Questions?

THE UNIVERSITY OF TEXAS AT AUSTIN

WHAT STARTS HERE CHANGES THE WORLD

# It's a Complex Problem

## LCA for $CO_2$-EOR have many path ways, products and sub products

Cradle-Grave Boundary

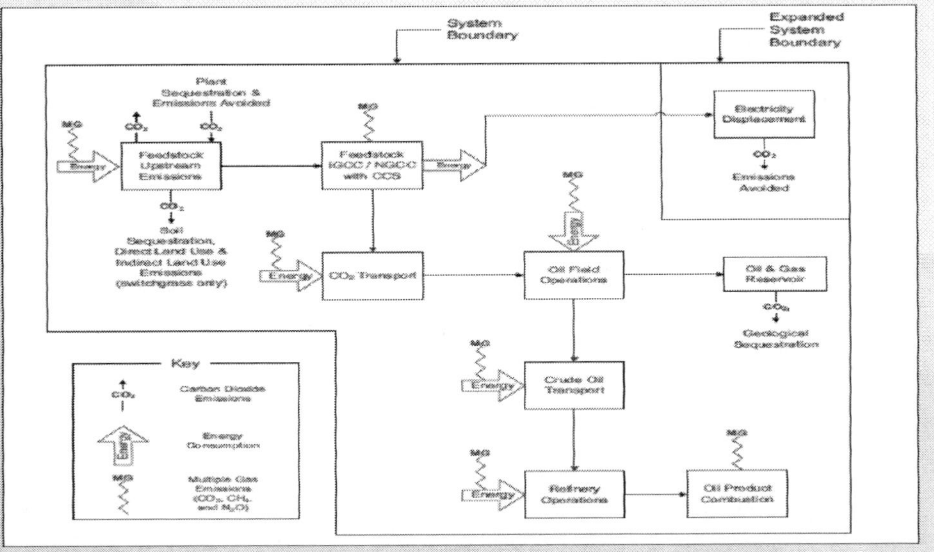


Jaramillio; et. Al., 2013, *Comparative LCI GHG emission in EOR Operations from diferents sources*

THE UNIVERSITY OF TEXAS AT AUSTIN

WHAT STARTS HERE CHANGES THE WORLD

# It's a Complex Problem

LCA for $CO_2$-EOR have many path ways, products and sub products

Gate-Gate Boundary

emissions

Energy

inputs

$CO_2$ makeup

emissions

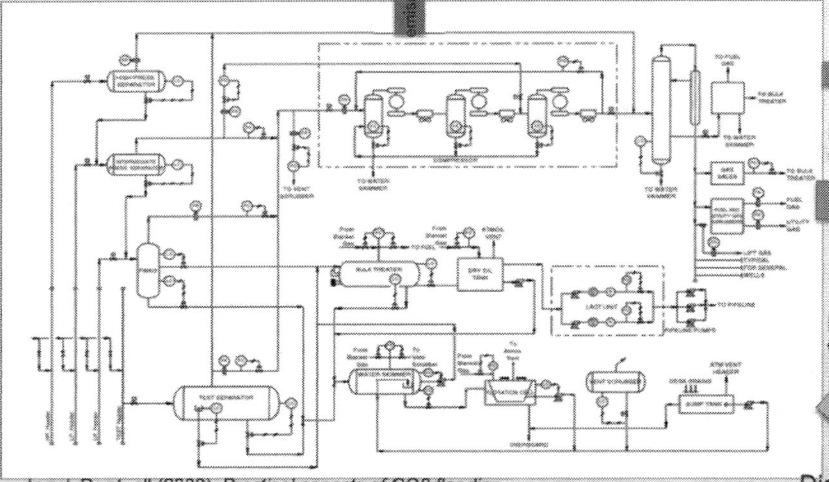

Jarrel, P.; et. all.(2002), *Practical aspects of CO2 flooding*

Crude Oil

sales

Hydrocarbon Gas

recycling

$CO_2$

Brine Water

storage

Disposal

THE UNIVERSITY OF TEXAS AT AUSTIN

WHAT STARTS HERE CHANGES THE WORLD

# My contribution:

**Central Objective:** Create model to analyze the relation between **energy consumption**, oil production, CO2 injection, GHG emissions and sequestration oriented to achieve the NCNO classification for CO2-EOR Operations and energy efficiency recommendations.

## Specific objective :

- Review the boundaries criteria
- Homogenize the functional units
- Clearly understand the CO2-EOR surface operations/emissions/energy efficiency
- Collaborate with other phases of the project
- Present advances of the model developed
- Help developing strategies that are conducive to achieving a NCNO classification.

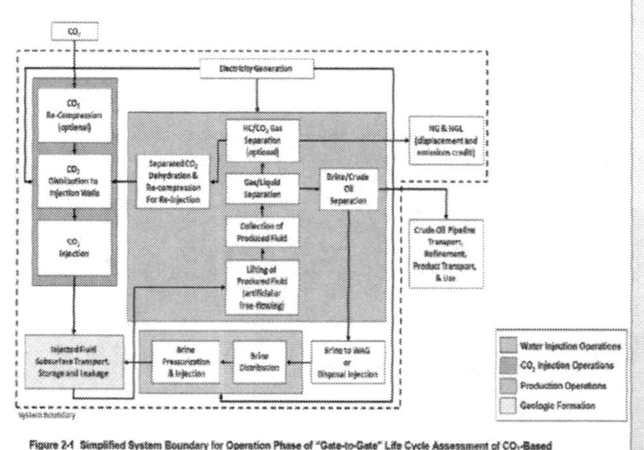

**Figure 2-1 Simplified System Boundary for Operation Phase of "Gate-to-Gate" Life Cycle Assessment of $CO_2$-Based Enhanced Oil Recovery. Dotted line indicates study boundary.**

**Dilmore, Robert M. (DOE/NETL). 2010.** *An Assessment of Gate-to-Gate Environmental Life Cycle Performance of Water-Alternating-Gas CO2-Enhanced Oil Recovery in the Permian Basin*, pp 206

24

THE UNIVERSITY OF TEXAS AT AUSTIN

WHAT STARTS HERE CHANGES THE WORLD

# Our Efforts

- ✓ Reviewed, process, classified and referred in the reports to DOE-NTEL large number of studies (60+)
- ✓ Selected the most consistent and commonly referred works
- ✓ Conversions and calculations
- ✓ Developed some schematics with all significant components for EOR and established the mass and/or energy flow between them
- ✓ Sought without much success a detailed real list of surface equipment and its operating conditions

THE UNIVERSITY OF TEXAS AT AUSTIN

WHAT STARTS HERE CHANGES THE WORLD

# System boundaries of previous studies

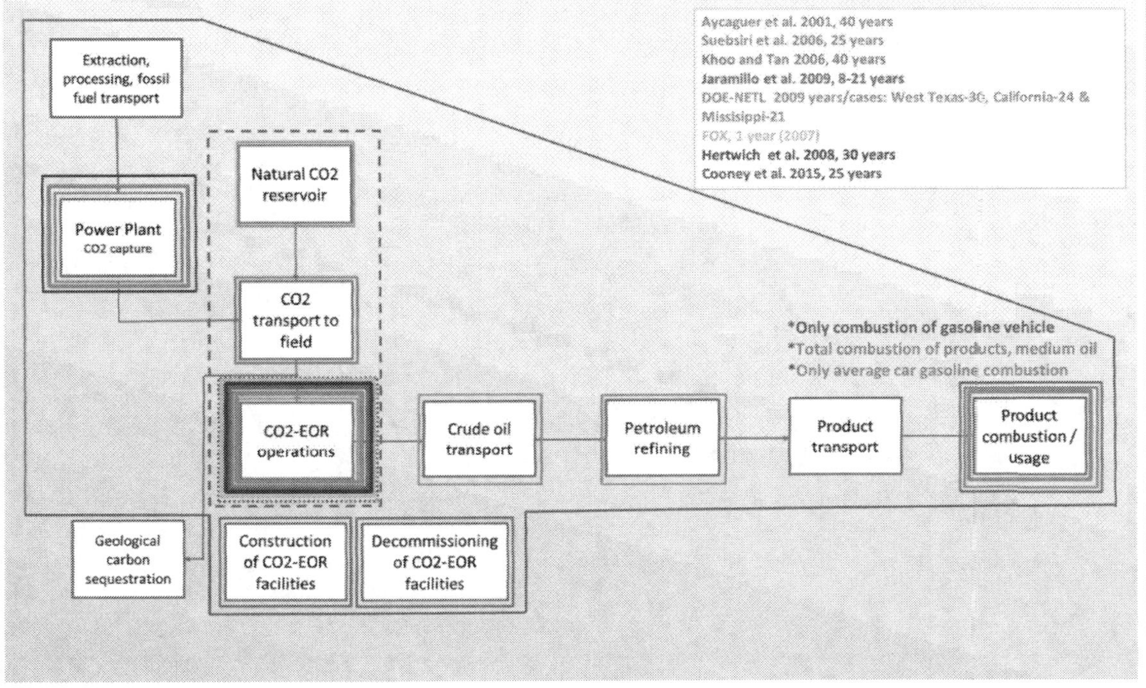

THE UNIVERSITY OF TEXAS AT AUSTIN — WHAT STARTS HERE CHANGES THE WORLD

# Summary

- Goings-on:
  - ✓ Selected the system boundaries relevant to NCNO classification
  - ✓ Identification of critical $CO_2$ emission components within the EOR site
  - ✓ Homogenize the functional units to determinate the parameter in our study
  - ✓ Looking for Cranfield $CO_2$-EOR electricity consumption
- Next Tasks:
  <u>In current Study</u>
  - Build a model for energy consumption of the $CO_2$-EOR operation
  - Start scenario analysis
  - identify and analyze significant relationships between **energy consumption**, oil production, CO2 injection, GHG emissions and sequestration
  - Link results from numerical simulations with energy consumption model
  - Help developing the strategies to achieve the NCNO classification for CO2-EOR Operations and energy efficiency recommendations

27

THE UNIVERSITY OF TEXAS AT AUSTIN — WHAT STARTS HERE CHANGES THE WORLD

# Conclusions

- Carbon balance of $CO_2$-EOR is sensitive to the system boundary.
- In a gate-to-gate life cycle analysis, the electricity consumption (purchased and generated) is responsible for almost all the emissions associated with the EOR operation, particularly at the $CO_2$ separation and compression processes.
- Each $CO_2$-EOR facility is unique. Different facility dissing and operational strategies, different energy requirements, performance and GHG emissions
- Electricity consumption data is critical to allow appropriate correlation in mass/energy flows. Not have this would lead to assume generalizations with very high uncertainties.
- Carbon balance is sensitive to CO2 flood performance (CO2 utilization rates).
- A universal methodology for NCNO classification will certainly benefit $CO_2$-EOR operations as there might be an economic impact if potential future regulations provide value to the emissions and/or storage of $CO_2$.

28

THE UNIVERSITY OF TEXAS AT AUSTIN — WHAT STARTS HERE CHANGES THE WORLD

# Future Objectives:

- Abstract that summarizes the conceptualization and first results of our model

- Draft conceptualization of a proposal for research:
  The economic Implications of:
  - CO2-EOR Operations with Classification NCNO (with VN)
  - Corrosion Behaviors in $CO_2$ Injection Wells (with AI)
  - Complement other studies

  Other topics of interest:
  - CCS Public acceptance (Japan)
  - $CO_2$ Pricing

- Start with contacts in L.A. Oil Companies managers, decision maker, academic and research institutions oriented to promote BEG research, cooperation and interchange interests. (goin-on)

THE UNIVERSITY OF TEXAS AT AUSTIN — WHAT STARTS HERE CHANGES THE WORLD

## Selection of system boundaries for NCNO classification: Cradle-to-Grave

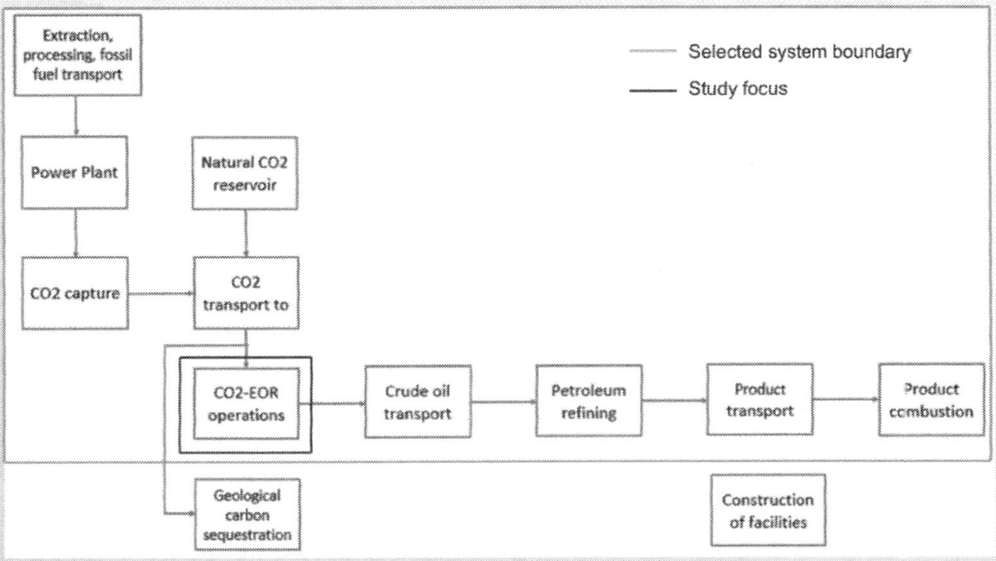

S

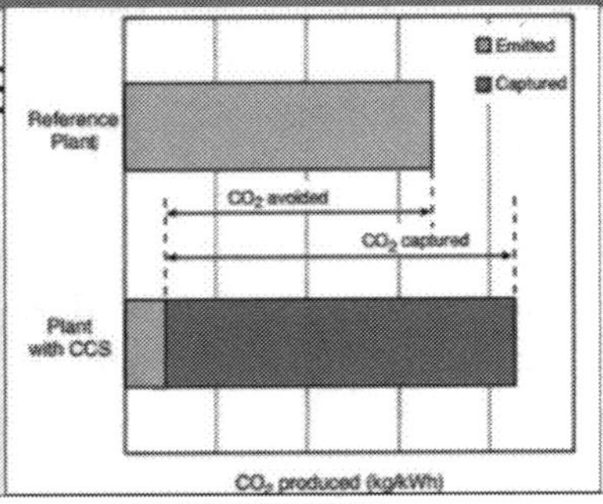


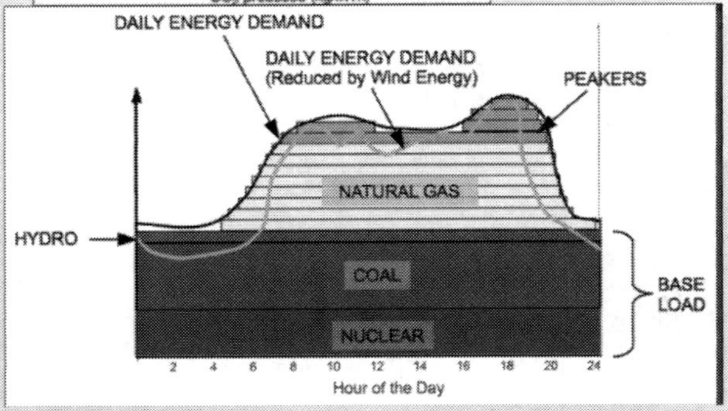


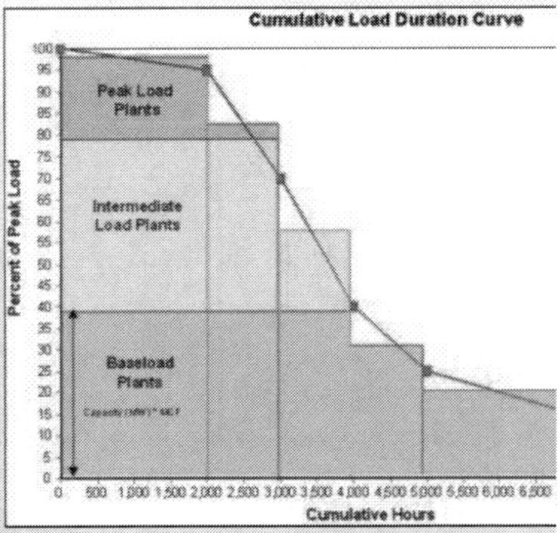


THE UNIVERSITY OF TEXAS AT AUSTIN
WHAT STARTS HERE CHANGES THE WORLD

# Strategic Power Plant Investment Planning under Fuel and Carbon Price Uncertainty by Ansgar Geiger 2010

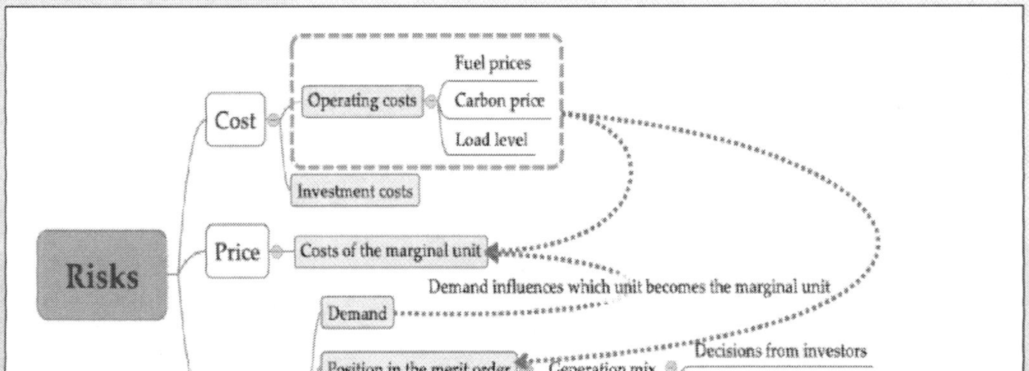

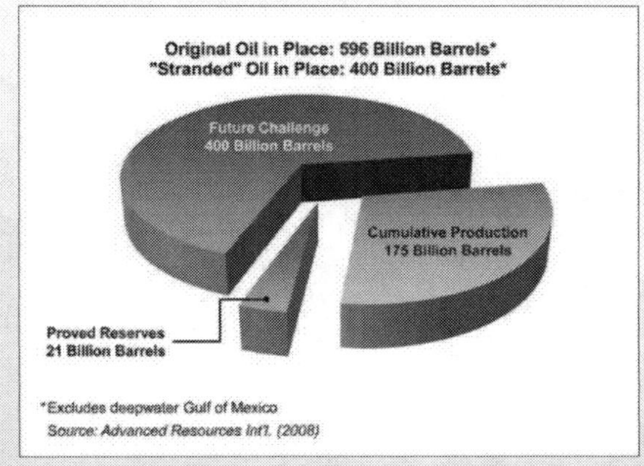


THE UNIVERSITY OF TEXAS AT AUSTIN
WHAT STARTS HERE CHANGES THE WORLD

# Idem befor (31)

THE UNIVERSITY OF TEXAS AT AUSTIN

WHAT STARTS HERE CHANGES THE WORLD

# Integrating Life Cycle Cost Analysis and LCA, InLCA: Selected Papers Gregory A. Norris

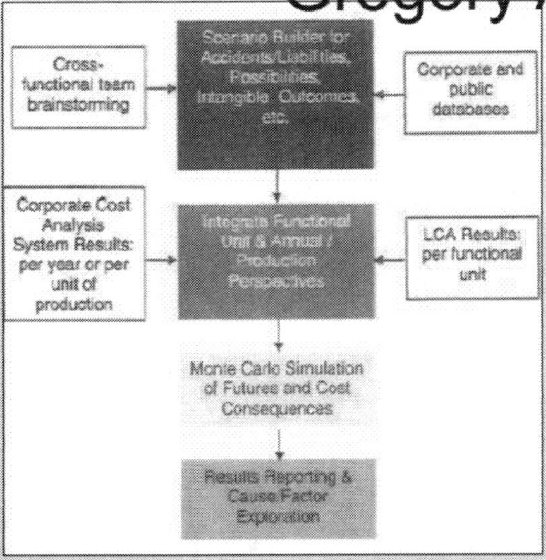


| Cost Type | Description |
|---|---|
| **Type 1:** Direct | Direct costs of capital investment, labor, raw mater costs. Includes both capital and O&M costs |
| **Type 2:** Indirect | Indirect costs not allocated to the product or proce: Includes both capital and O&M costs |
| **Type 3:** Contingent | Contingent costs such as fines and penalties, cost: liabilities |
| **Type 4:** Intangible | Difficult to measure costs, including consumer acc wellness, corporate image, community relations |
| **Type 5:** External | Costs borne by parties other than the company (e. |

THE UNIVERSITY OF TEXAS AT AUSTIN

WHAT STARTS HERE CHANGES THE WORLD

# Integrating Life Cycle Cost Analysis and LCA, InLCA: Selected Papers Gregory A. Norris

| Tool/Method | LCA | LCC |
|---|---|---|
| Purpose | Compare relative environmental performance of alternative product systems for meeting the same end-use function, from a broad, societal perspective | Determine cost-effectiveness of alternative investments and business decisions, from the perspective of an economic decision maker such as a manufacturing firm or a consumer |
| Activities which are considered part of the 'Life Cycle' | All processes causally connected to the physical life cycle of the product; including the entire pre-usage supply chain; use and the processes supplying use; end-of-life and the processes supplying end-of-life steps | Activities causing direct costs or benefits to the decision maker during the economic life of the investment, as a result of the investment |
| Flows considered | Pollutants, resources, and inter-process flows of materials and energy | Cost and benefit monetary flows directly impacting decision maker |
| Units for tracking flows | Primarily mass and energy; occasionally volume, other physical units | Monetary units (e.g., dollars, euro, etc.) |
| Time treatment and scope | The timing of processes and their release or consumption flows is traditionally ignored; impact assessment may address a fixed time window of impacts (e.g., 100-year time horizon for assessing global warming potentials) but future impacts are generally not discounted | Timing is critical. Present valuing (discounting) of costs and benefits. Specific time horizon scope is adopted, and any costs or benefits occurring outside that scope are ignored |



## CO2 Enhanced Oil Recovery Institute for 21st Century Energy | U.S. Chamber of Commerce U.S. Chamber of Commerce

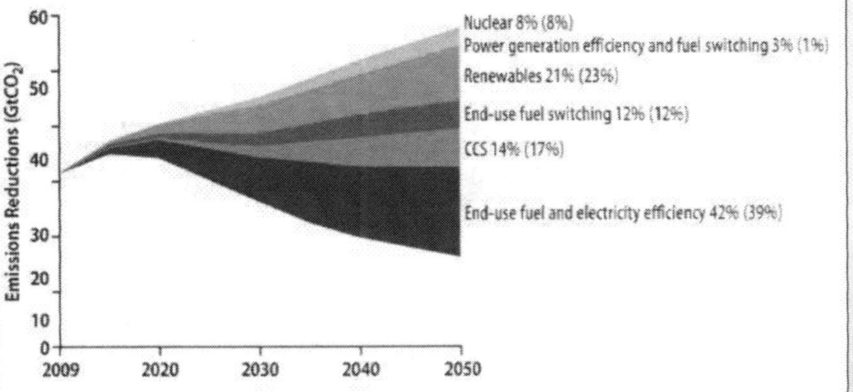


Note: Per IEA, the numbers in parentheses are shares in 2050. For example, 14% is the share of CCS in cumulative emission reductions through 2050, and 17% is the share of CCS in emission reductions in 2050, compared with the 6 °C Scenario.

(Source: 2013 IEA Global CCS Roadmap).

THE UNIVERSITY OF TEXAS AT AUSTIN

WHAT STARTS HERE CHANGES THE WORLD

# New unabated1 coal is not compatible with keeping global warming below 2°C

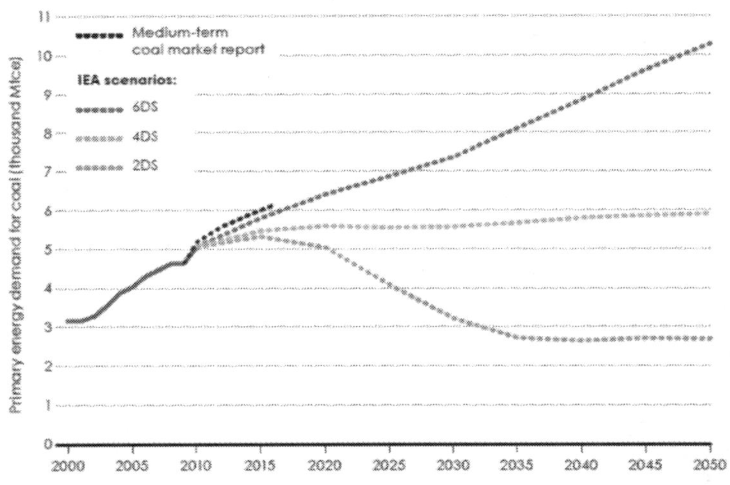


Figure 3: Primary energy demand for coal under IEA's medium-term coal market report and IEA Scenarios 2DS, 4DS and 6DS consistent with 2, 4 and 6 degrees warming above pre-industrial levels in the long term. Source: IEA/Gagné, 2012, adapted.

THE UNIVERSITY OF TEXAS AT AUSTIN

WHAT STARTS HERE CHANGES THE WORLD

# New unabated1 coal is not compatible with keeping global warming below 2°C

Statement by leading climate and energy scientists

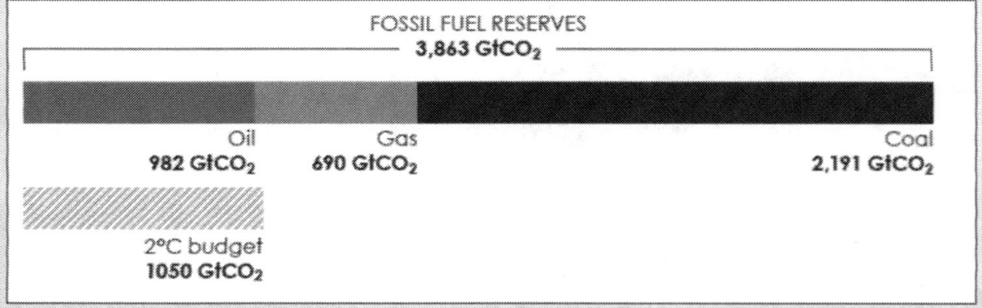

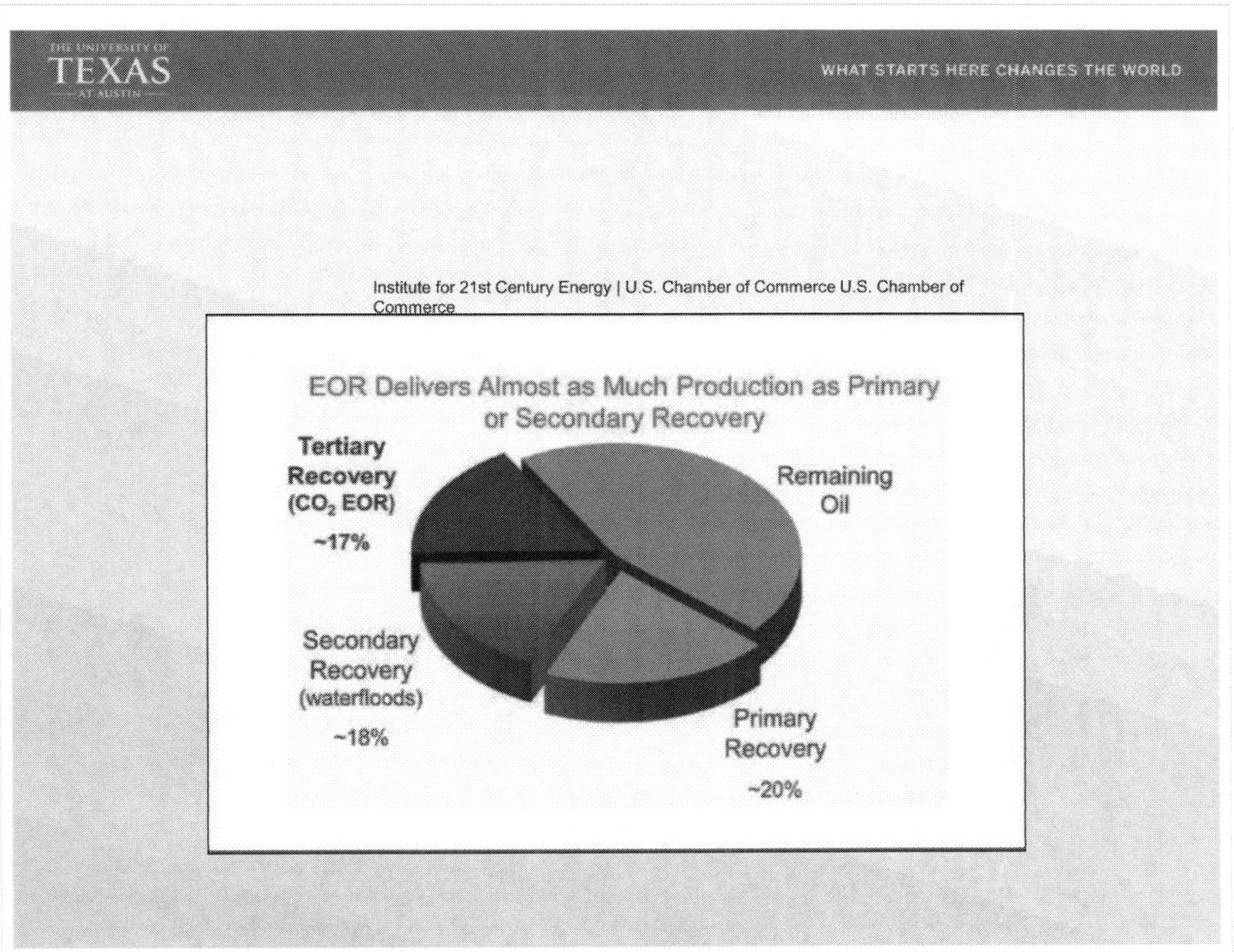
THE UNIVERSITY OF
TEXAS
AT AUSTIN
WHAT STARTS HERE CHANGES THE WORLD
Institute for 21st Century Energy | U.S. Chamber of Commerce U.S. Chamber of Commerce
EOR Delivers Almost as Much Production as Primary or Secondary Recovery
Tertiary Recovery ($CO_2$ EOR) ~17%
Remaining Oil
Secondary Recovery (waterfloods) ~18%
Primary Recovery ~20%

# Life Cycle Implications of Using $CO_2$-Based Fracturing Fluids as a Substitute for Slickwater

To reduce the carbon and water intensity of unconventional oil and gas

Andres Clarens, Anne Menefee, Buddy Wilkins, Tao Zhiyuan
University of Virginia

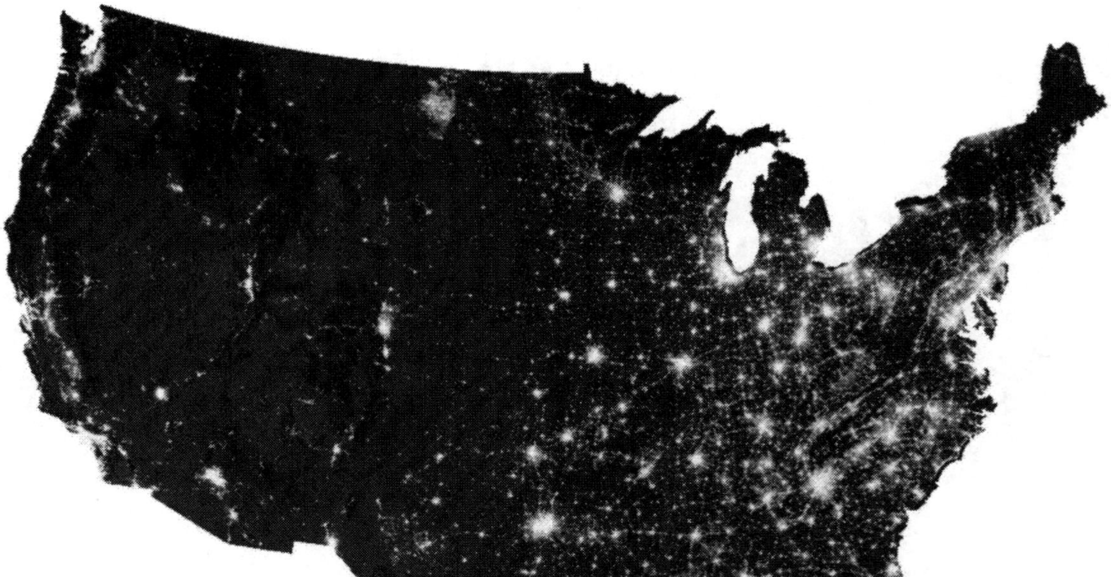

# LCA system boundaries

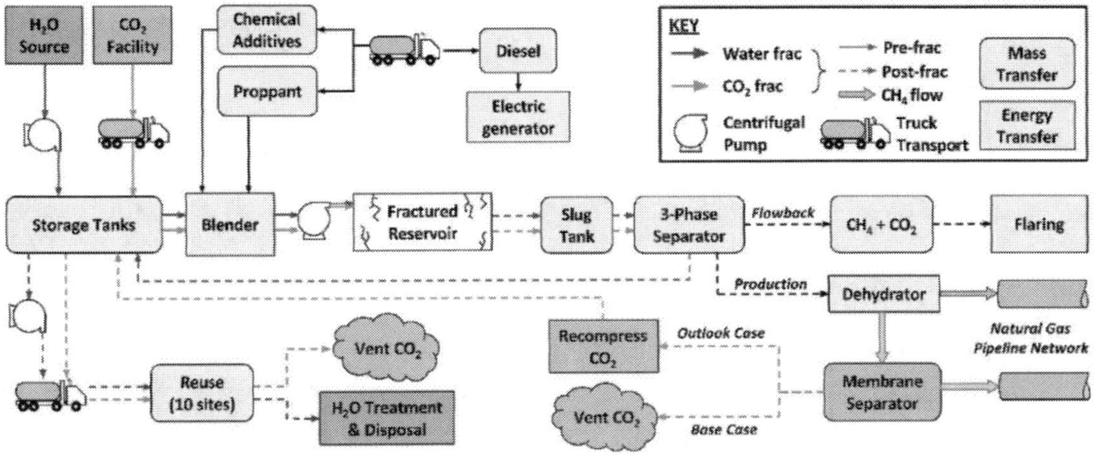


CMTC 2017
7.19.17

# $CO_2/CH_4$ flow profiles

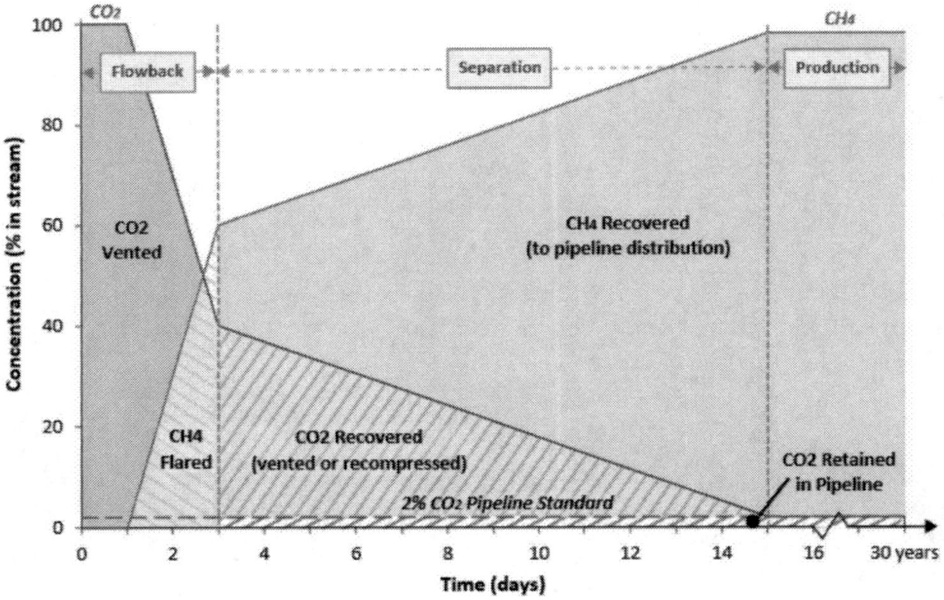


CMTC 2017
7.19.17

# Major inputs

| Input | Units | Water frac | $CO_2$ frac, base | $CO_2$ frac, outlook |
|---|---|---|---|---|
| Estimated ultimate recovery (EUR) | GJ | 4.98E+06 | 7.64E+06 | 2.67E+07 |
| Production increase factor (PIF) | % | -- | 50 (0-110) | 200 (50-1340) |
| Total fluid volume per well | $m^3$ | 9.09E+04 | 6.36E+04 | 6.04E+04 |
| Volume of gas flared | $m^3$ | 1.03E+04 | 1.68E+05 | 1.60E+05 |
| Flowback (% of injected fluid) | % | 7 (4-47) | 70 (50-90) | 50 (10-90) |
| Produced water volume per well | $m^3$ | 1.76E+04 | -- | -- |
| Truck trips per well | -- | 1206 | 3269 | 3036 |
| Diesel fuel consumption | $m^3$ | 626 | 985 | 950 |
| $CO_2$ compression energy (source) | $MJ/tCO^2$ | -- | 241 (72-285) | 241 (72-285) |

CMTC 2017
7.19.17

# results

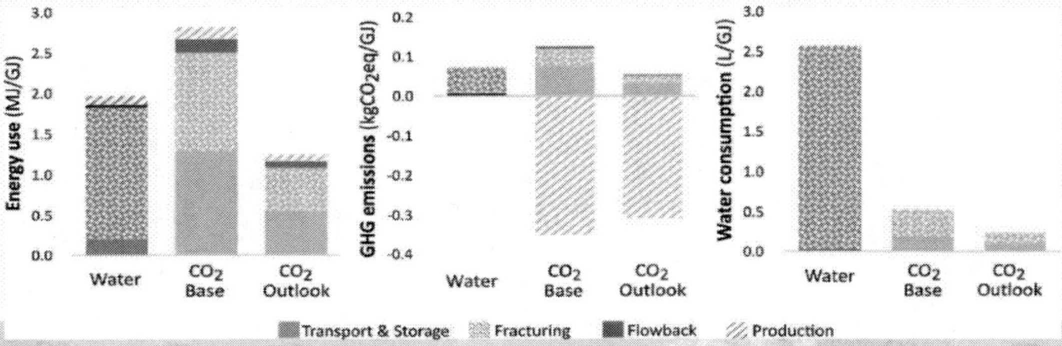


CMTC 2017
7.19.17

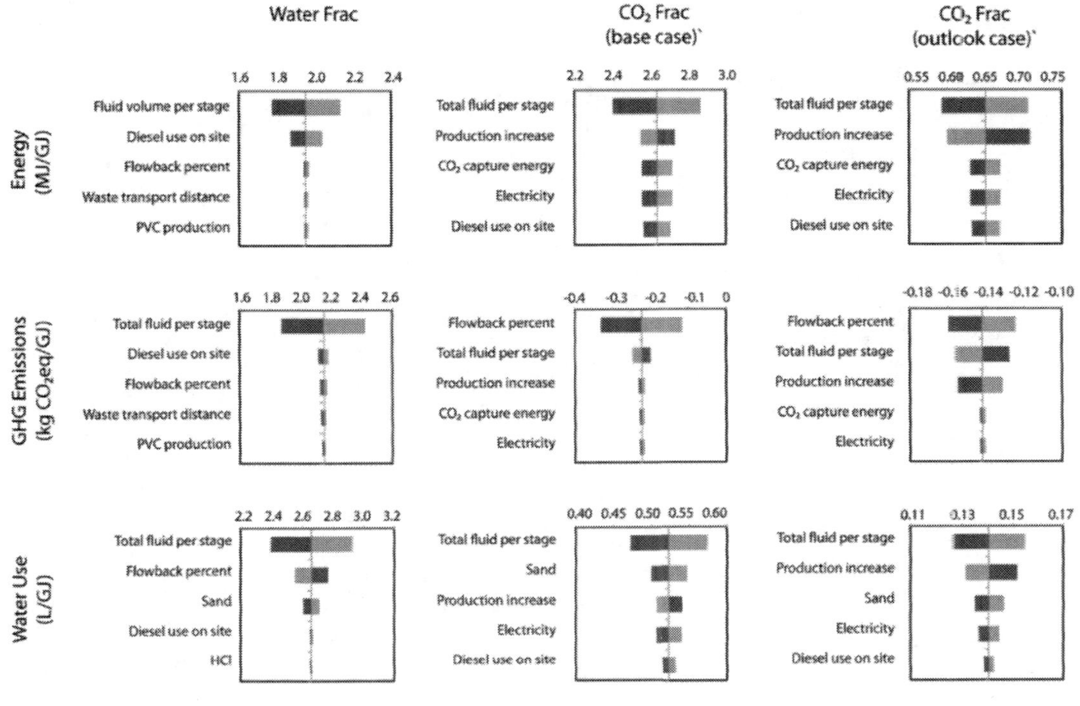


R Wilkins, AH Menefee, AF Clarens (2016)
Environmental science & technology 50 (23), 13134-13141

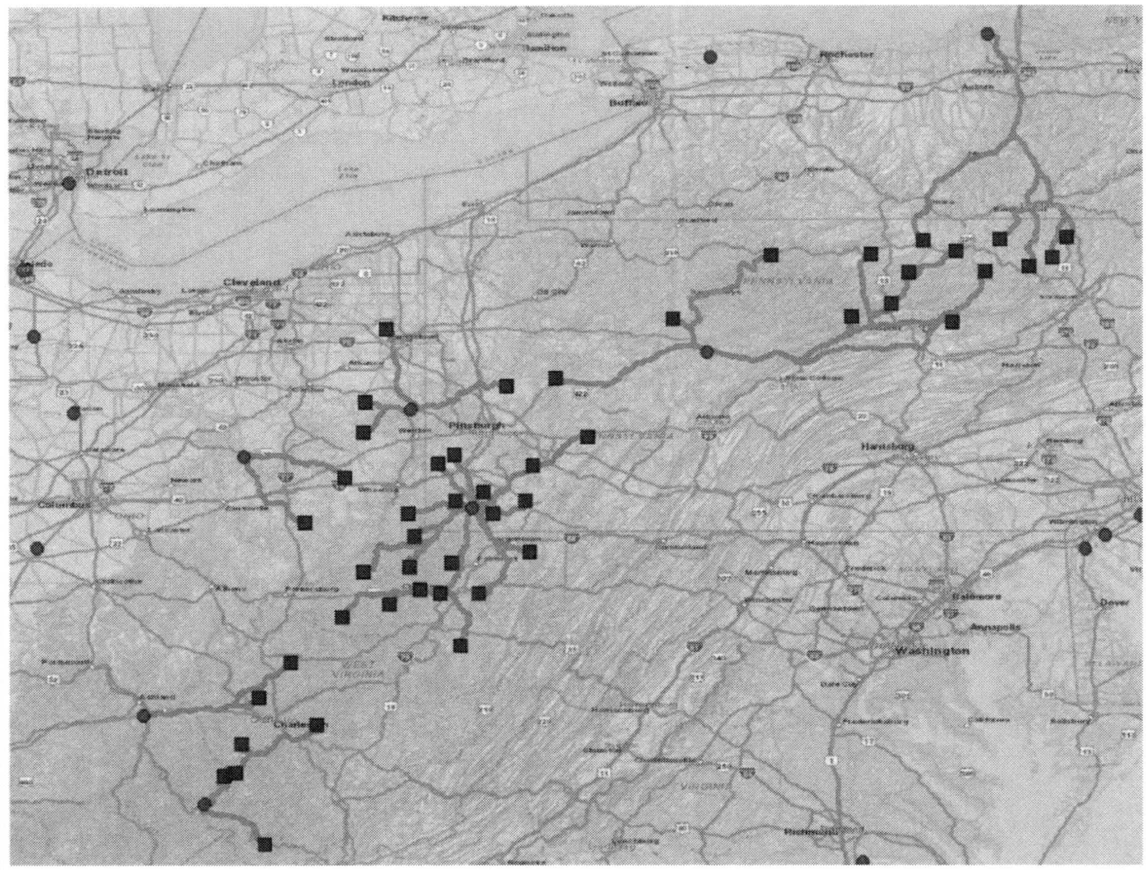

## sources of $CO_2$ (from currently capturable sources)

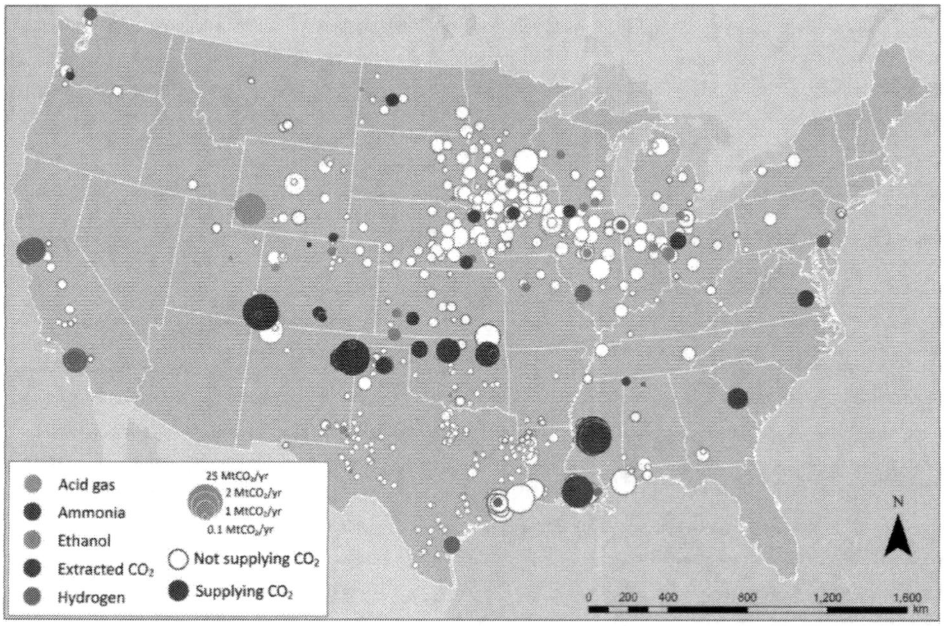


CMTC 2017
7.19.17

# sources of $CO_2$

CMTC 2017
7.19.17

# size distribution of the industry

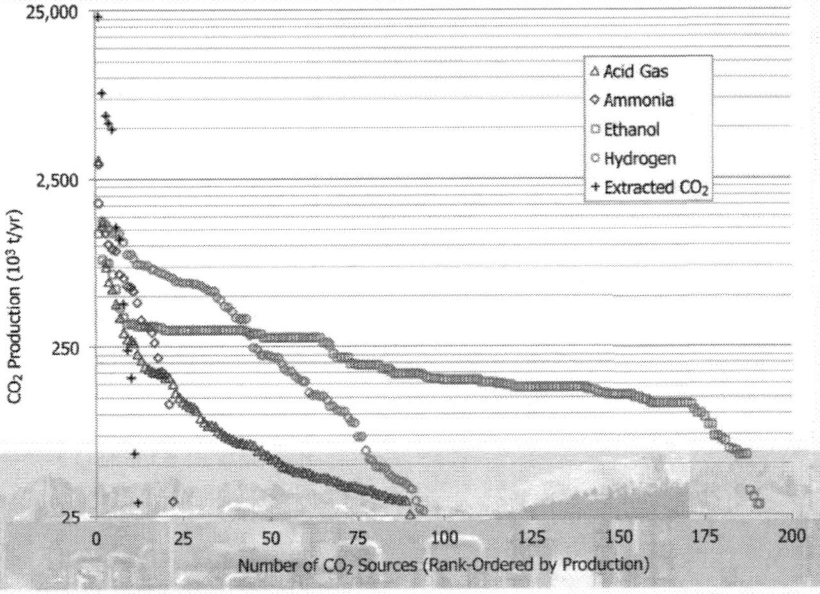


CMTC 2017
7.19.17

## $CO_2$ deserts

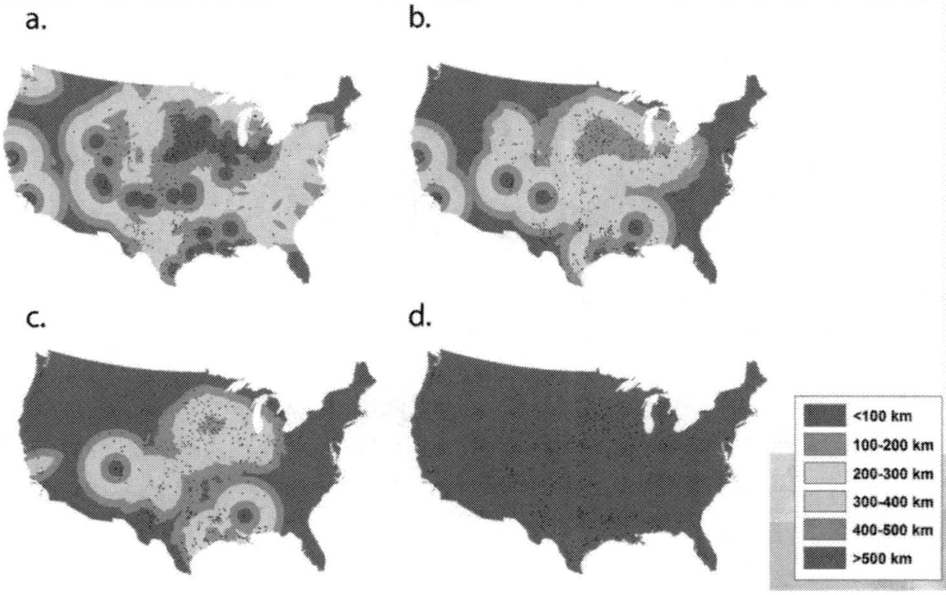


CMTC 2017
7.19.17

## regional implications of $CO_2$ availability

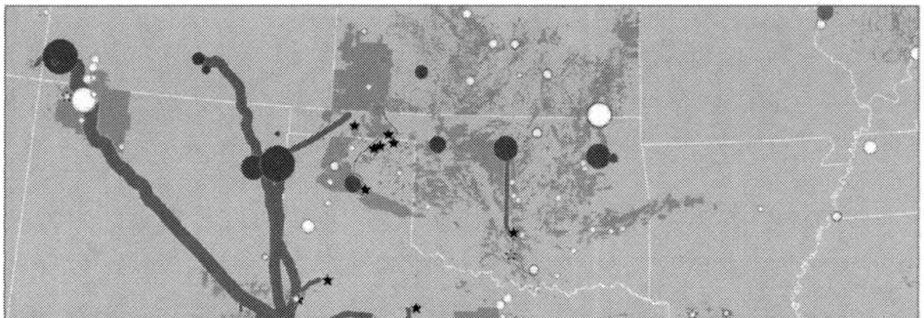


RS Middleton, AF Clarens, X Liu, JM Bielicki, JS Levine (2014)
Environ. Sci. Technol 48 (19), 11713-11720

CMTC 2017
7.19.17

# shale formations v. deep saline aquifers

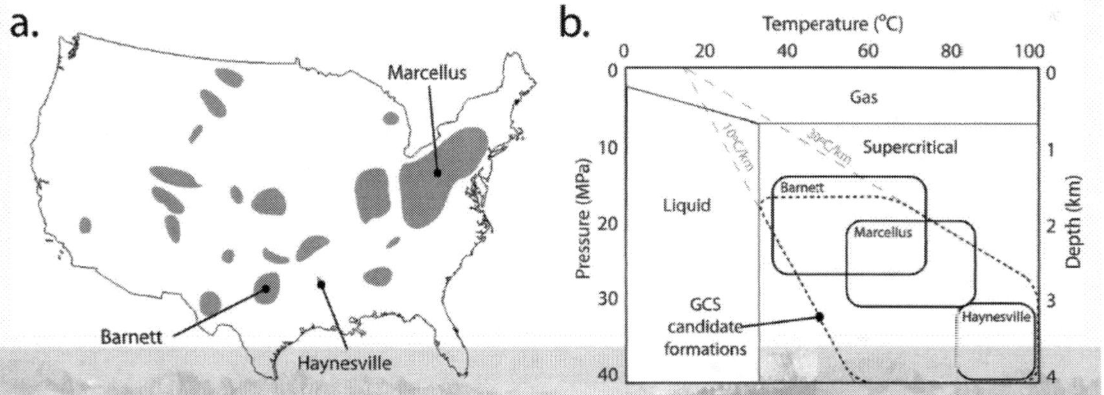


CMTC 2017
7.19.17

# only a fraction of methane in shale matrix is sorbed

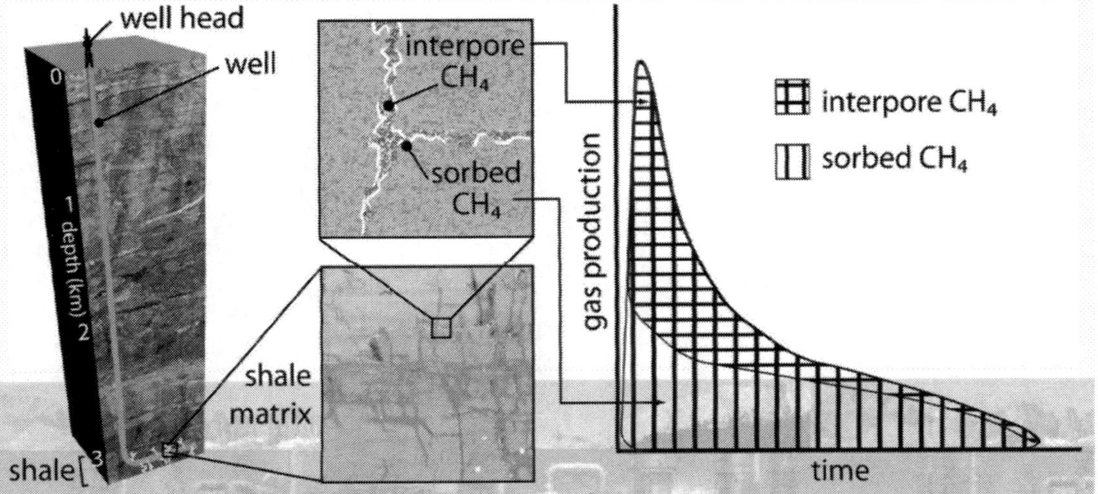


CMTC 2017
7.19.17

## sorption on shales a function of TOC

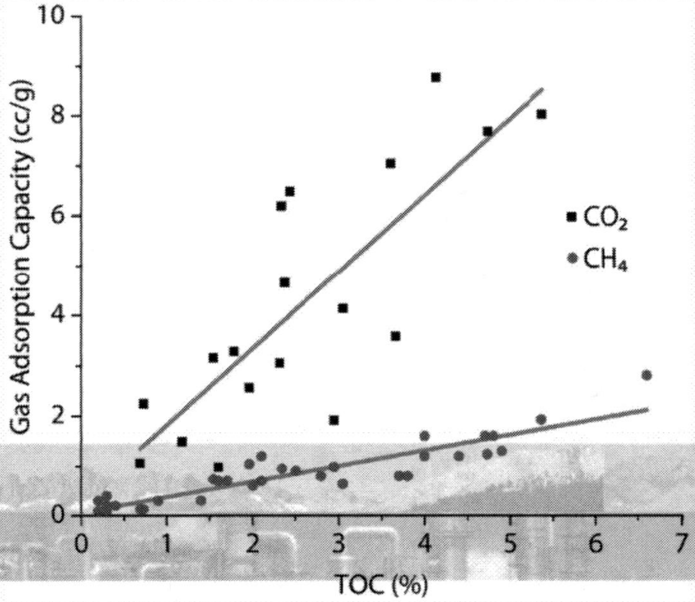


CMTC 2017
7.19.17

## unipore model for gas transport in kerogen

$$\frac{D}{r^2}\frac{\partial}{\partial r}(r^2\frac{\partial C}{\partial r}) = \frac{\partial C}{\partial t}$$

$$\frac{V_t}{V_\infty} = 1 - \frac{6}{\pi^2}\sum_{n=1}^{\infty}\frac{1}{n^2}\exp(\frac{-n^2 D\pi^2 t}{r_p^2})$$

$$D_{e,CH_4} = \frac{D_{CH_4}\pi^2}{r_p^2}$$

CMTC 2017
7.19.17

## model contains three independent parameters

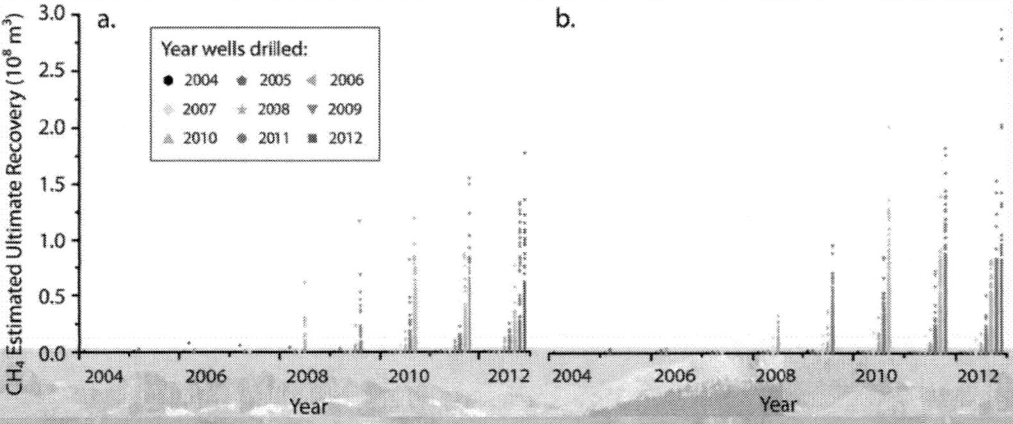


CMTC 2017
7.19.17

## model architecture

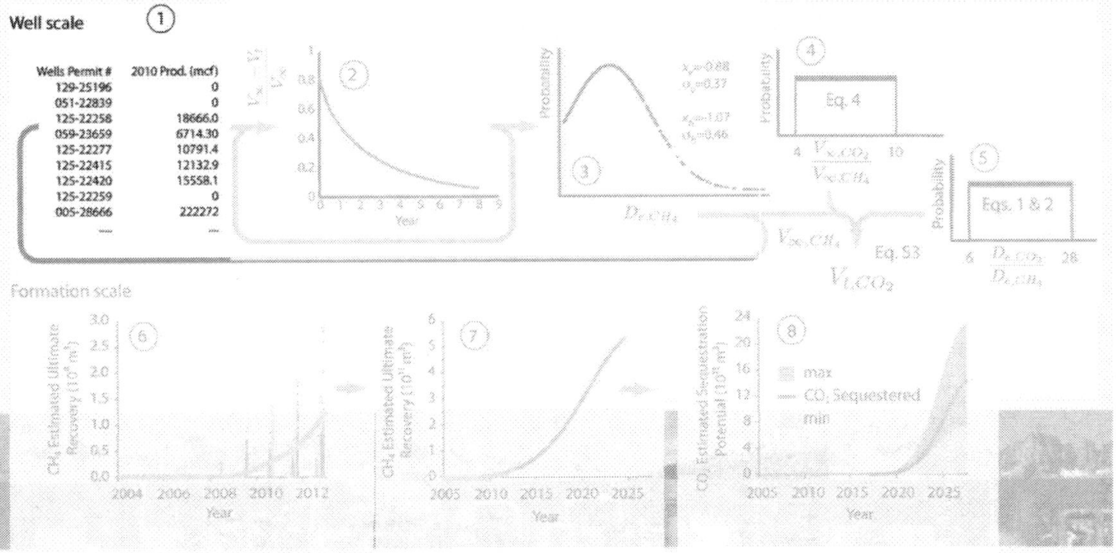


CMTC 2017
7.19.17

## model results

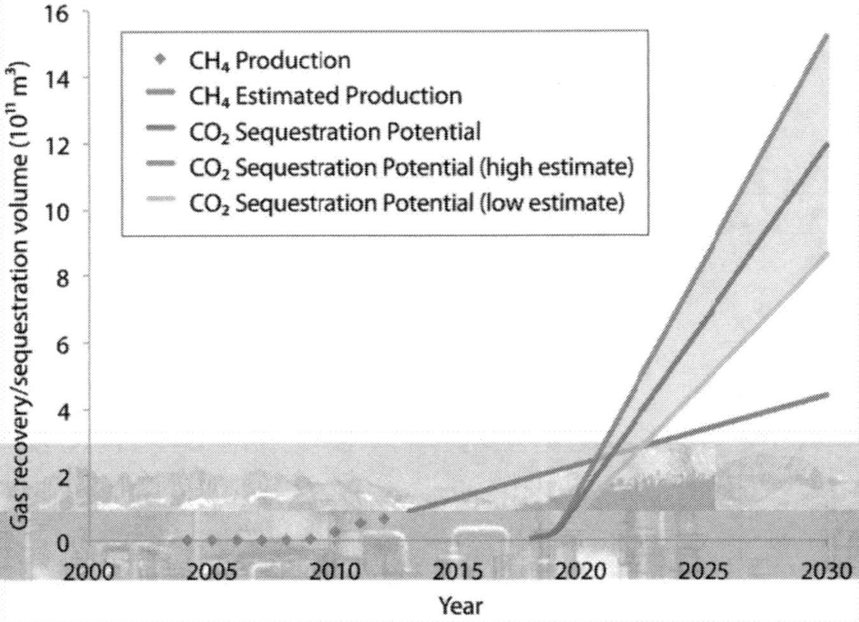


CMTC 2017
7.19.17

## sequestration capacity in context

2.2 Gt $CO_2$/yr

10.4-18.4 Gt $CO_2$ (2018-2030)

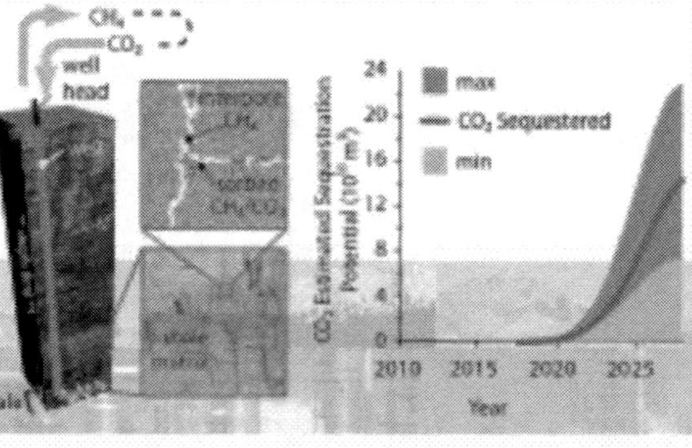


Z Tao, A Clarens (2013) Environmental science & technology 47 (19), 11318-11325

CMTC 2017
7.19.17

# considerable interest in this topic recently

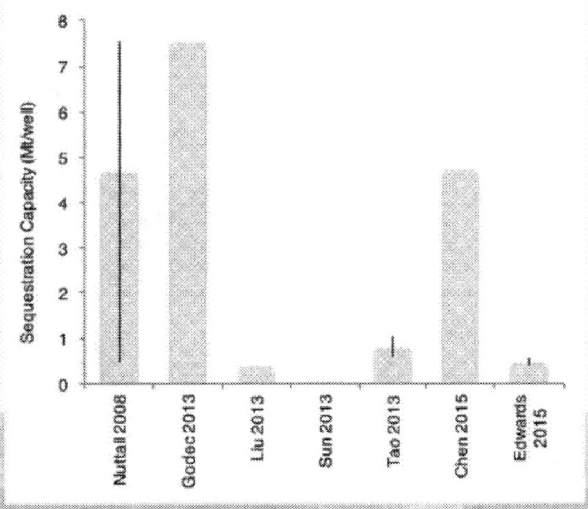


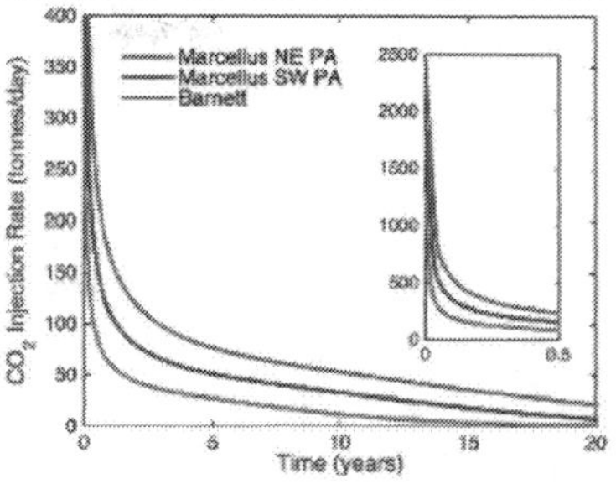


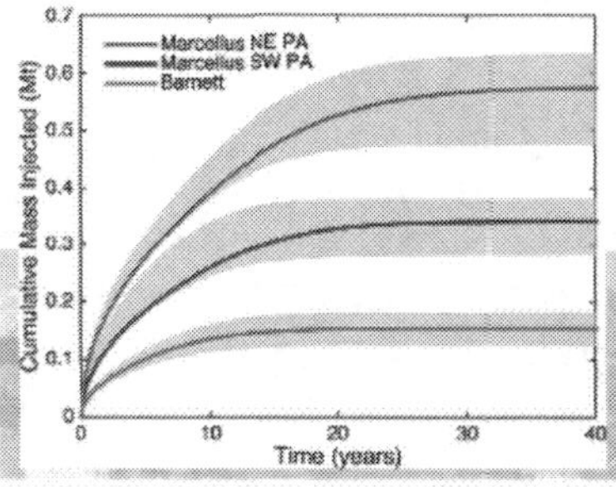


Edwards, R., et al. ES&T 49.15 (2015): 9222-9229.

CMTC 2017
7.19.17

# technoeconomic analysis

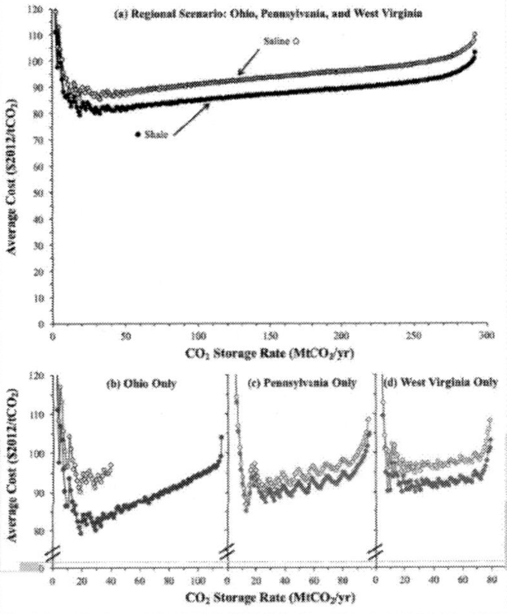


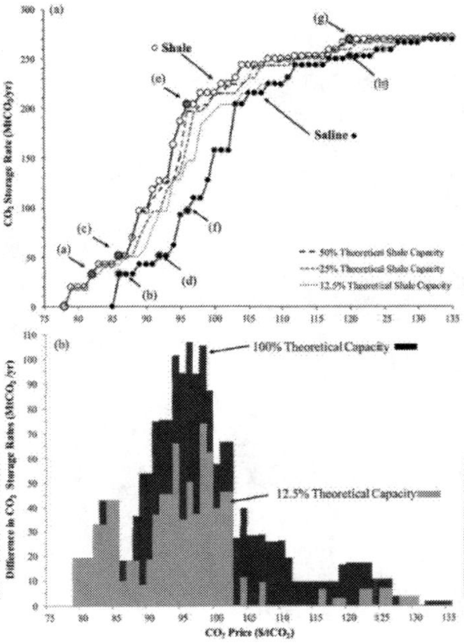


CMTC 2017
7.19.17

## summary

1. shale fracturing will create a large volume of pore space
   enhancing and preserving that pore space by avoiding water would be desirable

2. using $CO_2$ as a fracturing fluid could produce major emissions reductions
   but learning and optimization will only take place through trial and error

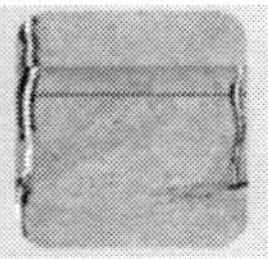

3. shale has a number of favorable characteristics for storage
   low permeability, high TOC, high pressure

4. the Marcellus shale in PA has significant capacity
   it could store between 10.4-18.4 Gigatonnes $CO_2$ before 2030

5. technoeconomics of the process for CCS are favorable
   storage in shales could reduce costs relative to greenfield saline sites by 5-10%

CMTC 2017
7.19.17

## acknowledgments

University of Virginia Engineering

NSF

CBET-
1134397
CBET-
1254839

CMTC 2017
7.19.17

CARBON MANAGEMENT TECHNOLOGY CONFERENCE

**CMTC-486622-MS**

# Surfactant and a Mixture of Surfactant and Nanoparticles Stabilized-$CO_2$/Brine Foam for Gas Mobility Control and Enhanced Oil Recovery

Z. A. AlYousef, M. A. Almobarky, and D. S. Schechter, Texas A&M University



## Abstract

Injecting carbon dioxide ($CO_2$) into oil reservoirs has the potential to enhance oil recovery (EOR) and mitigate climate change by storing $CO_2$ underground. Despite successes in using $CO_2$ to enhance oil recovery, mobility control remains a major challenge facing $CO_2$ injection projects. The objective of this work is to investigate the potential of using surfactant and a mixture of surfactant and nanoparticles (NPs) to generate foam to reduce gas mobility and enhanced oil recovery.

A newly developed anionic surfactant and a mixture of the surfactant and surface modified silica NPs were used to assess the ability of generating a stable foam at harsh reservoir conditions: sc-$CO_2$ and high temperature. Dynamic foam tests and coreflood experiments were conducted to evaluate foam stability and strength. To measure the mobility of injected fluids in sandstone rocks, the foam was generated by co-injection of sc-$CO_2$ and surfactant, as well as a mixture of surfactant and NPs at 90% quality. The coreflood experiments were conducted using non-fractured and fractured sandstone cores at 1550 psi and 50°C.

Surfactant alone and mixtures of surfactant and NPs were able to generate foam in porous media and reduce $CO_2$ mobility. The mobility reduction factor (MRF) for both cases was about 3.5 times higher than that of injecting $CO_2$ and brine at the same conditions. The coreflood experiments in non-fractured sandstone rocks showed that both surfactant and a mixture of surfactant and NPs were able to enhance oil recovery. The baseline experiment in the absence of surfactant resulted in a total recovery of 71.50% of the original oil in place (OOIP). Using surfactant brought the oil recovery to 76% of the OOIP. The addition of NPs to surfactant resulted in a higher oil recovery still, 80% of the OOIP. In fractured rocks, oil recoveries during secondary production mechanisms for the mixture, the surfactant alone, and sc-$CO_2$ alone were 12.62, 8.41 and 7.21% of the OOIP, respectively.

Large amounts of oil remain underground following primary and secondary oil production schemes. $CO_2$ has been widely used to enhance oil recovery. However, its high mobility might result in unfavorable and unsuccessful projects. The use of specially designed surfactants and the synergistic effect of surfactant and NPs may provide a solution to stabilize $CO_2$-brine foam at harsh reservoir conditions and, therefore, reduce gas mobility, consequently enhancing oil recovery.

## Introduction

The remaining oil underground following traditional recovery mechanisms is considerable (Hirasaki, Miller, and Puerto 2011). Typically, fields can produce about 45 to 50 % of the original oil in place (OOIP) following primary and secondary oil production mechanisms (Sandrea and Sandrea 2007). As a result, oil production is coming up short in meeting the ever increasing global energy demand (EIA, 2011). Enhanced oil recovery (EOR) techniques are needed to recover this huge amount of residual oil. $CO_2$ is used extensively to enhance oil recovery. Technically, it can promote swelling, reduce oil viscosity, vaporize, and extract portions of crude oil. Moreover, the easy solubility of $CO_2$ in oil makes it an ideal gas for EOR applications (Slobod and Koch, 1953, Enick, Holder, and Morsi 1988, Bayraktar and Kiran 2000). Despite the reported successes of $CO_2$ injection, a major challenge facing this technique is poor volumetric sweep efficiency. Major factors that contribute to this problem are the low density and viscosity of $CO_2$ relative to reservoir fluids, as well as reservoir heterogeneity such as high permeability and heavily fractured zones (Campbell and Orr 1985, Chakravarthy et al. 2004, Masalmeh et al. 2010). The high mobility of injected gas may lead to early breakthrough of gas, leaving most of the residual/ trapped oil untouched and increasing the gas to oil ratio (GOR). To solve the $CO_2$ injection issues, several approaches have been tested. The most often reported and applied approaches are: water alternating gas (WAG); generation of foams; and increasing gas viscosity by adding thickening agents (Christensen, Stenby, and Skauge 1998, Chkravarthy, et al. 2004, Enick 1998, Dalland and Hanssen 1996, Enick and Olsen 2012, Dandge and Heller 1987, Heller, J. 1994). The use of foam has the potential to reduce the gas mobility in petroleum reservoirs by increasing the gas apparent viscosity and reducing the gas relative permeability and, hence, improve the volumetric sweep efficiency (Falls et al. 1988, Kovscek and Radke 1994). However, the generation and stabilization of foam at reservoir conditions are major challenges. The major contributors to foam destabilization in porous media are: the harsh conditions such as reservoir temperature and salinity, surfactant adsorption to the rock, and the presence of crude oil (Mannhardt, Schramm, and Novosad 1993, Al-Hashim, H.S. et al. 1988, Figdore 1982, Grigg and Bai 2005).

Nanoparticles (NPs) have been used to stabilize $CO_2$/brine emulsions at reservoir conditions (Espinoza et al. 2010, Al Otaibi et al. 2013, Worthen et al. 2013). The use of specially designed surfactants and the synergistic effects of surfactant and NPs may help to stabilize $CO_2$/brine foams at harsh reservoir conditions and, therefore, reduce gas mobility, consequently enhancing oil recovery. For instance, Worthen et al. (2013) used non-modified silica NPs and caprylamidopropyl betaine (CAPB) surfactant. The mixture produced a stable and viscous $CO_2$-in-water foam when neither of these materials could stabilize foam individually at experimental conditions. Similarly, Singh et al. (2015) used fly ash powder and three types of surfactants: anionic, cationic and nonionic. In the presence of NPs, anionic and nonionic surfactants produced foam with smaller bubble size. In porous media, NPs and anionic surfactant produced a stable foam. Binks et al. (2015) reported a stable foam by mixing calcium carbonate ($CaCO_3$) particles and sodium stearoyl lactylate surfactant (SSL). Finally, Xue et al. (2016) found that mixing silica NPs and laurylamidopropyl betaine (LAPB) surfactant produced a viscous foam with small bubble sizes.

The objective of this study is to investigate foam strength using a newly developed anionic surfactant and the mixture of the surfactant and surface-modified silica NPs. Importantly, this study reports the $CO_2$ mobility reduction factor (MRF) and oil recovery factors as a result of using the surfactant and the mixture at 1550 psi and 50°C.

## Materials

The surfactant used in this study is a complex nanofluid (CNF) anionic surfactant. The NPs used are surface modified silica nanoparticles received in aqueous form from Nyacol Chemicals (DP 9711). The size of the particles was measured using Dynamic Light Scattering (DLS) and found to be 30 nm ± 1. Brine was prepared using deionized water (DI) (ASTM Type II, Lab Chem) and sodium chloride (99%, Cole-Parmer). The cores used in this study were non-fractured and fractured Bentheimer sandstone from Kocurek Industries. **Table 1** summarizes the properties of these cores. The oil used in this study was North Burbank Unit (NBU) oil with an average viscosity of 3.2 cp at 50°C.

**Table 1: Properties of Rock Samples**

| Sample # | Length (in) | Diameter (in) | Type of Rock | Porosity (%) | Pore Volume (ml) | Permeability (D) |
|---|---|---|---|---|---|---|
| 1 | 12 | 1 | Non-fractured | 21.76 | 33.61 | 1.50 |
| 2 | 12 | 1 | Non-fractured | 21.44 | 33.11 | 1.55 |
| 3 | 12 | 1 | Non-fractured | 21.20 | 32.74 | 1.72 |
| 4 | 12 | 1 | Non-fractured | 21.20 | 32.74 | 1.76 |
| 5 | 12 | 1 | Non-fractured | 21.84 | 33.74 | 1.77 |
| 6 | 12 | 0.96 | Fractured | 20.68 | 29.74 | - |
| 7 | 12 | 0.95 | Fractured | 19.90 | 27.74 | - |
| 8 | 12 | 0.95 | Fractured | 19.90 | 27.74 | - |

## Methodology

This study consists mainly of dynamic foam tests and coreflood experiments for $CO_2$, surfactant and the mixture of surfactant and silica NPs. The dynamic foam was generated using a coreflood apparatus, **figure 1**, and the $CO_2$ mobility was evaluated in rock samples at 1550 psi, 50°C, and 90% quality (the gas fractional flow in the co-injection process). At least five pore volumes (PVs) of 1 wt% brine were injected at 5 ft/day to ensure the sample was 100% saturated with brine. The BPR was set to be 1550 psi. The baseline experiment was conducted through a co-injection of sc-$CO_2$ and brine at 90% quality. For other experiments, the samples were pre-flushed with surfactant or a mixture of surfactant and NPs at 5 ft/day for 1 PV before starting the co-injection. Then, the co-injection of sc-$CO_2$ and surfactant/mixture was conducted also at 90% quality and the drop in pressure was recorded for each case. The same setup and at the same conditions, except that water was injected at 3 ft/d during waterflooding process, was used to conduct coreflood experiments to assess the ability of generated foam to reduce gas mobility and enhanced oil recovery. Non-fractured rocks were used to run the mobility tests while fractured and non-fractured rocks were used to conduct the coreflood experiments. For fractured rocks, the sample were initially 100 % saturated with crude oil. Fractures were created through the horizontal axis by cutting the rocks from the center.

During sample preparation, the diluted surfactant and NP solutions were stirred separately overnight to ensure homogeneity. The NPs were then added to the surfactant solution slowly, in stepwise fashion, to avoid aggregation of NPs. The size of NPs was measured before and after the mixing to verify that no extensive aggregation occurred during mixing. The concentration of surfactant and NPs used was 0.50 wt%. The brine was prepared with 1 wt% NaCl.

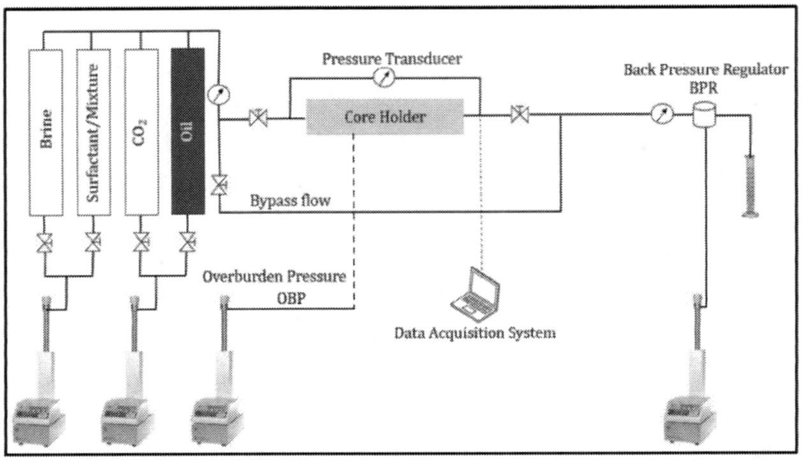


**Figure 1: Experimental Setup**

# Results and Discussion

## Dynamic Foam Tests

Comparisons here were based on recorded pressure drops across core samples and calculated MRF (steady state pressure drop/ steady state pressure drop for baseline experiment) for the three cases: baseline, surfactant, and mixture of surfactant and NPs. Rock sample #1 was used to conduct the baseline experiment. The results, as shown in **figure 2**, showed that the steady state pressure drop for the baseline experiment was about 0.29 psi. Bentheimer sample #2 was used to conduct the experiments in the absence and presence of NPs. In the absence of NPs, the foam behavior was excellent at the first PVs injected. After that, it had a sudden drop in the pressure values and it produced a foam with a steady state pressure drop of 0.88 psi, as shown in **figure 2**. The surfactant has the ability to reduce the $CO_2$-water IFT and generate foams, but the stability is challenging. In the presence of NPs, the behavior was similar to that in the absence of NPs. However, it had a lower foam generation ability in the first PVs injected. After 1.5 PVs of the co-injection process, as shown in **figure 3**, the mixture resulted in a slightly higher steady state pressure drop, 1 psi, than the surfactant case. This is an indication of the ability of NPs to produce a more stable foam in porous media. The permeability of the rocks used here was about 1.5 Darcy, so these reported values are still acceptable. The MRF values calculated for both the surfactant and mixture were found to be 3.04 and 3.45, respectively. This means that both the surfactant and the mixture were able to reduce the $CO_2$ relative permeability and increase the gas apparent viscosity, thereby reducing gas mobility.

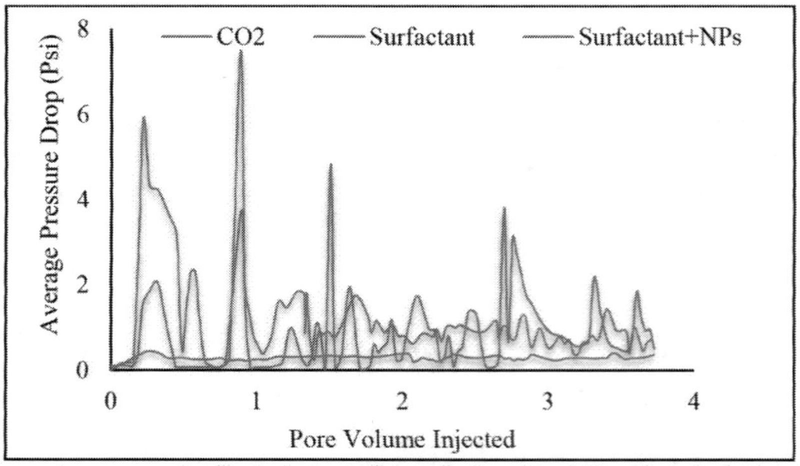


**Figure 2: Average pressure drop across the Bentheimer sandstone for baseline, 0.50 wt% surfactant and a mixture of 0.50 wt% surfactant and 0.50 wt% NPs at 50°C using $CO_2$**

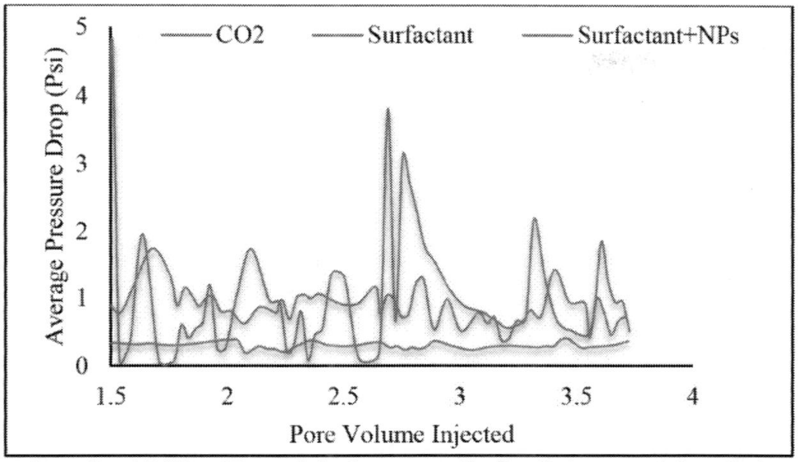

**Figure 3: Average pressure drop after 1.5 PVs of the co-injection across the Bentheimer sandstone for baseline, 0.50 wt% surfactant and a mixture of 0.50 wt% surfactant and 0.50 wt% NPs at 50°C using $CO_2$**

## Coreflood Experiments

Two sets of experiments were conducted to assess the ability of foam to enhance oil recovery, one in non-fractured rocks (3-5) and the other in fractured rocks (6-8).

### *Non-Fractured Rocks*

Coreflood experiments showed that both conditions, with and without NPs, improved oil recovery during foam injection processes, with higher recovery in the presence of NPs. **Figure 4** shows the results of coreflood experiments following waterflooding and $CO_2$ injection. Oil recovery following the waterflooding process was about 32.82% of the OOIP. At least 3.5 PVs of water were injected to ensure that no more oil could be recovered in this process and to diminish any capillary end effects that might exist. Then, $CO_2$ was injected at 5 ft/d and total oil recovery reached 71.50% of the OOIP. This means that $CO_2$ was able to produce about 38.68% of the OOIP and 57.58 of the remaining oil in place. The average pressure drop during $CO_2$ injection was about 0.36 psi.

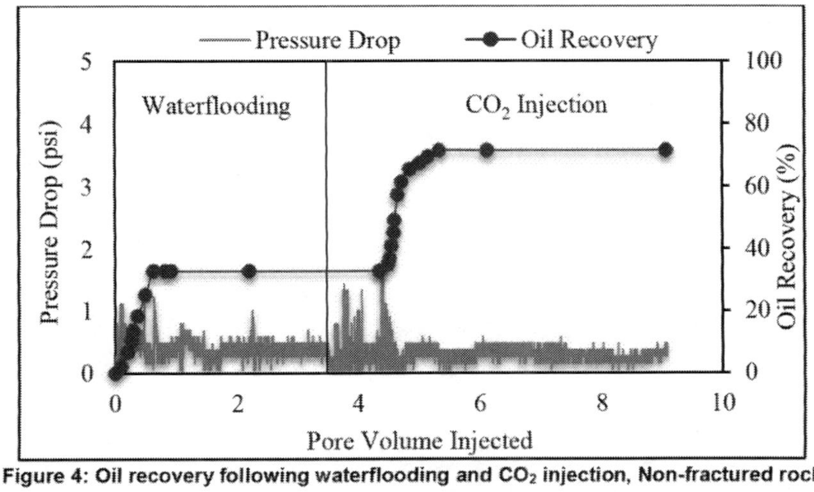

**Figure 4: Oil recovery following waterflooding and $CO_2$ injection, Non-fractured rock**

**Figure 5** shows the results of coreflood experiments when surfactant was used. Oil recovery following the waterflooding process was about 36.15 % of the OOIP. As before, at least 4 PVs of water were injected to ensure that no more oil could be recovered in this process and to diminish any capillary end effects that might exist. Then, 1 PV of surfactant was injected as a pre-flush step. The objective of this step was to minimize the adsorption that might occur during the co-injection processes. There was no significant amount of oil produced during the pre-flush step. The surfactant foam was able to produce about 39.90% of the OOIP and 62.50 % of the remaining oil in place. This brought the total oil recovery to around 76.06% of the OOIP. This is 4.56% higher than injecting $CO_2$ alone. The average pressure drop during the co-injection process of $CO_2$ and surfactant was about 0.71 psi. This is almost double that of injecting $CO_2$ alone.

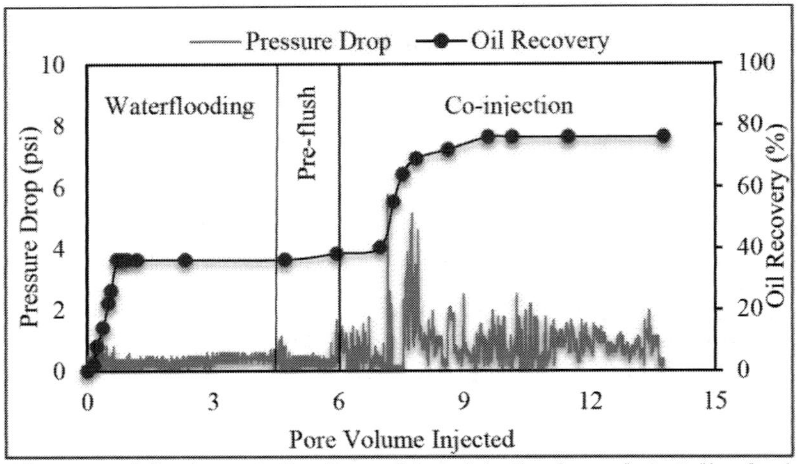


**Figure 5: Oil recovery following waterflooding and foam injection for surfactant, Non-fractured rock**

The next experiment, as shown in **figure 6**, was for the mixture of surfactant and NPs. The same procedures used in the previous experiment were used in this run. Oil recovery following the waterflooding process was about 35.73 % of the OOIP. The pre-flush with the mixture was not able to significantly recover any additional oil. During the co-injection processes, the mixture was able to produce about 44.33% of the OOIP and 68.97% of the remaining oil in place. The total oil recovery following the mixture foam process was around 80.05% of the OOIP. This is around 4% higher than the previous experiment where only surfactant was used and 8.55% higher than $CO_2$. The average pressure drop during the co-injection process of $CO_2$ and the mixture was about 1.16 psi. This is higher than both the surfactant and $CO_2$ cases.

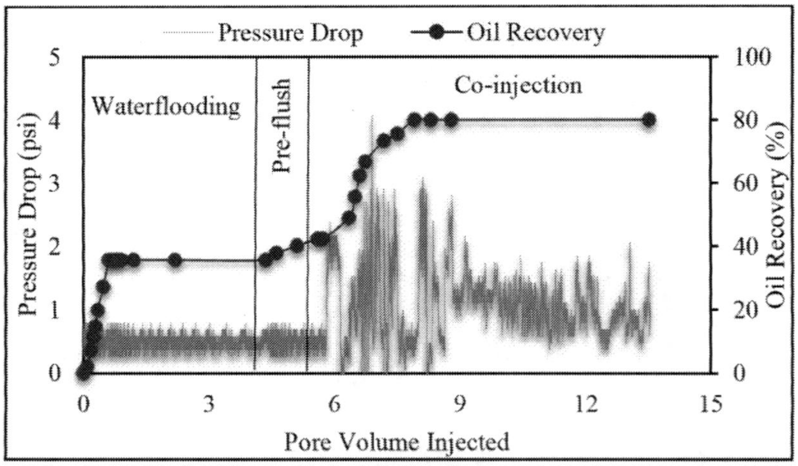


**Figure 6: Oil recovery following waterflooding and foam injection for a mixture of surfactant and NPs, Non-fractured rock**

A comparison between the three cases is presented in **figure** 7. The highest oil recovery was reported for the mixture while the lowest was for $CO_2$. The high oil recovery produced for $CO_2$ was because the experiment was conducted at or near the minimum miscibility pressure (MMP) of $CO_2$ in NBU oil. The higher oil recovery reported for surfactant compared to $CO_2$ demonstrates the ability of foam flooding to reduce gas mobility and enhance oil recovery. Also, the higher recovery of the mixture compared to that of surfactant demonstrates the ability of the presence of NP to further reduce gas mobility, improving the gas sweep efficiency and, therefore, recovering more oil.

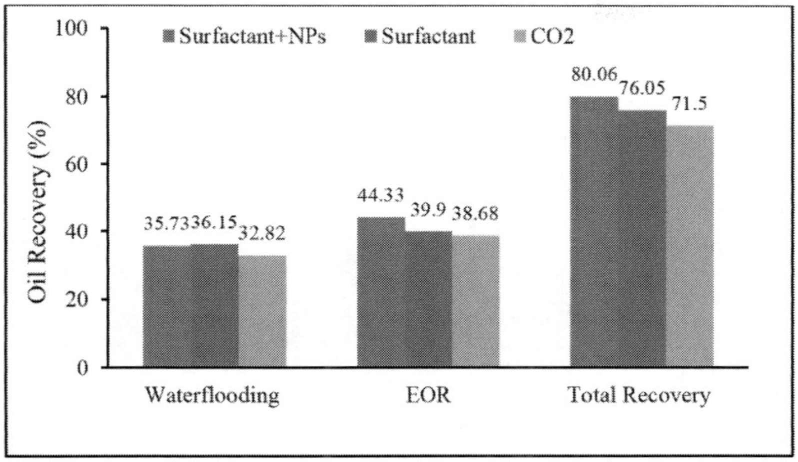


**Figure 7: Oil recovery following waterflooding, $CO_2$ and foam injection, Non-fractured rocks**

***Fractured Rocks***

Similar to the previous experiments, the results of coreflood experiments here on non-fractured rocks showed improved oil recovery during the foam injection processes, with higher recovery when NPs were used. **Figure 8** shows the results of coreflood experiments for the baseline case, surfactant and the mixture.

For the baseline experiment, the oil recovery following the waterflooding process was about 59.71% of the OOIP. At least 4 PVs of water were injected at 3 ft/d to ensure that no more oil could be recovered in this process and to diminish any capillary end effects that might exist. Then, $CO_2$ was injected at 5 ft/d and the total oil recovery reached 66.92% of the OOIP. This means that the $CO_2$ was able to produce about 7.21% of OOIP and 17.90% of the remaining oil in place.

For the surfactant case, the oil recovery following the waterflooding process was about 54.01% of the OOIP. At least 5.5 PVs of water were injected at 3 ft/d to ensure that no more oil could be recovered in this process and to diminish any capillary end effects that might exist. Then, 1 PV of surfactant was injected at 1.5 ft/d as a pre-flush step. There was no significant amount of oil produced during the pre-flush step. The co-injection process was conducted at 5 ft/d and 90% quality. The surfactant foam was able to produce about 8.41% of the OOIP and 18.28% of the remaining oil in place. This brought the total oil recovery to be around 62.42% of the OOIP. Even though the total oil recovery of $CO_2$ was higher than surfactant, the recovery factor during foam injection was higher than the $CO_2$ case. The surfactant produced 8.41% following waterflooding, whereas the $CO_2$ recovered 7.21% of the OOIP.

The next run, as shown in **figure 8**, was for the case where the mixture of surfactant and NPs was used. The same procedure as in the previous experiment was used in this run. The oil recovery following the waterflooding process was about 57.90% of the OOIP. A small amount of oil was produced during the pre-flush process. During the pre-flush and co-injection processes, the mixture was able to produce about 12.62% of the OOIP and 29.98% of the remaining oil in place. The total oil recovery following the mixture foam process was around 70.52% of the OOIP. This is around 8.10% higher than the previous experiment in which only surfactant was used and 3.60% higher than $CO_2$.

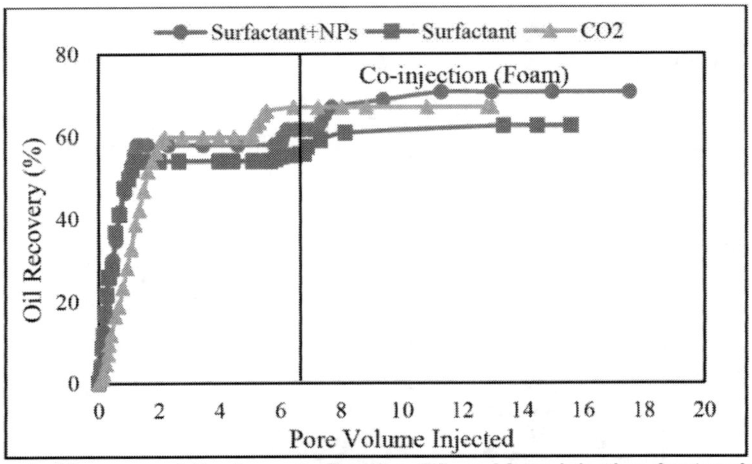


**Figure 8: Oil recovery following waterflooding, $CO_2$ and foam injection, fractured rock**

A comparison among the three cases is presented in **figure 9**. The highest oil recovery was reported for the mixture while the lowest was for surfactant. However, the results reported for surfactant compared to $CO_2$ are already discussed above. The high oil recovery reported for all cases was because the experiments were conducted at or near the minimum miscibility pressure (MMP) of $CO_2$ in NBU oil. Also, the rock samples were 100% saturated with oil. The higher oil recovery reported for surfactant compared to $CO_2$, at the secondary recovery scheme, demonstrates the ability of foam flooding to reduce gas mobility and, hence, to improve oil recovery. Also, the higher recovery of the mixture compared to that of surfactant and $CO_2$ demonstrates the ability of NPs to further reduce gas mobility, thus improving the sweep efficiency and, hence, recovering more oil. The summary of the performance of waterflooding and secondary recovery schemes can be found in **figure 9**.

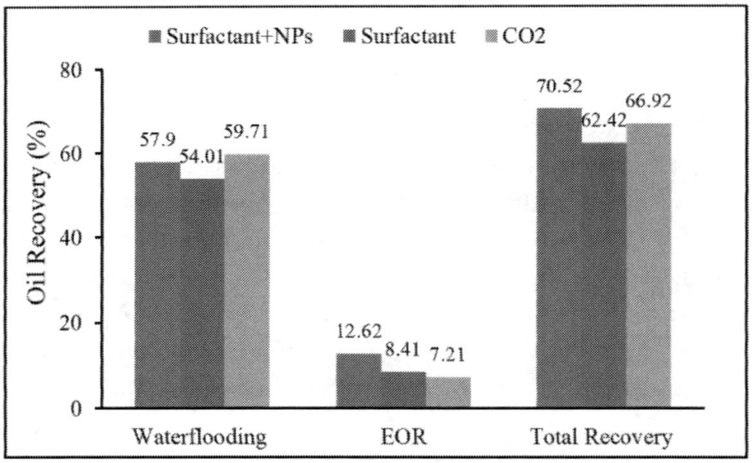


**Figure 9: Summary of coreflood experiments, fractured rocks**

## Conclusion

Anionic surfactant and a mixture of anionic surfactant and surface modified silica NPs were used in this study to assess the ability of the surfactant and the mixture to stabilize $CO_2$-brine foam at reservoir conditions. Dynamic foam tests were conducted to test the ability of surfactant and the mixture to generate foam in porous media and to reduce $CO_2$ mobility. Core flood experiments were performed in Bentheimer non-fractured and fractured sandstone rocks to examine the ability of generated foam to reduce gas mobility and to enhance oil recovery. Based on the results of dynamic foam tests and coreflood experiments:

- At harsh reservoir conditions, both surfactant and the mixture were able to reduce the sc-$CO_2$ mobility about 3-4 times.

- Using non-fractured rocks, the mixture of surfactant and NPs recovered about 80.05% of the OOIP. This is around 4% higher than surfactant and 8.55% higher than sc-$CO_2$.
- Using fractured rocks, the presence of NPs was able to improve the oil recovery compared to the surfactant and pure sc-$CO_2$ injection cases. The oil recoveries during secondary production mechanisms for $CO_2$, surfactant, and mixture were 7.21, 8.41 and 12.62% of the OOIP, respectively.

## Acknowledgment

The authors would like to acknowledge the chemical companies for providing the chemicals to conduct this study. We would like also to thank laboratory assistant, Rodolfo Marquez Vivas, for help with experiments.

## References

Al-Hashim, H. S., M. S. Celik, M. M. Oskay et al. 1988. Adsorption and Precipitation Behaviour of Petroleum Sulfonates from Saudi Arabian Limestone. *Journal of Petroleum Science and Engineering* **1** (4): 335-344.

Al Otaibi, Fawaz Mohammed, Sunil Lalchand Kokal, Yun Chang et al. 2013. Gelled Emulsion of $CO_2$-Water-Nanoparticles. SPE Annual Technical Conference and Exhibition, New Orleans, Louisiana, USA. SPE 166072.

Bayraktar, Zeynep, Erdogan Kiran. 2000. Miscibility, Phase Separation, and Volumetric Properties in Solutions of Poly(Dimethylsiloxane) in Supercritical Carbon Dioxide. *Journal of Applied Polymer Science* **75** (11): 1397-1403.

Binks, B. P., Campbell, S., Mashinchi, S., & Piatko, M. P. 2015. Dispersion Behavior and Aqueous Foams in Mixtures of a Vesicle-Forming Surfactant and Edible Nanoparticles. Langmuir, **31**(10): 2967–2978.

Campbell, Bruce T., Franklin M. Orr, Jr. 1985. Flow Visualization for $CO_2$/Crude-Oil Displacements. *Society of Petroleum Engineers Journal* **25** (5): 665-678.

Chakravarthy, D., V. Muralidharan, E. Putra et al. 2004. Application of X-Ray CT for Investigation of $CO_2$ and WAG Injection in Fractured Reservoirs. Canadian International Petroleum Conference, Calgary, Alberta. PETSOC-2004-232.

Christensen, J. R., E. H. Stenby, A. Skauge. 1998. Review of WAG Field Experience. International Petroleum Conference and Exhibition of Mexico, Villahermosa, Mexico. SPE-39883-MS.

Dalland, Mariann, Jan Erik Hanssen. 1997. Enhanced Foams for Efficient Gas Influx Control. International Symposium on Oilfield Chemistry, Houston, Texas. SPE-37217-MS.

Dandge, D. K., J. P. Heller. 1987. Polymers for Mobility Control in $CO_2$ Floods. SPE International Symposium on Oilfield Chemistry, San Antonio, Texas. SPE-16271-MS.

EIA. 2011. Annual Energy Outlook 2011 with Projections to 2035. *US Department of Energy.*

Enick, Robert M. 1998. A Literature Review of Attempts to Increase the Viscosity of Dense Carbon Dioxide. University of Pittsburgh, Pittsburgh, Pennsylvania.

Enick, Robert M., Gerald D. Holder, Badie I. Morsi. 1988. A Thermodynamic Correlation for the Minimum Miscibility Pressure in $CO_2$ Flooding of Petroleum Reservoirs. *SPE Reservoir Engineering* 3 (1): 81-92.

Enick, Robert Michael, David Kenneth Olsen, James Robert Ammer et al. 2012. Mobility and Conformance Control for $CO_2$ EOR via Thickeners, Foams, and Gels -- A Literature Review of 40 Years of Research and Pilot Tests. SPE Improved Oil Recovery Symposium, Tulsa, Oklahoma, USA. SPE-154122-MS.

Espinoza, David Alejandro, Federico Manuel Caldelas, Keith P. Johnston et al. 2010. Nanoparticle-Stabilized Supercritical $CO_2$ Foams for Potential Mobility Control Applications. SPE Improved Oil Recovery Symposium, Tulsa, Oklahoma, USA. SPE-129925-MS.

Falls, A. H., G. J. Hirasaki, T. W. Patzek et al. 1988. Development of a Mechanistic Foam Simulator: The Population Balance and Generation by Snap-Off. *SPE Reservoir Engineering* **3** (3): 884-892.

Figdore, Phillip Eugene. 1982. Adsorption of Surfactants on Kaolinite: NaCl versus $CaCl_2$ Salt Effects. *Journal of Colloid and Interface Science* **87** (2): 500-517.

Grigg, R. B., B. Bai. 2005. Sorption of Surfactant Used in $CO_2$ Flooding onto Five Minerals and Three Porous Media. SPE International Symposium on Oilfield Chemistry, The Woodlands, Texas. SPE-93100-MS.

Heller, J.P. 1994. Foams: Fundamentals and Applications in the Petroleum Industry. $CO_2$ Foams in Enhanced Oil Recovery, *ACS Advances in Chemistry* (Reprint) **242**: 201–234 . Washington DC.

Hirasaki, George, Clarence A. Miller, Maura Puerto. 2011. Recent Advances in Surfactant EOR. *SPE Journal* **16** (4): 889-907.

Kovscek, A. R., C. J. Radke. 1994. Fundamentals of Foam Transport in Porous Media. In *Foams: Fundamentals and Applications in the Petroleum Industry*, Chap. 3, 115-163. *Advances in Chemistry*, American Chemical Society.

Mannhardt, Karin, L. L. Schramm, J. J. Novosad. 1993. Effect of Rock Type and Brine Composition on Adsorption of Two Foam-Forming Surfactants. *SPE Advanced Technology Series* 1 (1): 212-218.

Masalmeh, Shehadeh K., Heiko Hillgartner, Rifaat Abdul-Munim Al-Mjeni et al. 2010. Simultaneous Injection of Miscible Gas and Polymer (SIMGAP) to Improve Oil Recovery and Sweep Efficiency from Layered Carbonate Reservoirs. SPE EOR Conference at Oil & Gas West Asia, Muscat, Oman. SPE-129645-MS.

Sandrea, Ivan, Rafael Sandrea. 2007. Global Oil Reserves - Recovery Factors Leaves Vast Target for EOR Technologies. *Oil & Gas Journal* **105** (41): 1-8.
Singh, Robin, Abhay Gupta, Kishore K. Mohanty et al. 2015. Fly Ash Nanoparticle-Stabilized $CO_2$-in-Water Foams for Gas Mobility Control Applications. SPE Annual Technical Conference and Exhibition, Houston, Texas, USA. SPE-175057-MS.
Slobod, R. L., H. A. Koch, Jr. 1953. High-Pressure Gas Injection- Mechanism of Recovery Increase. Drilling and Production Practice, New York, USA. API-53-082.
Worthen, Andrew J., Hitesh G. Bagaria, Yunshen Chen et al. 2013. Nanoparticle-Stabilized Carbon Dioxide-in-Water Foams with Fine Texture. *Journal of Colloid and Interface Science* **391**: 142-151.
Worthen, Andrew J., Steven L. Bryant, Chun Huh et al. 2013. Carbon Dioxide-in-Water Foams Stabilized with Nanoparticles and Surfactant Acting in Synergy. *AIChE Journal* **59** (9): 3490-3501.
Xue, Zheng, Andrew Worthen, Ali Qajar et al. 2016. Viscosity and Stability of Ultra-High Internal Phase $CO_2$-in-Water Foams Stabilized with Surfactants and Nanoparticles with or without Polyelectrolytes. *Journal of Colloid and Interface Science* **461**: 383-395.

# AUTHOR INDEX

# AUTHOR INDEX

**AIChE**
120 Wall Street, FL 23
New York, NY 10005-4020

ISBN 978-1-5108-4815-3

9 781510 848153

CURRAN ASSOCIATES INC.
proceedings.com

# Carbon Management Technology Conference (CMTC 2017)

Global CCUS Innovation Nexus

Houston, Texas, USA
17-20 July 2017

Volume 2 of 2

AIChE
The Global Home of Chemical Engineers

# Carbon Management Technology Conference (CMTC 2017)

## Global CCUS Innovation Nexus

Houston, Texas, USA
17-20 July 2017

Volume 2 of 2

ISBN: 978-1-5108-4815-3

**Printed from e-media with permission by:**

Curran Associates, Inc.
57 Morehouse Lane
Red Hook, NY 12571

**Some format issues inherent in the e-media version may also appear in this print version.**



Printed by Curran Associates, Inc. (2017)

For permission requests, please contact AIChE
at the address below.

AIChE
120 Wall Street, FL 23
New York, NY 10005-4020

Phone: (800) 242-4363
Fax: (203) 775-5177

www.aiche.org

**Additional copies of this publication are available from:**

Curran Associates, Inc.
57 Morehouse Lane
Red Hook, NY 12571 USA
Phone: 845-758-0400
Fax: 845-758-2633
Email: curran@proceedings.com
Web: www.proceedings.com

# TABLE OF CONTENTS

CMTC-486652-MS

# Field Testing of Cryogenic Carbon Capture

Aaron Sayre, Dave Frankman, Andrew Baxter, and Kyler Stitt, Sustainable Energy Solutions; Larry Baxter, Brigham Young University



## Abstract

Sustainable Energy Solutions has been developing Cryogenic Carbon Capture™ (CCC) since 2008. In that time two processes have been developed, the External Cooling Loop and Compressed Flue Gas Cryogenic Carbon Capture processes (CCC ECL™ and CCC CFG™ respectively).

The CCC ECL™ process has been scaled up to a 1TPD $CO_2$ system. In this process the flue gas is cooled by an external refrigerant loop. SES has tested CCC ECL™ on real flue gas slip streams from subbituminous coal, bituminous coal, biomass, natural gas, shredded tires, and municipal waste fuels at field sites that include utility power stations, heating plants, cement kilns, and pilot-scale research reactors. The $CO_2$ concentrations from these tests ranged from 5 to 22% on a dry basis. $CO_2$ capture ranged from 95-99+% during these tests. Several other condensable species were also captured including $NO_2$, $SO_2$ and PMxx at 95+%. NO was also captured at a modest rate.

The CCC CFG™ process has been scaled up to a .25 ton per day system. This system has been tested on real flue gas streams including subbituminous coal, bituminous coal and natural gas at field sites that include utility power stations, heating plants, and pilot-scale research reactors. $CO_2$ concentrations for these tests ranged from 5 to 15% on a dry basis. $CO_2$ capture ranged from 95-99+% during these tests. Several other condensable species were also captured including $NO_2$, $SO_2$ and PMxx at 95+%. NO was also captured at 90+%. Hg capture was also verified and the resulting effluent from CCC CFG™ was below a 1ppt concentration.

This paper will focus on discussion of the capabilities of CCC, the results of field testing and the future steps surrounding the development of this technology.

## Cryogenic Carbon Capture (CCC)

Since 2008 Sustainable Energy Solutions has developed a process called Cryogenic Carbon Capture (CCC). The CCC process is a true bolt-on technology that can easily be retrofit on virtually any stationary emission source without modification to the boiler, turbine, or steam piping. Using assumptions and modeling techniques from NETL base case 11 the Cryogenic Carbon Capture process is projected to cost

\$35/tonne $CO_2$ avoided (~\$30/tonne captured) but because of unique features of the CCC process this cost can be significantly reduced (Figure 1).

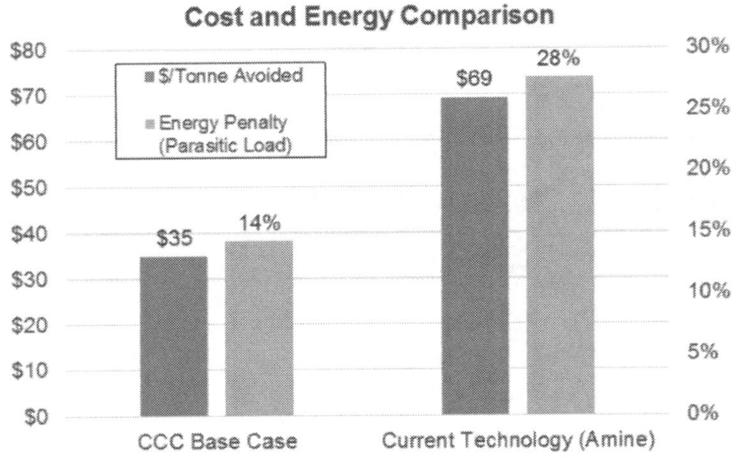


**Figure 1. Cost and Energy Comparison**

Leveraging existing power plant assets with minimal integration is the most cost-effective way to reduce $CO_2$ emissions from fossil-fueled power. CCC is an enabling technology that installs on existing infrastructure and greenfield infrastructure alike with no intrusive integration. Using existing infrastructure reduces costs below \$30/tonne $CO_2$ captured. The cost is reduced further as the CCC process removes $SO_2$, $NO_2$, Hg, and other regulated pollutants. This offsets the related emission-control costs. The CCC process can also provide efficient large-scale energy storage and demand response capabilities that minimize the parasitic load during peak demand and apply that parasitic load during non-peak demand or when intermittent renewable energy comes online. These unique features make it possible for Cryogenic Carbon Capture to be revenue positive in certain markets.

**How CCC™ Works**

The CCC process utilizes a large amount of recuperative heat exchange through commercially-available heat exchangers to cool the flue gas to the verge of $CO_2$ desublimation before it enters a proprietary desublimating heat exchanger which desublimates, or freezes, the $CO_2$ out of carrier gas stream. The solid $CO_2$ is separated from the carrier gas and melted for heat recovery then delivered as a high-pressure high-purity liquid. Because the $CO_2$ is compressed as a liquid rather than a gas the energy input for this step is minimal. The cold light gasses ($N_2$, $O_2$ and others) do not condense but they do return through the recuperator for energy recovery. This minimizes the cooling load on the desublimator.

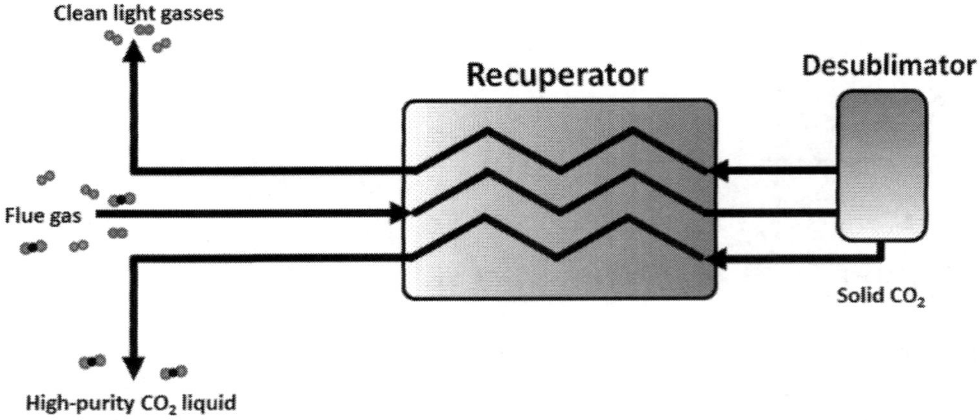


**Figure 2. How CCC Works**

A frequent mistake people make in energy calculations for this process is to treat the CCC process as a typical refrigeration process. They may look at the energy required to cool the gas from its warmest temperature to its coldest temperature while neglecting the energy that is recovered through recuperative heat exchange. In a typical refrigeration process a stream enters at one temperature and leaves at a lower temperature. In this process the gas is cooled to very low temperatures but most of the energy used to cool the gas is recovered through heat exchange and recuperation so the gas leaving is only slightly colder than the gas coming into it. The energy requirement for the CCC process is the energy required to make up for inherent losses through the process and drive the phase change of the $CO_2$. The diagram below (Figure 3) represents the conceptual thermal energy to cool the gas stream ($T_2$), the thermal energy recovered through the recuperator ($T_2$-$T_1$), and the thermal energy required to drive the process and make up losses ($T_1$).

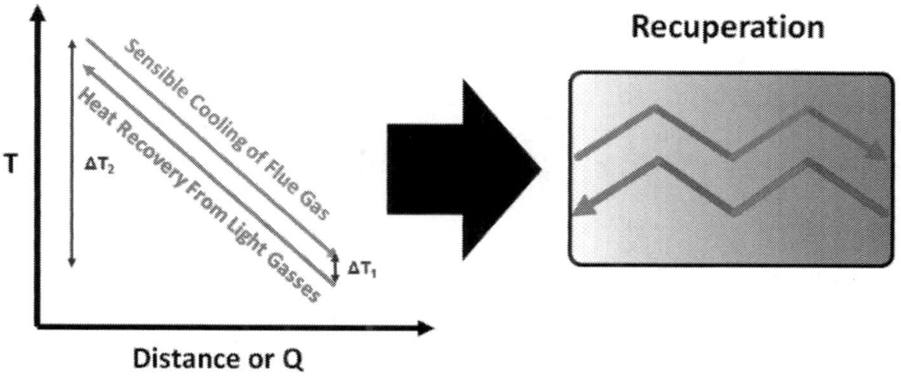


**Figure 3. Thermal Energy Recovery Through Recuperator**

Pressure loss through the system is an energy load that is not recoverable; it accounts for 10-15% of the total load of the system.

## CCC™ Demo Systems

In 2014 SES began field testing two demo systems to gather capture data on flue gas from real sources. One of these systems, called the External Cooling Loop or ECL system, was funded by the Wyoming Advanced Conversion Technology Task Force (Figure 4). This system processes a flue gas stream at about 1.5 bar absolute (about 7 psig). It uses an external refrigeration loop to provide cooling to the desublimator. The second demo system was funded through the Department of Energy's ARPA-E program (Figure 5). Instead of using an external cooling loop this process compresses the flue gas to about 6 bar absolute (about 87 psig) and then expands it to cool it directly; this is sometimes called a self-cooled system. Both demo systems were built in intermodal shipping containers to make them easy to transport from site to site.

**Figure 4. 1 tonne/day ECL™ demo system**

Figure 5. 0.25 tonne/day CFG™ demo system

## CCC™ Field Testing

### CCC-ECL Field Testing at BYU

In August of 2014 SES began testing the ECL at Brigham Young University's coal combustion research laboratory on their Burner Flow Reactor (BFR). Using the BFR allowed us to test a variety of coal and biomass mixtures. We tested Wyoming Black Thunder coal as well as other bituminous and subbituminous coal and natural gas. We also co-fired 90% (mass) Black Thunder Wyoming coal with 10% biomass (finely ground hardwood) and captured the $CO_2$ from the flue gas. Greater than 98% $CO_2$ was captured. This demonstrated the ability of CCC to produce a negative carbon emission in a realistic scenario. All $CO_2$ from the coal was captured along with the $CO_2$ from the carbon-neutral biomass creating a net-negative atmospheric $CO_2$ emission.

Another highlight from this testing was the measurement of particulate matter in the flue gas stream before and after the CCC™ process. Measurements were taken for particles between 10 µm and 2.5 µm with 98% overall reduction (Figure 7).

Figure 6. CCC-ECL™ System at BYU (left) Viewport on Burner Flow Reactor (right)

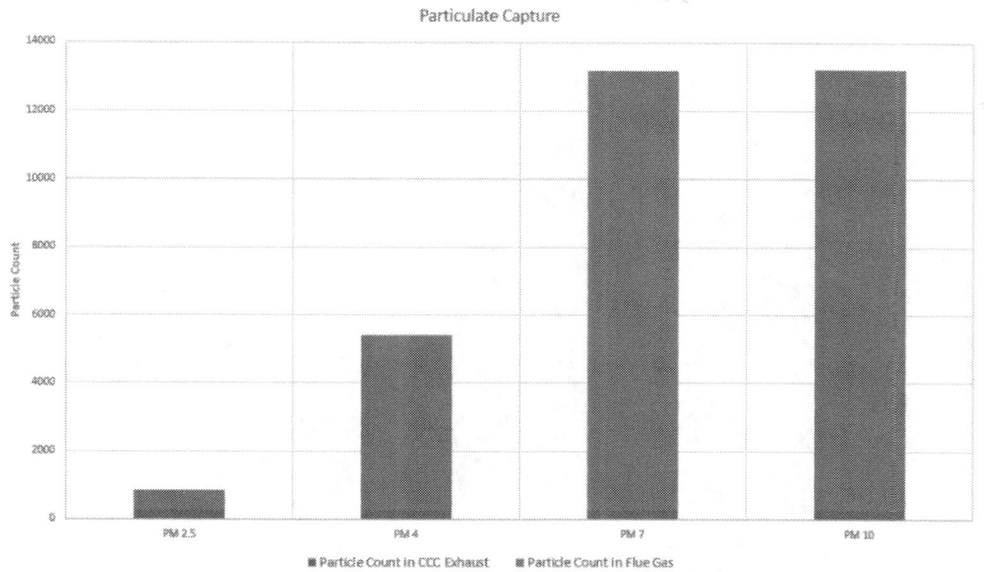


**Figure 7. Particulate Capture Data from Initial BYU Testing**

## CFG Shakedown Testing at BYU

In October of 2014 SES began testing the CFG system at the BYU heating plant. At the time of testing the BYU Heating Plant burned Utah Skyline coal before November 1 and natural gas after November 1 so we set up the system so we could test it on both fuels without changing the setup. Both tests went very well. After tests at the heating plant were completed we moved the skid across the parking lot and continued testing at the Burner Flow Reactor with various coal mixtures. In each case over 90% $CO_2$ capture was achieved with minimal upsets from source to source.

**Figure 8. CCC-CFG™ system being unloaded at BYU Heating Plant**

## CCC-CFG ™ Testing at a Coal-Fired Power Station in Wyoming

In late November Pacificorp hosted us for testing at one of their coal-fired power stations near Glenrock, Wyoming. At this site they burn a Wyoming Powder River Basin (PRB) coal. We were set up taking a slip stream from unit #3 through a port on the stack. While we were there we completed our first 12 and 24 hour tests with the CCC-CFG system and significantly improved the automation and controls in our process.

Figure 9. CCC-CFG testing at Pacificorp Power Station Near Glenrock, Wyoming

$CO_2$ inlet concentration varied during runs but outlet concentration was consistent and followed thermodynamically predicted concentrations. Figures 10 and 11 show inlet and outlet concentrations for three 8+ hour runs with moving averages. $CO_2$ spikes in outlet concentration on 19-Dec-14 are the result of an experimental procedure to mitigate fouling in the system.

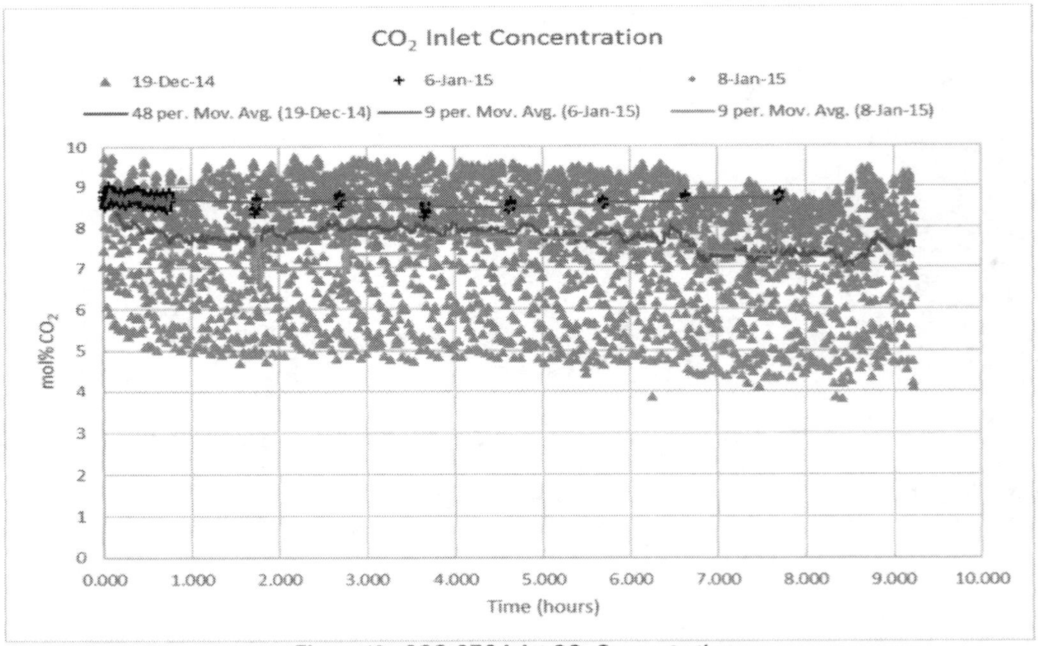


Figure 10. CCC-CFG Inlet $CO_2$ Concentrations

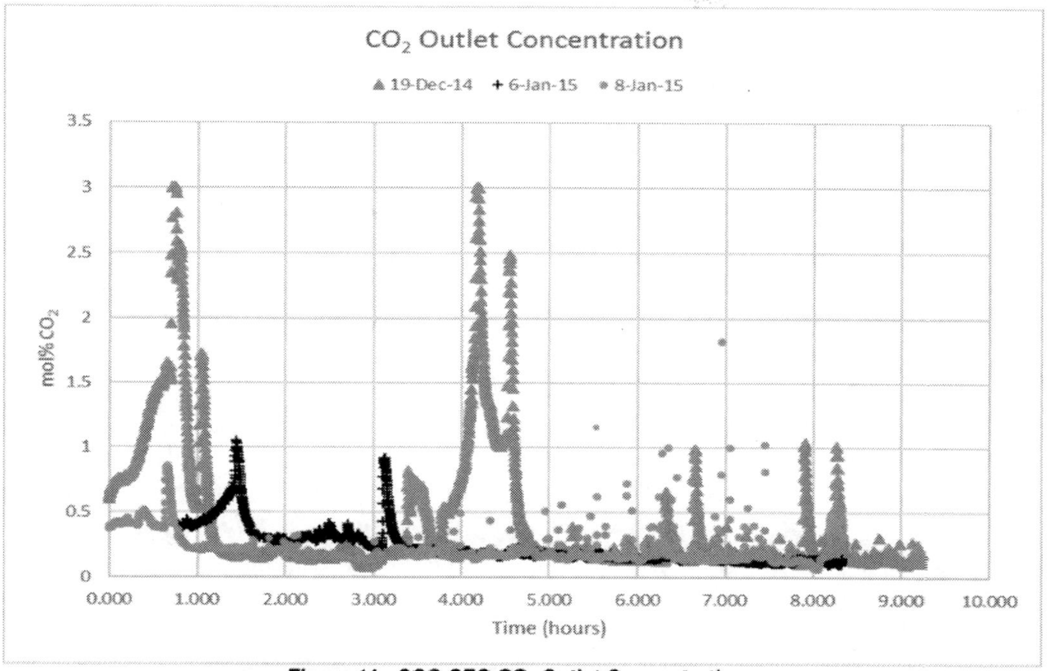


**Figure 11. CCC-CFG $CO_2$ Outlet Concentrations**

The CCC process captures pollutant species that condense at $CO_2$ desublimation temperatures. The capture of $CO_2$, $SO_2$, $NO_2$, and Hg have been measured by SES. The system also captured over 99% of the NO during a run in January.

| | | 1/6/2015 | 1/8/2015 |
|---|---|---|---|
| NO In | ppm | 57.825 | 23.941 |
| NO Out | ppm | 0.311 | 1.834 |
| NO Capture | % | 99.5% | 92.3% |

**Table 1. NO Capture Data at PacifiCorp Site**

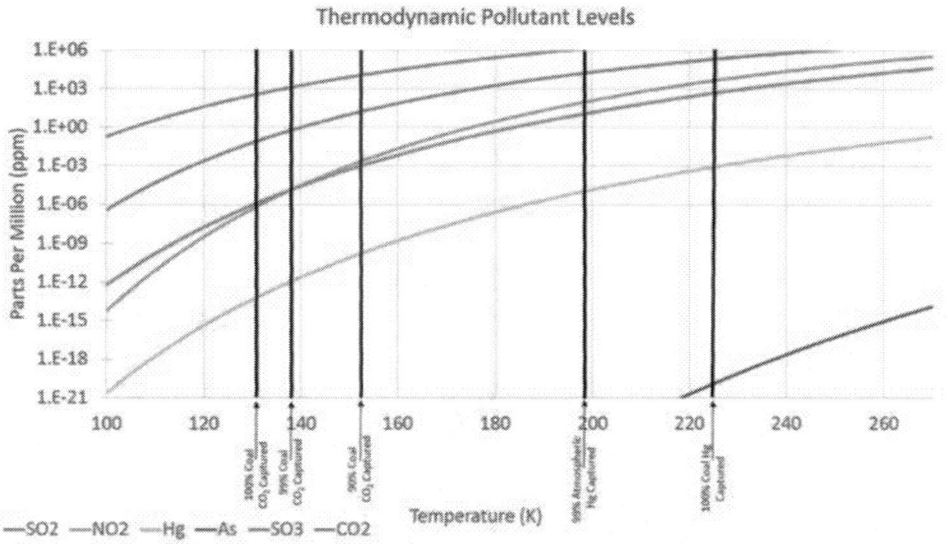


**Figure 12. Theoretical Pollutant Capture at Various Temperatures**

NO capture was somewhat unexpected when we we first measured it because it condenses at temperatures that are lower than the desublimation temperatures of $CO_2$. After further research, testing, and collaboration with industry experts we found that NO reacts with $O_2$ at low temperatures to form $NO_2$

which then condenses and is captured by the CCC process. This reaction is especially favorable at higher pressures in the CFG-CCC process but both CFG-CCC and ECL-CCC systems have demonstrated NO capture. Since our first NO capture observation we have run additional tests in the ECL system and additional laboratory experiments which repeatedly show NO capture in the CCC process.

During tests in Wyoming we contracted a 3rd party emissions testing company to measure Hg emissions before and after the CCC process. Hg emissions from the stack were recorded at 735 pptv or 5.77 μg/m$^3$. After the CCC unit, Hg emissions were below the detection limit of 1 pptv or 0.01 μg/m$^3$. This means that Hg levels after the CCC process are lower than atmospheric Hg levels.

**CCC-ECL Testing at a Coal-Fired Power Station in Wyoming**

In February of 2015 we brought the CCC-CFG back to Orem for additional closed-loop testing with simulated flue gas and we took the CCC-ECL system out to the same power station where we had previously tested the CCC-CFG unit. One of the highlights of these tests was demonstrating the demand response capability of the CCC-ECL. During non-peak demand the CCC-ECL process can liquify extra refrigerant then use that refrigerant at peak demand times while keeping the compressors and other refrigeration equipment turned off. This reduces the energy requirement of the process by about 85%.

Figure 13. Testing Energy Storage at SES Research Lab

We liquified extra natural gas and nitrogen as sample refrigerants during test runs and then turned off all powered equipment in the refrigeration loop and continued to run at >90% $CO_2$ capture with minimal load from the process. Burning the spent refrigerant gives an additional benefit because during peak demand times after the natural gas refrigerant has been vaporized to provide cooling it can be burned in a quick-response simple-cycle natural gas turbine to supplement power need. If the simple-cycle turbine is integrated into the power station, its flue gas could be reinjected into the boiler giving the response time of a simple-cycle turbine with the efficiency of a combined-cycle gas turbine.

Figure 14. Unloading the ECL system at the PacifiCorp Power Station in Wyoming

Figure 15. Testing the ECL on Unit #3 at the Power Station

## CCC-ECL Testing at a Commercial Cement Plant in Morgan, Utah

In June of 2015 a cement producer hosted our CCC-ECL system at their plant in Morgan, Utah. Cement plants produce $CO_2$ in higher concentrations than most sources because $CO_2$ is released in the cement production process and added to the $CO_2$ from the fuel used to heat the process. Consequently, cement plants can produce flue gas streams with upwards of 30% (mol) $CO_2$. In our tests $CO_2$ concentrations were typically around 22% (mol) and were sometimes higher. The plant we tested at set us up at a port near the bypass baghouse. In addition to burning coal at their site they also burn shredded tires and municipal waste. It provided a great opportunity for testing because we tested the CCC process on flue gas with high concentrations of $CO_2$ from unique fuel sources. The CCC process easily separates $CO_2$ in both low and high concentrations. Table 2 shows flue gas inlet and outlet concentrations from various runs as measured using a MKS fourier-transform infrared spectroscope (FTIR).

Figure 16. CCC-ECL System at the Cement Plant in Morgan, Utah

| Date | Temp | | | CO2 in | | | CO2 out | | | CO2 capture | | | Start | End | Total |
|---|---|---|---|---|---|---|---|---|---|---|---|---|---|---|---|
| | Avg | Min | Max | Avg | Min | Max | Avg | Min | Max | Avg | Min | Max | | | Hours |
| 6/12/2015 | -138.6 | -140.9 | -127.2 | 20.2 | 18.4 | 20.9 | 0.62 | 0.27 | 1.04 | 97.5 | 95.9 | 99.0 | 15:30 | 17:04 | 1.57 |
| 6/16/2015 | -137.8 | -142.9 | -120.7 | 16.9 | 13.9 | 18.1 | 0.63 | 0.15 | 1.45 | 96.9 | 93.0 | 99.1 | 9:44 | 12:25 | 2.68 |
| 6/17/2015 | -133.8 | -137.4 | -128.8 | 21.0 | 17.0 | 22.1 | 0.40 | 0.24 | 0.72 | 98.5 | 97.3 | 99.1 | 10:11 | 14:06 | 3.92 |
| 6/18/2015 | -135.7 | -140.5 | -122 | 19.2 | 16.8 | 20.8 | 0.58 | 0.22 | 1.54 | 97.5 | 93.4 | 99.0 | 12:10 | 16:25 | 4.25 |
| 6/19/2015 | -134.7 | -138.4 | -123.1 | 21.2 | 18.9 | 23.6 | 0.57 | 0.25 | 0.95 | 97.9 | 96.5 | 99.0 | 10:15 | 16:19 | 6.07 |
| 6/22/2015 | -132.7 | -137.6 | -111.2 | 21.1 | 19.3 | 22.4 | 0.74 | 0.29 | 1.69 | 97.2 | 93.4 | 98.9 | 12:31 | 16:45 | 4.23 |
| 6/23/2015 | -132.9 | -137.7 | -113.5 | 21.8 | 14.7 | 23.1 | 0.70 | 0.30 | 1.33 | 97.5 | 95.4 | 98.9 | 12:54 | 16:15 | 3.35 |
| 6/25/2015 | -127.7 | -135.7 | -114.4 | 20.7 | 18.2 | 21.5 | 0.57 | 0.25 | 1.39 | 97.8 | 94.8 | 98.9 | 8:53 | 15:28 | 6.58 |
| 6/30/2015 | -132.9 | -137.4 | -128.6 | 17.6 | 13.0 | 19.8 | 0.50 | 0.23 | 0.68 | 97.6 | 96.5 | 98.7 | 10:09 | 11:35 | 1.43 |
| 6/30/2015 | -127.0 | -136.6 | -122 | 20.1 | 7.3 | 21.6 | 0.73 | 0.29 | 0.95 | 97.1 | 95.1 | 97.7 | 13:18 | 13:36 | 0.30 |
| 6/30/2015 | -128.2 | -140.9 | -116.1 | 21.4 | 16.1 | 24.3 | 0.67 | 0.21 | 1.66 | 97.5 | 93.7 | 99.1 | 14:52 | 17:10 | 2.30 |
| 7/1/2015 | -128.1 | -135.4 | -113.6 | 20.4 | 16.5 | 22.8 | 0.61 | 0.25 | 2.31 | 97.6 | 90.6 | 99.0 | 11:16 | 20:33 | 9.28 |
| 7/2/2015 | -127.8 | -132.4 | -122.3 | 18.4 | 16.9 | 19.9 | 0.57 | 0.32 | 1.12 | 97.4 | 94.8 | 98.7 | 10:12 | 12:35 | 2.38 |
| 7/7/2015 | -130.1 | -136.3 | -120 | 21.6 | 19.9 | 22.5 | 0.65 | 0.25 | 1.16 | 97.6 | 95.8 | 99.0 | 11:35 | 13:23 | 1.80 |
| 7/8/2015 | -127.1 | -134.4 | -122.3 | 21.7 | 12.9 | 23.5 | 0.51 | 0.18 | 1.21 | 98.1 | 95.4 | 99.2 | 10:15 | 15:08 | 4.88 |
| 7/9/2015 | -126.5 | -131.3 | -124.9 | 19.9 | 18.4 | 20.6 | 0.62 | 0.35 | 0.76 | 97.5 | 96.9 | 98.5 | 9:50 | 13:40 | 3.83 |
| 7/10/2015 | -126.3 | -134.2 | -118.5 | 20.5 | 17.2 | 21.6 | 0.77 | 0.24 | 1.79 | 97.0 | 93.0 | 99.0 | 7:26 | 11:26 | 4.00 |
| 7/15/2015 | -128.7 | -136.5 | -111.2 | 21.0 | 18.3 | 23.3 | 0.76 | 0.34 | 1.69 | 97.1 | 93.3 | 98.7 | 10:27 | 19:26 | 8.98 |
| 7/16/2015 | -126.5 | -136.8 | -116.3 | 20.0 | 17.8 | 21.3 | 0.70 | 0.19 | 2.17 | 97.2 | 91.2 | 99.1 | 9:11 | 12:57 | 3.77 |
| 7/18/2015 | -125.3 | -136.9 | -114.8 | 19.7 | 17.2 | 21.6 | 0.81 | 0.18 | 1.54 | 96.7 | 93.5 | 99.1 | 12:24 | 13:21 | 0.95 |
| 7/21/2015 | -126.6 | -128 | -120.6 | 20.5 | 19.0 | 21.9 | 0.82 | 0.64 | 1.14 | 96.8 | 95.4 | 97.5 | 15:59 | 19:37 | 3.63 |

**Table 2. Run Data from Various Runs at The Cement Plant**

$CO_2$ outlet concentration can be predicted using the pressure and temperature in the desublimator. Figure 17 shows agreement between measured $CO_2$ capture data and predicted capture at the cement plant.

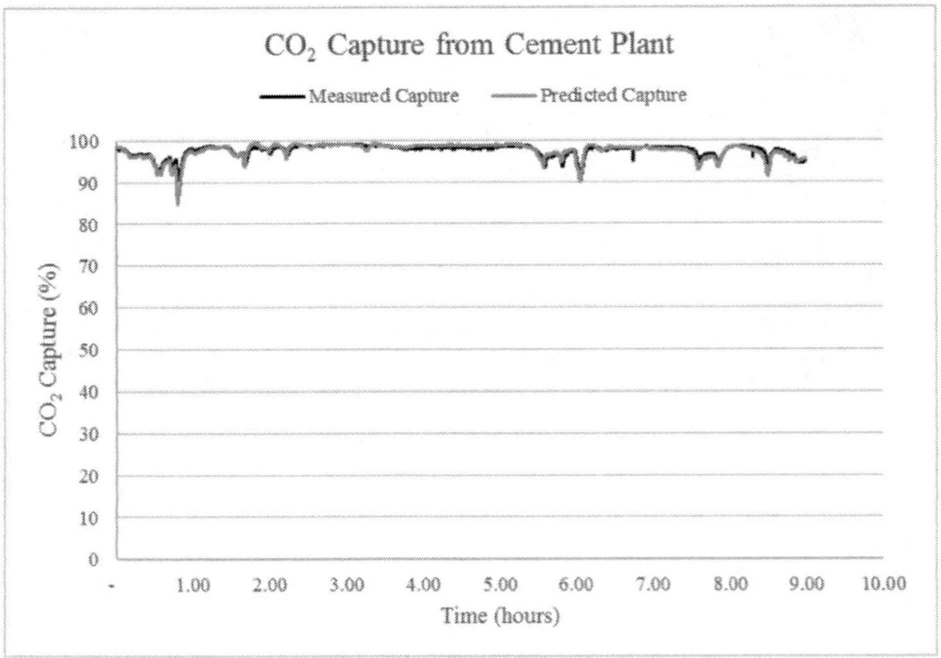


**Figure 17. Measured and Predicted CO2 Capture Data**

## Current Testing and Future Plans

In 2015 and 2016 SES demonstrated the CCC-CFG system at other sites in Utah and Wyoming. We continue to do closed-loop and unit-operations testing daily at our lab in Orem, Utah. These tests focus on improving process reliability and efficiency and developing intellectual property around the process.

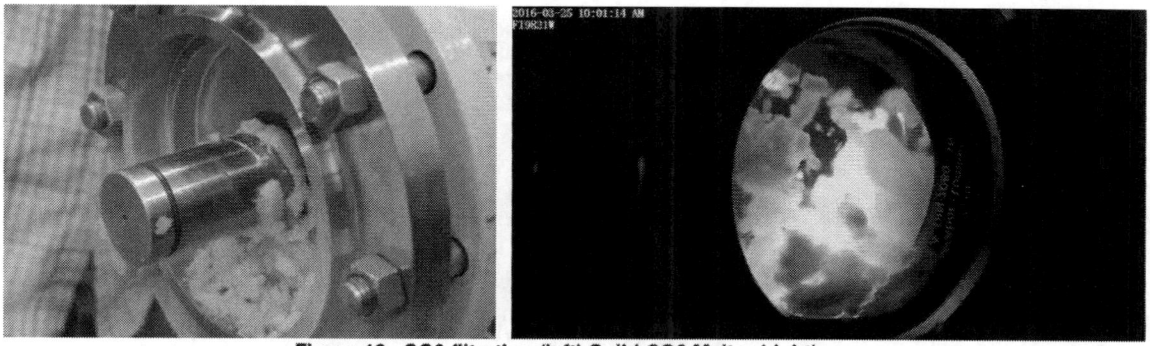

**Figure 18. CO2 filtration (left) Solid CO2 Melter (right)**

We are developing CCC technology for small commercial deployment. We are currently working under funding from NETL, Rocky Mountain Power, and others. Our current work will culminate in long-term testing and demonstration at a commercial coal-fired power station next year. We are interested in working with anyone interested in partnering or assisting in the scale up and deployment of this technology. Additional information can be found at https://sesinnovation.com/.

## Acknowledgements

This work was funded by Sustainable Energy Solutions LLC of Orem, Utah, the Advanced Research Projects Agency – Energy (ARPA-E), U.S. Department of Energy, under Award Number DE-AR0000101, the Climate Change and Emissions Management Corporation (CCEMC) of Alberta, Canada, and the Advanced Conversion Technologies Task Force in Laramie, Wyoming. Thank you to PacifiCorp and Holcim and their employees for their support in site preparation and testing at their facilities. Sustainable Energy Solutions owns the Cryogenic Carbon Capture (CCC) technology portfolio.

## Disclaimer

The information, data, or work presented herein was funded in part by an agency of the United States Government. Neither the United States Government nor any agency thereof, nor any of their employees, makes any warranty, express or implied, or assumes any legal liability or responsibility for the accuracy, completeness, or usefulness of any information, apparatus, product, or process disclosed, or represents that its use would not infringe privately owned rights. Reference herein to any specific commercial product, process, or service by trade name, trademark, manufacturer, or otherwise does not necessarily constitute or imply its endorsement, recommendation, or favoring by the United States Government or any agency thereof. The views and opinions of authors expressed herein do not necessarily state or reflect those of the United States Government or any agency thereof.

## Works Cited

US Department of Energy, N.E.T.L., Cost and Performance Baseline for Fossil Energy Plants Volume 1: Bituminous Coal and Natural Gas to Electricity. Revision 2a, September 2013.

CMTC-486671-MS

# $CO_2$-EOR in Fractured Ultra-Low Permeability Reservoirs: Problems and Remedial Measures

Xiaochun Liu[1], Liping Ma[1], Junling Tan[1], Tangying Yang[1], Xiaorong Li[1], Jirui Hou[2], Qi Wei[2], Hongda Hao[2], Zhaojie Song[2], Shitou Wang[1] and Weiyu Bi[1], (1) Research Institute of Oil and Gas Technology, Changqing Oilfield, Petrochina, Xi'an, China, (2) Research Institute of Enhanced Oil Recovery, China University of Petroleum (Beijing), Beijing, China



This paper was prepared for presentation at the Carbon Management Technology Conference held in Houston, Texas, USA, 17-20 July 2017.



## Abstract

H-3 Block is an ultra-low permeability reservoir in Changqing oil field, China which had been waterflooded from 2009 and was switched to $CO_2$ flooding in 2013 due to excess water production. However, the nature fractures in NE-SW region have resulted in early $CO_2$ breakthrough and poor production performance. The investigation of $CO_2$ production performance and the method to control $CO_2$ production becomes a key to continue $CO_2$-EOR project.

Outcrop cores are used to perform a series of $CO_2$ flooding experiments at reservoir conditions of pressure, temperature and formation water salinity. Permeability heterogeneity and injection pressure are considered as two variables to affect gas channeling characteristics. It is figured out that producing gas-oil ratio and components analysis of effluent could be used to judge gas channeling and timing to control $CO_2$ production for field use. Starch gel is developed to control $CO_2$ production within nature fractures to improve $CO_2$ swept volume in rock matrix. Ethylenediamine (EDA) is proposed to delay $CO_2$ production within high-permeability zones, and the application boundary as a function of permeability heterogeneity is determined.

Three production stages are clearly stated based on production performance and experimental observation, including gas-free production stage, oil/gas co-production stage, and gas channeling stage. A significant new finding is that oil/gas co-production stage contributes the most to oil recovery. And oil-$CO_2$ mass transfer zone, rather than free $CO_2$, reaches the outlet at this stage, which is proved by color of effluent and chromatographic analysis. For field cases, producing gas-oil ratio and components analysis of effluent at wellhead could help field engineers make a decision: keep producing with caution at oil-gas co-production stage or control the $CO_2$ production at gas channeling stage. The conformance improvement and the increase in injection pressure could remarkably enhance the oil recovery at oil/gas co-production stage. To delay gas channeling and extent oil/gas co-production stage, two-level gas channeling control is presented. A slug of starch gel is first injected to block fractures and then ethylenediamine is injected to react with in-situ $CO_2$ within high-permeability zone. The starch gel, acting as pure viscous fluid, would not leave contamination in rock matrix. And the viscous reactant of ethylenediamine and in-situ $CO_2$ could successfully tune injected $CO_2$ to flood low-permeability zone when permeability ratio is less than 100.

Poor sweep efficiency and immiscibility seriously limit field application of $CO_2$-EOR, especially in fracture reservoirs. The novelty of this study is to present a new approach for field engineers to judge $CO_2$ channeling and make the right decision at right time. Two-level gas channeling control is developed to improve $CO_2$ sweep efficiency and oil production performance in fractured reservoirs.

Keywords: $CO_2$-EOR, fractured sandstone reservoirs, gas channeling characteristic, $CO_2$ production control, laboratory experiment

## Introduction

There are many difficulties in ultra-low permeability reservoir water flooding development; the micro sweep efficiency is low, and the injection water pressure is high. So, $CO_2$ flooding has obvious advantages in the development of low permeability reservoir, and has been widely used in many domestic and foreign oil fields. However, because of gravity overriding and viscous fingering, the channeling of $CO_2$ is very serious. Moreover, with the increase of reservoir heterogeneity, gas can easily penetrate along the high permeable layer, and the effect of $CO_2$ flooding is obviously reduced. Therefore, gas channeling is a key problem to be solved during $CO_2$ flooding in ultra-low permeability reservoirs. Right understanding of $CO_2$ channeling characteristics has important guiding significance for improving $CO_2$ flooding in low permeability reservoirs.

Many scholars have done a lot of research on $CO_2$ channeling, but there is no unified understanding. The injection pressure of $CO_2$ and the heterogeneity of reservoir have a significant influence on gas channeling characteristics.

When the temperature is above 31℃ and the pressure is larger than 7.382MPa, $CO_2$ will be in a supercritical state. The viscosity is close to the gas, the density is close to the liquid, and it has a strong extraction and dissolution capacity. Further increasing in pressure, $CO_2$ and crude oil will be miscible. The increase of injection pressure is beneficial to extracting the formation of crude oil, and also to some extent influence $CO_2$ channeling characteristics. When the reservoir heterogeneity is very serious, $CO_2$ is advanced along the high permeability layer. Especially, when nature fractures are developed, $CO_2$ forms the gas channel easily along the fractures, greatly reducing the sweep efficiency. This will lead to a large amount of $CO_2$ gas in a futile cycle in the formation, influencing the ultimate recovery.

In order to further clarify the gas channeling law of $CO_2$ flooding in ultra-low permeability reservoirs, this paper analyzes the production characteristics of $CO_2$ flooding in ultra-low permeability reservoirs under the conditions of reservoir temperature and pressure. And taking $CO_2$ injection pressure and reservoir heterogeneity into account, their influence on the $CO_2$ flooding of ultra-low permeability reservoirs is studied. This paper was based on preliminary laboratory research of Hao et al. (2015). Two-stage gas channeling control is proposed and optimized. This study could provide an experimental guide for the prevention and control of $CO_2$ gas channeling in field case.

## Experiments

### Materials

- Cores: Natural outcrop cores, artificial cores (the permeability ratios are 5,10,15 and 20) and fractured core with the size of 30*4.5*4.5 cm are utilized in the 1-D flooding experiments; Two layers solid circular artificial cores (the permeability ratios are 10) with the size of Φ45*4.5 cm are utilized in the 3-D flooding experiments. Table 1 shows the basic data for the experimental cores.
- Gas: $CO_2$ (Purity≥99.00%, Beijing, China)
- Water: Simulating the formation of Huang 3 block in Changqing Oilfield, The salinity of the brine is 61.2 g/L and the water type is $CaCl_2$.
- Oil: The oil sample and are collected from the layer of Huang 3 block in Changqing Oilfield. The crude oil has a viscosity of 2.16 mPa·s and a density of 0.768 g/cm$^3$ at a temperature of 86.2°C
- Modified starch gel system: Modified starch (content, ≥99.00%, Beijing, China), monomer of acrylic amide (content, ≥98.00%, Beijing, China), crosslinking agent (content, ≥98.00%, Beijing, China), initiator (content, ≥99.00%, Beijing, China) and stabilizer (content, ≥98.00%, Beijing, China) were used to form the gel systems with brine.
- Polymer gel system (called PLS gel): Modified starch gel system: polymer (content, ≥99.00%, Qingdao, China), Additive (content, ≥99.00%, Qingdao,

China), crosslinking agent (content, ≥98.00%, Beijing, China), deoxidizer (content, ≥99.00%, Qingdao, China) and stabilizer (content, ≥98.00%, Qingdao, China) were used to form the gel systems with brine.

Table 1 Experimental core data

| Types of cores | Core number | Gas permeability ($\times10^{-3}\mu m^2$) | Permeability ratio | Porosity volume ($cm^3$) | Porosity (%) | Oil saturation (%) |
|---|---|---|---|---|---|---|
| Homogeneous | CQJZ-1 | 5 | 1 | 89 | 14.65 | 53.93 |
| | CQJZ-2 | | | 90 | 14.81 | 51.11 |
| | CQJZ-3 | | | 96 | 15.80 | 51.00 |
| | CQJZ-4 | | | 92 | 15.14 | 54.35 |
| | CQJZ-5 | | | 92 | 15.14 | 45.65 |
| Heterogeneous | CQFJZ-5 | 5/25 | 5 | 98 | 16.13 | 48.98 |
| | CQFJZ-10 | 5/50 | 10 | 130 | 21.40 | 42.31 |
| | CQFJZ-15 | 5/75 | 15 | 135 | 22.22 | 44.44 |
| | CQFJZ-20 | 25/500 | 20 | 128 | 21.07 | 35.94 |
| | CQLF | Fracture permeability /5 | - | 148 | 24.63 | 57.63 |

## Experimental equipments

The main experimental equipment used in the experiment include KDHW-II Incubator, Isco pump, Core Holder ($30\times4.5\times4.5cm^3$), Core Holder System (Φ 45*4.5cm), Piston Intermediate Container, Return Valve, D07-11C Gas flow meter, Pressure transmitter. Fig. 1 shows the set-up of this experiment.

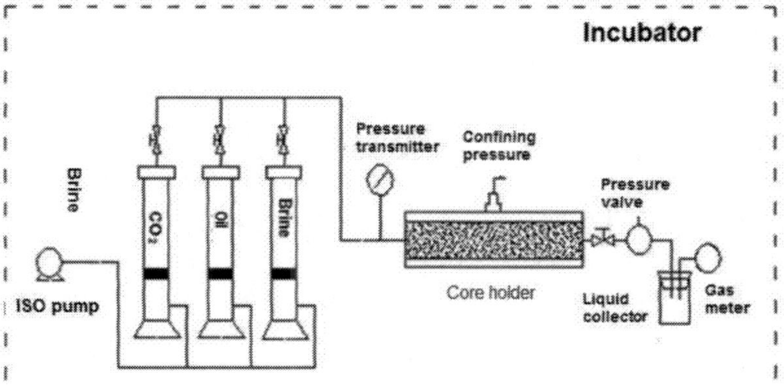


Fig.1 $CO_2$ flooding experimental set-up

## Experimental procedure

The experiment is divided into two major parts, one is to study gas channeling law with homogeneous and heterogeneous cores and the fracture model, the other is to optimize two-stage gas channeling control with fractured two layers solid circular artificial cores, which is used to improve displacement result. The law of gas channeling is studied from two aspects: the heterogeneity and $CO_2$ phase. By optimizing the two-stage blocking agent to screen out the best combination of blocking agent.

The bulk volumes of the cores are measured before oil displacement experiments. The cores are firstly evacuated and then saturated with the simulated water. The porosities are determined as the ratio of brine saturation volume to the bulk volume. The permeability of the cores is measured by changing water injection rate from 0.1 to 0.3 mL/min. Subsequently, the cores are displaced by simulated oil to reach the residual water saturation condition. Then the initial oil saturation is calculated using the injected oil volume and the pore volume.

In $CO_2$ phase scenarios, homogeneous are used in the experiments. The detailed sequence during continuous $CO_2$ flooding experiments is explained as follows. ①The production pressure setted in this scenarios range from 5 to 25 MPa. ②The phase change of $CO_2$ in the formation is controlled by changes in pressure, phase transition of $CO_2$ from gas to supercritical state (not including miscibility) to miscibility. ③The experiment is terminated when the producing gas/ oil ratio reaches 3000mL/mL, which is equal to 3000 $Sm^3/m^3$ when the $CO_2$ flooding is non-commercial in the oilfield.

In heterogeneity scenarios, homogeneous and heterogeneous cores and fractured are used in the experiments. The detailed sequence during continuous $CO_2$ flooding experiments is explained as follows. ①the permeability of the cores used in this scenarios range from 1(homogeneous core) to 20, and one fractured core is also used ②The production pressure is set to be 15 MPa, and the injection rate is 1 mL/min under formation pressure. ③The experiment is terminated when the producing gas/ oil ratio reaches 3000mL/mL, which is equal to 3000 $Sm^3/m^3$ when the $CO_2$ flooding is non-commercial in the oilfield.

In two-stage gas channeling control scenarios, fractured two layers solid circular artificial cores are used in the experiments. The detailed sequence during continuous $CO_2$ flooding experiments is explained as follows. ①The production pressure is set to be 15 MPa, and the injection rate is 1 mL/min under formation pressure. ②The starch gel or PLS gel is injected into the model when the producing gas/ oil ratio reaches 500mL/mL, Then closed for 24 hours waiting to form a gel. ③The EDA or foam is injected into the model when the producing gas/ oil ratio reaches 500mL/mL again.④The experiment is terminated when the producing gas/ oil ratio reaches 3000mL/mL, which is equal to 3000 $Sm^3/m^3$ when the $CO_2$ flooding is non-commercial in the oilfield.

## Results and Discussion

Fig. 2 depicts production dynamic curve of $CO_2$ flooding at 15MPa injection pressure. Three production stages are clearly stated based on production performance and experimental observation, including gas-free production stage, oil/gas co-production stage, and gas channeling stage. ①In gas-free production stage, single-phase oil is produced in the form of slugs. ②When the gas appears in the output side, $CO_2$

flooding enters into the oil/gas co-production stage, in which the oil is a continuous phase while the gas is a dispersed phase. ③When a large number of gas comes out of the core, $CO_2$ flooding enters into the gas channeling stage, in which the gas phase is continuous coming with a small amount of oil.

Carrying out $CO_2$ flooding experiment at different injection pressure or under heterogeneous conditions, the displacement characteristics are the same as that shown in the Fig. 2. Whether the gas injection pressure or heterogeneous conditions change, dynamic characteristic curve of $CO_2$ flooding will be influenced.

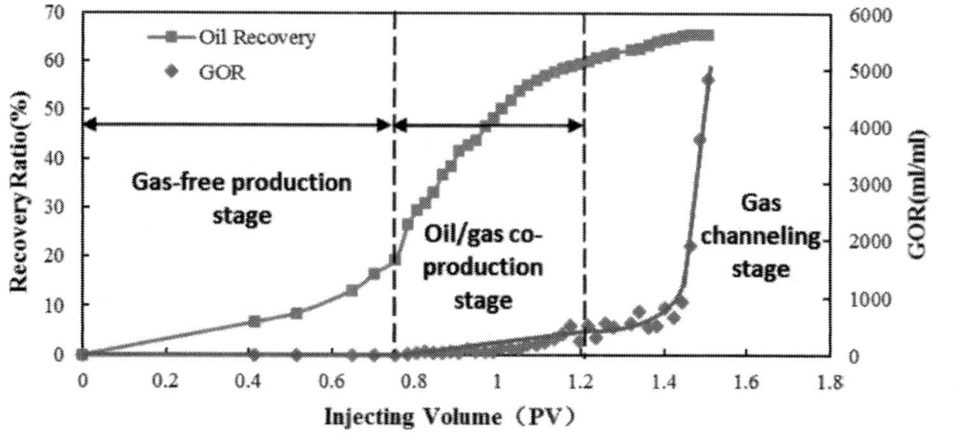


Fig.2 Producing gas oil ratio and recovery curve of supercritical $CO_2$ immiscible flooding in homogeneous core sample

## Effect of $CO_2$ phase behavior

Table 2 is the experimental results of continuous $CO_2$ flooding at different injection pressure in homogeneous cores, while Fig.3 depict the oil recovery of each stage with the injection pressure curve during continuous $CO_2$ flooding. In general, the total oil recovery of $CO_2$ flooding increases with the increase of injection pressure, but the degree of oil recovery in each stage is different. With the increase of injection pressure, the oil recovery of gas free production stage does not increase very much, which is mainly caused by the basic properties of the core. For ultra-low permeability reservoirs with similar properties, oil and the area of pure oil are similar in gas phase percolation curve, while the increase of the injection pressure slows the gas viscous fingering to a certain extent, which makes the displacement front move more uniform. Relatively, the change of $CO_2$ injection pressure will change oil and gas two-phase flow region to a large extent, and oil/gas co-production stage and gas channeling stage increased significantly at the macro level. When $CO_2$ is converted from a gas phase to a supercritical state, the ability of dissolve and diffuse is greatly enhanced, and the ability to extract and extract light hydrocarbons is also significantly improved, so the oil recovery of oil/gas co-production stage is increased significantly. When the injection pressure reaches above 20MPa, $CO_2$ and oil will form miscible phase, which can greatly enhance the oil recovery, and a certain amount of crude oil can be produced in the gas channeling stage at this condition.

Table 2 Experimental results of $CO_2$ phase effect on the gas channeling

| Core number | Injection Pressure (MPa) | Recovery degree (%) | | | Total recovery (%) | Injection volume (PV) | | | Total injection volume (PV) |
|---|---|---|---|---|---|---|---|---|---|
| | | gas-free production stage | oil/gas co-production stage | gas channeling stage | | gas-free production stage | oil/gas co-production stage | gas channeling stage | |
| CQJZ-1 | 5 | 18.13 | 9.58 | 1.14 | 28.85 | 0.75 | 0.24 | 0.11 | 1.10 |
| CQJZ-2 | 10 | 18.93 | 25.18 | 2.26 | 53.89 | 0.75 | 0.35 | 0.15 | 1.25 |
| CQJZ-3 | 15 | 19.41 | 40.63 | 5.57 | 65.61 | 0.75 | 0.47 | 0.29 | 1.51 |
| CQJZ-4 | 20 | 25.4 | 48.8 | 10.4 | 84.60 | 0.76 | 1.0 | 0.75 | 2.51 |
| CQJZ-5 | 25 | 26.47 | 51.2 | 11.25 | 88.81 | 0.77 | 1.1 | 0.82 | 2.69 |

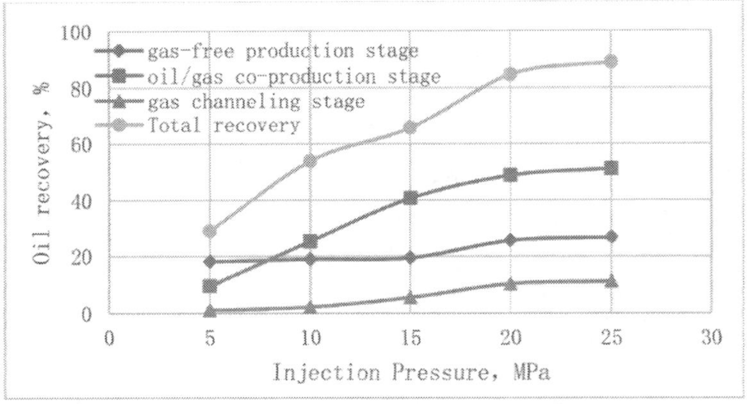


Fig.3 The recovery of each stage with the injection pressure curve

## Effect of reservoir heterogeneity

Table 3 is the experimental results of continuous $CO_2$ flooding at different permeability ratio in heterogeneous and homogeneous cores and fractured model, while Fig.4 depict the oil recovery of each stage with the heterogeneity during continuous $CO_2$ flooding. On one hand, with the increase of core heterogeneity, the total recovery rate was significantly lower in $CO_2$ flooding. The recovery of the homogeneous core was significantly higher than that of heterogeneous core and fractured cores, and with the enhance of reservoir heterogeneity, the total injection volume of $CO_2$ and the injection volume displacement in each stage decrease significantly. the recovery degree of gas-free production stage decreases with the increase of reservoir heterogeneity, but slightly lower. The recovery of gas-free production stage is mainly affected by the reservoir physical properties, flow area of pure oil of low permeability heterogeneous reservoir are similar, so there is the oil recovery of gas-free production stage of small decline. For the fractured reservoir, the oil recovery of gas-free production stage mainly originates from the oil in the cracks. On the other hand, Heterogeneity of the core has a great influence on oil/gas co-production stage and gas channeling stage. Compared with the homogeneous core, when the permeability ratio is 5, the recovery degree of oil/gas co-production stage of the heterogeneous core is obviously decreased. $CO_2$ is infiltrated along the hypertonic layer, and the injected gas cannot effectively spread to the low permeability reservoir. With the further increase of the permeability difference, the contradiction between the layers increases, and the phenomenon of $CO_2$ channeling becoming higher and higher. In summary, reservoir heterogeneity mainly affects the development of $CO_2$ flooding oil/gas co-production stage and gas channeling stage in ultra-low permeability reservoirs. How to improve reservoir heterogeneity effectively, and to prevent the injection of gas along the hypertonic layer or crack, which is the key point to improve the development of $CO_2$ flooding in ultra-low permeability reservoirs.

Table 3 Experimental results of heterogeneity effect on the gas channeling

| Core NO. | Permeability ratio | Recovery degree (%) | | | Total recovery (%) | Injection volume (PV) | | | Total injection volume (PV) |
|---|---|---|---|---|---|---|---|---|---|
| | | gas-free production stage | oil/gas co-production stage | gas channeling stage | | gas-free production stage | oil/gas co-production stage | gas channeling stage | |
| CQJZ-3 | 1 | 19.41 | 40.63 | 5.57 | 65.61 | 0.75 | 0.47 | 0.29 | 1.51 |
| CQFJZ-5 | 5 | 17.95 | 24.25 | 5.39 | 47.59 | 0.54 | 0.49 | 0.39 | 1.42 |
| CQFJZ-10 | 10 | 17.56 | 21.39 | 4.31 | 43.26 | 0.45 | 0.39 | 0.34 | 1.18 |
| CQFJZ-15 | 15 | 17.25 | 20.16 | 3.97 | 41.38 | 0.41 | 0.36 | 0.28 | 1.05 |
| CQFJZ-20 | 20 | 16.96 | 19.45 | 3.81 | 40.22 | 0.39 | 0.27 | 0.24 | 0.9 |
| CQLF | fracture | 14.31 | 15.74 | 1.32 | 31.37 | 0.25 | 0.2 | 0.15 | 0.6 |

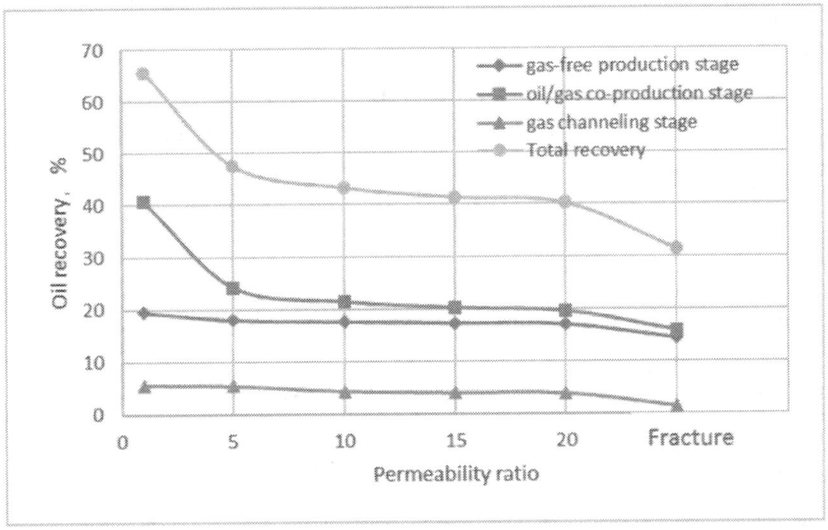


Fig.4 The recovery of each stage with the Permeability ratio curve

## Two-stage gas channeling control scenarios

Based on the previous study of the law of gas channeling, the method of $CO_2$ immiscible flooding was carried out in five-spot pattern model, and the plugging process and the process parameters were optimized and evaluated. Fig. 5 is the five-spot pattern model. Three kinds of sealing agent combinations were evaluated including starch gel + EDA combination, PLS gel + EDA combination and starch gel + foam t combination.

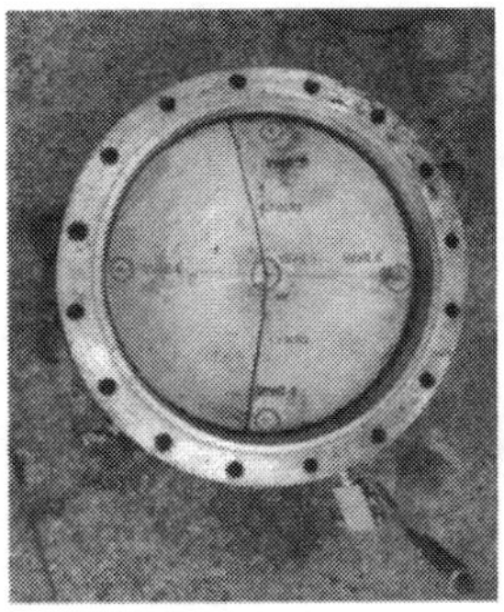

Fig.5 Five-spot pattern model

### Starch gel + EDA scenario

Fig. 6 depict the oil recovery and GOR of each production well during continuous $CO_2$ flooding. According to the change of production gas and oil ratio in the process of displacement and the different blocking measures, the flooding process can be divided into three stages: primary gas flooding stage, secondary gas flooding stage and tertiary gas flooding stage. Primary gas flooding stage: from the start of the gas injection to the beginning of the gel injection; Secondary gas flooding stage: from the gel injection into the beginning of EDA injection; Tertiary gas flooding stage: from the EDA injection to the end of the experiment.

The production performance curves of four wells are similar. The characteristic of primary gas flooding stage is described as follows: the GOR of gas free period is 0, during which the oil recovery is proportional to the $CO_2$ injection volume, while the

GOR increases rapidly as gas breakthrough, and the recovery curve tends to be flat, which means $CO_2$ flows along the fracture. The characteristic of secondary gas flooding stage is described as follows: the production of gasoline is maintained at a relatively low level for a period of time, indicating that the injected gel effectively blocks the cracks, forcing the injected $CO_2$ to turn toward the core matrix, and the recovery curve does not rise immediately after the injection of the gel, because the effective sealing of the gel has a certain delay. After a while the GOR rises rapidly, which is due to the gas flowing along the high permeability layer caused by matrix heterogeneity, the stage of the recovery curve changes from flat to upturned, and then tends to be flat again. The characteristic of tertiary gas flooding stage is described as follows: the recovery curve is first increased, indicating that the injected EDA is well improved in the heterogeneity of the core, and the production gas curve also declines at the same time. Then the recovery curve is flattened again, and the GOR is increased rapidly.

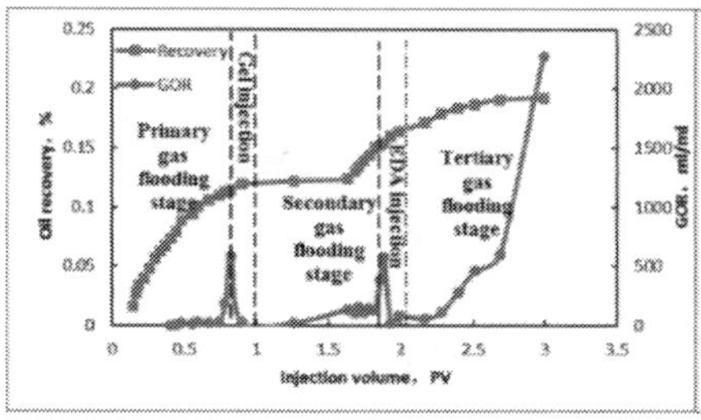


(a) well 1

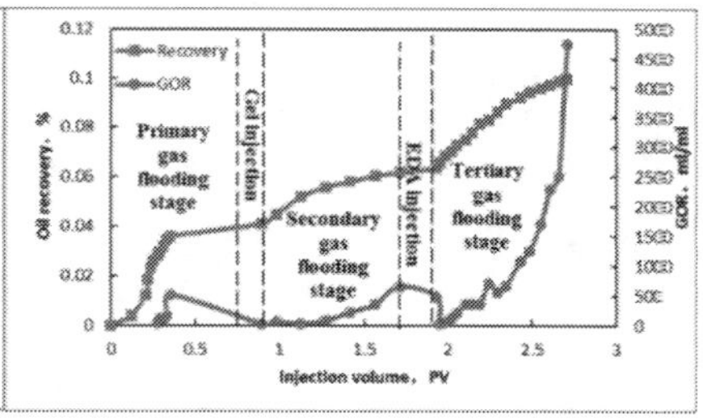


(b) well 2

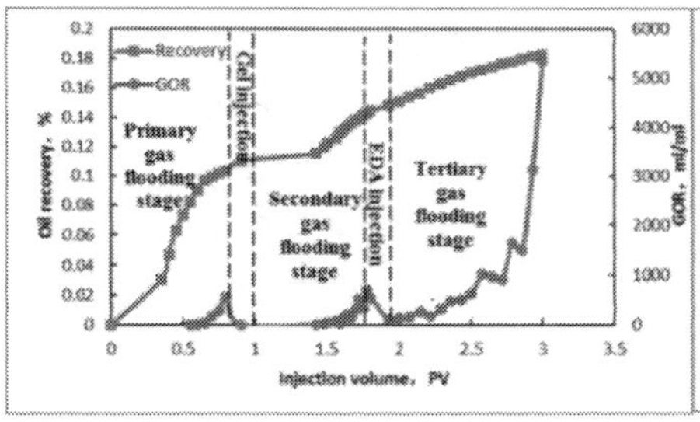


(c) well 3

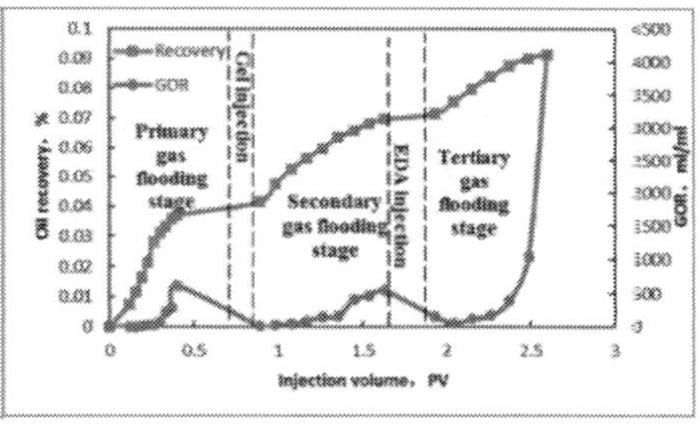


(d) well 4

Fig.6 The production dynamic curve of each well

Table 4 is the experimental results of continuous $CO_2$ flooding with starch gel + EDA in five-spot pattern model. Well 1 and well 3 are far crack wells, and oil recovery of them are 11.44% and 10.35% respectively during primary gas flooding stage. Well 2 and well 4 are near crack wells, which are affected by cracks. The oil recovery of them are 3.596% and 3.692% respectively during the same period. After primary gas flooding stage the cracks are blocked by the starch gel. The results show that the recovery of the four wells is maintained at the same level in the secondary gas flooding stage, the recovery of each well is about 3%, and the total recovery degree is 14.50% in this stage. Similarly, the subsequent injection of EDA improves the heterogeneity of the model to a certain extent, and constantly adjusts the injection profile of $CO_2$ to inhibit the viscosity of $CO_2$ in the model, the recovery of the tertiary gas flooding stages is 13.35%. Plugging system of starch gel + EDA enhances oil recovery which reaches 27.85%, and has a good adaptability to reservoir of fracture and serious heterogeneity.

Table 4 Experimental results of starch gel + EDA scenario

| recovery/% | primary gas flooding stage | secondary gas flooding stage | tertiary gas flooding stage | Total recovery |
|---|---|---|---|---|
| well 1 | 11.44 | 3.94 | 3.79 | 19.17 |
| well 2 | 3.596 | 2.769 | 3.616 | 9.981 |
| well 3 | 10.35 | 4.52 | 3.77 | 18.64 |
| well 4 | 3.692 | 3.27 | 2.17 | 9.132 |
| Total recovery | 29.078 | 14.499 | 13.346 | 56.923 |

**PLS gel + EDA scenario**

Fig. 6 depict the oil recovery and GOR of each production well during continuous $CO_2$ flooding. Table 5 is the experimental results of continuous $CO_2$ flooding with PLS gel + EDA in five-spot pattern model. Similarly, according to the change of production gas and oil ratio in the process of displacement and the different blocking measures, the flooding process can be divided into three stages.

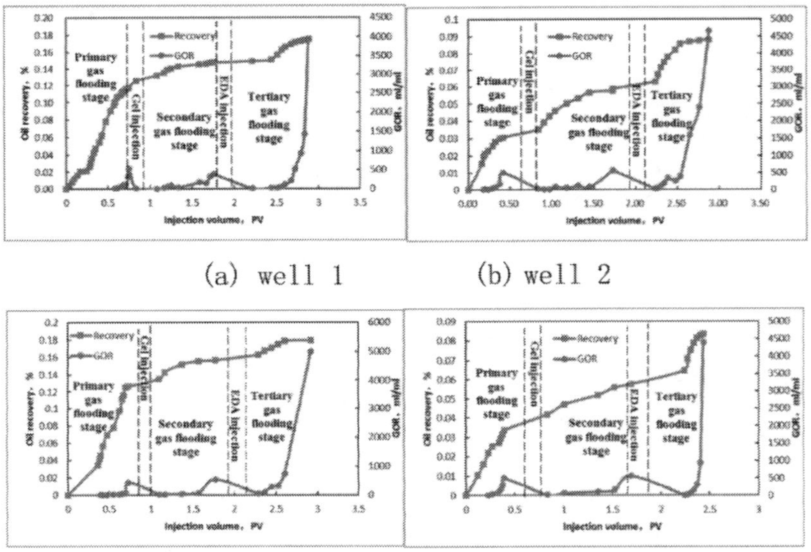


(a) well 1 (b) well 2

(c) well 3 (d) well 4

Fig.7 The production dynamic curve of each well

Table 5 Experimental results of PLS gel + EDA scenario

| recovery/% | primary gas flooding stage | secondary gas flooding stage | tertiary gas flooding stage | Total recovery |
|---|---|---|---|---|
| well 1 | 11.73 | 2.846 | 2.72 | 17.296 |
| well 2 | 3.078 | 2.782 | 2.957 | 8.817 |
| well 3 | 12.63 | 3.04 | 2.26 | 17.93 |

| | | | | |
|---|---|---|---|---|
| well 4 | 3.374 | 2.377 | 2.574 | 8.325 |
| Total recovery | 30.812 | 11.045 | 10.511 | 52.368 |

Table 6 Comparision of Starch + EDA with PLS +EDA

| recovery/% | primary gas flooding stage | secondary gas flooding stage | tertiary gas flooding stage | Total recovery |
|---|---|---|---|---|
| Starch gel+ EDA | 29.078 | 14.499 | 13.346 | 56.923 |
| PLS +EDA | 30.812 | 11.045 | 10.511 | 52.368 |

The trend of these four wells is roughly the same as above. The development of near-crack wells is far less than that of far-flung wells, but the production of four can be roughly maintained at the same level after the production measures are taken. Table 5 is comparison of oil recovery of starch gel + EDA system and PLS gel + EDA system at each stage. The recovery of tertiary gas flooding stage reaches 10.51%, which proves PLS + EDA to be adaptable to reservoir of fracture and serious heterogeneity too. However, compared with the starch gel, the recovery of secondary gas flooding stage of PLS gel + EDA is about 3.45% lower than that of starch gel +EDA. And the recovery of tertiary gas flooding stage of PLS gel + EDA is about 2.84% lower than that of starch gel +EDA. Although both cases are injected with EDA at tertiary gas flooding stage, the effect of former is better due to the strong ability of the starch gel to seal the crack. In summary, starch gel in is slightly better than PLS gel the sealing effect, and the sealing effect of the cracks will also directly affect the effect of EDA.

The reasons that starch gels are more resistant to cracks than PLS gels are described as follows. The viscosity modulus of the starch gel is 24.19Pa, and the elastic modulus is 1.335Pa, which is he rigid gel according to the standard, while the viscosity modulus of the PLS gel is 5.802Pa, and the elastic modulus is 0.494Pa, which is the medium strength gel according to the standard. So the starch gel is stronger than the PLS gel. The viscosity of Starch gel is lower than PLS gel at room temperature, the ability of starch gel injection is higher, which makes it more conducive to migration to the deep cracks. In summary, one stage of blocking agent recommended the use of starch gel.

**Starch gel + foam scenario**

Fig. 8 depicts the oil recovery and GOR of each production well during continuous $CO_2$ flooding. Table 7 is the experimental results of continuous $CO_2$ flooding with starch gel + foam in five-spot pattern model. By comparing the recovery of the difference stages of the starch gel + EDA and starch gel + foam system, it is shown that the difference in the degree of primary gas flooding in the two cases is due to the subtle differences in the experimental core. Because of the same two-stage flooding agent, the recovery of tertiary gas flooding stage is not much difference, and the recovery of the former is 3.12% higher than that of the latter, indicating that the foam is not as efficient as EDA in improving heterogeneous cores.

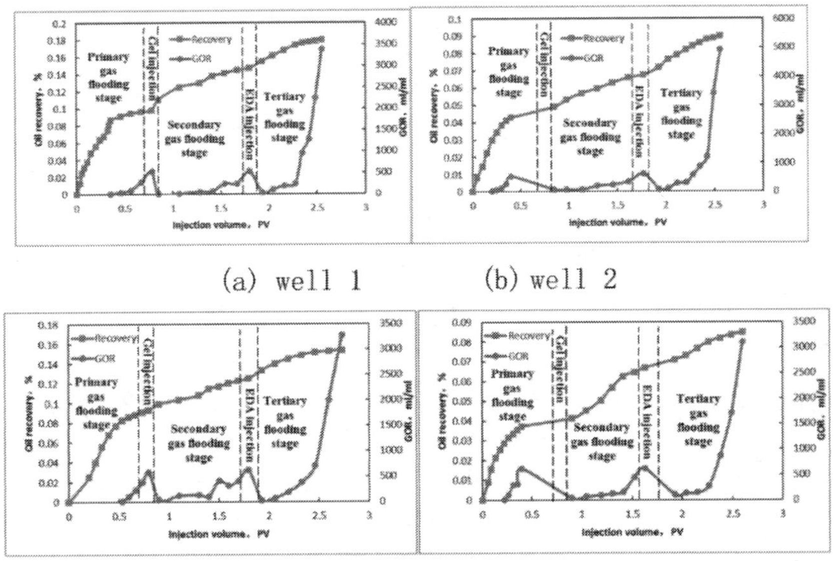


(a) well 1 (b) well 2

(c) well 3 (d) well 4

Fig.8 The production dynamic curve of each well

Table 7 Experimental results of Starch gel + foam combination

| recovery/% | primary gas flooding stage | secondary gas flooding stage | tertiary gas flooding stage | Total recovery |
|---|---|---|---|---|
| well 1 | 9.79 | 4.92 | 3.31 | 18.02 |
| well 2 | 4.29 | 2.48 | 2.23 | 9.00 |
| well 3 | 9.29 | 3.17 | 2.88 | 15.33 |
| well 4 | 3.71 | 3.17 | 1.81 | 8.69 |
| Total recovery | 27.08 | 13.73 | 10.23 | 51.04 |

EDA is stronger than foam in improving heterogeneous for the following reasons. First, the EDA is very low in viscosity and has good injective ability. It can migrate to the high permeability layer in the matrix, during which it reacts with $CO_2$ to form a yellow viscous substance, which can effectively block the high permeability layer, while the foam only can increase the power of the displacement process, forcing the gas into the low permeability layer, thus expanding swept volume. but this does not directly plug the effect of high permeability layer. Second, the reaction of EDA with $CO_2$ can deplete $CO_2$ in the formation of gas channel, greatly reducing the GOR, which is conducive to shielding the channeling channel, while the foam cannot be a permanent collection of carbon dioxide, once the foam burst $CO_2$ will be released out. Finally, Alkaline EDA, before being reacted with $CO_2$, can be used as an alkali to carry out the oil displacement, with excellent performance of alkali flooding, and after the reaction with carbon dioxide can be used as a profile control agent, with dual effects, which are the bubble cannot do.

## Conclusions

Heterogeneity and $CO_2$ phase behavior can greatly affect $CO_2$ flooding efficiency. The stages of the displacement process can help us to study the law of gas channeling more specifically. Two-level gas channeling control is an effective way to improve $CO_2$ sweep efficiency and oil production performance in fractured reservoirs. Some conclusions can be summarized as follows.

(1) Three production stages are clearly stated based on production performance and experimental observation, including gas-free production stage, oil/gas co-production stage, and gas channeling stage. A significant new finding is that oil/gas co-production stage contributes the most to oil recovery.

(2) With the increase of $CO_2$ injection pressure, the recovery degree of the gas free production stage, the oil/gas co-production stage and the gas channeling stage is gradually increased, and the effect of the injection pressure on oil recovery of the oil/gas co-production stage is especially significant. When the injected gas and the oil reaches near-miscible or miscible state, a certain amount of crude oil can still be produced during the gas channeling stage.

(3) With the increase of reservoir heterogeneity, the recovery degree of the gas free production stage, the oil/gas co-production stage and the gas channeling stage is gradually reduced, and the oil recovery of the oil/gas co-production is deeply affected by reservoir heterogeneity. With the increase of reservoir heterogeneity, $CO_2$ breakthroughs along the hypertonic layer or fractures, and the channeling phenomenon is increasingly significant.

(4) Through the optimization of the two-level gas channeling control system, it is recommended the use of high-strength starch gel as first blocking agent and EDA as secondary blocking agent. The combination of the two chemicals can be a good control of gas channeling, which can greatly enhance oil recovery.

## Acknowledgments

This work is sponsored by Changqing oil field project (16CY2-FW-003).

## References

1. Zhao, F., Hao, H., Hou, J., Hou, L., & Song, Z. (2015). CO2, mobility control and sweep efficiency improvement using starch gel or ethylenediamine in ultra-low permeability oil layers with different types of heterogeneity. Journal of Petroleum Science & Engineering, 133, 52-65.
2. Hao, H., Hou, J., Zhao, F., Song, Z., Hou, L., & Wang, Z. (2016). Gas channeling control during CO2, immiscible flooding in 3d radial flow model with complex fractures and heterogeneity. Journal of Petroleum Science & Engineering, 146, 890-901.
3. Yang, S. H., & Reed, R. L. (1989, January 1). Mobility Control Using CO2 Forms. Society of Petroleum Engineers. doi:10.2118/19689-MS

4. Cho, J., Kim, T. H., & Lee, K. S. (2016, March 21). Modeling of CO2 EOR Process Combined with Intermediate Hydrocarbon Solvents for Higher Recovery Efficiency. Society of Petroleum Engineers. doi:10.2118/179778-MS
5. Chen, S., Li, H., Yang, D., & Tontiwachwuthikul, P. (2010, October 1). Optimal Parametric Design for Water-Alternating-Gas (WAG) Process in a CO2-Miscible Flooding Reservoir. Society of Petroleum Engineers. doi:10.2118/141650-PA
6. Odi, U., & Gupta, A. (2010, January 1). Optimization and Design of Carbon Dioxide Flooding. Society of Petroleum Engineers. doi:10.2118/138684-MS
7. Wang, R., Lv, C. Y., Lun, Z., Yue, X., Zhao, R., & Wang, H. T. (2011, January 1). Study on Gas Channeling Characteristics and Suppression Methods in CO2 Flooding for Low Permeability Reservoirs. Society of Petroleum Engineers. doi:10.2118/142306-MS
8. Tan, S. H., Zhou, Z. P., Wu, Z. L., Qian, W. M., Liu, W., & Xiao, W. (2004). Production of attic oil in complex fault block reservoir by co2 gas drive. Petroleum Exploration & Development.
9. Spivak, A., & Chima, C. M. (1984, January 1). Mechanisms of Immiscible CO2 Injection in Heavy Oil Reservoirs, Wilmington Field, CA. Society of Petroleum Engineers. doi:10.2118/12667-MS
10. Simon, R., & Graue, D. J. (1965). Generalized correlations for predicting solubility, swelling and viscosity behavior of co2 crude oil systems. Journal of Petroleum Technology, 17(1), 102-106.

# Wyoming Reservoir Information Tool (WyRIT) $CO_2$-EOR & $CO_2$ Buffer storage opportunities

Carbon Management Technical Conference (CMTC)
Houston Texas

*Nick Jones – Project/Client Manager,*

*July 19th*

Enhanced Oil Recovery Institute
UNIVERSITY OF WYOMING

## EORI – Mission

***The mission of EORI is to facilitate a meaningful and measurable increase in recoverable reserves and production of oil and natural gas in Wyoming that may otherwise not be realized. Key to this is the effective and efficient transfer of relevant technology, information and knowledge to Wyoming producers. EORI believes that its mission is being met when producers consider EORI as a vital source of relevant technology, information, expertise and knowledge for Wyoming fields.***

Enhanced Oil Recovery Institute

# Enhanced and Improved Oil Recovery

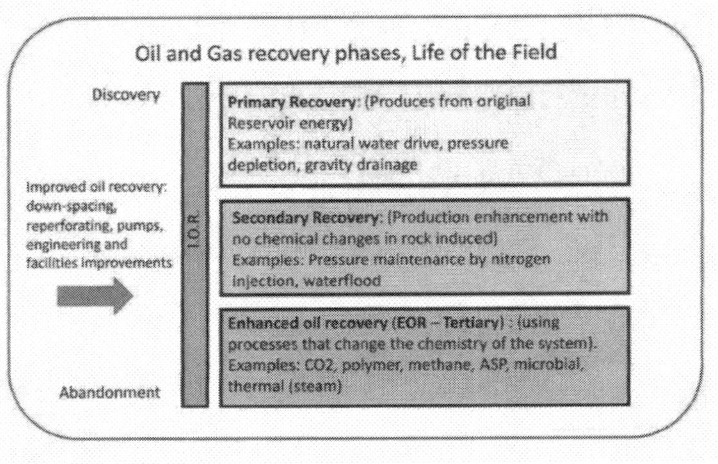


Enhanced Oil Recovery Institute

# Reported EOR Projects in Wyoming

- Chemical
  - Alkaline
  - Polymer
  - Surfactant
  - Combinations
- Gas
  - Nitrogen
  - Methane
  - Carbon Dioxide
- Thermal
  - In-Situ Combustion
  - Steam
- Mining
  - Gravity Drainage
- Combination
  - Chemical/Thermal
  - Chemical/Gas
  - Gas/Thermal

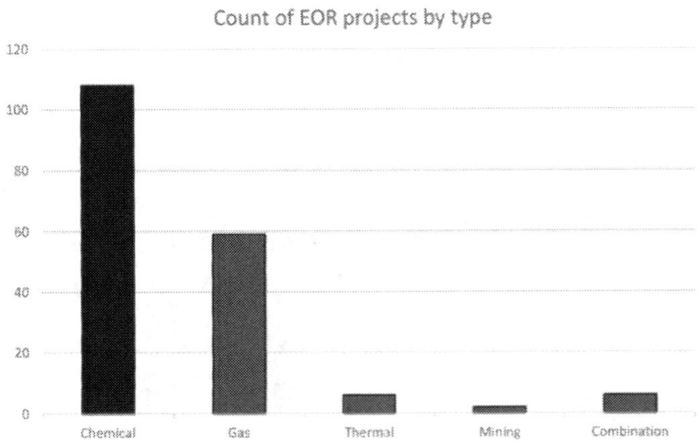


References: US Department of Energy & Wyoming Oil and Gas Conservation Commission

Enhanced Oil Recovery Institute

# EOR Projects by Reservoir

- Mesaverde
  - Teapot
  - Parkman
- Cody
  - Sussex
  - Shannon
- Carlile
  - Turner
- Frontier
  - Wall Creeks
- Muddy/Newcastle
- Lakota/Dakota
- Phosphoria
- Minnelusa
- Tensleep
- Madison

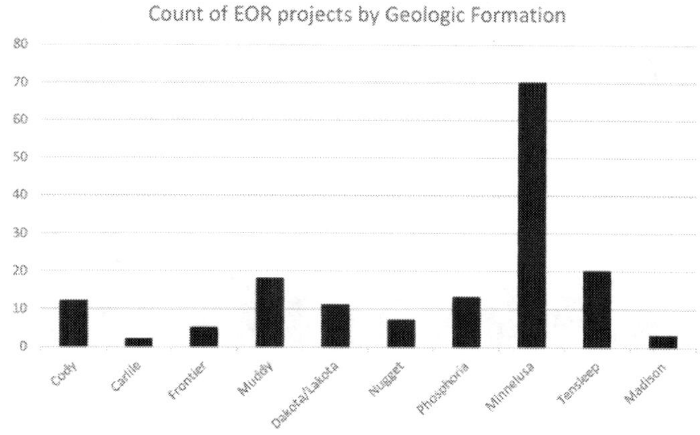


References: US Department of Energy & Wyoming Oil and Gas Conservation Commission

Enhanced Oil Recovery Institute

# Well count

- Mesaverde
  - Teapot
  - Parkman
- Cody
  - Sussex
  - Shannon
- Carlile
  - Codell
  - Turner
- Niobrara
- Frontier
  - Wall Creeks
- Muddy/Newcastle
- Lakota/Dakota
- Phosphoria
- Minnelusa
- Tensleep
- Madison

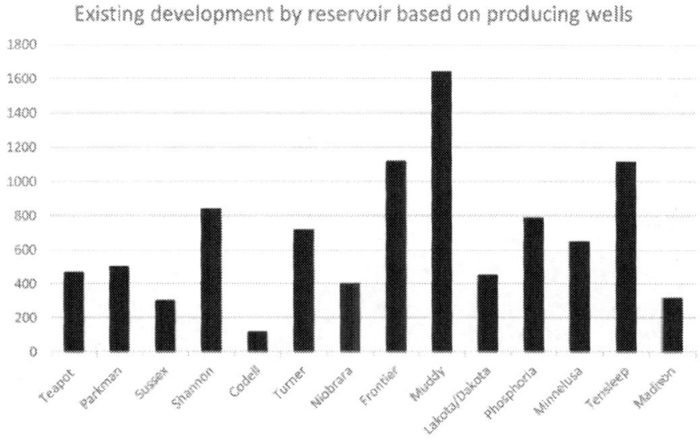


References: Wyoming Oil and Gas Conservation Commission

Enhanced Oil Recovery Institute

# 2016 Oil Production

- Mesaverde
  - Teapot
  - Parkman
- Cody
  - Sussex
  - Shannon
- Carlile
  - Codell
  - Turner
- Niobrara
- Frontier
  - Wall Creeks
- Muddy/Newcastle
- Lakota/Dakota
- Phosphoria
- Minnelusa
- Tensleep
- Madison

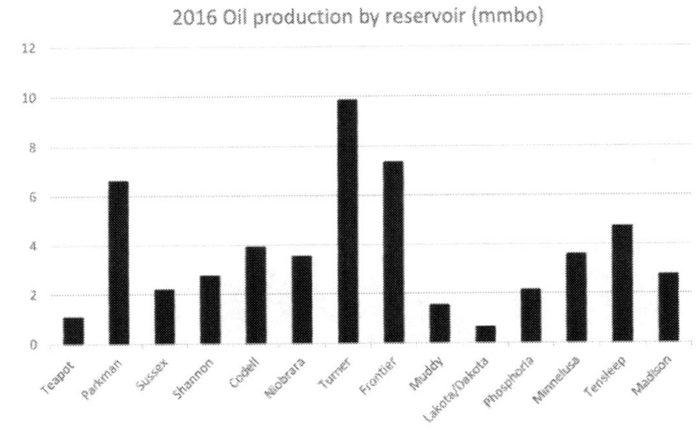


Together these Formations represent ~ 72% of total state oil production for 2016

References: Wyoming Oil and Gas Conservation Commission

Enhanced Oil Recovery Institute

# Unconventional Targets

- Lance
- Mesaverde
  - Teapot
  - Parkman
- Cody
  - Sussex
  - Shannon
- Niobrara
- Carlile
  - Codell
  - Turner
- Frontier
  - Wall Creek
- Mowry

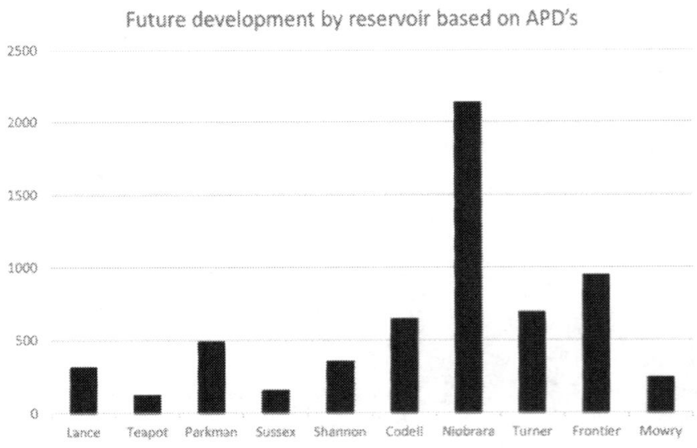


References: Wyoming Oil and Gas Conservation Commission

Enhanced Oil Recovery Institute

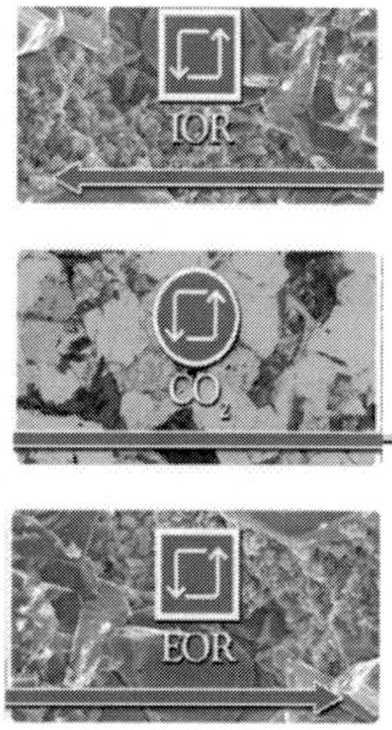


WyRIT

## Information Management

Enhanced Oil
Recovery Institute

# Collaboration is Key

- Enhanced Oil Recovery Institute
- Wyoming Geographic Information Science Center
- ***Wyoming Oil & Gas Conservation Commission
- ***Wyoming Geological Association
- ***Wyoming Pipeline Authority
- Wyoming State Geological Survey
- ***Seismic Exchange

***Key data providers



# Background

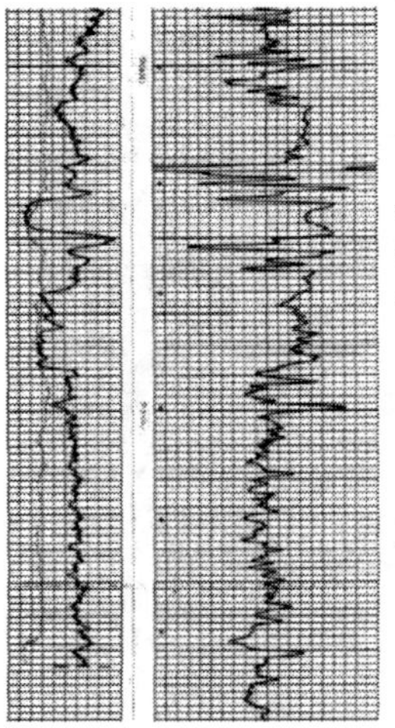

- Wyoming Oil & Gas Reservoir Information
  - Database development
  - Using relational database design
  - Geospatial perspective
  - Data sharing and collaboration
- Reservoir Screening
- Data Dissemination
- Web Application
  - Wyoming Reservoir Information Tool
    - WyRIT
      - End Users

| | | |
|---|---|---|
| Denver | 378 | 9.94% |
| Houston | 119 | 3.13% |
| Dallas | 111 | 2.92% |
| Cheyenne | 98 | 2.58% |
| Rock Springs | 78 | 2.05% |
| Casper | 66 | 1.74% |
| Boulder | 56 | 1.47% |
| (not set) | 54 | 1.42% |
| Oklahoma City | 39 | 1.03% |

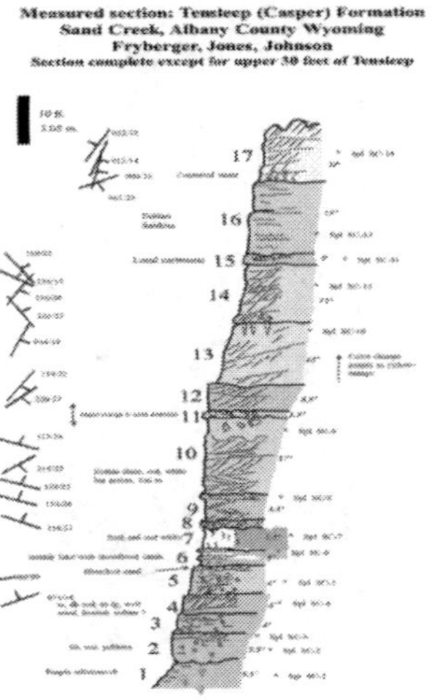

## Wyoming Reservoir Information Tool (WyRIT)

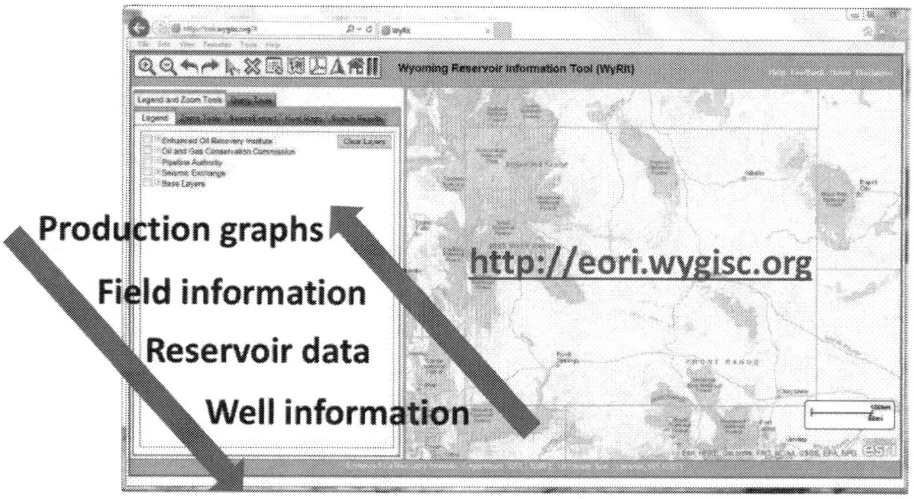


Enhanced Oil Recovery Institute

## Available layers and outputs

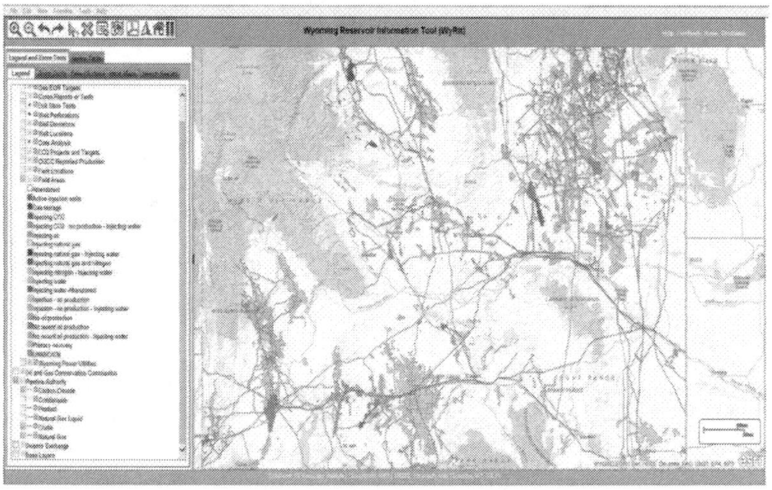


Statewide coverage

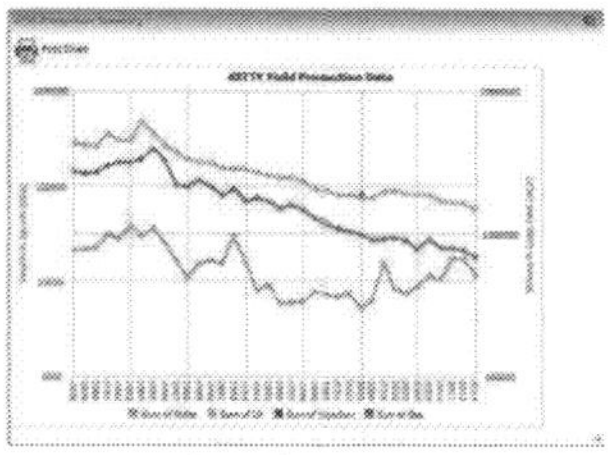

Annual Production graphs

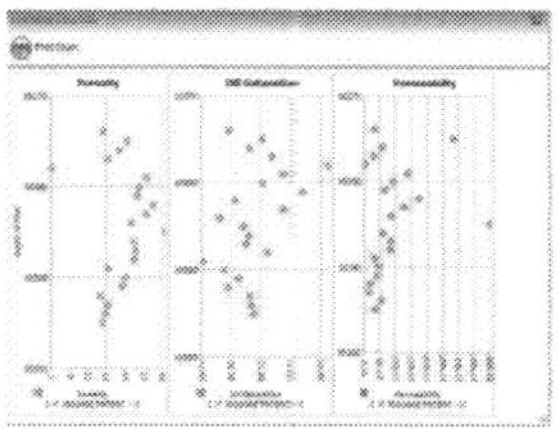

Core analysis charts (in press)

Enhanced Oil Recovery Institute

# Available links to external data sources

Seismic attributes and links to SEI

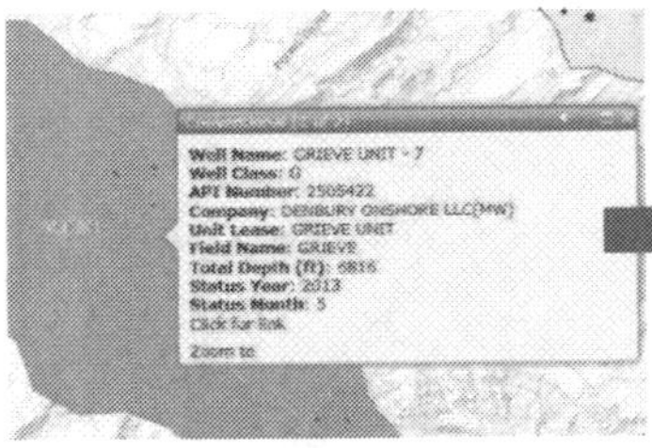


Well attributes and links to WOGCC

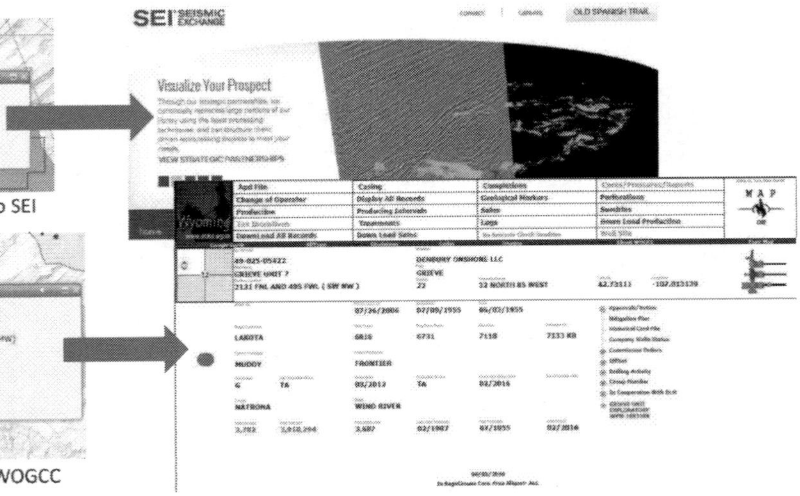






# Production maps

Oil

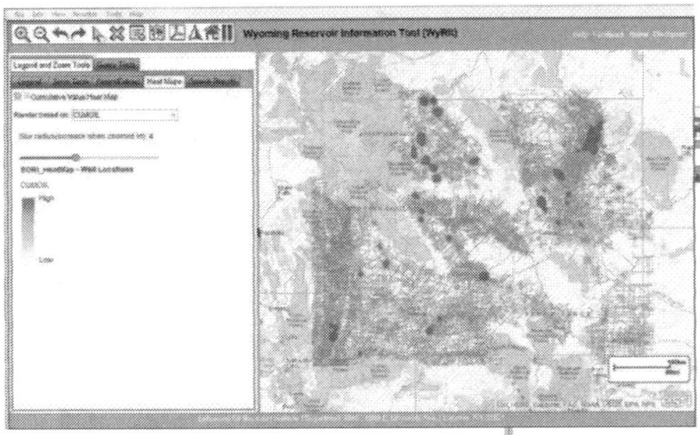


Gas

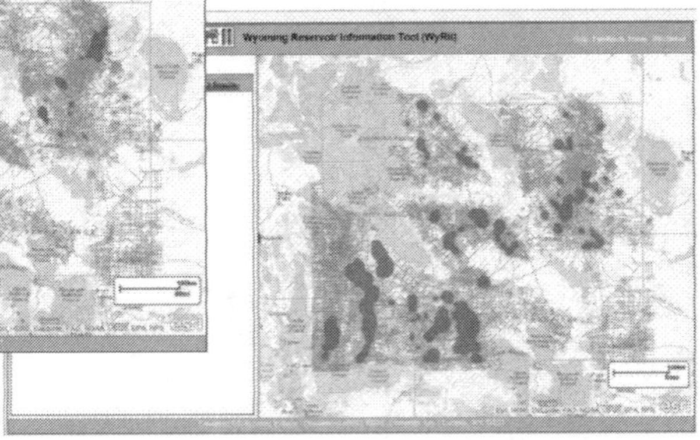

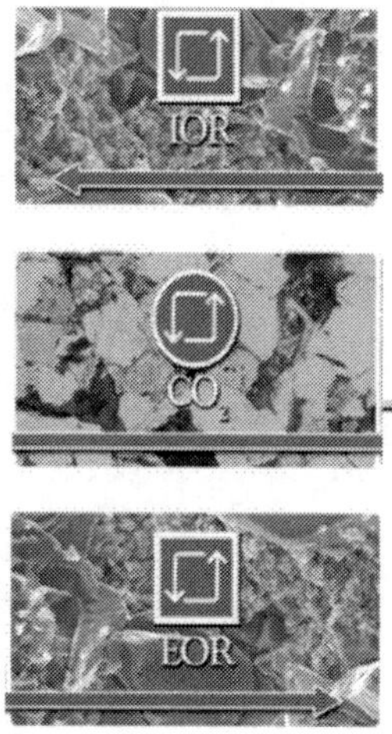


# $CO_2$-EOR & Buffer Storage

## Incidental $CO_2$ sequestration



Enhanced Oil Recovery Institute

# Challenges and Questions

**How much $CO_2$ is incidentally sequestered?**

- Reported $CO_2$ injection volumes - Check
- Reported natural gas production volumes – Check
- *Little to no available public information on composition of produced natural gas*
- $CO_2$ loss to the reservoir is estimated to be between 15 and 30% during the project.
- It is Estimated that 90 to 95% of injected $CO_2$ is permanently stored at the end of a $CO_2$-EOR project (Melzer et al)

- This is a common theme at $CO_2$ conferences



Enhanced Oil Recovery Institute

# Summary $CO_2$ Injection in Wyoming

| Year | Project | Incremental bo | Cumulative CO2 (mcf) | Cumulative CO2 (tons) |
|---|---|---|---|---|
| 1986 | Wertz | 24,150,949 | 203,167,188 | 11,620,842 |
| 1989 | Lost Soldier | 45,834,473 | 675,416,216 | 38,632,741 |
| 2008 | Beaver Creek | 8,825,906 | 168,613,153 | 9,644,406 |
| 2013 | Big Sand Draw | 916,792 | 39,838,773 | 2,278,715 |
| 2012 | Grieve | | 36,258,949 | 2,073,955 |
| 2003 | Patrick Draw | 17,079,074 | 334,713,367 | 19,145,076 |
| 2003 | Salt Creek | 25,762,143 | 2,838,584,423 | 162,362,548 |

**122,567,280 Incremental barrels of oil** **245,758,283 tons of CO2 in buffer storage**

*2.01 tons of CO2 per barrel of oil*

| Assumed Oil Price | | | Assumed cost | |
|---|---|---|---|---|
| $50.00 | $735,403,682 | state revenue | 2% | $184,318,713 cost of CO2 for EOR |
| | $5,392,960,336 | industry | of the price of oil | |



Enhanced Oil Recovery Institute

# $CO_2$ Demand Potential

- Bighorn Basin – Potential of 1 Billion barrels in $CO_2$ EOR (Yin et al)
- Potential $CO_2$ demand ~3 to 4.5 Tcf
    - Phosphoria
    - Tensleep
    - Madison
- Green River Basin
- Potential $CO_2$ demand ~0.8 to 1.3 Tcf
    - Almond
- Powder River Basin – Potential of 885 Million barrels in CO2 EOR (Branting and Whitman)
- Potential $CO_2$ demand ~4.7 to 7.1 Tcf
    - Frontier
    - Muddy
    - Minnelusa
- Sand Wash Basin
- Potential $CO_2$ demand ~0.08 to 0.1 Tcf
    - Tensleep
    - Madison
- Wind River Basin
- Potential $CO_2$ demand ~1.1 to 1.7 Tcf
    - Tensleep
    - Madison

## Operator willingness to implement

- Based on $CO_2$ availability
- Pipelines
- Field infrastructure costs
- It's a natural progression from secondary recovery
- This is a common theme at $CO_2$ conferences

**The top 180 $CO_2$-EOR Targets in Wyoming would require approximately 700 mmtons $CO_2$. An additional 150 mmtons would need to do the rest.**

Branting, J.K. and Whitman, L.D., 1992. The Feasibility of Using $CO_2$-EOR Techniques in the Powder River Basin. Society of Petroleum Engineers – SPE-24337 – MS.

Wo, S., Whitman, L.D., and Steidtmann. 2009. Estimates of Potential $CO_2$ demand for $CO_2$-EOR in Wyoming Basins. Society of Petroleum Engineers – SPE-122921 – MS.



Enhanced Oil Recovery Institute

# Candidate PRB Reservoirs suitable for $CO_2$-EOR

Enhanced Oil Recovery Institute

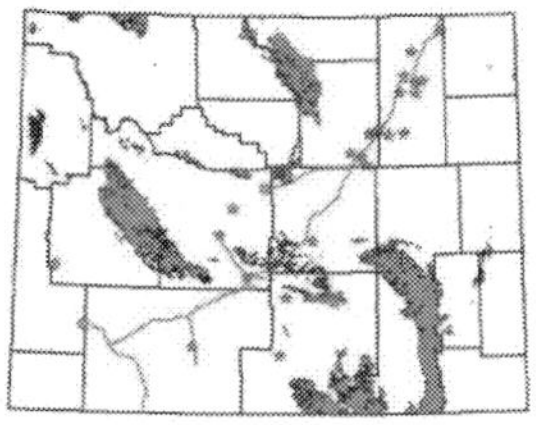

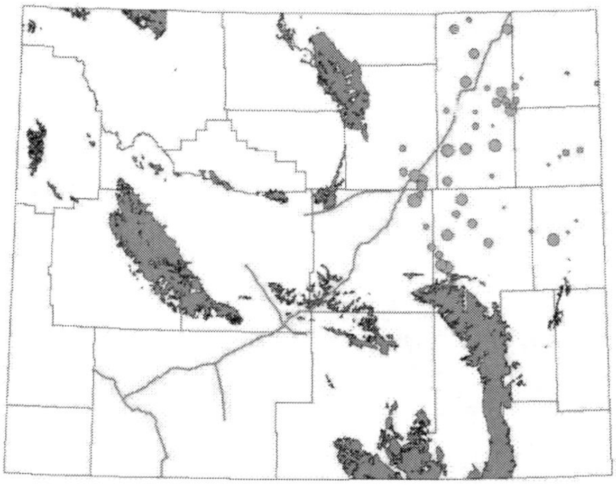

Criteria used to identify these reservoirs include:

- Depth between 2,200 and 12,000 feet
- Api oil gravity between 22 and 45
- Cumulative oil production greater than 1 mmbo

*Note – $CO_2$ needed was derived from estimated recovery factors, cumulative production, 2 bo per ton $CO_2$,

Tons $CO_2$ required = Estimated incremental production/2



# Enhanced and Improved Oil Recovery

WyGISC

## Acknowledgements and Credits

- Enhanced Oil Recovery Commission
- Wyoming Geographic Information Science Center
- Wyoming Oil & Gas Conservation Commission
- Wyoming Geological Association
- Melzer Consulting
- Wyoming State Geological Survey
- University of Wyoming School of Energy Resources

WOGCC

Wyoming PIPELINE AUTHORITY

School of Energy Resources

UNIVERSITY of WYOMING

# Thank You!

Nick Jones- Manager, Conventional Resources

NJones@uwyo.edu

Office 307-766-3284

www.uwyo.edu/eori

Enhanced Oil Recovery Institute
UNIVERSITY OF WYOMING

KAPSARC

مركز الملك عبدالله للدراسات والبحوث البترولية

King Abdullah Petroleum Studies and Research Center

# Could China meet CO2 emission reduction targets by CO2-EOR?

Sa'd Shannak

## Table of Content

## KAPSARC in Brief ...

King Abdullah Petroleum Studies and Research Center (KAPSARC)

- Independent, non-profit, research institution
- Focuses on energy economics, policy, technology, and the environment
- Global and local
- Has financial and administrative independence
- Located in Riyadh

KAPSARC 3

## KAPSARC Research Programs and Focus Areas

Research Programs (Teams):

| Energy Systems Modeling and Economics | Energy Markets, Fuels & Feedstocks | Energy Transitions, Econometrics & Macro | Human Geography of Energy |
|---|---|---|---|

Research Focus Area:

- **Allocating Energy Resources in an Economy**
- **Energy Flows and Market Impacts**
- **Social and Environmental Spillovers of Energy Policy Choices**
- **Energy Productivity**
- **Transportation; Modes, Fuels and Technologies**
- **Energy Transitions**

KAPSARC 4

## The Carbon Cycle

- There is a certain amount of carbon circulating through the environment.
- This amounts has been preserved for hundred of million of years;
- We are adding to the mix and change net amounts can be absorbed

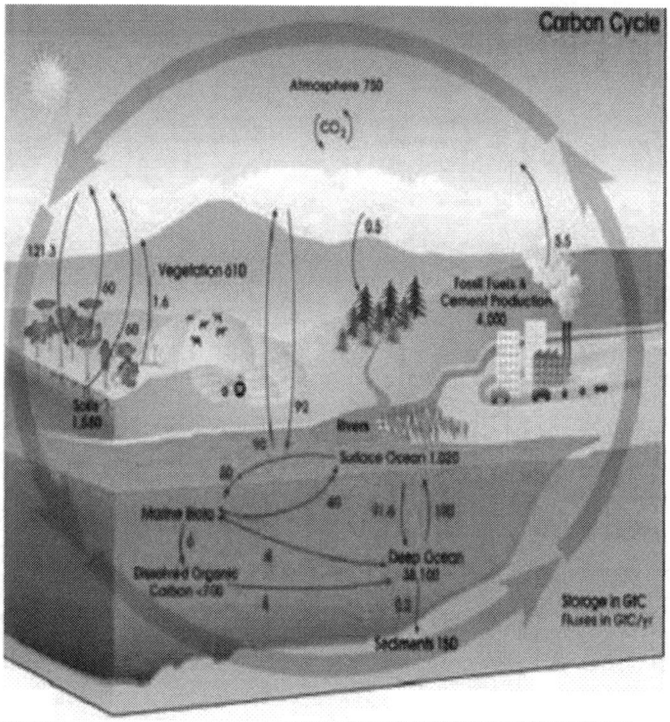




## Fossil Fuel CO2 emissions : Top emitters

- The top four emitters in 2015 covered 59% of global emissions China (29%), United States (15%), EU28 (10%), India (6%)

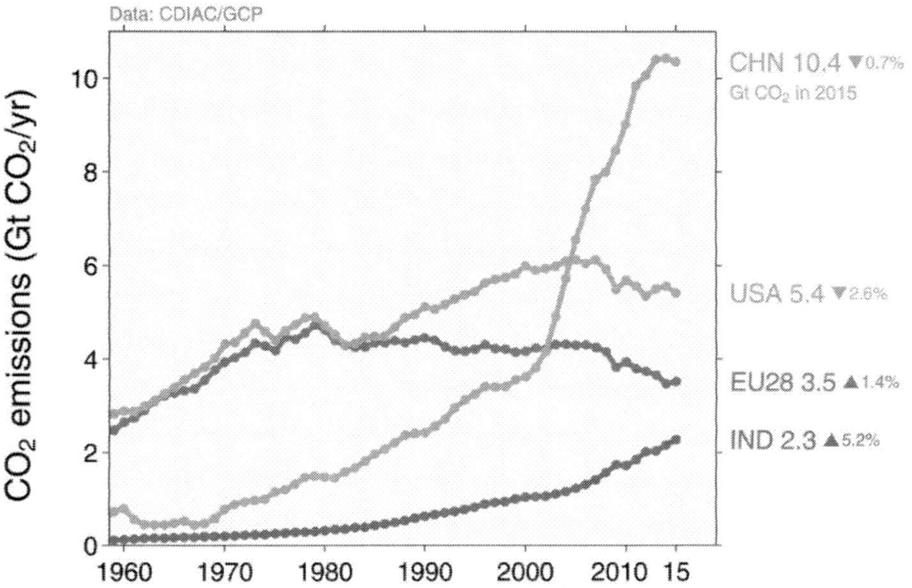




Source: CDIAC; Le Quéré et al 2016; Global Carbon Budget 2016

## Overview of Goal and Plans

- China at COP 21 has promised to:
  - Cut emissions from its coal power plants by 60% by 2020
  - Lower total carbon dioxide emissions per unit of GDP by 60% to 65% from the 2005 level by 2030
  - Start in 2017 its national emission trading system, covering key industry sectors

KAPSARC

7

## CCUS in China

- China current 13th Five-year plan (2016–2020) focuses on technology development for CCS R&D with an industry matching fund of around two billion

- CCUS roadmap targets during the Five-Year plan:
  - identify unique low-cost opportunities for CCS demonstration
  - launch at least two large-scale CCS demonstration projects, with an installed capacity to capture at least 2 million tons of CO2 per year.

- Two main challenges :
  - public funding and
  - knowledge about potential CO2 storage sites.

KAPSARC

8

## Opportunity

- $CO_2$ is injected into reservoirs to boost oil recovery and slow declining oil production
- Regulations pertaining EOR are generally clear
- Operators especially in the states have 30-years commercial experience
- CO2 storage can be attained along oil production

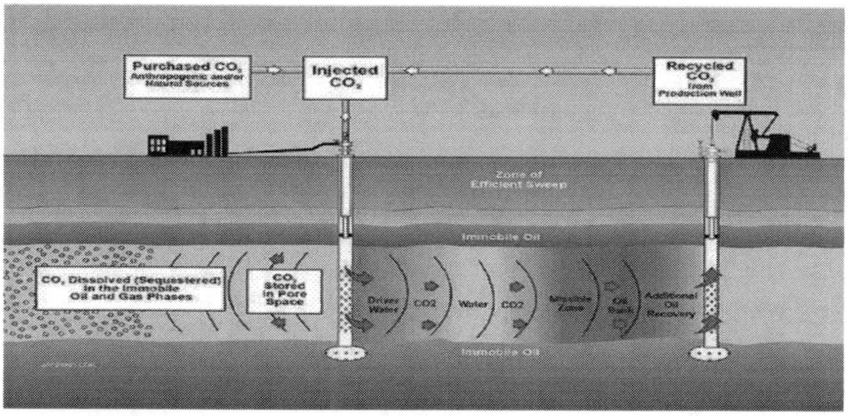




## Research Question

- What is the potential of CO2-EOR projects in China and its impact on electricity market ?

## Objectives

- Analyze the consequence of using CO2 capture for CO2-EOR on the Chinese electricity market.
- Develop an economic model to understand the supply side economics associated with CO2-EOR

## Motivations

- China ambition to reduce CO2 emissions
- Existing opportunities to use CO2-EOR for climate change purposes

# A framework for analyzing the CO2-EOR suitability

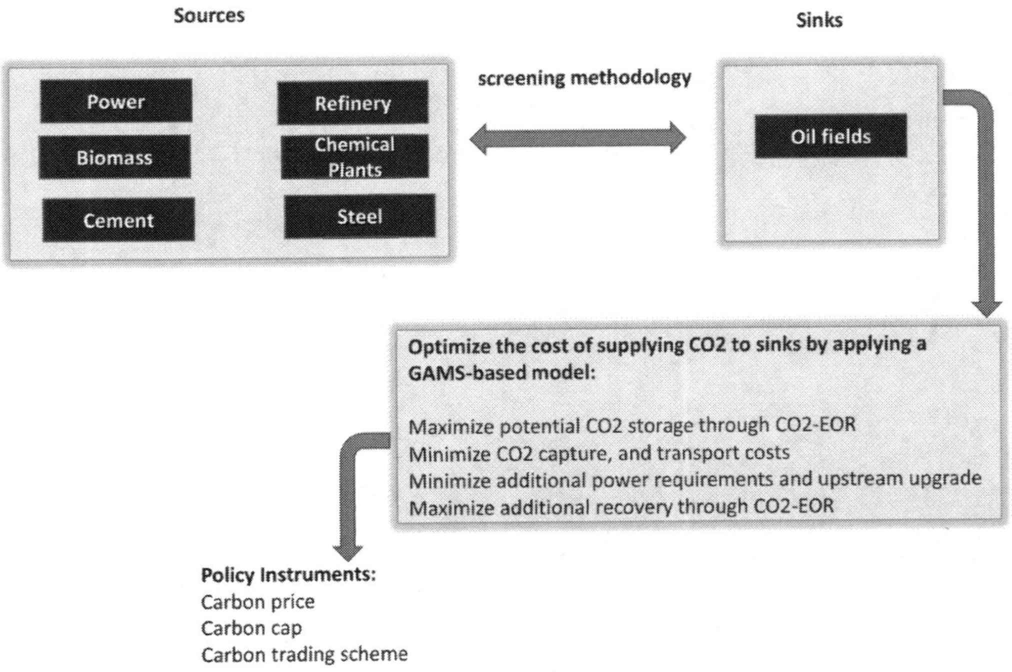




# Screening Criteria (sources)

- Considered so far screening criteria
  - Age of facility: Younger than 20 years for power plant
  - Status (operating and constructed )
  - Coal and natural gas
  - Total capacity >100 MW (Power Plant)
  - Distance to sink (max 500 KM)

# Sources

- Potential CO2 emission from each source was calculated by multiplying the fuel consumption by corresponding emission factor

| | Pounds of CO2 per million Btu | Heat rate (Btu per kWh) | Pounds of CO2 per kWh | Grams of CO2 per kWh |
|---|---|---|---|---|
| Coal | | | | |
| Bituminous | 205.691 | 10,080 | 2.07 | 938.9 |
| Subbituminous | 214.289 | 10,080 | 2.16 | 979.7 |
| Lignite | 215.392 | 10,080 | 2.17 | 984 |
| Natural gas | 116.999 | 10,408 | 1.22 | 553 |
| Distillate oil (No. 2) | 161.29 | 10,156 | 1.64 | 743 |
| Residual oil (No. 6) | 173.702 | 10,156 | 1.76 | 798 |



# Sources

- In total the combined database comprises 3275 sources (2361 power plants, 674 cement plants, 16 steel plants, 114 oil refinery plants, 60 biomass plants, and 23 coal to liquid plants)

- The total emission amounts estimated to be 5730 MtonCO2/year (almost 6 Gt/year) and the emission share as follows: 79%, 10%, 9%, 0.4%, 1%, and 1% for power, cement, steel, refinery, biomass and coal to liquid respectively

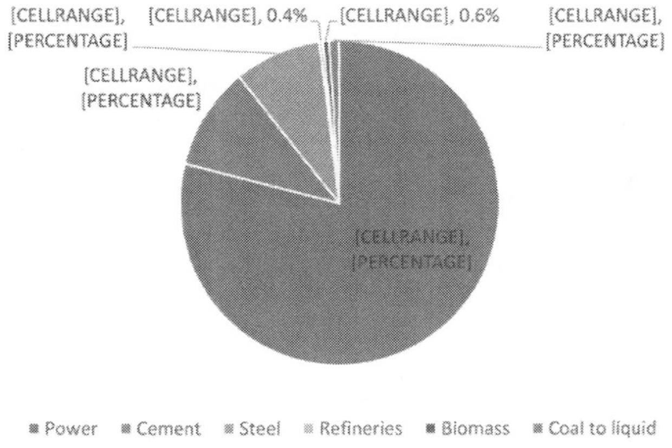

## Screening methodology of Sinks (oil fields)

| Scoring criteria | Miscible flood | Immiscible flood | Score |
|---|---|---|---|
| Remaining oil saturation | The intention of this criteria is to exclude fields that has come too far in its production life cycle. However, the criteria is redundant since this is captured by the recovery phase criteria | | 0 or 2 |
| Effective mobility | Reservoirs are assumed to require an effective mobility higher than 5 | N/A | 0, 1 or 2 |
| Viscosity | N/A | Minimum viscosity for Immiscible flood is set at 10 cP, i.e. viscosity must be greater than 10 cP | 0, 1 or 2 |
| Pure oil field | Gas/oil ratio (GOR) must be less than 25% | | 0, 1 or 2 |
| Drive mechanism | Fields with gas injection are unfit for CO2 EOR and receive a zero if known to have gas injection. Similarly, water or WAG injection has high potential and receive score 2 | | 0, 1 or 2 |
| Recovery phase score | Remaining oil in place must be higher than 30% of OOIP in order to make EOR economically viable, i.e. current recovery factor must be less than 70% (this was set at 80% in the report which was a typo) | | 0 or 2 |
| Total | The total score is calculated as the product of all the above | | 0, 8, 16 or 64 |

KAPSARC

15

## Screening methodology of Sinks (oilfields)

$$MMP = -329.558 + (7.727 * \text{MW} * 1.005\ \text{Temp}) - (4.377 * \text{MW})$$

Where MW is the mole weight of the C5+ component of the reservoir oil estimated as follows:

$$MW = (7864.9 \div API)^{\left(\frac{1}{10386}\right)}$$

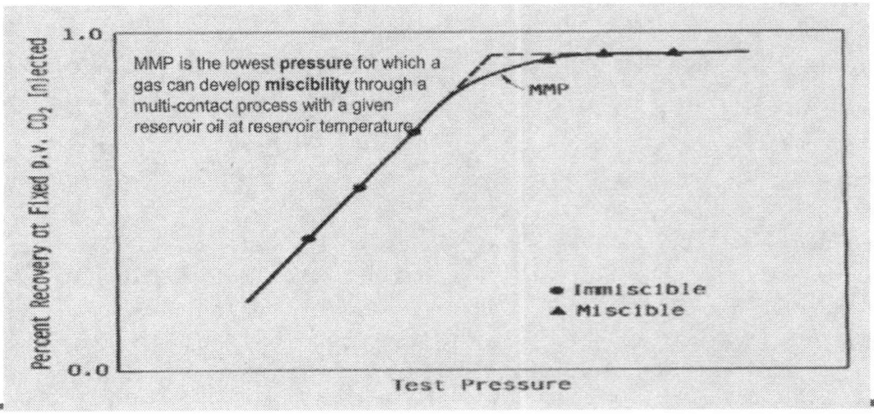


KAPSARC

16

## Oil field candidates suitable for CO2-EOR application

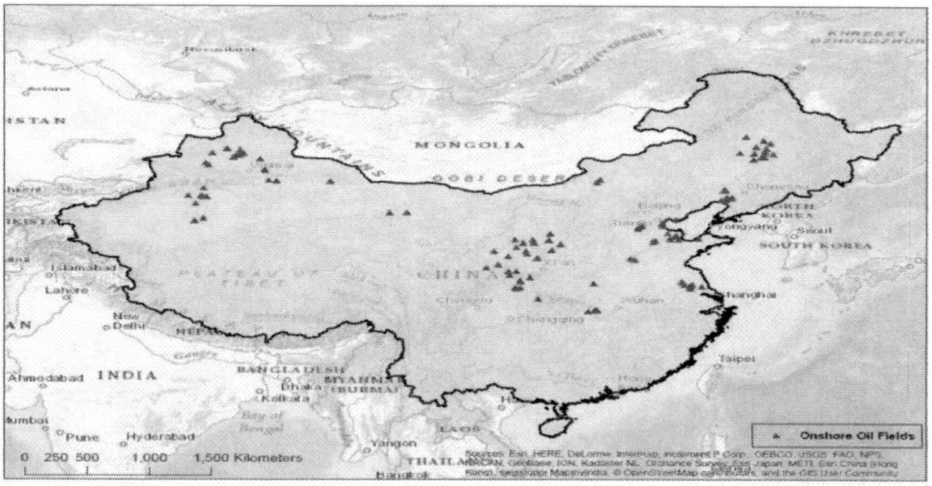


- 151 onshore oilfields for CO2-EOR flooding were selected and used in this study. It is worth noting that all offshore fields were excluded from initial screening as the CO2-EOR application can be exceptionally costly.



## Optimization model to analyze tradeoffs in developing CO2-EOR projects

- The objective function maximizes the revenues as follows:

$$\max = \sum (Q \times O_p) + (Q_c \times C_p)$$

Q is the additional oil recovery due to CO2-EOR in barrels
OP is the price of oil barrel in U.S dollars
Qc is the quantity of CO2 stored in the formation
Cp is the CO2 price

- Constraints:

CO2 demand for each CO2-EOR operation has been met and additional power requirements to capture CO2 is also met

# Optimization model to analyze tradeoffs in developing CO2-EOR projects

The model eventually works on maximizing the profit of oil revenues from total costs as shown below:

$$P = (Q \times O_p) + (Q_c \times C_p) - \{(D_{OS} \times S_s \times T) + (S_s \times C_C) + (U_c \times Q) + (S_S \times E_r \times A_{cc}) + (Q \times TX)\}$$

- DOS is the distance between oil filed and CO2 supply source,
- T is the transport rate,
- CC is the capture cost for different CO2 sources,
- Uc : is the upstream facility upgrade cost, and
- TX is the tax rate payable for the government



# CO2-EOR: Storage and Recovery (IEA Scenarios)

| Scenario Name | Description | Incremental Recovery, % OOIP | Utilisation, $tCO_2$/bbl |
|---|---|---|---|
| **Conventional** | Miscible WAG flood with vertical injector and producer wells in a "five spot" or similar pattern. Operational practices that seek to minimize $CO_2$ utilization. | 6.5 | 0.3 |
| **EOR+** | Miscible flooding following current best practices optimized for oil recovery. May also involve some "next" and "second generation" approaches that drive increased recovery and utilization. | 13 | 0.6 |
| **Maximum Storage** | Miscible flooding where injection is designed and operated with the explicit goal of increasing storage. Could include approached where water is removed from reservoir to increase available pore volume. | 13 | 0.9 |

## Sources to sinks: KAPSARC as a matchmaker

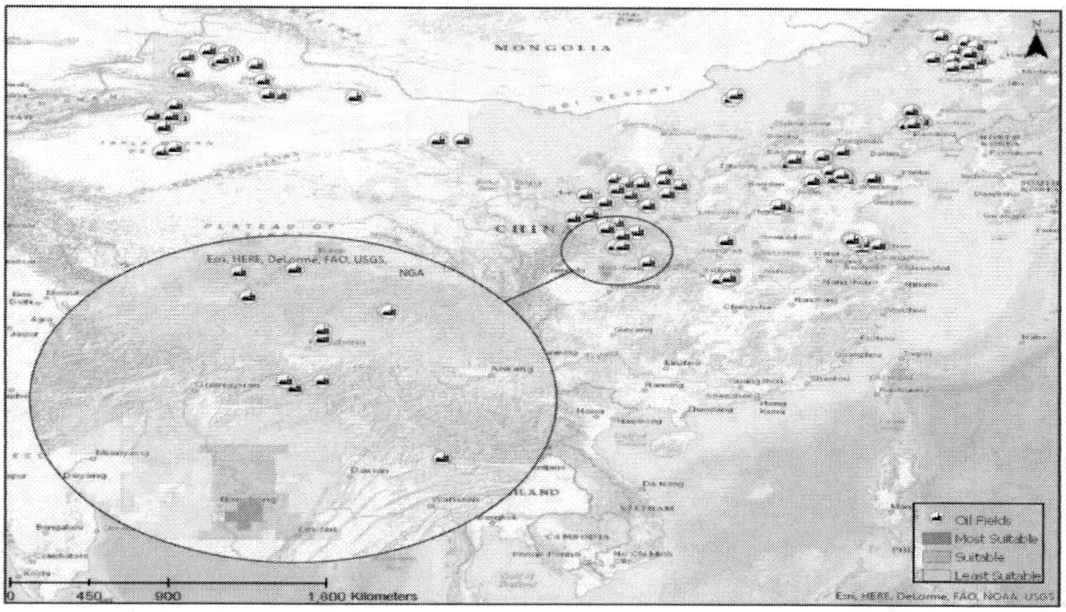




## Conclusions

Substantial opportunity for storage & oil recovery through CO2-EOR

31% emission reduction through CO2-EOR can be attained

Additional sources of emission besides power plants could provide an economic opportunity

Need for pilot projects to demonstrate the dual CO2-EOR objectives.

Design of proper policy instruments and incentives need to be in place

# Thank You

Sad.Shannak@kapsarc.org

KAPSARC King Abdullah Petroleum Studies and Research Center

23

# AQUISTORE

ptrc
Petroleum Technology
Research Centre

Aquistore®

**Kyle Worth, P.Eng, PMP**
**Aquistore Project Manager**
**Petroleum Technology Research Centre**

1

# PETROLEUM TECHNOLOGY RESEARCH CENTRE

ptrc
Petroleum Technology
Research Centre

- **Non-Profit Research & Development Company**
- **Collaborative partnership with Industry, Government and Research Organizations**
- **Extensive network of researchers worldwide**
- **Research associated with $CO_2$ management**
  - **IEAGHG Weyburn –Midale $CO_2$ Monitoring & Storage Project**
  - **Sask$CO_2$USER**
  - **Aquistore**

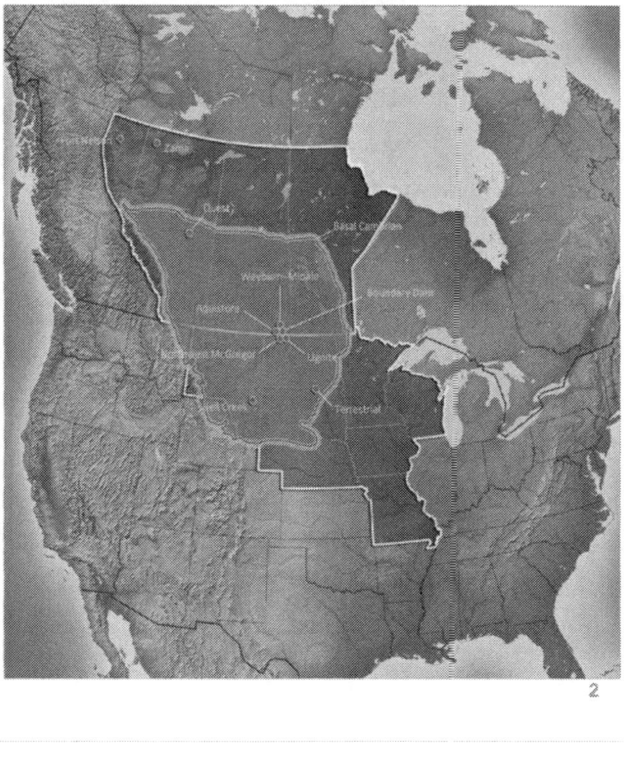

2

## WEYBURN ($CO_2$-EOR) AQUISTORE ($CO_2$ STORAGE)

ptrc Petroleum Technology Research Centre

- **Integrated $CO_2$-EOR and $CO_2$ Storage CCUS Project**
- **Practical application of observations, new data, and evidence is key**
- **Moving beyond pilot demonstration to commercially operating scale**
  - **Risk Assessment**
  - **Uncertainties**
  - **MMV**

3

## AQUISTORE: DEEP SALINE $CO_2$ STORAGE PROJECT

ptrc Petroleum Technology Research Centre

**107,000** CUMULATIVE TONNES OF $CO_2$ STORED

**0** INDUCED SEISMIC EVENTS

**2100** MAX INJECTION RATE TONNES/DAY

**30** MONITORING TECHNOLOGIES DEPLOYED

4

# Deepest Wells in Saskatchewan

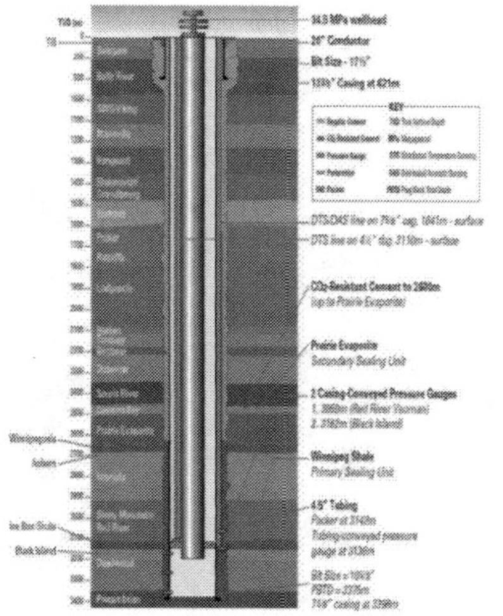


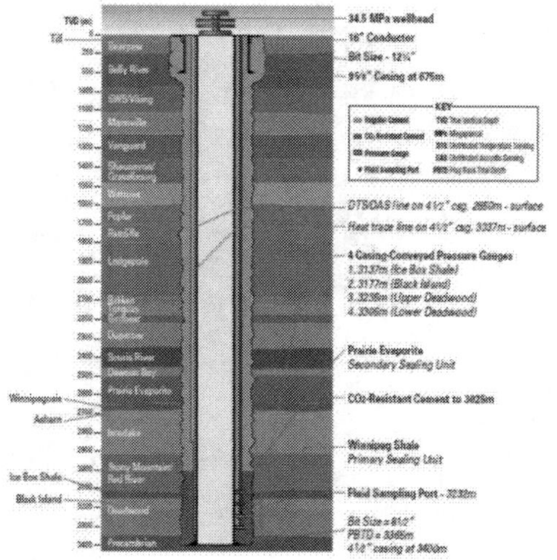


5

# 2015-2017 MMV Program

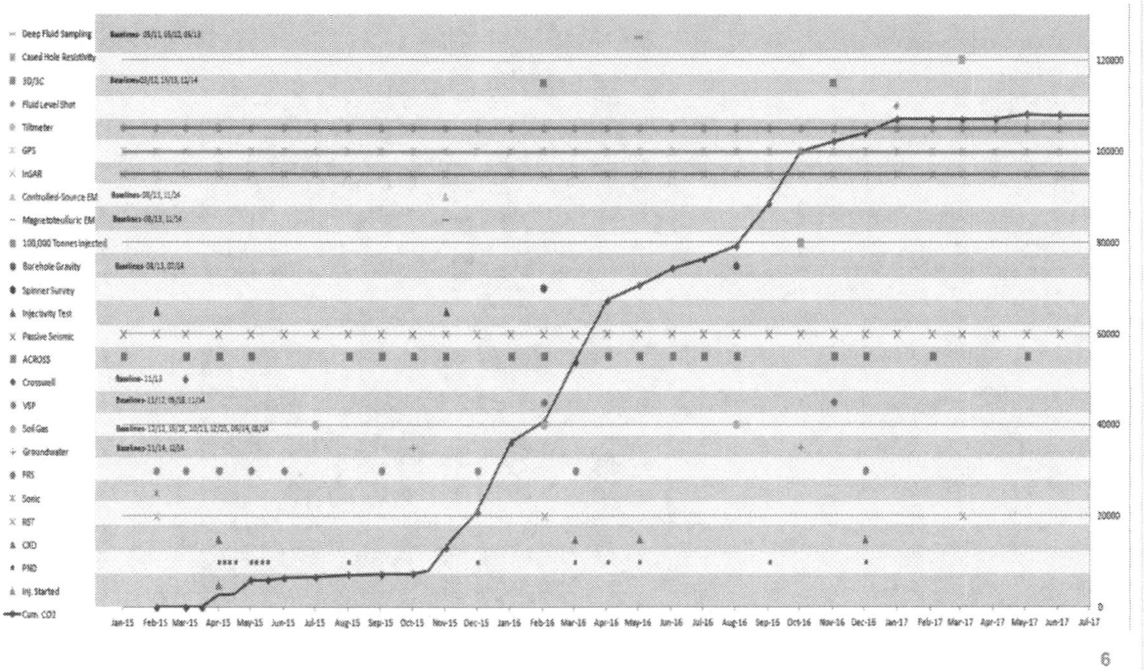


6

# MONITORING INSTALLATIONS

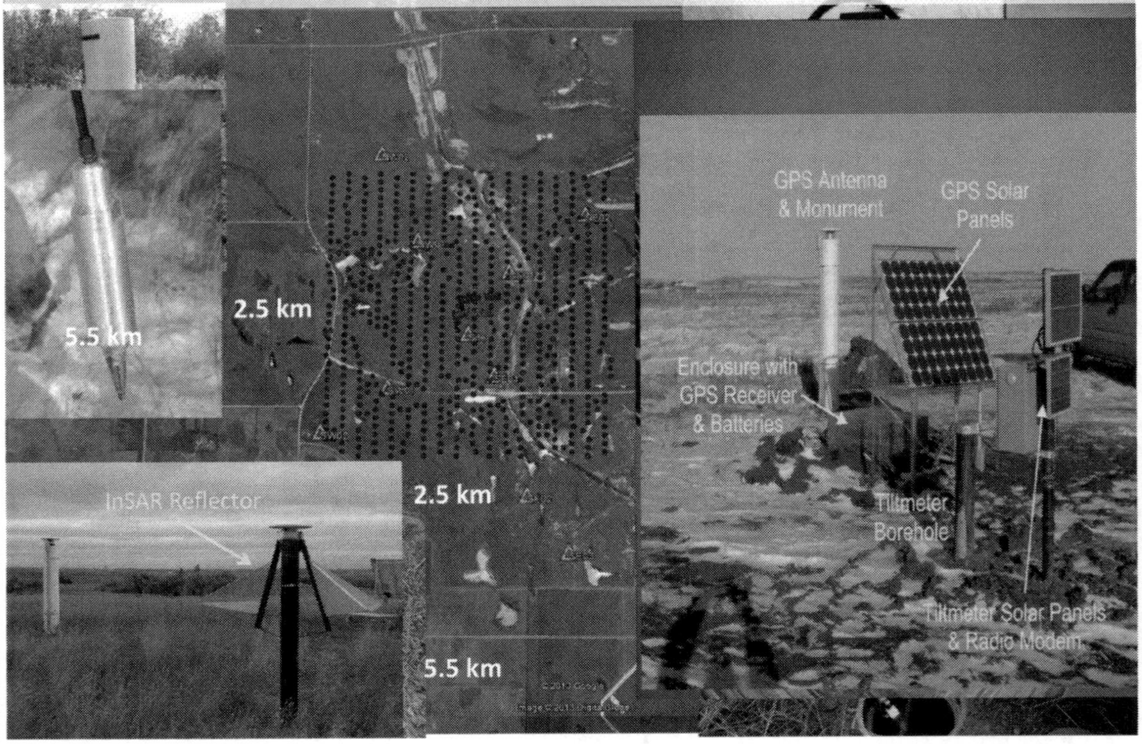


# 3D TIME-LAPSE SEISMIC

ptrc
Petroleum Technology Research Centre

**Baseline survey(March-2012)**
•3D Surface seismic
•1kg explosives @20m
•Buried permanent receivers @15m

**Pre $CO_2$ Monitoring 1 (May-2013)**
•3D Surface seismic

**Pre $CO_2$ Monitoring 2 (Nov-2013).**
•3D Surface seismic
•3D DAS VSP-I

$CO_2$ injection started (Apr 2015)

**Monitor 3 (Feb 2016)** 36 kT $CO_2$
•3D Surface seismic
•3D DAS VSP-II

**Monitor 4 (Nov 2016)** 100 kT $CO_2$
•3D Surface seismic
•3D DAS VSP-III

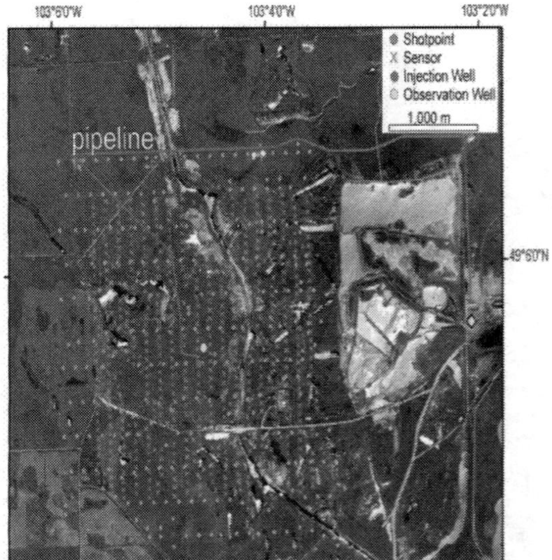


8

# PLUME EVOLUTION

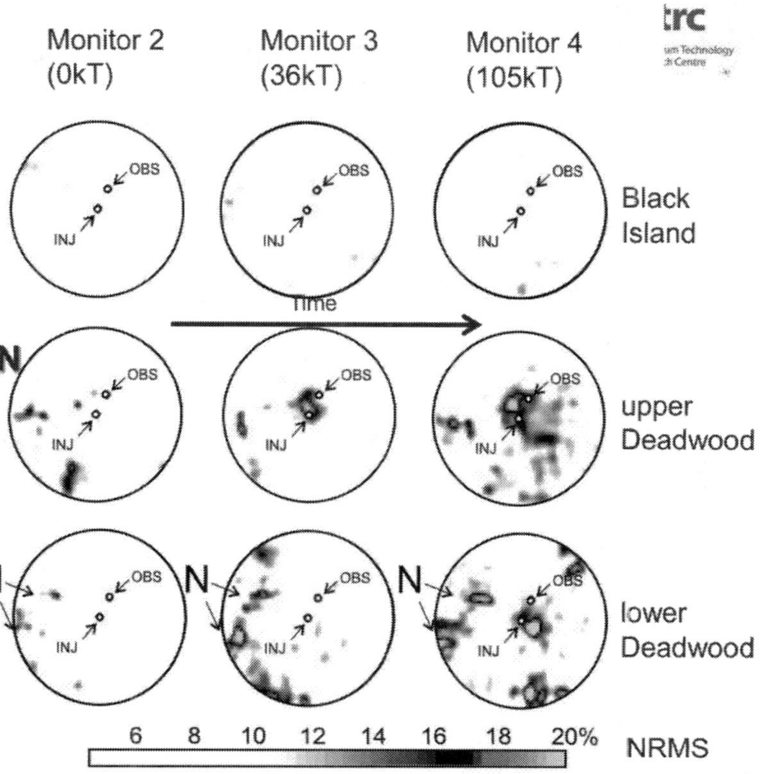


9

# OBSERVED SEISMIC VS. MODELLED $CO_2$ (105 KT INJECTED)

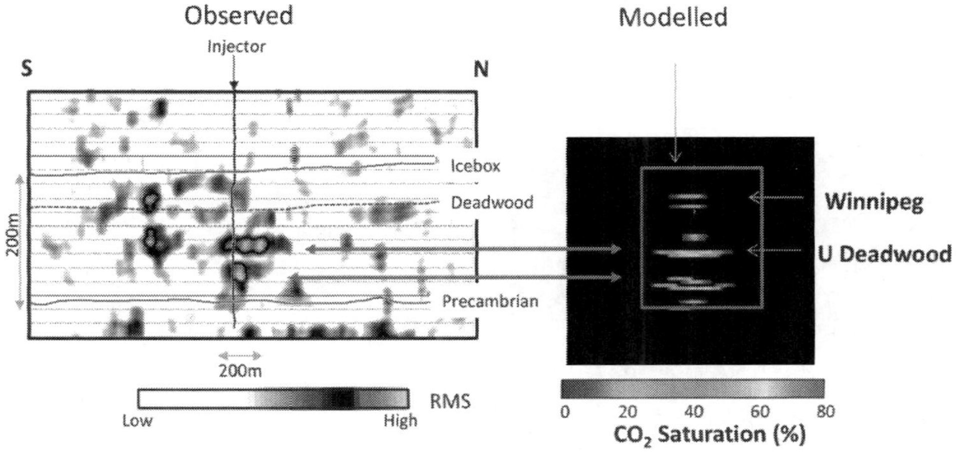


10

## Test Site for Casing Conveyed and Buried Fibre

ptrc Petroleum Technology Research Centre

### DAS (Distributed Acoustic System)

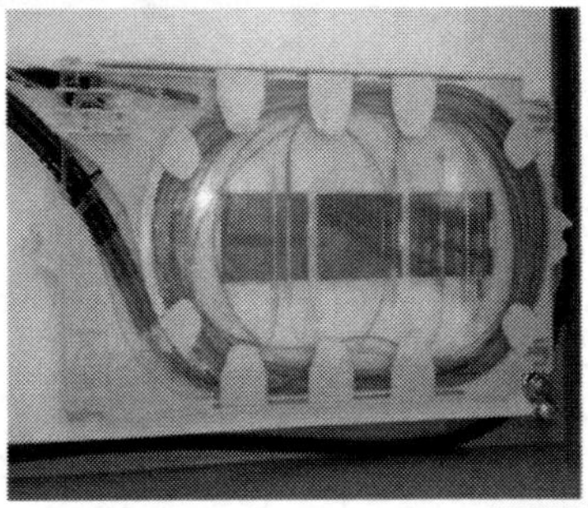

- Distributed acoustic sensing (DAS) technology for seismic monitoring could revolutionize seismic imaging.
- Employs a optical fibre as a replacement for traditional geophones.
- High quality 3D imaging methods that are lower in cost to deploy than traditional geophone surveys
- Further research and demonstration will confirm monitoring capabilities that surpass the current standard technologies.
- Can be used for natural resources extraction (oil, gas), underground storage ($CO_2$, wastewater) and in the potash and mining industries.

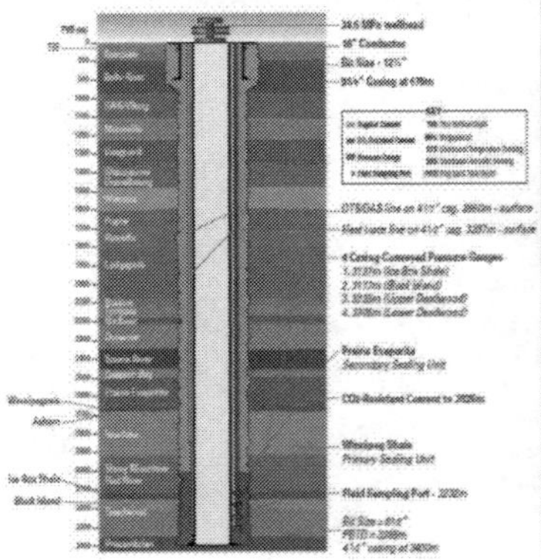

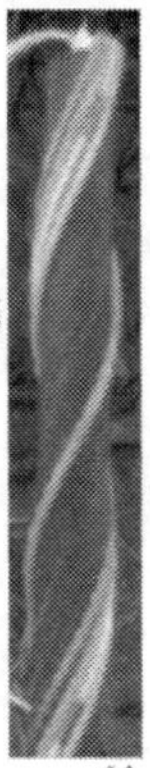

11

## Monitor I DAS VSP Difference

ptrc Petroleum Technology Research Centre

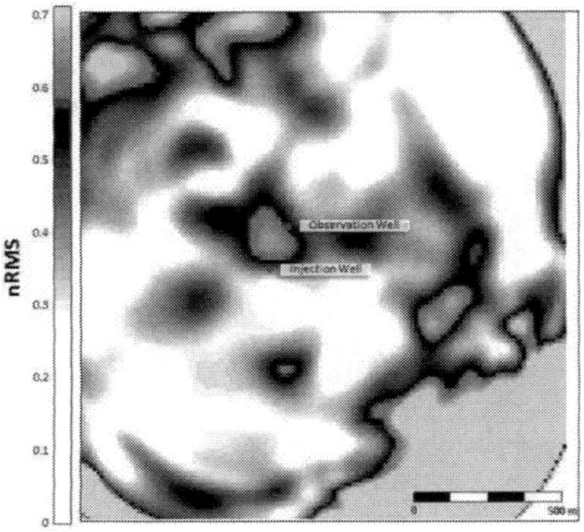


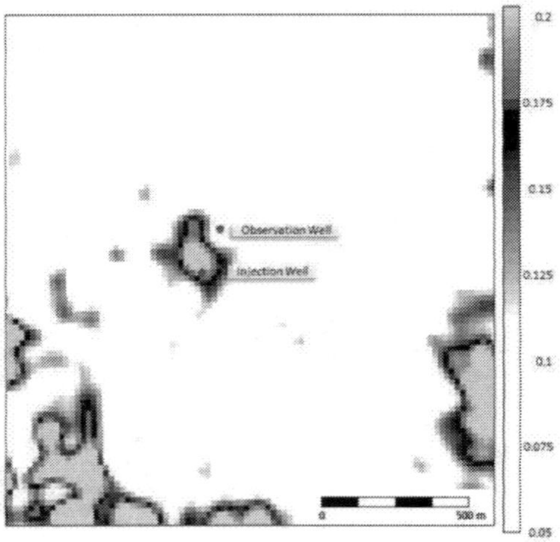


12

## ACROSS Source

- **ACROSS (Accurately Controlled and Routinely Operated Signal System)**
- **Fixed in cement at surface, can produce repeatable, extremely precise two-component seismic signal**
- **Used in collaboration with Aquistore's 630 buried geophones.**

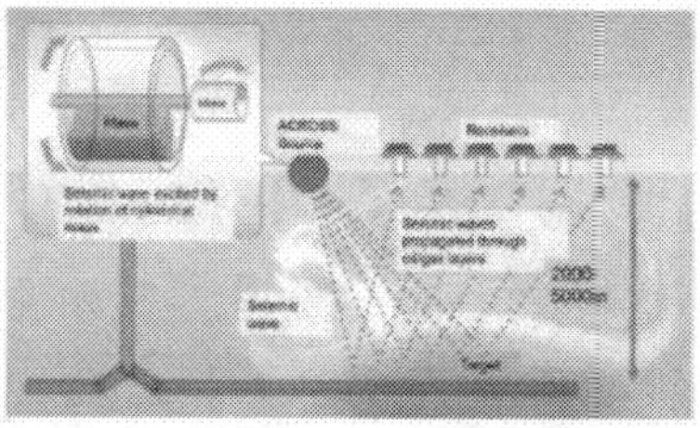

JOGMEC

13

## Induced Seismicity Monitoring

- No injection related seismicity (Mw >1) detected during first 2 years of operation.
- No smaller magnitude events (Mw> -3) recorded during 8 month deployment of more sensitive downhole system.
- Local events: mining blasts recorded.
- Regional and teleseismic events have been detected and recorded.

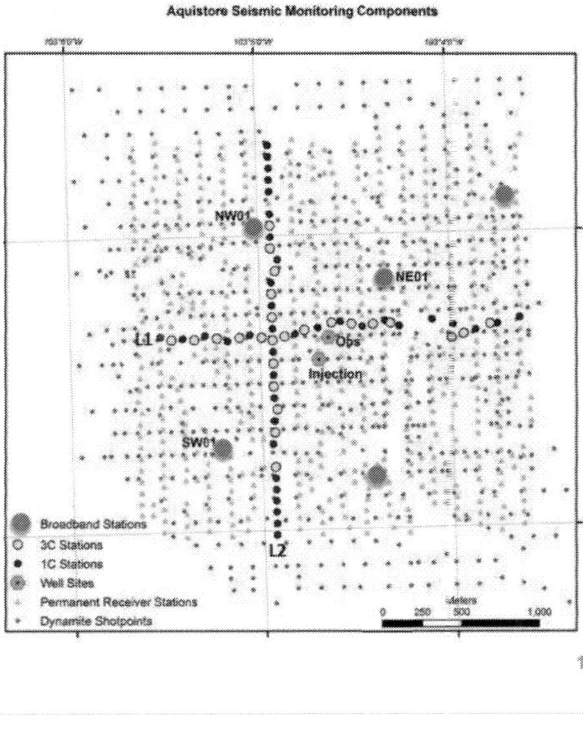


Bristol University Microseismicity ProjectS
BUMPS

14

## PND Logging Results – Differential Saturations from Σ

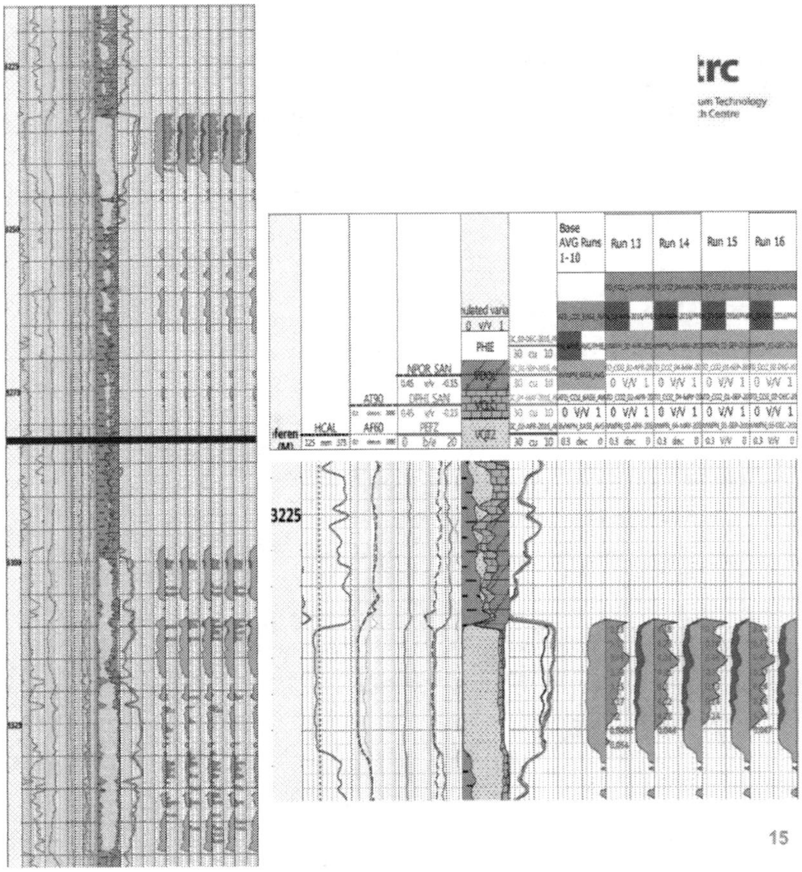


15

## Geomechanical Coupled Non-Isothermal Model

ptrc
Petroleum Technology Research Centre

- Cold $CO_2$ Injection
- Wellbore heat transmission
- Thermal effects in fluid/mineral reactions
- $CO_2$ thermo-physical behavior
- $CO_2$ injectivity
- $CO_2$ migration in cooled area
- Thermal fracturing
- $CO_2$ hydrates

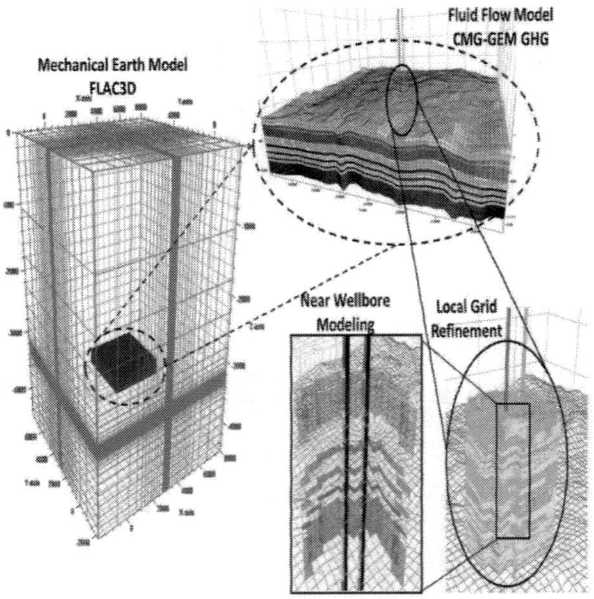


16

# 2017-2018

ptrc Petroleum Technology Research Centre

- Small quantity of $CO_2$ at significant depth imaged with permanent and DAS seismic surveys.
- Evaluate impacts of a 75°F drop in downhole temperature
- Incorporate non-isothermal parameters into future reservoir simulations.
- Evaluate how variable rates of injected $CO_2$ may have an impact on injection well: casing, tubing, packer and cement integrity.
- Evaluate the impact on near-wellbore and long-term injectivity due to salt precipitation phenomena.

*"The continued success of this Project will have incredible implications for reducing $CO_2$ emissions throughout the world."*

\- John Gale, IEA

17

# Thanks to Our Sponsors & In-Kind Collaborators

ptrc Petroleum Technology Research Centre

Sustainable Development Technology Canada · Enbridge · goGreen Saskatchewan · Schlumberger Carbon Services · SaskPower Powering the future · CO-OP · Natural Resources Canada · bp · EERC · PCOR Partnership · Chevron · Korea National Oil Corporation · UKCCS Research Centre · MGSC · GFZ Helmholtz Centre Potsdam · Berkeley Lab · JOGMEC · Scintrex A Division of LRS · Global CCS Institute · OYO · RITE · $CO_2$ Capture Project · SaskEnergy · INRS · St. Francis Xavier University · University of Alberta · University of Saskatchewan · University of Manitoba · Carleton University · University of Strathclyde Glasgow · University of Leeds · University of Bristol

For updates and news, follow us on Twitter:
@PTRC_SK

18

Center for Applied Energy Research

# Evaluating the Effect of Aqueous Calcium Chloride Desiccant on Cooling Tower Performance in a Heat-Integrated Post-Combustion $CO_2$ Capture System


**Amanda Warriner[1],**

**Heather Nikolic[1], Jonathan Pelgen[1], Kunlei Liu[1,2]**

*[1]Center for Applied Energy Research, University of Kentucky, 2540 Research Park Drive, Lexington, Kentucky 40511-8410, United States*

*[2]Department of Mechanical Engineering, University of Kentucky, Lexington, Kentucky 40506, United States*


*http://www.caer.uky.edu/powergen/home.shtml*

caer.uky.edu

CMTC 2017 July 17-20 , 2017

---

Center for Applied Energy Research

# University of Kentucky Center for Applied Energy Research

***Creating Technology for Tomorrow's Energy***

www.caer.uky.edu

caer.uky.edu

CMTC 2017 July 17-20 , 2017

UK Center for Applied Energy Research

caer.uky.edu

# Power Generation and Utility Fuels Group

## *Using Fossil Resources to Produce Clean Electricity*

| | | | | |
|---|---|---|---|---|
| Post-Combustion $CO_2$ Capture | Solvent Development | Chemical Looping | Water Treatment | Corrosion |
| Process Integration and Scale-Up | Process Controls | Electro-chemistry | Membrane Separations | Analytical Methods Development |

- About 30 researchers (engineers, scientists, technicians and students)
- 5 active projects sponsored by U.S. DOE and 2 active projects sponsored by industrial consortia
- 10-18 peer reviewed publications, annually
- 5-7 invention disclosures, annually
- 5-10 project proposal submitted, annually

CMTC 2017 July 17-20 , 2017

UK Center for Applied Energy Research

caer.uky.edu

# Carbon Capture Project Background

Primarily funded by U.S. DOE NETL, a 0.7 MWe $CO_2$ Capture System (CCS) was developed by the University of Kentucky's Center for Applied Energy Research (UKy- CAER), installed at Kentucky Utilities E.W. Brown Generation Station and evaluated for overall plant efficiency.

Title: Application of a Heat Integrated Post-Combustion $CO_2$ Capture System with Hitachi Advanced Solvent into Existing Coal-Fired Power Plant (DE-FE-0007395)

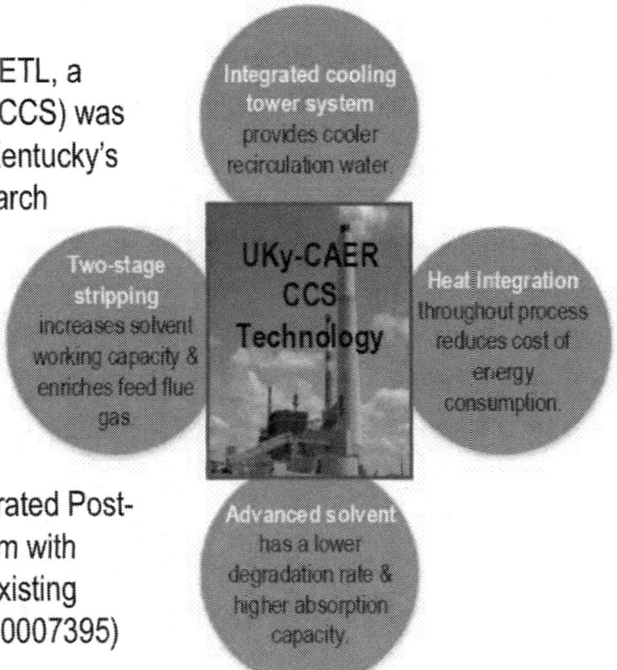


CMTC 2017 July 17-20 , 2017

Center for Applied Energy Research

caer.uky.edu

# Benefits Realized in UKy-CAER CCS Technology

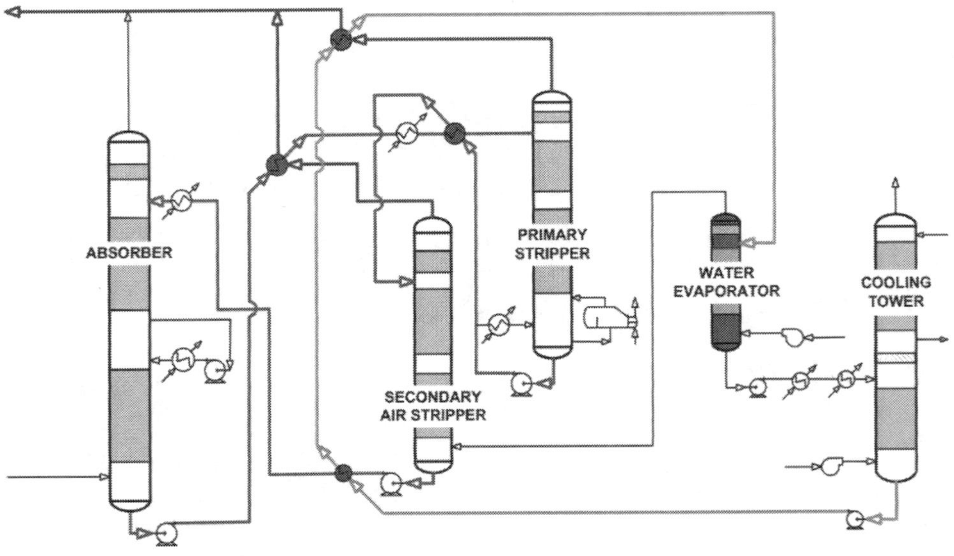


—— = Solvent; —— = Gas Return; —— = Liquid Desiccant

**BENEFITS: 1)** recovers waste energy, **2)** enhances $CO_2$ removal, **3)** improves commercial-scale turbine efficiency, **4)** reduces fresh water requirement

CMTC 2017 July 17-20, 2017

Center for Applied Energy Research

caer.uky.edu

# 0.7 MWe Carbon Capture System

**Application of a Heat Integrated Post-Combustion $CO_2$ Capture System with Hitachi Advanced Solvent into Existing Coal-Fired Power Plant (DE-FE-0007395)**

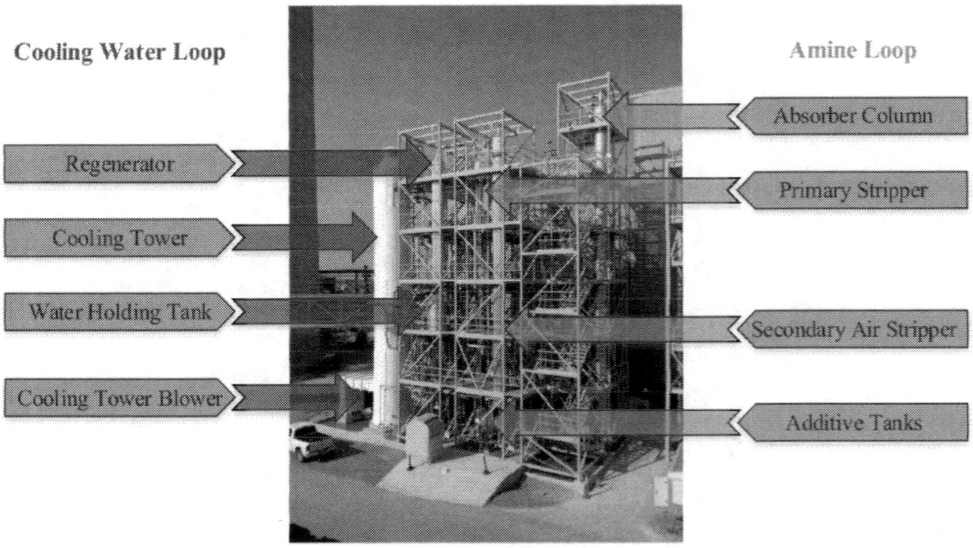


**Small Pilot Scale $CO_2$ Capture Unit at E.W. Brown Generating Station in Harrodsburg, KY**

CMTC 2017 July 17-20, 2017

Center for Applied Energy Research

caer.uky.edu

# Heat-integrated Liquid Desiccant Loop and Cooling Tower

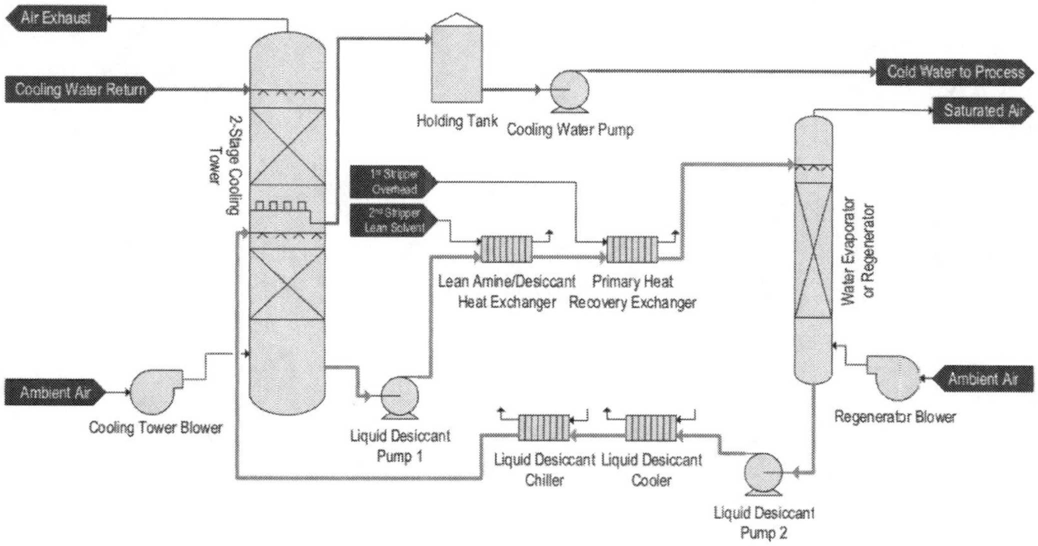


This process reduces the air relative humidity and wet bulb temperature ($T_{wb}$) in the top section, resulting in a lower cooling water temperature than otherwise plausible.

CMTC 2017 July 17-20 , 2017

Center for Applied Energy Research

caer.uky.edu

# Cooling Tower Preliminary Simulation

- Model suggested a $T_{wb}$ depression of 3-7 degrees may be observed
- Simulations showed increased performance of the liquid desiccant loop as the CTWref increases above 72°F, achieving as much as a 12 degree difference in the cooling water exit temperature over the reference design

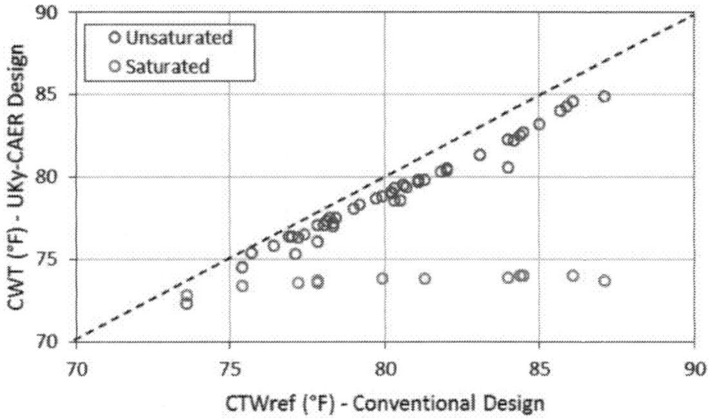


**Simulated Cooling Tower Water Exit Temperatures in August for the UKy-CAER Modified and Traditional Cooling Tower Designs**

Source: Version 8.2, Aspen Technology, Inc.

CMTC 2017 July 17-20 , 2017

Center for Applied Energy Research

# Sensitivity Study

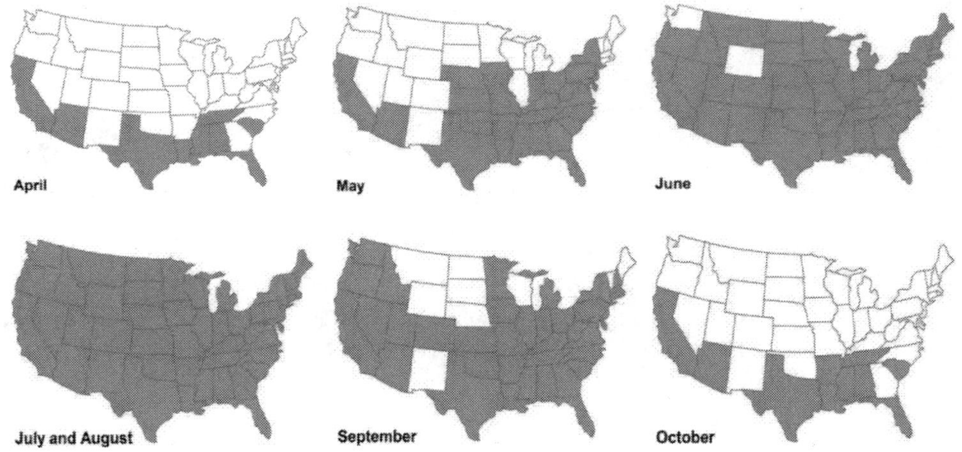


**States in the Continental United States with Average, Monthly Weather Conditions in which a Liquid Desiccant Air Drying System Will Provide Additional Cooling**

- ❑ Operation of a liquid desiccant air drying system will allow for increased efficiency in warm-weather months that are typical throughout the eastern and mid-western United States, and year-around in southern United States

Source: www.usclimatedata.com & www.currentresults.com

CMTC 2017 July 17-20 , 2017

caer.uky.edu

Center for Applied Energy Research

# Application Study

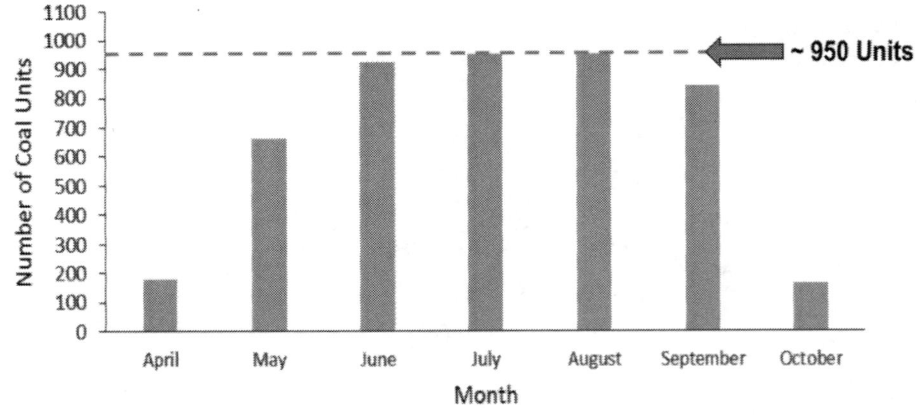


**Number of Coal-fired Units in the Continental United States with a CWTref Greater Than 72°F**

- ❑ From May-September, ~600-950 coal-fired units in the continental United States may benefit from the liquid desiccant air drying system

Source: www.eia.gov/coal/data.cfm

CMTC 2017 July 17-20 , 2017

caer.uky.edu

# Experimental Methods

**Elektronik EE33-J Probe Set:**

- ❑ Evaporates condensation in high humidity environments
- ❑ Protects the humidity sensor from calcium chloride contamination with PTFE SS filter

**Concept Validation:**

- ❑ Regenerator Air = 300 ACFM
- ❑ Cooling Tower Air = 15,000-28,000 SCFM
- ❑ Liquid Desiccant = 40-60 GPM
- ❑ Brine Concentration = 30-36 wt.%
- ❑ 1 Hour Periods of 90% $CO_2$ Capture

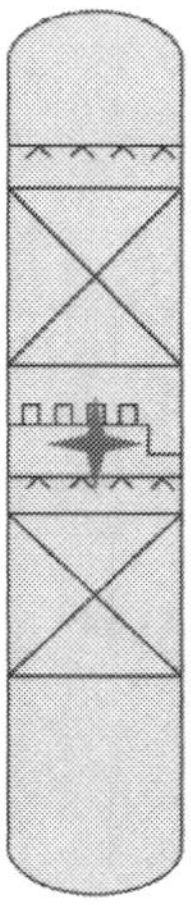

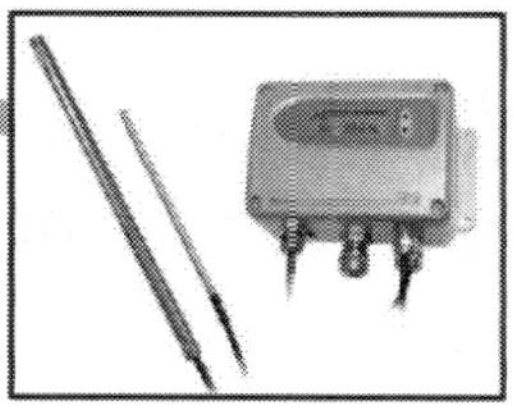





# Parametric Campaign Results

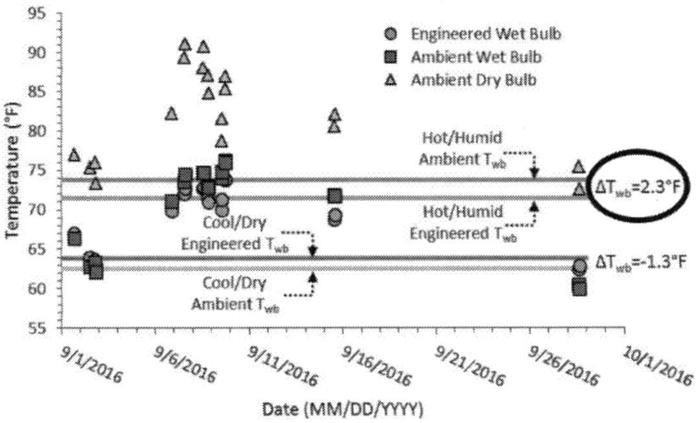


**Ambient $T_{db}$, Ambient $T_{wb}$, and Engineered $T_{wb}$ from September 2016 UKy-CAER Parametric Tests**

- ❑ $T_{wb}$ was reduced by ~2.3°F with desiccant application, confirming design effectiveness
- ❑ Humidity was found to increase after energizing the desiccant loop on relatively cool, dry days, suggesting desiccant solution evaporation rate may exceed the absorption rate

# Parametric Campaign Results

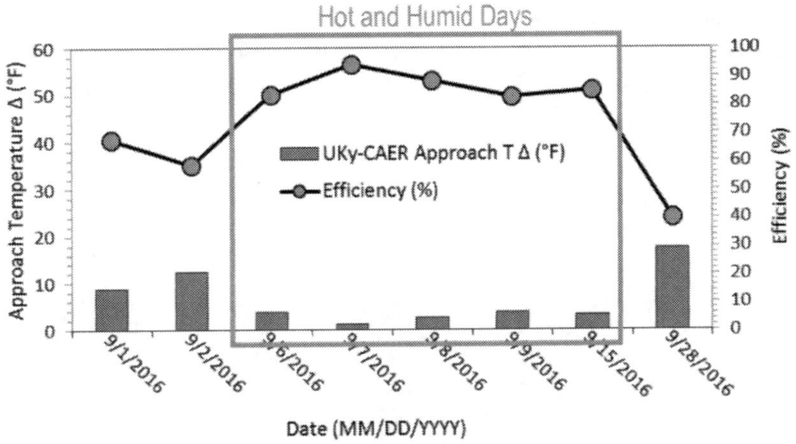


**Cooling Tower Approach Temperature Differences and Efficiencies from September 2016 UKy-CAER Parametric Tests**

- Cooling tower performance improved on hot and humid environment days
- Energy needed to drive the thermal regeneration of the brine was within the inventory of available waste heat



# Increased Stability of Cooling Water Exit Temperature

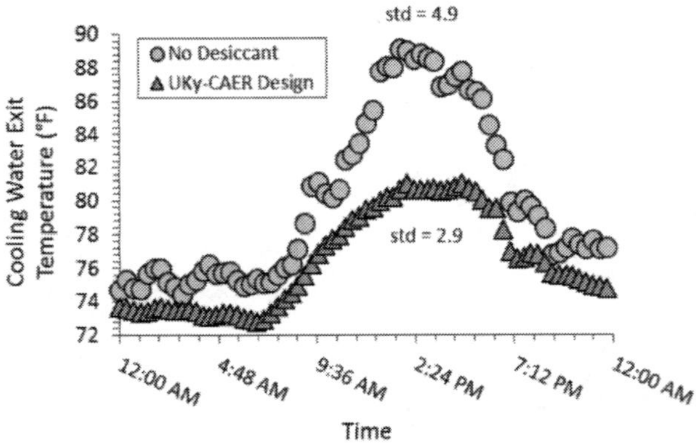


**Cooling Water Exit Temperature with Liquid Desiccant Application in Comparison to the Conventional Design**

- Absorption process may dampen temperature variability of inlet air, and thus, increase water exit temperature predictability
- Dehumidification process may help maintain the absorber temperature profile at the design condition
- Accurate estimations of turbine and condenser design specifications will result in a reliable and economic process

Center for Applied Energy Research

caer.uky.edu

# Conclusions

1) Recover Waste Energy → Captured heat to recover solvent capacity in 2-stage stripping process, maintaining 90% capture without process manipulation

2) Reduce the Air Wet Bulb Temperature → Observed wet bulb temperature drop, and competitive performance results on hot, humid days

3) Reduce Fresh Water Usage → Solvent concentration and process levels maintained with controlled thermal input via regeneration process

CMTC 2017 July 17-20 , 2017

Center for Applied Energy Research

caer.uky.edu

# Next Objectives

- Scale-up the unique UKy-CAER $CO_2$ Capture Process to 10 MWe post-combustion $CO_2$ capture system, to be located at LG&E Trimble County Power Plant near Bedford, KY
- Provide scale-up data, and design and operational information for commercial-scale deployment assessment

CMTC 2017 July 17-20 , 2017

Center for Applied Energy Research

caer.uky.edu


## Acknowledgements

- U.S. DOE NETL (DE-FE007395): Jose Figueroa, Lynn Brickett
- Carbon Management Research Group (CMRG)
- Louisville Gas & Electric (LG&E) and Kentucky Utilities (KU)
- Duke Energy
- Electric Power Research Institute (EPRI)
- Kentucky Department of Energy Development and Independence (KY-DEDI)
- E.W. Brown Generating Station
- David Link, Mahyar Ghorbanian, Jeff Fraley
- Len Goodpaster, Otto Hoffmann, Marshall Marcum, Andy Placido, Jesse Thompson


# Thank you!

CMTC 2017

July 17-20 , 2017

**SOUTHEAST OFFSHORE STORAGE RESOURCE ASSESSMENT (SOSRA)**
**PROJECT NUMBER: DE-FE0026086**

# OFFSHORE $CO_2$ STORAGE POTENTIAL OF THE EASTERN GULF OF MEXICO

Jack C. Pashin, Avinash Chandra, Paul Charbonneau, and Jenny Meng, Oklahoma State University
Denise J. Hills, Guohai Jin, and Marcella R. Redden, Geological Survey of Alabama

OKLAHOMA STATE UNIVERSITY
GEOLOGICAL SURVEY OF ALABAMA Founded 1848

SOUTHERN COMPANY *Energy to Serve Your World*
Advanced Resources International, Inc.
SOUTHERN STATES ENERGY BOARD
NETL

Carbon Management Technology Conference
Hilton Americas, Houston, TX
July 17-20, 2017

## SUMMARY – SOSRA

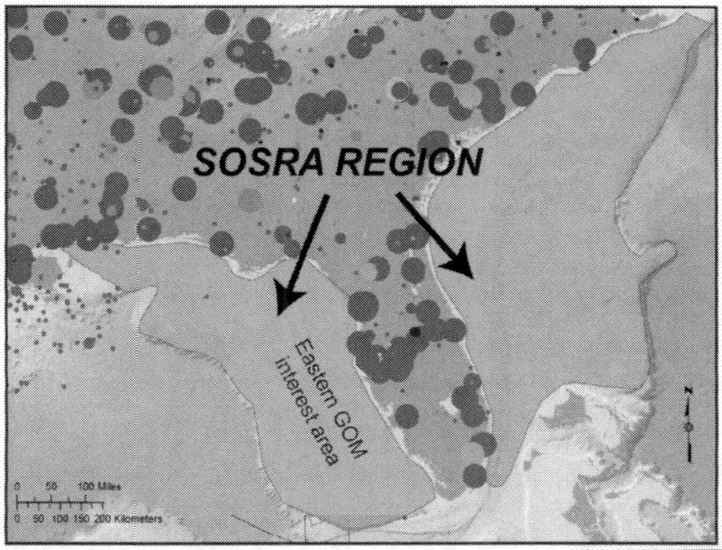

## Study Area and Subregions

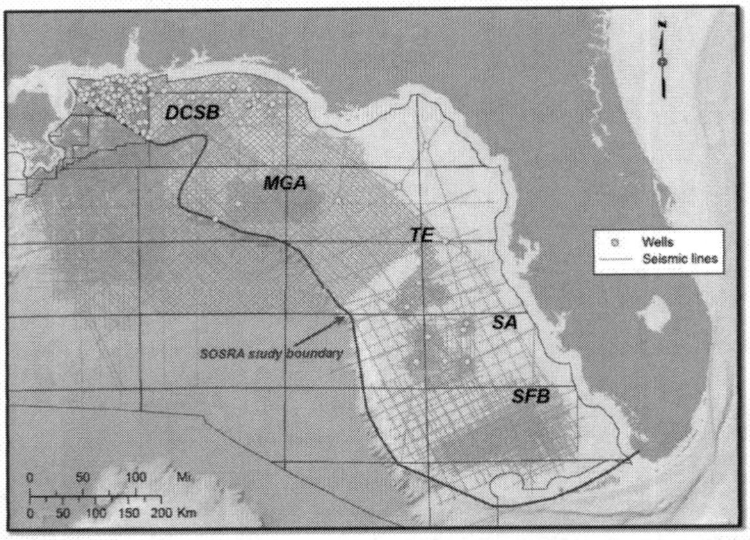


**DCSB** DeSoto Canyon Salt Basin

**MGA** Middle Ground Arch

**TE** Tampa Embayment

**SA** Sarasota Arch

**SFB** South Florida Basin


*This material is based upon work supported by the U.S. Department of Energy National Energy Technology Laboratory.*

*Cost share and research support are provided by the Project Partners and an Advisory Committee*


## Cretaceous Facies

*SOSRA REGION*

*NETL EOR TEST*
*SECARB ANTHROPOGENIC TEST*

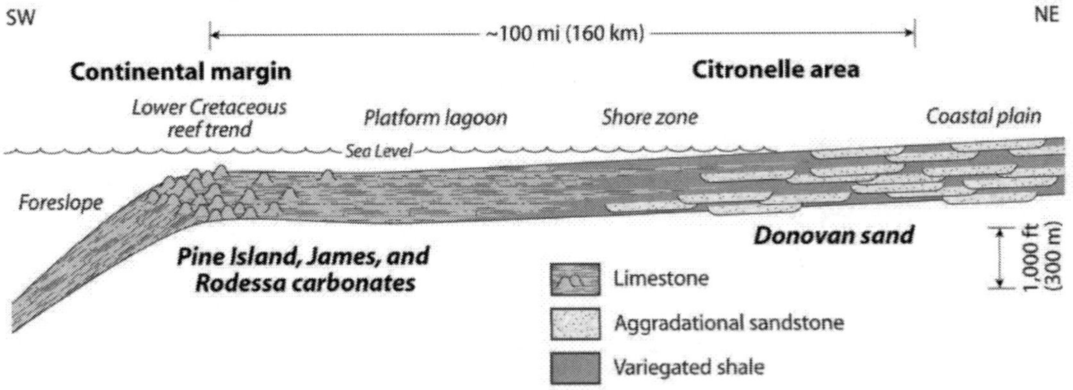


Pashin et al. (2014)

## Geothermal and Burial Data, DCSB

*Temperature-depth profile*

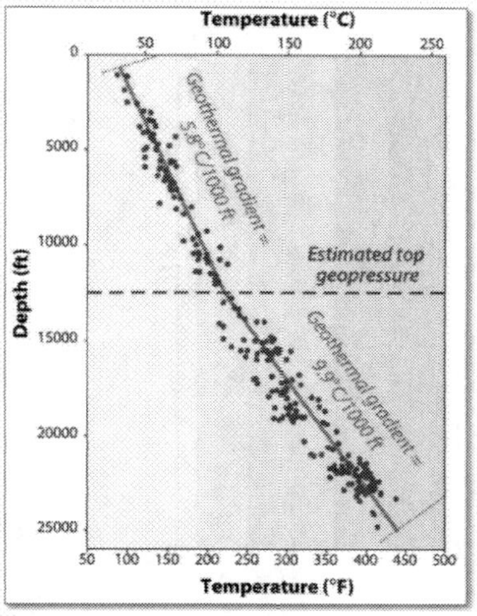


*Burial history curve*

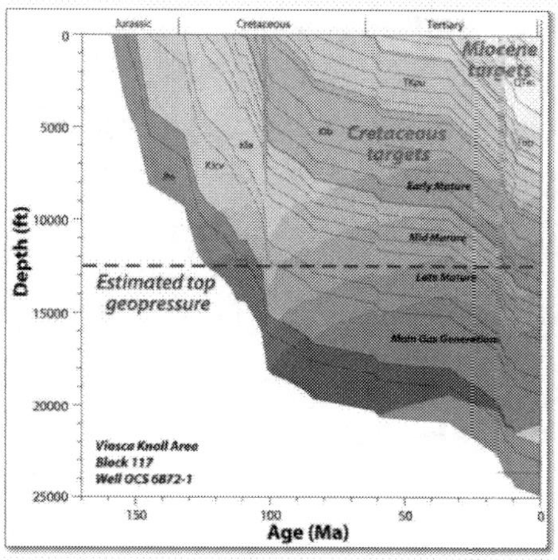


## Normal Brine, Pressure Gradients, Onshore Eastern Gulf

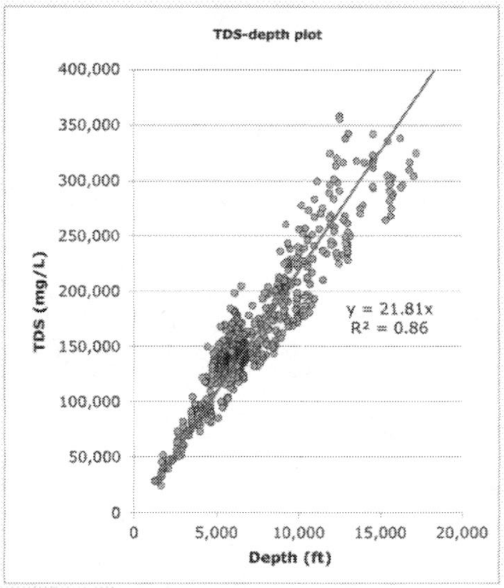


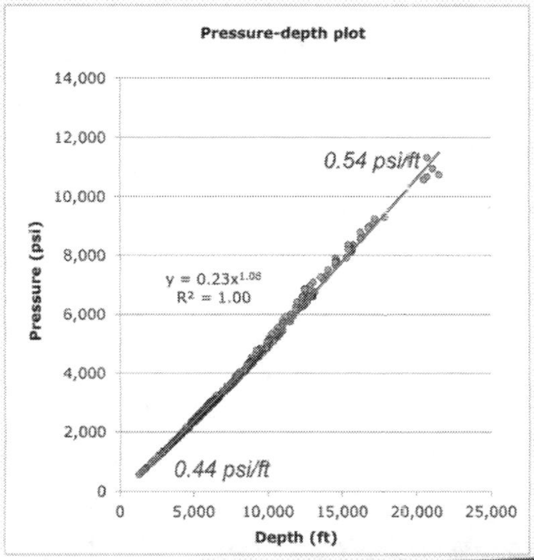

## Seismic Velocity Surveys

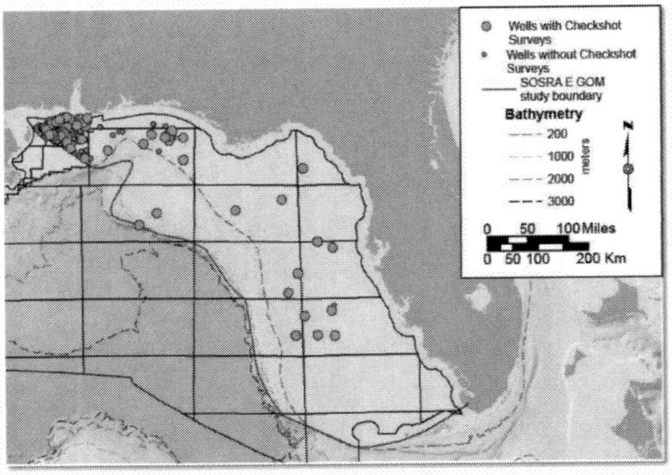


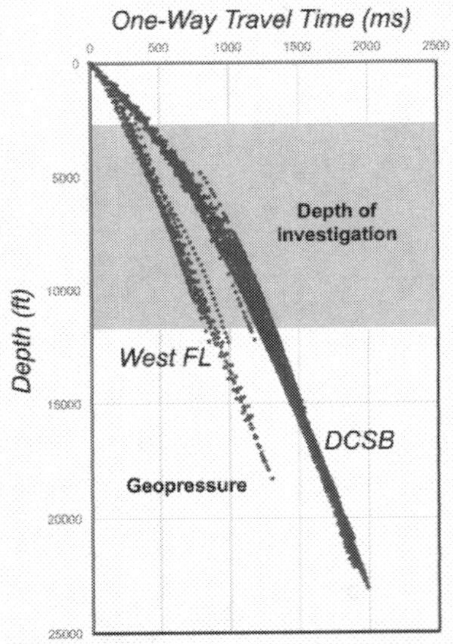


## DCSB Destin Dome

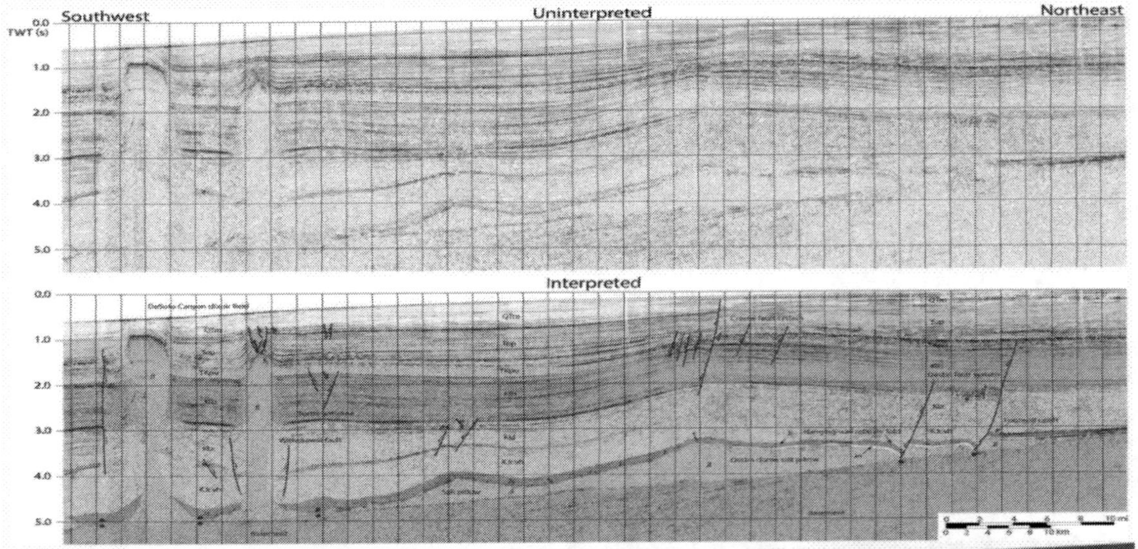

## DCSB Salt Roller Province

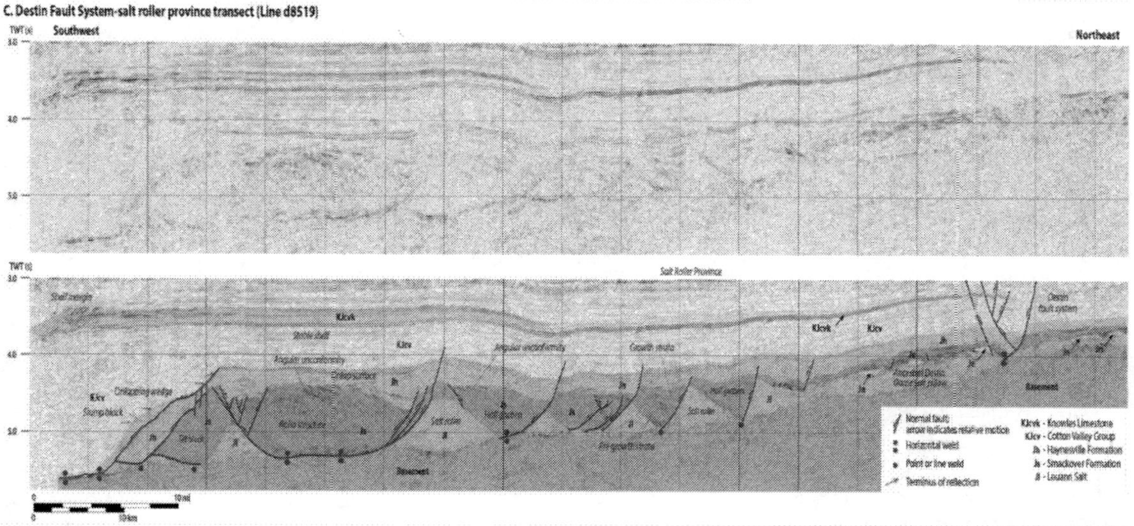


## Depth Converted Structural Cross Sections, DeSoto Canyon Salt Basin

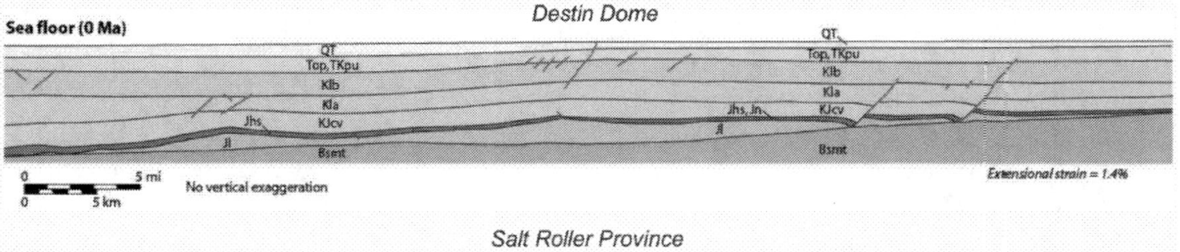


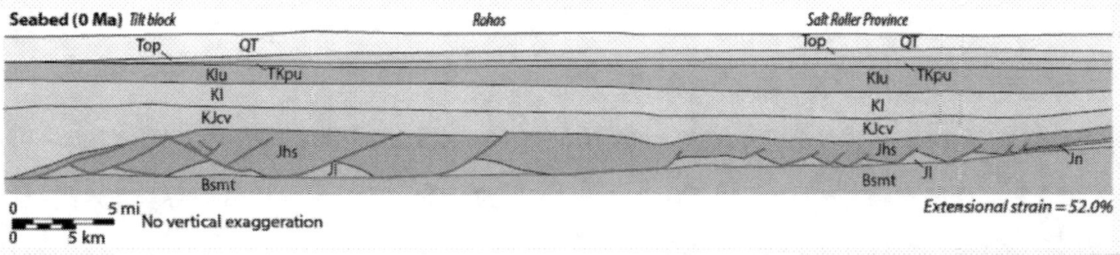

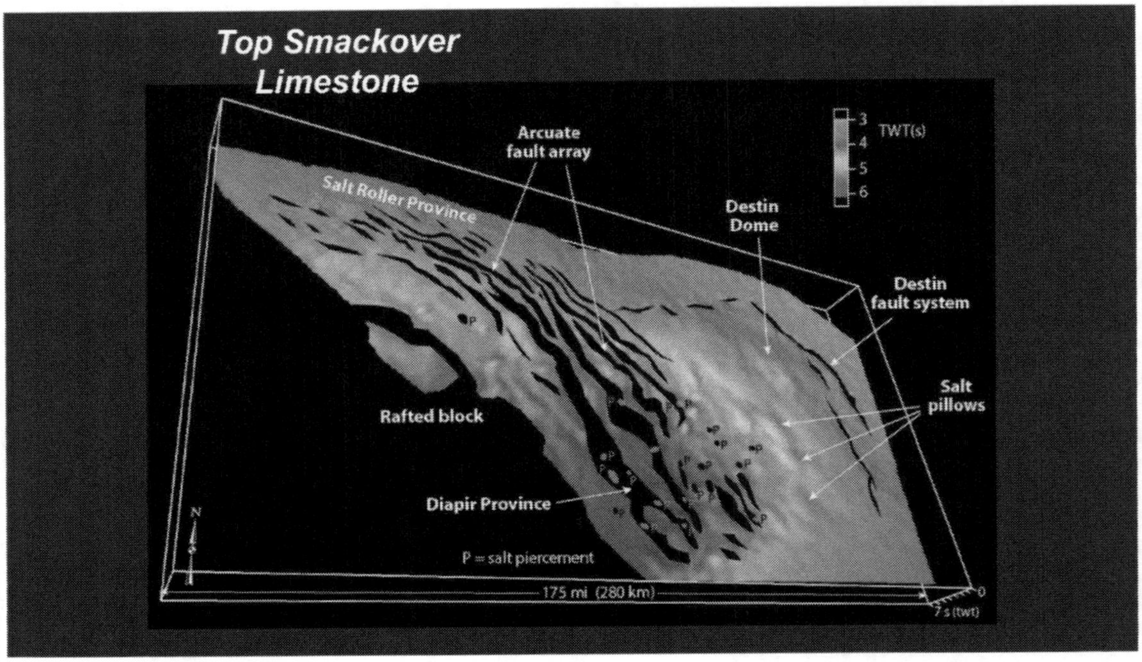

Top Smackover Limestone
Arcuate fault array
TWT(s)
3
4
5
6
Salt Roller Province
Destin Dome
Destin fault system
Salt pillows
Rafted block
Diapir Province
P = salt piercement
175 mi (280 km)
0
7 s (twt)
N


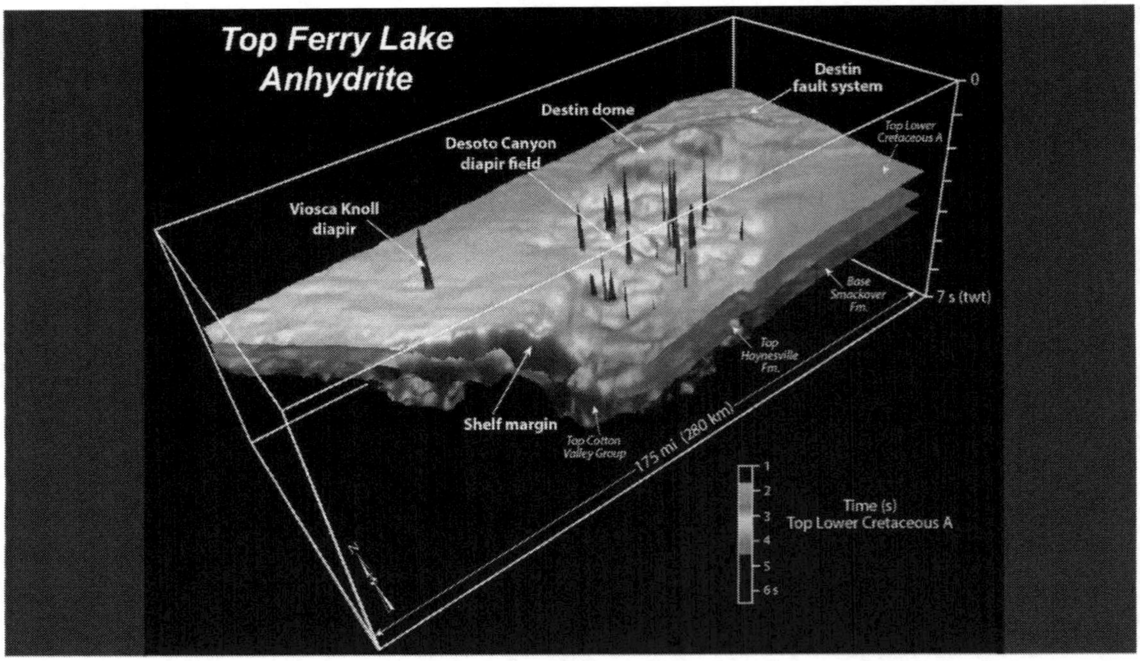

Top Ferry Lake Anhydrite
Destin fault system
Destin dome
Desoto Canyon diapir field
Viosca Knoll diapir
Top Lower Cretaceous A
0
7 s (twt)
Base Smackover Fm.
Top Haynesville Fm.
Shelf margin
Top Cotton Valley Group
175 mi (280 km)
1
2
3
4
5
6 s
Time (s)
Top Lower Cretaceous A
N

## West Florida Shelf Bathymetry

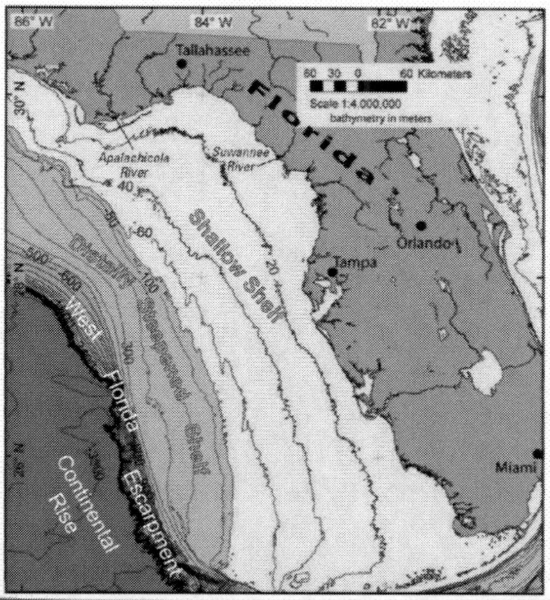


- Broad, shallow, region near shore (NE of 80 m contour).
- Distally steepening outer shelf leading to West Florida Escarpment.

## Florida – Lithologic Column

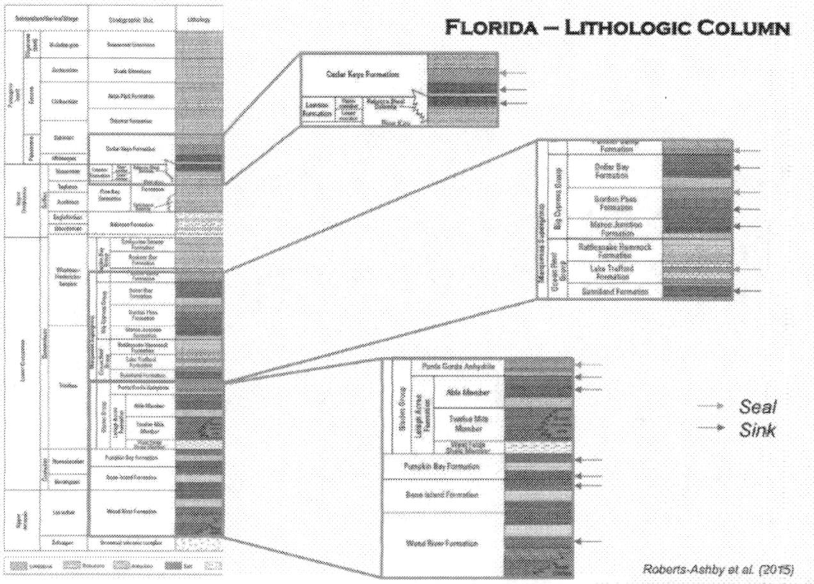


Roberts-Ashby et al. (2015)

## West Florida Shelf-Escarpment

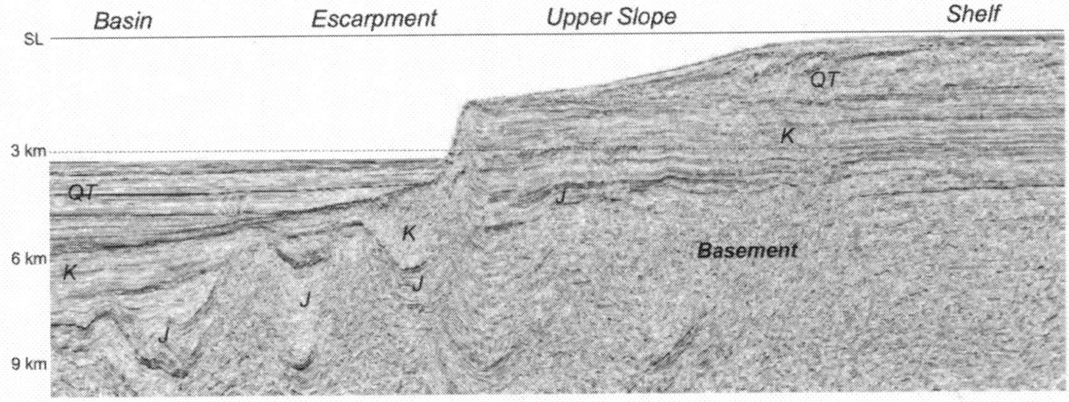


Roberts and Erickson (2009)

## Data Quality – West Florida

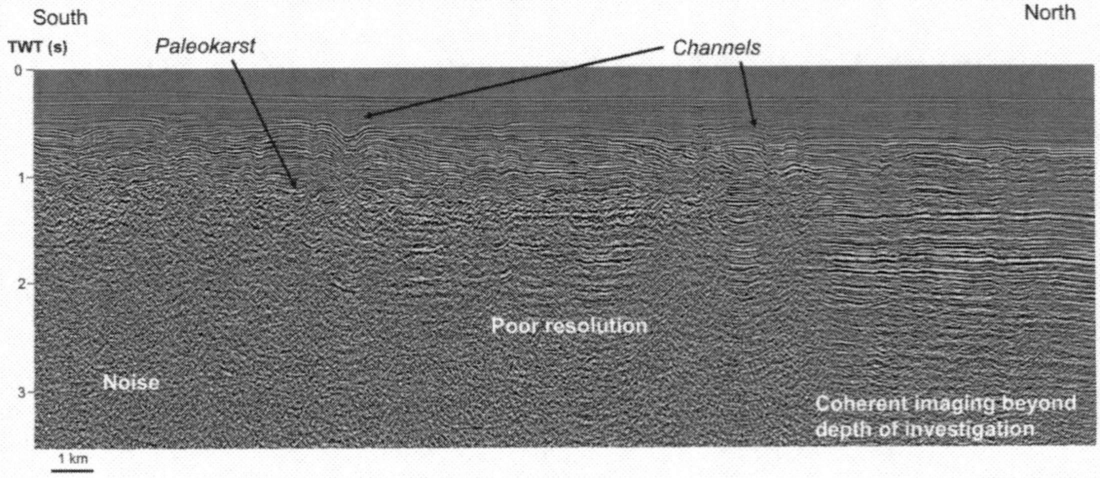

### DATA QUALITY – WEST FLORIDA

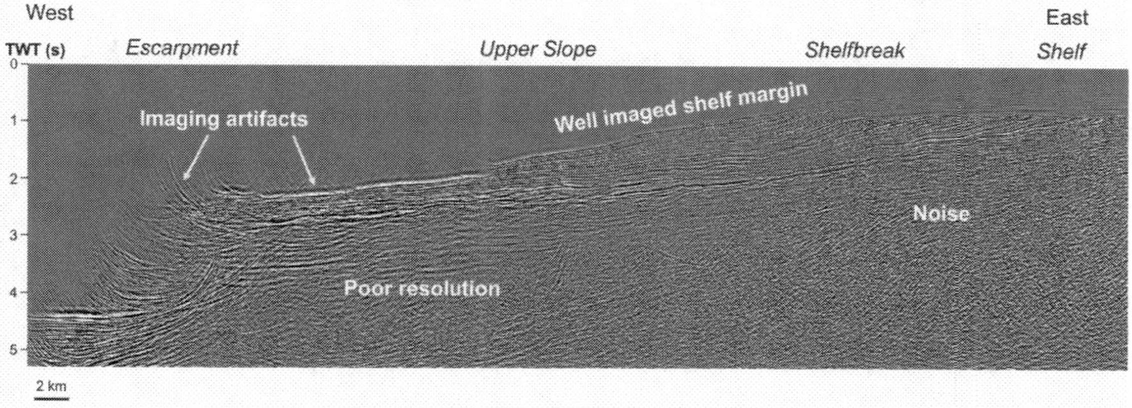


### DATA QUALITY – WEST FLORIDA

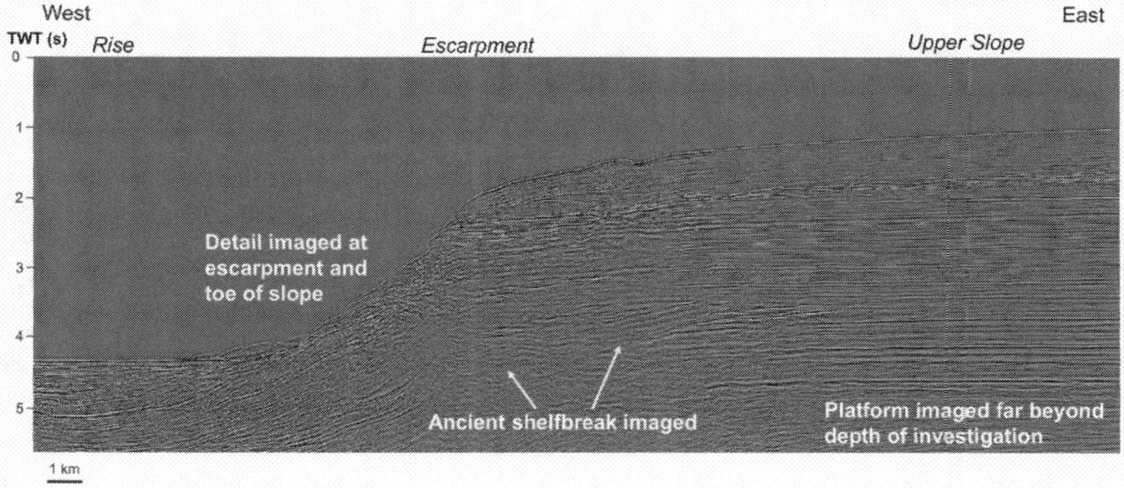

## Prospective EGOM Sinks

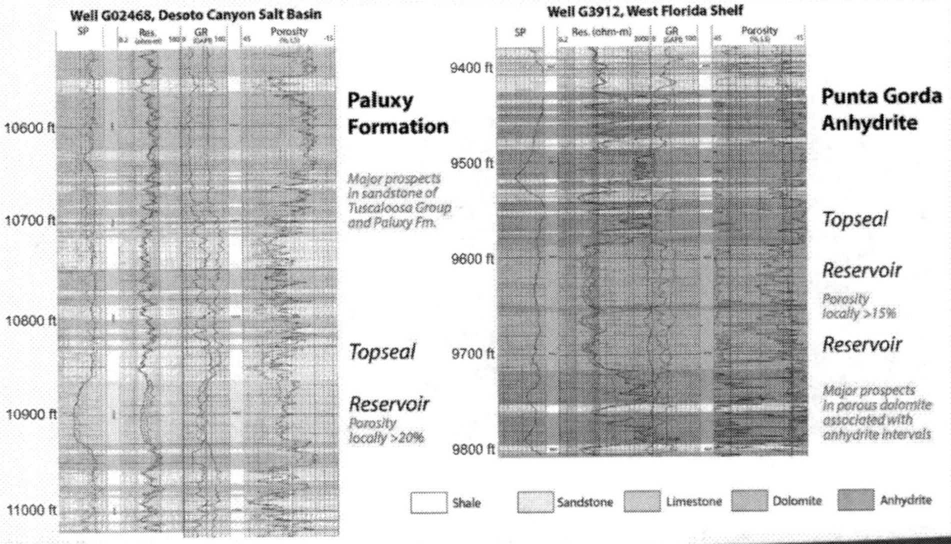


### Observations and Issues

- Large portfolio of potential sinks and seals in eastern Gulf of Mexico region.
- Seismic and well data being interpeted.
- Geopressure >12,000 ft; main storage prospects in Cretaceous-Miocene section.
- Multiple sandstone formations prospective in DeSoto Canyon Salt Basin; abundant mudrock and carbonate seals, including chalk.
- Relatively simple Cretaceous carbonate platform and distally steepened Cenozoic shelf in West Florida.
- Variable seismic quality in West Florida.
- Porous dolomite below anhydrite seals.

# Reducing Regeneration Energy and Capital Costs in an Advanced PCCC System

Jonathan V. Pelgen
University of Kentucky - Center for Applied Energy Research

**Power Generation and Utility Fuels Group**

http://www.caer.uky.edu/powergen/home.shtml

Award: **DE-FE0007395**

caer.uky.edu

CMTC-17 Houston, Texas　　July 17 -20, 2017

# University of Kentucky Center for Applied Energy Research

www.caer.uky.edu

caer.uky.edu

CMTC-17 Houston, Texas　　July 17 -20, 2017

Center for Applied Energy Research
caer.uky.edu

# Overview

1. **Background**
2. **Reduction in Regeneration Energy**
3. **Oxidative Degradation and Solvent Emissions**
4. **Corrosion Studies**
5. **Final Thoughts**
6. **Acknowledgements**

CMTC-17 Houston, Texas    July 17 -20, 2017

Center for Applied Energy Research
caer.uky.edu

# UKy-CAER CCS Project Overview

- 0.7 MWe (1300 ACFM) advanced post-combustion small pilot $CO_2$ capture
- Catch and release program
- Designed as a modular configuration
- Testing at Kentucky Utilities E.W. Brown Generating Station, Harrodsburg, KY, approximately 30 miles from UKy-CAER
- Includes several UKy-CAER developed technologies
- Three solvent testing campaigns (MEA baseline, advanced H3-1, and CAER-B3)

CMTC-17 Houston, Texas    July 17 -20, 2017

UK Center for Applied Energy Research

caer.uky.edu

# Process Flow Diagram

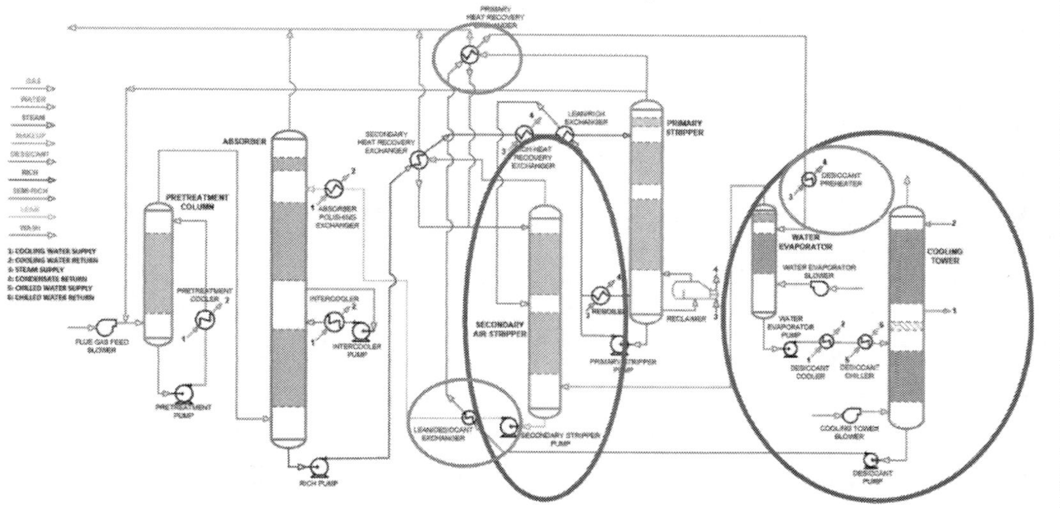


CMTC-17 Houston, Texas

July 17-20, 2017

UK Center for Applied Energy Research

caer.uky.edu

# Heat Integration: $CO_2$ Released in Secondary Air Stripper

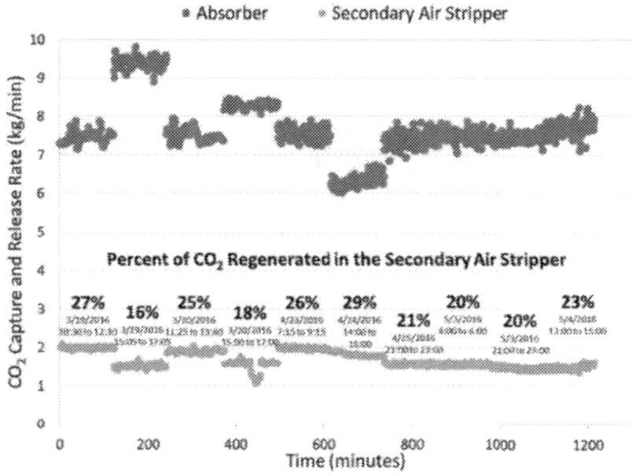


**The secondary stripper is effective in removing >15% of the $CO_2$ absorbed.**

CMTC-17 Houston, Texas

July 17-20, 2017

UK Center for Applied Energy Research

caer.uky.edu

# Heat Integration: Solvent Carbon Loadings

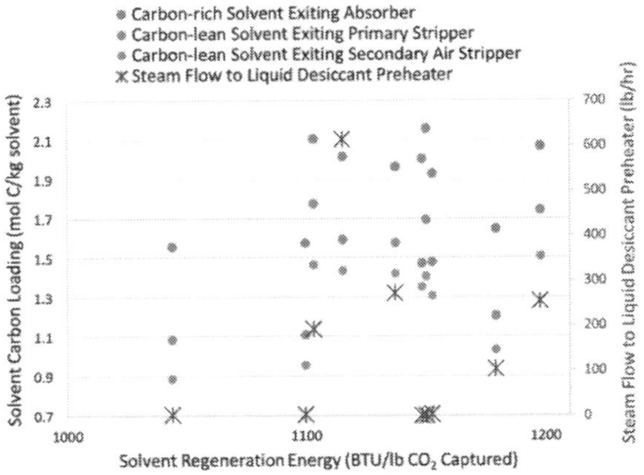


**No additional heat recovery required in the desiccant preheater for effective stripping $CO_2$ stripping.**

CMTC-17 Houston, Texas

July 17 -20, 2017

UK Center for Applied Energy Research

caer.uky.edu

# Regeneration Energy: Summary

## Experimental Results Compared to TEA

UKy-CAER process reduces the energy consumption

| Energy Consumption | |
|---|---|
| DOE Reference Case 10 | 1540 BTU/lb-$CO_2$ |
| UKy-CAER CCS process MEA case, according to TEA | 1340 BTU/lb-$CO_2$ |
| UKy-CAER CCS process MEA case, experimental long term | ~1350 BTU/lb-$CO_2$ |
| UKy-CAER CCS process H3-1 case, according to TEA | 937 BTU/lb-$CO_2$ |
| UKy-CAER CCS process H3-1 case, experimental campaign | ~1000 BTU/lb-$CO_2$ |

CMTC-17 Houston, Texas

July 17 -20, 2017

# MEA Campaign Results

L/G, blue
Inlet $CO_2$ Concentration (%), red
Primary Stripper Pressure (psia), green

All Conditions during MEA Campaign

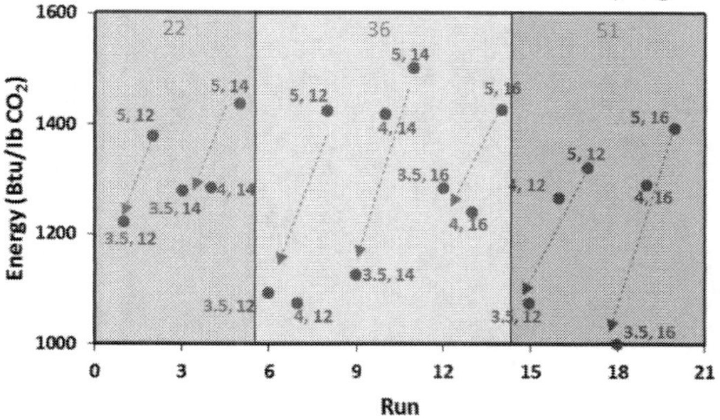


CMTC-17 Houston, Texas

July 17-20, 2017

Center for Applied Energy Research

caer.uky.edu

# Secondary Stripper Impact on Degradation

Similar oxidative degradation rates (formate) with conventional stripper and secondary air stripper from two separate solvent testing campaigns

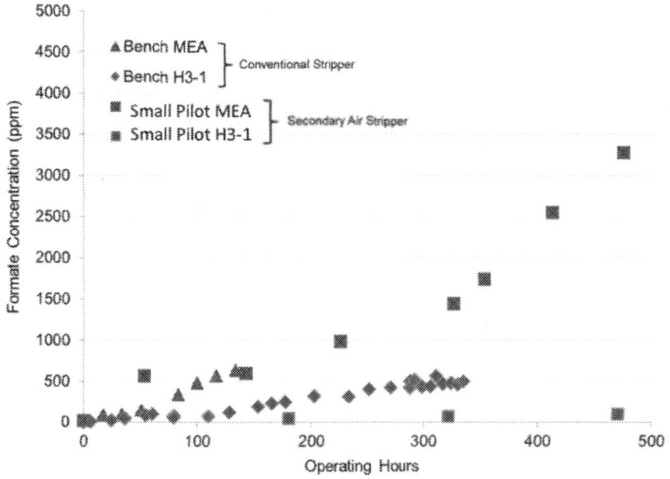


CMTC-17 Houston, Texas

July 17-20, 2017

Center for Applied Energy Research

caer.uky.edu

Center for Applied Energy Research

caer.uky.edu

# Ammonia Emissions vs Fe

Positive correlation between $NH_3$ emission and higher Fe in the solvent.

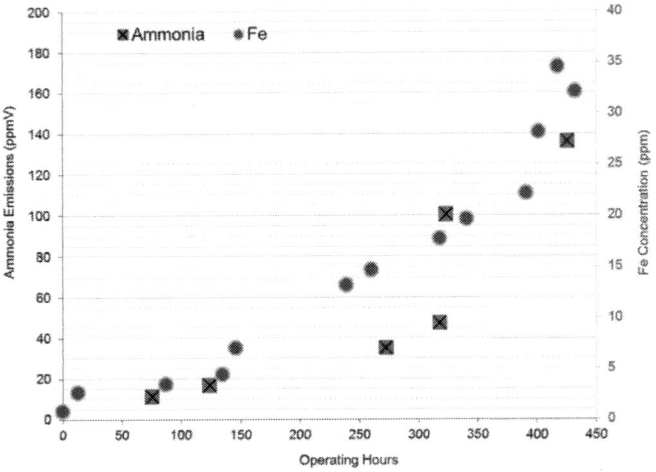


CMTC-17 Houston, Texas

July 17 -20, 2017

Center for Applied Energy Research

caer.uky.edu

# Ammonia Emissions vs Multi-Functional Additive

General increase in $NH_3$ emission with lower additive concentration in the solvent

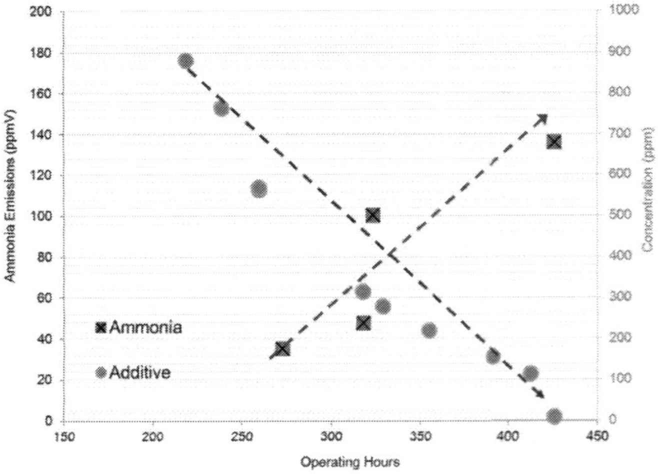


CMTC-17 Houston, Texas

July 17 -20, 2017

# Corrosion Studies

- Carbon steel is widely used in CCS applications.
- Is there any beneficial corrosion inhibition impact from anti-oxidation additives?
  - Is there a dual-use additive to reduce oxidative degradation and corrosion to reduce initial and long-term CAPEX?

## YES

Center for Applied Energy Research

caer.uky.edu

CMTC-17 Houston, Texas

July 17 -20, 2017

# Corrosion Sampling Locations

A = absorber
HR = hot-rich
CL = cold-lean
S = stripper

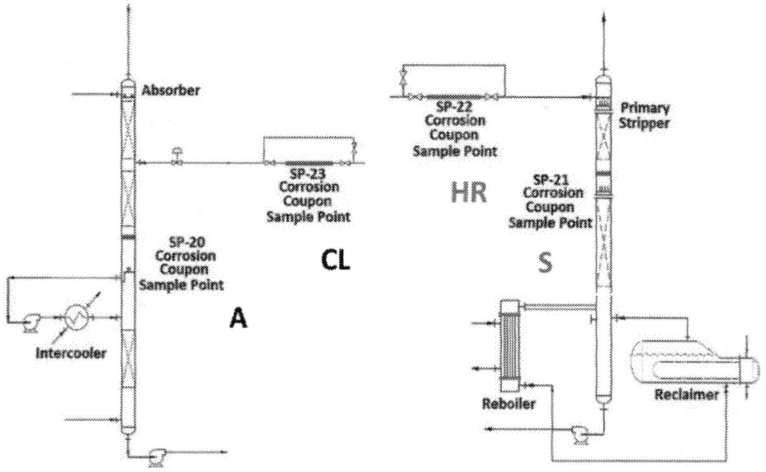


Center for Applied Energy Research

caer.uky.edu

CMTC-17 Houston, Texas

July 17 -20, 2017

Center for Applied Energy Research

caer.uky.edu

# Corrosion Studies: MEA Baseline

A = absorber
HR = hot-rich
CL = cold-lean
S = stripper

A106 carbon steel (CS) held up well during the MEA campaign in the absorber and cold-lean return piping.

Focus directed to making CS last longer in the hot-rich piping and stripper.

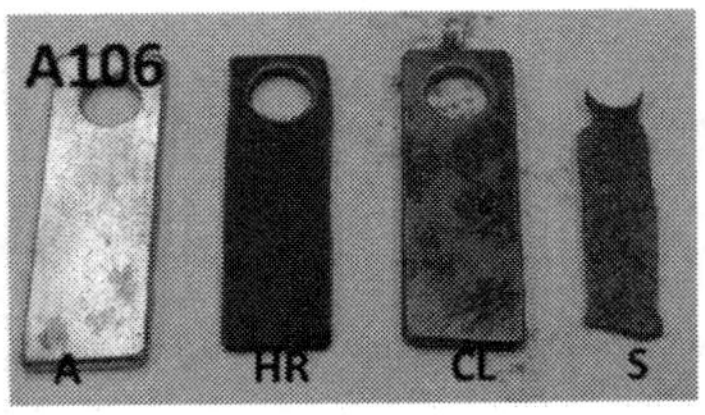


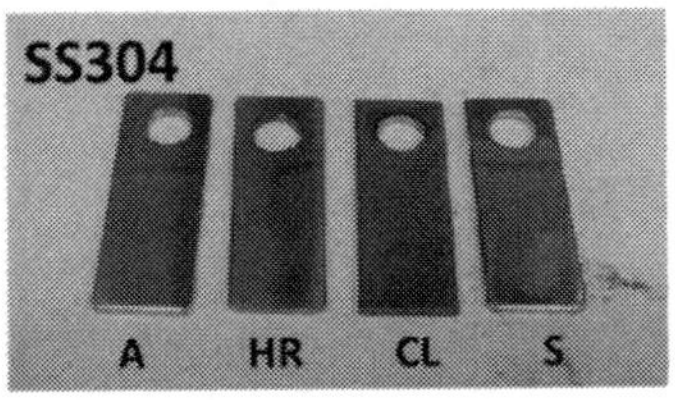


After approximately 430 run hours

CMTC-17 Houston, Texas

July 17 -20, 2017

Center for Applied Energy Research

caer.uky.edu

# Corrosion Samples

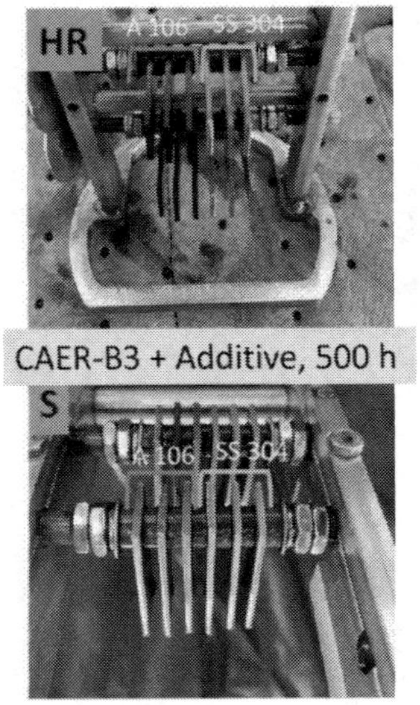


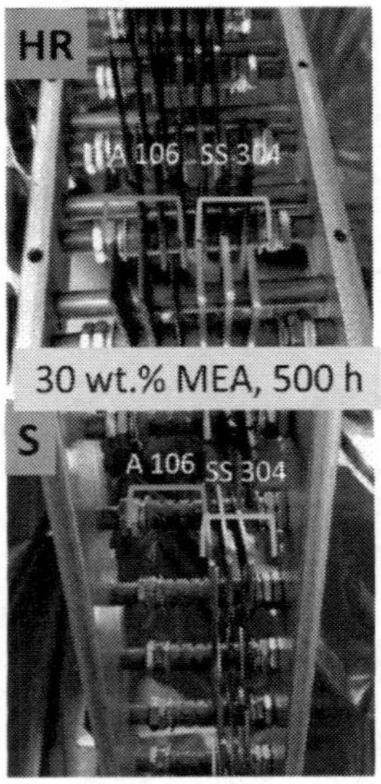


CMTC-17 Houston, Texas

July 17 -20, 2017

# Corrosion Rate Based on Total Run Time

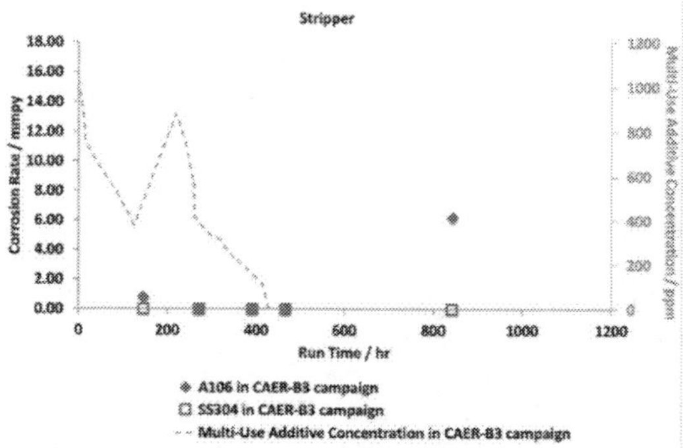


Temperature range in the stripper: 185–266 °F

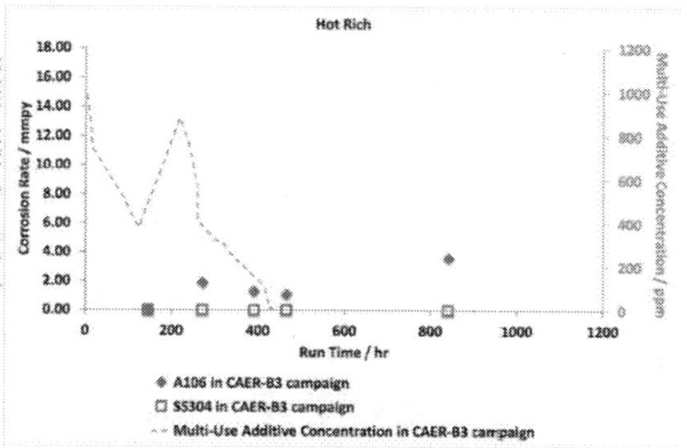


Temperature range in the hot rich piping: 185–230 °F

UK Center for Applied Energy Research
caer.uky.edu

CMTC-17 Houston, Texas

July 17-20, 2017

# Final Thoughts:

- The UKy-CAER process demonstrates a pathway that leads closer to the DOE's goal of 90% capture with no more than a 35% increase in the cost of electricity.
- The UKy-CAER process demonstrates the benefits of heat integration and two-stage stripping
- UKy-CAER's use of a multi-use additive can potentially reduce CAPEX due to corrosion and potentially reduce OPEX related to solvent degradation

UK Center for Applied Energy Research
caer.uky.edu

CMTC-17 Houston, Texas

July 17-20, 2017

# Acknowledgements

**José Figueroa, DOE NETL (DE-FE0007395)**
**CMRG Members**
**David Link, Mahyar Ghorbanian, Jeff Fraley - LG&E and KU**
**UKy-CAER Slipstream Team**

UK Center for Applied Energy Research
caer.uky.edu

CMTC-17 Houston, Texas
July 17 -20, 2017

# Thank you!

## 0.7 MWe Pilot Scale $CO_2$ Capture Project

KU E.W. Brown Generating Station

**Sponsored by:**

U.S. Department of Energy Office of Fossil Energy
National Energy Technology Laboratory
Kentucky Department of Energy Development and Independence
Carbon Management Research Group
University of Kentucky

UK Center for Applied Energy Research
caer.uky.edu

CMTC-17 Houston, Texas
July 17 -20, 2017

# Carbon Dioxide Storage Resource Assessment of the Mid-Atlantic Ocean Seaboard: Analysis of Geological and Geophysical Vintage Data

MARINA ROSSI

PHD STUDENT

ADVISOR: DR. NINO RIPEPI

1

## SOSRA

**Southeast Offshore Storage Resource Assessment**

- Founded by the Department of Energy
- Managed by the Southern States Energy Board (SSEB)
- SSEB appointed three planning area managers to each offshore region (defined by BOEM)
- Mid-Atlantic Seaboard Planning Area management was awarded to researchers at Virginia Tech's Virginia Center for Coal and Energy Research (VCCER)

2

## Mid-Atlantic: Team

- Virginia Tech
- Virginia Center for Coal and Energy Research (VCCER)
- VA DMME
- Advisors and Consultants

VirginiaTech

Virginia Center for Coal and Energy Research

DMME
Virginia Department of Mines Minerals and Energy

3

## SOSRA Mid-Atlantic Study Area

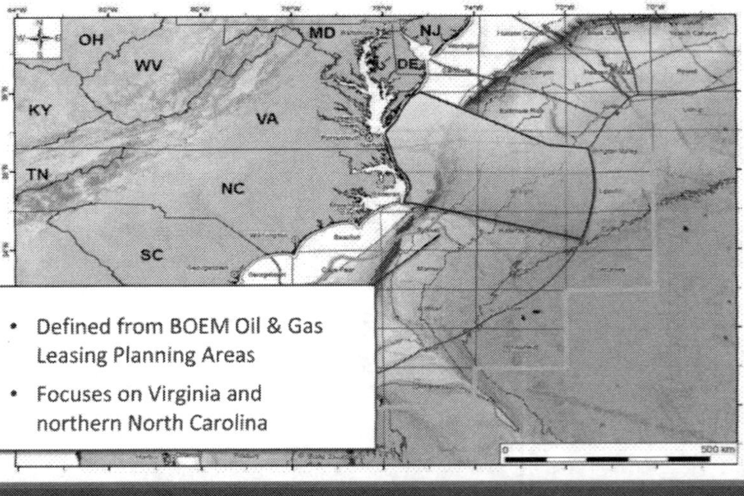


- Defined from BOEM Oil & Gas Leasing Planning Areas
- Focuses on Virginia and northern North Carolina

FUGRO

4

## Summary

- SOSRA's Objectives
- Phase I Objectives
- Identify Potential Geological Target for the Storage of CO2
- Data Collection
- Preliminary Results
- Conclusion

5

# SOSRA: Objectives

**Southeast Offshore Storage Resource Assessment**

- Provide a high quality prospective carbon dioxide storage resource assessment of the Atlantic ocean and the eastern part of the Gulf of Mexico
- Predicting storage capacity to within ± 30%
- Provide a pathway toward commercialization

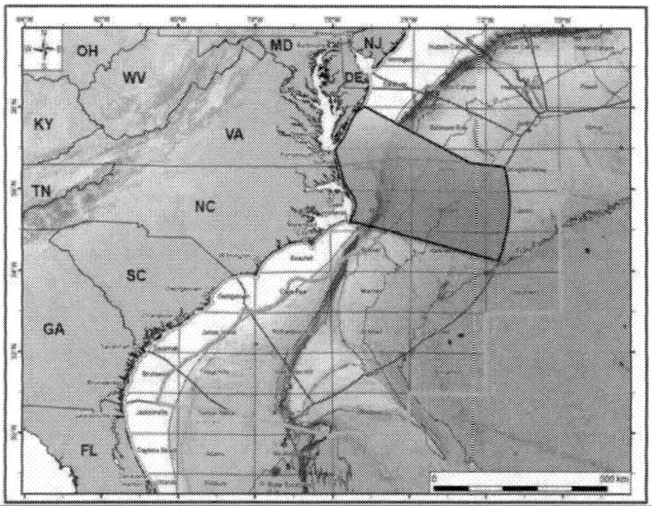


6

# Phase I: Objectives

- Data Collection
- Basic Geological Framework: Based on Publications
- Data Analysis
- ➢ Quality assessment
- ➢ Coverage assessment
- ➢ Well ties
- Seismic Interpretation



# Identification of Geological Targets

- **Reservoir Parameters**

➢High Porosity (up to 30%)

➢Permeability (1 to $10^{-5}$ Darcy)

(Ideally Sandstone and Limestone)

- **Presence of a Seal**

➢Low permeability layer above the porous rock

(Shale: $10^{-5}$ to $10^{-9}$ Darcy, Clay...)

- Thickness of Reservoir Large Enough

# Initial Geologic Characterization

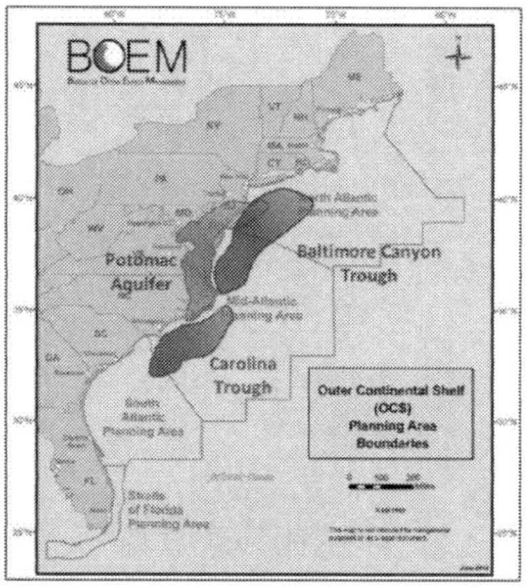


Modified from BOEM, USGS, GCCC, and Bayer and Milici,1987.

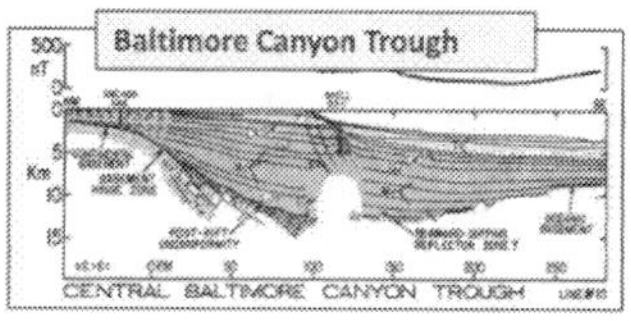


From Klitgord, 1988

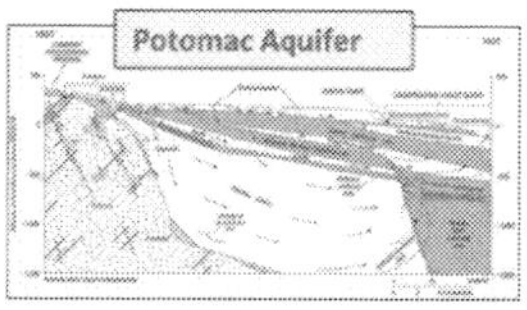


From USGS, 2013.

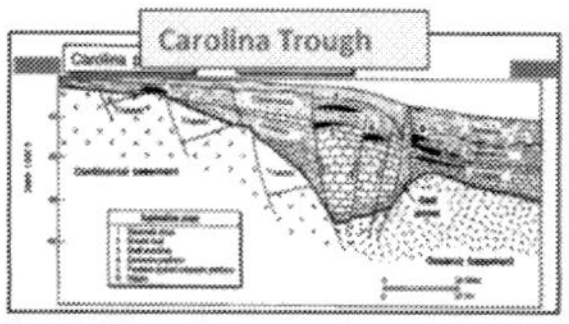


From Carpenter and Amato, 1992.



# Data Overview

**Wells**

- Atlantic Slope Project (1967): 13
- Atlantic Margin Coring (1976): 3
- Ocean Drilling Program (1987): 2
- Shell Oil and Gas Exploration (1984): 1

**30 Surveys: 2D multi-channel seismic (vintage)**

- Over 1,000 lines collected

**Publications**

**Main Issue: Lack of <u>Well Control</u>**

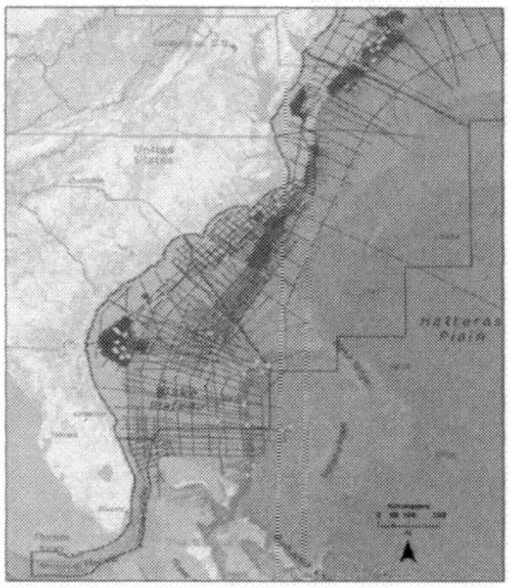

# Quality Assessment

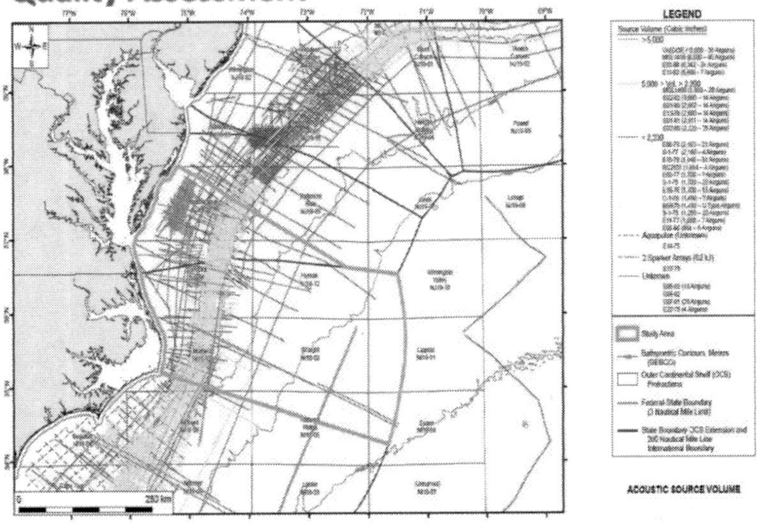


❖ **Acoustic Source Volume**
Much smaller than modern air gun volumes
Limits Depth Penetration

❖ **Streamer Length**
For most less than 4km
Can't go deeper than 4km and therefore not recommend for exploration purposes

❖ **CDP Fold**
For most 48 or less so S/N ratio not ideal

**This parameters decrease the quality of the legacy surveys compared to modern 2-D surveys**

11

# Coverage

Critical Parameter: Line Spacing between Dip line and Strike Line

Surveys divided into 3 categories based on their line-spacing:

- **Regional**: greater than 8 nautical miles (15km)

Gives basic overview of the basin of interest

- **Semi-Regional**: between 3 and 8 nautical miles (5 to 15km)

To identify potential area of interest within a basin

- **Exploration Scale**: less than 3 nautical miles (5km)

To characterize a potential reservoir

**30 Surveys provide a <u>good coverage</u> of the Mid-Atlantic region.**
**However, not enough wells in the region to establish good ties (1 well).**

**Seismic Line Spacing for Surveys Acquired in Virginia and Northern North Carolina Waters***

| Permit or Survey | Specific OCS Area Tabulated[1] | Survey Scope[2] | Seismic Data Type Available | Number of Dip Lines | Typical Spacing Between Dip Lines | Number of Strike Lines | Typical Spacing Between Strike Lines |
|---|---|---|---|---|---|---|---|
| All four 1973-1978 USGS surveys in study area | Lines extending into VA and Northern NC waters | Regional | Demux & Stacked | 6 | 22-27 nmi (40-50 km) | 2 | 92 nmi (170 km) |
| E14-75 | Cape Hatteras, NC to Northern SC | Semi-Regional | Stacked | 25 | 5-11 nmi (10-20 km) | Variable, 3 to 7 | 5-11 nmi (10-20 km) |
| E16-76 | Cape Hatteras, NC to MD | Semi-Regional | Stacked & Migrated | 35 | 2-8 nmi (4-15 km) | 6 | 5-14 nmi (10-25 km) |
| BGR 79 | Lines extending into VA waters | Regional | Stacked & Migrated | 0 | NA | 4 | 7-14 nmi (12-25 km) |
| E01-80 | Currituck Sound Protraction | Exploration | Stacked & Migrated | 18 | 1-2 nmi (2-3 km) | 6 | 1-3 nmi (2-5 km) |
| | Manteo Protraction | Exploration | Stacked & Migrated | 23 | 1-3 nmi (2-5 km) | 7 | 1.5-2 nmi (3-4 km) |
| E02-80 | Currituck Sound Protraction | Exploration | Stacked, Migrated & Depth | 8 | 1-3 nmi (2-5 km) | 2 | 4 nmi (7 km) |
| | Manteo Protraction | Exploration | Stack, Migrated & Depth | 22 | 1-3 nmi (2-5 km) | 2 | 5-11 nmi (10-20 km) |
| E01-81 | Northern NC and VA | Regional | Stacked & Migrated | 7 | 14 nmi (26 km) | 2 | 10 nmi (18 km) |
| E07-81 | Manteo Protraction | Exploration | Migrated & Depth | 11 | 1.5-3 nmi (2.5-5 km) | 1 | NA |
| E02-82 | Northern NC to Southern MD | Exploration | Stack, Migrated, Depth | 104 | 1 nmi (1.5 km) | 4 | 7 nmi (13 km) |
| E04-82 | Northern NC and VA | Regional | Migrated & Depth | 9 | 5-16 nmi (9-30 km) | 0 | NA |
| E11-82 | Norfolk Basin, VA | Exploration | Migrated | 33 | 1.5 nmi (2.5 km) | 11 | 2 nmi (3 km) |
| | Northern NC to Southern MD | Regional | Migrated | 5 | 14-34 nmi (25-63 km) | 3 | 4-15 nmi (7-27 km) |
| E05-83 | Northern NC slope | Exploration | Migrated & Depth | 6 | 1-3 nmi (2-5 km) | 0 | NA |
| E03-88 | Currituck Sound Protraction | Exploration | Migrated | 8 | 3 nmi (5 km) | 1 | NA |
| | Norfolk Basin, VA | Exploration | Stacked | 7 | 3 nmi (5 km) | 1 | NA |

12

## Well Ties

- Seismic lines passing as close as possible to the wells
- Most are in the Northern part of the Mid-Atlantic
- Starting point for interpretation
- Will allow calibration of geological model and insure consistency of the interpretation
- For each well collected:

One or more lines passes as close as 1km and at the most 3km.

| Wells | Permit | Line | Orientati | Type | Approximate distance from well |
|---|---|---|---|---|---|
| AMCOR 6006 | E02-77 | BP-102 | Dip | Depth | 3 km |
| AMCOR 6006 | E02-82 | pr92-225 | Dip | Migrated | 3 km |
| AMCOR 6006 | E14-75 | we-011-2 | Strike | Stack | 2 km |
| AMCOR 6006 | E14-75 | we-011-2 | Dip | Stack | 3 km |
| AMCOR 6007 | E04-82 | 18079 | Dip | Migrated or Depth | 2 km |
| AMCOR 6007 | E16-76 | ma-22 | Dip | Migrated | 3 km |
| AMCOR 6007 | E16-76 | ma-7-2 | Strike | Migrated | 1 km |
| ASP 10 | E01-81 | pp81-324b | Strike | Migrated | 1 km |
| ASP 10 | E02-82 | pr82-121 | Dip | Migrated or Depth | < 1 km |
| ASP 10 | E02-82 | pr82-122a | Dip | Migrated or Depth | < 1 km |
| ASP 13 | C-1-78 | 25 | Dip | Stack | 1.5 km |
| ASP 13 | E11-77 | MA-104 D/E | Strike | Migrated | < 1 km |
| ASP 14 | E11-77 | MA-1088 | Strike | Migrated | 1 km |
| ASP 14 | E11-77 | MA-129 | Dip | Migrated | 2 km |
| ASP 15 | C-1-78 | 25 | Strike | Stack | 1 km |
| ASP 15 | E11-77 | MA-112A | Strike | Migrated | 1 km |
| ASP 15 | E11-77 | MA-131 | Dip | Migrated or Depth | 2 km |
| ASP 22 | E01-81 | pp81-324b | Strike | Migrated | < 1 km |
| ASP 22 | E02-82 | pr82-124 | Dip | Migrated or Depth | < 1 km |
| ASP 23 | E02-82 | pr82-099 | Dip | Migrated or Depth | < 1 km |
| ASP 23 | E02-82 | pr82-230-1 | Strike | Migrated or Depth | 1.5 km |
| ASP 7 | E01-80 | 36-059 | Strike | Migrated | < 1 km |
| ASP 7 | E02-80 | sa1048 | Dip | Migrated | < 1 km |
| ASP 8 | E01-80 | 36-088 | Dip | Migrated | < 1 km |
| ASP 8 | E07-81 | csa81-12b | Strike | Migrated | < 1 km |
| ASP 8 | E07-81 | csa81-9 | Dip | Migrated | < 1 km |
| Shell 372-1 | E11-77 | MA-116A | Strike | Migrated | < 1 km |
| Shell 372-1 | E11-77 | MA-139 | Dip | Migrated or Depth | < 1 km |
| Shell 586-1 | BGR-79 | 203 | Strike | Migrated | 1.5 km |
| Shell 586-1 | E11-77 | MA-112B | Strike | Migrated | 1 km |
| Shell 586-1 | E11-77 | MA-151 | Dip | Migrated or Depth | 2.5 km |
| Shell 587-1 | E11-77 | MA-116A | Strike | Migrated | < 1 km |
| Shell 587-1 | E11-77 | MA-151 | Dip | Migrated or Depth | 1.5 km |
| Shell 93-1 | S-1-75 | 10 | Dip | Stack | 1 km |
| Shell 93-1 | E11-77 | MA-112B | Strike | Migrated | 1 km |
| Shell 93-1 | E11-77 | MA-191 | Dip | Migrated or Depth | 1 km |
| Tenneco 495- | E22-75 | 5044 | Oblique | Stack | < 1 km |
| Tenneco 495- | E11-77 | MA-153 | Dip | Migrated or Depth | 1 km |
| Tenneco 495- | S-1-77 | TD15A | Strike | Stack | < 1 km |



## Preliminary Results: Seismic Interpretation


- Klitgord, K.D., Poag, C.W., Schneider, C.M., and North, L., 1994, Geophysical database of the East Coast of the United States: northern Atlantic margin - cross sections and gridded database (Georges Bank Basin, Long Island Platform, and Baltimore Canyon Trough), U.S. Geological Survey Open-File Report 94-637, 190 p.

- Klitgord, K.D., and Schneider, C., 1994, Geophysical database of the East Coast of the United States: northern Atlantic margin - velocity analyses, U.S. Geological Survey Open-File Report 94-192, 74 p.

- Bayer, K.C., and Milici, R.C., 1987, Geology and petroleum potential of Mesozoic and Cenozoic rocks, offshore Virginia: Virginia Division of Mineral Resources Publication 73, pt. D, 111 p. and two plates.


Few surveys were public when these papers were published and therefore only the USGS regional lines have been used for the interpretation.

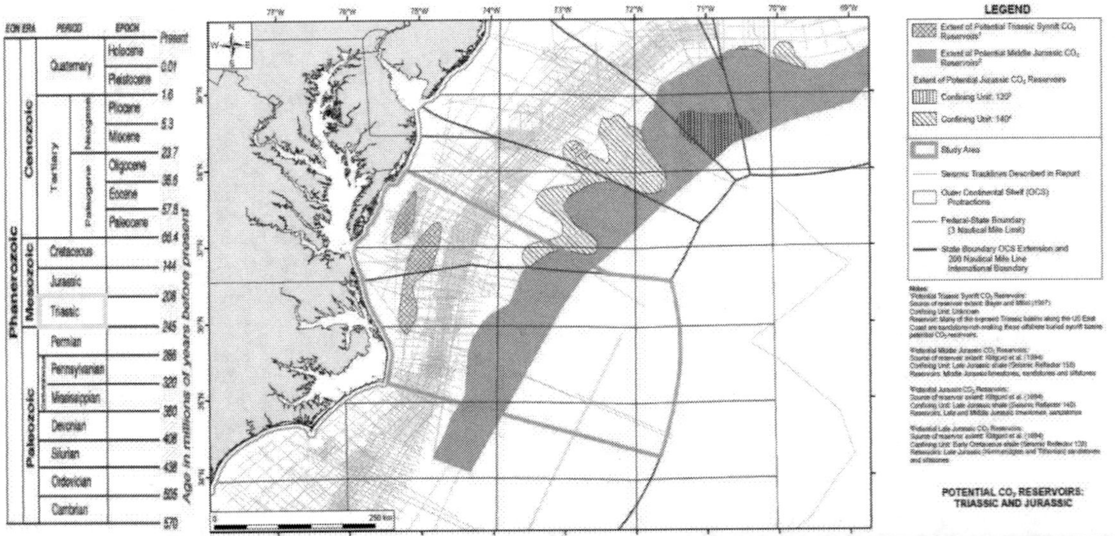
LEGEND
Study Area
Seismic Tracklines Described in Report
Outer Continental Shelf (OCS) Protractions
Federal-State Boundary (3 Nautical Mile Limit)
POTENTIAL $CO_2$ RESERVOIRS: TRIASSIC AND JURASSIC
Age in millions of years before present
Phanerozoic
Cenozoic
Mesozoic
Paleozoic
Quaternary
Tertiary
Cretaceous
Jurassic
Triassic
Permian
Pennsylvanian
Mississippian
Devonian
Silurian
Ordovician
Cambrian
Holocene
Pleistocene
Pliocene
Miocene
Oligocene
Eocene
Paleocene

15

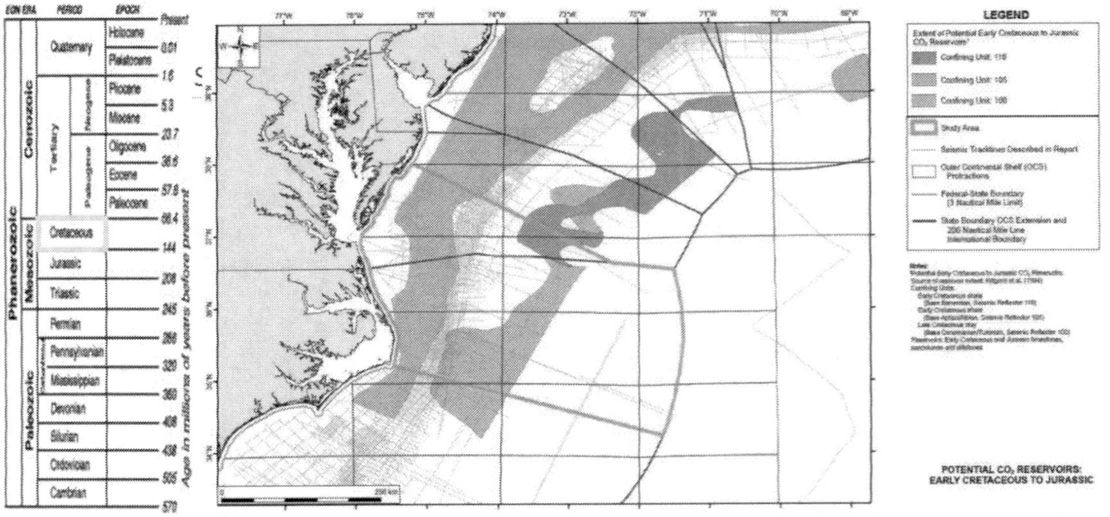
LEGEND
Study Area
Seismic Tracklines Described in Report
Outer Continental Shelf (OCS) Protractions
Federal-State Boundary (3 Nautical Mile Limit)
POTENTIAL $CO_2$ RESERVOIRS: EARLY CRETACEOUS TO JURASSIC
Age in millions of years before present
Phanerozoic
Cenozoic
Mesozoic
Paleozoic
Quaternary
Tertiary
Cretaceous
Jurassic
Triassic
Permian
Pennsylvanian
Mississippian
Devonian
Silurian
Ordovician
Cambrian
Holocene
Pleistocene
Pliocene
Miocene
Oligocene
Eocene
Paleocene

16

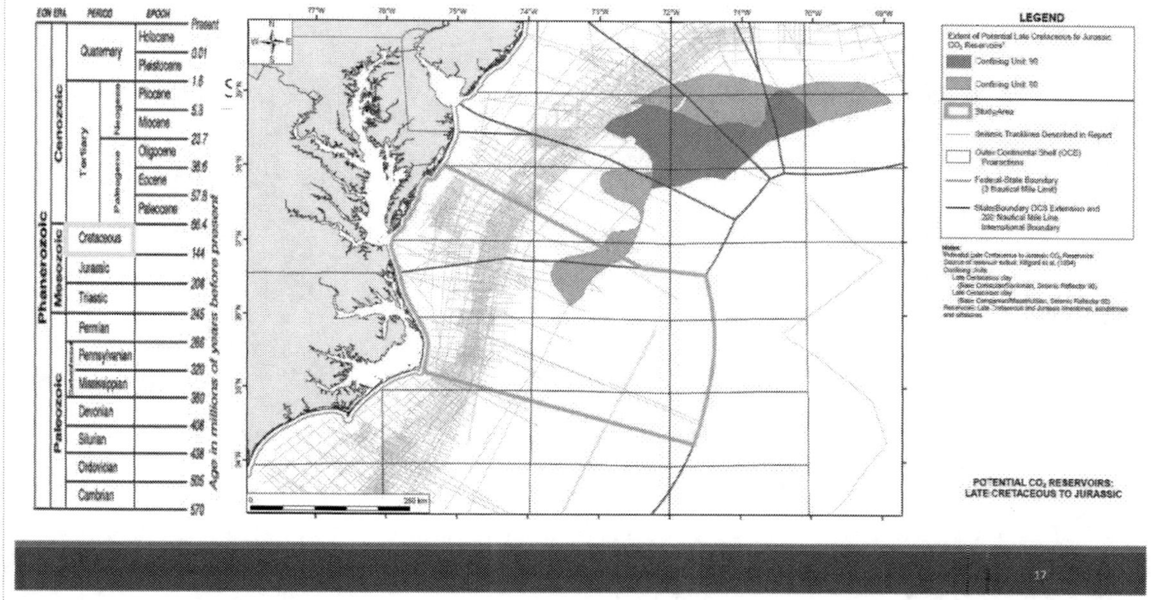
EON ERA
PERIOD
EPOCH
Present
Phanerozoic
Cenozoic
Mesozoic
Paleozoic
Quaternary
Tertiary
Holocene
Pleistocene
Pliocene
Miocene
Oligocene
Eocene
Paleocene
Cretaceous
Jurassic
Triassic
Permian
Pennsylvanian
Mississippian
Devonian
Silurian
Ordovician
Cambrian
Age in millions of years before present
LEGEND
POTENTIAL CO2 RESERVOIRS:
LATE CRETACEOUS TO JURASSIC
17

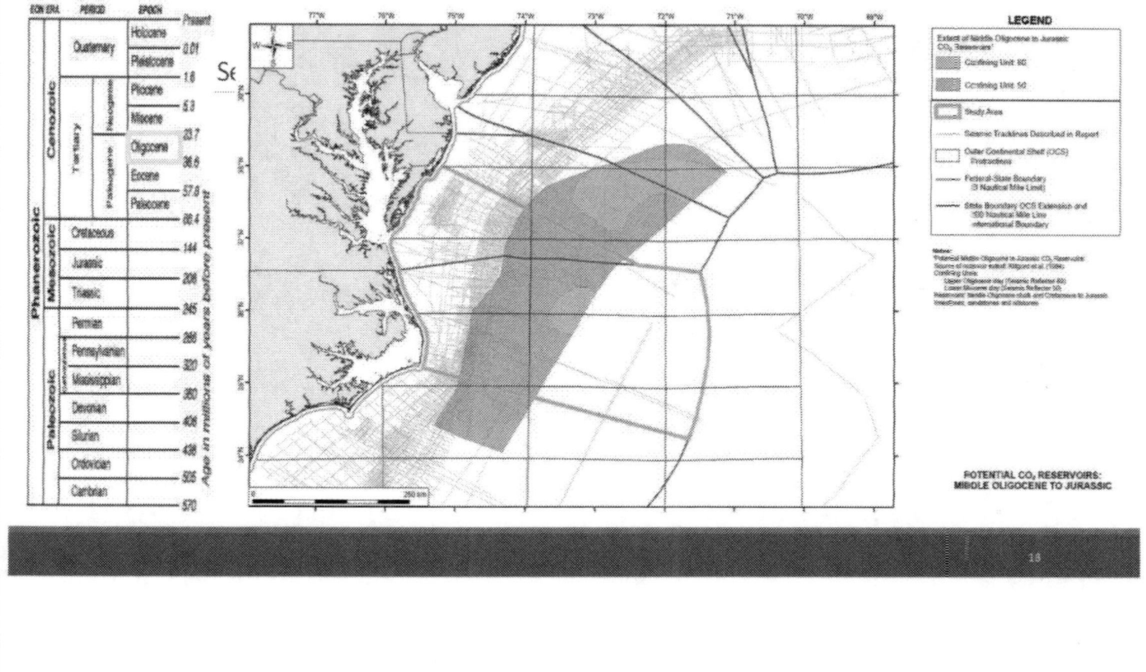
EON ERA
PERIOD
EPOCH
Present
Phanerozoic
Cenozoic
Mesozoic
Paleozoic
Quaternary
Tertiary
Holocene
Pleistocene
Pliocene
Miocene
Oligocene
Eocene
Paleocene
Cretaceous
Jurassic
Triassic
Permian
Pennsylvanian
Mississippian
Devonian
Silurian
Ordovician
Cambrian
Age in millions of years before present
LEGEND
POTENTIAL CO2 RESERVOIRS:
MIDDLE OLIGOCENE TO JURASSIC
18

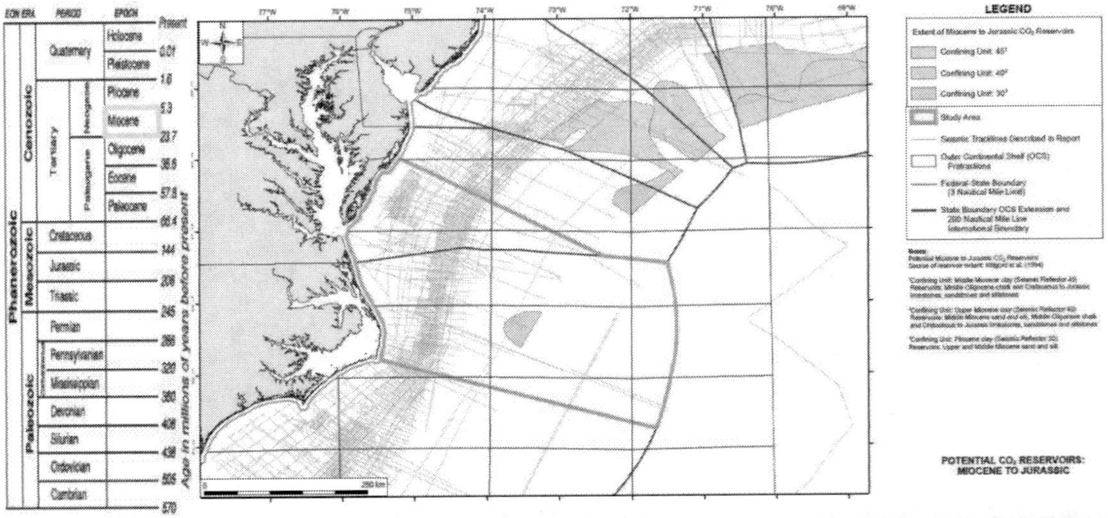


19

## Summary on Data Analysis

- Good Coverage of Seismic Lines in the Mid-Atlantic
- Poor Coverage of Wells in the Mid-Atlantic

- Quality variable of the Seismic Lines: Older lines tend to have a lower quality than younger ones
- Quality variable of the Well-Logs: Scanned paper copies provide poor quality data

20

# Summary of Initial Interpretation

- Starting Point: Interpretation of the lines closest to the wells selected
- Interpretation of the regional lines to establish basic framework of the Baltimore Canyon Trough and Carolina Trough
- Interpretation of the best quality data to identify formation target for $CO_2$ sequestration
- Use onshore wells to provide a potential correlation between onshore and offshore basins (Potomac Aquifer)



# Questions

# Project $ECO_2S$ Phase 2 CarbonSAFE Field Project

Prepared for:
**2017 Carbon Management Technology Conference**
**CMTC 2017**

Abstract # 497382

Prepared By:
**Vello Kuuskraa,** President
**Advanced Resources International, Inc.**

July 17-20, 2017
Houston, Texas





# DOE Disclaimer

This presentation is based upon work supported by the Department of Energy and was prepared as an account of work sponsored by an agency of the United States Government. Neither the United States Government nor any agency thereof, nor any of their employees, makes any warranty, express or implied, or assumes any legal liability or responsibility for the accuracy, completeness, or usefulness of any information, apparatus, product, or process disclosed, or represents that its use would not infringe privately owned rights. Reference herein to any specific commercial product, process, or service by trade name, trademark, manufacturer, or otherwise does not necessarily constitute or imply its endorsement, recommendation, or favoring by the United States Government or any agency thereof. The views and opinions of authors expressed herein do not necessarily state or reflect those of the United States Government or any agency thereof.

# The CarbonSAFE Project: Background

**In mid-2016, the U.S. Department of Energy/National Energy Technology Laboratory (DOE/NETL) issued a Funding Opportunity Announcement to develop an integrated CCS storage program ready for commercial operation by 2025.**

This program called CarbonSAFE (Carbon Storage Assurance Facility Enterprise) consisted of four distinct phases:

- Phase I. Pre-Feasibility Study
- Phase II. Storage Complex Feasibility Study
- Phase III. Site Characterization and Permitting
- Phase IV. Infrastructure Construction

A consortium of companies, led by Mississippi Power Company (MPC), submitted a proposal and were awarded a Phase II portion of CarbonSAFE. The project called $ECO_2S$ (Early $CO_2$ Storage) is located in Kemper County, Mississippi.





# Kemper County $CO_2$ Storage Complex

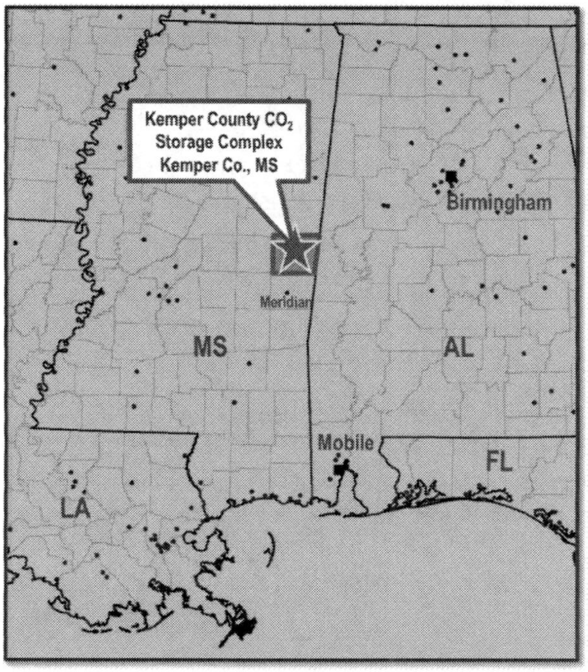


The Kemper County $CO_2$ Storage Complex is located in Kemper County, east-central Mississippi, north of Meridian, Mississippi.

# Background

**Prior Pre-Feasibility on the Kemper County $CO_2$ Storage Complex.** Mississippi Power Company (MPC), in conjunction with Southern Company Services (SCS), Advanced Resources and others had performed considerable pre-feasibility work on the Kemper County $CO_2$ Storage Complex. This enabled the project to qualify for the Phase II: Storage Complex Feasibility for CarbonSAFE.

The storage pre-feasibility work involved:

1. Performing a Geologic Evaluation of the Proposed $CO_2$ Storage Site
2. Establishing the Presence of Sufficient $CO_2$ Storage Capacity
3. Utilizing Data from Prior Deep Well Drilling
4. Performing Reservoir Modeling for the Areal Extent of the $CO_2$ Plume and a More Rigorous Estimate of $CO_2$ Storage Capacity
5. Establishing the Availability and Reliability of the $CO_2$ Source
6. Securing Surface Pore Space Rights, and
7. Conducting Initial Stakeholders Analysis and Other Pre-Feasibility Tasks

This presentation will discuss the pre-feasibility work conducted by the study team in preparation for proceeding with the Phase II. Storage Complex Feasibility Study that is now underway.





# 1. Performing Geologic Evaluation of the Proposed $CO_2$ Storage Site

The first pre-feasibility effort was a comprehensive, multi-disciplinary geologic evaluation by the Geological Survey of Alabama (GSA), with Dr. Jack Pashin as Principal Investigator, entitled, "Geological Evaluation of the Potential for $CO_2$ Sequestration in Kemper County, Mississippi."

This study identified two Lower Cretaceous strata, the Paluxy and the Washita-Fredericksburg sandstone saline formations, and one Upper Cretaceous interval, the Lower Tuscaloosa Massive Sand, as geologically favorable settings for storing $CO_2$ in Kemper County.

These three formations are overlain by the thick Tuscaloosa Marine Shale, a regionally extensive confining unit (seal).

The study used a 25 well data set, plus information from two previous drilled wells, to establish the geologic foundation for the $CO_2$ Storage Complex.

# Kemper County - Generalized Stratigraphy

| | | | | |
|---|---|---|---|---|
| Tertiary | Eocene | Lower Wilcox Group | Nanafolia Fm. | USDW |
| Tertiary | Paleocene | Midway Group | Naheola Fm | Potential USDW |
| Tertiary | Paleocene | Midway Group | Porter's Creek Clay | Regional Seal |
| Cretaceous | Upper | Selma Group | Predominately Chalk | Regional Seal |
| Cretaceous | Upper | Eutaw Fm. | | Potential USDW |
| Cretaceous | Upper | Tuscaloosa Group | Upper | Potential USDW & Water Supply |
| Cretaceous | Upper | Tuscaloosa Group | Marine Shale | Regional Seal |
| Cretaceous | Upper | Tuscaloosa Group | Lower & Massive Sand | ★ Potential Saline Fm. |
| Cretaceous | Lower | Washita- Fredericksburg | | ★ Saline Fm. |
| Cretaceous | Lower | Paluxy Fm. | | ★ Saline Fm. |
| *Paleozoic Unconformity* | | | | |

*(Lower Cretaceous Morringsport, Ferry Lake & Rodessa Fms missing @ Kemper Co.)*

Source: Pashin, J.C., D.J. Hills, D. C. Kopaska-Merkel, M.R. McIntyre, Geological Evaluation of the Potential for $CO_2$ Sequestration in Kemper County, Mississippi, Final Report, prepared for Southern Company Research and Environmental Affairs, June 1, 2008.

- **Tuscaloosa Massive Sand**
  - Depth: 3,000' to 3,252'
  - Porosity: 20%
  - Net Sand: 246'
- **Washita-Fredericksburg**
  - Depth: 3,252' to 4,225'
  - Porosity: 18%
  - Net Sand: 638'
- **Paluxy**
  - Depth: 4,225' to 4,808'
  - Porosity: 18%
  - Net Sand: 444'





# 2. Establishing the Presence of Sufficient $CO_2$ Storage Capacity

The GSA study provided a preliminary $CO_2$ storage capacity estimate (at 10% efficiency) of 4 to 5 MMmt per square mile (640 acres) for the Paluxy Sandstone implying a storage capacity of 200+ MMmt for the proposed $CO_2$ storage unit.

Subsequently, Advanced Resources performed additional geologic work and constructed a series of geological cross-sections for the $CO_2$ storage site. These cross-sections confirmed that all three of the saline formations - - the Paluxy, the Washita-Fredericksburg, and the Tuscaloosa Massive Sand - - are regionally extensive with considerable net sand thickness.

Using this additional information, the Project Team calculated a $CO_2$ storage capacity of 480 MMmt for the proposed $CO_2$ storage site using the volumetric DOE/NETL $CO_2$ storage capacity estimation methodology (at 10% efficiency):

- 170 MMmt for the Paluxy,
- 220 MMmt for the Washita-Fredericksburg, and
- 90 MMmt for the Tuscaloosa Massive Sand.

# Index Map of Kemper Co., MS Showing Paleozoic Structural Features, IGCC Site and Oil & Gas Exploratory Wells

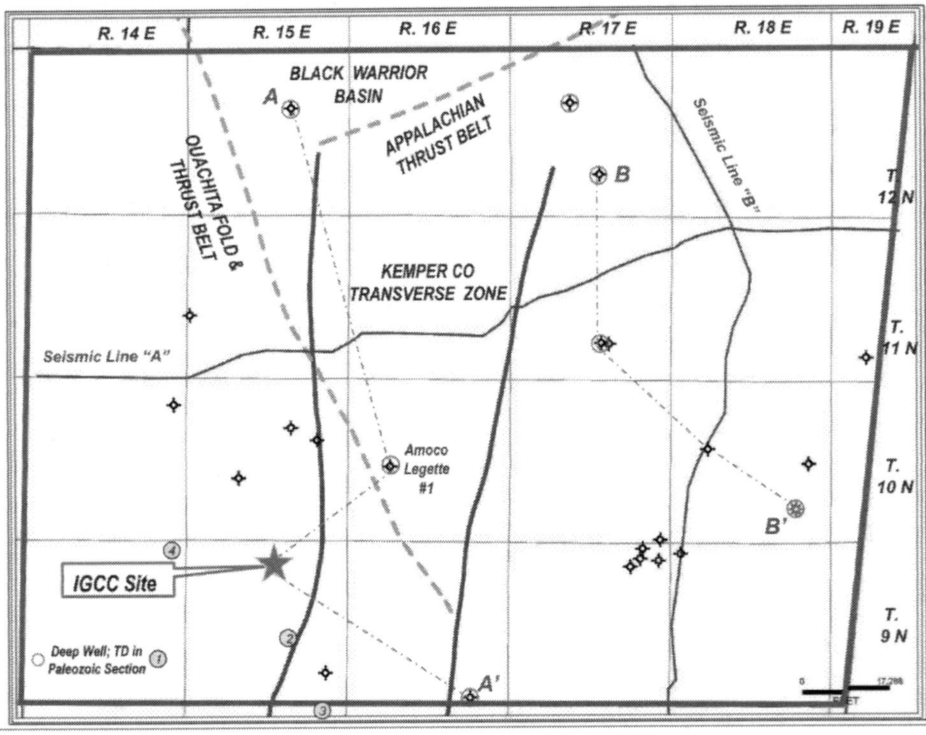






# Structure Cross-Section A-A'; Depth 1300' – 5500'

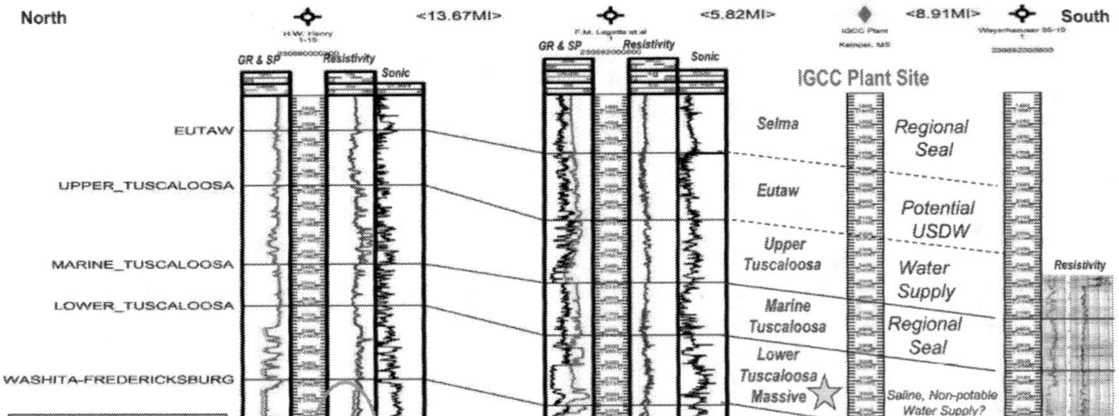

# 3. Utilizing Data from Prior Deep Well Drilling

To characterize the reservoir setting and establish a potential water source, MPC and SCS drilled two deep wells in the area:

- The first well was a small-diameter test hole drilled to 3,962 ft (below surface) through the Massive Sand formation of the Tuscaloosa Group. A geophysical log suite, involving gamma ray, multi-resistivity and spontaneous potential, was run in this well.
- The second well was a large-diameter water production well to 3,442 ft (below surface) with an 80 ft screened interval in the Massive Sand.

Once the second well was drilled and completed, pumping tests, up to 1,200 gallons per minute, were conducted to quantify the flow potential (product of permeability, pressure and sand thickness) of the Tuscaloosa Massive Sand.

Water samples from the Tuscaloosa Massive Sand indicated a Total Dissolved Solids (TDS) of 23,000 mg/L (ppm), establishing that the water is non-potable.





# 4. Performing Reservoir Modeling to Establish the Areal Extent of the $CO_2$ Plume and A More Rigorous Estimate of $CO_2$ Storage Capacity

Advanced Resources undertook reservoir modeling to: (1) better understand $CO_2$ injectivity, (2) calculate the areal extent of the $CO_2$ plume, and (3) provide a more rigorous estimate of $CO_2$ storage capacity.

The reservoir model used 43 distinct layers to represent the individual sands and shales of the three proposed $CO_2$ storage formations - - Tuscaloosa Massive Sand, Washita-Fredericksburg and Paluxy, including their interburdens and seals.

Reservoir modeling showed that the 30,000 acre proposed $CO_2$ storage site had an overall $CO_2$ storage capacity of 300 MMmt and could readily accommodate 3 MMmt/yr of $CO_2$ injection for 30 years.

| Name | | Rock Type | Depth (ft) | Thickness | Porosity (%) | Permeability (md) |
|---|---|---|---|---|---|---|
| Tuscaloosa Massive Sand | 1 | Sand | 3360 | 36 | 20 | 360 |
| | 2 | | 3396 | 37 | 20 | 360 |
| | 3 | | 3433 | 37 | 20 | 360 |
| | 4 | Interburden | 3470 | 65 | 7 | 0.002 |
| | 5 | Sand | 3535 | 38 | 20 | 360 |
| | 6 | | 3573 | 38 | 20 | 360 |
| | 7 | | 3611 | 39 | 20 | 360 |
| | 8 | Interburden | 3650 | 30 | 7 | 0.002 |
| | 9 | Sand | 3680 | 20 | 20 | 360 |
| Washita-Fredericksberg | 10 | Interburden | 3700 | 40 | 7 | 0.002 |
| | 11 | Sand | 3740 | 30 | 18 | 170 |
| | 12 | | 3770 | 30 | 18 | 170 |
| | 13 | Interburden | 3800 | 60 | 7 | 0.002 |
| | 14 | Sand | 3860 | 30 | 18 | 170 |
| | 15 | | 3890 | 30 | 18 | 170 |
| | 16 | Interburden | 3920 | 50 | 7 | 0.002 |
| | 17 | Sand | 3970 | 50 | 18 | 170 |
| | 18 | | 4020 | 50 | 18 | 170 |
| | 19 | | 4070 | 50 | 18 | 170 |
| | 20 | | 4120 | 50 | 18 | 170 |
| | 21 | | 4170 | 50 | 18 | 170 |
| | 22 | Interburden | 4220 | 20 | 7 | 0.002 |
| | 23 | Sand | 4240 | 30 | 18 | 170 |
| | 24 | | 4270 | 30 | 18 | 170 |
| | 25 | | 4300 | 30 | 18 | 170 |
| | 26 | Interburden | 4330 | 50 | 7 | 0.002 |
| | 27 | Sand | 4380 | 40 | 18 | 170 |
| | 28 | Interburden | 4420 | 160 | 7 | 0.002 |
| | 29 | Sand | 4580 | 20 | 18 | 170 |
| | 30 | Interburden | 4600 | 40 | 7 | 0.002 |
| Paluxy | 31 | Sand | 4640 | 33 | 18 | 170 |
| | 32 | | 4673 | 33 | 18 | 170 |
| | 33 | | 4706 | 34 | 18 | 170 |
| | 34 | Interburden | 4740 | 40 | 7 | 0.002 |
| | 35 | Sand | 4780 | 40 | 18 | 170 |
| | 36 | Interburden | 4820 | 10 | 7 | 0.002 |
| | 37 | Sand | 4830 | 10 | 18 | 170 |
| | 38 | Interburden | 4840 | 20 | 7 | 0.002 |
| | 39 | Sand | 4860 | 46 | 18 | 170 |
| | 40 | | 4906 | 46 | 18 | 170 |
| | 41 | | 4952 | 46 | 18 | 170 |
| | 42 | | 4998 | 46 | 18 | 170 |
| | 43 | | 5044 | 46 | 18 | 170 |

## Geologic Model of the $CO_2$ Storage Facility

- Numerous shale interburdens ("reservoir architecture") will help constrain vertical flow and $CO_2$ override within each formation.
- The reservoir pressure is estimated to be hydrostatic (1,844 psi) with a temperature of 130°F.
- The salinity of the brine is in excess of 20,000 ppm.
- Relative permeability data was obtained from the Citronelle $CO_2$ storage test.





## $CO_2$ Plume Model for Kemper County $CO_2$ Storage Complex

$CO_2$ Plume After 30 Years

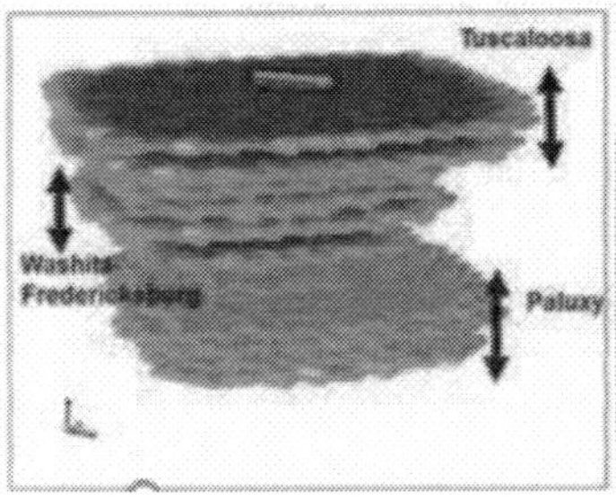


$CO_2$ Plume After 50 Years

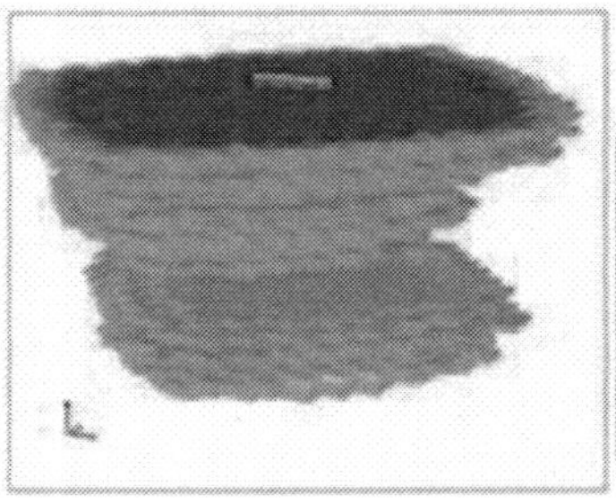

The reservoir model used 30 years of $CO_2$ injection at 3 MMmt/year and 20 years of shut-in to establish the maximum areal extent of the $CO_2$ plume.

$CO_2$ Plume After 50 Years

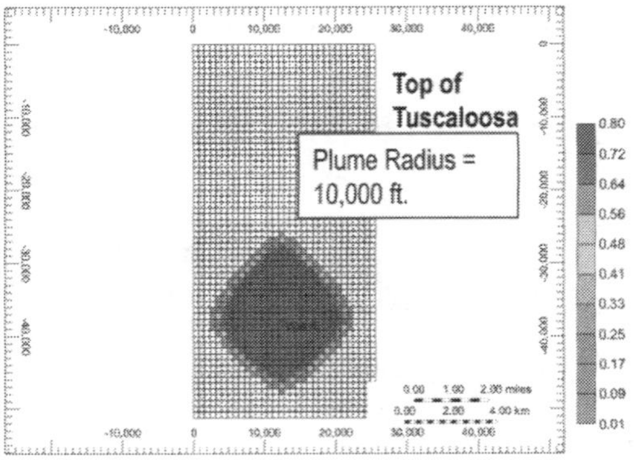

# 5. Establishing the Availability and Reliability of the $CO_2$ Source

Kemper energy facility, with $CO_2$ output of about 3 MMmt per year, is the intended source of the $CO_2$.

Source: Mississippi Power Company https://www.flickr.com/photos/mississippipower/27709115165/

A $141 million, 61-mile pipeline for transporting $CO_2$ already exists.

As such, the project only needs short distance gathering lines for connecting the $CO_2$ source with $CO_2$ injection wells.





# 6. Securing Surface and Pore Space Rights for the Kemper $CO_2$ Storage Complex

The proposed $CO_2$ storage area would encompass a ~30,000 acre area surrounding the Kemper energy facility. As part of defining and securing the proposed $CO_2$ storage site, MPC had performed the following work:

- Defined the protected and environmentally sensitive areas at and near the proposed well pad to avoid potential conflict.
- Successfully secured surface and pore space rights in the center of the $CO_2$ storage site.
- Examined and resolved any conflicts with lignite development and mineral rights in the $CO_2$ storage area.

# 7. Conducting Initial Stakeholder Analysis and Other Pre-Feasibility Tasks

MPC, as part of siting and constructing the Kemper energy facility, had already conducted extensive stakeholder analysis in the communities near the $CO_2$ storage site and within the expected areal extent of the $CO_2$ plume.

MPC, in conjunction with ARI, also prepared preliminary capital and operating costs for commercial-scale $CO_2$ storage at the Kemper County $CO_2$ Storage Complex.





# Current Status of ECO$_2$S

Currently, two deep wells have been drilled, logged and cored to provide more site specific reservoir characterization data.

These wells will subsequently become key $CO_2$ plume and pressure monitoring wells for the commercial phase of Project ECO$_2$S.

# Collection of Reservoir Characterization Data

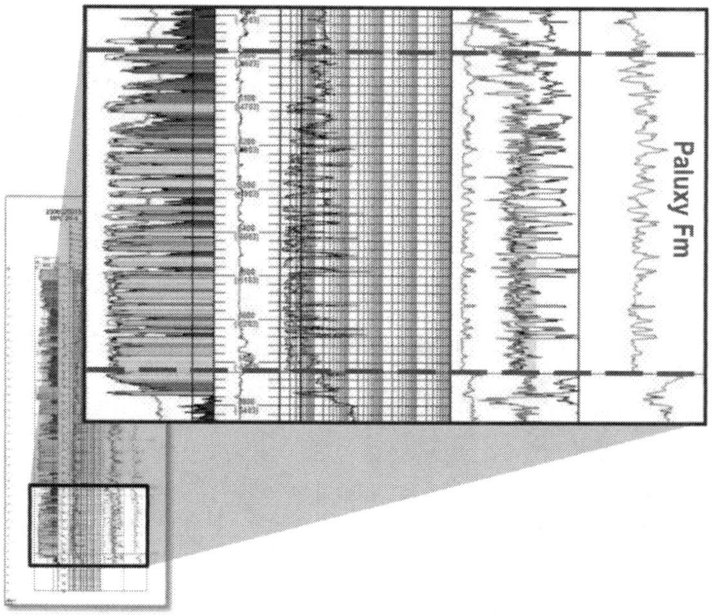


A suite of geophysical logs were run across the 3,160 foot reservoir interval, from 2,560' to 5,720'.

Obtaining core samples in the unconsolidated sand portions of the reservoir has been a challenge.

# Acknowledgements

The Project Team led by Southern States Energy Board, Mississippi Power Company and Southern Company Services, with technical support from Advanced Resources and a host of key subcontractors, acknowledge the valuable support provided by the U.S. DOE National Energy Technology Laboratory on this challenging and valuable CarbonSAFE field project.

Office Locations
**Washington, DC**
4501 Fairfax Drive, Suite 910
Arlington, VA 22203
Phone: (703) 528-8420

**Houston, TX**
11931 Wickchester Ln., Suite 200
Houston, TX 77043-4574
Phone: (281) 558-9200

**Knoxville, TN**
1210 Kenesaw Ave.
Suite 1210A
Knoxville, TN 37919-7736

## CMTC-502143-MS

# Applicability of CO2-EOR in Shale-Oil Reservoirs Using Diagnostic Plots

Dheiaa Alfarge, Iraqi Ministry of Oil, Missouri University of Science and Technology; Mingzhen Wei and Baojun Bai, Missouri University of Science and Technology



This paper was prepared for presentation at the Carbon Management Technology Conference held in Houston, Texas, USA, 17-20 July 2017.



## Abstract

Unconventional resources have played a significant role in changing oil industry plans recently. Shale formations in North America such as Bakken, Niobrara, and Eagle Ford have huge oil in place, 100-900 Billion barrels of recoverable oil in Bakken only. However, the predicted primary recovery is still below 10%. Therefore, seeking for techniques to enhance oil recovery in these complex plays is inevitable. In this paper, two engineering-reversed approaches have been integrated to investigate the feasibility of $CO_2$ huff-n-puff process in shale oil reservoirs. Firstly, a numerical simulation study was conducted to upscale the reported experimental-studies outcomes to the field conditions. As a result, different forward diagnostic plots have been generated from different combinations of $CO_2$ physical mechanisms with different shale-reservoirs conditions. Secondly, different backward diagnostic plots have been produced from the history match with $CO_2$ performances in fields' pilots which were performed in Bakken formation of North Dakota and Montana. Finally, fitting the backward with the forward diagnostic plots was used to report and diagnose some findings regarding the injected-$CO_2$ performance in field scale.

This study found that the porosity and permeability of natural fractures in shale reservoirs are significantly changed with production time, which in turn, led to a clear gap between $CO_2$ performances in lab-conditions versus to what happened in field pilots. As a result, although experimental studies reported that $CO_2$ molecular-diffusion mechanism has a significant impact on $CO_2$ performance to extract oils from shale cores, pilot tests performances indicated a poor role for this mechanism in field conditions. Therefore, the bare upscaling process for the oil recovery improvement and the $CO_2$-molecualr diffusion rate, which are obtained from $CO_2$ injection in lab-cores, to the field scale via numerical simulations needs to be reconsidered. In addition, this study found that kinetics of oil recovery process in productive areas and $CO_2$-diffusivity level are the keys to perform a successful $CO_2$-EOR project. Furthermore, general guidelines have been produced from this work to perform successful $CO_2$ projects in these complex plays. Finally, this paper provides a thorough idea about how $CO_2$ performance is different in field scale of shale oil reservoirs as in lab-scale conditions.

## Introduction

In current days, conventional oil and gas reservoirs are showing a clear trend of depletion and diminish in number. Therefore, seeking for unconventional reservoirs has been the target over the 20 years. Fortunately, the investment in these unconventional plays has been yet successful. Oil production from tight formations including shale plays has shared for more than 50% of total oil production in the US (Alfarge et al., 2017). Hoffman et al., (2016) reported that 4 million barrels per day as an increment in US-oil daily production comes from these unconventional oil reservoirs. From 2011 to 2014, Unconventional Liquid Rich (ULR) reservoirs contributed to all natural gas growth and nearly 92% of oil production growth in the US (Alfarge et al., 2017). Specifically, Bakken and Eagle Ford contributed more than 80% of total US oil production from these tight formations (Yu et al., 2016a). More recently, Bakken formation alone delivers close to 10% of the total US production with more than 1.1 million barrels per day (Alvarez et al, 2016). This revolution in oil and gas production happened mainly because shale oil reservoirs have been just increasingly developed due to the advancements in horizontal wells and hydraulic fracturing in last decade. Several studies have been conducted to estimate the recoverable oil in place in these complex formations indicating a large amount of oil in place. The available information refers to 100-900 Billion barrels in Bakken only. However, the predicted recovery from primary depletion could lead to 7% only of original oil in place (Clark, 2009). Furthermore, some investigators argued that the primary recovery factor is still in a range of 1-2 % in some of the plays in North America (Wang et al., 2016). For example, the North Dakota Council reported that "With today's best technology, it is predicted that 1-2% of the reserves can be recovered" (Sheng, 2015). The main problem during the development of unconventional reservoirs is how to sustain the hydrocarbon production rate, which also leads to low oil recovery factor. **Fig. 1** explains the typical trend for oil production in these complex plays. The producing wells usually start with high production rate initially; however, they show steep decline rate in first 3-5 years until they get leveled off at very low rate. According to Yu et al. (2014), the main reason beyond the quick decline in production rate is due to the fast depletion of natural fractures networks with slow recharging from matrix system, which is the major source of hydrocarbon. Therefore, oil recovery factor from primary depletion has been predicted typically to be less than 10% (LeFever et al. 2008; Clark, 2009; Alharthy et al., 2015; Kathel and Mohanty 2013, Wan et al., 2015; Alvarez et al, 2016).

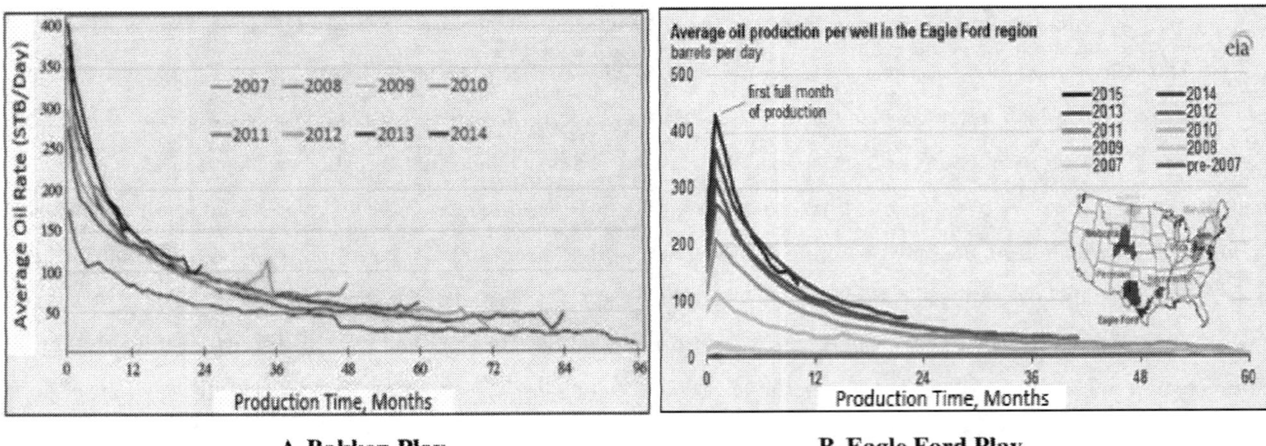


**A-Bakken Play** **B-Eagle Ford Play**

**Fig. 1: Average oil production per well in unconventional reservoirs (EIA, 2016)**

Since these reservoirs have huge original oil in place, any improvement in oil recovery factor would result in enormous produced oil volumes. Therefore, IOR methods have huge potential to be the major stirrer in these huge reserves.

Although IOR methods are well understood in conventional reservoirs, they are a new concept in unconventional ones. All basic logic-steps such as experimental works, simulation studies, and pilot tests for investigating the applicability of different IOR methods have just started over the last decade. Miscible gas injection has shown excellent results in conventional reservoirs with low permeability and light oils. Extending this approach to unconventional reservoirs including shale oil reservoirs in North America has been extensively investigated over the last decade. The gases which have been investigated are $CO_2$, $N_2$, and an enrich natural gases. However, the majority of studies focused on $CO_2$ due to different reasons. $CO_2$ can dissolve in shale oil easily, swells the oil and lowers its viscosity. $CO_2$ has a lower miscibility pressure with shale oil rather than other gases such as $N_2$ and $CH_4$ (Zhang et al., 2016). Moreover, the $CO_2$ performance in lab conditions was excellent in increasing oil recovery from shale cores as shown in **Fig. 2**. However, the minimum miscible pressure of $CO_2$ in these types of oil has a controversial range in between 2500 psi to 3300 psi. Furthermore, it has been reported that oil of these reservoirs has a low acid number which might give the hope to apply $CO_2$ injection successfully without asphaltenes precipitation problems (Kurtoglu et al., 2014). Unfortunately, the results of pilot-tests for $CO_2$-EOR, huff-n-puff protocol, which have been conducted in unconventional reservoirs of North America were disappointing (Hoffman et al., 2016). This gap in $CO_2$ performance between lab-conditions versus to what happened in field-scale suggests that there is something missing between microscopic-level and macroscopic-level in these plays. Most of the experimental studies reported that the molecular-diffusion mechanism for $CO_2$ is beyond the increment in oil recovery obtained in lab scale (Alfarge et al., 2017). Furthermore, most of the previous simulation studies relied on the lab-diffusivity level for these miscible gases to predict the expected oil increment on field scale (Alfarge et al., 2017). One of the main reasons for the poor-performance for $CO_2$ in the pilot tests might be due to the wrong prediction for $CO_2$ diffusion-mechanism in these types of reservoirs. The purpose of this study is to diagnose the reasons behind the gap in the $CO_2$ performance in lab-conditions versus to what happened in field-scale.

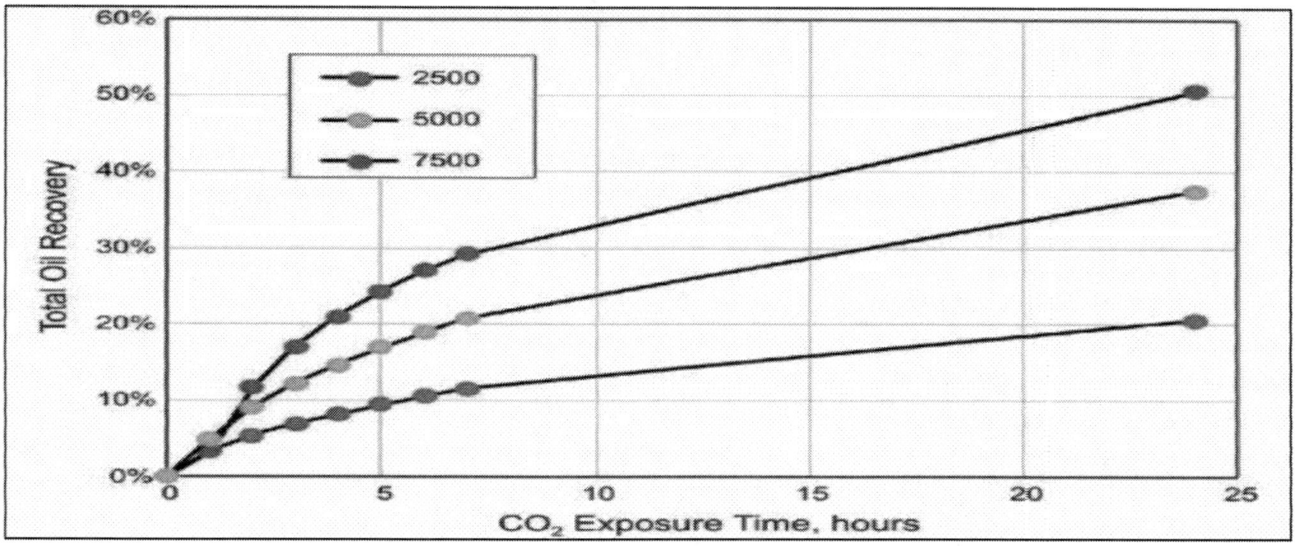


**Fig. 2: Sample for how much oil extracted from natural cores by $CO_2$ in lab conditions (Hawthorne et al., 2017)**

## Background

Starting with lab-work tools, the study of Song et al. (2013) conducted an experimental investigation to compare results from injecting $CO_2$ and water in cores from Bakken. They found that water flooding would enhance oil recovery better than immiscible $CO_2$ in Huff-n-Puff protocol. However, miscible and near miscible $CO_2$ Huff-n-Puff would achieve better performance than water flooding in enhancing oil recovery. Hawthorne et al., (2013) investigated the mechanism beyond increasing oil recovery by $CO_2$ injection in Bakken cores. They proved that diffusion mechanism is the main mechanism for $CO_2$ to increase oil recovery in these complex plays. However, to extract oil from shale matrix by $CO_2$, long times of exposure combined with large contact areas are required. Gamadi et al. (2014) conducted experimental work on shale cores from Mancos and Eagle Ford to investigate the potential of $CO_2$ injection in these reservoirs. Their laboratory results indicated that cyclic $CO_2$ injection could improve oil recovery from shale oil cores from 33% to 85% depending on the shale core type and other operating parameters. Alharthy et al., (2015) compared the performance of injecting different types of gases such $CO_2$, $C_1$-$C_2$ mixtures, and $N_2$ on enhancing oil recovery from Bakken cores experimentally. They concluded that the injecting gas, composed of $C_1$, $C_2$, $C_3$, and $C_4$, could produce nearly as much oil as $CO_2$ injection could which was 90% from several Middle Bakken cores and nearly 40% from Lower Bakken cores. Also, they found the counter-current mechanism is the main mechanism for these gases to recover oil from shale cores. Finally, Yu et al., (2016) investigated $N_2$ flooding process experimentally on Eagle Ford core plugs saturated with dead oil. They examined different flooding time range and different injection pressure on $N_2$ flooding performance. They found that more oil was produced with a longer flooding time and higher injection pressure. To sum up, $CO_2$ showed a good potential to extract oil from shale cores in experimental works (Jin et al., 2016).

The numerical simulation studies of Shuaib et al., (2009) and Wang et al., (2010) might be the early-published studies in this area. Those models showed that 10-20% of increment oil could be recovered by continuous gas flooding while 5-10% could be recovered by huff-n-puff gas protocol (Hoffman et al., 2016). Dong et al., (2013) reported a numerical study evaluating $CO_2$ injection performance for the Bakken interval in a sector of the Sanish Field. They came up with a scenario to increase $CO_2$ injectivity in that field by drilling more horizontal injection wells. This scenario predicted the possibility to inject 5000 Mscf/day at a maximum injection pressure of 8000 psi. From their simulation study, they found that using $CO_2$ injection method might increase oil recovery from 5% to 24% in that field. Xu et al., (2013) evaluated the reservoir performance of Elm Coulee field in Eastern Montana under $CO_2$ flooding with different hydraulic fracture orientations. They found that transverse fractures have higher oil recovery factor, but it has lower utilization value than longitudinal fractures due to breakthrough problems. Zhu et al., (2015) constructed a model in which gas could be injected into a hydraulic fracture along a horizontal well and the production process could occur from an adjacent fracture which has an intersection with the same well. They found a substantial improvement in oil recovery by injecting $CO_2$ in reservoirs with fluid flow from fracture to fracture. Pu et al., (2016) introduced a new model which considers capillarity and adsorption effect of the small pores for shale reservoirs. They found that using this model would simulate $CO_2$-EOR in unconventional reservoirs properly. Furthermore, capillarity consideration in modeling process would predict higher oil recovery by $CO_2$ injection than the cases which did not include capillarity property.

It is clear from the previous studies that $CO_2$ would have a great potential to enhance oil recovery in these poor-quality reservoirs. However, whether using $CO_2$ in Huff-n-Puff protocol or injecting $CO_2$ in flooding scenario is still a controversial argument. Due to the low permeability, conformance problems in these reservoirs, and the significant observed molecular-diffusion rate for $CO_2$ in lab conditions, most of the previous researchers prefer the $CO_2$ Huff-n-Puff on $CO_2$ flooding. Unfortunately, the results of pilot tests of $CO_2$ in the cyclic process were disappointing (Hoffman et al., 2016). Therefore, this study has been conducted for determining the reasons causing the gap for $CO_2$ performance in lab-conditions versus to what happened in field-scale.

**Molecular Diffusion**

Gravity drainage, physical diffusion, viscous flow, and capillary forces are the common forces which control the fluids flow in porous media. However, one force might eliminate the contributions of others depending on the reservoir properties and operating conditions. Molecular diffusion is defined as the movement of molecules caused by Brownian motion or composition gradient in a mixture of fluids (Mohebbinia et al., 2017). This type of flow would be the most dominated flow in fractured reservoirs with a low-permeability matrix when gravitational drainage is inefficient (Moorgate and Firoozabadi, 2013; Mohebbinia et al., 2017). It has been noticed and approved that gas injection is the most common EOR-process affected by calculations of molecular-diffusion considerations. Ignoring or specifying incorrect diffusion-rate during simulation process can lead to overestimate or underestimate the oil recovery caused by the injected gas. This happens not only due to the variance in miscibility-process between the injected-gas and formation-oil but also due to the path change of the injected gas species from fractures to the formation-matrix.

The Péclet number (Pe) is a class of dimensionless numbers which have been used to measure the relative importance of molecular diffusion flow to the convection flow. This number can be calculated as shown in **Eq. 1**. If Pe number is less than 1, diffusion is the dominant flow. However, if Pe is greater than 50, convection is the dominant flow. The dispersion flow is dominant when Pe in range 1 to 50 (Hoteit and Firoozabadi, 2009). **Fig. 3** explains the flow regimes according to Péclet number cutoffs.

$$\mathbf{Pe} = \frac{diffusion\ time}{convection\ time} = (L2/D)/(L/v) = Lv/D \tag{1}$$

Where v is the bulk velocity, L is a characteristic length, and D is the diffusion coefficient.

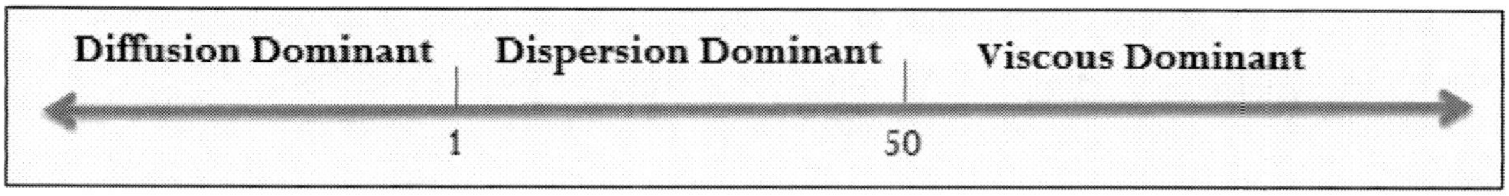


**Fig. 3: Flow Regimes According to Péclet Number Cutoffs**

## $CO_2$ Molecular-Diffusion Mechanism

Different mechanisms have been proposed for the injected $CO_2$ to improve oil recovery in unconventional reservoirs as shown in **Table 1**. However, since the matrix permeability in these unconventional reservoirs is in range (0.1 –0.00001 md), $CO_2$ would not be transported by convection flux from fracture to matrix (Yu et al., 2014). The main transportation method for $CO_2$ is happening due to the difference in the concentration gradient between $CO_2$ concentration in the injected gases and the target-oil. This process of transportation is subjected to Fick's law. Hawthorne et al., (2013) extensively investigated the $CO_2$ diffusion-mechanism in Bakken cores and proposed five conceptual-steps to explain it. These conceptual steps include: (1) $CO_2$ flows into and through the fractures, (2) unfractured rock matrix is exposed to $CO_2$ at fracture surfaces, (3) $CO_2$ permeates the rock driven by pressure, carrying some hydrocarbon inward; however, the oil is also swelling and extruding some oil out of the pores, (4) oil migrates to the bulk $CO_2$ in the fractures via swelling and reduced viscosity, and (5) as the $CO_2$ pressure gradient gets smaller, oil production is slowly driven by concentration-gradient diffusion from pores into the bulk $CO_2$ in the fractures.

**Table 1: The proposed $CO_2$ EOR mechanisms for improving oil recovery in unconventional reservoirs**

| $CO_2$ mechanism | Approach tool |
|---|---|
| 1-Diffusion | Lab |
| 2-Reduction in Capillary forces | Lab and simulation |
| 3-Repressurization | Lab |
| 4-Extraction | Lab |
| 5-Oil swelling and pressure maintenance | Lab and simulation |
| 6-Oil Viscosity reduction | Lab and simulation |
| 7-Combination of more than one mechanism from above | - |

Most of the previous experimental studies reported that $CO_2$ diffusion mechanism is beyond the increment in oil recovery obtained in lab conditions. Then, the observed increment in oil-recovery and/or the $CO_2$ diffusion-rate obtained in lab conditions were upscaled directly to field scale by using numerical simulation methods. This direct upscaling methodology might be so optimistic due to that the lab-cores have higher contact area and longer exposure time to $CO_2$ than what might happen in the real-conditions of unconventional reservoirs. As a result, both of previous simulation studies and experimental works might be too optimistic to predict a quick improvement in oil recovery from injecting $CO_2$ in these tight formations.

## Methodology

To determine the reasons causing the gap for $CO_2$ performance in lab conditions versus to what happened in the pilot tests, field scale conditions, we need to start with screening the parmeters which we are sure of them and the parameters which have some ambiguity. The parameters which are known in these pilot-tests are the following:

- PVT data, and oil composition properties
- Major Wells and stimulation criteria
- Rock properties
- $CO_2$ injection operating parameters (rate, pressure, time)
- Produced oil rate versus time
- $CO_2$ performance in lab conditions (most of the mechanisms and observations)

On the other hand, the parameters which have ambiguity in these pilot-tests are the following:

- $CO_2$ diffusivity level in field scale
- Natural fracture intensity, porosity, and permeability in field scale

According to the previous diagnose, the systematic methodology for this work is falling into two reversed scenarios. Firstly, a numerical simulation study was conducted to upscale the reported experimental-studies outcomes to the field conditions. As a result, different forward diagnostic plots have been generated from different combinations of $CO_2$ physical mechanisms with different shale-reservoirs conditions. Secondly, different backward diagnostic plots have been produced from the history match with $CO_2$ performances in fields' pilots which were performed in Bakken formation of North Dakota and Montana. Finally, fitting the backward with the forward diagnostic plots was used to report and diagnose some findings regarding the injected-$CO_2$ performance in field scale. **Fig. 4** shows the detailed methodology for this study. This study is the first numerical simulation study to integrate the two reversed approaches in this area of research. All of the previous numerical simulation studies either chose the direct upscale for lab observation or built merely a conceptual model.

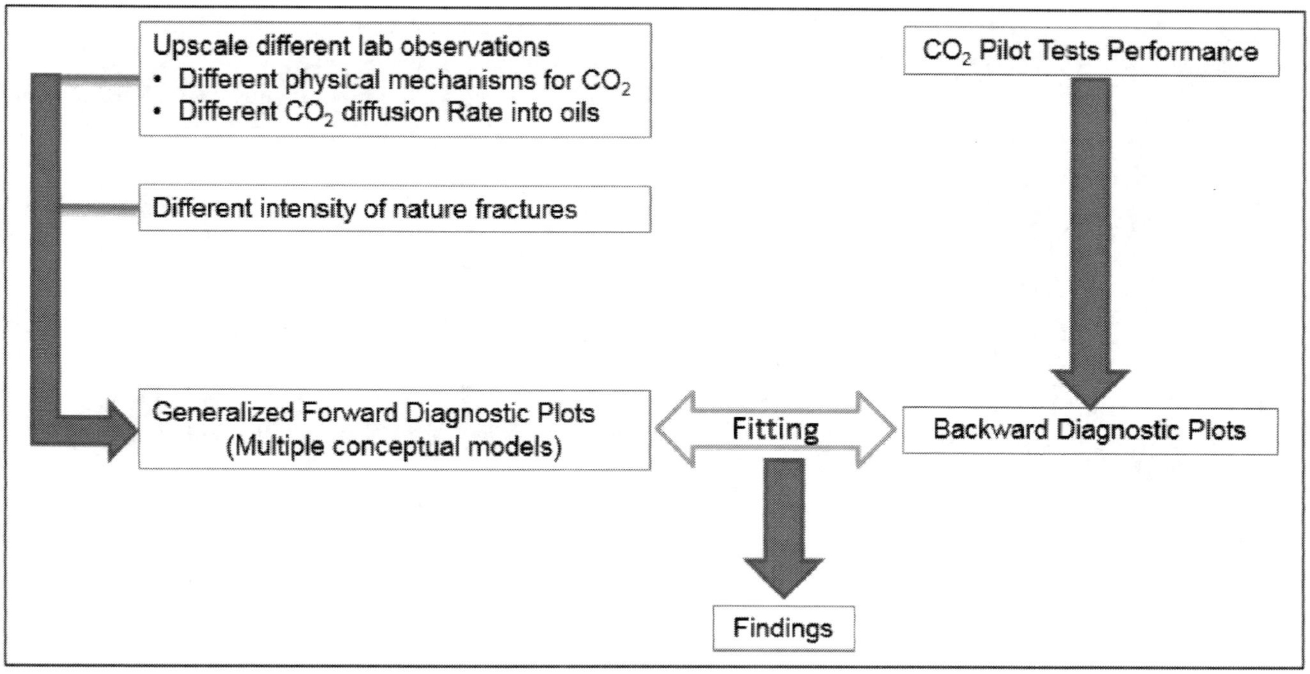


**Fig. 4: Systematic methodology for this study**

## Numerical Simulation

In this simulation study, the LS-LR-DK (logarithmically spaced, locally refined, and dual permeability) model is used. The LS-LR-DK method can accurately simulate the fluid flow in fractured shale-oil reservoirs. Also, an advanced general equation-of-state compositional simulator has been used to build an equation-of-state model. Then, both of models have been combined to simulate compositional effects of reservoir fluid during primary and enhanced oil recovery processes. Furthermore, implementation of a diffusion model in the LS-LR-DK (logarithmically spaced, locally refined, and dual permeability) model has been conducted. In this study, we tried to build a numerical model which has the typical fluid and rock properties of Bakken formation, one of the most productive unconventional formations in the US. In this model, we injected EOR-$CO_2$ in different scenarios as Huff-n-Puff protocol through hydraulically fractured well in Bakken formation. All the mechanisms which were proposed in **Table 1** have been considered in this model.

In this field case study, the production well was stimulated with 5 hydraulic fractures as shown in **Fig. 5**. The spacing between the hydraulic fractures is 200 ft. The simulation model includes two regions which are stimulated reservoir volume (SRV) and un-stimulated reservoir volume (USRV). The dimensions of the reservoir model are 2000 ft x 2000ft x42 ft, which corresponds to length, width, and thickness respectively. The dimensions of the fractured region are 5 fractures with half-length of 350 ft in J direction, width 0.001 ft in I direction, and fracture height of 42 ft in K direction. Fracture conductivity is 15 md.ft. The other model input parameters are shown in **Table 2**.

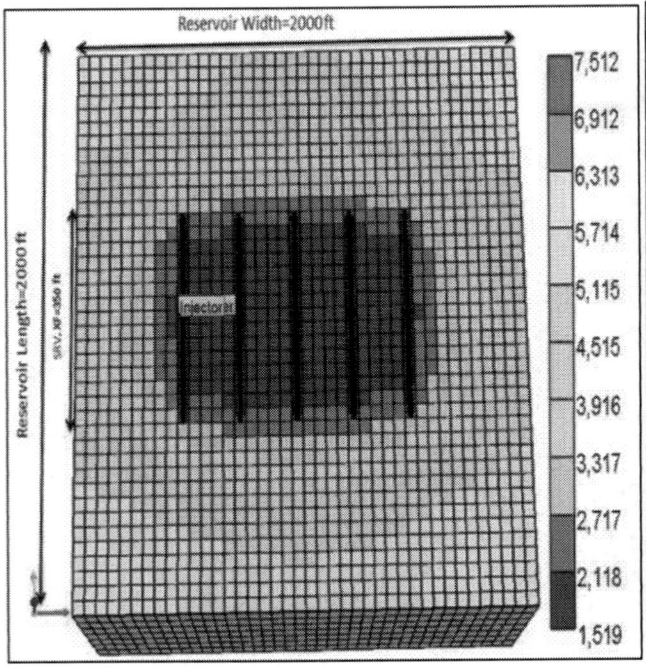


Figure 5a- Average Pressure in a depleted well in Bakken

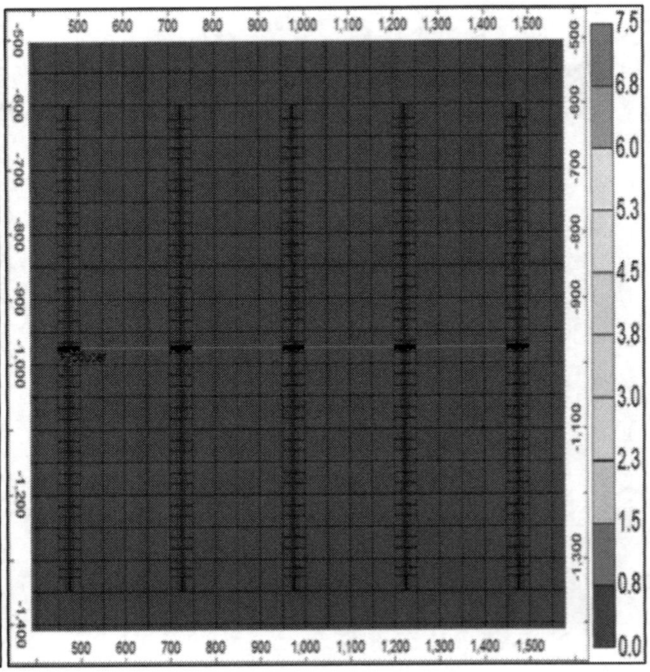


Figure 5b- A closed view for SRV of Production well

**Table 2: Model input parameters for the base case**

| Parameter | value | Unit |
|---|---|---|
| The model dimensions | 2000x2000x42 | ft |
| Production Time | 20 | year |
| Top of Reservoir | 8000 | ft |
| Reservoir Temperature | 240 | °F |
| Reservoir pressure | 7500 | psi |
| Initial Water saturation | 0.3 | value |
| Total compressibility | $1x10^{-6}$ | $psi^{-1}$ |
| Matrix permeability | 0.005 | mD |
| Matrix porosity | 0.085 | value |
| Horizontal Well length | 1000 | ft |
| Total number of fractures | 5 | value |
| Fracture conductivity | 15 | mD-ft |
| Fracture half-length | 250 | ft |
| Fracture Height | 42 | ft |

## Compositional Model for the Formation Fluids

The typical Bakken oil has been simulated in this study. The oil which was used in this model has 42 API°, 725

SCF/STB, and 1850 psi as oil gravity, gas oil ratio, and bubble point pressure respectively. It is known that compositional models are the most time-consumed models due to the number of components in a typical reservoir oil. In our model, we have 34 components so that would take a long time for the simulator to complete run one scenario. The common practice in numerical simulation for such situation is the careful lump of reservoir oil components into a short representative list of pseudo components. These pseudo components could be acceptable if they have matched with the laboratory–measured phase behavior data. The supplied data for reservoir oil needs to have a description of associated single carbon numbers and their fractions, saturation pressure test results, separator results, constant composition expansion test results, differential liberation test results, and swelling test results (CMG, 2016). All of these data can be used for tuning the EOS to match the fluid behavior. In our simulation, we lumped the original 34 components into 7 pseudo components as shown in **Table 3** by using WinProp-CMG. WinProp is an Equation-of-State (EOS)-based fluid behavior and PVT modeling package. In WinProp laboratory data for fluids can be imported and an EOS can be tuned to match its physical behavior. Fluid interactions can be then predicted and a fluid model can be also created for use in CMG software (CMG, 2016). **Table 4** presents the Peng-Robinson EOS fluid description and binary interaction coefficients of the Bakken crude oil with different gases. **Fig. 6** represents the two-phase envelope for Bakken oil which was generated by WinProp-CMG.

**Table 3: Compositional data for the Peng-Robinson EOS in the model oil**

| Component | Mole fraction | Critical pressure (atm) | Critical Temp. (K) | Acentric Factor | Molar Weight (g/gmole) |
|---|---|---|---|---|---|
| CO2 | 0 | 7.28E+01 | 3.04E+02 | 0.225 | 4.40E+01 |
| N2-CH4 | 0.2704 | 4.52E+01 | 1.90E+02 | 0.0084 | 1.62E+01 |
| C2H-NC4 | 0.2563 | 4.35E+01 | 4.12E+02 | 0.1481 | 4.48E+01 |
| IC5-CO7 | 0.127 | 3.77E+01 | 5.57E+02 | 0.2486 | 8.35E+01 |
| CO8-C12 | 0.2215 | 3.10E+01 | 6.68E+02 | 0.3279 | 1.21E+02 |
| C13-C19 | 0.074 | 1.93E+01 | 6.74E+02 | 0.5672 | 2.20E+02 |
| C20-C30 | 0.0508 | 1.54E+01 | 7.92E+02 | 0.9422 | 3.22E+02 |

**Table 4: Binary interaction coeficients for Bakken oil**

| Component | CO2 | N2-CH4 | C2H-NC4 | IC5-CO7 | CO8-C12 | C13-C19 | C20-C30 |
|---|---|---|---|---|---|---|---|
| CO2 | | | | | | | |
| N2-CH4 | 1.01E-01 | | | | | | |
| C2H-NC4 | 1.32E-01 | 1.30E-02 | | | | | |
| IC5-CO7 | 1.42E-01 | 3.58E-02 | 5.90E-03 | | | | |
| CO8-C12 | 1.50E-01 | 5.61E-02 | 1.60E-02 | 2.50E-03 | | | |
| C13-C19 | 1.50E-01 | 9.76E-02 | 4.24E-02 | 1.72E-02 | 6.70E-03 | | |
| C20-C30 | 1.50E-01 | 1.45E-01 | 7.79E-02 | 4.27E-02 | 2.51E-02 | 6.00E-03 | |

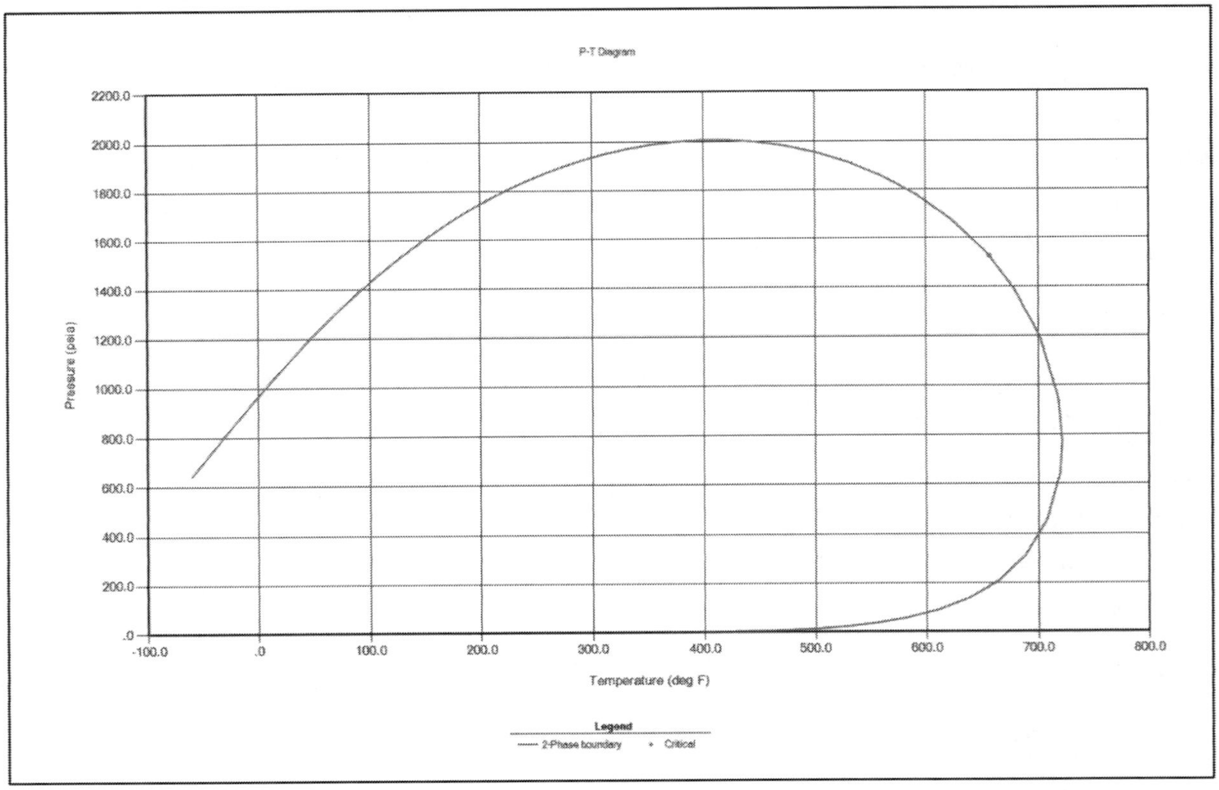


**Fig. 6: The two-phase envelope for Bakken oil which was generated by WinProp-CMG**

## Results and Discussion

Hoffman et al., (2016) reported seven pilot-tests in Bakken formation which was conducted in North Dakota and Montana. We are presenting just one pilot of them in this section. This pilot was indicated in his paper as pilot test#2. This pilot-test injected $CO_2$ as Huff-n-Puff process in Bakken formation, in Montana portion. They injected 1500-2000 Mscf/day of CO2 for 45 days at 2000-3000 psi. The soaking period was proposed to be 2 weeks. Then, the well was put back in the production process. The operating parameters for this pilot tests were suggested as shown in **Table 5**.

**Table 5: The operating parameters for Pilot test#2 conducted in Bakken formation**

| Scenario | Time |
|---|---|
| Primary depletion at BHP=1500psi | 9 Years |
| $CO_2$ Injection period (at rate of 1500 Mscf/day) | 2 months |
| Soaking | 14 days |
| Back for production | 10 years and 7.5 months |

Different proposed mechanisms for $CO_2$ to enhance oil recovery in shale oil reservoirs, which are shown in **Table 1,** have been combined with both of different intensity, porosity, and permeability for natural fractures and the operating parameters for pilot test#2 which are shown in **Table 5**. As a result, different diagnostic plots of these combinations have been generated. In this paper, we call these diagnostic plots as the forward diagnostic plots as shown in **Fig. 7**. In the meantime, we create a history match process for pilot test#2 as shown in **Fig. 7b.** We call these plots, history-match plots, as backward diagnostic plots. If we compare the forward diagnostic plots, **Fig. 7**, with the pilot test#2 performance which is shown in **Fig. 8a,** it is clear there are some diagnostic curves are close to what happened in that pilot test, however, other diagnostic curves are far away from what happened in the field. We concluded that the diagnostic plots which have a good match with the pilot test have two main criteria. These two main properties are as following:

(1) The pilot test is matching the solutions of low-effective diffusivity for the injected $CO_2$.

(2) The pilot test is matching the solutions which have natural fractures of changeable porosity and permeability as shown in **Fig. 8b.** In the beginning of the well life, the well performance is matching the solutions which have high porosity and permeability for the simulated natural fractures. However, the well performance would match the solutions which have lower porosity and permeability at a later time. The reasons beyond this behavior are similar to the reasons causing the permeability and porosity reduction with production time progress in both of shale gas and coal-bed methane reservoirs as shown in **Fig. 10**.

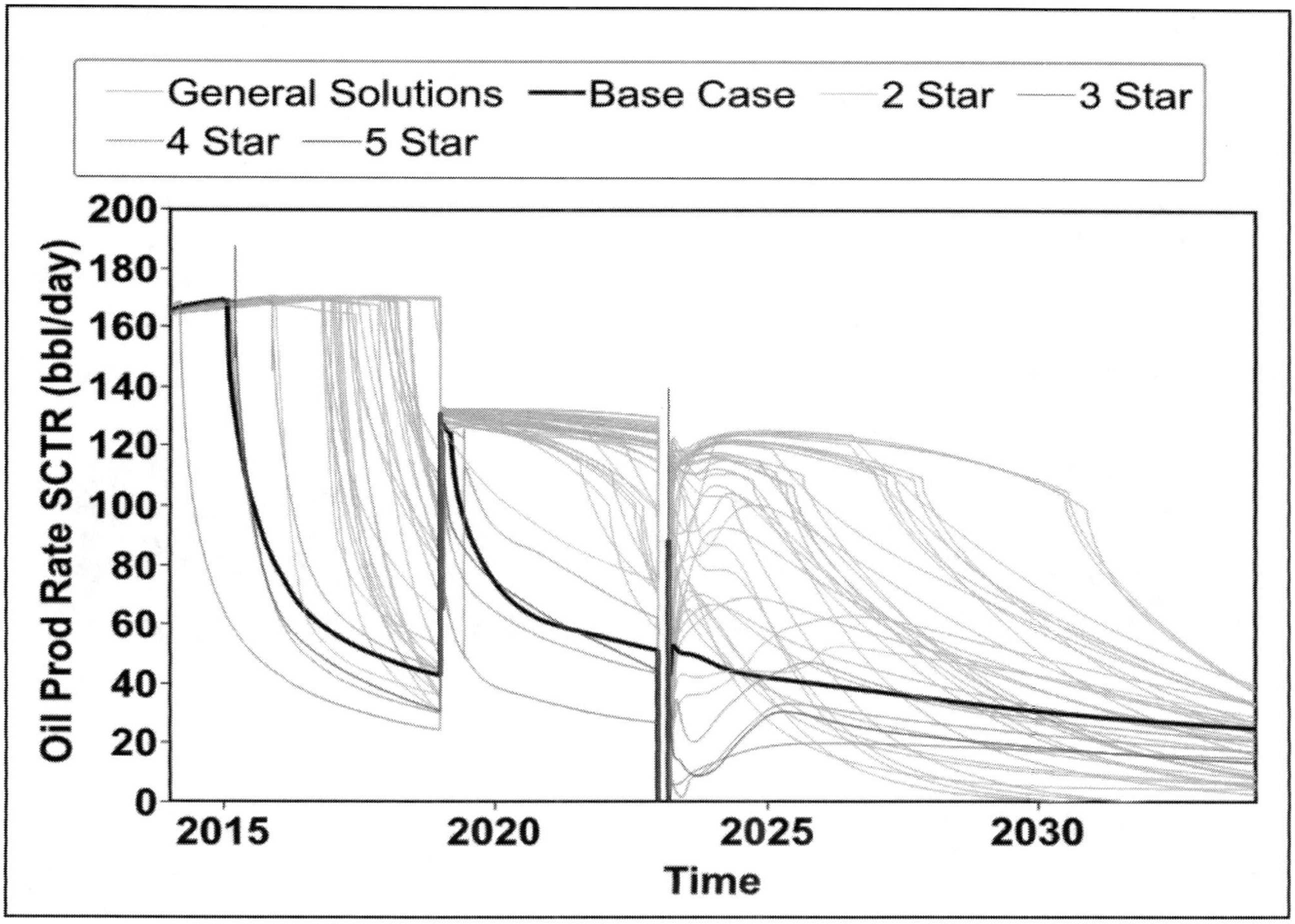


**Fig. 7: The simulated forward diagnostic plots**

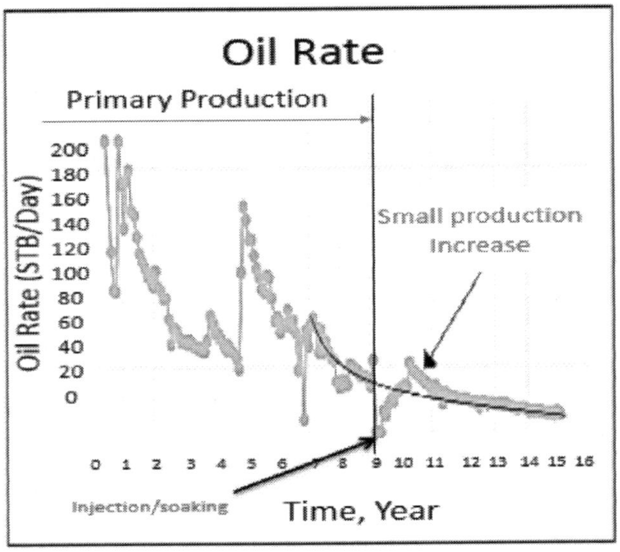


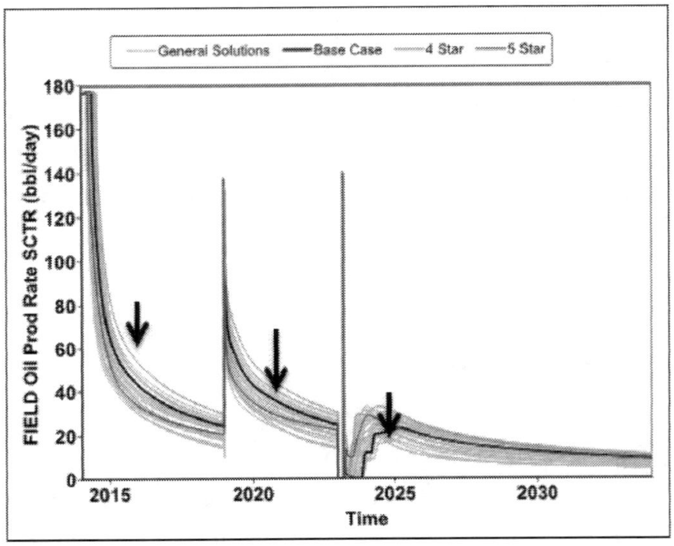


**Fig. 8a: $CO_2$ Pilot test#2 (Hoffman et al. 2016)** **Fig. 8b: History match from the simulated model**

We think that the previous two main characteristics for shale reservoirs are the critical points for $CO_2$-EOR success in shale reservoirs. In addition, these two main criteria are not fully considered in lab conditions which in turn led to this gap in $CO_2$ performance in lab conditions versus to what happened in field scale. In most of the reported experimental studies, small chips of natural cores were exposed to $CO_2$ for a long time under high pressure and temperature. Therefore, in lab scale, the contact area and exposure time between $CO_2$ and formation cores are much larger than what happens in the field scale. In the reported pilot test, there is no such long exposure time and large contact area. Therefore, $CO_2$ needs a good molar-diffusivity so it can invade the matrix-oil and extract the formation-oil by counter-current mechanism because the diffusion flow is the dominated flow in these types of reservoirs as shown in **Fig. 9.** Furthermore, it needs larger contact area between the injected $CO_2$ and the formation-oil. This can be done by performing $CO_2$ in early time of the production well life before natural fractures get closed.

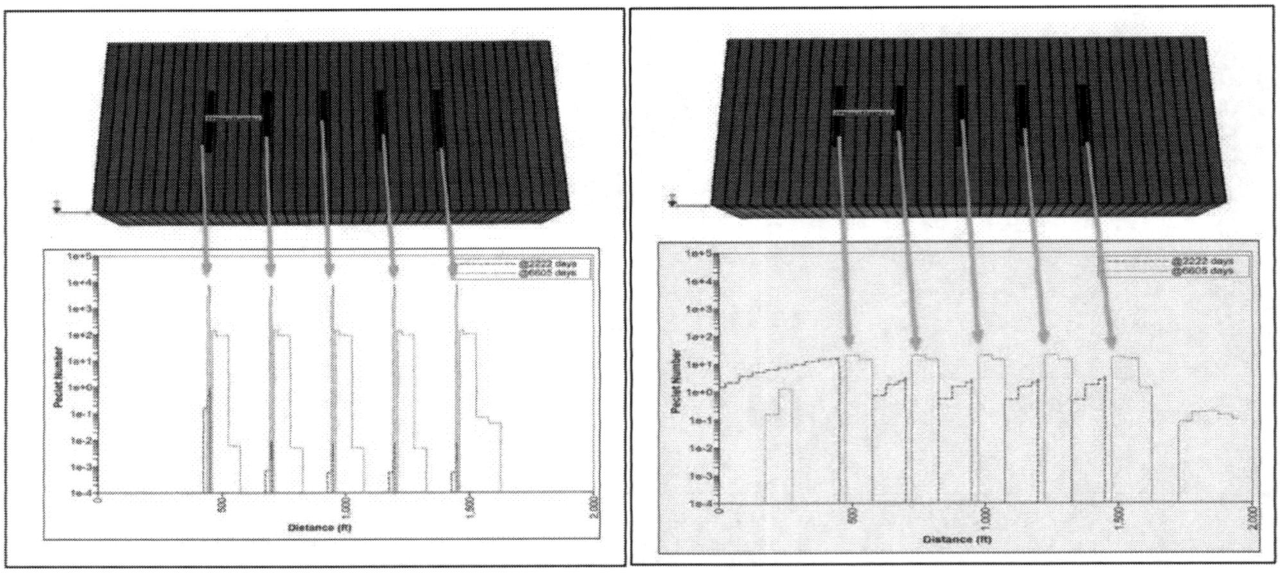

A-Gas Phase B-Oil Phase

**Figure 9: Péclet number distribution a long cross section in the matrix-model**

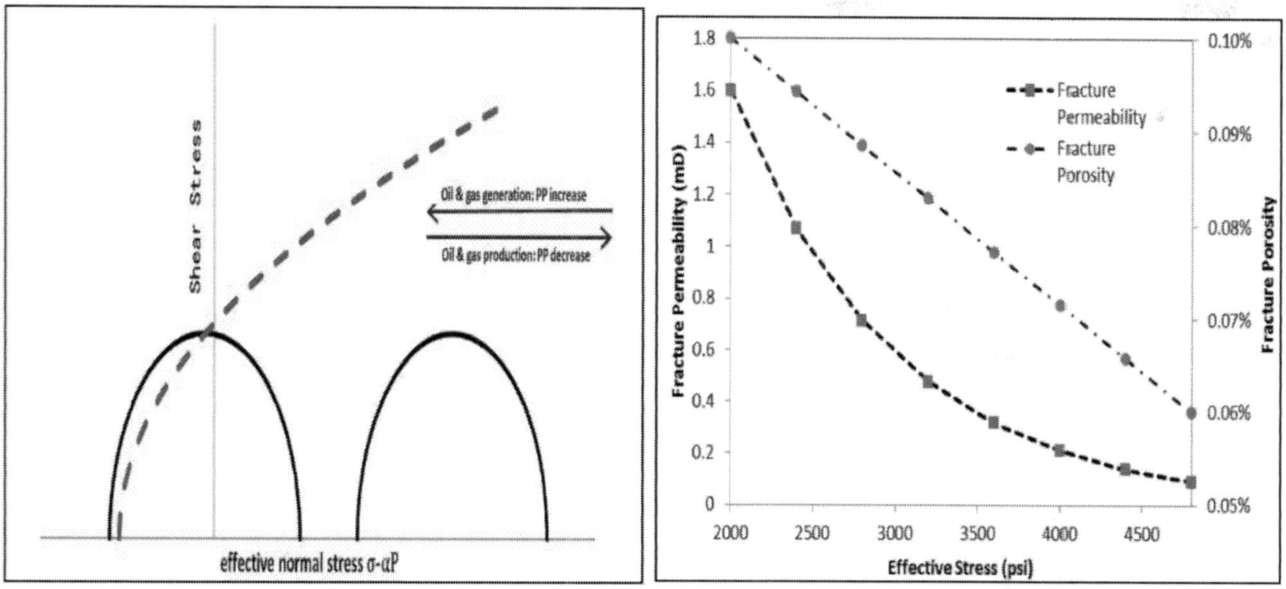


**Fig. 10: How and why permeability and porosity of natural fractures are changeable with time (Wang et al., 2015)**

To approve these speculations, we injected $CO_2$ CO2 into the same well of Bakken in two separated cases. In the first case, we injected $CO_2$ in the production well assuming there is no molecular-diffusion for $CO_2$ into formation-oil. However, in the second scenario, we injected $CO_2$ with a molecular-diffusion mechanism enabled. Two cycles of $CO_2$ Huff-n-Puff have been used for each case. The agenda and the time breakdown for both cases are shown in **Table 6**.

**Table 6: The agenda and time breakdown for both of two cases in CO2 Huff-n-Puff scenario**

| Scenario | Time |
|---|---|
| Primary depletion at BHP=1500 psi | 10 Years |
| CO2 Injection at rate of 500 Mscf/day (1st cycle) | 6 months |
| Soaking time | 3 months |
| Back for production | 4 years and 3 months |
| CO2 Injection at rate of 500 Mscf/day (2nd Cycle) | 6 months |
| Soaking | 3 months |
| Back for production | 4 years and 3 months |
| Total time for modeling | 20 years |

The results indicated that the $CO_2$ performance for without-molecular diffusion case did not provide a significant improvement in oil recovery or oil production rate from what was obtained at natural depletion production as shown in **Fig. 9**. If we look closely, we found that the enhancement in oil recovery due to $CO_2$ injection would not offset the loss in oil production which was happening due to the soaking and injection period. However, $CO_2$ with the molecular-diffusion case has improved the oil recovery and oil production in a significant way as shown in **Fig. 11**.

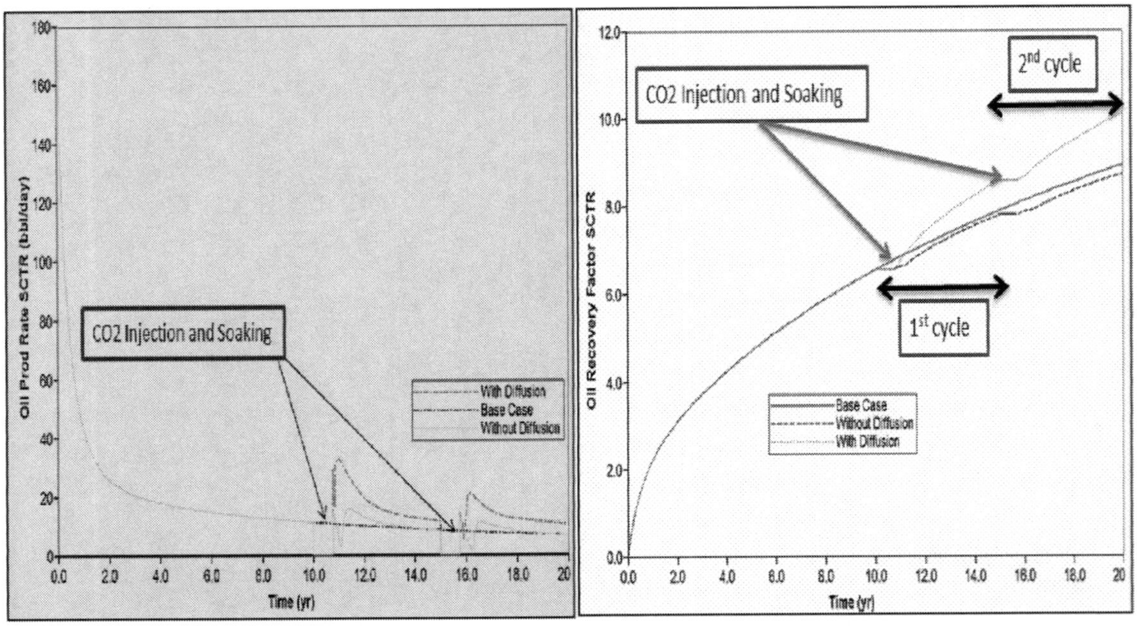


**Fig. 11: Effect of Molecular diffusion on CO2-EOR in Shale Oil reservoirs**

If we investigate the reasons beyond the role of molecular-diffusion mechanism on $CO_2$ performance to enhance oil recovery in these tight formations, we found this mechanism makes $CO_2$ penetrate deeper into the tight matrix, far away from hydraulic fractures as shown in **Fig. 12**. However, the case of $CO_2$ injection which does not have diffusion capacity makes the $CO_2$ penetrate just in the limited areas around the hydraulic fractures. Therefore, for the cases in which $CO_2$ penetrate deeper in the tight matrix, $CO_2$ would swell more volumes of oil, reduce oil viscosity, and finally produce larger quantities of oil by counter-current mechanism. On the other hand, the cases in which $CO_2$ has low molecular-diffusion rate would produce the injected-$CO_2$ back very soon. Therefore, producing the injected $CO_2$ back would put another hold on oil production due to slippage-effect making the enhancement in oil production for these types of reservoirs even worse. Since the diffusion mechanism for $CO_2$ has a significant effect to enhance oil recovery in shale oil reservoirs, any change in $CO_2$ diffusion rate would result in a clear change in oil recovery factor.

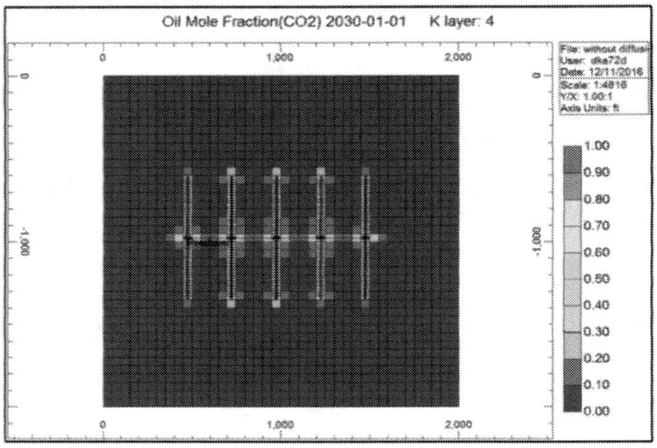


**Fig. 12a- CO2 Injection without-molecular diffusion**

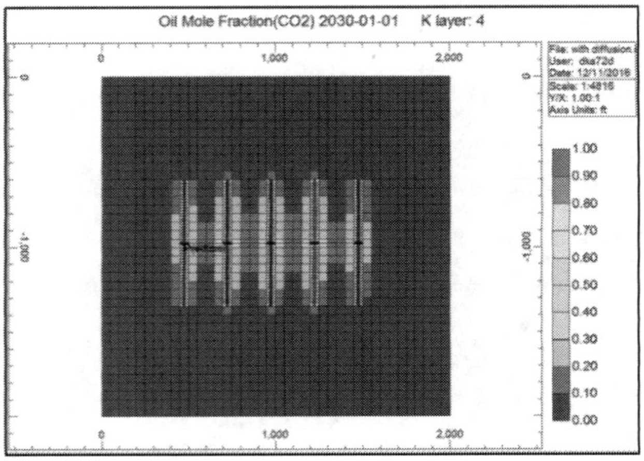


**Fig. 12b- CO2 Injection with Molecular Diffusion**

To investigate the effect of the CO2-EOR performing time on CO2 performance, we injected CO2 at a different time from the production-well life. In the first scenario, we injected CO2 after 5 years of production life. However, in the second scenario, we injected CO2 after 10 years from the production life as in the pilot-test case. The results confirmed the prediction which is that CO2 would perform better in the cases which have earlier CO2-EOR rather than the cases which have late CO2-EOR as shown in Fig. 13. This could be explained by the effective-stress principle which might be significantly important to control the permeability and porosity of natural fractures in shale oil reservoirs. As far as the CO2-EOR performed earlier, its performance would be better because the injected CO2 would find a good intensity of natural fracture which helps in enhancing its diffusivity into formation-oil. Another verification has been conducted to verify the effect of exposure time on CO2-molecular diffusion. This verification has been done by injecting CO2 in low-conductivity hydraulic fractures versus injecting CO2 in high-conductivity hydraulic fractures. The results indicated that CO2 would enhance oil recovery in low-conductivity fractures more than in high-conductivity fractures. The reason causing the difference in CO2-EOR performance according to the fracture conductivity is that the CO2 would be produced back in a faster way in high-conductivity fractures cases as shown in Fig. 14. The fast production for CO2 would downgrade the CO2 diffusivity into formation-oil, which in turn would reduce its performance to enhance production more oil. To sum up, as far as the kinetic of oil recovery process in productive areas does not overcome the CO2-diffusion rate, the CO2 would experience more exposure time with formation oil before its being produced back which is making CO2 successful to enhance oil recovery in these poor-quality reservoirs.

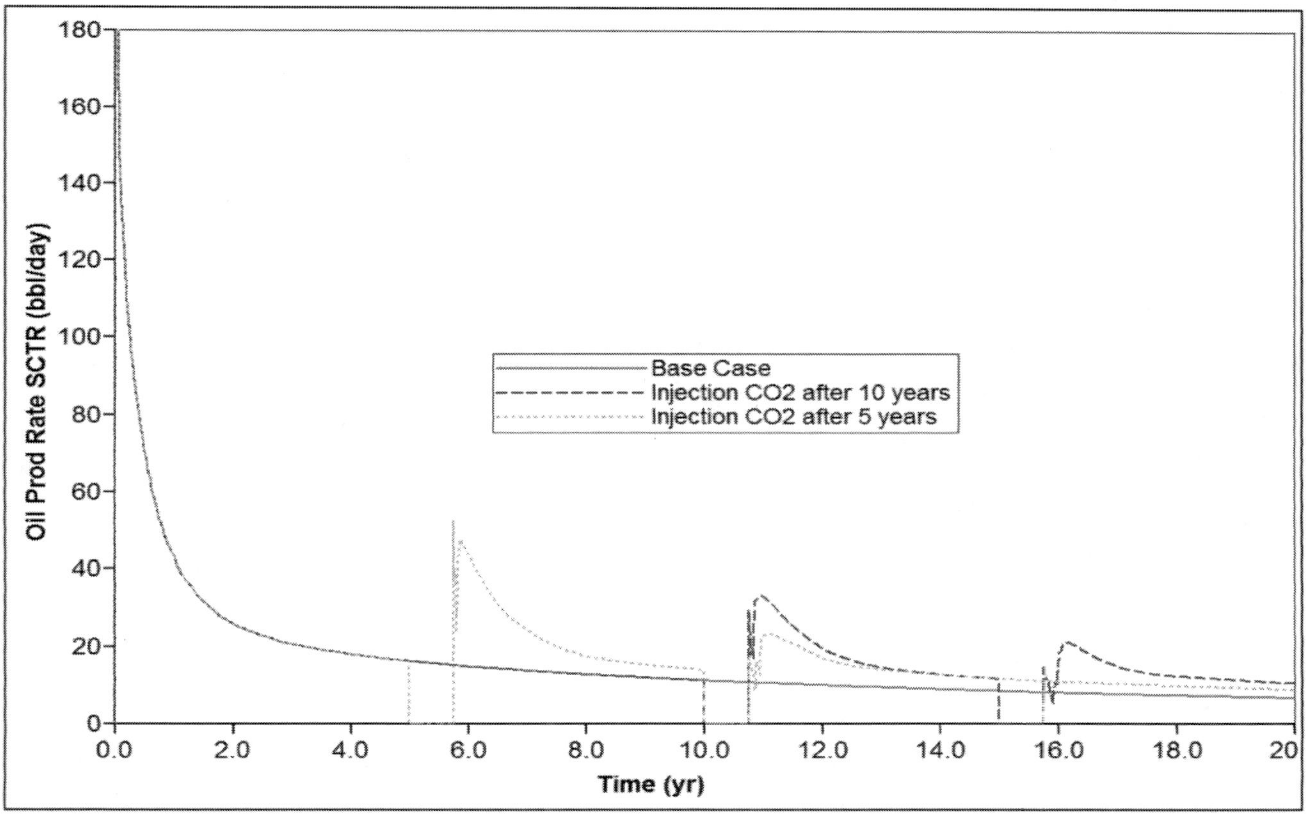


**Fig. 13: Effect of the performing time on CO2-EOR in shale oil reservoirs**

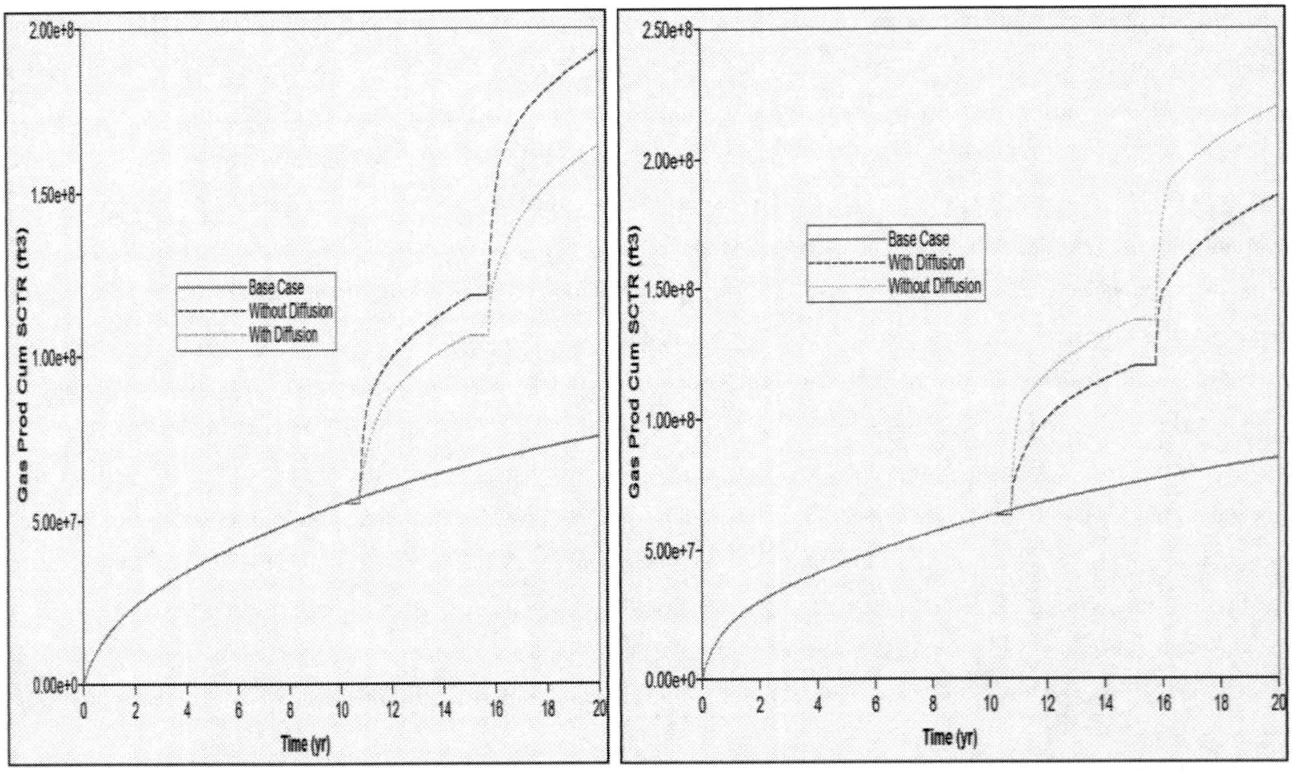


**5 md.ft** **15 md.ft**

**Fig. 14: Effect of fracture conductivity on $CO_2$-EOR performance**

## Conclusions

- Most of the previous experimental studies relied on $CO_2$ Molecular-diffusion mechanism to predict the potential success for $CO_2$-EOR in shale reservoirs.
- Upscaling this mechanism to the field scale via conventional simulation methods by using the same lab-obtained $CO_2$-diffusion rate is misleading.
- To be significant in the field scale, this mechanism requires having either of the kinetics for oil recovery process in the productive areas of these reservoirs to be too slow or the $CO_2$ diffusion rate in field conditions to be too fast.
- The results from the reported pilot-tests are matching with the low-diffusivity diagnostic plots.
- The intensity of natural fractures has the potential role for the successful CO2-EOR project. However, CO2-EOR projects need to be performed earlier to find opened natural-fractures which help in enhancing $CO_2$ performance in these complex reservoirs.

## References

Alfarge, D., Wei, M., and Bai, B., (2017). IOR Methods in Unconventional Reservoirs of North America: Comprehensive Review. SPE-185640-MS prepared for presentation at the SPE Western Regional Meeting held in Bakersfield, California, USA, 23-27 April 2017.

Alharthy, N., Teklu, T., Kazemi, H. et al. 2015. Enhanced Oil Recovery in LiquidRich Shale Reservoirs: Laboratory to Field. Society of Petroleum Engineers. DOI: 10.2118/175034MS.

Alvarez, J.O. and Schechter, D.S. 2016. Altering Wettability in Bakken Shale by Surfactant Additives and Potential of Improving Oil Recovery During Injection of Completion Fluids. Society of Petroleum Engineers. http:10.2118/SPE-179688-MS.

Clark, A. J., (2009). Determination of Recovery Factor in the Bakken Formation, Mountrail County, ND. Society of Petroleum Engineers. doi:10.2118/133719-STU.

Computer Modeling Group, GEM Manual. Https: www.CMG.Ca/ accessed 2017.

Dong, C., & Hoffman, B. T. (2013). Modeling Gas Injection into Shale Oil Reservoirs in the Sanish Field, North Dakota. Society of Petroleum Engineers. doi:10.1190/URTEC2013-185Energy Information Administration (EIA), (2015).

Gamadi, T. D., Sheng, J. J., Soliman, M. Y., Menouar, H., Watson, M. C., & Emadibaladehi, H. (2014). An Experimental Study of Cyclic CO2 Injection to Improve Shale Oil Recovery. Society of Petroleum Engineers. doi:10.2118/169142-MS

Hawthorne et al., (2017). Measured Crude Oil MMPs with Pure and Mixed CO2, Methane, and Ethane, and Their Relevance to Enhanced Oil Recovery from Middle Bakken and Bakken Shales. SPE-185072-MS paper was prepared for presentation at the SPE Unconventional Resources Conference held in Calgary, Alberta, Canada, 15-16 February 2017.

Hawthorne, S. B., Gorecki, C. D., Sorensen, J. A., Steadman, E. N., Harju, J. A., & Melzer, S. (2013). Hydrocarbon Mobilization Mechanisms from Upper, Middle, and Lower Bakken Reservoir Rocks Exposed to CO. Society of Petroleum Engineers.

Hoffman, B. T., and Evans J., (2016). Improved Oil Recovery IOR Pilot Projects in the Bakken Formation. SPE-180270-MS paper presented at the SPE Low Perm Symposium held in Denver, Colorado, USA, 5–6 May 2016.

Hoteit, H. and Firoozabadi, A. 2009. Numerical Modeling of Diffusion in Fractured Media for Gas-Injection and Recycling Schemes. SPE J. 14(02): 323–337. SPE-103292-PA. http://dx.doi.org/10.2118/103292-PA.

Jin, L., Sorensen, J.A., Hawthorne, S.B., Smith, S.A., Bosshart, N.W., Burton Kelly, M.E., Miller, D.J., Grabanski, C.B., and Harju, J.A., (2016). Improving oil transportability using CO2 in the Bakken System—a laboratorial investigation: Proceedings of the SPE International Conference & Exhibition on Formation Damage Control, SPE 178948, Lafayette, Louisiana, February 24–26.

Kathel, P. and Mohanty, K.K. 2013. EOR in Tight Oil Reservoirs through Wettability Alteration. Society of Petroleum Engineers. DOI: 10.2118/166281MS.

Kurtoglu, B., Kazemi, H., Rosen, R., Mickelson, W., & Kosanke, T. (2014). A Rock and Fluid Study of Middle Bakken Formation: Key to Enhanced Oil Recovery.Society of Petroleum Engineers. doi:10.2118/171668-MS.

LeFever, J., & Helms, L., 2008, Bakken Formation Reserve Estimates, North Dakota Geological Survey.

Mohebbinia et al., (2017). Molecular Diffusion Calculations in Simulation of Gasfloods in Fractured Reservoirs. SPE-182594-MS Paper presented at the SPE Reservoir Simulation Conference held in Montgomery, TX, USA, 20–22 February 2017.

Moortgat, J. and Firoozabadi, A. 2013. Fickian Diffusion in Discrete-Fractured Media from Chemical Potential Gradients and Comparison to Experiment. Energ Fuel27 (10): 5,793–5,805. http://dx.doi.org/10.1021/ef401141q.

Pu, H., and Li, Y., (2016). Novel Capillarity Quantification Method in IOR Process in Bakken Shale Oil Reservoirs. SPE-179533-MS presentated at the SPE Improved Oil Recovery Conference held in Tulsa, Oklahoma, USA, 11–13 April 2016

Schmidt, M., & Sekar, B. K. (2014). Innovative Unconventional 2EOR-A Light EOR an Unconventional TertiaryRecovery Approach to an Unconventional Bakken Reservoir in Southeast Saskatchewan.World Petroleum Congress.

Sheng, J.J. 2015. Enhanced oil recovery in shale reservoirs by gas injection. Journal of Natural Gas Science and Engineering 22: 252–259. http://dx.doi.org/10.1016/j.jngse.2014.12.002

Shoaib, S., & Hoffman, B. T. (2009). CO2 Flooding the Elm Coulee Field. Society of Petroleum Engineers. doi:10.2118/123176-MS.

Song, C., & Yang, D. (2013). Performance Evaluation of CO2 Huff-n-Puff Processes in Tight Oil Formations. Society of Petroleum Engineers. doi:10.2118/167217-MS.

Wan, T., & Sheng, J. (2015). Compositional Modelling of the Diffusion Effect on EOR Process in Fractured Shale-Oil Reservoirs by Gasflooding. Society of Petroleum Engineers. doi:10.2118/2014-1891403-PA.

Wang, D., Zhang, J., Butler, R., and Olatunji, K., (2016). Scaling Laboratory-Data Surfactant-Imbibition Rates to the Field in Fractured-

Shale Formations. Socienty of Petroleum Engineers. http://dx.doi.org/10.2118/178489-PA.

Wang et al., (2015). Geomechanics Coupling Simulation of Fracture Closure and Its Influence on Gas Production in Shale Gas Reservoirs. SPE-173222-MS paper presented at the SPE Reservoir Simulation Symposium held in Houston, Texas, USA, 23–25 February 2015.

Wang, X., Luo, P., Er, V., & Huang, S.-S. S. (2010). Assessment of CO2 Flooding Potential for Bakken Formation, Saskatchewan. Society of Petroleum Engineers. doi: 10.2118/137728-MS.

Wood, T., & Milne, B., (2011). Waterflood potential could unlock billions of barrels: Crescent Point Energy. http://www.investorvillage.com/uploads/44821/files/CPGdundee.pdf

Xu, T., & Hoffman, T. (2013). Hydraulic Fracture Orientation for Miscible Gas Injection EOR in Unconventional OilReservoirs. Society of Petroleum Engineers. doi:10.1190/URTEC2013-189

Yu et al., (2016b). Experimental Evaluation of Shale Oil Recovery from Eagle Ford Core Samples by Nitrogen Gas Flooding. SPE-179547-MS Paper presented at the SPE Improved Oil Recovery Conference held in Tulsa, Oklahoma, USA, 11–13 April 2016.

Yu, W., Lashgari, H., & Sepehrnoori, K. (2014). Simulation Study of CO2 Huff-n-Puff Process in Bakken Tight OilReservoirs. Society of Petroleum Engineers. doi:10.2118/169575-MS

Yu, Y., Li, L. and Sheng J., (2016a). Further Discuss the Roles of Soaking Time and Pressure Depletion Rate in Gas Huff-n-Puff Process in Fractured Liquid-Rich Shale Reservoirs. SPE-181471-MS paper presented in at the SPE Annual Technical Conference and Exhibition held in Dubai, UAE, 26-28 September 2016.

Zhang et al., (2015). Intergrated Method to Screen Tight Oil Reservoirs for CO2 Flooding. SPE-175969-MS paper presented at the SPE/CSUR Unconventional Resources Conference held in Calgary, Alberta, Canada, 20-22 October 2015.

Zhang, K., (2016). Experimental and Numerical Investigation of Oil Recovery from Bakken Formation by Miscible CO2 Injection. Paper SPE 184486 presented at the SPE international Student Paper Contest at the SPE Annual Technical Conference and Exhibition held in Dubai, UAE, 26-28 September 2016.

Zhu, P., Balhoff, M. T., & Mohanty, K. K. (2015). Simulation of Fracture-to-Fracture Gas Injection in an Oil-Rich Shale. Society of Petroleum Engineers. doi:10.2118/175131-MSSPE-180270-MS17

CMTC-502823-MS

# Performance Control Methods Applied in Heterogeneous Reservoirs for Improving $CO_2$ flooding Performance

Fenglan Zhao, Hongda Hao, Zhixing Wang, Jirui Hou, Zhongfeng Fu, Wenfeng Li, Peng Wang, Meng Zhang, and Guoyong Lu, China University of Petroleum (Beijing); and Guangzhong Lv, Institute of Exploration and Development, Shengli Oil Field, SINOPEC



This paper was prepared for presentation at the Carbon Management Technology Conference held in Houston, Texas, USA, 17-20 July 2017.



## Abstract

$CO_2$-EOR is an effective technology for reducing $CO_2$ emissions while enhancing oil recovery in ultra-low permeability reservoir, which has been performed in Shengli Oil Field, China since 2013 with cumulative $CO_2$ injection of 12588 t by 2016. However, the area heterogeneity of reservoir resulted in serious gas channeling and poor production performance. Performance control methods including sweeping area regulation, differential pressure control and real-time producing regulation were proposed to enlarge sweeping area and improve $CO_2$ utilization in areal heterogenous reservoir. 3D physical models of areal heterogeneity and five-spot pattern were utilized in the laboratory. Conventional $CO_2$ flooding, sweeping area regulation, differential production pressure control and real-time producing regulation were conducted respectively in the 3D models, and the flooding efficiency was evaluated through oil recovery increments and changes of performance curves. Corescale numerical modeling was also built to study the profile improvements of $CO_2$ flooding by the performance control methods.

Experimental and numerical simulation results showed that $CO_2$ was displaced unevenly in the areal heterogeneous reservoir, leaving plenty of oil remained in the relatively high and relatively low permeability area. The oil recovery of $CO_2$ flooding in areal heterogenous reservoir can be doubled by performance control methods of sweeping area regulation, differential pressure control or real-time producing regulation. The remaining oil in relatively low permeability area can be effectively displaced by sweeping area regulation, while both larger sweeping area and better $CO_2$ flooding profile can be achieved by differential pressure control and real-time producing regulation. Higher productivity of individual well can be obtained in the early and middle stage of $CO_2$ flooding by differential pressure control, while similar oil & gas production performance and longer displacement period of $CO_2$ injection can be achieved by real-time producing regulation. The performance improvement of $CO_2$ flooding by performance control methods provided a feasible technical strategy for enhancing oil recovery of areal heterogeneous reservoir in the oil field under the condition of a lower oil price.

**Keywords:** low permeability reservoir; $CO_2$ flooding; areal heterogeneity; performance control; sweeping area

## 1 Introduction

$CO_2$ flooding is considered to be a promising and effective process for enhancing oil recovery ($CO_2$-EOR) because it can not only increase oil production, but also reduce greenhouse gas emissions by sequestrating $CO_2$ in reservoirs. Field applications present $CO_2$ flooding can enhance oil recovery by 8-15% (Pyo et al., 2003; Bachu and Shaw, 2004). However, when $CO_2$ was injected into the formation, flooding performance was commonly affected by reservoir heterogeneity, and left large amount of oil remained in the reservoirs. Therefore, it is of great importance to improve $CO_2$ flooding performance in the heterogeneous reservoirs. Previous studies have been conducted for the purpose of maximizing the sweep efficiency of $CO_2$ flooding with different methods. Malik and Islam (2000) have reported that horizontal injection wells have proved to be efficient for $CO_2$ flooding processes to improve oil recovery and increase the $CO_2$ storage in the Weyburn Field of Canada. Shayeqi et al. (1996) have shown that $CO_2$ injection followed by $N_2$ slug could enhance oil sweep efficiency and oil recovery in the core flooding experiments. $CO_2$ water alternating gas (WAG) flooding is considered to be an effective method to improve sweep efficiency in both laboratory experiments and field pilot tests (Christensen et al.,1998; Heeremans et al., 2006; Elfeel et al., 2013a). The injected gas can enhance microscopic displacement efficiency, and the injected water can enhance the macroscopic sweep efficiency. However, the properties of formation fluids and reservoir conditions limit the wide application of WAG flooding (Dijke et al.,2006; Elfeel et al., 2013b). The modified WAGs, such as surfactant alternating gas (SAG) flooding and foam assisted water alternating gas (FAWAG) flooding, have been purposed to expand gas production time during $CO_2$ flooding (Spirov et al.,2012; Salehi et al.,2014). The utilize of foam can further reduce gas mobility, and then increase sweep efficiency and alleviate the effect of heterogeneity during $CO_2$ injection, however, the foam stability is seriously affect with the presence of crude oil (Falls et al.,1989; Farajzadeh et al., 2010). Recently, Telmadarreie et al. (2016) has reported that microbubbles of $CO_2$ with high performance injectivity demonstrated a great potential in enhancing oil recovery in heterogeneous porous media. Zhao et al. (2015) has proposed a profile control method by the reaction products between ethylenediamine and $CO_2$, and Hao et al. (2016) proposed a two-stage gas channeling control method by high-strength gel and ethylenediamine to improve $CO_2$ flooding efficiency in ultra-low permeability fractured reservoirs.

As described above, the displacement efficiency can be greatly improved with the utilize of chemical agents during $CO_2$ flooding, while economic and effective methods for enhancing oil recovery still need to be evolved under the condition of a lower oil price. $CO_2$ flooding has been performed in Shengli Oil Field, China since 2013 with cumulative $CO_2$ injection of 12588t by 2016. After serious gas channeling and poor production performance, performance control methods including sweeping area regulation, differential production pressure control and real-time producing regulation are introduced to improve the performance of $CO_2$ flooding in areal heterogeneous reservoir through 3D experimental and corescale numerical simulations.

## 2 Experiments

### 2.1 Materials

The oil sample and formation water were collected from one block of Shengli Oil Field, China. The density of the formation oil is 0.7463 g/cm$^3$, the viscosity is 1.20 mPa·s, and the gas/oil ratio is 45.6 m$^3$/m$^3$. The oil sample mainly contains light or medium components, and the minimum miscible pressure (MMP) measured by slim tube test is 31.6 MPa under the formation conditions (44.87 MPa, 60 °C). The salinity of the formation water is 29884 mg/L and the water type is $CaCl_2$. The injected $CO_2$ with purity of 99.99 mol% is from Beijing, China.

The 3D heterogeneous models used in the experiments were artificial fabricated by quartz sand, clay and proxy resin. The diameter of the physical model is 400mm, and the thickness is 45mm (As shown in Fig.1). The model was areal heterogeneous with relatively high permeability of $50\times10^{-3}\mu m^2$, and relatively low permeability of $10\times10^{-3}\mu m^2$. An injection-production unit of 5-spot pattern was designed in the 3D physical model. $CO_2$ injection well (I-1) was located in the center of the model, three production wells (P-1, P-2 and P-3) in relatively high permeability area, and one production well (P-4) in relatively low permeability area. The well spacing between injector and producer was 190mm. Fig.2 is the coreholder of 3D physical model (Jiangsu, China) with maximum operation pressure of 15MPa.

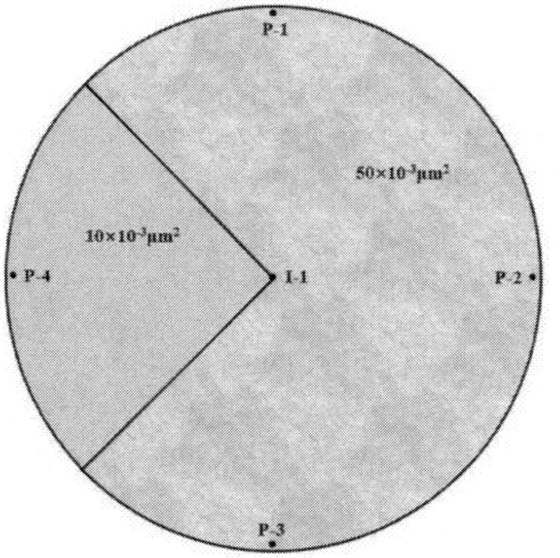


**Fig.1 Picture of 3D physical model with areal heterogeneity**

**Fig.2 Corehorder of 3D physical model**

## 2.2 Experimental setup

The experimental setup consisted of five sub-systems: injection system, displacement system, production system, temperature control system and data acquisition system (As shown in Fig.3). In the injection system, the formation water, formation oil and $CO_2$ were stored in transfer cylinders and then injected into the core sample by a constant pressure and rate pump (HAS-100HSB, Jiangsu, China). The injection rate of $CO_2$ was controlled by a gas flow meter (D07-11C, Beijing, China). 3D heterogeneous models were placed in the 3D radial flow coreholder in the displacement system. In the production system, backpressure regulators (BPRs) were used to control the production pressure. The produced oil was recorded by graduated tubes, and the gas was measured by gas flow meters (CS200, Beijing, China). The thermostat was used to set the experimental temperature as the reservoir temperature, and the displacement pressure was recorded by the data acquisition system.

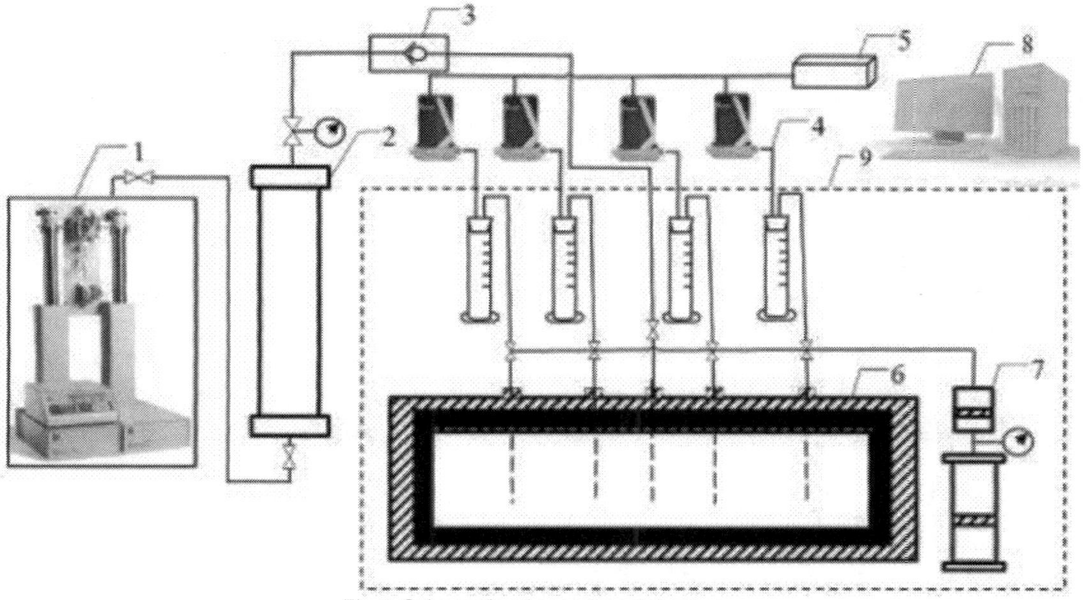


**Fig.3 Schematic of experimental setup.**

1-Constant pressure and rate pump; 2-Transfer cylinder; 3-One-way valve; 4-Gas-oil separators; 5-Gas flow meters; 6-Radial flow coreholder; 7-Backpressure regulators (BPRs); 8-Data acquisition system; 9-Thermostat.

### 2.3 Experimental procedure

Several 3D displacement experiments were designed to improve the performance of $CO_2$ flooding by performance control methods in the laboratory, and the experimental procedures are detailed as follows:

First, the 3D physical model and the wells were fabricated and pre-located according to the designing. Then, epoxy resins were coated on the surface of the core to avoid corruption of $CO_2$. The bulk volume of the core was measured before displacement experiment. The core was firstly evacuated and then saturated with formation water. The porosity was determined as the ratio of brine saturation volume to the bulk volume. The permeability of the core was measured by changing water injection rate from 0.1 to 0.3 mL/min. Subsequently, the core was displaced by formation oil to reach the residual water saturation condition. The initial oil saturation was calculated as the ratio of saturated oil volume to the pore volume.

In conventional $CO_2$ flooding experiment, gas was injected into 3D physical model without any performance control method during the whole experimental process. The detailed sequence is explained as follows. ① The 3D physical model with areal heterogeneity was placed into 3D coreholder, and the formation temperature was set as 60°C. ② Same production pressure of P-1, P-2, P-3 and P-4 was set as 8 MPa by BPRs to maintain the formation pressure and the supercritical condition of $CO_2$. ③ Conventional $CO_2$ flooding was conducted by injecting $CO_2$ through I-1, and the injection rate was set as 1.5 mL/min. All the producers with the same backpressure were opened simultaneously. ④ The experiment was terminated when the producing gas/oil ratio (GOR) reached 3000 $m^3/m^3$. The displacement pressure, oil production and gas production were measured during the experiment.

In sweeping area regulation experiment, relatively low permeability area was designed to be firstly displaced to enlarge $CO_2$ sweeping area. The detailed is explained as follows. ① The 3D physical model was placed into 3D coreholder. The formation temperature was set as 60°C, and same production pressure of P-1, P-2, P-3 and P-4 was set as 8 MPa. ② $CO_2$ was injected through I-1 with the injection rate of 1.5 mL/min. P-4 located in relatively low permeability area was firstly opened, while P-1, P-2, and P-3 located in relatively high permeability area were closed. ③ As the displacement of $CO_2$ flooding, P-1, P-2, and P-3 were then opened when the producing GOR of P-4 reached 1000 $m^3/m^3$. ④ The experiment was terminated when the producing GOR reached 3000 $m^3/m^3$. The displacement pressure, oil production and gas production were measured during the experiment.

In differential pressure control experiment, different production pressure was set by BPRs in different permeability area. The detailed sequence is explained as follows. ① The 3D physical model was placed into 3D coreholder, and the formation temperature was set as 60°C. ② The BPR pressures of P-1, P-2 and P-3 were set as 8.2, 8.3 and 8.2 MPa, and P-4 was set as 8 MPa. ③ $CO_2$ was injected through I-1 with the injection rate of 1.5 mL/min, and all the producers were opened simultaneously. ④ The experiment was terminated when the producing GOR reached 3000 $m^3/m^3$. The displacement pressure, oil production and gas production were measured during the experiment.

In real-time producing regulation experiment, the production pressure was real-time regulated by BPRs to achieve similar oil and gas production performance during $CO_2$ flooding. The detailed sequence is explained as follows. ① The 3D physical model was placed into 3D coreholder, and the formation temperature was set as 60°C. ② The BPR pressures of P-1, P-2 and P-3 were initially set as 8.2, 8.3 and 8.2 MPa, and P-4 was initially set as 8 MPa. ③ $CO_2$ was injected through I-1 with the injection rate of 1.5 mL/min, and all the producers were opened simultaneously. ④ When $CO_2$ was breakthrough in the producers, the production pressure was real-time regulated by changing BPRs to achieve similar gas and oil production performance. ⑤ The experiment was terminated when the producing GOR reached 3000 $m^3/m^3$. The displacement pressure, oil production and gas production were measured during the experiment.

# 3 Results and discussion

## 3.1 Oil recovery increment by $CO_2$ performance control flooding

$CO_2$ flooding efficiency was seriously affected by the heterogeneity of reservoir. In order to improve the flooding efficiency and enhance oil recovery of $CO_2$ flooding, performance control methods including sweeping area regulation, differential production pressure control and real-time producing regulation were studied in the experiments to improve the performance of $CO_2$ flooding in areal heterogeneous reservoir. The experimental results of performance control methods for $CO_2$ flooding are shown in Table 1. The oil recovery of conventional $CO_2$ flooding without performance control was 25.85%, while the recovery of $CO_2$ flooding with performance control methods by sweeping area regulation, differential production pressure control and real-time producing regulation were 42.50%、45.56% and 49.25%, respectively. The oil recovered by performance control methods was almost twice as the oil recovered without performance control during $CO_2$ flooding.

**Table 1 Experimental results of $CO_2$ performance control flooding**

| | Producer No. | Permeability ($\times10^{-3}\mu m^2$) | Performance control methods | | | |
|---|---|---|---|---|---|---|
| | | | Conventional $CO_2$ flooding | Sweeping area regulation | Differential pressure control | Real-time producing regulation |
| Oil recovery | P-1 | 50 | 6.13 | 5.98 | 12.46 | 11.99 |
| | P-2 | | 12.18 | 10.00 | 13.52 | 13.39 |
| | P-3 | | 5.52 | 7.75 | 11.22 | 11.58 |
| | P-4 | 10 | 2.02 | 18.77 | 9.22 | 12.29 |
| | Total | / | 25.85 | 42.50 | 46.42 | 49.25 |

Oil recovery of each individual well in production control flooding experiments is shown in Fig.4. In conventional $CO_2$ flooding without production control treatment, the oil recovery was mostly achieved by the producers of P-1, P-2, and P-3 located in the relatively high permeability area, especially P-2, and the oil recovery of P-4 located in relatively low permeability area was only 2.02% after producing GOR reached $3000m^3/m^3$. $CO_2$ was mostly injected into the relatively high permeability area, especially along P-2 direction, and then left plenty of crude oil remained in both relatively high permeability area and relatively low permeability area.

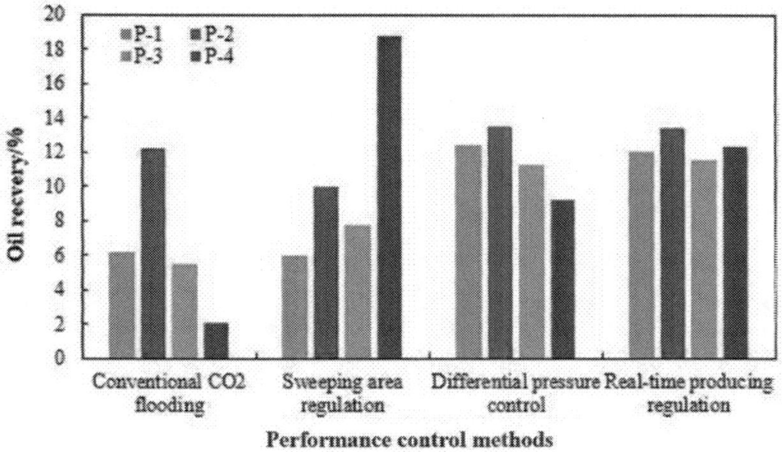


**Fig.4 Oil recovery of individual well in $CO_2$ performance control flooding**

In order to effectively displace the oil in relatively low permeability area, sweeping area regulation experiments was designed to firstly displace relatively low permeability area, and then the relatively high permeability area. With the sweeping area regulation, more than 16.65% OOIP was recovered compared of the conventional $CO_2$ flooding. Through sweeping area regulation, the relatively low permeability area was effectively swept, and the oil recovery of P-4 located in relatively low permeability area contributed significantly to the total recovery of the injection-production unit. Comparatively, differential pressure control and real-time producing regulation achieved higher total oil recovery. The oil produced by each individual well was more balanced, which means that the displacement front of injected $CO_2$ was more even in the 5-spot injection-production unit. The production control treatments of differential pressure control and real-time producing regulation can not only enlarge $CO_2$ sweeping area, but also achieve better flooding profile of $CO_2$ in areal heterogeneous reservoir.

### 3.2 Performance analysis of $CO_2$ performance control flooding

The performance of $CO_2$ flooding in the injection-production unit was further analyzed to discuss the mechanisms of enhancing oil recovery by performance control methods for $CO_2$ flooding in areal heterogeneous reservoir. Cumulative oil production of each individual well in $CO_2$ performance control flooding is shown in Fig.5. In conventional $CO_2$ flooding as shown in Fig.5(a), the oil production of each individual well showed great difference without performance control methods. P-2 located in relatively high permeability area with the best petrophysical condition achieved the highest oil production, while the oil productions of P-1 and P-3 were similar which were both affected by the relatively low permeability area. However, lowest productivity of P-4 lead to the lowest oil production with little $CO_2$ injected into the relatively low permeability area.

In sweeping area regulation experiment as shown in Fig.5(b), P-4 achieved the highest oil production because $CO_2$ was firstly injected into relatively low permeability area. However, fluid diversion occurred immediately when P-1, P-2 and P-3 opened. $CO_2$ was then mostly injected into relatively high permeability area, leading to the very low productivity of P-4 again. Although the relatively low permeability area can be effectively swept by sweeping area regulation, the oil production of individual well still remained uneven in the injection-production unit during $CO_2$ flooding process.

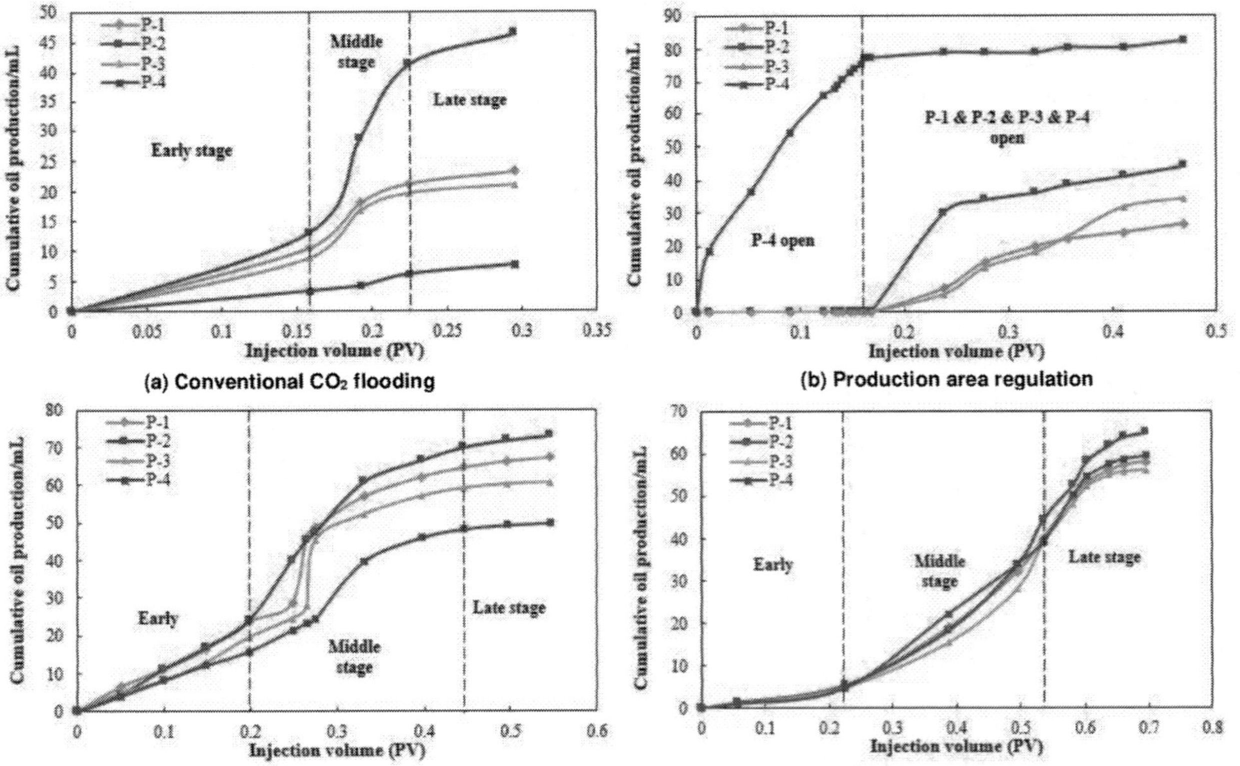


(c) Differential pressure control (d) Real-time producing regulation

Fig.5 Cumulative oil production of individual well in $CO_2$ performance control flooding

Comparatively, similar productivity of individual well can be achieved by production pressure control treatments and real-time producing regulation during $CO_2$ flooding in areal heterogenous model as shown in Fig.5(c) and Fig.5(d). It is worth notice that $CO_2$ flooding process can be subdivide into three stages: early stage with no gas producing, middle stage with gas and oil producing simultaneously, and late stage with plenty of gas breakthrough in the producers. In convetional $CO_2$ flooding, cumulative oil production of individual well increased greatly in the middle stage of the process, and the middle stage can be effectively prolonged by differential pressure control. However, the cumulative oil production increased greatly in the late stage by real-time producing regulation. In other words, although the individual-well productivity of real-time producing regulation was less than the productivity of differential pressure control in the early and middle stage of $CO_2$ flooding, highest oil recovery could still be achieved with more oil produced from the late stage of real-time producing regulation.

The displacement period of $CO_2$ flooding can also be significantly prolonged by performance control methods. Table 2 is the gas production performance of each individual well in performance control experiments. The total $CO_2$ injection volume of conventional $CO_2$ flooding, sweeping area regulation, production pressure control, and real-time producing regulation was 0.28, 0.47, 0.55 and 0.69 PV, respectively. In conventional $CO_2$ flooding, $CO_2$ breakthrough with injectional volume of 0.16 PV, and channeled with injectional volume of 0.28 PV when production GOR reached 3000 $m^3/m^3$. And no gas was produced through P-4 located in relatively low permeability area. In sweeping area regulation, $CO_2$ breakthrough firstly in P-4 with injection volume of 0.11 PV, then breakthrough in P-1, P-2 and P-3 with injection volume of 0.21 PV. And when $CO_2$ channeled in P-1, P-2 and P-3, no gas was produced through P-4 with the gas diversion to relatively high permeability area. In differential pressure control and real-time producing regulation, $CO_2$ was produced almost the same time both in the relatively high permeability area and the relatively low permeability area, which also reflect the more even displacement by the differential pressure control and real-time producing regulation.

**Table 2 Gas production of individual well in $CO_2$ performance control flooding**

| | Performance control methods | Relatively higher permeability area | | | Relatively lower permeability area |
|---|---|---|---|---|---|
| | | P-1 | P-2 | P-3 | P-4 |
| Time of $CO_2$ breakthrough /PV | Conventional $CO_2$ flooding | 0.16 | 0.15 | 0.16 | - |
| | Sweeping area regulation | 0.21 | 0.20 | 0.21 | 0.11 |
| | Differential pressure control | 0.19 | 0.19 | 0.19 | 0.20 |
| | Real-time producing regulation | 0.22 | 0.22 | 0.22 | 0.22 |
| Time of producing GOR>3000$m^3/m^3$ /PV | Conventional $CO_2$ flooding | 0.28 | 0.27 | 0.28 | - |
| | Sweeping area regulation | 0.47 | 0.46 | 0.47 | - |
| | Differential pressure control | 0.55 | 0.55 | 0.55 | 0.55 |
| | Real-time producing regulation | 0.69 | 0.69 | 0.69 | 0.69 |

## 3.3 Corescale numerical simulation of $CO_2$ performance control flooding

Although the mechanisms of enhancing oil recovery by performance control methods for $CO_2$ flooding in areal heterogeneous reservoir were analyzed above, $CO_2$ displacement front cannot be observed directly in the laboratory experiments. In order to further study the profile improvements of $CO_2$ flooding by the performance control methods, corescale numerical modeling was built using CMG-GEM Simulator. The diameter of the model was 400mm, and the thickness was 45mm, which was the same as the 3D physical model. An injection-production unit of 5-spot pattern was also designed in the numerical model, with P-1, P-2, P-3 located in relatively high permeability area, and P-4 located in relatively low permeability area. The well spacing between injector and producer is 190mm. The oil viscosity is 1.20 mPa·s, and the initial oil saturation is 0.50. Other parameters of the corescale numerical model are summarized in Table 3.

**Table 3 Parameters of the corescale numerical model**

| Parameter | Value | Parameter | Value |
|---|---|---|---|
| Reservoir depth (m) | 3189.65 | Average porosity | 0.18 |
| WOC (m) | 3189.675 | *K* of higher permeability area ($\times10^{-3}\mu m^2$) | 50 |
| Reservoir temperature (K) | 333.15 | *K* of lower permeability area($\times10^{-3}\mu m^2$) | 10 |
| Initial formation pressure (psi) | 1160.30 | Permeability ratio $V_k$ | 5 |

Similar to the 3D laboratory experiments, conventional $CO_2$ flooding, sweeping area regulation, differential pressure control and real-time producing regulation were also conducted respectively in the corescale numerical models. The gas saturation distributions of $CO_2$ performance control simulations are shown in Fig.6. Fig.6 visually impressed the flooding profile of $CO_2$ in the areal heterogeneous model, which was in accordance with the 3D experimental results. In conventional $CO_2$ flooding without performance control methods, the injected gas mostly displaced along relatively high permeability area, and only part of relatively low permeability area near the injection well (I-0) was swept when the producing GOR reached 3000m$^3$/m$^3$ (as shown in Fig.6(a)). In sweeping area regulation, both the relatively high permeability area and the relatively low permeability area were swept by $CO_2$. The relatively low permeability area was firstly swept when P-4 was opened, and the relatively high permeability area was then swept after P-1, P-2 and P-3 opened (as shown in Fig.6(b)). Compared with conventional $CO_2$ flooding and sweeping area regulation, the relatively high permeability area and relatively low permeability area can be swept simultaneously by differential pressure control and real-time producing regulation. With differential pressure control and real-time producing regulation, the displacement front of $CO_2$ tended to be more even, and larger sweeping area can be achieved when the producing GOR reached 3000 m$^3$/m$^3$ (as shown in Fig.6(c) and Fig.6(d)).

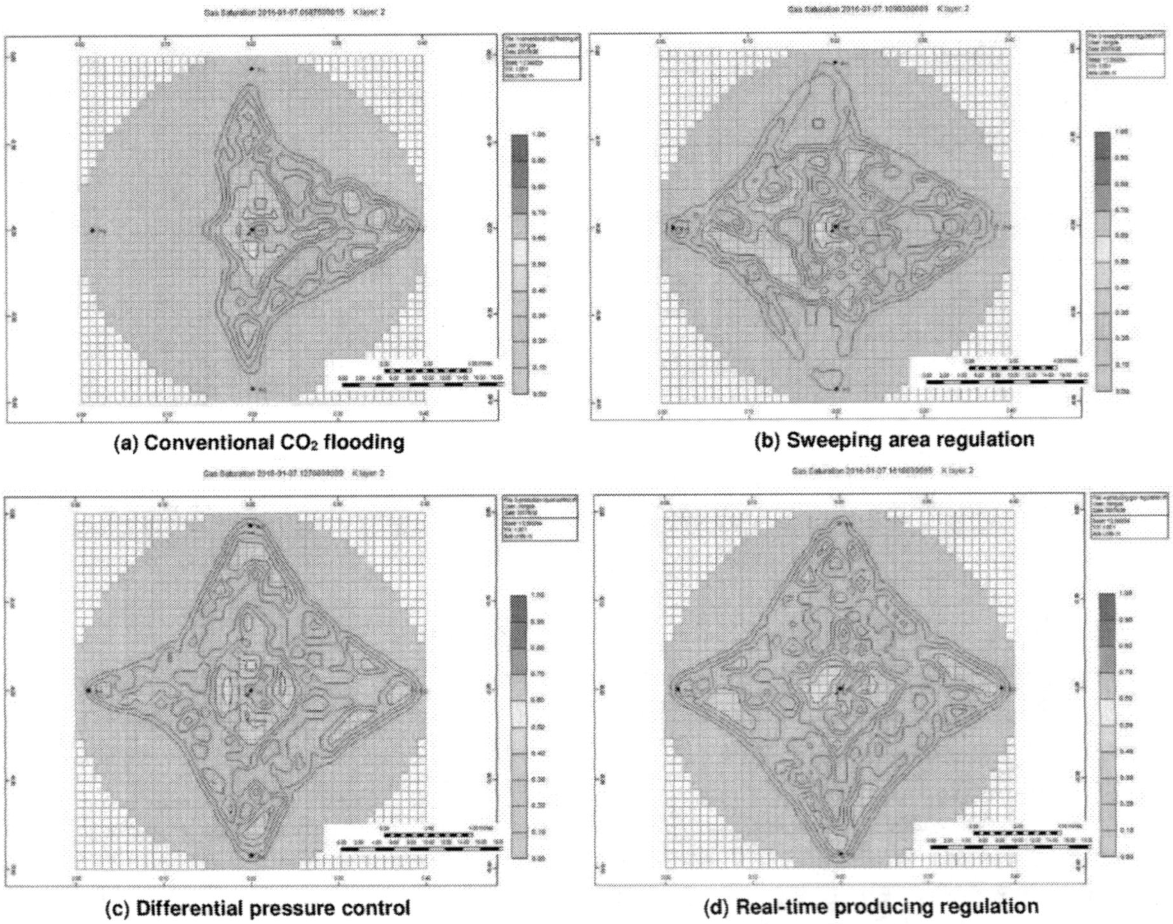

(a) Conventional $CO_2$ flooding (b) Sweeping area regulation

(c) Differential pressure control (d) Real-time producing regulation

**Fig.6 Gas saturation distributions of $CO_2$ performance control simulations**

The numerical results of $CO_2$ performance control flooding were consistent with the 3D laboratory experiments. As a summarize, $CO_2$ flooding performance of areal heterogeneous reservoir can be improved by simply performance control methods such as sweeping area regulation, differential pressure control or real-time producing regulation, which provided a feasible technical strategy for enhanced oil recovery of the well group in the oil field under the condition of a lower oil price.

## 4 Conclusion

The oil recovery of areal heterogeneous reservoir could be significantly enhanced by $CO_2$ perofrmance control flooding. Through the experimental and numerical simulation of performance control methods, some conclusions can be summarized as follows.

(1) $CO_2$ was displaced unevenly in the areal heterogeneous reservoir, leaving plenty of oil remained in the relatively high and relatively low permeability area. Through performance control methods of sweeping area regulation, differential pressure control or real-time producing regulation, the oil recovery can be doubled during $CO_2$ flooding process.

(2) The relatively low permeability area can be effectively displaced by prerformance control method of sweeping area regulation, which contributed significantly to the total oil recovery of the injection-production unit.

(3) Through performance control methods of differential pressure control and real-time producing regulation, not only the sweeping area can be effectively enlarged, but also $CO_2$ profile improvement can be achieved.

(4) Through performance control method of differential pressure control, higher productivity of individual well can be obtained in the early and middle stage of $CO_2$ flooding. While, similar oil & gas production performance and longer displacement period of $CO_2$ injection can be achieved by real-time producing regulation.

(5) The performance improvement of $CO_2$ flooding by performance control methods provided a feasible technical strategy for enhancing oil recovery of areal heterogeneous reservoir in the oil field under the condition of a lower oil price.

## Acknowledgements

The project is supported by *China National Major Technology Project (2017ZX05009-004)*. The authors want to acknowledge all the colleagues of *Institute of Enhanced Oil Recovery, China University of Petroleum (Beijing), China* and the support of *Institute of Exploration and Development, Shengli Oil Field, SINOPEC, China.*

## References

Bachu S., Shaw J., 2004. Estimation of oil recovery and $CO_2$ storage capacity in $CO_2$ EOR incorporating the effect of underlying aquifers. In: SPE/DOE Symposium on Improved Oil Recovery, Tulsa, Oklahoma, USA.

Christensen J., Stenb E., Skauge A.,1998. Review of WAG field experience. In: International Petroleum Conference and Exhibition of Mexico, Villahermosa, Mexico.

Dijke M., Sorbie K., Sohrabi M., et al., 2006. Simulation of WAG floods in an oil-wet micro model using a 2-D pore-scale network model. Journal of Petroleum Science and Engineering, 52, 71–86.

Elfeel M., Al-Dhahli A., Geiger S., et al., 2013a. Multi-Scale simulation of WAG flooding in naturally fractured reservoirs. In: EAGE Annual Conference & Exhibition incorporating SPE Europec, London, UK.

Elfeel M., Al-Dhahli A., Jiang Z., et al., 2013b. Effect of rock and wettability heterogeneity on the efficiency of WAG flooding in carbonate reservoirs. In: SPE Reservoir Characterization and Simulation Conference and Exhibition, Abu Dhabi, UAE.

Falls A., Muster, J., Ratulowski J., 1989. The apparent viscosity of foams in homogeneous bead packs. SPE Reservoir Engineering, 4 (02), 155-164.

Farajzadeh R., Andrianov A., Zitha P., 2010. Investigation of immiscible and miscible foam for enhancing oi recovery. Industrial & Engineering Chemistry Research, 49 (4), 1910-1919.

Hao H., Hou J., Zhao F., et al. 2016. Gas channeling control during $CO_2$ immiscible flooding in 3D radial flow model with complex fracture sand heterogeneity. Journal of Petroleum Science and Engineering, 146, 890-901.

Heeremans J., Esmaiel T., Van Kruijsdijk P., et al., 2006. Feasibility study of WAG injection in naturally fractured reservoirs. In: SPE/DOE Symposium on Improved Oil Recovery, Tulsa, Oklahoma, USA.

Malik Q., Islam M., 2000. $CO_2$ injection in the Weyburn Field of Canada: optimization of enhanced oil recovery and greenhouse gas storage with horizontal wells. In: SPE/DOE Improved Oil Recovery Symposium, Tulsa, Oklahoma, USA.

Pyo K., Damian-Diaz N., Powell M., et al., 2003. $CO_2$ flooding in Joffre Viking Pool. In: Canadian International Petroleum Conference, Calgary, Alberta, Canada.

Salehi M., Safarzadeh M., Sahraei E., et al., 2014. Comparison of oil removal in surfactant alternating gas with water alternating gas, water flooding and gas flooding in secondary oil recovery process. Journal of Petroleum Science and Engineering, 120,

86–93.
Shayegi S., Jin Z., Schenewerk P., et al., 1996. Improved cyclic stimulation using gas mixtures. In: SPE Annual Technical Conference and Exhibition, Denver, Colorado, USA.
Spirov P., Rudyk S., Khan A., 2012. Foam assisted WAG, snorre revisit with new foam screening model. In: North Africa Technical Conference and Exhibition, Cairo, Egypt.
Telmadarreie A., Doda A., Trivedi J., et al., 2016. $CO_2$ microbubbles – a potential fluid for enhanced oil recovery: bulk and porous media studies. Journal of Petroleum Science and Engineering, 138, 160–173.
Zhao F., Zhang L., Hou J., et al., 2014. Profile improvement during $CO_2$ flooding in ultra-low permeability reservoirs. Petroleum Science, 11 (2), 279-286.

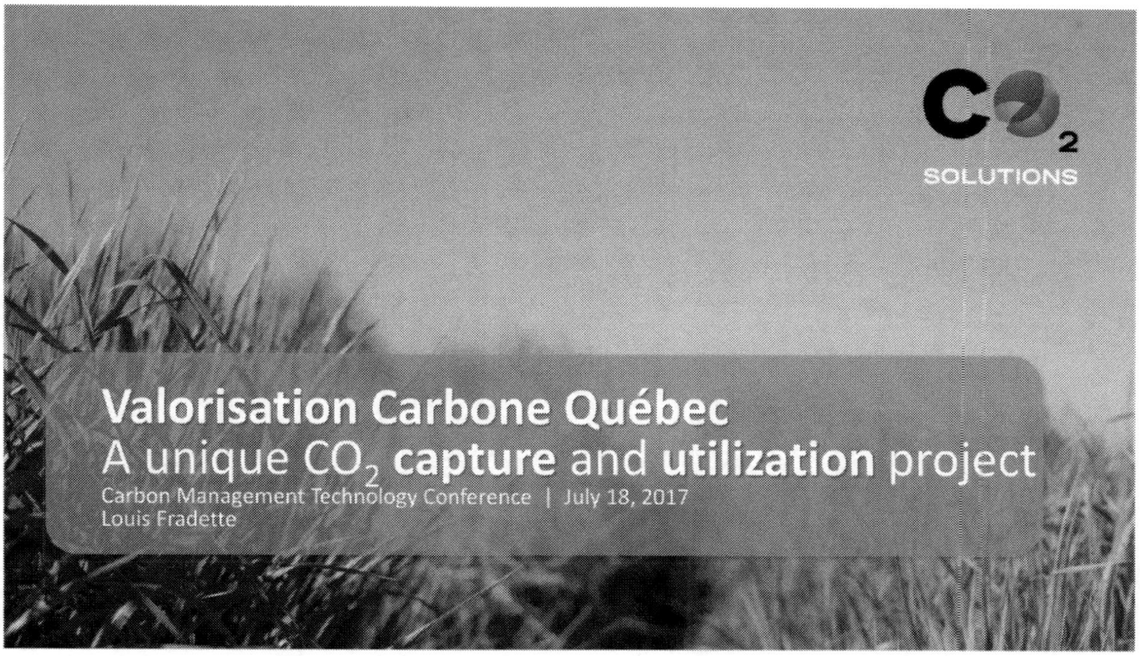


## Forward Looking Statements

- All statements in this presentation that are other than statements of historical facts are forward-looking statements which contain our current expectations about our future results. Forward-looking statements involve numerous risks and uncertainties. We have attempted to identify any forward-looking statements by using words such as "anticipates", "believes", "could", "expects", "intends", "may", "should" and other similar expressions.

- Although we believe that the expectations reflected in all of our forward-looking statements are reasonable, we can give no assurance that such expectations will prove to be correct. A number of factors may affect our future results and may cause those results to differ materially from those indicated in any forward-looking statements made by us or on our behalf. Such factors include our early stage of technology development; our need for capital to finance necessary research and product development; our ability to attract and retain key employees and strategic partners; our ability to achieve and maintain profitability; fluctuations in the trading price and volume of our stock; competition from other providers of similar products and services; and other unanticipated future events and conditions. For further information concerning risks and uncertainties that may affect our future results, please review the disclosures as may be contained from time to time in our filings with SEDAR. Other than as required by applicable securities laws, we undertake no obligation to publicly update or revise any of our forward-looking statements, whether as a result of changed circumstances, new information, future events, or for any other reason occurring after the date of this presentation. This presentation does not constitute an offer to sell or solicitation of an offer to buy securities in any jurisdiction.

C$O_2$ SOLUTIONS

## WHO ARE WE?

- **Canadian company**, head office in Québec City
- 23 employees, high tech (7 PhD's)
- CSI is the provider of a **patented** (80+ patents) **$CO_2$ capture technology** based on the use of *Carbonic Anhydrase* (CA)
  - ✓ Can capture any level of $CO_2$ emitted from stationary sources, at any rate
  - ✓ Delivers extremely pure $CO_2$, dry or wet
- The process offers the **lowest capture cost** among existing technologies
- The process is the only **commercially available** capture process that doesn't generate **toxic by-products**
- Potential applications:
  - ✓ Direct utilization of captured $CO_2$ as industrial gas
  - ✓ Sequestration of $CO_2$ for carbon mitigation
- Publicly traded on TSX-V (CST)

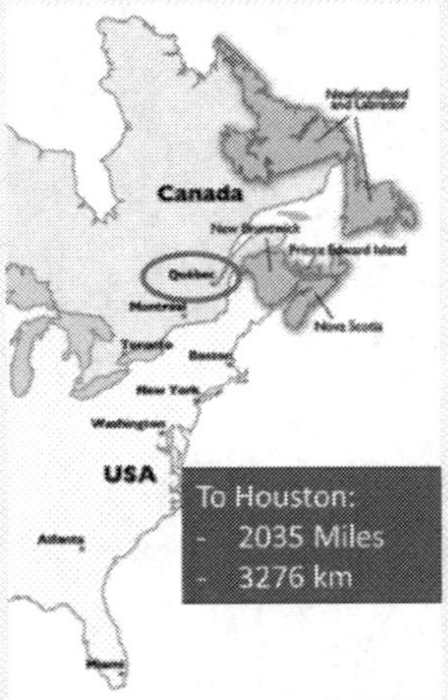


$CO_2$ SOLUTIONS

3

## THE TECHNOLOGY

- Biomimicking
  - ✓ **Industrial lung** using the **enzyme** *Carbonic Anhydrase* (CA)
- The process

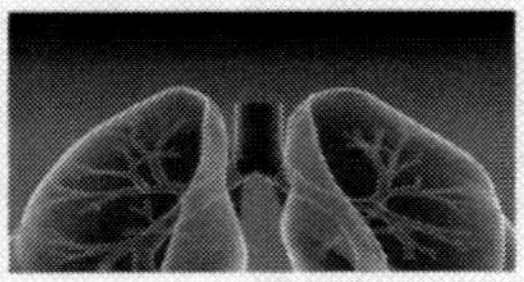

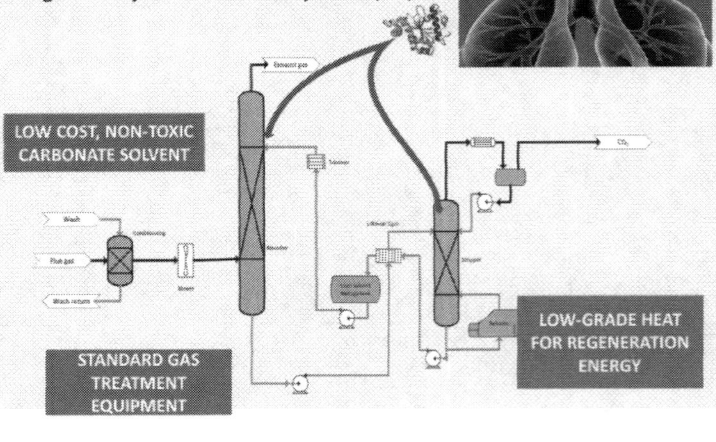


$CO_2$ SOLUTIONS

4

## Robust Patent Portfolio

**BROAD PATENT PORTFOLIO** FOR USE OF **CARBONIC ANHYDRASE** IN CANADA, U.S., EU, CHINA, AUSTRALIA AND OTHER MARKETS

**46** ISSUED **37** PENDING*

vs. 13 and 19 respectively in 2008

**SOLVENTS**
Amines
Carbonates
Amino Acids
Combinations

**PROCESSES**
Packed Tower
Spray Scrubber
Bubble Column
Universal

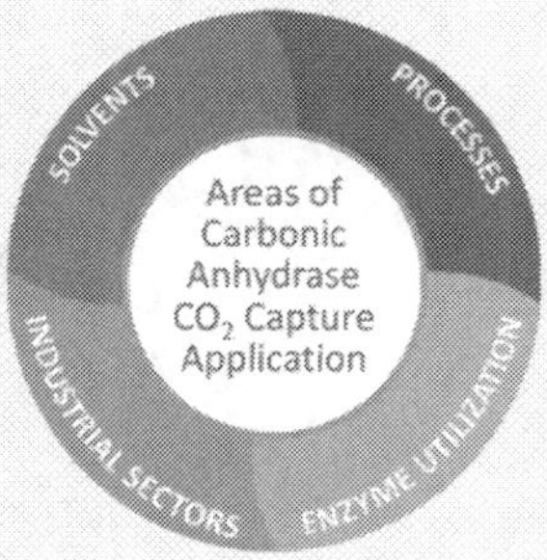


**INDUSTRIAL SECTORS**
Power
Steam
'Drop-in' Applications

**ENZYME UTILIZATION**
Soluble
Particle-Based
Analogs

CO2 SOLUTIONS

*as at June 30, 2015

## Commercial Progress – Capture and Utilization

**30 tpd project, Can$7.4 M**
- Preliminary engineering completed.
- Detailed engineering underway.
- To be commissioned in 2018.

**Up to 300 tpd, Can$30 M**
- In negotiatiion with industrial and financial partners.

| <2014 | 2014-2015 | 2015 | 2016+ |
|---|---|---|---|
| Lab and Bench-Scale: 0.5 tpd | Pilot: 1 tpd | Demonstration: 10 tpd | Commercial: 10-300 tpd; $37M CAPEX in Canada |

CO2 SOLUTIONS

## Capture Technology Demonstrated

- 5 times less enzyme than the design basis
- $CO_2$ produced is at 99.95%+ purity
- No solvent degradation observed
- No solvent makeup required
- No toxic waste products generated
    - Solvent sent to the municipal sewer
- Operation with day shift operators only
- 3rd party validation of performance (Tetratech Inc.)

$CO_2$ SOLUTIONS

7

## CURRENT COMMERCIAL PROJECT – 30 TPD SCALE

- Toundra Greenhouses (ST) – Resolute Forest Products (RFP)
- Commercial Contract
    - ✓ August 2016
- Detailed engineering underway

**SERRES TOUNDRA (ST)**
Saint-Félicien, QC

- Capture of up to 30tpd $CO_2$ from Resolute Forest Products' (RFP) pulp mill
- Utilize mill's low-grade / waste heat for process
- Reduce $CO_2$ output form mill
- Share carbon credits
- Supply of captured $CO_2$ to neighboring ST greenhouse complex through binding offtake agreement

**Full $CO_2$ value chain enabled by $CO_2$ Solutions' technology**

## Current $CO_2$ Utilization Project – 30 TPD Scale

- Startup in Q3 of 2018

9

## The Valorisation Carbone Québec Project

- The province of Québec has a 100% renewable power portfolio

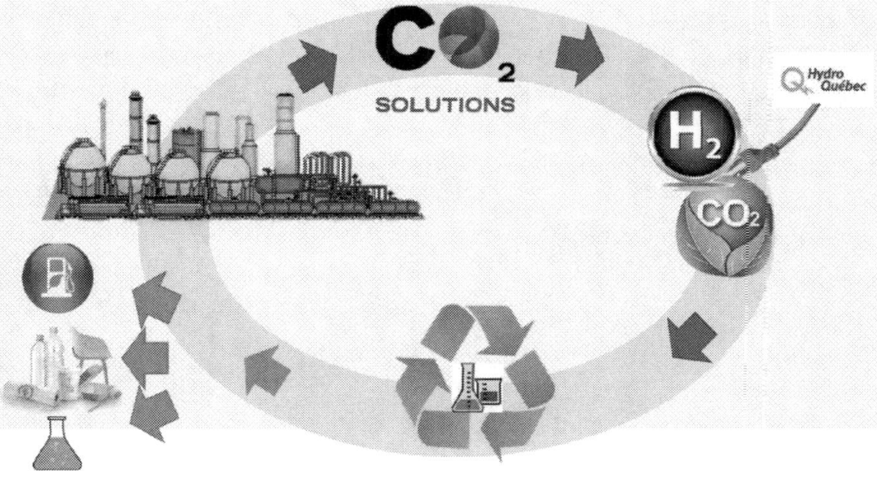


10

## VCQ Overview – A Unique Partnership

- Led by $CO_2$ Solutions in partnership with Laval University and Hatch
- \$15 million grant from the Government of Québec (Green Fund) / \$5 million in-kind contribution from the partners
- Goals
  - ✓ Industrially demonstrate (tpd scale) processes for $CO_2$ utilization into value-added products coupled to $CO_2$ Solutions' capture technology
  - ✓ Help industry reduce GHG emissions without damaging its ability to compete on economic terms
- Reviewing 100+ potential $CO_2$ utilization processes for demo
- Particular interest in processes that can use Québec's renewable/hydro electricity
- Parachem site selected as the VCQ demonstration site

CO2 SOLUTIONS
Québec
Fonds vert
UNIVERSITÉ LAVAL
HATCH

CO2 SOLUTIONS

## VCQ Overview – Two Parallel Tracks

- Demonstration
  - Quickly setup capture and utilization demonstration, before end of 2017, combining:
    - The 10 tpd capture unit from $CO_2$ Solutions
    - Attach at least one $CO_2$ utilization unit to demonstrate
      - Technical feasibility
      - Economic viability
      - Generate data to support both
- Development
  - Assemble a multidisciplinary scientific team to accelerate the development of processes close to maturity
  - Solve issues related to scale-up
  - Establish development path to allow for demonstration before March 2019

CO2 SOLUTIONS

12

## VCQ Parachem Site in Montréal East

- Parachem (Former Gulf Refinery)
  - ✓ 51% owned by Suncor Energy;
  - ✓ Producer of paraxylene
- Available at site
  - ✓ Large site (6 000 sq. ft.)
  - ✓ $H_2$
  - ✓ Water: cooling, fresh, hot, and treatment
  - ✓ Compressed air
  - ✓ Offices, operators
  - ✓ Electrical power

## VCQ – CSI $CO_2$ Capture Technology Improvements

- Stripping
  - ✓ Reduce energy consumption further
  - ✓ Determine best-suited equipment and configuration
- Advance the application of high-intensity contactors
  - ✓ Test RPB at 10 tpd scale
  - ✓ Design container-sized capture unit
  - ✓ Accumulate process experience from many months of operation

14

## A THOROUGH REVIEW OF UTILIZATION TECHNOLOGIES

- $100^{+}$ technologies reviewed (Laval University)
- Comprehensive list of criteria, double rating

**Technology**
- Conversion type (Bio, chemical, hybrid)
- Efficiency of $CO_2$ conversion
- Number of reactions involved (complexity)
- Catalyst used (cost, lifespan)
- Product (output)

**Process**
- Capacity demonstrated so far (tpd)
- Complexity of the separation process required
- Process conditions (P, T, ...)
- Feeds required (H2, CH4, ...)
- Energy involved (GJ/tonne $CO_2$)
- Energy type/quality (electricity, stream, ...)
- Conversion costs using inputs (\$/tonne $CO_2$)

CO2 SOLUTIONS

15

## VCQ - $CO_2$ Utilization Product Families

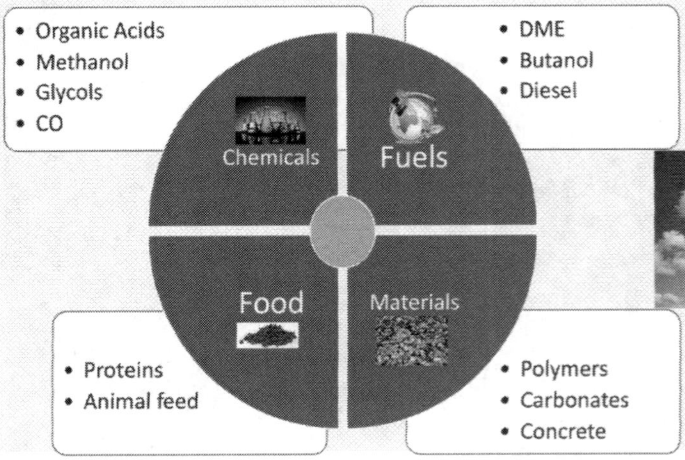

CO2 SOLUTIONS

# ACETIC ACID – FIRST UTILIZATION TECHNOLOGY **SELECTED**

- Industrial ecology in Montreal East
  - ✓ Complete **Poly Ester Chain** already exists
  - ✓ Addition of $CO_2$ capture and utilization into Acetic Acid strengthens existing relationships

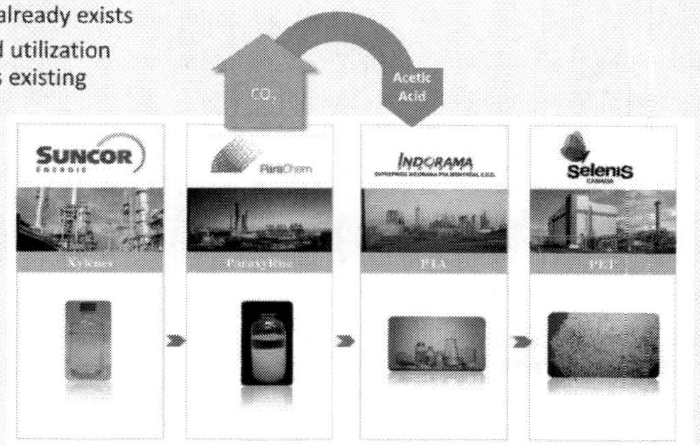


CO₂ SOLUTIONS

17

# $CO_2$ Utilization – Routes and Impacts on Investments – Some Learnings

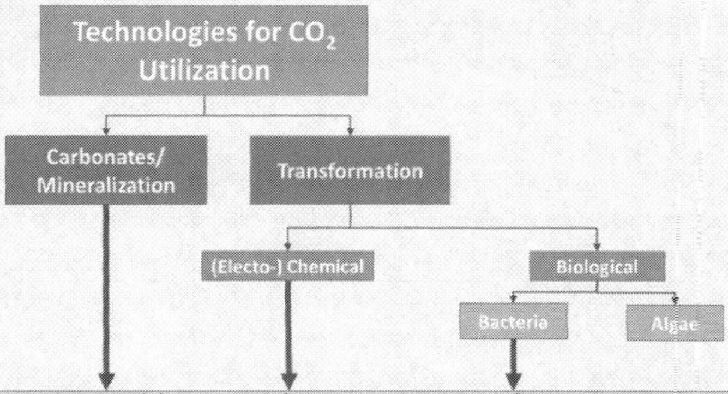


| | | | |
|---|---|---|---|
| Scale | Commercial | Lab++ | Lab+ |
| $CO_2$ uptake (/wt) | <10% | 50-80% | 50-100% |
| Electricity | - | 80% of the cost | 50% of the cost |
| CAPEX to convert 20 tpd | <$5M | >$20M | >$20M |

CO₂ SOLUTIONS

## VCQ – Timeline Summary

- Since January 2017
  - ✓ Signature
  - ✓ Hatch Inc. officially named as the engineering partner
  - ✓ Laval University technology review
  - ✓ Scientific advisory committee launched
  - ✓ Acetic acid confirmed as first $CO_2$ utilization product
  - ✓ Completed relocation of $CO_2$ capture unit to Parachem site
  - ✓ Agreement under review with three more $CO_2$ utilization technologies
- August
  - ✓ Capture unit restarts at Parachem
- October
  - ✓ First $CO_2$ utilization unit installed and running
  - ✓ Agreements with up to three more $CO_2$ utilization technologies

CO₂ SOLUTIONS

19

## CSI Summary

- Process performance confirmed
  - ✓ Enzyme stable in process
  - ✓ No toxic waste generation
  - ✓ No aerosol issues
  - ✓ No solvent degradation
  - ✓ Solvent
    - Can be eliminated in standard water treatment plant
    - Can be converted into fertilizers when spent
- Capture cost confirmed
  - ✓ **<30$/tonne @ 1250 tpd scale**
- Commercialization on its way
  - ✓ 30 tpd project is on its way, engineering completed
  - ✓ Up to 300 tpd project in Alberta – at negotiation stage
  - ✓ VCQ has started and will provide access to the most advanced $CO_2$ utlization technologies in the coming months

CO₂ SOLUTIONS

## VCQ – Seeking $CO_2$ Utilization Techs and Strategic Partners

- Fully paid demonstration site
- Providers of $CO_2$ utilization technologies get access to demonstration site
  - ✓ Funds available to support technology evolution
- Operation for up to 24 months to develop and mature technologies
  - ✓ Full access to performance evaluation for each technology, at all levels of maturity
- Access to first class engineering and expertise in $H_2$ production
- Privileged relationship with providers of $CO_2$ utilization technologies
  - ✓ Strategic positioning in the carbon market
- Visibility as leaders in the field
- **Still seeking additional strategic partners and technologies for demonstration**

CO2 SOLUTIONS

# CMTC-502866-MS

# Enhanced Oil Recovery – The History of $CO_2$ Conventional WAG Injection techniques developed from Lab in the 1950's to 2017

David Merchant, Merchant Consulting



## Abstract

During the past 45 years, $CO_2$ flood technology for Enhanced Oil Recovery projects evolved from a partially understood process filled with uncertainties to a process based on proven technology and experience. Many questions involved with $CO_2$ flooding have been thoroughly analyzed and answered. This knowledge is currently being used by a limited number of companies that actually know how to design, implement, and manage a $CO_2$ flood for long term profit. The purpose of this report is to help disseminate this knowledge to operating companies interested in EOR flooding or to $CO_2$ Sequestration Communities interested in storing $CO_2$ in EOR projects.

In 2015, Merchant Consulting published CMTC-440075-MS "Life beyond 80 – A look at Conventional WAG Recovery beyond 80% HCPV Injection in $CO_2$ Tertiary Floods". The primary objective of the report was to target all 10 $CO_2$ Recovery Methods used today including "Conventional WAG Techniques" which have been used in over 90% of all the Enhanced Oil Recovery projects implemented to date. These include projects in the Permian Basin in Texas, Colorado, Oklahoma, and Wyoming. The paper presents answers to the question "What is life after 80% HCPV Injected?" And "What effect does life after 80% HCPV have on Tertiary Oil Recovery, $CO_2$ Utilization and $CO_2$ Retention in different producing formations?" Results of this study show Tertiary Oil Recovery can be as high as 26% OOIP when slug sizes exceed 190% HCPV injected.

**Conventional WAG History in $CO_2$ Tertiary Oil Projects:**

To achieve $CO_2$ Injection beyond 80% HCPV Injection requires proper $CO_2$ WAG Management. The purpose of this report is to provide both the EOR and $CO_2$ Sequestration Communities an understanding of the "History of Conventional WAG" and how it has changed from first introduced in the Lab in the 1950's, to how it was implemented and developed in the 1980's by the Major Oil Companies in the Permian Basin, and how Conventional WAG is being managed today in the field.

# Introduction – $CO_2$ Flooding

## $CO_2$ Flood History

The Permian Basin has had over 85+ years of oil production history and produced over 30 billion barrels of oil and represents the 3rd largest petroleum producing area in the United States after the Gulf of Mexico and Alaska. In 1972, $CO_2$ injection was first introduced commercially in the Sacroc Field. Today, over 1.3 billion barrels have been produced with $CO_2$ and accounts for over 250,000 BOPD from over 17,000 production wells and 14,000 $CO_2$ injection wells, through 4,500 miles of pipeline.

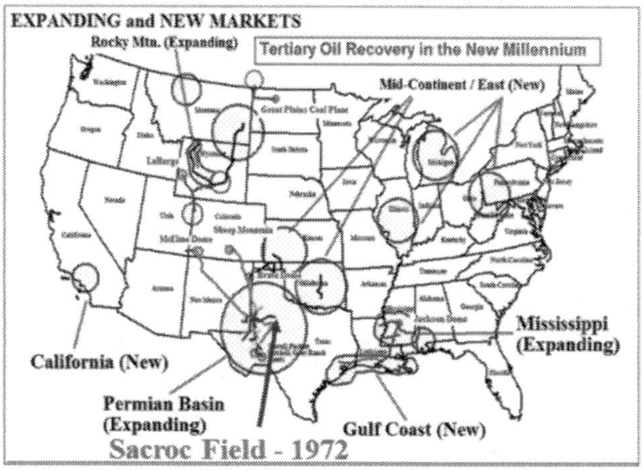


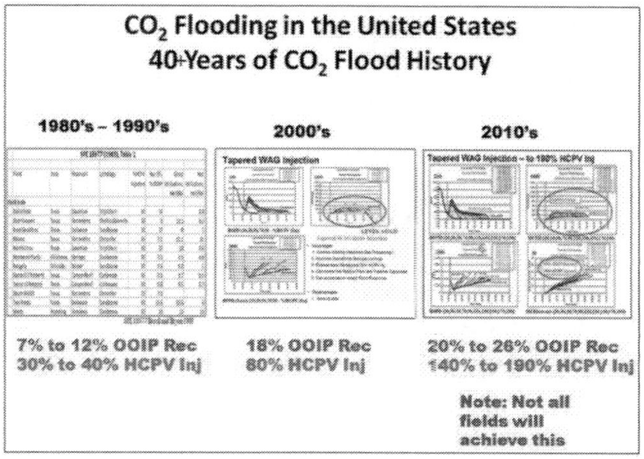


## What makes $CO_2$ Tertiary Oil Recovery Work?

The reason why $CO_2$ works is simple. The $CO_2$ acts as a solvent when injected into the reservoir and swells the Residual Oil (oil left after water flood) and reduces its viscosity. This causes the Residual Oil to swell and become mobile and be produced.

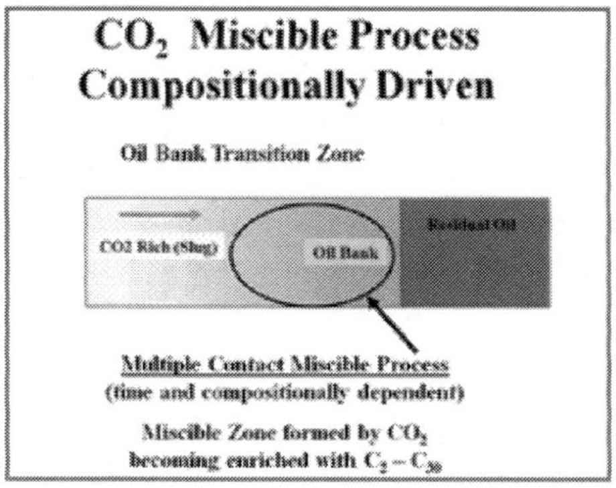


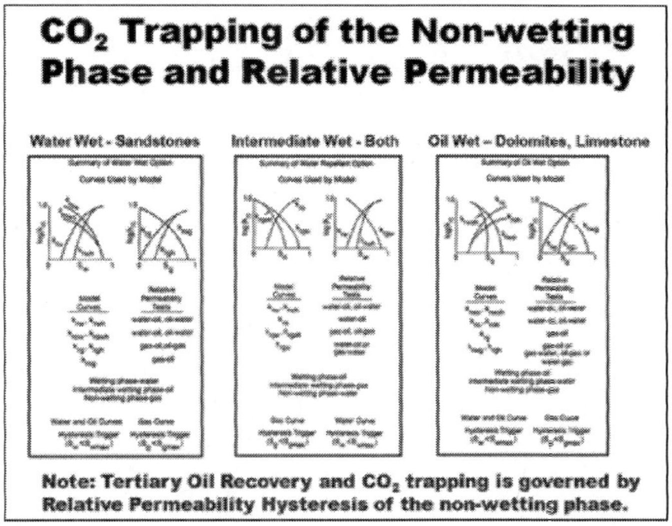


The amount of Tertiary Oil Recovered and the amount of $CO_2$ Trapped or Sequestered is dependent on the rock type. In addition to mobilizing residual oil saturation, a portion of the $CO_2$ becomes trapped. In the Petrophysical world, this is known as Relative Permeability Hysteresis or Trapping of the Non-wetting Phase. In this case $CO_2$ is the non-wetting phase. As demonstrated above, this phenomenon occurs in Water Wet Rocks (Sandstones), Intermediate Wet Rocks (Sandstones or Carbonates), and in Oil Wet Rocks (Dolomites and Limestone formations).

# $CO_2$ Tertiary Recovery Methods

## Introduction

$CO_2$ Tertiary Recovery Processes to date encompass "Ten" Recovery Methods. Four of these methods are used in the Permian Basin. Conventional WAG Techniques have been used in over 90+% of the $CO_2$ floods around the World. The Seminole field example presented in this report is an example of a field operated under Conventional WAG that targets both the Main Pay and ROZ zones.

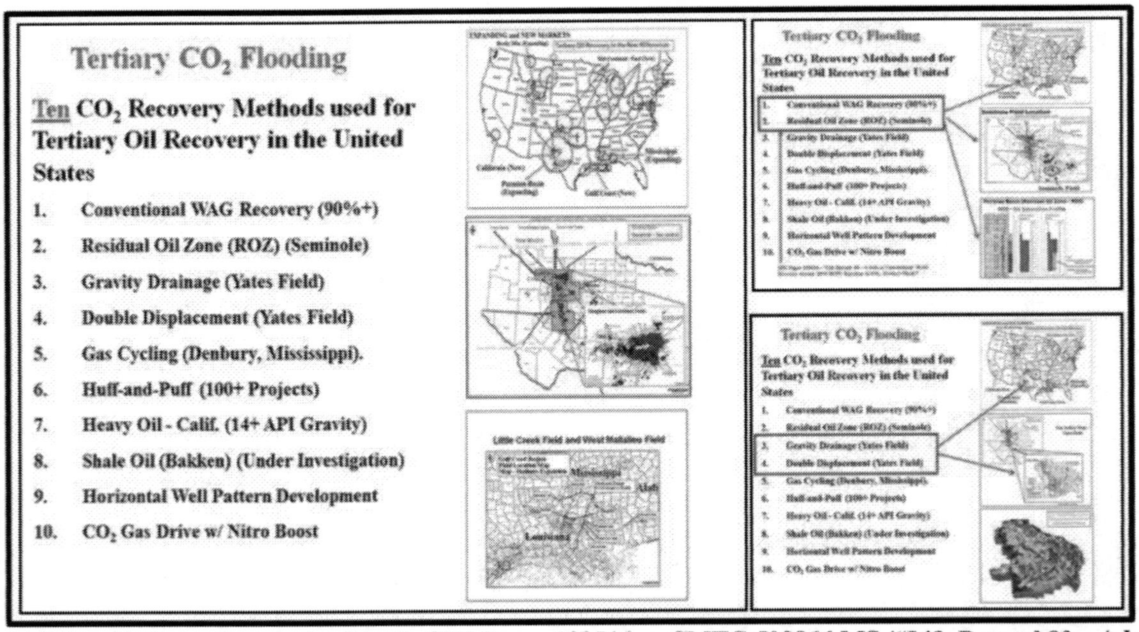


**Additional information can be found: SPE Paper 139516 or CMTC-502866-MS ("Life Beyond 80 – A look at Conventional WAG Recovery beyond 80% HCPV Injection in $CO_2$ Tertiary Floods")**

## Carbon Sequestration Options: Ten Recovery Methods for EOR

Most Enhanced Oil Recovery Projects use one of the following ten operating methods: Conventional WAG Recovery, ROZ Recovery, Gravity-stabilized Recovery, Double Displacement, Gas-cycling Huff-and-Puff, Heavy Oil-California, Shale Oil (Bakken, Wolfcamp), Horizontal Well Pattern Development, and $CO_2$ Gas Drive w/ Nitro Boost. The primary difference between methods depends on the reservoir geology and well pattern configuration. In Conventional $CO_2$ floods, typical of West Texas, the formations are basically flat (Ramp Sequence), low perm, the fields are developed on pattern spacing (e.g. 5-spot patterns, 9-spot patterns, or Chickenwire patterns), and Conventional WAG Operating schemes are used to control mobility and $CO_2$ flood response. In conventional WAG operations, the objective is to minimize the amount of $CO_2$ purchased ($CO_2$ stored in Sequestration projects), which is typically in the range of range of 30%-50% of the total HCPV $CO_2$ injected. In un-conventional Gravity-Stabilized and Double Displacement case histories, Flue Gas, $CO_2$, Lean Gas or $N_2$ is usually injected in the top of the structure and oil is produced from the bottom. More $CO_2$ can be sequestered than conventional WAG operations. As much as 80% of the total pore volume can be displaced with $CO_2$. In Gas-cycling projects, typical of projects operated by Denbury in Mississippi, $CO_2$ is cycled through the formation. As much as 6 pore-volumes of $CO_2$ are injected to recover 18% OOIP. In Huff-and-Puff operations, the $CO_2$ is injected into and produced from the same well. The objective is to mobilize tertiary oil in the near vicinity of the well-bore, and then produce the $CO_2$ and tertiary oil back. In California the Wilmington field had three $CO_2$ pilots in 14 API Gravity Crude. Recovery was comparable with West Texas Operations. Shale Oil with $CO_2$ is still under investigation. Horizontal Well $CO_2$ floods are operational in Aneth field in Utah and Weyburn field in Canada. The Gravity Drainage case with added Nitro-Boost was developed for rate acceleration cases where nitrogen follows $CO_2$ Injection.

# $CO_2$ Storage - Mis-conceptions about EOR

Over the years, there have been many mis-conceptions and half truths developed about EOR and its ability to store $CO_2$.

The **first mis-conception** about $CO_2$ flooding deals with ranges of Tertiary Oil Recovery reported over the years. In the 1990's, most all of the major oil companies reported tertiary oil recovery in the range of 10%-12% OOIP with a 30%-40% HCPV $CO_2$ slug size. By 2000, this number grew to 18% OOIP recovery with an 80% slug size. Today, expected recoveries in certain fields are expected to exceed 26% OOIP with 190% HCPV $CO_2$ injected. However, NOT all fields will achieve this level of oil recovery. Each $CO_2$ project must be judged on its own merits.

The **second mis-conception** about $CO_2$ flooding deals with the amount of $CO_2$ Storage in EOR operations. Normal EOR operations have always stored or trapped $CO_2$ in the reservoir. Tertiary recovery is a displacement process. For the EOR process to work, $CO_2$ must be cycled through the reservoir, similar to water flood operations. The $CO_2$ acts as a solvent to swell the remaining oil left after water flood and decreases the oil's viscosity; thus, allowing the tertiary oil to flow. Three phase relative permeability dictates the amount of $CO_2$ trapped in the formation due to phase trapping. From existing $CO_2$ flood evaluations, this equates to the amount of $CO_2$ purchased and represents approximately 30%-40% of the Hydrocarbon-Pore-Volume. With extended WAG with $CO_2$ volumes exceeding 125% HCPV, the storage capacity could increase to 50%-60% HCPV, but involves large volumes of $CO_2$ through-put with little incremental gain in oil recovery.

The **third mis-conception** deals with throughput Injection Rate and Injection Pressure (Water Injection or $CO_2$ Injection). All reservoirs have a Maximum Through-put Rate that can be injected into the reservoir. The maximum amounts of $CO_2$ that can be stored in a reservoir are based on a number of factors. These include: The total number of production and injection wells available, pattern configuration, current reservoir pressure, fracture pressure, injection flow capacity (permeability-thickness (kh) of each injection well), and relative permeability. Most water flood operations operate at or near fracture pressure. The same is true for $CO_2$ floods. The objective is to provide maximum $CO_2$ throughput rate across the reservoir at a level that maintains good seal integrity.

The **fourth mis-conception** about $CO_2$ flooding deals with the Total $CO_2$ Storage Capacity of a reservoir. To date, all $CO_2$ floods operate with the intent to minimize the amount of $CO_2$ purchased. Under these conditions, $CO_2$ is purchased up-front and cycled through the reservoir until the revenue from the oil and hc-gas production doesn't have the ability to support the cost to recover the oil (abandonment). This date has been very elusive to determine. Of all the $CO_2$ projects operated to date, most are still operating today. A limited few projects have been shut-in due to a low oil price environment, but could be returned back to production if given the right economic conditions.

The **fifth mis-conception** about $CO_2$ flooding deals with $CO_2$ Storage Timing. Is it at the beginning? Does it occur during EOR $CO_2$ operations? Or, is it at the end of EOR operations? For those advocating a switch from EOR to $CO_2$ Storage two-thirds though or near the end of the EOR project should think through the physics. You can't fool Mother Natue. The amount of $CO_2$ **TRAPPED**, **STORED**, OR **SEQUESTERED** is dependent on the Composition of the Residual Oil Saturation and Phase Trapping of the Non-Wetting Phase, which occurs throughout the life of $CO_2$ injection, with 75% of $CO_2$ storage occurring during the first one-third of the project's life. After this initial period, most of the $CO_2$ that is injected is re-cycle $CO_2$ from the $CO_2$ recovery plant. $CO_2$ purchases near the end of the life of an EOR project are minimal.

The **last mis-conception** deals with achieving Total $CO_2$ Storage Potential of the reservoir. Currently there are ten methods used by the industry to recover oil with $CO_2$. All target the remaining tertiary oil. None target the **water**. The Eleventh Recovery Method which targets $CO_2$ storage removes both the **$CO_2$** and **Water** from the reservoir. In a World where there is an infinite amount of $CO_2$ available for storage, the remaining water in the reservoir becomes a commodity target along with the oil. Under this scenario, the objective will be to displace as much of the remaining oil plus water from the reservoir and replace it with $CO_2$. Since seal integrity will always be maintained below formation parting fracture pressure, the additional storage capacity provided by removing both the oil and water would far exceed EOR storage operations alone.

# Conventional WAG Techniques

Conventional WAG Techniques have been used in over 90% of the $CO_2$ Floods implemented today. They are designed to work well in reservoirs that are developed on pattern spacing. These include: Five-spot, Nine-spot, and Chickenwire pattern development. The objective is to "Level Load" produced gas production to a $CO_2$ Recovery Plant inlet gas rate, which extends field life and recovers more tertiary oil.

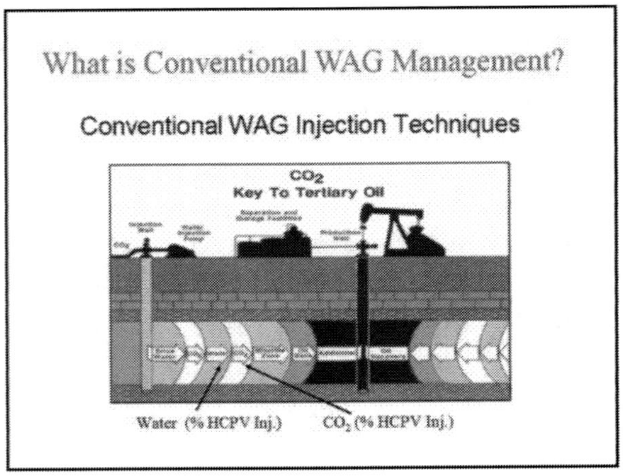


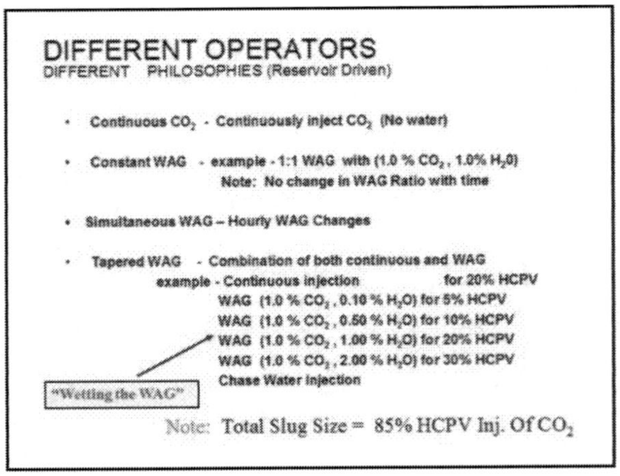


Today, most operators have adopted the Tapered WAG approach to optimize WAG management.

### Conventional WAG Recovery

Conventional WAG Operating Methods in the Permian Basin fall into one of four categories:

1. Continuous $CO_2$ Injection
2. Constant WAG Injection
3. Tapered or Hybrid WAG Injection
4. Simultaneous $CO_2$ Injection (Limited use)

### What is WAG Management?

Through the 1980's and 1990's, Amoco, Shell, Arco, Exxon, Mobil, and Texaco committed significant manpower to evaluate the feasibility of full field scale $CO_2$ flooding in the Permian Basin. Before the initiation of field scale floods, many pilots were drilled and much reservoir simulation was conducted to understand the $CO_2$ flooding process. The results are presented below.

All Conventional WAG Injection Projects have one thing in common. $CO_2$ is injected into the reservoir and the produced recycle $CO_2$ must be re-injected back into the reservoir to maximize oil recovery. This was first demonstrated by Caudle and Dyes in 1958 when water was added to $CO_2$ to decrease solvent mobility [Turek, 1,102]. As $CO_2$ technology was transferred from the lab to the field, most all of the Major Oil Companies in the 1970's and early 1980's adopted the use of Constant Water-Alternating-Gas (WAG) Injection based on the theory that alternate gas water injection is necessary to maintain mobility control and maximize oil recovery. During the late 1990's, Tapered and Hybrid WAG Operations were adopted to improve the overall recovery process [2,3,4,5,6,104,105]. During the WAG process, $CO_2$ and Water are injected into the reservoir in alternating $CO_2$ and water slug sizes time periods. For Constant WAG operating schemes, the half cycle slug size is typically fixed for example at 1.0% HCPV $CO_2$ for the Gas Cycle and 1.0% $H_2O$ for water. For Tapered Wag projects, WAG Ratios change with time. "Wetting the WAG" or increasing water half cycle volume with time improves conformance by slowing the gas in the fast zones. The water half cycle can be increased or decreased to to improve overall conformance or adjusted to "Level Load" gas production to a Plant inlet rate.

# Conventional WAG History

**Lab Experiments in the 1950's and 1960's**

Conventional **W**ater-**A**lternating-**G**as (**WAG**) was originally discovered in the 1950's when Caudle and Dyes introduced water in Lab experiments to decrease gas mobility. During the 1960's over 150 small scale Miscible and Immiscible projects were implement across the United States by labs and oil companies interested in developing Enhanced Oil Recovery for additional recovery of oil beyond Waterflood operations.

**Permian and Rocky Mountain Basins History of "WAG"**

**1950's and 1960's**

1950's – 150 Small Scale Miscible and Immiscible Projects
1958 - Lab Experiments Caudle and Dyes introduced water to decrease solvent mobility

**1970's**

1972 – Sacroc Unit (Kelly Snyder field) First Commercial Large Scale Project
1972-1979 North Cross, Two Freds, Maljamar, others

**1980's Large Scale Permian Basin and Rocky Mountain Expansions**

Slaughter field three projects plus Wasson ODC Unit – Amoco's Tapered WAG
Wasson field Denver Unit – Shell's Continuous Injection area plus DUWAG
Wasson field Willard Unit - Arco's Area Wide WAG went to Individual Patterns
Seminole field – Amerada Hess's Constant WAG scheme
Rangely field – Chevron's Wide Area WAG scheme
Bairoil (Lost Soldier and Wertz field ) – Amoco's Tapered WAG

**1990's, 2000's, and 2010's**

Most new $CO_2$ flood projects adopt the "Tapered WAG" approach to WAG Management

**SPE 18977 (1989), Table 1**

| Field | State | Reservoir | Lithology | %HCPV Injected | Rec Eff, % OOIP | Gross Utilization, Mcf/Bbl | Net Utilization, Mcf/Bbl |
|---|---|---|---|---|---|---|---|
| **Field Scale** | | | | | | | |
| Dollarhide | Texas | Devonian | Trip Chert | 30 | 14 | | 2.4 |
| East Vacuum | Texas | San Andres | Oolitic Dolomite | 30 | 8 | 11.1 | 6.3 |
| Ford Geraldine | Texas | Delaware | Sandstone | 30 | 17 | 9 | 5 |
| Means | Texas | San Andres | Dolomite | 55 | 7.1 | 15.2 | 11 |
| North Cross | Texas | Devonian | Trip Chert | 40 | 22 | 18 | 7.8 |
| Northeast Purdy | Oklahoma | Springer | Sandstone | 30 | 7.5 | 6.5 | 4.6 |
| Rangely | Colorado | Weber | Sandstone | 30 | 7.5 | 9.2 | 5 |
| Sacroc (17 Pattern) | Texas | Canyon Reef | Carbonate | 30 | 7.5 | 9.7 | 6.5 |
| Sacroc (4 Pattern) | Texas | Canyon Reef | Carbonate | 30 | 9.8 | 9.5 | 3.2 |
| South Welch | Texas | San Andres | Dolomite | 25 | 7.6 | | |
| Two Freds | Texas | Delaware | Sandstone | 40 | 15.6 | 15.6 | 8 |
| Wertz | Wyoming | Tensleep | Sandstone | 60 | 10 | 13 | 10 |

SPE 18977 Brock and Bryon 1989

**$CO_2$ Flooding in 1970's**
**First Commercial Scale $CO_2$ Flooding in 1970's**

Chevron, who operated the Sacroc Unit in the Permian Basin, was the first field to inject $CO_2$ as a commercial project in1972. Chevron injected $CO_2$ over the total unit with less than 15% in any one pattern and sold the property in 1992 to Pennzoil. Pennzoil ramped up $CO_2$ Injection and implemented a 300 acre pilot in the middle of the field of which 1/3 of the total field's production out of 55,000 acres was being produced from the pilot area. In 1998, Pennzoil sold the field to Devon and Devon then sold the field to Kinder Morgan in 2000. Kinder Morgan then ramped-up $CO_2$ injection to over 900 MMSCFIPD. The result was a Tertiary oil response growth from 9,600 BOPD to over 30,000 BOPD.

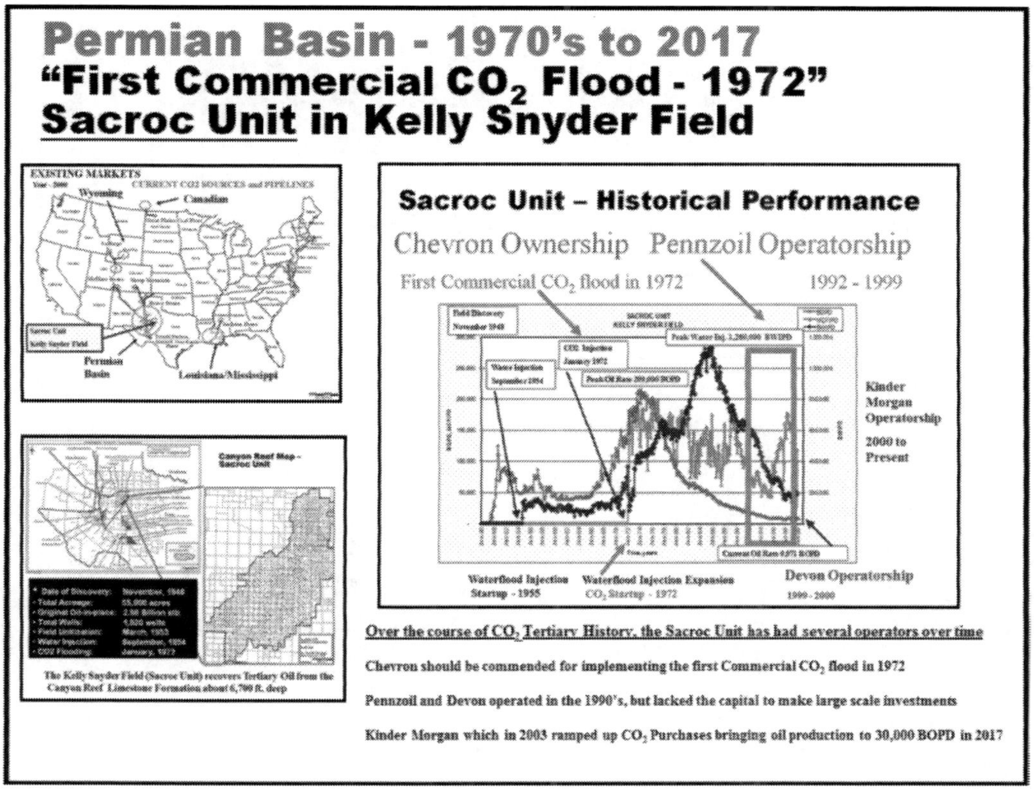

Chevron should be commended for implementing the first Commercial $CO_2$ flood in the United States, but only sprinkled $CO_2$ over the total reservoir limit.

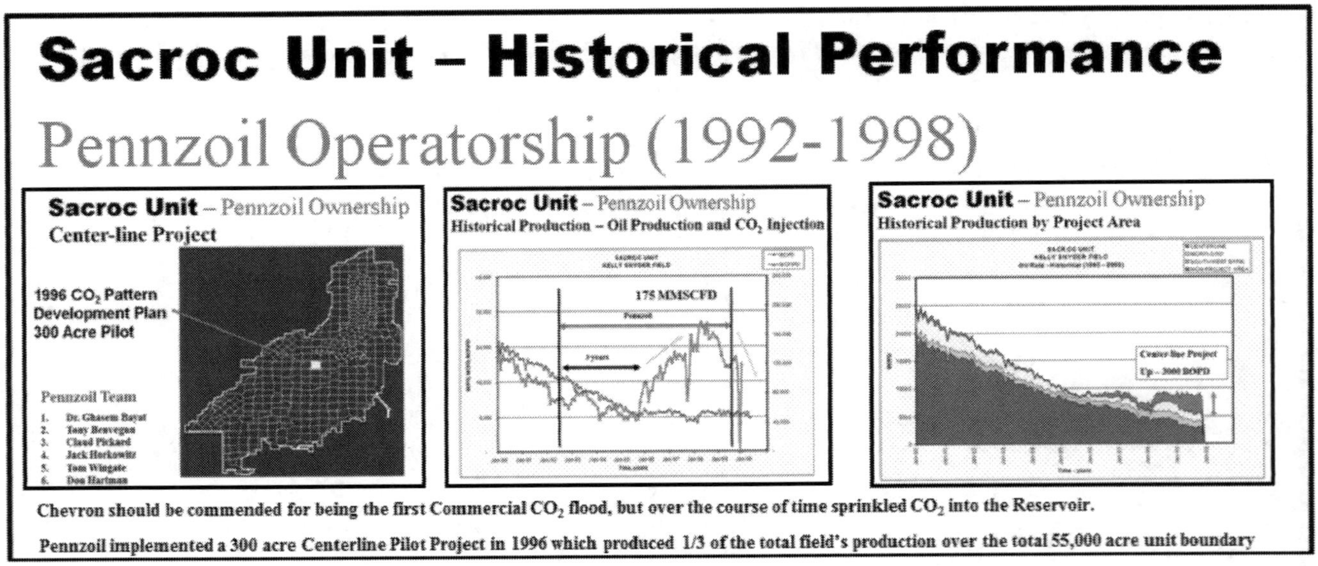


Pennzoil in 1996 implemented a 300 acre Centerline Pilot Project which produced one third of the fields 9,600 BOPD production. (Ref: 106) In 1998, Pennzoil sold the field to Devon who later sold the field to Kinder Morgan in 2000.

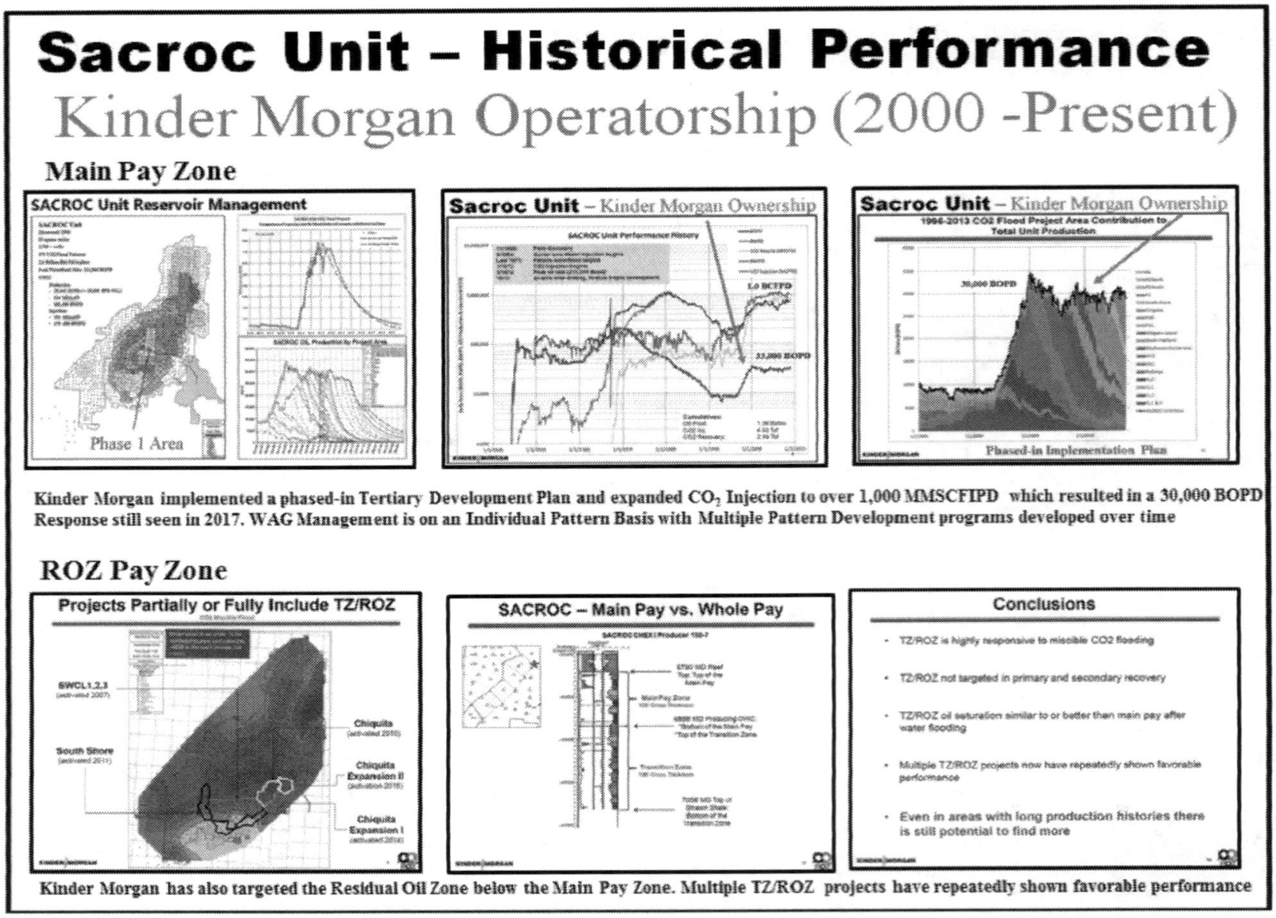


In 2000, Kinder Morgan implemented a multi-phase Tertiary Development Plan that increased $CO_2$ Injection to over 800 MMSCFIPD, which resulted in a 30,000 BOPD Response still seen in 2017.

# $CO_2$ Flooding in 1980's

In the 1980's, major oil companies including Shell, Amoco, Texaco, Arco, Phillips, Oxy, and Exxon implemented $CO_2$ projects in many of the large fields located across the Permian Basin, Rockies, and Gulf Coast Regions. As reported by the major oil companies, expected tertiary recovery was typically in the 8% to 12% OOIP range. Optimum slug size for most projects was projected to be in the 30% to 40% HCPV $CO_2$ injection range.

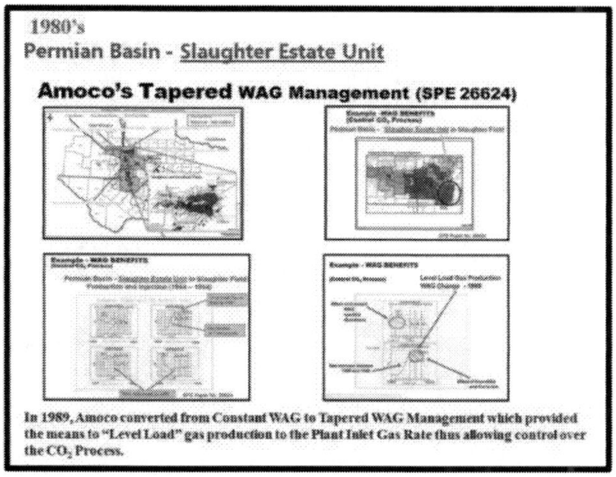


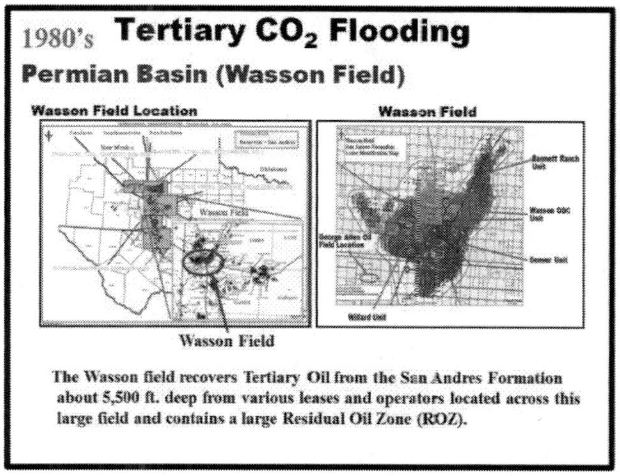


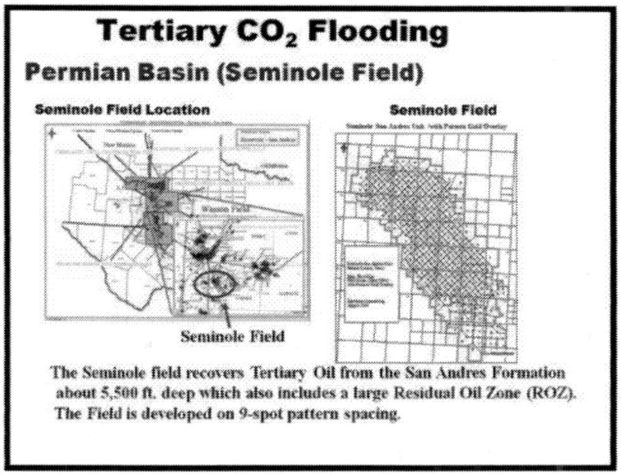


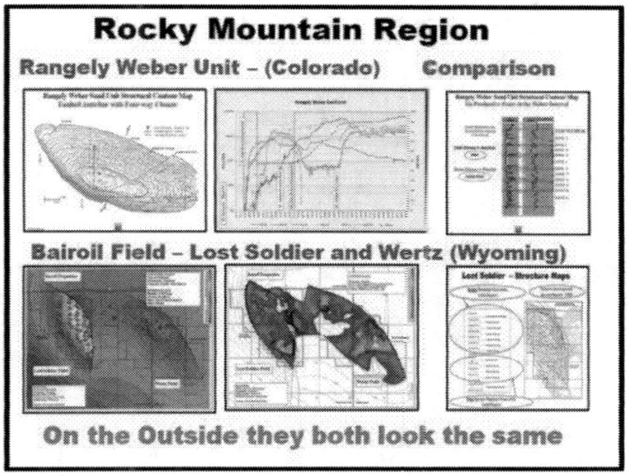


In the 1980's, Major Oil Companies also implemented different types of $CO_2$ Recovery Methods to both experiment and determine which process works best. Each will be discussed below.

## Permian and Rocky Mountain Basins History of "WAG"

**1980's Large Scale Permian Basin and Rocky Mountain Expansions**

Slaughter field three projects plus Wasson ODC Unit – Amoco's Tapered WAG
Wasson field Denver Unit – Shell's Continuous Injection area plus DUWAG
Wasson field Willard Unit - Arco's Area Wide WAG went to Individual Patterns
Seminole field – Amerada Hess's Constant WAG scheme
Rangely field – Chevron's Wide Area WAG scheme
Bairoil (Lost Soldier and Wertz field ) – Amoco's Tapered WAG

## $CO_2$ Flooding in 1980's (Amoco – Tapered WAG)

Amoco initiated $CO_2$ injection into four floods in the mid 1980's. These included: Slaughter Estate Unit, Central Mallet Unit, and Frazier Unit in Slaughter field plus Wasson ODC in Wasson field. All initially were operated under a constant WAG. During this time frame, extensive modelling was conducted to determine a method that could optimize the WAG process and extend oil recovery. (Ref: 107) The result was "Tapered WAG" which provided the means to "Level Load" total gas production from the field to a maximum "Plant Inlet Gas Rate". In 1989, Amoco implemented this new technology into all four projects, achieving total success.

As this technology improved, advancements in $CO_2$ flood design have been implemented into more improved reservoir management practices. In the 1990's, Constant WAG was being replaced with Tapered WAG operations and tertiary oil recovery increased to 18% of a field's Original Oil-in-Place. Today, most fields implemented during the 1980's have surpassed the 40% HCPV injection, advancing to or passing earlier HCPV $CO_2$ predictions.

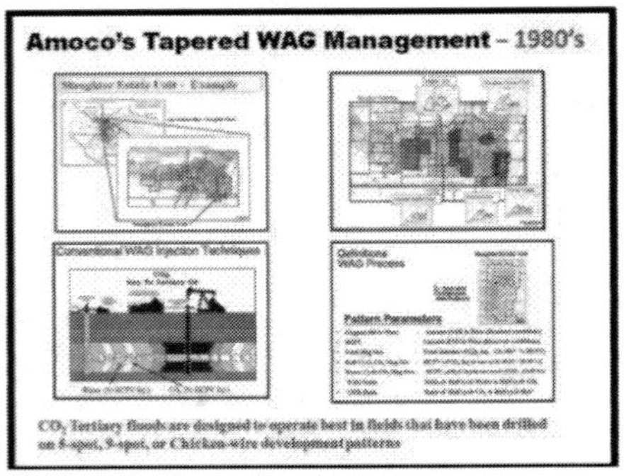


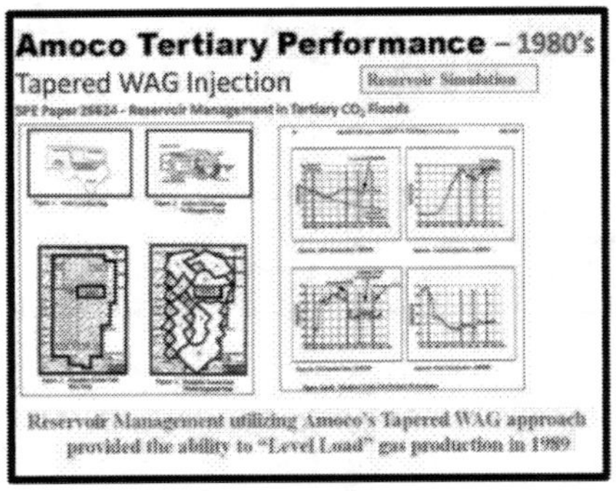


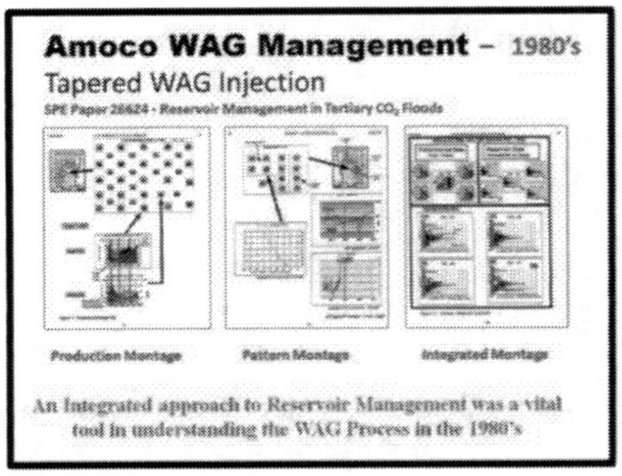


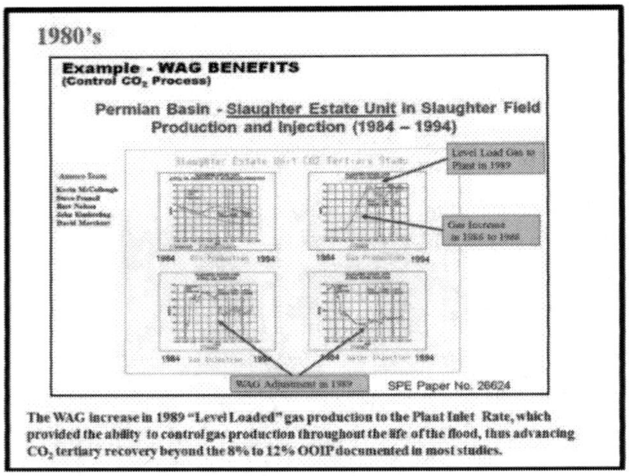


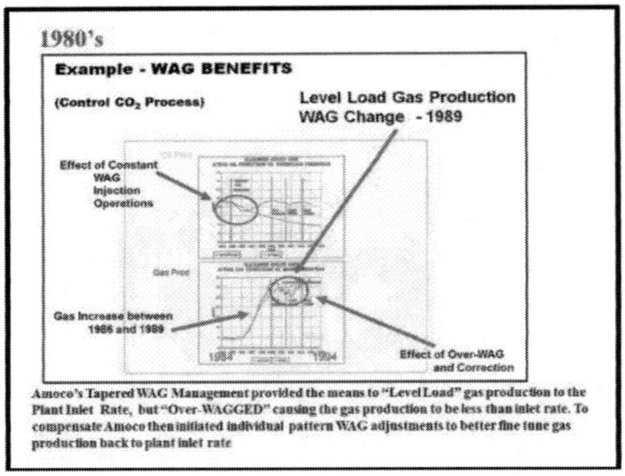


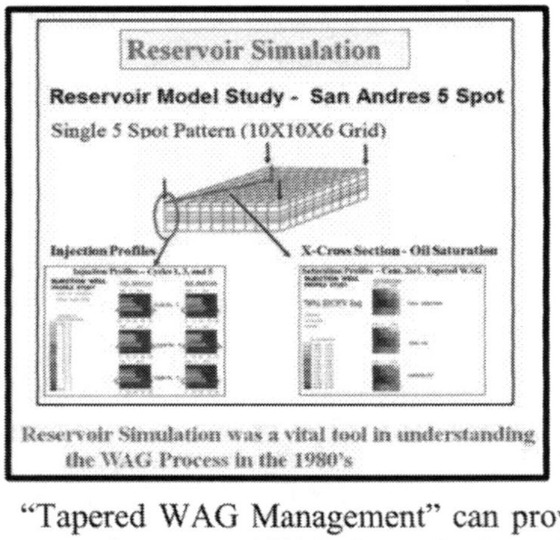


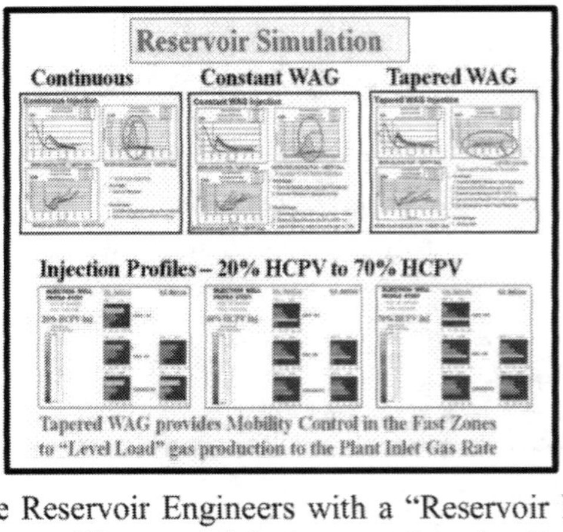


Slaughter Field
Bibliography References: 19 thru 29

"Tapered WAG Management" can provide Reservoir Engineers with a "Reservoir Management Tool" that can properly manage WAG, but only through proper "Reservoir Management Practices"

# $CO_2$ Flooding in 1980's Wasson Field (Shell, Arco, and Amoco)

Shell, in the Denver Unit of Wasson field took a different approach to understanding $CO_2$ management. It first divided the Denver Unit up into three project areas. The main area was a Continuous Injection project where they targeted 4% HCPV/per year, the second project area was a Constant WAG injection area (DUWAG Area) where they tartegeted 2% HCPV/per year $CO_2$ and 2% HCPV/per year Water. The final injection area in the western part of the unit was poorer quality reservoir and reminents of a gas cap. As technology improved, advancements in $CO_2$ flood design were implemented into reservoir management practices by Shell through the 1990's. The nine-spot inverted line drive configuration was changed to more of a direct line drive after conformance issues were identified in the field based on performance and pattern adjustments could better control conformance issues.

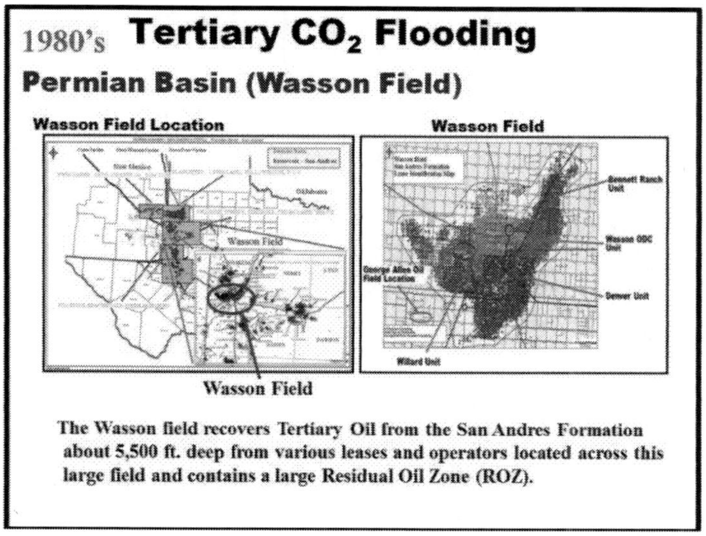


Arco, in the Willard Unit first implemented an area wide switching program based on making large area switches. The reservoir responed with some injectors taking 4% HCPV where others may have had 1% HCPV injected. Due to pattern imbalance issues, Arco then switched to an individual pattern WAG management program.

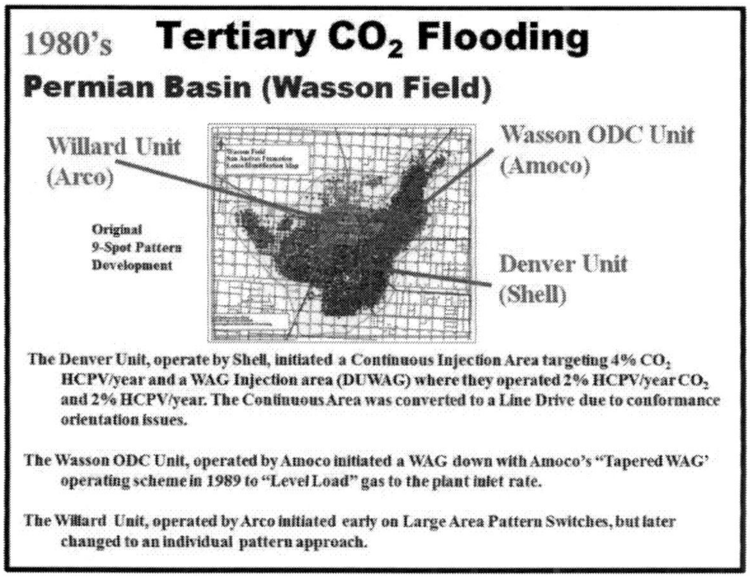


Wasson Field
Bibliography References:
7,8,9,10,11,18,12,13,14,15,16

Amoco, in 1989, at the same they implemented Tapered WAG in Slaughter field, implemented a Tapered WAG at Wasson ODC Unit to "Level Load" gas production to its Ryan Holmes Plant Gas Rate Limit. As a result, the benefit of modifying to a Tapered WAG extended Tertiary Oil Recovery way beyond all of Amoco's original model predictions.

## $CO_2$ Flooding in 1980's (Amerada Hess – Seminole San Andres Unit – Constant WAG)

The best example of $CO_2$ Tertiary Recovery is the Seminole San Andres Unit in the Permian Basin. The field has undergone Primary, Secondary Waterflood, $CO_2$ Tertiary operations in the Main Pay, and in 1996 initiated $CO_2$ Tertiary operations in the Residual Oil Zone (ROZ).

The Seminole field was discovered in 1936 with **water injection operations** initiated in 1971. Under Primary Recovery the field would have only recovered 12.8% of its Original Oil-in-Place (OOIP). With Secondary Water flood operations, the field would have recovered 42.7% of its Original Oil-in-Place.

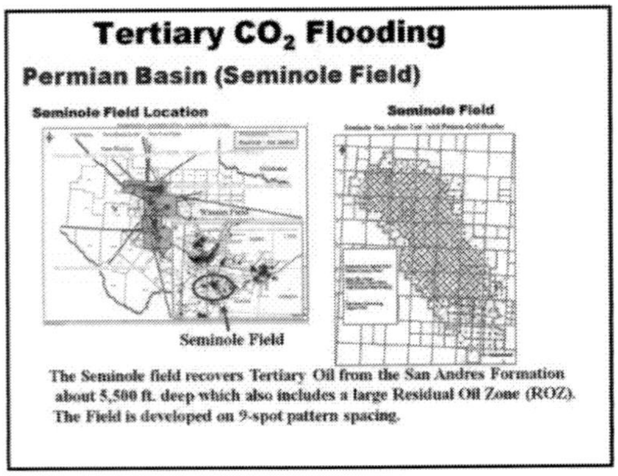


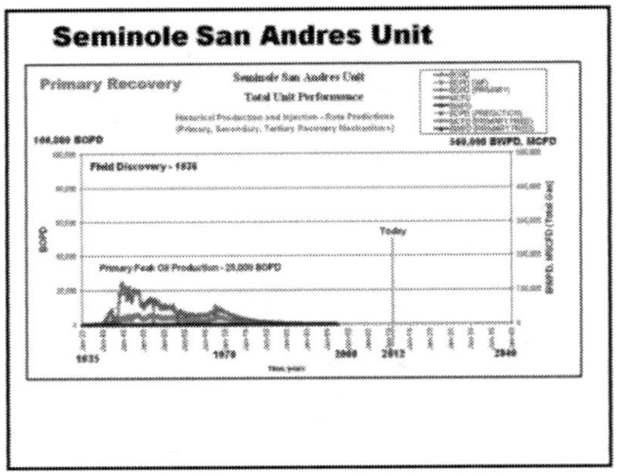


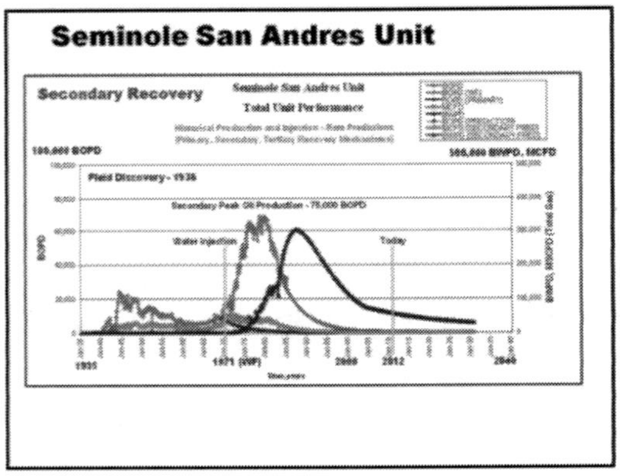


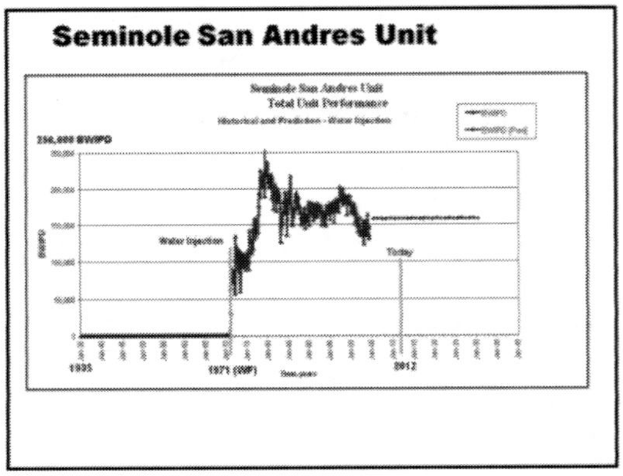


**Tertiary $CO_2$ Injection operations** into the Main Pay Zone was initiated in 1983. Under $CO_2$ Injection the field's Main Pay Section is expected to recover an additional 24% of its Original Oil-in-Place (OOIP).

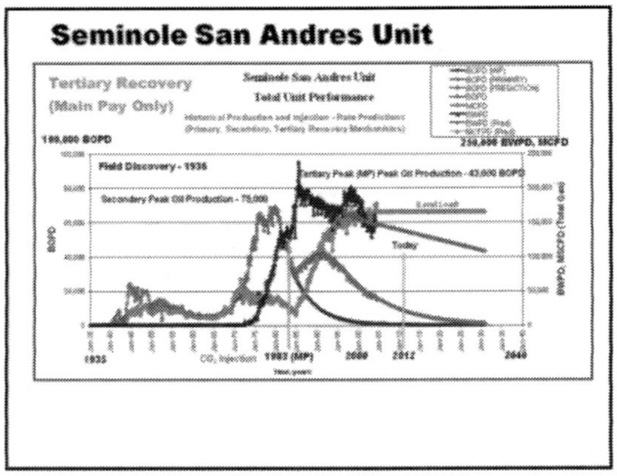


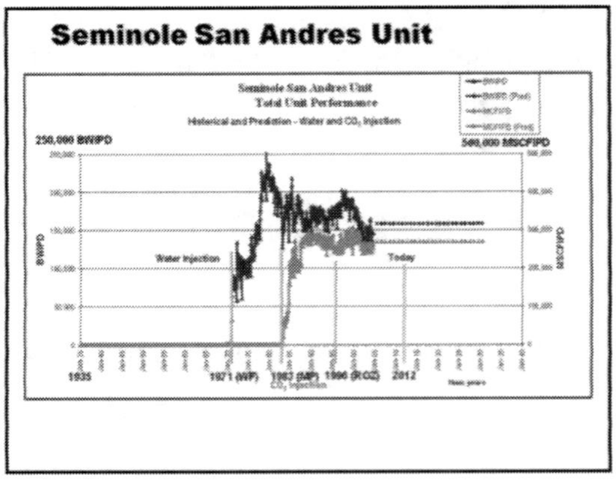


| Seminole Unit |
|---|
| Bibliography References: 58 thru 65 |

## Seminole San Andres Unit – (Main Pay Oil Zone Example)

Primary, Secondary, and $CO_2$ Tertiary Main Pay Oil Recovery as a percent of Original Oil-in-Place (OOIP) are shown below. Under Primary Operations, the field would have recovered 12.8% of its OOIP before abandoning operations. Water flood operations would have increased oil recovery to 42.7% OOIP (Primary plus Secondary). Tertiary operations with $CO_2$ in the Main Pay zone would have increased oil production by 24% OOIP. Total Primary+Secondary+Tertiary (MP) = 66.6% OOIP.

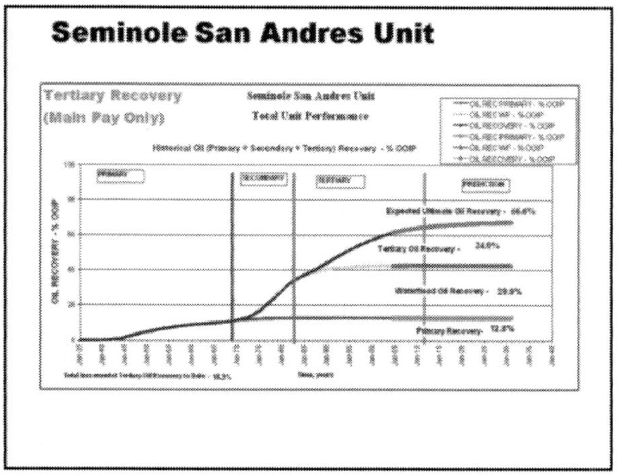


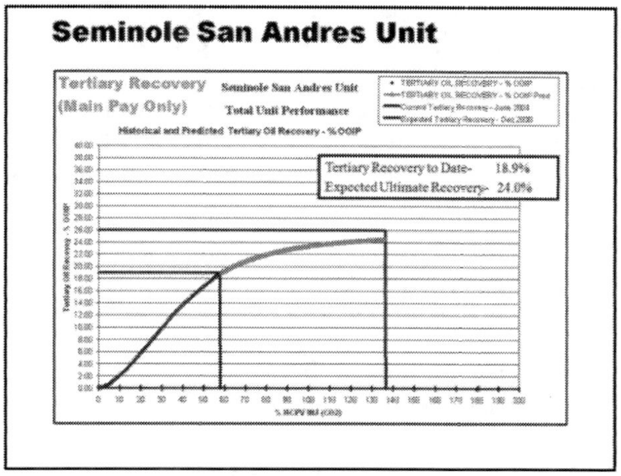


## Seminole San Andres Unit – (Main Pay – Amerada Hess WAG Management)

The Main Pay Zone was developed on Inverted a Nine-spot Pattern development. In 1983, the field initiated $CO_2$ injection into a phased-in pattern development across the field. By 1987, 95% of the $CO_2$ flood had been completed south to the town of Seminole. Amerada Hess's WAG Management was based on a 2:1 WAG Ratio with 3.0% HCPV $CO_2$ Half cycles and 6.0% HCPV Water Half Cycle. The field has maintained or increased the water to better control gas breakthrough over time.

## Seminole San Andres Unit – (Residual Oil Zone (ROZ) Example)

A Residual Oil Zone (ROZ) is created when oil within the original oil column migrates away from the field over geologic time creating a ROZ interval. In addition to the Main Pay Zone, the Seminole field contains a very large Residual Oil Zone (ROZ). The size of the ROZ is about the same size at the residual oil remaining in the Main Pay. $CO_2$ Injection commenced in 1996 into the ROZ. Injection into the ROZ will extend field life beyond the 2050's.

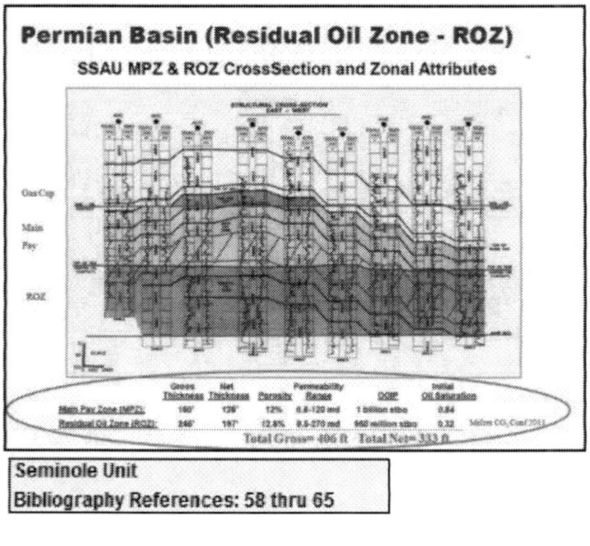


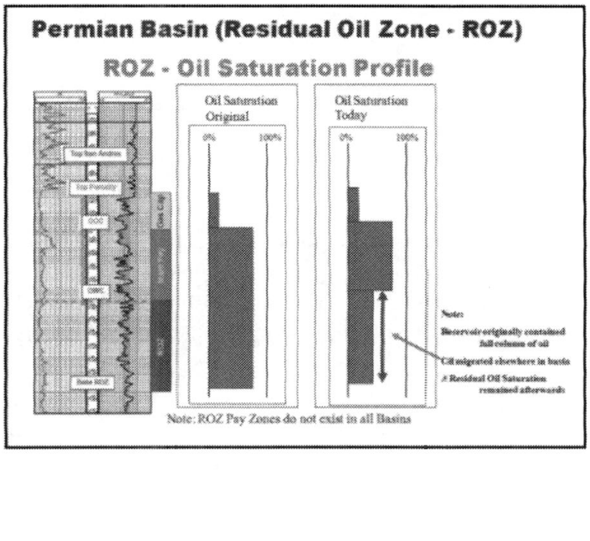


| Seminole Unit |
|---|
| Bibliography References: 58 thru 65 |

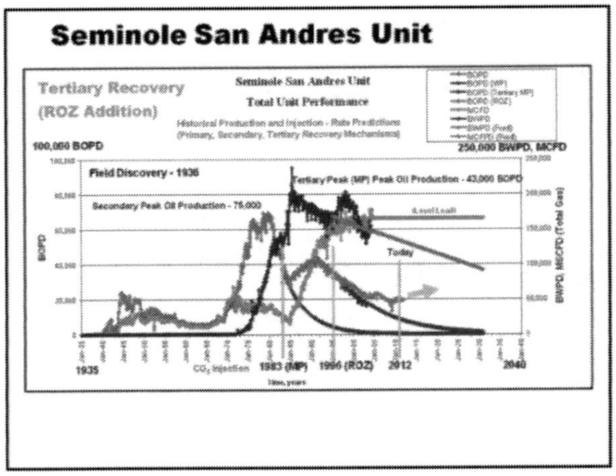


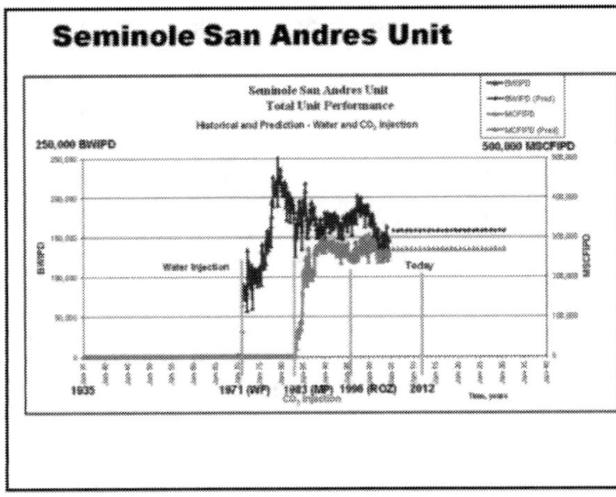


The Residual Oil Zone is responding well to $CO_2$ Injection. Since the ROZ reserves have never been tied to any Basin Study Estimates, these reserves are "NEW" bookable reserves. Future Performance is based on ROZ reservoir characterization unique to the Seminole field. Tertiary Performance is expected to be similar to that of the Main Pay or better, which depends on the quality of the ROZ pay section.

## Residual Oil Zone (ROZ) Enhanced Oil Recovery Potential

Compared to other types of oil recovery mechanisms, $CO_2$ is the best method designed to recover tertiary oil from Residual Oil (ROZ) Zones. Residual Oil Zones have been identified across the World. The Permian basin is the first region to identify it, characterize it, and now exploit it. Advanced Resources International has estimated the ROZ in the Permian Basin could contain as much as 30 Billion barrels of recoverable tertiary oil from the ROZ. The Wyoming Enhanced Oil Recovery Institute has estimated the state of Wyoming may contain as much as 800 to 1,200 million barrels of ROZ potential in the Big Horn Basin of Wyoming, not counting potential in other basins in the region.

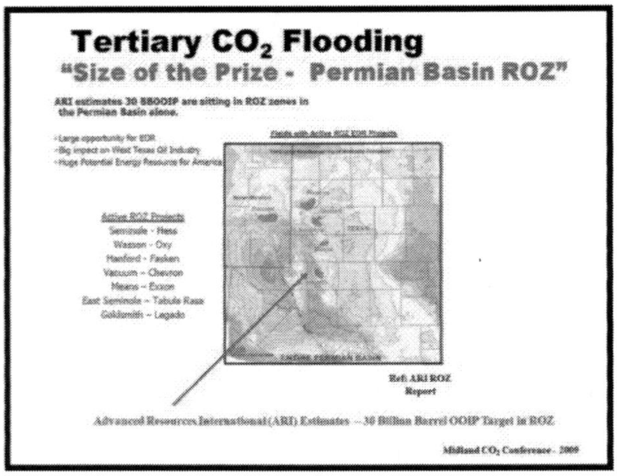


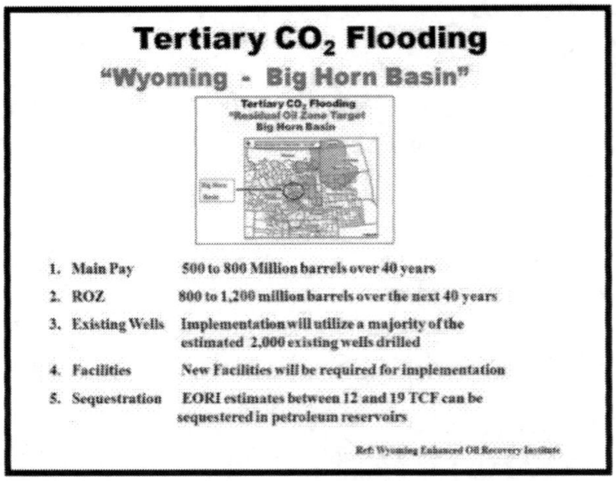


Anthropogenic $CO_2$ through $CO_2$ sequestration could expand this target to areas of the World that contain ROZ zones, but lacked the $CO_2$ to make the projects economically attractive.

# $CO_2$ Flooding in the 2000's (New Millennium)

In the 21st Century, the benefit of controlling the $CO_2$ WAG process has extended oil recovery to new levels. As previously shown, the Seminole Main Pay $CO_2$ flood in the Permian Basin is expected to recover 24% of its Original Oil-in-Place with a 140% HCPV injection of $CO_2$, not counting the additional recovery from the ROZ.

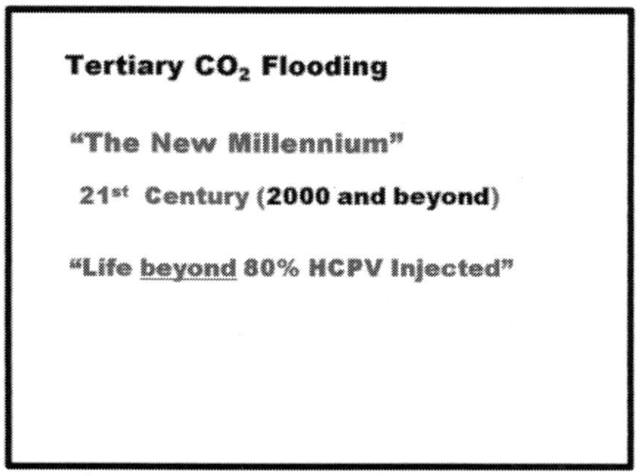

However, not all $CO_2$ floods are alike; Tertiary Oil Recovery from sandstone reservoirs are different from recovery from carbonate reefs just as un-fractured reservoirs are different from fractured reservoirs.

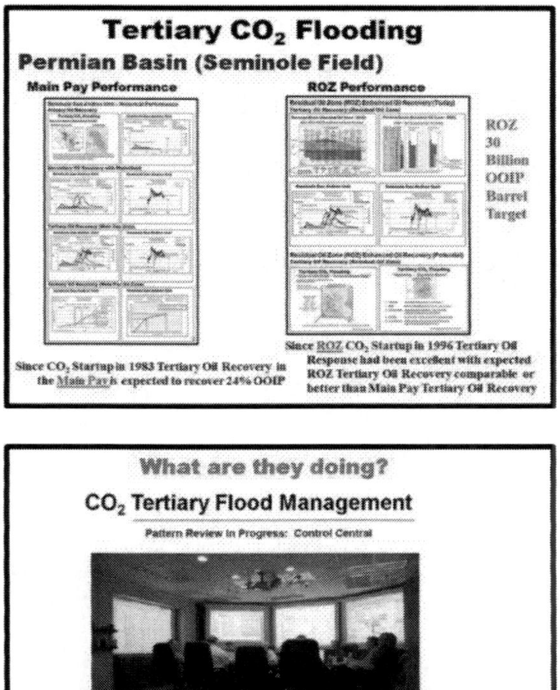


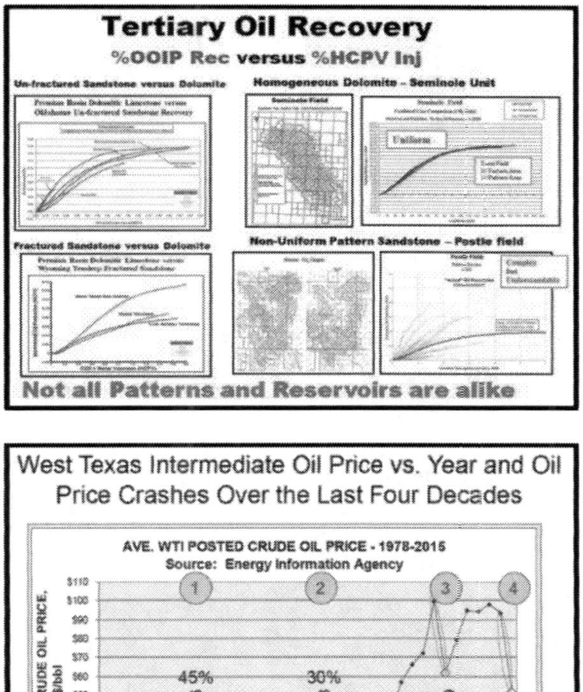


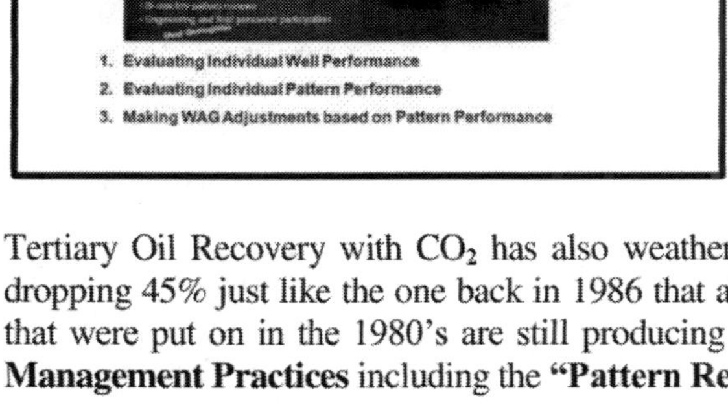


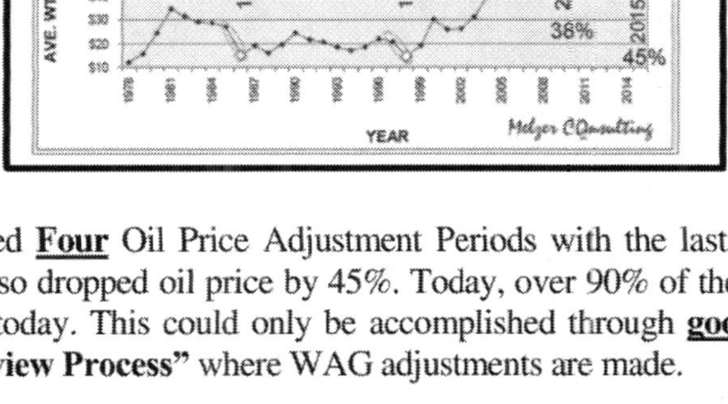


Ref. 108

Tertiary Oil Recovery with $CO_2$ has also weathered **Four** Oil Price Adjustment Periods with the last one in 2015 dropping 45% just like the one back in 1986 that also dropped oil price by 45%. Today, over 90% of the $CO_2$ Floods that were put on in the 1980's are still producing today. This could only be accomplished through **good Reservoir Management Practices** including the **"Pattern Review Process"** where WAG adjustments are made.

**$CO_2$ Flooding in the Rocky Mountains (WAG Management Decisions (The Good, the Bad, and the Ugly)**

Different Operators in the 1980's developed different WAG philosophies depending on the reservoir pattern configuration, formation dip, reservoir pressure, miscibility pressure, etc. As technology improved, advancements in $CO_2$ flood design were implemented into reservoir management practices to improve overall performance. However, companies should be aware that **NOT** all reservoirs are alike even though they were formed in the same geologic time period with similar producing characteristics.

The example shown below illustrates the fact that not all reservoirs are alike. The Rangely Weber Unit in Wyoming produces from the Weber Sand which is similar in geologic age to Bairoil's Tensleep Reservoir producing section. In addition, $CO_2$ Tertiary WAG performance at 60% $CO_2$ HCPV Injected recovers about 9% OOIP in both fields.

**NOT ALL FORMATIONS ARE THE SAME**

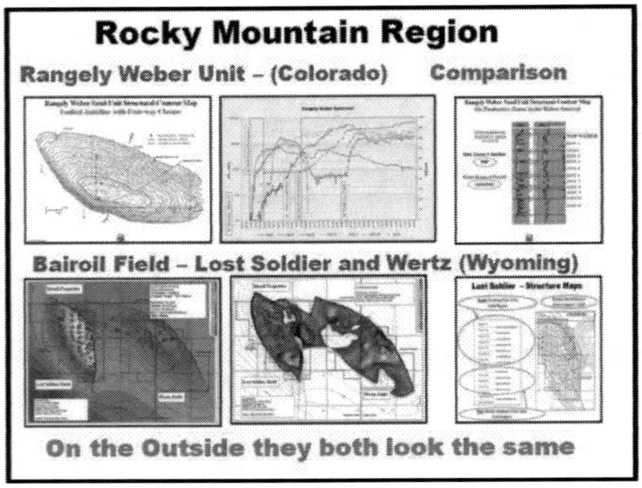


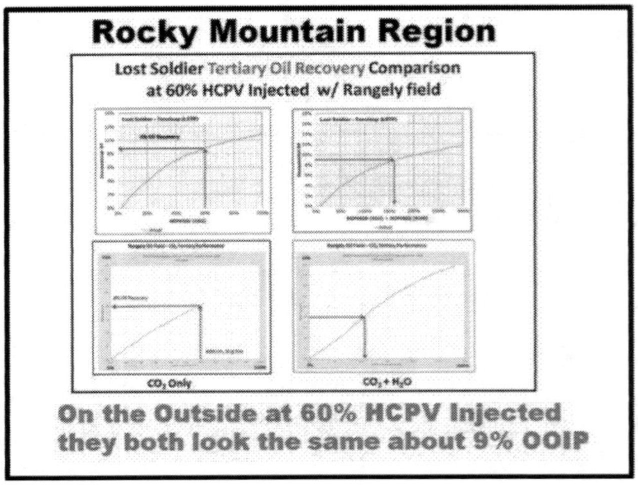


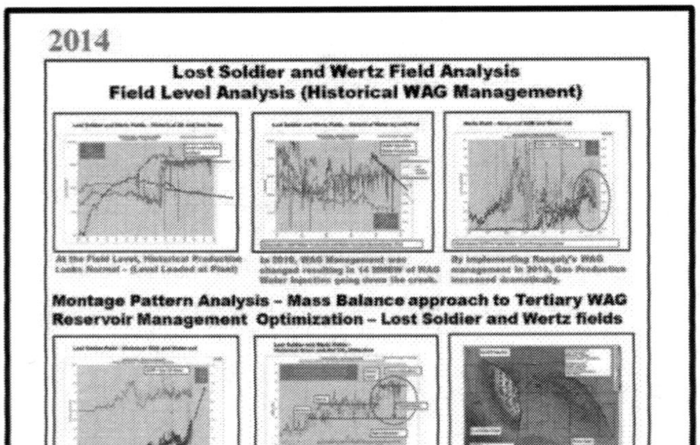


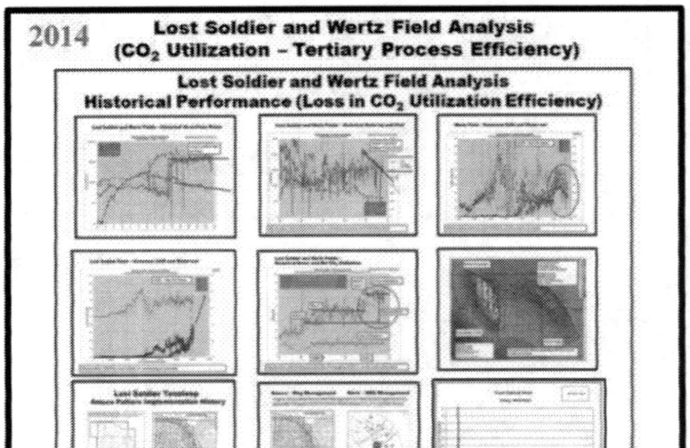


In 2010, Merit Company, who operated the field, modified Amoco's five-spot pattern 9:1 WAG scheme to Rangely's 1:1, 2:1, and 3:1 WAG scheme by dividing the unit up into thirds. From 2010 to 2014, 14 million barrels of water went down the creek that was supposed to be used for WAG Management. The overall effect gassed out 37% of the wells in the Lost Soldier field that deteriated Amoco's Gross $CO_2$ Utilization WAG efficiency rating of 13 MSCF/BO to over 40 MSCF/BO under Merit's operatorship.

**Fortunately BAD WAG Practices can be fixed…..**

# Life beyond 80% HCPV- (Tertiary Oil Recovery, $CO_2$ Gross and Net Utilization)

Since 1972, over 130+ Commercial $CO_2$ floods have been operated in the United States with the majority of these projects still active today. During this time, Engineers have reported a wide range of Tertiary Oil Recovery, Gross Utilization and Net Utilization values at varying stages of maturity. The question becomes "What would Operators report on these $CO_2$ floods today?" And "What would Oil Recovery, Gross Utilization and Net Utilization look like under Extended $CO_2$ Slug volumes?" The answer to those questions depends on Reservoir Type. Data from these $CO_2$ projects from various SPE and DOE reports are listed below by formation type:

**Table 2. - San Andres and Grayburg Formation – Dolomite**

| | | | | Current Economic | | | Life beyond 80% HCPV Inj | | | |
|---|---|---|---|---|---|---|---|---|---|---|
| State | Formation | Field | Lease | Tertiary Recovery (%OOIP) | Gross $CO_2$ Utilization (MSCF/STB) | Net $CO_2$ Utilization (MSCF/STB) | Tertiary Recovery (%OOIP) | Gross CO2 Utilization (MSCF/STB) | Net CO2 Utilization (MSCF/STB) | Bibliography References |
| Texas | San Andres | Wasson | Denver Unit | 15 to 20 | 10 to 15 | 5 to 10 | 20 to 25 | 5 to 10 | 1 to 5 | 7,8,9,10 |
| Texas | San Andres | Wasson | Denver Unit-ROZ | 10 to 15 | 15 to 20 | 5 to 10 | 15 to 20 | 10 to 15 | 5 to 10 | 11,18 |
| Texas | San Andres | Wasson | Wasson ODC Unit | 15 to 20 | 10 to 15 | 5 to 10 | 20 to 25 | 5 to 10 | 1 to 5 | 12 |
| Texas | San Andres | Wasson | Cornell Unit | 15 to 20 | 10 to 15 | 5 to 10 | 20 to 25 | 5 to 10 | 1 to 5 | 13 |
| Texas | San Andres | Wasson | Bennett Ranch Unit | 15 to 20 | 10 to 15 | 5 to 10 | 20 to 25 | 5 to 10 | 1 to 5 | 14,15 |
| Texas | San Andres | Wasson | Willard Unit | 15 to 20 | 10 to 15 | 5 to 10 | 20 to 25 | 5 to 10 | 1 to 5 | 16 |
| Texas | San Andres | Seminole | Seminole Unit-Main Pay | 20 to 25 | 10 to 15 | 5 to 10 | 25 to 30 | 5 to 10 | 1 to 5 | 17 |
| Texas | San Andres | Seminole | Seminole Unit-ROZ | 10 to 15 | 15 to 20 | 5 to 10 | 15 to 20 | 10 to 15 | 5 to 10 | 11,18 |
| Texas | San Andres | Slaughter | Slaughter Estate Unit | 15 to 20 | 15 to 20 | 5 to 10 | 20 to 25 | 10 to 15 | 5 to 10 | 19 thru 24 |
| Texas | San Andres | Slaughter | Frazier Unit | 15 to 20 | 15 to 20 | 5 to 10 | 20 to 25 | 10 to 15 | 5 to 10 | |
| Texas | San Andres | Slaughter | Central Mallet Unit | 15 to 20 | 15 to 20 | 5 to 10 | 20 to 25 | 10 to 15 | 5 to 10 | 25 |
| Texas | San Andres | Slaughter | Slaughter Sundown | 15 to 20 | 15 to 20 | 5 to 10 | 20 to 25 | 10 to 15 | 5 to 10 | 26, 27,28 |
| Texas | San Andres | Slaughter | Mobil Mallet Unit | 15 to 20 | 15 to 20 | 5 to 10 | 20 to 25 | 10 to 15 | 5 to 10 | 29 |
| Texas | San Andres | Slaughter | H T Boyd Unit | 15 to 20 | 15 to 20 | 5 to 10 | 20 to 25 | 10 to 15 | 5 to 10 | |
| Texas | San Andres | Slaughter | Igoe Smith | 15 to 20 | 15 to 20 | 5 to 10 | 20 to 25 | 10 to 15 | 5 to 10 | |
| Texas | San Andres | Slaughter | Alex Slaughter Estate | 15 to 20 | 15 to 20 | 5 to 10 | 20 to 25 | 10 to 15 | 5 to 10 | |
| Texas | San Andres | Levelland | Levelland | 5 to 10 | 15 to 20 | 5 to 10 | 10 to 15 | 15 to 20 | 5 to 10 | 30,31,32 |
| Texas | San Andres | Means (San Andres) | Means (San Andres) | 10 to 15 | 15 to 20 | 5 to 10 | 20 to 25 | 10 to 15 | 5 to 10 | 33,34 |
| Texas | San Andres | Goldsmith | Goldsmith | 10 to 15 | 15 to 20 | 5 to 10 | 20 to 25 | 10 to 15 | 5 to 10 | 35,36 |
| Texas | San Andres | Hanford | Hanford | 10 to 15 | 15 to 20 | 5 to 10 | 20 to 25 | 10 to 15 | 5 to 10 | 37 |
| New Mexico | Grayburg San Andres | Vacuum | East Vacuum | 10 to 15 | 15 to 20 | 5 to 10 | 15 to 20 | 10 to 15 | 5 to 10 | 38,39 |
| New Mexico | Grayburg San Andres | Vacuum | Central Vacuum | 10 to 15 | 15 to 20 | 5 to 10 | 15 to 20 | 10 to 15 | 5 to 10 | 40 |
| New Mexico | Grayburg San Andres | Maljamar | Maljamar Unit | 10 to 15 | 15 to 20 | 5 to 10 | 15 to 20 | 10 to 15 | 5 to 10 | 41,42 |
| New Mexico | Grayburg San Andres | North Hobbs | North Hobbs | 10 to 15 | 15 to 20 | 5 to 10 | 15 to 20 | 10 to 15 | 5 to 10 | |
| Texas | San Andres | East Penwell (SA) Unit | East Penwell (SA) Unit | 5 to 10 | 15 to 20 | 5 to 10 | 10 to 15 | 15 to 20 | 5 to 10 | |
| Texas | Grayburg | Mabee | Mabee | 10 to 15 | 15 to 20 | 5 to 10 | 15 to 20 | 10 to 15 | 5 to 10 | 43 |
| Texas | San Andres | GMK South | GMK South | 5 to 10 | 15 to 20 | 5 to 10 | 10 to 15 | 15 to 20 | 5 to 10 | 44 |
| Texas | San Andres | Adair San Andres Unit | Adair San Andres Unit | 5 to 10 | 15 to 20 | 5 to 10 | 10 to 15 | 15 to 20 | 5 to 10 | |
| Texas | San Andres | Cedar Lake | Cedar Lake | 5 to 10 | 15 to 20 | 5 to 10 | 5 to 10 | 15 to 20 | 5 to 10 | |
| Texas | San Andres | Welch | West Welch | 5 to 10 | 15 to 20 | 5 to 10 | 10 to 15 | 10 to 15 | 5 to 10 | 45 |
| Texas | San Andres | Welch | South Welch | 5 to 10 | 15 to 20 | 5 to 10 | 10 to 15 | 10 to 15 | 5 to 10 | 46 |
| Texas | Grayburg San Andres | Cowden | North Cowden Demo. | 5 to 10 | 15 to 20 | 5 to 10 | 5 to 10 | 15 to 20 | 5 to 10 | |
| Texas | San Andres | Cowden | South Cowden | 5 to 10 | 15 to 20 | 5 to 10 | 5 to 10 | 15 to 20 | 5 to 10 | 47, 48 |

**Table 3. - Clearfork Formation – Limestone (Tight - Low Permeability)**

| | | | | Current Economic | | | Life beyond 80% HCPV Inj | | | |
|---|---|---|---|---|---|---|---|---|---|---|
| State | Formation | Field | Lease | Tertiary Recovery (%OOIP) | Gross $CO_2$ Utilization (MSCF/STB) | Net $CO_2$ Utilization (MSCF/STB) | Tertiary Recovery (%OOIP) | Gross $CO_2$ Utilization (MSCF/STB) | Net $CO_2$ Utilization (MSCF/STB) | Bibliography References |
| Texas | Clearfork | Anton Irish | Anton Irish | 5 to 10 | 10 to 15 | 5 to 10 | 10 to 15 | 10 to 15 | 5 to 10 | 49 |
| Texas | Clearfork | Wasson (South) | Wasson (South) | 1 to 5 | 10 to 15 | 5 to 10 | 1 to 5 | 10 to 15 | 5 to 10 | 50 |

**Table 4. - Devonian Formation – Tripolitic Chert**

| | | | | Current Economic | | | Life beyond 80% HCPV Inj | | | |
|---|---|---|---|---|---|---|---|---|---|---|
| State | Formation | Field | Lease | Tertiary Recovery (%OOIP) | Gross $CO_2$ Utilization (MSCF/STB) | Net $CO_2$ Utilization (MSCF/STB) | Tertiary Recovery (%OOIP) | Gross $CO_2$ Utilization (MSCF/STB) | Net $CO_2$ Utilization (MSCF/STB) | Bibliography References |
| Texas | Devonian | Dollarhide | Dollarhide Devonian Unit | 10 to 15 | 5 to 10 | 1 to 5 | 15 to 20 | 5 to 10 | 1 to 5 | 51 thru 54 |
| New Mexico | Devonian | Dollarhide | West Dollarhide Unit | 10 to 15 | 5 to 10 | 1 to 5 | 15 to 20 | 5 to 10 | 1 to 5 | |
| Texas | Devonian | Dollarhide | North Dollarhide Unit | 5 to 10 | 5 to 10 | 1 to 5 | 10 to 15 | 5 to 10 | 1 to 5 | 55 |
| Texas | Devonian | North Cross | N. Cross-Devonian Unit | 35 to 40 | 5 to 10 | 1 to 5 | 40 to 45 | 5 to 10 | 1 to 5 | 56-57 |

**Table 5. - Canyon Reef Formation – Karsted Limestone (High Permeability)**

| | | | | Current Economic | | | Life beyond 80% HCPV Inj | | | |
|---|---|---|---|---|---|---|---|---|---|---|
| State | Formation | Field | Lease | Tertiary Recovery (%OOIP) | Gross $CO_2$ Utilization (MSCF/STB) | Net $CO_2$ Utilization (MSCF/STB) | Tertiary Recovery (%OOIP) | Gross $CO_2$ Utilization (MSCF/STB) | Net $CO_2$ Utilization (MSCF/STB) | Bibliography References |
| Texas | Canyon Reef | Kelly Snyder (Sacroc) | Sacroc Unit | 10 to 15 | 10 to 15 | 5 to 10 | 15 to 20 | 10 to 15 | 5 to 10 | 58 thru 65 |
| Texas | Canyon Reef | Salt Creek | Salt Creek | 10 to 15 | 10 to 15 | 5 to 10 | 10 to 15 | 10 to 15 | 5 to 10 | 66,67 |
| Texas | Canyon Reef | Sharon Ridge | Sharon Ridge | 5 to 10 | 10 to 15 | 5 to 10 | 10 to 15 | 10 to 15 | 5 to 10 | 68 |
| Texas | Canyon Reef | Cogdell | Cogdell | 5 to 10 | 10 to 15 | 5 to 10 | 10 to 15 | 10 to 15 | 5 to 10 | 69 |

**Table 6. - Strawn, Morrow, Delaware, Springer, Marmaton, and Yates (Fluivial Deltaic, Point Bar, Turbidite)**

| | | | | Current Economic | | | Life beyond 80% HCPV Inj | | | |
|---|---|---|---|---|---|---|---|---|---|---|
| State | Formation | Field | Lease | Tertiary Recovery (%OOIP) | Gross $CO_2$ Utilization (MSCF/STB) | Net $CO_2$ Utilization (MSCF/STB) | Tertiary Recovery (%OOIP) | Gross $CO_2$ Utilization (MSCF/STB) | Net $CO_2$ Utilization (MSCF/STB) | Bibliography References |
| Texas | Strawn | Katz | Katz | 10 to 15 | 10 to 15 | 5 to 10 | 15 to 20 | 10 to 15 | 5 to 10 | 70 |
| Oklahoma | Morrow | Postle | Postle | 10 to 15 | 10 to 15 | 5 to 10 | 15 to 20 | 10 to 15 | 1 to 5 | 71,72 |
| Oklahoma | Morrow | Camrick | Camrick | 10 to 15 | 10 to 15 | 5 to 10 | 10 to 15 | 10 to 15 | 5 to 10 | |
| Texas | Delaware Ramsey | Twofreds | Twofreds | 10 to 15 | 10 to 15 | 5 to 10 | 15 to 20 | 10 to 15 | 5 to 10 | 73,74,75 |
| Texas | Delaware Ramsey | Ford Geraldine | Ford Geraldine | 10 to 15 | 10 to 15 | 5 to 10 | 15 to 20 | 10 to 15 | 5 to 10 | 76,77 |
| Texas | Delaware Ramsey | El Mar | El Mar | 10 to 15 | 10 to 15 | 5 to 10 | 10 to 15 | 10 to 15 | 5 to 10 | |
| Texas | Delaware Ramsey | East Ford | East Ford | 10 to 15 | 10 to 15 | 5 to 10 | 10 to 15 | 10 to 15 | 5 to 10 | |
| Texas | Springer | Northeast Purdy | Northeast Purdy | 10 to 15 | 10 to 15 | 5 to 10 | 10 to 15 | 10 to 15 | 5 to 10 | 78 |
| Texas | Marmaton | Hansford Marmaton | Hansford Marmaton | 10 to 15 | 10 to 15 | 5 to 10 | 10 to 15 | 10 to 15 | 5 to 10 | 79 |
| Texas | Yates | North Ward Estes | North Ward Estes | 5 to 10 | 10 to 15 | 1 to 5 | 10 to 15 | 10 to 15 | 1 to 5 | 80 |
| California | Stevens MMB | Elk Hills | Elk Hills MMB | 10 to 20 | 10 to 15 | 5 to 10 | 20 to 30 | 10 to 15 | 5 to 10 | 81 |

**Table 7. -Tensleep, Mesaverde Almond, Weber, Sprayberry (Fractured Sandstone)**

| | | | | Current Economic | | | Life beyond 80% HCPV Inj | | | |
|---|---|---|---|---|---|---|---|---|---|---|
| State | Formation | Field | Lease | Tertiary Recovery (%OOIP) | Gross $CO_2$ Utilization (MSCF/STB) | Net $CO_2$ Utilization (MSCF/STB) | Tertiary Recovery (%OOIP) | Gross $CO_2$ Utilization (MSCF/STB) | Net $CO_2$ Utilization (MSCF/STB) | Bibliography References |
| Wyoming | Tensleep | Wertz | Wertz | 5 to 10 | 10 to 15 | 5 to 10 | 10 to 15 | 10 to 15 | 5 to 10 | 82 |
| Wyoming | Tensleep | Lost Soldier | Lost Soldier | 5 to 10 | 10 to 15 | 5 to 10 | 10 to 15 | 10 to 15 | 5 to 10 | 83 thru 86 |
| Wyoming | Mesaverde Almond | Patrick Draw Monell | Patrick Draw Monell | 5 to 10 | 10 to 15 | 5 to 10 | 10 to 15 | 10 to 15 | 5 to 10 | |
| Colorado | Weber SS | Rangely | Rangely Weber Unit | 5 to 10 | 5 to 10 | 1 to 5 | 10 to 15 | 5 to 10 | 1 to 5 | 87 thru 92 |
| Texas | Sprayberry | Sprayberry | Sprayberry | 0 to 2 | 10 to 15 | 5 to 10 | 0 to 2 | 10 to 15 | 5 to 10 | 93,94 |

**Table 8. - Heavy Oil**

| | | | | Current Economic | | | Life beyond 80% HCPV Inj | | | |
|---|---|---|---|---|---|---|---|---|---|---|
| State | Formation | Field | Lease | Tertiary Recovery (%OOIP) | Gross $CO_2$ Utilization (MSCF/STB) | Net $CO_2$ Utilization (MSCF/STB) | Tertiary Recovery (%OOIP) | Gross $CO_2$ Utilization (MSCF/STB) | Net $CO_2$ Utilization (MSCF/STB) | Bibliography References |
| California | Ranger | Wilmington (Onshore) | FB1, FB3, FB5 | 10 to 15 | 15 to 20 | 5 to 10 | 10 to 15 | 15 to 20 | 5 to 10 | 95 thru 98 |
| Arkansas | Ozan (Meaken Sand) | Lick Creek | Lick Creek Unit | 10 to 15 | 15 to 20 | 5 to 10 | 10 to 15 | 15 to 20 | 5 to 10 | 99, 100 |

# Conclusion

With more than forty-five years of successful enhanced oil recovery (EOR) projects in the Permian Basin (Texas), Mississippi, Wyoming, Colorado, California, Oklahoma, and several countries worldwide, carbon dioxide $CO_2$ flooding is a proven method for extending field life. $CO_2$ acts as a solvent to overcome forces that trap oil in tiny rock pores, helping sweep immobile oil left after primary or secondary recovery operations. Generally, $CO_2$ is not miscible at first contact with reservoir oils, but miscibility can be developed in reservoirs above or near the Minimum Miscibility Pressure (MMP). $CO_2$ can attain miscibility through a multiple-contact process that vaporizes or extracts both intermediate and higher molecular weight hydrocarbons from the reservoir oil. The $CO_2$ phase picks up many intermediate hydrocarbon components from the oil, swells the oil, and reduces oil viscosity, making it mobile to move through the rock.

Advances in technology and reservoir understanding have made detailed evaluation of potential EOR candidates obtainable within months, not years. In addition, improved reservoir management and innovative investment plans have significantly reduced risks and increased rewards. Many of the original questions about $CO_2$ flooding involved the displacement efficiency of the process, how $CO_2$ would interact with the oil, and how much oil could be recovered. Many of these questions have been answered with better reservoir management tools. Not all fields are good candidates for $CO_2$ Tertiary Recovery. A reservoir must contain certain characteristics for a $CO_2$ flood to be successful. In the past, it was thought the oil must be found at depths sufficient to allow for high pressures, so that $CO_2$ and oil develop total miscibility. This is not necessary correct. Most $CO_2$ floods operate at reservoir pressures that are above their minimum miscibility pressure. But today, it is not uncommon to find $CO_2$ projects that operate below or near the minimum miscibility pressure. The $CO_2$ still produces tertiary oil. The process is not as efficient as that operated above the minimum miscibility pressure. Most historical $CO_2$ floods have targeted reservoirs that have a gravity of 25 API units or greater, but low API Gravity reservoirs are also targets. For example: Wilmington field in California produces a14 API Gravity crude from the Ranger formation. Three pilots were conducted in Fault Blocks I, III, and V. Eventhough economic performance was reported poor, mostly due to an inadequate

$CO_2$ source and low oil price enviornment, a good number of wells increased oil rate from 30 BOPD to over 300 BOPD after $CO_2$ was injected. $CO_2$ has the ability to affect the full $C_2$ through $C_{30+}$ compositional range. Whereas, Nitrogen, and in some cases flue gas injection, will only extract the lighter components ($C_2$ through $C_6$). In addition, a high percentage of intermediate hydrocarbons in the oil composition can be beneficial in making the overall recovery process more efficient. If these occur naturally in the oil, then the oil will probably contain a low value of Minimum Miscibility Pressure (MMP). If the oil has a high MMP, then additions such as propane, butane, condensate, or other types of hydrocarbons can be added to the $CO_2$ injection stream to lower the minimum miscibility pressure and improve overall oil recovery.

# References

## Conventional WAG $CO_2$ Flood History


1. Turek, E.A., Christopher, C.A., Stein, M.H., Merchant, D.H., "History of WAG Development" by Turek, Prepared for BP under Amoco's Learning Program, Internal Report, October, 2000.
2. Merchant, D.H., "GWR and Slug Size Optimization of the WAG Process", Prepared for Amoco Production Company, Permian Basin Business Unit, Internal Report, File: JWA-539.41-111, Model Study 1986-1987, Report Published: January 1988.
3. Merchant, D.H., "Injection Well Profiling of the $CO_2$ Recovery Process", Prepared for Amoco Production Company, Permian Basin Business Unit, Internal Report, File: DDF-539.41-1586, Model Study 1987-1988, Report Published: August 1988.
4. Merchant, D.H., "EOR/ $CO_2$ Storage Optimization and Economics 2004 CCP Program ", Prepared for BP Alternative Energy under the Carbon and Capture Program CCP, An Integrated, Collaborative Technology Development Project for Next Generation $CO_2$ Separation, Capture and Geologic Sequstration, Subcontract C010: EOR/CO2 Storage Optimization and Economics – 2004 Program, Date Published: December 2004.
5. Merchant, D.H., "Enhanced Oil Recovery Methods – Conventional $CO_2$ WAG Injection", Prepared for BP Alternative Energy under the Carbon and Capture Program CCP1, An Integrated, Collaborative Technology Development Project for Next Generation $CO_2$ Separation, Capture and Geologic Sequstration, Subcontract C010: EOR/$CO_2$ Storage Optimization and Economics – 2004 Program, Date Published: October 2005.
6. Merchant, D.H., "Comparisons of Conventional $CO_2$ WAG InjectionTechniques used in the Permian Basin", presented at the 15th Annual $CO_2$ Flooding Conference, Midland, Texas, and December 10-11, 2009.
7. Tanner, C.S., Baxley, P.T., CHI, J.G., and Miller W.C., "Production Performance of the Wasson Denver Unit $CO_2$ Flood," SPE 24156, SPE/DOE Eighth Symposium on Enhanced Oil Recovery, Tulsa, Oklahoma, April 1992.
8. Kittridge, M.G., "Quantitative $CO_2$ Monitoring, Denver Unit, Wasson (San Andres field), SPE 24644, presented at the 67th Annual Technical Conference and Exhibition of SPE held in Washington DC, October, 1992.
9. Hsu, C-F, Morell, J.I., Falls, A.H., "Field-Scale $CO_2$ Flood Simulations and Their Impact on the Performance of the Wasson Denver Unit", SPE 29116, presented at 13th SPE Symposium on Reservoir Simulation held in San Antonio, Texas, 12-15 February 1995.
10. Coalmer, M.S., Hsu, C.F., Cooper, J.C., Ward, J.B., Voto, J.G., Way, K.F., and Valdez, R., "Reservoir Characterization and Development Plan of the Wasson San Andres Denver Unit Gas Cap", SPE 56549-MS, presented at the Annual Technical Conference and Exhibition of SPE held in Houston, Texas, October 3-6, 1999.
11. Koperna, G.J., Melzer, L.S., Kuuskraa, V.A., "Recovery of Oil Resources From the Residual and Transitional Oil Zones of the Permian Basin", SPE 102972-MS, presented at the Annual Technical Conference and Exhibition of SPE held in San Antonio, Texas, September 24-27, 2006.
12. Bullock, G.W., Wood, T.B, Konecki, M.L., "A Brief History of the Wasson EOR Project", Wasson ODC, SPE 17754-PA, SPE Production Engineering Journal, Volume 5, Number 3, Pages 338-344, August, 1990.
13. Todd, M.R., Cobb, W.M., McCarter, E.D., "$CO_2$ Flood Performance Evaluation for the Cornell Unit, Wasson San Andres Field", SPE 10292-PA, Journal of Petroleum Technology, Volume 34, Number 10, Pages 2271-2282, October 1982.
14. Hsu, C-F, Fitzgerald, M.A., Musallam, S.C., McCray, T.L., Purvis, W.D., "Design and Implementation of a Grass-Roots $CO_2$ Project for the Bennett Ranch Unit", SPE 35188, presented at the Permian Basin Oil and Gas Recovery Conference, Midland, Texas, March 27-29, 1996.
15. Altura Company, "Wasson Field, Bennett Ranch Unit, Wasson ODC Unit, and Denver Unit $CO_2$ Performance Update, Yoakum and Gaines County, Texas", presented at the Annual $CO_2$ Flooding Conference, Midland, Texas, December 7-8, 1999.
16. Johnston, J.W., "A Review of the Willard (San Andres) Unit $CO_2$ Injection Project", SPE 6388-MS, presented at the Permian Basin Oil and Gas Recovery Conference, Midland, Texas, March 10-11, 1977.
17. Wang, F.P., Lucia, F.J., Kerans, C., "Integrated Reservoir Characterization Study of a Carbonate Ramp Reservoir: Seminole San Andres Unit, Gaines County, Texas", SPE 36515-PA, SPE Reservoir Evaluation and Engineering Journal, Volume 1, Number 2, Pages 105-113, April 1998.

18. Melzer, L.S., Koperna, G.J., Kuuskraa, V.A., "The Origin and Resource Potential of Residual Oil Zones", SPE 102964-MS, presented at the Annual Technical Conference and Exhibition of SPE held in San Antonio, Texas, September 24-27, 2006.
19. Rowe, H.G., York, S.D., and Ader, J.C., "Slaughter Estate Unit Tertiary Pilot Performance", SPE 9796, presented at the 1981 SPE/DOE Enhanced Oil Recovery Symposium, Tulsa, Oklahoma, April 5-8, 1981.
20. Ader, J.C. and Stein, M.H., "Slaughter Estate Unit Tertiary Miscible Gas Pilot Reservoir Description," SPE 10727, Oil and Gas Journal, May, 1984.
21. Stein, M.H., Frey, D.D., Walker, R.D., and Pariani G.J., "Slaughter Estate Unit $CO_2$ Flood: Comparison between Pilot and Field-Scale Performance", SPE 13975, JPT, September, 1992.
22. Pariani, G. J., McCofloch, K. A., Warden, SJ-., and Edens, D.R., "An Approach To Optimize Economics in a West Texas C02 Flood," JPT, Sept., 1992.
23. Merchant D.H. and Thakur, S.C.,"Reservoir Management in Tertiary $CO_2$ Floods", SPE 26624, presented at the 68th Annual Technical Conference and Exhibition, Houston, Texas, October 1994.
24. Drennon, M.D., Kelm, C.H., and Whittington, H.M., "A Method for Appraising the Feasibility of Field-Scale $CO_2$ Mscible Flooding," SPE 9323, 1980 SPE SPE Annual Technical Conference and Exhibition, Dallas, Sept 21-24, 1980.
25. Honnert, M., Creel, P., Tate, R., and Everett, D., "Five Years of Ongoing Conformance Work in the Central Mallet Unit $CO_2$ Flood in West Texas Yields Improved Economics for Operator", SPE 101701-MS, presented at First International Oil Conference and Exhibition in Mexico, August 31st to September 2nd, 2006.
26. Guillot, S.N., "Horizontal Well Applications in a Miscible $CO_2$ Flood, Sundown Slaughter Estate Unit, Hockley County, Texas", ," SPE 30742, SPE SPE Annual Technical Conference and Exhibition, Dallas, October 22-25, 1995.
27. Folger, L.K., "Improved $CO_2$ Flood Predictions Using 3D Geologic Description and Simulation on the Sundown Slaughter Unit", SPE 35410, presented at SPE/DOE Tenth Symposium on Improved Oil Recovery, Tulsa, Oklahoma, April 21-24, 1996.
28. Folger, L.K., Guillot, S.N, "A Case Study of the Development of the Sundown Slaughter Unit $CO_2$ Flood Hockley County, Texas", SPE 35189, presented at the Permian Basin Oil and Gas Recovery Conference, Midland, Texas, March 27-29, 1996.
29. Kuo, M.C.T., Dulaney, J.P, Deer, M.W., Evans, B.S., Granquist, M.R., "Optimization of Waterflood Performance and $CO_2$ Flood Design Using a Modelling Approach, Mallet Unit, Slaughter Field", SPE 20377, presented at the 65TH Annual Technical Conference and Exhibition held in New Orleans, LA, September 23-26, 1990.
30. Henry, R.L., Feather, G.L., Smith, L.R., Fussell, D.D.,"Utilization of Composition Observation Wells in a West Texas $CO_2$ Pilot Flood", Levelland Unit, SPE 9786-MS, presented at SPE/DOE Symposium on Improved Oil Recovery, Tulsa, Oklahoma, April 5-8, 1981.
31. Yellig, W., "Carbon Dioxide Displacement of a West Texas Reservoir Oil", Levelland, SPE 9785-PA, SPE Oil and Gas Journal, Volume 22, Number 6, Pages 805-815, December, 1982.
32. Macon, R.B., Design and Operation of the Levelland Unit $CO_2$ Injection Facility", SPE 8410-MS, presented at the Annual Technical Conference and Exhibition held in Las Vegas, Nevada, September 23-26, 1979.
33. Magruder, J.B., Stiles, L.H. and Yelverton, T.D., "A Review of the Means San Andres Full-Scale $CO_2$ Tertiary Project", SPE/DOE 17349, presented at the SPE/DOE Enhanced Oil Recovery Symposium, Tulsa, Oklahoma, April 17-20, 1988.
34. Hambly, E.,"Means San Andres Unit – A History of $CO_2$ Flood Performance and Management", presented at the Annual $CO_2$ Flooding Conference, Midland, Texas, December 7-8, 1999.
35. Jasek, D.E., Frank, J.R., Smith, D.J., "Goldsmith San Andres Unit $CO_2$ Pilot – Design, Implementation, and Early Performance", SPE 48945-MS, presented at the Annual Technical Conference and Exhibition held in New Orleans, LA, September 27-30, 1998.
36. Merchant, D.H., "Goldsmith Unit – Tertiary Performance Evaluation – $CO_2$ Prediction", developed for Shell $CO_2$ Company and Chevron USA, Internal Report, January, 2000.
37. Merritt, M.B. and Groce, J.F., "A Case History of the Hanford San Andres Miscible $CO_2$ Project", SPE/DOE 20229, presented at the SPE/DOE Seventh Symposium on Enhanced Oil Recovery, Tulsa, Oklahoma, April 22-25, 1990.
38. Brownlee, M.H. and Sugg, L.A., "East Vacuum Grayburg-San Andres Unit $CO_2$ Injection Project Developmnet and Results to Date", SPE 16721, presented at the 62nd Annual Technical Conference and Exhibition, Dallas, Texas, Sept 27-30, 1987.
39. Martin, F.D. and J.E. Stevens, Harpole, K.J., "$CO_2$-Foam Field Test at East Vacuum Grayburg/San Andres Unit", SPE 27786, presented at the SPE/DOE Tenth Symposium on Improved Oil Recovery, Tulsa, Oklahoma, April 17-20, 1995.
40. Boomer, R.J., "Central Vacuum Unit $CO_2$ Performance Overview (SE New Mexico)", presented at the Annual $CO_2$ Flooding Conference, Midland, Texas, December 7-8, 2001.

41. Pittaway, K.R., Albright, J.C. and Hoover, J.W., "The Maljamar Carbon Dioxide Pilot: Review and Results", SPE/DOE 14940, presented at the SPE/DOE Fifth Symposium on Enhanced Oil Recovery, Tulsa, Oklahoma, April 20-23, 1986.
42. Moore, J.S. and Clark, G.C., "History Match of the Maljamar $CO_2$ Pilot Performance", SPE/DOE 17323, presented at the SPE/DOE Enhanced Oil Recovery Symposium, Tulsa, Oklahoma, April 17-20, 1988.
43. Roper, M.K., Cheng, C.T., Varnon, J.E., Pope, G.A. Sepehmoori, K., "Interpretation of a $CO_2$ WAG Injectivity Test in the San Andres Formation Using a Compositional Simulator", Mabee Field, SPE 24163-MS, presented at the SPE/DOE Seventh Symposium on Improved Oil Recovery, Tulsa, Oklahoma, April 22-24, 1992.
44. Jaramillo, A.R.,"Utilization of a Black-Oil Simulator as a Monitor of Waterlood Operations in a San Andres Reservoir", SPE 19046-MS, GMK Unit, Unsolicited, This document was submitted for consideration for publication but included in SPE library, 1989.
45. Taylor, A.R. Hinterlong, G.H., Kumar, K.H., "West Welch $CO_2$ Flood Simulation with an Equation of State and Mixed Wettability", SPE 39808, presented at the Permian Basin Oil and Gas Recovery Conference, Midland, Texas, March 23-26, 1998.
46. Keeling, R.J., "$CO_2$ Miscible Flooding Evaluation of the South Welch Unit, Welch San Andres Field", SPE/DOE 12664, presented at the SPE/DOE Fouth Symposium on Enhanced Oil Recovery, Tulsa, Oklahoma, April 15-18, 1984.
47. Hallenbeck, L.D., Harpole, K.J., Sistrunk, G.T., and Wier, D.R., "Innovative Approach to $CO_2$ Project Development Holds Promise for Improving $CO_2$ Flood Economics in Smaller Fields Nearing Abandonment ", South Cowden Field Example, SPE 28334, presented at the 69th Annual Technical Conference and Exhibition held in New Orleans, LA, September 25-28, 1994.
48. Wegener, D.C., Harpole, K.J., "Determination of Relative Permeability and Trapped Gas Saturation for Predictions of WAG Performance in the South Cowden $CO_2$ Flood", SPE/DOE 35429, presented at the SPE/DOE Tenth Symposium on Enhanced Oil Recovery, Tulsa, Oklahoma, April 21-24, 1996.
49. Oxy Permian, "Anton Irish Clearfork $CO_2$ Flood Case History" presented at the Annual $CO_2$ Flooding Conference, Midland, Texas, December 7-8, 1999.
50. Burbank, D.E., "Early $CO_2$ Flood Experience at the South Wasson Clearfork Unit", SPE/DOE 24160, presented at the SPE/DOE Eighth Symposium on Improved Oil Recovery, Tulsa, Oklahoma, April 22-24, 1992.
51. Poole, E.S. "Evaluation and Implementation of $CO_2$ Injection at the Dollarhide Devonian Unit", SPE 1722, presented at the SPE Permian Basin Oil and Gas Recovery Conference, Midland, Texas, March 1988.
52. Poole, E.S., "Evaluation and Implementation of $CO_2$ Injection at the Dollarhide Devonian Unit", SPE 17277, presented at the Permian Basin Oil and Gas Recovery Conference, Midland, Texas, March 10-11, 1988.
53. Lin, E.C., Poole, E.S., "Numerical Evaluation of Single-Slug, WAG, and Hybrid $CO_2$ Injection Processes, Dollarhide Devonian Unit, Andrews County Texas", SPE 20098, presented at the Permian Basin Oil and Gas Recovery Conference, Midland, Texas, March 1990.
54. Bellavance, J.F.R, "Dollarhide Devonian $CO_2$ Flood: Project Performance Review 10 years later", SPE 35190, presented at the Permian Basin Oil and Gas Recovery Conference, Midland, Texas, March 27-29, 1996.
55. Kovarik, J.J., Prasad, R.K., and Waddell, Wade, and Watts, "North Dollarhide (Devonian) Unit: Reservoir Characterization and $CO_2$ Feasibility Study", SPE 27678, presented at the Permian Basin Oil and Gas Recovery Conference, Midland, Texas, March 1994.
56. Pontius, S.B. and Tham, M.J., "North Cross (Devonian) Unit $CO_2$ Flood – Review of Flood Performance and Numerical Simulation Model", Journal of Petroleum Technology (December 1988), pp 1706-1714.
57. Mizenko, G.J., "North Cross (Devonian) Unit $CO_2$ Flood: Status Report", SPE/DOE 24210, presented at the SPE/DOE Eighth Symposium on Enhanced Oil Recovery, Tulsa, Oklahoma, April 22-24, 1992.
58. Crameik, T.D., Plassey, J.A., "Carbon Dioxide Injection Project Sacroc Unit, Scurry County, Texas", SPE 72-D001, presented at the Annual Meeting Papers, Division of Production, Houston, Texas, March 6-8, 1972.
59. Graue, D.J., Blevins, T.R., "Sacroc Tertiary $CO_2$ Pilot Project", SPE 7090-MS, presented at the SPE Symposium on Improved Methods of Recovery, Tulsa, Oklahoma, April 16-17, 1978.
60. Kane, A.V., "Performance Review of a Large-Scale $CO_2$-WAG Enhanced Recovery Project, Sacroc Kelly-Snyder Field," SPE 7091-PA, Journal of Petroleum Technology, Volume 31, Number 2, Pages 217-231, February 1979.
61. Gill, T.E., "Ten Years of Handling $CO_2$ for Sacroc Unit", SPE 11162-MS, presented at the Annual Technical Conference and Exhibition of Society of Petroleum Engineers, New Orleans, LA, Septermber 26-29, 1982.
62. Langston, M.V., Hoadley, S.F., Young, D.N., "Definitive $CO_2$ Flooding Response in the Sacroc Unit", SPE 17321-MS, presented at the SPE/DOE Enhanced Oil Recovery Symposium, Tulsa, Oklahoma, April 17-20, 1988.
63. Hawkins, J.T., Benvegnu, A.J., Wingate, T.P., McKameie, J.D., Pickard, C.D., Altum, J.T., "Sacroc Unit $CO_2$ Flood: Multidisciplinary Team Improves Reservoir Management and Decreases Operating Costs", SPE Reservoir Engineering, August 1996.
64. Bayat, M.G., Pickard, C.D., Benvegnu, A.J., Wingate, T.P., Larkin, R.,"Linking Reservoir Characteristics and Recovery Processes at Sacroc", Second Annual Subsurface Fluid Control Symposium, Houston, Texas, September, 1996.

65. Gonzalez, R., Schepers, K., Reeves, S., Eslinger, E., Back, T., "Integrated Clustering/Geostatistical/Evolutionary Strategies Approach for 3D Reservoir Characteization and Assisted History Matching in a Complex Carbonate Reservoir, Sacroc Unit, Permian Basin", SPE 113978-MS, presented at the SPE/DOE Enhanced Oil Recovery Symposium, Tulsa, Oklahoma, April 20-230, 2008.
66. Genetti, D.B., Whitaker, C.A., Smith, D.P., Price, L.M., "Applying Improved Recovery Processes and Effective Reservoir Management to Maximize Oil Recovery at Salt Creek", SPE 81458-MS, presented at the Middle East Oil Show, Bahrain, June 9-12, 2003.
67. Bishop, D.L., Williams, M.E., Gardner, S.E., Smith, D.P., Cochrane, T.C., "Vertical Conformance in a Mature Carbonate $CO_2$ Flood: Salt Creek Field Unit, Texas", SPE 88720-MS, presented at the Abu Dhabi International Conference and Exhibition, Abu Dhabi, United Arab Emirates, October 10-13, 2004.
68. Brinkman, F.P., Kane, T.V, McCullough, R.R., Miertschin, J.W., "Use of Full-Field Simulation to Design a Miscible $CO_2$ Flood", Sharon Ridge Field, SPE 56882-PA, SPE Reservoir Evaluation and Engineering Journal, Volume 2, Number 3, pages 230-237,June, 1999.
69. Brown, G., Carvalho, B., Wray, A., Sensa, Smith, D., Tooms, M., Pennell, S., "Monitoring Alternating $CO_2$ and Water Injection and its Effect on Production in a Carbonate Reservoir using Permanent Fiber-Optic Distributed Temperature Systems", Cogdell Field, SPE 90248-MS, presented at the Annual Technical Conference and Exhibition of Society of Petroleum Engineers, Houston, Texas, Septermber 26-29, 2004.
70. Smith, D. and Kelly, T., "Katz (Strawn) Field $CO_2$ Project – Permian Basin, Texas", presented at the 15th Annual $CO_2$ Flooding Conference, Midland, Texas, December 10-11, 2009.
71. Roy, M.B., Tucker, C.W., Lakey, C.J., Cloud, W.B., "Waterflood Redevelopment Prior to Future Tertiary Attempts – A Case History", SPE 8460 MS, Postle, presented at the SPE Oklahoma City Regional Meeting, Oklahoma City, Oklahoma, February 21-22, 1977.
72. Wehner, S.C., "A $CO_2$ EOR Update from "No Man's Land", Challenges and Successes – Postle Field, Oklahoma", presented at the 15th Annual $CO_2$ Flooding Conference, Midland, Texas, December, 10-11, 2009.
73. Kirpatrick, R.K., Flanders, W.A., and Depauw, R.M., "Performance of the Twofreds $CO_2$ Injection Project", SPE 14439, presented at the SPE/DOE Fouth Symposium on Enhanced Oil Recovery, Tulsa, Oklahoma, April 15-18, 1984.
74. Flanders, W.A., and Depauw, R.M., "Update Case History: Performance of the Twofreds Tertiary $CO_2$ Project", SPE 26614-MS, presented at the Annual Technical Conference and Exhibition of Society of Petroleum Engineers, Houston, Texas, October 3-6, 1993.
75. Flanders, W.A., and McGinnis, R.A., "$CO_2$ EOR Economics for Small-to-Medium-Size Fields", SPE 26391-MS, presented at the Annual Technical Conference and Exhibition of Society of Petroleum Engineers, Houston, Texas, October 3-6, 1993.
76. Pittaway, K.R. and Runyan, E.E, "The Ford Geraldine Unit $CO_2$ Flood: Operating History", SPE 17278, presented at the SPE Permian Basin Oil and Gas Recovery Conference, Midland, Texas, March 10-11, 1988.
77. Phillips, J.L., McPherson, J.L., and Leibrecht, R.J., "$CO_2$ Flood: Design and Initial Operations, Ford Jeraldine (Delaware Sand) Unit", SPE 12197, presented at the 58th Annual Technical Conference and Exhibition held in San Francisco, CA, October 5-8, 1983.
78. Fox, M.J., Simlote, V.N., Stark, K.L., and Brinlee, L.D., "Review of $CO_2$ Flood, Springer "A" Sand, NE Purdy Unit", SPE/DOE 14938, presented at the SPE/DOE Fifth Symposium on Enhanced Oil Recovery, Tulsa, Oklahoma, April 20-23, 1986.
79. Flanders, W.A., Stanberry, W.A., and Martinez, M., "Review of $CO_2$ Performance of the Hansford Marmaton Unit", SPE/DOE 12327, presented at the SPE/DOE Enhanced Oil Recovery Symposium, Tulsa, Oklahoma, April 17-20, 1988.
80. Ring, J.N., "An Overview of the North Ward Estes $CO_2$ Flood", SPE 30729, presented at the SPE Annual Technical Conference and Exhibition, Dallas, Texas, October 22-25, 1995.
81. Oxy Elk Hills,"Enhanced Recovery from Mature Fields - An Operators Perspective", presented at RPSEA Problem Identification Forum, University of California – School of Engineering, November 29, 2006.
82. Kleinsteiber, S.W., "The Wertz Tensleep $CO_2$ Flood: A Review of the Engineering Design and Initial Performance", SPE 18067, presented at the 63rd Annual Technical Conference and Exhibition of Society of Petroleum Engineers, Houston, Texas, October 2-5, 1988.
83. Brokmeyer, R.J. and Borling, D.C., "Lost Soldier Tensleep $CO_2$ Tertiary Project – Performance Case History – Bairoil, Wyoming", SPE 35191, presented at the Permian Basin Oil and Gas Recovery Conference, Midland, Texas, March 27-29, 1996.
84. Merit Energy Company, "Baroil Operations Center – $CO_2$ Tertiary Recovery Review Lost Soldier and Wertz Fields, Sweetwater and Carbon County, Wyoming", Fourth Annual Wyoming Natural Gas Fair, September 28, 2000.
85. Eves, K.E. and Nevarez, J.J., "Update of Lost Soldier/Wertz Floods – Living in a Constrained $CO_2$ Environment", presented at the 15th Annual CO2 Flooding Conference, Midland, Texas, December 10-11, 2009.

86. Kulkarni, M. M., Chen, Hung-Lung, Brummert, A.C., "$CO_2$ IOR Evaluation for the U.S. Rocky Mountain Assets", SPE 113297-MS, presented at the SPE/DOE Symposium on Enhanced Oil Recovery, Tulsa, Oklahoma, April 20-23, 2008.
87. Larson, W.K., "Rangeley Weber Sand $CO_2$ Project: A Case History", presented at the Denver SPE MiniSymposium, November 19, 1987.
88. Wackowski, R.K., Stevens, C.E., Masoner, L.O., Attanucci, V., Larson, J.L., Aslesen, K.S., "Applying a Rigorous Decision Analysis Methodology to Optimization of a Tertiary Recovery Project: Rangely Weber Sand Unit, Colorado", SPE Oil and Gas Economics, Finance, and Management Conference held in London, England, 28-29, April 1992.
89. Wakowski, R.K. and Masoner, L.O., "Rangely Weber Sand Unit $CO_2$ Project Update: Operating History", SPE/DOE 27755, presented at the SPE/DOE Ninth Symposium on Improved Oil Recovery, Tulsa, Oklahoma, April 17-20, 1994.
90. Fulbright, G.D, Hild, G.P., Horf, T.A., Myers, F.S., O'Toole, F.S., Wakowski, R.K., "Evolution of Conformance Improvement Efforts in a Major $CO_2$ WAG Injection Project, Rangly Field, Colorado", SPE/DOE 35361, presented at SPE/DOE Tenth Symposium on Improved Oil Recovery, Tulsa, Oklahoma, April 21-24, 1996.
91. Masoner, L.O., Abidi, H.R., Hild, G.P., "Diagnosing $CO_2$ Flood Performance Using Actual Performance Data", Rangely Field – Colorado, SPE/DOE 35363, presented at SPE/DOE Tenth Symposium on Improved Oil Recovery, Tulsa, Oklahoma, April 21-24, 1996.
92. Wackowski, R.K., "Rangely Weber Sand Unit $CO_2$ Project Update", Powerpnt talk presented at the 5th Annual $CO_2$ Flooding Conference, Midland, Texas, December 10-11, 1999.
93. Schechter, D.S., McDonald, P., Sheffield, T., Baker, R. "Integration of Laboratory and Field Data for Development of a $CO_2$ Pilot in the Naturally Fractued Spraberry Trend Area, Permian Basin, Texas", SPE 36657, presented at the Annual Technical Conference and Exhibition held in Denver, CO., October 6-9, 1996.
94. Schechter, D.S., McDonald, P., Sheffield, T., Baker, R. "Reservoir Characterization and $CO_2$ Pilot Design in the Naturally Fractured Spraberry Trend Area, Permian Basin, Texas", SPE 35469, presented at the Permian Basin Oil and Gas Recovery Conference, Midland, Texas, March 27-29, 1996.
95. Spivak, A., Garrison, W.H., Nguyen, J.P, "Review of an Immiscible $CO_2$ Project, Tar Zone, Fault Block V, Wilmington Field, California", SPE 17407-PA, SPE Reservoir Engineering Journal, Volume 5, Number 2, Pages 155-162, May, 1990.
96. Saner, W.B., Patton, J.T., "$CO_2$ Recovery of Heavy Oil: Field Test", SPE 12082, presented at the Annual Technical Conference and Exhibition held in San Francisco, California, October 5-8, 1983.
97. Sankur, V., Creek, J.L., DiJutio, S.S, Emanuel, A.S., "A Laboratory Study of Wilmington Tar Zone $CO_2$ Injection Project", SPE 12751, presented at the California Regional Meeting held at Long Beach, California, Apr 17-23, 1984.
98. Jeschke, P.A., Schoeling, L., Hemmings, J., "$CO_2$ Flood Potential of California Oil Reservoirs and Possible $CO_2$ Sources", SPE 63305, presented at the SPE/AAPG Western Regional Meeting held in Long Beach, California, June 19-23, 2000.
99. Reid, T.B., Robinson, H.J., "Lick Creek Meakin Sand Unit Immiscible $CO_2$/Waterflood Project", SPE 9795, presented at SPE/DOE Tenth Symposium on Improved Oil Recovery, Tulsa, Oklahoma, April 6-7, 1981.
100. Moffitt P.D., Zornes, D.R., "Lick Creek Meakin Sand Unit Immiscible $CO_2$/Waterflood Project", presented at the 67th Annual Technical Conference and Exhibition held in Washington DC, October 4-7, 1992.
101. Brock, W.R. and Bryan, L.A.: "Summary Results of $CO_2$ EOR Field Tests, 1972-1987," paper SPE 18977 presented at the 1989 SPE Joint Rocky Mountain Regional/Low Permeability Reservoir Symposium and Exehibition, Denver, 6-8 March, 1989
102. Caudle, B.H. and Dyes, A.B.: "Improving Miscible Displacement by Gas-Water Injection" Trans., AIME (1958) 213, 281.
103. Hadlow, R.E.: "Update of Industry Experience with $CO_2$ Injection," paper SPE 24928 presented at the 1992 SPE Annual Technical Conference and Exhibition, Washington, DC, 4-7 October.
104. Grigg, R.B. and Schechter, D.S.: "State of the Industry in $CO_2$ Floods," paper SPE 38849 presented at the 1997 SPE Annual Technical Conference and Exhibition, San Antonio,Texas 5-8 October, 1997
105. John D. Rogers,*SPE, and Reid B. Grigg, SPE, New Mexico Petroleum Recovery Reseach Center.: "A Litature Analysis of the WAG Injectivity Abnormalities in the CO2 Process" SPE Report (SPE 73830) was revised for publication, first presented at 2000 SPE/DOE IOR Symposium, Tulsa Apr 3-5 2000
106. Pennzoil's Sacroc Development Team (1994-1998): Dr. Ghasem Bayat, Tony Benvegnu, Claud Pickard, Jack Horkowitz, Tom Wingate, Don Hartman; Team was Responsible for Sacroc's Centerline Project plus two other pilot programs that was responsible for 1/3 of Sacroc's total 9,600 BOPD Production through this time period.
107. Amoco's Permian Basin Original Model Development Group: Jack Aulick(Manager), Reservoir Modellers: Kevin McCollough, Steve Pennell, Burt Nelson, Gary Pariani, Dennis Edens, John Kimberling, and David Merchant, Operations: Perry Jarrell; Team was responsible for developing Amoco's Tapered WAG Operating scheme.
108. Steve Melzer, Midland 2016 $CO_2$ Conference, "Plot of Historical Oil Price Crash Periods from the 1970's through 2016", Date: December 2016.

# Acknowledgements

I would like to express my gratitude to the members of the Carbon Management Technology Conference (CMTC) for allowing me the opportunity to be part of this $CO_2$ Conference. I would especially like to thank George Koperna of Advanced Resources International and Jose Figueroa for their support in making this effort possible.

David H. Merchant
Merchant Consulting
Magnolia, Texas

# CO2CRC Storage and Capture Demonstration

Abdul Qader, PhD, FAIE, FIEAust
Capture Manager

CO2CRC
BUILDING A LOW EMISSIONS FUTURE

CMTC
19 July 2017

## Who is CO2CRC?

- CO2CRC is a world leading carbon capture & storage (CCS) research organisation that works closely with global industry to reduce large scale greenhouse gas emissions through:
  - Investigating carbon capture, transport and storage technologies
  - Examining the costs and drivers across the CCUS chain
  - Conducting research, development and demonstration across the CCUS value chain
  - Providing efficient collaboration amongst 150 researchers in capture and storage
  - Building experience in CCUS technology by addressing knowledge gaps

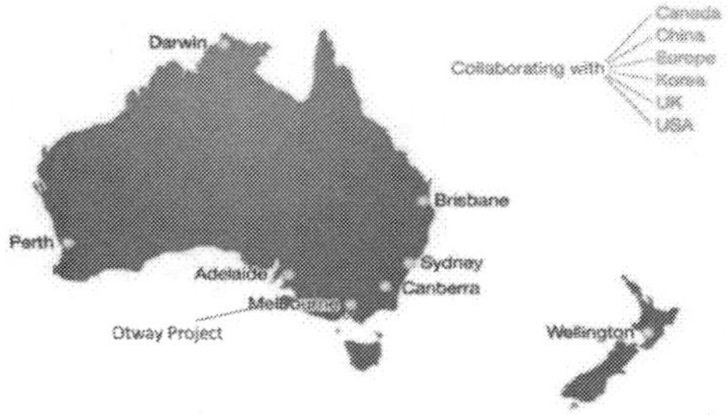


32 research and industry partners globally

CO2CRC

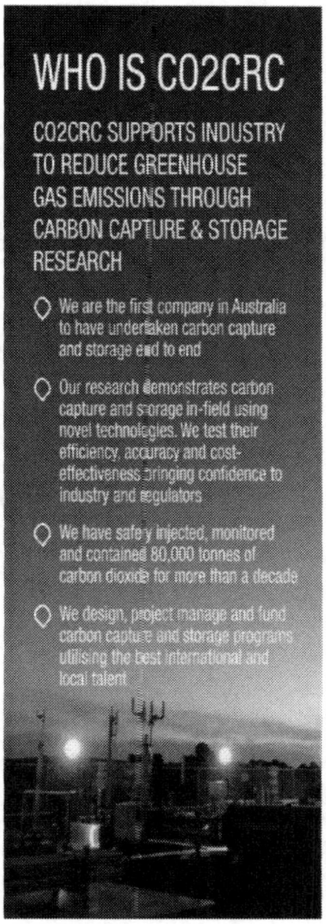

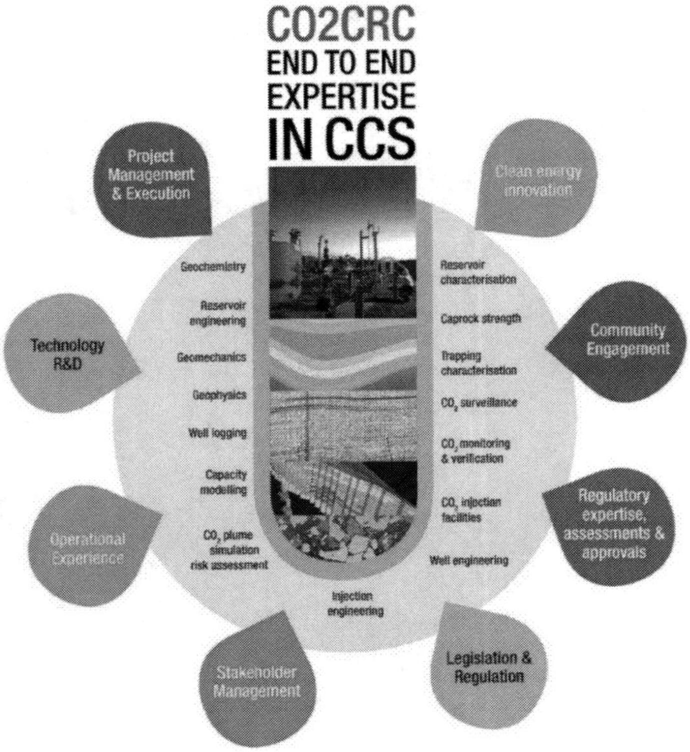


3

# Australia's Integrated CCS activities

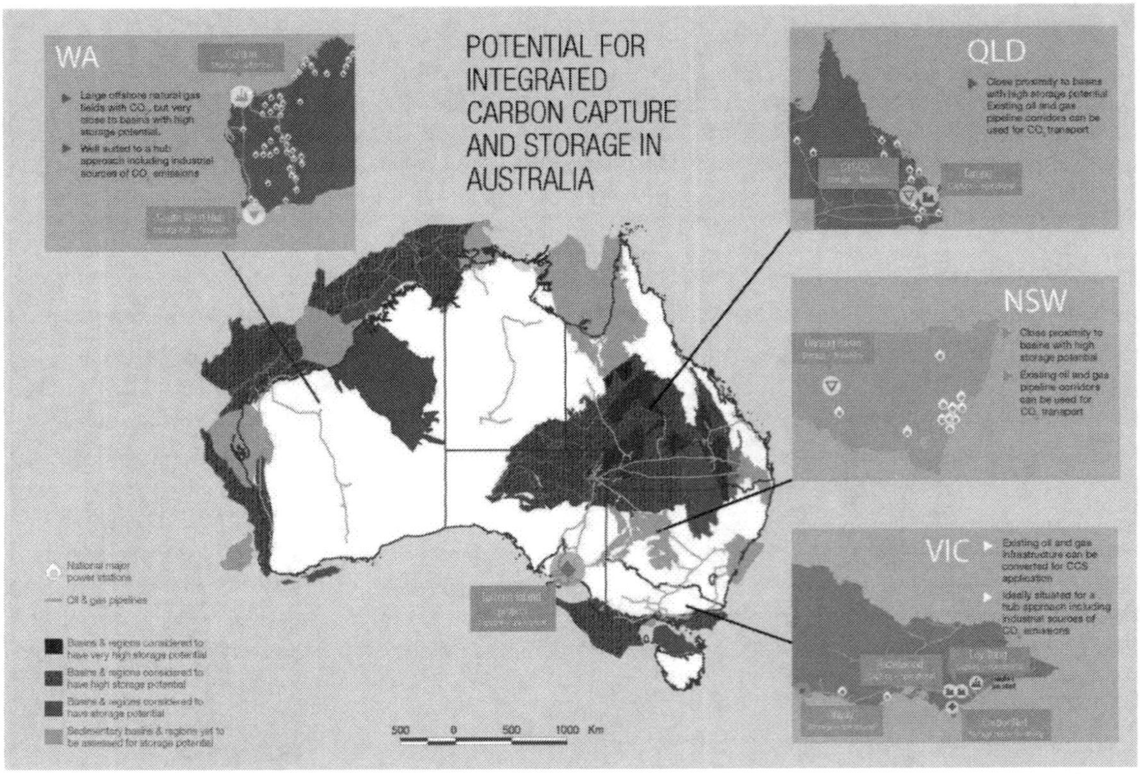

# How is Carbon Dioxide Stored Geologically?

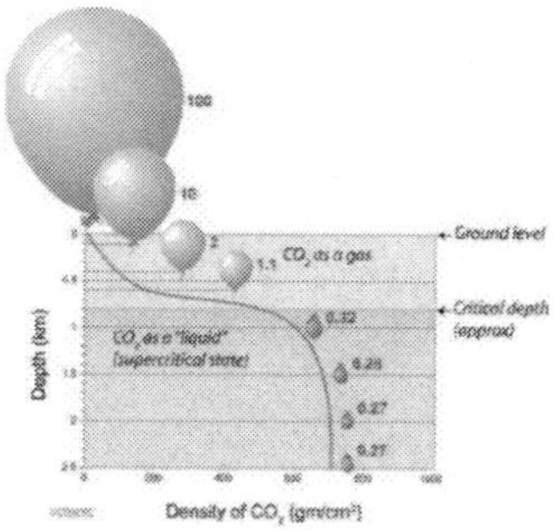


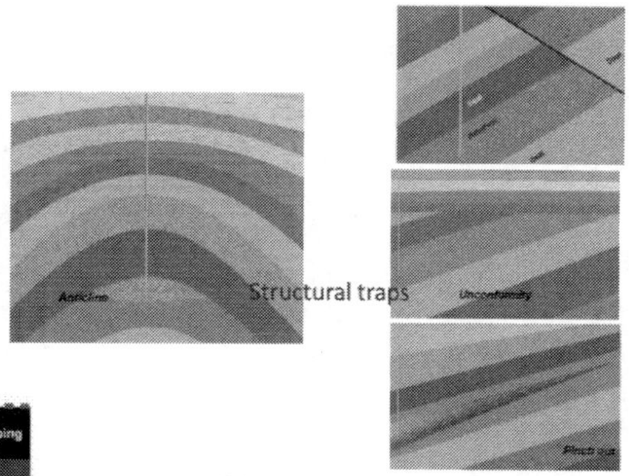


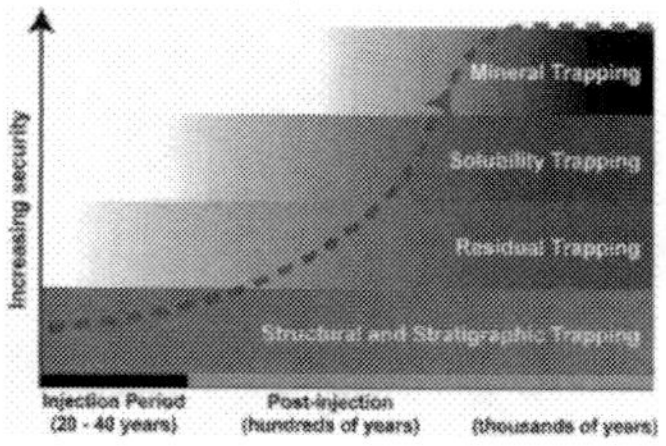


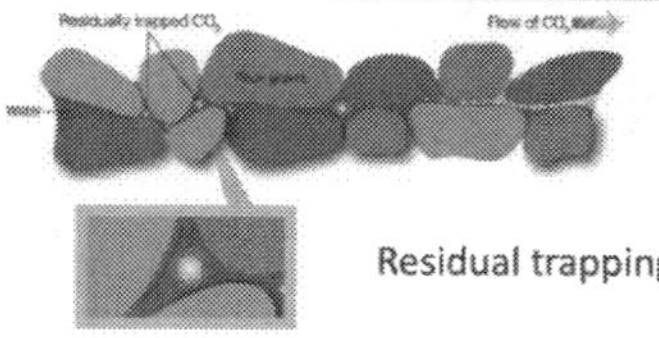

Residual trapping

5

# Otway Site from above

Proposed new well locations

# CO2CRC Otway Storage Projects

- Stage 1: 2004 – 2009 (Concept)
  - ✓ Demonstrated safe transport, injection and storage of $CO_2$ into a depleted gas reservoir
- Stage 2: 2009 – 2019 (Risk Reduction)
  - ✓ Demonstrate safe injection and monitoring into a saline formation
- Stage 3: 2016 – 2028 (Cost Effective Storage Management and Subsurface Monitoring)
  - ✓ Demonstrate safe, reliable and cost-effective subsurface monitoring of $CO_2$

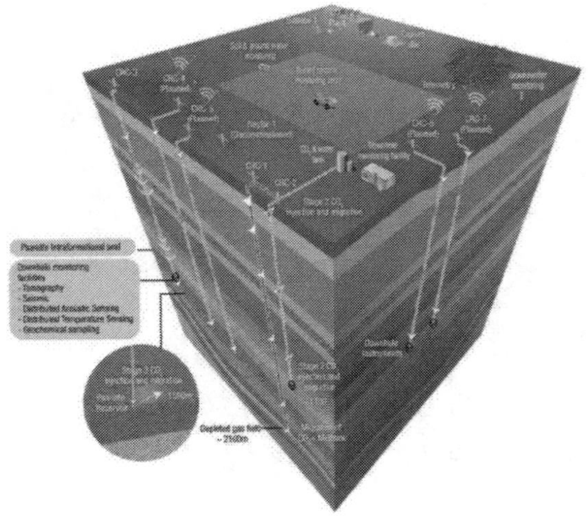

7

# Otway Stage 2

*Demonstrate that $CO_2$ storage can be safely conducted at scale within a Saline Formation*

Appraisal
- ✓ 2A :Drill CRC-2
- ✓ 2B: Measure parameters affecting residual and dissolution trapping in a saline formation
- ✓ 2B Extension: interactions with impurities & well test refinement

Operation
- 2C: Spatially track injected $CO_2$ in a saline formation
  - ✓ Minimum detection limit
  - ✓ Migration behaviour
  - Stabilisation (on track for 2018)

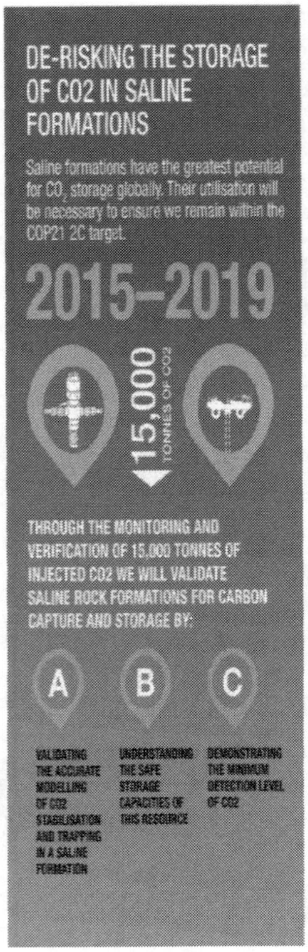

# Otway Stage 2B – Conclusions

- Residual trapping is a key mechanism for the storage of carbon dioxide in saline formations
- The test measures field-scale residual trapping using a single well configuration and six different methods – pressure, temperature, pulsed neutron logging, noble gas tracers, reactive tracers and a dissolution test.
- The field results enable us to evaluate the effectiveness of each method, and to recommend how such a test could be improved.

9

# CO2CRC Otway Storage Project Stage 2C

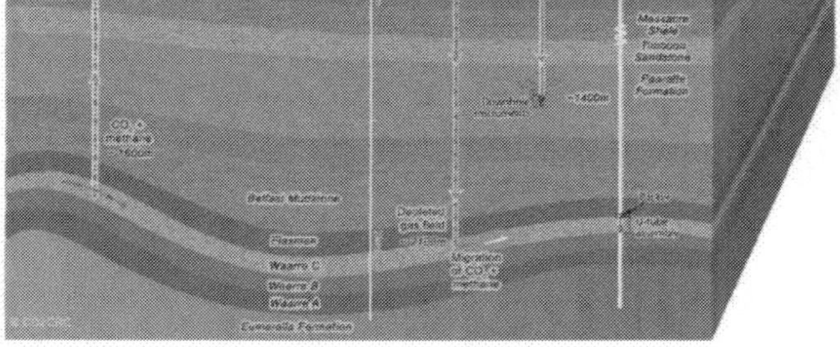


DE-RISKING THE STORAGE OF CO2 IN SALINE FORMATIONS

Saline formations have the greatest potential for $CO_2$ storage globally. Their utilisation will be necessary to ensure we remain within the COP21 2C target

2015–2019

15,000 TONNES OF CO2

THROUGH THE MONITORING AND VERIFICATION OF 15,000 TONNES OF INJECTED CO2 WE WILL VALIDATE SALINE ROCK FORMATIONS FOR CARBON CAPTURE AND STORAGE BY:

A — VALIDATING THE ACCURATE MODELLING OF CO2 STABILISATION AND TRAPPING IN A SALINE FORMATION

B — UNDERSTANDING THE SAFE STORAGE CAPACITIES OF THIS RESOURCE

C — DEMONSTRATING THE MINIMUM DETECTION LEVEL OF CO2

# Seismic Array

- *Installation of 1km seismic array.*
- *900 monitoring devices – geophones*
- *Over 30km of fibre optic cable*
- *Receivers buried 4 metres underground*
- *Lines 100 metres apart, receivers 15 metres apart*
- *Installed in early 2015*

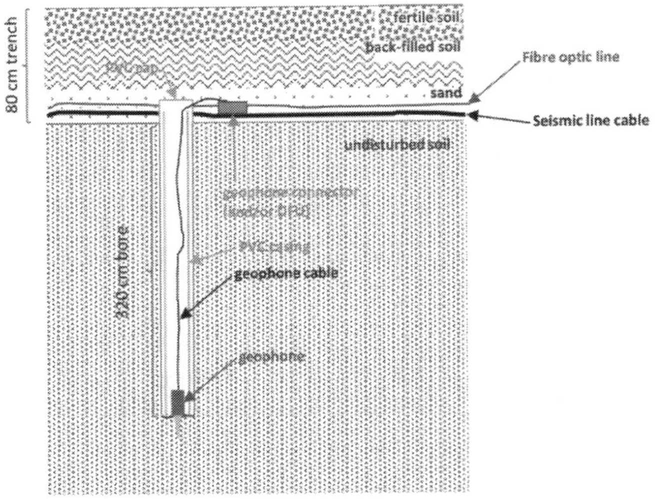


CO2CRC

11

# Geophone and fiber array installation

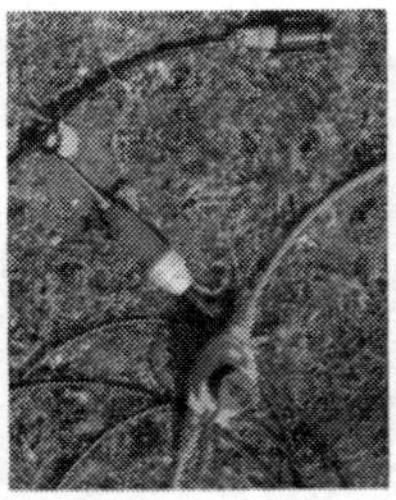

Geophones

Deployed in trench

Fibre optic cable

CO2CRC

12

# Otway Stage 2C Preliminary results

1. 15,000 tonnes of $CO_2$ safely injected into a saline formation, migrating as predicted
2. Seismic and pressure monitoring resolution is beyond expectations
3. Minimum detection levels of supercritical $CO_2$ are being evaluated through the direct observation
4. The combination of these results, with ongoing regular monitoring through to 2019, will de-risk the injection, monitoring and trapping of CO2 in a saline formation giving confidence to technology users and regulators.

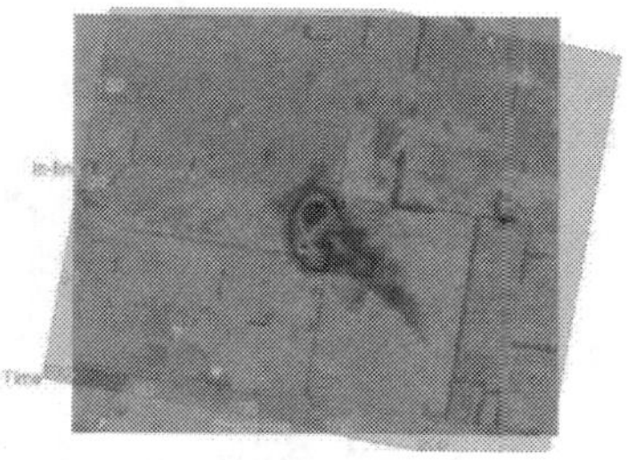

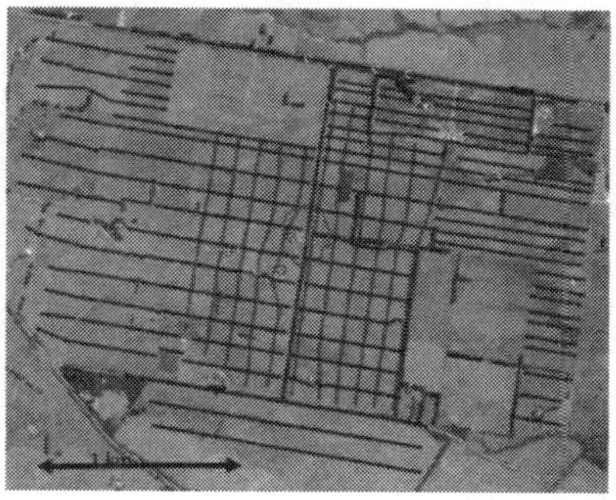

13

# Otway Stage 2 – 4D change still image

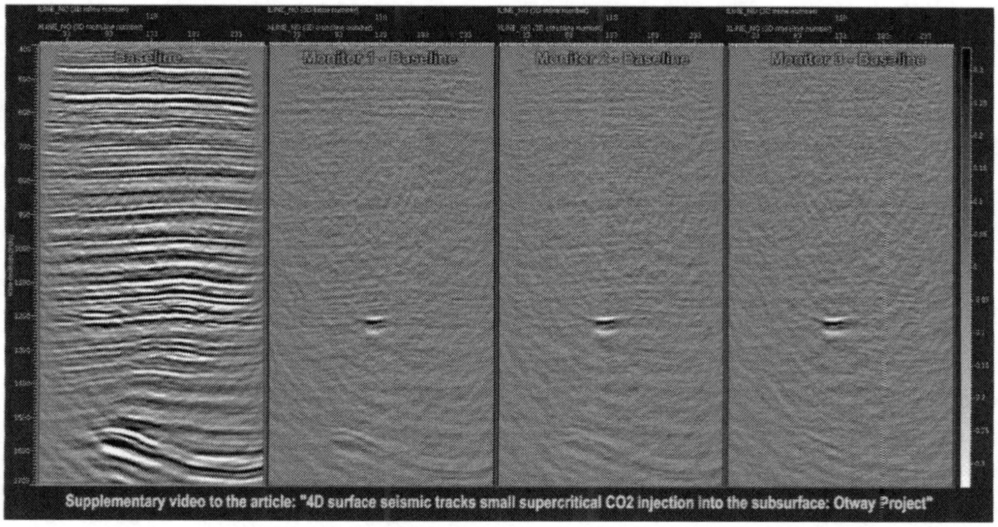

## Energy(M1 – B)

## Energy(M2 – B)

# Energy(M3 – B)

# Otway Storage Project Stage 2C

# Highlights on Storage

**Exceptional Datasets**

- Atmospheric (continuous monitoring)
- Soil Gas
- Seismic
- Reservoir fluids

**Field Validation**

- Validation of models in actual field tests
- Fully permitted and operational site
- Availability of $CO_2$
- Globally unique advantage

**Developing Novel & Innovative Concepts & Technology**

- Single Well Test for reservoir characterisation
- Several generations of U-Tube systems deployed
- Fibre Optics for reservoir interrogation and well diagnostics
- Innovative well instrumentation

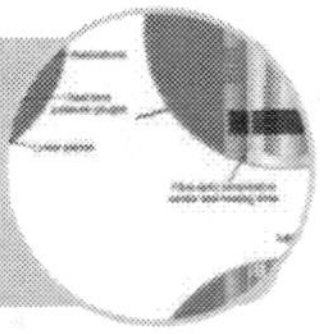

**First of its kind experiments**

- First $CO_2$ storage demonstration project in Australia
- Reducing risk and uncertainty
- Advising regulators and project proponents in Australia and world wide
- A go-to project

CO2CRC

19

# Success Factors for the CO2CRC Otway Project

1. A locally based Site operator who is available all year round at short notice.
2. A locally based Community Liaison Officer who is contactable always.
3. The availability of gas on call from Buttress-1 at no further cost.
4. The proximity and availability of pure $CO_2$ gas from Boggy Creek at market cost.
5. Supportive land owner community.
6. Availability of formations and structures in the sub surface that allow safe injection of product.
7. A surface facility that has been built (right standard and superior materials) to last a long time.
8. A workable regulatory setup for R&D type work. Approvals to inject both $CO_2$ and Buttress-1 gas and produce formation water.
9. Project expertise availability as and when required.
10. Monitoring
11. Demonstrating it can be done under Australian conditions

CO2CRC

20

## We have the social licence to operate

21

## What does our future activities look like?

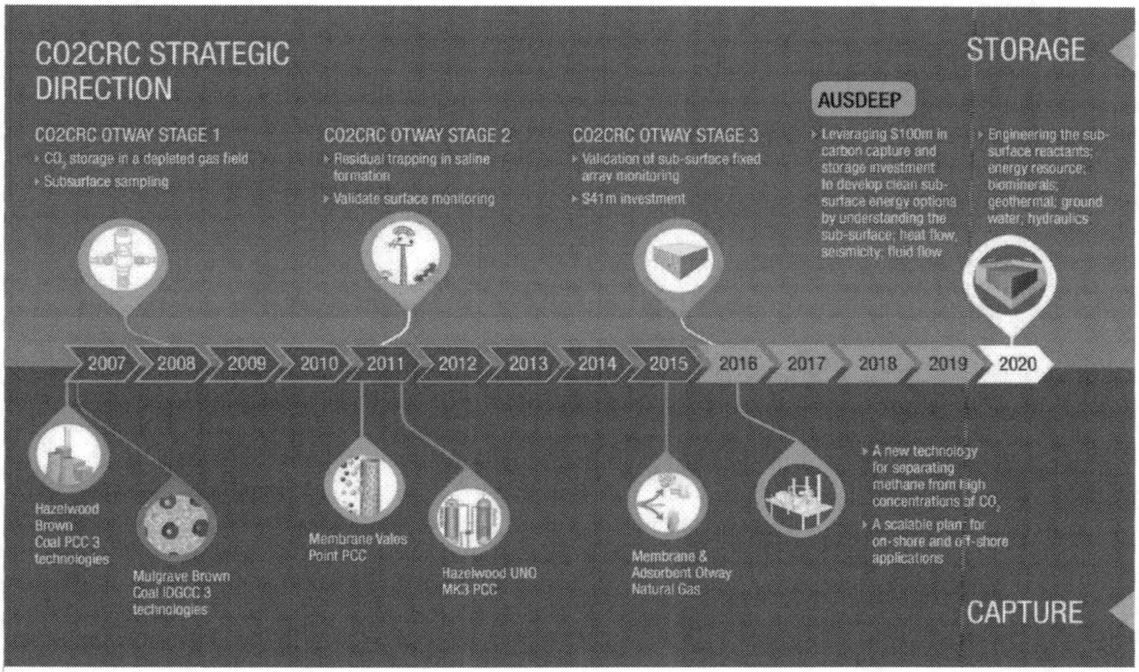


22

## Implications of results to date

- *CO2CRC is delivering a broadly applicable procedure to predict, monitor, verify & assure $CO_2$ migration & trapping in saline formations*
- *Otway 2C is providing in-field evidence for the use of saline formations as a suitable $CO_2$ storage targets*
  - Validation of $CO_2$ behaviour is achievable, and
  - $CO_2$ behaviour in saline formations is predictable
- *The project outcomes provide a clear foundation to future operators and regulators to the leakage assurance threshold in the subsurface*
  - The minimum technical limit for a surface-based risk-targeted seismic monitoring system has been determined

23

# Stage 3 $CO_2$ Storage Management & Subsurface M&V $49.5 million project

1. We will develop a high-resolution, **real-time monitoring** capability. The technologies will provide an early warning solution for industry and regulators.
2. We will provide non invasive monitoring techniques that will be acceptable for community and regulators.
3. The project will provide a prospectus of technologies and workflows that can be used to define costs in commercial monitoring projects.
4. The project will evolve technology from benchtop application to in-field validation, which is being driven by current and prospective operator need.

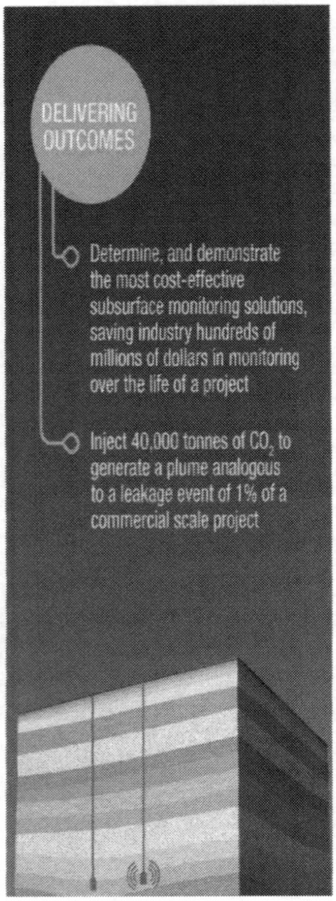

## Otway site specific

# Otway Storage Project Stage 3 Concept

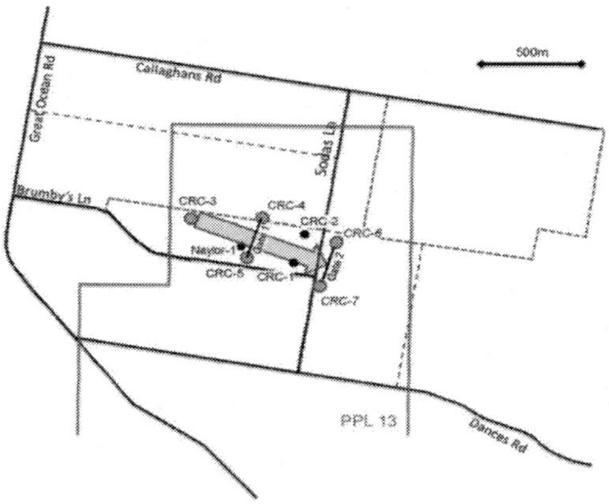


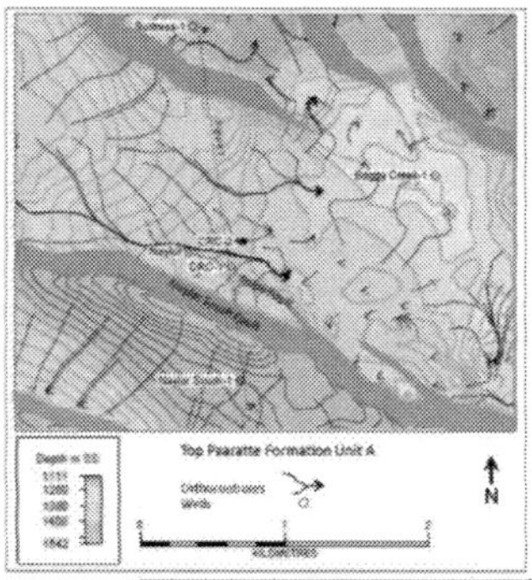


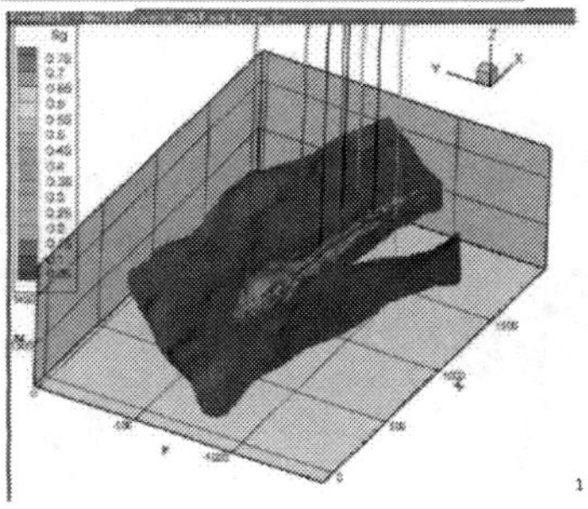

# Otway Stage 3 Project

Project Timeline

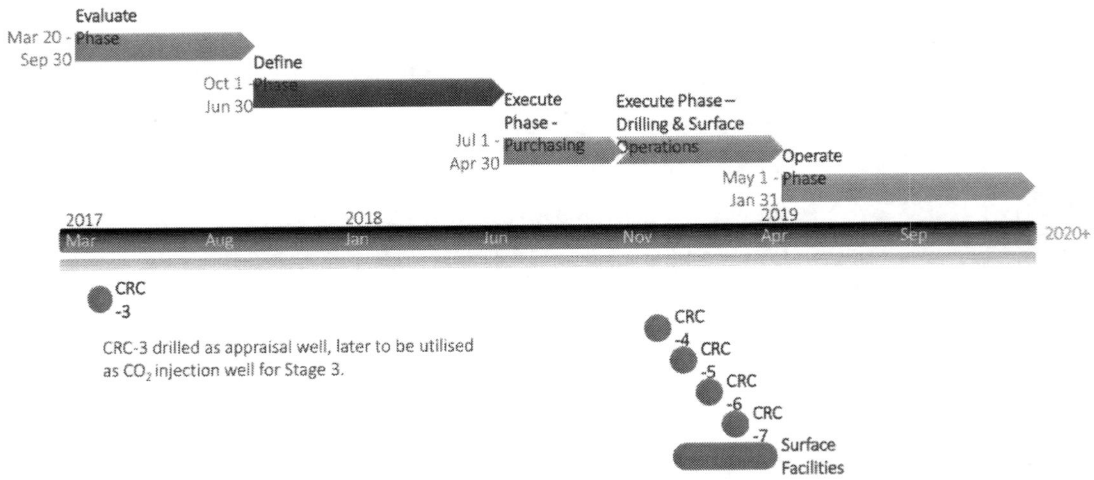


27

## Field Capture Projects

- First application of 3 different capture technologies in parallel and their comparison for large scale implementation
- Attained capability of scaling up of capture plants from concept till large scale (TRL level 1 till 9) with economic feasibility

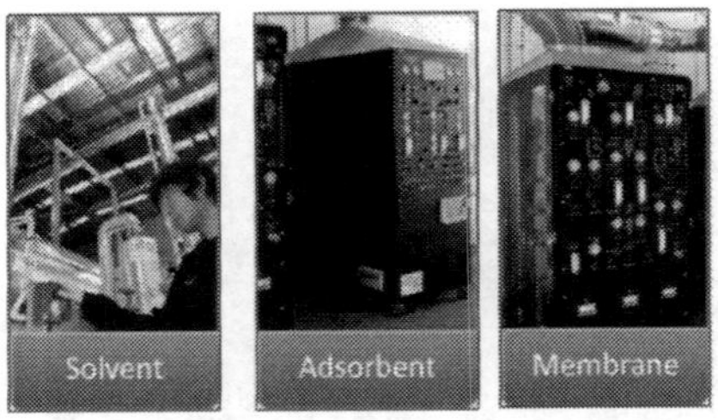


3 pre-combustion technologies at HRL, VIC : 2007-2011, ETIS/BCIA funded

3 post-combustion technologies at Hazelwood, VIC : 2007-2011 ETIS/BCIA funded

BCIA funded UNO MK 3 post-combustion at Hazelwood, VIC : 2011-2014

Membrane plant at Delta's Vales Point, NSW : ANLEC R&D funded

28

29

# Current and Future Capture Program Overview

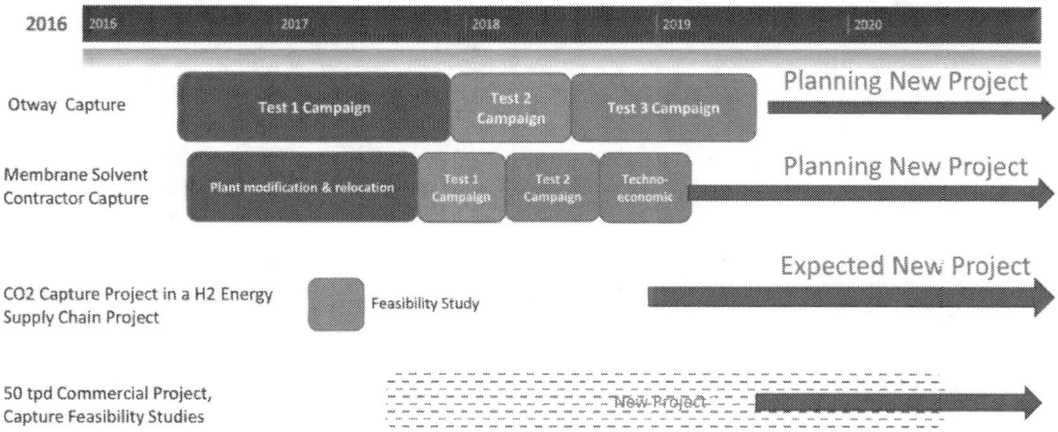


30

# Otway Capture Project

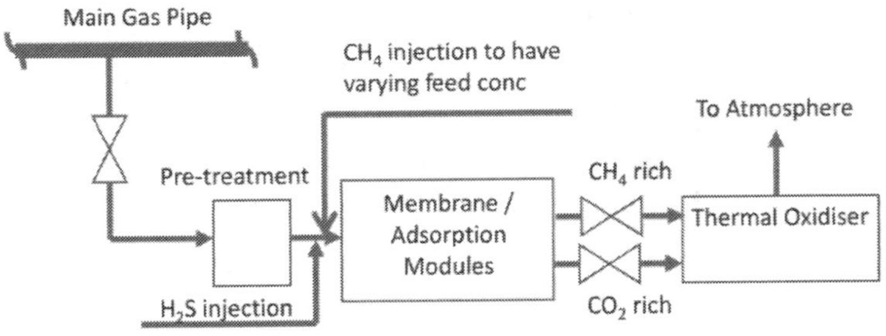

## Project Aims

- To develop cost effective, compact technologies to capture $CO_2$ mainly from high $CO_2$ content wells.
- To test new capture materials (membranes and adsorbents) and develop new capture processes over a range of adjusted $CO_2$ concentrations

## Main Features

- Feed Pressure: 80 to 30 bar
- Feed $CO_2$ Concentration: 5% to 80% achieved by $CH_4$ addition (existing is ~80% $CH_4$ and 19% $CO_2$)
- $H_2S$ addition for impurities effect tests
- Rig on a common skid delivered on site for installation and commissioning (flow diagram shown left) 2016
- Adsorbent – 4 kg, Current use: Silica (S1), Second stage: Z1; Membranes – 3 different modules
- Opportunities for vendors to conduct long term testing of capture technologies for CO2/CH4 separation at a fully instrumented site

CO2CRC

31

CO2CRC

UNSW

THE UNIVERSITY OF MELBOURNE

32

## Membrane Gas-Liquid Contactors

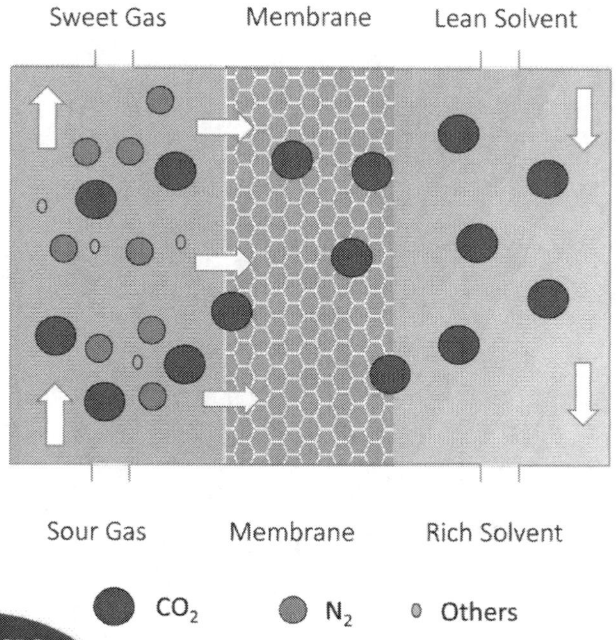


- Avoidance of hydrodynamic issues.
- Equipment size reduced by 60 – 75%.
- Corrosion is significantly reduced.
- Around 40% reduction in operating costs.
- Around 35 – 40% savings in capital costs.
- Footprint requirements reduced by 40%.

33

## Government, Industry and Research Partners

34

CO2CRC

35

## Otway Site Operations

### Appraisal (CRC 3) Well Operations

- ✓Rig operations lasted 30 days
- ✓Well was drilled to 1,667m depth
- ✓Executed FE Plan and gathered high quality dataset
- ✓Captured detailed lessons learnt
- ✓Maintained social licence to operate
- ✓Provided a pathway forward for Stage 3
- ✓ Achieved success in challenging wellbore operations

CO2CRC

CO2CRC

36

# CRC-3 Appraisal Well

Recovered Formation Evaluation Dataset

**Wellsite Formation Evaluation Program**

**Wireline Logging**

Open Hole Logs
- Survey, GR, Res, Sonic, Neu, Den, FMI, ECS, CMR
- Cased Hole Logs
- CBL (Full bore)
- Completion Mapper

**Coring**

Targeting

Primary Seal

Primary & Secondary Storage intervals

Recovered 127m, 3" core, 97% recovery

**Well Injection Tests**

Injection Operations

PS2: 100T injection

PS1: 100T injection

Data Recovered

D/H Pressure in DST string

D/H Pressure in CRC-2

Surface Pressures

**Seismic Surveys**

DAS during cementing

Zero Offset & Walk Away VSP

Recording iDAS and Geophone data for comparison

**Post well Evaluation**

**Drilling & Wireline Data**

Petrophysical Wireline Analysis

Image Processing

Wellbore Integrity

Drilling Data

EOWR

M&V Well Design

**Core Analysis**

Core Analysis

Routine

Special

Digital

MICP

Palanology

Hylogging

CO2CRC

BUILDING A LOW EMISSIONS FUTURE

37

Project ECO2S Phase 2 CarbonSAFE Field Project

# Project $ECO_2S$: Geologic Characterization and Preliminary Storage Capacity Estimates

Prepared for:
**2017 Carbon Management Technology Conference**

Prepared By:
**David Riestenberg, George Koperna,** Advanced Resources International, Inc.

**Richard Esposito,** Southern Company

**Jack Pashin**, Oklahoma State University

July 17-20, 2017
Houston, Texas

Advanced Resources International, Inc.

Project ECO2S Phase 2 CarbonSAFE Field Project

# Disclaimer

This presentation is based upon work supported by the Department of Energy National Energy Technology Laboratory under award number DE-FE0029465 and was prepared as an account of work sponsored by an agency of the United States Government. Neither the United States Government nor any agency thereof, nor any of their employees, makes any warranty, express or implied, or assumes any legal liability or responsibility for the accuracy, completeness, or usefulness of any information, apparatus, product, or process disclosed, or represents that its use would not infringe privately owned rights. Reference herein to any specific commercial product, process, or service by trade name, trademark, manufacturer, or otherwise does not necessarily constitute or imply its endorsement, recommendation, or favoring by the United States Government or any agency thereof. The views and opinions of authors expressed herein do not necessarily state or reflect those of the United States Government or any agency thereof.

NETL | Southern States Energy Board | Southern Company | Mississippi Power

Advanced Resources International, Inc.

2

# Presentation Outline

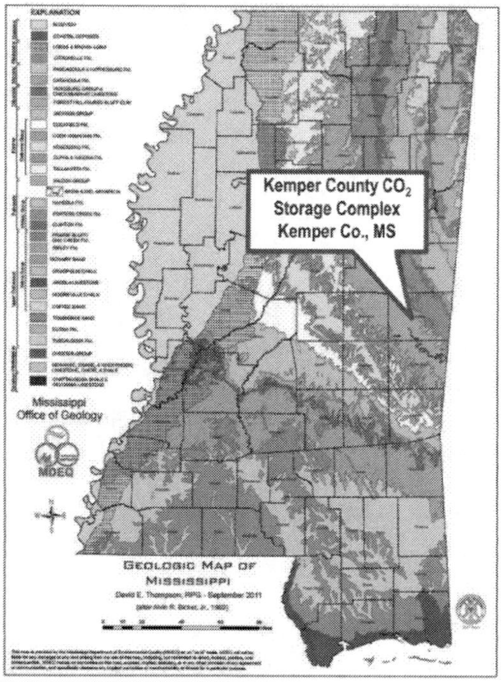


I. Project $ECO_2S$ Introduction

II. Geologic Characterization Plan

III. Results to Date

IV. Preliminary Storage Capacity

V. Conclusions





# Project $ECO_2S$ Introduction

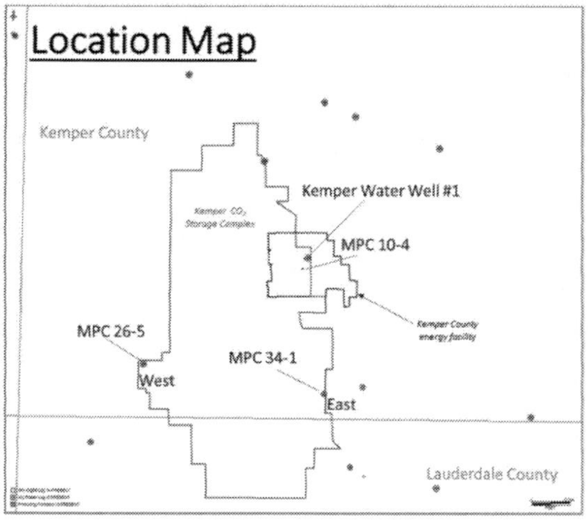

MPC has established an area of interest exceeding 30,000 acres around the Kemper County energy facility and has begun acquisition of surface and mineral rights.

Project $ECO_2S$, a DOE-supported CarbonSAFE program, will pursue key advances in $CO_2$ storage knowledge and technology, including **optimizing $CO_2$ storage efficiency**, **modeling the fate of injected $CO_2$** and **establishing residual $CO_2$ saturations**. In addition, Project $ECO_2S$ will involve "real-life" experiences, issues, and challenges of **scaling-up from its regional, pre-feasibility assessment of $CO_2$ storage** to **establish a site-specific, commercial-scale $CO_2$ storage facility**, including **capturing the "lessons learned"** in making this transition.

# Kemper Storage Complex Stratigraphy

| | | | | |
|---|---|---|---|---|
| Tertiary | Eocene | Lower Wilcox Group | Nanafolia Fm. | Lignite/USDW |
| | Paleocene | Midway Group | Naheola Fm | Potential USDW |
| | | | Porter's Creek Clay | Regional Seal |
| Cretaceous | Upper | Selma Group | Predominately Chalk | Regional Seal |
| | | Eutaw Fm. | | Potential USDW |
| | | Tuscaloosa Group | Upper | Potential USDW |
| | | | Marine Shale | Regional Seal |
| | | | Lower Tusc. Massive Sand ☆ | Potential Saline |
| | Lower | Washita- Fredericksburg ☆ | | Saline |
| | | Paluxy Fm. ☆ | | Saline |
| *Paleozoic Unconformity Ouachita Facies* | | | | |

- Three Cretaceous storage clastic units with high porosity:
  - Lower Tuscaloosa Grp (massive sand)
  - Washita-Fredericksburg interval
  - Paluxy formation
- Three prominent caprocks (reservoir seals):
  - Tuscaloosa marine shale
  - Shale interval at top of the Washita-Fredericksburg
  - Shale interval at base of Washita-Fredericksburg
- Shallow seals in the Selma and Midway Groups

Source: Pashin, J.C., D.J. Hills, D. C. Kopaska-Merkel, M.R. McIntyre, Geological Evaluation of the Potential for $CO_2$ Sequestration in Kemper County, Mississippi, Final Report, prepared for Southern Company Research and Environmental Affairs, June 1, 2008.





# Initial Geologic Assessment

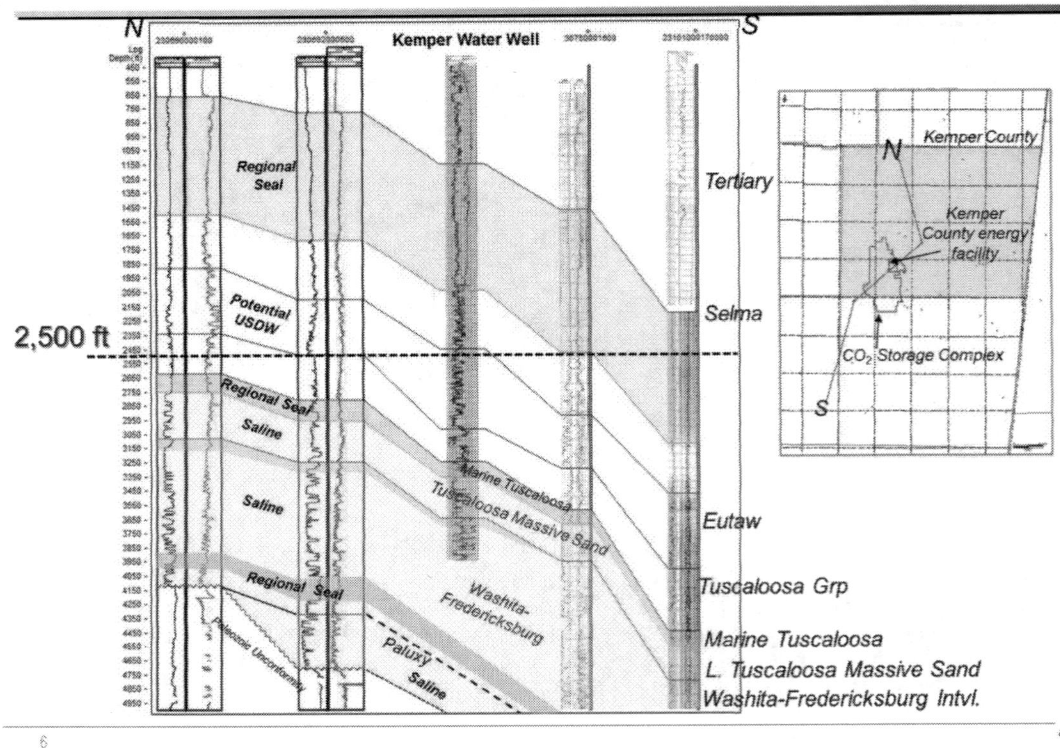

# Kemper Storage Complex Deep Structure

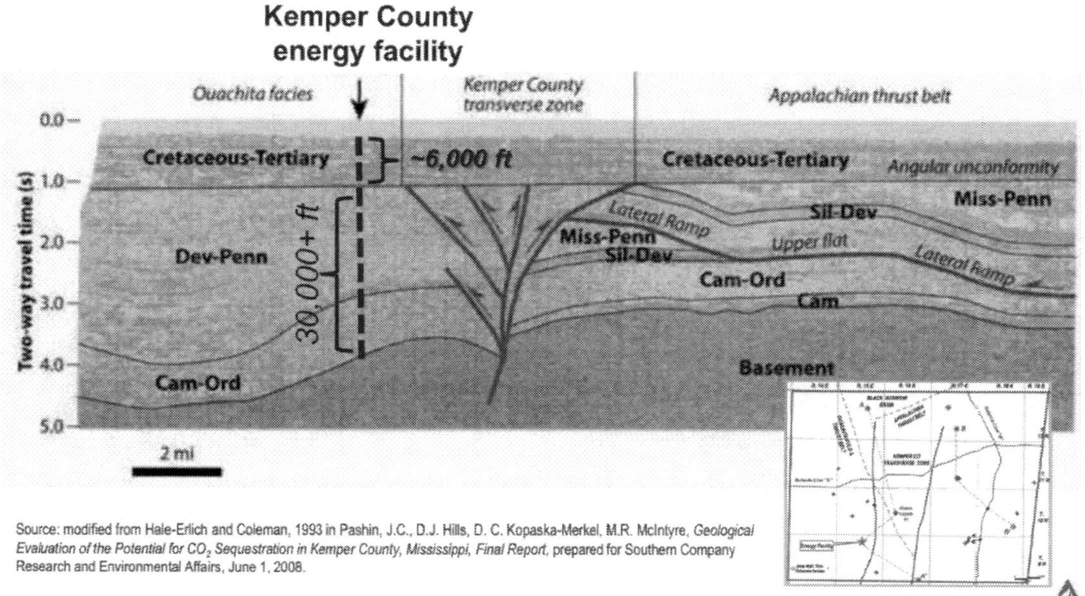


Source: modified from Hale-Erlich and Coleman, 1993 in Pashin, J.C., D.J. Hills, D. C. Kopaska-Merkel, M.R. McIntyre, *Geological Evaluation of the Potential for $CO_2$ Sequestration in Kemper County, Mississippi, Final Report*, prepared for Southern Company Research and Environmental Affairs, June 1, 2008.





# Initial Geologic Assessment

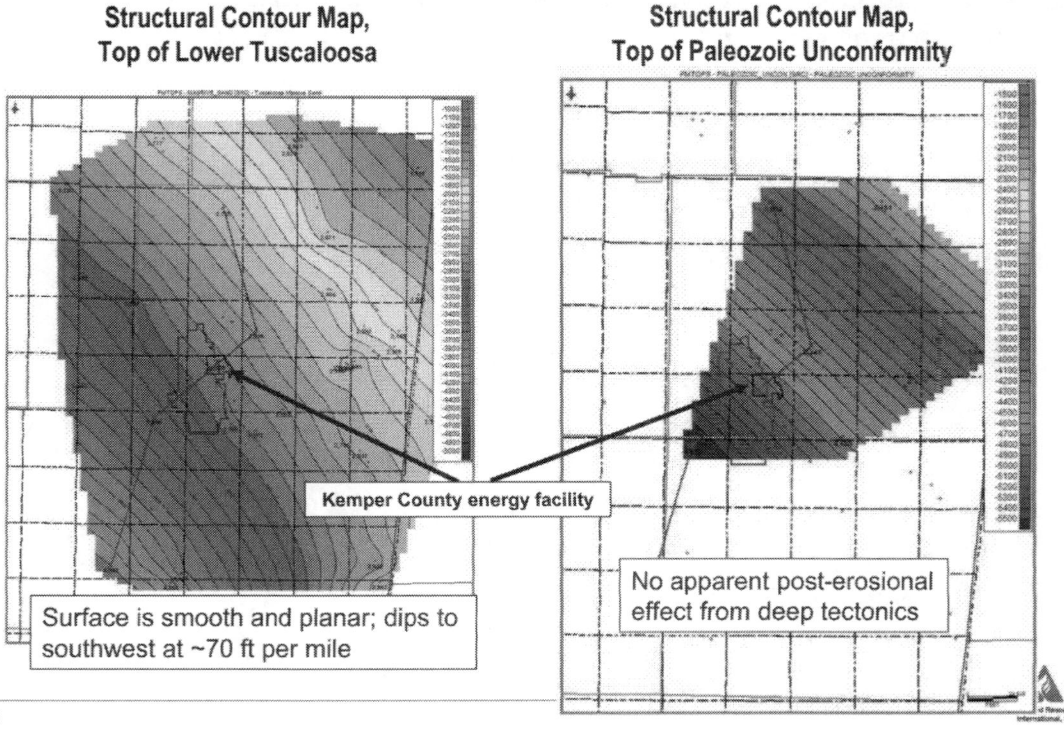

# Project $ECO_2S$ Organization Chart

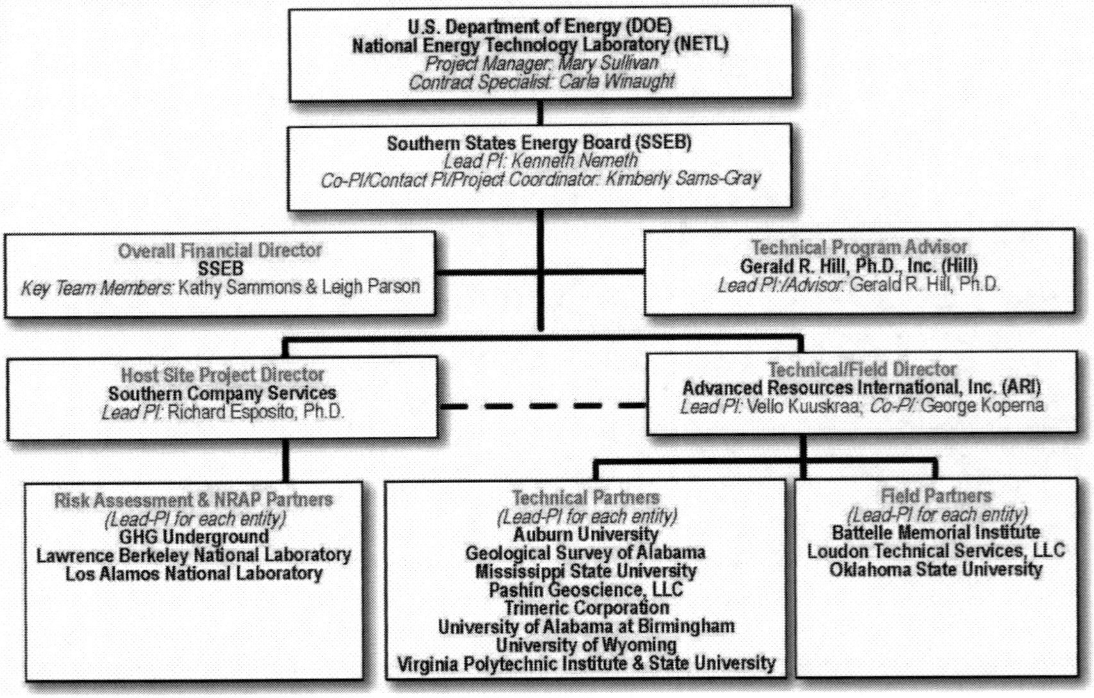


Project ECO2S Phase 2 CarbonSAFE Field Project

# $ECO_2S$ Geologic Studies

- Confirm storage reservoir volumetric properties; develop dataset on flow properties
  - Geophysical log response
  - Petrophysical properties observed in core
  - Advanced core tests, including rel-perm, CT scans under steady-state flow
- Caprock studies including
  - Threshold pressure tests, minimum capillary displacement pressure
  - Clay mineralogy
- Describe depositional facies, rock types, mineralogy, facies and environments of deposition for storage reservoirs and caprocks
- Develop a conceptual geologic model honoring interpreted depositional style
- Develop initial rock mechanics model
- Extend evaluation to regional framework
- Fluid-rock interactions
- Evaluation of existing 2D data, Identify any structural concerns

# $ECO_2S$ Geologic Data Gathering

- Drill three wells to gather drilling performance data, whole and sidewall core, logs
- Openhole Logs
  - Triple combo (caliper, array induction, gamma ray, density porosity, neutron porosity, spontaneous potential, photoelectric)
  - Combined magnetic resonance (CMR)
  - Formation micro imager (FMI)
  - Dipole sonic (mechanical properties)
  - Rotary sidewall cores
- Whole core of reservoir and caprock intervals
- Evaluation of existing 2D seismic





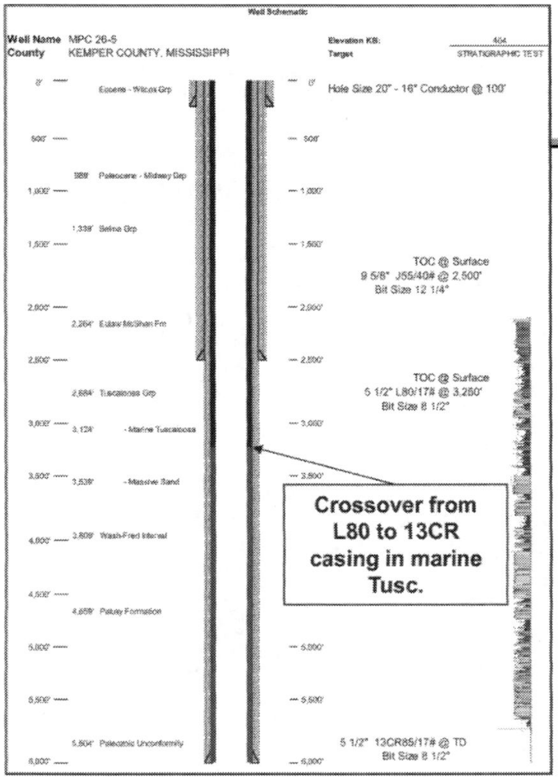


# Well Design

- Crossover from carbon steel to chrome casing in marine Tuscaloosa
- Surface and long string casing cemented to surface

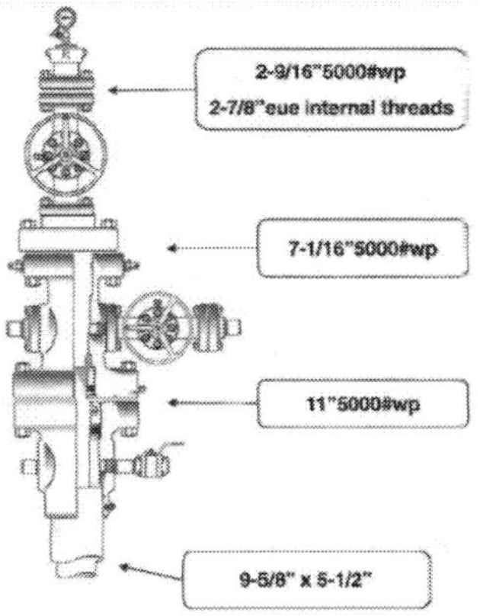

# $ECO_2S$ Field Status

- First project well, the MPC 26-5, spud in May
  - 17 days from spud to TD including two core points
- Second well, the MPC 34-1, spud in June
  - 14 days from spud to TD including two core points
- Third well, the MPC 10-4, will be spud in early August

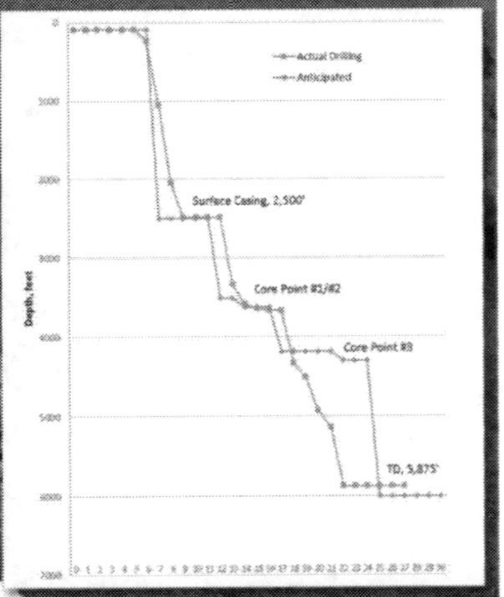


## MPC 26-5 triple combo and mud log



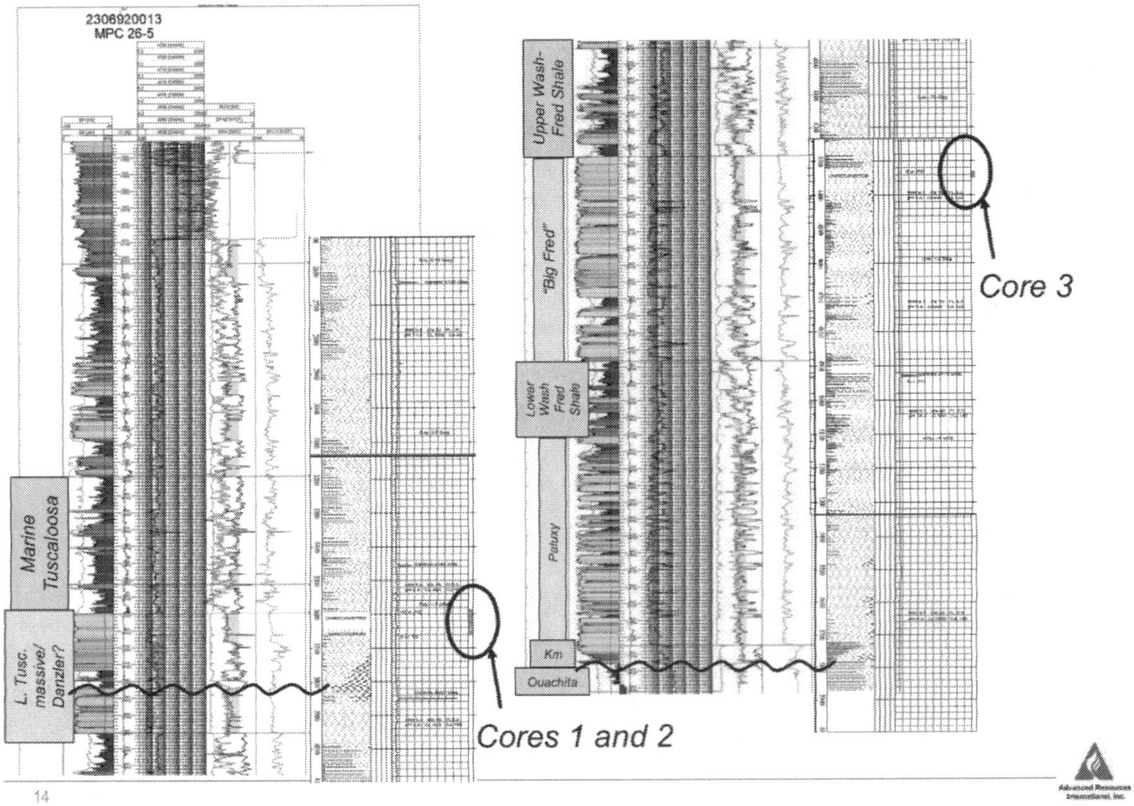

# MPC 26-5 Coring Results

| Core 1 (shale above L.T. massive) | Core 2 (L.T. massive) | Core 3 (Wash-Fred) |
|---|---|---|
| • 3,587 – 3,643 ft<br>• Cored 56ft, Recovered 4ft<br>➢ *Gray-brown and red-brown shale* | ▪ 3,645 – 3,662 ft<br>▪ Cored 17ft, Recovered 10.5ft<br>▪ Recovered Portion:<br>➢ Gray to gray-brown shale<br>➢ Medium to fine grained sandstone | • 4,331 – 4,349 ft<br>• Cored 18ft, Recovered 4.3ft<br>• Recovered Portion<br>➢ Medium to fine grained sandstone |





# Core Pictures MPC 26-5

Core 2 Lower Tuscaloosa massive – very poorly indurated sandstone, well caked

Core 3 Wash-Fred – less indurated than Tuscaloosa core

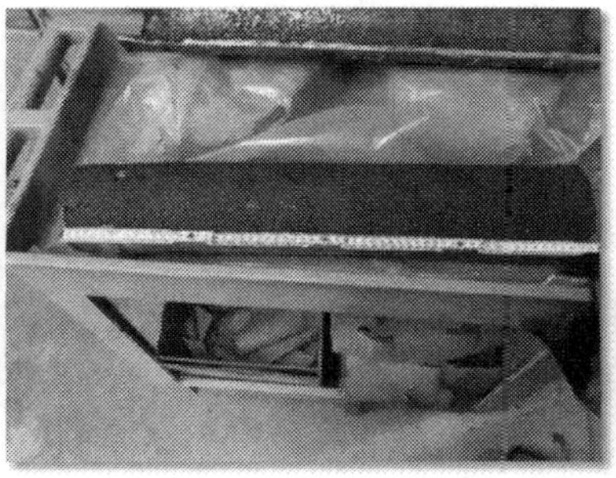

# Storage Complex Reservoir Continuity

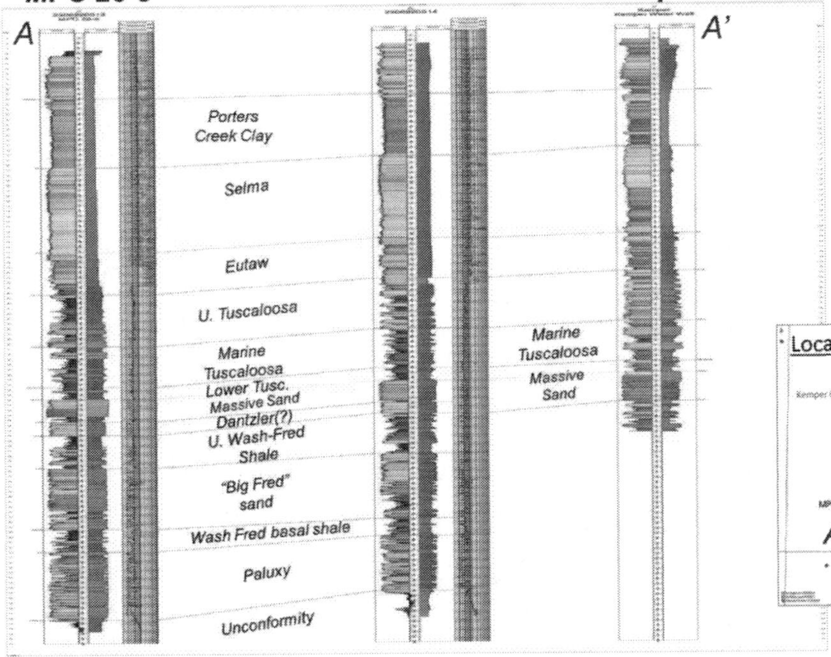


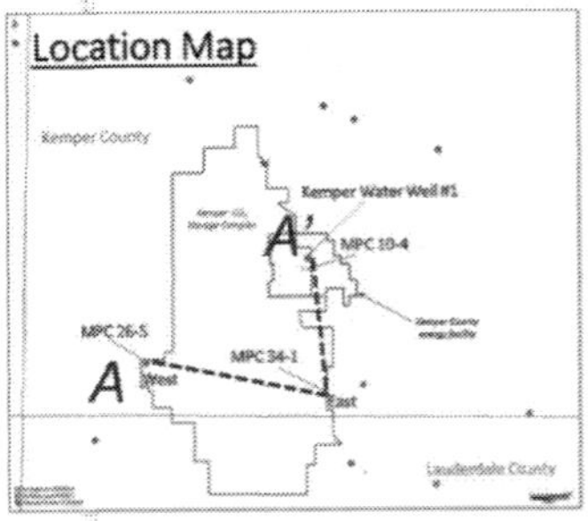






# Existing 2D Seismic

- 30 miles of existing 2D seismic was acquired to evaluate structure, regional stratigraphy

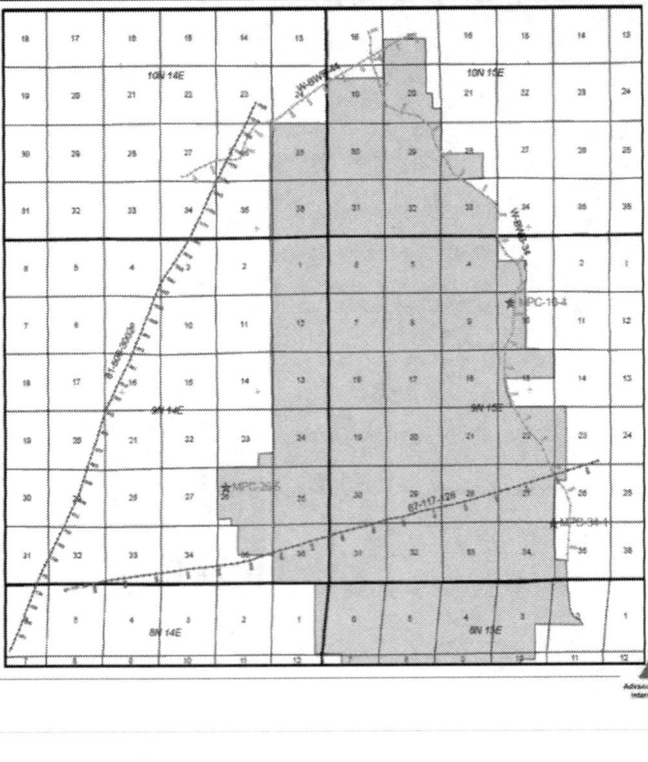

# Kemper Storage Complex Capacity

Net thickness* and porosities** from MPC 26-5

| Reservoir | Net Pay (ft) | Porosity |
|---|---|---|
| L. Tusc. Massive | 162 | 28% |
| Wash.-Fred. | 630 | 28% |
| Paluxy | 370 | 26% |
| **TOTAL** | **1,162** | **27%** |

- Calculate $CO_2$ storage capacity at 100% pore volume utilization for 30,000 acres (approximate Kemper Storage Complex area)
- Apply DOE capacity estimate approach with *site specific**** saline formation efficiency factors for clastics of 3.1% (P10), 6.1% (P50) and 10% (P90) (Goodman et al., 2011)

** shale volume less than 20% using gamma ray index*
*** log density porosity*
**** site specific efficiency factors assume that the net/gross area and net/gross thickness terms are fixed at the P90 level*

Advanced Resources International, Inc.



Project ECO2S Phase 2 CarbonSAFE Field Project

# Kemper Storage Complex Capacity

| Formation | 100% Storage Capacity (MMte)*** | P10 (3.1%) Storage Capacity (MMte)*** | P50 (6.1%) Storage Capacity (MMte)*** | P90 (10%) Storage Capacity (MMte)*** |
|---|---|---|---|---|
| Tusc. Massive Sand | 760 | 20 | 50 | 80 |
| Wash-Fred | 3,140 | 100 | 190 | 310 |
| Paluxy | 1,830 | 60 | 110 | 180 |
| **Total** | **5,720** | **180** | **350** | **570** |

** Assume 0.43 psi/ft hydraulic pressure gradient*
*** from IPCC 2005 Annex Chart*
****million metric tonnes*

Advanced Resources International, Inc.

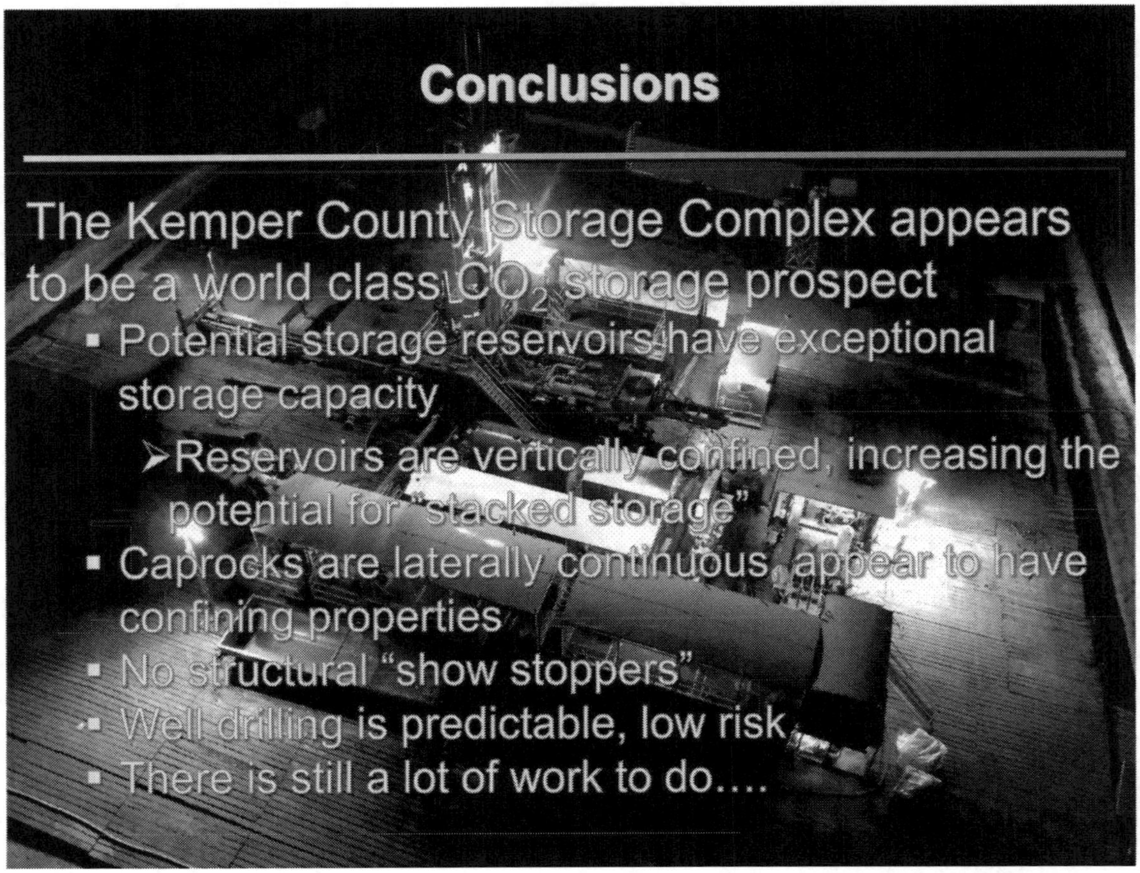

# Conclusions

The Kemper County Storage Complex appears to be a world class $CO_2$ storage prospect

- Potential storage reservoirs have exceptional storage capacity
  - Reservoirs are vertically confined, increasing the potential for "stacked storage"
- Caprocks are laterally continuous, appear to have confining properties
- No structural "show stoppers"
- Well drilling is predictable, low risk
- There is still a lot of work to do....



# Acknowledgements

The Project Team led by Southern States Energy Board, Mississippi Power Company and Southern Company Services, with technical support from Advanced Resources and a host of key subcontractors, acknowledge the valuable support provided by the U.S. DOE National Energy Technology Laboratory on this CarbonSAFE field project.

Office Locations
**Washington, DC**
4501 Fairfax Drive, Suite 910
Arlington, VA 22203
Phone: (703) 528-8420

**Houston, TX**
11931 Wickchester Ln., Suite 200
Houston, TX 77043-4574
Phone: (281) 558-9200

**Knoxville, TN**
1210 Kenesaw Ave.
Suite 1210A
Knoxville, TN 37919-7736





# Project ECO$_2$S Characterization Objectives

1. Demonstrate that the subsurface at the storage complex can store commercial volumes of $CO_2$ safely and permanently within the saline reservoir system,
2. Rigorously establish and optimize the $CO_2$ storage capacity of the storage complex, including the areal extent of the $CO_2$ plume,
3. Confirm the viability of each of the reservoir seals as a long-term, reliable confining system for the $CO_2$ storage site,
4. Undertake rock mechanics and geomechanical modeling to define the potential for induced seismicity at the $CO_2$ storage site,
5. Conduct a comprehensive risk assessment utilizing reservoir modeling and the NETL-sponsored integrated assessment models.

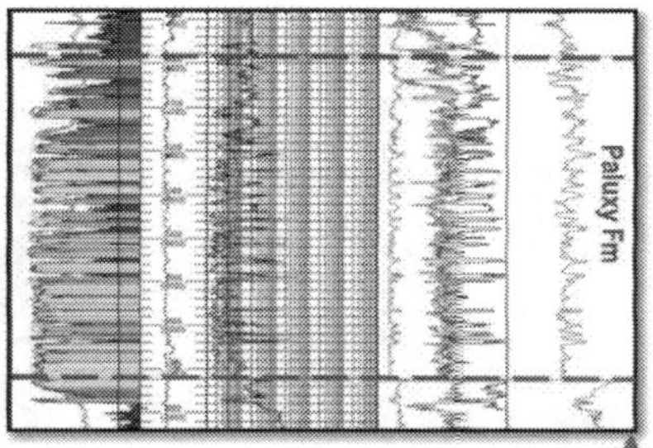

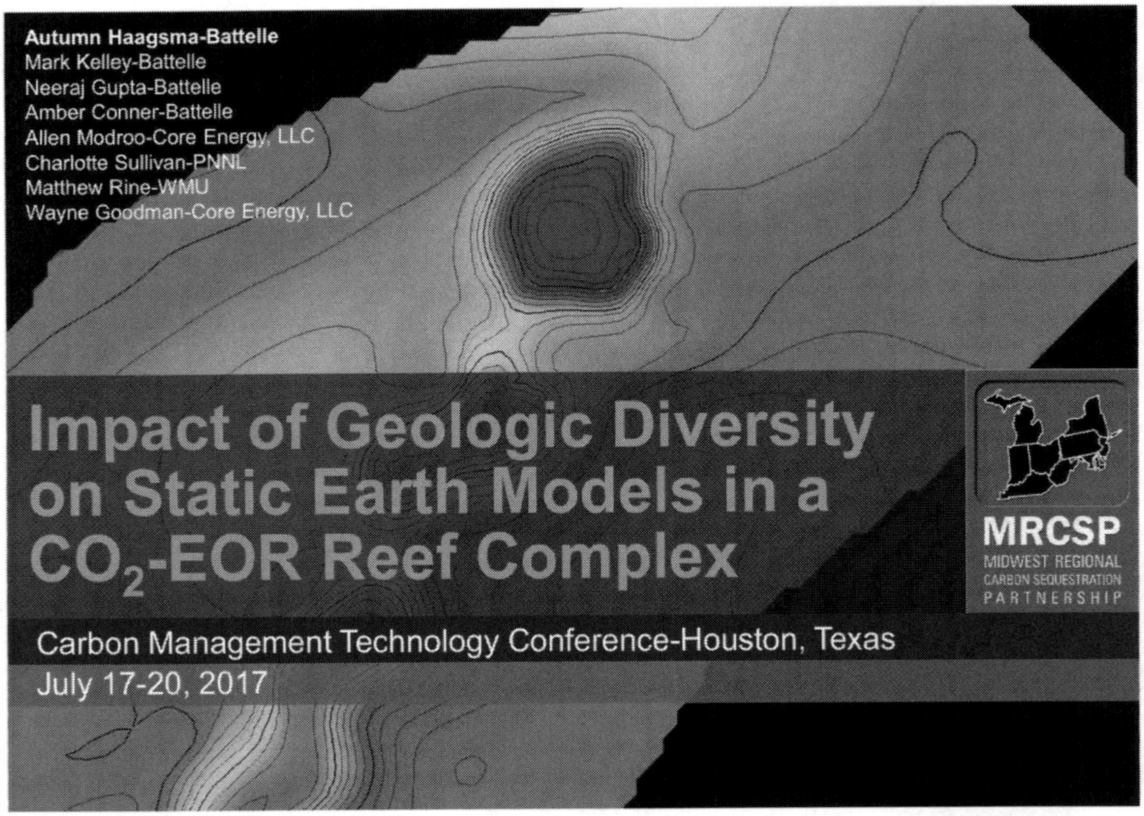


**Autumn Haagsma-Battelle**
Mark Kelley-Battelle
Neeraj Gupta-Battelle
Amber Conner-Battelle
Allen Modroo-Core Energy, LLC
Charlotte Sullivan-PNNL
Matthew Rine-WMU
Wayne Goodman-Core Energy, LLC


# Impact of Geologic Diversity on Static Earth Models in a $CO_2$-EOR Reef Complex

MRCSP
MIDWEST REGIONAL CARBON SEQUESTRATION PARTNERSHIP

Carbon Management Technology Conference-Houston, Texas
July 17-20, 2017

1 DOE/NETL Coop. Agreement # DE-FC26-0NT42589 BATTELLE

## Acknowledgements


***Battelle's MRCSP team members for work shown here***

**DOE/NETL** has worked with us and our partners to structure a program that adds to the knowledge base and extends the state-of-the-art research.

**Core Energy, LLC** our host site and $CO_2$ supplier for 10 years of collaboration under Phase II and Phase III

**Western Michigan University** has worked closely with us to develop geologic interpretations and static earth models and provided access to the core database


CORE ENERGY, LLC

Michigan Basin Core Research Laboratory

BATTELLE
It can be done™

Pacific Northwest National Laboratory
Operated by Battelle for the U.S. Department of Energy

NATIONAL ENERGY TECHNOLOGY LABORATORY

2 MRCSP BATTELLE

# Presentation Outline

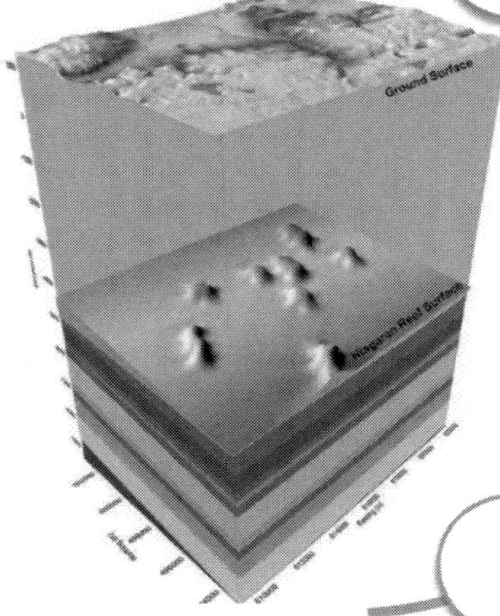

- Project Background
- Geologic Characterization
- Conceptual Model
- Static Earth Modeling
- Key Findings/Conclusions



# MRCSP Project Overview: Goals and Objectives

- Primary goal: To execute a large-scale $CO_2$ injection test to evaluate best practices and technologies required to implement carbon sequestration
- Objectives are to advance operational, monitoring, and modeling techniques needed to:
  - Develop and validate reservoir models
  - Address public concerns such as leakage and storage security
  - Address other topics such as cost effectiveness and CCUS practicability

# MRCSP Large-Scale Test Site

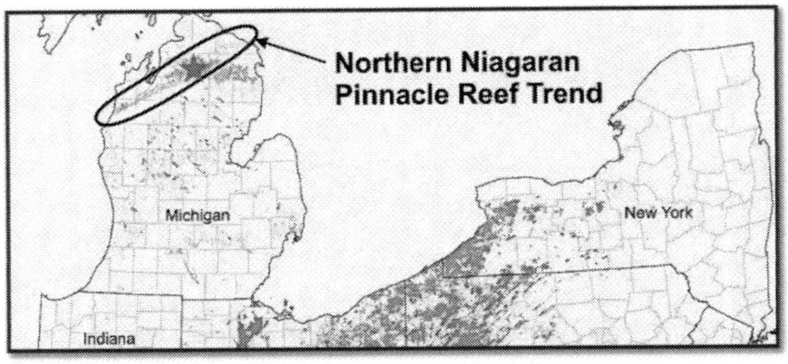


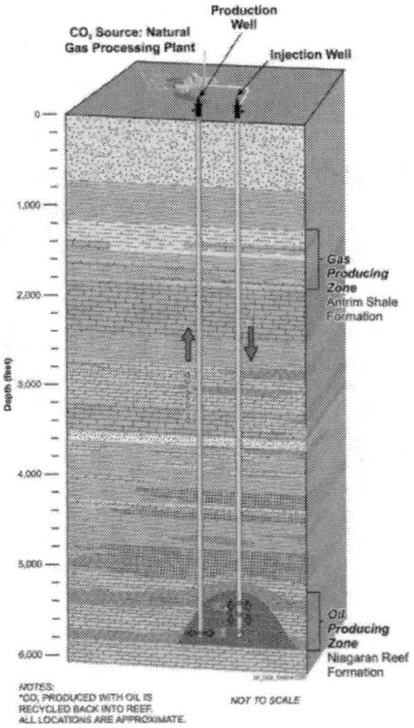


- **Location**: Otsego County, Michigan
- **Source of $CO_2$**:
  - Local natural gas processing plant
- **Reservoir Type**:
  - Closely spaced, highly compartmentalized oil & gas fields located in the Northern Michigan's Niagaran Reef Trend

5

MRCSP BATTELLE

# MRCSP Large-Scale Test Site

3D Representation of Study Area

Map of Study Area

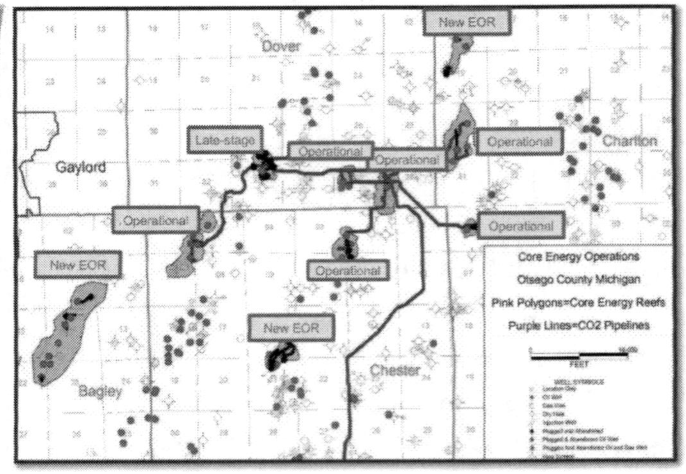


6

MRCSP BATTELLE

# Compiled Dataset

- Working closely with Core Energy, LLC and WMU
- >4200 wells in study area
  - Many shallow (<3000 ft)
  - Many missed the reefs
- 60+ sq miles of 3D seismic
- 2 2D lines
- ~ 20 Whole core
  - Many missed the reefs
- **Recently drilled 4 new wells and collected advanced data**

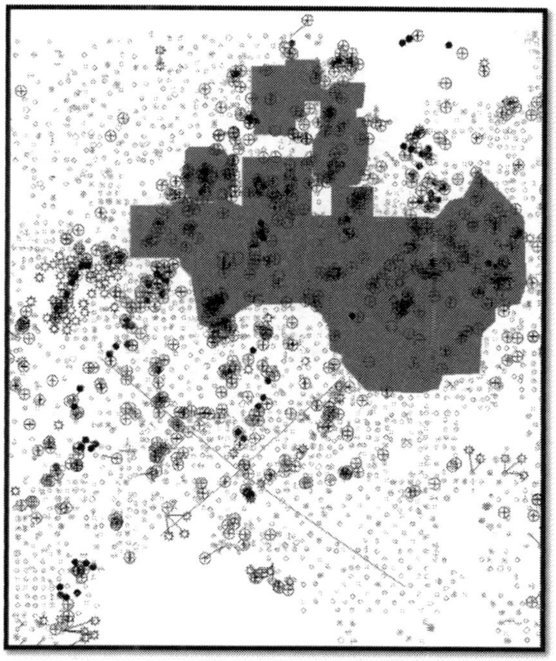

7

MRCSP BATTELLE

# Reef Building and Diagenesis Leads to Complex Geology

- Silurian-age, Niagaran group reef
- Sealed by overlying Salina group
  - Carbonates and evaporites

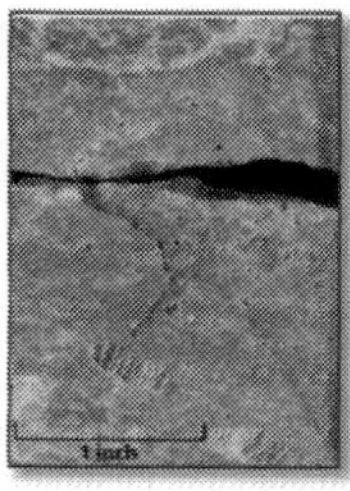

Bioherm

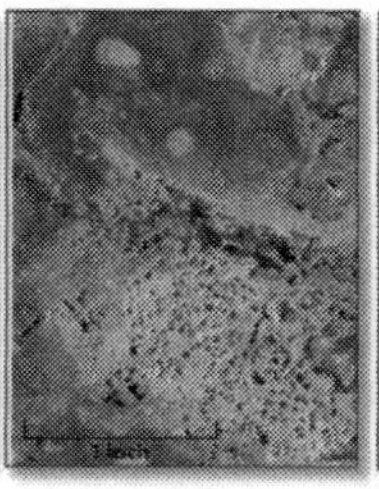

Reef Core

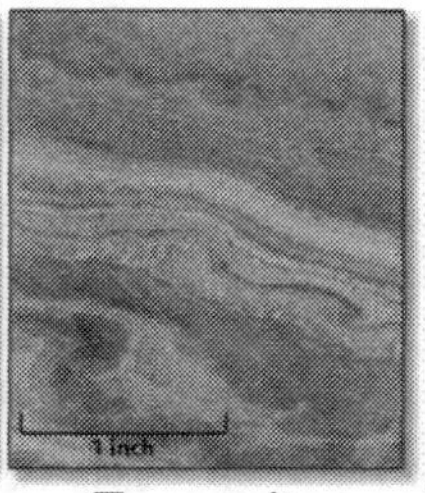

Evaporite

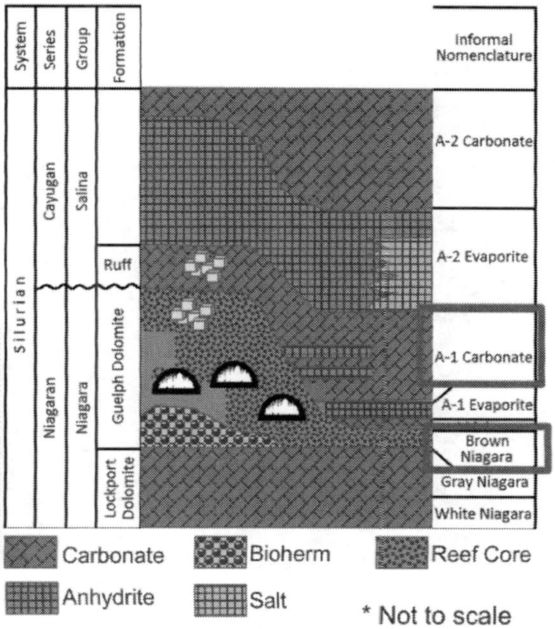

* Not to scale

8

MRCSP BATTELLE

# Current Reefs of Study are Diverse

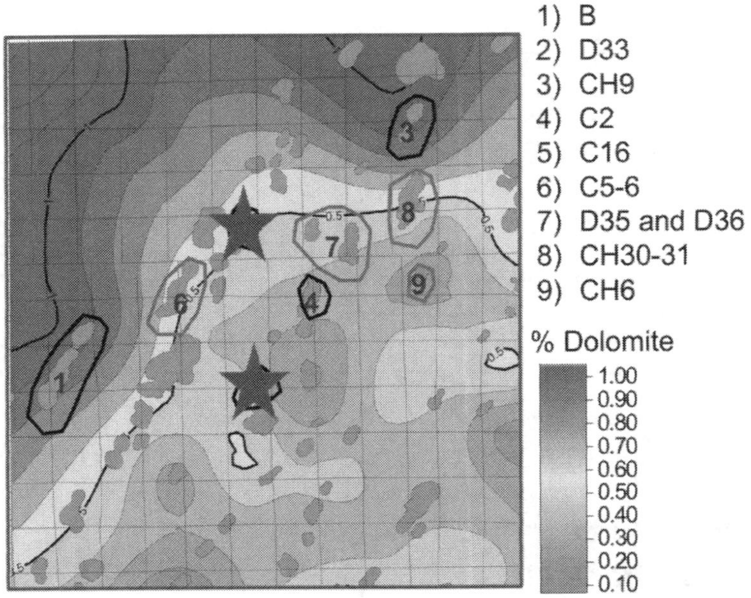


9

MRCSP BATTELLE

# Developing Workflows for Static Earth Models

- Building comprehensive database
- Collaboration with WMU and Core Energy, LLC
- Simple and efficient workflows
  - Characterize reefs
  - Justify decisions
  - Build geologic models
  - Scale from site to regional

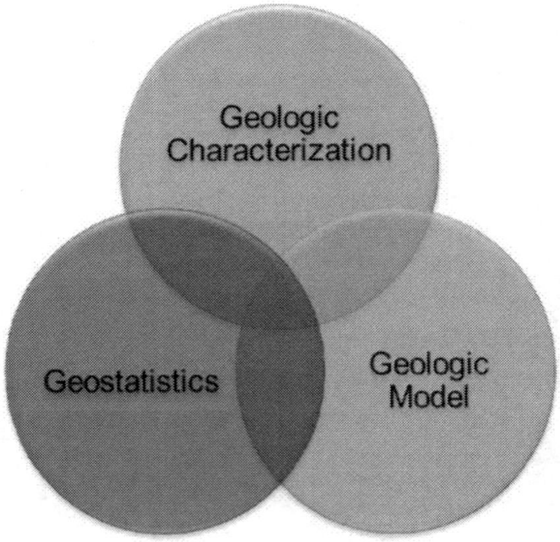


10

MRCSP BATTELLE

# Basic Log Analysis to Identify Porosity and Lithology

D33

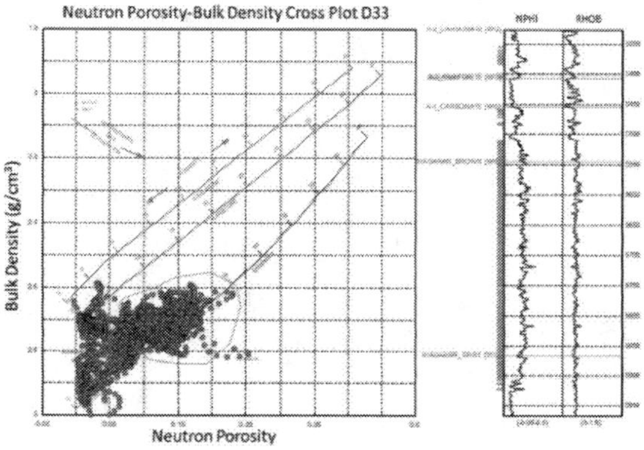


C16

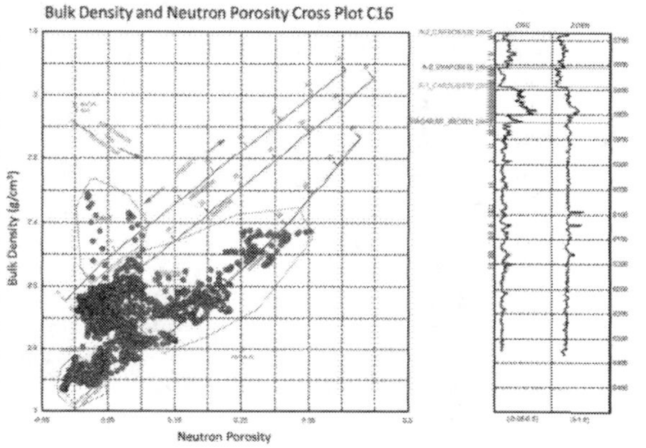


11

MRCSP BATTELLE

# Advanced Log Analysis to Identify Changes in Texture and Structures

D33

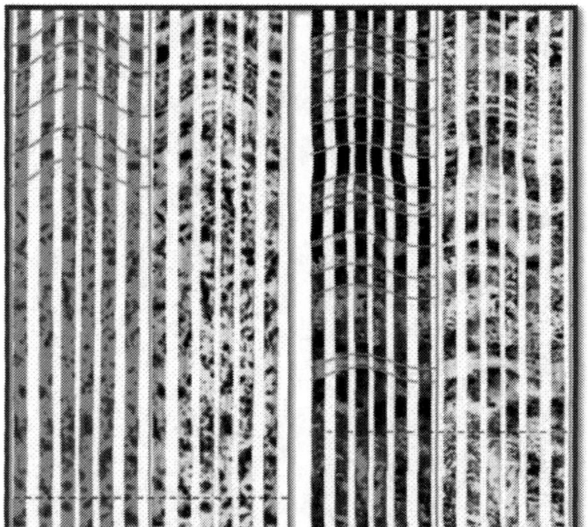

C16

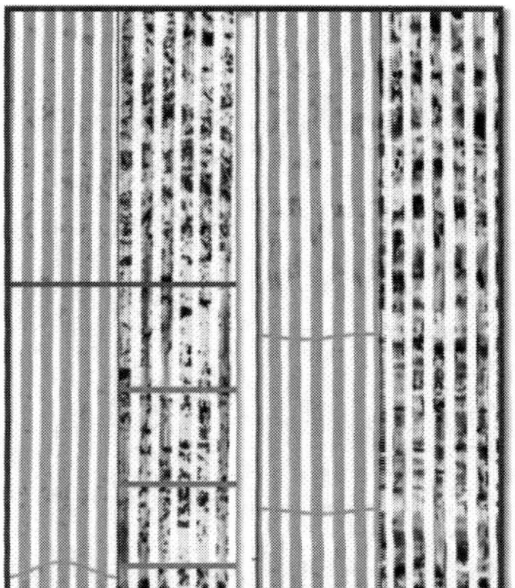

12

MRCSP BATTELLE

# Whole Core Analysis to Analyze Reservoir and Caprock Properties

D33

C16

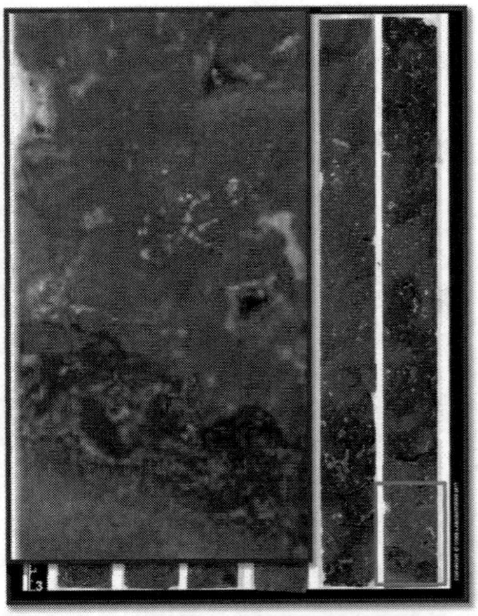

13

MRCSP BATTELLE

# Whole Core Analysis to Analyze Reservoir and Caprock Properties

D33

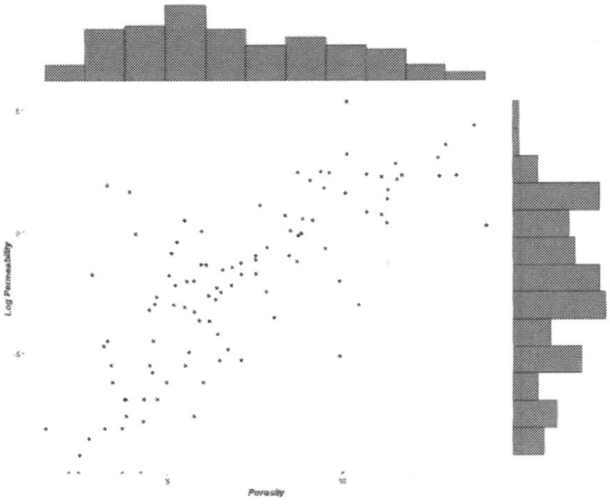


C16

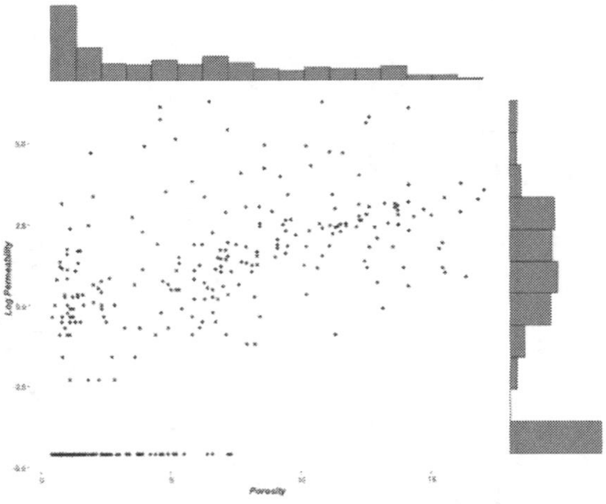


*New data not yet available from new wells

14

MRCSP BATTELLE

# 3D Seismic to Determine Reef Geometry

D33

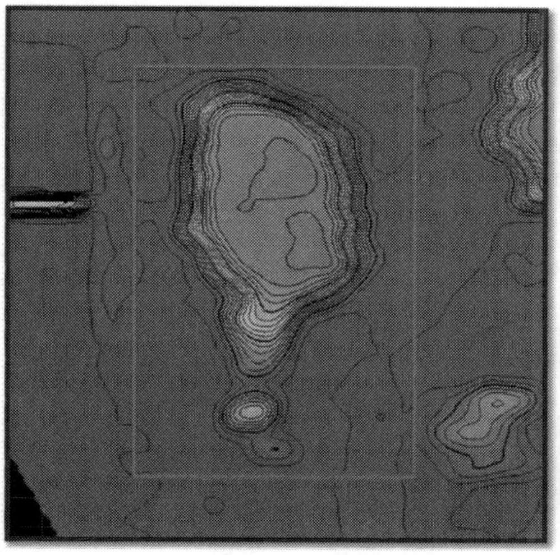

C16

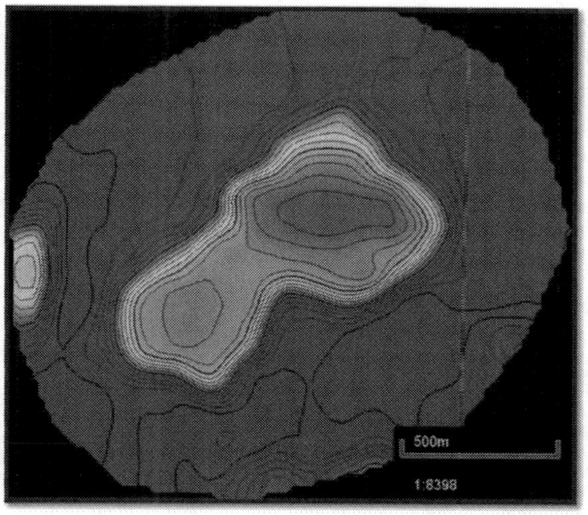


15

MRCSP BATTELLE

# Integration of Data to Develop Conceptual Models- D33

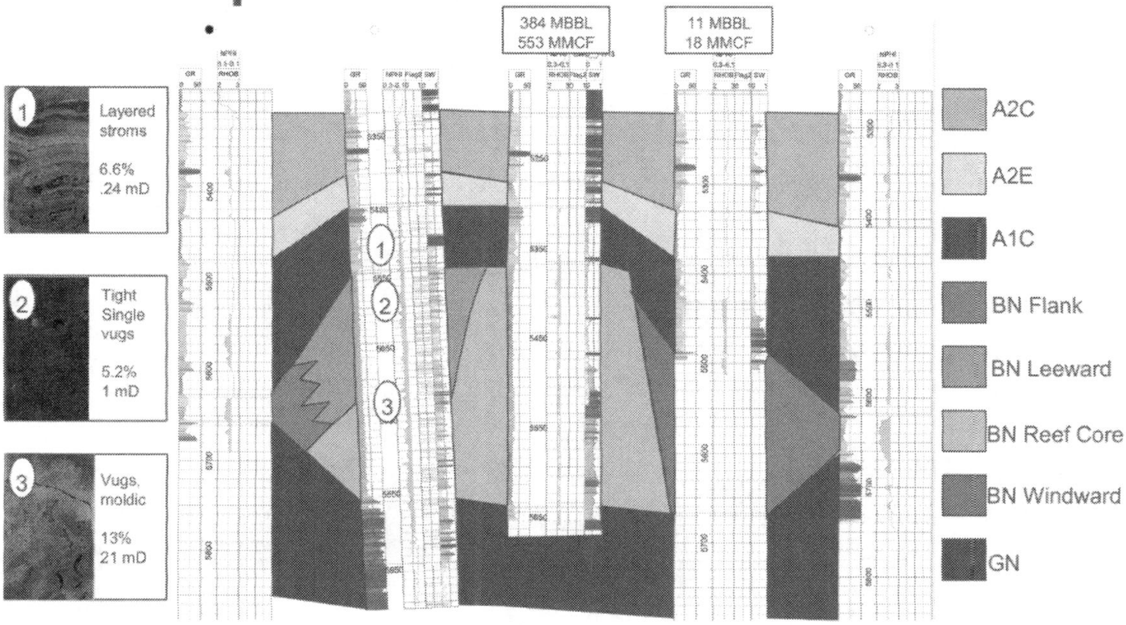


16

MRCSP BATTELLE

## Integration of Data to Develop Conceptual Models- C16

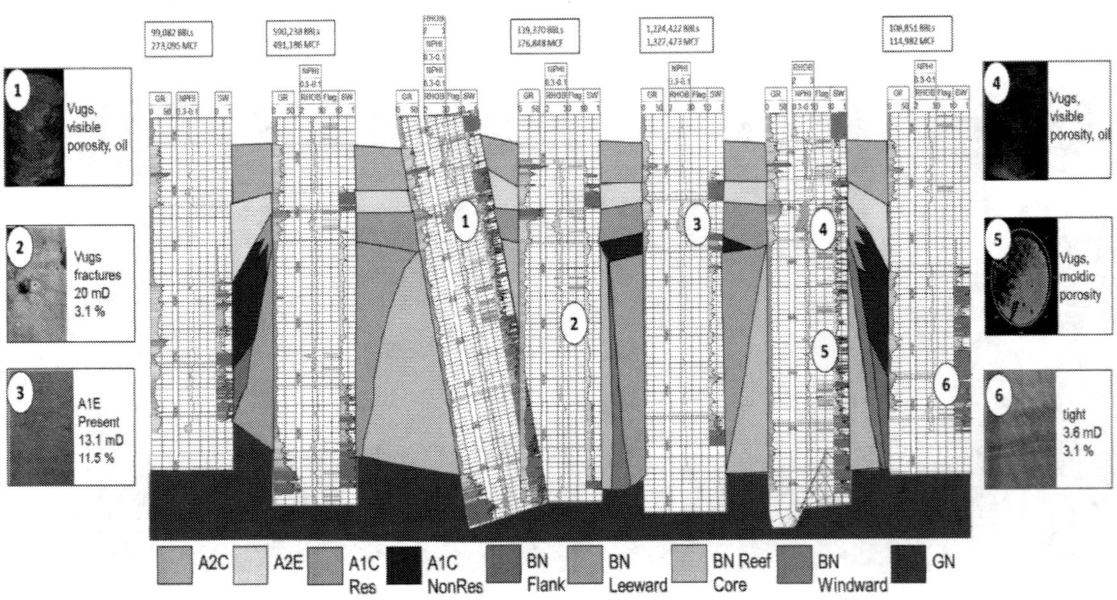


17

MRCSP BATTELLE

## Static Earth Modeling

- Goals
  - Develop efficient workflow
  - Incorporate WMUs work
  - Suitable for dynamic
- Combine simple structure model with advanced model
- Keep interpretations consistent across the team

**Analyze Wireline Logs**
- Pick sequence and formation tops
- Pick Facies
- Compute Petrophysics

**Build Depositional Model**
- Determine windward and leeward flanks
- Determine reservoir and non reservoir zones
- Create boundaries around zones
- Assess influence of salt-filled porosity, if any

**Generate Surfaces**
- Gray Niagaran, reservoir reef, non reservoir reef, A1, A2, etc.
- Force surfaces to pinch-out along boundaries and eliminate cross overs
- Incorporate seismic for geometry

**Generate Structure Model**
- Use appropriate grid size based on well spacing
- Define horizons, zones, and layers

**Generate Segments**
- Define segments from depositional model
- Build segments using clipped or contoured surface tops and bases
- Merge segment into one unit

**Generate Property Model**
- Scale up well logs
- Populate properties through the model using segments as filters to control distributions
- Upscale model

18

MRCSP BATTELLE

# Application of Conceptual Model to Guide SEM

D33

C16

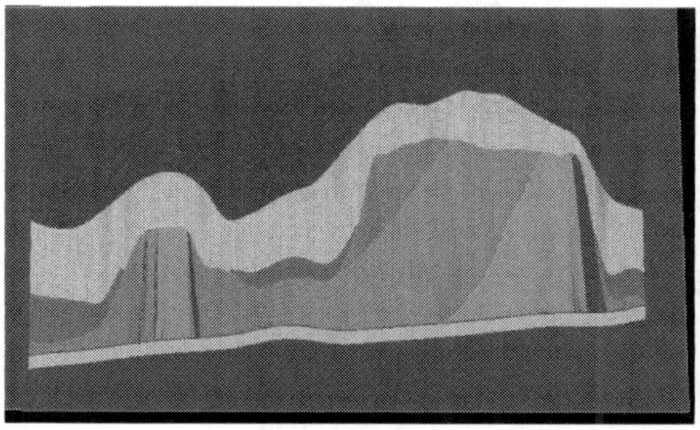

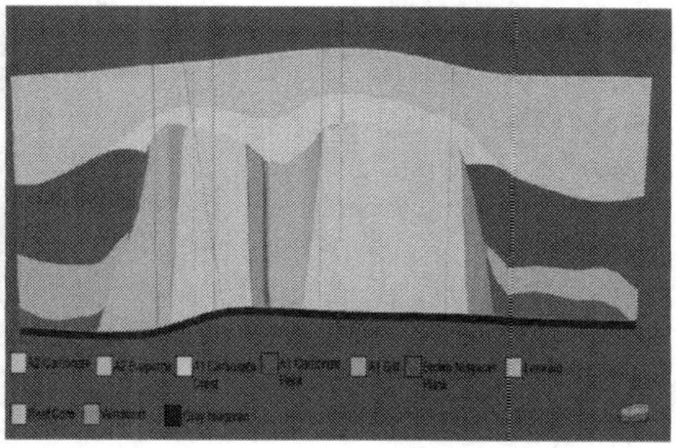

19

MRCSP BATTELLE

# Porosity Distribution is Significantly Different between Reefs

D33

C16

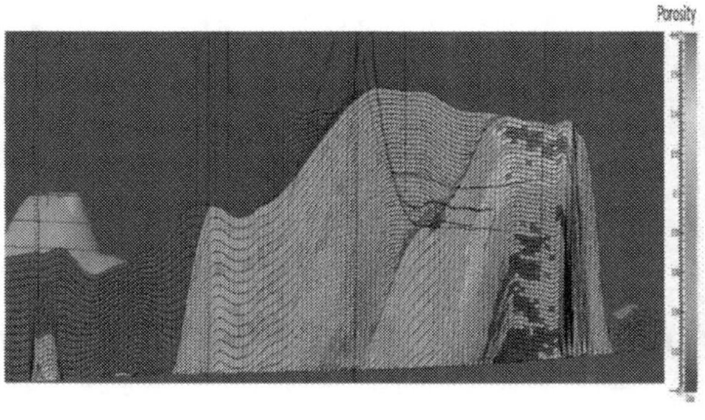


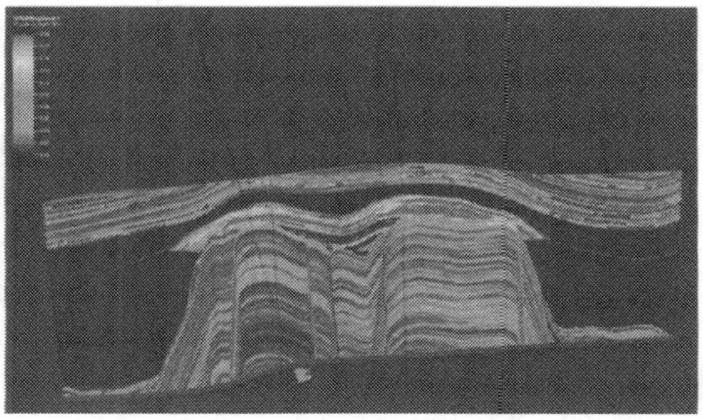

20

MRCSP BATTELLE

# Diversity of Niagaran Reefs

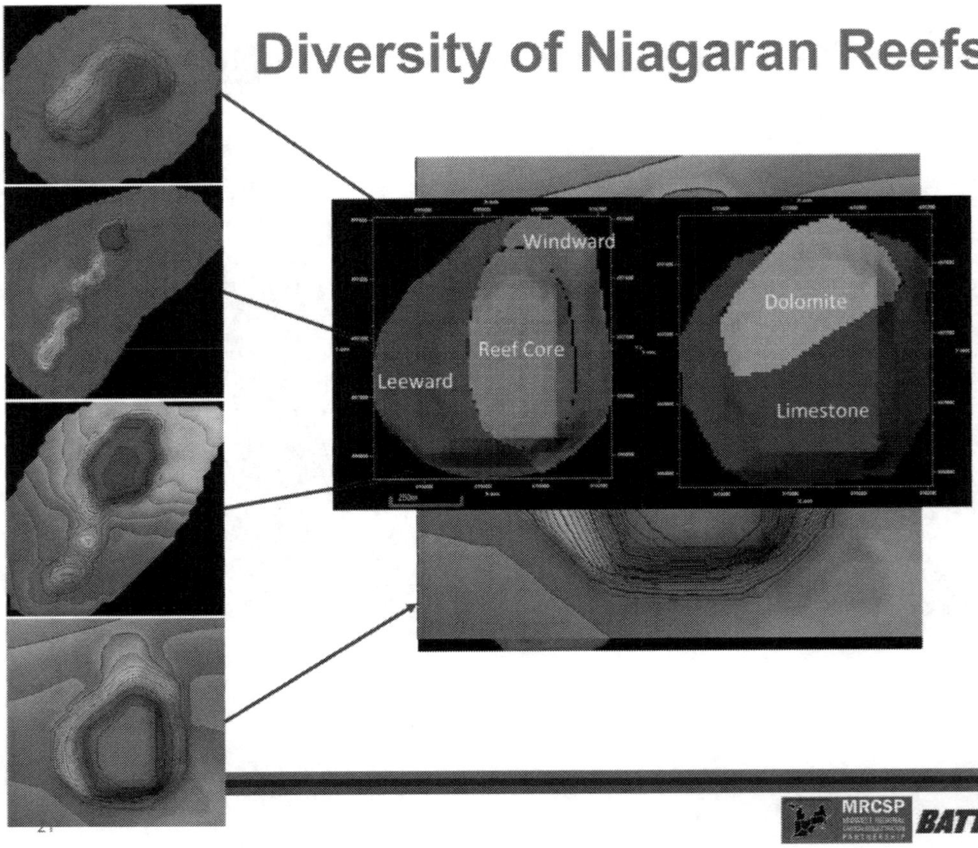




MRCSP BATTELLE

# Conclusions

- Standard analyses provide initial geologic interpretations
- Integration of data types is needed to fully understand geologic diversity
- Development of conceptual models are crucial to guide SEMs
- Geologic variability between reefs influences the reservoir position and distribution

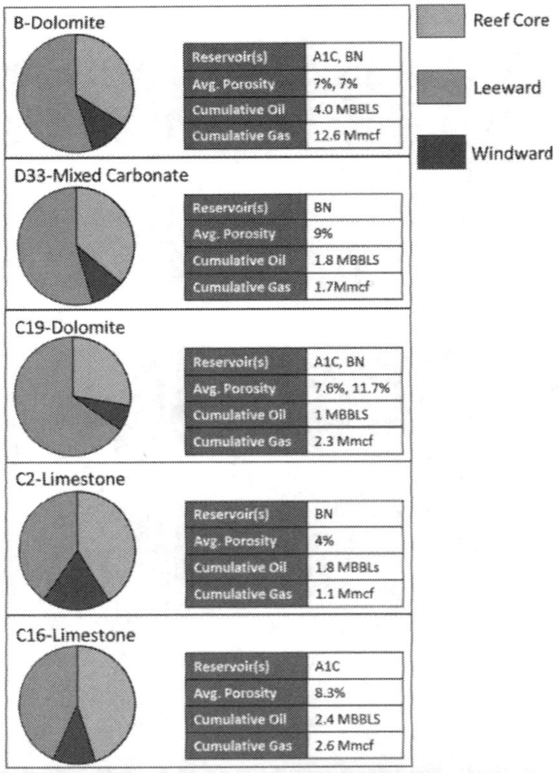


B-Dolomite

| | |
|---|---|
| Reservoir(s) | A1C, BN |
| Avg. Porosity | 7%, 7% |
| Cumulative Oil | 4.0 MBBLS |
| Cumulative Gas | 12.6 Mmcf |

D33-Mixed Carbonate

| | |
|---|---|
| Reservoir(s) | BN |
| Avg. Porosity | 9% |
| Cumulative Oil | 1.8 MBBLS |
| Cumulative Gas | 1.7Mmcf |

C19-Dolomite

| | |
|---|---|
| Reservoir(s) | A1C, BN |
| Avg. Porosity | 7.6%, 11.7% |
| Cumulative Oil | 1 MBBLS |
| Cumulative Gas | 2.3 Mmcf |

C2-Limestone

| | |
|---|---|
| Reservoir(s) | BN |
| Avg. Porosity | 4% |
| Cumulative Oil | 1.8 MBBLs |
| Cumulative Gas | 1.1 Mmcf |

C16-Limestone

| | |
|---|---|
| Reservoir(s) | A1C |
| Avg. Porosity | 8.3% |
| Cumulative Oil | 2.4 MBBLS |
| Cumulative Gas | 2.6 Mmcf |



MRCSP BATTELLE

# Expanding to Regional Characterization and Capacity Estimations

- Continued collaboration between Battelle, Core Energy, and WMU
- Expand datasets to northern reef trend (NRT)
- Identify trends and/or predictors for capacity

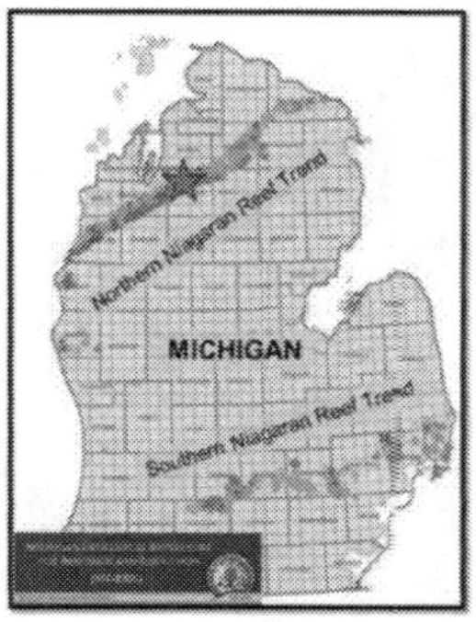


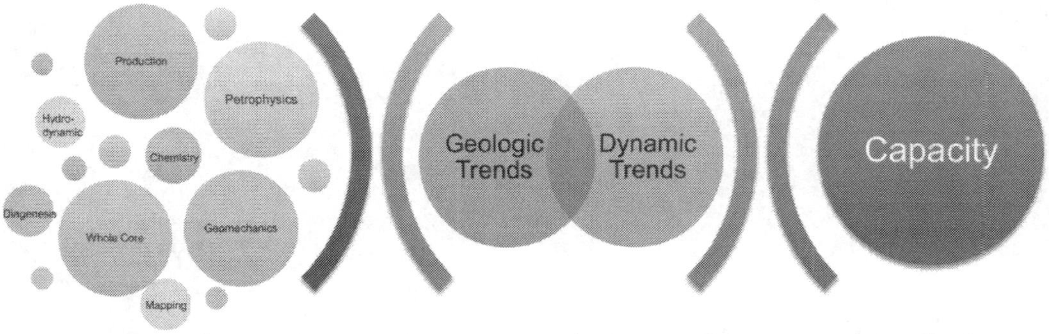

## Questions?

Autumn Haagsma
haagsma@battelle.org

EERC
Critical Challenges.
Practical Solutions.

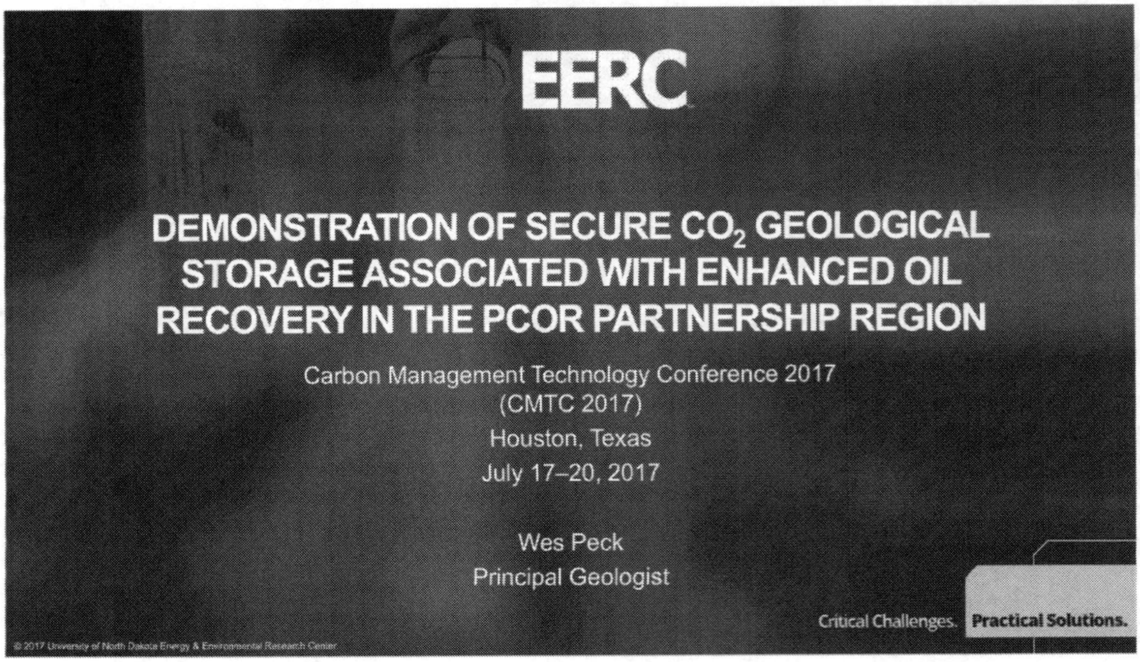
EERC
DEMONSTRATION OF SECURE $CO_2$ GEOLOGICAL STORAGE ASSOCIATED WITH ENHANCED OIL RECOVERY IN THE PCOR PARTNERSHIP REGION
Carbon Management Technology Conference 2017
(CMTC 2017)
Houston, Texas
July 17–20, 2017
Wes Peck
Principal Geologist
Critical Challenges. Practical Solutions.
© 2017 University of North Dakota Energy & Environmental Research Center

## PLAINS $CO_2$ REDUCTION (PCOR) PARTNERSHIP

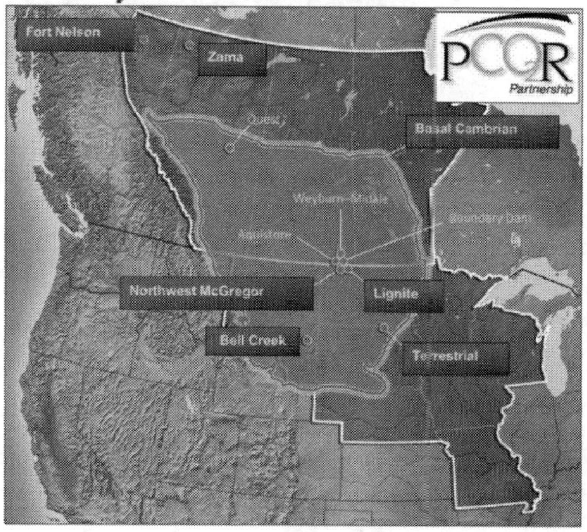


- Region includes:
  - Nine states.
  - Four Canadian provinces.
  - Over 3.6 million $km^2$.
- Several completed field projects.
- Over 3 million tons of $CO_2$ stored and monitored in association with $CO_2$ enhanced oil recovery (EOR).
- More than 120 partners.

EERC | U.S. DEPARTMENT OF ENERGY | NATIONAL ENERGY TECHNOLOGY LABORATORY | PCOR Partnership

Critical Challenges. **Practical Solutions.**

## PCOR PARTNERSHIP

PCOR Partnership 2003 – Present

## PCOR PARTNERSHIP OBJECTIVES

- Safely and permanently achieve $CO_2$ storage on a commercial scale.
- Establish a relationship between the $CO_2$ EOR process and long-term storage of $CO_2$.
- Establish monitoring, verification, and accounting (MVA) methods to effectively monitor $CO_2$ storage.
- Use commercial oil/gas practices as the backbone of MVA strategies, and augment with additional cost-effective techniques.
- Share lessons learned for the benefit of similar projects across the region.

## $CO_2$ EOR

- **A great near-term storage option:**
  - Over 40 years of handling and injecting large volumes of $CO_2$.
  - Much of the infrastructure already in place.
  - Storage cost can be offset by income from EOR.

**"Greener" than conventionally produced oil:**

- Existing EOR operations are already storing $CO_2$.
- Nearly every tonne of $CO_2$ purchased is eventually stored.

# COMPARING $CO_2$ EOR TO "REGULAR" OIL

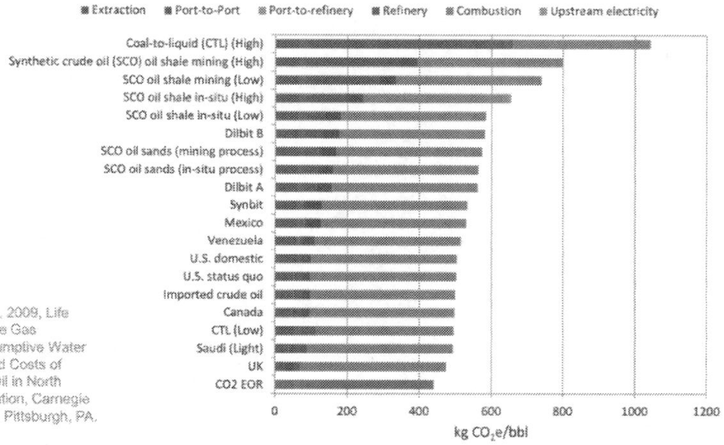


Adapted from:
Mangmeechai, A., 2009, Life Cycle Greenhouse Gas Emissions, Consumptive Water Use and Levelized Costs of Unconventional Oil in North America: Dissertation, Carnegie Mellon University, Pittsburgh, PA.

EERC | U.S. DEPARTMENT OF ENERGY | NATIONAL ENERGY TECHNOLOGY LABORATORY | PCOR    7    Critical Challenges. **Practical Solutions.**

# JOURNAL ARTICLE

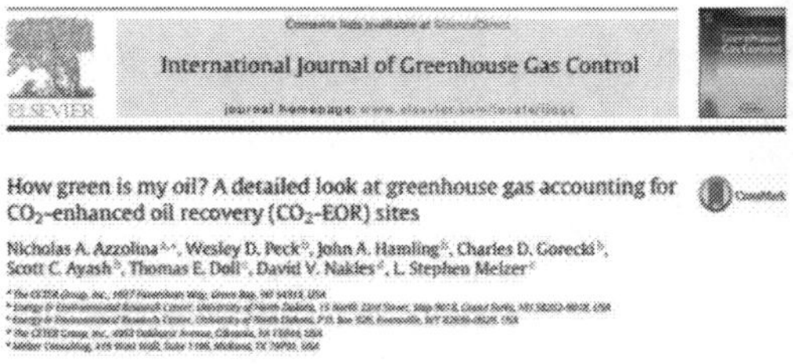

International Journal of Greenhouse Gas Control

How green is my oil? A detailed look at greenhouse gas accounting for $CO_2$-enhanced oil recovery ($CO_2$-EOR) sites

Nicholas A. Azzolina, Wesley D. Peck, John A. Hamling, Charles D. Gorecki, Scott C. Ayash, Thomas E. Doll, David V. Nakles, L. Stephen Melzer

http://www.sciencedirect.com/science/article/pii/S1750583616302985

The spreadsheet $CO_2$ EOR life cycle analysis model is available on the PCOR Partnership public Web site!

http://www.undeerc.org/pcor/technicalpublications/CO2-EOR-Life-Cycle-Analysis.aspx

# BELL CREEK

- The Bell Creek oil field is operated by Denbury Onshore LLC.
- $CO_2$ is sourced from ConocoPhillips' Lost Cabin and ExxonMobil's Shute Creek gas-processing plants.
- The Energy & Environmental Research Center is studying $CO_2$ storage associated with commercial $CO_2$ EOR.

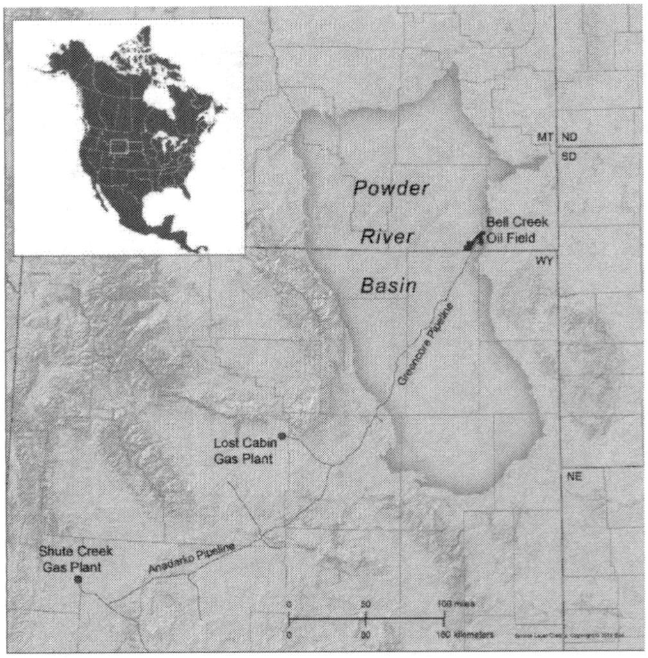


EERC | U.S. DEPARTMENT OF ENERGY | NATIONAL ENERGY TECHNOLOGY LABORATORY | PCOR

# FIELD DEVELOPMENT

- **Primary production and waterflooding produced ~37.5% original oil in place (OOIP).**
- **Estimated 40–50 million incremental bbl of oil.**
- **Estimated 12.7 million tonnes of $CO_2$ stored.**

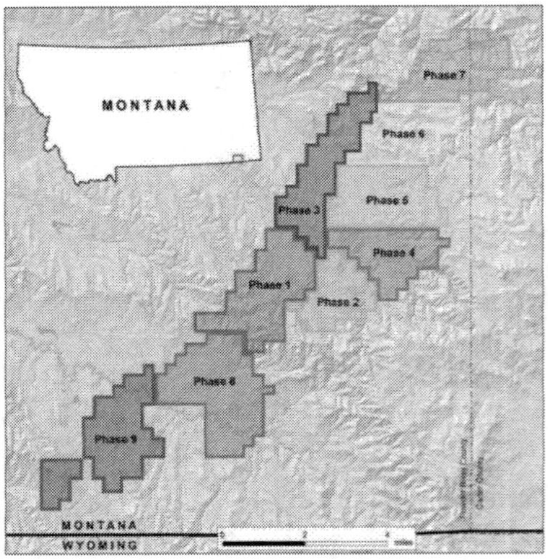


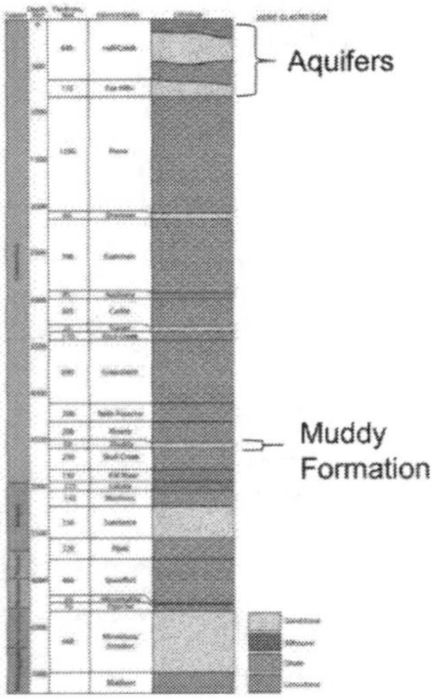


EERC | U.S. DEPARTMENT OF ENERGY | NATIONAL ENERGY TECHNOLOGY LABORATORY | PCOR



Critical Challenges. **Practical Solutions.**

# $CO_2$ INJECTION

**As of March 2017**

- Oil Produced: ~3.7 million barrels
  *(source: Montana Board of Oil & Gas [MBOG] database)*
- $CO_2$ Stored: ~3.7 million tonnes
  *(source: Denbury)*

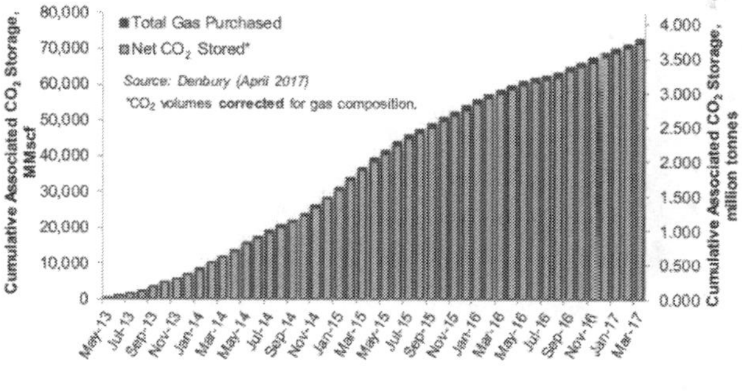


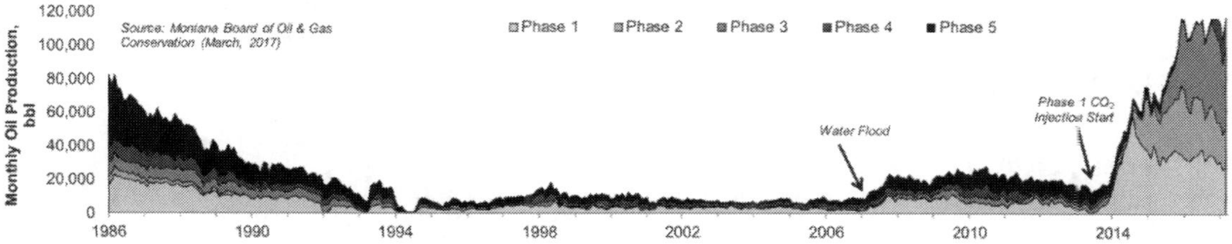




# ADAPTIVE MANAGEMENT APPROACH

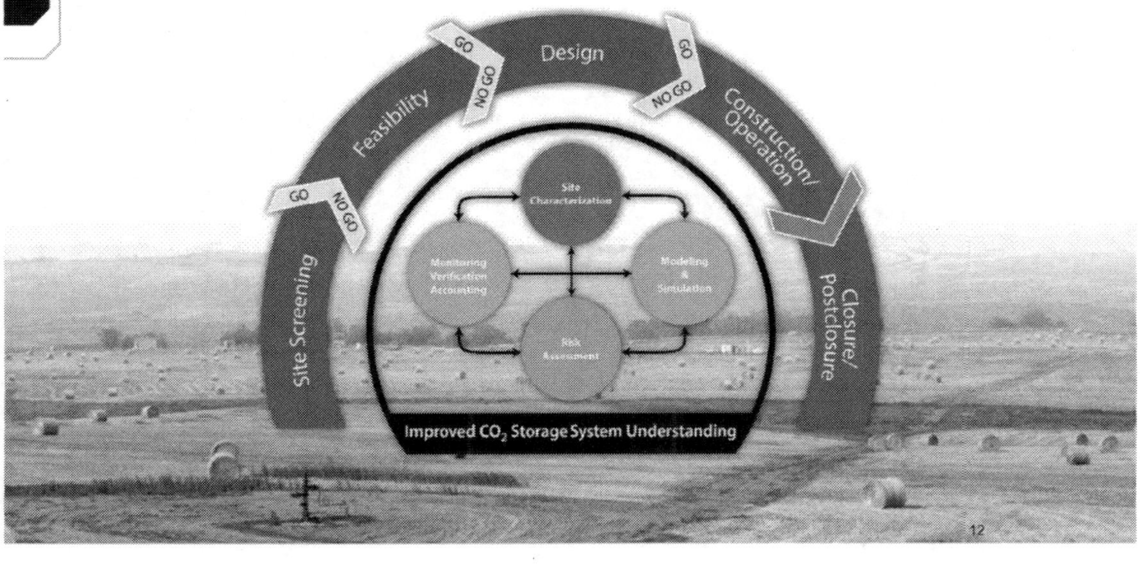

# MVA

- 16 techniques
- 1.5 years of preinjection monitoring
- 3+ years of operational monitoring

Demonstrate and validate monitoring techniques and their associated economics to inform viable MVA strategies for commercial-scale carbon capture and storage (CCS).

Building off of the backbone of commercial operations data.

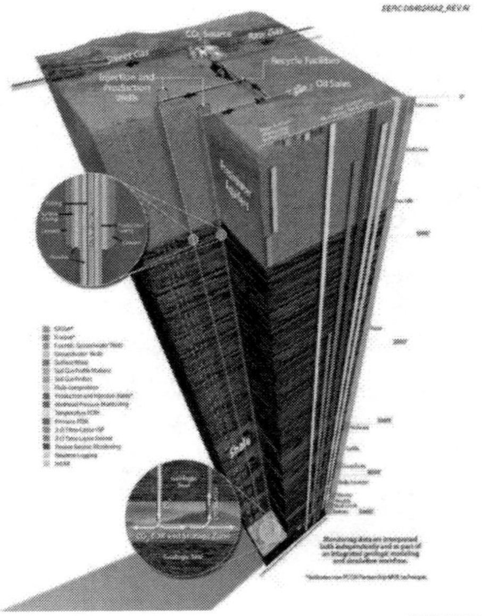

EERC | U.S. DEPARTMENT OF ENERGY | NATIONAL ENERGY TECHNOLOGY LABORATORY | PCOR Partnership

13

Critical Challenges. **Practical Solutions.**

# PULSED-NEUTRON LOG (PNL) CAMPAIGN

- **Seven PNL campaigns:**
  - **45 wells (92 total logs) logged to date:**
    - ♦ **45 baseline**
    - ♦ **47 repeat**

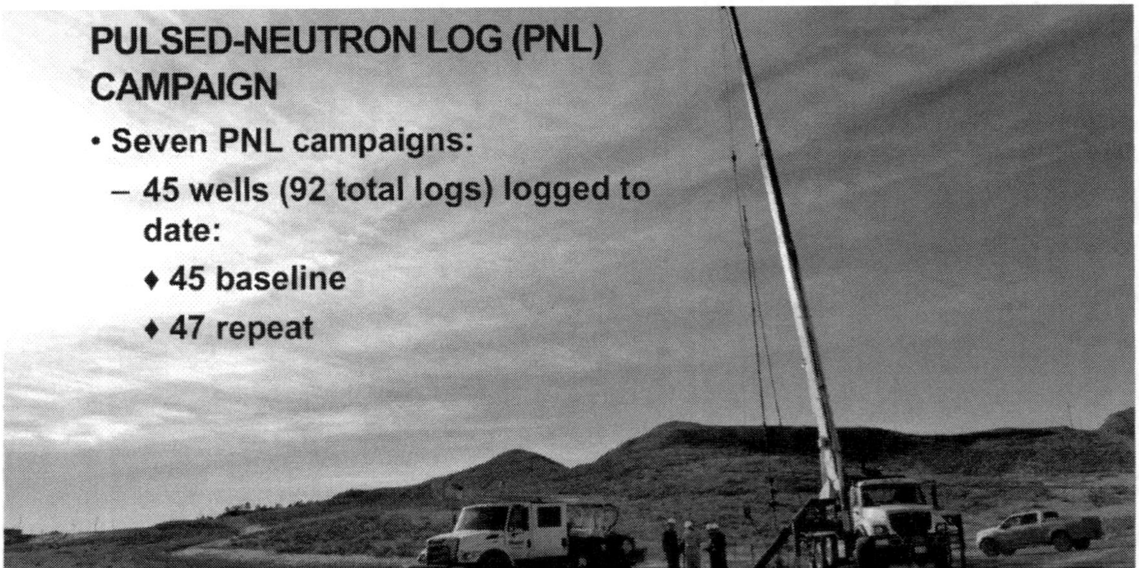

## MVA FOR MODEL VALIDATION – PULSED-NEUTRON LOGGING

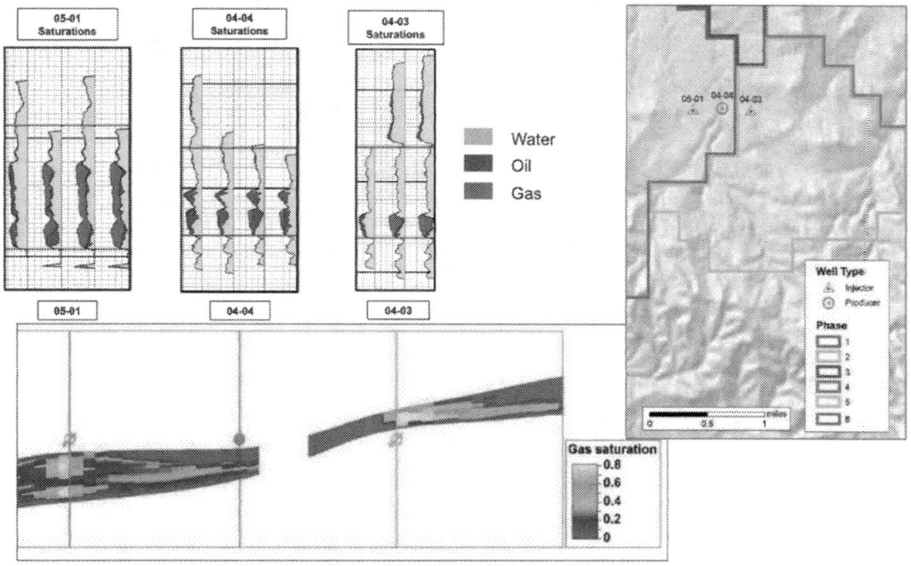


## BASELINE 3-D SEISMIC SURVEY

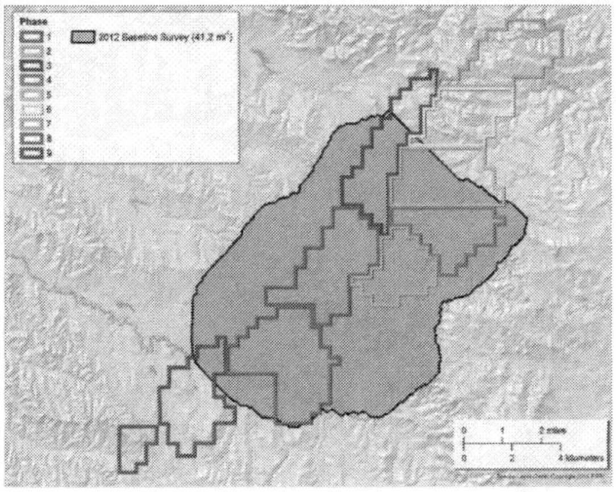


- 104-km Baseline Survey (August 2012)

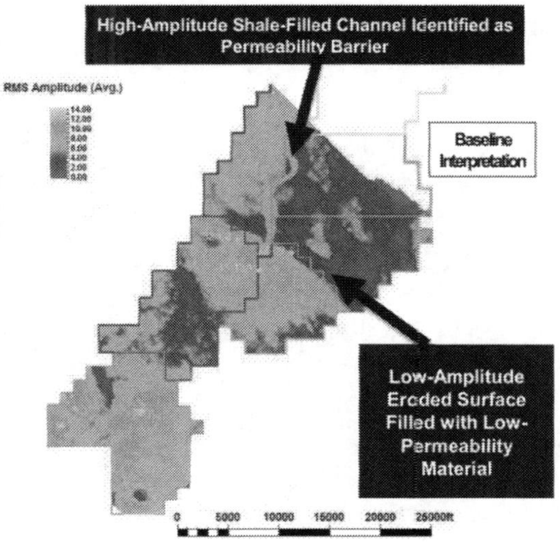


18

# SIMULATION-GUIDED MVA

- Simulation was used to predict location and saturation of $CO_2$.
- 2-D seismic line used to confirm ability of seismic to detect $CO_2$ in the reservoir.
- Results supported decision to conduct large 3-D survey.

Predictive Simulation Results ($CO_2$ plumes)

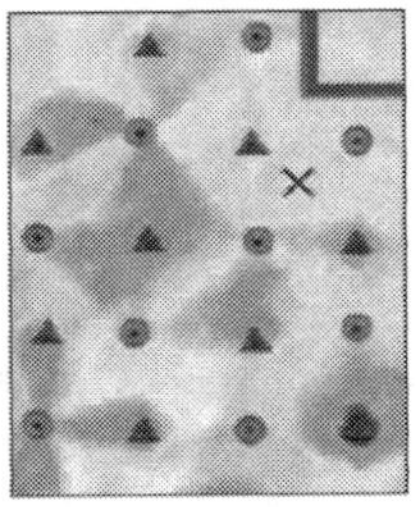

Seismic Line Overlaying Simulation

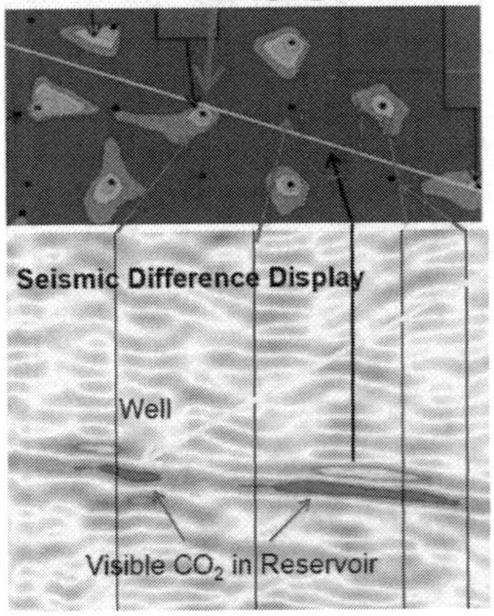




# FIRST REPEAT 3-D SURVEY

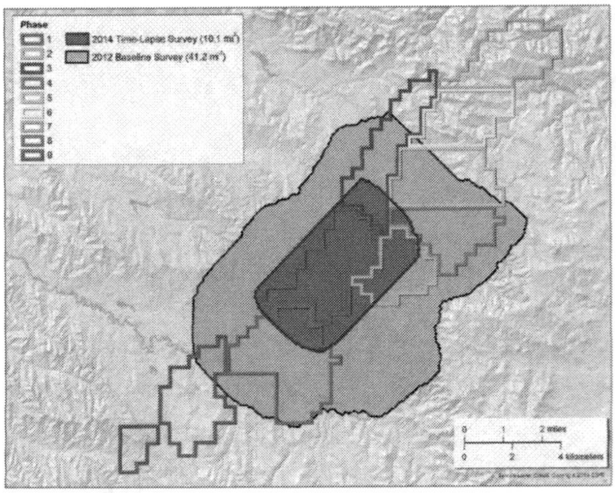

| Phase | Start of $CO_2$ Injection | Estimated Associated $CO_2$ Storage (Oct 2014), Mt |
|---|---|---|
| 1 | May 2013 | 1.04 |
| 2 | Dec 2013 | .166 |

Calculated using MBOG data.

- 26-km² repeat (October 2014).
- ~1.2 Mt $CO_2$ stored in monitored area at the time of survey.

# FIRST REPEAT 4-D DIFFERENCE INTERPRETATION (2012–2014)

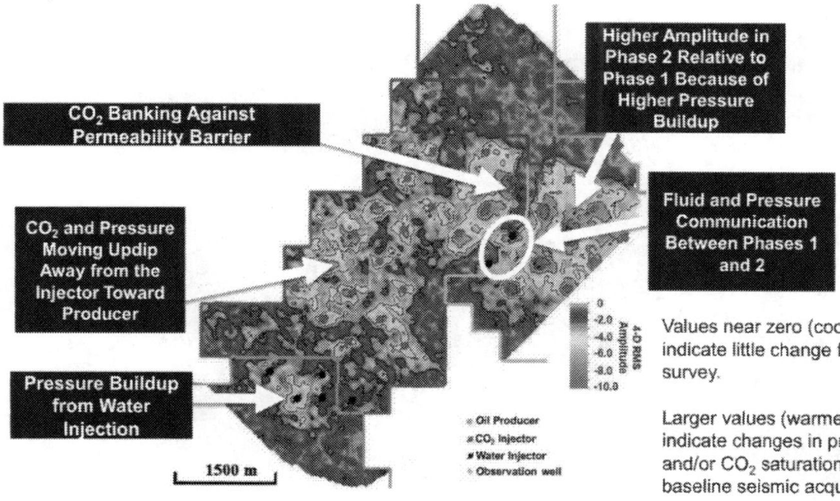


Values near zero (cooler colors) indicate little change from baseline survey.

Larger values (warmer colors) indicate changes in pressure and/or $CO_2$ saturation since baseline seismic acquisition.

# ADDITIONAL OBSERVATIONS

An additional 4-D seismic analysis (survey acquired in 2015) shows suspected fluid communication between the Phase 1 and 3 areas…

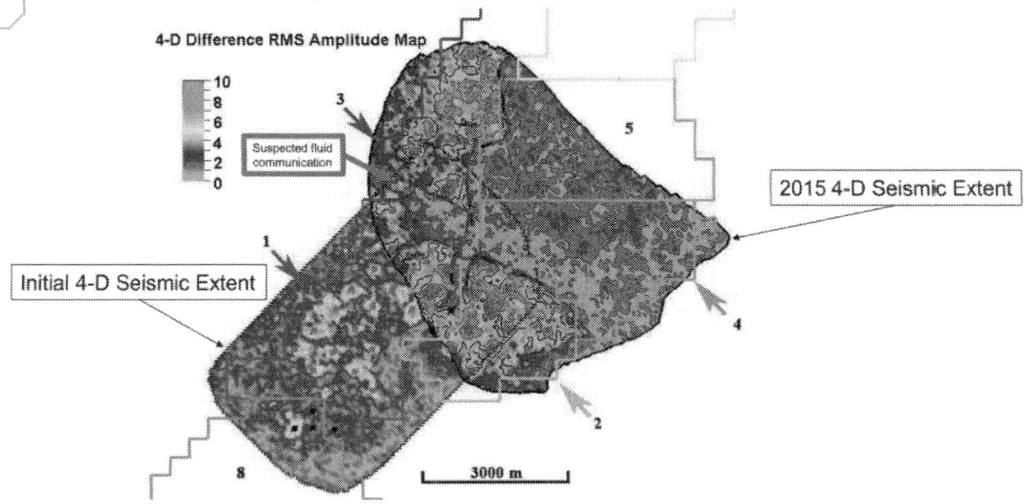

# WHAT DOES IT MEAN?

What does the change in amplitude response in 4-D investigations mean?

- The difference is due to comingled effects of changes in pressure and $CO_2$ saturation, which are difficult to distinguish from one another.
- However, initial investigation of pulsed-neutron logs acquired contemporaneously with the 2015 seismic survey indicate an average $S_{CO2}$ of 3%–4% over an interval of 20 feet is detectable (at a minimum) in Bell Creek 4-D seismic.

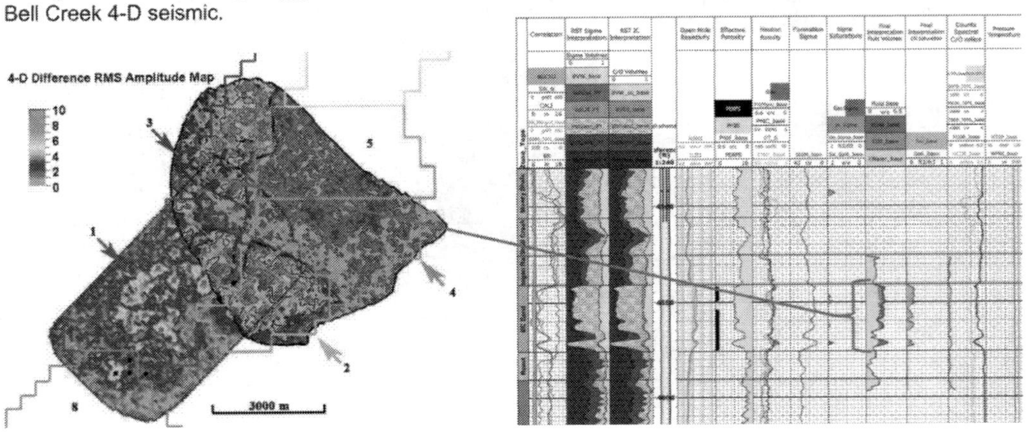


# IMPORTANT LEARNINGS FROM THE 4-D SEISMIC INVESTIGATION

- Permeability barriers (preventing fluid communication and pressure dissipation) have been illuminated where they were masked previously in the baseline seismic.
- Cross-phase fluid communication has been identified.
- $CO_2$ accumulating updip along the western edge of the N–S permeability barrier is visible.
- $CO_2$ and pressure plumes associated with injection wells are clearly shown.
- Improved insight into the reservoir's interwell heterogeneity.
  - Yielded important details necessary to adapt static models, enable better history matching, and increase accuracy in predictive simulations.
- Provided actionable information for the engineers operating the field while demonstrating associated CO2 storage incidental to EOR.

# CONTACT INFORMATION

**Energy & Environmental Research Center**
University of North Dakota
15 North 23rd Street, Stop 9018
Grand Forks, ND 58202-9018

**www.undeerc.org**
701.777.5195 (phone)
701.777.5181 (fax)

**Wes Peck**
**Principal Geologist**
wpeck@undeerc.org

EERC

THANK YOU!

Critical Challenges. **Practical Solutions.**

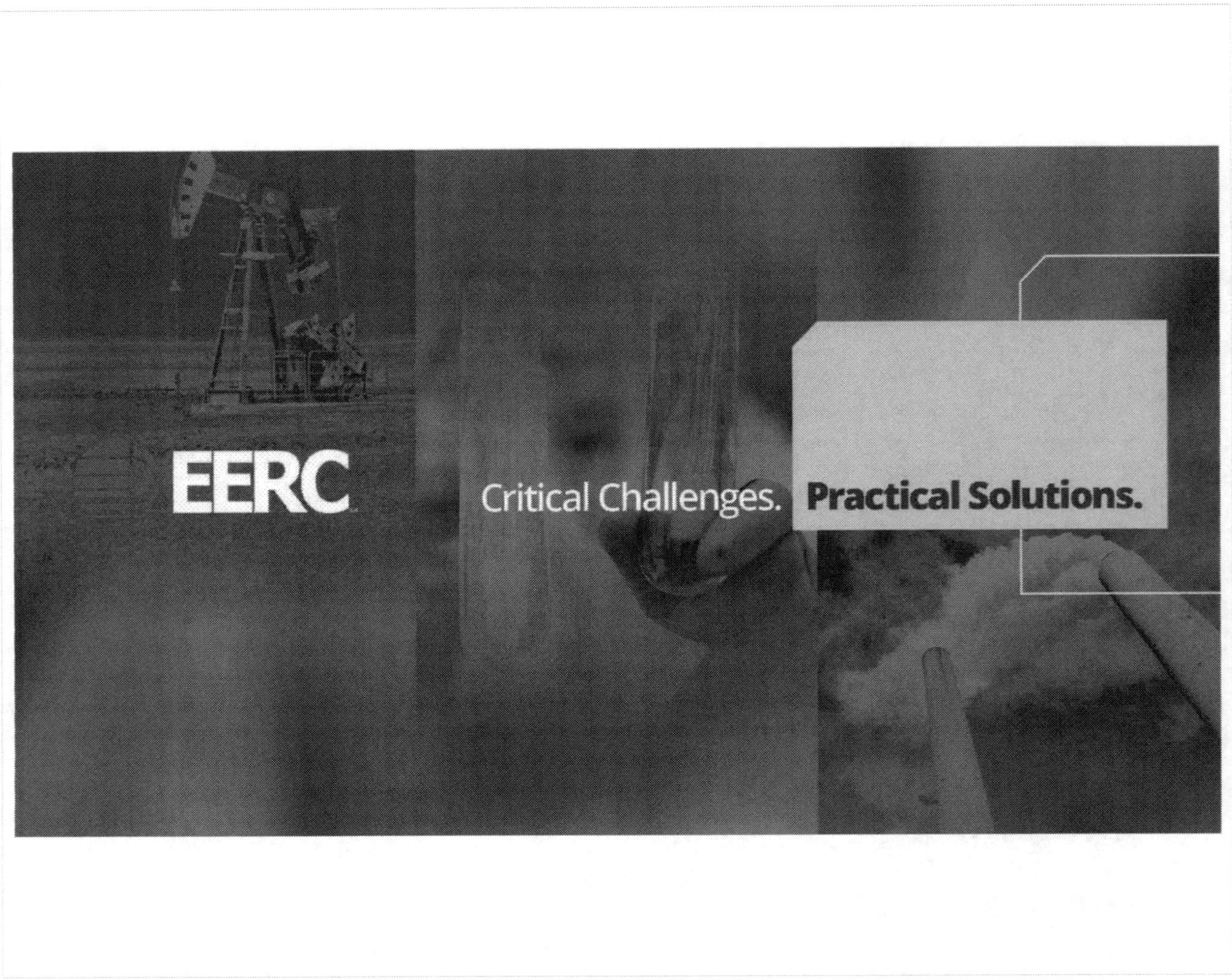
EERC
Critical Challenges. Practical Solutions.

ENHANCED OIL RECOVERY INSTITUTE

*SCHOOL OF ENERGY RESOURCES*

4th CMT Conference

Session 2 EOR 1 – July 18th, 2017 Houston, TX


**Benjamin R. Cook, PhD**

Economist/Visiting Professor

Enhanced Oil Recovery Institute &

UW College of Business


UNIVERSITY OF WYOMING

2

# The Mission of EORI

*EORI Headquarters*
*Business Innovation Center*
*Casper, WY*

The Enhanced Oil Recovery Institute (EORI) was created pursuant to Wyoming Statute to facilitate research programs intended to increase recoverable reserves and the production of oil and natural gas in Wyoming. The Institute achieves its statutory purpose by facilitating the transfer of relevant technology, information and knowledge to operators producing Wyoming reserves, and by promoting research and technology transfers in conventional and unconventional oil and gas reservoirs.

SER + EORI

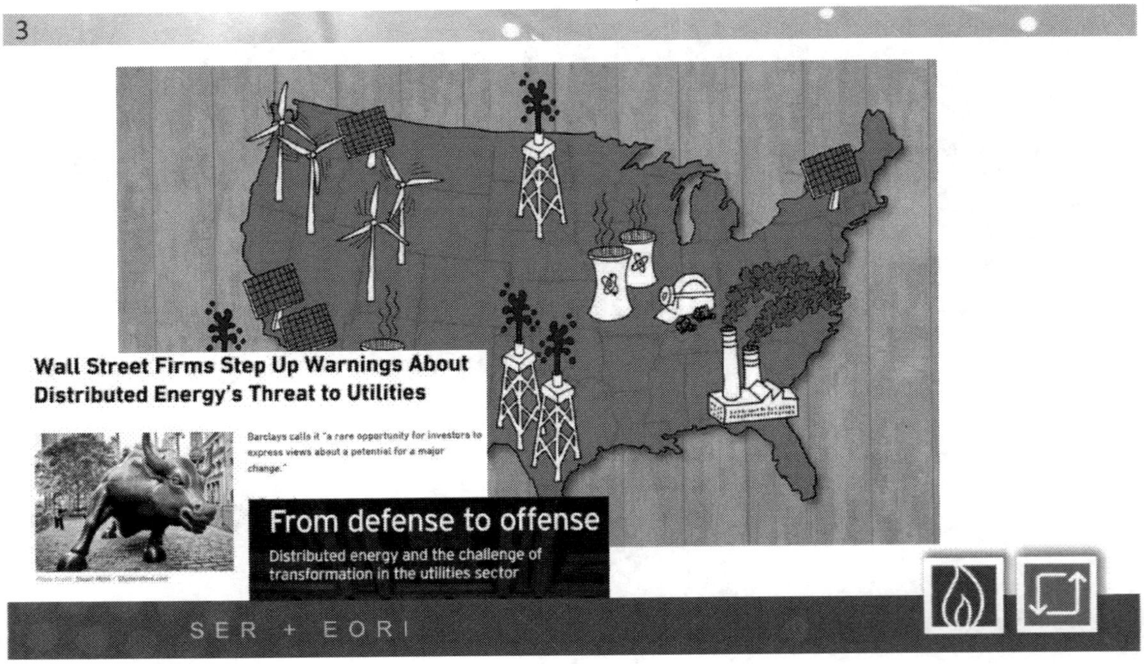
3
Wall Street Firms Step Up Warnings About Distributed Energy's Threat to Utilities
Barclays calls it "a rare opportunity for investors to express views about a potential for a major change."
From defense to offense
Distributed energy and the challenge of transformation in the utilities sector
SER + EORI

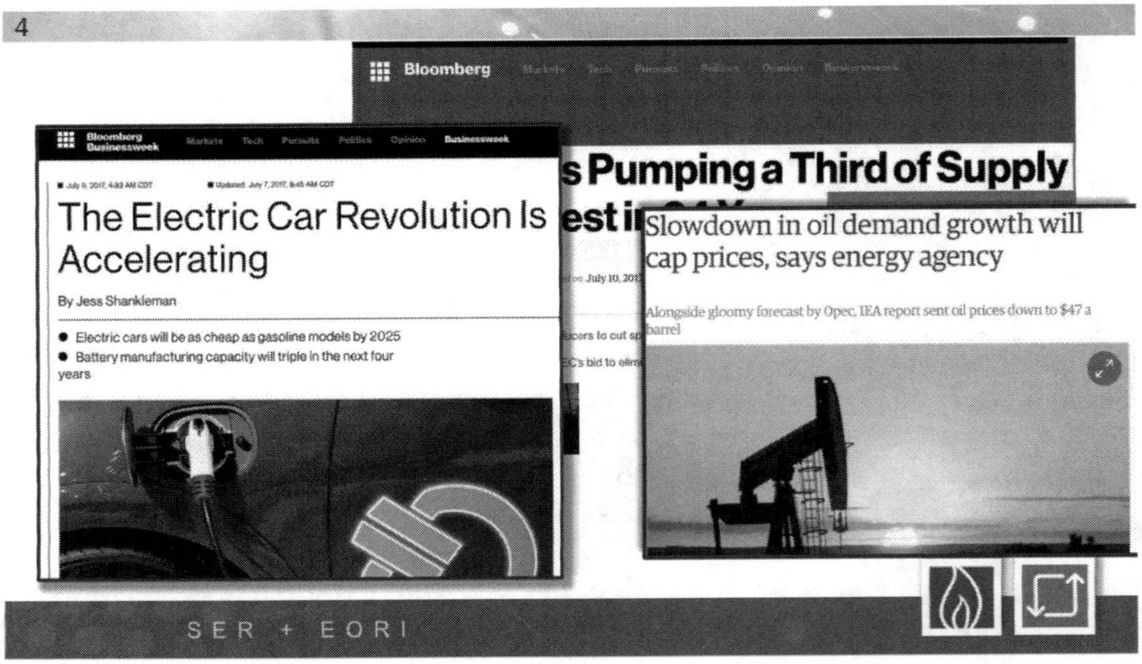
4
Bloomberg
s Pumping a Third of Supply
Bloomberg Businessweek
The Electric Car Revolution Is Accelerating
By Jess Shankleman
Electric cars will be as cheap as gasoline models by 2025
Battery manufacturing capacity will triple in the next four years
Slowdown in oil demand growth will cap prices, says energy agency
Alongside gloomy forecast by Opec, IEA report sent oil prices down to $47 a barrel
SER + EORI

5

# Electricity in the Oil Patch

**Major Share of Lifting Cost**

Especially for conventional oil fields and water-floods, the electricity cost of pumping fluid to the surface can be deciding factor in when to shut-in a well.

SER + EORI

6

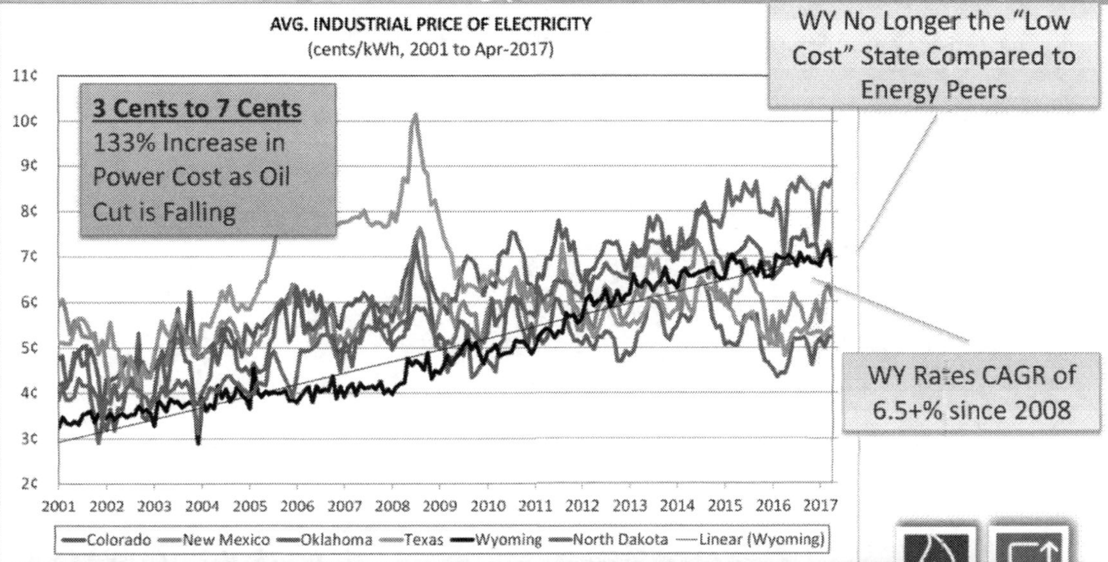


SER + EORI

7

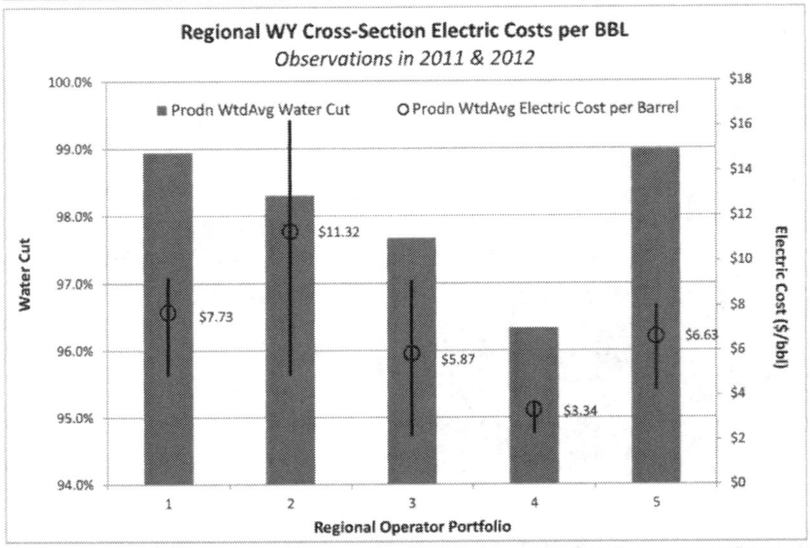


The Big Horn Basin has many productive waterfloods, but they make a bunch of water for each barrel of oil

SER + EORI

8

# A Weight Average (Pretty) WY Oil Well

- **A production weighted average WY oil well in July 2016 would be one producing 114 bpd of 35 deg. API oil and 604 bpd of water at a depth of 7,966'**
  - Would require 39,000 – 54,000 kWhs per month
  - 468,000 – 648,000 kWhs per year
  - Average U.S. home in 2014 using 10,932 kWhs per year.
  - This well uses the equivalent power of 43 – 59 homes
  - 9,950 wells ~= New Hampshire

SER + EORI

9

# …and the Cost $/bbl

- **According to EIA.gov the average industrial rate in WY is around 7 cents per kilowatt-hour.**
  - Well electricity of $2,700 - $3,780 per month
  - Average of $39,060 per year
  - WY average worker pay in 2014 was $46,500
  - At 16% oil, 84% water the electricity only comes out to about $1/bbl-oil
  - 25% of WY oil fields, including 12 of the top 30 are 95% water-cut or more
  - In these cases, the electric costs may be $2.70 - $14.60 or more per bbl

SER + EORI

10

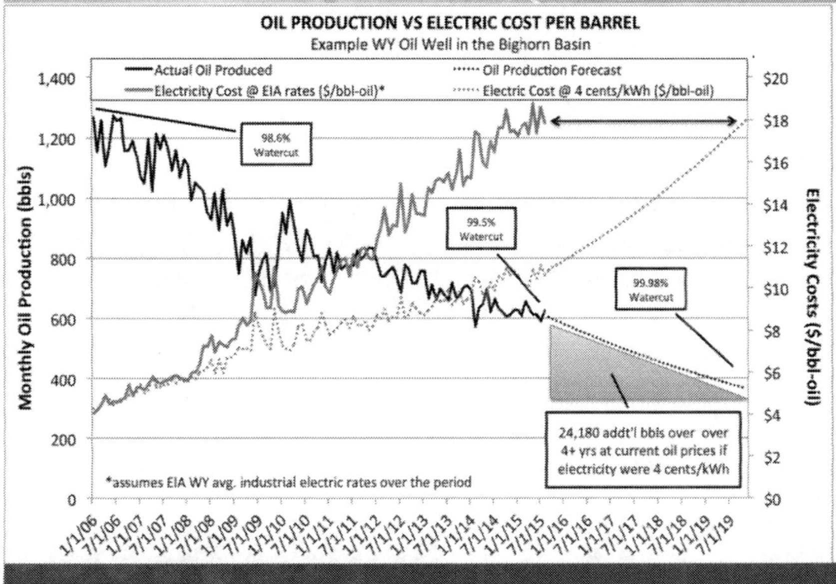


Assuming that the economic limit on electricity costs is $18/bbl-oil….

…this well would continue producing 4+ years at 4 cents versus 7 cents/kWh.

SER + EORI

11

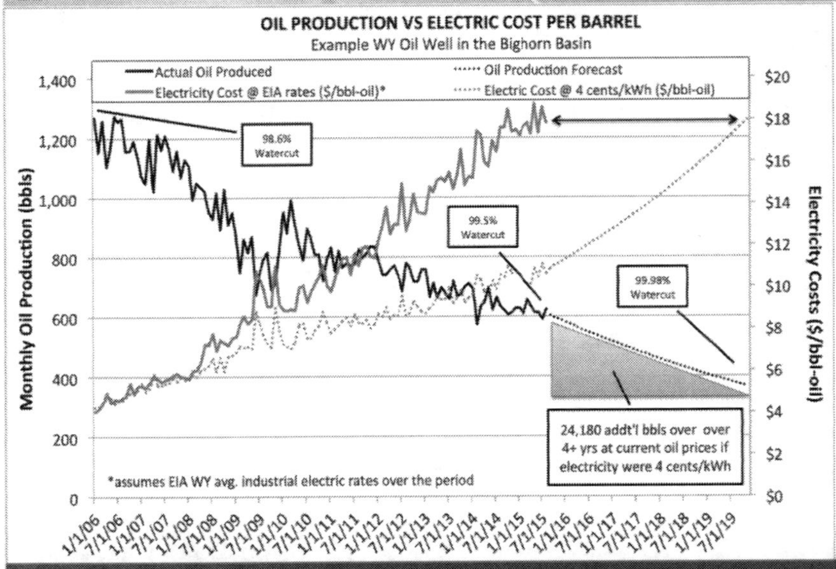


| Watercut | Electric $/bbl-oil |
|---|---|
| 70.0% | $0.49 |
| 75.0% | $0.59 |
| 80.0% | $0.75 |
| 85.0% | $1.01 |
| **90.0%** | **$1.52** |
| 95.0% | $3.03 |
| 96.0% | $3.79 |
| 97.0% | $5.11 |
| 98.0% | $7.77 |
| 99.0% | $15.53 |
| 99.5% | $31.07 |

SER + EORI

12

# What if Utility Dropped the Rate?

- **Assume that the oil operator was somehow able to negotiate for a lower electric rate on this well rather than shut-in?**

| Rate Schedule | Additional Prod. Months | Additional Oil Produced (bbls) | Additional Energy Used (MWhs) | Additional Electric Payments |
|---|---|---|---|---|
| EIA Rates... | - | 0 | 0 | 0 |
| then 6 cents | 10 | 5,765 | 1,639 | $98,000 |
| then 5 cents | 28 | 14,788 | 4,589 | $229,000 |
| then 4 cents | 51 | 24,179 | 8,360 | $334,000 |
| then 3 cents | 79 | 33,037 | 12,950 | $388,000 |

"Ceteris Paribus" -> flat oil prices & $18 electric cost shut down.

SER + EORI

13

$$Rates = \frac{Revenue\ Requirement}{Sales\ (MWhs)}$$

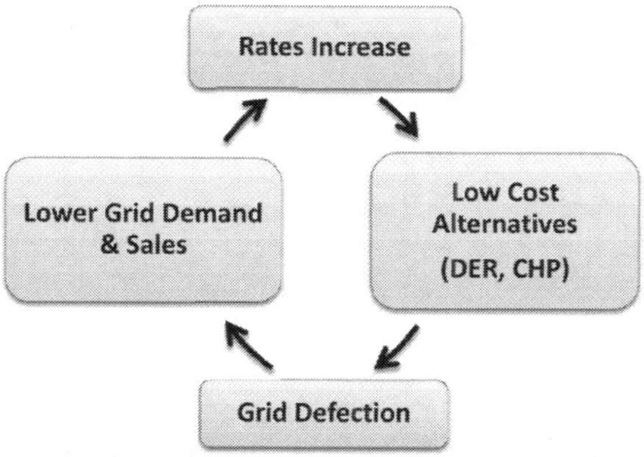

SER + EORI

14

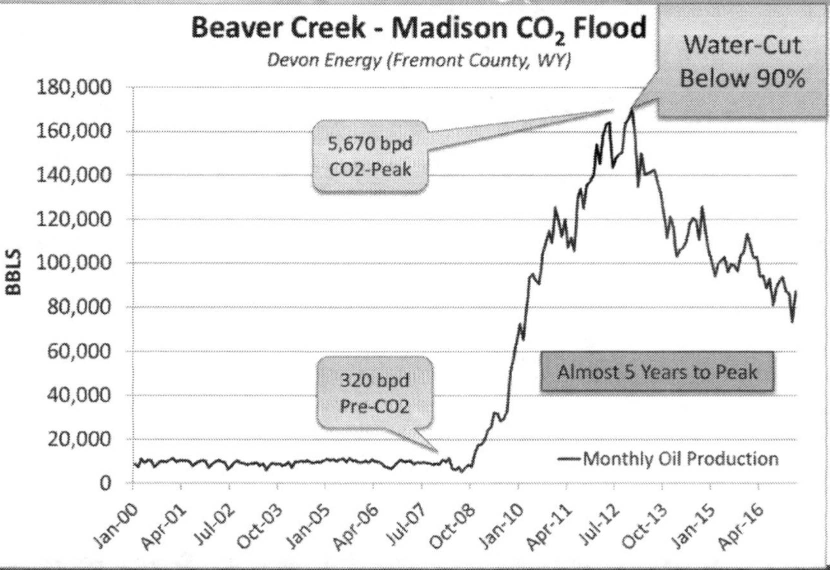


**Technically Feasible**

+

**Economically Feasible**

+

**$CO_2$ Available**

SER + EORI

15

# Costs of Compression at Recylce

| $CO_2$ Volume (mmcfd) | HP of Compression | KWs (90% uptime) | MWhs/y ear | Annual Electric Cost ($60/MWh) |
|---|---|---|---|---|
| 5 | 504 | 338 | 2,964 | $177,855 |
| 10 | 1,007 | 676 | 5,923 | $355,358 |
| 20 | 2,011 | 1,350 | 11,828 | $709,657 |
| **30** | **3,018** | **2,026** | **17,750** | **$1,065,016** |
| 40 | 4,022 | 2,700 | 23,655 | $1,419,315 |
| 50 | 5,029 | 3,376 | 29,578 | $1,774,673 |
| 60 | 6,033 | 4,051 | 35,483 | $2,128,972 |
| 70 | 7,040 | 4,727 | 41,406 | $2,484,330 |
| 80 | 8,044 | 5,401 | 47,310 | $2,838,630 |
| 90 | 9,051 | 6,077 | 53,233 | $3,193,988 |
| 100 | 10,055 | 6,751 | 59,138 | $3,548,287 |

SER + EORI

16

# Capital Budgeting (Allocation)

| Regulated Generation/Utility | Upstream Oil & Gas |
|---|---|
| Integrated Resource Planning: "reliable and least-cost electric service to all of our customers while addressing the substantial risks and uncertainties inherent in the electric utility business."--PacifiCorp | Growing or replacing reserves (economically recoverable hydrocarbons) PLUS growing or replacing current production (improved oil recovery, new drilling or exploration, acquisitions) |
| Serving customer load, reliability and regulatory compliance. | Serving investors |
| Regulated return (8-10%) | Maximize return |
| Payback over decades | Usually the sooner the better |

SER + EORI

17

## Preferences of Utilities versus O&G

| Regulated Generation/Utility | Upstream Oil & Gas | Policy/Market Options? |
|---|---|---|
| Fixed price of $CO_2$ versus a variable price. Exposing customers to oil price risk in the electricity cost… | $CO_2$ prices tied to the oil price. Fixed prices create margin risk in low oil price environment. | Price Stabilization Policy… Utilities get constant price, oil operators pay variable price. |
| Interruptible $CO_2$ due to different markets. | Reliable/constant streams of $CO_2$. | $CO_2$ pipeline networks and diversification of sources. |
| 20-40 year time horizon. | Shut-down if oil prices are too low. Reduce $CO_2$ purchases over time as recycle volumes increase. | Diversification of sinks, and incentives for storage. |

SER + EORI

18

## Do Oil Operators Have Electric Grid Alternatives?

1. Efficiency and Alternative Pumps
   - Operators switching to Permanent Magnet ESPs and reducing power costs
   - ....extends the life of the wells and fields, at comparable pump costs

2. On-site Distributed Energy Resources (DER)
   - Natural Gas Reciprocating Engines at the wellhead…
   - ....are they cheaper than grid power?

SER + EORI

19

# What about NG Engine Generators?

Electrifying the Oilfield:
The Comparative Economics of Grid Power and Onsite Gas Generators

- **Comparative Economics of Grid vs NG Generators**
  - "Is grid-based line power still the lowest cost source of electricity for new oil developments in light of decreased leasing rates for natural gas reciprocating generators?"
- **Especially for Greenfields….**
  **Right-of-Ways, Distribution Lines, Maintenance, Transmission Upgrades**
- http://bit.ly/genvgrid

S E R + E O R I

20

# Six States

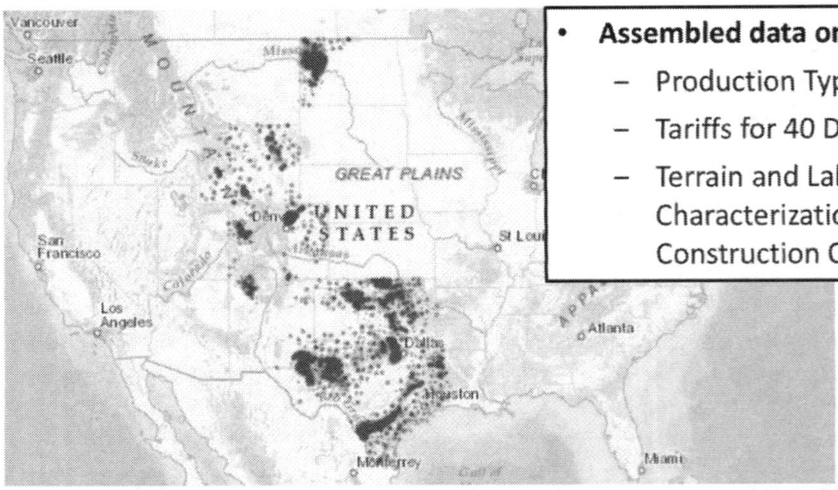


- **Assembled data on 93 counties**
  - Production Type Curves
  - Tariffs for 40 Different Utilities
  - Terrain and Labor Cost Characterization for Line Power Construction Costs

S E R + E O R I

21

## Line Construction & Tariffs

[TABLE 1 – Estimated Distribution Line Construction Costs]

| Construction Component | Colorado | North Dakota | New Mexico | Oklahoma | Texas | Wyoming |
|---|---|---|---|---|---|---|
| Poles & Conductor | $26.5k | $26.5k | $26.5k | $26.5k | $26.5k | $26.5k |
| Labor & Vehicles | $15.7k | $14.2k | $13k | $11.8k | $11.9k | $14.5k |
| ROW, Engineering & Terrain Adjustment | $10.6k - $38k | $19.5k - $44k | $12.5k - $20.1k | $10.2k - $22.7k | $10.3k - $42k | $19.5k - $44.4k |
| **Total Cost per Mile** | **$52.9k - $80.3k** | **$60.3k - $84.9k** | **$52k - $59.6k** | **$48.6k - $61k** | **$48.7k - $80.4k** | **$60.6k - $85.5k** |

[TABLE 3 – Summary of Utility Tariff's for a 2MW Customer]

| Tariff Component | Colorado | North Dakota | New Mexico | Oklahoma | Texas | Wyoming |
|---|---|---|---|---|---|---|
| Fixed Charge | $86.25 - $1,300 | $25.74 - $350 | $25.00 - $500 | $10.00 - $287.17 | $21.00 - $239.48 | $42.25 - $810 |
| Demand Charge($/peak-kW) | $0.00 - $16.46 | $9.29 - $16.74 | $1.49 - $12.30 | $0.00 – $6.45 | $2.13 - $16.50 | $0.00 - $29.47 |
| Energy Usage ($/kWh) | $0.034 - $0.145 | $0.04 - $0.06 | $0.038 - $0.066 | $0.043 - $0.073 | $0.025 - $0.076 | $0.031 - $0.119 |
| **Total Rate per Month ($/kWh)** | **$0.062 - $0.16** | **$0.067 - $0.094** | **$0.065 - $0.075** | **$0.065 - $0.075** | **$0.035 - $0.114** | **$0.044 - $0.131** |

SER + EORI

22

## Genset Costs

[TABLE 4 – Monthly Generator Lease Rates]

| Genset Engine | Prime Rating (kW) | Low Pricing ($/kW) | Average Pricing ($/kW) | High Pricing ($/kW) | Peak Pricing ($/kW) |
|---|---|---|---|---|---|
| 8LNA | 70 | $18.10 | $20.69 | $23.27 | $47.14 |
| 8LT | 125 | $18.99 | $21.70 | $24.42 | $47.20 |
| 11LT | 170 | $18.49 | $21.13 | $23.77 | $47.65 |
| 14LT | 225 | $18.86 | $21.55 | $24.25 | $42.22 |
| 22LT | 350 | $21.01 | $24.01 | $27.01 | $41.43 |

SER + EORI

23

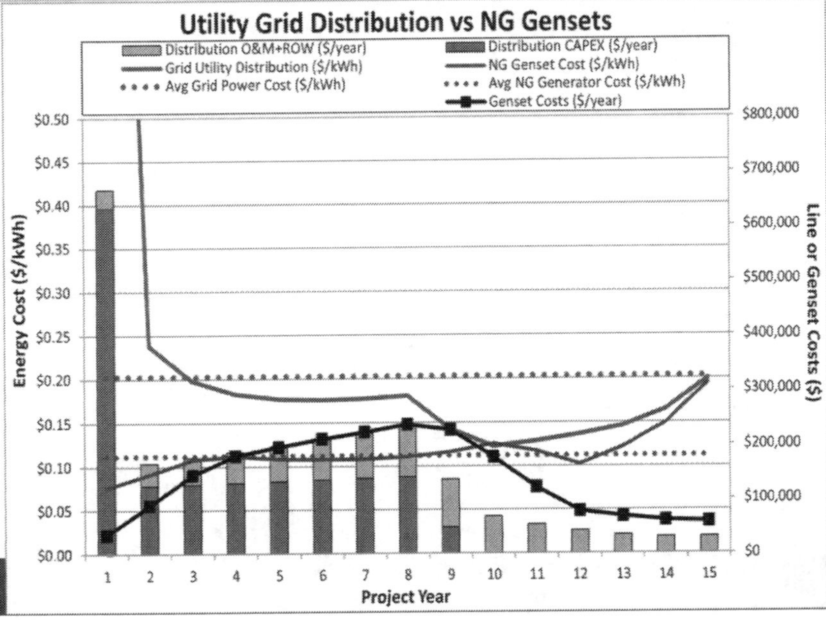


## Example Output

24

### [TABLE 5 – Example Results]

| 50 Well Colorado-Niobrara Development Example | Grid Power (Weld County-CO) | NG Generator |
|---|---|---|
| Est. Total Oil Production | 7,813,688 | 7,813,688 |
| Total Lease-Life kWhs | 18,699,306 | 18,699,306 |
| Utility Bill or Generator Leases | $1,340,010 | $2,146,602 |
| *Energy Cost per kWh* | *$0.07166* | *$0.11480* |
| Line Power CAPEX plus O&M | $2,451,086 | |
| Total Electricity Costs | $3,791,097 | $2,146,602 |
| *Total Cost per kWh* | *$0.20274* | *$0.11480* |
| Cost per BBL of Oil | *$0.48519* | *$0.27472* |

## Example Output

SER + EORI

## Results

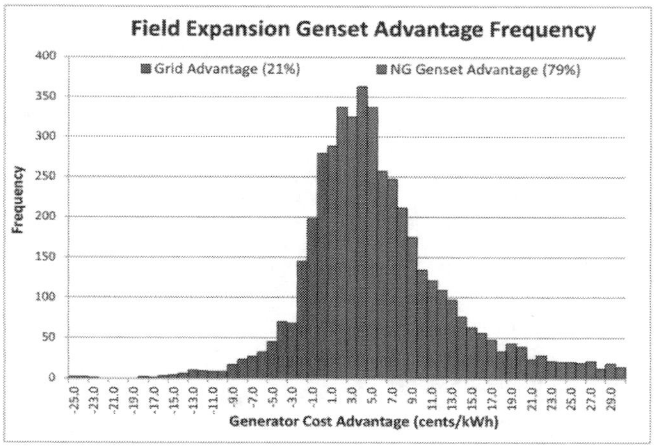


- ~170,000 unique scenarios outcomes across counties and utilities
- NG Gensets have a cost advantage in 66% of cases.
- If you exclude the Infill Drilling cases gensets are favored in 80% or more of cases considered.
- Closer spacing, more wells, and higher production type curves shift advantage to the grid.
- Greenfields, trunk distance, larger spacing, fewer wells, and lower production type curves favor gensets.

SER + EORI

ENHANCED OIL RECOVERY INSTITUTE

SCHOOL OF ENERGY RESOURCES

# Thank You!

UNIVERSITY OF WYOMING

27

# Economically Feasible?

## Costs Vary by Project Size

- Upgrading wells/facilities
- Infill drilling
- $CO_2$ recycling & compression
- Major or spur $CO_2$ pipelines
- $CO_2$ flow lines
- Monitoring stations
- $CO_2$ purchases

## Other Considerations

- Long term projects & contracts
- Payback in years, not months
- Flood last decades, peak in 4-5 years
- Proximity to $CO_2$

### Example Scoping Economics (small)

61.8 MMbbls Original Oil in Place
Total CAPEX of $28.7 million
Addt'l OPEX of $1.4 million
Payback 5-6 years @ $50/oil

- 15 Existing Wells plus 5 New Wells
  ~$12 million
- **<5 MMcfd** of $CO_2$ injection
  ~$6 million for $CO_2$ plant
  plus $500k in OPEX
- 4-inch 30-mile pipeline
  ~$10.5 million for pipeline
- $CO_2$ Price = $0.50 + 2% of WTI
  $1.50/mcf ($29/tonne) @ $50/oil

S E R + E O R I

28

**PROJECTED $CO_2$, EOR OPERATIONS, AND $CO_2$ SOURCES: 2020**

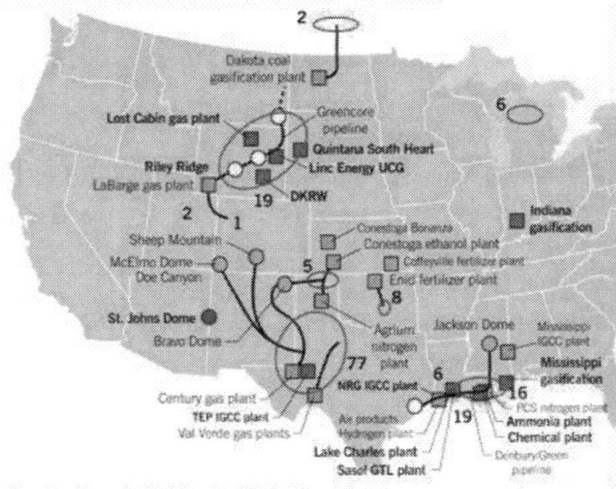


Source: Advanced Resources International Inc. based on OGJ EOR/Heavy Oil Survey 2014 and other sources.

FIG. 4

| Oil production, 2020 | |
|---|---|
| $CO_2$-EOR projects | 147 |
| Oil production, 1,000 b/d | 638 |
| **$CO_2$ supplies, 2020** | |
| Number of sources | 30 |
| • Natural | 6 |
| • Industrial | 24 |
| $CO_2$ supply, MMcfd | 6.5 |
| • Natural | 3.4 |
| • Industrial | 3.1 |

147 Number of $CO_2$-EOR projects
● Natural $CO_2$ source
■ Industrial $CO_2$ source
— $CO_2$ pipeline
····· $CO_2$ proposed pipeline

# Is the $CO_2$ Available?

Proximity to the $CO_2$ source can be a huge driver of costs. Assuming you get the permits pipelines cost $80-100k per inch-mile.

S E R + E O R I

29

# Miscible $CO_2$ - Technically Feasible

| Property | Recommendation |
|---|---|
| **Crude Oil** | |
| Gravity, Deg. API | >22 (and probably <45) |
| Viscosity, cp | <10 |
| Composition | High Cut of $C_5$-$C_{12}$ |
| **Reservoir** | |
| Oil Saturation, % PV | >20 |
| Type of Formation | Sandstone/Carbonate |
| Average Permeability | Just need good injection rates |
| Depth & Temperature | Sufficient for Injection Pressure > MMP |
| >40 API | >2,500 ft |
| 32 – 40 API | >2,800 ft |
| 28 – 32 API | >3,300 ft |
| 22 – 28 API | >4,000 ft |
| Fracture Pressure | Fracture Pressure > Injection Pressure |

*Source:* "EOR Screening Criteria Revisited – Part 2", Taber et.al, SPE Reservoir Engineering, Aug. 1997

Failing to reach Minimum Miscibility Pressure (MMP) can slash incremental oil recovery by 50% or more.

SER + EORI

30

# Economically Feasible?

## Costs Vary by Project Size

- Upgrading wells/facilities
- Infill drilling
- $CO_2$ recycling & compression
- Major or spur $CO_2$ pipelines
- $CO_2$ flow lines
- Monitoring stations
- $CO_2$ purchases

## Other Considerations

- Long term projects & contracts
- Payback in years, not months
- Flood last decades, peak in 4-5 years
- Proximity to $CO_2$

### Example Scoping Economics (small)

61.8 MMbbls Original Oil in Place

Total CAPEX of $28.7 million
Addt'l OPEX of $1.4 million
Payback 5-6 years @ $50/oil

- 15 Existing Wells plus 5 New Wells
  ~$12 million
- **<5 MMcfd** of $CO_2$ injection
  ~$6 million for $CO_2$ plant
  plus $500k in OPEX
- 4-inch 30-mile pipeline
  ~$10.5 million for pipeline
- $CO_2$ Price = $0.50 + 2% of WTI
  $1.50/mcf ($29/tonne) @ $50/oil

SER + EORI

31

# Mind the Gap – Industrial Capture vs EOR Pricing

| Pricing (Metric Tonnes) | |
|---|---|
| Metric Tonne | MCF |
| $10 | $0.52 |
| $15 | $0.78 |
| $20 | $1.04 |
| $25 | $1.30 |
| $30 | $1.56 |
| $35 | $1.82 |
| $40 | $2.08 |
| $45 | $2.34 |
| $50 | $2.60 |
| $55 | $2.86 |
| $60 | $3.12 |
| $65 | $3.38 |
| $70 | $3.64 |
| $75 | $3.90 |
| $80 | $4.16 |

| Pricing vs WTI Oil Price | | | |
|---|---|---|---|
| WTI | $0.50+1% | $0.5+2% | $0.5+2.5% |
| $20 | $0.70 | $0.90 | $1.00 |
| $30 | $0.80 | $1.10 | $1.25 |
| $40 | $0.90 | $1.30 | $1.50 |
| $50 | $1.00 | $1.50 | $1.75 |
| $60 | $1.10 | $1.70 | $2.00 |
| $70 | $1.20 | $1.90 | $2.25 |
| $80 | $1.30 | $2.10 | $2.50 |
| $90 | $1.40 | $2.30 | $2.75 |
| $100 | $1.50 | $2.50 | $3.00 |
| $110 | $1.60 | $2.70 | $3.25 |
| $120 | $1.70 | $2.90 | $3.50 |
| $130 | $1.80 | $3.10 | $3.75 |
| $140 | $1.90 | $3.30 | $4.00 |
| $150 | $2.00 | $3.50 | $4.25 |
| $160 | $2.10 | $3.70 | $4.50 |

**"Commodity Price"**
cost of atmospheric $CO_2$ at the source

**"Delivered At Field Price"**
= commodity price
+ processing/compression
+ pipeline/truck transport

**"Oil Price Tie"**
delivered $CO_2$ price is often tied to the WTI oil price
= $0.50 + 1-2.5% of WTI

*thanks for the idea Keith @ Chaparral Energy

SER + EORI

32

# Other Policy Initiatives

NEORI & Great Plains Institute

- Broad coalition of companies, NGOs and unions supporting $CO_2$-EOR.
- Major initiative is on extending, reforming and strengthening Section 45Q tax credit.

Other Policy Ideas

- Authorize private activity bonds (PABs) for $CO_2$ capture projects.
- Authorize MLPs for $CO_2$ projects.
- Price stabilization program.
- Incorporate $CO_2$ pipelines into national infrastructure agenda.
- Give CCUS plants more preferential ranking in electric power dispatch.

SER + EORI

33

# Units & Pricing

| Volume per Day | | | |
|---|---|---|---|
| Metric Tonnes | Short Tons | MCF/d | MMCF/d |
| 1 | 0.91 | 19 | 0.02 |
| 10 | 9.09 | 192 | 0.19 |
| 100 | 90.91 | 1,923 | 1.92 |
| 500 | 454.55 | 9,615 | 9.62 |
| 1,000 | 909.09 | 19,231 | 19.23 |
| 2,500 | 2,272.73 | 48,077 | 48.08 |
| 5,000 | 4,545.45 | 96,154 | 96.15 |
| 10,000 | 9,090.91 | 192,308 | 192.31 |

| Volume per Year | | | |
|---|---|---|---|
| Metric Tonnes | Short Tons | MCF/yr | MMCF/yr |
| 365 | 332 | 7,019 | 7 |
| 3,650 | 3,318 | 70,192 | 70 |
| 36,500 | 33,182 | 701,923 | 702 |
| 182,500 | 165,909 | 3,509,615 | 3,510 |
| 365,000 | 331,818 | 7,019,231 | 7,019 |
| 912,500 | 829,545 | 17,548,077 | 17,548 |
| 1,825,000 | 1,659,091 | 35,096,154 | 35,096 |
| 3,650,000 | 3,318,182 | 70,192,308 | 70,192 |

| Pricing (Short Tons) | |
|---|---|
| Short Ton | MCF |
| $10 | $0.57 |
| $15 | $0.86 |
| $20 | $1.14 |
| $25 | $1.43 |
| $30 | $1.72 |
| $35 | $2.00 |
| $40 | $2.29 |
| $45 | $2.57 |
| $50 | $2.86 |
| $55 | $3.15 |
| $60 | $3.43 |
| $65 | $3.72 |
| $70 | $4.00 |
| $75 | $4.29 |
| $80 | $4.58 |

| Pricing (Metric Tonnes) | |
|---|---|
| Metric Tonne | MCF |
| $10 | $0.52 |
| $15 | $0.78 |
| $20 | $1.04 |
| $25 | $1.30 |
| $30 | $1.56 |
| $35 | $1.82 |
| $40 | $2.08 |
| $45 | $2.34 |
| $50 | $2.60 |
| $55 | $2.86 |
| $60 | $3.12 |
| $65 | $3.38 |
| $70 | $3.64 |
| $75 | $3.90 |
| $80 | $4.16 |

*thanks for the idea Keith @ Chaparral Energy

SER + EORI

# CMTC-502924-MS

# Pressure Transient Analysis for Characterization of Lateral and Vertical Leakage through Faults

Mojtaba Mosaheb and Mehdi Zeidouni, Louisiana State University



This paper was prepared for presentation at the Carbon Management Technology Conference held in Houston, Texas, USA, 17-20 July 2017.



## Abstract

A fault is a potential pathway for fluid leakage, which can contaminate underground water resources. In addition, fault leakage can affect hydrocarbon production. This study aims to develop a type-curve-based methodology to characterize a fault both laterally and vertically using pressure transient analysis. We develop an analytical model to assess the pressure perturbations corresponding to production/injection from/into a reservoir with a leaky fault. Displacement of layers during the fault displacement may cause alteration of the reservoir properties across the fault. This alteration is accounted for by considering different properties on the two sides of the fault. The reservoir is divided into two regions separated by the fault, which are in hydraulic communication with one another and with the overlying/underlying permeable layers. The governing system of differential equations and corresponding boundary conditions are solved using Fourier and Laplace transforms. At early times of the fault leakage, the recorded well pressure changes are mostly affected by the fault properties and the effects of resistance from the upper zone emerge later. In this model, we neglect the resistance to leakage flow caused by the overlying zone to focus on the pressure changes at early times of the fault leakage. We show that these assumptions are valid to arrive at correct fault characteristics. For fault characterization, type curves are presented in terms of dimensionless vertical and horizontal conductivities of the fault. A computational optimization method is used in combination with type curves to fully characterize the reservoir-fault system. Results show that the characterization method is useful to estimate the fault vertical and lateral conductivities.

Keywords: Analytical model, Characterization, Fault leakage, Pressure transient

## Introduction

A fault can cause a discontinuity in formation permeability distorting the fluid flow and may act as a conduit to both lateral and vertical fluid flow. Fault zone permeability may be enhanced or reduced depending on the forces that cause the displacement of the layers and slip location through the fault plane (Sibson, 1977). Faults are generally composed of a core zone surrounded by a damaged zone. The permeability of the core zone is commonly low because this is the location of the slip that the original rock is ground. Compared to the core zone, the surrounding damaged zone's permeability may be enhanced due to possible fractures (Sibson, 1977; Caine et al., 1996). Analogous to the lateral permeability, the vertical permeability of a fault may be enhanced and the fault may be a vertical flow conduit (Maslia and Prowell, 1990). As an example, Bense and Person (2006) investigated the sealing and conductance behaviors of the Baton Rouge Fault in southern Louisiana and showed that the permeability of the fault is enhanced through the fault plane. They also found that the fault permeability is considerably low perpendicular to the fault plane. The existence of faults in underground formations can cause inter-formational fluid migration, particularly affecting water resources (Huntoon and Lundy, 1979). Stoessell and Prochaska (2005) showed that brine from deep saline aquifers migrated upward along the Baton Rouge Fault by several hundreds of meters.

Injection into deep saline aquifers is an industrial method to dispose of unwanted fluids. For instance, $CO_2$ storage is a method to cut $CO_2$ atmospheric emissions as a means to mitigate climate change. The injected $CO_2$ can leak through a nearby fault to shallower aquifers and contaminate the underground water resources (IPCC, 2005). The quality of the underground water resources may change due to $CO_2$ leakage through local high-permeability pathways such as permeable faults or abandoned wells (Apps et al., 2010). During underground storage of natural gas, migration of injected gas from the target zone through the faults can reduce the accessible gas for reproduction and contaminate underground water resources (Thomas and Benson, 2005). It is of crucial importance to characterize the potential leakage pathways including faults to manage the negative consequences.

In this paper, we investigate characterization of leaky faults based on the pressure transient analysis (PTA).

PTA is an applicable technique to characterize the hydraulic characteristics of the reservoirs based on mathematical modeling of fluid flow in porous media.

Analytical models are especially useful because of their independence of time and space discretization, capability for quick implementation, providing an explicit relationship between the system properties/measurements, and less complex models requiring fewer input of data. Several mathematical methods were used to present analytical solutions for fluid flow in a system including a fault as a discontinuity. Integral transforms (Bixel et al., 1963), Laplace-Fourier transformation (Ambastha et al., 1989), and Green's function (Raghavan, 2010) are examples of the mathematical solution methods. Analytical approaches to modeling a faulted system are reviewed in the following.

Bixel et al. (1963) investigated the transient pressure behavior of a well located near a fault and proposed an analytical solution based on the integral transforms. Stewart and Gupta (1984) investigated interference testing in a reservoir including a non-sealing fault and introduced drawdown type curves using numerical simulation approach. The pressure discontinuity across the fault was shown by Yaxley (1987). He obtained an analytical solution for pressure transient behavior of a vertical non-sealing barrier. The reservoir properties of two sides of the fault were assumed identical. Ambastha et al. (1989) presented an analytical solution with different reservoir properties at two sides of the fault. They demonstrated that the results of the interference tests are influenced by the property contrasts of the composite system and the location of the observation well. Rahman et al. (2003) presented an analytical solution to the transient flow problem of a system of two regions with a finite conductivity fault. The solution accounts for the transient flow within the fault. They concluded that the effect of transient flow in the fault can be negligible.

Modeling of the vertical leakage through the fault has received attention more recently. Shan et al. (1995) incorporated the effect of vertical leakage to an upper permeable zone in the suggested analytical model. However, they ignored the pressure discontinuity through the fault in the injection layer. Zeidouni (2012) presented analytical solutions of two-layer and multi-layer systems, which demonstrated the pressure discontinuity through the fault in presence of vertical leakage. Zeidouni (2016) extended the multi-layer solution for a leaky fault by fully accounting for the lateral resistance of the fault in all layers. Many other works have been done to address fluid leakage from a target zone (Pruess, 2005; Ebigbo et al., 2007; Birkholzer et al., 2009; Humez et al., 2011; Zeidouni and Pooladi-Darvish, 2012; Shakiba and Hosseini, 2016; Mosaheb and Zeidouni, 2017a, b).

The main goal of this study is to present type curves for fault characterization and demonstrate how they can be used to determine fault lateral and vertical conductivities. In obtaining the analytical solution, we build on an existing analytical solution by Zeidouni (2012). we neglect the resistance to flow caused by the overlying/underlying formations to reduce the number of parameters required to characterize the system. A fault usually intersects a sequence of adjacent permeable layers, which can be represented, by a single thick zone with high flow capacity (permeability×thickness). The superposition of the numerous overlying/underlying layers may show no resistance to vertical flow, especially at early times of leakage. The validity of this assumption is investigated as described in the next paragraph. In addition, unlike Zeidouni (2012, 2016), the flow capacities in the injection zone are considered different at two sides of the fault, allowing for possible layer juxtaposition.

In the following, we first present the physical system followed by corresponding analytical model. The analytical solution is verified against numerical simulation results. The analytical solution is cast in the form of type curves that can be readily applied to characterize lateral and vertical conductivity of the fault. Due to the large number of dimensionless groups obtained by the analytical solution, the system cannot be fully characterized using the type curves alone. We present a computational optimization method in combination with type curves to fully characterize the reservoir-fault system. Our approach is based on estimating the dimensionless parameters that describe the hydraulic properties of the fault and the altered region on the other side of the fault. Finally, we apply our proposed method to two example problems.

## Analytical model

For the physical system to be modeled, the target zone is separated into two regions (region 1 and region 2) by a vertical planar fault. Region 1 is on the side of the fault where the active well is located and region 2 is on the opposite side of the fault (Fig. 1). Because of possible displacement of the layers on the two sides of the fault plane, the thickness and reservoir properties of region 2 may be different from those of region 1. The reservoir properties are homogeneous and isotropic at each region. The $y$-axis is horizontal and perpendicular to the $x$-axis. The active well is assumed to be a line source/sink and is perforated over the whole thickness of the reservoir at $x=a$ and $y=0$ in the target zone. The reservoir is initially saturated with a single-phase fluid and the injected fluid is the same as the initial fluid. Both the upper zone and the target zone are infinite at both sides of the $x$ and $y$ axes. The fault plane is perpendicular to the $x$-axis, and it is located at $x=0$. The fault allows flow communications between region 1, region 2, and the upper zone. The horizontal and vertical permeabilities of the fault are considered constant.

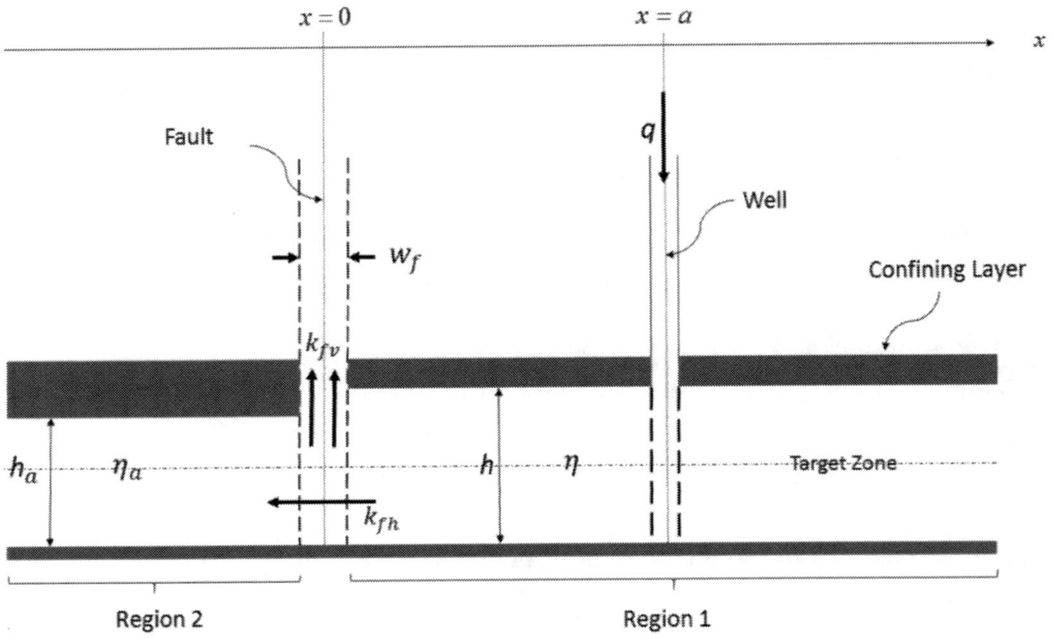


**Fig. 1 Schematic representation of the physical model**

The diffusivity equations for regions 1 and 2 make a system of two linear differential equations. The pressure change of the upper zone is negligible because the flow capacity of that zone is assumed to be large. Equations (1)-(10) represent diffusivity equations of regions 1 and 2 and the corresponding initial and boundary conditions:

Region 1:

$$\frac{\partial^2 \Delta P_1}{\partial x^2}+\frac{\partial^2 \Delta P_1}{\partial y^2}+\frac{q\mu}{kh}\delta(x-a)\delta(y)=\frac{1}{\eta}\frac{\partial \Delta P_1}{\partial t} \tag{1}$$

$$\Delta P_1(x,y,0)=0 \tag{2}$$

$$\Delta P_1(x,\pm\infty,t)=0 \tag{3}$$

$$\Delta P_1(+\infty,y,t)=0 \tag{4}$$

$$\frac{kh}{\mu}\frac{\partial \Delta P_1(0,y,t)}{\partial x}=\frac{k_{fh}h}{\mu w_f}(\Delta P_1(0,y,t)-\Delta P_2(0,y,t))+\frac{k_{fv}w_f}{2\mu L}(\Delta P_1(0,y,t)) \tag{5}$$

Region 2:

$$\frac{\partial^2 \Delta P_2}{\partial x^2}+\frac{\partial^2 \Delta P_2}{\partial y^2}=\frac{1}{\eta_a}\frac{\partial \Delta P_2}{\partial t} \tag{6}$$

$$\Delta P_2(x,y,0)=0 \tag{7}$$

$$\Delta P_2(x,\pm\infty,t)=0 \tag{8}$$

$$\Delta P_2(-\infty,y,t)=0 \tag{9}$$

$$\frac{k_a h_a}{\mu}\frac{\partial \Delta P_2(0,y,t)}{\partial x}+\frac{k_{fv} w_f}{2\mu L}(\Delta P_2(0,y,t))=\frac{k_{fh} h}{\mu w_f}(\Delta P_1(0,y,t)-\Delta P_2(0,y,t)) \tag{10}$$

where $L$ is half thickness of the above zone. The system of partial differential equations and the corresponding initial and boundary conditions (Equations (1)-(10)) are simplified to a system of ordinary differential equations using a combination of Laplace and Fourier transforms. Equations (11)-(12) state the sequence of applying Laplace and Fourier transforms to time and space domains:

$$\overline{\Delta P}(x,y,s)=\mathcal{L}\,[\Delta P(x,y,t)]=\int_0^{\infty}\Delta P(x,y,t)e^{-st}dt \tag{11}$$

$$\overline{\overline{\Delta P}}(x,\omega,s)=\mathcal{F}\,[\overline{\Delta P}(x,y,s)]=\int_{-\infty}^{+\infty}\overline{\Delta P}(x,y,s)e^{i.\omega.y}dy \tag{12}$$

where $s$ and $\omega$ are Laplace and Fourier transform dummy variables, respectively. Equations (13)-(14) provide the final solution of the pressure distribution at region 1 and region 2, respectively in dimensionless form (the solution details are given in the Appendix). The Laplace-Fourier domain is shown by = on the pressure (Equation (12) ) and dimensionless pressure (Equations (13)-(14) ).

$$\overline{\overline{P}}_{D1}=\frac{1}{2sA_1}\left(e^{-A_1|x_D-1|}+\frac{A_1\alpha-A_1\alpha_u-2\alpha\alpha_u-\alpha_u^{\,2}+A_1A_2T_D-A_2T_D\alpha-A_2T_D\alpha_u)}{A_1\alpha+A_1\alpha_u+2\alpha\alpha_u+2\alpha_u+A_1A_2T_D+A_2T_D\alpha+A_2T_D\alpha_u}e^{-A_1(x_D+1)}\right) \tag{13}$$

$$\overline{\overline{P}}_{D2}=\frac{\alpha}{s}\times\left(\frac{1}{A_1\alpha+A_1\alpha_u+2\alpha\alpha_u+2\alpha_u+A_1A_2T_D+A_2T_D\alpha+A_2T_D\alpha_u}e^{(-A_1+A_2x_D)}\right) \tag{14}$$

where:

$$P_{D1}=\frac{kh}{q\mu}\Delta P_1\,,\ P_{D2}=\frac{kh}{q\mu}\Delta P_2\,,\ t_D=\frac{\eta t}{a^2}\,,\ x_D=\frac{x}{a}\,,\ y_D=\frac{y}{a} \tag{15}$$

$$\alpha=\left(\frac{k_{fh}(h+h_a)}{2w_f}\right)\Big/\left(\frac{kh}{a}\right),\ \alpha_u=\left(\frac{k_{fv}w_f}{2L}\right)\Big/\left(\frac{kh}{a}\right) \tag{16}$$

$$T_D=\frac{k_a h_a}{kh}\,,\ \eta_D=\frac{\eta_a}{\eta} \tag{17}$$

$$A_1^{\,2}=\omega^2+s\,,\ A_2^{\,2}=\omega^2+\frac{s}{\eta_D} \tag{18}$$

where $w_f$, $k_{fv}$, and $k_{fh}$ are the fault width and vertical and horizontal permeability of the fault, respectively. $\alpha$ and $\alpha_u$ are dimensionless horizontal and vertical conductivities of the fault and $t_D$ is dimensionless time. $T_D$ and $\eta_D$ are flow capacity and diffusivity ratios, respectively. The solution must be inverted from the Laplace-Fourier domain to time-space domain. Analytical inversion of the solution to a closed-form solution in the time-space domain is impossible. Therefore, numerical Laplace and Fourier inversion methods are used to obtain the solution in time-space domain. Stehfest algorithm (Stehfest, 1970) is used for Laplace inversion and the Inverse Discrete Fourier Transform (IDFT) is used for Fourier inversion.

## Verification of the analytical model

The analytical solution is verified by comparison of spatial and temporal variation of the pressure with the numerical simulation results. In dimensional terms (corresponding to the dimensional values), the width of the fault is 0.1 m, the porosity is 0.2, total compressibility factor is $1 \times 10^{-6}$ 1/kPa, and injection rate is 0.005 $m^3/s$. The reservoir thickness and permeability of region 1 are 10 m and 10 mD, respectively. Fault lateral and vertical permeabilities are 0.01 mD and 5000 mD, respectively. The system specifications are given in Table 1. Fig. 2 exhibits the pressure distribution on the line drawn through the well perpendicular to the fault plane. The lateral pressure discontinuity is visible across the fault, which shows that the lateral permeability of the fault is less than the reservoir permeability. Fig. 2 illustrates that the analytical solution is in good agreement with the numerical simulation at both regions 1 and 2. If there is no lateral leakage, the pressure in region 2 remains constant at initial pressure. Therefore, the pressure gradient along region 2 shown in Fig. 2 (from $x$=-100 to $x$=0) is a sign of lateral leakage through the fault.

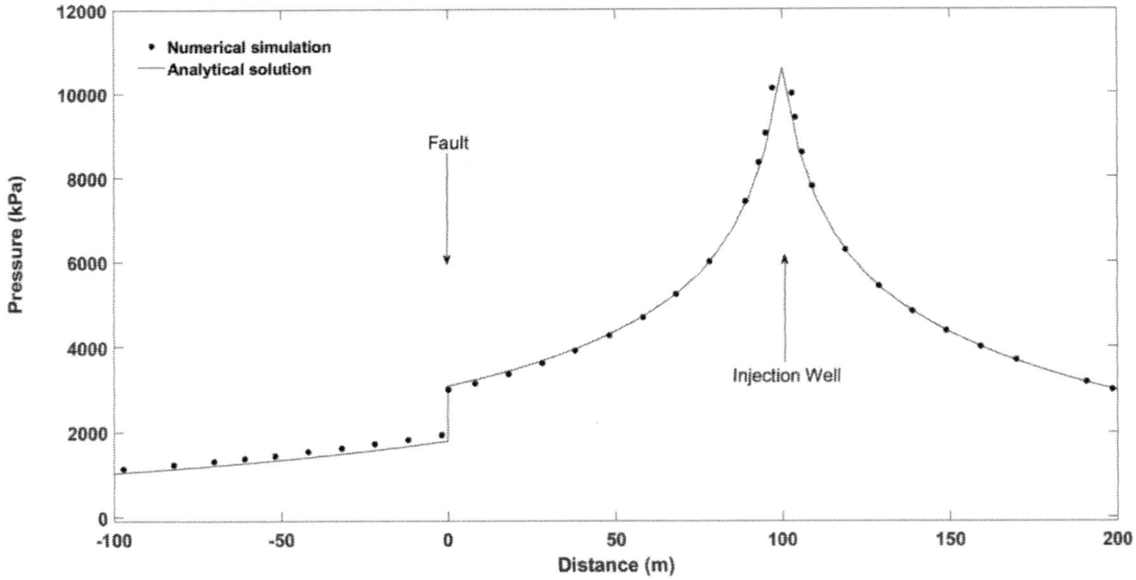


**Fig. 2 Validation of the spatial pressure distribution in the injection zone after 10 days.**

Fig. 3 illustrates good agreements of pressure and logarithmic pressure derivative (referred to as derivative hereafter) between analytical and numerical simulation. The verification of the pressure derivative is important because the characterization method to be presented later is based on the pressure derivative curves.

**Table 1. Reservoir properties**

| Parameter | Value | Parameter | Value |
|---|---|---|---|
| Effective width of the fault (m) | 0.1 | Viscosity (cp) | 0.5 |
| Lateral permeability of the fault (mD) | 0.01 | Injection rate ($m^3.s^{-1}$) | 0.005 |
| Vertical permeability of the fault (mD) | 5000 | Total compressibility (1/kPa) | $10^{-6}$ |
| Target zone thickness (m) | 10 | fault vertical conductivity , $\alpha_u$ | 0.12 |
| Fault-well Distance (m) | 100 | fault lateral conductivity , $\alpha$ | 1 |
| Permeability of region 1 (mD) | 10 | Diffusivity ratio , $\eta_D$ | 1 |
| Permeability of region 2 (mD) | 10 | Flow capacity ratio , $T_D$ | 1 |
| Porosity (fraction) | 0.2 | | |

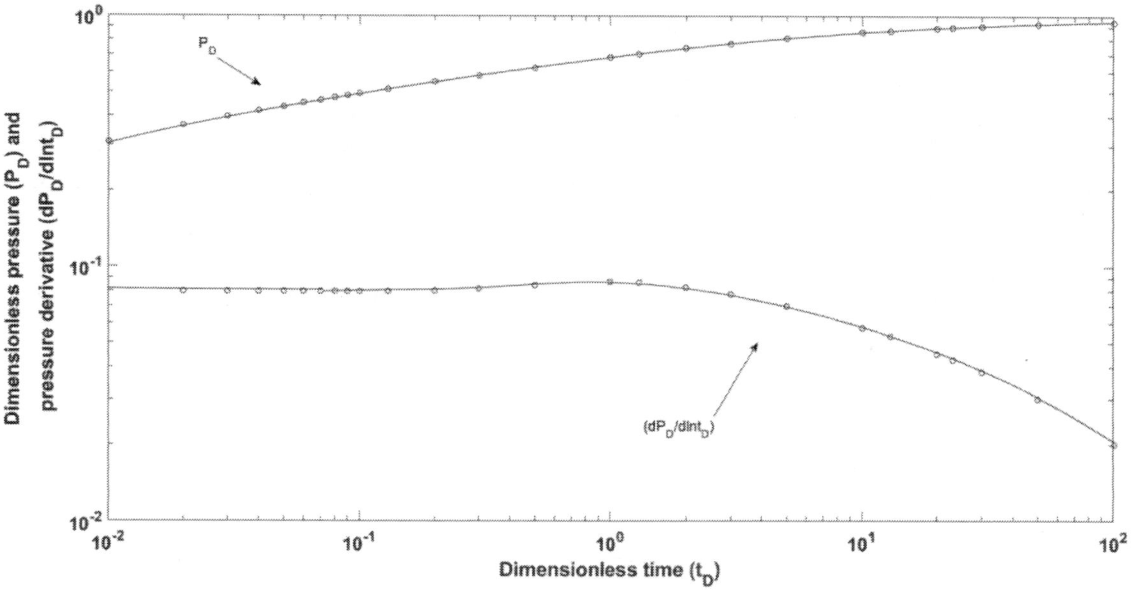


**Fig. 3: Validation of temporal variation of well pressure and (logarithmic) pressure derivative with respect to time using numerical simulation**

## Fault characterization

In this section, type curves are introduced for characterization of the leaky fault. Type curves are generalized and adjusted in terms of dimensionless variables. Based on the analytical solution, the reservoir-fault system can be characterized by four groups: Fault lateral and vertical conductivities, reservoir flow capacity ratio, and diffusivity ratio. These four groups are represented by dimensionless parameters: $\alpha$, $\alpha_u$, $T_D$, and $\eta_D$, respectively. The characterization procedure is presented based on the type curves.

## Type curves

Fig. 4 illustrates the type curves corresponding to the leaky fault system in terms of fault lateral ($\alpha$) and vertical ($\alpha_u$) conductivities for $T_D = \eta_D = 1$. We refer to type curves of Fig. 4 as the base type curves. The type curves are grouped for various values of $\alpha_u$. At each group, $\alpha$ varies from zero to infinity. The negative unit slope line of the logarithmic derivative curves in Fig. 4 shows the early time of the vertical leakage. The negative unit slope line would occur before appearance of flow resistance from the overlying/underlying zone. Using these type curves, hydraulic characteristics of the fault can be estimated with the well pressure data before detecting any resistance from the overlying/underlying zones.

A well pressure data corresponding to unique values of $\alpha$ and $\alpha_u$ would match with a unique type curve. In finding the matching type curve, $\alpha_u$ is easier to obtain from late time pressure data while parameter $\alpha$ is easier to estimate using early time data. Implementing this technique of using the early time and the late time data separately can help in resolving the choice of intermediate values of $\alpha$ and $\alpha_u$. In other words, for two close values of $\alpha_u$, the two corresponding groups of type curves may be very close to one another at late time, but the two curves should deviate from each other at early time curvature making it easy to recognize which one belongs to which group. In addition, two curves with too close values of $\alpha$ and different values of $\alpha_u$ may be close to each other at the early time. Then, their late time deviation can be used to determine the correct type curve. In short, there is a unique type curve for each combination of $\alpha_u$ and $\alpha$ which should be easy to be distinguished from the other curves. In addition, the point when deviation from the radial flow (the zero-slope derivative line prior to reaching the fault) would commence are useful to find the best matching type curve.

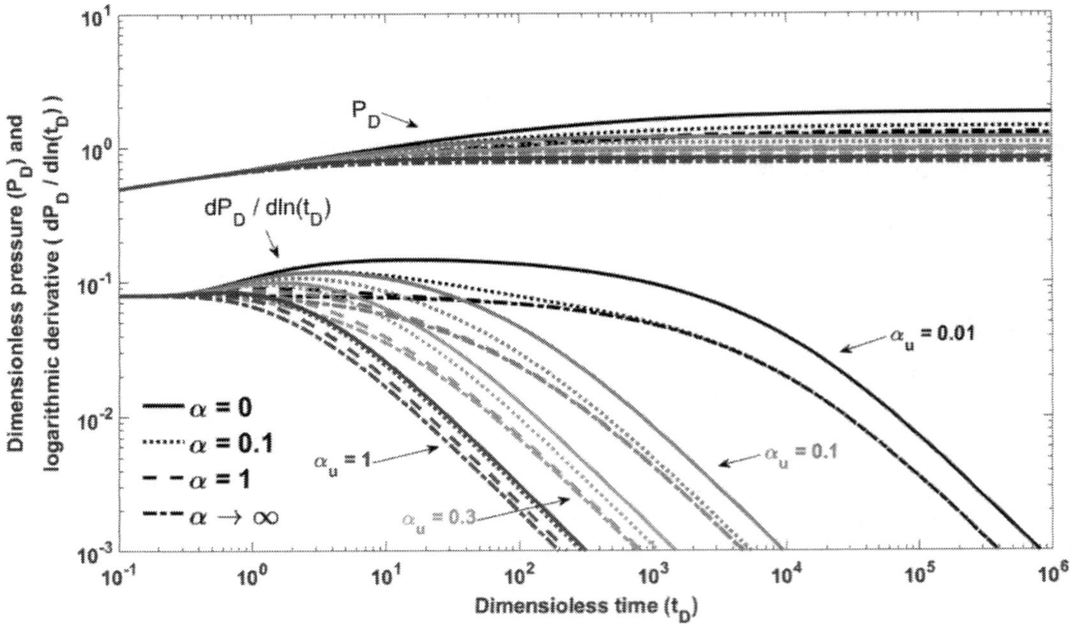


**Fig. 4: Type curves of the leaky fault system considering $T_D$=1 and $\eta_D$=1**

In order to use these type curves, the well pressure data should be plotted versus time in the same scale as the type curves on a transparent plot. The resulting plot should be next moved horizontally and vertically (without rotation) to find the best match with the type curves to estimate $\alpha$ and $\alpha_u$. In addition to $\alpha$ and $\alpha_u$, region 1 permeability and fault-well distance can also be evaluated based on the radial flow prior to reaching the fault. By selecting an arbitrary match point, we get $(P_D)_M$, $(\Delta P)_M$, $(t)_M$, and $(t_D)_M$ of that match point to calculate region 1 permeability ($k$) and fault-well distance ($a$) based on Equations (19) and (20).

$$k = \frac{q\mu}{h}\left(\frac{P_D}{\Delta P}\right)_M \tag{19}$$

$$a = \sqrt{\eta\left(\frac{t}{t_D}\right)_M} \tag{20}$$

One observation from these type curves is that 2 to 3 log-cycles of data may be required for identification of the fault characteristics. This implies that if the fault is felt after 1 hr test, 10 to 100 hrs of test may be required to enable characterizing of the fault. The type curves can be extended to determine $T_D$ and $\eta_D$ as well. Fig. 5 illustrates the effect of $T_D$ and $\eta_D$ on the pressure and pressure derivative for a fixed value of $\alpha$ (=0.1). The effect of $T_D$ and $\eta_D$ is most visible at the lowest values of $\alpha_u$. Similar patterns are achieved for different values of $\alpha$.

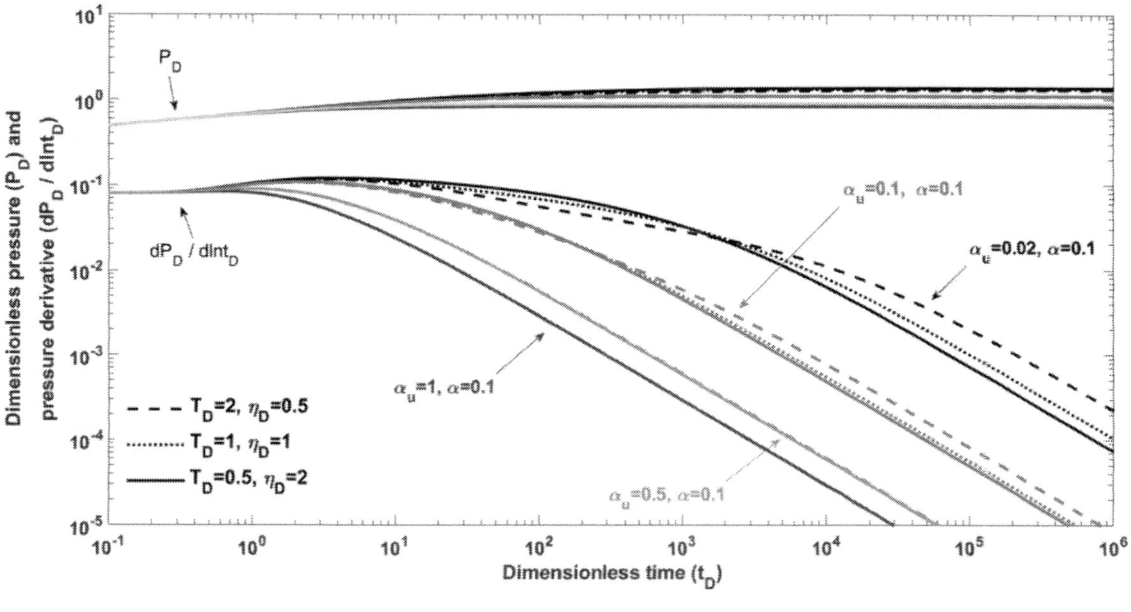


**Fig. 5 Effect of $T_D$ and $\eta_D$ on the type curves**

Fig. 5 shows that by increasing $\alpha_u$, the sensitivity of the pressure derivative curve to the parameters $T_D$

and $\eta_D$ will decrease. In the following, we introduce a procedure to estimate the dimensionless parameters. First, initial values of $\alpha$ and $\alpha_u$ are estimated by the base type curves. Next, an optimization method is applied to find the accurate values of the dimensionless parameters using the initial values obtained from type curve analysis. We use the MATLAB built-in optimization tool given by the *fmincon* optimization function (Matlab, 2015). This function is based on a sequential quadratic programming, which is a gradient-based algorithm. Here, fmincon optimization is applied to minimize the square of the difference between well pressure data and the mathematical solution by changing the four dimensionless parameters.

## Characterization procedure

In characterization procedure, first, we estimate the initial values of $\alpha$ and $\alpha_u$ from the type curves. Next, we calculate the values of permeability and fault-well distances using Equations (19) and (20). In the optimization process, we use the pressure data to modify the estimated initial value of $\alpha$ and $\alpha_u$. Next, we fix $T_D$ and $\eta_D$. This step is done by the late time data because $T_D$ and $\eta_D$ have a negligible effect on the early time data. The initial values of $T_D$ and $\eta_D$ can be considered equal to one. This sequential procedure improves the optimization process compared to optimizing all four dimensionless parameters simultaneously using the whole pressure data. The characterization procedure is summarized in Fig. 6.

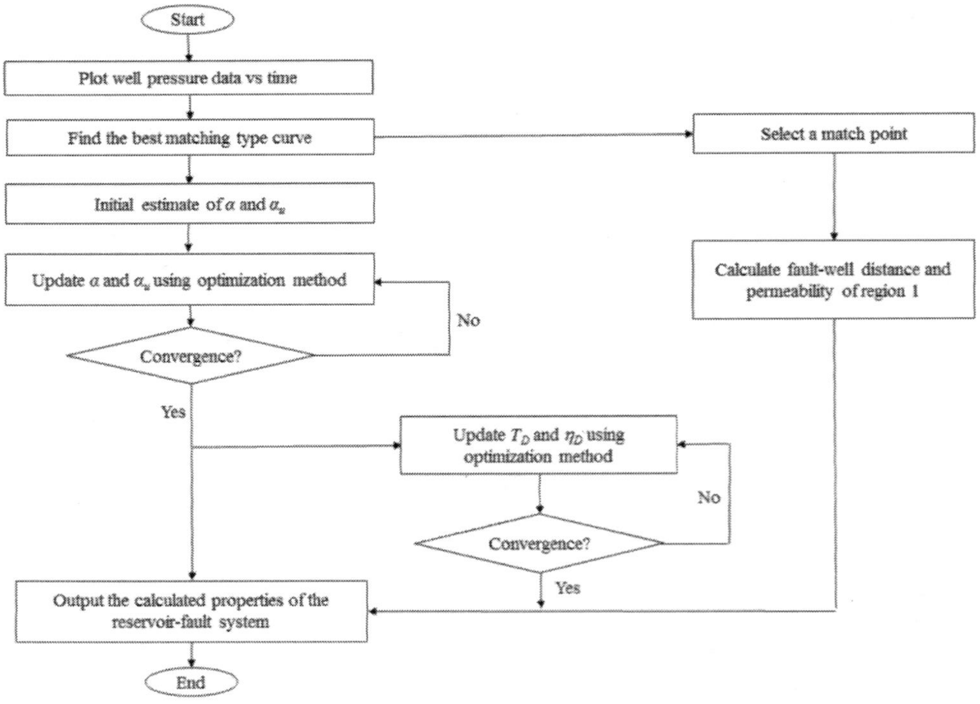


**Fig. 6 Characterization procedure of the leaky fault system**

## Results and discussion

In this part, two examples are presented to estimate fault characteristics by the type-curve method using the procedure described in the previous section.

### Example 1: No alteration of reservoir properties across the fault

In this example, we analyze the well pressure data of an injection well near a fault. The permeability, porosity, total compressibility, and thickness of the target zone at both sides of the fault are respectively 100 mD, 0.1, $10^{-6}$ 1/kPa, and 10 m. The flow capacity of the upper zone is large enough compared to the injection zone. The injection rate is 0.005 $m^3.s^{-1}$ and fluid viscosity is 0.5 cp. Fig. 7 shows the synthetic well pressure data and pressure derivative of the first example in which $\alpha$ and $\alpha_u$ are considered 1 and 0.3, respectively. These values correspond to 62 D and 0.1 mD for vertical and horizontal permeabilities of the fault considering a 0.1-m wide fault. Fluid leakage through the fault can be inferred from the derivative curve. If the fault is sealing, the pressure derivative becomes zero-slope after departing from the zero-slope corresponding to initial radial flow (Fig. 8). The late time horizontal line in Fig. 8 exhibits the pressure response of a sealing fault in which $\alpha$ and $\alpha_u$ are equal to zero.

The pressure data and the derivative are required on a plot with the same scale as the type curves for type-curve matching. Fig. 9 shows the type curve that best agrees with the pressure data. Corresponding values of $\alpha$ and $\alpha_u$ are estimated for the fault. The estimated values of $\alpha$ and $\alpha_u$ are 1 and 0.3, which are the true values used in generating our synthetic data. By selecting an arbitrary match point and using equations (19) and (20), the permeability of region 1 and the fault-well distance can be calculated as below:

$$(P_D)_M = 0.08 \ , \ (\Delta P)_M = 199 \ kPa \ , \ (t_D)_M = 0.07 \ , \ (t)_M = 1000 \ s \tag{21}$$

$$k = \frac{q\mu}{h}\left(\frac{P_D}{\Delta P}\right)_M = 100.5 \text{ mD} \tag{22}$$

$$a = \sqrt{\eta\left(t/t_D\right)_M} = 100.2 \ \text{ m} \tag{23}$$

These values are in close agreement with the correct values ($k$=100 mD, $a$=100 m).

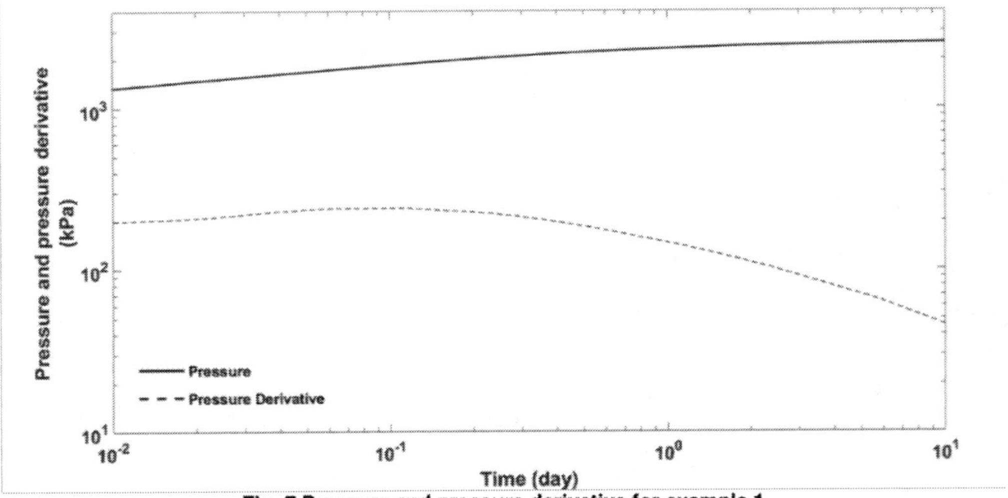


**Fig. 7 Pressure and pressure derivative for example 1**

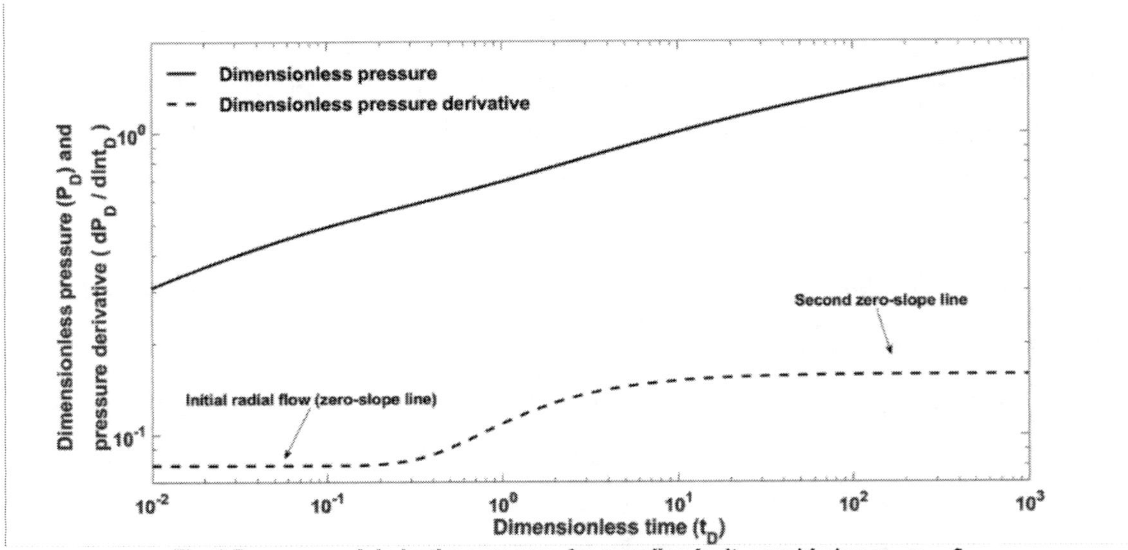


**Fig. 8 Pressure and derivative response for a sealing fault, considering $\alpha = \alpha_u = 0$.**

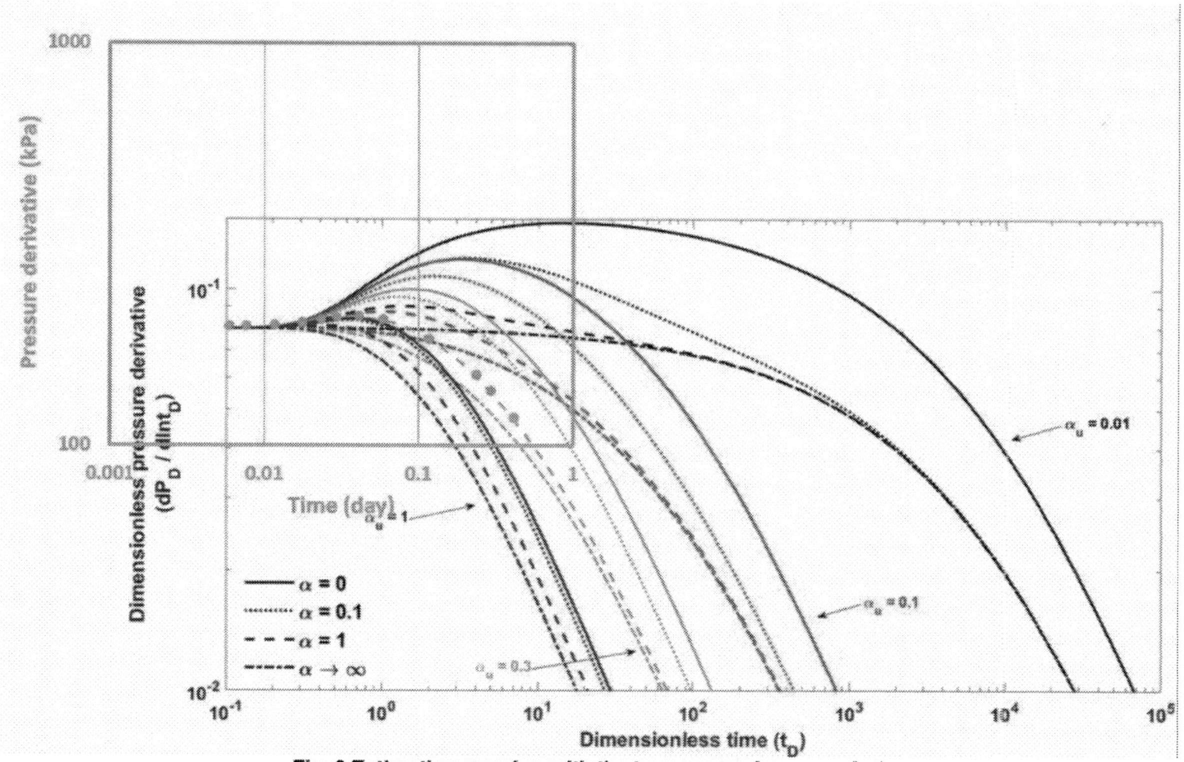


**Fig. 9 Estimating $\alpha$ and $\alpha_u$ with the type curves for example 1**

**Example 2: Alteration of reservoir properties across the fault**

Fig. 10 shows the synthetic pressure data of an injection well near a fault in which $\alpha$, $\alpha_u$, $T_D$, and $\eta_D$ are considered 0.02, 0.5, 0.1, and 0.1 respectively. The flow capacity of the upper zone is large in proportion to the injection zone. In this example, we characterize the leaky fault while the reservoir properties of region 2 may not be identical to those of region 1. The injection rate is 0.005 $m^3.s^{-1}$, the viscosity is 0.5 cp, the porosity is 0.1, total compressibility is $10^{-6}$ 1/kPa, and reservoir thickness is 20 m. First, we use the base type curves to find the initial values for $\alpha$ and $\alpha_u$. In this case, the pressure data curve may not accurately match with the type curves but we try to find the best match. The best match of the pressure data with the type curves is illustrated in Fig. 11. The estimated initial values of $\alpha$ and $\alpha_u$ are 0.2 and 0.03, respectively.

When $T_D \neq 1$ and/or $\eta_D \neq 1$, the effects of $T_D$ and $\eta_D$ on the pressure derivative curve appear after the arrival of the pressure pulse to the fault. The variations of $T_D$ and $\eta_D$ will not affect the early time pressure data. Thus, similar to example 1, we can select an arbitrary match point to estimate the fault-well distance and permeability of region 1 (Equations (24)-(26)).

$$(P_D)_M = 0.098 \ , \ (t_D)_M = 1 \ , \ \ (t_D)_M = 1 \ , \ (t)_M = 10022 \ \text{ s} \tag{24}$$

$$k = \frac{q\mu}{h}\left(\frac{P_D}{\Delta P}\right)_M = 49.4\ \text{mD} \tag{25}$$

$$a = \sqrt{\eta\left(t/t_D\right)_M} = 100.1\ \text{m} \tag{26}$$

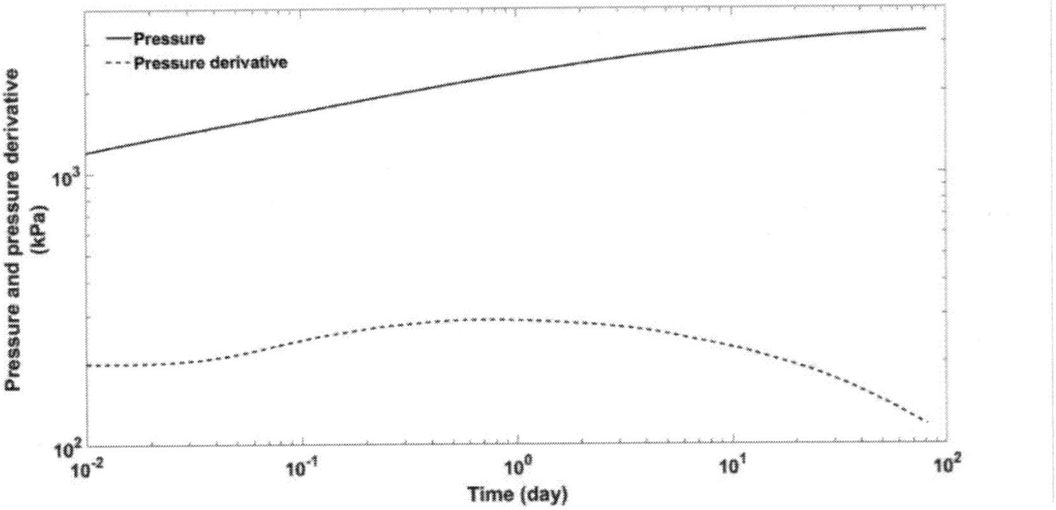


Fig. 10 Pressure and derivative data for example 2

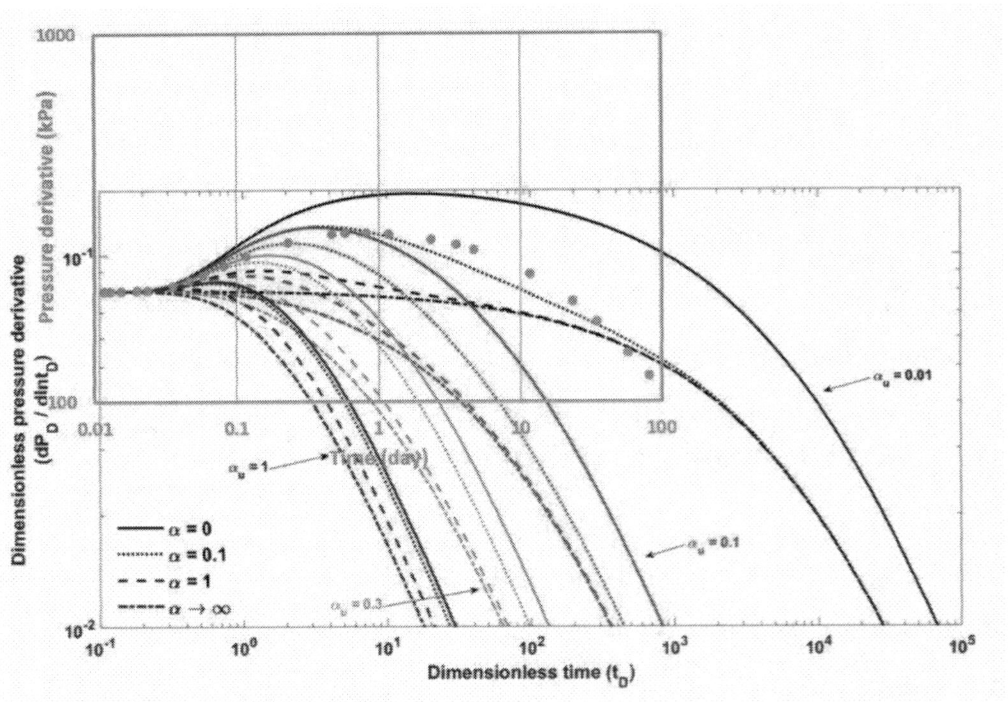


Fig. 11 Estimation of $a$ and $a_u$ with the base type curves for example 2

Next, the optimization method is used to find accurate values of the dimensionless parameters. We set the initial values of $T_D$ and $\eta_D$ equal to one. The calculated permeability and fault-well distance ($k$ and $a$)

are used to convert the well pressure data to dimensionless numbers. We apply optimization function to find the best match between the dimensionless pressure data and the analytical solution by modifying the four dimensionless parameters. Using the optimization method, the estimated values of the dimensionless parameters are $\alpha$=0.5, $\alpha_u$=0.02, $T_D$=0.1 and $\eta_D$=0.29. The estimated values of fault conductivities ($\alpha$ and $\alpha_u$) are true values corresponding to synthetic data. Generally, effects of fault conductivities on the pressure derivative are more than effects of diffusivity and flow capacity ratios in this model. Because the model is focused on the early times of fault leakage.

## Conclusions

In this study, we preseted a type-curve based method to characterize a vertically and laterally leaking fault using pressure data. In order to develop the type curves, we developed an analytical solution. The reservoir properties on the two sides of the fault are considered non-identical. The analytical solution was verified against the numerical simulation results. In this model, we neglect the resistance to flow by overlying/underlying permeable layers connected to the reservoir by the fault. We investigated the validity of this assumption by assigning different flow capacities to the overlying/underlying zones. This assumption is valid before reaching to the effects of the adjacent zone. The model was cast in the form of type curves to characterize lateral and vertical leakage through the fault. We used the optimization method in combination with the type curves to identify $\alpha$, $\alpha_u$, $T_D$, and $\eta_D$. The type curves were used to find the initial values of unknown parameters. Applying the optimization method, very accurate values of $\alpha_u$ and $\alpha$ were estimated by modifying the initial values. As $\alpha_u$ increased, the sensitivity of the pressure response to $T_D$ and $\eta_D$ was reduced. However, for low values of $\alpha_u$, the well pressure data were considerably sensitive to $T_D$ and $\eta_D$. The sensitivity of this range of pressure data on flow capacity and diffusivity ratios ($T_D$ and $\eta_D$) is less than that to the fault vertical and lateral conductivities. This work shows that the pressure data prior to reaching the flow resistance of the overlying/underlying zone can be enough to estimate hydraulic characteristics of the leaky fault.

## Acknowledgement

This study was partially supported by the Louisiana Board of Regents under contract #43950.

## Appendix: Mathematical solution

We present the system of differential equations in dimensionless form for generality and convenience. The system of differential equations of regions 1 and 2 and the corresponding boundary conditions (Equations (1)-(10)) are converted to dimensionless form using Equations (15)-(17).

Region 1:

$$\frac{\partial^2 P_{D1}}{\partial x_D^2}+\frac{\partial^2 P_{D1}}{\partial y_D^2}+\delta(x_D-1)\delta(y_D)=\frac{\partial P_{D1}}{\partial t_D} \tag{A1}$$

$$P_{D1}(x_D,y_D,0)=0 \tag{A2}$$

$$P_{D1}(x_D,\pm\infty,0)=0 \tag{A3}$$

$$P_{D1}(+\infty,y_D,t_D)=0 \tag{A4}$$

$$\frac{\partial P_{D1}(0,y_D,t_D)}{\partial x_D}=\alpha(P_{D1}(0,y_D,t_D)-P_{D2}(0,y_D,t_D))+\alpha_u(P_{D1}(0,y_D,t_D)) \tag{A5}$$

Region 2:

$$\frac{\partial^2 P_{D2}}{\partial x_D^2}+\frac{\partial^2 P_{D2}}{\partial y_D^2}=\frac{1}{\eta_D}\frac{\partial P_{D2}}{\partial t_D} \tag{A6}$$

$$P_{D2}(x_D,y_D,0)=0 \tag{A7}$$

$$P_{D2}(x_D,\pm\infty,t_D)=0 \tag{A8}$$

$$P_{D2}(-\infty,y_D,t_D)=0 \tag{A9}$$

$$T_D\frac{\partial P_{D2}(0,y,t)}{\partial x_D}+\alpha_u(P_{D2}(0,y,t))=\alpha(P_{D1}(0,y,t)-P_{D2}(0,y,t)) \tag{A10}$$

To solve the system of differential equations, the method of Laplace and Fourier transforms is used. Applying the transforms, the system of partial differential equations in time-space domain is converted into a system of ordinary differential equations in Laplace-Fourier domain that can be solved. Then, the solutions should be converted back to the time-space domain using Laplace-Fourier inverse transforms. The Laplace transform of the function $P_D$ is defined as below:

$$\overline{P_D}(x_D,y_D,s)=\mathcal{L}[P_D(x_D,y_D,t_D)]=\int_0^\infty P_D(x_D,y_D,t_D)e^{-st_D}dt_D \tag{A11}$$

where $s$ is the Laplace transform variable. First, the Laplace transform is applied into the time domain.

Region 1:

$$\frac{\partial^2 \overline{P}_{D1}}{\partial x_D^2} + \frac{\partial^2 \overline{P}_{D1}}{\partial y_D^2} + \frac{1}{s}\delta(x_D - 1)\delta(y_D) = s\overline{P}_{D1} \tag{A12}$$

$$\overline{P}_{D1}(x_D, \pm\infty, s) = 0 \tag{A13}$$

$$\overline{P}_{D1}(+\infty, y_D, s) = 0 \tag{A14}$$

$$\frac{\partial \overline{P}_{D1}(0, y_D, s)}{\partial x_D} = \alpha(\overline{P}_{D1}(0, y_D, s) - \overline{P}_{D2}(0, y_D, s)) + \alpha_u(\overline{P}_{D1}(0, y_D, s)) \tag{A15}$$

Region 2:

$$\frac{\partial^2 \overline{P}_{D2}}{\partial x_D^2} + \frac{\partial^2 \overline{P}_{D2}}{\partial y_D^2} = \frac{s}{\eta_D}\overline{P}_{D2} \tag{A16}$$

$$\overline{P}_{D2}(x_D, \pm\infty, s) = 0 \tag{A17}$$

$$\overline{P}_{D2}(-\infty, y_D, s) = 0 \tag{A18}$$

$$T_D \frac{\partial \overline{P}_{D2}(0, y_D, s)}{\partial x_D} + \alpha_u(\overline{P}_{D2}(0, y_D, s)) = \alpha(\overline{P}_{D1}(0, y_D, s) - \overline{P}_{D2}(0, y_D, s)) \tag{A19}$$

The Fourier transform of function $\overline{P_D}$ is defined as below:

$$\overline{\overline{P_D}}(x_D, \omega, s) = \mathcal{F}[\overline{P_D}(x_D, y_D, s)] = \int_{-\infty}^{+\infty} \overline{P_D}(x_D, y_D, s)e^{i.\omega.t}dt \tag{A20}$$

where $\omega$ is the Fourier parameter. Second, the Fourier transform is applied into the space domain ($y_D$) and the ordinary system of differential equations and corresponding boundary conditions in the Laplace-Fourier domain is obtained:

Region 1:

$$\frac{d^2 \overline{\overline{P}}_{D1}}{dx_D^2} - A_1^2 \overline{\overline{P}}_{D1} = -\frac{1}{s}\delta(x_D - 1) \tag{A21}$$

$$\overline{\overline{P}}_{D1}(+\infty) = 0 \tag{A22}$$

$$\frac{\partial \overline{\overline{P}}_{D1}(0)}{\partial x_D} = \alpha(\overline{\overline{P}}_{D1}(0) - \overline{\overline{P}}_{D2}(0)) + \alpha_u(\overline{\overline{P}}_{D1}(0)) \tag{A23}$$

Region 2:

$$\frac{d^2\overline{\overline{P}}_{D2}}{dx_D^2} - A_2{}^2\overline{\overline{P}}_{D2} = 0 \tag{A24}$$

$$\overline{\overline{P}}_{D2}(-\infty) = 0 \tag{A25}$$

$$T_D\frac{\partial\overline{\overline{P}}_{D2}(0)}{\partial x_D} + \alpha_u(\overline{\overline{P}}_{D2}(0)) = \alpha(\overline{\overline{P}}_{D1}(0) - \overline{\overline{P}}_{D2}(0)) \tag{A26}$$

where $A_1=\omega^2+s$ and $A_2=\omega^2+s/\eta_D$. The solution in Laplace-Fourier domain is as below.

Region 1:

$$\overline{\overline{P}}_{D1} = \frac{1}{2sA_1}\left(e^{-A_1|x_D-1|}\right) + C_1 e^{-A_1 x_D} \tag{A27}$$

$$C_0A_1 - C_1A_1 = \alpha(C_0 + C_1 - C_2) + \alpha_u(C_0 + C_1) \tag{A28}$$

Region 2:

$$\overline{\overline{P}}_{D2} = C_2 e^{A_2 x_D} \tag{A29}$$

$$T_D C_2 A_2 + C_2\alpha_u = \alpha(C_0 + C_1 - C_2) \tag{A30}$$

where $C_0 = (1/(2sA_1))e^{-A_1}$. $C_1$ and $C_2$ are unknown constants, which must be calculated by combining the equations (A28) and (A30). The solutions for dimensionless pressure of regions 1 and 2 in Laplace-Fourier domain are:

$$\overline{\overline{P}}_{D1} = \frac{1}{2sA_1}\left(e^{-A_1|x_D-1|} + \frac{A_1\alpha + A_1\alpha_u - 2\alpha\alpha_u - \alpha_u{}^2 + A_1A_2T_D - A_2T_D\alpha - A_2T_D\alpha_u}{A_1\alpha + A_1\alpha_u + 2\alpha\alpha_u + \alpha_u{}^2 + A_1A_2T_D + A_2T_D\alpha + A_2T_D\alpha_u} e^{-A_1(1+x_D)}\right) \tag{A31}$$

$$\overline{\overline{P}}_{D2} = \frac{\alpha}{s}\times\left(\frac{1}{A_1\alpha + A_1\alpha_u + 2\alpha\alpha_u + \alpha_u{}^2 + A_1A_2T_D + A_2T_D\alpha + A_2T_D\alpha_u} e^{(-A_1+A_2x_D)}\right) \tag{A32}$$

## References


Ambastha, A. K., Mcleroy, P. G. & Grader, A. S. 1989. Effects of a partially communicating fault in a composite reservoir on transient pressure testing. SPE Formation Evaluation, 4 (02), 210-218.

Apps, J. A., Zheng, L., Zhang, Y., Xu, T. & Birkholzer, J. T. 2010. Evaluation of Potential Changes in Groundwater Quality in Response to CO2 Leakage from Deep Geologic Storage. Transport in Porous Media, 82 (1), 215-246.

Bense, V. F. & Person, M. A. 2006. Faults as conduit-barrier systems to fluid flow in siliciclastic sedimentary aquifers. Water Resources Research, 42 (5).

Birkholzer, J. T., Zhou, Q. & Tsang, C.-F. 2009. Large-scale impact of CO 2 storage in deep saline aquifers: a sensitivity study on pressure response in stratified systems. International Journal of Greenhouse Gas Control, 3 (2), 181-194.

Bixel, H. C., Larkin, B. K. & Vanpoollen, H. K. 1963. Effect of Linear Discontinuities on Pressure Build-up and Drawdown Behavior. Transactions of the Society of Petroleum Engineers of Aime, 228 (8), 885-895.

Caine, J. S., Evans, J. P. & Forster, C. B. 1996. Fault zone architecture and permeability structure. Geology, 24 (11), 1025-1028.

Ebigbo, A., Class, H. & Helmig, R. 2007. CO2 leakage through an abandoned well: problem-oriented benchmarks. Computational Geosciences, 11 (2), 103-115.

Humez, P., Audigane, P., Lions, J., Chiaberge, C. & Bellenfant, G. 2011. Modeling of CO 2 leakage up through an abandoned well from deep saline aquifer to shallow fresh groundwaters. Transport in Porous Media, 90 (1), 153-181.

Huntoon, P. & Lundy, D. A. 1979. Fracture-Controlled Ground-Water Circulation and Well Siting in the Vicinity of Laramie. Ground Water, 17(5) (5), 463-469.

IPCC 2005. Special Report on Carbon Dioxide Capture and Storage: Prepared by Working Group III of the Intergovernmental Panel on Climate Change. Cambridge University Press.

Maslia, M. L. & Prowell, D. C. 1990. Effect of Faults on Fluid-Flow and Chloride Contamination in a Carbonate Aquifer System. Journal of Hydrology, 115 (1-4), 1-49.

Matlab, 2015. 8.6. The MathWorks, Inc., Natick, Massachusetts, United States.

Mosaheb, M. & Zeidouni, M. Above-Zone Pressure Response to Distinguish Between Fault and Caprock Leakage. SPE Western Regional Meeting, 2017a. Society of Petroleum Engineers.

Mosaheb, M. & Zeidouni, M. Pressure Transient Analysis for Leaky Well Characterization and its Identification From Leaky Fault. SPE Health, Safety, Security, Environment, & Social Responsibility Conference-North America, 2017b. Society of Petroleum Engineers.

Pruess, K. 2005. Numerical studies of fluid leakage from a geologic disposal reservoir for CO2 show self-limiting feedback between fluid flow and heat transfer. Geophysical Research Letters, 32 (14).

Raghavan, R. 2010. A composite system with a planar interface. Journal of Petroleum Science and Engineering, 70(3), 229-234.

Rahman, N. M., Miller, M. D. & Mattar, L. 2003. Analytical solution to the transient-flow problems for a well located near a finite-conductivity fault in composite reservoirs. In SPE Annual Technical Conference and Exhibition.

Shakiba, M. & Hosseini, S. A. 2016. Detection and characterization of CO 2 leakage by multi-well pulse testing and diffusivity tomography maps. International Journal of Greenhouse Gas Control, 54, 15-28.

Shan, C., Javandel, I. & Witherspoon, P. A. 1995. Characterization of leaky faults: Study of water flow in aquifer-fault-aquifer systems. Water Resources Research, 31(12), 2897-2904.

Sibson, R. 1977. Fault rocks and fault mechanisms. Journal of the Geological Society, 133 (3), 191-213

Stehfest, H. 1970. Algorithm 368: Numerical inversion of Laplace transforms [D5]. Communications of the ACM, 13 (1), 47-49.

Stewart, G. & Gupta, A. 1984. The interpretation of interference tests in a reservoir with sealing and partially communicating faults. European Petroleum Conference.

Stoessell, R. K. & Prochaska, L. C. 2005. Chemical evidence for migration of deep formation fluids into shallow aquifers in south Louisiana. Gulf Coast Association of Geological Sciences Transactions, 794-808.

Thomas, D. C. & Benson, S. M. 2005. Carbon dioxide capture for storage in deep geologic formations, Elsevier.

Yaxley, L. M. 1987. Effect of a partially communicating fault on transient pressure behavior. SPE Formation Evaluation, 2(04), 590-598.

Zeidouni, M. 2012. Analytical model of leakage through fault to overlying formations. Water Resources Research, 48.

Zeidouni, M. 2016. Semi-Analytical Model of Pressure Perturbations Induced by Fault Leakage in Multilayer System. Journal of Hydrologic Engineering, 21 (6).

Zeidouni, M. & Pooladi-Darvish, M. 2012. Leakage characterization through above-zone pressure monitoring: 1—Inversion approach. Journal of Petroleum Science and Engineering, 98, 95-106.

**Carbon Management Technology Conference**

*Global CCUS Innovation Nexus*

# ECO₂NOMICS™ THAT MAKE SENSE

**VeloxoTherm™ $CO_2$ Capture Technology**
**Rapid Cycle Thermal Swing Adsorption Process**

July 18, 2017 | 1:50 PM

inventys

*Hilton Americas Houston*

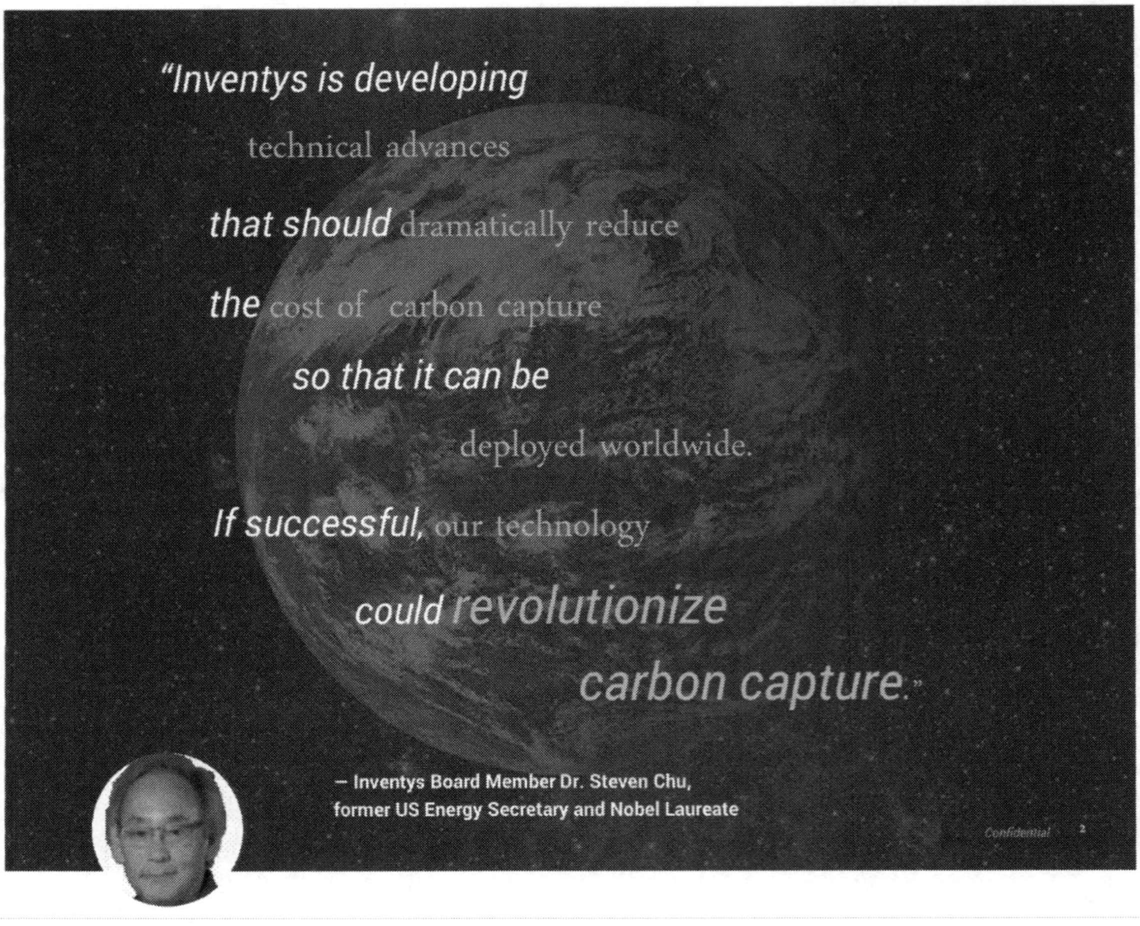

# E$CO_2$nomics That Make Sense

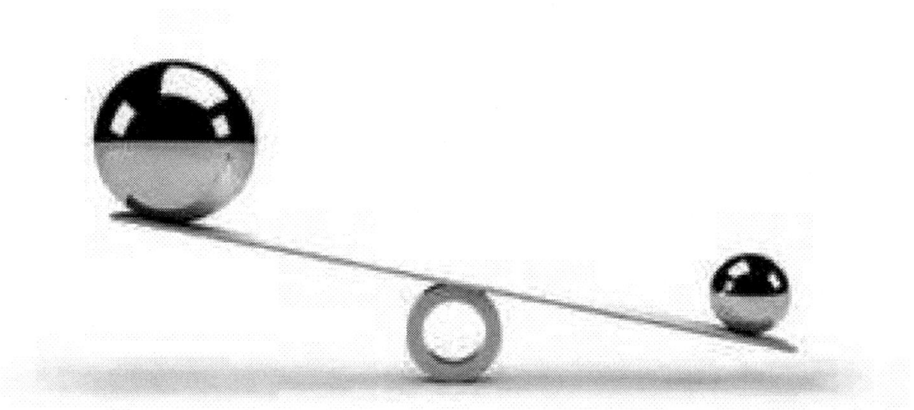

**$60-90/tonne**
*Solvent-Based*
**Liquid**

**<$30/tonne**
*Structured Adsorbent*
**Solid**

inventys

Confidential 3

# Inventys Story

**COST EFFECTIVE**
*1/3 the capital cost and 50% lower in total capture cost of traditional solvent*

**COMPACT**
*3-steps rapid cycle process (adsorb, regenerate, & cool) in one single compact unit*

**HIGH PERFORMANCE**
*Nanomaterials with high surface area per unit volume and no chemical degradation (filling)*

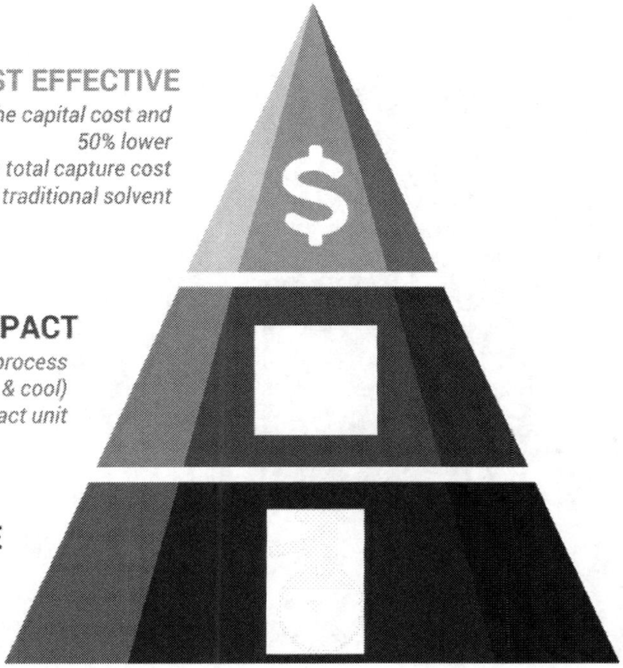


inventys

Confidential 4

# Small Size & Compact Design

*~4,800 TPD $CO_2$ CAPTURE PLANT*

Solvent-based Liquid System

Structured Adsorbent Solid System

**inventys**

Confidential 5

# Key Performance Indicators

| Performance Objectives | Bench Scale Testing (2016) | Mark I Beds Target (2017) | Mark II Beds Target (2018) |
|---|---|---|---|
| **Product $CO_2$ Purity** *(%; dry basis)* | 92 | 85–95 | 90–95 |
| **Recovery** *(%; product $CO_2$/feed $CO_2$)* | 80 | 75–85 | 85–90 |
| **Productivity** *(TPD $CO_2$/$m^3$ adsorbent)* | ✓ 11 | 9–11 | 10–12 |
| **Steam Ratio** *(kg/kg; steam/product $CO_2$)* | 1.5 | 1.3–1.7 | ➢ 0.8–1.2 |

✓ **High Productivity → Compact Solution → Low Capital Cost**

➢ **Low Steam Ratio → Low Regeneration Energy → Low Operating Cost**

**inventys**

Confidential 6

# Core Competencies

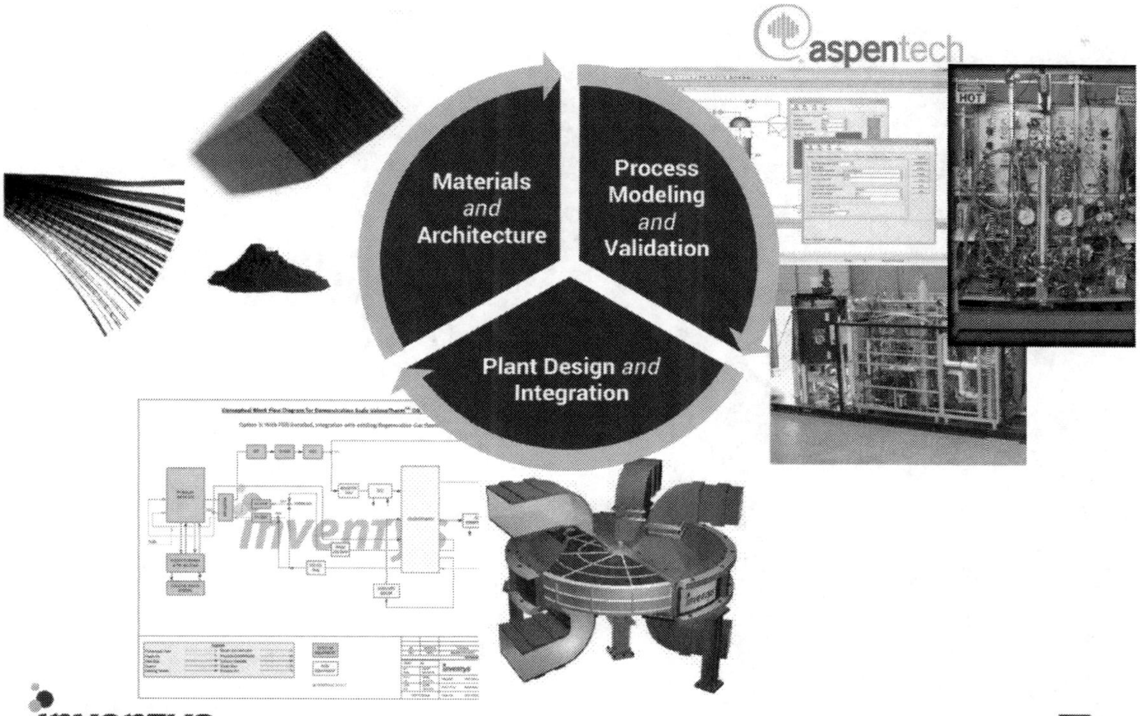


inventys

Confidential 7

# VeloxoTherm™ Process is Fast

*RC-TSA: RAPID CYCLE TEMPERATURE SWING ADSORPTION*

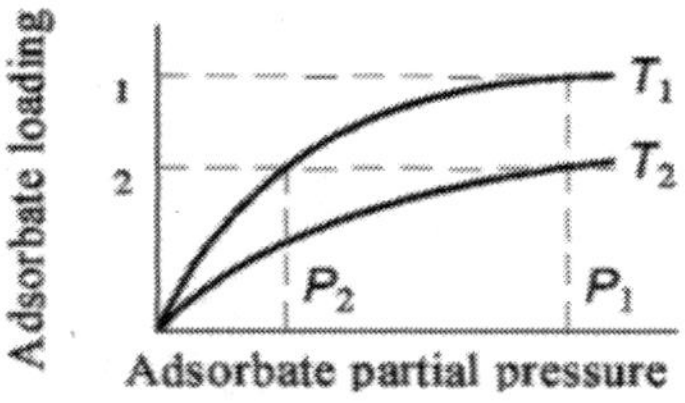


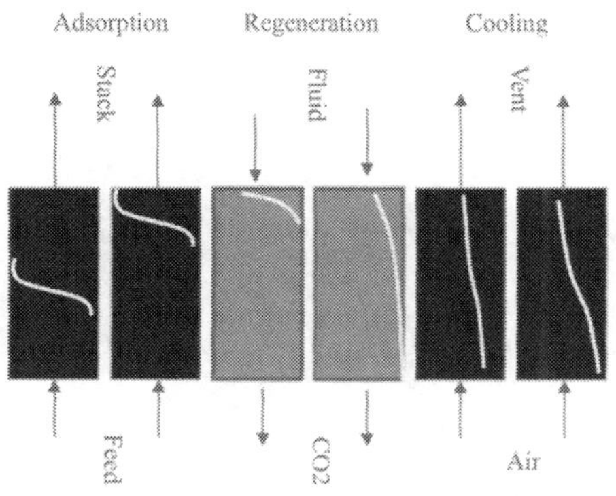


- » **Utilize cyclic gas concentration profile**
- » **Utilize cyclic heat temperature profile**

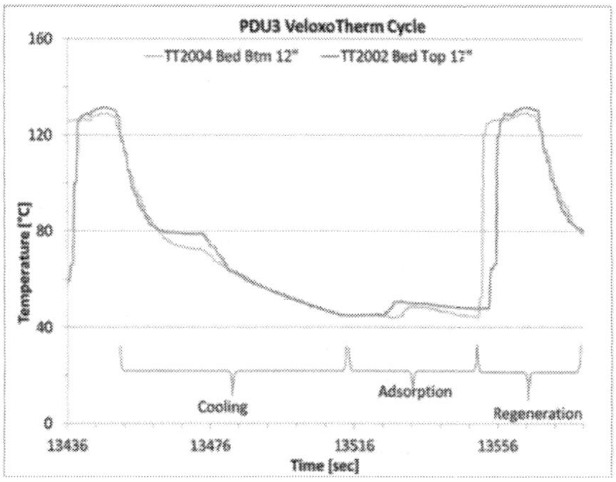


T=60-120 Seconds

inventys

Confidential 9

# Core Technology

Packaged Structured Solid Adsorbent Bed

Structured Adsorbent

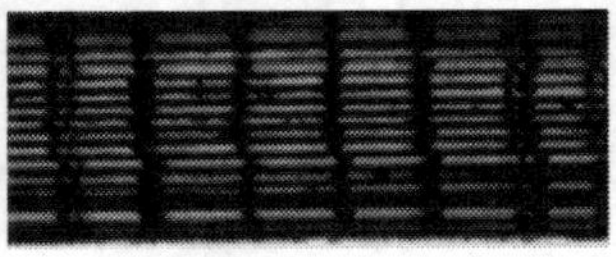

Adsorbent Sheet or Laminate

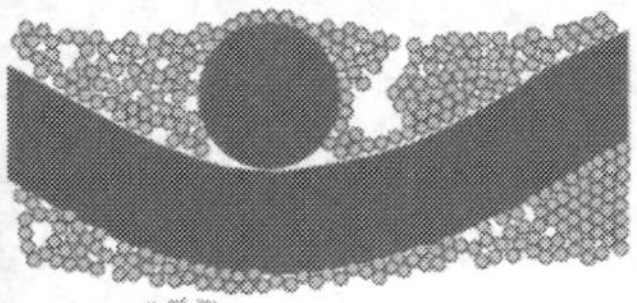

Solid Adsorbents

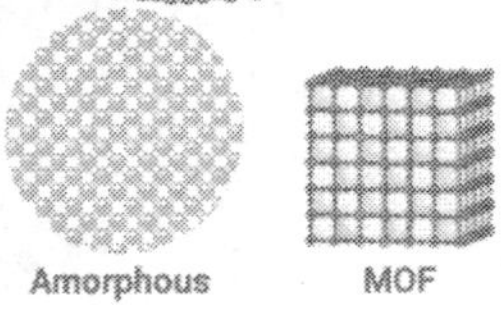


Porous Nano-Material

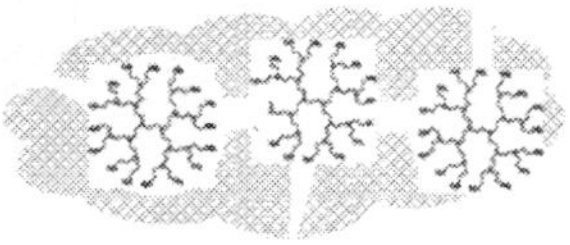

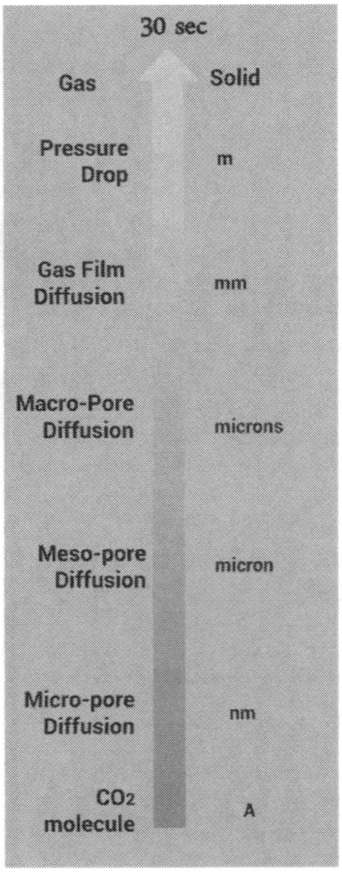


# Structured Adsorbents

*HOW THEY WORK*

» **Inventys' Structured Adsorbents enable critical advantages:**

- **Effective Hydrodynamics** *with very low pressure drop*
- **Fast Transport Phenomena** *with high specific surface area*
- **Higher Durability** *with immobilized adsorbent*
- **Fast Kinetics** *and mass transfer with short diffusion paths*
- **Fast and Controlled Heat Transfer** *with anisotropic properties*
- **No Fluidization** *with immobilized mechanica function*
- **Superior Cyclic Performance** *based on kine separations*
- **High Heat and Mass Transport** *due to short diffusion paths*
- **Independently Engineered Thermal Propert** *of adsorbent*
- **Tailored Kinetic Selectivity** *structures desig*
- **Tailored Void Fraction** *and controlled void r*
- **Tailored Packing Densities** *in adsorbent sh*

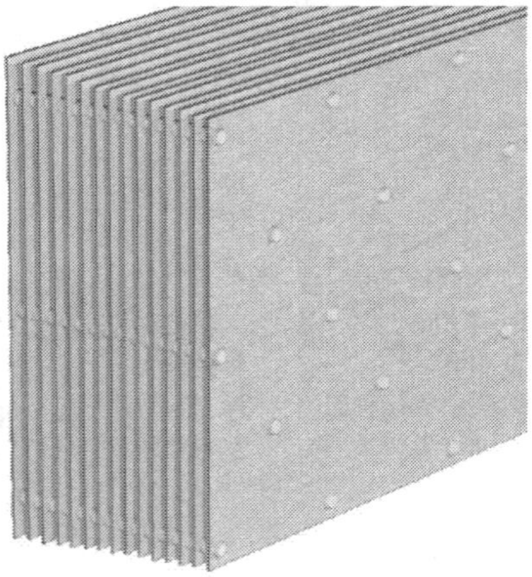

inventys

Confidential 30

# Structured Adsorbents

*HOW THEY WORK*

## » Carbon capture requires:

- *Handling large volumes of flue gas at low pressures.*
- *Low pressure drop at high superficial gas velocities required.*
- *High surface area to be able handle the energy in/out with minimum losses.*

| Sorbent Property | Granular | Structures |
|---|---|---|
| Sorbent Configuration | *Packed Bed* | *Spaced Sheets* |
| Characteristic Dimension(s) | *0.7 mm* | *0.1 mm* |
| Specific Surface Area [$m^2/m^3$] | *5,400* | *10,000* |
| Mass Transfer Coefficient [$s^{-1}$] | *287* | *1629* |
| Superficial Gas Velocity [cm/s] | *280* | *280* |
| Pressure Drop (Pa) | *2,000* | *110* |

inventys



# Structured Adsorbents

*FROM POWDER TO BED*

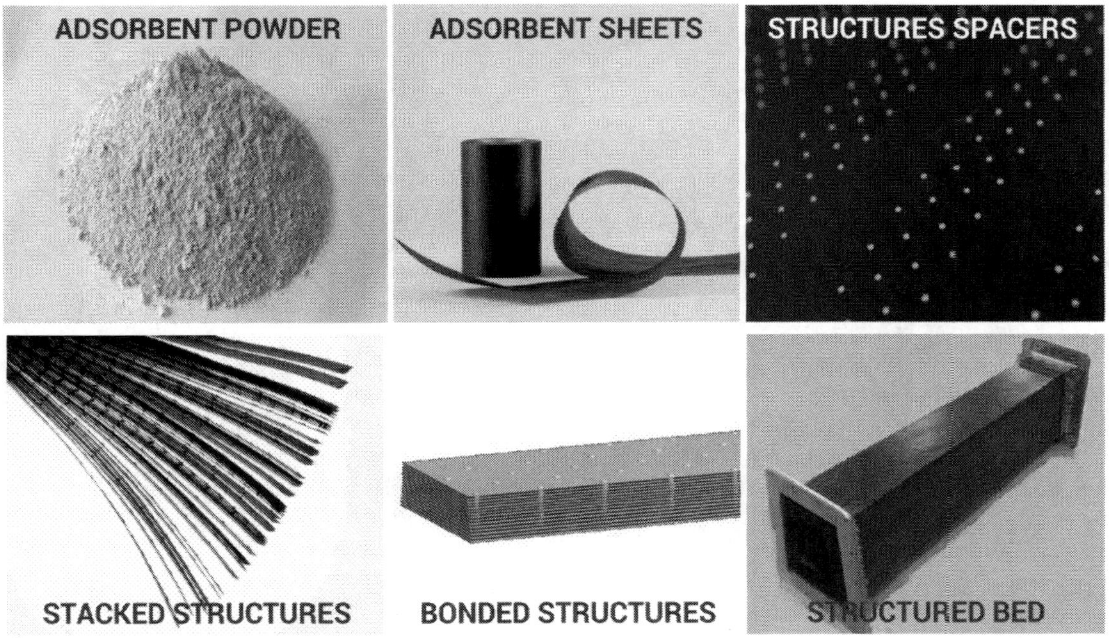


inventys

# Materials & Cycle Development

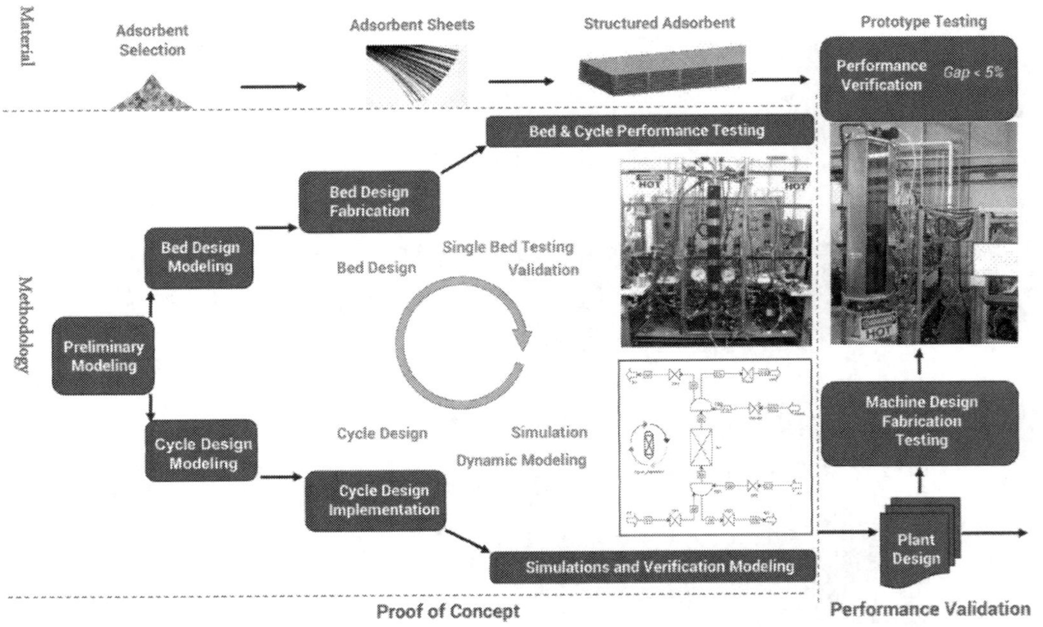


**Modeling Concept Design → Fabrication → Experimentation → Verification → Detailed Design**

inventys

Confidential 13

# Rapid Cycle Temperature Swing Adsorption

*THREE SIMPLE STEPS*

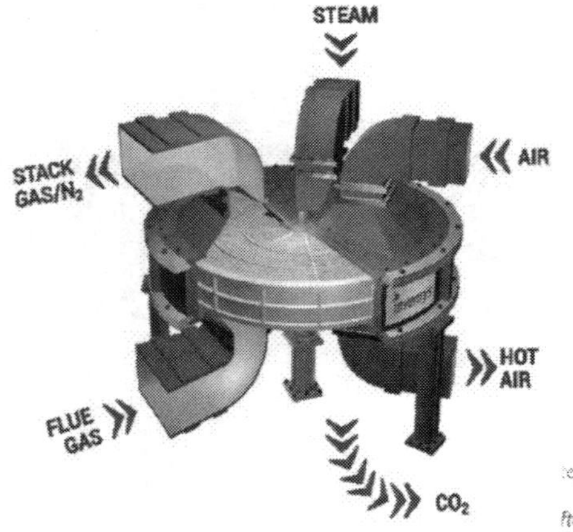


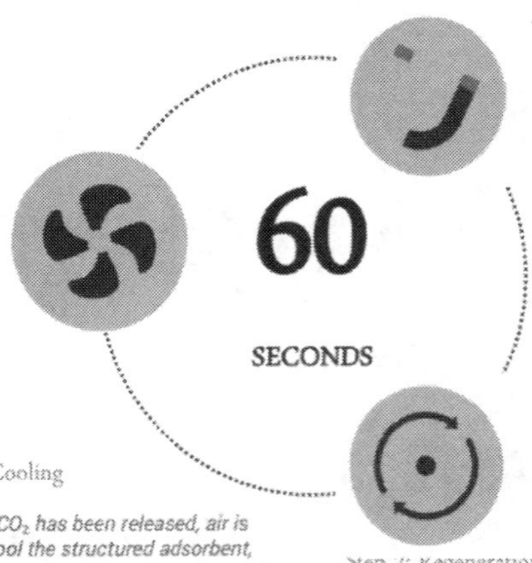


Step 1: Adsorption

*As flue gas passes through the VeloxoTherm™ Adsorbent Structure, $CO_2$ clings to the adsorbent while the other gases pass through.*

Step 2: Regeneration

*After the structured adsorbent becomes saturated with $CO_2$, it is regenerated. Low pressure steam is used to release the $CO_2$ from the adsorbent.*

Step 3: Cooling

*After the $CO_2$ has been released, air is used to cool the structured adsorbent, preparing it for the adsorption step and the process is started over again.*

inventys

Confidential 14

# Complex Cycle in Simple Hardware

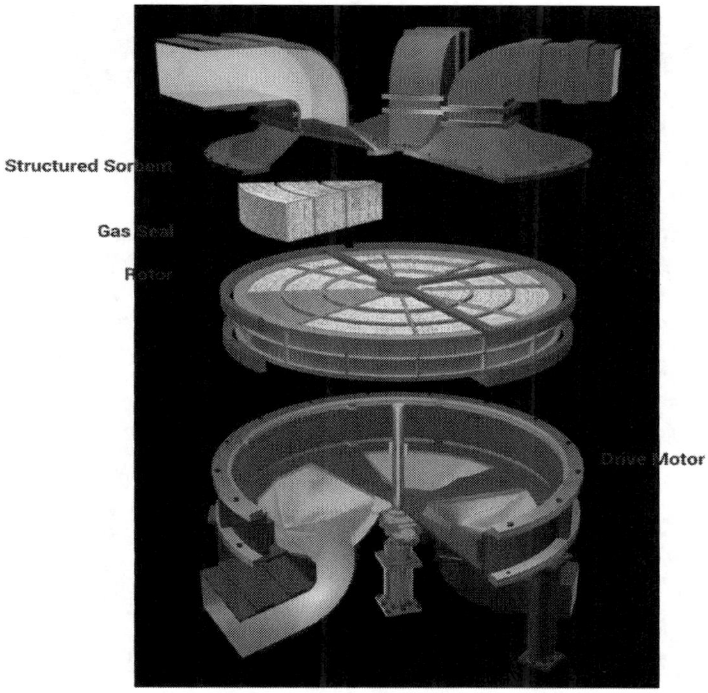


inventys

Confidential 15

# Process Cycle Development

- Adsorbent Thermodynamic Properties (isotherms, δH)
- Structures Transport Properties (δP, D, k, Cp, etc.)
- Dynamic Simulation (ADSIM Custom Modeling)
- Process Cycle Testing (VeloxoTherm Test Station)
- Cycle Design Model Verification (Test = Model)

inventys

Confidential 16

# Dynamic Simulations

» **Proprietary process models; Built on commercial software**
- *TSA Cycle with proprietary custom code*
- *Finite Element Method*
- *Solid & gas phase material energy balance*
- *Momentum mass and heat transfer*
- *Customized for parallel plate geometry*

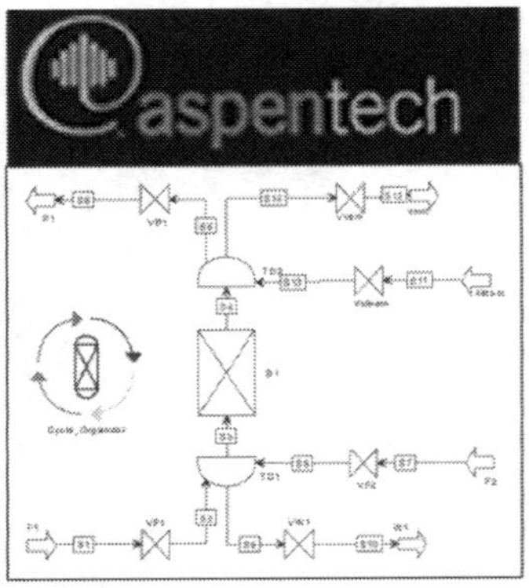


$$-\varepsilon_i E_{zk}\frac{\partial^2 c_k}{\partial z^2} - \varepsilon_i E_{rk}\frac{1}{r}\frac{\partial}{\partial r}\left(r\frac{\partial c_k}{\partial r}\right) + \frac{\partial(v_g c_g)}{\partial z} + \varepsilon_B\frac{\partial c_k}{\partial t} + J_k = 0$$

$$-k_{ga}\varepsilon_i\frac{\partial^2 T_g}{\partial z^2} + C_{vg}v_g\rho_g\frac{\partial T_g}{\partial z} + \varepsilon_B C_{vg}\rho_g\frac{\partial T_g}{\partial t} + P\frac{\partial v_g}{\partial z} + HTCa_p(T_g - T_s) = 0$$

$$-k_{sa}\frac{\partial^2 T_s}{\partial z^2} - k_{sr}\frac{1}{r}\frac{\partial}{\partial r}\left(\frac{1}{r}\frac{\partial T_s}{\partial r}\right) + \rho_s C_{ps}\frac{\partial T_s}{\partial t} + \rho_s\sum_{i=1}^{n}(C_{pai}w_i)\frac{\partial T_s}{\partial t} + \rho_s\sum_{i=1}^{n}\left(\Delta H_i\frac{\partial w_i}{\partial t}\right) - HTCa_p(T_g - T_s) = 0$$

inventys

Confidential 17

# Process Model Verification

*Simulated Adsorbent Bed Gas Compositions vs Test*

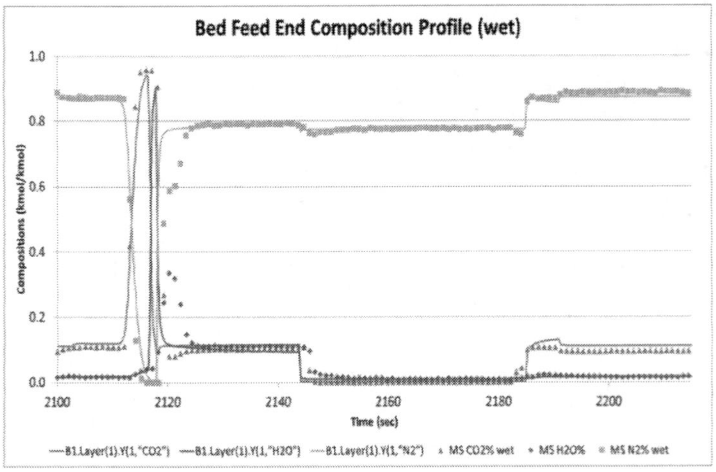


**Dynamic Simulation Finite Element Modeling INPUTS:**
- *Equilibrium Isotherms*
- *Kinetic Diffusion Rates*
- *Mass Transfer Coefficient*
- *Adsorbent Bed Properties*
- *Bed Geometry*
- *Cycle Steps and Durations*

Simulation N2
Simulation CO2
◆ Test CO2
■ Test N2

inventys

Confidential 18

# Cost Analysis Methodology

- *Simulation and Verification Modeling*
- *Identify Key Performance Indicators*
- *Process Flow Diagrams and Major Equipment List*
- *Equipment Cost calculated based on budgetary quotes where possible or APEA*
- *Class IV TEA developed using Factored Estimate (AACE methodology)*

inventys



# Cost Model Definition

*CAPITAL COST ESTIMATION*

*Using factor methodology - based on AACE International 16R-90 with user variations based on cost factors from recently conducted FEED studies.*

- **Purchased Equipment Cost (PEC)**
- **Bare Erected Costs (BEC):** *PEC, supporting facilities, materials, bulks/commodities and direct and indirect labor expense*
- **Total Plant Cost (TPC):** *BEC + engineering/construction management/home office and contractor premiums, allowances and freight, process and project contingencies.*
- **Total Overnight Cost (TOC):** *TPC + pre-production costs, inventory capital, financing costs and other owner costs (where applicable).*

inventys

# Block Flow Diagram

*~10,000 TPD $CO_2$ CAPTURE AND COMPRESSION PLANT*

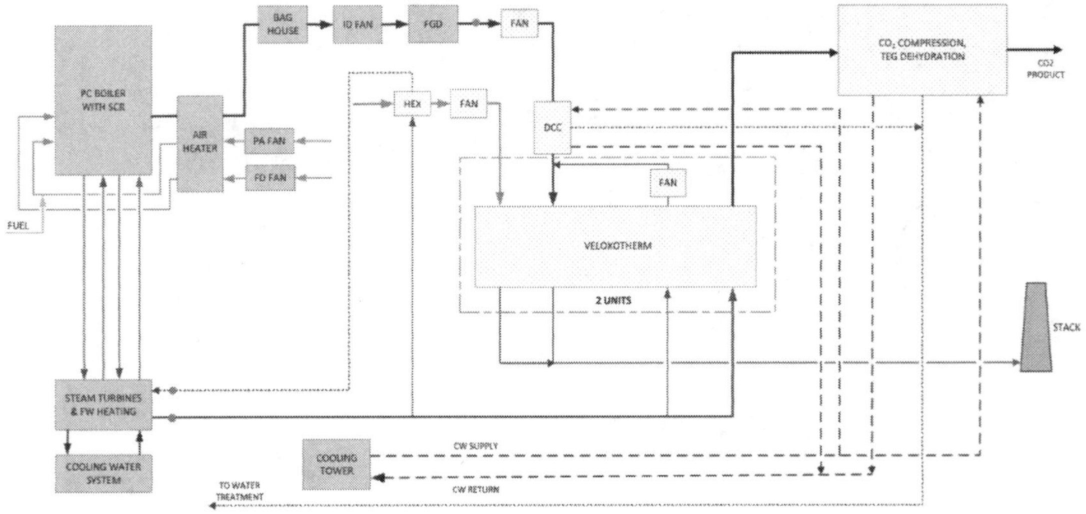


*Simple and compact process resulting in low CAPEX*



# Coal Flue Gas Carbon Capture Cost

## » Process design basis and performance targets

| Design Basis | | |
|---|---|---|
| *Coal Flue Gas $CO_2$ Concentration* | *%v/v dry*<br>*%v/v wet* | *15.0*<br>*12.8* |
| *$CO_2$ Capture Capacity* | *TPD* | *9,583* |
| *$CO_2$ Capture Efficiency/Recovery* | *%* | *90* |
| *$CO_2$ Product Purity* | *%v/v (dry)* | *95* |
| *$CO_2$ Product Pressure* | *psia* | *2215* |
| *Plant Capacity Factor* | *%* | *85* |

| Performance Targets | | |
|---|---|---|
| *Steam Ratio* | *kg/kg* | *1.5:1* |
| *Max Pressure Drop Per Adsorbent Pass* | *kPa* | *10* |
| *Adsorbent Productivity* | *TPD $CO_2/m^3$* | *11* |
| *Auxiliary Flow to Fresh Feed Ratio* | *mol/mol (dry)* | *~1.1:1* |
| *$O_2$ Product Purity* | *%v/v* | *Less than 0.1%* |

# Cost Model Results Summary

| RESULTS ($USD) | | |
|---|---|---|
| *Total Cost of Capture* | *\$/T $CO_2$* | *33* |
| *Total Overnight Costs* | *\$MM* | *288* |
| *Steam Energy Requirement** | *GJ/T $CO_2$* | *4.0* |
| *Auxiliary Heating** | *GJ/T $CO_2$* | *0.5* |
| *Capture Plant BoP Energy Requirement* | *MWe* | *16.3* |
| *Compression Energy Requirement* | *MWe* | *47.4* |
| *Steam Ratio* | *kg/kg $CO_2$* | *1.5:1* |

**Based on enthalpy of steam at take-off conditions – heat integration and pinch analysis to be conducted*

*The scope of this analysis, and associated capital and operating cost estimation, considers all unit operations, equipment and utilities implied below:*

- *Gross power losses due to steam extraction at LP steam turbines*
- *Net parasitic losses due to increase in auxiliary power demand*
- *Raw make-up water (evaporative and stack losses)*

inventys

Confidential 23

# Cost Model Results

*CAPITAL COST COMPONENTS*

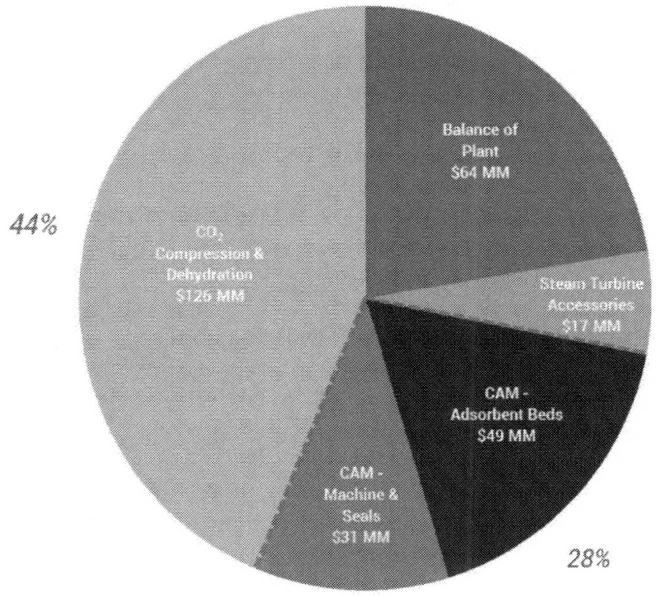


*Total Capital <\$290 MM USD*

inventys

Confidential 24

# Cost Model Results

*COST OF $CO_2$ CAPTURE AND COMPRESSION*

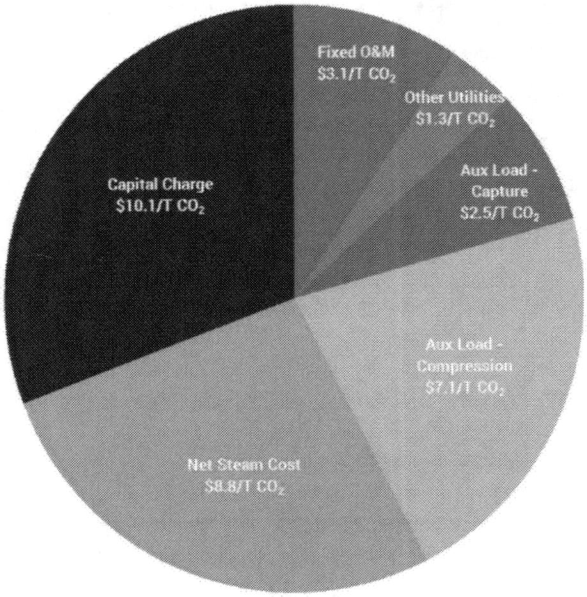


*Total Cost of Capture and Compression <$33USD/MT $CO_2$*

inventys

Confidential 25

# Steam Ratio Sensitivity

*COST OF $CO_2$ CAPTURE AND COMPRESSION*
*STEAM RATIO 1:1 kg/kg $CO_2$*

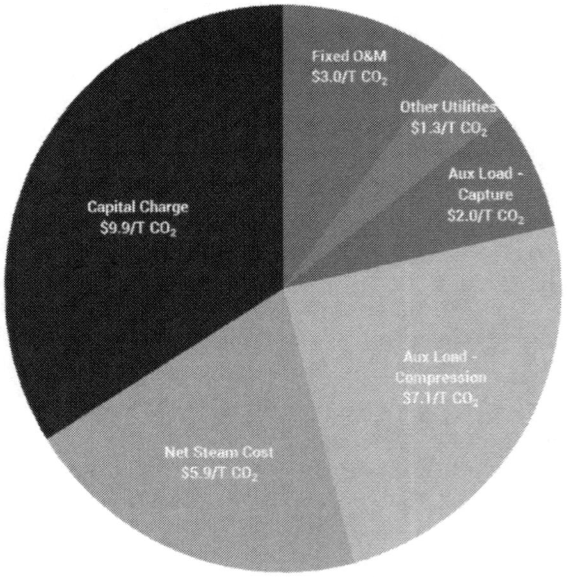


*Total Cost of Capture and Compression <$30USD/MT $CO_2$*

inventys

Confidential 26

# Field Demo Plant

*LOCATED AT PIKES PEAK SOUTH LLOYD THERMAL PROJECT*

inventys



# Product Development Roadmap

| | | | |
|---|---|---|---|
| Mark I | Successfully developed and tested | Detailed TEA completed 3-fold reduction in capital cost | $33 USD* |
| Mark II | Significant reduction in energy cost, utilizing tailored adsorbent | New architecture and adsorbent properties within low-cost embodiment | $25 USD* |
| Mark III | Adsorbent development program | Advances in adsorbent chemistry to utilize waste heat | $15 USD* |

*Lifecycle cost per tonne of Captured and Compressed $CO_2$

inventys

# Commercialization Path

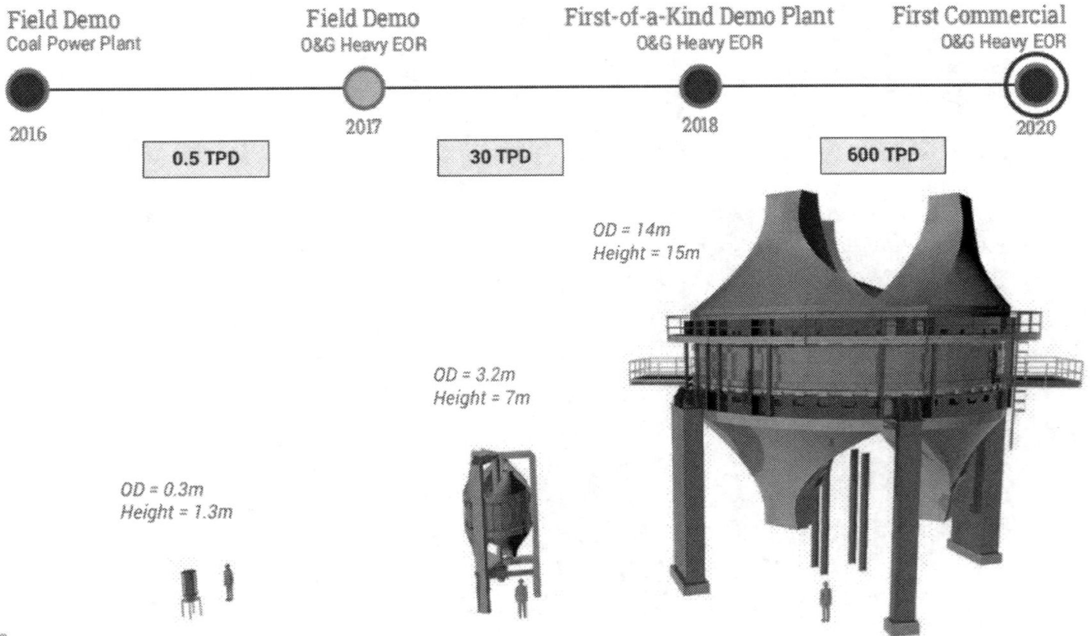


inventys

Confidential 29

# Strong Partners for Execution

*INDUSTRIAL, GOVERNMENT, & FINANCIAL PARTNERSHIPS*

| | |
|---|---|
| Customer Relationships | nrg; Husky Energy; OXY; SUNCOR ENERGY |
| Project Investors | NRC-CNRC; U.S. DEPARTMENT OF ENERGY; energy technologies institute; SUSTAINABLE DEVELOPMENT TECHNOLOGY CANADA™; CCEMC Climate Change & Emissions Management Corporation; Western Economic Diversification Canada |
| Equity Investors | CHRYSALIX ENERGY VENTURE CAPITAL; Chevron; THE RODA GROUP; MITSUI GLOBAL INVESTMENT |

inventys

Confidential 30

# APPENDIX

Soheil Khiavi
Inventor & Co-Founder, Chief Technology Officer,
*Inventys*
*8528 Glenlyon Pkwy #143, Burnaby, BC V5J 0B6*
*Cell: 1 604 818 5218*
soheil.khiavi@inventysinc.com

inventys

EPRI | ELECTRIC POWER RESEARCH INSTITUTE

# Techno-economic analysis of extraction, treatment, and distribution of brines as a part of pressure management operations for $CO_2$ storage in saline aquifers

**Joseph Swisher**, Robert Trautz, Jeffery Preece, Richard Rhudy, Richard Breckinridge, Abhoyjit Bhown

*Electric Power Research Institute*

Michael DiFilippo

*Consultant*

Carbon Management Technology Conference 2017

Houston, TX

July 19, 2017



# Acknowledgement and Disclaimer

NATIONAL ENERGY TECHNOLOGY LABORATORY

Acknowledgment: "This material is based upon work supported by the Department of Energy under Award Number DE-FE0026140."

Disclaimer: "This presentation was prepared as an account of work sponsored by an agency of the United States Government. Neither the United States Government nor any agency thereof, nor any of their employees, makes any warranty, express or implied, or assumes any legal liability or responsibility for the accuracy, completeness, or usefulness of any information, apparatus, product, or process disclosed, or represents that its use would not infringe privately owned rights. Reference herein to any specific commercial product, process, or service by trade name, trademark, manufacturer, or otherwise does not necessarily constitute or imply its endorsement, recommendation, or favoring by the United States Government or any agency thereof. The views and opinions of authors expressed herein do not necessarily state or reflect those of the United States Government or any agency thereof."





EPRI | ELECTRIC POWER RESEARCH INSTITUTE

# Brine extraction may be an important component of a comprehensive pressure management plan

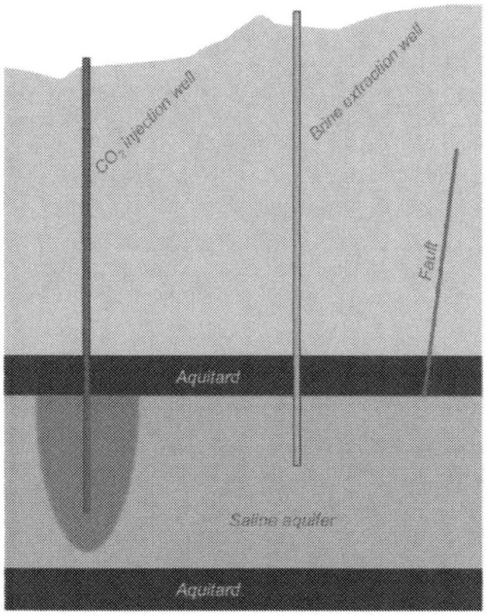


Reasons to extract brine as part of storage reservoir management

- Protect confining units from fracturing
- Steer the plume of CO2
- Improve injectivity or capacity
- Protect faults





EPRI | ELECTRIC POWER RESEARCH INSTITUTE

# We wanted to study what it would take to treat different brines at the quantities potentially necessary for long-term operation

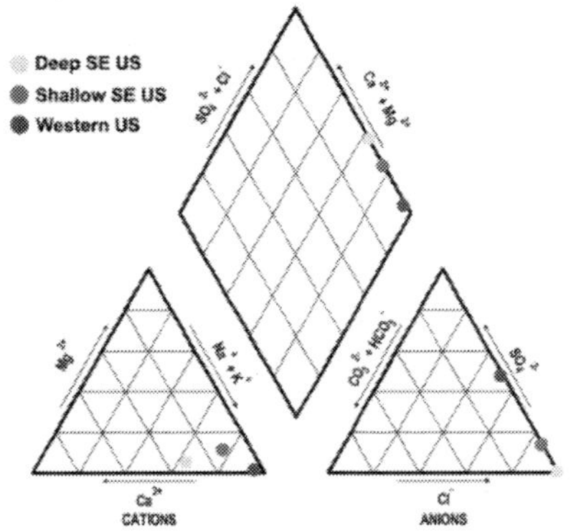


| Brine | TDS (mg/L) |
|---|---|
| Shallow Southeastern US | 29,118 |
| Deep Southeastern US | 165,883 |
| Western US | 30,553 |

- Our analysis was based on storing about 3,600,000 metric tons $CO_2$/year (41% of a 1000 MWe coal station's emissions)
- Three brines from formations of different geologic ages were selected as targets for treatment.
- We considered three scenarios for each brine: brine would be extracted at a rate of 1-to-1, 1-to-2, or 1-to-5 with respect to $CO_2$ (mass basis).





EPRI | ELECTRIC POWER RESEARCH INSTITUTE

# We selected several commercial or near-commercial desalination technologies to develop a menu of process options

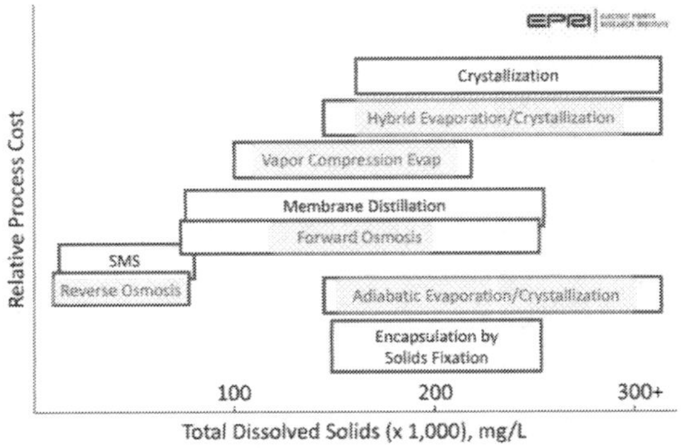


| Process | Code | Uses electricity | Uses natural gas |
|---|---|---|---|
| Reverse Osmosis | RO | Yes | No |
| Vapor Compression Evaporation | VCE | Yes | No |
| Forward Osmosis | FO | Yes | Yes |
| Hybrid evaporation/crystallization | HEC | Yes | Yes |
| Adiabatic evaporation/crystalization | AEC | Yes | Yes |







# Techno-economic analysis of extracting, treating, and distributing brine

We determined the power consumption of providing artificial lift to the brine.

The project was assumed to operate for 30 years.

*Well drawdowns under different permeability assumptions*

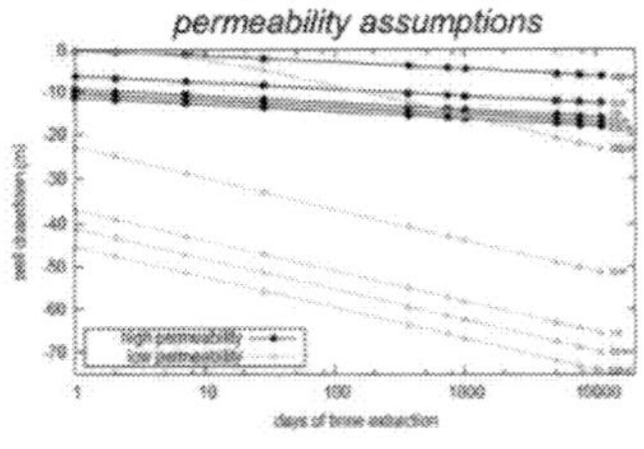

Different process combinations were considered. The final selection were all capable of recovering >88% of the brine as distillate.

*Final process combinations considered in this study*

| | | |
|---|---|---|
| RO → VCE | VCE | VCE → HEC |
| FO → HEC | FO → AEC | VCE → AEC |

The distillate was distributed via a 24 km pipeline to a nominal municipal customer.

We also considered different energy price scenarios, reflecting potential higher prices in a carbon-constrained market.

| Scenario | Low energy price | High energy price |
|---|---|---|
| Electricity | $0.089/kWh | $0.161/kWh |
| Natural gas | $4.00/MMBtu | $8.00/MMBtu |

# The treatment process is the dominant cost in all scenarios

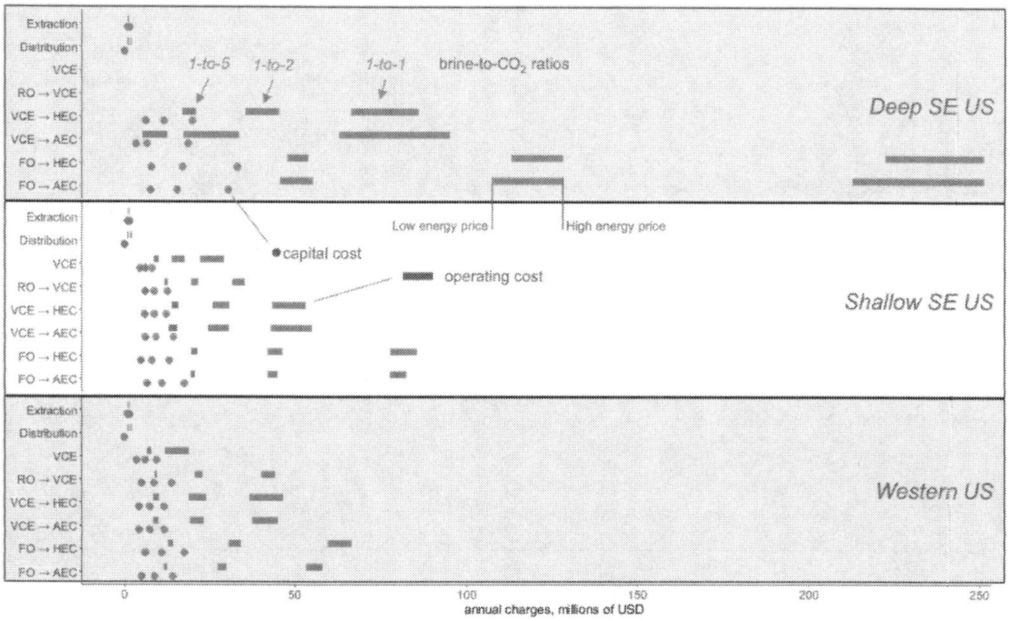






EPRI | ELECTRIC POWER RESEARCH INSTITUTE

# High hardness in a brine contributes to a high annual cost of operation for pre-treatment

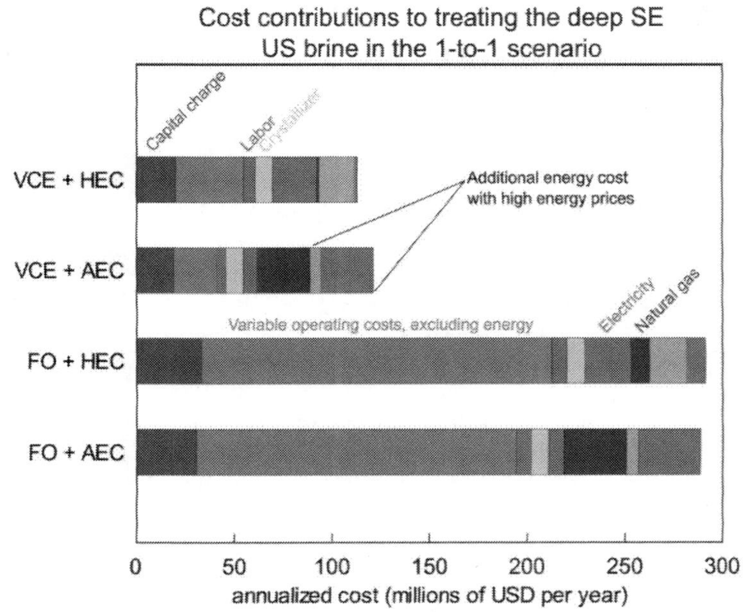






EPRI | ELECTRIC POWER RESEARCH INSTITUTE

## The overall process of extraction, treatment, and distribution can add significant costs to storing $CO_2$

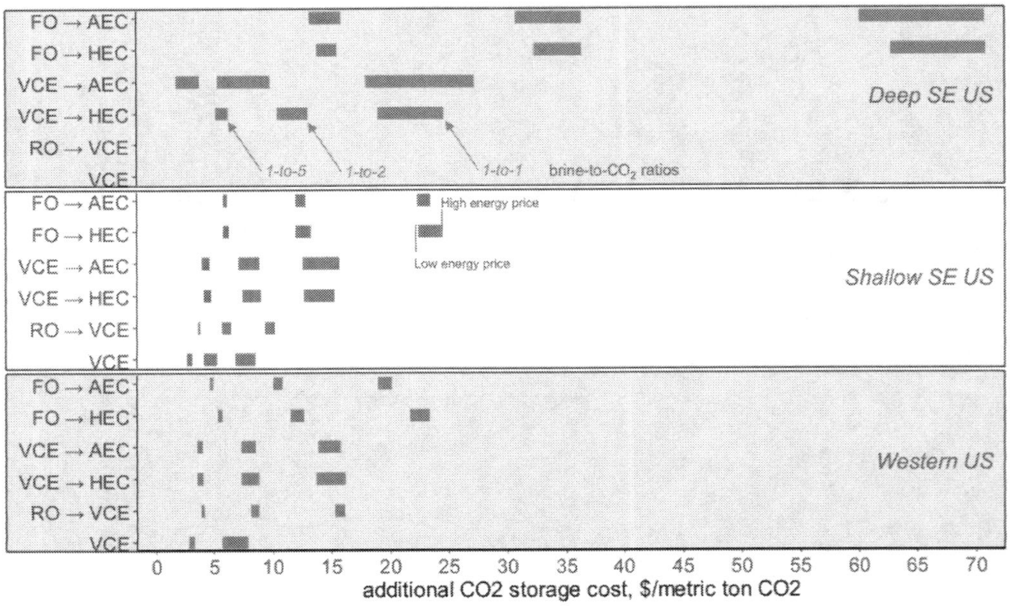






EPRI | ELECTRIC POWER RESEARCH INSTITUTE

## Conclusion

Brine extraction and treatment can add significant cost to a $CO_2$ storage project.

Understanding the brine chemistry and the potential need for extraction is an important part of evaluating any potential storage reservoir.





EPRI | ELECTRIC POWER RESEARCH INSTITUTE

# Together…Shaping the Future of Electricity

11



EPRI | ELECTRIC POWER RESEARCH INSTITUTE

# SHOCKWAVECO2

*Ultra low-cost skid-based shockwave technology to turn waste CO2 into carbon products*

*Vertical prototype of CO2 system in TX*

*Shockwave-tech clean coal plant in China*

# Simple Instruction Manual

1. Connect Shockwave CO2 system to CO2-rich stack gas
2. Input excess trona or Nahcolite brine, carbonate, or hydroxide liquid
3. Push the “ON” button (open compressed air valve + turn on liquid reagent pump)
4. When pH of liquid in tank drops to 12.0 or 9.0 respectively, pump carbonate or bicarbonate product for further processing (drying, packaging, or heating)

*Make <u>more product</u> using <u>excess or wasted resource</u> for little cost!*

2

# 5 System Requirements

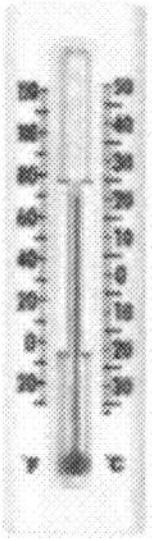

1. $CO_2$ stack gas conditioned to 75°-85°F (23°-30°C)

2. Chemical liquids to absorb $CO_2$

3. 110v electrical outlet

4. Holding tank for processed $CO_2$ product

5. Commercial air compressor or stack gas compressor

3

# What is Shockwave Technology?

Efficient chemical mixer and contactor for mass/energy transfer using compressed fluid (air, steam, or stack gas)

Benefits:

- High speed liquid vaporizer
- Collision energy transfer
- Robust mixer in forced reaction zone
- Controllable gas expansion

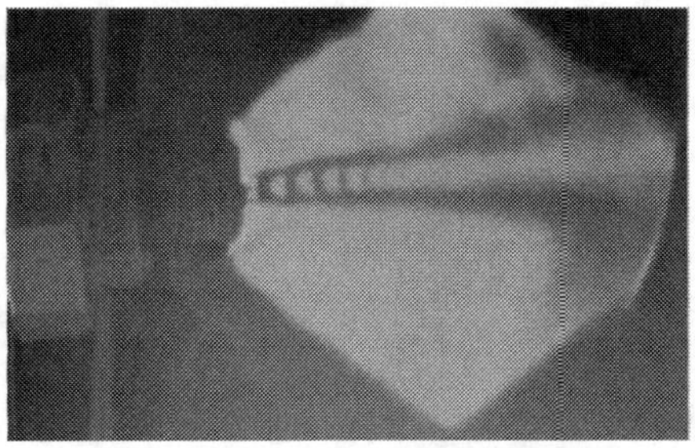

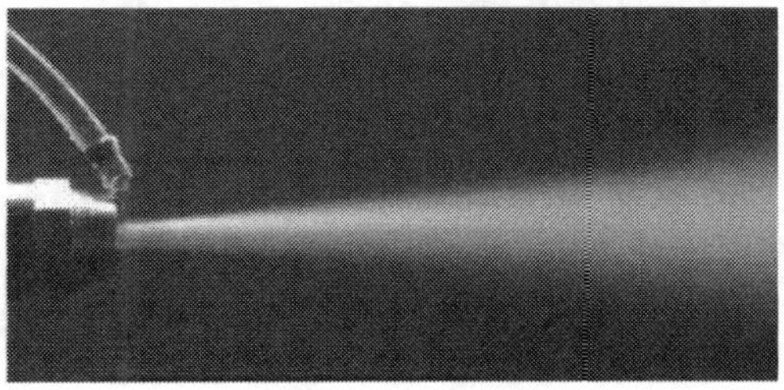

4

# How is Shockwave Used?

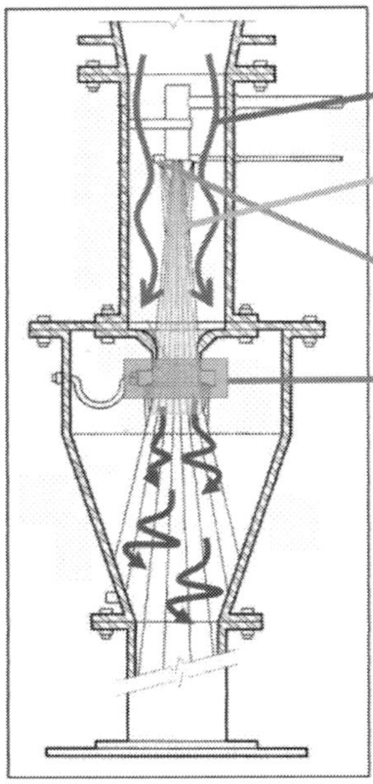

- Stack gas flows through pipe
- Shockwave turns into high speed air blender and collides and mixes entire stack gas flow
- Inject chemicals into shock mixer
- Chemicals are vaporized and forced to mix with entire stack gas using shockwave inside confined shocktube (primary mixer)
- Gas expands in designed mixing pattern in production tanks (secondary mixer)

5

# Well-Published Technology

6

# Why is shockwave "ultra low-cost"?

CAPEX:

- Shocktube systems are simple and compact
- Common materials with no pressure buildup or heat input
- Modular configuration with add-on components
- Skid is the size of a pickup truck

*Shockwave system is so easy and convenient, it captures CO2 and produces product with no special equipment!*

OPEX:

- 100psi compressed air or stack gas at 3.5#/min for shock nozzle
- Shockwave needs to be 10% of the mass of the stack gas to react the entire stack gas
- 110v liquid chemical pump
- Each shocktube uses $5/hr of total electricity
- No internal moving parts, minimal caking, little downtime

7

# What carbon products does the system make?

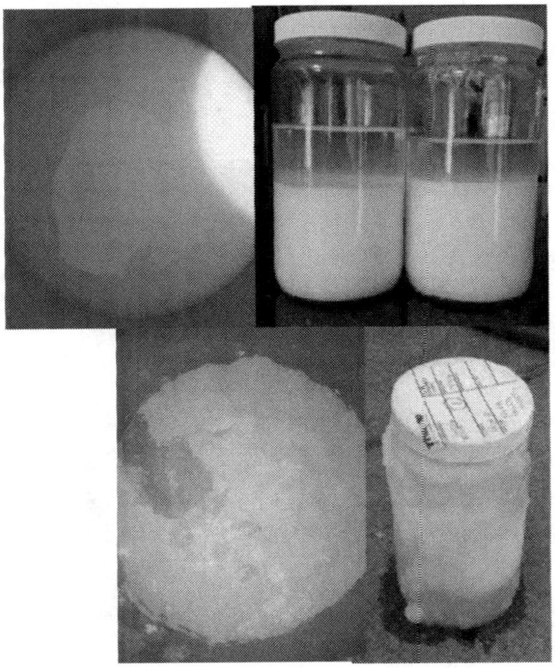

If using sodium hydroxide, then sodium carbonate and/or bicarbonate is produced

If using sodium carbonate, then sodium bicarbonate is produced

If using other chemicals (i.e. K, Mg, Ca, amine, etc.), then K2CO3, KHCO3, MgCO3, Mg(HCO3)2, etc. is produced under correct conditions

If heating bicarbonate, then 99.9% CO2 is produced for process application

8

# Process Flow for Bicarb

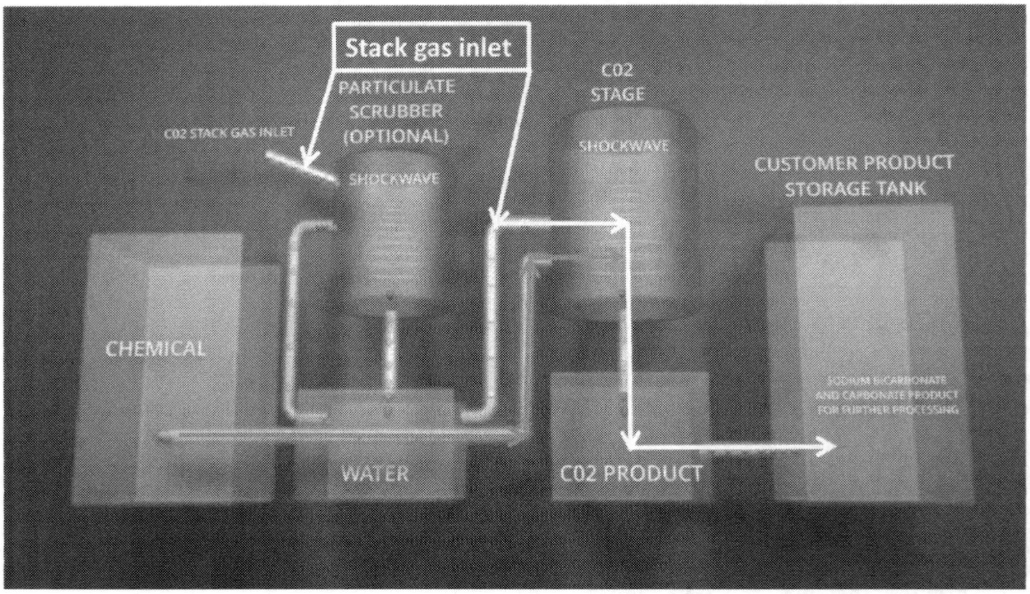


9

# Process Flow for Sustainable CO2

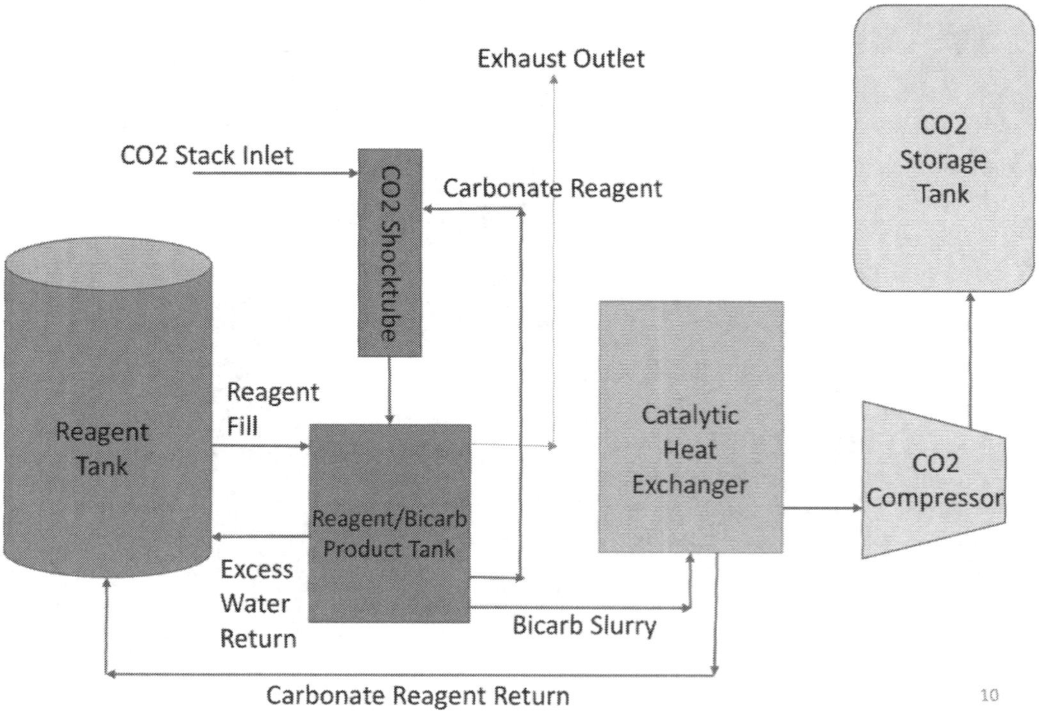


10

# Base Model System

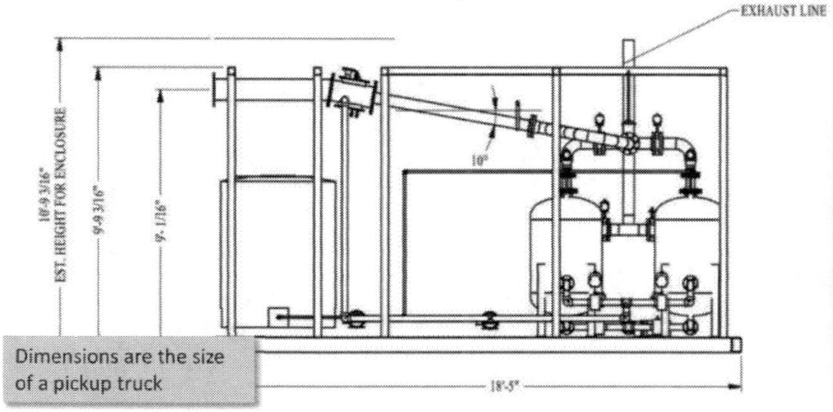


Dimensions are the size of a pickup truck

- 1x proprietary CO2 processing shocktube
- 2x custom alternating production tanks
- 450 gallon onboard reagent holding tank
- Pumps, meters, gauges, and gas analyzers
- External connections for "plug 'n play"
- Self cleaning cycle
- Climate controlled enclosure
- Programmed logic controls, basic automation, online data monitoring

11

# Add-on Configurations

Available Now

- Second onboard CO2 shocktube – doubles output volume
- Particulate scrubber – removes 99.9% PM material

In Development

- Stack gas compressor – radically increases efficiency instead of using compressed air
- CO2 release heat exchanger – produces pure CO2 from bicarbonate for process applications
- CO2 release catalyst – accelerates desorption of CO2 from bicarbonate, halving the heat requirement to produce concentrated CO2
- CO2 compressor – liquefies CO2 for process applications

Bicarbonate cyclone and rotary dryer

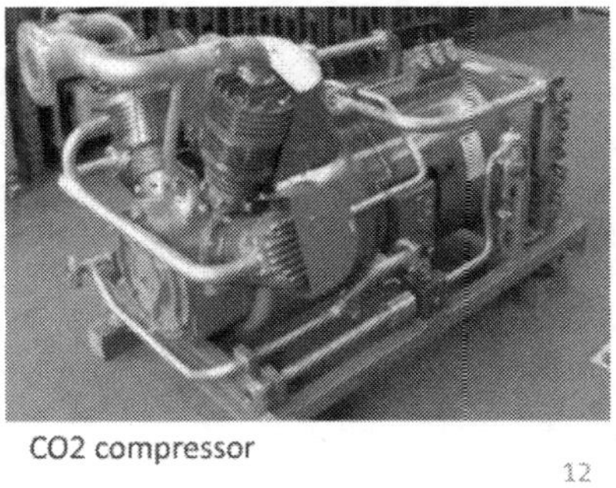

CO2 compressor

12

# Stack Gas Compressor Upgrade

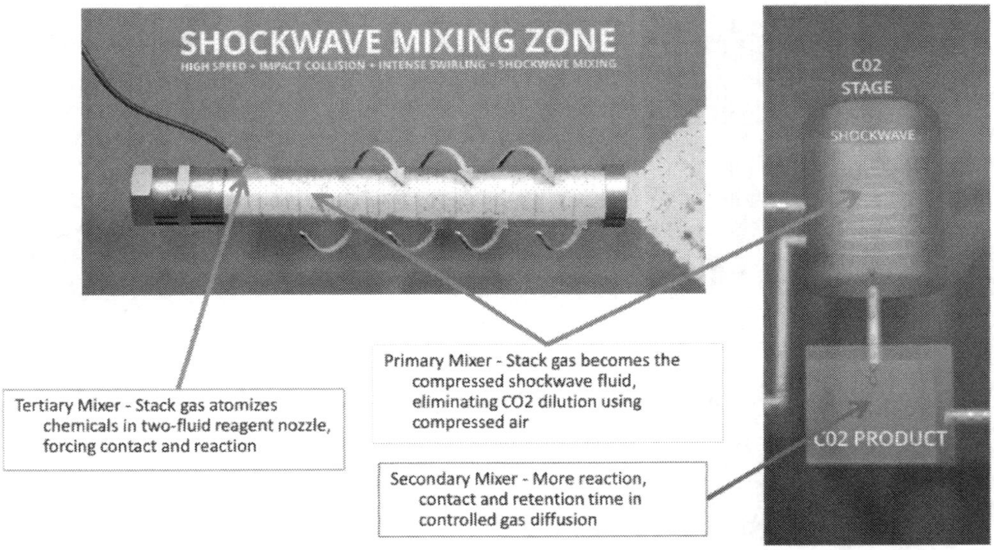


*3 forced reaction points and 2 mixers for 1 energy input = very high efficiency*

13

# Catalyst Upgrade

- Working with a highly recognized research team with a proven catalyst that accelerates desorption of CO2
- Catalyst is highly stable and has infinite reusability
- Can improve the release of CO2 from bicarbonate up to 2,500% at only 150°F with no moisture buildup
    - With a heat exchanger, this catalyst produces concentrated CO2 from bicarbonate at negligible cost

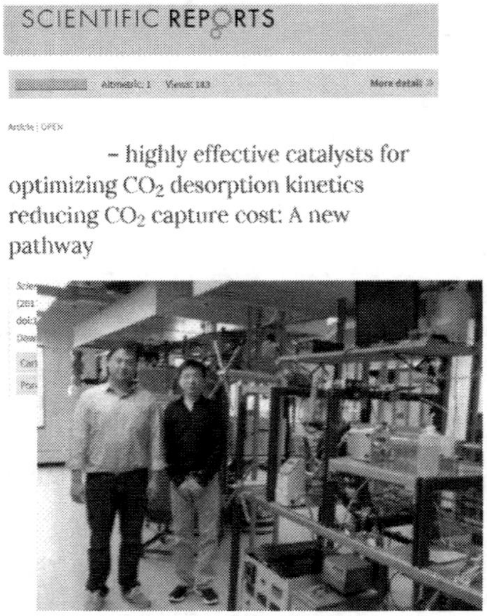


Brian Tang standing next to catalyst inventor at his research lab

14

# Potassium Reagent Substitution

- Using potassium carbonate will drastically improve $CO_2$ capture efficiency, enabling stricter $CO_2$ scrubbing applications
  - Better solubility
  - Better absorption and reaction rates
  - Higher heat tolerance

- Able to capture and produce more and cheaper $CO_2$ at scale, especially when paired with catalyst
- Ideal applications include the production of $CO_2$ from pre-combustion $H_2$ and natural gas scrubbing

15

# Economics and Efficiency Using Sodium

Each shocktube produces 3 to 5 tons/day (or more) of product depending on:

- Flow rate and (assisted) pressure of stack gas
- Chemical concentration of reagent liquid
- Richness of $CO_2$ in stack gas stream

- Targeting 70% $CO_2$ capture efficiency
- Projected bicarbonate production cost ≈$24 - $40/ton
- Projected 99.9% $CO_2$ production using catalyst and heat exchanger ≈$48-80/ton
- Each shocktube only requires $5/hr of electricity

Potential to increase production to 5 – 10 tons/day ($12-24/ton) of bicarbonate with mechanical improvements and add-ons. $CO_2$ production would then be $24-48/ton.

16

# Proposed Cost Structure to Customers

Base model

- $150,000 upfront payment
- 3 year lease at $3,000/month
- 5 year lease at $2,500/month

Add-ons

- Second CO2 shocktube with tanks for $50,000 + $1,500/month
- PM scrubber for $35,000 + $1,000/month

Optional Maintenance

- $2,000/month maintenance plan for 24/7 call support and 1 service visit per month

*Over 20 customers have already said, "YES, they will buy if we can supply CO2." In fact, no customer has ever said "NO," even if our cost to produce CO2 is similar or slightly higher than their current purchase price.*

*Price and terms are flexible for the right customer and partner to help develop our system to maximize bicarbonate/CO2 production.*

17

# Scalability

Shockwave technology on 35MW coal-fired boiler in China

1MWe system in Dallas, TX

18

# Highly Motivated CO2 Customers

Food (making, preservation, preparation) ≈\$500M in U.S.

Refrigerated Transportation >\$20B in U.S.

Greenhouse Agriculture <\$20M in U.S.

Industrial Refrigeration <\$10M in U.S.

Enhanced Oil Recovery >\$40B in U.S.

CO2 Blast Cleaning <\$10M in U.S.

19

# Technology Summary

- We are NOT the highest performance, high cost system capturing CO2 from large-scale sources trying to meet emission regulations
- We ARE the lowest cost, good performance, most affordable and practical system in compact, portable footprint for commercial and industrial customers
- We CAN improve performance and product yield through add-ons, using multiple systems, and further development
- If you use CO2 and want to recycle your CO2 to save money, you should want to try us out.

20

# Ideal Customer *and* Partner Characteristics

## Initial Customer

- Has appetite for adopting new technology
- Wants and monetizes carbonate or bicarbonate product
- Has insight and expertise in producing and processing bicarbonate

TRONOX
CHURCH & DWIGHT CO., INC.
SOLVAY
FMC
NATURAL SODA
TATA CHEMICALS

## Partner

- Understands bicarbonate as an intermediate to produce CO2
- Can help develop efficient bicarbonate refining and CO2 release processes
- Sees the vision and potential market for a low-cost portable CO2 production system

AIR LIQUIDE
Linde
PRAXAIR
MATHESON

21

# Contact Info

*Live system demo available starting summer 2017. For more info, please call or email.*

SHOCKWAVECO2


**Brian Tang, CEO**
Brian.tang@shockwaveco2.com
**917-497-8342**
**411 McKinney Pkwy**
**McKinney, TX 75071**


22

## Disclaimer

This presentation was prepared for the 2017 Carbon Management Technologies Conference. The Presentation contains forward-looking statements within the meaning of the Securities Act and the Securities Exchange Act of 1934, as amended. These forward-looking statements may include, without limitation, statements concerning the operating plans and objectives of the Company, statements concerning the actions of governmental and regulatory authorities, competition to the Company, the market for the services of the kind to be provided by the Company, commercial arrangements the Company intends to enter into or projections of income and expenses of the Company. A number of factors that could cause actual results to differ materially from the assumptions, results, actions or outcomes anticipated in the forward-looking statements contained in this Presentation. Consequently, there can be no assurance that any results, actions or outcomes anticipated in any such forward-looking statement will be realized.

 

## Who is Shenhua?

niE

- **State-owned enterprise**
  - Founded 1995 ... Fortune 200 last 7 years
  - Vertically integrated energy company ...
    ... mining, power, chemicals, transport
    ... #1 in coal in China (10+% share)
    ... Top 5 in power (12+% share, 83 GW total)

中国神华
CHINA SHENHUA

- **Clean energy roadmap ... "1245" strategy**
  - Ultra low emissions (ULE) technologies
  - Water and reuse treatment
  - Green mining
  - Renewables (6+ GW wind)

## Shenhua assets map

niE

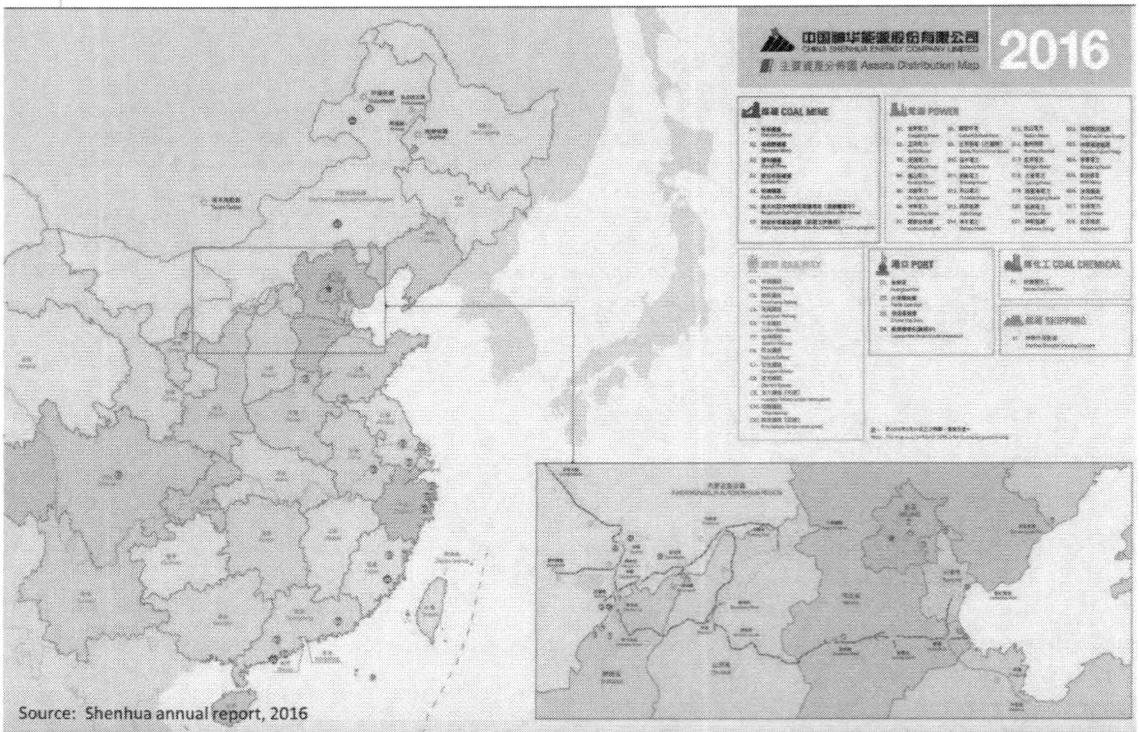


Source: Shenhua annual report, 2016

## Who is NICE?

nice

- **Background**
  - National Institute of Clean and low-carbon Energy
  - Founded in 2009 ... corporate R&D lab for Shenhua group
  - Mission ... to become a world class R&D institute supporting Shenhua's transition to clean and low carbon energy supplier
  - 490+ researchers
  - Sites: Beijing, China; Mountain View, CA; Schwäbisch Hall, Germany
- **Mission driven R&D ... platforms**

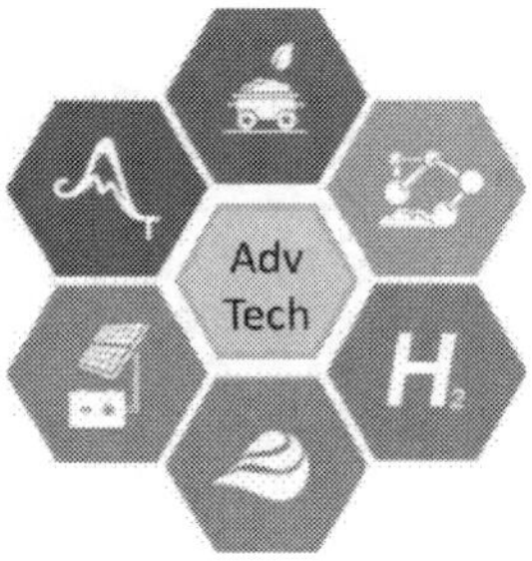


**Advanced Technologies**
- emissions mitigation
- engineering innovation
- strategic growth

 

# $CO_2$ management landscape in China

## Policy context

nie

**COP15 Copenhagen (2009)**
- 40-45% reduction in emissions vs 2005 by 2020

**COP21 Paris commitments (2015)**
- Peak emissions by 2030
- 60-65% reduction in emissions vs 2005 by 2030
- Increase non-fossil share to 20% by 2030

2009 | 2011-2015 | 2016-2020 | 2021-2025 | 2025-2030

**12th Five year plan (2011-2015)**
Energy intensity (per unit GDP): -16% vs 2010
Actual = 18.2%
Carbon intensity (per unit GDP): -17% vs 2010
Actual = 20%
Share of non-fossil energy: 11.4%
Actual = 12%
"Gradually establish a carbon trade market"

**13th Five year plan (2016-2020)**
Energy intensity: -15% vs 2015
Carbon intensity: -18% vs 2015
Share of non-fossil energy: 15%
National trading market: July 2017

 7

## Emissions trading markets – Phase 1

nie

*2013* *2014*

1 | 2 3 4 5 | 6 | 7

**1. Shenzhen**
July 18, 2013
**30 MM** $tCO_2$/yr
635 companies

**2. Shanghai**
Nov 26, 2013
**150 MM** $tCO_2$/yr
200 companies

**3. Beijing**
Nov 28, 2013
**70 MM** $tCO_2$/yr
490 companies

**4. Guangdong**
Dec 19, 2013
**350 MM** $tCO_2$/yr
202 companies

**5. Tianjin**
Dec 26, 2013
**150 MM** $tCO_2$/yr
114 companies

**6. Hubei**
Apr 4, 2014
**120 MM** $tCO_2$/yr
138 companies

**7. Chongqing**
Jun 19, 2014
**100 MM** $tCO_2$/yr
242 companies

**970 MM $tCO_2$/yr**

*Power*
*O&G*
*Steel*
*Cement*
*Chemical*

- Phase 2: National market (2017-2020)
- Phase 3: National market with reductions (2020-2030)
- Phase 4: International linkages (2030+)

X. Zhao et al. Ren Sust Energy Rev. 59, 1229 (2016)

 8

## Emissions trading markets – Historical activity

Shenzhen (~ 30 RMB, 2017)

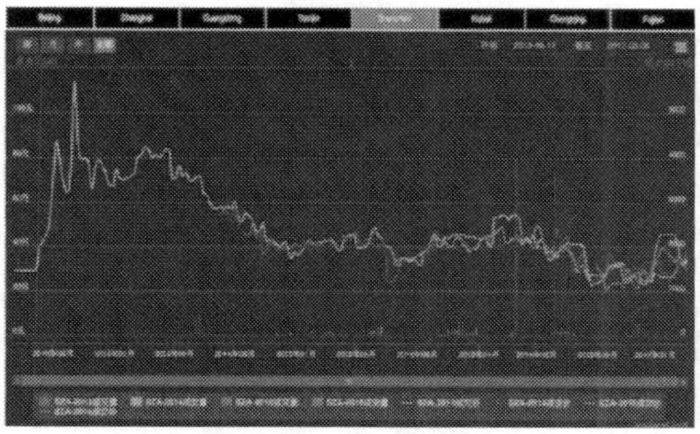

Beijing (~ 50 RMB, 2017)

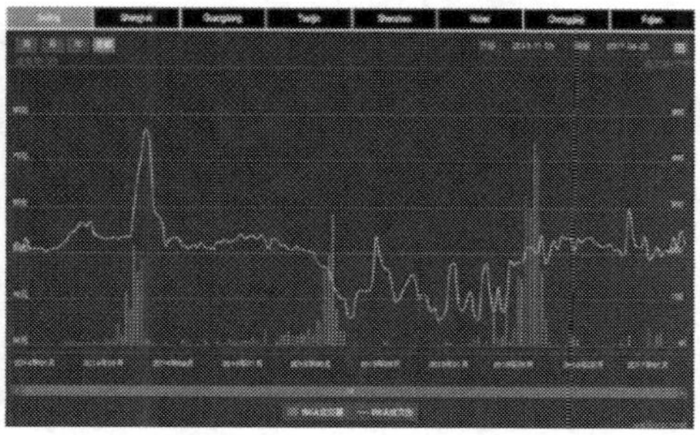

Shanghai (~ 40 RMB, 2017)

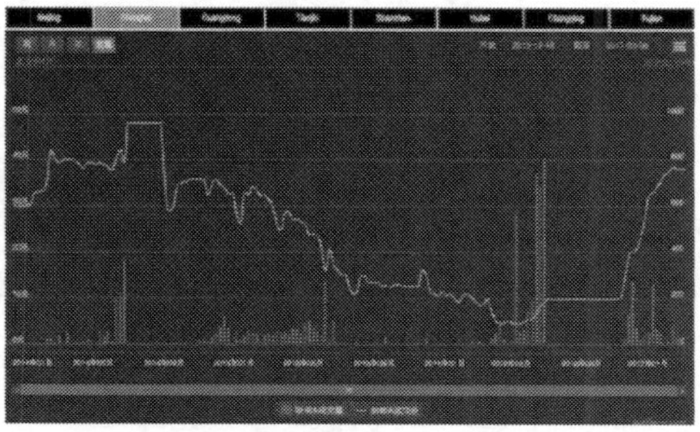

http://www.tanpaifang.com/tanhangqing/

By end of October 2014,
- Cumulative trading volume: 13.8 MM ton
  6M t in Hubei; 2M t in Beijing
  >1 M t in Shanghai, Guangdong, Tianjin
- Total turnover value = 500 MM RMB

However, lessons remain to be learned about:
- Allocations ... legacy vs market
- Market operations ... liquidity, information

X. Zhao et al. Ren Sust Energy Rev. 59, 1229 (2016)

 

# Shenhua $CO_2$ management strategy

## Shenhua $CO_2$ sources

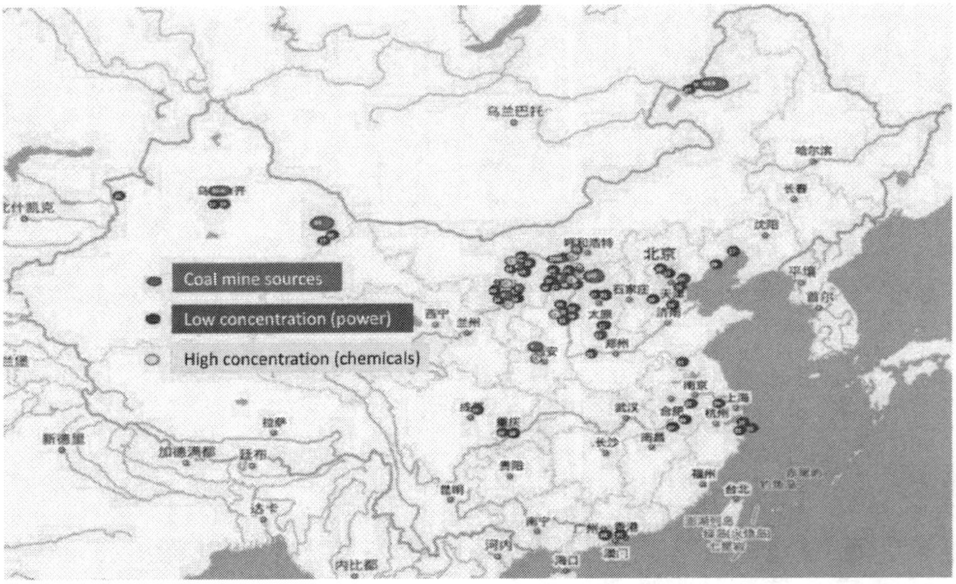


 

## Shenhua $CO_2$ strategy – Key elements

**1. Standards, controls and trading**
- Assessment, monitoring, and controls
- Participate in carbon trading markets

**2. Efficiency projects on established plant/capex**
- Close excess capacity
- Upgrade systems

**3. Renewables**
- Currently 6+ GW wind

**4. Advanced R&D and international cooperation**

 

## Carbon management within Shenhua Group

### Shenhua Group

Environmental Protection — Science & Technology — NICE

**NICE**
Advanced RD&D to reach 2030 targets

Provide Shenhua with real technical options to affordably reduce $CO_2$ emissions by 65%

**Power**
- Capture demo 1 (solvent, 2018)
- Capture demo 2 (solvent, 2020)

**Chemicals**
- Ordos storage project monitoring (Injection complete; site transferred to NICE)
- $CO_2$ capture and sales (ongoing)

**Mining**
- CMM (ongoing)

**Trading**
- Carbon Markets (2017)

Group-wide $CO_2$ reduction efforts (2016-2020)
- Closure of excess capacity (mines, low efficiency power generation, coking)
- Energy efficiency (boilers, machinery, mining)
- Renewables generation (wind, solar)



## Shenhua $CO_2$ footprint & reductions

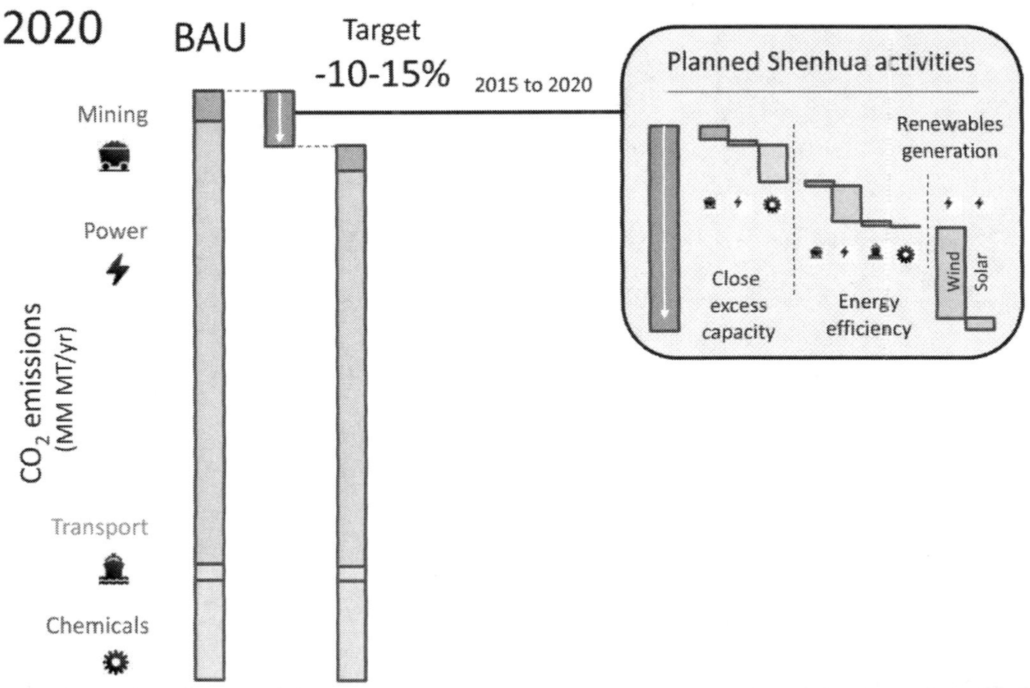

## Shenhua $CO_2$ footprint & reductions

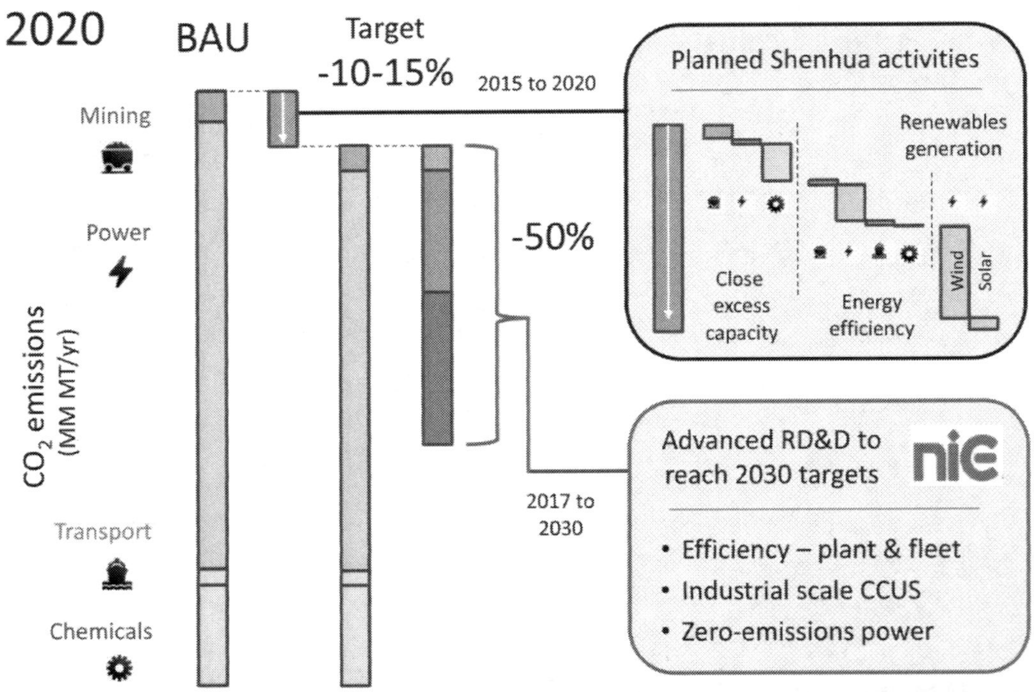




## NICE RD&D strategy for Carbon management

### Improve fleet-level efficiency.

- Technology insertion ... boosting efficiency at individual sites.
- System optimization ... marginal savings from fleet evolution.

### Accelerate affordable industrial-scale CCUS.

- Lower cost capture ... establish cost benefits and operability of Gen 2 tech under China-specific operating conditions
- Storage and utilization ... how do we ramp up to 100+ MM tpa scale?

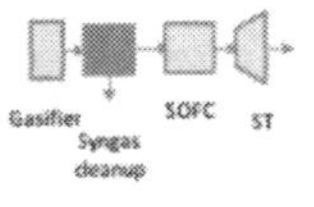


### Demonstrate concepts for zero-emissions power.

- "Transformational" power cycles ... design, enablers, pilots
- Renewables integration with fossil ... practical options in China

## Accelerating affordable CCUS at scale

The aspirational goal of the NICE CCUS RD&D effort is to provide Shenhua with the technical capacity to affordably implement CCUS at 100+MM tpa scale by 2030.

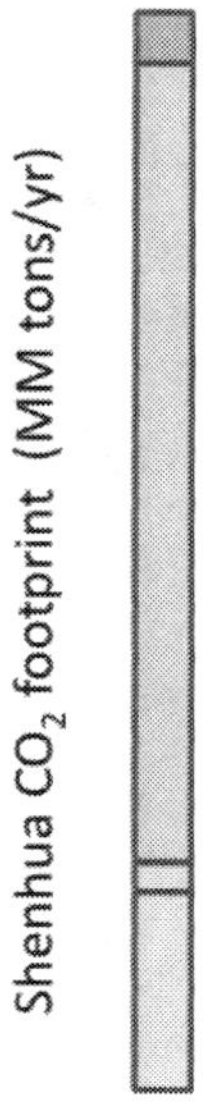


### RD&D focus areas

1. Enabling capabilities
   - Technoeconomic assessment tools
   - Carbon market dynamics

2. $CO_2$ capture
   - Power generation (existing)
   - Power generation (new builds)
   - Coal-to-chemicals (precombustion)

3. $CO_2$ disposition
   - Geological storage
   - Value-added utilization

 

# NICE RD&D
# 1. Enabling capabilities

## Snapshot: Technoeconomic analysis

## Survey of reported $CO_2$ capture costs

nie

China $CO_2$ capture costs
~ $34/ton

| Reference | Year | RMB/ton | $/ton |
|---|---|---|---|
| AE 87 3347 | 2010 | 162 | 25* |
| EP 4 1869 | 2011 | 203 | 31.2* |
| EP 4 1878 | 2011 | 206 | 31.7* |
| JCP 112 4123 | 2016 | 220 | 33.8* |
| E 58 117 | 2013 | 247** | 39.44 |
| JCP 139 612 | 2016 | 286* | 44 |

* Assumes RMB:USD = 6.5
** Assumes RMB:USD = 6.25

US $CO_2$ capture costs
~ $58/ton

| Reference | Year | $/ton |
|---|---|---|
| DOE Bituminous Baseline Report | 2011 | 58.2 |

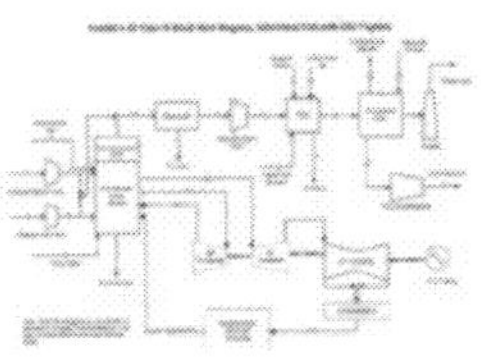

Why is there a difference?
What are the key drivers?

 

## Baseline plant – Key assumptions

nie

Reference plants:
US – DOE bituminous coal
(Cases 11 & 12, 2011)
China – Shenhua bituminous coal

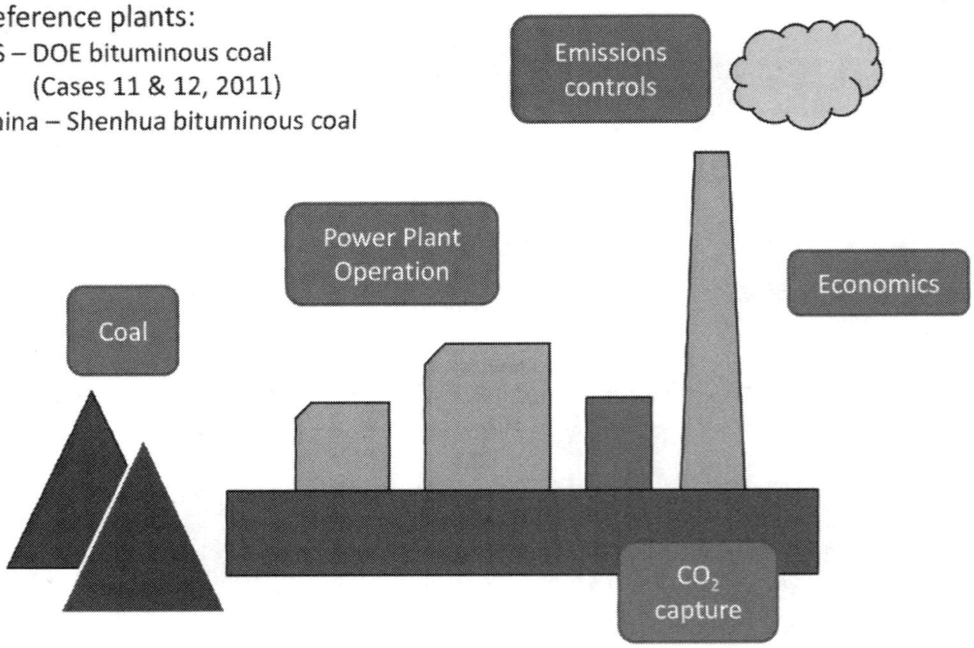


 

## Costs of electricity – US and China baselines

nic

Preliminary results

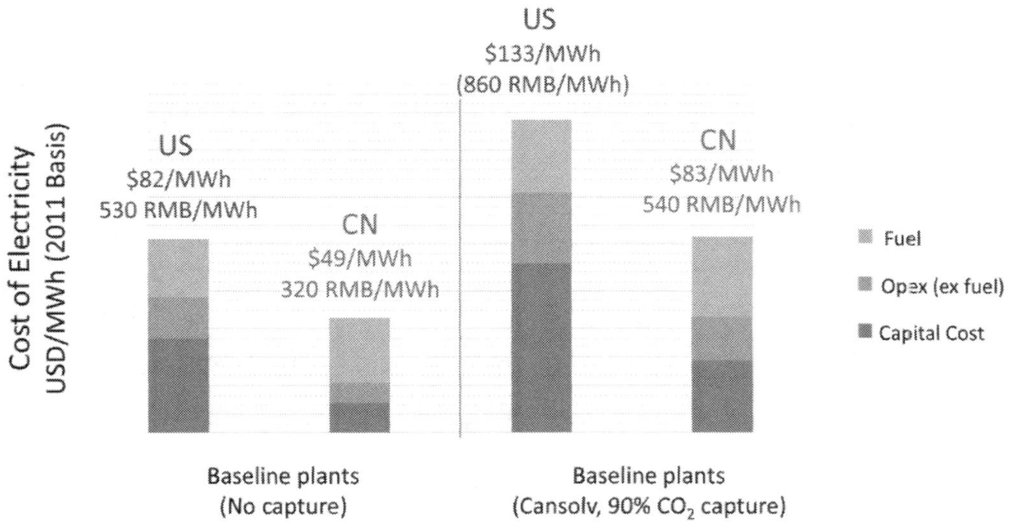


 21

## Cost walk

nic

Preliminary results

### Drivers for $CO_2$ capture cost differences

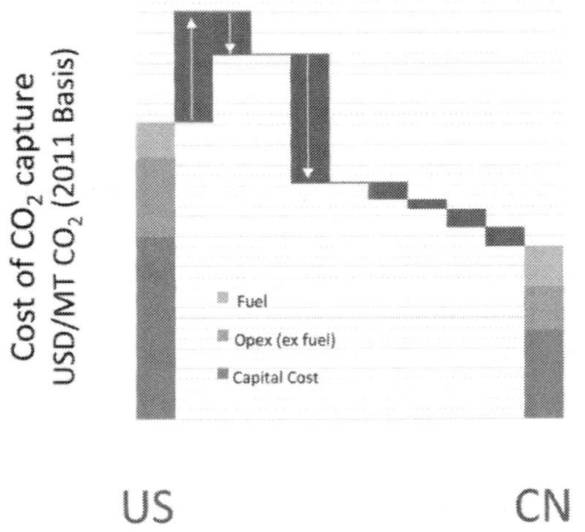


**Key drivers**

1. Capacity factor
2. Plant size
3. Coal consumption
4. Capex
5. Coal price
6. Labor
7. Fixed opex
8. Variable opex
9. Other economic assumptions

 22

## Summary and Next steps ... Enablers

nice

**Baseline China plant has been defined**

Internal methodology to convert "US costs" into "China costs"

Core design and economic assumptions

Performance and costing ... sensitivity analysis in progress

**Next steps and collaboration opportunities**

Validate internal results and publish a "China coal power base case"

Track and understand impacts of carbon trading markets

Engage global community on cost impacts of China-specific factors

 23

## NICE RD&D
## 2. $CO_2$ capture

**Snapshot: Gen 2 capture tech for power generation**

## Paths to affordable capture

nie

| Gen 1 technology | Gen 2 technology |
|---|---|
| (amine solvents) | (alternate solvents, membranes, sorbents) |
| Cost maturation | Scale-up and validation |

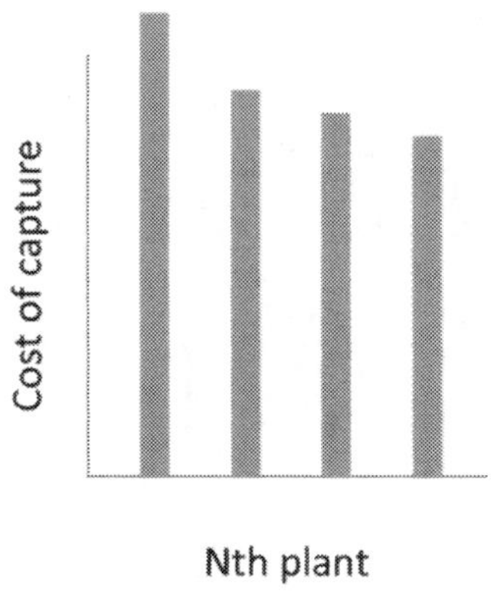


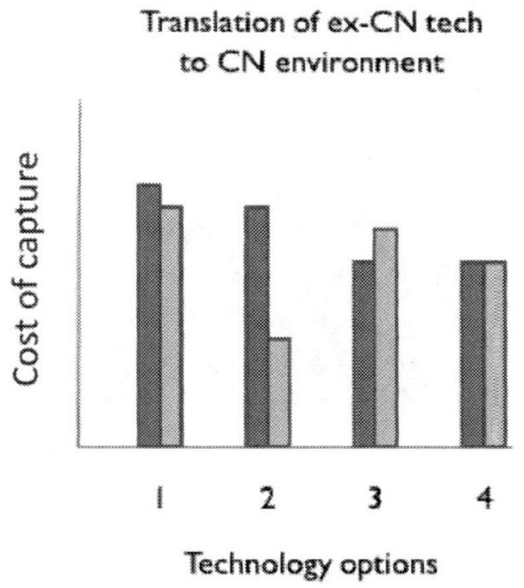


 25

## Operational differences – China vs US

nie

1. **Capacity factor/load following**
   - Turndown can cause $CO_2$ concentration fluctuations
   - Effect of ramping on performance and economics

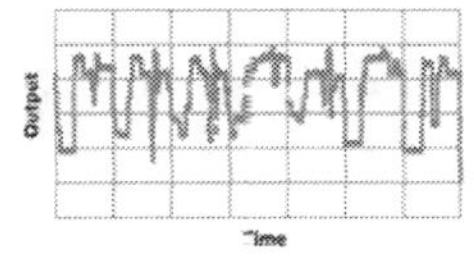


2. **Capture rate**
   - NG equivalence ... 60-65% capture
   - Fleet-level trade-offs for partial capture

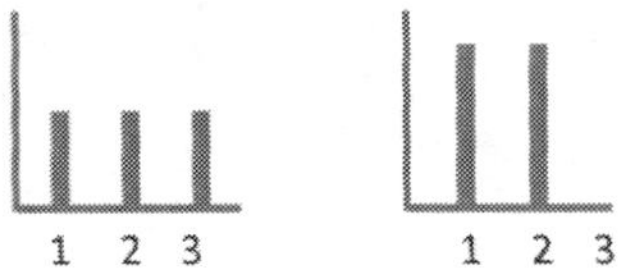


3. **Other**
   - Emissions controls ... sensitivity to contaminants and dynamics
   - $CO_2$ product specs ... pressure, purity, off-take rates

 26

## Gen 2 technology – Screening criteria

nie

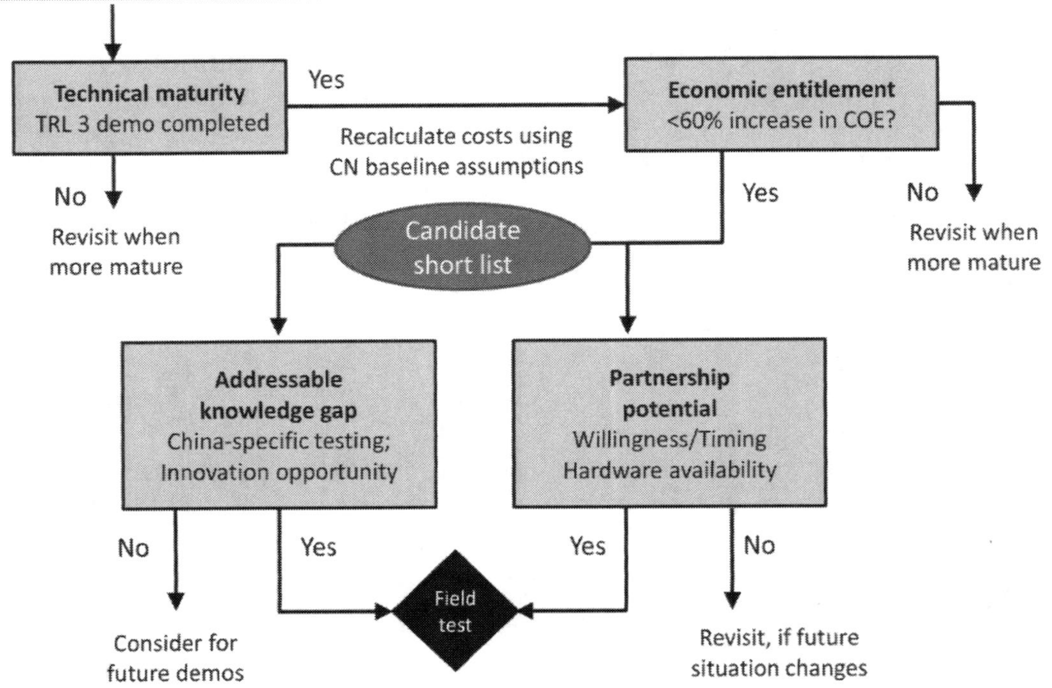


 27

## Slipstream test facility

nie

**Jiangyou site** (Sichuan province)

- 4 subcritical boiler units:
  - 2 x 300 + 2 x 330 MW
- Emissions controls: SCR, FGD
- Capture slipstream
  - 0.1 $MW_e$ (400 to 500 kg/hr)
  - Modeled after NCCC bench-scale testing platform
- Design and construction in progress ...
  Qualification testing starts 4Q2017

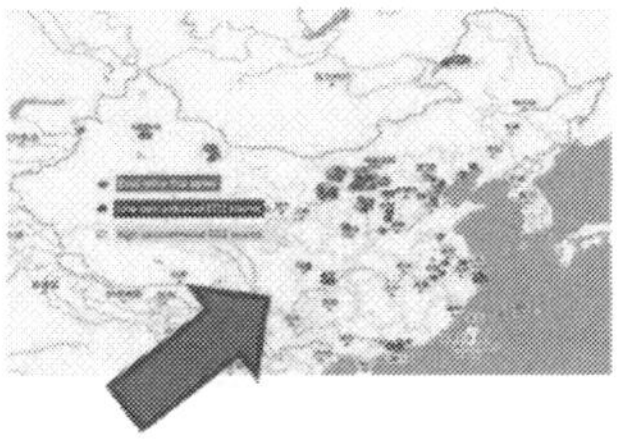

Thanks to NCCC for helping us get to this point quickly.

 28

## Summary and Next steps ... Capture

**Slipstream evaluation of Gen 2 technologies in progress**

Technology screening criteria:

- Economic potential (<60% increase in COE)
- Addressable knowledge gaps

**Next steps and collaboration opportunities**

Support maturation of Gen 1 and Gen 2 capture technologies

- Focus on operability and economic entitlement
- Open to new technology options

Develop technical and commercial roadmap for implementation within Shenhua power generation and coal-to-chemicals business units

 29

# NICE RD&D
# 3. $CO_2$ utilization and sequestration

## Snapshot: Geological storage demo at Ordos

## Shenhua Ordos demonstration project

- China's first geological storage project
- Active from 2011-2014
- Low permeability saline aquifer
- Total $CO_2$ injected: 300,000 MT
- NICE role: Monitoring

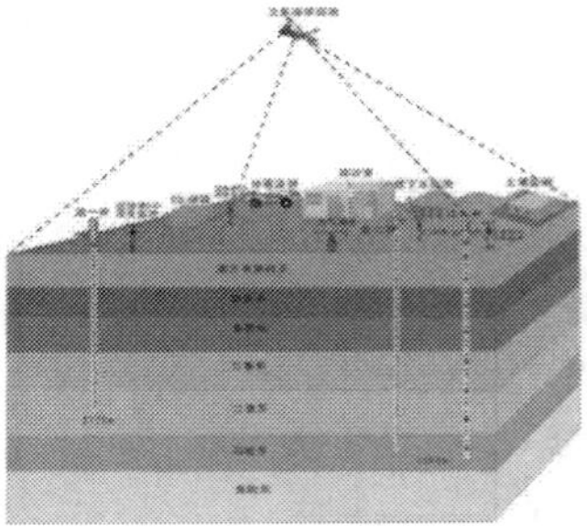

 

## $CO_2$ source

- **Coal to liquids (CTL) initiative**
  - Significant national initiative in 1990s-2000s
  - Reduces energy security risks around oil imports
  - Utilization of low S, low ash coal reserves

- **Ordos DCL project**
  - Direct coal liquefaction ... high T hydrogenation
  - Phase 1 Commercial operation 2011
    ... 1 MM tpa liquids product (3.4 MM tpa coal)
  - $CO_2$ footprint
    ... 3.6 MM tpa from coal to $H_2$
    ... 0.7 MM tpa from power and steam
  - $CO_2$ for storage demonstration
    ... DCL capture ... 80% purity
    ... purification ... 95% purity

 

## Ordos project ... Process map & current status

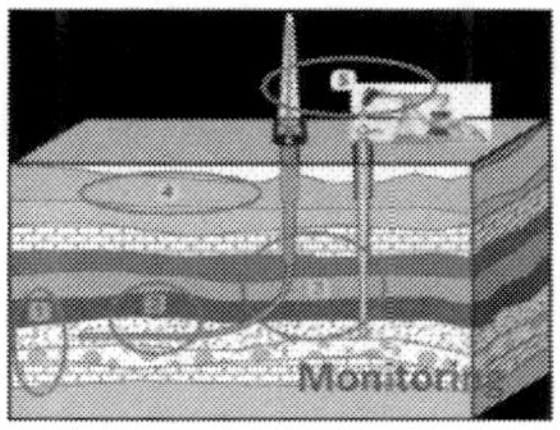

Long-term monitoring responsibility transferred to NICE

- Plume modeling
- Risk-based approaches

 

## Summary and Next steps

**Closing out Ordos project**

Continuing monitoring for long-term safety and $CO_2$ assurance

Review lessons learned to inform future sequestration projects

Transition focus to EOR partnerships and other utilization opportunities

**Next steps and collaboration opportunities**

Engage international community to stay current on storage developments

Ramp up efforts in utilization (EOR, others)

 

## Conclusion

Globally, much progress is being made towards industrial CCS.

NICE is pursuing RD&D to assemble the technical capacity for Shenhua to do affordable CCS at 100+MM tpa scale by 2030.

We are open to learning from others, sharing our lessons learned, and partnering to develop technologies suited to the China landscape.

 

# MOVING FORWARD

## Law and Policy Panel, CMTC 2017

Mary Ellen Ternes, BEChE, JD, Earth & Water Law, LLC
Charles McConnell, BSChE, MBA, Rice University
Frederick Eames, JD, Hunton & Williams
Victor B. Flatt, JD, University of Houston
E. Dennis Griffith, BSChE, MSChE, Granherne

# OUTLINE

- Utilization should be the focus
- Many policy changes: Energy, Environmental, Economic
- What are the subjects of these current reversals and how will it work?
  - Paris Agreement – What is it?
    - Individual US Commitments
  - Regulatory Reform Realities –
    - Legislative Process:
      - New Rules: Congressional Review Act
      - Older Rules: Regulatory Process, with or without statutory revision
    - Regulatory Process:
      - How: Administrative Procedures Act
        - Notice and comment rulemaking
        - Judicial Review
      - What: 2-1; Best Available Science and Endangerment Finding >> CAA New Source Review?
    - Litigation and Market Response: States, NGO's, citizen suits, corporate governance and business reputation
- Impact on current projects
- Opportunities and challenges

## BUT FIRST...

We've seen large scale swings before.

For calibration purposes:

| President | Terms | Clean Air Act |
|---|---|---|
| **Reagan** | 1981-1989 | **1988 – EPA applies New Source Review (NSR) strictly to Electric Generating Units (EGUs)** – WEPCO; NSR recognized as not intended to allow significant life extension projects |
| **Bush 1** | 1989-1993 | **1990** - Clean Air Act Amendments (no specific NSR revisions)<br>**1992** – EPA adopts WEPCO Rule for EGUs |
| **Clinton** | 1993-2001 | **1996** – EPA proposes NSR reform rules<br>**1997** – NSR Enforcement Initiative – EGUs<br>**1999 – EPA sues 7 EGU companies for NSR violations** |
| **Bush II** | 2001-2009 | **2001** – ***EPA/DOE conduct 90-day review of NSR impact on energy industry; USAG reviews NSR enforcement actions for CAA consistency.***<br>**2002** – EPA promulgates NSR reform rules: revised baseline actual; actual to projected actual; Plant-wide Applicability Limits (PALs); **Clean Units; and Pollution Control Projects (PCP).**<br>**2002** – EPA proposes NSR reform for EGUs, including **Equipment Replacement Provision (ERP)**<br>**2003** – EPA promulgates ERP<br>**2005-2006** – **DC Circuit vacates Clean Unit, PCP and ERP revisions** |

# OUTLINE

- Utilization should be the focus
- Many policy changes: Energy, Environmental, Economic
- What are the subjects of these current reversals and how will it work?
  - **Paris Agreement – What is it?**
    - **Individual US Commitments**
  - Regulatory Reform Realities –
    - Legislative Process:
      - New Rules: Congressional Review Act
      - Older Rules: Regulatory Process, with or without statutory revision
    - Regulatory Process:
      - How: Administrative Procedures Act
        - Notice and comment rulemaking
        - Judicial Review
      - What: 2-1; Best Available Science and Endangerment Finding >> CAA New Source Review?
    - Litigation and Market Response: States, NGO's, citizen suits, corporate governance and business reputation
- Impact on current projects
- Opportunities and challenges

# PARIS AGREEMENT

- Comes into force if nations representing 55% of global GHG emissions ratify – Came into Force October 2016
- Kyoto governs to 2020. Paris to govern post-2020.
- Pivots from top down multinational agreement with overarching targets to bottom up submission of national emissions reduction plans without global reduction target.
- Establishes commitments by all parties to make "indicated nationally determined contributions" (INDCs) based on voluntary domestic measures to reduce emissions – includes developing countries.
- New INDCs every five years, with expectation they will ratchet up targets every five years.
- National emission targets under INDCs are not "binding" and enforceable, transparency and the five year review cycle are intended to drive performance
- Cannot withdraw for 3 years – 1 year notice

*Use of markets:*

At this time, the United States does not intend to utilize international market mechanisms to implement its 2025 target.

*Domestic laws, regulations, and measures relevant to implementation:*

Several U.S. laws, as well as existing and proposed regulations thereunder, are relevant to the implementation of the U.S. target, including the Clean Air Act (42 U.S.C. §7401 et seq.), the Energy Policy Act (42 U.S.C. §13201 et seq.), and the Energy Independence and Security Act (42 U.S.C. § 17001 et seq.).

Since 2009, the United States has completed the following regulatory actions:

- Under the Clean Air Act, the United States Department of Transportation and the United States Environmental Protection Agency adopted fuel economy standards for light-duty vehicles for model years 2012-2025 and for heavy-duty vehicles for model years 2014-2018.
- Under the Energy Policy Act and the Energy Independence and Security Act, the United States Department of Energy has finalized multiple measures addressing buildings sector emissions including energy conservation standards for 29 categories

of appliances and equipment as well as a building code determination for commercial buildings.

- Under the Clean Air Act, the United States Environmental Protection Agency has approved the use of specific alternatives to high-GWP HFCs in certain applications through the Significant New Alternatives Policy program.

At this time:

- Under the Clean Air Act, the United States Environmental Protection Agency is moving to finalize by summer 2015 regulations to cut carbon pollution from new and existing power plants.
- Under the Clean Air Act, the United States Department of Transportation and the United States Environmental Protection Agency are moving to promulgate post-2018 fuel economy standards for heavy-duty vehicles.
- Under the Clean Air Act, the United States Environmental Protection Agency is developing standards to address methane emissions from landfills and the oil and gas sector.
- Under the Clean Air Act, the United States Environmental Protection Agency is moving to reduce the use and emissions of high-GWP HFCs through the Significant New Alternatives Policy program.
- Under the Energy Policy Act and the Energy Independence and Security Act, the United States Department of Energy is continuing to reduce buildings sector emissions including by promulgating energy conservation standards for a broad range of appliances and equipment, as well as a building code determination for residential buildings.

In addition, since 2008 the United States has reduced greenhouse gas emissions from Federal Government operations by 17 percent and, under Executive Order 13693 issued on March 25th 2015, has set a new target to reduce these emissions 40 percent below 2005 levels by 2025.

# OUTLINE

- Utilization should be the focus
- Many policy changes: Energy, Environmental, Economic
- What are the subjects of these current reversals and how will it work?
  - Paris Agreement – What is it?
    - Individual US Commitments
  - **Regulatory Reform Realities –**
    - **Legislative Process:**
      - **New Rules: Congressional Review Act**
      - **Older Rules: Regulatory Process, with or without statutory revision**
    - Regulatory Process:
      - How: Administrative Procedures Act
        - Notice and comment rulemaking
        - Judicial Review
      - What: 2-1; Best Available Science and Endangerment Finding >> CAA New Source Review?
    - Litigation and Market Response: States, NGO's, citizen suits, corporate governance and business reputation
- Impact on current projects
- Opportunities and challenges

# CONGRESSIONAL REVIEW ACT

## UP TO 2001 (1); 2017 (14)

**28. Congressional Review of Agency Rulemaking**
**[5 U.S.C. 801, 802, 804]**

The following excerpts of chapter 8 of title 5, United States Code, do not contain privileged procedures for the consideration of a measure in the House. They are depicted here because they constitute Rules of the House and potentially affect the legislative process. Detailed procedures for the consideration in the Senate of a joint resolution disapproving an agency rule may be found in the statute (5 U.S.C. 802).

SEC. 801. CONGRESSIONAL REVIEW.

(a)(1)(A) Before a rule can take effect, the Federal agency promulgating such rule shall submit to each House of the Congress and to the Comptroller General a report containing—

(i) a copy of the rule;

(ii) a concise general statement relating to the rule, including whether it is a major rule; and

(iii) the proposed effective date of the rule.

(B) On the date of the submission of the report under subparagraph (A), the Federal agency promulgating the rule shall submit to the Comptroller General and make available to each House of Congress—

[1247]

STATUTORY LEGISLATIVE PROCEDURES

(4) Except for a major rule, a rule shall take effect as otherwise provided by law after submission to Congress under paragraph (1).

(5) Notwithstanding paragraph (3), the effective date of a rule shall not be delayed by operation of this chapter beyond the date on which either House of Congress votes to reject a joint resolution of disapproval under section 802.

(b)(1) A rule shall not take effect (or continue), if the Congress enacts a joint resolution of disapproval, described under section 802, of the rule.

(2) A rule that does not take effect (or does not continue) under paragraph (1) may not be reissued in substantially the same form, and a new rule that is substantially the same as such a rule may not be issued, unless the reissued or new rule is specifically authorized by a law enacted after the date of the joint resolution disapproving the original rule.

SEC. 802. CONGRESSIONAL DISAPPROVAL PROCEDURE.

(a) For purposes of this section, the term "joint resolution" means only a joint resolution introduced in the period beginning on the date on which the report referred to in section 801(a)(1)(A) is received by Congress and ending 60 days thereafter (excluding days either House of Congress is adjourned for more than 3 days during a session of Congress), the matter after the resolving clause of which is as follows: "That Congress disapproves the rule submitted by the _____ relating to _____, and such rule shall have no force or effect." (The blank spaces being appropriately filled in).

[1250]

# OUTLINE

- Utilization should be the focus
- Many policy changes: Energy, Environmental, Economic
- What are the subjects of these current reversals and how will it work?
  - Paris Agreement – What is it?
    - Individual US Commitments
  - **Regulatory Reform Realities –**
    - Legislative Process:
      - New Rules: Congressional Review Act
      - Older Rules: Regulatory Process, with or without statutory revision
    - **Regulatory Process:**
      - **How: Administrative Procedures Act**
        - **Notice and comment rulemaking**
        - **Judicial Review**
      - **What: 2-1; Best Available Science and Endangerment Finding >> CAA New Source Review?**
    - Litigation and Market Response: States, NGO's, citizen suits, corporate governance and business reputation
- Impact on current projects
- Opportunities and challenges

# ADMINISTRATIVE PROCEDURES ACT

**§ 706. Scope of review**

To the extent necessary to decision and when presented, the reviewing court shall decide all relevant questions of law, interpret constitutional and statutory provisions, and determine the meaning or applicability of the terms of an agency action. The reviewing court shall—

(1) compel agency action unlawfully withheld or unreasonably delayed; and

(2) hold unlawful and set aside agency action, findings, and conclusions found to be—

(A) arbitrary, capricious, an abuse of discretion, or otherwise not in accordance with law;

(B) contrary to constitutional right, power, privilege, or immunity;

(C) in excess of statutory jurisdiction, authority, or limitations, or short of statutory right;

(D) without observance of procedure required by law;

(E) unsupported by substantial evidence in a case subject to sections 556 and 557 of this title or otherwise reviewed on the record of an agency hearing provided by statute; or

(F) unwarranted by the facts to the extent that the facts are subject to trial de novo by the reviewing court.

In making the foregoing determinations, the court shall review the whole record or those parts of it cited by a party, and due account shall be taken of the rule of prejudicial error.

(Pub. L. 89-554, Sept. 6, 1966, 80 Stat. 393.)

# CCUS?

All regulatory changes must comply with the **Administrative Procedures Act**:

- General:
  - Regulatory Reform Task Forces
  - Removing Two Rules for every new One Rule
  - Imposing new "Best Available Science" requirements
- Revising Clean Air Act Regulation
  - Clean Power Plan and NSPS 111(b)
  - Oil and Gas Methane Rules
  - ***Endangerment Finding? Supports Application of New Source Review***
  - ***New Source Review? Provides authority for imposing CCS requirements as Best Available Control Technology (BACT) for major stationary GHG "Anyway" Sources***

# OUTLINE

- Utilization should be the focus
- Many policy changes: Energy, Environmental, Economic
- What are the subjects of these current reversals?
  - Paris Agreement – What is it?
    - Individual US Commitments re CCUS forcing
  - Regulatory Reform Realities –
    - Legislative Process:
      - New Rules: Congressional Review Act
      - Older Rules: Regulatory Process, with or without statutory revision
    - Regulatory Process:
      - How: Administrative Procedures Act
        - Notice and comment rulemaking
        - Judicial Review
      - What: 2-1; Best Available Science and Endangerment Finding >> CAA New Source Review?
    - **Litigation and Market Response: States, NGO's, citizen suits, corporate governance and business reputation**
- Impact on current projects
- Opportunities and challenges

# CCUS?

- Litigation in rulemaking, enforcement and innovative common law claims
  - California and other state activism
  - Delegated States
  - Non-Governmental Organizations (NGO) environmental advocacy
  - Industry Groups
  - Citizen suits

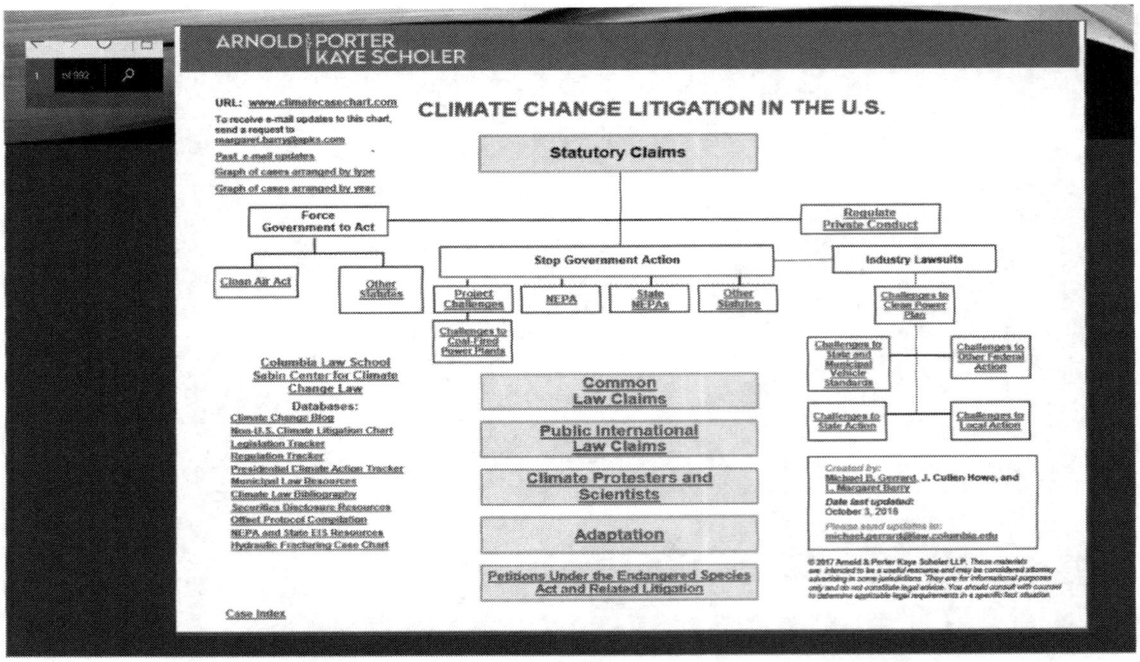

# CCUS?

- Corporate Governance
  - Disclosure requirements?
    - Materiality regardless of SEC enforcement
    - AG Lawsuits
  - Shareholder initiatives: petitions, lawsuits
  - Market driven global sustainability initiatives and business reputation
    - Financial Stability Board (FSB) Task Force on Climate-Related Financial Disclosures (TCFD),
    - Power Forward 3.0,
    - Annual Sustainable Innovation Forums,
    - Global Reporting Initiative,
    - Carbon Disclosure Project,
    - ISO 20400 Sustainable Procurement Guidance

# OUTLINE

- Utilization should be the focus
- Many policy changes: Energy, Environmental, Economic
- What are the subjects of these current reversals?
  - Paris Agreement – What is it?
    - Individual US Commitments re CCUS forcing
  - Regulatory Reform Realities –
    - Legislative Process:
      - New Rules: Congressional Review Act
      - Older Rules: Regulatory Process, with or without statutory revision
    - Regulatory Process:
      - How: Administrative Procedures Act
        - Notice and comment rulemaking
        - Judicial Review
      - What: 2-1; Best Available Science and Endangerment Finding >> CAA New Source Review?
    - Litigation and Market Response: States, NGO's, citizen suits, corporate governance and business reputation
- **Impact on current projects**
- **Opportunities and challenges**

LanzaTech

**The Road to Awesome**
**Jennifer Holmgren**
**CEO LanzaTech**

LanzaTech

## A Few Predictions

| | |
|---|---|
| **1876** | "This 'telephone' has too many shortcomings to be seriously considered as a means of communication" *Western Union Memo* |
| **1895** | "Heavier-than-air flying machines are impossible" *Lord Kelvin, President Royal Society* |
| **1920** | "The wireless music box (radio) has no imaginable commercial value" *David Sarnoffs Associates in response to his urgings for investments in the radio* |
| **1943** | "I think there's a world market for maybe five computers" *Thomas Watson, Chairman IBM* |
| **1949** | "Computer in the future may weigh no more than 1.5 tons" *Popular Mechanics forecasting the relentless march of science* |
| **1977** | "There is no reason anyone would want a computer in their home" *Ken Olson, President, Chairman and Founder of Digital Equipment* |
| **1981** | "640K ought to be enough computer memory for anyone" *Bill Gates* |

LanzaTech 2

**Predictions are simply extrapolations of the past…**

*…innovation expands the 'art of the possible'*

**…today's 'unimaginable' is tomorrow's 'conventional wisdom.'**

LanzaTech

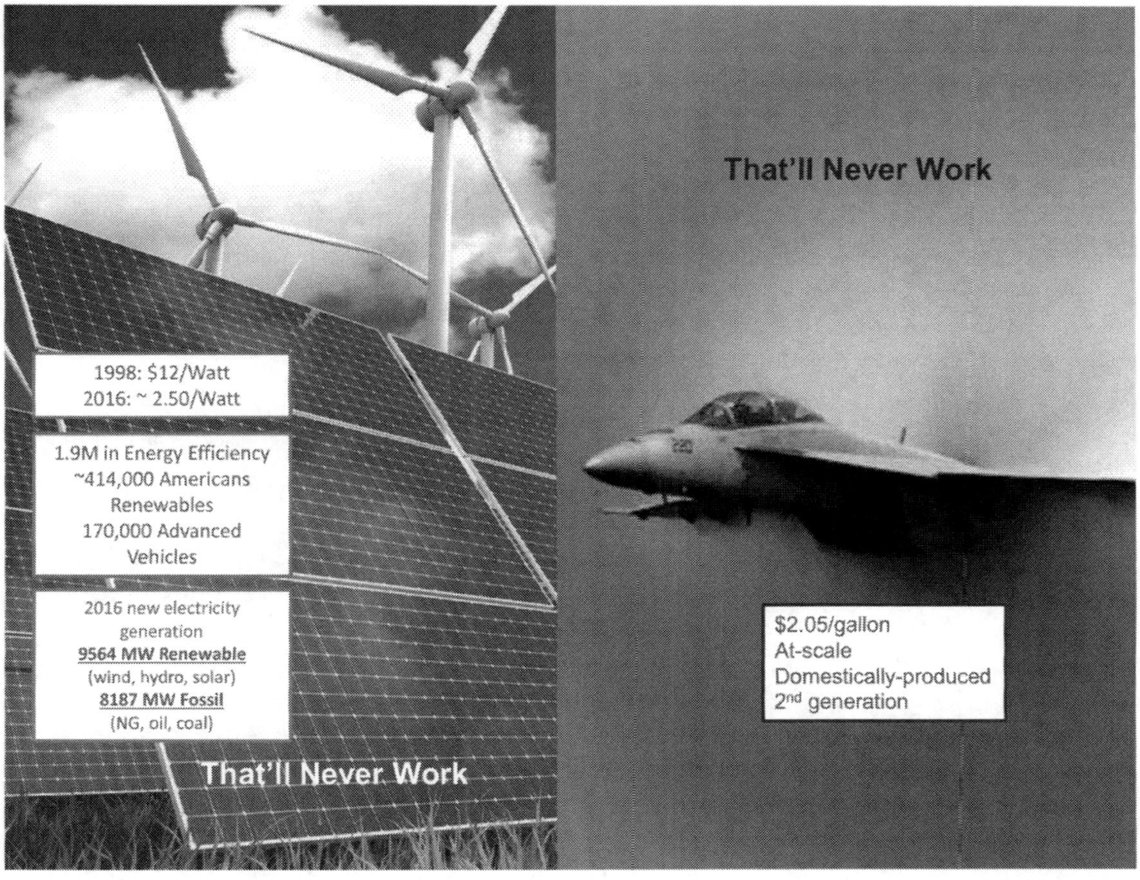

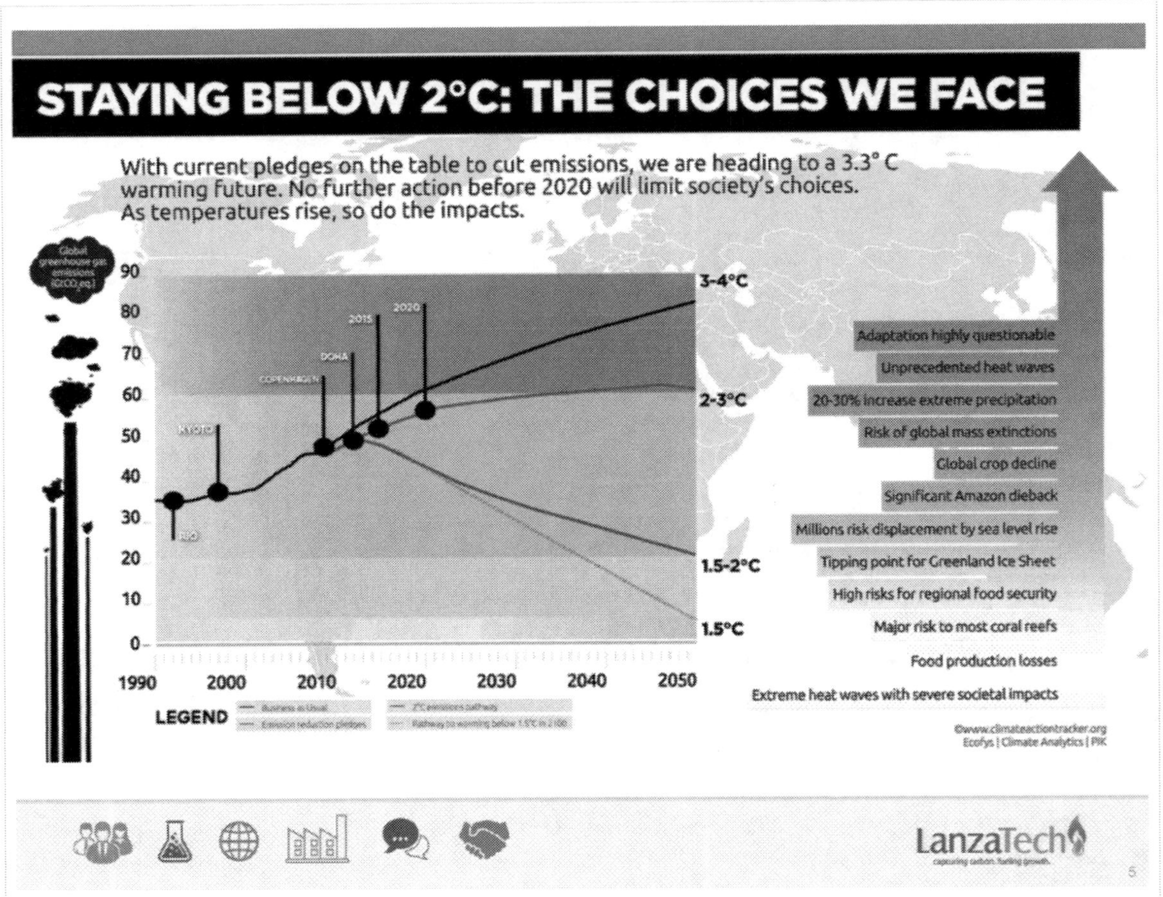
STAYING BELOW 2°C: THE CHOICES WE FACE
With current pledges on the table to cut emissions, we are heading to a 3.3° C warming future. No further action before 2020 will limit society's choices. As temperatures rise, so do the impacts.
90
80
70
60
50
40
30
20
10
0
1990
2000
2010
2020
2030
2040
2050
3-4°C
2-3°C
1.5-2°C
1.5°C
LEGEND
Adaptation highly questionable
Unprecedented heat waves
20-30% increase extreme precipitation
Risk of global mass extinctions
Global crop decline
Significant Amazon dieback
Millions risk displacement by sea level rise
Tipping point for Greenland Ice Sheet
High risks for regional food security
Major risk to most coral reefs
Food production losses
Extreme heat waves with severe societal impacts
LanzaTech
5

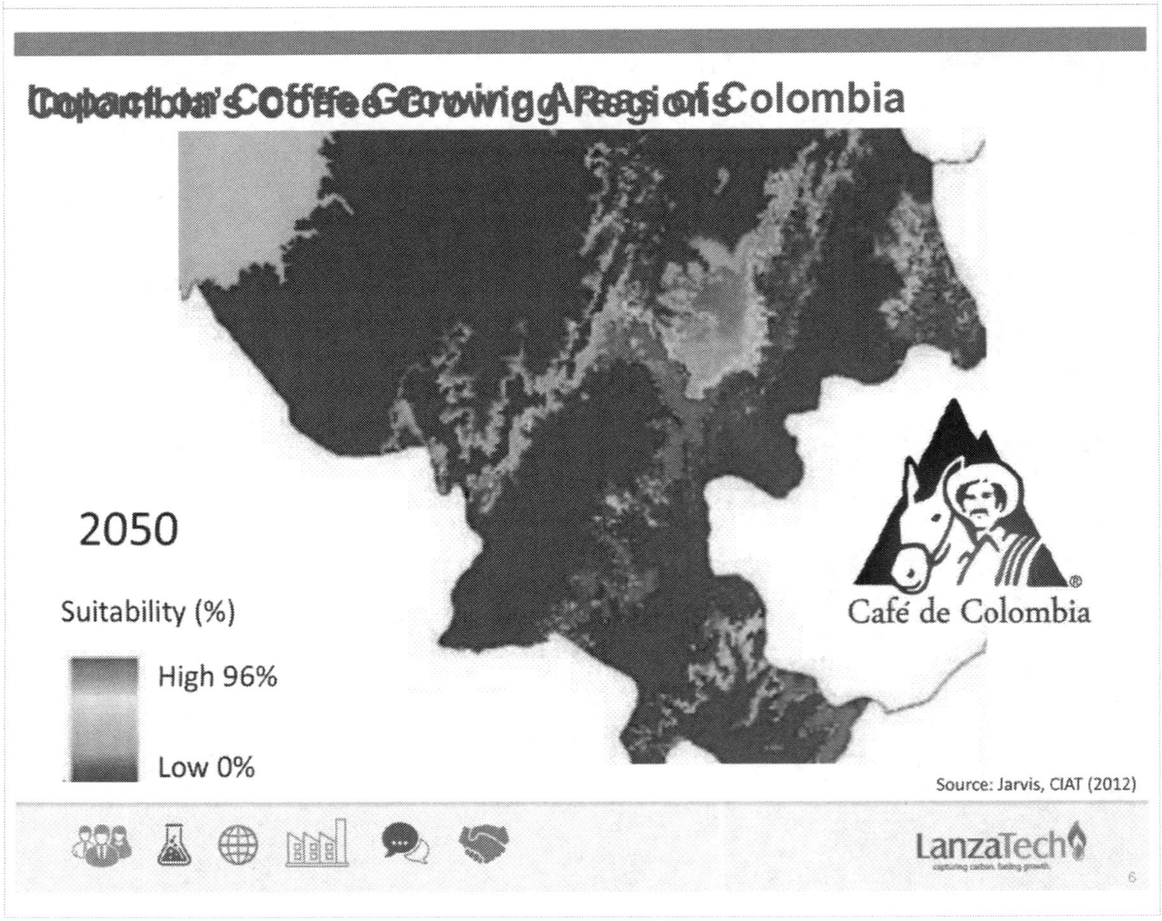
2050
Suitability (%)
High 96%
Low 0%
Café de Colombia
Source: Jarvis, CIAT (2012)
LanzaTech
6

Energy can be Carbon Free

Aviation Fuel needs Carbon

## Carbon: Problematic When Combusted

### Outdoor Air Pollution Impact

CHINA: 1.2M premature deaths in China in 2010
Or 25M healthy years of life from the population.

USA: 200,000 early deaths every year
(Equal to those who die from diabetes each year. )

*The New York Times*, April 1, 2013, sec. World / Asia Pacific,
quoting from the *Global Burden of Disease Study 2010*, The Lancet, December 2012.
Shindell, D. T. et al. (2016) Climate and health impacts of US emissions reductions consistent with 2C, Nature Climate Change, doi:10.1038/nclimate2935.

C

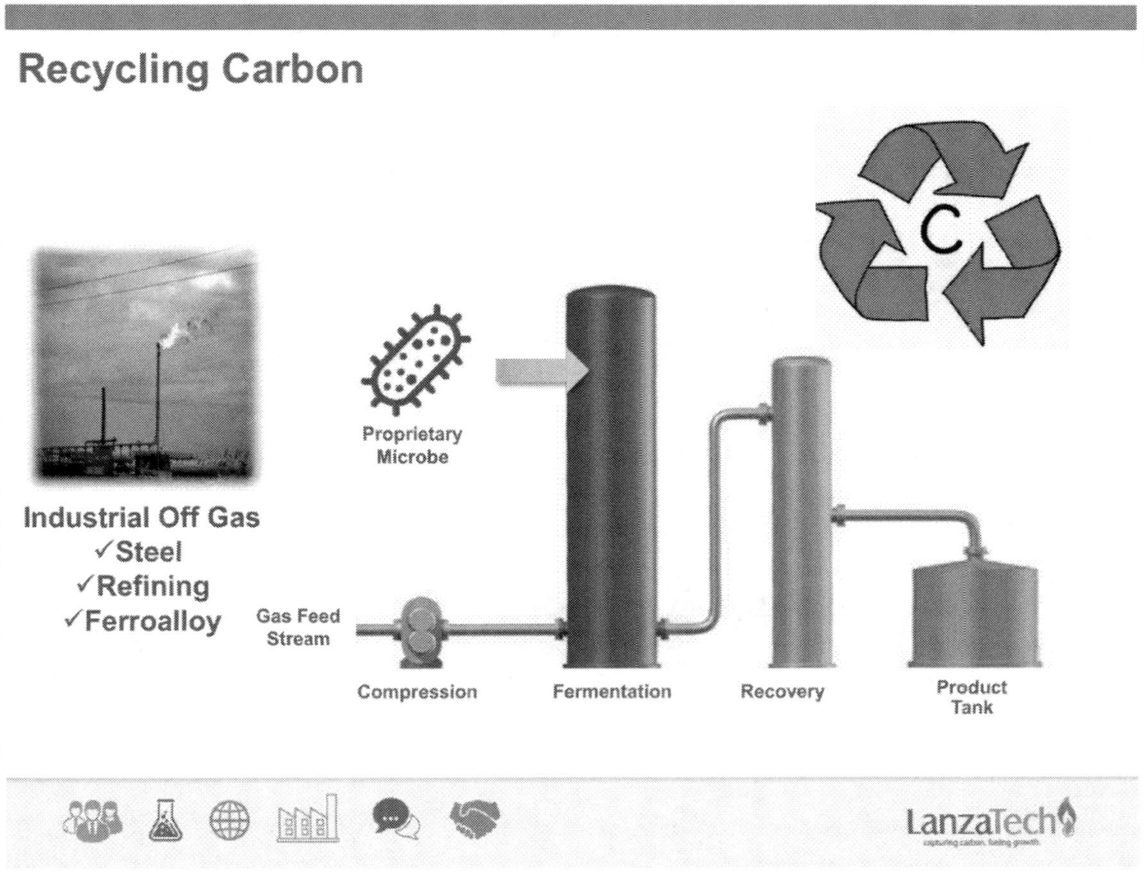
Recycling Carbon
C
Industrial Off Gas
✓Steel
✓Refining
✓Ferroalloy
Proprietary Microbe
Gas Feed Stream
Compression
Fermentation
Recovery
Product Tank
LanzaTech

## Commercial Scale Facilities

首钢朗泽
Shougang LanzaTech

**Caofeidian, China**
**16M gallons/year**
**2017**

ArcelorMittal

Steelanol

**Gent, Belgium**
**21M gallons/year**
**2018**

LanzaTech
capturing carbon. fueling growth.

## Commercial Scale Beyond Steel

इंडियनऑयल

IndianOil

**Haryana, India**
**13M gallons/year**
**Refinery offgas**
**2019**

AEMETIS

**Modesto, California**
**8M gallons/year**
**Biomass Syngas**
**2018**

LanzaTech

capturing carbon. fueling growth.

## Ethanol-blend Fuels Reduce Particulate Emissions

**30-40% reduction PM**
**E0:E20 fuel blends**

Storey et al., SAE Int. J. Fuels Lubr. 3(2):2010

LanzaTech

capturing carbon. fueling growth.

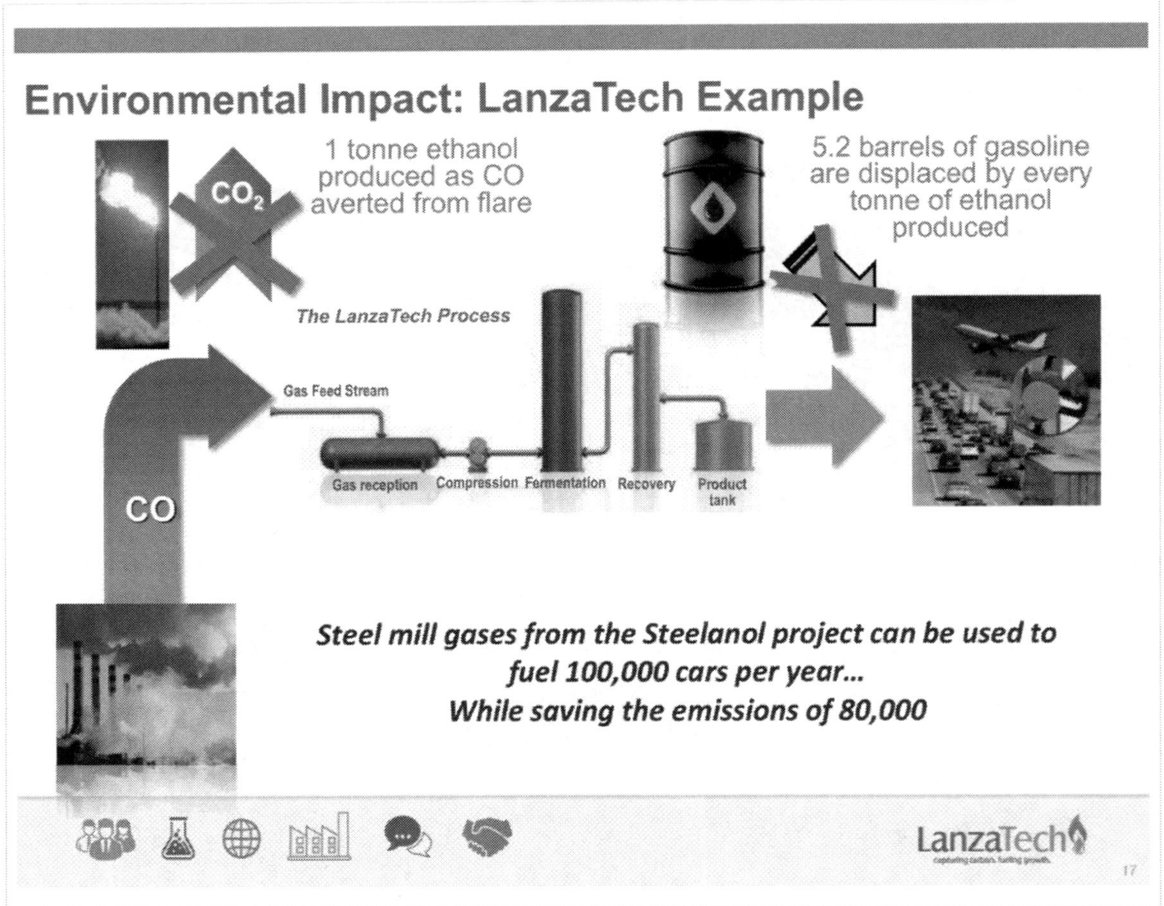
Environmental Impact: LanzaTech Example
1 tonne ethanol produced as CO averted from flare
$CO_2$
5.2 barrels of gasoline are displaced by every tonne of ethanol produced
The LanzaTech Process
Gas Feed Stream
Gas reception
Compression
Fermentation
Recovery
Product tank
CO
Steel mill gases from the Steelanol project can be used to fuel 100,000 cars per year...
While saving the emissions of 80,000
LanzaTech
17

Electric Mobility for Road Transport
Transition is underway for road transport
$35,000 USD
200miles/charge
Current LanzaTech Commercial Projects
Equivalent ~300k EVs on Road/Year
LanzaTech

Aviation Fuel needs Carbon

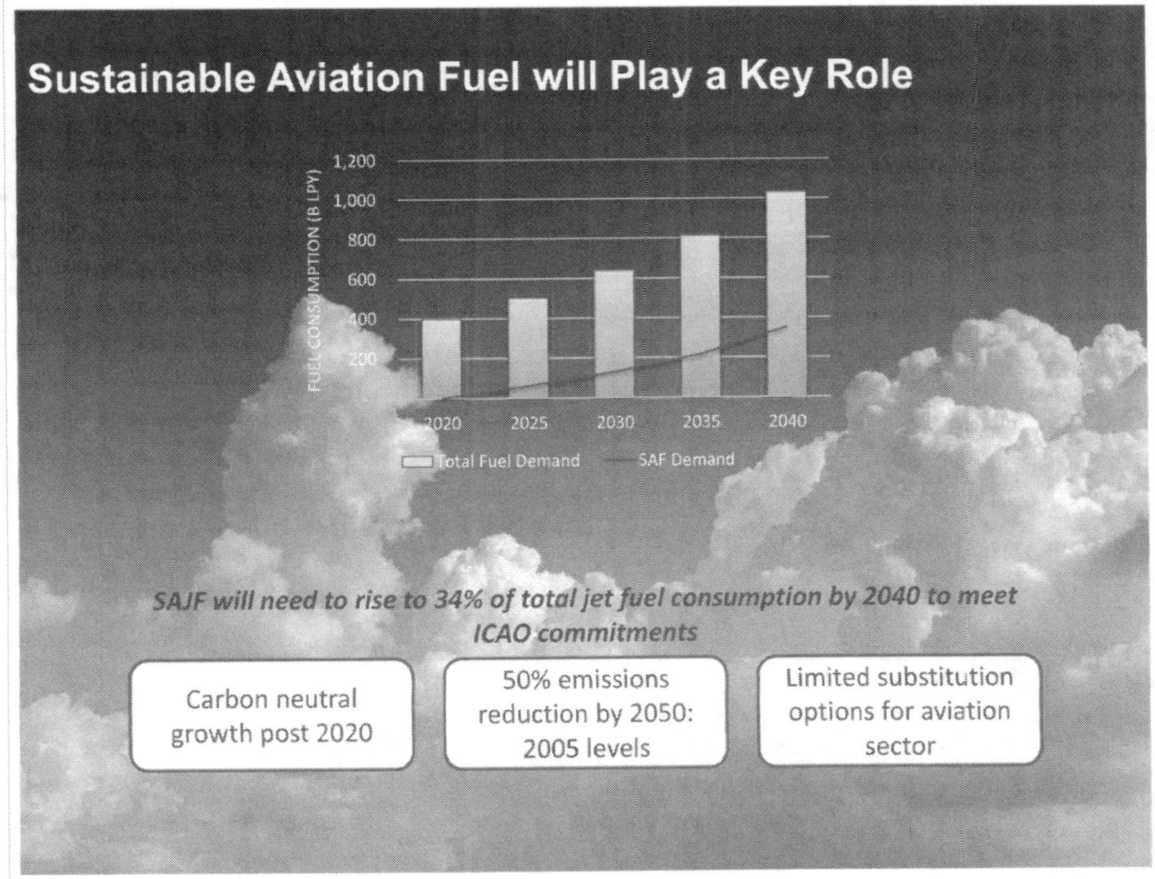
Sustainable Aviation Fuel will Play a Key Role
1,200
1,000
800
600
400
200
FUEL CONSUMPTION (B LPY)
2020
2025
2030
2035
2040
Total Fuel Demand
SAF Demand
SAJF will need to rise to 34% of total jet fuel consumption by 2040 to meet ICAO commitments
Carbon neutral growth post 2020
50% emissions reduction by 2050: 2005 levels
Limited substitution options for aviation sector

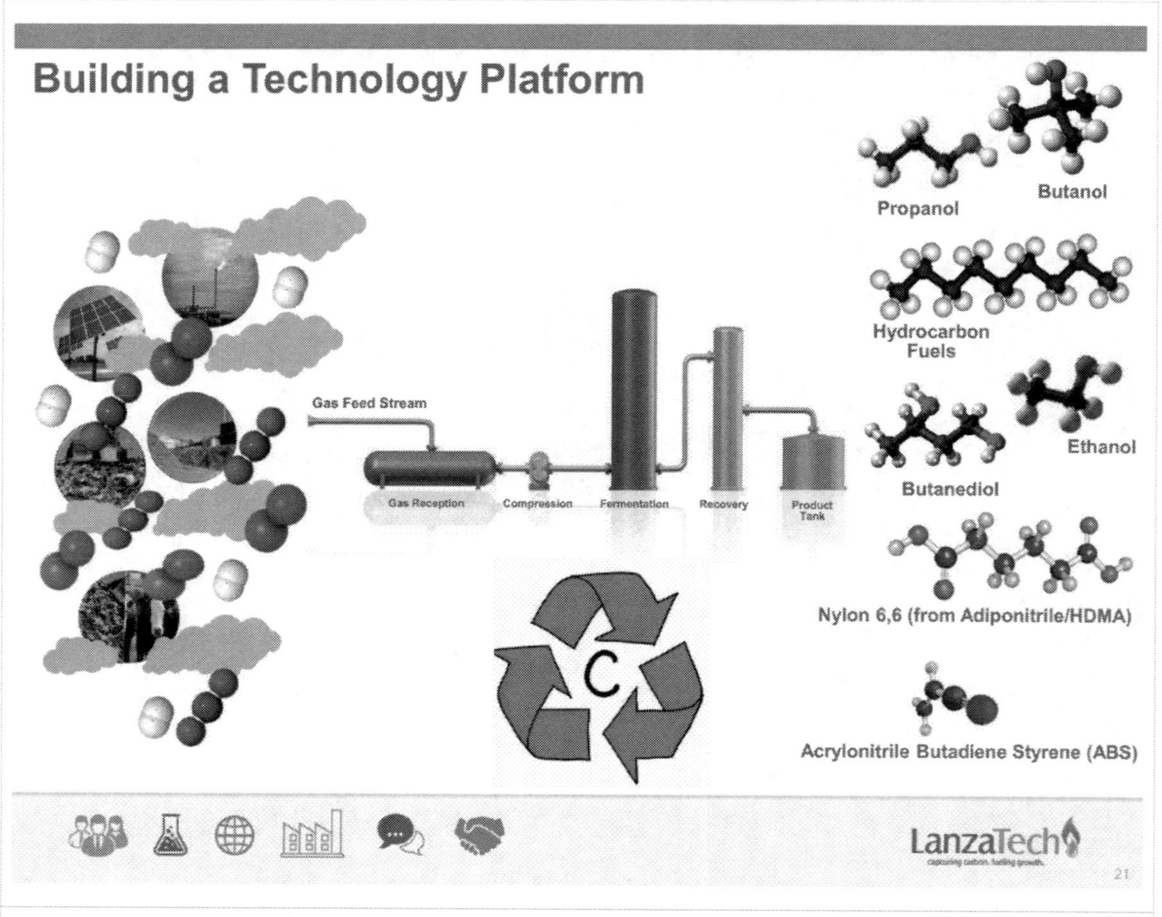
Building a Technology Platform
Propanol
Butanol
Hydrocarbon Fuels
Ethanol
Butanediol
Nylon 6,6 (from Adiponitrile/HDMA)
Acrylonitrile Butadiene Styrene (ABS)
Gas Feed Stream
Gas Reception
Compression
Fermentation
Recovery
Product Tank
C
LanzaTech
capturing carbon. fueling growth.
21

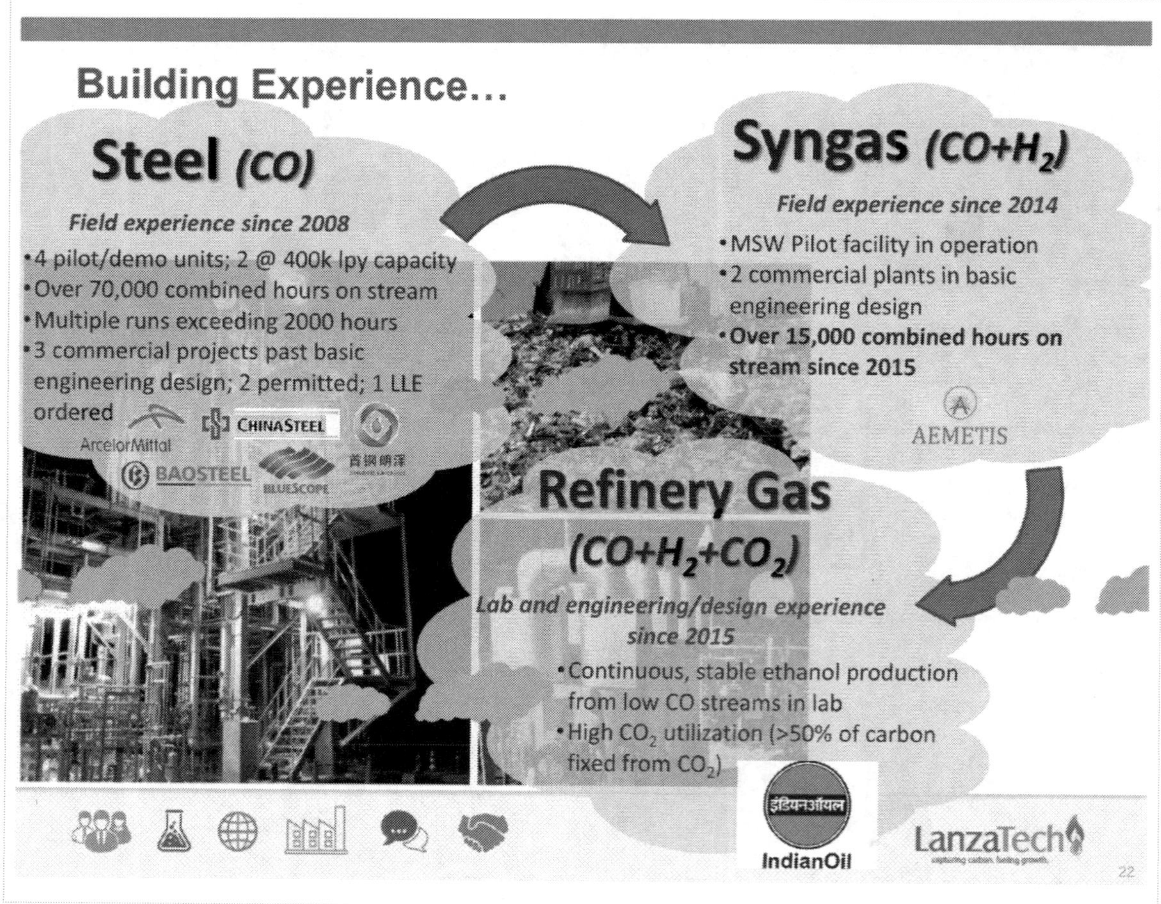
Building Experience…
Steel (CO)
Field experience since 2008
•4 pilot/demo units; 2 @ 400k lpy capacity
•Over 70,000 combined hours on stream
•Multiple runs exceeding 2000 hours
•3 commercial projects past basic engineering design; 2 permitted; 1 LLE ordered
ArcelorMittal
CHINASTEEL
BAOSTEEL
BLUESCOPE
Syngas (CO+H2)
Field experience since 2014
•MSW Pilot facility in operation
•2 commercial plants in basic engineering design
•Over 15,000 combined hours on stream since 2015
AEMETIS
Refinery Gas
(CO+H2+CO2)
Lab and engineering/design experience since 2015
•Continuous, stable ethanol production from low CO streams in lab
•High CO2 utilization (>50% of carbon fixed from CO2)
इंडियनऑयल
IndianOil
LanzaTech
capturing carbon. fueling growth.
22

## Stoichiometry of Ethanol Production with Different Gas Compositions

| Gas | Reaction | $CO:H_2$ ratio | |
|---|---|---|---|
| CO | $6\ CO + 3\ H_2O \rightarrow$ <u>EtOH</u> $+ 4\ CO_2$ | 1:0 | **Steel Mill Gas** Demonstrated at scale |
| $CO + H_2$ | $3\ CO + 3\ H_2 \rightarrow$ <u>EtOH</u> $+ CO_2$ | 1:1 | **Syngas (e.g. MSW)** Demonstrated at scale |
| $CO + H_2$ | $4\ H_2 + 2\ CO \rightarrow$ <u>EtOH</u> $+ H_2O$ | 1:2 | |
| $CO + H_2 + CO_2$ | $5\ H_2 + CO + CO_2 \rightarrow$ <u>EtOH</u> $+ 2\ H_2O$ | 1:5 | **High $H_2$ Syngas** Demonstrated at pilot, allows $CO_2$ fixing in products |
| $CO + H_2 + CO_2$ | $5.4\ H_2 + 1.4\ CO_2 + 0.6\ CO + 0.6\ H_2O \rightarrow$ <u>EtOH</u> $+ 3\ H_2O$ | 1:9 | |
| $CO_2 + H_2$ | $4\ H_2 + 2\ CO_2 \rightarrow$ <u>Acetate</u> $+ 2\ H_2O$ | | Primarily **Acetate** production |

LanzaTech
capturing carbon. fueling growth.

## Building a Technology Platform

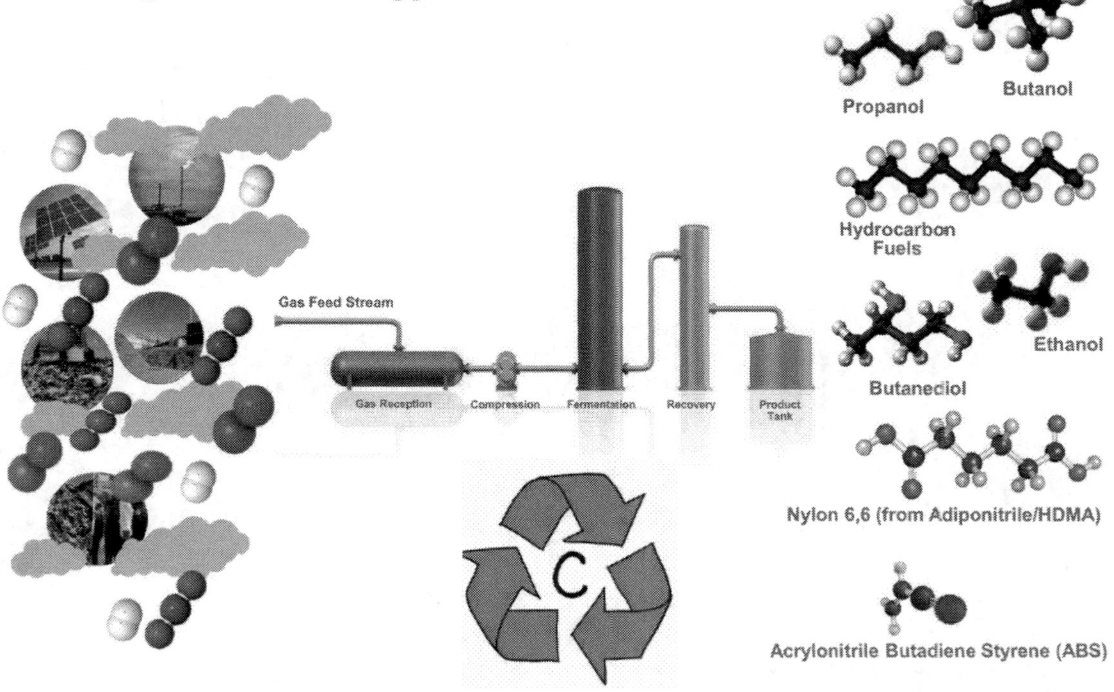


LanzaTech
capturing carbon. fueling growth.

24

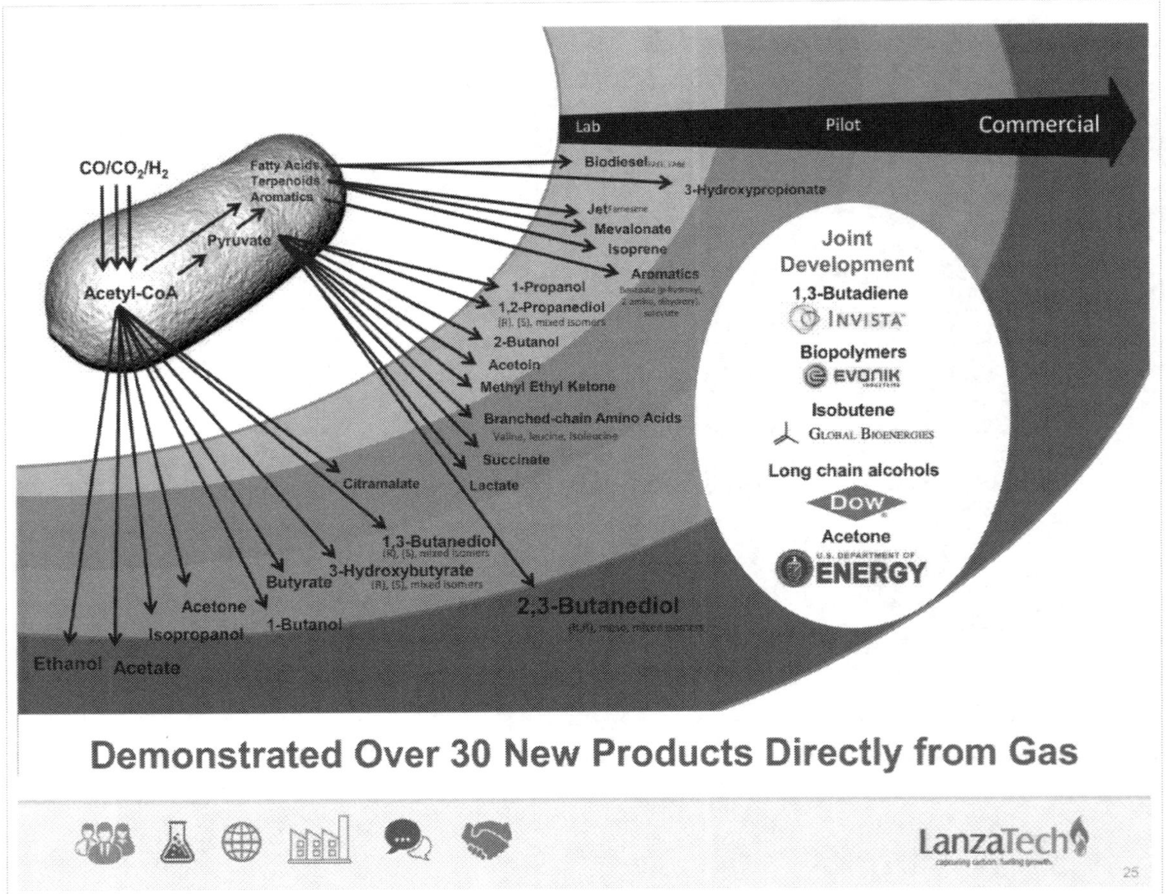
Lab
Pilot
Commercial
CO/CO2/H2
Fatty Acids
Terpenoids
Aromatics
Pyruvate
Acetyl-CoA
Biodiesel
3-Hydroxypropionate
Jet
Mevalonate
Isoprene
Aromatics
1-Propanol
1,2-Propanediol
(R), (S), mixed isomers
2-Butanol
Acetoin
Methyl Ethyl Ketone
Branched-chain Amino Acids
Valine, leucine, isoleucine
Succinate
Citramalate
Lactate
1,3-Butanediol
(R), (S), mixed isomers
3-Hydroxybutyrate
(R), (S), mixed isomers
Butyrate
Acetone
1-Butanol
Isopropanol
Ethanol
Acetate
2,3-Butanediol
Joint Development
1,3-Butadiene
INVISTA
Biopolymers
EVONIK
Isobutene
GLOBAL BIOENERGIES
Long chain alcohols
DOW
Acetone
U.S. DEPARTMENT OF ENERGY
Demonstrated Over 30 New Products Directly from Gas
LanzaTech
capturing carbon. fueling growth.
25

"hardware"
Microbe 2.0
✓new product molecu
✓Same operating
conditions
DISRUPTION = 1) Rapid Reaction to
Market Fluctuations 2) Feedstock ≠ Commodity

## Carbon: Opportunities Today

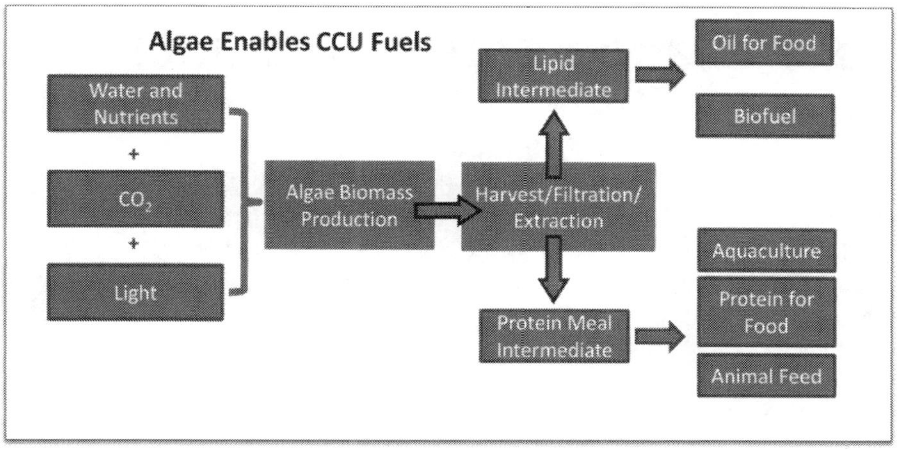


LanzaTech

## Crossing the Valley of Death

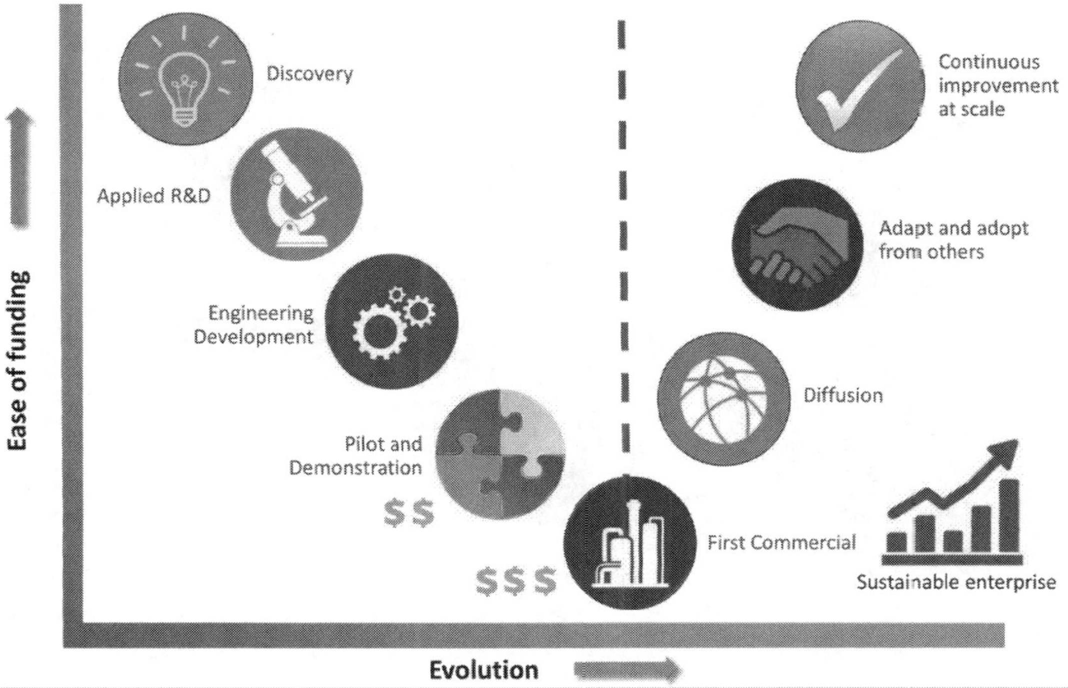


LanzaTech

**"I run in the darkness long before I ever dance under the lights"**

*-Muhammad Ali*

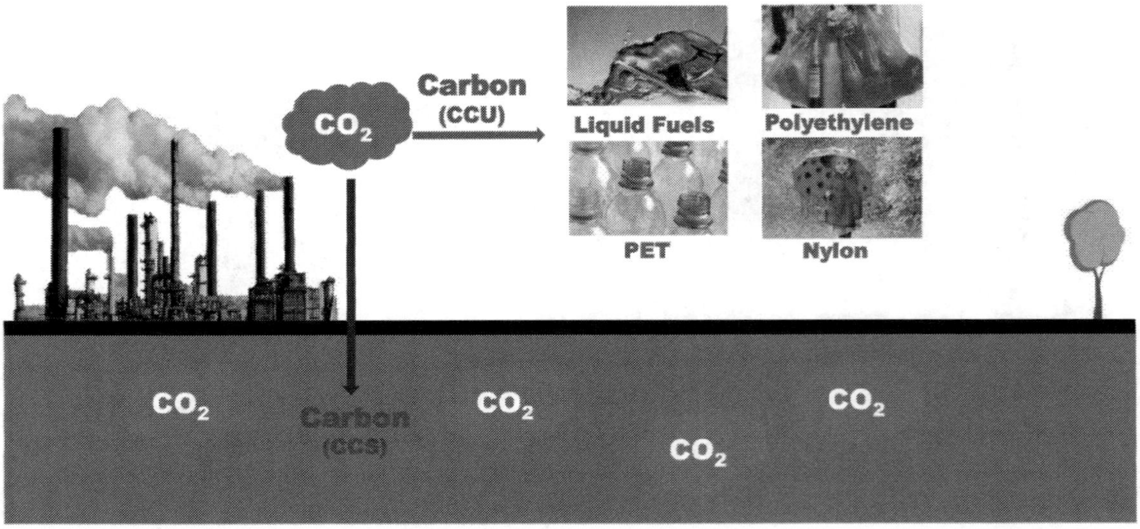


**Capturing Carbon. Generating Revenue. Reducing Emissions.**

**NOW**

**Carbon Management Technology Conference**
**Houston, Texas**
**July 18, 2017**

Robert M. Sturtz

VP of Business Development

World Fuel Services

# Aviation is a Ready Made Market for Bio-Fuels.

- Highly concentrated demand nodes
- Highly concentrated users
- Aviation has no alternative source of energy
- Economic benefits
- No changes needed for airplanes/airports/pipelines
- Environmental benefits

# Aviation Taking Action on $CO_2$

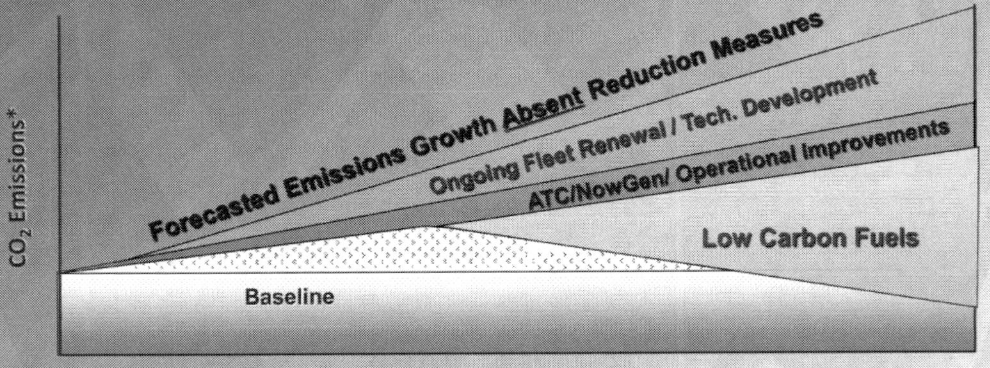


Carbon-Neutral Growth and Reduction Timeline 2050

- United worldwide – global sectoral approach
- Sustainable bio-fuels a key part of the solution

5

## Low-Carbon Fuels Critical to Voluntary Environmental Commitments of Worldwide Aviation Industry

1. Improve fuel efficiency by 1.5% p.a. in 2009-2020
2. Achieve carbon-neutral growth ("CNG") from *2020*
3. Reduce sector emissions by 50% from 2005 to *2050*

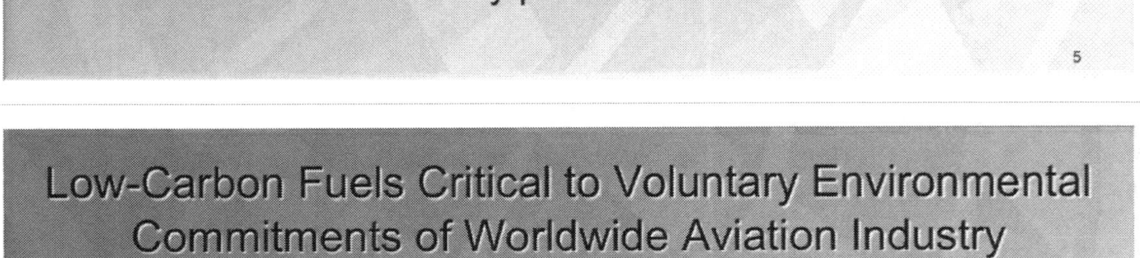

www.airlines.org Air Transport Association 6

## Airlines Take Their Environmental Responsibility Seriously

- Committed to Action on four Pillars
  - Technology including cleaner Fuels
  - Operations
  - Infrastructure
  - Basket of measures
    - Includes recent Agreements under ICAO that will lead to a cost for Carbon

www.airlines.org

Air Transport Association

## Airline Criteria for Alternative Fuels

- Reliable (Must meet ASTM Fuel Specifications)
- "Drop-In" Fuel
- Cost Competitive
- Environmentally better than conventional jet on an LCA basis

# Airline Off-take Agreements

| | | |
|---|---|---|
| • AltAir | 10 MG | United, WFS |
| • Fulcrum BioEnergy | 46.5 MG | United, Cathay |
| • Red Rocks | 6 MG | FedEx, SW |
| • SG Preston | 10MG | Jet Blue |
| • Gevo | 8 MG | Luthansa |
| • Amyris/Total | 1 MG | Cathay, Airbus |

9

## AltAir Project

- Off-take Agreements executed in 2012
- Repurposed the Paramount, CA Refinery
- Converts Beef Tallow to RD and SAJF
- Began Production in March 2016
- 1st Commercial production of SAJF in the World
- 1st commercial production of RD in CA
- Supplies united, Gulfstream, KLM, UPS, US Navy, and LAX ground vehicles.

CAAFI Sponsors
From across the industry
AIA
AEROSPACE INDUSTRIES ASSOCIATION
FEDERAL AVIATION ADMINISTRATION
CAAFI®
ACI
NORTH AMERICA
AIRPORTS COUNCIL INTERNATIONAL
Airlines for America™
We Connect the World
6 September 2017
11
CAAFI®

How CAAFI works
Bringing interested parties together*
20 Airlines, Military, Airport orgs.
15 Aircraft, Engine, Subsystem OEM's
15 U.S. States / State Univ's
CAAFI®
19 Countries / 5 Continents
30 U.S. Government Offices
Airlines for America™
We Connect the World
54 fuel producers
300 Global Sponsor/Stakeholder Attendees
180 Different Organizations, 32 EXPO Participants
6 September 2017
12
CAAFI®
* Attendance at 2011 Annual General Meeting

# CAAFI Work Teams

## Building the foundation to enable commercialization

**Research & Development**

***Enabling Multiple "Drop-in" Solutions***

**Certification / Qualification**

***Fostering ASTM D7566 Approval***

**Environmental**

***GHG LCA, PM2.5 Quantification, Sustainability***

**Business**

***Facilitating Deployment, Investment***

6 September 2017 13

# The Great Green Fleet Exercise

# CCU: Part II

## How Advances in Biological and Other Recycling Technologies are Changing the Carbon Equation

Carbon Management Technology Conference
July 18, 2017

**Matt Carr**
Executive Director
mcarr@algaebiomass.org

AB ALGAE BIOMASS ORGANIZATION

AB ALGAE BIOMASS ORGANIZATION

### PLATINUM MEMBERS

bioprocessalgae
ACCELERGY CORPORATION
WSGR Wilson Sonsini Goodrich & Rosati PROFESSIONAL CORPORATION

### GOLD MEMBERS

ALGENOL
LanzaTech
SYNTHETIC GENOMICS
subitec
cellana
TerraVia
MATRIX GENETICS
JOULE
TRITON ALGAE INNOVATIONS
KELLER AND HECKMAN LLP SERVING BUSINESS THROUGH LAW AND SCIENCE
CarlsonHighland Certified Public Accountants & Consultants

### SILVER MEMBERS

EARTHRISE
SCHOTT glass made of ideas
APPLIED CHEMICAL TECHNOLOGY The World Leader in Process Equipment Design
DIVERSIFIED TECHNOLOGIES, INC.
CLEARAS WATER RECOVERY
heliae
MBIO
FedEx Express
Sapphire Energy
NESTE OIL
DSM BRIGHT SCIENCE. BRIGHTER LIVING.
QUALITAS HEALTH
evodos
NCMA
PHITEC SUSTAINING EFFICIENCY
TEXAS A&M AGRILIFE RESEARCH
TASA
GLOBAL ALGAE INNOVATIONS
ZIVO BIOSCIENCE

ABO
ALGAE BIOMASS ORGANIZATION

# The Challenge

- Reduce GHG emissions from fossil power generation and other industrial sources...

**<u>Economically</u>**

Algae Biomass Organization | **www.algaebiomass.org**

ABO
ALGAE BIOMASS ORGANIZATION

# The Biological Capture and Utilization Opportunity

**<u>Algae & Other Autotrophs</u>**

- Earth's original carbon mitigation strategy
- $CO_2$ eating machines:
  - 100 acres of algae consume as much $CO_2$ as 2,000,000 trees
- Exceptionally fast-growing
  - Mature in days vs. months/years
- Factory for production of oils, proteins, other high-value products

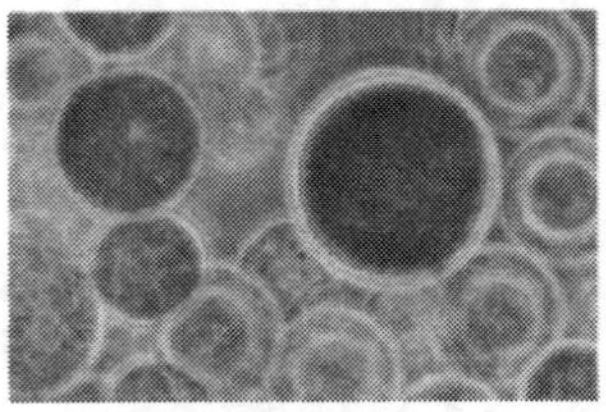

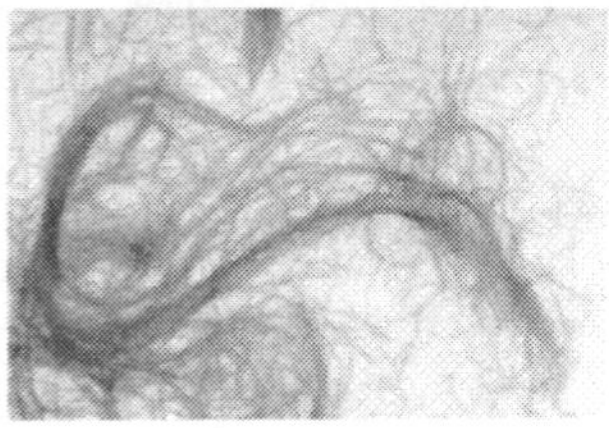

Algae Biomass Organization | www.algaebiomass.org

ALGAE BIOMASS ORGANIZATION

## The Biological Capture and Utilization Opportunity

- Can produce …
  **50-100 gallons biofuel per ton $CO_2$**
- Product value from algae conversion:
  > $100 / ton $CO_2$
- Today: Algae developers <u>buy</u> $CO_2$
  (at $40+ / ton) as a feedstock
- Industrial $CO_2$ Challenge + Algae =
  **Opportunity**



ALGAE BIOMASS ORGANIZATION

## Co-Location Potential

- **2016 DOE/ORNL Analysis*:**
  **<u>Substantial U.S. Potential for Algae CCU</u>:**

❖ 140,000 sq. miles suitable for open pond production (≈**75,000 farms** )

❖ Even with highly conservative assumptions, using existing technology:
  - More than 500 viable point sources
  - Potential to use **>200 million tons CO2 / year**

❖ Much larger quantities possible beyond 2030, especially with CO2 price/regulation

*2016 Billion-Ton Report *https://bioenergykdf.net/billionton2016/overview*

Algae Biomass Organization | www.algaebiomass.org

# Algae Chapter Highlights

## **COAL** - Saline Strain, Current Productivity

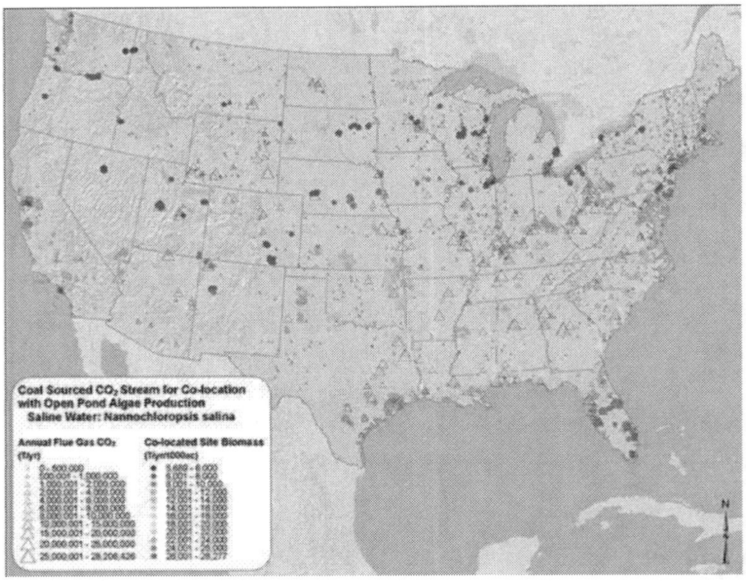


Algae Biomass Organization | www.algaebiomass.org

# Algae Chapter Highlights

## **NATURAL GAS** - Saline Strain, Current Productivity

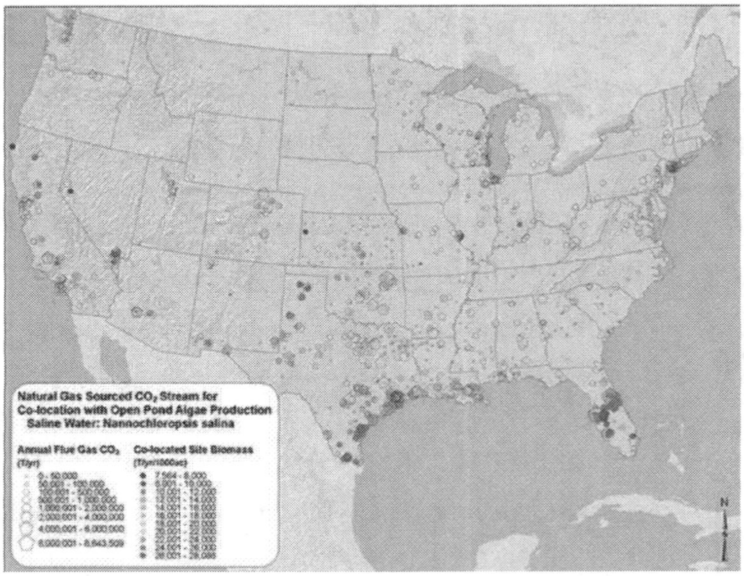


Algae Biomass Organization | www.algaebiomass.org

# Algae Chapter Highlights

## ETHANOL - Saline Strain, Current Productivity

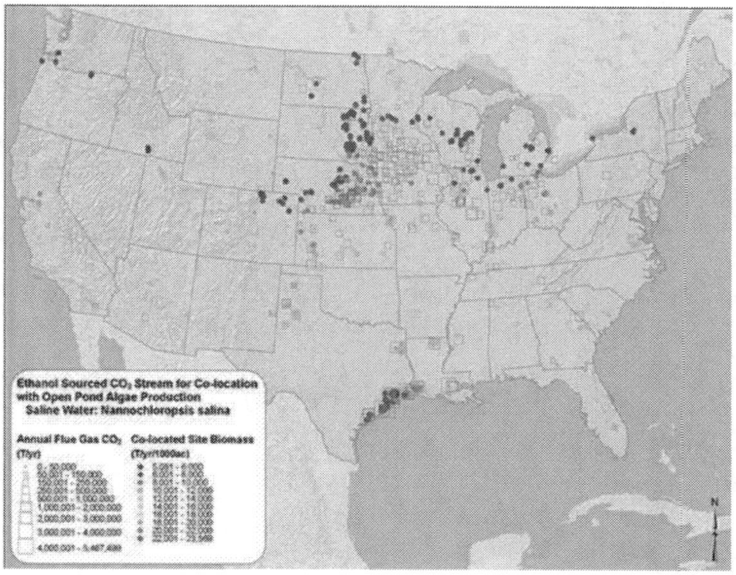


Algae Biomass Organization | www.algaebiomass.org

# Technology Progress

## ➢ Cost of Production

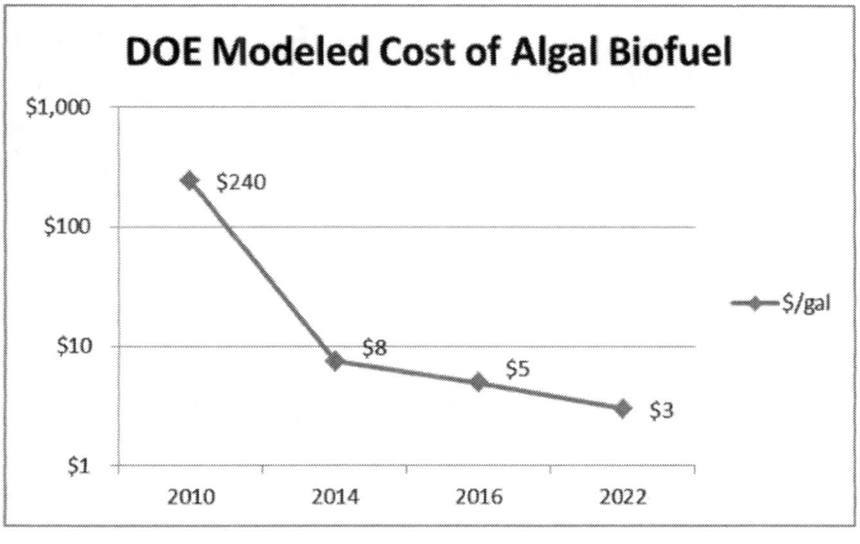


Algae Biomass Organization | www.algaebiomass.org

# Technology Progress

- DOE-funded integrated algae biorefinery commercial demonstration projects in:
  - Florida (Algenol)
  - New Mexico (Sapphire)
  - Iowa (BioProcess Algae)

- Pilot / demonstrations with flue gas from variety of industrial sources:
  - Coal, other power plants
  - Cement, steel
  - Ethanol

Algae Biomass Organization | www.algaebiomass.org

# Beneficial Re-use of Carbon Using Microalgae University of Kentucky / Duke Energy

- **Pilot project at Duke's East Bend Coal-powered Generating Station**
- **Recently awarded $1 million DOE Fossil Energy Grant**

CAER Center for Applied Energy Research | UK UNIVERSITY OF KENTUCKY | DUKE ENERGY

# 33-Acre Kauai Algae Facility

GLOBAL ALGAE INNOVATIONS

Algal Biomass Organization | www.algalbiomass.org

Distribution Statement "A" (Approved for Public Release, Distribution Unlimited)

ALGAE BIOMASS ORGANIZATION

# State of Algae Carbon Utilization Technology

## ➢ BioProcess Algae

- 5-acre commercial demonstration in Shenandoah, IA, co-located with Green Plains **corn ethanol plant**
- Continuous operation since 2009 utilizing waste CO2, waste heat, waste nutrients
- Markets: Fuel, fish/animal feed

bioprocessalgae

Valuing Carbon

Algae Biomass Organization | www.algaebiomass.org

## 3 Commercial Projects Underway

Strong Pipeline and Committed Partners

*Commercial Production 2017*

*1 M barrels/year in the ground*

LanzaTech

AB
ALGAE BIOMASS ORGANIZATION

## Big Market Opportunities

- **Aquaculture Feed**
  - **130 million tons, $9B / yr**

TerraVia DSM cellana ADM

- **Livestock Feed**
  - **1B tons, $370B / yr**
- **Fertilizer**
  - **180 million tons, $90B /yr**

ACCELERGY heliae

Source: Bloomberg New Energy Finance Research Note 5 June 2015

Algae Biomass Organization | www.algaebiomass.org

ALGAE BIOMASS ORGANIZATION

# Product Innovation & Diversification

## ➢ Soil Amendments and Biofertilizers

*200 Acre Algae Biofertilizer Demonstration – Arizona Farm*

ACCELERGY

- Algae cultivated in simple PBR's adjacent to the melon field
- Algae distributed through drip irrigation system

Del Monte Quality

| Yosemite Melons | |
|---|---|
| Germination | 2X |
| Early Growth | 2.5X |
| Plant Size & Density | 3X |
| Early Melon Maturity | 2X |
| Mature Melon Size | 1.4X |
| Mature Melon Weight | 1.5X |
| Taste (blind panel test) | 14:1 |



ALGAE BIOMASS ORGANIZATION

# Globalization

## ➢ Strategic Partnerships / Globalization

❖ **TerraVia / Bunge / BioMar – Aquaculture Feed**

TerraVia BUNGE BioMar

❖ **LanzaTech / ArcelorMittal / Virgin Atlantic – Steel Emissions to Aviation Biofuel**

LanzaTech ArcelorMittal virgin atlantic

Algae Biomass Organization | www.algaebiomass.org

## Other CCU Markets & Technologies

- **Concrete**
  - **Blue Planet** - Biomimetic mineralization into aggregates (43 billion ton / $350 billion global market) and other materials
  - **CarbonCure** – Sequestration in cement
- **Transportation Fuels**
  - 2 billion ton market
- **Performance Materials**
  - Graphene

Algae Biomass Organization | www.algaebiomass.org

## Policy Imperative: Parity with CCS

- **Federal CCU Tax Credit**
  - New Senate bill (FUTURE Act) provides $35 / ton for carbon captured and *utilized via photosynthetic, chemosynthetic or other conversion pathways.*
  - House bill of record limited to CCS
- **Carbon Utilization R&D**
  - Increased from $2 million in FY2015 to $10 million in FY2016-17
  - Still 1/10 funding level for capture, storage

Algae Biomass Organization | www.algaebiomass.org

## Policy Outlook

- CCU Well Aligned with Trump Priorities
  - Manufacturing
  - Infrastructure
  - Rural Revitalization

**JOBS!**

Algae Biomass Organization | www.algaebiomass.org

## 2017 Algae Biomass Summit

**Salt Lake City, Utah**

October 29 – November 1, 2017

Algae Biomass Organization | www.algaebiomass.org

BRIDGESTONE
Your Journey, Our Passion
Greer Tidwell
Carbon Management Technology
July 18, 2017
Houston, Texas
Copyright 2016 Bridgestone Americas

メガマック
MEGA×MEGA×MEGA

***Successful factory life***

REQUIRES TURNING

***NEIGHBORS INTO FANS***

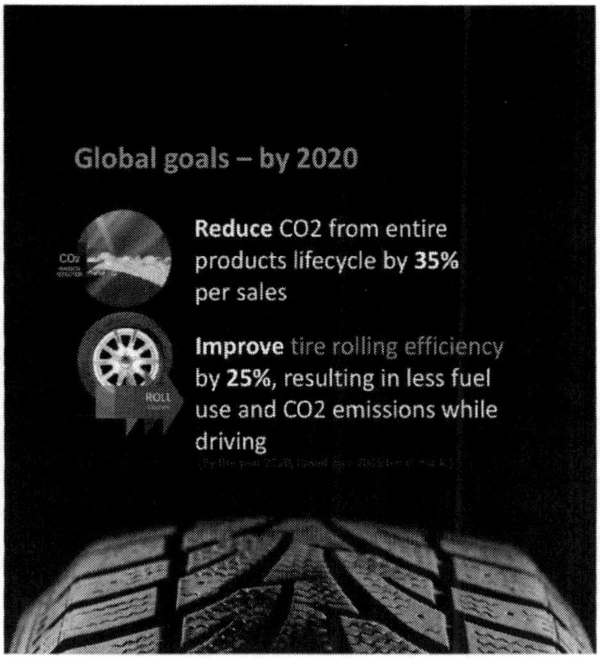


## Environmental Mission: Reducing C02 Emissions

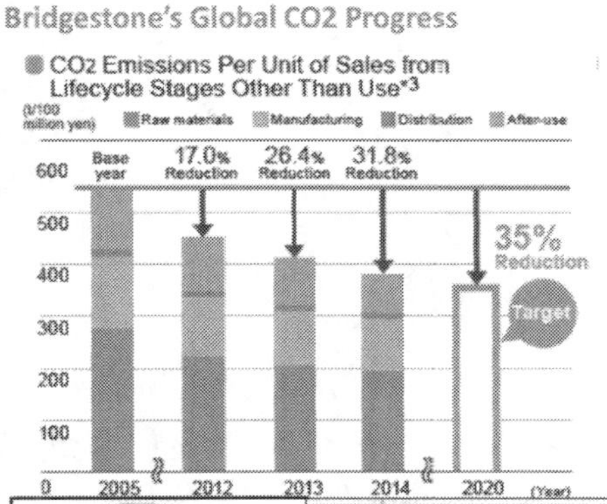

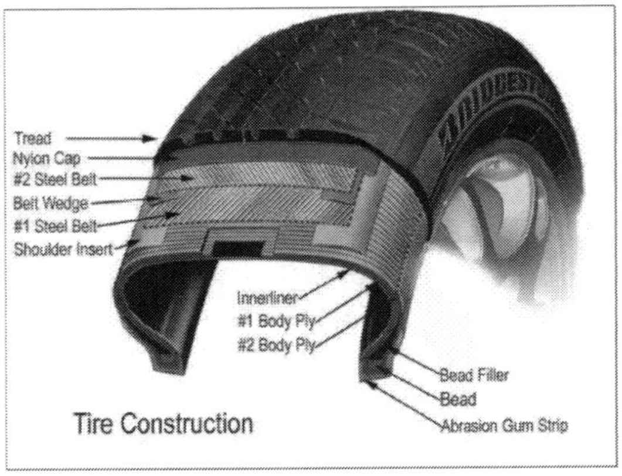

## 100% Sustainable Material Concept Tire 2012 Paris Auto Show

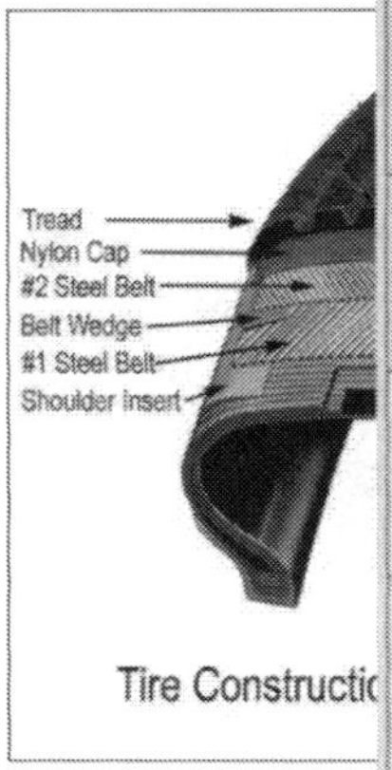


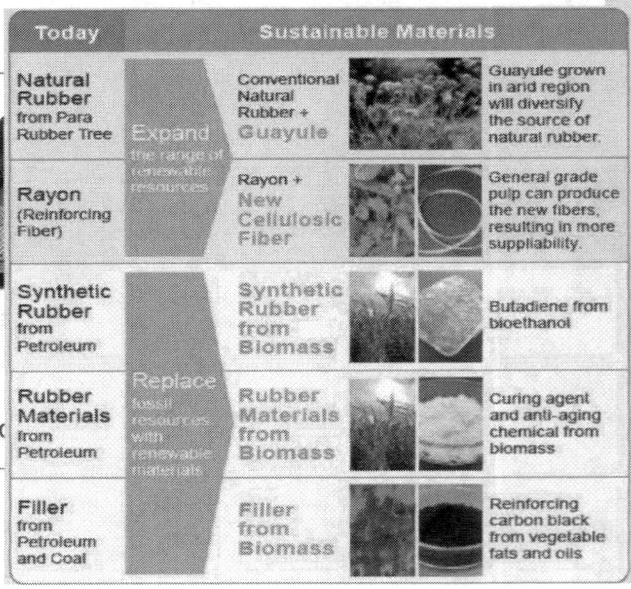

| Today | Sustainable Materials | | |
|---|---|---|---|
| Natural Rubber from Para Rubber Tree | Expand the range of renewable resources | Conventional Natural Rubber + Guayule | Guayule grown in arid region will diversify the source of natural rubber. |
| Rayon (Reinforcing Fiber) | | Rayon + New Cellulosic Fiber | General grade pulp can produce the new fibers, resulting in more suppliability. |
| Synthetic Rubber from Petroleum | Replace fossil resources with renewable materials | Synthetic Rubber from Biomass | Butadiene from bioethanol |
| Rubber Materials from Petroleum | | Rubber Materials from Biomass | Curing agent and anti-aging chemical from biomass |
| Filler from Petroleum and Coal | | Filler from Biomass | Reinforcing carbon black from vegetable fats and oils |

# Bridgestone GUAYULE R&D  MESA, ARIZONA

Current source
*Hevea brasiliensis*

# Nationwide: PROGRESS   Tire Recycling and Stockpile Reduction

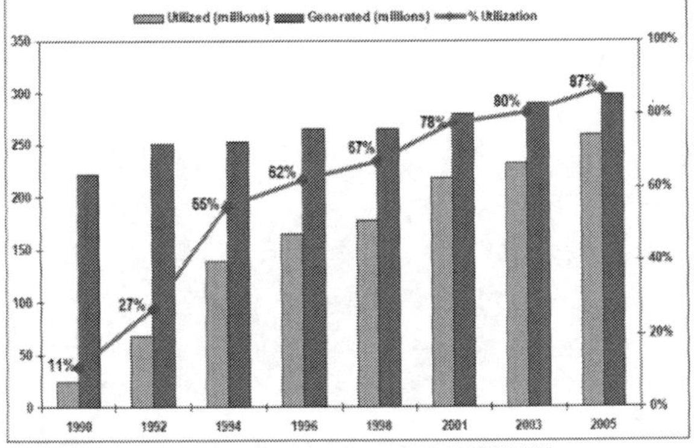


**Recycling U. S. Used Tires**
**THEN 10%  NOW >90%**

Source: Rubber Manufacturers Association

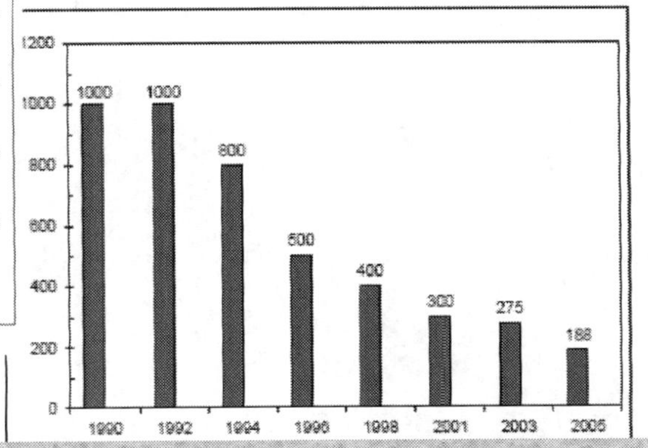


**Cleaning up U. S.**
**Used Tire Stockpiles, Now <75M**

"What do you do with all those scrap tires?"

BRIDGESTONE

TIRES4WARD

PUTTING OLD TIRES TO GOOD USE

Vision: WASTE FREE Tire Industry

- ✓ 100% of used tires returned to our stores get a beneficial next use.
- ✓ Provide Free scrap tire recycling for river cleanups
  - ✓ http://www.bridgestoneamericas.com/en/corporate-social-responsibility/environment-and-sustainability/tires4ward-request

>150,000 Tires recycled from 600+ events

*Your Journey, Our Passion*

GLOBAL
CCS
INSTITUTE

## The Current Narrative

E&E NEWS
Will CCS Ever Work?

HUFFPOST
**Donald Trump Promised 'Clean Coal,' But It Doesn't Exist**

The New York Times
**Companies Struggle to Make Carbon Capture Viable**

SCIENTIFIC AMERICAN
*Carbon Capture May Be Too Expensive to Combat Climate Change*

MORNING CONSULT
**Congress, White House Drag Feet on Support for Carbon Capture Expansion**

Bloomberg
Will Trump Make This $7 Billion Clean-Coal Plant Irrelevant?

Forbes
**Carbon Capture: An Expensive Option For Reducing U.S. $CO_2$ Emissions**

FINANCIAL TIMES
Carbon capture and storage — too little, too late, too expensive

Mashable
**Michael Bloomberg calls 'BS' on clean coal technology**

## Getting from Facts to Facilities

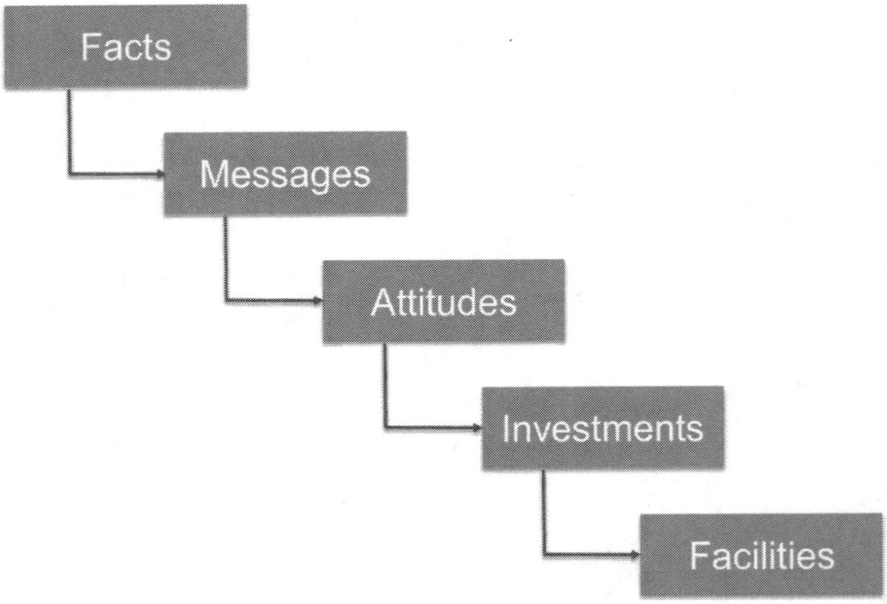


## The Language We Use

***It is better to be <u>understood</u> than to be comprehensive***

## Keep it Simple

SECU

CCUS

WECU

## Congressional Survey – Does Carbon Capture Work?

- **Language** – "CCS" is not an easily identified term
- **Cost** – widely held belief that CCS is too expensive
- **Climate change**
  - extends use of fossil fuel
  - Important option to address climate change
- **Safety** – risks not well understood

## Positive Language, Supported by Facts

**P**roven
**A**ffordable
**V**ersatile
**E**ssential

## Proven

**17 large-scale facilities**
**35+ million tonnes/yr**

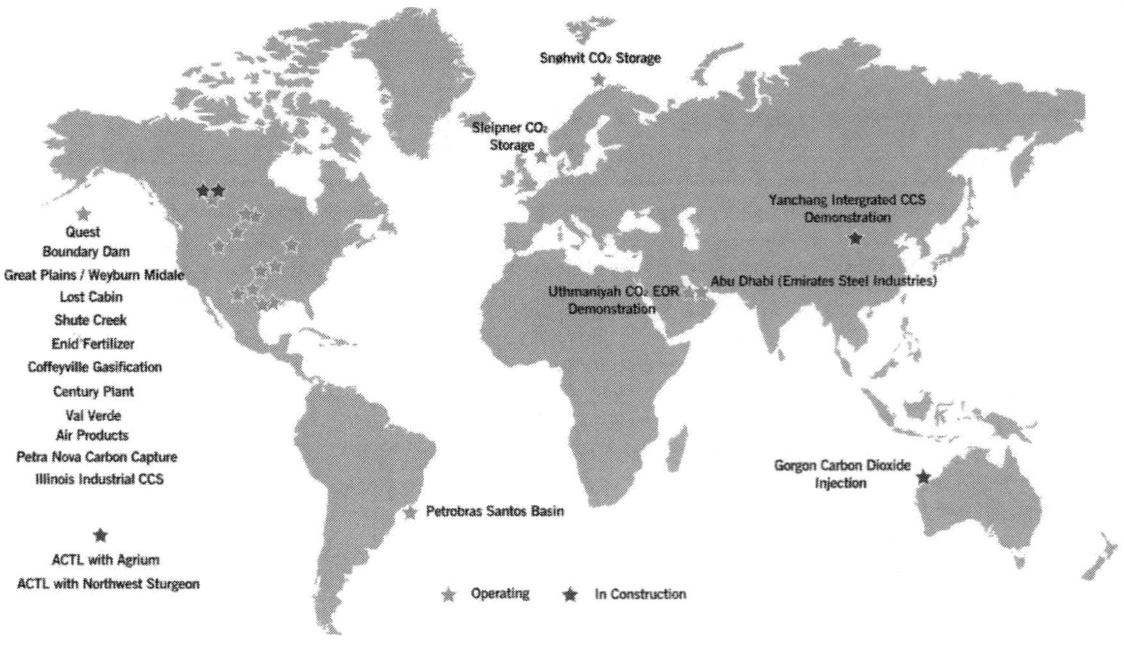

## Affordable

**Without CCS, cost of mitigation more than doubles**

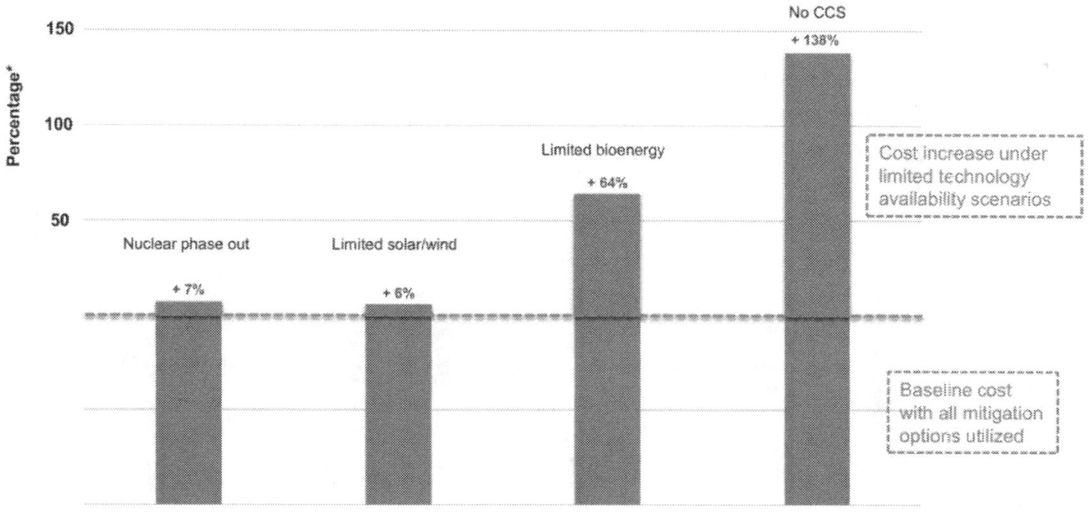


*Percentage increase in total discounted mitigation costs (2015-2100) relative to default technology assumptions – median estimate

**Source**: *IPCC Fifth Assessment Synthesis Report, Summary for Policymakers, November 2014.*

## Versatile

**CCS can be used across many industries**

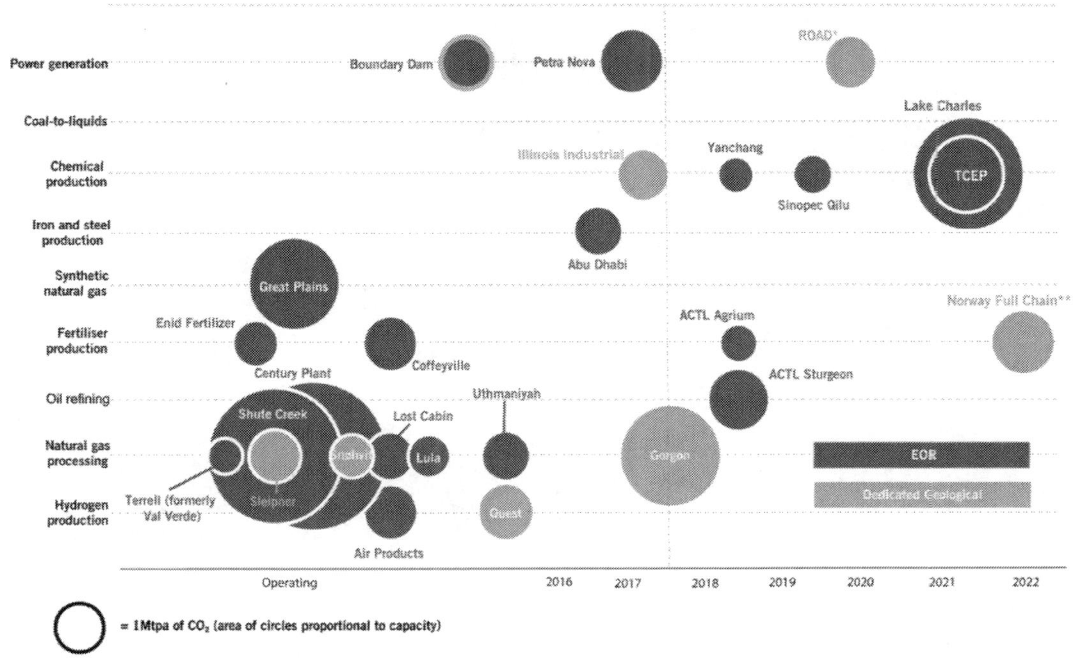

## Essential

**Numerous authorities say we can't achieve 2DS without it**

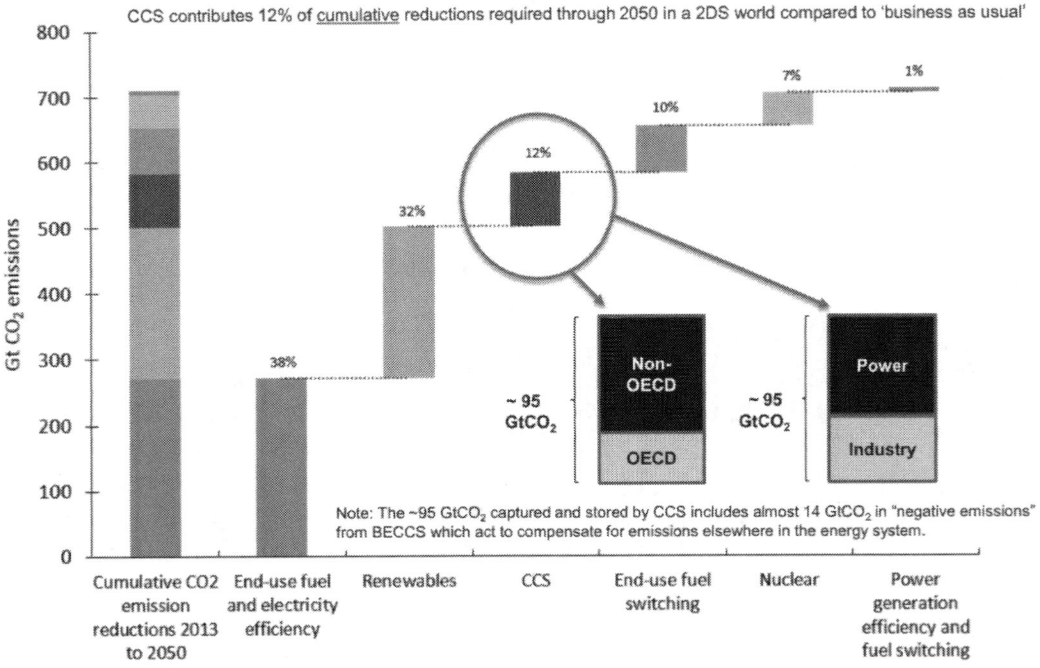


**Source:** *IEA, Energy Technology Perspectives (2016)*

## Join the Underground

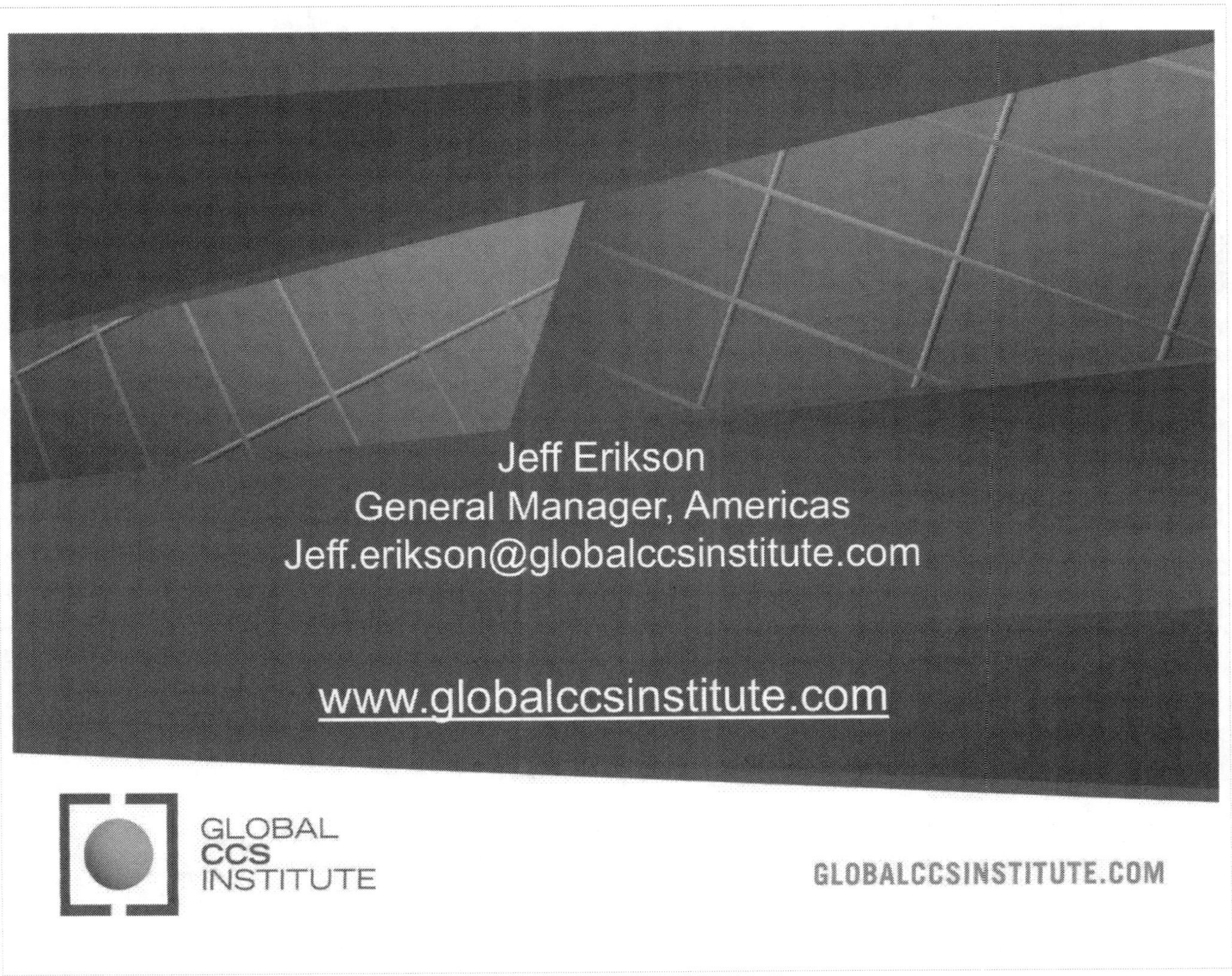
Jeff Erikson
General Manager, Americas
Jeff.erikson@globalccsinstitute.com
www.globalccsinstitute.com
GLOBAL
CCS
INSTITUTE
GLOBALCCSINSTITUTE.COM

# CHARTING A COURSE FOR COAL'S FUTURE

**Janet Gellici, National Coal Council**
Carbon Management Technology Conference
Houston, TX – July 18, 2017

## National Coal Council

Celebrating 33 years ~ 1984|2017

**The National Coal Council provides advice and recommendations to the Secretary of Energy on general policy matters relating to coal and the coal industry.**

**NCC is a Federal Advisory Committee organized under FACA legislation**

## National Coal Council

NATIONAL COAL COUNCIL

**Members**
Appointed to serve
by Secretary of Energy
125-150 members

- **Industry –**
  coal suppliers, utility & industrial consumers & coal transportation
- **Support Services –**
  engineering firms, vendors, consultants & attorneys
- **Academics**
- **NGOs –**
  environmental & trade association reps
- **Government –**
  PUC & state energy officials

**Reports**
~ 35 reports prepared by NCC
at no cost to DOE

**Extensive Range of Topics**

Advanced Coal Technologies
Coal Exports
Coal Conversion
Utility Deregulation
Climate & Clean Air Regulations
Building New Coal Power Plants
Industrial Coal Use
CCUS for EOR
Value of Existing Coal Fleet
Fossil Forward: CCS Technologies
Policy Parity for CCS
$CO_2$ Building Blocks: $CO_2$ Utilization

## Presentation Overview

- Trump Administration Priorities
- Energy Industry Priorities
- What's Ahead

NATIONAL COAL COUNCIL

NATIONAL COAL COUNCIL

# Administration Priorities

- Pro Development
  - *Economic/Job Growth*
  - *Infrastructure Development*
- Pro Market
  - *More Markets, Less Mandates*
  - *Regulatory Reform*
- Pro America
  - *Balance of Trade*
  - *Exports*

NATIONAL COAL COUNCIL

# Administration Energy Priorities

- American Energy Dominance
  - *Energy Resource Development*
  - *Regulatory Reform*
  - *Exports*

"... we're going to be an exporter
... We will be dominant.
We will export American energy all over the world."
President Trump – June 29th, 2017

## DOE Initiatives

NATIONAL COALCOUNCIL

- Energy Resource Development
  - *New offshore oil & gas leasing*
- Regulatory Reform
  - *RFI Regulatory Burden Reduction*
  - *Capacity Market Impacts Study*
- Exports/Balance of Trade
  - *$6.9 million Rare Earth Elements*
  - *Increased natural gas exports*
  - *New petro pipeline to Mexico*

"Becoming energy dominant means that we are getting government out of the way so that we can share our energy wealth with developing nations."

Secretary Perry, Secretary Zinke, Administrator Pruitt
June 26th, 2017 - Washington Times Op Ed

## Presentation Overview

- Trump Administration Priorities
- Energy Industry Priorities
- What's Ahead

NATIONAL COALCOUNCIL

## Optimize/Preserve Existing Coal Fleet

- Energy Security
- Economic/Job Growth
- Infrastructure Development

NATIONAL COALCOUNCIL

## Advance New Markets for Coal

- Job Creation/Economic Growth
- Balance of Trade/Exports
- Energy Security

NATIONAL COALCOUNCIL

## RDD&D for Advanced Coal Technologies

- Energy Security
- Infrastructure Development
- Economic/Job Growth

NATIONAL COAL COUNCIL

## National Coal Council Advanced Coal Technologies Subcommittee

- Technology Development & Policy Go Hand-in-Hand
- Technology Priorities
- Continued Government Support for RDD&D

NATIONAL COAL COUNCIL

## Advance Export Markets for Coal

- Balance of Trade/Exports
- Job Creation/Economic Growth
- Energy Dominance

NATIONAL COAL COUNCIL

## Advance Industrial Markets for Coal

- Infrastructure Development
- Economic Growth/Job Creation
- National Security

NATIONAL COAL COUNCIL

## Presentation Overview

- Trump Administration Priorities
- Energy Industry Priorities
- What's Ahead

NATIONAL COAL COUNCIL

## What's Ahead

- Budget FY2018
- Pending Infrastructure Plan
- Regulatory Reforms Initiative
- Tax Reform
- State vs. Federal Jurisdiction
- Paris Agreement Renegotiation
- Congressional Action

NATIONAL COAL COUNCIL

»»FOSSIL FORWARD
LEVELING THE PLAYING FIELD
Policy Parity for Carbon Capture and Storage Technologies
CO2 BUILDING BLOCKS
NATIONAL COALCOUNCIL
Janet Gellici, CEO
jgellici@NCC1.org
www.nationalcoalcouncil.org

# CARBON FARMING

## ASSOCIATED WITH SOLAR ENERGY DISSIPATION FOR REACTIVATING AND INCREASING GLOBAL PHOTOSYNTESIS ON ARID AND DESERT LANDS FOR ATMOSPHERE'S CARBON SEQUESTRATION AND FOR ADAPTING AGRICULTURE TO RECENT CLIMATE CHANGE

**Ferdinando Tessarolo**

Presented at CMTC 2017
Houston, TX
July 19, 2017

**Arid and desert lands are the result of an excess of solar energy and lack of rain and recently desertification is becoming the effect of climate change.**

CARBON MANAGEMENT TECHNOLOGY CONFERENCE

Los Monegros Desert, Spain

Los Monegros Desert, Spain

Los Monegros Desert, Spain

Los Monegros Desert, Spain

Los Monegros Desert, Spain

Los Monegros Desert, Spain

More than 35,000,000 km2 are deserts or arid land.

1) In case of lack of irrigation water a practical way to reduce aridity and desertification is getting rid of some of the solar energy. Of course a better result can be obtained by the combination of solar energy dissipation plus a bit of irrigation water. The application of solar energy dissipation reduces at least more than 50% the need of irrigation water in summer time. The dissipation of solar energy is a simple way to adapt arid and desert lands to climate change and at the same time it is an easy way to take CO2 down from the atmosphere and reduce the effect of climate change due to CO2 emissions.

## Carbon Farming Definition

By definition Carbon Farming means: agriculture dedicated to the take down and fixing atmospheric CO2 in the soil. Carbon farming associated with solar energy dissipation means agricultural dedicated to take CO2 down from the atmosphere in arid and desert lands and sinks CO2 into the soil. There for Carbon Farming Associated with Solar Energy Dissipation fulfills many objects. In one hand it is the best way to reactivate arid and desert lands for grass and plants colonization which reactivate photosynthesis for taking down carbon emissions from the atmosphere and, on the other hand, it is an excellent way to reactivate agriculture for increasing agriculture production in arid or desert lands. Of course CO2 coming from combustion is the same as carbon coming from the microorganisms' digestion or animals' breath. Of course plants will sequestrate CO2 from atmosphere without being able to see the difference between CO2 coming from combustion or carbon coming from microorganisms' digestion or animals' breath. The main object of Carbon Farming Associated with Solar Energy Dissipation is reducing the excess of CO2 concentration in the atmosphere by removing part of the CO2 specially the excess of CO2 of dispersed emissions, as in the cases of the CO2 emissions of all types of transport or CO2 emissions from factories or CO2 emission of coal or other fossil fuel power plants.

The combination of Carbon Farming with Solar Energy Dissipation could be the best instrument for reducing the evaporation of moister in the soil. Doing so we could be able to: 1) increase agriculture production and 2) take down atmosphere's CO2 and 3) mitigate climate change.

# Working Principle

The installation of parallel solar energy dissipation panels is an easy compatible and sustainable way to catch the solar energy before the solar energy reaches and heats up the soil. With less solar energy reaching the soil there will be less evaporation of soil's moisture. Each panel captures some solar radiation and produces two shades during the day, there will be one shade in the morning and the other in the afternoon. In the morning each panel produces a shade to the west and in the afternoon it produces a shade to the east. The result is two cross shades protecting the lands from excess of solar radiation. Additionally, the panels can protect the land from wind, doing so panels will also be able to reduce soil erosion and some soil's moister evaporation. Moisture will always be the best fertilizer of all in arid lands. At the end of the day our Carbon Farming Associated with Solar Energy Dissipation will be able to facilitate: 1) plant's growing, 2) better living conditions for microorganisms colonizing the land underground, and 3) better life conditions for insects colonizing the soil above the ground. All of them will, bit by bit, be able to improve the fertility of the normally sterile soil.

Result in the Los Monegros Desert.

In Los Monegros Rain falls from October to April are not more than 250 mm, and wind will be blowing for more than 100 days in the same period. After 8 years of trial project, in the **Los Monegros Desert**, Spain, we could see that biomass production has increased around 300 to 400% compared with similar biomass' production in arid lands cultivated in a traditional way. Obviously biomass production also depends on fertilization. It has been proved that protected lands can produce between 1 to 2 kilos of biomass per m2/year, which corresponds to 10 to 20 ton of atmospheric $CO_2$ sequestration/hectare/year.Obviously agriculture production depends on the rainfall, so there may be years with a bit more than 1 kilo biomass production and others with 2 kilos of biomass production per m2. Normally the difference in biomass production between protected land and land without protection is around 1 kg which corresponding to around 15 to 20 tons of $CO_2$/ hectare/year. The Biomass growing under solar energy dissipation panels' protection corresponds to solid $CO_2$ taken down from the atmosphere; said biomass is made of cereal, straw and roots.

## CO2 and microorganisms

Atmosphere's $CO_2$ sequestration by microorganisms has never been taken into account among the positive effects of solar energy dissipation. Microorganisms also participate in atmosphere's carbon sequestration by exchanging soil's minerals for exudates from plants. Exudates are juices rich in sugars and others organic elements. The exchange process of exudates from plants to microorganisms in exchange for minerals from microorganisms to plants allows plants growing better and facilitates microorganisms colonizing the soil. Because microorganisms can't do photosynthesis they need to receive plants' exudates for their development. Microorganisms are essential for improving the structure and fertility of soil. It has been noticed that year after year, after the application of solar energy dissipation, there is a great increase of the biomass production in the land protected by solar energy dissipation panels.

Additionally it has been found out that, in the case of durum wheat, there is an important increase of protein content which is usually close to 16% versus 12 or 13% protein of the durum wheat from open field irrigated or not irrigation land. In protected land we also got durum wheat with more than 17% proteins. 16% protein is an indicator of the high quality of the durum wheat such durum wheat is specially indicated for producing semolina for pasta and cuscus for high quality of food for humans.

## Weather influence

A large solar energy dissipation panel installation, 10.000 hectares for instance, will probably be able to produce local positive climate change as it is able to keep soil temperature lower than the temperature of land without panels and, at the same, because it is able to produce convective worm air movements, which, due to the V effect, will probably, during a storm, be able to induce more rain.

CARBON MANAGEMENT TECHNOLOGY CONFERENCE

**Cost of installation**

The cost of solar energy dissipation panels depends on the intensity of solar energy dissipation per hectare that is being looked for. The cost also depends on the result that is wanted to be achieved. In Spain the cost of the materials to build 2 meters high solar energy dissipating panels is around 7 to 8 E/per meter (one meter = to 3.28 feet), which means that the cost of materials per hectare (1 hectare = to 2.47 acres) can be between 8000 E for panels forming 10 m wide corridors or 12000 E for panels forming 7 m wide corridors. For the time being we only mention to the cost of materials used to build the panels as our panels have been built manually. We consider that the cost of installing panels by machine would be between 500 to 1000 E/ hectare. If shading fabric is used to build the panels shade fabric should be changed every 12 to 14 years depending on the intensity of the solar energy and the quality of the solar fabric. If others materials were used the initial cost for the panels could be higher than the cost of panels made with shading fabric and wood supports but the lifetime of the panels could be much longer. We have also noted that plants prefer solar energy dissipating panels to be black rather than green. Renewing the shade fabric cost around 4 E per meter. In the construction of our panels we have used wooden supports. The manufacturer of the wooden supports says supports have about 20 years guaranty. Cheaper and more durable supports such as cement supports could be used instead.

## Financial amortization

Cereal's production in arid land increases to about 4 ton/per hectare/year if solar energy dissipation panels protect the land. The benefit for selling the cereal should be taken under account for the cost of amortization of the panels. In the process of financial amortization of the cost of the panels it should be necessary to take into account the benefit of the sequestration of the atmospheric CO2. Unfortunately at present there are not laws to compel CO2 emitters to pay to take down CO2 emissions. CO2 emission should be organized like cities have organized the garbage's pickup. Climate change will be seriously taken under control only when emitters shall be compel by law to pay to store or to sequestrate their own emissions. The solar energy dissipation panels will be a business only when there will be a law establishing the fare price for CO2 emission real sequestration. For the time been the only thing we can do to take CO2 down is carrying on testing the application if solar energy dissipation. In this world, always looking for more and more energy, so far almost nothing has been done to reduce CO2 emissions. Instead there is a CO2 stock exchange to exchange CO2 emissions rights between rich emitters and poor countries. The CO2 right exchange market has never been able to taken any CO2 down from the atmosphere.

CARBON MANAGEMENT TECHNOLOGY CONFERENCE

Geo storage risks

Geo storage is the most known method for $CO_2$ emissions reduction. Frequently $CO_2$ geo storage is used for improving oil fracking. $CO_2$ together with others chemical products are pumped into old oil wells to increase oil production. Nobody knows if geo storage in old oil field will give some surprise in the future such as chemical leakage or $CO_2$ leakage. Leakages are the main risks on $CO_2$ geo stores practices (geo storage requires high pressure, there for high temperature at 2000 m deep could gasifies and push $CO_2$ to find a way out from the geo store and make $CO_2$ to return back to the atmosphere. In addition, it is difficult to find a natural chemical compound, such as calcium oxide, also known as quicklime, at sufficient geological depth to be able make chemical reaction with $CO_2$ to keep it ($CO_2$) in a geologically stable state such as calcium carbonate for long time. Among others risks, geological storage may have an additional risk. As it has happen in the case of natural gas geo storage CASTOR project ($CH_4$ natural gas geo storage) off the coast of Catalonia near Tarragona or the Mediterranean coast of Spain, http://politica.elpais.com/politica/2017/05/03 /portal/1493802085_594860.html. In 2013, the high-pressure geo storage has produced medium-intensity earthquakes onshore the coast of Tarragona, Spain. After those earthquakes CASTOR gas project has been suspended. In our opinion photosynthesis is to be considered one of the best stable way to take $CO_2$ down in the form of biomass in the first place and turning biomass into BIOCHAR later on. Dispersing BIOCHAR on the land is a simple way to sink $CO_2$ in land for long time.

**crop comparison**

**Agriculture $CO_2$ emissions**

It has been demonstrated that traditional agriculture is a net emitter of $CO_2$ as all lands dedicated to agriculture abandon the natural $CO_2$ cycle. The natural $CO_2$ cycle starts by taking $CO_2$ down in the form of biomass which is partially broken down by microorganisms in the form of organic matter and $CO_2$. Lands used for agriculture is progressively losing the organic matter that has been accumulated on the soil during millions of years. Agriculture practices, by the effect of: microorganisms, fertilizers and solar radiations, gasify more organic matter in $CO_2$ than the organic matter is being replaced by new biomass. In order to maintain the present current global carbon cycle and increase global photosynthesis, it is necessary to:

1) maintain the area already available for the carbon cycle (for instance we should stop colonizing other forest lands for agricultural use), 2) we should reduce forest fires, 3) we could use Solar Energy Dissipation on arid and desert lands for special agriculture together with rational use of the little irrigation water if available, 4) use some of the land used for agriculture today for producing pine trees or eucalyptus trees for the paper industries instead of chopping down natural pine or eucalyptus forest for the paper industries. For the above, for rehabilitation arid lands, in case of lack of irrigation water, we could need the following process: a) break or tilting arid or desert soil as deep as possible, 60 cm deep will be enough, breaking soil should be necessary in order establish cracks to let moisture penetrates deep into the soil, (desert land is almost always as hard as rock. b) provide natural mineral fertilizers and manure or other compatible fertilizers, c) installing parallel rows of solar energy dissipating panels to reduce evaporation of soil moisture and evapotranspiration stress of plants, all of them will be a possible compatible and sustainable solution for arid and/or desert lands without irrigation water or with little irrigation water available, d) it is necessary to leave straw or biomass on the soil after harvesting, e) leave the crushed straw on the soil will be necessary during the first years to in order to protect the soil from the vertical sun radiations during central hours of the day. f) After a sufficient lay of biomass covers and protects the soil from excess of solar radiation during the central hours of the day, some biomass can be taken away and be transformed into **BIOCHAR**. The transformation of biomass in **BIOCHAR** is necessary to avoid microorganisms to gasify most biomass in CO2. Microorganisms' bio digestion of the biomass and organic matter degrades organic matter and biomass by retransforming biomass into CO2 and release it back to the atmosphere. g) **BIOCHAR** should be used to improve the structure of the same soil which has produced the biomass, in this way, among others reasons, **BIOCHAR** will be able to turn soil a bit looser than the original hard soil and work as a shelter for microorganisms during drought seasons. h) Applies Carbon Farming Associated with Solar Energy Dissipation. **It is everybody's responsibility and interest to take care of the environment of the only planet we have.**

**Private transport's emissions.**

Every km we drive with our cars cost probably 0,50 E (Cost of car+ insurance+ garage etcetera etcetera) the government also spend taxpayer's found to build road, pay police and pay cost of roads's maintenance, so the real cost for driving a car will be a lot more than 50 cent/km. Suppose a driver drives an average of 25.000 km per year and release 4 tons of CO2/year the cost of taking that CO2 down will be around less that 2 cents per km.

**Industrial and Freight transport emissions**

Carbon Farming Associated with Solar energy dissipation could take down as much as all type of industrial and transport's CO2 emission released in the atmosphere.

**Domestic garbage emissions**

**Today many millions tons of domestic garbage are stored in garbage damp. Garbage damp are environmental risky and expensive to take care of. Garbage could be used as fuel for energy production and CO2 could be taken down in arid or desert lands, protected by solar energy dissipation. The same amount of CO2 released by garbage combustion can be taken down even in an arid land located far away from the place where the emission have taken place. A trash bottle of a soft drinks can emit around 200 gr. of CO2. If CO2 compensation for taking CO2 down from the atmosphere were 200 Euros per ton the cost of taking a soft drink trash bottle' CO2 should be around 4 Cents per bottle. There are many millions of hectares of arid and desert lands now abandoned waiting for Carbon Farming Associated with Solar Energy Dissipation to increase globe's photosynthesis to take CO2 down from the atmosphere. Carbon Farming associated with solar energy dissipation will be able to sink atmosphere's CO2 into the soil and will also be able to produce food, or other commodities, in those abandon arid or desert lands.**

# Climate change will carry on even with zero emissions.

It must be understood that if CO2 emissions from fossil fuel combustion were to be reduced to zero all at once today, climate change will continue due to more than 400 PPM of CO2 already in the atmosphere. In addition natural global photosynthesis is decreasing quickly day by day due to the colonization of forest lands and natural pasture for agriculture and industrial business. All form of traditional agriculture and forestry for paper production push lands out from the natural land's cycle of CO2 (Photosynthesis, biomass production and biomass break down). It is commonly accepted that bio fuels, from corn or sugar cane or others agriculture commodities, are better fuel than fossil fuel because they are consider climate friendly due to be of zero emission. This fact is true only if CO2 cycle, organic matter gasification and global natural photosynthesis reduction, are not being taken under account. In fact all lands, dedicated to traditional agriculture and the use of forest for paper, are losing organic matter day by day by natural elements' effects like microorganisms, fertilizers, etcetera which break down and gasify organic matter in form of CO2. When land is used for agriculture CO2 is pushed auto of land and get accumulated in the atmosphere increasing the GHG effect. So, even if agriculture emissions are set to zero, the present concentration of CO2 in the atmosphere will remain around 400 PPM, there for climate change will continue to increase even if combustion and agriculture emissions are set to zero. And again even if forests fire will be reduced to zero climate change will carry on its course for the 400 PPM of CO2 already in the air. Others GHG like CH4 from land and marine PERMAFROST (CH4 is six or seven times more GHG than CO2) will carry on melting and massively liberated in the atmosphere for the present global high temperature. So even if others GHG emission will stop today climate change will go on for the 400 PPM of CO2 already in the air. And again even if all what has been mentioning before will magically be set to zero today polar and glacier melting will carry on melting due to present global air temperature as less snow covering the planet means there

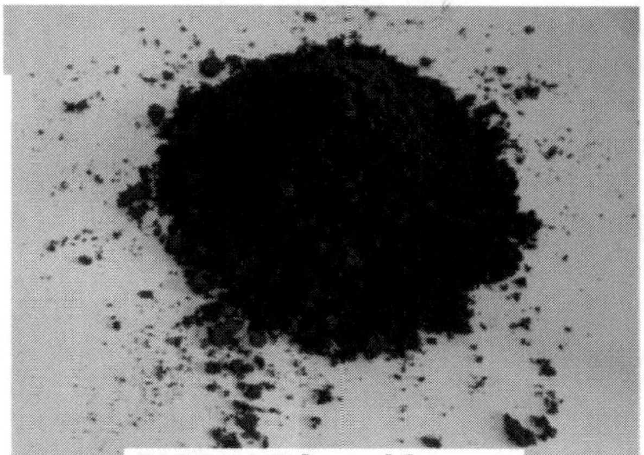
BIOCHAR from biomass

will be less sun rays reflection to the space so climate change will carry on. So, because of the 400 PPM of CO2 is already in the atmosphere, climate change will carry on even if CO2 emissions, agriculture CO2 and CH4 emission will be set to zero today. At the end of the day it is clear that, no matter if we were able to reduce all CO2 emission and other anthropological GHG direct or indirect emission to zero today, climate change will carry on indefinitely. So it is imperative to start reducing the CO2 concentration in the air now without delay.

Therefore the combination of Carbon Farming Associated with the Dissipation of Solar Energy for protecting arid or desert lands seems to be the only way to reduce CO2 in the atmosphere below 400 PPM.

BIOCHAR is biomass reduced to char by anaerobic combustion. It has been demonstrated that BIOCHAR lasts very long time in the soil.

https://en.wikipedia.org/wiki/Terra_preta

Concentrated CO2 emissions can be used for fracking or geostoring

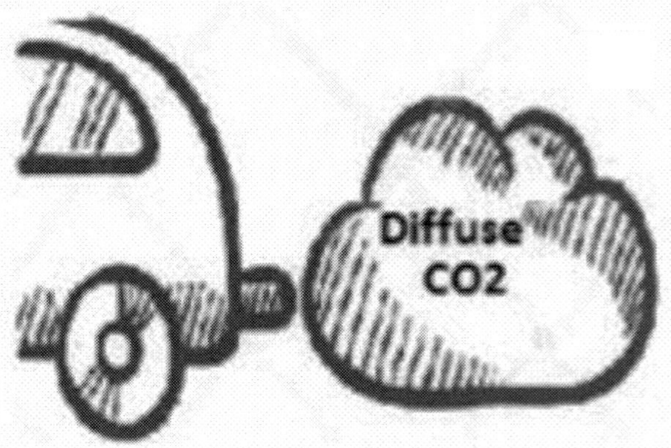


Car's emission can be taken down for 200 or 300 USD a Year

can of aluminum Less than 1 cent

2 litter bottle of soft drink: 4 cents

Wine bottle 5 cents

All the others items will cost between less than a cent to 2 cents

# Thank You

For more information contact
Ferdinando Tessarolo at
carbonfarming@yahoo.com

CARBON MANAGEMENT TECHNOLOGY CONFERENCE

# Carbon Farming Associated with Solar Energy Dissipation

for increasing global photosynthesis in arid and desert land for taking $CO_2$ down from atmosphere and sink it in the soil.

Desert land

Protected land

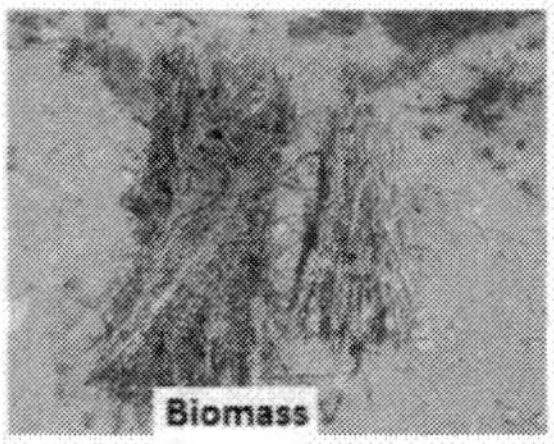

Biomass

Wheat

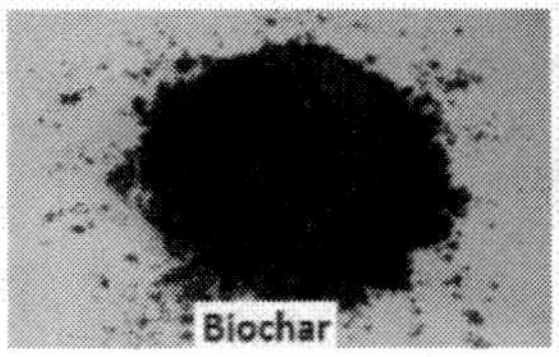

Biochar

US desert and arid lands

**Carbon farming associated with solar energy dissipation reduces plant's evapotranspiration and the evaporation of moister in the soil for facilitating biomass production in arid and in desert lands. The increases of global photosynthesis is the best way to take $CO_2$ down from the atmosphere. Biomass can be reduced in BIOCHAR and Biochar can be put back in the same land where the biomass has been produced in the first place. BIOCHAR can stay stable in the land for hundreds protected by solar energy dissipation panels can produce more than 15 tons of biomass per hectare per year. 10 tons of biomass is equivalent to around 18 ton of $CO_2$ taken down from the atmosphere. So even if the solar energy dissipation panels are taken away the BIOCHAR will stay stable in the land for hundreds of years.**

Concentrated $CO_2$ emissions can be used for fracking or geostoring

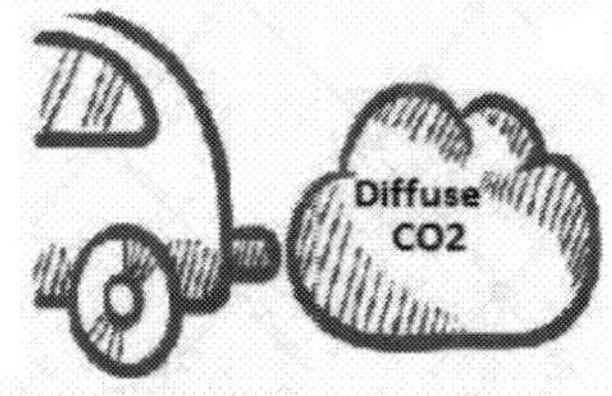


Car's emission can be taken down for 200 or 300 USD a Year

## Patent for sale

**Carbonfarming@yahoo.com**

**If the compensation for diffuse emissions were 200 USD a ton, the compensation for taking down the emissions of the following items could be:**

can of aluminum Less than 1 cent

2 litter bottle of soft drink: 4 cents

Wine bottle 5 cents

All the others items will cost between less than a cent to 2 cents

**All diffuse emissions can be taken down by carbon farming associated with solar energy dissipation on arid and desert lands for a sustainable little compensations.**

for more information contact

**carbonfarming@yahoo.com**

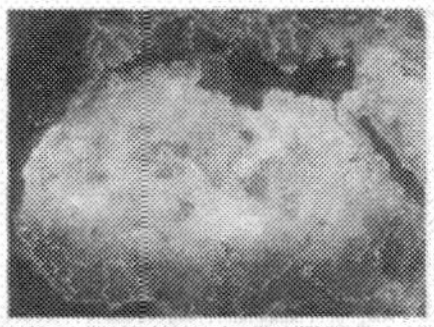

Africa desert and arid lands

**Carbon farming associated with solar energy dissipation increases global photosynthesis and global biodiversity in arid and desert lands**

# Carbon Farming Associated with Solar Energy Dissipation

for increasing global photosynthesis in arid and desert land for taking $CO_2$ down from atmosphere and sink it in the soil.

Arid or desert lands have no plants on them. No plants mean no photosynthesis and no photosynthesis means no carbon sequestration above the ground level nor carbon sequestration below the ground level. So drought means sterile land, poverty and climate change. If there is no irrigation water available the only thing we can do to compensate the excess of solar energy is dissipating some of the solar energy.

Desert Land

Desert land with solar energy protection

Parallel lines of solar energy dissipating panels will be able to reduce the temperature of the ground and there for reduce moister evaporation in the corridor of land between the panels. The panels will also be able to reduce plant's evapotranspitation stress and increase the amount of biomass that can grow in protected land. More biomass means more atmosphere carbon sequestration and more crop.

The difference of dry biomass production between protected land and unprotected land is normally 1kg of biomass per m2 which means between 15 to 20 tons of carbon sequestration from the atmosphere $CO_2$. At the same time crop will increase up to around 4 ton of cereal per hectare.

**Patent for sale**
**for more information contact**
**carbonfarming@yahoo.com**

Wheat

Carbon Farming Associated with Solar Energy dissipation increases global photosynthesis in arid and desert land for taking $CO_2$ down from atmosphere and sink it in the form of BIOCHAR in the soil. Solar Energy Dissipation panels reduce the evaporation of moister in the soil and plant's evapotranspiration. Doing so we can: 1) increase agriculture production in arid and desert land, 2) take down $CO_2$ from the atmosphere, and 3) mitigate climate change.

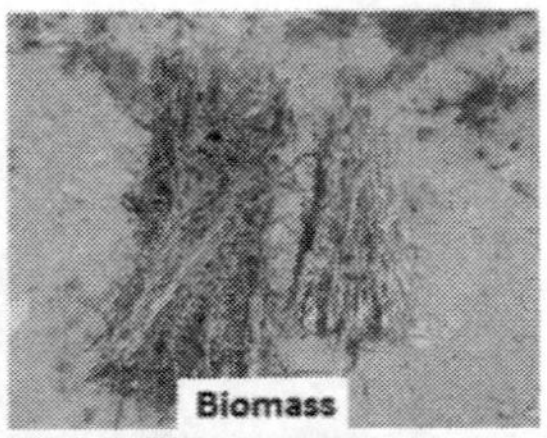

Biomass

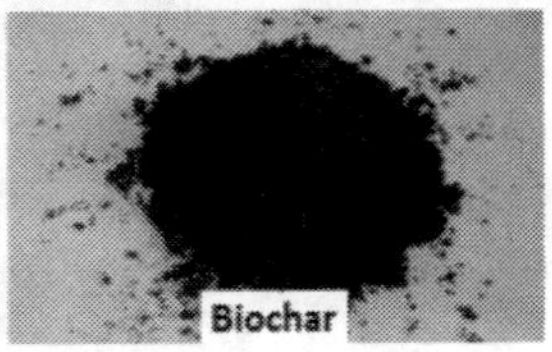

Biochar

## Climate change will carry on even with zero emissions.

**It must be understood that if $CO_2$ emissions from fossil fuel were to be reduced to zero all at once today, climate change will continue due to others GHG emission and for more than 400 PPM of $CO_2$ already in the atmosphere. If also agriculture $CO_2$ emissions will be reduced to zero today climate change will carry on for the 400 PPM of $CO_2$ already in the air and for the others GHG emissions. And again even if also forests fire will be reduced to zero climate change will carry on its course for the others GHG emissions and for the GHG already in the air. And even if also natural gas emission from land and marine PERMAFROST will be reduced to zero climate change will carry on for the GHG already in the air. And again even if all what has been mentioned before will magically be set to zero today polar and glacier melting will carry on due to present climate disorder. So, because a lot of GHG are already in the atmosphere, climate change will carry on even if all GHG emissions will be set to zero today. At the end of the day it is clear that, no matter if we were able to reduce all GHG emission to zero today, climate change will carry on indefinitely. So it is imperative to set to zero the GHG emissions and start reducing the $CO_2$ concentration in the air now without delay.**

Sahara desert

Therefore the combination of Carbon Farming with Solar Energy Dissipation protecting arid and desert lands seems to be the only way to reduce $CO_2$ in the atmosphere below 400 PPM.

America deserts

HUNTON & WILLIAMS

# Trump Administration Policy Developments and CCUS

Carbon Management Technology Conference
Houston, Texas

**Fred Eames**
Partner, Hunton & Williams LLP
July 18, 2017

## Introduction

HUNTON & WILLIAMS

**Hunton & Williams LLP**

- Founded in 1901, Richmond, VA
- 700+ lawyers, 19 offices, 6 countries
- Dallas, Houston, Austin
- Strong focus on energy/environment

**Personal**

- Partner, 2005 - present
- Decade experience on Capitol Hill
- Counsel, House Committee on Energy & Commerce
- Represent utilities, oil and gas, other energy clients on energy/environment
- Founded CCS Alliance in 2008
- Technical Chair and Lead Author, 2015 NCC report “Leveling the Playing Field: Policy Parity for Carbon Capture and Storage Technologies

HUNTON& WILLIAMS

If you don't read the newspaper, you're uninformed. If you read the newspaper, you're misinformed.
- Mark Twain

Far and away the best prize that life has to offer is the chance to work hard at work worth doing.
- Theodore Roosevelt



## Policy Factors Influencing CCUS

HUNTON& WILLIAMS

- Energy Policy
  - Support for coal and fossil
- Environmental Policy
  - Climate: Re-establishing legal and evidential foundations
- Economic Policy
  - Shrink government
  - Reduce spending and regulation
  - Minimize incentives and government distortion of markets
  - Grow the economy

**So, regarding CCUS, how do we increase government support for coal while eliminating climate mandates, reducing R&D spending, and stopping market distortions?**

## What CCUS Needs from Government

HUNTON & WILLIAMS

Consistent government financial support sufficient to sustain investment decisions

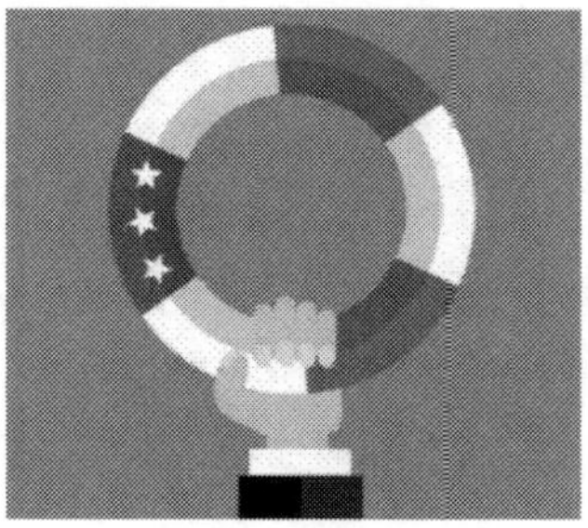

Minimize regulatory barriers

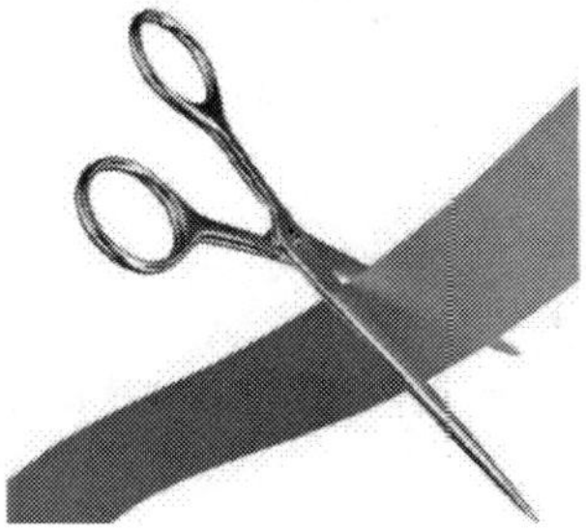

July 26, 2017

5

## President Trump on Coal

HUNTON & WILLIAMS

"The miners told me about the attacks on their jobs and their livelihoods. They told me about the efforts to shut down their mines, their communities, and their very way of life. I made them this promise: We will put our miners back to work. . . . Today, I'm taking bold action to follow through on that promise. My administration is putting an end to the war on coal. We're going to have clean coal -- really clean coal. With today's executive action, I am taking historic steps to lift the restrictions on American energy, to reverse government intrusion, and to cancel job-killing regulations."
- March 28

July 26, 2017

6

## Energy Policy

HUNTON & WILLIAMS

- Actions supporting fossil energy – coal in particular
  - New theme: Energy dominance
  - Stream Protection Rule revoked – CRA resolution signed February 16
  - Removal of regulatory barriers – Executive Order 13783, Promoting Energy Independence and Economic Growth, March 28
  - DOI coal leasing moratorium lifted, March 29
  - DOE grid study to support baseload energy, April 14
    - "Over the last few years . . . grid experts have expressed concerns about the erosion of critical baseload resources . . . highlight[ing] the diminished diversity of our nation's electric generation mix, and what that could mean for baseload power and grid resilience. This has resulted in part from regulatory burdens introduced by previous administrations that were designed to decrease coal-fired power generation. Such policies have destroyed jobs and economic growth, and they threaten to undercut the performance of the grid well into the future."
    - "The results of this analysis will help the federal government formulate sound policies to protect the nation's electric grid. In establishing these policies, the Trump Administration will be guided by the principles of reliability, resiliency, affordability, and fuel assurance – principles that underpin a thriving economy."
- Political appointments

July 26, 2017 7

## President Trump on Climate Change

HUNTON & WILLIAMS

- "As someone who cares deeply about the environment, which I do, I cannot in good conscience support a deal that punishes the United States -- which is what it does — the world's leader in environmental protection, while imposing no meaningful obligations on the world's leading polluters. [We will] begin negotiations to reenter either the Paris Accord or a really entirely new transaction on terms that are fair to the United States, its businesses, its workers, its people, its taxpayers."
  - June 1

July 26, 2017 8

## Environmental (Climate) Policy

HUNTON & WILLIAMS

- Actions on climate policy
  - Paris Accord – U.S. will withdraw
  - Clean Power Plan/111(b) rules – EPA reviewing, CPP litigation stayed
  - Social Cost of Carbon – Interagency Work Group disbanded, documents withdrawn
  - "Red Team" Exercise – EPA Administrator Pruitt announces evaluation of climate science
    - "The public is largely unaware of the intense debates within climate science. . . . Consensus statements necessarily conceal judgment calls and debates and so feed the 'settled,' 'hoax,' and 'don't know' memes that plague the political dialogue around climate change."

      - Steve Koonin, Wall Street Journal, April 21
  - SEC climate risk disclosure guidance – enforcement suspended
  - To be determined: Endangerment finding, GHG reporting rule



## Key Environmental Issues for CCUS

HUNTON & WILLIAMS

*From the National Coal Council's 2015 Report "Leveling the Playing Field: Policy Parity for Carbon Capture and Storage Technologies"*

- New Source Review
- Defining secure geological storage
- Siting
  - NEPA, ESA, CWA, MBTA, etc.
- Misunderstanding of the growing – not declining – global use of fossil fuels

## Economic Policy

HUNTON & WILLIAMS

### Trump Administration on Economic Priorities

- "[W]e looked at this budget through the eyes of the people who are actually paying the bills. . . . [The President's budget] begins to reduce the size of the debt relative to the size of the economy in year one. . . . Everything that we do in this administration . . . the focus is sustained, 3 percent economic growth."

  - OMB Director Mick Mulvaney, May 23

### Key proposals and actions

- Proposed FY 2018 budget
  - DOE energy programs budget cut from \$11.3 B to \$7.3 B
  - CCS-related funding cut by nearly 85%
  - House Appropriations budget - \$9.6 B; CCS funding increased
  - Senate – Bill to be released this week
- Disaffection with funding for commercialization; focus on early stage R&D
- \$1 trillion infrastructure bill – contents tbd

July 26, 2017 11

## Economic Policy

HUNTON & WILLIAMS

### Key proposals and actions

- Tax reform
  - Principles: wage and job growth, simplicity and fairness, "service-first" IRS
  - Structure: reduce corporate and individual rates, paid for by border adjustment tax and elimination of special tax incentives
  - Key for CCUS: increasing the Section 45Q credit; extending PABs to CCS
    - Heitkamp-Whitehouse-Capito bill reintroduced July 12

| 45Q tax credit | Current | Proposed |
|---|---|---|
| CCUS | \$10/ton | \$35/ton |
| CCS | \$20/ton | \$50/ton |
| Limit | 75 million tons | None |
| Threshold | 500,000 tpy | 100,000 tpy |
| Secure storage | IRS guidance | No change |

    - H.R. 2010 (Cramer – R-ND) – Would define secure storage as Subpart UU compliance
  - Carbon tax?

July 26, 2017 12

## Where Does This Leave CCUS?

HUNTON& WILLIAMS

Key issues:

- CCUS = energy dominance, with the additional benefit of addressing carbon emissions
- DOE R&D budget for CCUS – defend and enhance
- Increase the Section 45Q tax credit, cultivate support for other incentives
- "Institutionalist" political appointees need to win internal battles over budgeting and incentives

Key State issue:

- Will States support CCUS projects?

Outlook summary:

- The glass is more than half full



HUNTON& WILLIAMS

**Fred Eames**
Hunton & Williams LLP
2200 Pennsylvania Avenue, NW
Washington, DC 20037
202.955.1500
feames@hunton.com
https://www.linkedin.com/in/frederick-eames-2b3215a/

Carbon Management Technology Conference 2017, Houston, TX

# Toward Negative Emissions on Electric Grids With High Penetrations of Intermittent Renewables


Robert H. Williams, Senior Research Scientist Emeritus
Andlinger Center for Energy and the Environment
Princeton University


18 July 2017

PRINCETON UNIVERSITY

andlinger center
for energy+the environment

## BECCS as Carbon Mitigation Option

- The 2015 Paris Climate Accord, since ratified by 153 countries, seeks to limit the rise in global temperature to 2 °C [**2DS future**] and 1.5 °C if possible.
- According to the IPCC's 5AR:
  - "Many models cannot reach concentrations of about 450 ppm $CO_{2e}$ by 2100 [roughly equivalent to a 2DS future] in the absence of CCS," and
  - "Many models could not limit *likely* warming to below 2°C if bioenergy, CCS, and their combination (BECCS) are limited *(high confidence)*."
- In *Energy Technology Perspectives 2017* the IEA charted a possible path for global energy to 2060 consistent with limiting global warming to 1.75 °C—a path that gives prominence to BECCS—e.g., photosynthetic $CO_2$ storage in geological media at a rate of ~ 300 Mt/y by 2030 (15% of total global CCS) and ~ 4900 Mt/y by 2060 (45% of total global CCS).
- But progress has been slow in advancing BECCS technologies.
- Proposal to accelerate BECCS development:
  - Launch BECCS in market via coprocessing coal and biomass:
    - Benefits to BECCS: lower average feedstock costs/economies of scale compared to "pure" BECCS;
    - Benefit to coal: path whereby coal in US might be better able to compete with low-cost natural gas under a carbon policy…by exploiting economic benefit of negative emissions.
  - Carry out coal/biomass coprocessing with CCS with aim of decarbonizing balancing capacity on electric grids with high penetrations of intermittent renewables—a high priority in quest for a low carbon energy future.

## Timeliness of BECCS Market Launch via Coal/Biomass Coprocessing

- There is a reasonable chance that in 2017 **S.1460** (*Energy and Natural Resources Act of 2017*) will be enacted—passed Senate, 20 April 2016, but needs to be reconciled with House version.
- *Inter alia*, **S.1460** authorizes FE to spend $22 million/year for 5 years (2018-2022) for support of FEED studies for net-negative $CO_2$ emissions projects. The term 'net-negative carbon dioxide emissions project' means a project employing technologies for thermochemical co-conversion of coal and biomass with CCS that:
  - The Secretary of Energy determines can provide electricity, fuels, or chemicals with net-negative carbon dioxide emissions from production and consumption of the end products, while removing atmospheric $CO_2$;
  - Will proceed initially through a large-scale pilot project for which front-end engineering will be performed for bituminous, subbituminous, and lignite coals; and
  - Through which each use of coal will be combined with the use of a regionally indigenous form of biomass energy, provided on a renewable basis, that is sufficient in quantity to allow for net-negative emissions of carbon dioxide (in combination with a carbon capture system), while avoiding impacts on food production activities.
- Co-sponsored by Senators Murkowski (R-AK) and Cantwell (D-WA), **S.1460** has broad bipartisan support.
- The 'net-negative $CO_2$ emissions project' provision of **S.1460** was developed by Senator Manchin (D-WV).
- The 'net-negative carbon dioxide emissions project' provision is consistent with President Trump's promise to find ways to advance coal.
- The seminal Sanchez and Kammen (2016) article highlights the strategic importance of BECCS market launch via coal/biomass coprocessing.

## Toward Reliable Grid Power with High Penetrations of Intermittent Renewable Electricity (IRE)

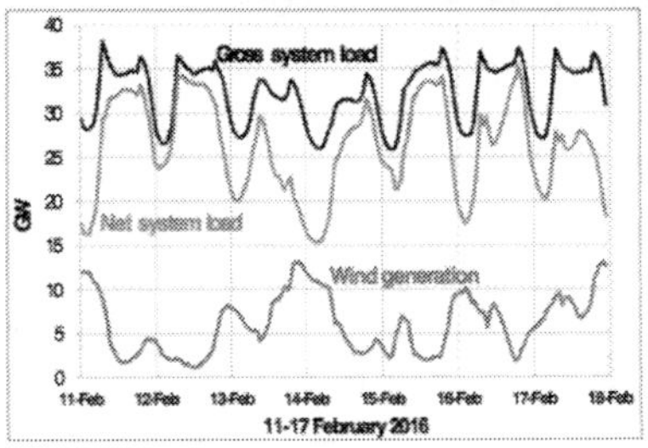


7 days of wind power and electric load for ERCOT grid (Texas)

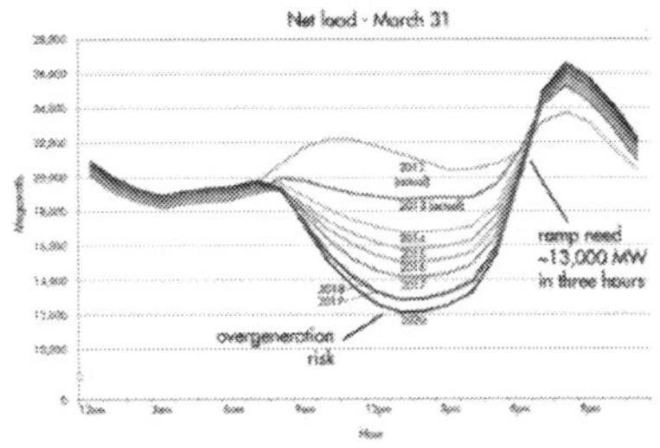


"Duck Curve" for CAISO grid (California)

- Ongoing technological revolution in IRE technology, with attendant remarkable cost reductions, is likely to continue for decades.
- Fast-ramping balancing capacity (BC) (some combination of backup capacity and storage capacity) is needed to ensure reliable grid power at high IRE grid penetrations.
- BC is often provided in US today by mix of natural gas (NG)-fired gas turbine combined cycle (GTCC) and combustion turbine (CT) units, but…

## Toward CCS for Balancing Capacity at High IRE Penetrations

- Paris climate goals ➔ BC must eventually (post-2030) be decarbonized—e.g., via $CO_2$ capture and sequestration (CCS)—with $CO_2$ storage via EOR, in deep saline formations, etc.
- "Early mover" CCS projects should be deployed during next decade to gain experience and get underway technology cost buydown (TCB) process ("learning by doing").
- CCS energy systems are capital intensive ➔ require continuous (baseload) operation.
- At high IRE penetrations, BC plants must be operated at low capacity factors (CFs): 30% to 50% (not baseload!).
- At even relatively modest IRE grid penetrations, roles for baseload power will be limited as illustrated by the situation in Germany:

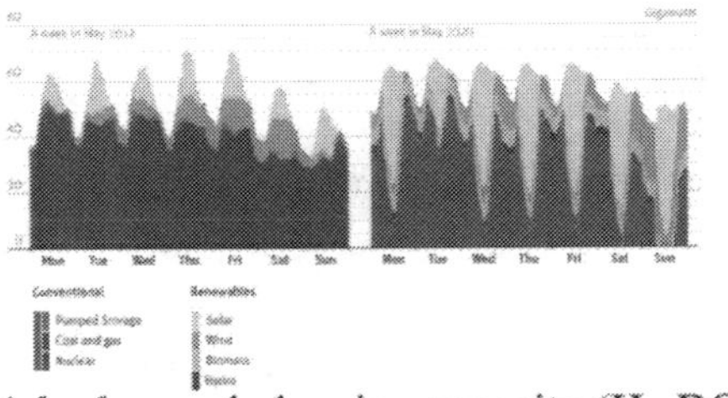

These curves show the Germany electricity load for 1 week in May in 2012 (left, when IRE accounted for 13% of electricity generation) and for 1 week in May in 2020 (right, when IRE is expected to account for 29% of electricity generation). **It is striking that with < 30% IRE in 2020 there is "no room" on German grid for baseload power.**

- A hydrogen balancing capacity ($H_2$-BC) approach for addressing this challenge is discussed.

## Firming Up Electric Grids at High IRE Penetrations with $H_2$-Balancing Capacity

- Three elements of a $H_2$-BC system [see, e.g., Davison (2009), ETI (2015)]:
  - $H_2$ is produced in baseload (~ 90% CF) plants with CCS from natural gas (NG), coal, biomass, or coal/biomass or NG/biomass;
  - $H_2$ is consumed in fast-ramping BC units [e.g., CT, GTCC, proton exchange membrane (PEM) fuel cells, or compressed air energy storage (CAES) units] that operate at low CFs;
  - $H_2$ is stored in buffer underground storage systems to enable decoupling baseload $H_2$ production from highly variable $H_2$ consumption by BC units.
- The concept "works" because underground $H_2$ storage part of system expected to be inexpensive [see, e.g., Davison (2009), ETI (2015)].
- Present focus:
  - $H_2$ with CCS via cogasification of coal + torrefied corn stover—a promising approach for enabling coal to compete with natural gas in a carbon-constrained world based on near-term technologies
    - Making $H_2$ from coal is commercial technology—hundreds of Chinese plants that make fertilizer from coal;
    - Corn stover is an abundant biomass supply that can be exploited in near term;
    - Torrefaction process for corn stover is ready to be demonstrated at commercial scale;
    - Cogasification of coal/torrefied biomass is technically feasible and ready for commercial-scale demonstration.
  - $H_2$ so produced can be used in CT, GTCC, PEM fuel cells or CAES units.
  - It is assumed (pessimistically) that captured $CO_2$ is stored in deep saline formations.
  - In near term, underground buffer storage of $H_2$ is feasible only in salt caverns.

## Underground Storage of $H_2$

| | Clemens (USA) | Moss Bluff (USA) | Teesside (UK) |
|---|---|---|---|
| Geology | Domal salt | Domal salt | Bedded salt |
| Operator | Conoco Phillips | Praxair | Sabic Petroleum |
| Stored fluid | Hydrogen | Hydrogen | Hydrogen |
| Commissioned | 1983 | 2007 | ~1972 |
| Volume [m³] | 580,000 | 566,000 | 3 x 70,000 |
| Reference depth [m] | 930 | > 822 | 350 |
| Pressure range [bar] | 70 - 135 | 55 - 152 | ~ 45 |
| Possible working gas capacity $H_2$ Mio [kg] | 2.56 | 3.72 | 0.83 |

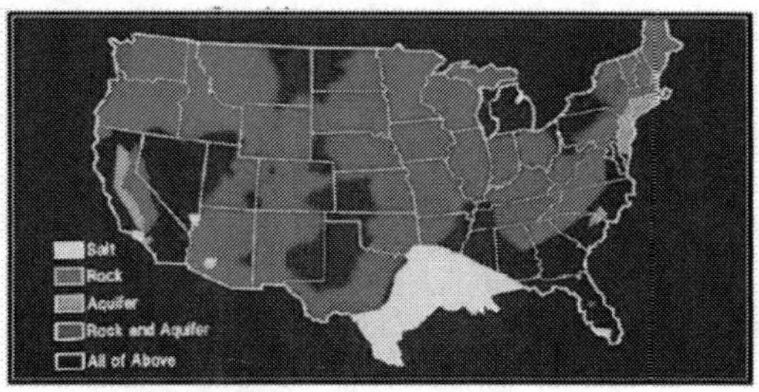


- Since 1972 $H_2$ has been stored in 3 caverns in bedded salt in UK (at Teeside).
- $H_2$ is also being stored in 2 caverns in salt domes on Texas Gulf Coast.
- Because of limited geographical availability of these salt formations in the US, underground storage of $H_2$ in porous media (aquifers and depleted HC fields) and rock caverns are also desirable; with these additional options the $H_2$-BC strategy could be pursued in most parts of US.
- In recent years, H2STORE and HyINTEGER programs in Germany have supported R&D on $H_2$ storage in porous media (Pudlo et al., 2013; 2016; Pfeiffer et al., 2016; Pfeiffer et al., 2017; Energy Storage Funding Initiative, 2017).
- Little US R&D on $H_2$ storage in porous media has been carried out recently, although this was to have become a key targeted R&D activity under the 2016 interagency, EERE-led *$H_2$ at Scale Initiative,* which envisioned major roles for $H_2$ as necessary to realize the Obama Administration's goal of deep reductions in US GHG emissions by midcentury.

## Levelized cost of $H_2$ vs GHG Emissions Price for Alternative Options for Making $H_2$ w/CCS (via Co-gasification of Coal + Torrefied Corn Stover)—with Comparisons to NG Price

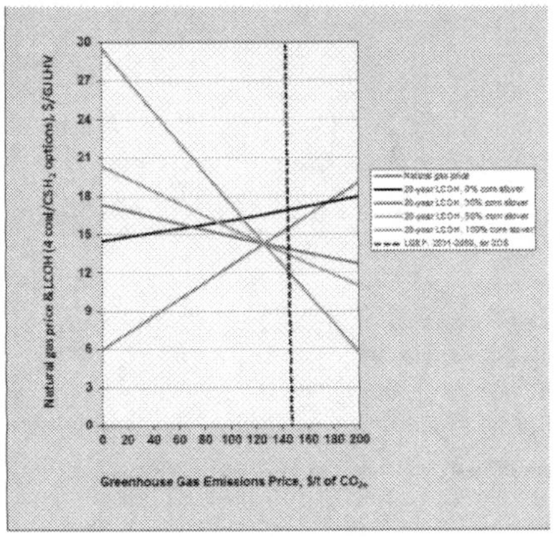


- Preliminary cost estimates neglecting $H_2$ storage costs (NOAK plants); $O_2$-blown quench gasifier; 93% $CO_2$ capture, $CO_2$ storage in deep saline formations (see Appendix for details).
- All options using corn stover (CS) consume 1 x $10^6$ dry t/y.

| | Prices in absence of GEP, $/GJ HHV |
|---|---|
| Natural gas | 5.4 |
| Coal | 2.1 |
| Corn stover | 3.9 |
| Torrefied corn stover | 6.9 |
| For H2 from: | **GHG emissions** (relative to NG emissions) |
| **100% Coal** | **+ 0.18** |
| 70% coal + 30% corn stover | - 0.31 |
| 50% coal + 50% corn stover | - 0.68 |
| 100% corn stover | - 1.8 |

- For negative emission options, LCOH falls with GHG emissions price.
- For 100% corn stover: $H_2$ capacity 0.3X that for 30% corn stover; average feedstock cost 2X more.
- LCOHs for all CS options < NG price (including emissions charge) for GHG emissions prices > $137/t $CO_{2e}$.
- Post 2030, this GHG emissions price < $CO_2$ price consistent with realization of 2DS for global energy future.

## $CO_2$ Price Trajectories (solid curves) for Alternative Atmospheric $CO_2$ Stabilization Scenarios & Corresponding Levelized $CO_2$ Prices, 2031-2050 (dashed curves)

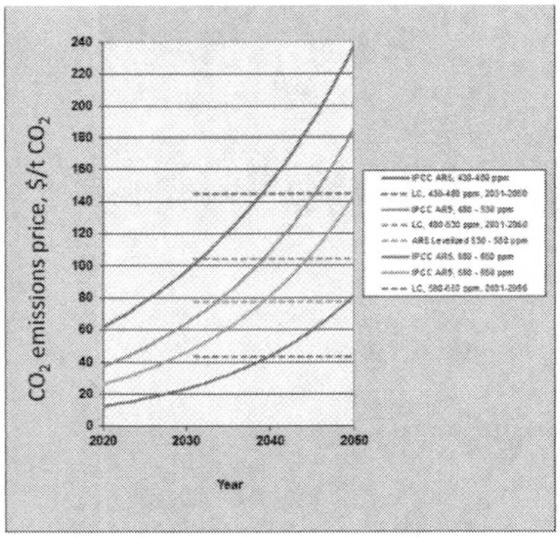


- At left are $CO_2$ emission price trajectories (solid curves) consistent with 4 alternative scenarios for stabilizing atmospheric $CO_2$ concentrations for these emissions strategies, according to IPCC's AR5.
- Solid red curve roughly consistent with limiting global warming to 2 °C (2DS global energy future).
- Dashed red curve = levelized cost of $CO_2$ emissions, 2031-2050, = $145/t $CO_2$ (dashed vertical **black curve** on previous graph).

## Policy Recommendations for Advancng the Coal/Biomass $H_2$-BC Concept

- DOE should support detailed pre-feasibility/feasibility studies of coal/biomass $H_2$-BC concept for providing reliable negative-emissions electricity on electric grids with high penetrations of IRE—as well as R&D on key component technologies.
  - If those studies sustain the preliminary findings presented here that the coal/biomass $H_2$-BC concept is promising under a strong carbon mitigation policy, and
  - If the final (reconciled) version of **S.1460** contains the 'net-negative $CO_2$ emissions project' provision.

  then DOE should consider supporting, via the **S.1460** authorization, at least 1 FEED study for advancing the concept. This strategy ought to be embraced by the Trump Administration, because:
  - Pres. Trump is committed to advancing coal's role in the US energy economy, and
  - A U.S. program focused on net-negative $CO_2$ emissions technology development would be seen as "proof positive" of Sec. Perry's assertion that departing the Paris Climate Accord does not mean the U.S. will abandon its technology leadership in vital areas that can contribute solutions for mitigation.
- In parallel, DOE should support systems analyses exploring natural gas-based balancing capacity options for providing reliable zero or negative-emissions electricity on electric grids with high penetrations of IRE—one aim of which would be a better understanding of the prospective competition between coal and natural gas in providing low, zero or negative emitting balancing capacity.
- Finally, high priority should be given to finding ways to fix electricity markets that have been "broken" by the IRE revolution (Larson et al., 2017)—e.g., these markets do not provide investors adequate incentives to build new capacity needed to ensure a reliable grid electricity supply even though that capacity will be idle most of the time at high grid penetrations of IRE.

## Is a Viable Long Term Future for Coal Possible in US Power Market?

- Abandoning the Paris Climate Accord and halting incentives for renewables will do little to create a Coal Renaissance in the US power market because of fierce competition from abundant gas at prospective natural gas prices.
- Substantial US power market roles for coal in the presence of an eventual strong US carbon mitigation policy (likely in ~ 4-8 years) are not possible without CCS.
- But high grid penetrations of IRE with attendant greatly reduced demand for baseload grid power make CCS very challenging economically for coal CCS technologies, because these very capital-intensive systems must be operated at high (baseload) capacity factors to be competitive.
- This capital intensity challenge can be addressed effectively by pursuing CCS via the $H_2$-BC concept—which makes it possible to carry out CCS with a system part ($H_2$ production) that operates at a high (90%) capacity factor while the much less capital-intensive part of the system (flexible balancing capacity) operates at low capacity factors.
- Deploying coal-based $H_2$-BC systems in the US is not likely to be sufficient to enable coal to compete with natural gas because unrealistically high GHG emission prices would be required to realize breakeven between coal-derived $H_2$ and natural gas.
- However, there are reasonably good prospects that, for IRE-intensive electric grids, coal/biomass $H_2$-BC systems characterized by strong negative GHG emission rates would be able to compete as balancing capacity against natural gas-fired GTCC and CT options at GHG emissions prices consistent with a 2DS global energy future.
- Ascertaining how successful coal/biomass $H_2$-BC systems might be in competing against advanced low, zero, or negative emitting natural gas balancing capacity options (e.g., natural gas fired Allam cycles, natural gas-fired $H_2$-BC systems, and natural gas/biomass-fired $H_2$-BC systems) on IRE-intensive grids should be a priority focus of future energy systems studies.

## References


Davison, John, "Electricity systems with near-zero emissions of $CO_2$ based on wind energy and coal gasification with CCS and hydrogen storage," *International Journal of Greenhouse Gas Control*, **3:** 683–692, 2009.

Energy Storage Funding Initiative, "HyINTEGER: Long-term underground storage of hydrogen," German Federal Ministry for Economic Affairs and Energy and Federal Ministry of Education and Research, 18 April 2017. (Website accessed 11 May 2017: http://forschung-energiespeicher.info/en/projektschau/gesamtliste/projekt-einzelansicht/127/Wasserstoff_langfristig_unterirdisch

_speichern/.)

ETI (Energy Technologies Institute), *The Role of Hydrogen Storage in a Clean Responsive Power System*, Holywell Building, Holywell Way, Loughborough LE11 3UZ, United Kingdom, www.eti.co.uk, 2015.

Kreutz, Tom, Eric Larson, and Bob Williams, "Understanding Challenges with Intermittent Renewable Electricity Expansion," contribution of the Energy Systems Analysis Group (ESAG), Andlinger Center for Energy and the Environment, to the *2017 Annual Report of the Carbon Mitigation Initiative of Princeton University*, April 2017, http://cmi.princeton.edu/annual_reports

/2016/integration/index.php#cio_2

Pfeiffer, W.T., S.A., Hagrey, D. Köhn, W. Rabbel, and S. Bauer, "Porous media hydrogen storage at a synthetic, heterogeneous field site: numerical simulation of storage operation and geophysical monitoring," *Environmental Earth Sciences*, 75(16): 1177, 2016. doi:10.1007/s12665-016-5958-x.

Pfeiffer, W.T., C. Beyer, and S. Bauer. (2017). Hydrogen storage in a heterogeneous sandstone formation: dimensioning and induced hydraulic effects. *Petroleum Geoscience*, petgeo2016-2050.

Pudlo, D., L. Ganzer, S. Henkel, M. Kuhn, A. Liebscher, M. De Lucia, M. Panfilov, P. Pilz, V. Reitenback, D. Albrecht, H. Wurdemann, and R. Gaupp, "The H2STORE Project: Hydrogen Underground Storage – A Feasible Way in Storing Electrical Power in Geological Media?," in *Clean Energy Systems in the Subsurface: Production, Storage, and Conversion*, M.Z. Hou, *et al.* (eds), Springer Verlag, Berlin, 2013, pp. 395-412.

Sanchez, D.L., and D.M. Kammen, "A commercialization strategy for carbon-negative energy," *Nature-Energy*, 11 January 2016 https://www.nature.com/articles/nenergy20152

## Appendix: Documentation for Calculations Presented in Slide # 8

The final version of this appendix will have a detailed discussion of assumptions leading to the construction of the cost calculations presented in Slide 8. At present, only relevant citations are indicated.

### Citations for Appendix


Chiesa, P., S. Consonni, T. Kreutz, and R. Williams, "Co-production of hydrogen, electricity and $CO_2$ from coal with commercially ready technology. Part A: Performance and Cost," *International Journal of Hydrogen Energy*, **20**: 747-767, 2005.

EIA (Energy Information Administration), *Annual Energy Outlook 2017*, January 2017.

Kreutz. T., R. Williams, S. Consonni, and P. Chiesa, "Co-production of hydrogen, electricity and $CO_2$ from coal with commercially ready technology. Part B: Economic analysis," *International Journal of Hydrogen Energy*, **20**: 769-784, 2005.

Larson, E., G. Fiorese, G. Liu, R. Williams, T. Kreutz, and S. Consonni, "Co-production of synfuels and electricity from coal + biomass with zero net greenhouse gas emissions: an Illinois case study," *Energy and Environmental Science*, **3**: 28-42, 2010.

NETL (National Energy Technology Laboratory), *Assessment of $H_2$ Production with $CO_2$ Capture*, Volume 1: Baseline State-of-the-Art Plants, DOE/NETL-2010/1434, 30 August 2010.

NETL (National Energy Technology Laboratory), *Cost and Performance Baseline for Fossil Energy Plants Volume 1a: Bituminous Coal (PC) and Natural Gas to Electricity*, Revision 3, DOE/NETL-2015-1723, 6 July 2015.

Svanberg, M., I. Ollafson, J. Floden, and A. Nordin, "Analyzing Biomass Torrefaction Supply Chain Costs," *Biosource Technology*, **142**: 287-296. 2013.

# Utilizing systems analysis to inform carbon capture technology development

Dale Keairns[1], Tim Fout[2], Robert Stevens[2], Travis Shultz[2], Richard Newby[3], Alex Zoelle[4], Mark Woods[3]

Carbon Management Technology Conference
Houston, Texas, July 17-20, 2017

NATIONAL ENERGY TECHNOLOGY LABORATORY

U.S. DEPARTMENT OF ENERGY

[1] Deloitte Consulting LLP, [2] US DOE National Energy Technology Laboratory, [3] KeyLogic, [4] Leidos

## Illustrating the role of systems analysis

NATIONAL ENERGY TECHNOLOGY LABORATORY

- **Boundaries for the systems analysis**
  - Technology Focus: Economic carbon management
  - Technology State: Pre-commercial
  - Application: Fossil electric power generation
- **Question for systems analysis**
  - What are the priority data needs
  - What scale to invest

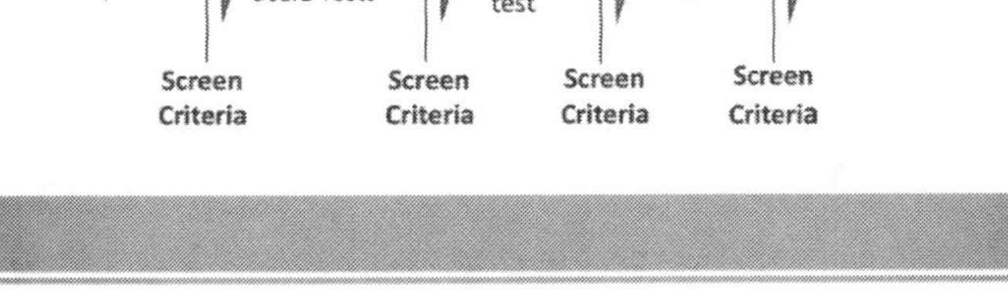


U.S. DEPARTMENT OF ENERGY 2

# Supporting technology development



**Develop Reference Plant Incorporating Innovative Technology**

- Review and assess innovative technology test data and concepts
- Model innovative technology
- Model integrated system using Baseline plant data and assumptions

**Baseline Report – PC Capture Plant**

- Bituminous coal
- 550 MW greenfield plant
- Midwestern U.S. ISO conditions
- Base load
- Amine-absorbent $CO_2$ capture
- Plant performance and cost

U.S. DEPARTMENT OF ENERGY 3

# Supporting technology development

NATIONAL ENERGY TECHNOLOGY LABORATORY

**Develop Reference Plant Incorporating Innovative Technology**

- Review and assess innovative technology test data and concepts
- Model innovative technology
- Model integrated system using Baseline plant data and assumptions

**Innovative Carbon Management Technology Studies Include**

- PC post-combustion capture
- IGCC pre-combustion capture
- Chemical looping technologies
- Supercritical $CO_2$ cycles
- Solid oxide fuel cells

**Baseline Report – PC Capture Plant**

- Bituminous coal
- 550 MW greenfield plant
- Midwestern U.S. ISO conditions
- Base load
- Amine-absorbent $CO_2$ capture
- Plant performance and cost

U.S. DEPARTMENT OF ENERGY 4

## Supporting technology development

NATIONAL ENERGY TECHNOLOGY LABORATORY

**Develop Reference Plant Incorporating Innovative Technology**

- Review and assess innovative technology test data and concepts
- Model innovative technology
- Model integrated system using Baseline plant data and assumptions

**Design and Operating Parameter Sensitivity Studies**

- Determine effect of parameters on performance and cost

**Results**

- Inform basis for establishing technology goals
- Inform priorities for technology development

**Baseline Report – PC Capture Plant**

- Bituminous coal
- 550 MW greenfield plant
- Midwestern U.S. ISO conditions
- Base load
- Amine-absorbent $CO_2$ capture
- Plant performance and cost

U.S. DEPARTMENT OF ENERGY

5

## Supporting technology development

NATIONAL ENERGY TECHNOLOGY LABORATORY

**Develop Reference Plant Incorporating Innovative Technology**

- Review and assess innovative technology test data and concepts
- Model innovative technology
- Model integrated system using Baseline plant data and assumptions

**Design and Operating Parameter Sensitivity Studies**

- Determine effect of parameters on performance and cost

**Results**

- Inform basis for establishing technology goals
- Inform priorities for technology development

**Baseline Report – PC Capture Plant**

- Bituminous coal
- 550 MW greenfield plant
- Midwestern U.S. ISO conditions
- Base load
- Amine-absorbent $CO_2$ capture
- Plant performance and cost

**Methodology to Assess Development Status of Post-Combustion $CO_2$ Separation Technologies***

- Provides metric that quantifies performance and cost gaps relative to a desired COE goal

U.S. DEPARTMENT OF ENERGY

* solvents, sorbents, adsorbents, membranes, phase change separation technologies

6

## Supporting technology development

NATIONAL ENERGY TECHNOLOGY LABORATORY

**Develop Reference Plant Incorporating Innovative Technology**

- Review and assess innovative technology test data and concepts
- Model innovative technology
- Model integrated system using Baseline plant data and assumptions

**Design and Operating Parameter Sensitivity Studies**

- Determine effect of parameters on performance and cost

**Results**

- Inform basis for establishing technology goals
- Inform priorities for technology development

**Adapt Analysis for Alternative Applications**

- New plant, retrofit
- Baseload, cyclic operation
- Plant scale
- Water constraints
- Alternative system boundaries

**Baseline Report – PC Capture Plant**

- Bituminous coal
- 550 MW greenfield plant
- Midwestern U.S. ISO conditions
- Base load
- Amine-absorbent $CO_2$ capture
- Plant performance and cost

**Methodology to Assess Development Status of Post-Combustion $CO_2$ Separation Technologies***

- Provides metric that quantifies performance and cost gaps relative to a desired COE goal

U.S. DEPARTMENT OF ENERGY

* solvents, sorbents, adsorbents, membranes, phase change separation technologies

7

## Chemical Looping Combustion (CLC)

NATIONAL ENERGY TECHNOLOGY LABORATORY

- **Chemical Looping Combustion Concepts**
  - Combustion, chemical looping with oxygen uncoupling (CLOU)
  - Candidate Oxygen carriers: Ni, Mn, Cu, Fe, Ca
  - Fuel options: coal, natural gas, syngas, biomass, petcoke
  - Atmospheric and pressurized
- **International studies: U.S., Sweden, Korea, Spain, China, UK, Canada**
- **Experimental testing**
  - Multiple international academic testing limited to bench scale up to 150 kWt
  - 3 MWt GE-Alstom CFB facility

U.S. DEPARTMENT OF ENERGY

8

# Chemical Looping Combustion Circulating Fluid Bed Concept

NATIONAL ENERGY TECHNOLOGY LABORATORY

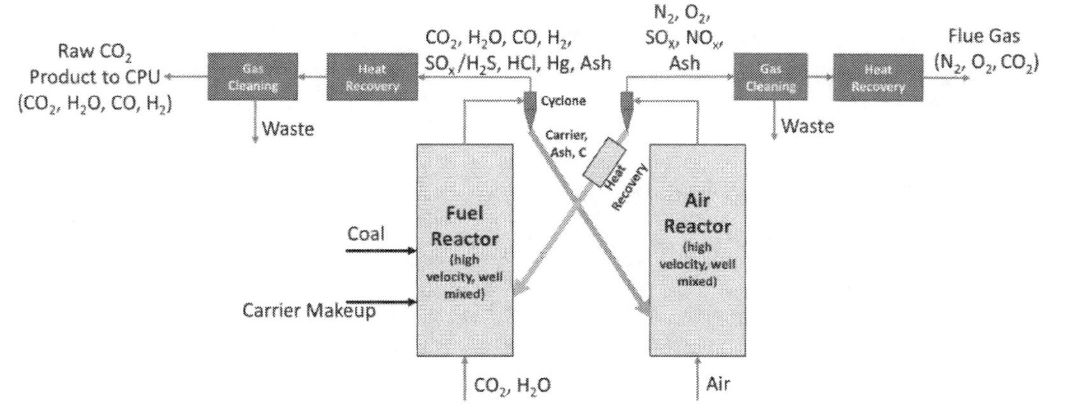


*Source: NETL*

U.S. DEPARTMENT OF ENERGY 9

# Utilizing systems analysis to inform CLC development priorities

NATIONAL ENERGY TECHNOLOGY LABORATORY

**Chemical Looping Combustion (CLC) Reference Plant**

- Review and assess CLC test data and concepts
- Model chemical looping process
- Model integrated system using Baseline plant data and assumptions

**Baseline Report – PC Capture Plant**

- Bituminous coal
- 550 MW greenfield plant
- Midwestern U.S. ISO conditions
- Base load
- Amine-absorbent $CO_2$ capture
- Performance and cost data base

**CLC Reference Plant – Design and Operating Parameter Sensitivity Studies**

- Oxygen carrier: conversion/circulation rate
- Oxygen carrier makeup
- Oxygen carrier makeup cost
- Fuel reactor carbon gasification conversion
- Component costs (e.g., char separator)
- Use of $CO_2$ purification unit (CPU)
- Fuel and air reactor temperatures
- Fuel and air reactor gas velocities
- Air reactor outlet gas oxygen content

**Process Model Simulation and Sensitivity Case Study Results**

- Inform basis for establishing technology goals
- Inform priorities for technology development

U.S. DEPARTMENT OF ENERGY 10

## CLC CFB Reference Plant Performance and Cost

NATIONAL ENERGY TECHNOLOGY LABORATORY

| Oxygen carrier type | $Fe_2O_3$ | $CaSO_4$ | BBR Case 12 |
|---|---|---|---|
| Plant Capacity (MW) | 550 | 550 | 550 |
| Plant Efficiency (%, HHV) | 32.8 | 31.9 | 28.4 |
| Carbon Capture Efficiency (%) | 95.8 | 85.9 | 90 |
| $CO_2$ Product Purity (mole% $CO_2$) | 99.0 | 99.95 | 100 |
| Total Plant Cost ($/kW) | 2,631 | 2,975 | 3,563 |
| Cost of Electricity ($/kWh) first-year w/o T&S | 126.4 | 117.5 | 137.3 |

U.S. DEPARTMENT OF ENERGY 11

## CLC CFB Reference Plant Capital Cost

NATIONAL ENERGY TECHNOLOGY LABORATORY

| Oxygen carrier type | $Fe_2O_3$ | $CaSO_4$ | BBR Case 12 |
|---|---|---|---|
| Plant Efficiency (%, HHV) | 32.8 | 31.9 | 28.4 |
| Total Plant Cost ($/kW) | 2,631 | 2,975 | 3,563 |
| CLC System Cost* (%) | 25.3 | 23.4 | |
| CPU Cost (%) | 5.7 | 12.2 | |
| Reducer Reactor Cost (% of CLC System Cost) | < 2 | < 2 | |

* **Chemical Looping Combustion System Components**: reducer reactor, cyclones, piping; oxidizer reactor, cyclones, piping; char-carrier separator; HRSGs; FD/ID fans

U.S. DEPARTMENT OF ENERGY 12

# Systems analysis messages for CFB concept development

NATIONAL ENERGY TECHNOLOGY LABORATORY

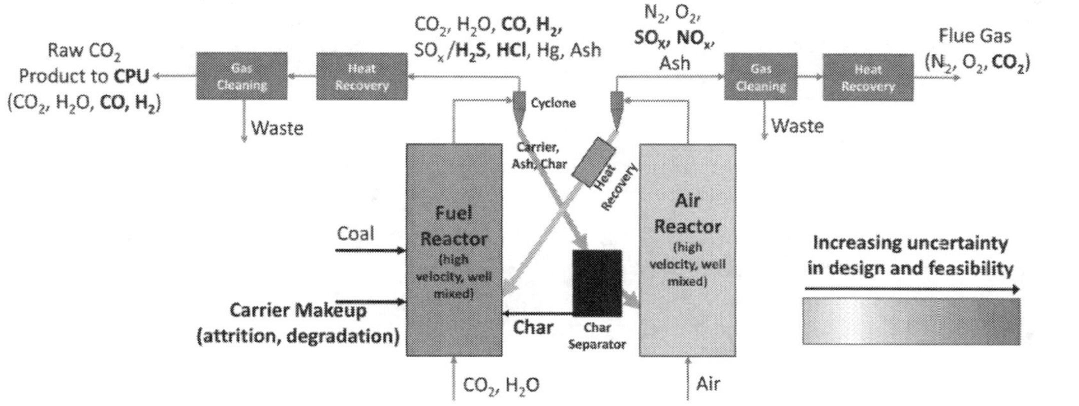


*Source: NETL*



# CLC Assessment Results Perspective

NATIONAL ENERGY TECHNOLOGY LABORATORY

- The assessment is founded on the premise that the CLC power plant operability and availability are comparable to that of the conventional PC power plant and do not limit the technology's ultimate feasibility – experimental basis is required to support this premise
- The "reference-case" process simulations show the possibility that CLC could provide significant performance and cost advantages over conventional PC power plants using conventional, amine-based $CO_2$ capture technology
- The fuel reactor is the key developmental component in the CLC power plant
- Steady-state test data is required that is representative of the integrated CLC reactors, or of individual CLC reactors, suitable for scale-up to commercial capacity

## CLC Assessment Results Perspective (cont'd)

NATIONAL ENERGY TECHNOLOGY LABORATORY

- Oxygen carrier cost and life are key parameters for achieving a low cost system
- The cost of the fuel reactor will represent only a small portion of the CLC power plant total cost and its development focus should be on achieving acceptable operability and reliability, and not on compact vessel design
- Carrier/char separation is required for circulating bed concepts: separation performance, reliable and within cost constraints
- The $CO_2$ specification is an important parameter and affects system cost
- The analyses provide understanding of risk: the combined effects of a few of the uncertain sensitivity parameters can result in the technology not being competitive; balance-of-plant equipment (e.g., heat recovery, gas cleaning) specifications, conceptual designs, and cost estimates are required

U.S. DEPARTMENT OF ENERGY 15

## Representation of Post-Combustion Capture

NATIONAL ENERGY TECHNOLOGY LABORATORY

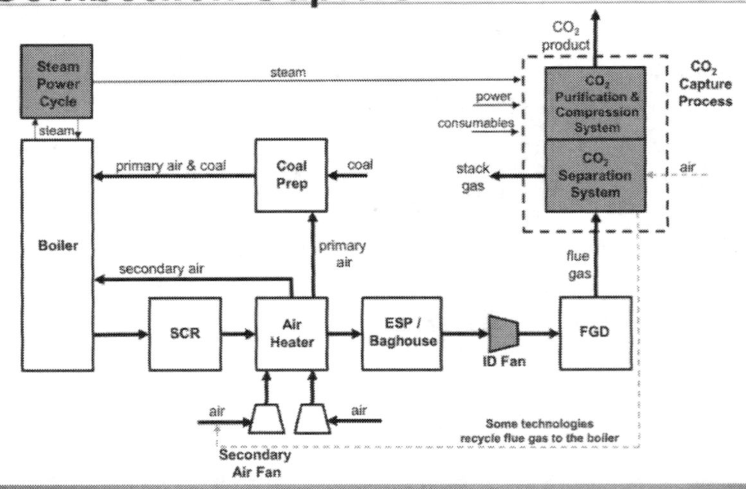


U.S. DEPARTMENT OF ENERGY 16

# Representation of Post-Combustion Capture



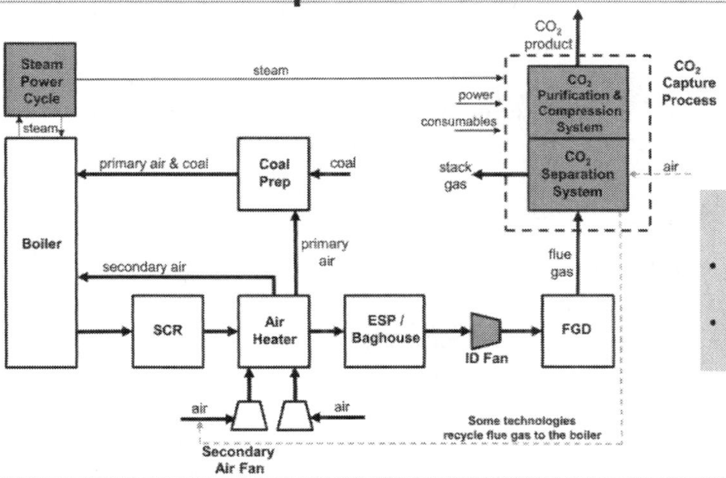


**Membrane Option**

- Eliminates interface with steam cycle
- No circulating medium



# Utilizing systems analysis to inform membrane development priorities

NATIONAL ENERGY TECHNOLOGY LABORATORY

**Post-Combustion Membrane Capture Reference Plant**

- Review and assess membrane test data and concepts
- Model membrane process
- Model integrated system using Baseline plant data and assumptions

**Baseline Report – PC Capture Plant**

- Bituminous coal
- 550 MW greenfield plant
- Midwestern U.S. ISO conditions
- Base load
- Amine-absorbent $CO_2$ capture
- Performance and cost data base

**Membrane Design and Operating Parameter Sensitivity Studies**

- $CO_2$ permeance
- Gas constituents selectivity ($CO_2$, $N_2$, $O_2$, $H_2O$, $SO_2$)
- Membrane thickness
- Membrane module design and performance (capacity, pressure drop, mass transfer)
- Membrane flow configuration (counter-current, cross-flow, co-current)
- Process configuration (flue gas pressure, single flue gas membrane, staged)
- Membrane cost and life

**Process Model Simulation and Sensitivity Case Study Results**

- Inform basis for establishing technology goals
- Inform priorities for technology development

# Membrane Guidance Illustrations

NATIONAL ENERGY TECHNOLOGY LABORATORY

- **Membrane-based $CO_2$ capture has the potential to provide performance and cost benefits over conventional, solvent-based technologies if specific membrane characteristics can be achieved**
- **Water vapor must be considered in membrane process modeling efforts due to its high permeability and tendency to be condensed between process stages – typically not included as a gas stream test constituent**
- **For a low-pressure, single flue gas membrane configuration, research focused on advanced, high-selectivity (>50 to 200) membranes is not a priority, since there is limited benefit in COE reduction**
- **Membrane-based $CO_2$ capture processes that apply flue gas pressurization are not a priority, since there is limited cost advantage over low-pressure processes**
- **All membrane-based $CO_2$ capture processes will require raw $CO_2$-stream purification (use of a CPU rather than conventional $CO_2$ compression) to meet $CO_2$ product gas purity specifications, with the $O_2$ specification being the most stringent**

U.S. DEPARTMENT OF ENERGY 19

# Assess the status of technology development with respect to performance and cost metrics*

NATIONAL ENERGY TECHNOLOGY LABORATORY

## Objectives:

- Identify the minimal requisite parameters that should be measured by a technology developer through laboratory investigations to effectively support post-combustion $CO_2$ capture system studies
  - ✓Applicable for systems analysis: pre-screening, screening, commercial techno-economic analysis (TEA)
- Create a tool to test new experimental results and updated equipment performance and cost estimates
  - ✓Applicable for post-combustion capture technologies: solvents, sorbents, adsorbents, membranes, phase change

U.S. DEPARTMENT OF ENERGY

*T. Fout, R. Stevens, D. Miller, R. Newby, D. Kearns, M. Matuszewski, D. Bhattacharyya; Guidelines for parameter measurements in laboratory-scale post-combustion research efforts, Proceedings of the 2016 NETL $CO_2$ Capture Technology Meeting, August 8, 2016 (www.netl.doe.gov/events/conference-proceedings)

20

## Approach

NATIONAL ENERGY TECHNOLOGY LABORATORY

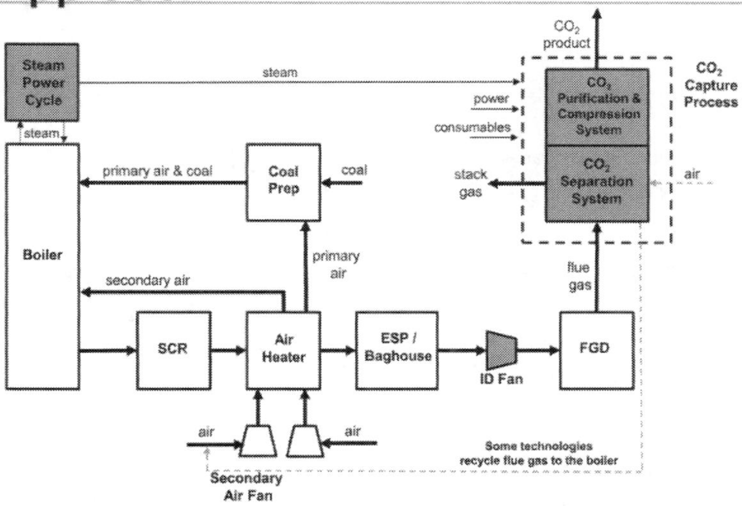


Develop two parameters that characterize the capture system economic potential

- Power consumption impact
- COE impact

ENERGY 21

## Characteristic Parameters

NATIONAL ENERGY TECHNOLOGY LABORATORY

- **Power Factor: Total $CO_2$ capture process power-impact on reference plant/(tonne/hr $CO_2$), dependent on:**
  1. $CO_2$ separation system auxiliary power
  2. $CO_2$ separation system heating duty and heating temperature
  3. $CO_2$ separation system cooling water duty
  4. Raw $CO_2$ gas composition and pressure
- **COE Factor: Total $CO_2$ capture process cost-impact on reference plant/(tonne/hr $CO_2$), dependent on:**
  1. $CO_2$ separation system materials consumption rates and materials replacement prices
  2. $CO_2$ separation system capital investment for process equipment

ENERGY 22

## Methodology Plot

NATIONAL ENERGY TECHNOLOGY LABORATORY

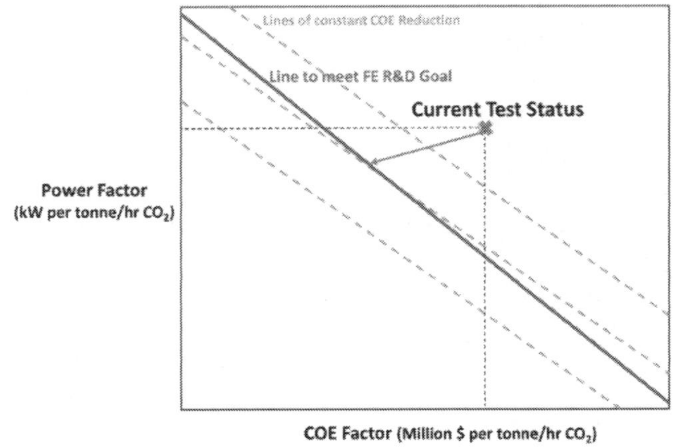


The acceptable Power Factor and COE Factor ranges can be translated into the range of test parameter goals for the technology



## Concluding Message


NATIONAL ENERGY TECHNOLOGY LABORATORY


- **Important tasks for development of innovative carbon management technology development include:**
  - Understand the application system boundaries, needs, and constraints
  - Utilize systems analysis to inform the priority process design, operation, and cost parameters that affect system performance and cost
  - Focus development on obtaining data to address the priority parameters
  - Implement testing at the appropriate scale to obtain relevant data
- **An approach and methodology has been developed and implemented to guide development of innovative carbon management technology**

Carbon Management Technology Conference 2017

INTERNATIONAL
**CCS KNOWLEDGE**
CENTRE

Presented By: Mike Monea

INTERNATIONAL
**CCS KNOWLEDGE**
CENTRE

| | |
|---|---|
| **Organization** | A Not for Profit Corporation with an Independent Board of Directors |
| **Sponsors** | BHP Billiton – Funding Sponsor ($20 million over 5 years)<br>SaskPower – Knowledge Sponsor |
| **Vision** | Projects undertaken at the Knowledge Centre will help inform stakeholders regarding "real world" considerations in the use of CCUS and advance the practical knowledge of CCUS as a viable solution.<br>• Promote open exchange of information<br>• No technology ownership<br>• No membership costs |

INTERNATIONAL
**CCS KNOWLEDGE**
CENTRE

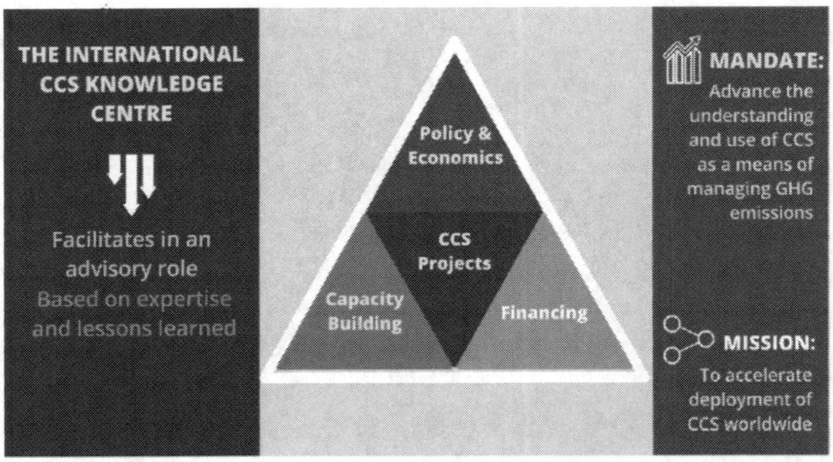


## YEAR 1 – OPERATING PLAN PROGRESS

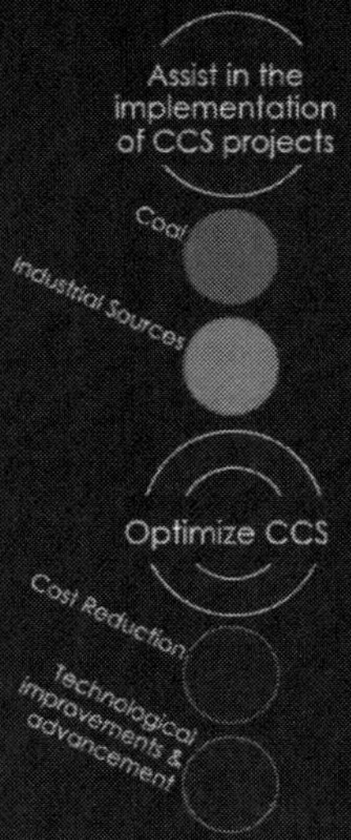


| Major Goal | Initiative | *Pathways Created* |
|---|---|---|
| A. Advance CCS in China | National CCUS Centre<br>Commercial Demonstration Project | • MOUs signed on Knowledge Sharing and Capacity Building<br>• CEM side event<br>• MOUs to be signed<br>• Facilitating subnational commitments<br>• Federal Joint Declaration update |
| B. Breaking Down CCS Barriers | Workshop Series - TBD | • October Symposium<br>• Links with the Asia Development Bank & CSLF<br>• GLOBE Series Workshops<br>• Creating Training Modules<br>• Modelling of CCS costs |
| C. Collaborate on industrial CCS applications | CCS Cement Project – Quebec<br>Industrial Applications on Steel - TBD | • Research in progress<br>• EIT paper to release<br>• Symposium panel with industrial CCS movers<br>Expanding to natural gas, cement, iron and steel / metallurgical coal, and fertilizers<br>• Still to pursue – Quebec and cement association |
| A. Collaborative Global Technical Community | CCS Cost Reduction and Optimization through Testing<br>Advanced Chemistry Laboratory to Support Test Facilities | This is now one larger project.<br>• Amine Degradation Alliance potential<br>• Analysis of linked international labs for lessons learned on amine reactions to date<br>• Rapid testing<br>• Create open source data from generic models<br>• Drives down the cost of amines |
| B. Modelling CCS Possibilities | Creation of Generic Models | |
| C. Breaking Down CCS Barriers | Series of Workshops - TBD | • Workshop on Amines with TNO – TBD<br>• Attending Mission innovation technical workshop on CCS |

INTERNATIONAL
**CCS KNOWLEDGE**
CENTRE

## Experiences of Full Chain Commercial CCUS

INTERNATIONAL
CCS KNOWLEDGE
CENTRE

## Regulations in Canada

CLEANER

- **1100** t/GWh = Lignite Coal Plant
- **550** = Current Natural Gas Plant
- **420** = Canadian Regulations on Power Plant
- **400** = New Natural Gas Plant
- **120-140** = CCS on Boundary Dam 3

INTERNATIONAL
CCS KNOWLEDGE
CENTRE

## BOUNDARY DAM CCS PERFORMANCE

- 1/2 $NO_x$
- 92% PM10
- 70% PM2.5
- 100% $SO_2$
- 90% $CO_2$

INTERNATIONAL
CCS KNOWLEDGE
CENTRE

## Canada's Commercial CCS: Securing Off-Takers

## CO2 EOR

Sale of CO2 to oil company for EOR

Sale of sulphuric acid, ex. fertilizer

Sale of fly ash for concrete production

INTERNATIONAL
CCS KNOWLEDGE
CENTRE

## WHAT WE KNOW

- Canada has committed to develop and deploy clean technologies.
- China and India are <u>not</u> on the path to eliminate coal.
  (IEA, 2015)
- CCS is a valuable tool in the toolbox – alongside solutions like renewable and energy efficiency.
- Without CCS, global GHG mitigation costs increase 138%.
  (IPCC, 2014)
- There are hardly any non-CCS pathways to 1.5°C.
  (GCCSI, 2016)

INTERNATIONAL
**CCS KNOWLEDGE**
CENTRE

## SUMMARY

- Saskatchewan has proven results on CCS
- Knowledge Center can go beyond borders to reduce GHGs at a global level
- Knowledge from Saskatchewan's CCS activities can be used to help achieve Canada's emission reduction targets
- This knowledge can be used to accelerate CCS deployment globally
- Open sharing of information a key to long term success of CCS world wide.

## Thank You

INTERNATIONAL
**CCS KNOWLEDGE**
CENTRE

For more information please visit our website at:

CCSKnowledge.com

or contact us by email: info@ccsknowledge.com

# National Carbon Capture Center Overview

Frank Morton – Director Technology Development

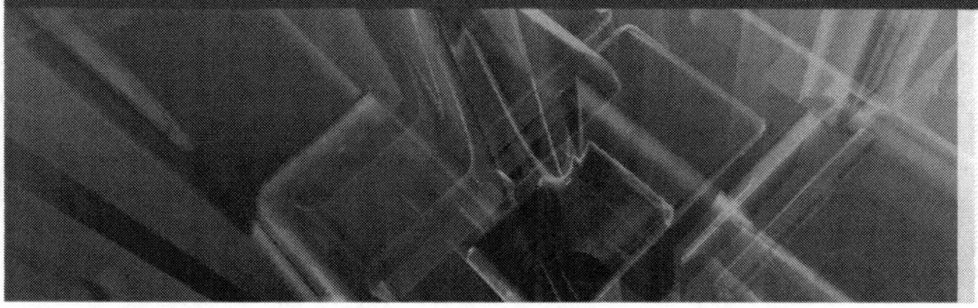

NATIONAL CARBON CAPTURE CENTER

1

## National Carbon Capture Center

*Offering a **world-class neutral test** facility and highly specialized staff to **accelerate the commercialization** of advanced technologies and enable coal-based power plants to achieve **near-zero emissions (low-cost $CO_2$)**.*

- Over **97,000 test hours** since founding in 2008
- Technology developers from **U.S. and six other countries**
- **First coal-derived gas testing** of solid oxide fuel cells and certain solvents, membranes and enzymes
- On-site **scale-ups** and **process enhancements** for 10 technologies
  - Scale-ups for testing at larger sites for five solvents
  - Scale-up to commercial operation for one solvent

Timeline: June 1, 2014 – May 31, 2019

U.S. DEPARTMENT OF ENERGY

NATIONAL ENERGY TECHNOLOGY LABORATORY

AEP AMERICAN ELECTRIC POWER | DUKE ENERGY | CLOUD PEAK ENERGY | EPRI ELECTRIC POWER RESEARCH INSTITUTE | Southern Company

2

# International Test Center Network

- 13 Members
- Share pubic knowledge with carbon capture test facilities.
  - Facility operations
  - Facility funding
  - Safety
  - Analytical techniques
- Collaborate on one technical item per year.
  - Amine carry-over and measurement techniques
  - Support advanced simulations and model development with a focus on reducing capital and operating cost and minimizing scale-up risks

***Share $CO_2$ Capture Knowledge***
*to encourage global collaboration and accelerate technology development of cost-effective carbon capture processes*

1

## International Test Center Network Members

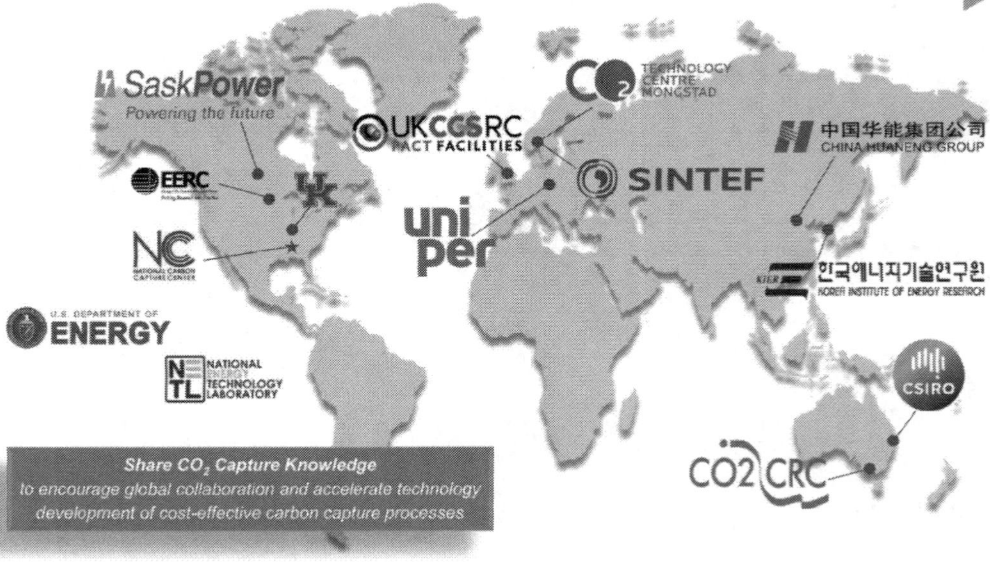


2

## What the Project Provides

- **Cost-efficient test site** with **infrastructure** for numerous technology developers
- **Real-world conditions** with coal-derived flue gas and syngas
- Flexible capability for testing at **multiple scales** and **on-site scale-ups**
- Expert **technical staff for** design, installation and testing support
- High-quality **data acquisition** and gas/liquid **sampling and analysis**

## Technology Development Process

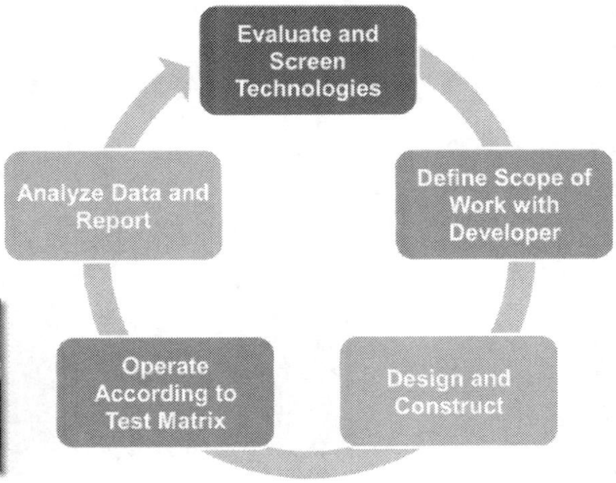

## PC4 Bench-Scale

- Simultaneous operation of up to five developers' test units
- Slipstream Solvent Test Unit (SSTU) for solvents in early development
- SSTU also used for solvent emissions studies and emission mitigation processes
- Flue gas/utilities and gas analysis systems operating independently of PC4 pilot-scale area

5

## PC4 Pilot-Scale

- Simultaneous operation of developer test units and Pilot Solvent Test Unit (PSTU)
- PSTU offers flexible operation to match developers' planned commercial configuration
- PSTU also supports solvent emissions and degradation studies

6

## Successful Testing and Partnerships

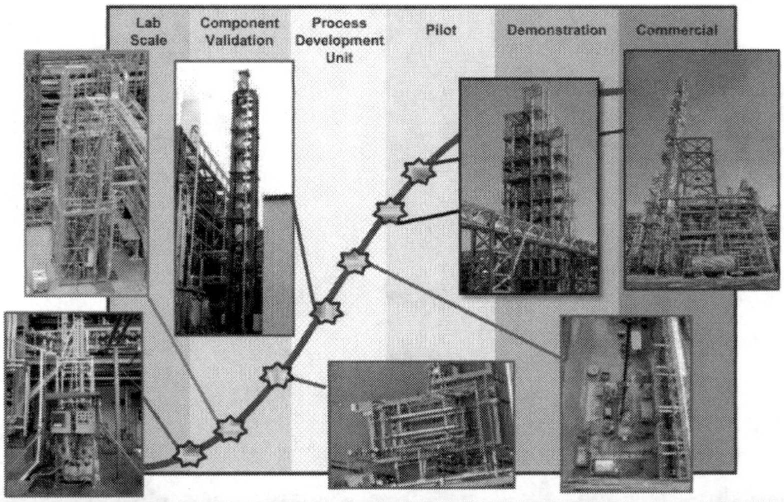


7

## NCCC International Collaboration

- Test facility – extensive support at no charge
- Co-funding partner – active participation in R&D
- Facilitate test partnerships
- International Collaboration – testing from 7 countries and ITCN
- Policy support – Organizations attempting to influence policy and improve carbon capture messaging need technical substance

NATIONAL CARBON CAPTURE CENTER

8

More information

www.nationalcarboncapturecenter.com

https://twitter.com/NCarbonCaptureC

9

UKCCSRC
PACT FACILITIES

# UKCCSRC PACT CORE FACILITIES SHEFFIELD

Jon Gibbins
Director, UK CCS Research Centre
Professor of Power Plant Engineering and Carbon Capture
University of Sheffield
www.ukccsrc.ac.uk
www.pact.ac.uk
j.gibbins@sheffield.ac.uk

Mohamed Pourkashanian
Director UKCCSRC PACT
Director University of Sheffield Energy 2050

**The UKCCSRC is supported by the Engineering and Physical Sciences Research Council as part of the Research Councils UK Energy Programme**

July 2017

EPSRC
Pioneering research and skills

# About UKCCSRC PACT

UKCCSRC PACT FACILITIES

- **UKCCSRC P**ilot-scale **A**dvanced **C**apture **T**echnology facilities
  - Funded by: BEIS (formerly DECC) and EPSRC
  - Cranfield, Edinburgh, Imperial, Leeds, Nottingham, Sheffield
  - Member of International CCS Test Centre Network (for UK)
- **Scope:** Specialist national facilities for research in advanced fossil-fuel energy, bioenergy and carbon capture technologies
  - Comprehensive range of pilot-scale facilities
  - Supporting specialist research and analytical facilities
  - Leading academic expertise
- **Aim:** Support and catalyse industrial and academic R&D to accelerate the development and commercialisation of novel low carbon technologies
- **Objectives**
  - Bridge gap between bench-scale R&D and industrial pilot trials
  - Provide shared access to industry and academia

Department for Business, Energy & Industrial Strategy

EPSRC Engineering and Physical Sciences Research Council

UKCCS RESEARCH CENTRE

Cranfield UNIVERSITY | THE UNIVERSITY of EDINBURGH | Imperial College London | UNIVERSITY OF LEEDS | The University of Nottingham | The University Of Sheffield.

## Fundamental research needed to increase Commercial Readiness as well as TRL

UKCCSRC PACT FACILITIES

For technologies to mature to be a "bankable asset class" this gap must also be bridged. Nuclear, Wind, Solar, Marine are all TRL9 and are still benefitting from fundamental research.

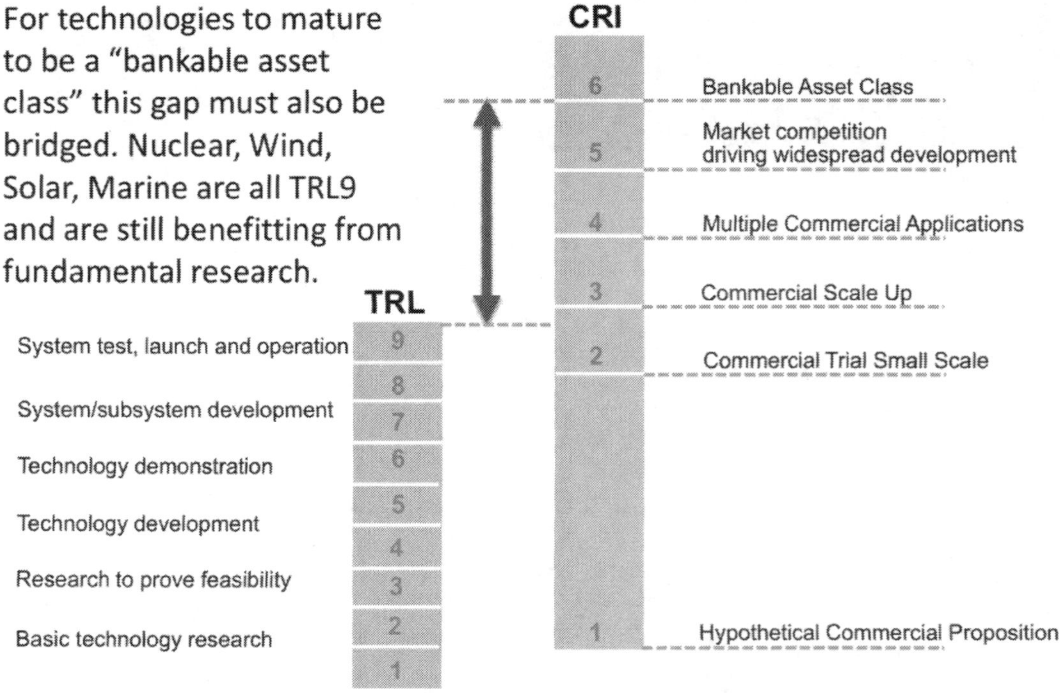


Faraday discussion, 2016, http://pubs.rsc.org/en/content/articlelanding/2016/fd/c6fd00046k#!divAbstract, based on http://arena.gov.au/files/2014/02/Commercial-Readiness-Index.pdf

UKCCSRC PACT FACILITIES

# PACT Core Facilities

## Coal/Biomass, NG-CCGT , Biofuel, $CO_2$ Capture Plant

## Large –Scale Combustion Test Facilities

UKCCSRC PACT FACILITIES

# PACT Core Facility: Overview

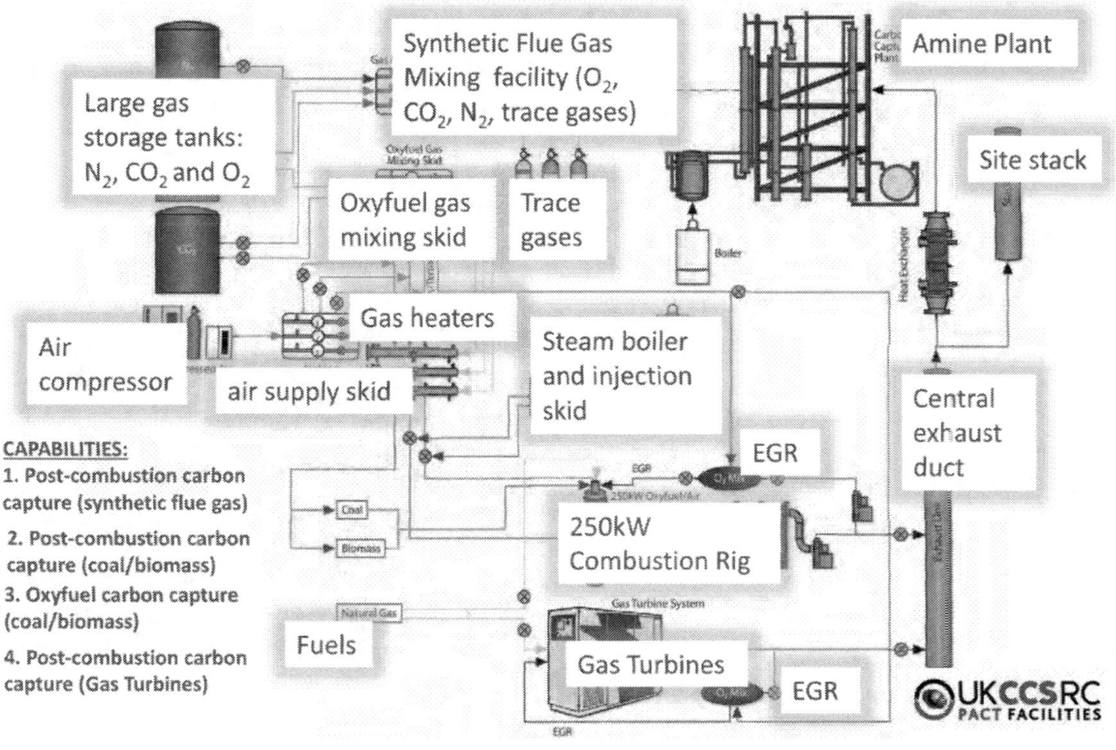


CAPABILITIES:
1. Post-combustion carbon capture (synthetic flue gas)
2. Post-combustion carbon capture (coal/biomass)
3. Oxyfuel carbon capture (coal/biomass)
4. Post-combustion carbon capture (Gas Turbines)

UKCCSRC PACT FACILITIES

UKCCSRC PACT FACILITIES

## Solvent- based Carbon Capture Plant

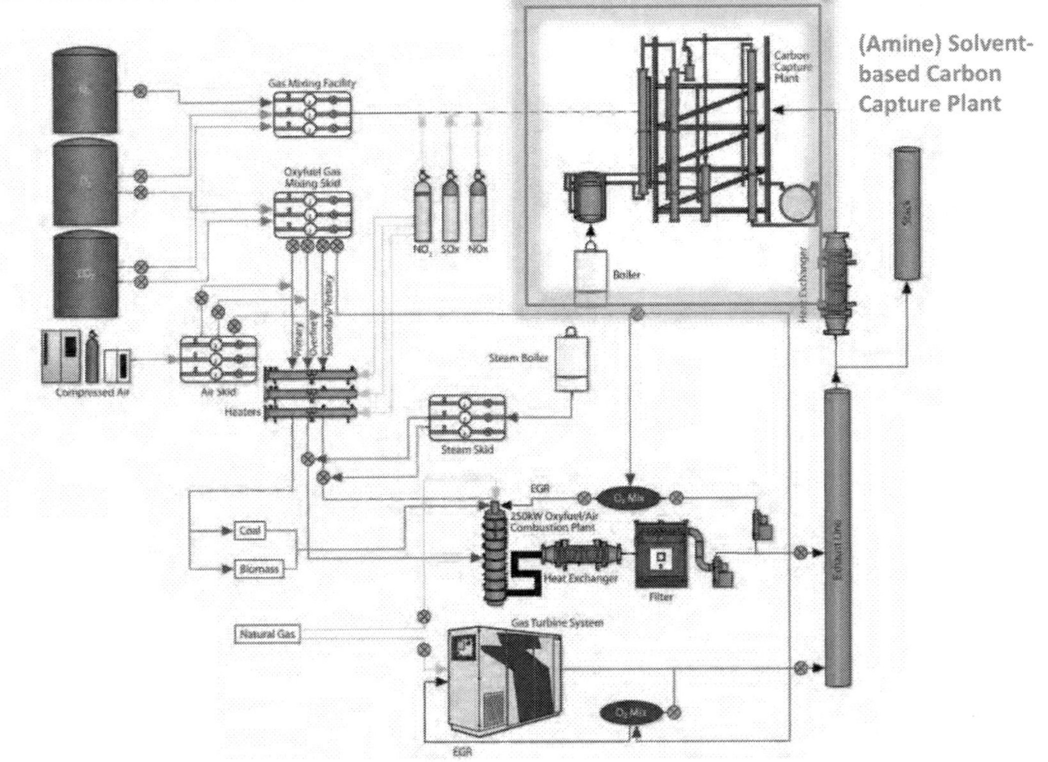


(Amine) Solvent-based Carbon Capture Plant

UKCCSRC PACT FACILITIES

## Solvent- based Carbon Capture Plant

### Overview

- ❑ 8m/300mm Absorber and Desorber columns
  - 2x 3m packed sections
  - Random/structured packing
- ❑ Integrated FGD (carbonate) wash system for removal of SOx from coal flue gas
- ❑ Flue gas treated: 210 $Nm^3/h$; eqiv. to 150kW coal flue gas
- ❑ Removes 1 tonne of $CO_2$ per day (MEA) with over 98% purity
- ❑ Solvent sampling on absorber and desorber
- ❑ Material corrosion testing sites
- ❑ Trace gas injection capability
- ❑ Analytical capability
  - Gas composition
  - Temperature monitoring
  - Pressure monitoring (e.g. foaming)

## Synthetic Flue/Process Gas Facility

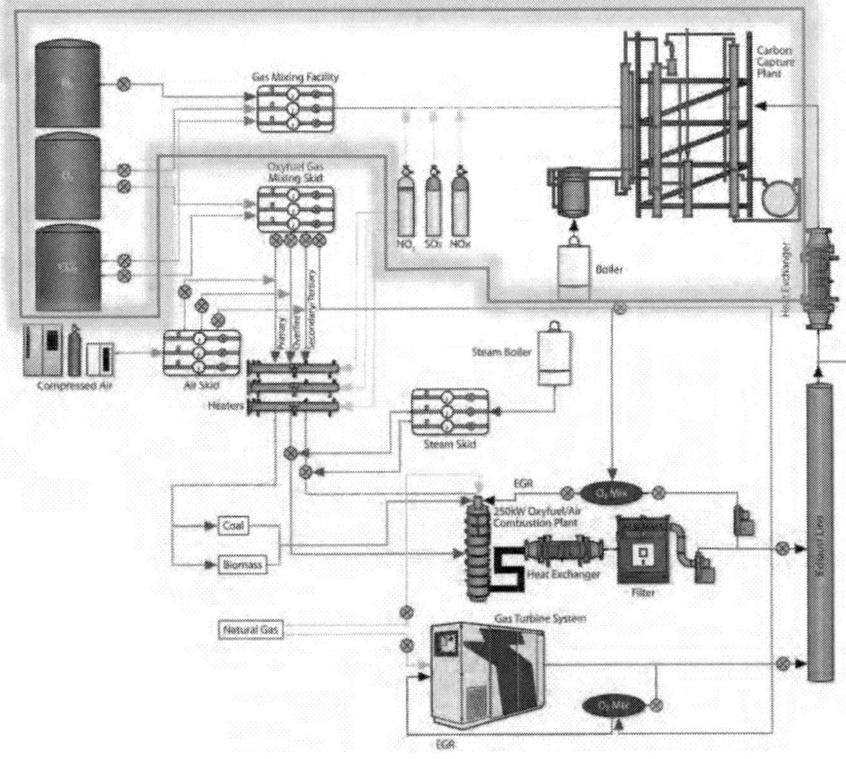


Synthetic Flue /Process Gas System

## Synthetic Flue/Process Gas Facility

### Overview

- Three gas metering and mixing lines, fed from $O_2$, $CO_2$ and $N_2$ storage tanks
- Complemented by trace gas injection NOx and SOx, other trace gasses
- Generate simulated flue/process gases
- Connected directly to the Solvent-based Carbon Capture Plant

UKCCSRC PACT FACILITIES

# 250kW Air Combustion Plant

## Overview

- ~250kWth, 4.5m high; 0.9m radius, cylindrical, down-fired rig with 8 sections
- Fuel: Coal, Biomass, Co-firing, Gas (primarily preheating)
- 2 x (interchangeable) coal/biomass burners - scaled from Doosan Power Systems commercial low-NOx burners
- Dedicated, high precession air metering skid
- Flue gas candle filter ( >99% ash removal);

- Furnace pressure (negative) balanced by exhaust fan
- Temperature and flow monitored water cooling system for the combustion rig, flue gas duct and heat exchanger.
- SCADA operating system with internet monitoring

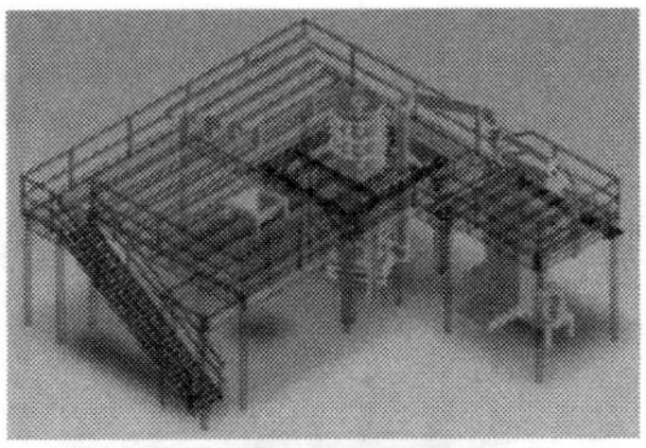

Biomass burner

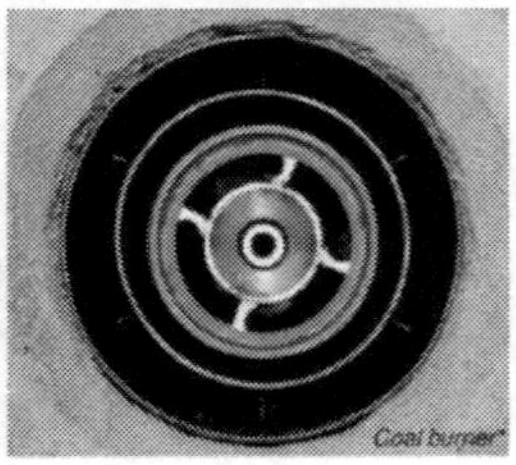

Coal burner*

UKCCSRC PACT FACILITIES

# PACT Core Facility: Layout

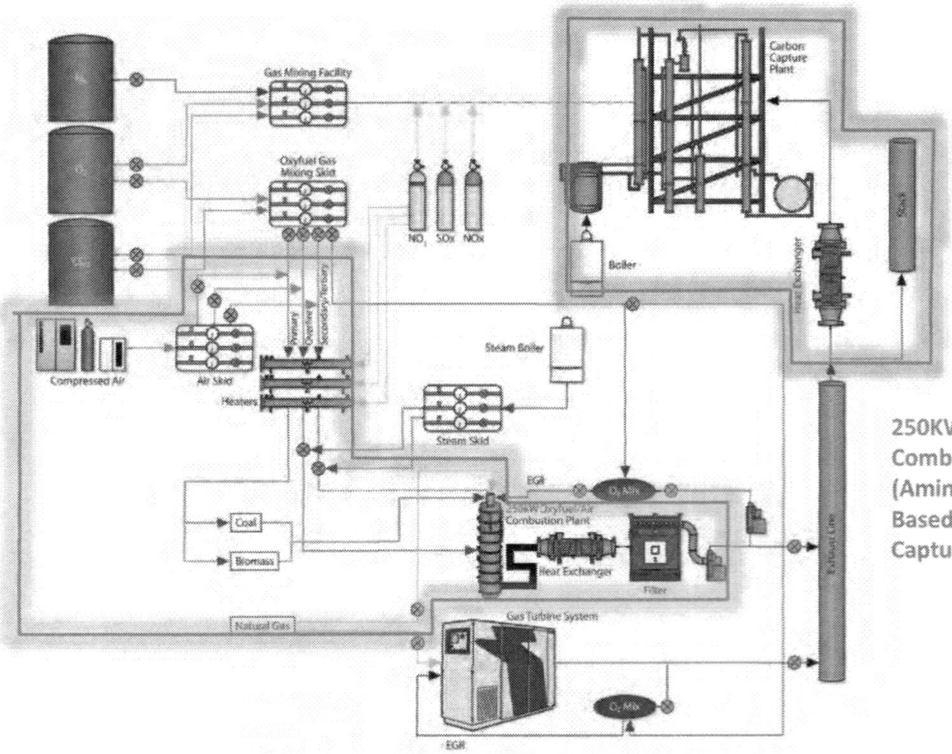


250KW Air Combustion Plant + (Amine) Solvent Based Carbon Capture Plant =

UKCCSRC PACT FACILITIES

## PACT Core Facility: Layout

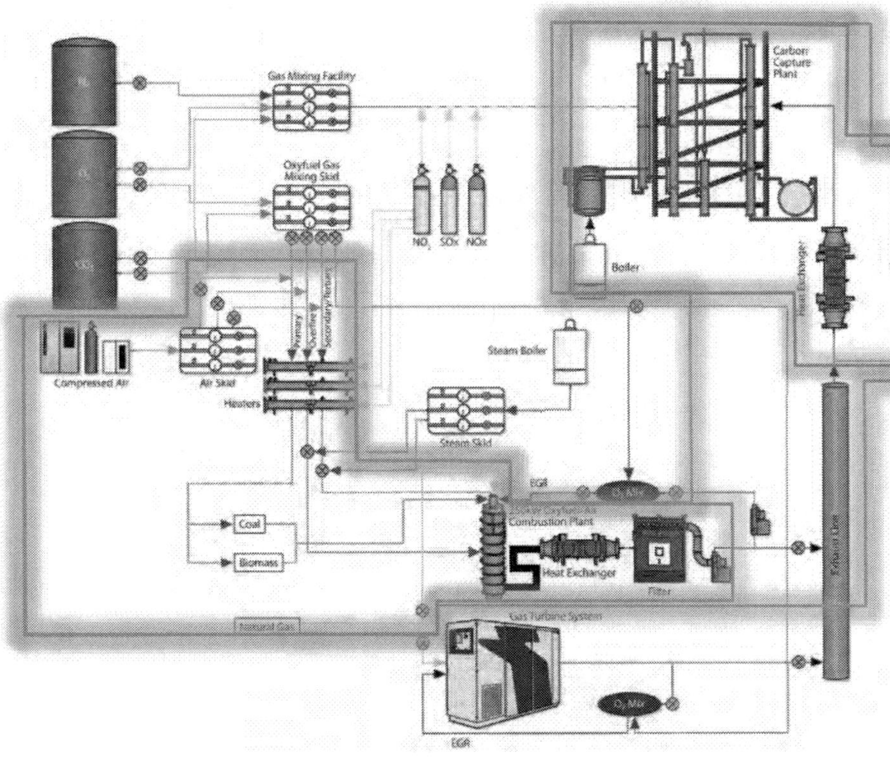


**250KW Air Combustion Plant + (Amine) Solvent Based Carbon Capture Plant = Post Combustion Capture** from Coal, Biomass, Co-firing, Gas Systems:

UKCCSRC PACT FACILITIES

## 250kW Oxyfuel Combustion Plant

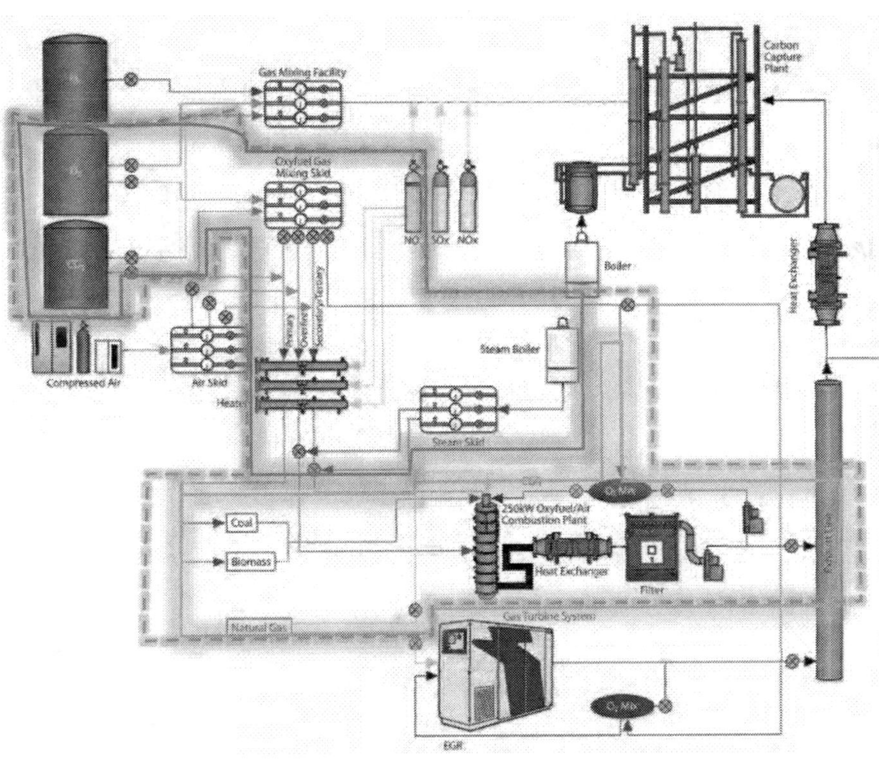


**Oxyfuel capture** from Coal, Biomass, Co-firing, Gas systems: (primarily) **Real oxyfuel mode+** (primarily) **Synthetic oxyfuel mode**

UKCCSRC PACT FACILITIES

# 250kW Oxyfuel Combustion Plant

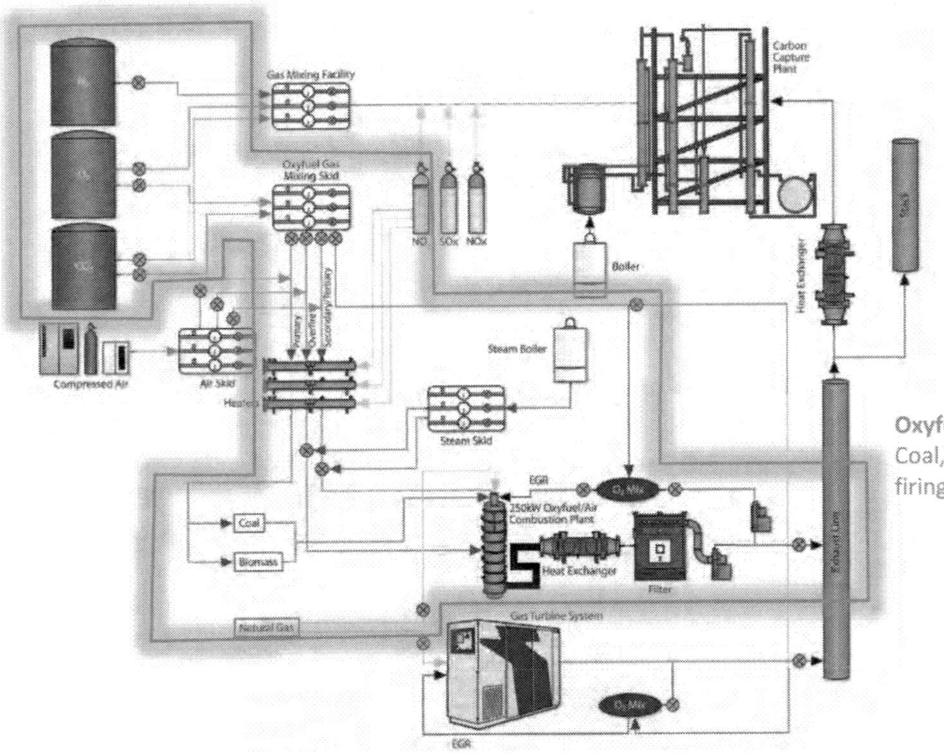


**Oxyfuel capture** from Coal, Biomass, Co-firing, Gas systems:

UKCCSRC PACT FACILITIES

# 250kW Oxyfuel Combustion Plant

## Example Applications

- ❑ Oxyfuel combustion R&D for coal, biomass or co-firing using a synthetic mixture of dry or wet $CO_2/O_2$ or wet flue gas recycle
- ❑ Fuel and process testing and optimisation;
- ❑ Integrated system modelling, for optimising Air Separation Unit (ASU) operation, combustion system control, and simulating the effect of different fuels on the combustion process
- ❑ System modelling and optimisation for flame visualisation and analysis, and latest modelling software for combustion system design, development and optimisation.

## Pilot-Scale Integrated Experimental Facilities for BIO-Cap Project

## Flame Imaging System

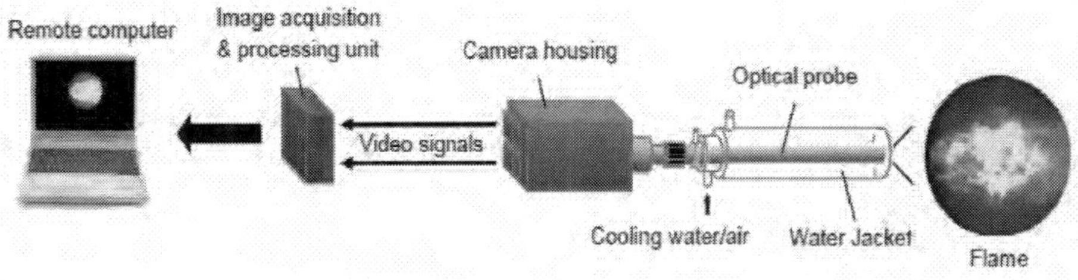


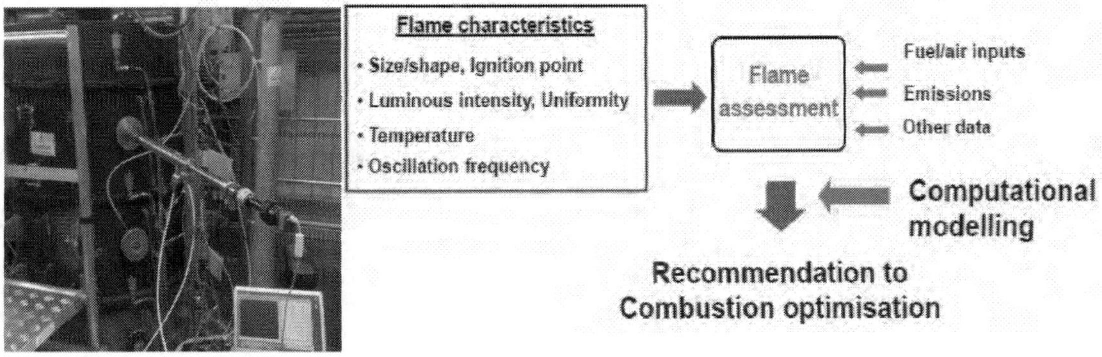


Imaging system has been developed by the University of Kent (Dr Gang Lu & Prof Yong Yan)

UKCCSRC PACT FACILITIES

# Results

Experimental calculation of the oscillation frequency - COAL

Original videos

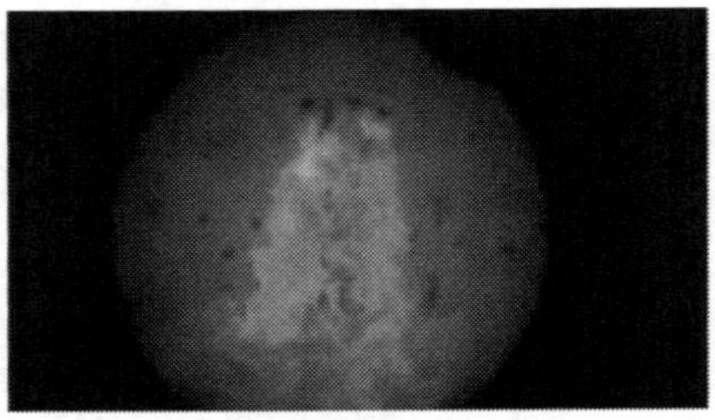

Air

Oxy24

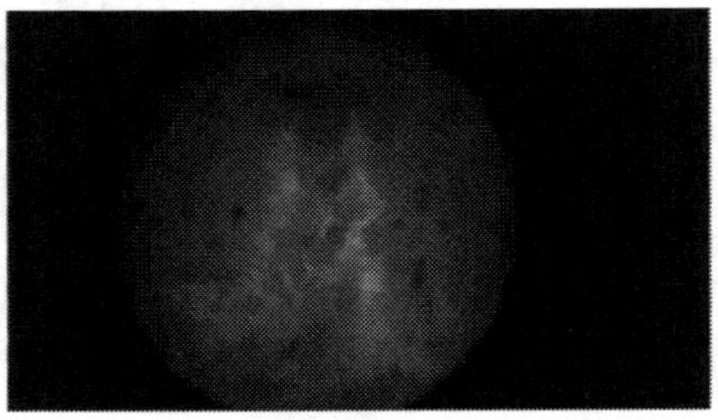

Oxy27

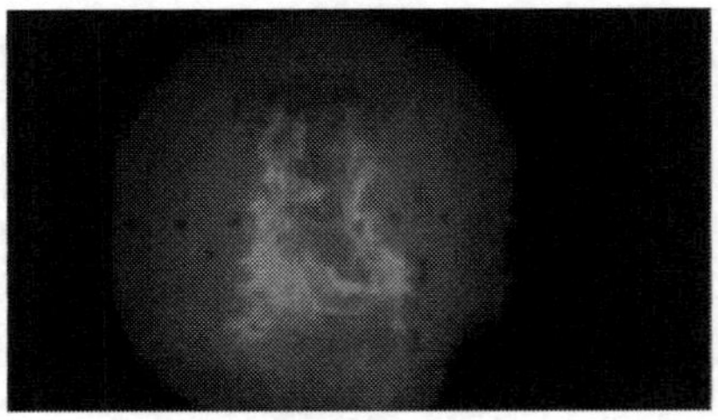

Oxy30



UKCCSRC PACT FACILITIES

# Results

Experimental calculation of the oscillation frequency

COAL

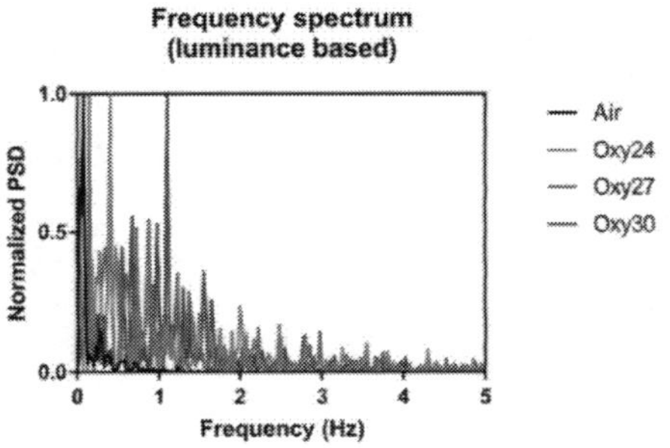


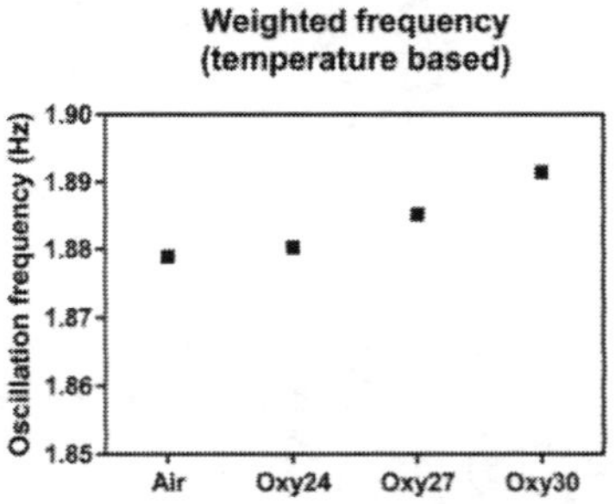


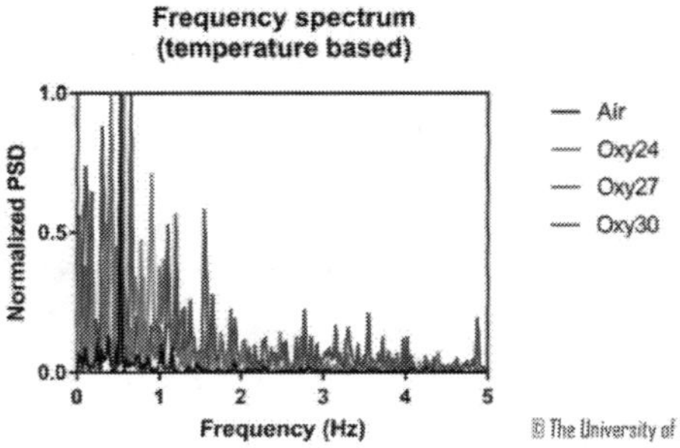


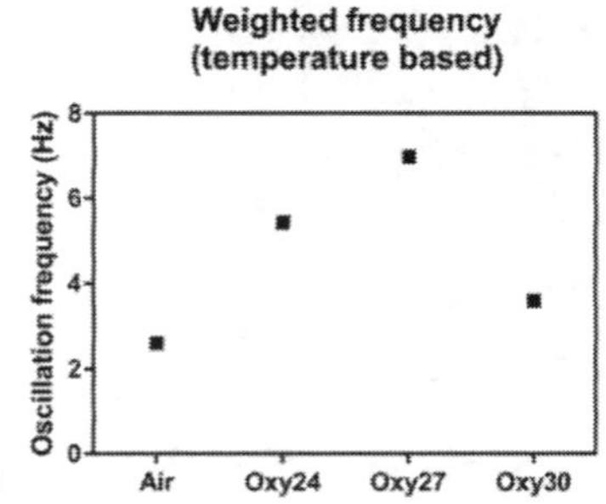

## Gas Turbine System

### Overview

- Two Turbec T100 Microturbines
- Consume 330kW of Natural gas
- Fuel: Natural gas, biogas, syngas, diesel, kerosene, methanol, LPC
- Generation 100kWe and 150kWth
- Overall efficiency up to 77% (33% electrical)

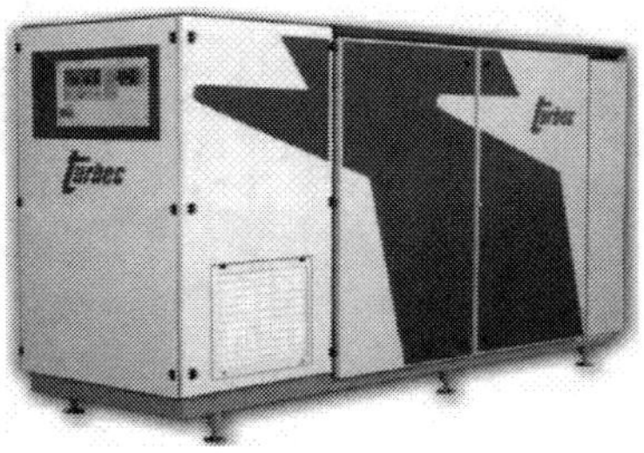

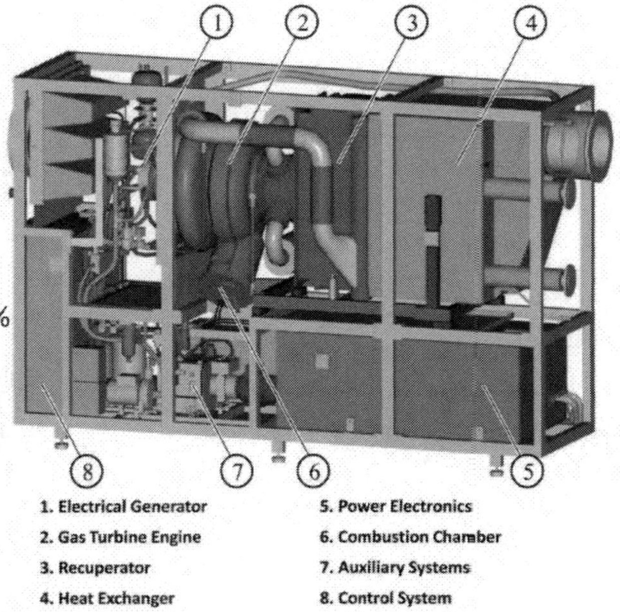


## Gas Turbine System

### System description

- **compressor** – radial centrifugal compressor compresses ambient air before sending to recuperator
- **recuperator** – preheats the compressed combustion air with the heat from the flue gases
- **combustor** – a lean combustion environment ensures low NOx, CO and hydrocarbon emissions
- **Turbine** – the hot, pressurised gas expands through the turbine to drive the turbine and the compressor and generator, which are all on the same shaft
- **Flue gas heat exchanger** – uses the hot flue gas to heat water

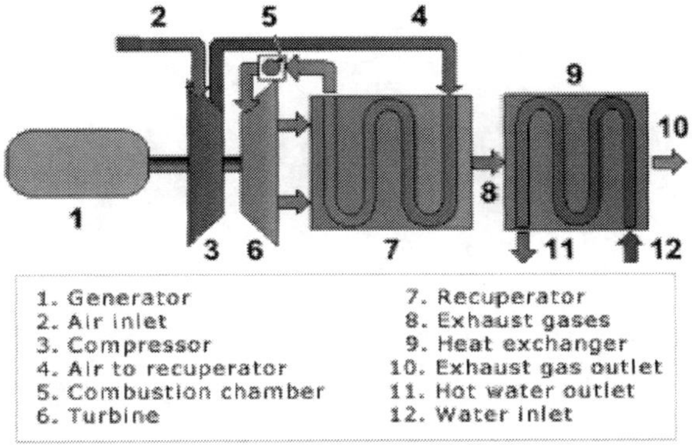

UKCCSRC PACT FACILITIES

# Gas Turbine System

## Analytical Facilities

- ❑ Flow rates, temperatures, dew points and pressure measurements throughout the system.
- ❑ Electrical and thermal power measurement
- ❑ Combustion Gas Analysis
  - Horiba VA-3000 Analyser I: Model VA-3002 for CO and NOx analysis
  - Horiba VA-3000 Analyser II: Model VA-3113 for CO2, O2 and SO2 analysis
  - Signal 3000HM Heated FID for total hydrocarbon analysis
- ❑ Particulate Spectrometer (Cambustion DMS500 Fast Particulate Spectrometer)
  - Classification by particle electrical mobility
  - Online analysis of particle mass, number and size spectra
  - range (5 – 1000nm)

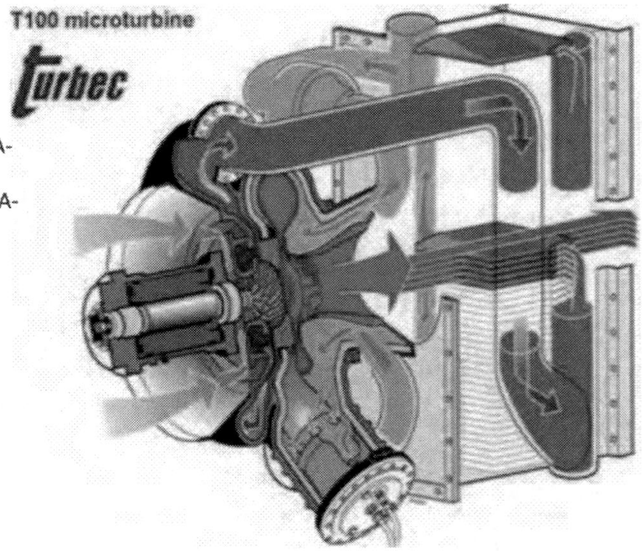


UKCCSRC PACT FACILITIES

# Gas Turbine System

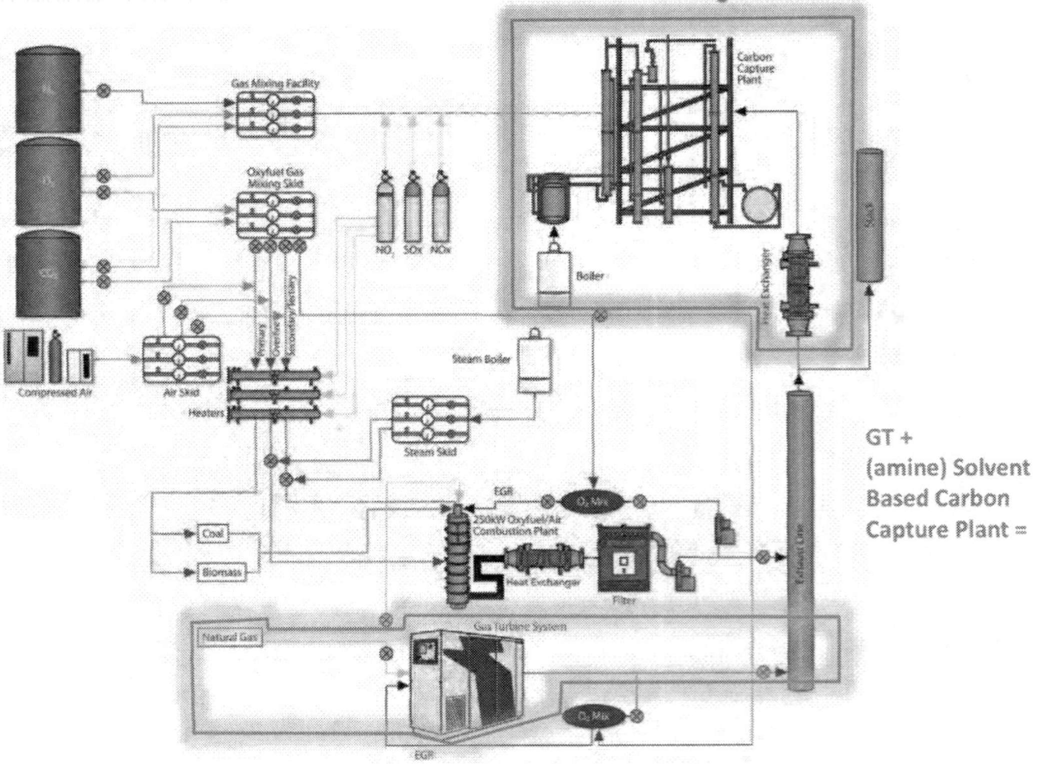


GT + (amine) Solvent Based Carbon Capture Plant =

UKCCSRC PACT FACILITIES

# Gas Turbine System

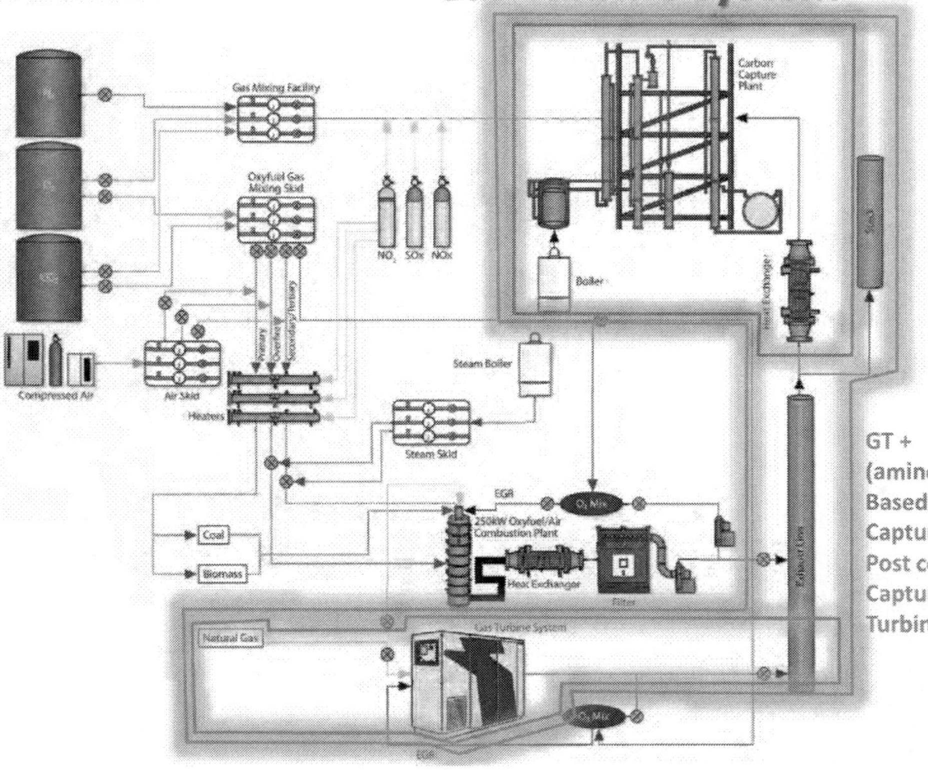


GT +
(amine) Solvent
Based Carbon
Capture Plant =
Post combustion
Capture from Gas
Turbine System

UKCCSRC PACT FACILITIES

**250 kW plant**

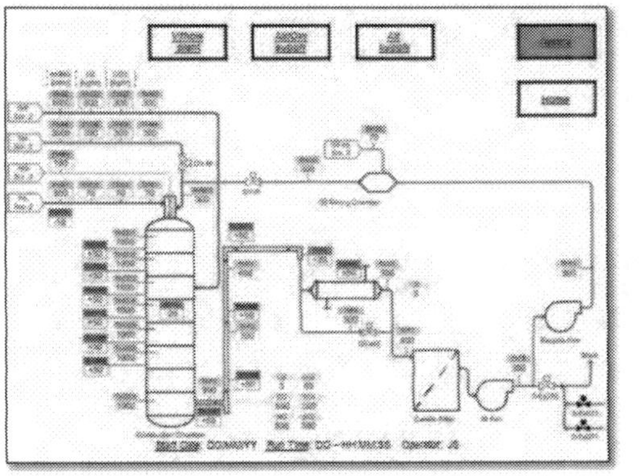

**Gas Turbine**

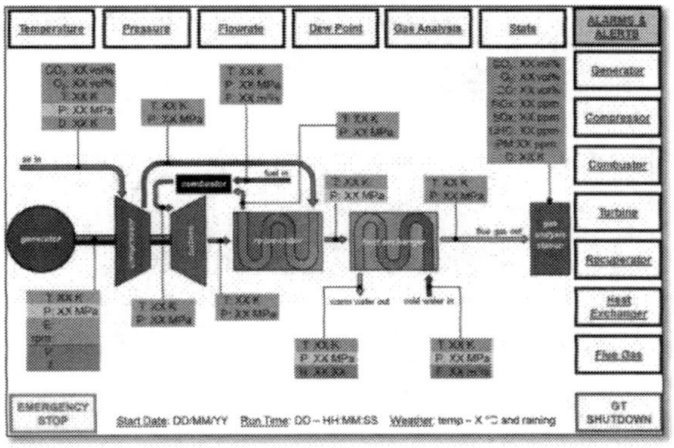


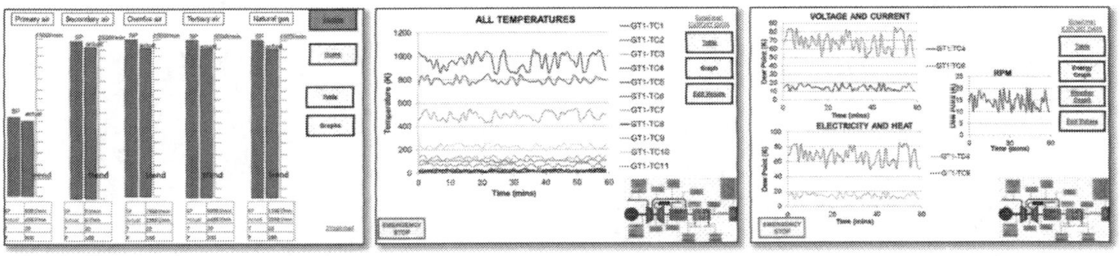

## Analytical Facilities: Labs

UKCCSRC PACT FACILITIES

Analytical labs

- ☐ Unique CEM mobile laboratory for solid-state detector based ICP-OES (SUWIC)
- ☐ Cambustion DMS500 Fast particulate analyser
- ☐ CHNS/O Elemental Analyser
- ☐ GC MS and TG-MS
- ☐ Thermogravimetric Analyser and TG-MS
- ☐ FT-IR and TG-IR
- ☐ Portable SERVOFLEX MiniMP gas analysers ($CO_2$ and $O_2$)

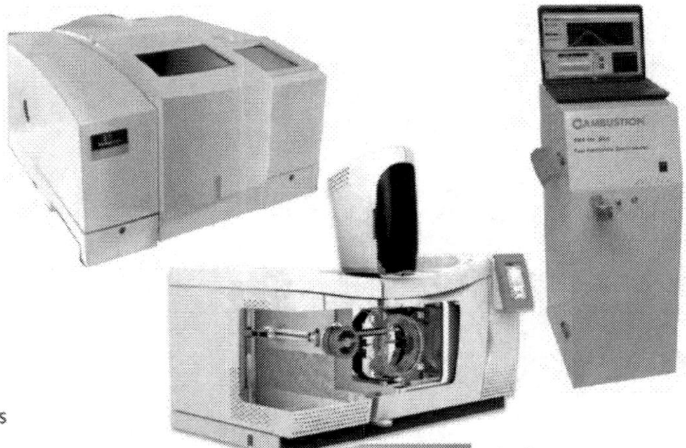

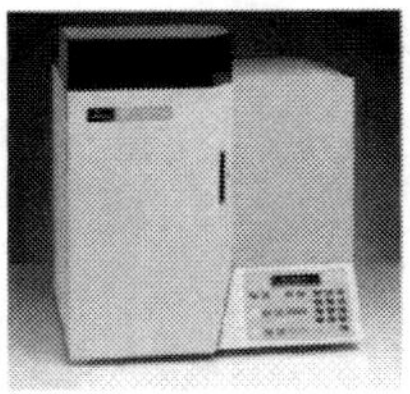

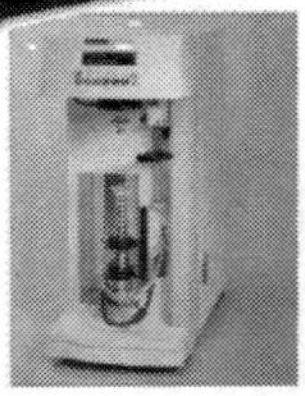

## Analytical Facilities: online monitoring

UKCCSRC PACT FACILITIES

- ☐ Gas analysis systems for both 250kW plant and the gas turbine.
- ☐ Particle size analyser
- ☐ Continuous Emissions Monitoring Laboratory (CEML) mobile laboratory, (Inductively Coupled Plasma) for monitoring metallic emissions from thermal processes;
  - ☐ Real-time, online diagnostics
  - ☐ Simultaneous multi-metal analysis

Center for Applied Energy Research

# Carbon Management Research at University of Kentucky Center for Applied Energy Research (UKy-CAER)

**Jesse Thompson**

University of Kentucky, Center for Applied Energy Research

Lexington, KY 40511

http://www.caer.uky.edu/powergen/home.shtml

caer.uky.edu

---

Center for Applied Energy Research

## University of Kentucky Center for Applied Energy Research

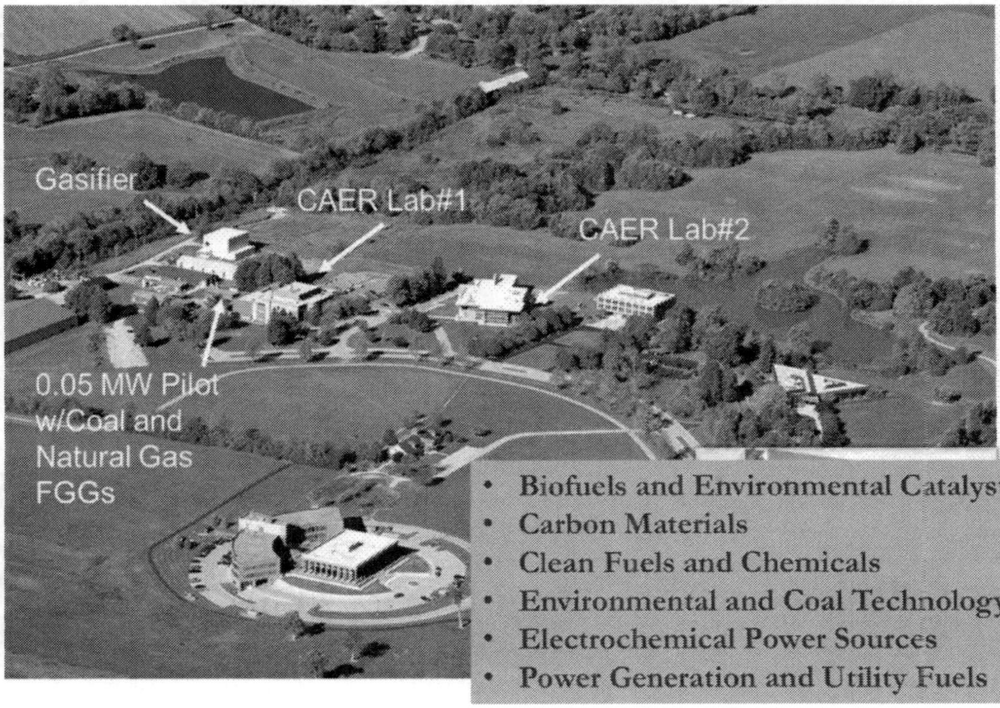


- Biofuels and Environmental Catalysts
- Carbon Materials
- Clean Fuels and Chemicals
- Environmental and Coal Technology
- Electrochemical Power Sources
- Power Generation and Utility Fuels

caer.uky.edu

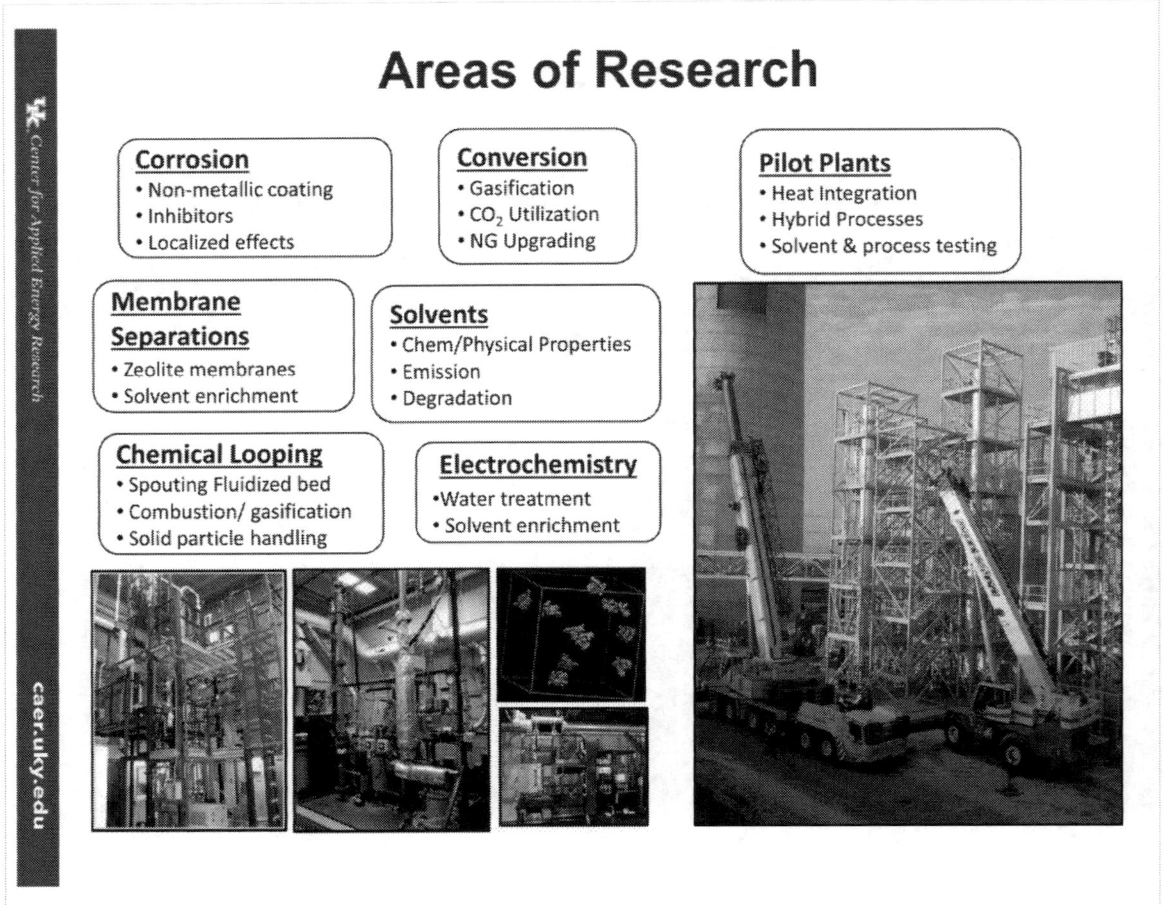
Areas of Research
Corrosion
• Non-metallic coating
• Inhibitors
• Localized effects
Conversion
• Gasification
• $CO_2$ Utilization
• NG Upgrading
Pilot Plants
• Heat Integration
• Hybrid Processes
• Solvent & process testing
Membrane Separations
• Zeolite membranes
• Solvent enrichment
Solvents
• Chem/Physical Properties
• Emission
• Degradation
Chemical Looping
• Spouting Fluidized bed
• Combustion/ gasification
• Solid particle handling
Electrochemistry
•Water treatment
• Solvent enrichment
UK Center for Applied Energy Research
caer.uky.edu

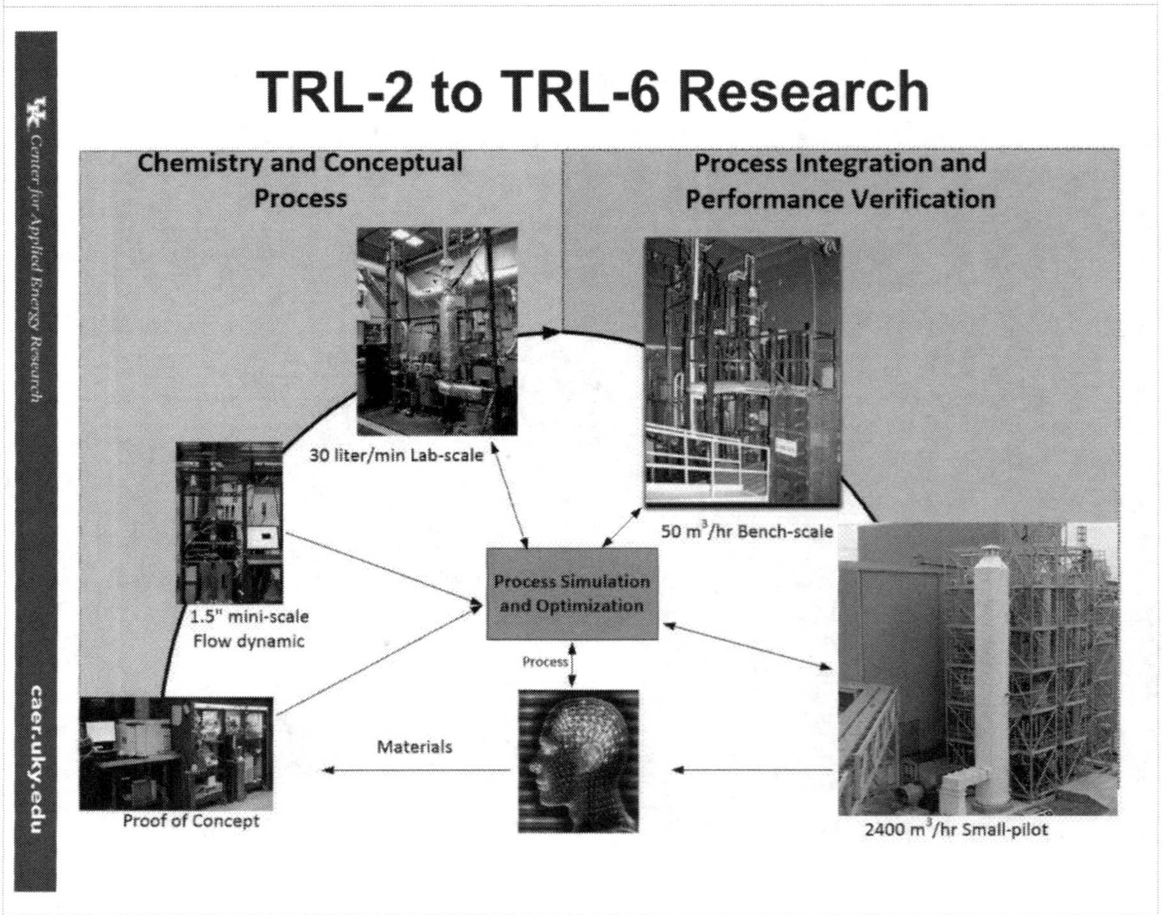
TRL-2 to TRL-6 Research
Chemistry and Conceptual Process
Process Integration and Performance Verification
30 liter/min Lab-scale
50 $m^3$/hr Bench-scale
1.5" mini-scale
Flow dynamic
Process Simulation and Optimization
Process
Materials
Proof of Concept
2400 $m^3$/hr Small-pilot
UK Center for Applied Energy Research
caer.uky.edu

# Small Pilot (1400 scfm) $CO_2$ Capture

- Unique UKy-CAER $CO_2$ Capture Process
  - **Heat integrated cooling tower** and liquid desiccant air drying system for decreased energy consumption
  - **Two-stage stripping** for increased solvent working capacity
- Continuous gas stream composition monitoring (**CEMS**)
- Continuous **steam usage** measurement
- Continuous **energy consumption** measurement
- Comprehensive liquid sampling ports for **solvent quality analyses**
- Comprehensive **gas sampling** ports for emissions and degradation analyses
- Liquid/gas **column profile** sample ports for absorber profile validation
- **Corrosion coupon** testing locations for material evaluation
- On-site **analytical laboratory** for routine liquid sample analysis

Phase 2 Modifications: Install $CO_2$ pre-concentration membrane and emission reduction systems

UK Center for Applied Energy Research

caer.uky.edu

# 0.1MWth Post-Combustion $CO_2$ Capture

**Coal-fired FGG and NG boiler**

**Flue Gas $CO_2$ Pre-Concentrating Membrane**

**13.7 cfm Unit**

**Gas Sampling Equipment**

**Emissions and degradation testing to address EH&S concerns**

- Integrated $CO_2$ capture process
- Solvent performance evaluation under real conditions with coal-derived flue gas
- Degradation studies for solvent degradation and make-up rates
- Flexibility for integration of other technologies (e.g., membranes) and process/method development (e.g., gas sampling)

UK Center for Applied Energy Research

caer.uky.edu

Center for Applied Energy Research

caer.uky.edu

# Technology Progression

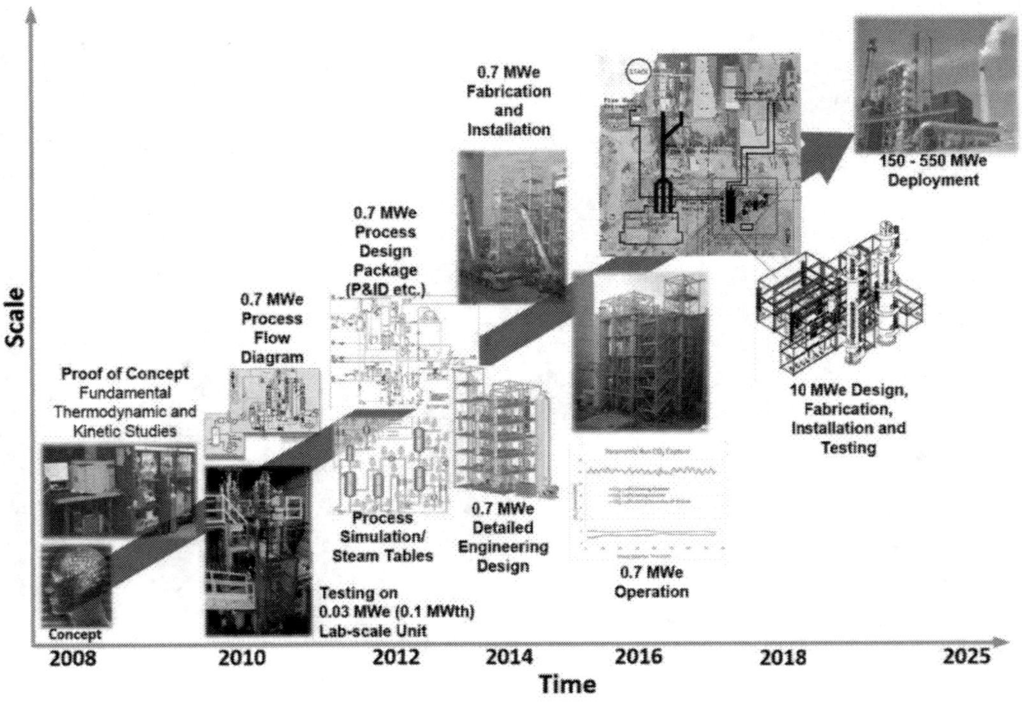


Center for Applied Energy Research

caer.uky.edu

# Gasification and Acid Gas Cleanup

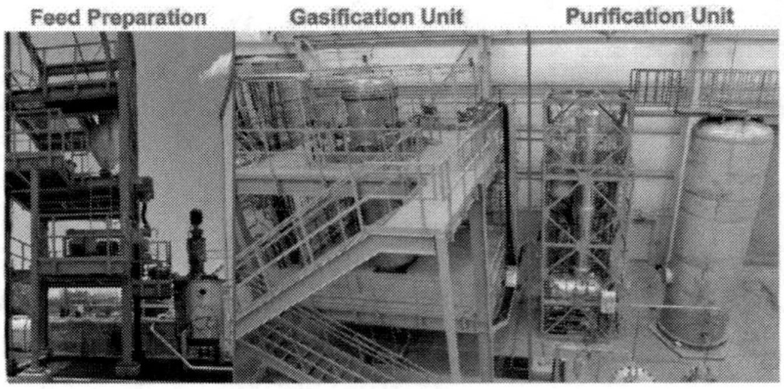


**Gasifier**

Coal and Coal/Biomass Integrated Feed Preparation

Entrained flow, oxygen blown, molten slag, OMB gasifier

1 ton dry coal/day

160 lb/hour syngas production

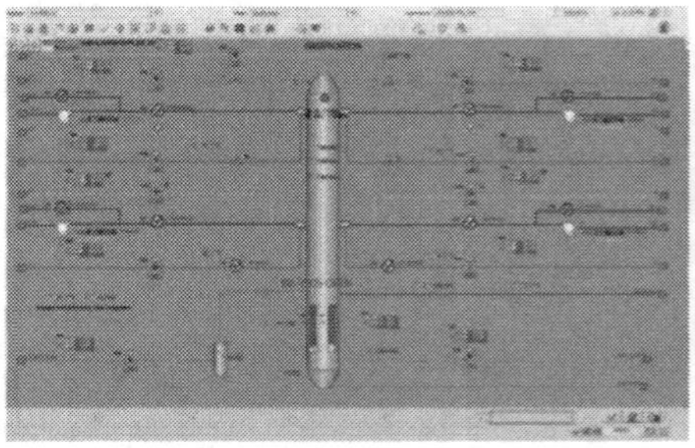

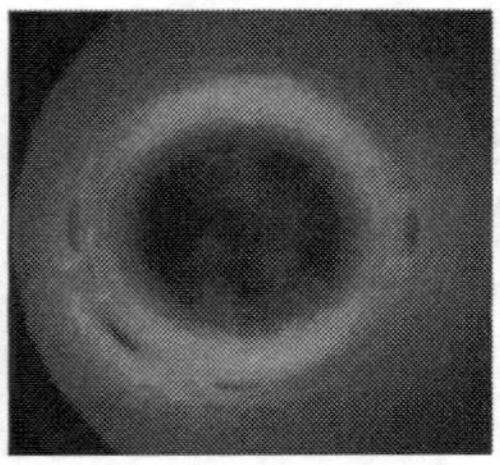

**Acid Gas**

Aqueous Amine system with absorber and stripper

Pressurized system (400 psi)

Sulfur treatment on both rejected and clean gas (<1ppmv sulfur)

Center for Applied Energy Research

caer.uky.edu

# WGS and F-T Module

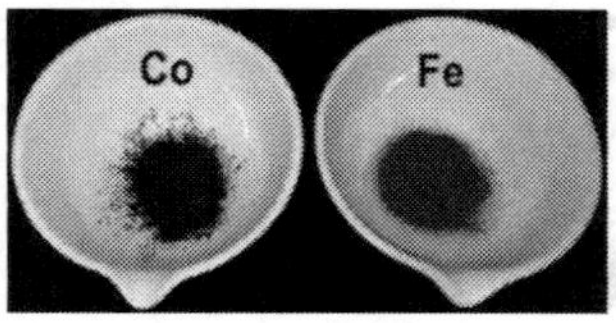


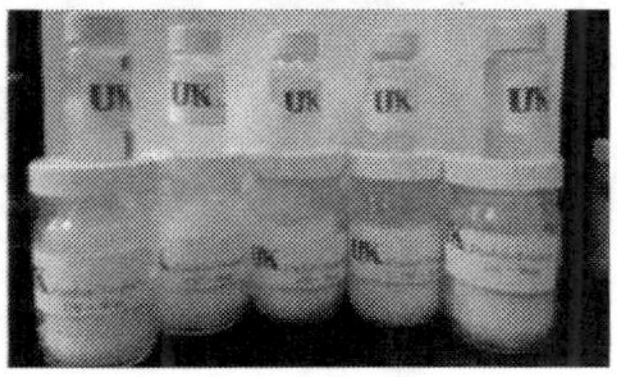

- Chart Energy microchannel reactor with production capacity of 1bbl./day
- FT Reactor is aluminum heat exchanger inside a steel kettle
- Capable of utilizing either Fe or Co based catalysts
- Wide ranging fuel production

Center for Applied Energy Research

caer.uky.edu

# Facilities and Apparatus: Lab-Bench Scale

Center for Applied Energy Research

caer.uky.edu

# Equipment and Instrumentation for Screening, Evaluation and Physical/Chemical Property Measurements

| Equipment/Instrumentation | Type | Capacity | Function |
|---|---|---|---|
| Vapor-Liquid Equilibrium (VLE) with GC (high pressure) | Parr reactor | 400 mL/charge | Determine chemical/physical properties such as equilibrium $CO_2$ pressure, capacity and heat of absorption |
| High pressure TGA | fixed bed | up to 10 gram/load | Evaluate sorbent stability, $CO_2$ absorption rate, carbon capacity, Vapor-Solid Equilibrium |
| Netzsch TG/QMS/DSC System | fixed bed | up to 30 gram/load | Evaluate sorbent stability, $CO_2$ absorption rate, carbon capacity, Vapor-Solid Equilibrium, heat of absorption. |
| Electrochemical Corrosion Cell | stirred reactor | 200 mL/charge | Corrosion |
| Traditional Corrosion Cell | equilibrium cell | 2 liter/run | Corrosion |
| Bench-Scale Mini-Scrubber | solvent-based, simulated flue gas | 1.1 kg/hr gas feed rate | Evaluate foaming tendency, batch-mode $CO_2$ absorption for mass transfer measurement |
| Bench-Scale Mini-Stripper | random packing | 20 mL/min liquid feed rate | Determine batch-mode $CO_2$ desorption rate |
| Wetted Wall Column (WWC) | - | 2 L/run | Determine mass transfer/kinetic data |
| Vapor-liquid equilibrium (VLE) apparatus (Low pressure) | - | 10 mL/charge | Determine Chemical/physical properties such as equilibrium $CO_2$ pressure, capacity and heat of absorption |
| Breakthrough Solvent Evaluation Apparatus | - | 10 mL/charge | Determine kinetic data/equilibrium $CO_2$ pressure, capacity |
| Stop Flow Reactor | - | 20 mL/charge | Reaction rate |
| Pilot-scale Extruder system with material preparation | thermo-compression | 20 kg/day product | Catalyst/solid sorbent production |
| Freeze Granulator and Freeze Dryer | slurry | 2 kg/day | Catalyst/solid sorbent production |
| EPA Method 5 Standard Gas Sampling Equipment | Apex Instruments (2009) | | Monitoring emissions of particulate matter (PM) and solvent degradation compounds for secondary environmental monitoring including EPA stack emission methods |

Center for Applied Energy Research

caer.uky.edu

# Learning from Scale-up Studies

## Amine Degradation

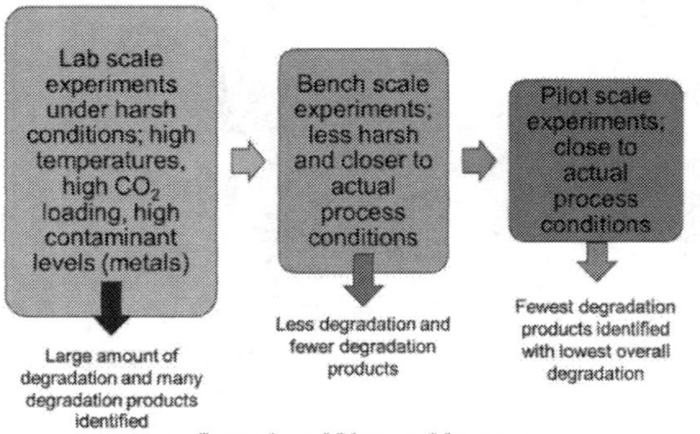


## Ammonia Emissions vs Fe

Positive correlation between $NH_3$ emission and higher Fe in the solvent.

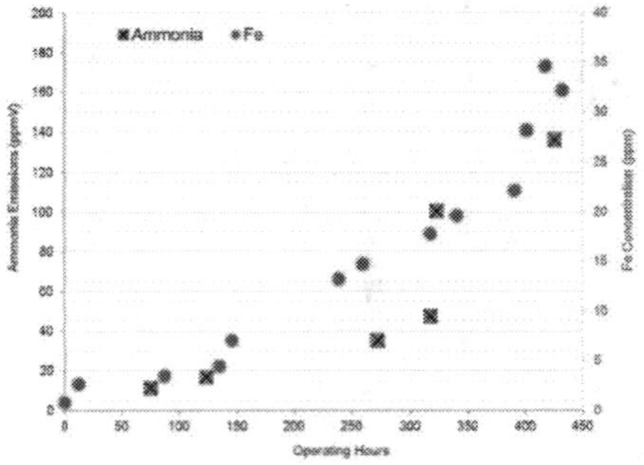


## Service Water Usage

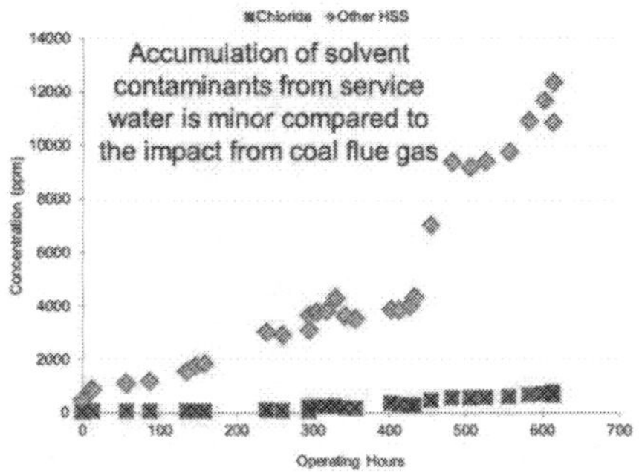


## Column Distribution / Channel Flow

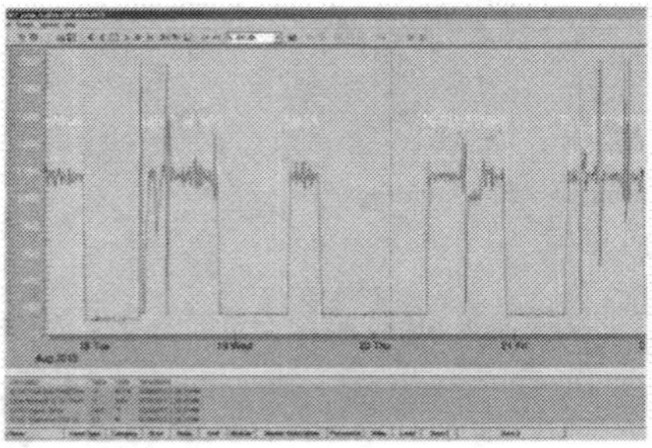

Center for Applied Energy Research

caer.uky.edu

# UKy-CAER Highlights

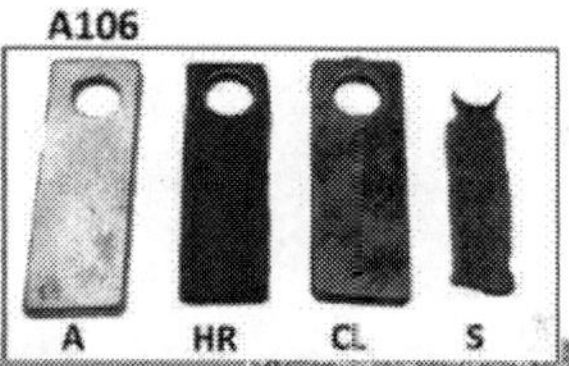


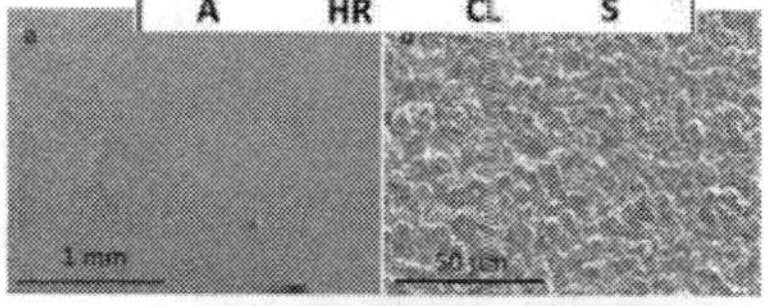


## Corrosion

"Use of Carbon Steel for Construction of Post-combustion $CO_2$ Capture Facilities: A Pilot-Scale Corrosion Study". *I&EC*, 2017, 56, 4792-4803.

"Understanding the Corrosion of $CO_2$-loaded 2-amino-2-methyl-1-propanol solutions assisted by thermodynamic modeling". *IJGGC* 2016, 54, 211-218.

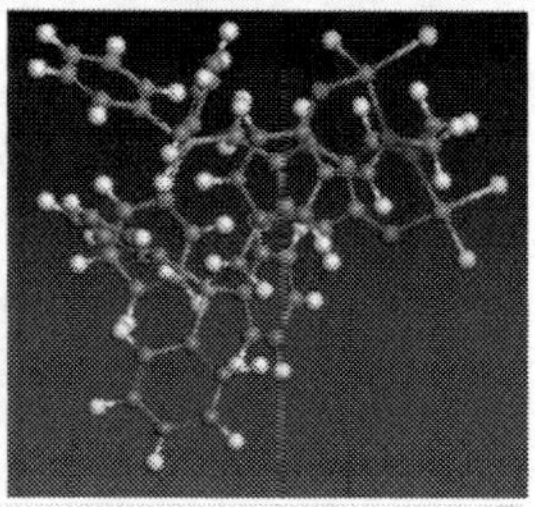

## Solvent Development

"Enhanced carbon capture through incorporation of surfactant additives". *I&EC*, 2016, 55, 7456-7461.

"Enhancements in mass transfer for carbon capture solvents part I: Homogeneous catalyst". *IJGGC*, 2017, 63, 249-259.

"Enhancements in mass transfer for carbon capture solvents part II: Micron-sized solid particles". *IJGGC*, 2017, 61, 138-145.

"Application of Surface Tension Model for Prediction of Interfacial Speciation of $CO_2$-Loaded Aqueous Solutions of Monoethanolamine". *I&EC*, 2017, 56, 5747-5755.

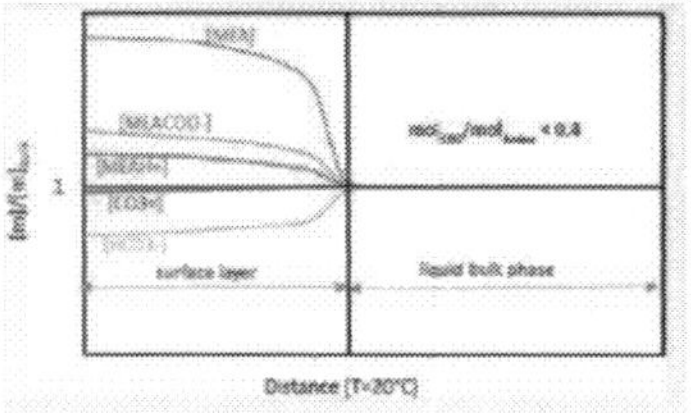


Center for Applied Energy Research

caer.uky.edu

# UKy-CAER Highlights

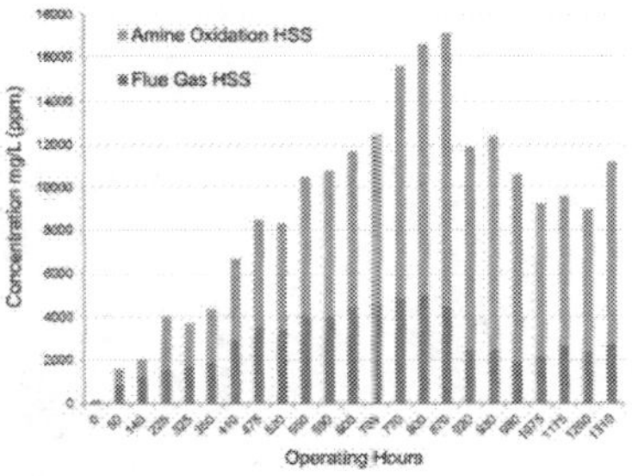


## Solvent Degradation

"Pilot testing of a heat integrated 0.7 MWe $CO_2$ capture system with two-stage air-stripping: Amine degradation and metal accumulation". *IJGGC*, 2017, 64, 23-33.

"Pilot testing of a heat integrated 0.7 MWe $CO_2$ capture system with two-stage air-stripping: Emissions". 2017, *IJGGC*, In review.

"Thermal degradation rate of AMP to cyclic DMOZD: Experiments and kinetic modeling". *I&EC*, 2016, 55, 9586-9593.

"Thermal degradation rate of 2-amino-2-methyl-1-propanol to cyclic 4,4-Dimethyl-1,3-oxazolidin-2-one; Mechanistic aspects and kinetics investigation". *I&EC*, 2017, In review.

"Selective removal of nitrosamines from a model amine carbon capture waterwash using modified solid sorbents". *ES&T*, 2017, In revision.

"Determining the Henry's volatility coefficient of nitrosamines in $CO_2$ capture solvents". *IJGGC*, 2017, Submitted.

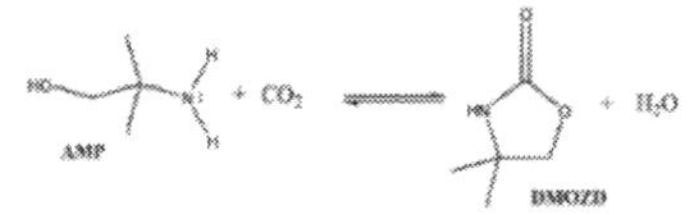


**Figure 1.** AMP thermal degradation, followed by cyclization and dehydration to form 4,4-dimethyl-1,3-oxazolidin-2-one (DMOZD).

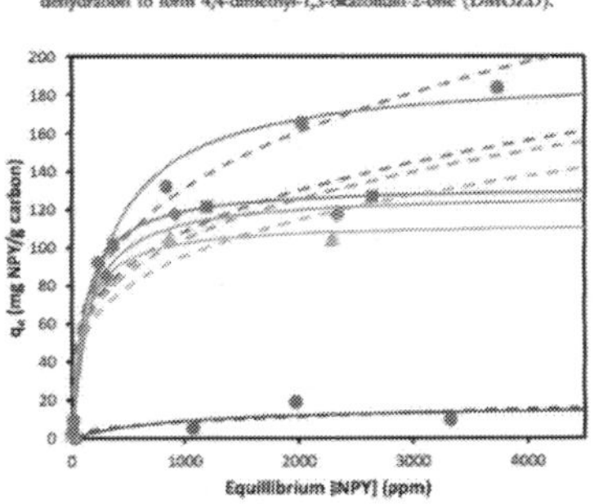

Center for Applied Energy Research

caer.uky.edu

# Acknowledgements


**Lynn Brickett and José Figueroa, DOE NETL**

**CMRG Members: LG&E and KU, EPRI, Duke**

# CO2CRC Capture Demonstration

Abdul Qader, PhD, FAIE, FIEAust
Capture Manager

CO2CRC
BUILDING A LOW EMISSIONS FUTURE

ITCN Panel in CMTC
19 July 2017

## Who is CO2CRC?

- CO2CRC is a world leading carbon capture & storage (CCS) research organisation that works closely with global industry to reduce large scale greenhouse gas emissions through:
  - Investigating carbon capture, transport and storage technologies
  - Examining the costs and drivers across the CCUS chain
  - Conducting research, development and demonstration across the CCUS value chain
  - Providing efficient collaboration amongst 150 researchers in capture and storage
  - Building experience in CCUS technology by addressing knowledge gaps

32 research and industry partners globally

# Australia's Integrated CCS activities

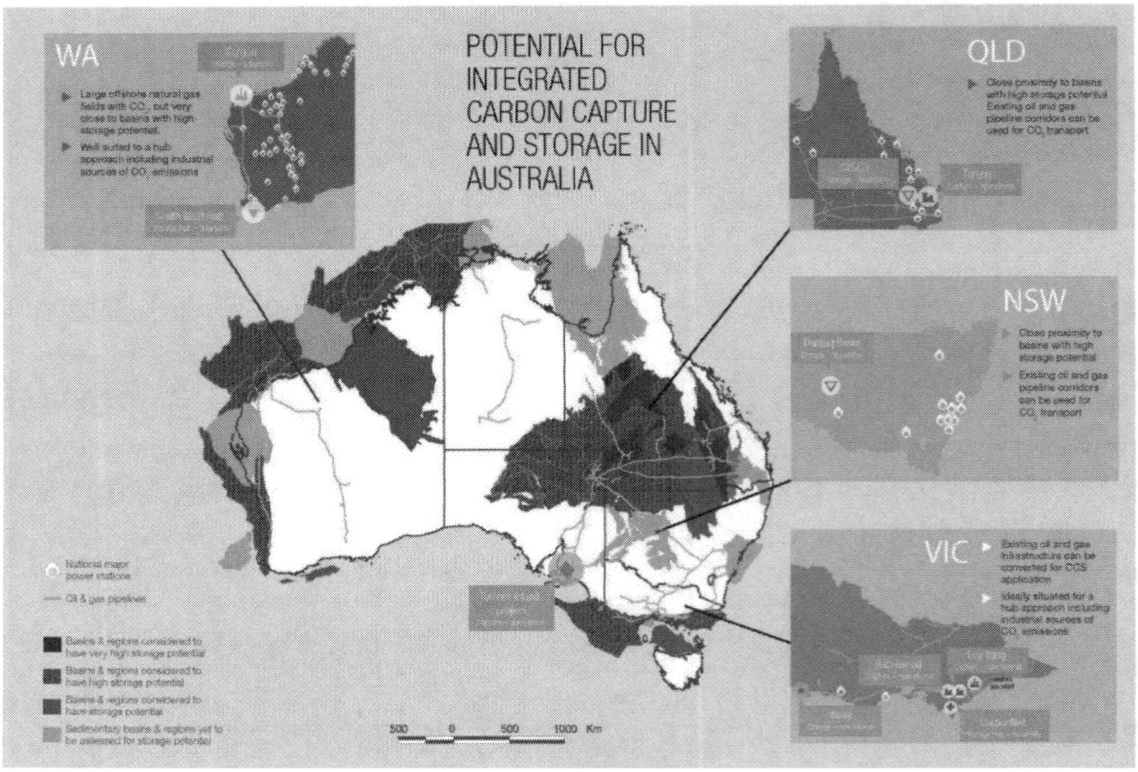

## Field Capture Projects

- First application of 3 different capture technologies in parallel and their comparison for large scale implementation
- Attained capability of scaling up of capture plants from concept till large scale (TRL level 1 till 9) with economic feasibility

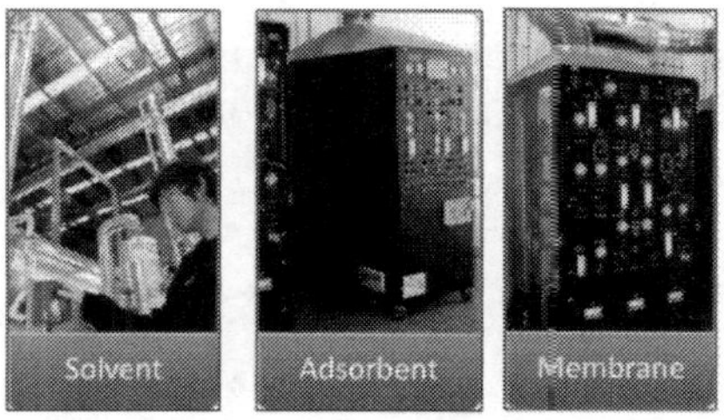


3 pre-combustion technologies at HRL, VIC : 2007-2011, ETIS/BCIA funded

3 post-combustion technologies at Hazelwood, VIC : 2007-2011 ETIS/BCIA funded

BCIA funded UNO MK 3 post-combustion at Hazelwood, VIC : 2011-2014

Membrane plant at Delta's Vales Point, NSW : ANLEC R&D funded

CO2CRC

4

CO2CRC

5

# Otway Capture Project

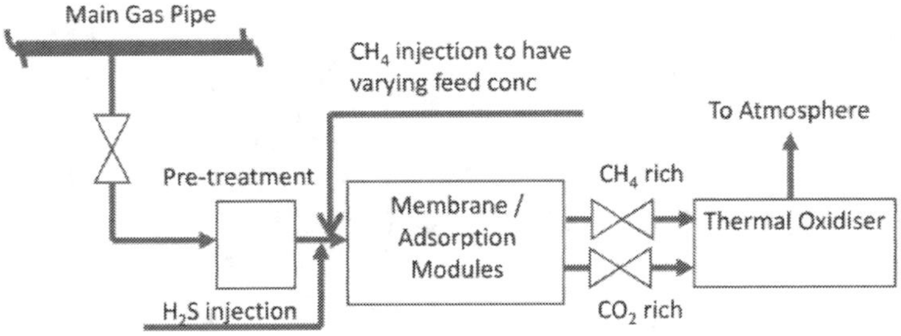

## Project Aims

- To develop cost effective, compact technologies to capture $CO_2$ mainly from high $CO_2$ content wells.
- To test new capture materials (membranes and adsorbents) and develop new capture processes over a range of adjusted $CO_2$ concentrations

## Main Features

- Feed Pressure: 80 to 30 bar
- Feed $CO_2$ Concentration: 5% to 80% achieved by $CH_4$ addition (existing is ~80% $CH_4$ and 19% $CO_2$)
- $H_2S$ addition for impurities effect tests
- Rig on a common skid delivered on site for installation and commissioning (flow diagram shown left) 2016
- Adsorbent – 4 kg, Current use: Silica (S1), Second stage: Z1; Membranes – 3 different modules
- Opportunities for vendors to conduct long term testing of capture technologies for CO2/CH4 separation at a fully instrumented site

CO2CRC

6

# Otway Capture Installation and Commissioning

## Hollow Fiber Modules prepared for Otway

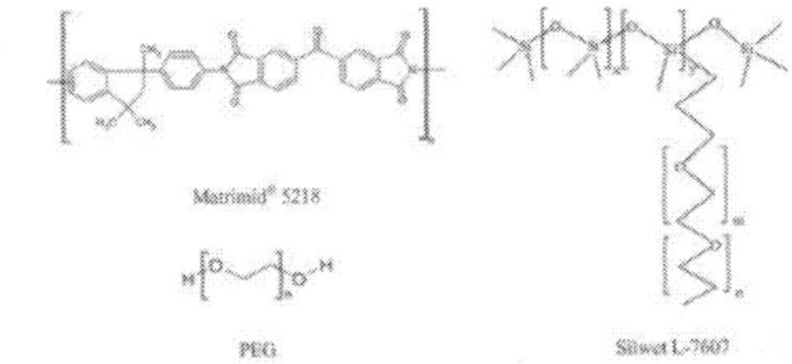


Addition of PEO-PDMS additive (Silwet-7607) could improve plasticization resistance and long term stable performance – so composite also be used

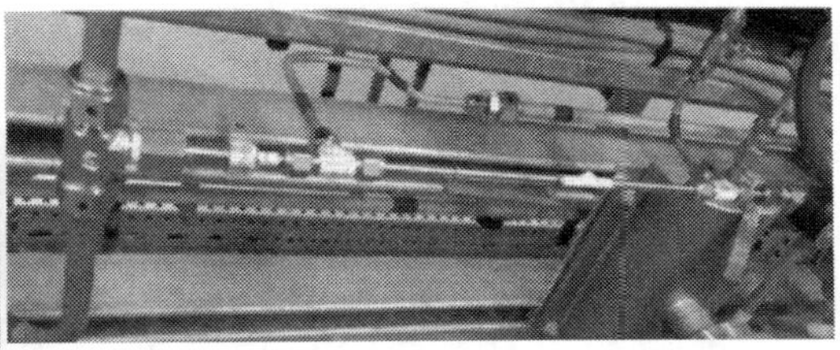

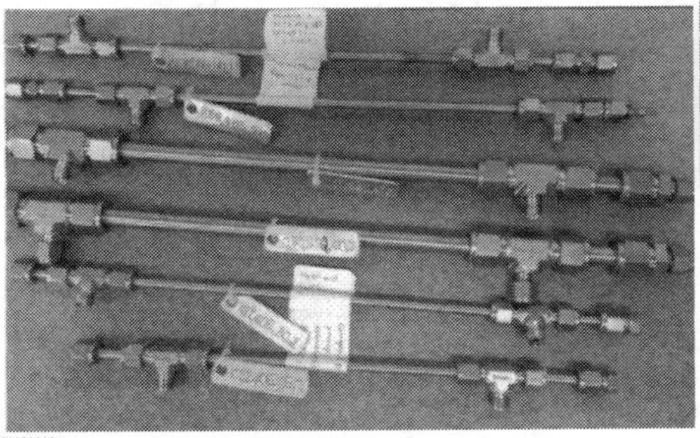

modules: -half inch module housing, 40 fibers, 40 cm fiber length inside the modules
moudle: - ¼ inch module housing, 5 fibers, 30 cm fiber length

## Membrane Gas-Liquid Contactors

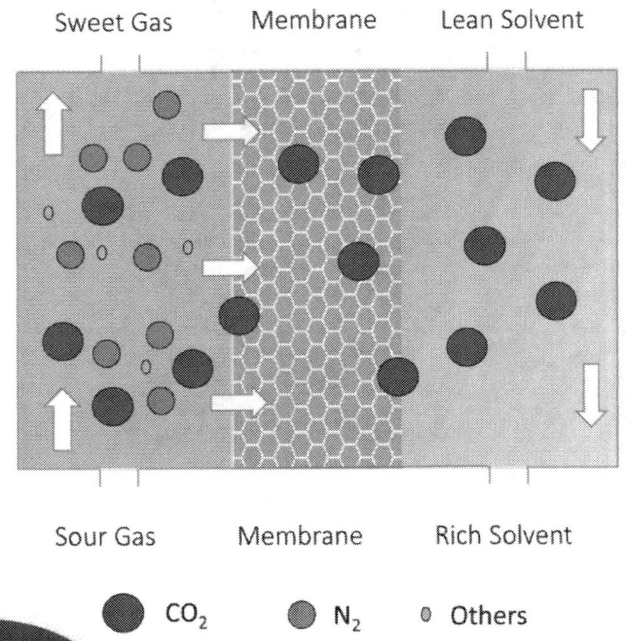


- Avoidance of hydrodynamic issues.
- Equipment size reduced by 60 – 75%.
- Corrosion is significantly reduced.
- Around 40% reduction in operating costs.
- Around 35 – 40% savings in capital costs.
- Footprint requirements reduced by 40%.

THE UNIVERSITY OF MELBOURNE

10

## Overall Concept Design for Membrane Contactors

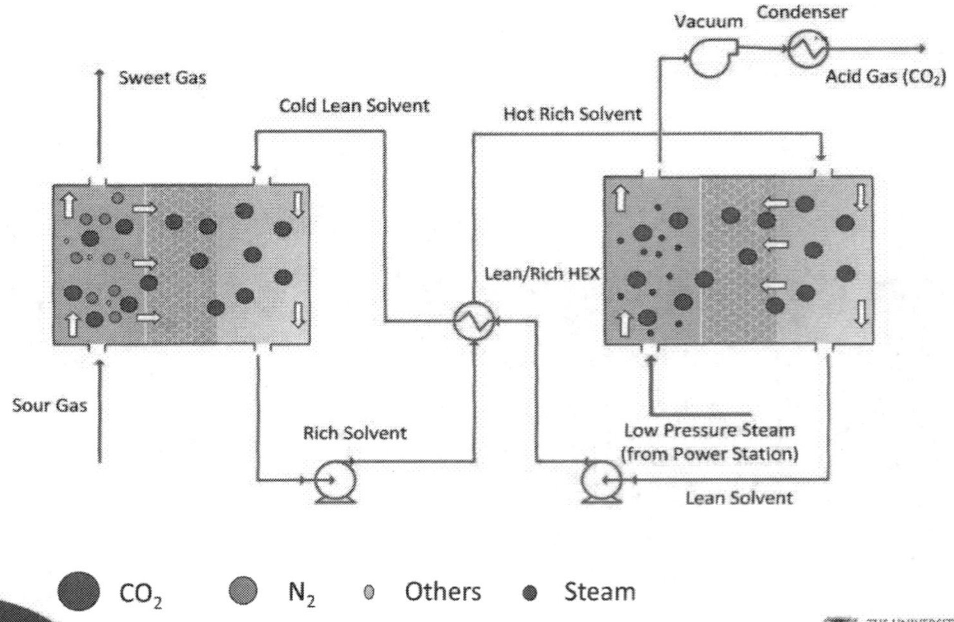


THE UNIVERSITY OF MELBOURNE

11

# International Collaboration and New Capture Ventures

- We are a member of ITCN – a special web page is available under NCCC web link (www.nationalcarboncapturecenter.com)
- A feasibility study on greenhouse gas reduction with a steel manufacturer in Australia is currently being discussed.
- A commercial venture for 50 tpd food grade $CO_2$
- A CCUS Hub Study for a Latrobe Valley power station is underway with global participants

12

# Current and Future Capture Program Overview

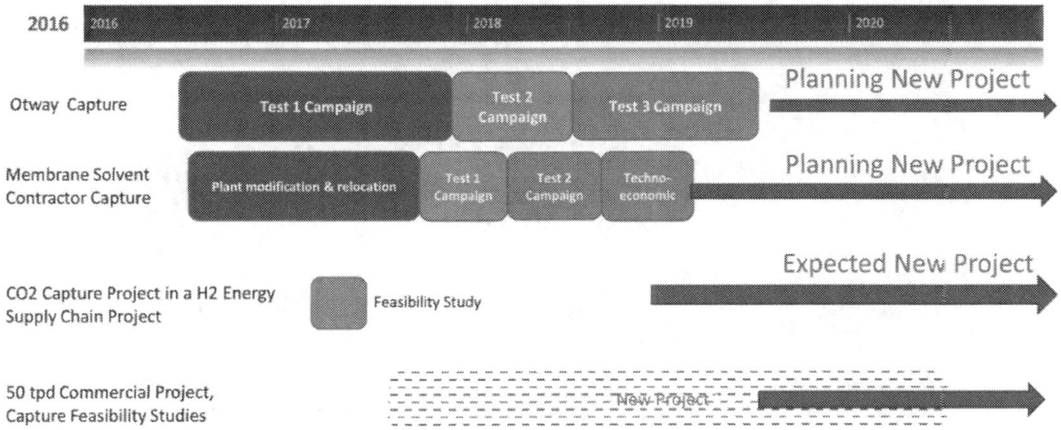


We are a member of ITCN – a special web page is available under NCCC web link (www.nationalcarboncapturecenter.com)

13

# Government, Industry and Research Partners

CO2CRC

14

CO2CRC

15

# International Collaboration on CCUS R&D

Frank Morton – Director Technology Development

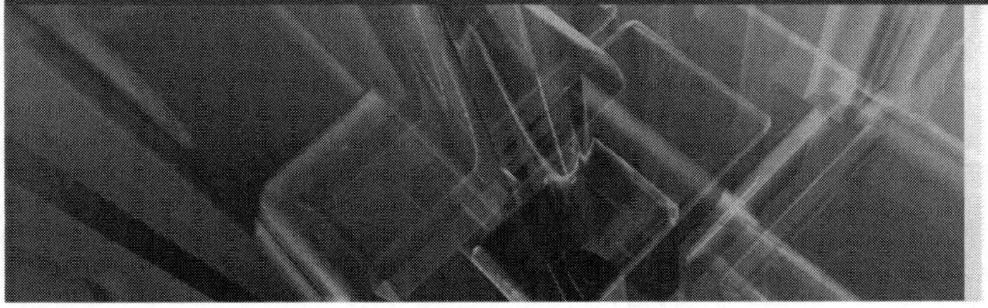

NATIONAL CARBON CAPTURE CENTER

1

## International Collaboration for the Development of CCUS R&D

- Overview of the NCCC
  - Establish cost-effective test facility which protects IP
- International activity
  - NCCC test collaborations
  - ITCN
- Path forward
- Importance of operating facilities
  - Policy, messaging, funding commitment from industry and gov't

NATIONAL CARBON CAPTURE CENTER

2

## National Carbon Capture Center

*Offering a **world-class neutral test** facility and highly specialized staff to **accelerate the commercialization** of advanced technologies and enable coal-based power plants to achieve **near-zero emissions (low-cost $CO_2$)**.*

- Over **97,000 test hours** since founding in 2008
- Technology developers from **U.S. and six other countries**
- **First coal-derived gas testing** of solid oxide fuel cells and certain solvents, membranes and enzymes
- On-site **scale-ups** and **process enhancements** for 10 technologies
  - Scale-ups for testing at larger sites for five solvents
  - Scale-up to commercial operation for one solvent

Timeline: June 1, 2014 – May 31, 2019

U.S. DEPARTMENT OF ENERGY

NATIONAL ENERGY TECHNOLOGY LABORATORY

Southern Company

CLEARPATH · AEP AMERICAN ELECTRIC POWER · CLOUD PEAK ENERGY · DUKE ENERGY · EPRI ELECTRIC POWER RESEARCH INSTITUTE

3

## What the Project Provides

- **Cost-efficient test site** with **infrastructure** for numerous technology developers
- **Real-world conditions** with coal-derived flue gas and syngas
- Flexible capability for testing at **multiple scales** and **on-site scale-ups**
- Expert **technical staff for** design, installation and testing support
- High-quality **data acquisition** and gas/liquid **sampling and analysis**

4

## PC4 Bench-Scale

- Simultaneous operation of up to five developers' test units
- Slipstream Solvent Test Unit (SSTU) for solvents in early development
- SSTU also used for solvent emissions studies and emission mitigation processes
- Flue gas/utilities and gas analysis systems operating independently of PC4 pilot-scale area

5

## PC4 Pilot-Scale

- Simultaneous operation of developer test units and Pilot Solvent Test Unit (PSTU)
- PSTU offers flexible operation to match developers' planned commercial configuration
- PSTU also supports solvent emissions and degradation studies

6

## NCCC International Collaboration

**Carbon capture R&D is an international issue that requires international collaboration**

- Support DOE goal of international cooperation
- Multiple paths for involvement; partners, developers, network members, consulting services and workshops
- Ease of collaboration since intellectual property is not being shared
- Broad effort China, India, Middle East, Korea, Japan, EU, Australia, Canada, Norway

7

## China

- Shenhua - NICE
  - MOU for knowledge sharing
  - Exploring test collaboration
- China Partner 1
  - Research group partnered with major power company
  - Signed MOU 2015
  - Negotiating contract for collaboration
- Huaneng – Clean Energy Research Institute
  - Joined ITCN
  - Exploring test collaboration
- Initial Meetings
  - Sinopec, Datang, CPI, Tsinghua Univ, Zhejiang Univ, Guodian, CATF, Wuhuan

Post-Combustion

Signing Ceremony
GDCCUS Facility

8

## India

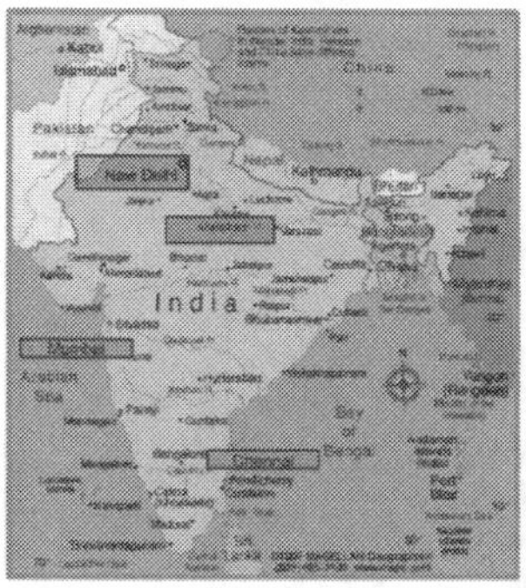

CCSi Test at Solvay Chemical in India

- Could become largest emitter of $CO_2$
- CCSi – Tested at the NCCC
- TIFAC within the Dept of Science and Tech is coordinating collaboration with government agencies
- In discussion with major oil & gas and power companies for knowledge sharing.



## Middle East

- Growing interest in EOR
- Extensive, well-funded green movement in the Middle East
- Countries want to participate in advanced power projects internationally
- UAE - Advanced negotiation with ADNOC on NCCC membership.

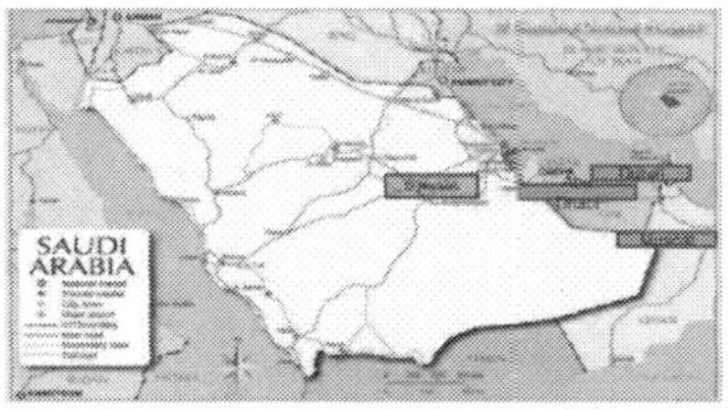

Emirates Steel - CCUS

Minister of Oil - Oman

## International Test Center Network Members

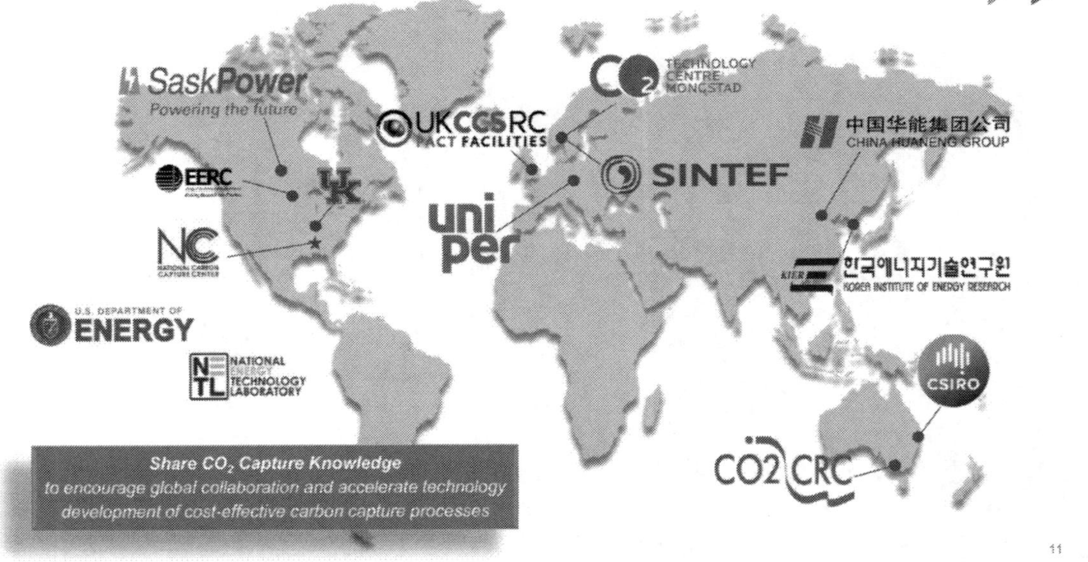




# International Test Center Network

***Share $CO_2$ Capture Knowledge***
*to encourage global collaboration and accelerate technology development of cost-effective carbon capture processes*

- Share pubic knowledge with carbon capture test facilities.
  - Facility operations
  - Facility funding
  - Safety
  - Analytical techniques
- Collaborate on one technical item per year.
  - Amine carry-over and measurement techniques
  - Support advanced simulations and model development with a focus on reducing capital and operating cost and minimizing scale-up risks

U.S. DEPARTMENT OF ENERGY

NATIONAL ENERGY TECHNOLOGY LABORATORY

EPRI ELECTRIC POWER RESEARCH INSTITUTE

Southern Company

AMERICAN ELECTRIC POWER BOUNDLESS ENERGY

CLOUD PEAK ENERGY

DUKE ENERGY

CLEARPATH

fcmorton@southernco.com

www.nationalcarboncapturecenter.com

https://twitter.com/NCarbonCaptureC

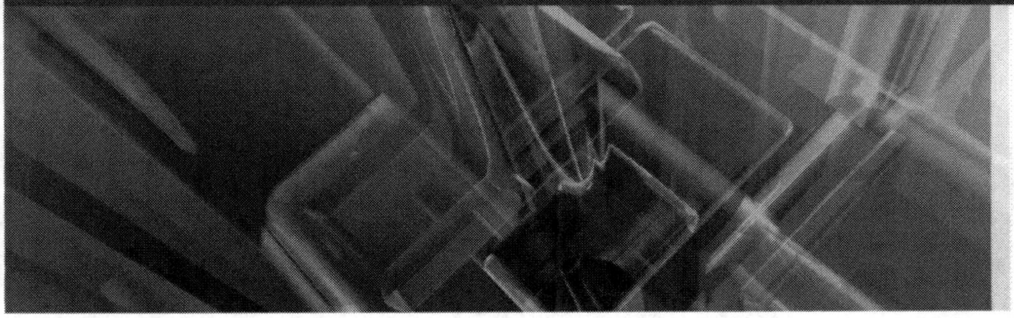

NC NATIONAL CARBON CAPTURE CENTER

13

CARBON MANAGEMENT

# CCUS, Wyoming, & Non-OECD Economies: ISO Standards as the common denominator

SPE International

SME Society for Mining, Metallurgy & Exploration

*Carbon Management Technology Conference*

*July 20, 2017*

AIChE

---

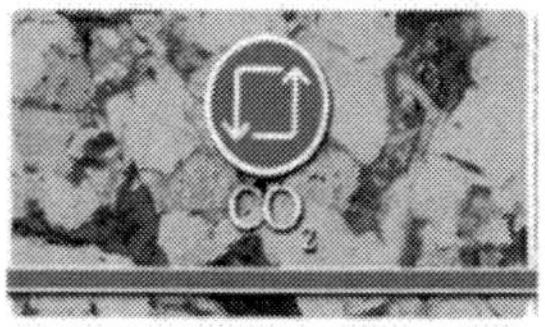

1. ISO Standards & CCUS
2. Wyoming & $CO_2$ Demand
3. Non-OECD Economies

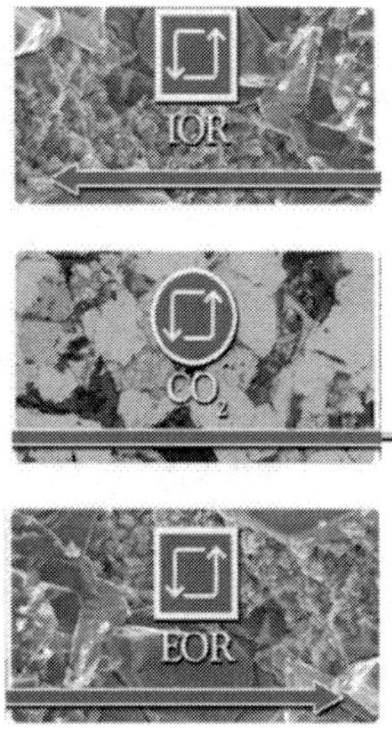


# 1: ISO Standards & CCUS

## What are Standards?

- Consensus based
- Designed as a rule, guideline or definition
- Revisable & updateable
- Voluntary
- Standards must fit to purpose:
  - *Prescriptive based*
  - *Objectives based*
  - *Performance based*
  - *Principles based*
  - *Hybrids*

International Organization for Standardization

ANSI American National Standards Institute

CSA Group Groupe CSA

ISO 14000

Enhanced Oil Recovery Institute

## Why Standards?

- Because they are not laws…
  - *Standards & regulations can work together*
- Not Mandated ∴ typically initiated by industry…
  - *And ∴ better received & used by industry as part of the process*
- Demonstrate regulatory compliance
- Streamline the regulatory process
- Harmonize across jurisdictions

Enhanced Oil Recovery Institute

## ISO Standards Development

- ISO does not write standards or influence technical content – manage the process
- Technical Committees (TC) of International Experts write standards
- P-Member countries approve standards
- Nations adopt ISO standards

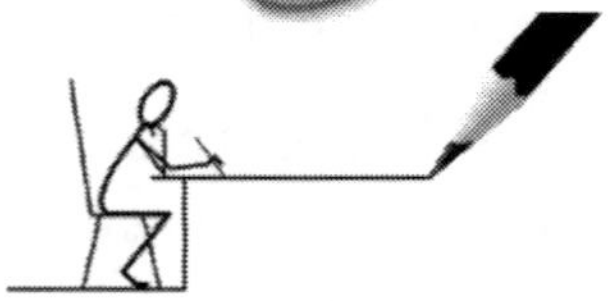

ANSI

Enhanced Oil Recovery Institute

# ISO Standards Development

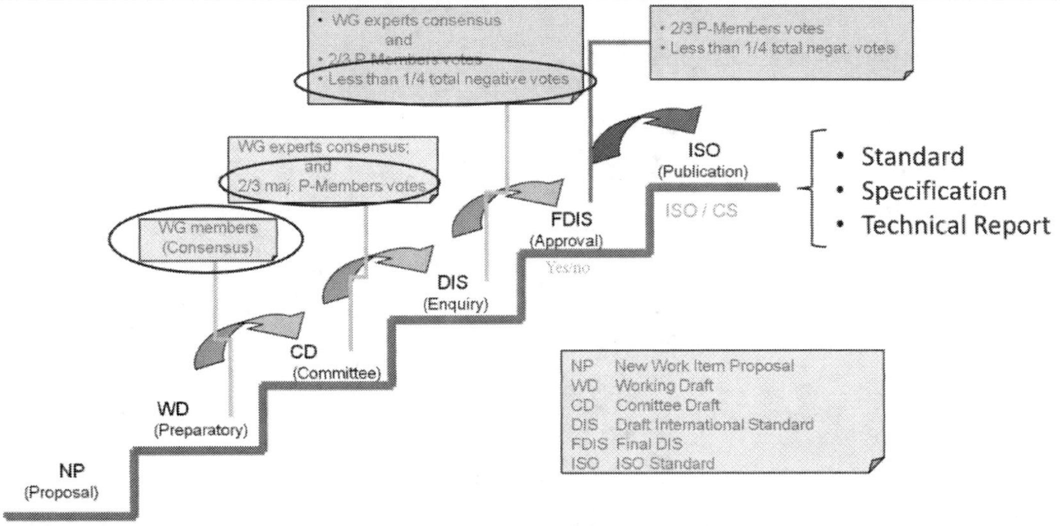


Enhanced Oil Recovery Institute

# ISO/TC 265 Carbon dioxide capture, transportation, and geological storage

Standardization of design, construction, operation, environmental planning and management, risk management, quantification, monitoring and verification, and related activities in the field of carbon dioxide capture, transportation, and geological storage

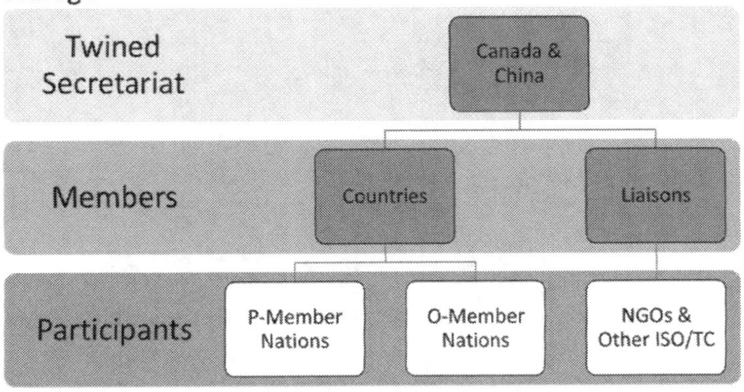


Enhanced Oil Recovery Institute

## ISO TC 265 – P-Members

Participating Countries (18):

| | |
|---|---|
| Australia | Norway |
| Canada | Saudi Arabia |
| China | South Africa |
| France | Spain |
| Germany | Sweden |
| India | Switzerland |
| Japan | United Kingdom |
| Malaysia | Republic of Korea |
| Netherlands | United States (ANSI) |

- ✓ *Voting Members*
- ✓ *Guaranteed International Expert Participation on all WGs*

9

Enhanced Oil Recovery Institute

## ISO TC 265 – O-Members

Observing Countries (10):

| | |
|---|---|
| Argentina | Italy |
| Czech Rep. | New Zealand |
| Egypt | Qatar |
| Finland | Serbia |
| Iran | Sri Lanka |

- ✓ Non-voting Members
- ✓ *May request* International Expert Participation on all WGs
- ✓ May upgrade to P-Member at any time

10

Enhanced Oil Recovery Institute

## ISO TC 265 – Liaisons

- ISO TC207 Environmental Management
- ISO TC67 Petroleum and Natural Gas
- CEN/TC 234 Gas Infrastructure
- Carbon Sequestration Leadership Forum (CSLF)
- European Industrial Gases Association (EIGA)
- International Energy Association (IEA)
- IEAGHG
- CO2 GeoNet
- World Resources Institute (WRI)

- ✓ Non-voting Members
- ✓ *Guaranteed* International Expert Participation on all WGs



Enhanced Oil Recovery Institute

## ISO TC 265 – Working Group Structure

**Cat Herding**

The challenges and rewards of managing online scientific communities

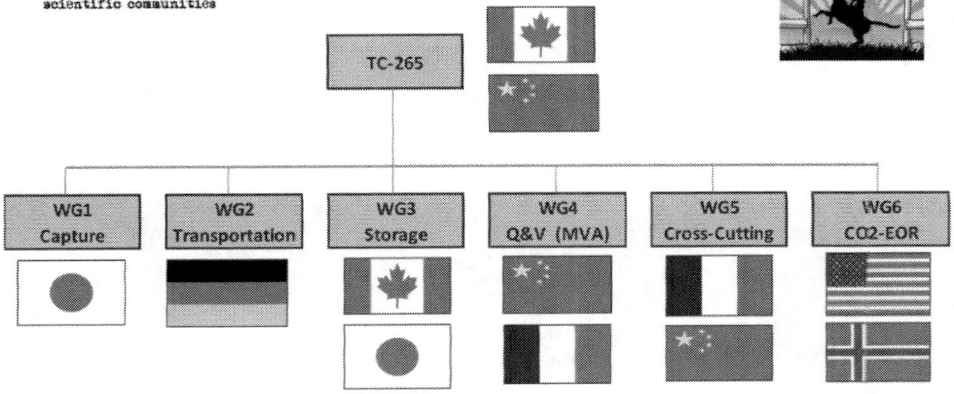




Enhanced Oil Recovery Institute

# WG1: Capture

Scope:

- Pre-, post-, & oxyfuel combustion capture
- Dehydration, compression & pumping
- Liquefaction, installation, operation, maintenance
- Quality of $CO_2$ streams
- Industrial processes
- Separation, purification
- Monitoring, management systems
- Plant retrofitting

Technical Report:

ISO TR 27912 Carbon dioxide capture systems, technologies, equipment and processes for power and industry

Drafting a Standard:

ISO CD 27919 Performance Evaluation Methods PCC Power Plant



Enhanced Oil Recovery Institute

# WG2: Transportation

Pipeline transportation systems boundaries:

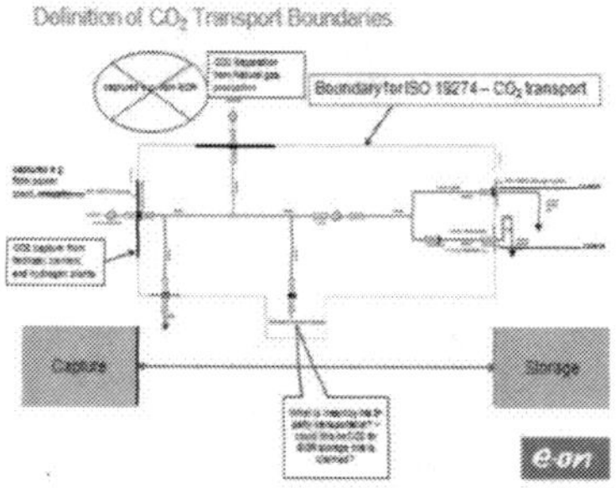


- Pipelines not currently covered by existing ISO/TC-67 standards
- Health, safety and environment (HSE) aspects specific to transport
- Materials of Construction

International Standard:

ISO 27913 Carbon dioxide capture, transportation and geologic storage – Pipeline transportations systems



Enhanced Oil Recovery Institute

## WG3: Storage

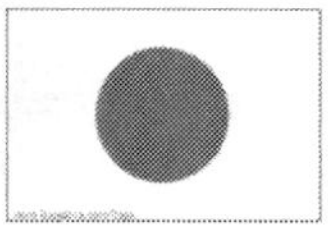

Geological storage of carbon dioxide
Canada (Onshore) & Japan (Offshore):

- Z-741-12 as seed document
- Site selection
- Site characterization
- Risk assessment & management
- Well construction
- Closure
- Post-closure

Z741-12
Geological storage of carbon dioxide
CSA Group
ANSI

Draft International Standard:
ISO DIS 27914 Geological Storage



## WG4: Quantification & Verification

Quantification & Verification Methodology; Led by China, with support from France:

- Project boundary & leakage
- $CO_2$ quantification
- Monitoring and reporting
- Third party verification
- Life Cycle Analysis

Draft Technical Report:
ISO DTR 27915 Quantification & Verification
Drafting a Standard:
ISO Ballot 27920 Quantification & Verification

# WG5: Crosscutting Issues

Crosscutting Issues; Led by France, with support from China:

- Terminology
- Definitions
- System Integration
- Public Participation & Engagement
- Mixing of gas streams from different sources

Drafting a Standard:
ISO Ballot 27920 $CO_2$ Stream Composition
Draft International Standard:
ISO DIS 27917 Vocabulary

Enhanced Oil Recovery Institute

# WG5: Crosscutting Issues

Example of harmonizing cross-cutting terms among WGs: $CO_2$ stream:

*"a stream consisting overwhelmingly of carbon dioxide"*

**WG5**: a stream consisting overwhelmingly of carbon dioxide

**WG2:** stream consisting overwhelmingly of carbon dioxide ***with a limited fraction of other chemical substances***

**WG3:** a stream of carbon dioxide ***that has been captured from an emission source (e.g., a fossil fuel power plant) and meets applicable regulatory requirements for $CO_2$ storage***

***Note: It may include any incidental associated substances derived from the source materials or the capture process, added as a result of commingling for transportation, added to the stream to enable or improve the injection process and/or trace substances added to assist in $CO_2$ migration detection.***

Enhanced Oil Recovery Institute

## WG6: $CO_2$-EOR

Carbon Dioxide Storage using EOR; led by USA, with support from Norway:

- Operating oil field environments
- Known reservoir & pore space management
- Manage known lateral stratigraphic traps in the target formation
- Coordination with WGs1-5

Drafting a Standard (2nd):
ISO CD 27916 Carbon dioxide storage using Enhanced Oil Recovery (EOR)
- *Lack of international experience in EOR*
- *No closure*

19 Enhanced Oil Recovery Institute

## ISO/TC 265 – Set of Standards & Technical Reports

Used individually or as an integrated full Life Cycle set:

- Capture
- Transport
- Storage (onshore & off)
- Quantification & Verification
- Crosscutting Issues
- $CO_2$-EOR

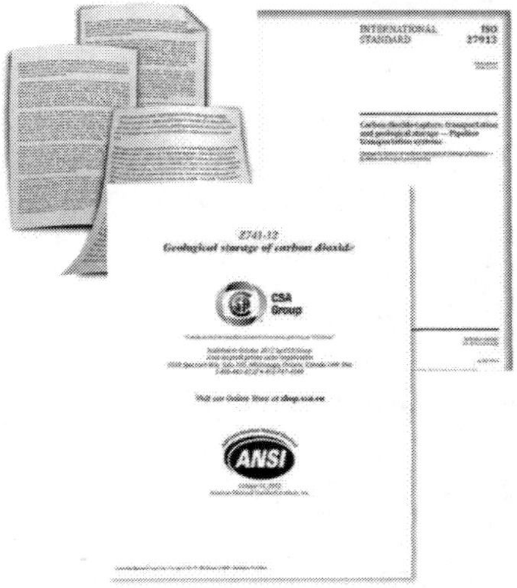


20 Enhanced Oil Recovery Institute

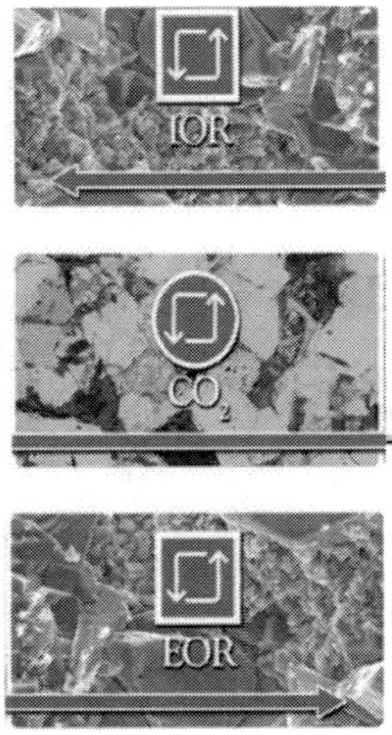


## 2: Wyoming & $CO_2$ Demand

## What is Enhanced or Improved Oil Recovery?

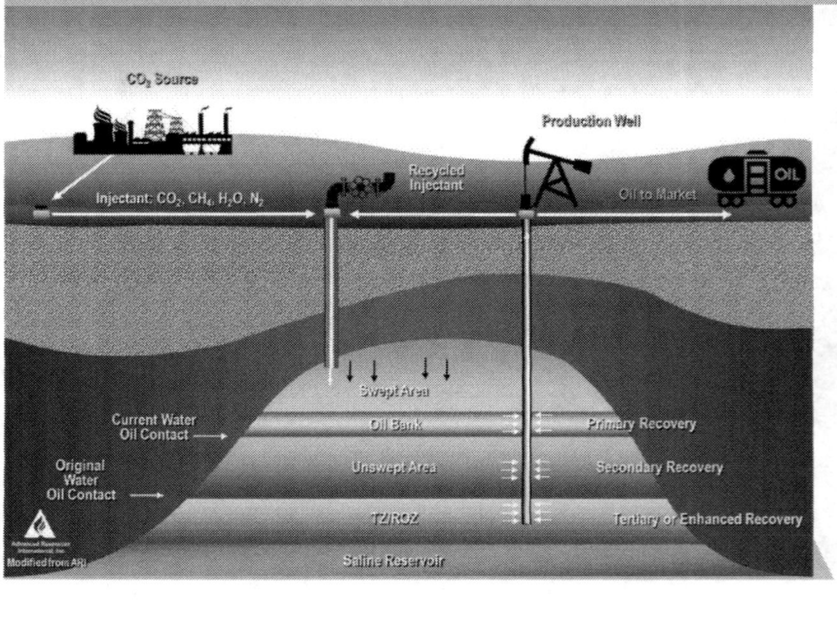

Enhanced Oil Recovery Institute

# $CO_2$ & WY – Potential

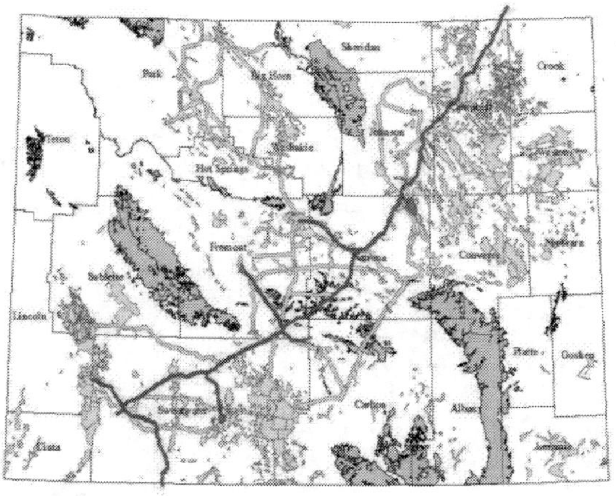


Based on $85/bbl. oil price, Onshore, L-48 per ARI, 2013

**WY Potential $CO_2$-EOR**

- Potential 1.8 Billion bbls
- Technically recoverable 1.1b bbls
- Economically recoverable 644MM bbls

**Wyoming $CO_2$ Injection (1991 – 2016)**

| Year | Project | Incremental bo | Cumulative CO2 (mcf) | Cumulative CO2 (tons) |
|---|---|---|---|---|
| 1986 | Wertz | 24,150,949 | 203,167,188 | 11,620,842 |
| 1989 | Lost Soldier | 45,834,473 | 675,416,216 | 38,632,741 |
| 2008 | Beaver Creek | 8,825,908 | 168,613,153 | 9,644,406 |
| 2013 | Big Sand Draw | 916,792 | 39,838,773 | 2,278,715 |
| 2012 | Grieve | -2,059 | 36,258,949 | 2,073,955 |
| 2003 | Patrick Draw | 17,079,074 | 334,713,367 | 19,145,076 |
| 2003 | Salt Creek | 25,762,143 | 2,838,584,423 | 162,362,548 |
| | | **122,567,280** | | **245,758,283** |
| | | | | *2.1 tons of CO2 per bo* |

½ = 3x

Enhanced Oil Recovery Institute

# $CO_2$ & WY – Demand

Bighorn Basin – Potential of 1 Billion barrels in $CO_2$ EOR[1,3]

- **Potential $CO_2$ demand ~3 to 4.5 Tcf**
  - Phosphoria
  - Tensleep
  - Madison

Powder River Basin – Potential of 885 Million barrels in $CO_2$ EOR[2,3]

- **Potential $CO_2$ demand ~4.7 to 7.1 Tcf**
  - Frontier
  - Muddy
  - Minnelusa

Green River Basin

- **Potential $CO_2$ demand ~0.8 to 1.3 Tcf**
  - Almond

Sand Wash Basin

- **Potential $CO_2$ demand ~0.08 to 0.1 Tcf**
  - Tensleep
  - Madison

Wind River Basin

- **Potential $CO_2$ demand ~1.1 to 1.7 Tcf**
  - Tensleep
  - Madison

**Big Horn & PRB**

- Potential $CO_2$ demand ~ 7.7 to 11.6 Tcf

**Realization of this production requires:**

- ***$CO_2$ availability***
- $CO_2$ Pipeline & Infrastructure
- Operator progression from Secondary Recovery to $CO_2$-EOR

24 1: Yin et al, 2014, 2: Branting & Whitman, 1992, 3: Wo, Whitman, & Steidtman, 2009

Enhanced Oil Recovery Institute

# $CO_2$-EOR + CCUS = SYNERGY: Examples of Success

**Sask's Powers Boundary Dam**

- First of it's kind – first fully operational CCUS at commercial scale
- Retrofit of Production unit 3 at the Boundary Dam coal-fired power station near Estevan
- Provides 1 Mtpa $CO_2$ supply to the Weyburn–Midale CO2-EOR projects
- $CO_2$ capture on line in 2014

Source: http://saskpowerccs.com

**Petra Nova Carbon Capture**

- 90% capture from a 240 MW flue gas slipstream of the 610 MW (net) pulverized coal-fired generating unit
- 50/50 joint venture between NRG Energy and JX Nippon Oil & Gas Exploration
- Operational in 2016 generating 1.4 Mtpa of $CO_2$ to EOR market

Source: /http://www.nrg.com/generation/projects/petra-nova/

25

Enhanced Oil Recovery Institute

# WYOMING's NextGen CCUS Facility

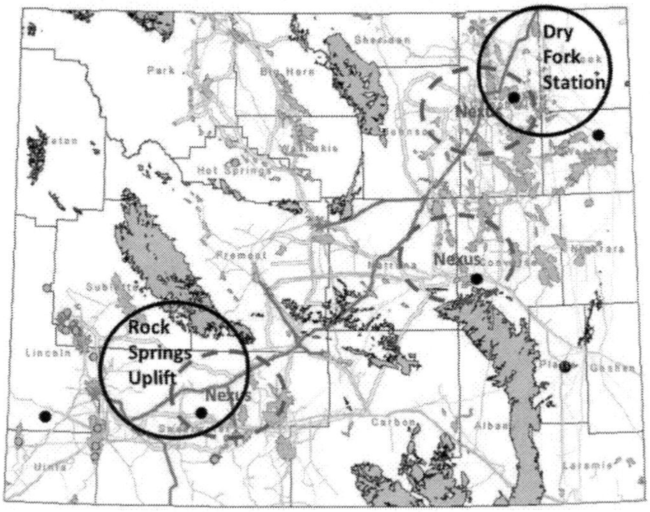


Perfect Synergy:

Overlay/Heat Map:

- CFPP
- $CO_2$ – EOR Infrastructure & Projects
- UW CarbonSAFE Awards

26

Enhanced Oil Recovery Institute

# Western Governors Association $CO_2$-EOR Initiative

**FIGURE ES-2: $CO_2$-EOR State Deployment Work Group – Participating States**

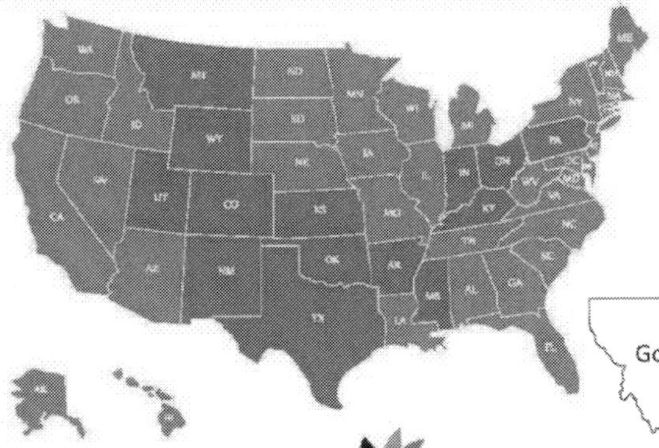

■ Participating States
■ Non-participating States

NATIONAL COAL COUNCIL

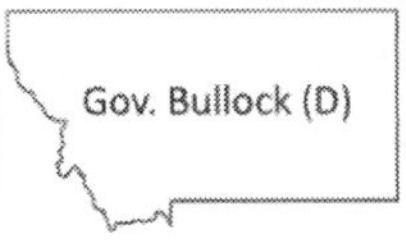

27

Enhanced Oil Recovery Institute

# Wyoming's CCUS Team

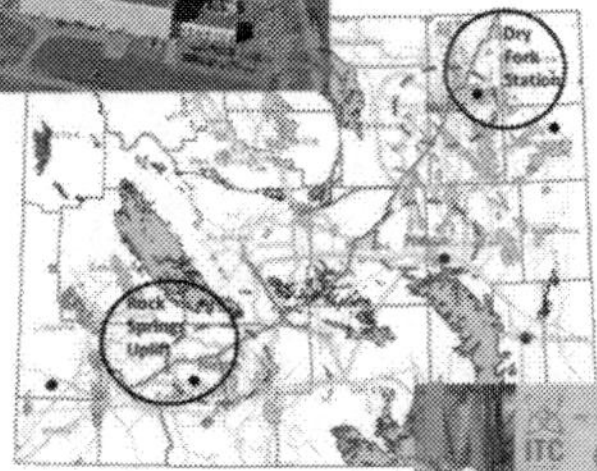

University of Wyoming

School of Energy Resources, University of Wyoming

Carbon Management Institute, University of Wyoming

Enhanced Oil Recovery Institute, University of Wyoming

PAW

ITC Wyoming Integrated Test Center

ENDOW Wyoming

Wyoming Infrastructure Authority

WMA Wyoming Mining Association

Wyoming Business Council

Western Research Institute

28

Enhanced Oil Recovery Institute

## 3: Non-OECD Economies

## Coal/Energy & Poverty: Energy is Good

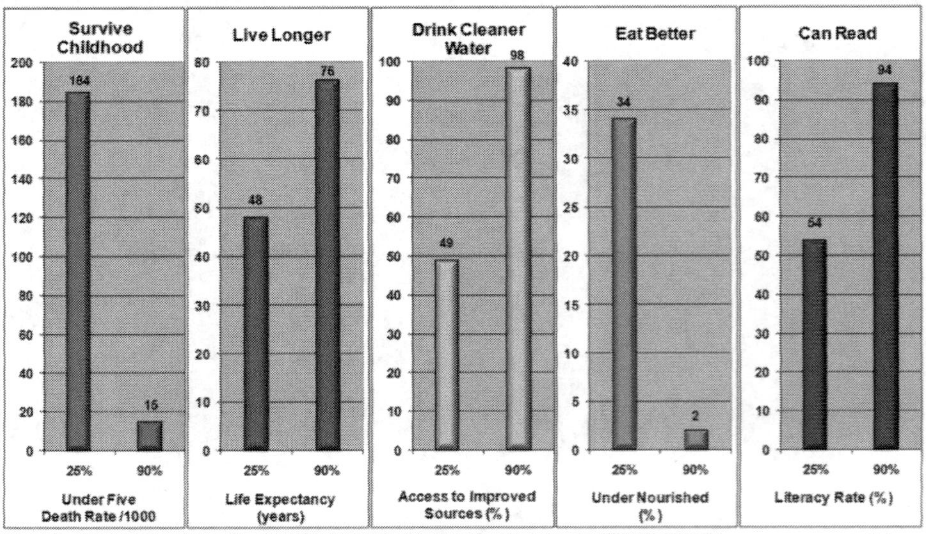


Source: Dr. Frank Clemente, Penn State University

Enhanced Oil Recovery Institute

# Non-OECD Energy Demand – Pick your trend line…

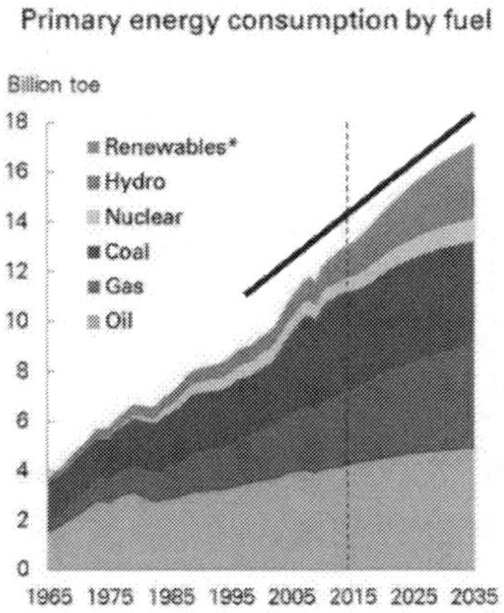


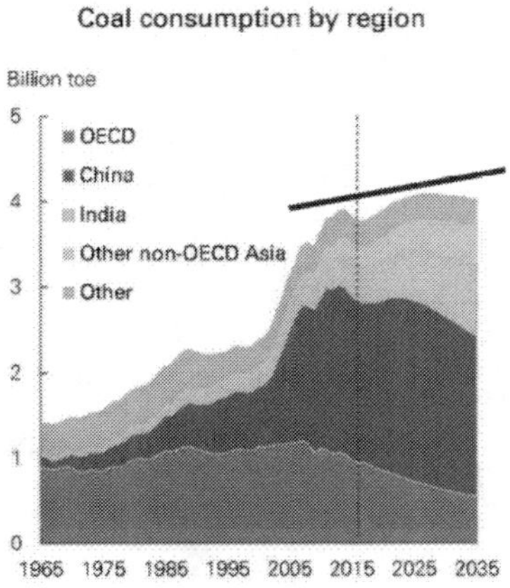


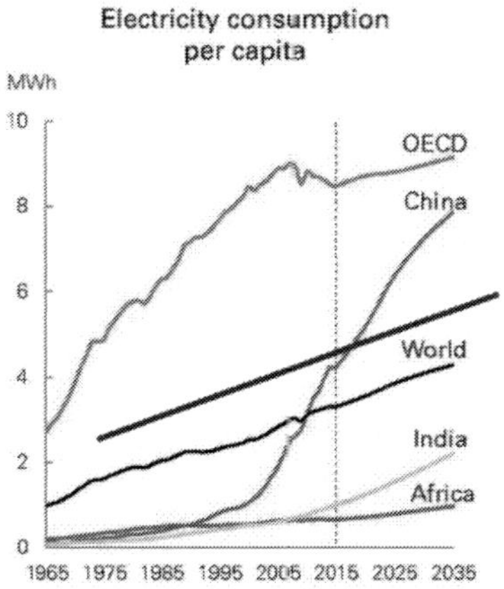


11 Source: BP Global Energy Outlook, 2017

Enhanced Oil Recovery Institute

# Why CCUS & Why ISO?

- How does this help advance CCUS?
- David Greeson relative to a 0.5% efficiency increase in capture process, *"who cares?"*
- Chuck McConnell described the *"let's get serious"* perspective derived from many who asked *"so what?"*
- CODE FOR: WHY DO CCUS & WHY ISO?

12

Enhanced Oil Recovery Institute

# Real-world applications – It Does Matter

SaskPower

**Mike Monea, President & CEO**
*International CCS Knowledge Center*

"Standards, smart local and ***global standards, are essential*** to the timely advancement of the technologies and equipment that will be necessary to make safe reliable power with the capture of emissions from hydrocarbon fueled power plants."

RICE

**Chuck McConnell, Executive Director**
*Rice University E & E Initiative*

"Get technology in ***places in the world*** where it can improve their lives and increase their energy security."

Enhanced Oil Recovery Institute

# Regulatory Confusion = Regulations

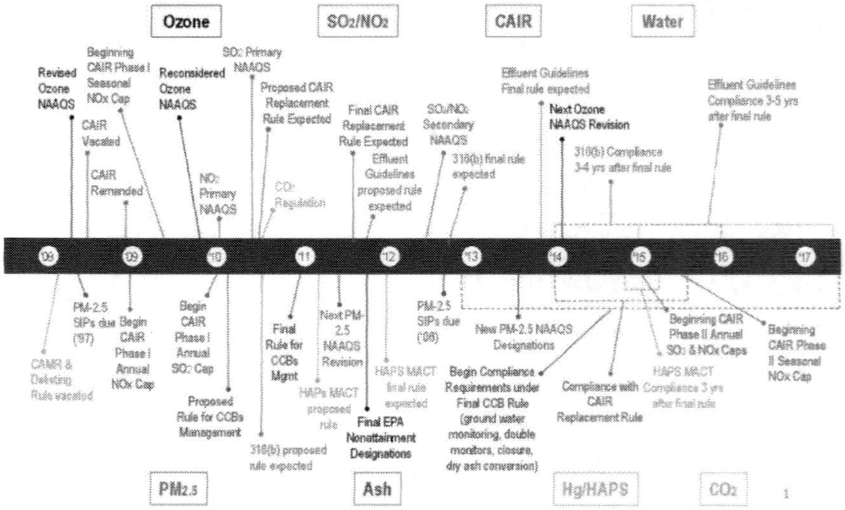


<u>$CO_2$ Issues:</u>

- GHG Subpart RR & UU
- UIC Class II
- UIC Class VI
- EPA MRV Plan
- 45Q

Source: Edison Electric Institute and Dick Winschel, CONSOL Energy

Enhanced Oil Recovery Institute

## Lack of...

OECD
BETTER POLICIES FOR BETTER LIVES

APEC
Asia-Pacific Economic Cooperation

United Nations
Framework Convention on Climate Change

...Regulatory Framework:

- Malaysia
- Argentina
- Iran
- Brazil
- Egypt

...Industry Experience:

- Saudi Arabia
- Mexico

Enhanced Oil Recovery Institute

## It matters...US ISO Technical Advisory Group Members

SOUTHERN COMPANY

nrg

Dominion

Denbury

OGCI
OIL AND GAS CLIMATE INITIATIVE

U.S. DEPARTMENT OF ENERGY

Advanced Resources International, Inc.

USGS
science for a changing world

Battelle
*The Business of Innovation*

Enhanced Oil Recovery Institute

## CCUS, Wyoming, and Non-OECD Economies

37

Enhanced Oil Recovery Institute

## Thank You – Question & Comments

Dr. Steven Carpenter, Director

**Wyoming Enhanced Oil Recovery Institute**

steven.carpenter@uwyo.edu

O: 307-315-6442

C: 513-460-0360

www.uwyo.edu/eori

Enhanced Oil Recovery Institute
UNIVERSITY OF WYOMING

nrg

## Carbon Capture at Petra Nova

July 2017

nrg | JX

PETRA NOVA Carbon Capture

# Petra Nova – so far so good

DELIVERED

On spec
On time
On budget

### Achieved COD on Dec. 29, 2016

$CO_2$

Over 700,000 tons of $CO_2$ captured as of July 17th

Oil

Producing more than 4,000 barrels of oil per day

PETRA NOVA Carbon Capture

# Petra Nova Project Overview

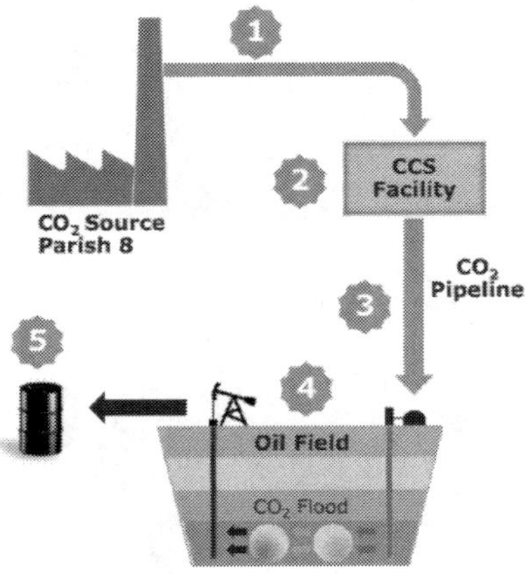


## Five Projects in One

1. **Diverting the flue gas** from an existing facility (Parish Unit 8)
2. **Processing flue gas** in a carbon capture system to strip out the $CO_2$
3. **Transport $CO_2$** to a nearby oil field.
4. **$CO_2$-EOR** operation to produce otherwise unrecoverable oil
5. **Transport and sell oil** – marketing, selling, and transporting the oil

Oil revenues pay for the entire project

No impact on power plant or its costs

PETRA NOVA Carbon Capture

# Carbon Capture System Site Layout

Petra Nova Carbon Capture Site

3

nrg | JX
PETRA NOVA Carbon Capture

# Oilfield Facilities Recapture and Inject $CO_2$

## West Ranch Field Central Facilities

- **200 new wells to be drilled (over 100 now complete)**
- **2 central processing facilities to separate oil-$CO_2$-water**
- **All produced $CO_2$ and water is reinjected into the formation**
- **University of Texas managing monitoring**

4

nrg | JX
PETRA NOVA Carbon Capture

# Commercial Structure

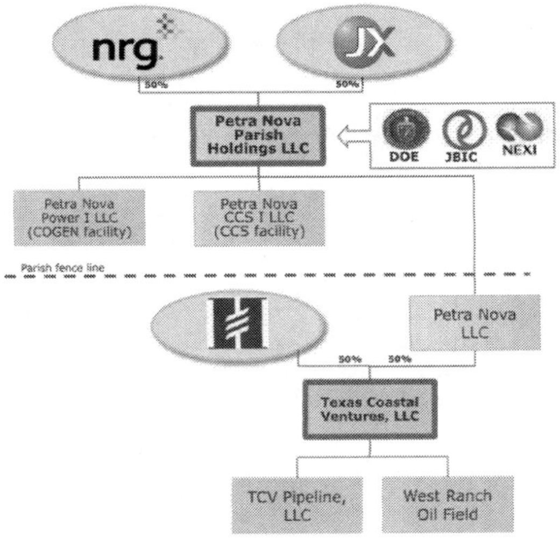


## Our Partners

JX Holdings is a leading integrated energy, resources, and materials company

NRG Energy, Inc. is the largest independent power company in the US

Hilcorp Energy is one of the largest privately-held oil and natural gas E&P companies in the US

JBIC and NEXI combined on a $250 million project loan.

US DOE awarded $190 MM grant funded through Clean Coal Power Initiative

5

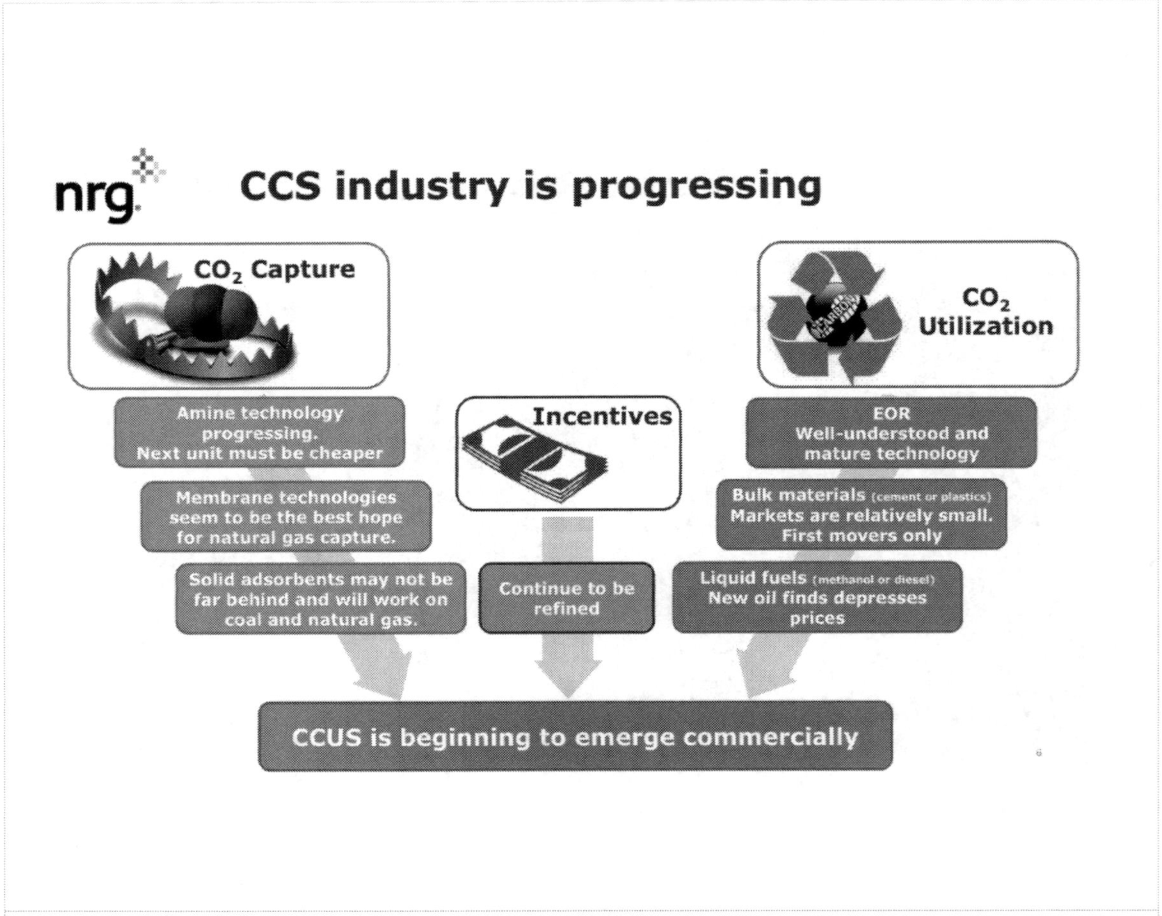
nrg
CCS industry is progressing
CO2 Capture
CO2 Utilization
Amine technology progressing. Next unit must be cheaper
Membrane technologies seem to be the best hope for natural gas capture.
Solid adsorbents may not be far behind and will work on coal and natural gas.
Incentives
Continue to be refined
EOR Well-understood and mature technology
Bulk materials (cement or plastics) Markets are relatively small. First movers only
Liquid fuels (methanol or diesel) New oil finds depresses prices
CCUS is beginning to emerge commercially
6

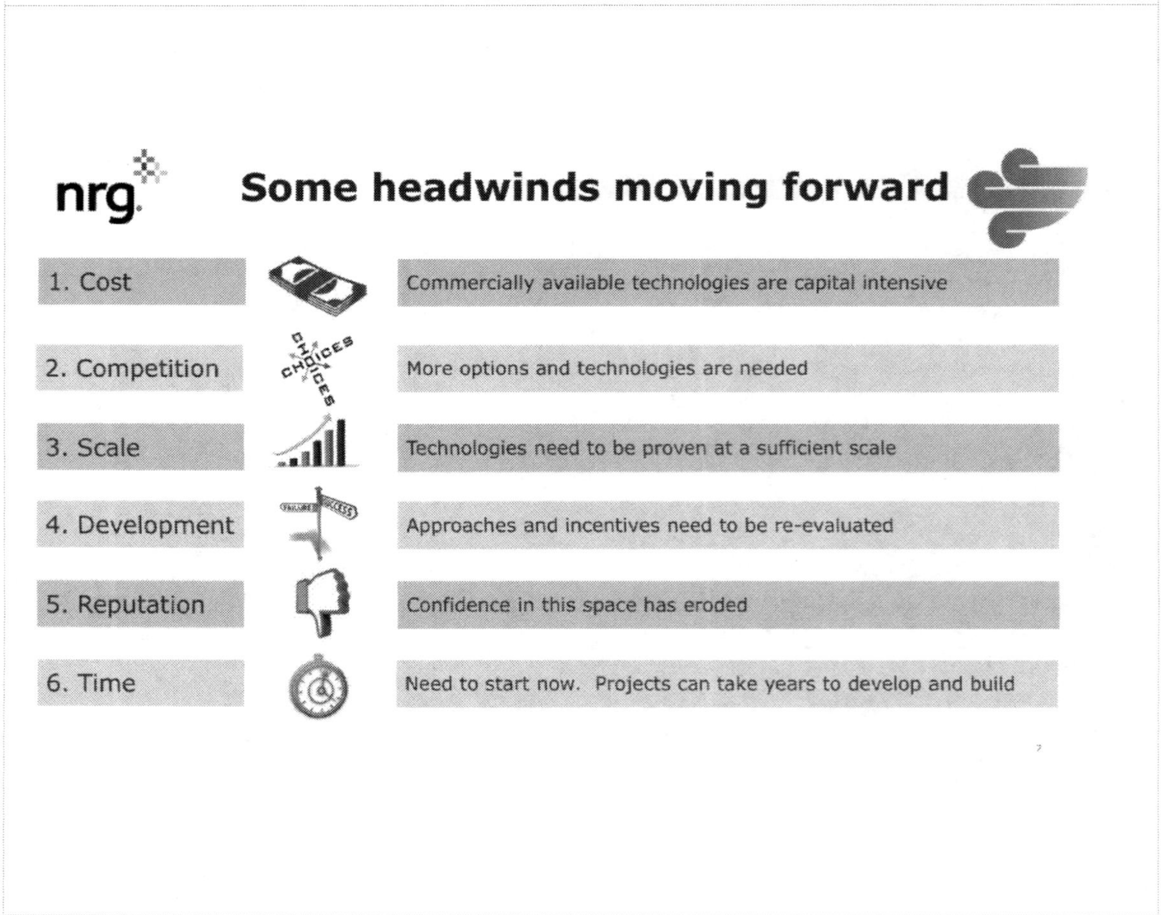
nrg
Some headwinds moving forward
1. Cost
Commercially available technologies are capital intensive
2. Competition
CHOICES
More options and technologies are needed
3. Scale
Technologies need to be proven at a sufficient scale
4. Development
FAILURE
SUCCESS
Approaches and incentives need to be re-evaluated
5. Reputation
Confidence in this space has eroded
6. Time
Need to start now. Projects can take years to develop and build
7

nrg

# Path to success – improving economics

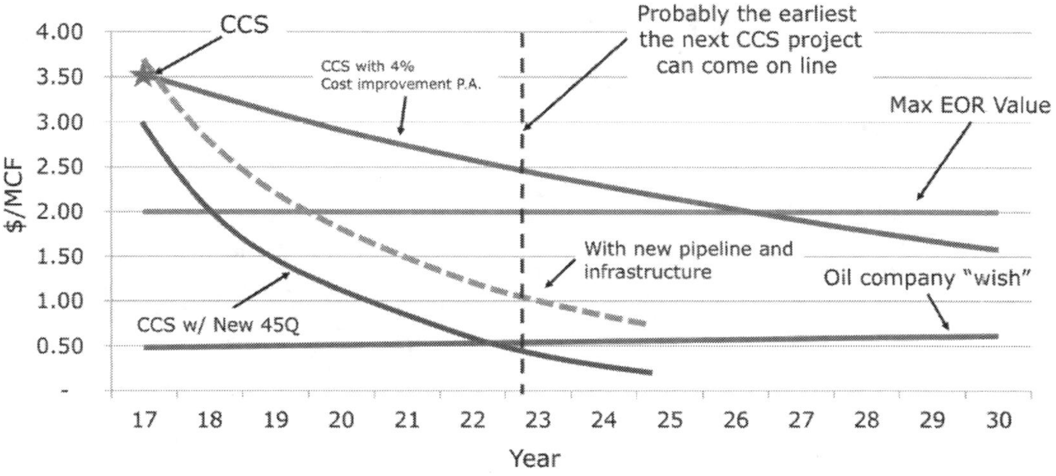


8

nrg | JX

PETRA NOVA Carbon Capture

# Interest is high right now

2017

30+ tours

8 speaking engagements

30+ articles written

When Petra Nova is operating, the coal-fired Parish Unit 8 has the same carbon intensity as a gas-fired power plant.

**International Interest:**

- Australia*
- Canada*
- Chile
- China
- Colombia
- India*
- Korea*
- Kuwait
- Mexico
- Norway*
- Pakistan*
- Singapore*
- Thailand*
- U.K.*

* Have sent delegations to visit the project.

9

## nrg Near Term – How to repeat?

**Requirements for successful CCS projects:**

- Technology evaluation
- Engineering and design management
- Pipeline development
- Commercial structuring and $CO_2$ sales
- Acquisition and management of oil field ownership and interface
- Project finance
- Government grant application and administration
- Environmental study management
- Permitting and licensing
- Construction management
- Communications and messaging
- Operational experience

With experience of bringing Petra Nova online on-budget and on-schedule, NRG can help!

10

## nrg JX Celebrate Success!

PETRA NOVA Carbon Capture

11

# Establishing $CO_2$ Utilization, Storage and Pipeline Systems for Oil Fields in Shallow and Deep Waters of the Gulf of Mexico

Prepared for:
**2017 Carbon Management Technology Conference**
**CMTC 2017**

**Session 11 – Offshore Storage**

Prepared By:
**Vello Kuuskraa**, President
**Advanced Resources International, Inc.**

July 19, 2017
Houston, Texas



Establishing $CO_2$ Utilization, Storage and Pipeline Systems for Oil Fields in Shallow and Deep Waters of the Gulf of Mexico

## Presentation Outline

| | |
|---|---|
| | **Executive Summary** |
| 1 | Eastern and East Central GOM Shallow Water $CO_2$ Pipeline System |
| 2 | Eastern and East Central GOM Deepwater $CO_2$ Pipeline System |
| 3 | Next Steps |

# Executive Summary

**The large oil fields in the Eastern and East Central GOM offshore offer significant opportunities for productively utilizing and storing $CO_2$ while helping increase domestic oil production and Federal revenues.**

- The 52 moderate to large oil fields in the shallow Federal waters of the GOM will need a three $CO_2$ pipeline system to deliver 40 million metric tons per year (2.1 Bcfd) for use and storage by $CO_2$ enhanced oil recovery,
- The 63 large oil fields in the deep Federal waters of the GOM will need a three $CO_2$ pipeline system to deliver 57 million metric tons per year (2.9 Bcfd) for use and storage by $CO_2$ enhanced oil recovery,
- The capital costs for these six $CO_2$ pipelines is estimated at nearly \$6 billion,
- The Federal royalties provided from the recovery of 7.3 billion barrels by $CO_2$ enhanced oil recovery (assuming all 7.3 B bbls would be economically viable to develop) would equal \$104 billion dollars, assuming an \$80 per barrel (WTI) oil price, as estimated by EIA for Year 2025.

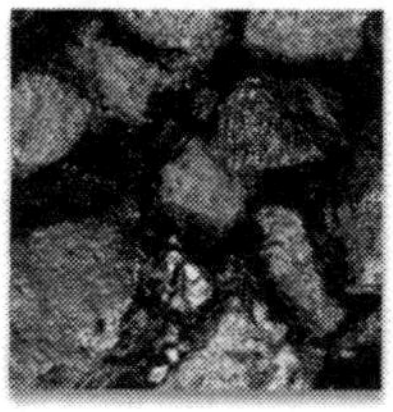

# 1. Eastern and East Central GOM Shallow Water $CO_2$ Pipeline System

# Pipeline Systems for Delivering $CO_2$ to Shallow Water Eastern and East Central GOM Oil Fields

**The Eastern and East Central portions of the Gulf of Mexico (GOM) shallow water offshore hold 52 moderate- to large-size oil fields. These oil fields contain 8.2 billion barrels of original oil reserves, with 97% of these original reserves already produced.**

**We started our study by targeting the largest 29 of these oil fields, each with original oil reserves of over 100 million barrels.**

We plotted the location of each of these 29 oil fields, estimated their technically viable oil recovery potential and $CO_2$ injection requirements, and then designed three $CO_2$ pipelines that would connect these 29 oil fields with $CO_2$ supply from onshore Louisiana and Mississippi.

- Eastern GOM $CO_2$ Pipeline System
- East Central GOM $CO_2$ Pipeline System #1
- East Central GOM $CO_2$ Pipeline System #2

Finally, we prepared an initial estimate of the capital costs of installing these three $CO_2$ pipeline systems in the shallow waters of Eastern and East Central GOM.



Establishing $CO_2$ Utilization, Storage and Pipeline Systems for Oil Fields in Shallow and Deep Waters of the Gulf of Mexico

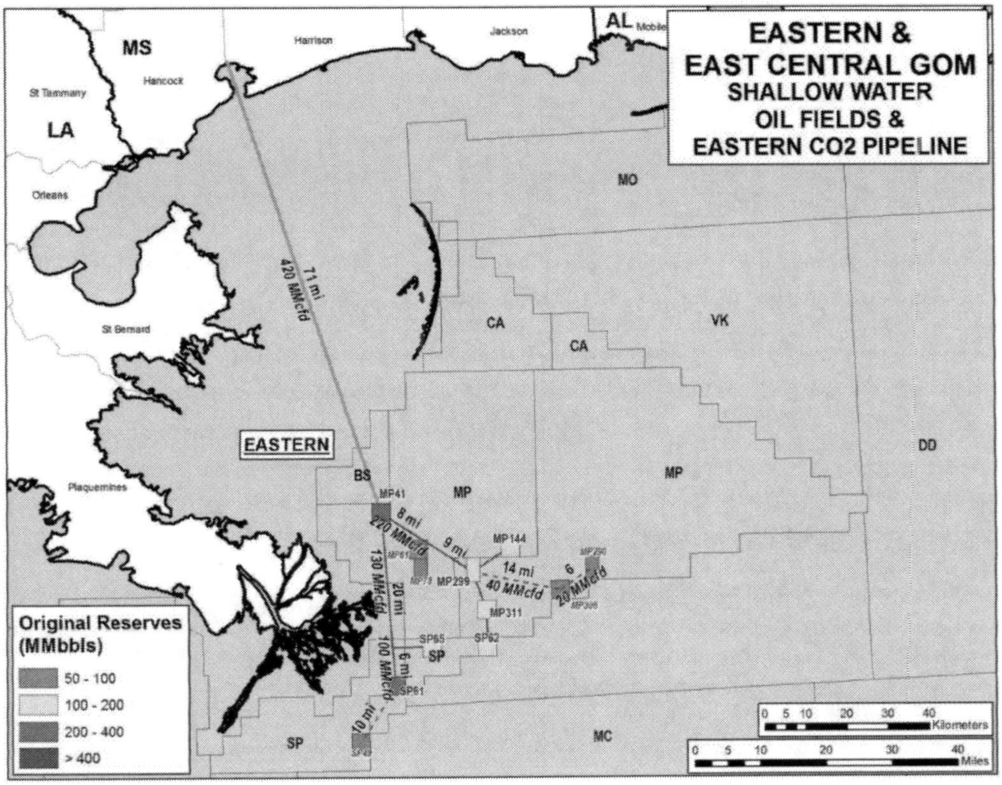

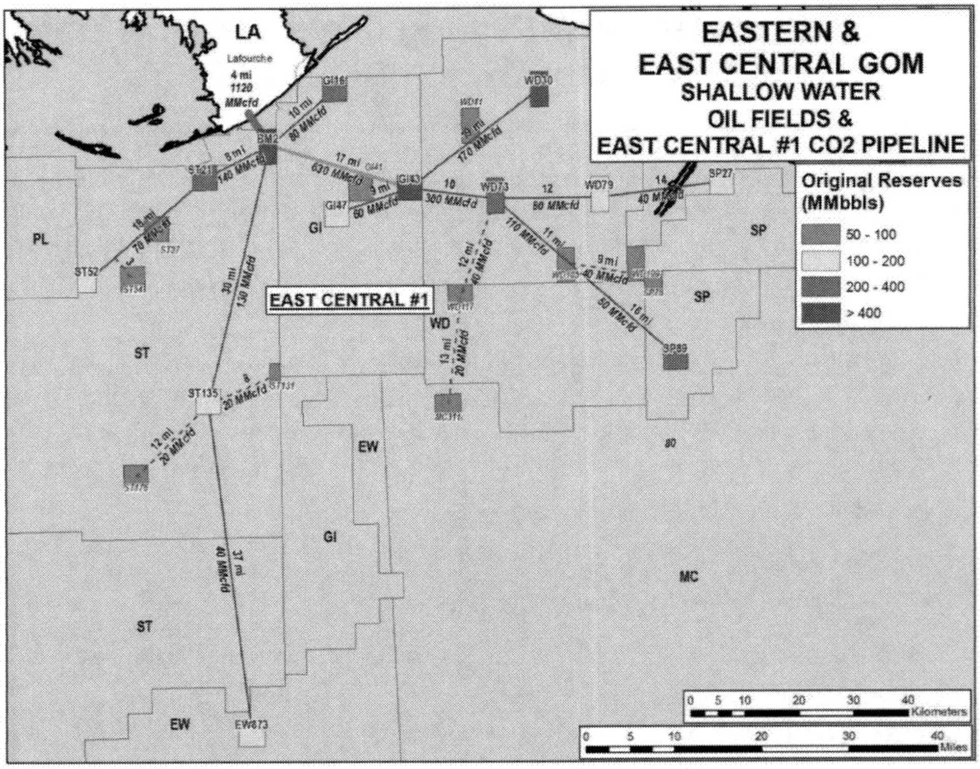
EASTERN & EAST CENTRAL GOM
SHALLOW WATER
OIL FIELDS &
EAST CENTRAL #1 CO2 PIPELINE
Original Reserves (MMbbls)
50 - 100
100 - 200
200 - 400
> 400
EAST CENTRAL #1
LA
Lafourche
4 mi
1120 MMcfd
PL
ST
GI
WD
SP
EW
MC
Kilometers
Miles





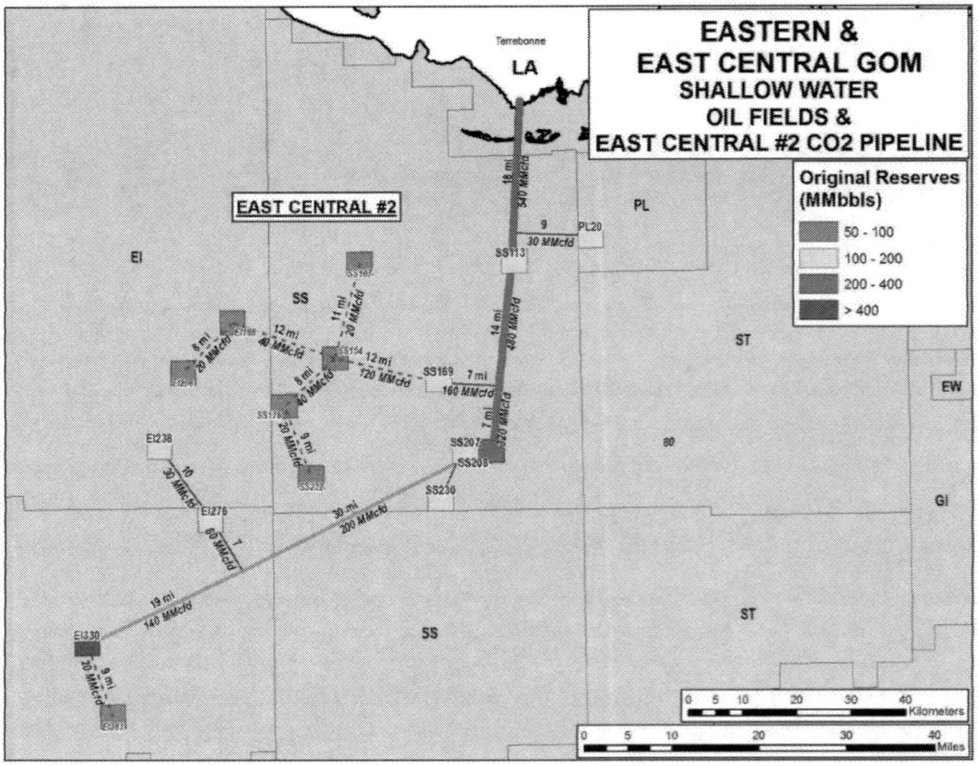
EASTERN & EAST CENTRAL GOM
SHALLOW WATER
OIL FIELDS &
EAST CENTRAL #2 CO2 PIPELINE
Original Reserves (MMbbls)
50 - 100
100 - 200
200 - 400
> 400
EAST CENTRAL #2
Terrebonne
LA
PL
EI
SS
ST
EW
GI
Kilometers
Miles

# Final Eastern and East Central GOM Shallow Water $CO_2$ Pipeline System

**Including both the 29 "targeted" oil fields and the 23 "opportunity" oil fields, the three pipelines would deliver about 2.1 Bcfd of $CO_2$ (40 million tons per year) to 52 shallow water GOM oil fields.**

| Pipeline System | No. of Fields | $CO_2$-EOR Oil Recovery | Total $CO_2$ Requirements | | $CO_2$ Requirements | |
|---|---|---|---|---|---|---|
| | (#) | (MMB) | (Bcf) | (MMmt) | (MMcfd) | (MMmt/yr) |
| Eastern | 12 | 610 | 6,110 | 323 | 420 | 8.1 |
| East Central #1 | | | | | | |
| • Part 1 | 6 | 520 | 5,220 | 276 | 360 | 6.9 |
| • Part 2 | 7 | 650 | 6,530 | 346 | 450 | 8.7 |
| • Part 3 | 6 | 240 | 2,380 | 126 | 160 | 3.2 |
| • Part 4 | 5 | 220 | 2,150 | 114 | 150 | 2.8 |
| Total | 24 | 1,630 | 16,280 | 862 | 1,120 | 21.6 |
| East Central #2 | | | | | | |
| • Part 1 | 12 | 500 | 5,040 | 267 | 350 | 6.8 |
| • Part 2 | 4 | 290 | 2,840 | 150 | 190 | 3.7 |
| Total | 16 | 790 | 7,880 | 417 | 540 | 10.5 |
| Total | 52 | 3,030 | 30,270 | 1,602 | 2,080 | 40.2 |

JAF2017_028.XLS

- The Eastern $CO_2$ pipeline system remains modest in size, transporting 420 MMcfd (8.1 MMmt/yr) of $CO_2$.
- The East Central #1 $CO_2$ pipeline system remains the dominant $CO_2$ system, with 1,120 MMcfd (21.6 MMmt/yr) of capacity.
- The East Central #2 $CO_2$ pipeline system would have 540 MMcfd (10.5 MMmt/yr) of transportation capacity.





# Eastern and East Central GOM Shallow Water $CO_2$ Pipeline Investment Costs

**Our estimate of installing the three $CO_2$ pipelines in the shallow waters of the Eastern and East Central Gulf of Mexico is about $1.7 billion.**

| Pipeline System | $CO_2$ Requirements | | Pipeline Requirements | Capital Costs* |
|---|---|---|---|---|
| | (MMcfd) | (MMmt/yr) | (in-mi) | ($MM) |
| Eastern | 420 | 8.1 | 2,478 | $600 |
| East Central #1 | 1,120 | 21.6 | 2,512 | $600 |
| East Central #2 | 540 | 10.5 | 2,074 | $500 |
| Total | 2,080 | 40.2 | 7,064 | $1,700 |

*We assume $240,000 per inch-mile for shallow water.

JAF2017_028.XLS

Opportunities for lowering these costs would involve further optimizing the $CO_2$ pipeline systems and selectively using existing, empty offshore natural gas pipelines, where possible.

# 2. Eastern and East Central GOM Deepwater $CO_2$ Pipeline System



Establishing $CO_2$ Utilization, Storage and Pipeline Systems for Oil Fields in Shallow and Deep Waters of the Gulf of Mexico

# Pipeline Systems for Delivering $CO_2$ to Deepwater Eastern and East Central GOM Oil Fields

**The 63 large deepwater Eastern and East Central GOM oil fields, concentrated in Green Canyon and Mississippi Canyon, offer considerable potential for $CO_2$ storage and technically viable oil recovery using $CO_2$-EOR.**

- These 63 large deepwater oil fields contain 8.6 billion barrels of original oil reserves, with about half of these original reserves already produced.
- We plotted the location of these 63 large deepwater oil fields and 26 Production Complexes (central platforms). We then estimated their oil recovery potential and $CO_2$ requirements for $CO_2$-EOR.
- We established seven major $CO_2$ Hubs served by three Deepwater $CO_2$ Pipeline Systems to link these 63 large oil fields with $CO_2$ supplies delivered from onshore Louisiana, Mississippi and Alabama.

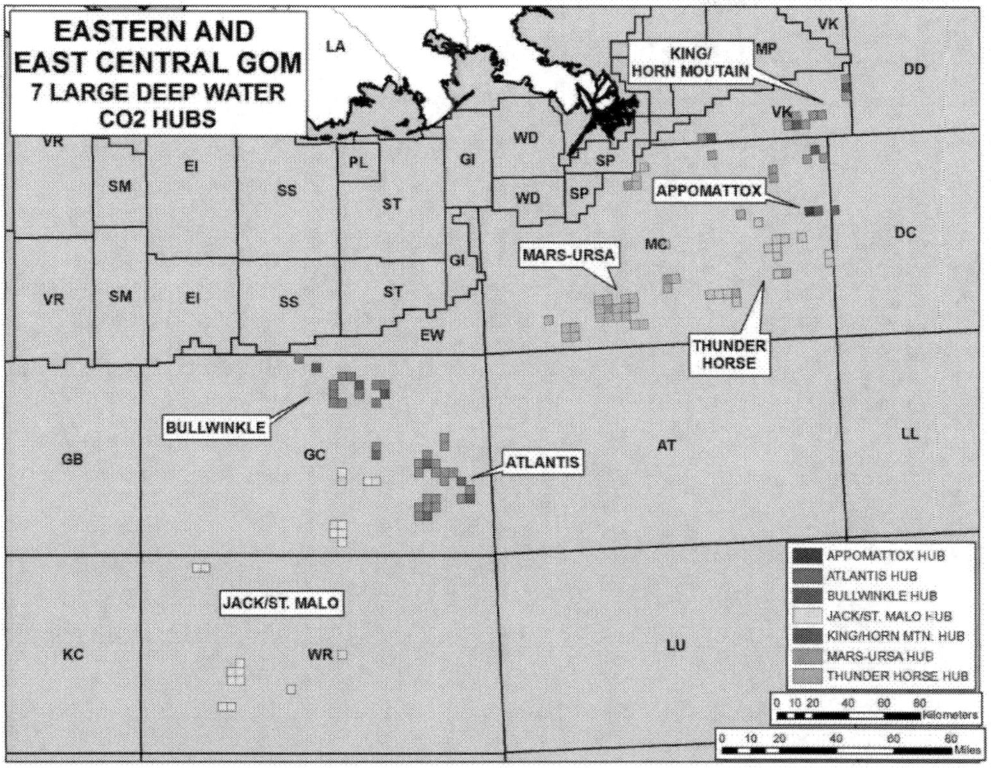






# Appomattox $CO_2$ Hub and Production Complexes

The Appomattox $CO_2$ Hub and Production Complex is linked to the Appomattox (MC391) and Vicksburg (DC353) oil fields located in about 7,300 feet of water depth. Vicksburg was discovered by Shell in 2013 with Appomattox discovered, also by Shell, in 2016. Both fields are projected to be placed on production in 2020.

Subsequently, Shell announced the Appomattox deepwater platform would also provide tieback opportunities for the nearby Gettysburg, Ryberg and other prospects.

Shell has proposed building a 24-inch pipeline, called the Mattox Pipeline, to transport crude oil from the Appomattox host platform to a shallow water offshore structure and from there to the onshore using existing infrastructure.

# Appomattox $CO_2$ Hub

Source: OGJ On-Line, 07/01/2015

The Appomattox deepwater semi-submersible production platform will be located in 7,200 feet of water, 80 miles south of Mobile, Alabama. It will contain six drill centers, 15 producing sub-sea wells and five water injection wells.

The production platform is scheduled to be placed on-stream in 2020.





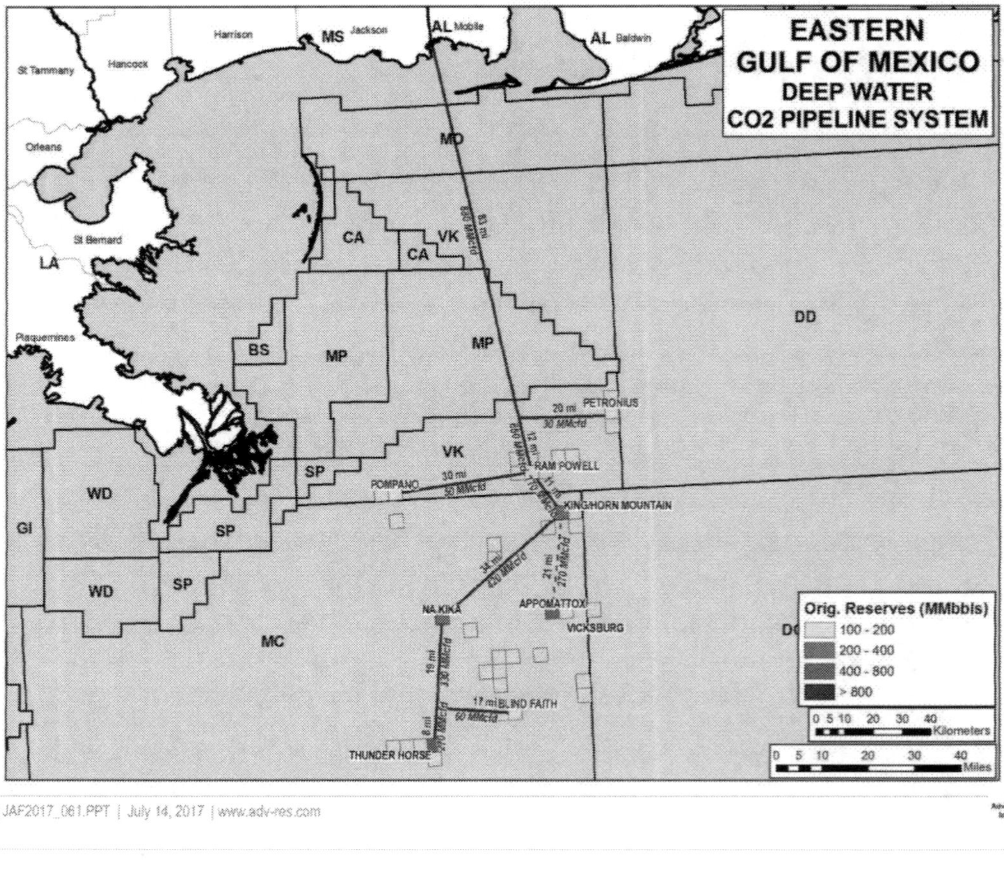

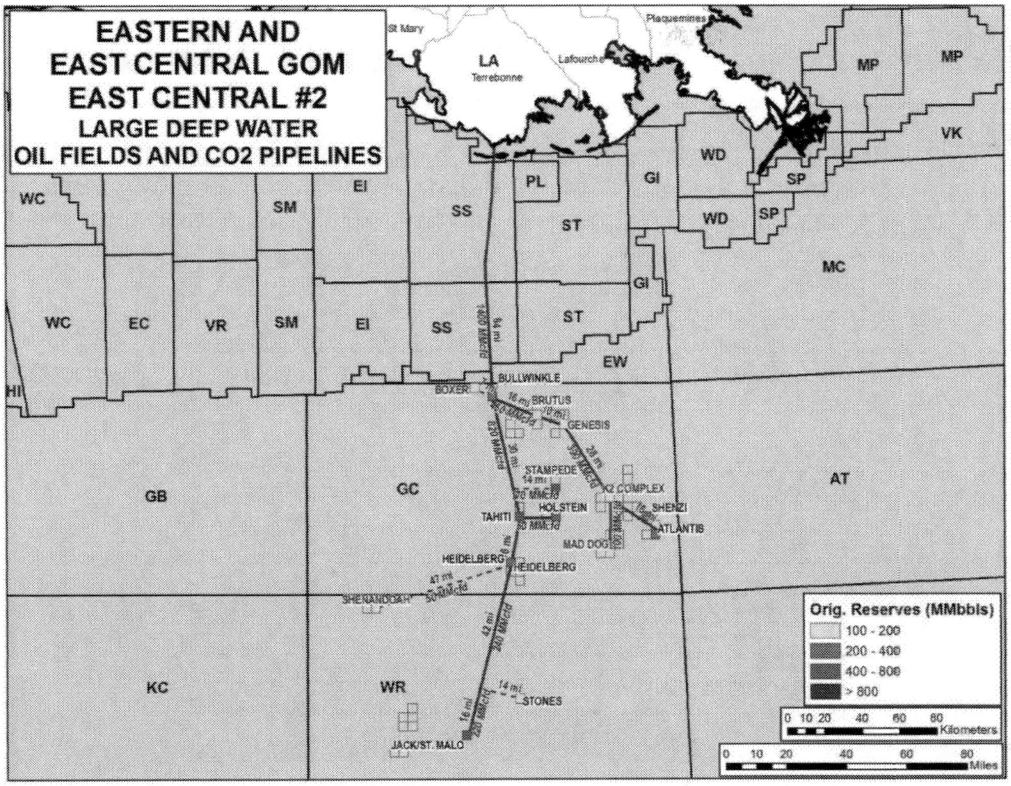
EASTERN AND
EAST CENTRAL GOM
EAST CENTRAL #2
LARGE DEEP WATER
OIL FIELDS AND CO2 PIPELINES
St Mary
LA
Terrebonne
Lafourche
Plaquemines
MP
MP
VK
WD
SP
WD
SP
GI
PL
WC
SM
EI
SS
ST
MC
GI
WC
EC
VR
SM
EI
SS
ST
EW
HI
BULLWINKLE
BOXER
BRUTUS
GENESIS
STAMPEDE
K2 COMPLEX
SHENZI
ATLANTIS
TAHITI
HOLSTEIN
MAD DOG
HEIDELBERG
HEIDELBERG
SHENANDOAH
STONES
JACK/ST. MALO
GB
GC
AT
KC
WR
Orig. Reserves (MMbbls)
100 - 200
200 - 400
400 - 800
> 800
Kilometers
Miles





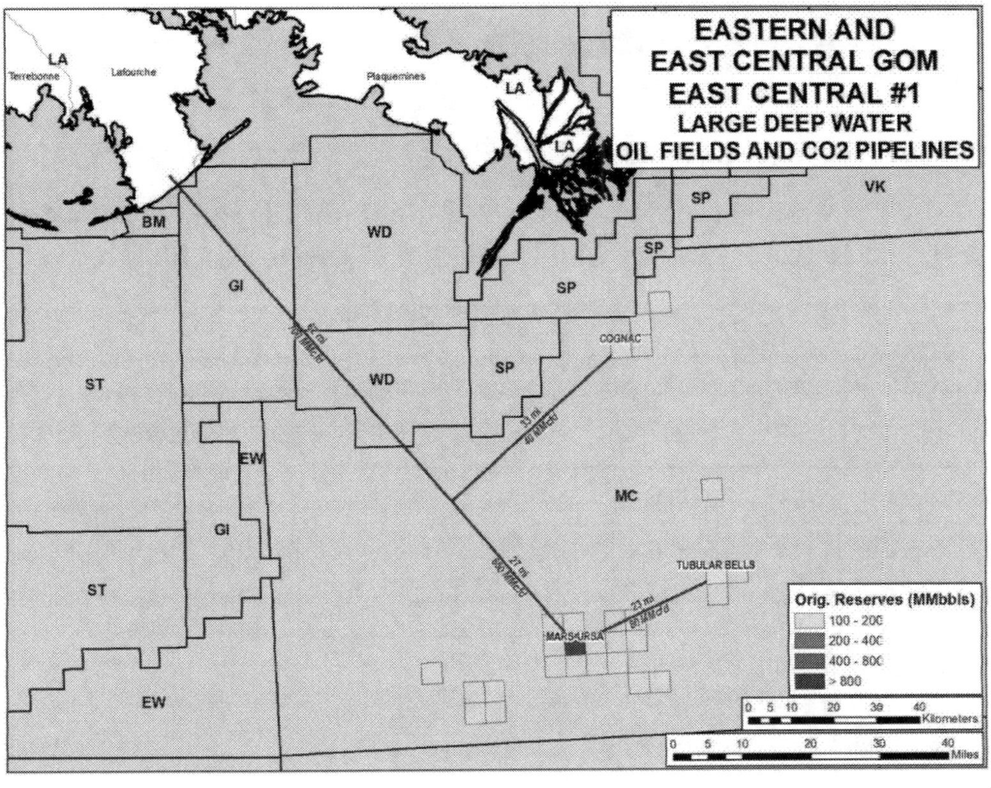
EASTERN AND
EAST CENTRAL GOM
EAST CENTRAL #1
LARGE DEEP WATER
OIL FIELDS AND CO2 PIPELINES
LA
Terrebonne
Lafourche
Plaquemines
LA
LA
BM
WD
GI
SP
VK
SP
SP
ST
WD
SP
COGNAC
EW
MC
GI
ST
TUBULAR BELLS
MARS-URSA
EW
Orig. Reserves (MMbbls)
100 - 200
200 - 400
400 - 800
> 800
Kilometers
Miles

# Eastern and East Central GOM Deepwater $CO_2$ Pipeline Systems

**We estimate technically feasible oil recovery of 4.3 billion barrels, an ultimate $CO_2$ demand of 43 Tcf (2,300 million metric tons), and $CO_2$ flows of 2.9 Bcfd (57 MMmt/yr) for the Deepwater GOM $CO_2$ Pipeline System.**

| Pipeline System | $CO_2$ Hub | $CO_2$ EOR* | $CO_2$ Demand** | | $CO_2$ Flow | |
|---|---|---|---|---|---|---|
| | | (MMbbls) | (Bcf) | (MMmt) | (MMcfd) | (MMmt/y) |
| Eastern | | | | | | |
| | King/Horn Mt. | 280 | 2,770 | 147 | 190 | 3.7 |
| | Thunder Horse | 600 | 6,040 | 319 | 420 | 8.1 |
| | Appomattox | 400 | 3,980 | 211 | 270 | 5.2 |
| **Eastern Sub-Total** | | **1,280** | **12,790** | **677** | **880** | **17.0** |
| **East Central #1** | | | | | | |
| | Mars-Ursa | **1,030** | **10,270** | **543** | **700** | **13.5** |
| **East Central #2** | | | | | | |
| | Bullwinkle | 270 | 2,730 | 144 | 190 | 3.7 |
| | Atlantis | 770 | 7,650 | 405 | 520 | 10.0 |
| | Jack/St. Malo | 950 | 9,520 | 504 | 650 | 12.6 |
| **East Central #2 Sub-Total** | | **1,990** | **19,900** | **1,053** | **1,360** | **26.3** |
| **Total** | | **4,300** | **42,960** | **2,273** | **2,940** | **56.8** |

5/24/2017

*Technically viable oil recovery is estimated at 15% of OOIP.
**Technically viable $CO_2$ requirements are estimated using 10 Mcf of $CO_2$ per barrel of oil recovery.





# Eastern and East Central GOM Deepwater $CO_2$ Pipeline Investment Costs

**Our overall estimate of installing the three $CO_2$ pipeline systems in the deep waters of the Gulf of Mexico is about $4.1 billion.**

| Pipeline System | $CO_2$ Requirements | | Pipeline Requirements | Capital Costs* |
|---|---|---|---|---|
| | (MMcfd) | (MMmt/yr) | (inch-miles) | ($MM) |
| **Eastern** | 880 | 17.0 | 4,862 | $1,310 |
| **East Central #1** | 700 | 13.5 | 3,048 | $820 |
| **East Central #2** | 1,360 | 26.3 | 7,166 | $1,980 |
| **Total** | **2,940** | **56.8** | **15,076** | **$4,110** |

JAF2017_029.XLS

Opportunities for lowering these costs would involve further optimizing the $CO_2$ pipeline system.

*We assume capital costs of $240,000 per inch-mile for shallow water and $300,000 per inch-mile for deepwater.

# Summary of Findings

**Our prefeasibility study of the Eastern and East Central Gulf of Mexico $CO_2$ Pipeline System provides the following findings:**

1. **Time Urgency for the $CO_2$ Pipeline System**. The majority of the large shallow water oil fields are close to abandonment as are a number (18) of the large deepwater oil fields having produced 93% to 98% of their original reserves. Once these fields are abandoned and their platforms removed, the feasibility of conducting $CO_2$-EOR and storing $CO_2$ in these oil fields become much more challenging and costly.
2. **The Eastern and East Central GOM $CO_2$ Pipeline System**. The six GOM $CO_2$ pipelines would facilitate the implementation of $CO_2$-EOR and $CO_2$ storage in 115 large GOM oil fields.
   - Technically feasible oil recovery of 7.3 billion barrels,
   - $CO_2$ demand and storage of 73 Tcf (3,870 million metric tons),
   - $CO_2$ delivery (and storage) of 5 Bcf per day, equal to nearly 100 million metric tons per year, over a 40 year time period, and
   - Capital costs of $5.8 billion dollars for an all new $CO_2$ pipeline system.



Establishing $CO_2$ Utilization, Storage and Pipeline Systems for Oil Fields in Shallow and Deep Waters of the Gulf of Mexico

# Summary of Findings (Cont'd)

3. **Pipeline Delivery Costs for $CO_2$.** Using a capital cost of $1.7 billion, a 14% capital charge, and an O&M charge of 5% of capital and with annual delivery of 40 million metric tons, the $CO_2$ transportation costs for the GOM Shallow Water $CO_2$ Pipeline System would be about $8 per metric ton.

   Using a capital cost of $4.1 billion, a 14% capital charge, and an O&M charge of 5% of capital and with annual delivery of nearly 57 million metric tons, the $CO_2$ transportation costs for the GOM Deepwater $CO_2$ Pipeline System would be about $14 per metric ton.
4. **GOM Deepwater $CO_2$ Pipeline Benefits.** With potential oil recovery of 7.3 billion barrels, an oil price of $80 per barrel (EIA AEO 2017 projected oil price (WTI) for Year 2025, in real 2016 dollars), and a royalty rate of 16.7% to 18.75%, the Federal Government would receive in excess of $100 billion dollars of royalty revenues from the oil produced using the GOM $CO_2$ Pipeline System (assuming all of the technically recoverable oil is developed).

# 3. Next Steps

**As part of our existing Scope of Work, we plan to address two additional topics.**

- $CO_2$ utilization/storage, oil recovery and $CO_2$ pipeline systems for eastern Louisiana, Mississippi and Alabama state waters.
- The potential and capacity of storing $CO_2$ in the depleted gas fields of the Eastern and East Central GOM, including state, shallow Federal, and deep Federal waters.

A critical topic, although not within our Scope of Work, would be addressing the question - - *How much of the $CO_2$ demand and potential for oil recovery in the GOM offshore would be economically viable to pursue?*

Office Locations
**Washington, DC**
4501 Fairfax Drive, Suite 910
Arlington, VA 22203
Phone: (703) 528-8420

**Houston, TX**
11931 Wickchester Ln., Suite 200
Houston, TX 77043-4574
Phone: (281) 558-9200

**Knoxville, TN**
1210 Kenesaw Ave.
Suite 1210A
Knoxville, TN 37919-7736

*** Formatting Issue - Best Available Paper/Graphic ***

# NET Power

Truly Clean, Cheaper Energy

CARBON MANAGEMENT

July 19, 2017



NETPOWER

# Taking Stock

*** Formatting Issue - Best Available Paper/Graphic ***

NETPOWER

# Where We Are

July 19, 2017



NETPOWER

## It's Getting Hotter

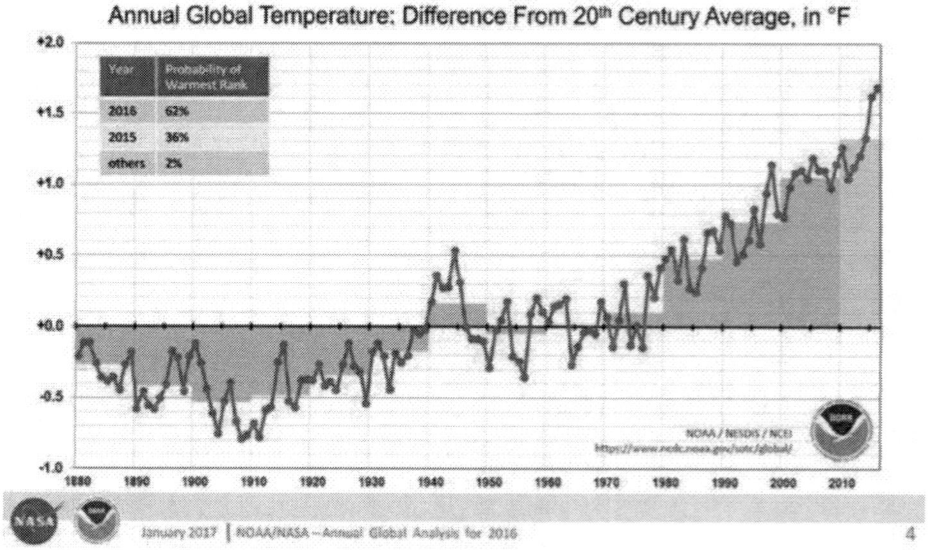


July 19, 2017

*** Formatting Issue - Best Available Paper/Graphic ***

NETPOWER

## Greenhouse gases are part of the problem

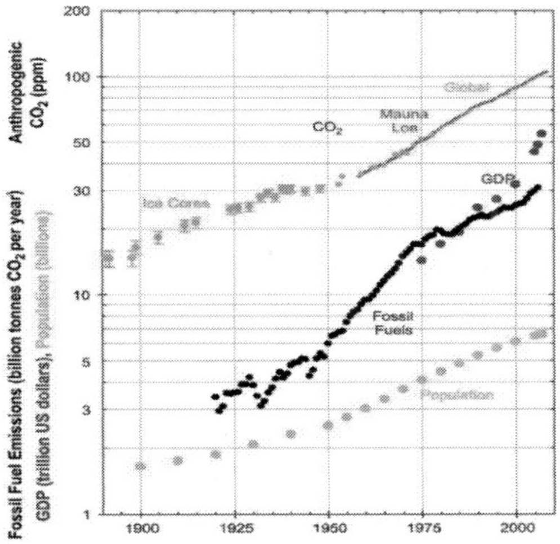


**... and greenhouse gases are a function of**

- *fugitive* carbon
- from our energy sources (among other places!)

**... which, in turn, are a function of**

- GDP
- Population

July 19, 2017

The information contained in this material is confidential and contains intellectual property of 8 Rivers Capital, LLC and its affiliates.

NETPOWER

## What to do?

July 19, 2017

The information contained in this material is confidential and contains intellectual property of 8 Rivers Capital, LLC and its affiliates.

***** Formatting Issue - Best Available Paper/Graphic *****

NETPOWER

# The future as seen by some smart people

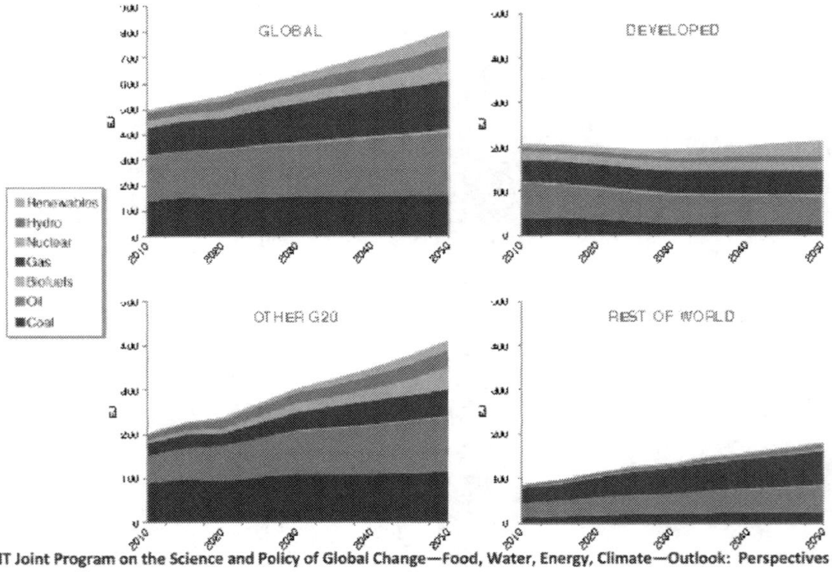


**MIT Joint Program on the Science and Policy of Global Change—Food, Water, Energy, Climate—Outlook: Perspectives from 2016**

July 19, 2017



NETPOWER

# Renewables are promising, but cannot do it alone

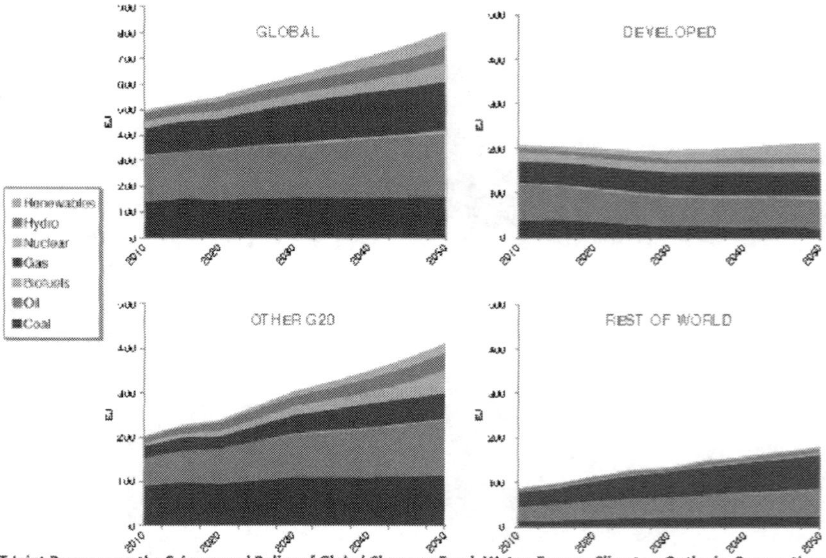


**MIT Joint Program on the Science and Policy of Global Change—Food, Water, Energy, Climate—Outlook: Perspectives from 2016**

July 19, 2017

The information contained in this material is confidential and contains intellectual property of 8 Rivers Capital, LLC and its affiliates.

NETPOWER

# Why can't renewables do it?

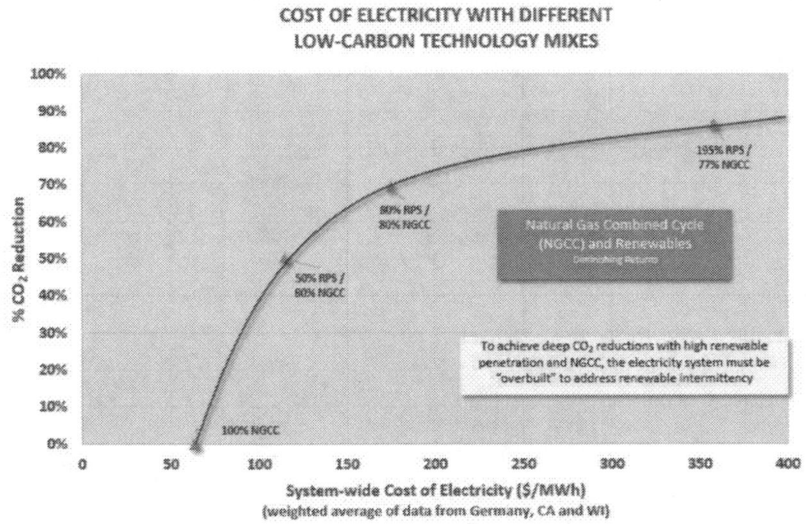


Data obtained from: Brick, S., and Thernstrom, S., Renewables and decarbonization: Studies of California, Wisconsin, and Germany, *The Electricity Journal*, 2016, 29, 6-12.

July 19, 2017



NETPOWER

# Nuclear is a great, but it also cannot do it alone

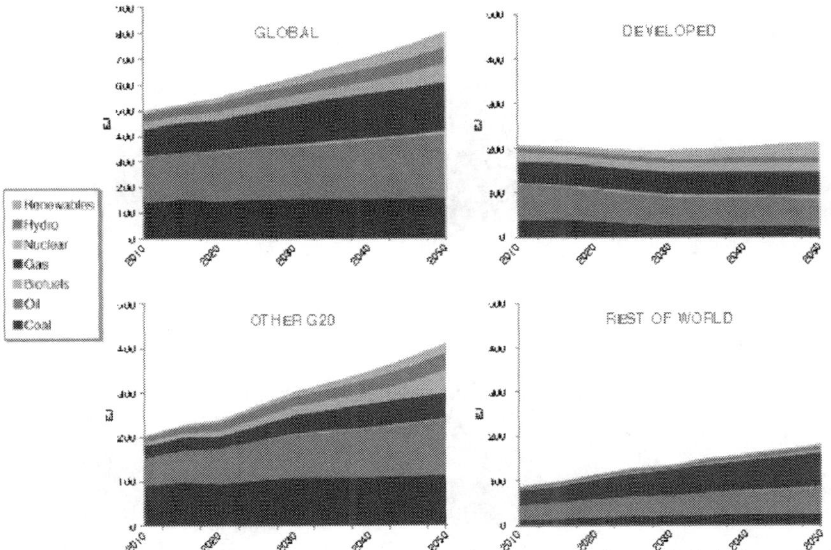


MIT Joint Program on the Science and Policy of Global Change—Food, Water, Energy, Climate—Outlook: Perspectives from 2016

July 19, 2017

*** Formatting Issue - Best Available Paper/Graphic ***

NETPOWER

# As a result, fossil fuels will be the core of the future

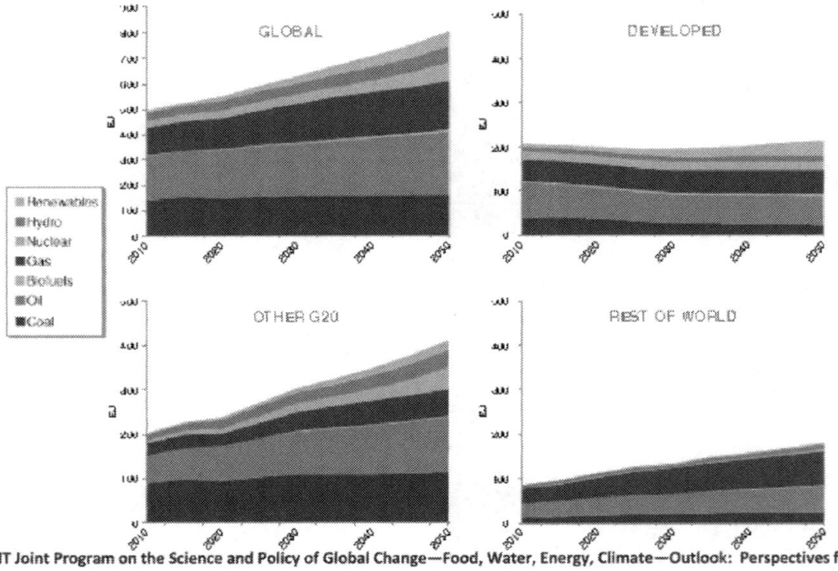


MIT Joint Program on the Science and Policy of Global Change—Food, Water, Energy, Climate—Outlook: Perspectives from 2016

July 19, 2017

The information contained in this material is confidential and contains intellectual property of 8 Rivers Capital, LLC and its affiliates.

NETPOWER

# And, if we don't fix fugitive carbon, we don't solve the problem

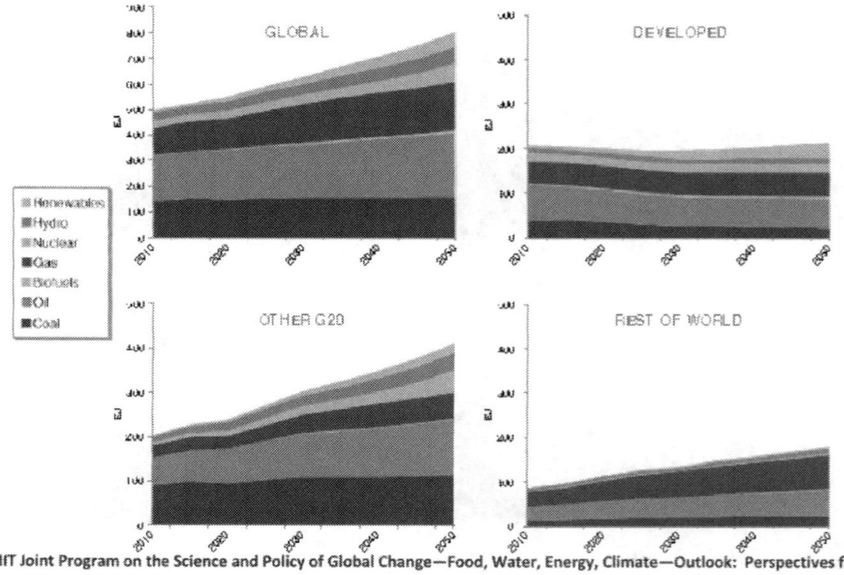


MIT Joint Program on the Science and Policy of Global Change—Food, Water, Energy, Climate—Outlook: Perspectives from 2016

July 19, 2017

The information contained in this material is confidential and contains intellectual property of 8 Rivers Capital, LLC and its affiliates.

NETPOWER

# We Have a Strong Team

July 19, 2017



NETPOWER

Capture

TOSHIBA nrg bp ExxonMobil aramco nie CBI MITSUBISHI HEAVY INDUSTRIES Southern Company Air Liquide Exelon NETPOWER inventys AIR PRODUCTS

Use

OXY LanzaTech JX NOVOMER Denbury bp Husky Solidia Technologies 8RIVERS aramco

Implementation, policy and research

U.S. DEPARTMENT OF ENERGY NETL GLOBAL CCS INSTITUTE nrg cosia CARBON XPRIZE Reimagine $CO_2$ UKCCS RESEARCH CENTRE NATIONALCOALCOUNCIL RICE Advanced Resources International, Inc. CCP The International CCS Knowledge Centre Enhanced Oil Recovery Institute UNIVERSITY OF WYOMING SME Society for Mining, Metallurgy & Exploration AIChE The Global Home of Chemical Engineers SPE Society of Petroleum Engineers 8RIVERS

July 2017

*** Formatting Issue - Best Available Paper/Graphic ***

NETPOWER

# And We Are Making Things Happen

July 19, 2017

The information contained in this material is confidential and contains intellectual property of 8 Rivers Capital, LLC and its affiliates.

NETPOWER

July 19, 2017

The information contained in this material is confidential and contains intellectual property of 8 Rivers Capital, LLC and its affiliates.

*** Formatting Issue - Best Available Paper/Graphic ***

NETPOWER

July 19, 2017



NETPOWER

July 19, 2017

***** Formatting Issue - Best Available Paper/Graphic *****

NETPOWER

July 19, 2017

The information contained in this material is confidential and contains intellectual property of 8 Rivers Capital, LLC and its affiliates.

NETPOWER

July 19, 2017

The information contained in this material is confidential and contains intellectual property of 8 Rivers Capital, LLC and its affiliates.

*** Formatting Issue - Best Available Paper/Graphic ***

NETPOWER

July 19, 2017

The information contained in this material is confidential and contains intellectual property of 8 Rivers Capital, LLC and its affiliates.

NETPOWER

July 19, 2017

The information contained in this material is confidential and contains intellectual property of 8 Rivers Capital, LLC and its affiliates.

*** Formatting Issue - Best Available Paper/Graphic ***

NETPOWER

July 19, 2017

The information contained in this material is confidential and contains intellectual property of 8 Rivers Capital, LLC and its affiliates.

NETPOWER

July 19, 2017

The information contained in this material is confidential and contains intellectual property of 8 Rivers Capital, LLC and its affiliates.

***** Formatting Issue - Best Available Paper/Graphic *****

NETPOWER

July 19, 2017



NETPOWER

July 19, 2017

The information contained in this material is confidential and contains intellectual property of 8 Rivers Capital, LLC and its affiliates.

*** Formatting Issue - Best Available Paper/Graphic ***

NETPOWER

July 19, 2017

The information contained in this material is confidential and contains intellectual property of 8 Rivers Capital, LLC and its affiliates.

NETPOWER

July 19, 2017

The information contained in this material is confidential and contains intellectual property of 8 Rivers Capital, LLC and its affiliates.

*** Formatting Issue - Best Available Paper/Graphic ***

NETPOWER

July 19, 2017

The information contained in this material is confidential and contains intellectual property of 8 Rivers Capital, LLC and its affiliates.

NETPOWER

And we know what to do with the $CO_2$

July 19, 2017

The information contained in this material is confidential and contains intellectual property of 8 Rivers Capital, LLC and its affiliates.

*** Formatting Issue - Best Available Paper/Graphic ***

NETPOWER

# $CO_2$ Use: EOR, Huge Opportunity

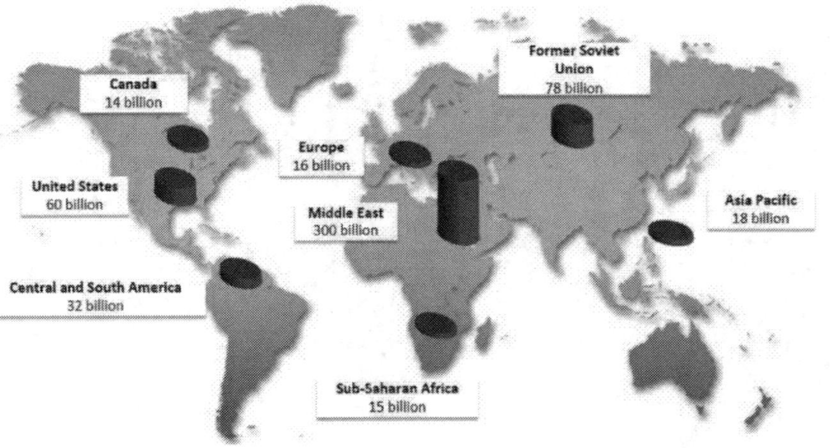


**Global perspective: 1.1B tons of $CO_2$ required to recover 470 billion barrels of oil**

July 19, 2017

The information contained in this material is confidential and contains intellectual property of 8 Rivers Capital, LLC and its affiliates.

NETPOWER

# $CO_2$ Use: Sequestration, 2,600 billion metric tons[1]

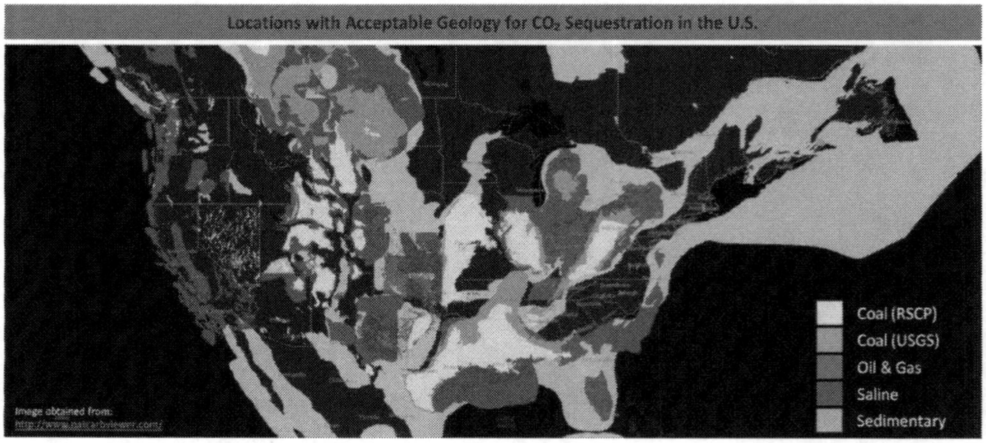


- **The federal 45Q tax credit puts CCS on a level playing field**
- **21 states have incentives and policies in place the provide value or preferential treatment to NET Power for capturing $CO_2$**

1. DOE & NETL Carbon Storage Atlas (Atlas V), 5th edition, 2015

July 19, 2017

The information contained in this material is confidential and contains intellectual property of 8 Rivers Capital, LLC and its affiliates.

*** Formatting Issue - Best Available Paper/Graphic ***



# $CO_2$ Use: Conversion to value added products

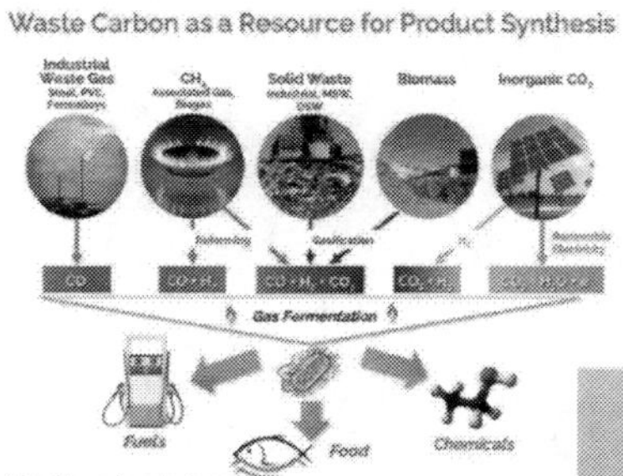


http://www.lanzatech.com

Wow

*** Formatting Issue - Best Available Paper/Graphic ***

NETPOWER

# What NET Power is Doing

July 19, 2017



NETPOWER

## NET Power

- **Electricity from natural gas**
- **Costs competitive** with existing natural gas power plants
- **Captures or eliminates** substantially all of the carbon and non-carbon atmospheric emissions *without any additional cost*
- **Water not required** (at a small reduction in efficiency)
- **Can use inexpensive fuels** such as acid gas, sour gas, associated gas, and produced gas
- **Also produces** important gases, including $N_2$, $O_2$ and Ar

July 19, 2017

NETPOWER

# The supercritical $CO_2$ Allam Cycle is simple

- **Historically, $CO_2$ capture has been expensive, whether using air to combust or oxy-combustion**
  - Air combustion
    - $\underbrace{8N_2 + 2O_2}_{air} + CH_4 \rightarrow \underbrace{8N_2 + CO_2}_{\substack{expensive\ to\\ separate}} + 2H_2O$
  - Oxy-combustion
    - $\underbrace{2O_2}_{\substack{expensive\\ to\ produce}} + CH_4 \rightarrow CO_2 + 2H_2O$
- **The Allam Cycle makes oxy-combustion economic by:**
  - Relying on a more efficient core power cycle
  - Recycling heat within the system to reduce $O_2$ and $CH_4$ consumption, and associated costs of the ASU

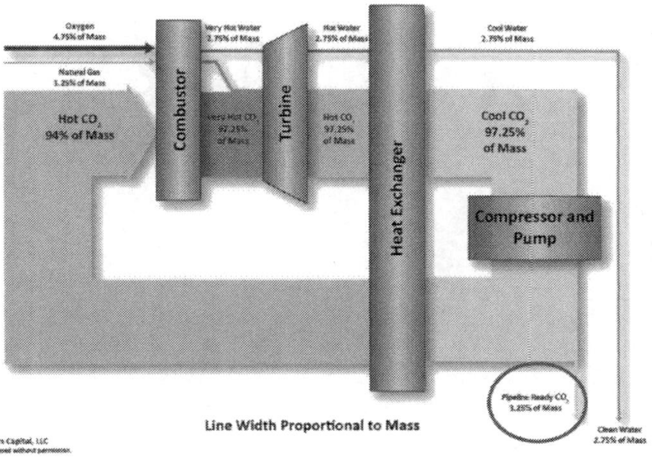






NETPOWER

# NET Power is about more than electricity

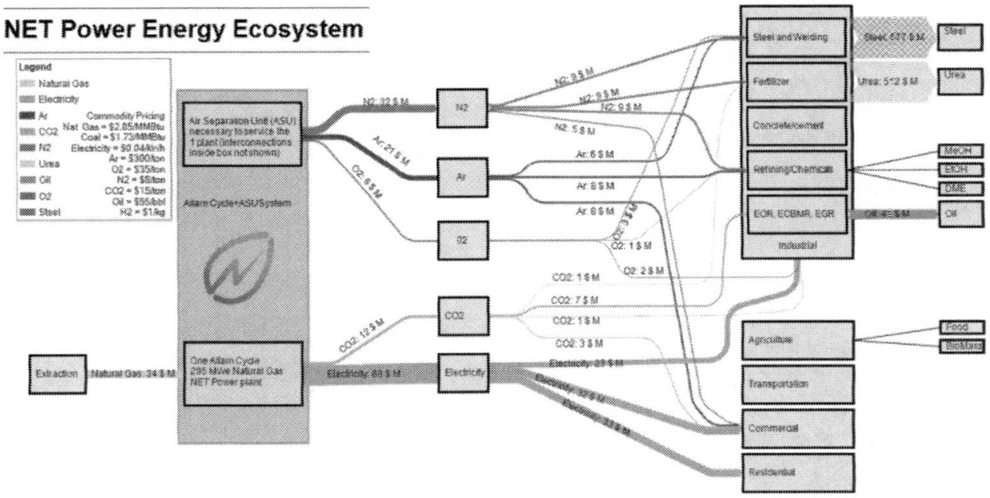

*** Formatting Issue - Best Available Paper/Graphic ***

NETPOWER

# NET Power partners with renewables to avoid diminishing returns

COST OF ELECTRICITY WITH DIFFERENT LOW-CARBON TECHNOLOGY MIXES

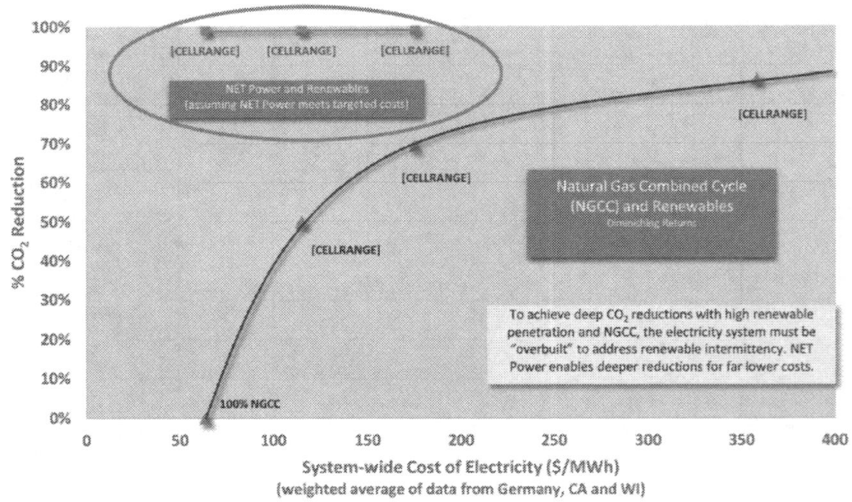


Data obtained from: Brick, S., and Thernstrom, S., Renewables and decarbonization: Studies of California, Wisconsin, and Germany, *The Electricity Journal*, 2016, 29, 6-12.

July 19, 2017



NETPOWER

# Total global addressable market: $749 billion

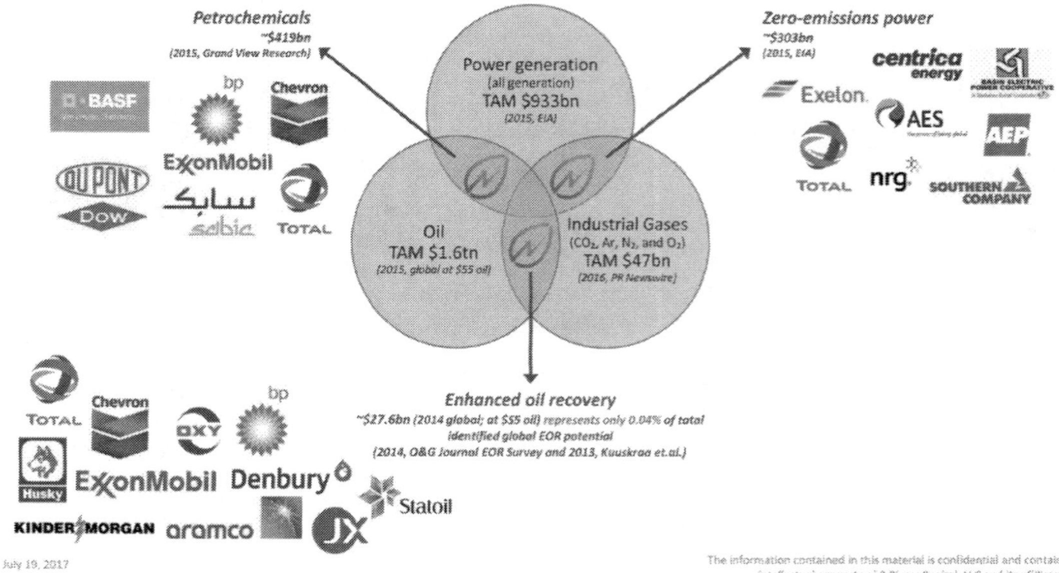


July 19, 2017

*** Formatting Issue - Best Available Paper/Graphic ***

NETPOWER

# NET Power's effective heat rate (LHV)

Note: Heat rates are net of additional ASU parasitics for capture of specialty gases. Assumes gas costs are offset at $4.44/MMBtu LHV





NETPOWER

# NET Power is clean, low cost and efficient

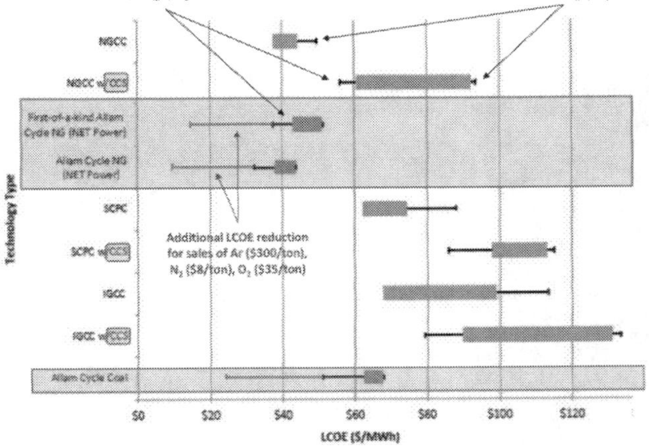


1. LCOE calculated using EPRI methodology
2. Assumes natural gas at $2.85/MMBTU and coal at $1.73/MMBTU
3. Every move of $1 in natural gas moves LCOE $6
4. Cost ranges represent range of data combined from: EIA (2013), Parsons Brinkerhoff (2013); Black & Veatch (2012); DOE NETL (2012)

**NET Power and Combined Cycle: Efficiency at ISO Conditions**

| | HHV | | LHV | |
|---|---|---|---|---|
| Energy Components | F-Class US NGCC Plant (0% CC)* | NET Power NG Plant (100% CC) | F-Class US NGCC Plant (0% CC)* | NET Power NG Plant (100% CC) |
| Gross Turbine Output | 51.06% (Compressors mechanically coupled) | 74.65% | 58.7% (Compressors mechanically coupled) | 82.7% |
| $CO_2$ Compressor Power | | -10.47% | | -11.6% |
| Plant Parasitic Auxiliary Power | -0.86% | -11.01% | -1.2% | -12.2% |
| Net Efficiency | 50.20% | 53.17% | 57.5% | 58.9% |

*Performance data from NETL Cost and Performance Baseline Report, 2013

| Parasitic Load Provides Opportunity for Efficiency Improvement | |
|---|---|
| ASU | 91.8% |
| NG Compressor | 8.2% |

NETPOWER

# NET Power demonstration plant timeline

**3/2016**
Groundbreaking

**12/2016**
Toshiba turbine arrived on site

**1/2017**
Engineering 99% complete; construction over 50% complete

**Q4 2017**
Full first-fire testing full-scale combustor

**Q1 2018**
Connect to grid





NETPOWER

# Toshiba turbine on site

- **50MWth turbine is complete.**
- **Actually sized as 200 MWth, but uses only 1 combustor vs 4**
- **500 MWth commercial turbine uses same technology**

July 19, 2017

The information contained in this material is confidential and contains intellectual property of 8 Rivers Capital, LLC and its affiliates.

*** Formatting Issue - Best Available Paper/Graphic ***

NETPOWER

## NET Power's demonstration plant is 95% complete

50 MWth Demonstration Plant





NETPOWER

## Commercial combustor complete with demonstration

Demonstration (50 MWth)

Commercial (500 MWth) = 10 x demonstration (50 MWth)

Image of Toshiba's NET Power combustor.

Image meant to be a visual aid for relative combustor-turbine configuration, not actual NET Power combustor configuration.

*** Formatting Issue - Best Available Paper/Graphic ***

NETPOWER

# NET Power 1st commercial plant timeline

| 6/2013 | 3/2016 | 9/2016 | 1/2017 | 1Q 2018 | 2Q 2018 | 1Q 2021 |
|---|---|---|---|---|---|---|
| Original Pre-FEED / Toshiba basic design | Start building Demo Plant | Pre-FEED initiated with demo lessons learned, performance optimization | Toshiba Kick-off in Tokyo | FEED for 1st commercial customer secured | 1st commercial plant MOU / order secured | Targeting first Commercial start-up |

July 19, 2017



NETPOWER

# World Class Partners

CBI

Exelon

8RIVERS

## In the press

Gas Turbine World Forbes Science AAAS Bloomberg Business The New York Times Vox Elsevier

July 19, 2017

NETPOWER

# Together we will solve this

July 19, 2017

The information contained in this material is confidential and contains intellectual property of 8 Rivers Capital, LLC and its affiliates.

NETPOWER

# Appendix

July 19, 2017

The information contained in this material is confidential and contains intellectual property of 8 Rivers Capital, LLC and its affiliates.

NETPOWER

# Allam Cycle Natural Gas Platform

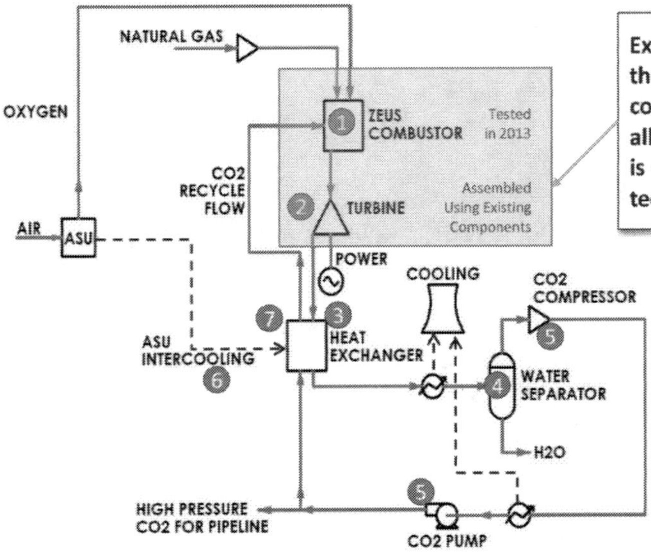


Except for these components, all equipment is mature technology

1. Fuel Combustion
2. $CO_2$ Turbine
3. Heat Rejection
4. Water Separation
5. Compression and Pumping
6. Additional Heat Input
7. Heat Recuperation

- 58.9% (LHV) net efficiency, with capture of >97% of carbon
- Oxy-fuel, closed-loop, $CO_2$ working fluid
- 200-400 bar; 6-12 pressure ratio
- $CO_2$ and water are the only byproducts
- All components, other than combustor and turbine, currently available
- 740H now in commercial use at two facilities

July 11, 2017



NETPOWER

# The NET Power Advantage - The Allam Cycle

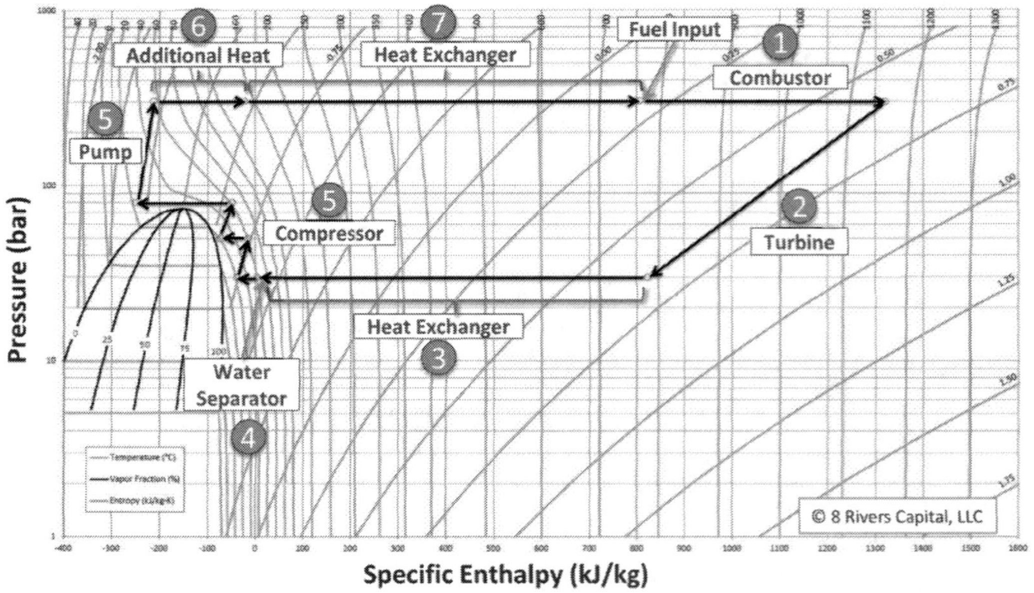


July 19, 2017

*** Formatting Issue - Best Available Paper/Graphic ***

NETPOWER

# NET Power has *higher energy density* than big projects

- Compared to IGCC, bulks are small driver of costs:

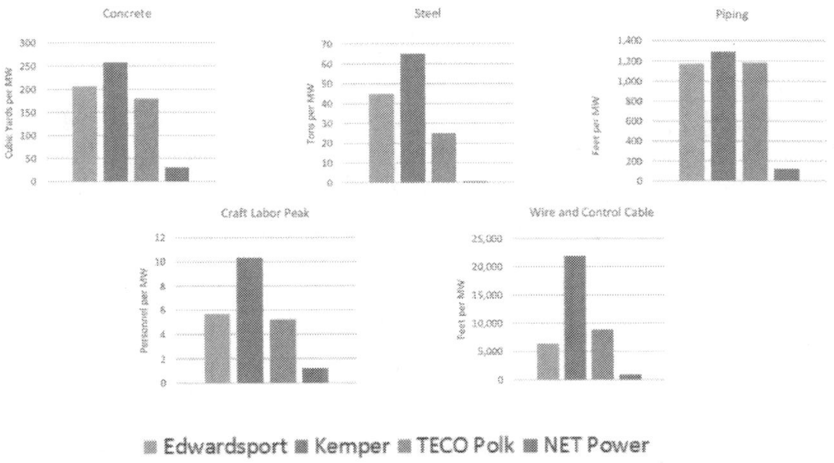


July 19, 2017



NETPOWER

# Globally, NET Power $CO_2$ will change the oil stack

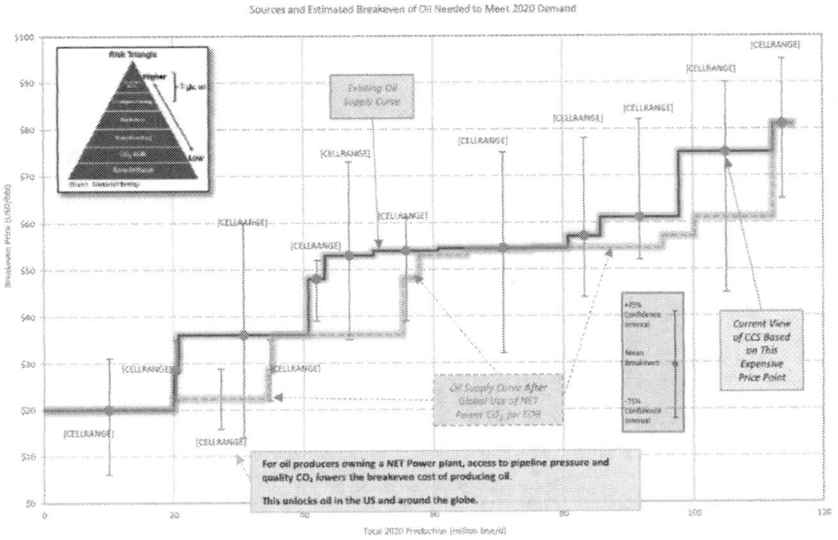


- Shutdown of tight oil/high cost plays highlights EOR as a low-cost opportunity for growth from existing fields
- NET Power further improves the economics of EOR and will significantly expand $CO_2$ supplies for producers
- NET Power solves the challenge of carbon capture adoption by achieving lower cost and higher efficiency than current non-carbon capture power generation options

*Source: Rystad Energy (2014); 8 Rivers Capital (2015)*

July 19, 2017

*** Formatting Issue - Best Available Paper/Graphic ***

NETPOWER

# Market opportunity for NET Power turbine sales

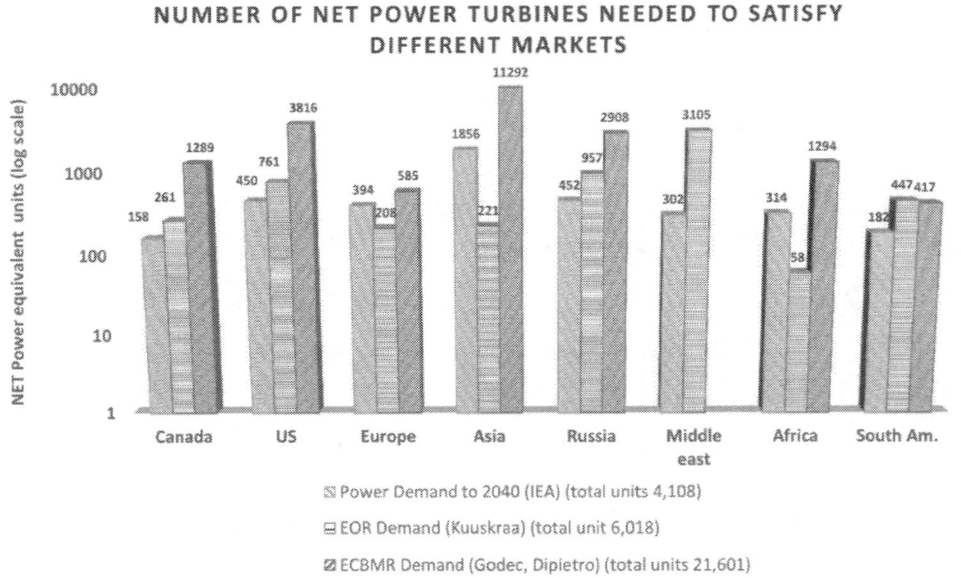


July 19, 2017



NETPOWER

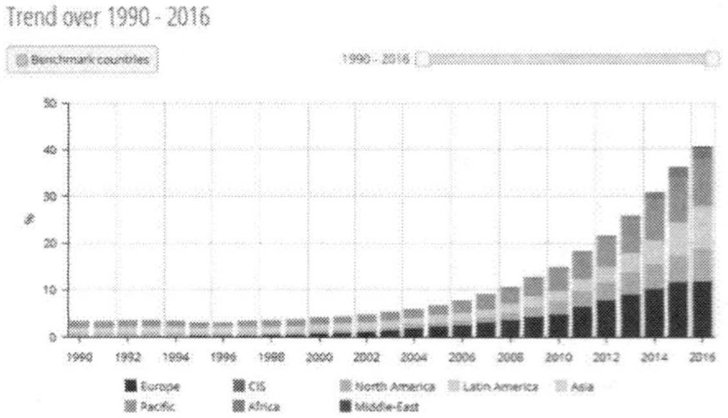


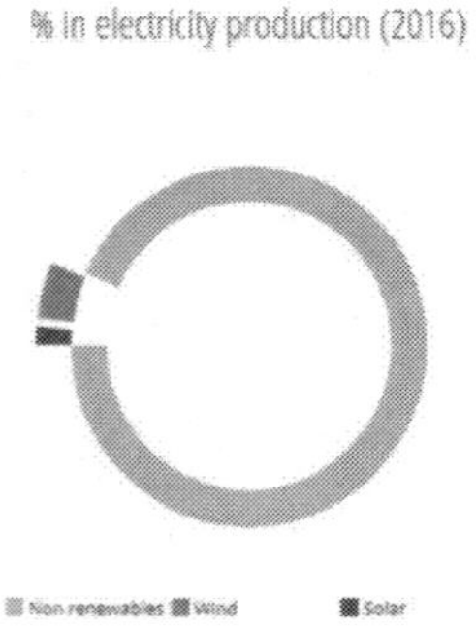


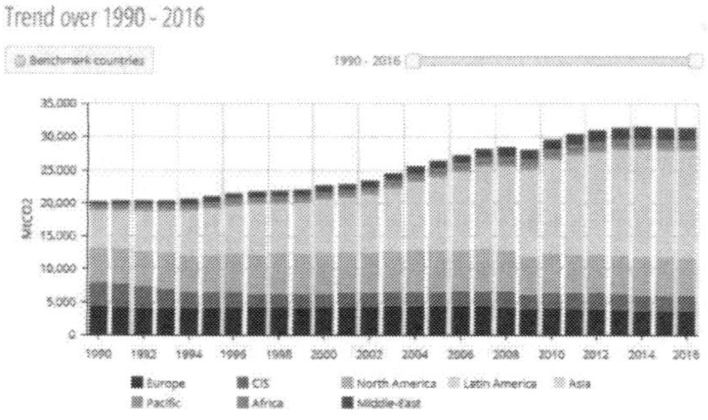

Source: https://yearbook.enerdata.net/renewables/wind-solar-share-electricity-production.html

July 19, 2017

NETPOWER

## Power Mix – G20 (2000 and 2016)

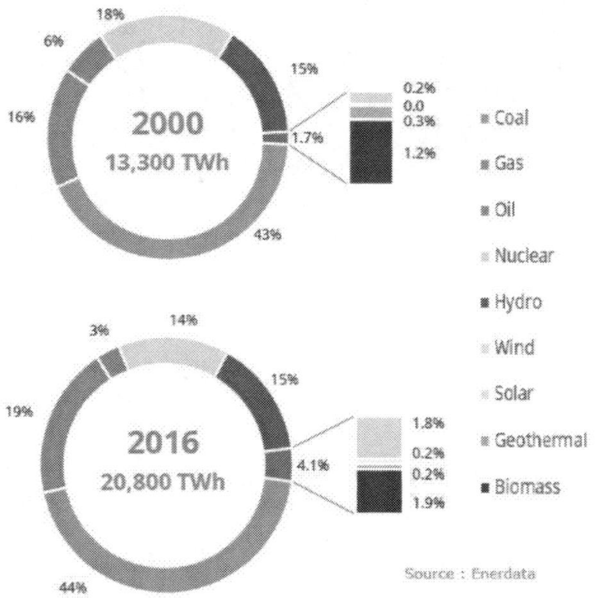


**Even with all the huge news on renewable growth fossil still dominates and has taken a larger share of the mix!**





# NET Power

+1 (919) 667-1800
www.NETPower.com / www.8Rivers.com

2

# CANADA'S LEADING INTEGRATED ENERGY COMPANY

**$81B** ENTERPRISE VALUE[1] MARCH 31, 2017

**35+** YEARS 2P RESERVE LIFE INDEX[2] AS AT DEC. 31, 2016

**725** MBOEPD 99% OIL PRODUCTION Q1 2017

**462** MBPD REFINING CAPACITY

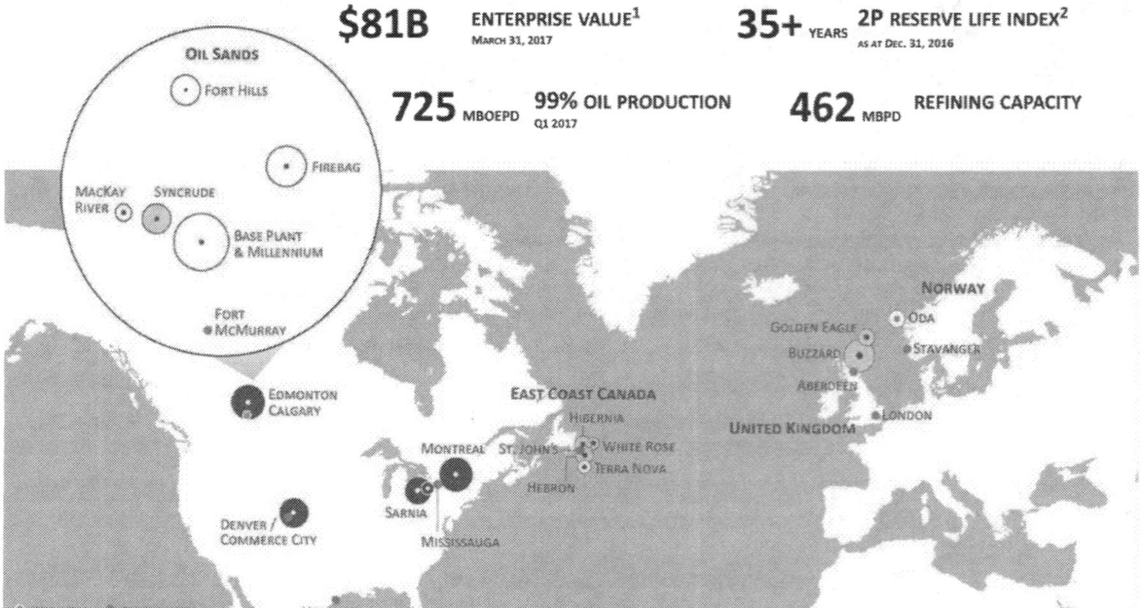


1, 2 SEE SLIDE NOTES AND ADVISORIES.

SUNCOR

# Long History of Sustainable Development

| 1995 | | 1998 | | | 2010 | | | 2016 |
|---|---|---|---|---|---|---|---|---|
| First sustainability report | Introduction of seven-point climate action plan | Flu gas De-sulphurization | Founding Member of GRI | Signed Ceres principles | Reclaimed pond 1 Wapisiw lookout | CEO participation in EcoFiscal Commission | Call for a price on carbon. Support Alberta climate leadership | Commit to transparent carbon reporting and launch new GHG goal |

**Suncor has a long, well established track record of being a leader on climate change issues**

"*Climate change is happening. Doing Nothing isn't an option*"- Steve Williams, CEO Suncor

- Suncor believes that climate change is a real and growing global challenge and that human activity, including the burning of fossil fuels, is contributing to increased concentrations of GHG emissions in the atmosphere.
- Suncor has acknowledged that we are in the early stages of an energy transition and we also believe hydrocarbons are going to be part of the energy mix for many years to come.

3

SUNCOR

See *Slide Notes* and *Advisories*.

4

# Environmental leadership: aggressive goal setting 2009-15

## Environmental performance goals
**2015 versus 2007 baseline**

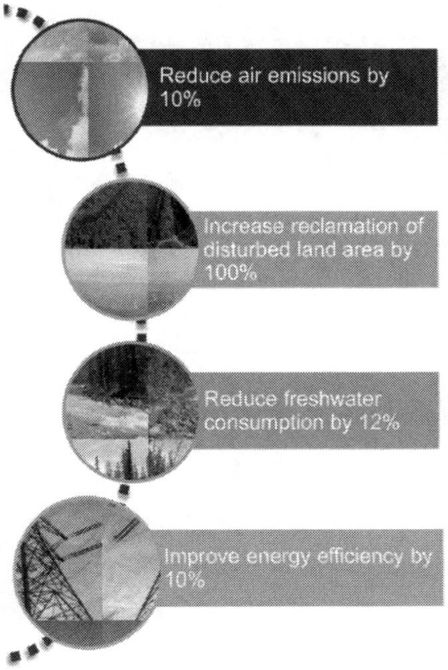


## Environmental performance results
**2015 versus 2007 baseline**

Reduced air emissions by 36%

Increased reclamation of disturbed land by 176%

Reduced freshwater consumption by 27%

Improved energy efficiency by 9%

SUNCOR

See *Slide Notes* and *Advisories*.

## GHG Goal to Lower Emissions Intensity 30% by 2030

| 2030 GHG Goal | | | |
|---|---|---|---|
| Reduce the total emission intensity of the production of our oil and petroleum products by 30% by 2030 | | | |
| **Energy efficiency** | **Transition to lower carbon fuels and power** | **Invest in Technology and Innovation** | **Participate in the low carbon future** |
| • Facility Energy management plans<br>• Improve reliability | • Fuel switching<br>• Cleaner fuels and low carbon electricity (e.g. cogeneration | • Low carbon bitumen extraction technologies<br>• Partial upgrading and crude oil decarbonization<br>• Low carbon heat and power<br>• Carbon capture, utilization and storage | • Low carbon power (cogeneration / solar / wind)<br>• Renewable fuels and biofuels<br>• Adapt to future transportation energy system |

SUNCOR

## Alberta Climate Leadership plan

**Phase-out coal** generated electricity by: **2030**

**Implement** price on GHG's **$50 / tonne by 2022**

Oil Sands Emissions Cap of **100 Mt**

**45%** reduction in methane by **2025**

- The Climate Leadership plan will replace the current regulation which is called the Specified Gas Emitters Regulation (SGER). SGER has been in effect since July 2007.
- http://alberta.ca/climate/leadership-plan.cfm

6

SUNCOR

## Suncor's approach to technology: collaboration is key

Leveraging a world of external capability and focusing on a path to deployment

### Venture capital

Selective investments into external venture capital funds

**Role:** Access companies for investment, partnership and leading edge knowledge and innovation trends

Emerald

### Direct strategic investments

Investments in young and growing companies in exchange for equity

**Role:** Support development of and access to close-to-commercial technologies

Lanzatech / Benefuel

### Technology partnerships

Partnerships with external innovation companies to develop/commercialize technologies

**Role:** Advance and test technology and commercialize within Suncor

EASE

### Industry partnerships

Alliance sharing expertise, risk, and technology in the Oil Sands (eg. COSIA, Evok Innovations)

**Role:** Advance and accelerate industry performance.

### Academic partnerships

Collaborations with universities to advance technology research

**Role:** Support and direct research into basic and applied science that support/informs strategy

### Innovation challenges

Competitions where external innovators submit solutions (eg. NRG COSIA Carbon XPRIZE)

**Role:** Tool for identifying technology partners

**"Win-win" approach with entrepreneurs**

- Enabled by an open and collaborative approach
  - Flexible with commercial structures and IP

7

## Suncor's Approach to technology: Path to Deployment

The process of moving from idea to commercialization

idea → concept → pilot → commercialization

Begin with the end in mind – commercial deployment

Large systemic innovations are delivered through integrated programs

Programs require as much design and effort as the invention

Understand the constraints in a path to deployment

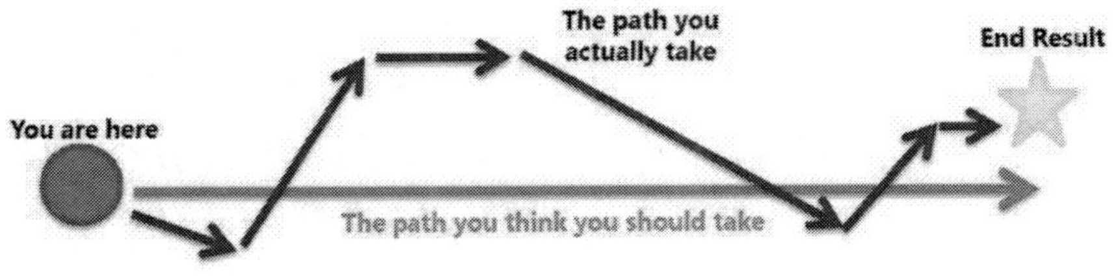


8

## In Situ Technology video

SUNCOR

Information as at December 31, 2016. Actual results may vary.



## Carbon Capture Challenge at Next SAGD Facility

**Typical 33,000 BPD facility**

- Six gas fired once through steram generators (1600 GJ/h)
- 2200 metric tonnes / day of CO2e
- ~10-15 hectares (+laydown). Largest stack: 30 m high

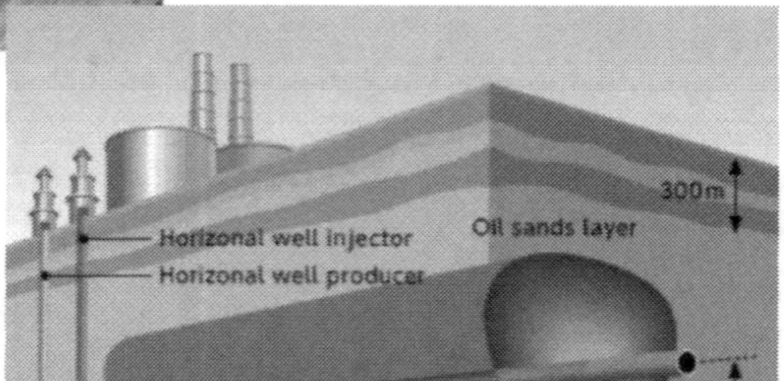


**Challenges:**

- Capital Cost
- Footprint/ Height
- Use
  - EOR within SAGD?
  - Storage in local formation

Information as at December 31, 2016. Actual results may vary.

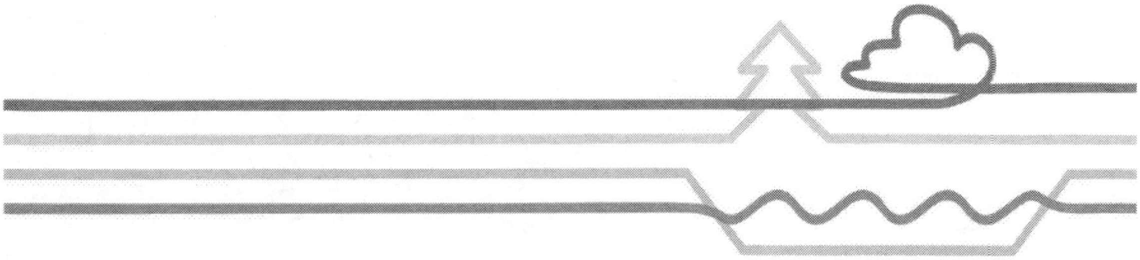

# All about **COSIA**

Canada's Oil Sands Innovation Alliance

## Our **Vision**

To enable responsible and sustainable growth of Canada's oil sands while delivering accelerated improvement in environmental performance through collaborative action and innovation.

Canadian Natural

bp

Syncrude

Suncor

cenovus ENERGY

We are cosia

nexen

Teck

Imperial

ConocoPhillips

devon

## GHG **Aspiration**

We will strive to...

"Produce oil with lower greenhouse gas emissions than other sources of oil."

## 2016 GHG **Project Portfolio**

### PROJECT PORTFOLIO

**154** Contributed technologies
(18 obtained in 2016)

**$208M** Cost to develop technologies
($21.5 million in 2016)

**12** Current (active) projects
(7 obtained for 2016)

**$15M** Cost for current projects
($5.6 million in 2016)

**42** Completed projects
(18 completed in 2016)

**$25M** Cost for completed technologies
($21.5 million in 2016)

### ACTIVE PROJECT PORTFOLIO BY 'D' PHASE

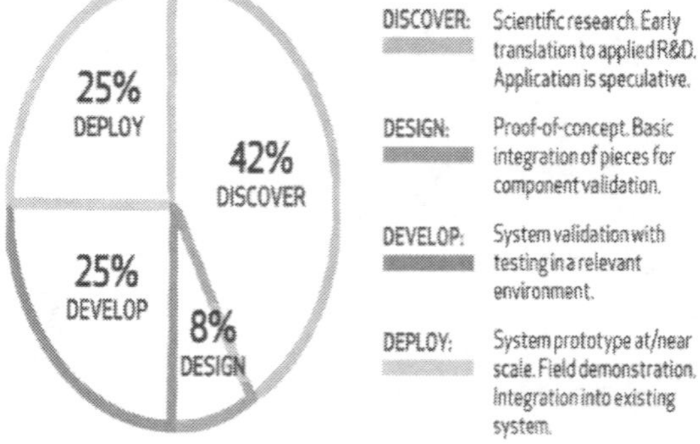

# GHG **Opportunities Areas & Gaps**

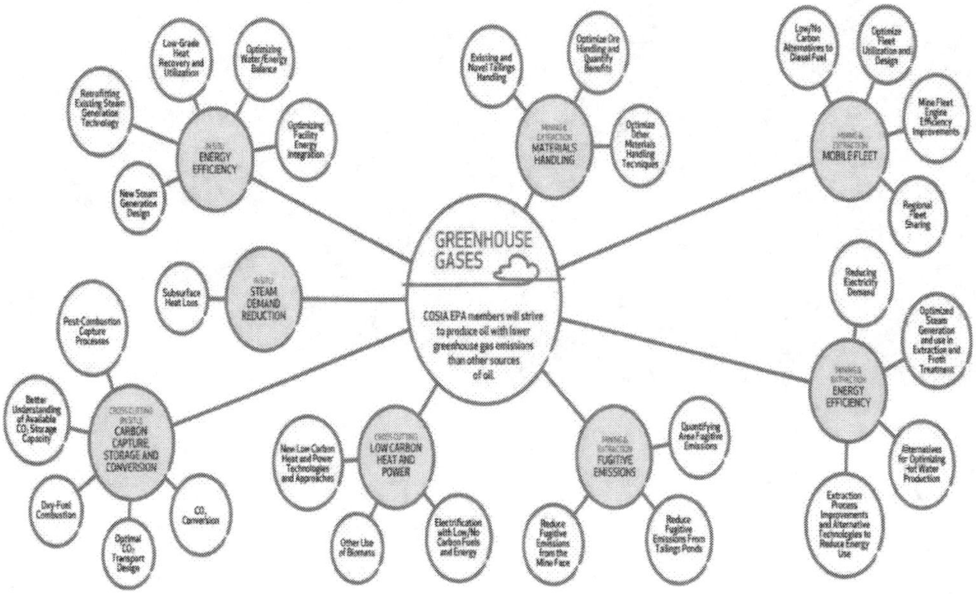


# GHG **Challenges**

### On-going GHG Challenges

- Direct Hot Water Production for an Oil Sands Mining & Extraction Process
- New High Efficiency Boiler
- Higher Value Use of Low Grade Heat
- Enriched Combustion Air
- Natural Gas Decarbonization
- Water and Energy Recovery
- New Heat Exchanger
- Pressure Let Down

### New Challenges

- Quantification of Area Fugitive Emissions
- Post Combustion $CO_2$ Capture from Natural Gas Combustion Flue Gas

## COSIA CHALLENGE

Mobilizing the world's minds and resources to improve environmental performance.

## **XPRIZE** Competition Timelines: Good things happening !

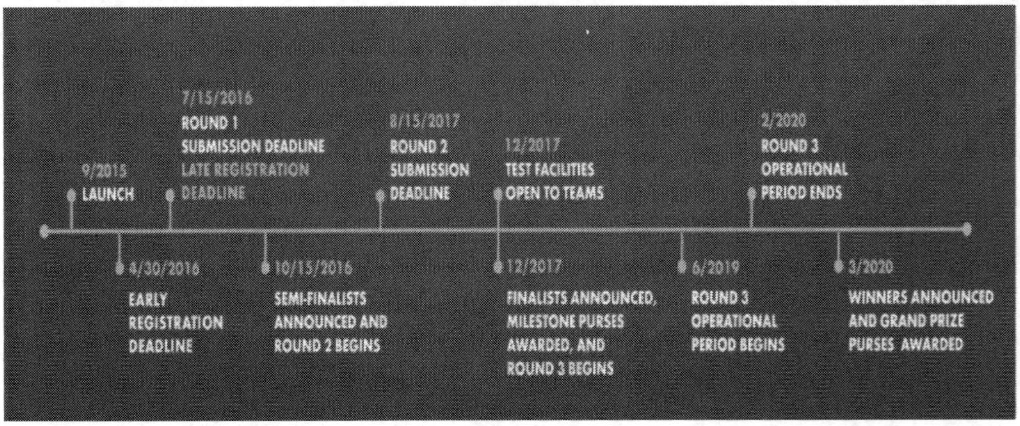


*Learn more at www.cosia.ca/carbon-xprize*

## Alberta Carbon Conversion Technology Centre

Innovation Infrastructure

## Vision: Alberta Carbon Conversion Technology Centre

- Create an innovation space to bridge the gap between lab and commercial scale
- Support and accelerate carbon conversion technology development to reduce emissions
- Bring great minds together to spark innovative ideas

## ACCTC: Specifications

- Five testing bays ~2,200 square meters (24,000 square feet) each
- Flue gas stream containing between 2 and 5 metric tons of $CO_2$ per day for each bay.
- Supply of electricity 600 V 3 phase 2MVA
- Supply of fresh water 5-20 m3/day
- Natural gas supply 400-1500 Sm3/day
- 400 bbl waste tank for each bay

## ACCTC: Timeline

2016 | 2017 | 2018 | 2019 | 2020

**Nov 2016**
Stack tie ins complete

**Feb 2017**
Detailed engineering complete

**Sept 2017**
Construction complete

**Nov 2017**
Ready for operations

**Feb 2018**
NRG COSIA Carbon XPRIZE teams move in

**Q2 2020**
Facility available for other technology testing

## **ACCTC Open for Testing** After NRG COSIA Carbon XPRIZE

### Technology Centre

- **Capitalize on technology centre successes** — create new industries and revenue streams for Alberta.
- Owned and operated by InnoTech Alberta, a subsidiary of Alberta Innovates

CMTC 2017

# An update on the state of CCS in the UK

Jon Gibbins
Director, UK CCS Research Centre
Professor of Power Plant Engineering and Carbon Capture
University of Sheffield
www.ukccsrc.ac.uk
j.gibbins@sheffield.ac.uk

The UKCCSRC is supported by the Engineering and Physical Sciences Research Council as part of the Research Councils UK Energy Programme

EPSRC Pioneering research and skills

# The Climate Problem

## IPCC Climate Change 2013 'The Physical Science Basis'

http://www.ipcc.ch/report/ar5/wg1/

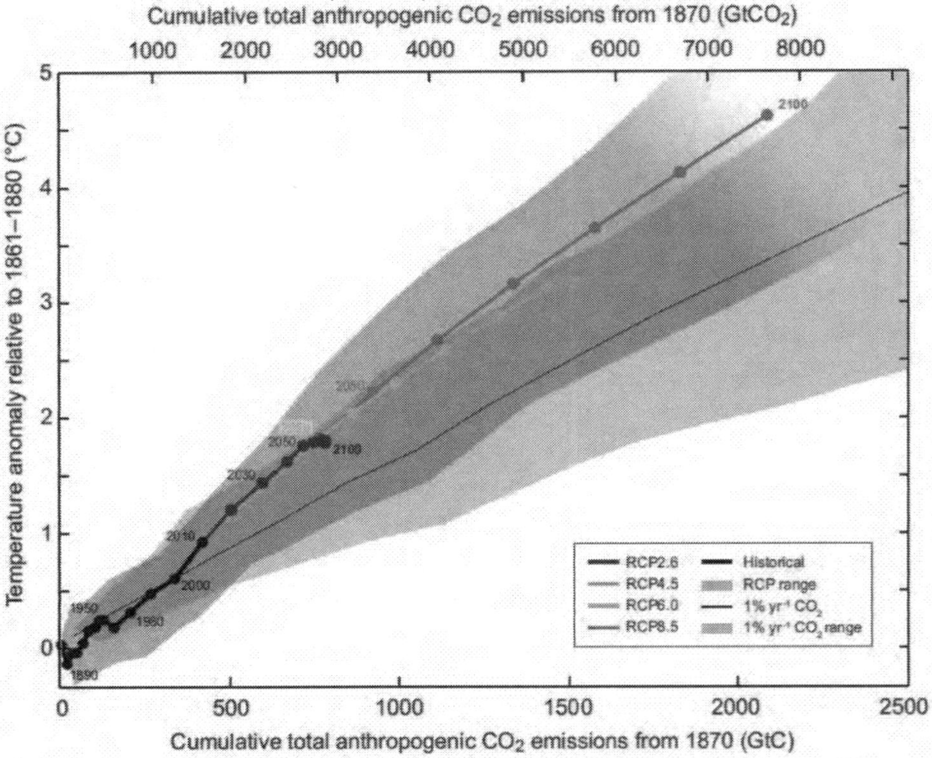


## The Paris Agreement

http://unfccc.int/paris_agreement/items/9485.php

**UNITED NATIONS**
**2015**

**Article 4**

1. In order to achieve the long-term temperature goal set out in Article 2, Parties aim to reach global peaking of greenhouse gas emissions as soon as possible, recognizing that peaking will take longer for developing country Parties, and to undertake rapid reductions thereafter in accordance with best available science, so as to achieve a balance between anthropogenic emissions by sources and removals by sinks of greenhouse gases in the second half of this century, on the basis of equity, and in the context of sustainable development and efforts to eradicate poverty.

**This implies Carbon Capture and Storage on all fossil fuel use, plus minimising other anthropogenic emissions e.g. from food production.**

## What do we need to achieve?

The prime climate objective is not to end the use of fossil fuels.

The prime objective is to develop and deploy 100% CCS in time to cap cumulative emissions of carbon at a safe level.

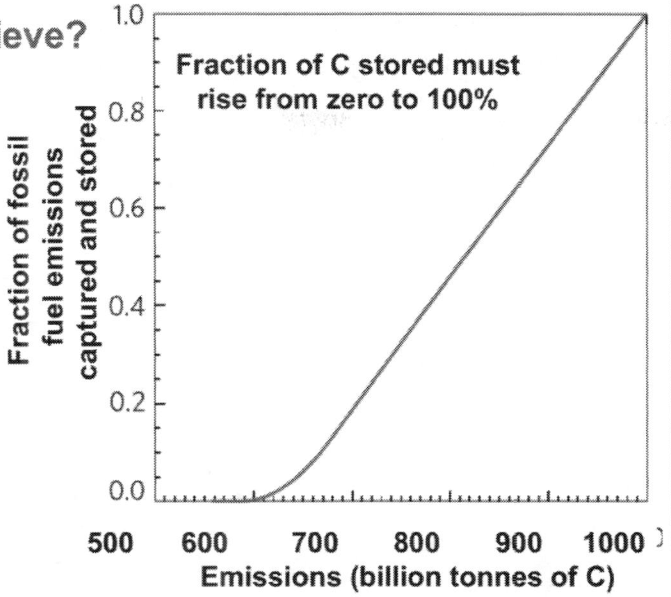


**Myles R. Allen, David J. Frame & Charles F. Mason, The case for mandatory sequestration, Nature Geoscience 2, 813 - 814 (2009), doi:10.1038/ngeo709**

$CO_2$ EOR and other applications with partial overall capture should be seen as a stage in a path from zero $CO_2$ capture to 100% CCS.

They can be a move in the right direction from where we are now – emitting 100% of fossil carbon to atmosphere.

The key factor is the extent to which technologies and/or projects can readily be adapted to get higher fractions of $CO_2$ stored.

## Key indicators to track current progress and future ambition of the Paris Agreement

UKCCS RESEARCH CENTRE

Peters, G. P. et al. , Nature Climate Change, 2017.

... without large-scale CCS deployment, most models cannot produce emission pathways consistent with the 2°C goal. ..... a **globally coordinated effort is needed to accelerate progress,** better understand the technological risks, and address social acceptability.

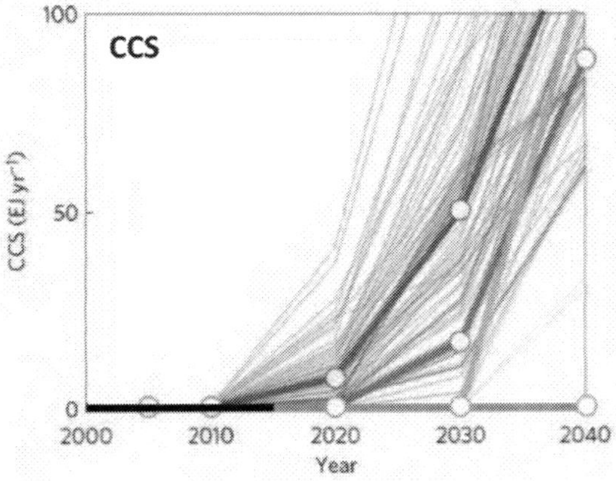


**Historical trends and future pathways to 2040, assuming actions start in 2010 (blue), 2020 (red), 2030 (yellow)**

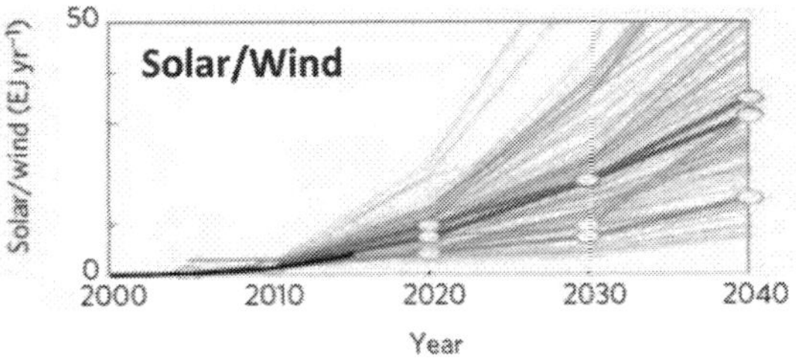

**Achieving net-zero emissions through the reframing of UK national targets in the post-Paris Agreement era,**
Pye et al, Nature Energy, 6 March 2017.

Total remaining global $CO_2$ budget for 66% chance of 2°C from the IPCC AR5 is 590 - 1240Gt$CO_2$. UK shares by equity (per capita basis) and inertia (current total emissions basis) imply emissions trajectories.

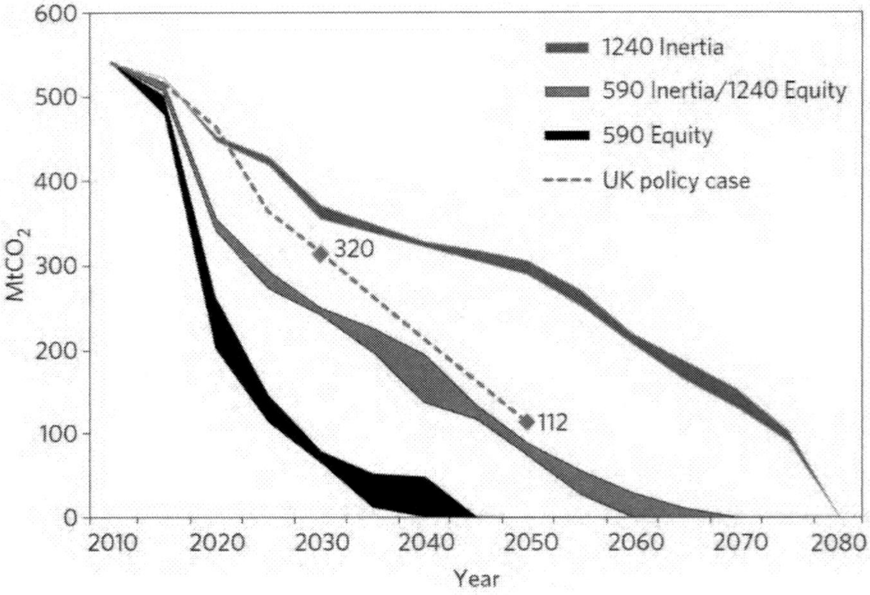


# CCS in the UK

## National Importance for UK Decarbonisation and CCS

UKCCS RESEARCH CENTRE

24 January 2017, Westminster Hall Debate

The Parliamentary Under-Secretary of State for Business, Energy and Industrial Strategy, Jesse Norman

CCS has a wide range of potential applications in which it could contribute to the reduction of carbon in our environment. Those include not merely decarbonising **heating** and **transport**, but providing a pathway for low-carbon **hydrogen** and producing **negative emissions** when **biomass is combined with CCS** in power generation. ....

It has been rightly noted that it has the potential to help **energy-intensive industries** in this country to remain competitive. ......

**The Government absolutely believe that CCS has a potential role** in long-term decarbonisation, but it must be affordable. ...... we are taking the time to look hard at CCS to ... find a cost-effective pathway.

https://hansard.parliament.uk/commons/2017-01-24/debates/242BBBF1-396C-463D-8861-CCC547E011D0/UKDecarbonisationAndCarbonCaptureAndStorage

## Committee on Climate Change

UKCCS RESEARCH CENTRE

Report 'UK climate action following the Paris Agreement', October 2016.

Carbon capture and storage is very important given its potential to reduce emissions across heavy industry and the power sector, open up new decarbonisation pathways (e.g. based on hydrogen) and remove $CO_2$ when coupled to bioenergy. **Estimates by the Committee and by the ETI indicate that the costs of meeting the UK's 2050 target could almost double without CCS.**

## Energy Technology Institute

Report 'Carbon capture and storage: Building the UK carbon capture and storage sector by 2030', 2015

**A complete failure to deploy CCS would imply close to a doubling of the annual cost of carbon abatement to the UK economy from circa 1% to 2% of GDP by 2050 (or roughly an extra £1000 on annual average household bills for energy and transport services).**

**Figure 2.** Assessment of current policies against the cost-effective path to meet carbon budgets and the 2050 target

**Meeting Carbon Budgets - 2016 Progress Report to Parliament Committee on Climate Change, June 2016**

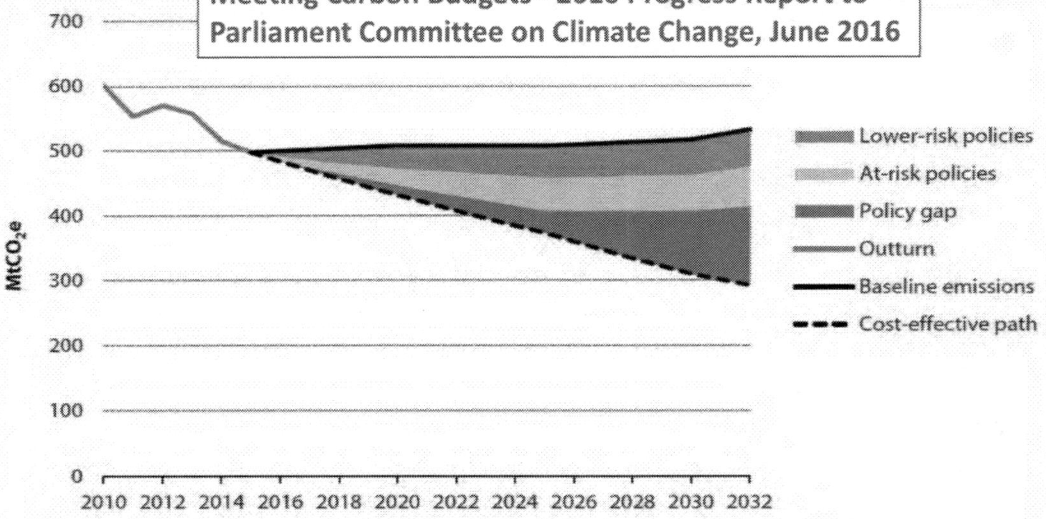


**Source:** DECC (2015) *Updated energy and emissions projections*; CCC analysis.
**Notes:** 'Lower-risk policies' (green) are those that aim to address known barriers and have sufficient funding and ambition to deliver with reasonable confidence. 'At-risk policies' (amber) either lack sufficient funding, do not address known barriers or have important design elements still to be confirmed. No funded policies exist to close the 'policy gap' (red), even though the Committee's scenarios identify abatement options to do so that are on the lowest cost path to meet the carbon budgets and the 2050 target. 'Baseline emissions' is the likely path of emissions in the absence of policy effort.

# Committee on Climate Change

## May/June 2016: A strategic approach for developing CCS in the UK

UKCCS RESEARCH CENTRE

https://www.theccc.org.uk/wp-content/uploads/2016/07/Poyry_-_A_Strategic_Approach_For_Developing_CCS_in_the_UK.pdf
https://www.theccc.org.uk/publication/meeting-carbon-budgets-2016-progress-report-to-parliament/

'**Sufficient scale of targeted roll-out**: 4-7 GW of power CCS and 3-5 Mt captured $CO_2$ from industrial plants by 2035 ..... to commercialise CCS and facilitate subsequent wide-scale deployment.'

'**An initial focus on one or two strategic clusters**: clusters in areas of industrial activity around storage sites that have been identified and successfully characterised.'

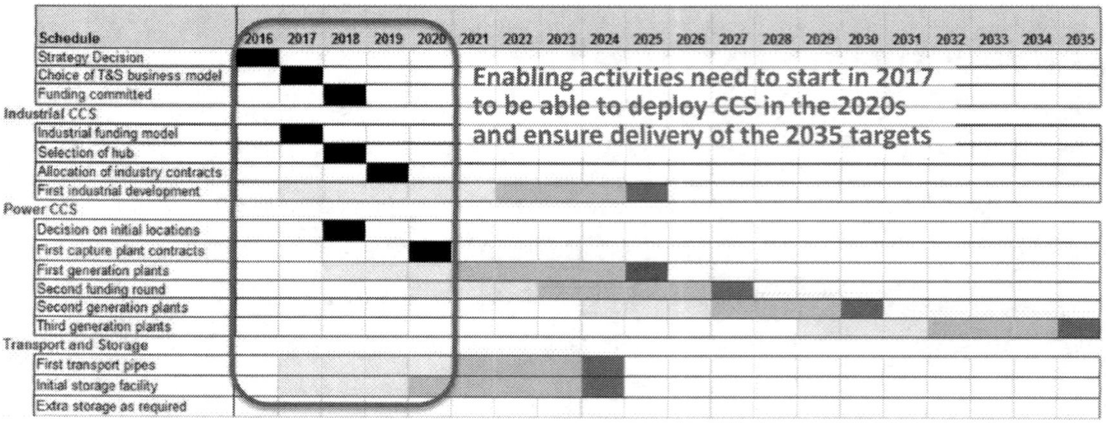

## UK Committee on Climate Change
## Greenhouse gas removal

- *The UK 2050 target to reduce emissions at least 80% from 1990 levels (i.e. less than around 160Mt$CO_2$e/yr) is challenging and requires significant action across the economy, but can be met in various ways using currently known technologies.*
- *Our UK scenarios to 2050 include up to 67 Mt$CO_2$/yr removals from three GGR options: afforestation, BECCS and wood in construction. ……. BECCS could become capable of reducing emissions at a comparable cost to other technologies in the 2030s. This would require the Government to implement an effective new approach to CCS development and development of sustainable bioenergy supplies without locking these into alternative uses. Our scenarios include up to 47 Mt$CO_2$/yr removed by BECCS while generating energy.*

https://www.theccc.org.uk/wp-content/uploads/2016/10/UK-climate-action-following-the-Paris-Agreement-Committee-on-Climate-Change-October-2016.pdf

## UK CCS 2008/2009 → UK CCS 2016/2017

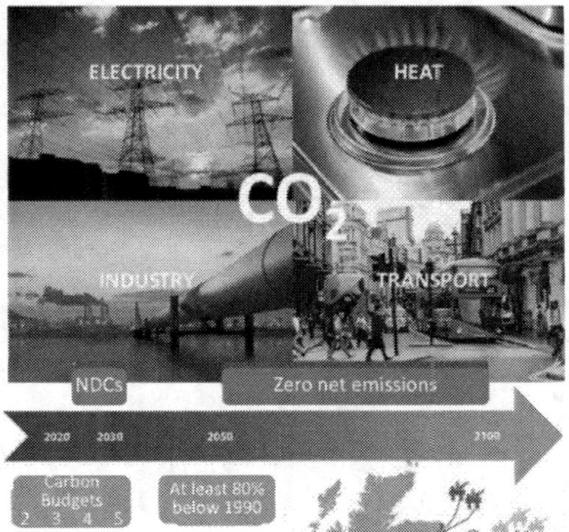

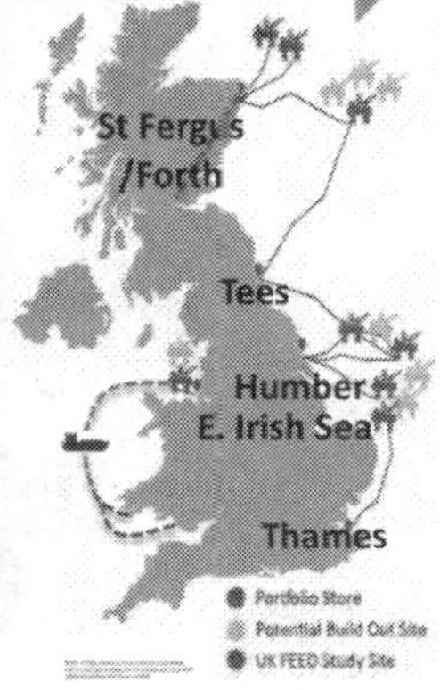

UKCCS RESEARCH CENTRE

October 11,2009

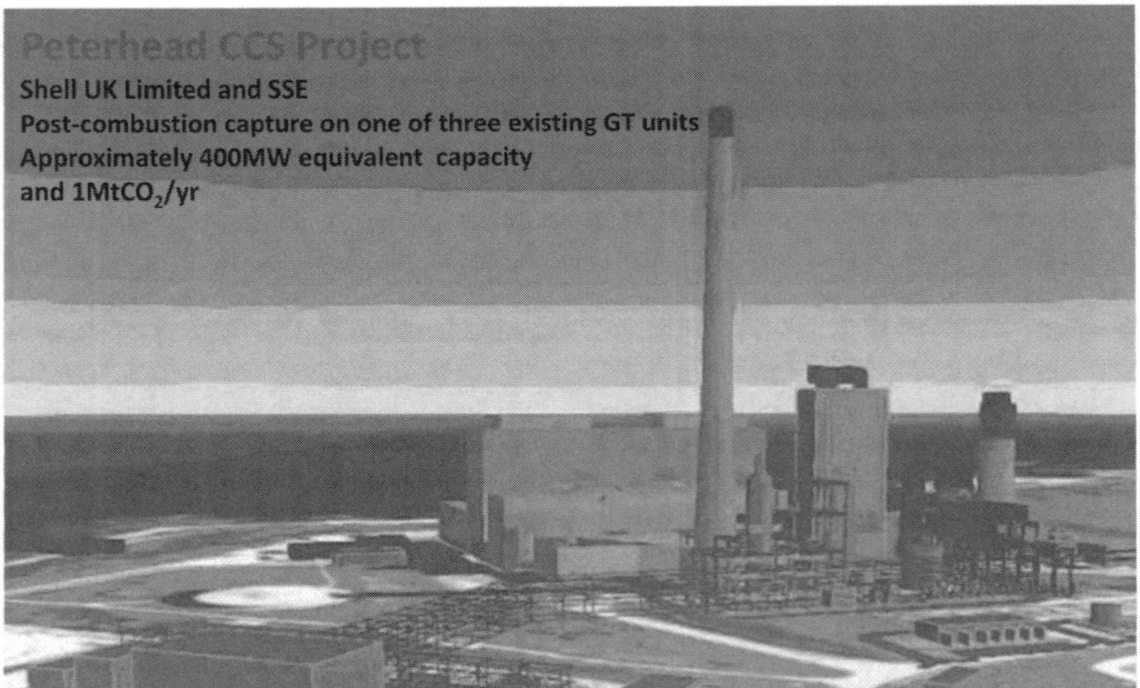

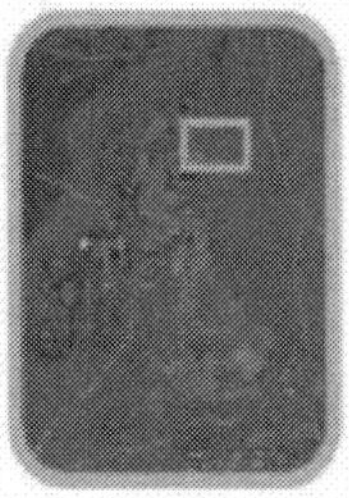

## Peterhead Geography

Jeremy Carey, Technology Manager, SSE, *CCS Deployment in SSE Peterhead and Beyond…*, IPA / UKCCSC CCS Conference 1st September 2011

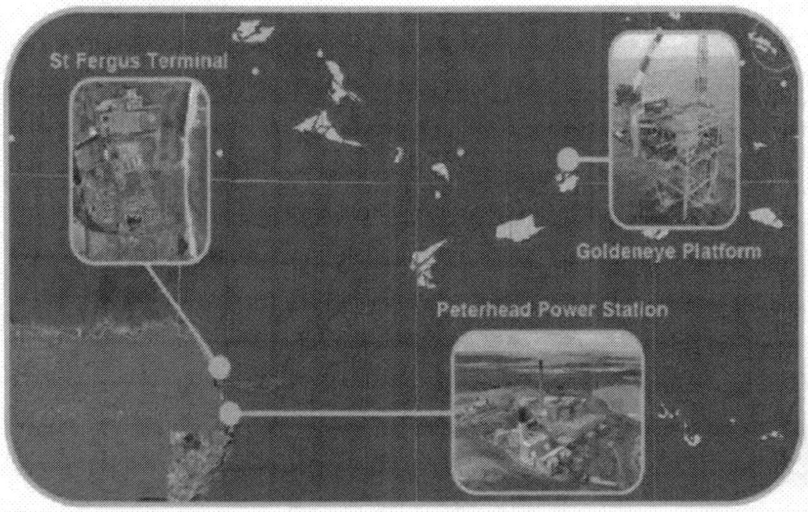


An indicative illustration of the White Rose Carbon Capture and Storage Project

ALSTOM drax BOC CAPTURE POWER

- New standalone power plant at the existing Drax Power Station site near Selby,
- State-of-the-art coal-fired power plant with the potential to co-fire biomass.
- 426MWe (gross) oxyfuel power and carbon capture and storage
- 90% of all $CO_2$ emissions captured
- Capturing approximately 2 million tonnes of $CO_2$ per year
- Anchor project for Yorkshire $CO_2$ transportation and storage network

http://www.whiteroseccs.co.uk

## CCS 'demonstrations' in retrospect

UK CCS RESEARCH CENTRE

### The target for CCS in the UK:

- Has to achieve multiple >10 $MtCO_2$/yr projects for meaningful impact on national emissions
- And deploy CCS across the UK to achieve 2050 and subsequent targets

### CCS demonstration programme so far:

- Has stipulated small ~300MW power plant projects processing 1-2 $MtCO_2$/yr (inherited from coal plant approach)
- No clear plans have existed for further deployment to use T&S fully
- Unit costs for sub-scale 'demonstrations' are therefore very high
- And a solution to the main problem of building and successfully operating >10 $MtCO_2$/yr T&S networks with multiple inputs would not be demonstrated

### Future CCS developments:

- Need scale to be cost-effective, suggested role for government in implementing T&S infrastructure (e.g Oxburgh Report*)
- National strategy for interlinked regional clusters, including shipping
- Multiple $CO_2$ sources: power, industry, hydrogen, BECCS
- Technology needs to be capable of net zero emissions

* http://www.ccsassociation.org/news-and-events/reports-and-publications/parliamentary-advisory-group-on-ccs-report/

http://www.eti.co.uk/library/taking-stock-of-uk-co2-storage

energy technologies institute

## Taking Stock

- Peterhead – project looked landable – simple chain.
- White Rose – assigning risks across several chain companies was too difficult.
- 3 CO2 stores have been appraised , plus 5 partially appraised - 8 sites totalling 1600MT - enough for many decades, at "10 GW" - 40MT/a+ CO2.
- Top risk of leakage is abandoned oil and gas wells.
- Transport and storage costs are £8-16/te – about £7/MWh for a gas fired power plant.
- Data is publically available on BEIS and ETI websites.
- Storage accessible to the South East and East England arguably the largest, and best.

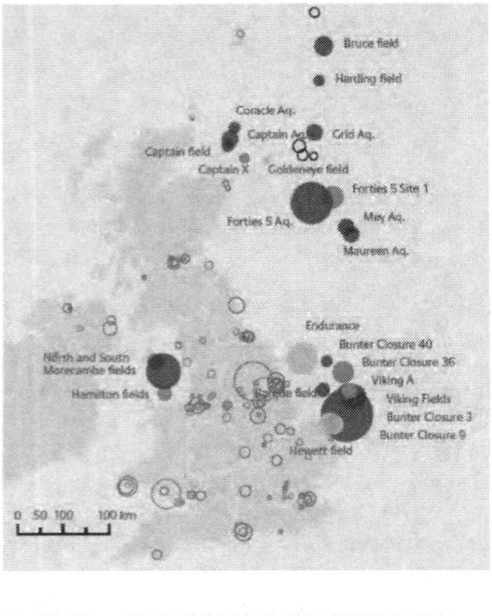




See also https://ukccsrc.ac.uk/about/delivering-cost-effective-ccs-2020s-new-start

http://www.eti.co.uk/library/taking-stock-of-uk-co2-storage

energy technologies institute

## UK has a wide range of offshore storage sites

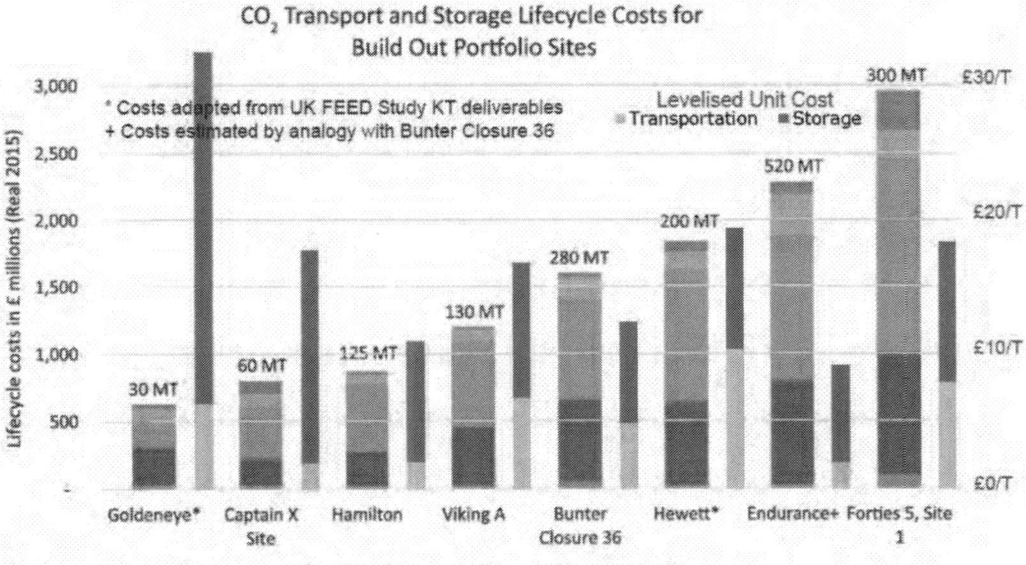




See also https://ukccsrc.ac.uk/about/delivering-cost-effective-ccs-2020s-new-start

http://www.eti.co.uk/library/taking-stock-of-uk-co2-storage

energy technologies institute

## ETI – Taking stock of UK $CO_2$ storage

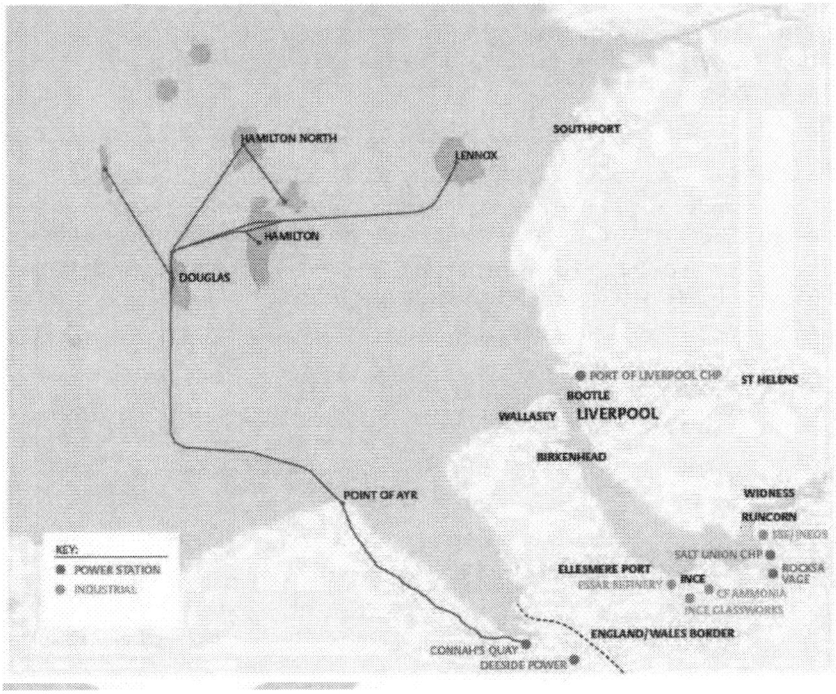

http://www.eti.co.uk/library/taking-stock-of-uk-co2-storage

energy technologies institute

## ETI – Taking stock of UK $CO_2$ storage

Example of key economic indices for different development options.

| CASE | Source | $CO_2$ Flow | Duration | Store | Up Front T&S Capex | Levelised T&S | Levelised Capture +T&S | Levelised Power Cost |
|---|---|---|---|---|---|---|---|---|
| | | MT/a | Years | | £M | £/Te | £/Te | £/MWh |
| Base Case – 2.6 GWe | Power | 7.5 | 25 | Endurance | 592 | 11 | 58 | 78 |
| Case A Ammonia Unit | Industrial | 0.33 | 8 | Hamilton | 324 | 256 | 300 | |
| Case B Single GT | Power | 1.5 | 25 | Hamilton | 255 | 33 | 63 | 95 |
| Case C Expansion of Case B | Power | 4.5 | 25 | Hamilton | 390 | 17 | 62 | 82 |
| Case D Add Ammonia Unit to Case C | Industrial | 0.33 | 20 | Hamilton | 56 | 37 | 80 | |

NAO

https://www.nao.org.uk/report/carbon-capture-and-storage-the-second-competition-for-government-support/

National Audit Office

UKCCS RESEARCH CENTRE

**Key findings** - The role of CCS

CCS could make a significant contribution to decarbonising the economy, but there are challenges which increase the costs to deploy it in the UK. CCS has the potential to contribute to the decarbonisation of the power, industrial, transport and heating sectors. Together these make up around 83% of the UK's $CO_2$ emissions.

### Recommendations

In developing its next phase of supporting CCS, the Department should:

a. **Maximise the potential value from the competition by incorporating the lessons it and the key stakeholders have learned into any new CCS strategy.**

b. **Ensure it understands, from the outset, the position of CCS developers and their ability or willingness to carry certain risks and applies this in its approach.**

c. **Assess options for how it can make early projects more affordable to taxpayers and consumers.**

d. **Agree early with HM Treasury any affordability constraints.**

More generally, the Department should:

e. **Work with HM Treasury to establish and use a consistent way of measuring the value of investments in different generating technologies that enable meaningful comparisons.**

f. **Regularly revisit its commercial strategy and the value-for-money case in light of the evolving understanding of the delivery environment and market conditions.**

g. **Consider the possible consequences of, and its risk appetite for, scenarios that are outside its central forecast or expectation when it develops a new project or programme.**

## CCS Cost Reduction Taskforce

Final Report, May 2013

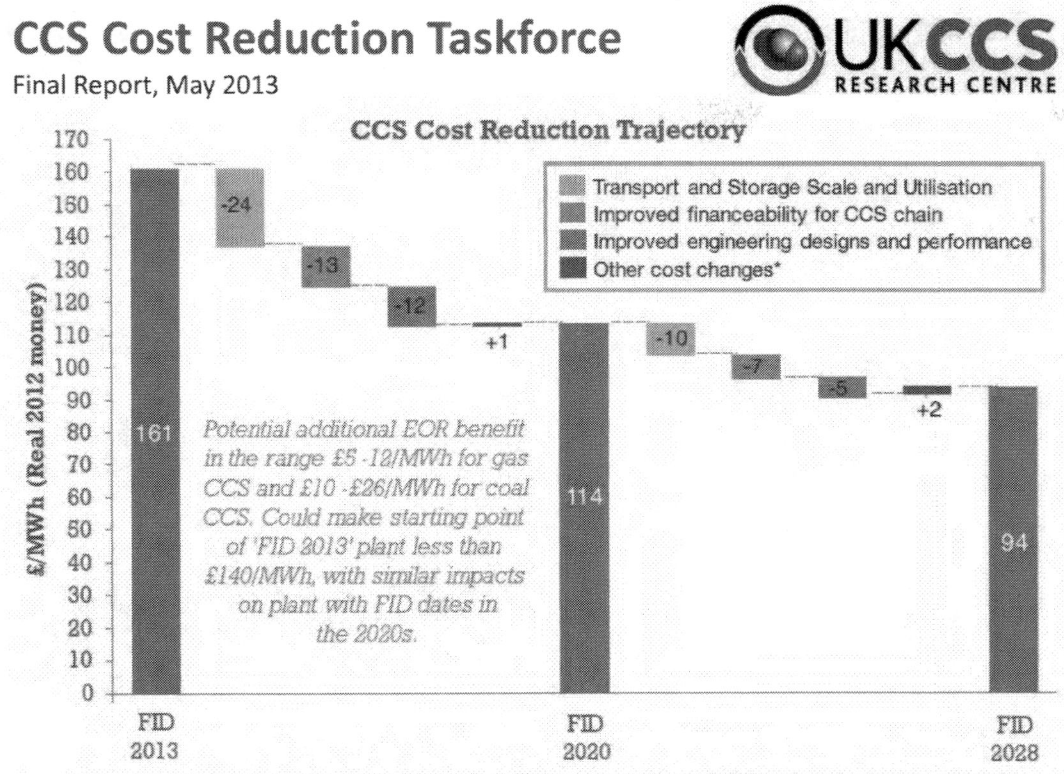


**Note: Shows average costs across technologies. *E.G. Increasing $CO_2$ price, falling storage abandonment costs**

https://www.gov.uk/government/uploads/system/uploads/attachment_data/file/201021/CCS_Cost_Reduction_Taskforce_-_Final_Report_-_May_2013.pdf

## Fundamental research needed to increase Commercial Readiness as well as TRL

For technologies to mature to be a "bankable asset class" this gap must also be bridged. Nuclear, Wind, Solar, Marine are all TRL9 and are still benefitting from fundamental research.

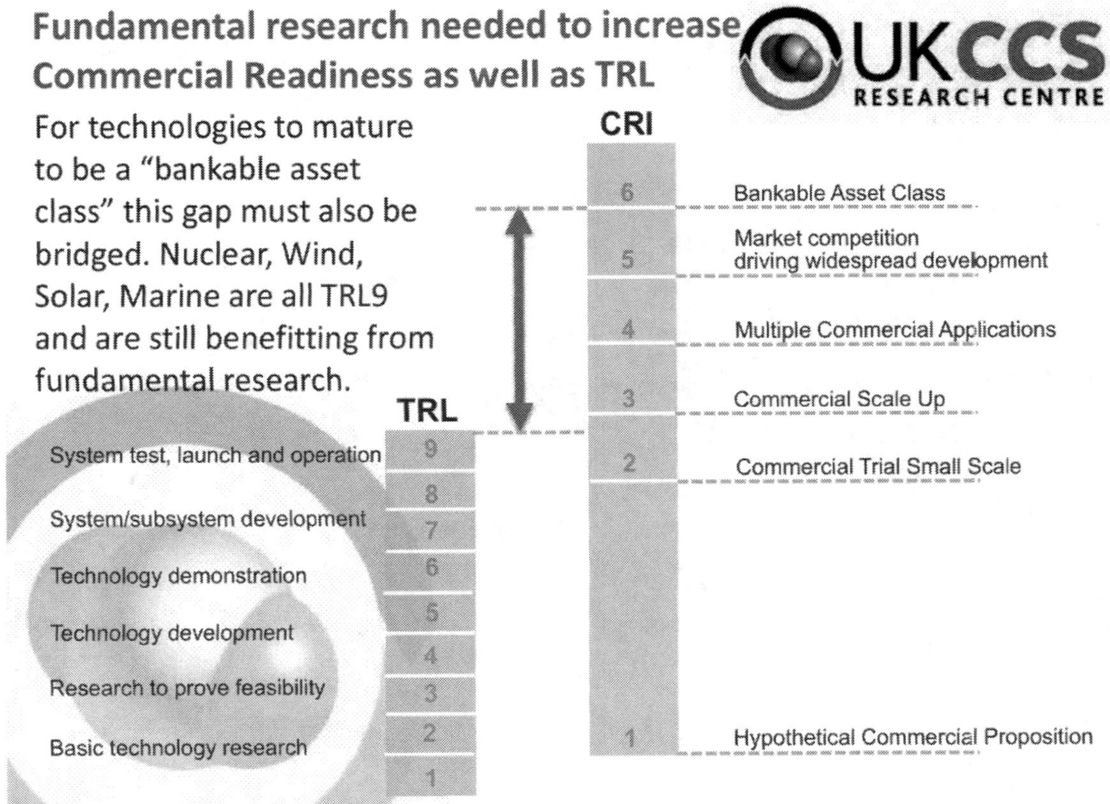


Faraday discussion, 2016, http://pubs.rsc.org/en/content/articlelanding/2016/fd/c6fd00046k#!divAbstract, based on http://arena.gov.au/files/2014/02/Commercial-Readiness-Index.pdf

# Potential Role of Hydrogen in the UK Energy System

Energy Research Partnership, October 2016, http://erpuk.org/wp-content/uploads/2016/10/ERP-Hydrogen-report-Oct-2016.pdf

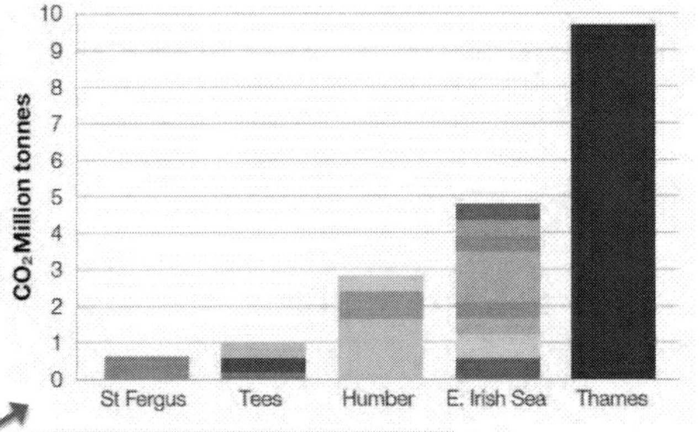


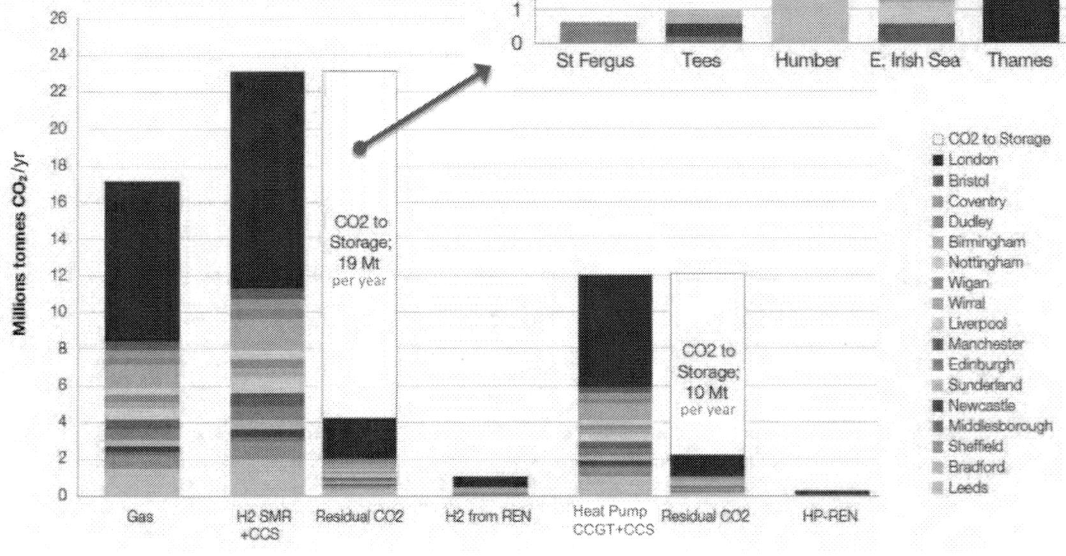


## LOCAL AUTHORITY CARBON DIOXIDE EMISSIONS ESTIMATES 2014

Statistical Release: National Statistics, 30 June 2016

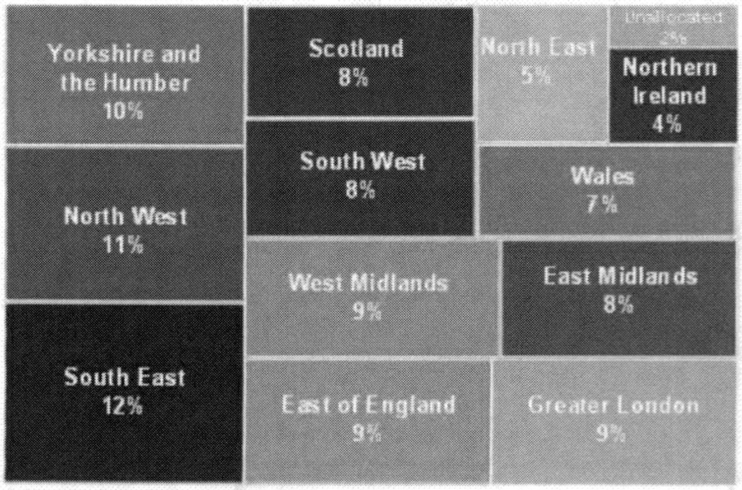


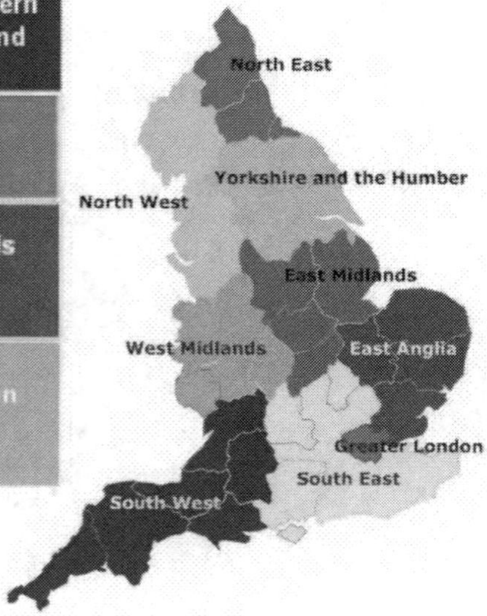


http://www.britaingallery.com/england_regions.php

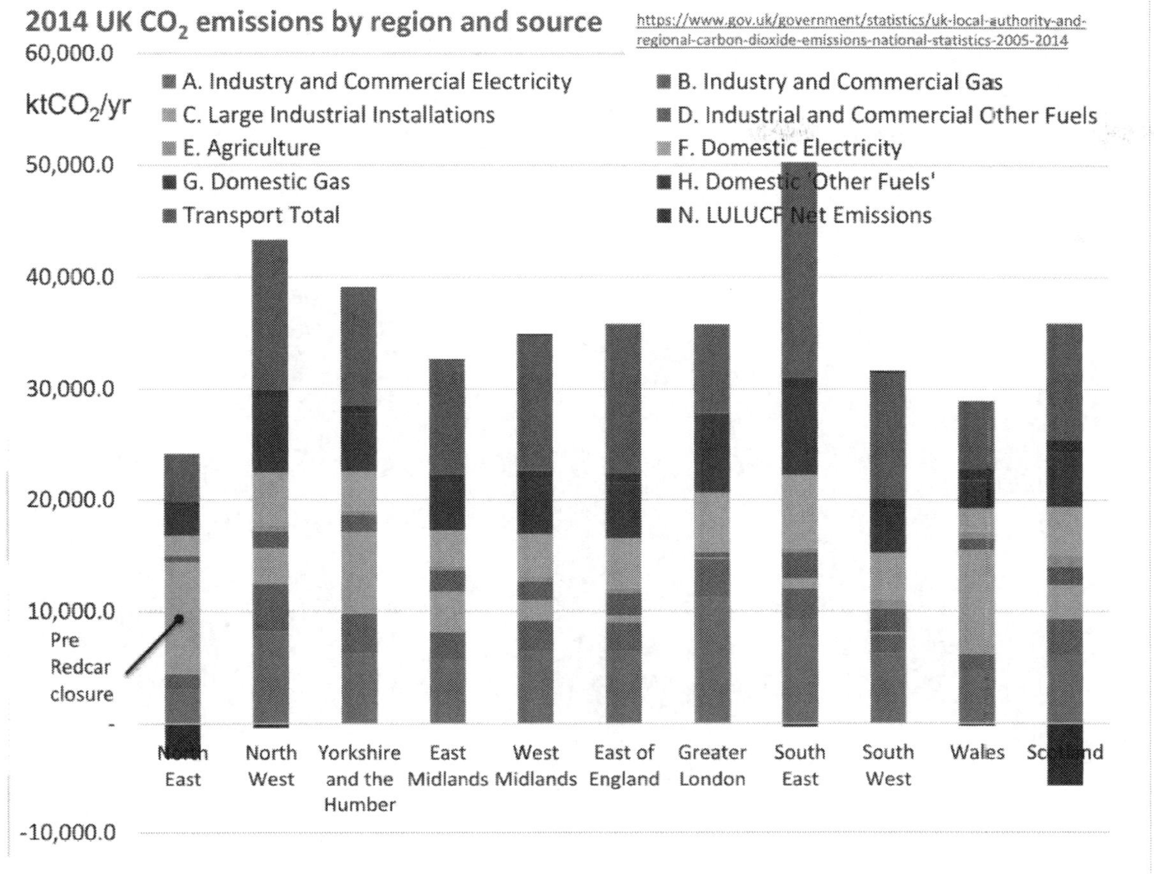


## DECC & BIS – March 2015 - Industrial Decarbonisation and Energy Efficiency Roadmaps to 2050

https://www.gov.uk/government/publications/industrial-decarbonisation-and-energy-efficiency-roadmaps-to-2050

A series of eight reports that assess the potential for a low-carbon future across the most heat-intensive industrial sectors in the UK.

The roadmap project aimed to:

· Improve understanding of the emissions-abatement potential of individual industrial sectors, the relative costs of alternative abatement options and the related business environment including investment decisions, barriers and issues of competitiveness;

· Establish a shared evidence base to inform future policy, and identify strategic conclusions and potential next steps to help deliver cost-effective decarbonisation in the medium to long term (over the period from 2020 to 2050).

**Significant direct CCS in four sectors by 2050 less than 20 $MtCO_2$/yr:**
**Chemicals (6), Iron & Steel (10 – but predates Redcar closure), Cement (2.5), Oil Refining (2.5).**

**Indirect CCS through decarbonised electricity and hydrogen in all sectors.**

# $CO_2$ from hydrogen by region

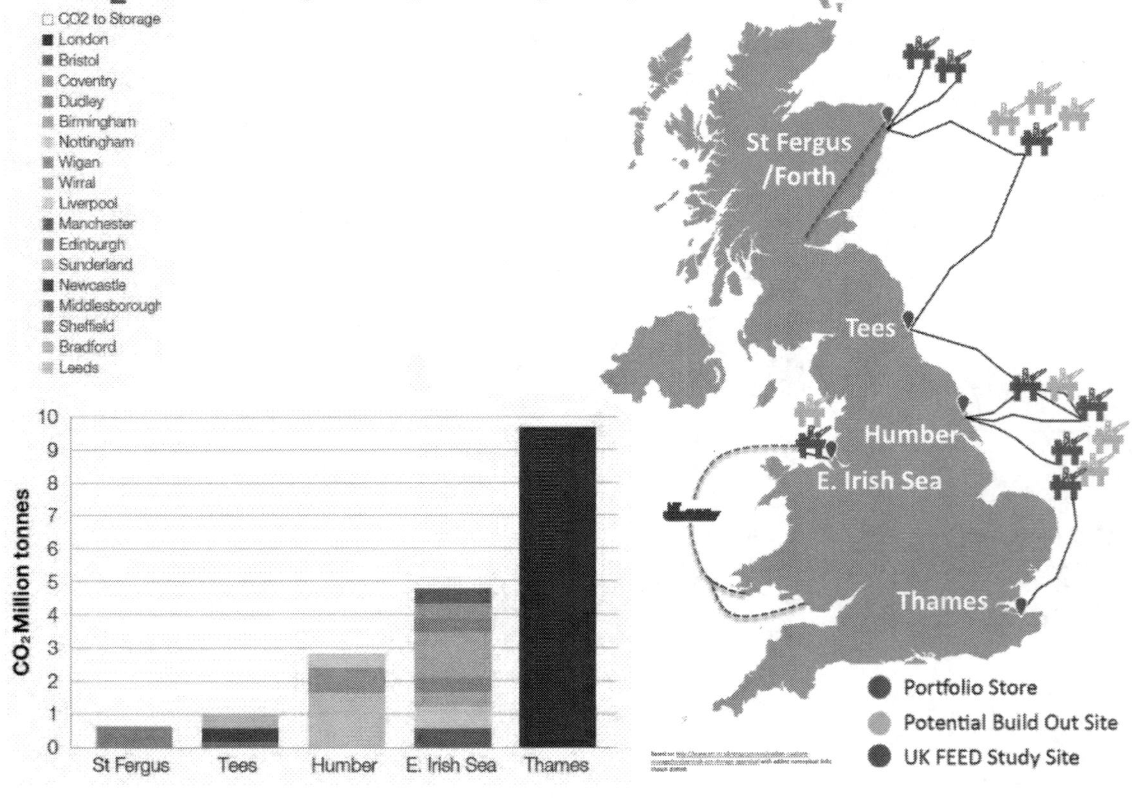


## H21 Leeds City Gate System

http://www.northerngasnetworks.co.uk/document/h21-leeds-city-gate/

UKCCS RESEARCH CENTRE

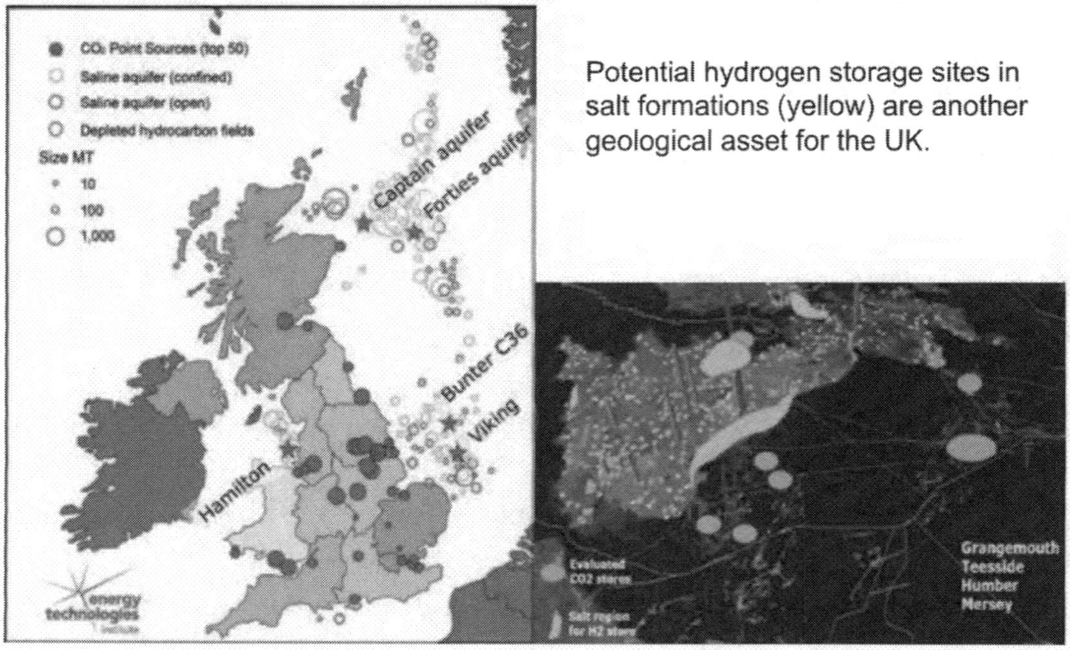


Potential hydrogen storage sites in salt formations (yellow) are another geological asset for the UK.

Image 7.3. *UK CCS availability*

## Type C Concepts:
## World's Largest Bilobe-Liquid Gas Storage Tanks

From: Gas Carriers: Arrangements & Characteristics, Rich Delpizzo, Manager, Global Gas Solutions, Presentation to Marine Chemists, Las Vegas, NV, July 2014.

9,686 m³ bilobe Type C LNG tanks building at Sinopacific for Denmark's Evergas

ABS Source: Maritime Propulsion, Feb 2014

## Independent Tanks: Type C – Bilobe

- Bilobe tanks being considered for 20-30,000 m³ size ships

From: Gas Carriers: Arrangements & Characteristics, Rich Delpizzo, Manager, Global Gas Solutions, Presentation to Marine Chemists, Las Vegas, NV, July 2014.

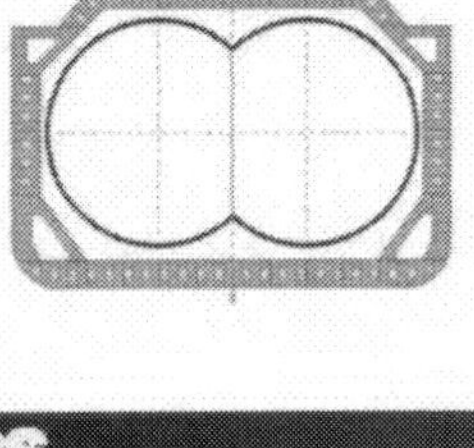

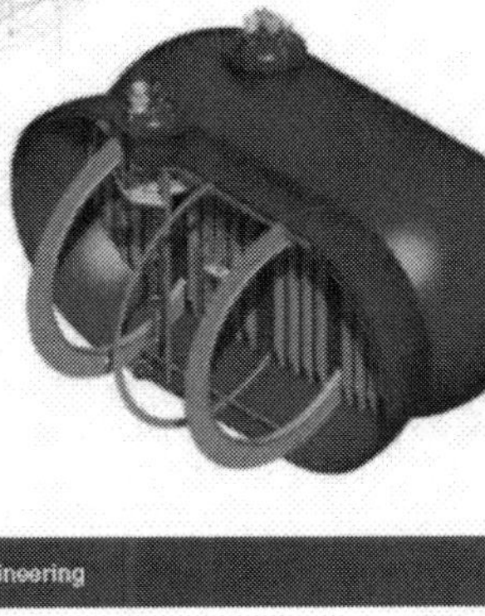

ABS Source: TGE Marine Gas Engineering

# UKCCSRC 2017-2022

## UKCCSRC 2017 Strategic Mission

**Help ensure that CCS will play an effective role in reducing net $CO_2$ emissions while securing affordable and controllable electricity supplies, low carbon heat and competitive industries for the UK.**

- **Lower costs and integration:** cost reductions for energy systems, industry and the wider economy.
- **Benefits of CCS in energy and economic systems:** inform policy and planning by government and by research funders, industry and other stakeholders.
- **Capacity to support delivery:** maintain an effective UK CCS community to deliver societal benefits to 2050 and beyond.
- **Cohesion, strategy, coordination, collaboration:** be a focal point for setting strategies, coordinating research and collaboration, nationally and globally.
- **Be an indispensable knowledge partner** in the UK and globally.

## UKCCSRC 2017 components

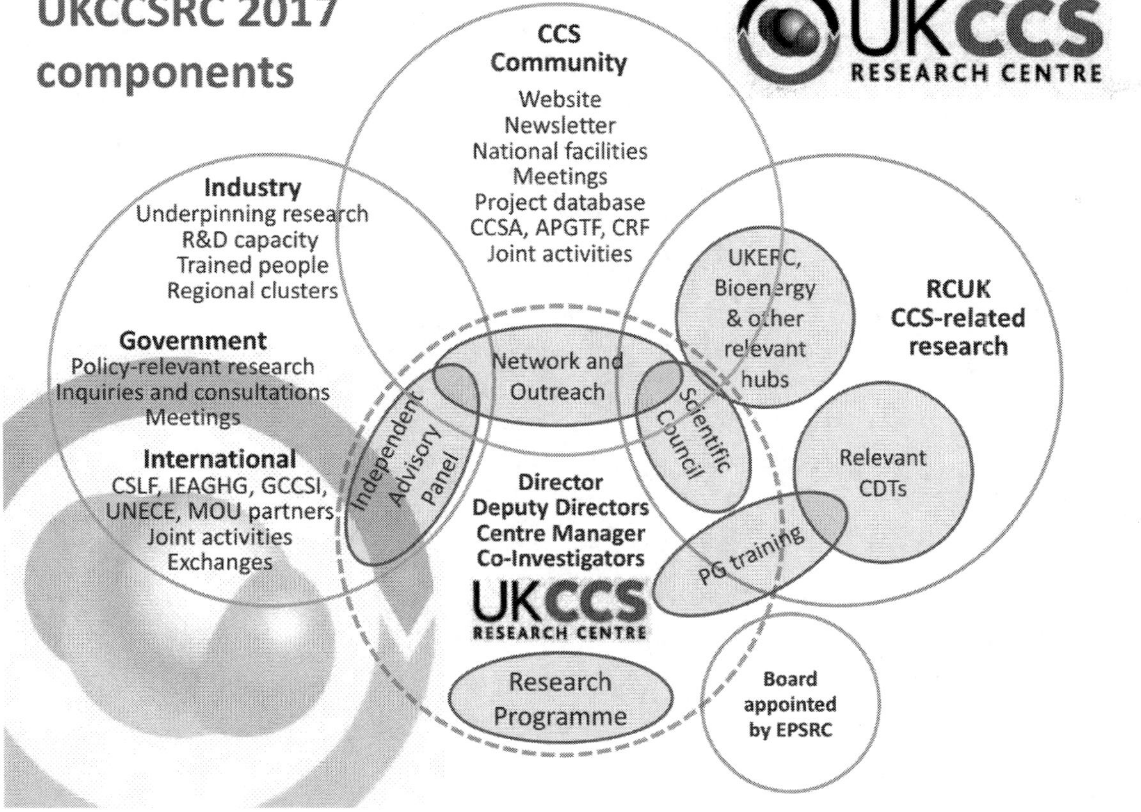


## Research challenges in CCS

## Part 1 - Have built up and be able to maintain the necessary research capacity

- Responsive and sustained strategy from funders
- Broad and deep research community
- Advanced facilities
- Established cooperation with UK industry and other stakeholders
- Links to key international partners for collaboration and delivery

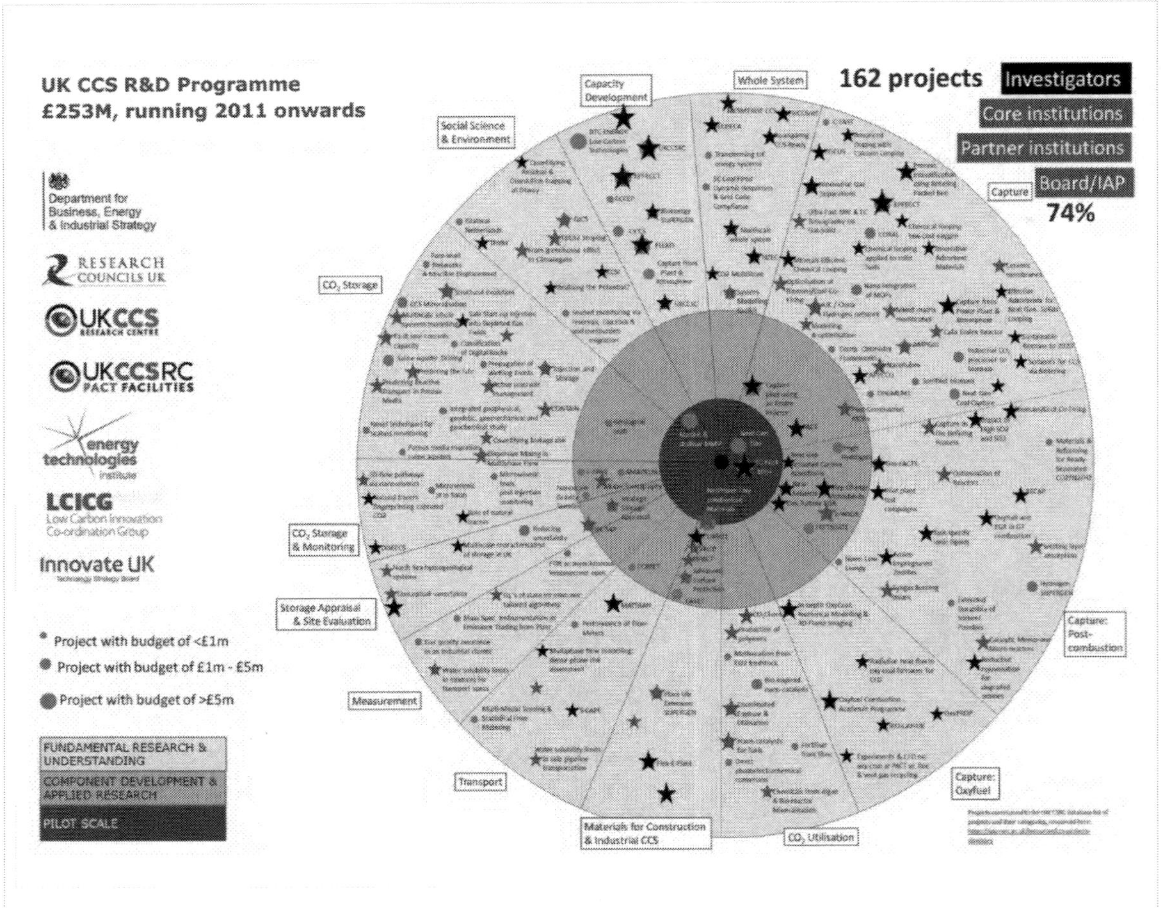
UK CCS R&D Programme
£253M, running 2011 onwards
162 projects
Investigators
Core institutions
Partner institutions
Board/IAP
74%
Department for Business, Energy & Industrial Strategy
RESEARCH COUNCILS UK
UKCCS RESEARCH CENTRE
UKCCSRC PACT FACILITIES
energy technologies institute
LCICG
Low Carbon Innovation Co-ordination Group
Innovate UK
Project with budget of <£1m
Project with budget of £1m - £5m
Project with budget of >£5m
FUNDAMENTAL RESEARCH & UNDERSTANDING
COMPONENT DEVELOPMENT & APPLIED RESEARCH
PILOT SCALE
Capacity Development
Whole System
Social Science & Environment
Capture
CO2 Storage
CO2 Storage & Monitoring
Storage Appraisal & Site Evaluation
Measurement
Transport
Materials for Construction & Industrial CCS
CO2 Utilisation
Capture: Oxyfuel
Capture: Post-combustion

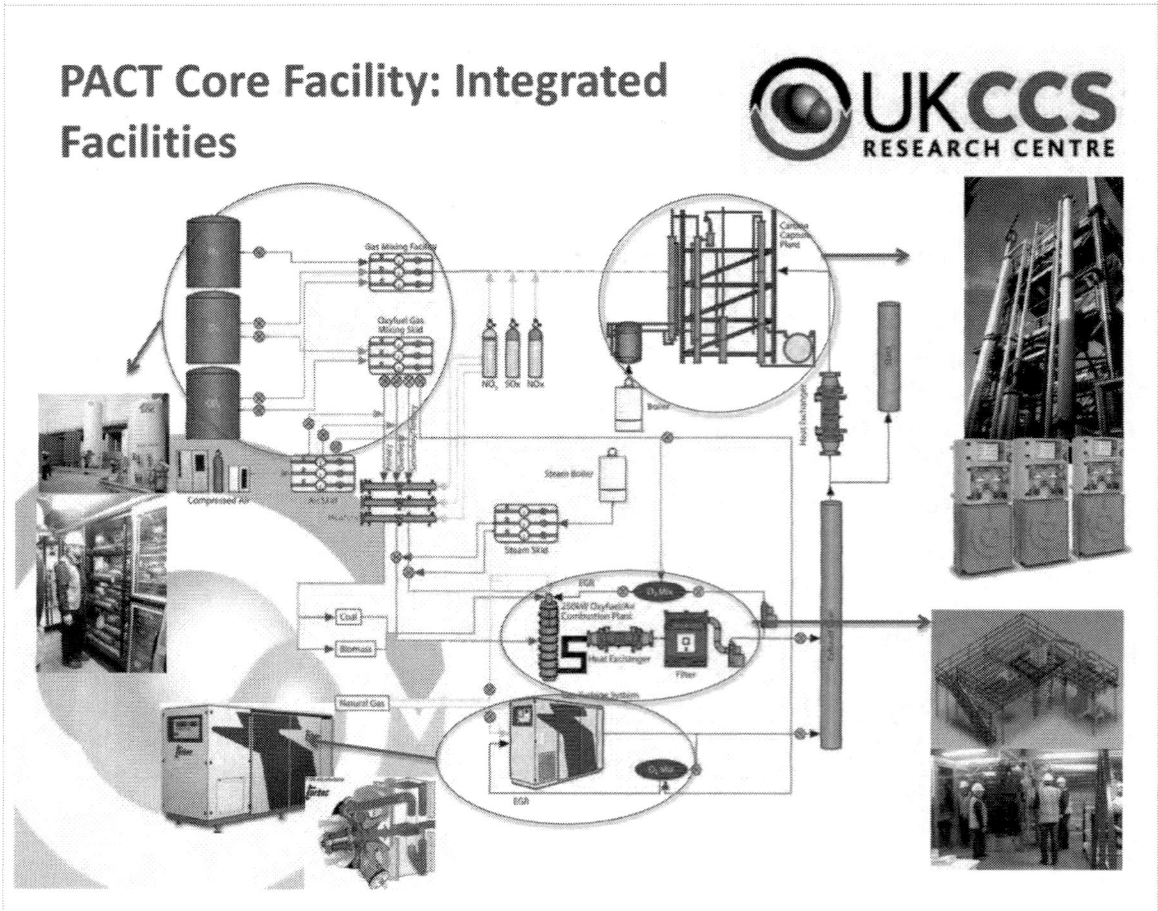
PACT Core Facility: Integrated Facilities
UKCCS
RESEARCH CENTRE
Gas Mixing Facility
Oxyfuel Gas Mixing Skid
Carbon Capture Plant
Boiler
Heat Exchanger
Steam Boiler
Steam Skid
Compressed Air
Coal
Biomass
Natural Gas
EGR
Heat Exchanger
Filter

## Research challenges in CCS

## Part 2 - Identify and help deliver cost-effective pathways to deploy CCS across the whole energy system

- Urgent need to transform CCS system concepts ....
- ... and develop the underpinning science to deliver cost-reduction,
- to engage with policy, regional and industry stakeholders
- ... and support new commercial, social and technical approaches.

## New CCS system concepts and some areas for cost reduction research

**Cross-cutting issues**

- CCS role in net-zero emission energy systems and pathways
- BECCS interactions with biomass and land use constraints
- Social, policy and commercial viability for CCS delivery models

**Combined power, industry and hydrogen clusters**

- Novel capture technologies e.g. solids and membranes, for a range of applications
- Gas to power with capture – post-combustion and oxyfuel
- Gas to hydrogen with capture – SMR with CCS and novel routes
- BECCS and other GGR technologies
- Synergies between cluster components (including CCU)

**Transport**

- Shipping as well as pipeline

**Storage**

- Using wells and pore space more effectively
- Reliable long-term modelling for liability management
- Offshore EOR
- UK test site

## Greenhouse gas removal technologies required before 2050 - and limited biomass available

UK CCS RESEARCH CENTRE

### UK Committee on Climate Change

https://www.theccc.org.uk/wp-content/uploads/2016/10/UK-climate-action-following-the-Paris-Agreement-Committee-on-Climate-Change-October-2016.pdf

- *The UK 2050 target to reduce emissions at least 80% from 1990 levels (i.e. less than around 160Mt$CO_2$e/yr) is challenging and requires significant action across the economy, but can be met in various ways using currently known technologies.*
- *Our UK scenarios to 2050 include up to 67 Mt$CO_2$/yr removals from three GGR options: afforestation, BECCS and wood in construction. ....... BECCS could become capable of reducing emissions at a comparable cost to other technologies in the 2030s. This would require the Government to implement an effective new approach to CCS development and development of sustainable bioenergy supplies without locking these into alternative uses. Our scenarios include up to 47 Mt$CO_2$/yr removed by BECCS while generating energy.*

**NERC GGR programme just started – World First!**

http://www.nerc.ac.uk/research/funded/programmes/ggr/

## Status of CCS

UK CCS RESEARCH CENTRE

- CCS has been transformed – its roles in the **net-zero emissions energy system** need to be researched and understood
- Social, policy and commercial viability is essential – Government, including the Treasury, must be on board
- Research communities need to develop and maintain capacity for cost-reduction research
- Plus the linkages to identify the right questions and deliver the answers to the users
- With support, fundamental research can help to deliver options to reduce costs before deployment in the 2020's
- And continue to reduce costs for the rest of the century

# AUTHOR INDEX

# AUTHOR INDEX

**AIChE**
120 Wall Street, FL 23
New York, NY 10005-4020

ISBN 978-1-5108-4815-3

9 781510 848153

CURRAN ASSOCIATES INC.
proceedings.com